SOME PHYSICAL PROPERTIES

Air (dry, at 20°C and 1 atm)

Density	1.21 kg/m^3
Specific heat capacity at constant pressure	$1010 \text{ J/kg} \cdot \text{K}$
Ratio of specific heat capacities	1.40
Speed of sound	343 m/s
Electrical breakdown strength	$3 \times 10^6 \text{ V/m}$
Effective molar mass	0.0289 kg/mol

Water

Density	1000 kg/m^3
Speed of sound	1460 m/s
Specific heat capacity at constant pressure	$4190 \text{ J/kg} \cdot \text{K}$
Heat of fusion (0°C)	333 kJ/kg
Heat of vaporization (100°C)	2260 kJ/kg
Index of refraction ($\lambda = 589$ nm)	1.33
Molar mass	0.0180 kg/mol

Earth

Mass	$5.98 \times 10^{24} \text{ kg}$
Mean radius	$6.37 \times 10^6 \text{ m}$
Free fall acceleration at the Earth's surface	9.81 m/s^2
Standard atmosphere	$1.01 \times 10^5 \text{ Pa}$
Period of satellite at 100 km altitude	86.3 min
Radius of the geosynchronous orbit	42,200 km
Escape speed	11.2 km/s
Magnetic dipole moment	$8.0 \times 10^{22} \text{ A} \cdot \text{m}^2$
Mean electric field at surface	150 V/m, down

Distance to:

Moon	$3.82 \times 10^8 \text{ m}$
Sun	$1.50 \times 10^{11} \text{ m}$
Nearest star	$4.04 \times 10^{16} \text{ m}$
Galactic center	$2.2 \times 10^{20} \text{ m}$
Andromeda galaxy	$2.1 \times 10^{22} \text{ m}$
Edge of the observable universe	$\sim 10^{26} \text{ m}$

SUPPLEMENTS

PHYSICS, FOURTH EDITION is accompanied
by a complete supplementary package.

STUDY GUIDE (A Student's Companion to Physics)

J. RICHARD CHRISTMAN
U.S. Coast Guard Academy

Provides self-tests for conceptual understanding and problem
solving.

SOLUTIONS MANUAL

EDWARD DERRINGH
Wentworth Institute of Technology

Provides approximately 302 of the solutions to textbook problems.

LABORATORY PHYSICS, SECOND EDITION

HARRY F. MEINERS
Rensselaer Polytechnic Institute
WALTER EPPENSTEIN
Rensselaer Polytechnic Institute
KENNETH MOORE
Rensselaer Polytechnic Institute
RALPH A. OLIVA
Texas Instruments, Inc.

This laboratory manual offers a clear introduction to procedures
and instrumentation, including errors, graphing, apparatus
handling, calculators, and computers, in addition to over 70
different experiments grouped by topic.

FOR THE INSTRUCTOR

A complete supplementary package of teaching and learning
materials is available for instructors. Contact your local Wiley
representative for further information.

VOLUME ONE

PHYSICS

FOURTH EDITION

Books by D. Halliday, R. Resnick, and K. Krane

Physics, Volume 1, Fourth Edition
Physics, Volume 2, Fourth Edition
Physics, Volume 2, Fourth Edition, Extended

Books by D. Halliday and R. Resnick

Fundamentals of Physics, Third Edition
Fundamentals of Physics, Third Edition, Extended

Books by R. Resnick

Introduction to Special Relativity

Books by Robert Eisberg and Robert Resnick

Quantum Physics of Atoms, Molecules, Solids, Nuclei, and Particles, Second Edition

Books by Kenneth S. Krane

Modern Physics
Introductory Nuclear Physics

VOLUME ONE

PHYSICS

FOURTH EDITION

ROBERT RESNICK

Professor of Physics
Rensselaer Polytechnic Institute

DAVID HALLIDAY

Professor of Physics, Emeritus
University of Pittsburgh

KENNETH S. KRANE

Professor of Physics
Oregon State University

JOHN WILEY & SONS, INC.

New York • *Chichester* • *Brisbane* • *Toronto* • *Singapore*

Acquisitions Editor *Clifford Mills*
Marketing Manager *Catherine Faduska*
Production Manager *Joe Ford*
Production Supervisor *Lucille Buonocore*
Manufacturing Manager *Lorraine Fumoso*
Copy Editing Manager *Deborah Herbert*
Photo Researcher *Jennifer Atkins*
Photo Research Manager *Stella Kupferberg*
Illustration *John Balbalis*
Text Design *Karin Gerdes Kincheloe*
Cover Design Direction *Karin Gerdes Kincheloe*
Cover Design *Lee Goldstein*
Cover Illustration *Roy Wiemann*

Library of Congress Cataloging-in-Publication Data

Halliday, David, 1916–
    Physics / David Halliday, Robert Resnick, Kenneth S. Krane. – – 4th ed.
      p.   cm.
    Includes index.
    ISBN 0-471-80458-4 (lib. bdg. : v. 1)
    1. Physics.   I. Resnick, Robert, 1923–   . II. Krane, Kenneth S.
III. Title.
QC21.2.H355 1992   91-35885
530– –dc20   CIP

Printed and bound by **Von Hoffmann Press, Inc.**

10 9 8 7 6

PREFACE TO VOLUME 1

The first edition of *Physics for Students of Science and Engineering* appeared in 1960; the most recent edition (the third), called simply *Physics,* was published in 1977. The present fourth edition (1992) marks the addition of a new coauthor for the text.

The text has been updated to include new developments in physics and in its pedagogy. Based in part on our reading of the literature on these subjects, in part on the comments from numerous users of past editions, and in part on the advice of a dedicated group of reviewers of the manuscript of this edition, we have made a number of changes.

1. Energy is treated in a coherent way throughout the text, beginning with the work-energy theorem and continuing with thermodynamics. For example, we consistently calculate work as that done *on* a system, thus using the same sign convention for work in both mechanics and thermodynamics. Attention to such details helps the student to discern the common concepts that permeate different areas of physics.

2. Special relativity, which was treated as a Supplementary Topic in the previous edition, is integrated throughout the text. Two chapters are devoted to special relativity: one (in Volume 1) follows mechanical waves and another (in Volume 2) follows electromagnetic waves. Topics related to special relativity (for instance, relative motion, frames of reference, momentum, and energy) are treated throughout the text in chapters on kinematics, mechanics, and electromagnetism. This approach reflects our view that special relativity should be treated as part of classical physics. However, for those instructors who wish to delay special relativity until the end of the course, the material is set off in separate sections that can easily be skipped on the first reading.

3. Changes in the ordering of topics from the third edition include the interchange of Chapters 2 and 3, so that one-dimensional kinematics now precedes vectors; the consolidation of all material on angular momentum into Chapter 13 (where it follows rotational kinematics and dynamics, thus making our presentation of rotational motion more nearly parallel to that of translational motion); and a reordering and substantial rewriting of the chapters on thermodynamics, emphasizing its statistical aspects and giving the subject a more "modern" flavor.

4. In response to requests from users, several new classical topics have been added to Volume 1; these include dimensional analysis, drag forces, elasticity, surface tension, viscosity, and musical acoustics.

5. Modern applications have been "sprinkled" throughout the text: for instance, quantization of energy and angular momentum, decays of nuclei and elementary particles, chaos theory, general relativity, and quantum statistics. These are not intended to be a coherent treatment of modern physics (which is available in the additional eight chapters of the extended version of Volume 2), but instead to indicate to the student the boundaries of classical physics and the relationships between classical and modern physics.

6. We have substantially increased the number of end-of-chapter problems relative to the previous edition of Volume 1: there are now 1519 problems compared with 958 previously, an increase of 59 percent. The number of end-of-chapter questions has been similarly increased from 614 to 821 (34%). We have tried to maintain the quality and diversity of problems that have been the hallmark of previous editions of this text.

7. The number of worked examples in Volume 1 has been increased from 135 to 183 (36%). The true increase in the number of worked examples (now called sample problems) is greater than this estimate, because the previous edition occasionally introduced new topics by means of worked examples. This edition eliminates that practice; new material is presented only in the exposition of the text, and the sample problems serve only as exercises in its application.

8. Computational techniques are introduced through several worked examples and through a variety of end-of-chapter computer projects. Some program listings are given in an appendix to encourage students to adapt those methods to other applications.

9. We have increased and updated the references to articles in the literature that appear as footnotes throughout the text. Some references (often to articles in popular magazines such as *Scientific American*) are intended to broaden the student's background through interesting applications of a topic. In other cases, often involving items of pedagogic importance to which we wish to call the attention of students as well as instructors, we make reference to articles in journals such as the *American Journal of Physics* or *The Physics Teacher*.

10. The illustrations have been completely redone and their number in Volume 1 has been increased by nearly a factor of 2, from 463 to 899. We have added color to many of the drawings where the additional color enhances the clarity or the pedagogy.

11. Many of the derivations, proofs, and arguments of the previous edition have been tightened up, and any assumptions or approximations have been clarified. We have thereby improved the rigor of the text without necessarily raising its level. We are concerned about indicating to students the limit of validity of a particular argument and encouraging students to consider questions such as: Does a particular result apply always or only sometimes? What happens as we go toward the quantum or the relativistic limit?

Although we have made some efforts to eliminate material from the previous edition, the additions mentioned above contribute to a text of increasing length. *It should be emphasized that few (if any) instructors will want to follow the entire text from start to finish.* We have worked to develop a text that offers a rigorous and complete introduction to physics, but the instructor is able to follow many alternate pathways through the text. The instructor who wishes to treat fewer topics in greater depth (currently called the "less is more" approach) will be able to select from among these pathways. Some sections are explicitly labeled "optional" (and are printed in smaller type), indicating that they can be skipped without loss of continuity. Depending on the course design, other sections or even entire chapters can be skipped or treated lightly. The Instructor's Guide, available as a companion volume, offers suggestions for abbreviating the coverage. In such circumstances, the curious student who desires further study can be encouraged independently to approach the omitted topics, thereby gaining a broader view of the subject. The instructor is thus provided with a wide choice of which particular reduced set of topics to cover in a course of any given length. For instructors who wish a fuller coverage, such as in courses for physics majors or honors students or in courses of length greater than one year, this text provides the additional material needed for a challenging and comprehensive experience. We hope the text will be considered a road map through physics; many roads, scenic or direct, can be taken, and all roads need not be utilized on the first journey. The eager traveler may be encouraged to return to the map to explore areas missed on previous journeys.

The text is available as separate volumes: Volume 1 (Chapters 1 to 26) covers kinematics, mechanics, and thermodynamics, and Volume 2 (Chapters 27 to 48) covers electromagnetism and optics. An extended version of Volume 2 (Chapters 27 to 56) is available with eight additional chapters which present an introduction to quantum physics and some of its applications. The following supplements are available:

Study Guide	Solutions Manual
Laboratory Manual	Instructor's Guide

A textbook contains far more contributions to the elucidation of a subject than those made by the authors alone. We have been fortunate to have the assistance of Edward Derringh (Wentworth Institute of Technology) in preparing the problem sets and J. Richard Christman (U. S. Coast Guard Academy) in preparing the Instructor's Guide and the computer projects. We have benefited from the chapter-by-chapter comments and criticisms of a dedicated team of reviewers:

Robert P. Bauman (University of Alabama)
Truman D. Black (University of Texas, Arlington)
Edmond Brown (Rensselaer Polytechnic Institute)
J. Richard Christman (U. S. Coast Guard Academy)
Sumner Davis (University of California, Berkeley)
Roger Freedman (University of California, Santa Barbara)
James B. Gerhart (University of Washington)
Richard Thompson (University of Southern California)
David Wallach (Pennsylvania State University)
Roald K. Wangsness (University of Arizona)

We are deeply indebted to these individuals for their substantial contributions to this project.

We are grateful to the staff of John Wiley & Sons for their outstanding cooperation and support, including physics editor Cliff Mills, editorial program assistant Cathy Donovan, marketing manager Cathy Faduska, illustrator John Balbalis, editorial supervisor Deborah Herbert, designer Karin Kincheloe, production supervisor Lucille Buonocore, photo researcher Jennifer Atkins, and copy editor Christina Della Bartolomea. Word processing of the manuscript for this edition was superbly done by Christina Godfrey.

September 1991

DAVID HALLIDAY
Seattle, Washington

ROBERT RESNICK
Rensselaer Polytechnic Institute
Troy, New York 12180-3590

KENNETH S. KRANE
Oregon State University
Corvallis, Oregon 97331

CONTENTS

CHAPTER 26
ENTROPY AND THE SECOND
LAW OF THERMODYNAMICS 571

APPENDICES

VOLUME ONE

PHYSICS

FOURTH EDITION

CHAPTER 1

MEASUREMENT

Despite the mathematical beauty of some of its most complex and abstract theories, including those of elementary particles and general relativity, physics is above all an experimental science. It is therefore critical that those who make precise measurements be able to agree on standards in which to express the results of those measurements, so that they can be communicated from one laboratory to another and verified. In this chapter we begin our study of physics by introducing some of the basic units of physical quantities and the standards that have been accepted for their measurement. We consider the proper way to express the results of calculations and measurements, including the appropriate dimensions and number of significant figures. We discuss and illustrate the importance of paying attention to the dimensions of the quantities that appear in our equations. Later in the text, other basic units and many derived units are introduced as they are needed.

1-1 THE PHYSICAL QUANTITIES, STANDARDS, AND UNITS

The building blocks of physics are the quantities that we use to express the laws of physics. Among these are length, mass, time, force, speed, density, resistivity, temperature, luminous intensity, magnetic field strength, and many more. Many of these words, such as length and force, are part of our everyday vocabulary. You might say, for example: "I will go to any *length* to help you as long as you do not *force* me to do so." In physics, however, we must not be misled by the everyday meanings of these words. The precise scientific definitions of length and force have no connection at all with the uses of these words in the quoted sentence.

We can define an algebraic quantity, for instance, *L* for length, any way we choose, and we can assume it is exactly known. However, when we try to assign a unit to a particular value of that quantity, we run into the difficulty of establishing a *standard,* so that those who have need of comparing one length with another will agree on the units of measurement. At one time, the basic unit of length was the yard, determined by the size of the king's waistline. You can easily see the problems with such a standard: it is hardly *accessible* to those who need to calibrate their own secondary standards, and it is not *invariable* to change with the passage of time.

Fortunately, it is not necessary to define and agree on standards for every physical quantity. Some elementary quantities may be easier to establish as standards, and more complex quantities can often be expressed in terms of the elementary units. *Length* and *time,* for example, were for many years among the most precisely measurable physical quantities and were generally accepted as standards. *Speed,* on the other hand, was less precisely measurable and therefore was treated as a derived unit (speed = length/time). Today, however, measurements of the speed of light have reached a precision beyond that of the former standard of length; we still treat length as a fundamental unit, but the standard for its measurement is now derived from the standards of speed and time.

The basic problem is therefore to choose the smallest possible number of physical quantities as fundamental and to agree on standards for their measurement. These standards should be both accessible and invariable, which may be difficult to satisfy simultaneously. If the standard kilogram, for instance, is to be an invariable object, it must be *in*accessible and must be kept isolated beyond the effects of handling and corrosion.

Agreement on standards has been accomplished through a series of international meetings of the General Conference on Weights and Measures beginning in 1889; the 19th meeting was held in 1991. Once a standard has been accepted, such as the *second* as a unit of *time,* then we can apply the unit to a vast range of measurements, from the lifetime of the proton (greater than 10^{40} seconds) to the lifetime of the least stable particles that can be produced in our laboratories (about 10^{-23} second). When we express such a value as 10^{40} in units of seconds, what we mean is that the *ratio* between the lifetime of the proton and the time interval that is arbitrarily defined as the standard second is 10^{40}. To accomplish such a measurement, we must have a way of comparing laboratory measuring instruments with the standard. Many of these comparisons are indirect, for no single measuring instrument is capable of operating precisely over 40 orders of magnitude. Nevertheless, it is essential to the progress of science that, when a researcher records a particular time interval with a laboratory instrument, the reading can in some way be connected to a calibration based on the standard second.

The quest for more precise or accessible standards is itself an important scientific pursuit, involving physicists and other researchers in laboratories throughout the world. In the United States, laboratories of the National Institute of Standards and Technology (formerly the National Bureau of Standards) are devoted to maintaining, developing, and testing standards for basic researchers as well as for scientists and engineers in industry. Improvements in our standards in recent years have been dramatic: since the first edition of this textbook (1960), the precision of the standard second has improved by more than a factor of 1000.

1-2 THE INTERNATIONAL SYSTEM OF UNITS*

The General Conference on Weights and Measures, at meetings during the period 1954–1971, selected as base units the seven quantities displayed in Table 1. This is the basis of the International System of Units, abbreviated SI from the French *Le Système International d'Unités.*

Throughout the book we give many examples of SI derived units, such as speed, force, and electric resistance, that follow from Table 1. For example, the SI unit of force, called the *newton* (abbreviation N), is defined in terms of the SI base units as

$$1 \text{ N} = 1 \text{ kg} \cdot \text{m/s}^2$$

as we shall make clear in Chapter 5.

* See "SI: The International System of Units," by Robert A. Nelson (American Association of Physics Teachers, 1981). The "official" U.S. guide to the SI system can be found in Special Publication 330 of the National Bureau of Standards (1986 edition).

TABLE 1 SI BASE UNITS

Quantity	SI Unit Name	Symbol
Time	second	s
Length	meter	m
Mass	kilogram	kg
Amount of substance	mole	mol
Thermodynamic temperature	kelvin	K
Electric current	ampere	A
Luminous intensity	candela	cd

If we express physical properties such as the output of a power plant or the time interval between two nuclear events in SI units, we often find very large or very small numbers. For convenience, the General Conference on Weights and Measures, at meetings during the period 1960–1975, recommended the prefixes shown in Table 2. Thus we can write the output of a typical electrical power plant, 1.3×10^9 watts, as 1.3 gigawatts or 1.3 GW. Similarly, we can write a time interval of the size often encountered in nuclear physics, 2.35×10^{-9} seconds, as 2.35 nanoseconds or 2.35 ns. Prefixes for factors greater than unity have Greek roots, and those for factors less than unity have Latin roots (except femto and atto, which have Danish roots).

To fortify Table 1 we need seven sets of operational procedures that tell us how to produce the seven SI base units in the laboratory. We explore those for time, length, and mass in the next three sections.

Two other major systems of units compete with the International System (SI). One is the Gaussian system, in terms of which much of the literature of physics is expressed. We do not use this system in this book. Appendix G gives conversion factors to SI units.

The second is the British system, still in daily use in the United States. The basic units, in mechanics, are length (the foot), force (the pound), and time (the second). Again Appendix G gives conversion factors to SI units. We use SI units in this book, but we sometimes give the British equivalents, to help those who are unaccustomed to SI units to acquire more familiarity with them. In only three countries [Myanmar (Burma), Liberia, and the United States] is a system other than SI used as the accepted national standard of measurement.

Sample Problem 1 Any physical quantity can be multiplied by 1 without changing its value. For example, 1 min = 60 s, so 1 = 60 s/1 min; similarly, 1 ft = 12 in., so 1 = 1 ft/12 in. Using appropriate conversion factors, find (*a*) the speed in meters per second equivalent to 55 miles per hour, and (*b*) the volume in cubic centimeters of a tank that holds 16 gallons of gasoline.

TABLE 2 SI PREFIXES[a]

Factor	Prefix	Symbol	Factor	Prefix	Symbol
10^{18}	exa-	E	10^{-1}	deci-	d
10^{15}	peta-	P	10^{-2}	**centi-**	c
10^{12}	tera-	T	10^{-3}	**milli-**	m
10^{9}	**giga-**	G	10^{-6}	**micro-**	μ
10^{6}	**mega-**	M	10^{-9}	**nano-**	n
10^{3}	**kilo-**	k	10^{-12}	**pico-**	p
10^{2}	hecto-	h	10^{-15}	femto-	f
10^{1}	deka-	da	10^{-18}	atto-	a

[a] In all cases, the first syllable is accented, as in na′-no-me′-ter. Prefixes commonly used in this book are shown in boldfaced type.

Solution (a) For our conversion factors, we need (see Appendix G) 1 mi = 1609 m (so that 1 = 1609 m/1 mi) and 1 h = 3600 s (so 1 = 1 h/3600 s). Thus

$$\text{speed} = 55 \ \frac{\text{mi}}{\text{h}} \times \frac{1609 \text{ m}}{1 \text{ mi}} \times \frac{1 \text{ h}}{3600 \text{ s}} = 25 \text{ m/s}.$$

(b) One fluid gallon is 231 cubic inches, and 1 in. = 2.54 cm. Thus

$$\text{volume} = 16 \ \text{gal} \times \frac{231 \text{ in.}^3}{1 \text{ gal}} \times \left(\frac{2.54 \text{ cm}}{1 \text{ in.}} \right)^3 = 6.1 \times 10^4 \text{ cm}^3.$$

Note in these two calculations how the unit conversion factors are inserted so that the unwanted units appear in one numerator and one denominator, and thus cancel.

1-3 THE STANDARD OF TIME*

The measurement of time has two aspects. For civil and for some scientific purposes we want to know the time of day so that we can order events in sequence. In most scientific work we want to know how long an event lasts (the time interval). Thus any time standard must be able to answer the questions "At what time does it occur?" and "How long does it last?" Table 3 shows the range of time intervals that can be measured. They vary by a factor of about 10^{63}.

We can use any phenomenon that repeats itself as a measure of time. The measurement consists of counting the repetitions, including the fractions thereof. We could use an oscillating pendulum, a mass–spring system, or a quartz crystal, for example. Of the many repetitive phe-

---

* For a history of timekeeping, see *Revolution in Time: Clocks and the Making of the Modern World,* by David S. Landes (Harvard University Press, 1983). Recent developments in precise timekeeping are discussed in "Precise Measurement of Time," by Norman F. Ramsey, *American Scientist,* January–February 1988, p. 42. An account of different systems for reporting time can be found in "Time and the Amateur Astronomer," by Alan M. MacRobert, *Sky and Telescope,* April 1989, p. 378.

TABLE 3 SOME MEASURED TIME INTERVALS[a]

Time Interval	Seconds
Lifetime of proton	$> 10^{40}$
Half-life of double beta decay of ^{82}Se	3×10^{27}
Age of universe	5×10^{17}
Age of pyramid of Cheops	1×10^{11}
Human life expectancy (U.S.A.)	2×10^{9}
Time of Earth's orbit around the Sun (1 year)	3×10^{7}
Time of Earth's rotation about its axis (1 day)	9×10^{4}
Period of typical low-orbit Earth satellite	5×10^{3}
Time between normal heartbeats	8×10^{-1}
Period of concert-A tuning fork	2×10^{-3}
Period of oscillation of 3-cm microwaves	1×10^{-10}
Typical period of rotation of a molecule	1×10^{-12}
Shortest light pulse produced (1990)	6×10^{-15}
Lifetime of least stable particles	$< 10^{-23}$

[a] Approximate values.

nomena in nature the rotation of the Earth on its axis, which determines the length of the day, was used as a time standard for centuries. One (mean solar) second was defined to be 1/86,400 of a (mean solar) day.

Quartz crystal clocks based on the electrically sustained periodic vibrations of a quartz crystal serve well as secondary time standards. A quartz clock can be calibrated against the rotating Earth by astronomical observations and used to measure time in the laboratory. The best of these have kept time for a year with a maximum accumulated error of 5 μs, but even this precision is not sufficient for modern science and technology.

To meet the need for a better time standard, atomic clocks have been developed in several countries. Figure 1 shows such a clock, based on a characteristic frequency of the microwave radiation emitted by atoms of the element cesium. This clock, maintained at the National Institute of Standards and Technology, forms the basis in this country for Coordinated Universal Time (UTC), for which time signals are available by shortwave radio (stations WWV and WWVH) and by telephone.

Figure 2 shows, by comparison with a cesium clock, variations in the rate of rotation of the Earth over a 4-year

Figure 1   Cesium atomic frequency standard No. NBS-6 at the National Institute of Standards and Technology in Boulder, Colorado. This is the primary standard for the unit of time in the United States. Dial (303) 499-7111 to calibrate your watch against the standard. Dial (900) 410-8463 for Naval Observatory time signals.

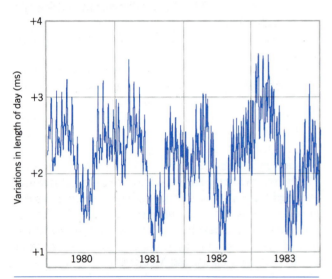

Figure 2   The variation in the length of the day over a 4-year period. Note that the vertical scale is only 3 ms = 0.003 s. See "The Earth's Rotation Rate," by John Wahr, *American Scientist,* January–February 1985, p. 41.

period. These data show what a poor time standard the Earth's rotation provides for precise work. The variations that we see in Fig. 2 can be ascribed to tidal effects caused by the Moon and seasonal variations in the atmospheric winds.

The second based on the cesium clock was adopted as the international standard by the 13th General Conference on Weights and Measures in 1967. The following definition was given:

One second is the time occupied by 9,192,631,770 vibrations of the radiation (of a specified wavelength) emitted by a cesium atom.

Two modern cesium clocks could run for 300,000 years before their readings would differ by more than 1 s. Hydrogen maser clocks have achieved the incredible precision of 1 s in 30,000,000 years. Clocks based on a single trapped atom may be able to improve on this precision by as much as 3 orders of magnitude. Figure 3 shows the impressive record of improvements in timekeeping that have occurred over the past 300 years or so, starting with the pendulum clock, invented by Christian Huygens in 1656, and ending with today's hydrogen maser.

1-4   THE STANDARD OF LENGTH*

The first international standard of length was a bar of a platinum–iridium alloy called the standard meter, which was kept at the International Bureau of Weights and Measures near Paris. The distance between two fine lines engraved near the ends of the bar, when the bar was held at a temperature of 0°C and supported mechanically in a prescribed way, was defined to be one meter. Historically, the meter was intended to be one ten-millionth of the distance from the north pole to the equator along the meridian line through Paris. However, accurate measure-

---

* See "The New Definition of the Meter," by P. Giacomo, *American Journal of Physics,* July 1984, p. 607.

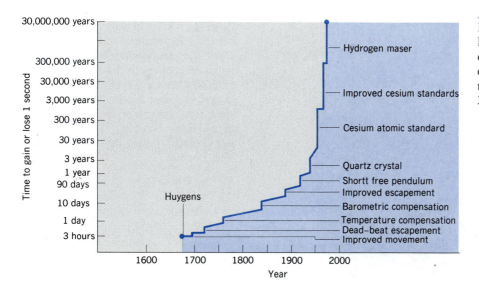

Figure 3 The improvement in time-keeping over the centuries. Early pendulum clocks gained or lost a second every few hours; present hydrogen maser clocks would do so only after 30,000,000 years.

ments showed that the standard meter bar differs slightly (about 0.023%) from this value.

Because the standard meter is not very accessible, accurate master copies of it were made and sent to standardizing laboratories throughout the world. These secondary standards were used to calibrate other, still more accessible, measuring rods. Thus, until recently, every measuring rod or device derived its authority from the standard meter through a complicated chain of comparisons using microscopes and dividing engines. Since 1959 this statement had also been true for the yard, whose legal definition in the United States was adopted in that year to be

$$1 \text{ yard} = 0.9144 \text{ meter} \quad \text{(exactly)}$$

which is equivalent to

$$1 \text{ inch} = 2.54 \text{ centimeters} \quad \text{(exactly)}.$$

The accuracy with which the necessary intercomparisons of length can be made by the technique of comparing fine scratches using a microscope is no longer satisfactory for modern science and technology. A more precise and reproducible standard of length was obtained when the American physicist Albert A. Michelson in 1893 compared the length of the standard meter with the wavelength of the red light emitted by atoms of cadmium. Michelson carefully measured the length of the meter bar and found that the standard meter was equal to 1,553,163.5 of those wavelengths. Identical cadmium lamps could easily be obtained in any laboratory, and thus Michelson found a way for scientists around the world to have a precise standard of length without relying on the standard meter bar.

Despite this technological advance, the metal bar remained the official standard until 1960, when the 11th General Conference on Weights and Measures adopted an atomic standard for the meter. The wavelength in vacuum of a certain orange-red light emitted by atoms of a particular isotope of krypton*, ^{86}Kr, in electrical discharge was chosen (see Fig. 4). Specifically, one meter was defined to be 1,650,763.73 wavelengths of this light. With the ability to make length measurements to a fraction of a wavelength, scientists could use this new standard to make comparisons of lengths to a precision below 1 part in 10^9.

The choice of an atomic standard offers advantages other than increased precision in length measurements. The ^{86}Kr atoms are available everywhere, are identical, and emit light of the same wavelength. The particular wavelength chosen is uniquely characteristic of ^{86}Kr and is sharply defined. The isotope can readily be obtained in pure form.

By 1983, the demands for higher precision had reached such a point that even the ^{86}Kr standard could not meet them and in that year a bold step was taken. The meter was redefined as the distance traveled by a light wave in a specified time interval. In the words of the 17th General Conference on Weights and Measures:

The meter is the length of the path traveled by light in vacuum during a time interval of 1/299,792,458 of a second.

This is equivalent to saying that the speed of light c is now *defined* as

$$c = 299,792,458 \text{ m/s} \quad \text{(exactly)}.$$

* The superscript 86 in ^{86}Kr gives the *mass number* (the number of protons plus neutrons in the nucleus) of this isotope of krypton. Naturally occurring krypton gas contains isotopes with mass numbers 78, 80, 82, 83, 84, and 86. The wavelength of the chosen radiation will differ in these different isotopes by about 1 part in 10^5, which is unacceptably large compared with the precision of the standard, about 1 part in 10^9. In the case of the cesium clock, there is only one naturally occurring isotope of cesium, which has mass number 133.

Figure 4   A krypton lamp at the National Physical Laboratories, Teddington, England. The glass capillary in the apparatus on the left contains the ^{86}Kr gas, which emits light when excited by an electric current. The lamp is inserted in the cryostat at right, where it is kept at the temperature of liquid nitrogen ($-210°$C). The light is viewed through the small porthole in the cryostat.

This new definition of the meter was necessary because measurements of the speed of light had become so precise that the reproducibility of the ^{86}Kr meter itself became the limiting factor. In view of this, it then made sense to adopt the speed of light as a defined quantity and to use it along with the precisely defined standard of time (the second) to redefine the meter.

Table 4 shows the range of measured lengths that can be compared with the standard.

TABLE 4   SOME MEASURED LENGTHS[a]

Length	Meters
Distance to farthest observed quasar	2×10^{26}
Distance to the Andromeda galaxy	2×10^{22}
Radius of our galaxy	6×10^{19}
Distance to the nearest star (Proxima Centauri)	4×10^{16}
Mean orbit radius for most distant planet (Pluto)	6×10^{12}
Radius of the Sun	7×10^{8}
Radius of the Earth	6×10^{6}
Height of Mt. Everest	9×10^{3}
Height of a typical person	2×10^{0}
Thickness of a page in this book	1×10^{-4}
Size of a typical virus	1×10^{-6}
Radius of a hydrogen atom	5×10^{-11}
Effective radius of a proton	1×10^{-15}

[a] Approximate values.

Sample Problem 2   A light-year is a measure of length (not a measure of time) equal to the distance that light travels in 1 year. Compute the conversion factor between light-years and meters, and find the distance to the star Proxima Centauri (4.0×10^{16} m) in light-years.

Solution   The conversion factor from years to seconds is

$$1 \text{ y} = 1 \text{ y} \times \frac{365.25 \text{ d}}{1 \text{ y}} \times \frac{24 \text{ h}}{1 \text{ d}} \times \frac{60 \text{ min}}{1 \text{ h}} \times \frac{60 \text{ s}}{1 \text{ min}}$$
$$= 3.16 \times 10^7 \text{ s} .$$

The speed of light is, to three significant figures, 3.00×10^8 m/s. Thus in 1 year, light travels a distance of

$$(3.00 \times 10^8 \text{ m/s}) (3.16 \times 10^7 \text{ s}) = 9.48 \times 10^{15} \text{ m},$$

so that

$$1 \text{ light-year} = 9.48 \times 10^{15} \text{ m}.$$

The distance to Proxima Centauri is

$$(4.0 \times 10^{16} \text{ m}) \times \frac{1 \text{ light-year}}{9.48 \times 10^{15} \text{ m}} = 4.2 \text{ light-years}.$$

Light from Proxima Centauri thus takes about 4.2 years to travel to Earth.

1-5 THE STANDARD OF MASS

The SI standard of mass is a platinum–iridium cylinder kept at the International Bureau of Weights and Measures and assigned, by international agreement, a mass of 1 kilogram. Secondary standards are sent to standardizing laboratories in other countries and the masses of other bodies can be found by an equal-arm balance technique to a precision of 1 part in 10^8.

The U.S. copy of the international standard of mass, known as Prototype Kilogram No. 20, is housed in a vault at the National Institute of Standards and Technology (see Fig. 5). It is removed no more than once a year for checking the values of tertiary standards. Since 1889 Prototype No. 20 has been taken to France twice for recomparison with the master kilogram. When it is removed from the vault two people are always present, one to carry the kilogram in a pair of forceps, the second to catch the kilogram if the first person should fall.

Table 5 shows some measured masses. Note that they vary by a factor of about 10^{83}. Most masses have been measured in terms of the standard kilogram by indirect methods. For example, we can measure the mass of the Earth (see Section 16-3) by measuring in the laboratory the gravitational force of attraction between two lead spheres and comparing it with the attraction of the Earth for a known mass. The masses of the spheres must be known by direct comparison with the standard kilogram.

On the atomic scale we have a second standard of mass, which is not an SI unit. It is the mass of the ^{12}C atom which, by international agreement, has been assigned an atomic mass of 12 unified atomic mass units (abbreviation u), exactly and by definition. We can find the masses of other atoms to considerable accuracy by using a mass spectrometer (Fig. 6; see also Section 34-2). Table 6 shows some selected atomic masses, including the estimated uncertainties of measurement. We need a second standard of mass because present laboratory techniques permit us to compare atomic masses with each other to greater precision than we can presently compare them with the standard kilogram. However, development of an atomic mass standard to replace the standard kilogram is well under way. The relationship between the present atomic standard and the primary standard is approximately

$$1 \text{ u} = 1.661 \times 10^{-27} \text{ kg}.$$

A related SI unit is the *mole,* which measures the quantity of a substance. One mole of ^{12}C atoms has a mass of exactly 12 grams and contains a number of atoms numerically equal to the Avogadro constant N_A:

$$N_A = 6.0221367 \times 10^{23} \text{ per mole.}$$

This is an experimentally determined number, with an uncertainty of about one part in a million. One mole of any other substance contains the same number of elementary entities (atoms, molecules, or whatever). Thus 1 mole of helium gas contains N_A atoms of He, 1 mole of oxygen contains N_A molecules of O_2, and 1 mole of water contains N_A molecules of H_2O.

To relate an atomic unit of mass to a bulk unit, it is necessary to use the Avogadro constant. Replacing the standard kilogram with an atomic standard will require an improvement of at least two orders of magnitude in the precision of the measured value of N_A to obtain masses with precisions of 1 part in 10^8.

Figure 5 The National Standard Prototype Kilogram No. 20, resting in its double bell jar at the U.S. National Institute of Standards and Technology.

TABLE 5 SOME MEASURED MASSES[a]

Object	Kilograms
Known universe (estimate)	10^{53}
Our galaxy	2×10^{43}
Sun	2×10^{30}
Earth	6×10^{24}
Moon	7×10^{22}
Ocean liner	7×10^{7}
Elephant	4×10^{3}
Person	6×10^{1}
Grape	3×10^{-3}
Speck of dust	7×10^{-10}
Virus	1×10^{-15}
Penicillin molecule	5×10^{-17}
Uranium atom	4×10^{-26}
Proton	2×10^{-27}
Electron	9×10^{-31}

[a] Approximate values.

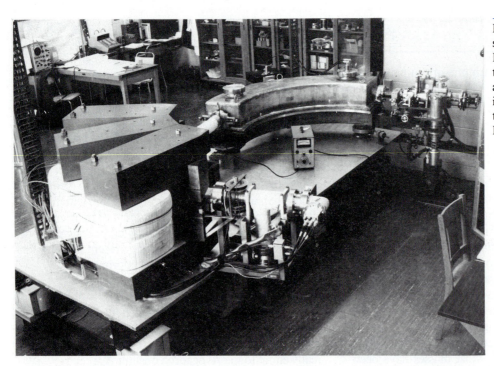

Figure 6 A high-resolution mass spectrometer at the University of Manitoba. Instruments of this type are used to obtain precise atomic masses such as those listed in Table 6. Work in this laboratory is supported by the National Research Council of Canada.

TABLE 6 SOME MEASURED ATOMIC MASSES

Isotope	Mass (u)	Uncertainty (u)
^{1}H	1.00782504	0.00000001
^{12}C	12.00000000	(exact)
^{64}Cu	63.9297656	0.0000017
^{102}Ag	101.91195	0.00012
^{137}Cs	136.907073	0.000006
^{190}Pt	189.959917	0.000007
^{238}Pu	238.0495546	0.0000024

1-6 PRECISION AND SIGNIFICANT FIGURES

As we improve the quality of our measuring instruments and the sophistication of our techniques, we can carry out experiments at ever increasing levels of precision; that is, we can extend the measured results to more and more *significant figures* and correspondingly reduce the *experimental uncertainty* of the result. Both the number of significant figures and the uncertainty tell something about our estimate of the precision of the result. That is, the result $x = 3$ m implies that we know less about x than the value $x = 3.14159$ m. When we declare $x = 3$ m, we mean that we are reasonably certain that x lies between 2 m and 4 m, while expressing x as 3.14159 m means that x probably lies between 3.14158 m and 3.14160 m. If you express x as 3 m when in fact you really believe that x is 3.14159 m, you are withholding information that might be important. On the other hand, if you express $x = 3.14159$ m when you really have no basis for know-

ing anything other than $x = 3$ m, you are being somewhat dishonest by claiming to have more information than you really do. Attention to significant figures is important when presenting the results of measurements and calculations, and it is equally as wrong to include too many as too few.

There are a few simple rules to follow in deciding how many significant figures to keep:

Rule 1 Counting from the left and ignoring leading zeros, keep all digits up to the first doubtful one. That is, $x = 3$ m has only one significant figure, and expressing this value as $x = 0.003$ km does not change the number of significant figures. If we instead wrote $x = 3.0$ m (or, equivalently, $x = 0.0030$ km), we would imply that we know the value of x to two significant figures. In particular, don't write down all 9 or 10 digits of your calculator display if they are not justified by the precision of the input data! Most calculations in this text are done with two or three significant figures.

Be careful about ambiguous notations: $x = 300$ m does not indicate whether there are one, two, or three significant figures; we don't know whether the zeros are carrying information or merely serving as place holders. Instead, we should write $x = 3 \times 10^2$ or 3.0×10^2 or 3.00×10^2 to specify the precision more clearly.

Rule 2 When multiplying or dividing, keep a number of significant figures in the product or quotient no greater than the number of significant figures in the least precise of the factors. Thus

$$2.3 \times 3.14159 = 7.2.$$

A bit of good judgment is occasionally necessary when applying this rule:

$$9.8 \times 1.03 = 10.1$$

because, even though 9.8 technically has only two significant figures, it is very close to being a number with three significant figures. The product should therefore be expressed with three significant figures.

Rule 3 In adding or subtracting, the least significant digit of the sum or difference occupies the same relative position as the least significant digit of the quantities being added or subtracted. In this case the *number* of significant figures is not important; it is the *position* that matters. For example, suppose we wish to find the total mass of three objects as follows:

$$
\begin{array}{ll}
103.\mathbf{9} & \text{kg} \\
2.\mathbf{1}0 & \text{kg} \\
0.31\mathbf{9} & \text{kg} \\
\hline
106.\mathbf{3}19 & \text{or} \quad 106.\mathbf{3} \text{ kg}
\end{array}
$$

The least significant or first doubtful digit is shown in **boldface**. By rule 1, we should include only one doubtful digit; thus the result should be expressed as 106.3 kg, for if the "3" is doubtful, then the following "19" gives no information and is useless.

Sample Problem 3 You wish to weigh your pet cat, but all you have available is an ordinary home platform scale. It is a digital scale, which displays your weight in a whole number of pounds. You therefore use the following scheme: you determine your own weight to be 119 pounds, and then holding the cat you find your combined weight to be 128 pounds. What is the fractional or percentage uncertainty in your weight and in the weight of your cat?

Solution The least significant digit is the units digit, and so your weight is uncertain by about one pound. That is, your scale would read 119 lb for any weight between 118.5 and 119.5 lb. The fractional uncertainty is therefore

$$\frac{1 \text{ lb}}{119 \text{ lb}} = 0.008 \quad \text{or} \quad 0.8\%.$$

The weight of the cat is 128 lb − 119 lb = 9 lb. However, the uncertainty in the cat's weight is still about 1 lb, and so the fractional uncertainty is

$$\frac{1 \text{ lb}}{9 \text{ lb}} = 0.11 = 11\%.$$

Although the *absolute* uncertainty in your weight and the cat's weight is the same (1 lb), the *relative* uncertainty in your weight is an order of magnitude smaller than the relative uncertainty in the cat's weight. If you tried to weigh a 1-lb kitten by this method, the relative uncertainty in its weight would be 100%. This illustrates a commonly occurring danger in the subtraction of two

numbers that are nearly equal: the relative or percentage uncertainty in the difference can be very large.

1-7 DIMENSIONAL ANALYSIS

Associated with every measured or calculated quantity is a *dimension*. For example, both the absorption of sound by an enclosure and the probability for nuclear reactions to occur have the dimensions of an area. The units in which the quantities are expressed do not affect the dimension of the quantities: an area is still an area whether it is expressed in m² or ft² or acres or sabins (sound absorption) or barns (nuclear reactions).

Just as we defined our measurement standards earlier in this chapter as fundamental quantities, we can choose a set of fundamental dimensions based on independent measurement standards. For mechanical quantities, mass, length, and time are elementary and independent, so they can serve as fundamental dimensions. They are represented respectively by M, L, and T.

Any equation must be *dimensionally consistent;* that is, the dimensions on both sides must be the same. Attention to dimensions can often keep you from making errors in writing equations. For example, the distance x covered in a time t by an object starting from rest and moving subject to a constant acceleration a will be shown in the next chapter to be $x = \frac{1}{2}at^2$. Acceleration is measured in units such as m/s². We use square brackets [ ] to denote "the dimension of," so that $[x] = \mathrm{L}$ or $[t] = \mathrm{T}$. It follows that $[a] = \mathrm{L/T^2}$ or $\mathrm{LT^{-2}}$. Keeping the units, and therefore the dimension, of acceleration in mind, you will therefore never be tempted to write $x = \frac{1}{2}at$ or $x = \frac{1}{2}at^3$.

The analysis of dimensions can often help in working out equations. The following two sample problems illustrate this procedure.

Sample Problem 4 To keep an object moving in a circle at constant speed requires a force called the "centripetal force." (Circular motion is discussed in Chapter 4.) Do a dimensional analysis of the centripetal force.

Solution We begin by asking "On which mechanical variables could the centripetal force F depend?" The moving object has only three properties that are likely to be important: its mass m, its speed v, and the radius r of its circular path. The centripetal force F must therefore be given, apart from any dimensionless constants, by an equation of the form

$$F \propto m^a v^b r^c$$

where the symbol $\propto$ means "is proportional to," and where a, b, and c are numerical exponents to be determined from analyzing the dimensions. As we wrote in Section 1-2 (and as we shall discuss in Chapter 5), force has units of kg·m/s², and therefore its dimensions are $[F] = \text{MLT}^{-2}$. We can therefore write the centripetal force equation in terms of dimensions as

$$[F] = [m^a]\, [v^b]\, [r^c]$$

$$\text{MLT}^{-2} = \text{M}^a\, (\text{L/T})^b \text{L}^c$$

$$= \text{M}^a \text{L}^{b+c} \text{T}^{-b}.$$

Dimensional consistency means that the fundamental dimensions must be the same on each side. Thus, equating the exponents,

exponents of M: $a = 1$;

exponents of T: $b = 2$;

exponents of L: $b + c = 1$ so $c = -1$.

The resulting expression is

$$F \propto \frac{mv^2}{r}.$$

The actual expression for centripetal force, derived from Newton's laws and the geometry of circular motion, is $F = mv^2/r$. The dimensional analysis gives us the exact dependence on the mechanical variables! This is really a happy accident, because dimensional analysis can't tell us anything about constants that do not have dimensions. In this case the constant happens to be 1.

Sample Problem 5   An important milestone in the evolution of the universe just after the Big Bang is the Planck time t_P, the value of which depends on three fundamental constants: (1) the speed of light (the fundamental constant of relativity), $c = 3.00 \times 10^8$ m/s; (2) Newton's gravitational constant (the fundamental constant of gravity), $G = 6.67 \times 10^{-11}$ m³/s²·kg; and (3) Planck's constant (the fundamental constant of quantum mechanics), $h = 6.63 \times 10^{-34}$ kg·m²/s. Based on a dimensional analysis, find the value of the Planck time.

Solution   Using the units given for the three constants, we can obtain their dimensions:

$$[c] = [\text{m/s}] = \text{LT}^{-1}$$

$$[G] = [\text{m}^3/\text{s}^2 \cdot \text{kg}] = \text{L}^3\text{T}^{-2}\text{M}^{-1}$$

$$[h] = [\text{kg} \cdot \text{m}^2/\text{s}] = \text{ML}^2\text{T}^{-1}$$

Let the Planck time depend on these constants as

$$t_P \propto c^i G^j h^k,$$

where i, j, and k are exponents to be determined. The dimensions of this expression are

$$[t_P] = [c^i]\, [G^j]\, [h^k]$$

$$T = (\text{LT}^{-1})^i\, (\text{L}^3\text{T}^{-2}\text{M}^{-1})^j\, (\text{ML}^2\text{T}^{-1})^k$$

$$= \text{L}^{i+3j+2k}\text{T}^{-i-2j-k}\text{M}^{-j+k}.$$

Equating powers on both sides gives

exponents of L: $0 = i + 3j + 2k$

exponents of T: $1 = -i - 2j - k$

exponents of M: $0 = -j + k$

and solving these three equations for the three unknowns, we find

$$i = -\tfrac{5}{2}, \qquad j = \tfrac{1}{2}, \qquad k = \tfrac{1}{2}.$$

Thus

$$t_P \propto c^{-5/2} G^{1/2} h^{1/2}$$

$$= \sqrt{\frac{Gh}{c^5}} = \sqrt{\frac{(6.67 \times 10^{-11}\ \text{m}^3/\text{s}^2 \cdot \text{kg})(6.63 \times 10^{-34}\ \text{kg} \cdot \text{m}^2/\text{s})}{(3.00 \times 10^8\ \text{m/s})^5}}$$

$$= 1.35 \times 10^{-43}\ \text{s}.$$

As commonly defined, the Planck time differs from this value by a factor of $(2\pi)^{-1/2}$. Such dimensionless factors cannot be found by this technique.

In similar fashion, we can determine the Planck length and the Planck mass, which also have very fundamental interpretations (see Problems 41 and 42).

QUESTIONS

1. How would you criticize this statement: "Once you have picked a standard, by the very meaning of 'standard' it is invariable"?

2. List characteristics other than accessibility and invariability that you would consider desirable for a physical standard.

3. Can you imagine a system of base units (Table 1) in which time was not included?

4. Of the seven base units listed in Table 1, only one — the kilogram — has a prefix (see Table 2). Would it be wise to redefine the mass of that platinum–iridium cylinder at the International Bureau of Weights and Measures as 1 g rather than 1 kg?

5. What does the prefix "micro-" signify in the words "microwave oven"? It has been proposed that food that has been irradiated by gamma rays to lengthen its shelf life be marked "picowaved." What do you suppose that means?

6. Many capable investigators, on the evidence, believe in the reality of extrasensory perception. Assuming that ESP is

indeed a fact of nature, what physical quantity or quantities would you seek to define to describe this phenomenon quantitatively?

7. According to a point of view adopted by some physicists and philosophers, if we cannot describe procedures for determining a physical quantity, we say that the quantity is undetectable and should be given up as having no physical reality. Not all scientists accept this view. What in your opinion are the merits and drawbacks of this point of view?

8. Name several repetitive phenomena occurring in nature that could serve as reasonable time standards.

9. You could define "1 second" to be one pulse beat of the current president of the American Association of Physics Teachers. Galileo used his pulse as a timing device in some of his work. Why is a definition based on the atomic clock better?

10. What criteria should be satisfied by a good clock?

11. From what you know about pendulums, cite the drawbacks to using the period of a pendulum as a time standard.

12. On June 30, 1981, the minute extending from 10:59 to 11:00 a.m. was arbitrarily lengthened to contain 61 s. The last day of 1989 also was lengthened by 1 s. Such a *leap second* is occasionally introduced to compensate for the fact that, as measured by our atomic time standard, the Earth's rotation rate is slowly decreasing. Why is it desirable to readjust our clocks in this way?

13. A radio station advertises "at 89.5 on your FM dial." What does this number mean?

14. Why are there no SI base units for area or volume?

15. The meter was originally intended to be one ten-millionth of the meridian line from the north pole to the equator that passes through Paris. This definition disagrees with the standard meter bar by 0.023%. Does this mean that the standard meter bar is inaccurate to this extent?

16. Can length be measured along a curved line? If so, how?

17. When the meter bar was taken to be the standard of length,

its temperature was specified. Can length be called a fundamental property if another physical quantity, such as temperature, must be specified in choosing a standard?

18. In redefining the meter in terms of the speed of light, why did the delegates to the 1983 General Conference on Weights and Measures not simplify matters by defining the speed of light to be 3×10^8 m/s exactly? For that matter, why did they not define it to be 1 m/s exactly? Were both of these possibilities open to them? If so, why did they reject them?

19. Suggest a way to measure (a) the radius of the Earth, (b) the distance between the Sun and the Earth, and (c) the radius of the Sun.

20. Suggest a way to measure (a) the thickness of a sheet of paper, (b) the thickness of a soap bubble film, and (c) the diameter of an atom.

21. If someone told you that every dimension of every object had shrunk to half its former value overnight, how could you refute this statement?

22. Is the current standard kilogram of mass accessible, invariable, reproducible, and indestructible? Does it have simplicity for comparison purposes? Would an atomic standard be better in any respect? Why don't we adopt an atomic standard, as we do for length and time?

23. Why do we find it useful to have two standards of mass, the kilogram and the ^{12}C atom?

24. How does one obtain the relation between the masses of the standard kilogram and the mass of the ^{12}C atom?

25. Suggest practical ways by which one could determine the masses of the various objects listed in Table 5.

26. Suggest objects whose masses would fall in the wide range in Table 5 between that of an ocean liner and the Moon and estimate their masses.

27. Critics of the metric system often cloud the issue by saying things such as: "Instead of buying 1 lb of butter you will have to ask for 0.454 kg of butter." The implication is that life would be more complicated. How might you refute this?

PROBLEMS

Section 1-2 The International System of Units

1. Use the prefixes in Table 2 and express (a) 10^6 phones; (b) 10^{-6} phones; (c) 10^1 cards; (d) 10^9 lows; (e) 10^{12} bulls; (f) 10^{-1} mates; (g) 10^{-2} pedes; (h) 10^{-9} Nannettes; (i) 10^{-12} boos; (j) 10^{-18} boys; (k) 2×10^2 withits; (l) 2×10^3 mockingbirds. Now that you have the idea, invent a few more similar expressions. (See p. 61 of *A Random Walk in Science,* compiled by R. L. Weber; Crane, Russak & Co., New York, 1974.)

2. Some of the prefixes of the SI units have crept into everyday language. (a) What is the weekly equivalent of an annual salary of 36K (= 36 k$)? (b) A lottery awards 10 megabucks as the top prize, payable over 20 years. How much is re-

ceived in each monthly check? (c) The hard disk of a computer has a capacity of 30 MB (= 30 megabytes). At 8 bytes/word, how many words can it store? In computerese, *kilo* means 1024 (= 2^{10}), not 1000.

Section 1-3 The Standard of Time

3. Enrico Fermi once pointed out that a standard lecture period (50 min) is close to 1 microcentury. How long is a microcentury in minutes, and what is the percent difference from Fermi's approximation?

4. New York and Los Angeles are about 3000 mi apart; the time difference between these two cities is 3 h. Calculate the circumference of the Earth.

5. A convenient substitution for the number of seconds in a year is $\pi \times 10^7$. To within what percentage error is this correct?

6. Shortly after the French Revolution, as part of their introduction of the metric system, the revolutionary National Convention made an attempt to introduce decimal time. In this plan, which was not successful, the day—starting at midnight—was divided into 10 decimal hours consisting of 100 decimal minutes each. The hands of a surviving decimal pocket watch are stopped at 8 decimal hours, 22.8 decimal minutes. What time is it? See Fig. 7.

Figure 7   Problem 6.

7. (*a*) A unit of time sometimes used in microscopic physics is the *shake*. One shake equals 10^{-8} s. Are there more shakes in a second than there are seconds in a year? (*b*) Humans have existed for about 10^6 years, whereas the universe is about 10^{10} years old. If the age of the universe is taken to be 1 day, for how many seconds have humans existed?

8. In two *different* track meets, the winners of the mile race ran their races in 3 min 58.05 s and 3 min 58.20 s. In order to conclude that the runner with the shorter time was indeed faster, what is the maximum tolerable error, in feet, in laying out the distances?

9. A certain pendulum clock (with a 12-h dial) happens to gain 1 min/day. After setting the clock to the correct time, how long must one wait until it again indicates the correct time?

10. Five clocks are being tested in a laboratory. Exactly at noon, as determined by the WWV time signal, on the successive days of a week the clocks read as follows:

How would you arrange these five clocks in the order of their relative value as good timekeepers? Justify your choice.

11. The age of the universe is about 5×10^{17} s; the shortest light pulse produced in a laboratory (1990) lasted for only 6×10^{-15} s (see Table 3). Identify a physically meaningful time interval approximately halfway between these two on a logarithmic scale.

12. Assuming that the length of the day uniformly increases by 0.001 s in a century, calculate the cumulative effect on the measure of time over 20 centuries. Such a slowing down of the Earth's rotation is indicated by observations of the occurrences of solar eclipses during this period.

13. The time it takes the Moon to return to a given position as seen against the background of fixed stars, 27.3 days, is called a *sidereal* month. The time interval between identical phases of the Moon is called a *lunar* month. The lunar month is longer than a sidereal month. Why and by how much?

Section 1-4 The Standard of Length

14. Your French pen pal Pierre writes to say that he is 1.9 m tall. What is his height in British units?

15. (*a*) In track meets both 100 yards and 100 meters are used as distances for dashes. Which is longer? By how many meters is it longer? By how many feet? (*b*) Track and field records are kept for the mile and the so-called metric mile (1500 meters). Compare these distances.

16. The stability of the cesium clock used as an atomic time standard is such that two cesium clocks would gain or lose 1 s with respect to each other in about 300,000 y. If this same precision were applied to the distance between New York and San Francisco (2572 mi), by how much would successive measurements of this distance tend to differ?

17. Antarctica is roughly semicircular in shape with a radius of 2000 km. The average thickness of the ice cover is 3000 m. How many cubic centimeters of ice does Antarctica contain? (Ignore the curvature of the Earth.)

18. A unit of area, often used in expressing areas of land, is the *hectare*, defined as 10^4 m^2. An open-pit coal mine consumes 77 hectares of land, down to a depth of 26 m, each year. What volume of earth, in cubic kilometers, is removed in this time?

19. The Earth is approximately a sphere of radius 6.37×10^6 m. (*a*) What is its circumference in kilometers? (*b*) What is its surface area in square kilometers? (*c*) What is its volume in cubic kilometers?

20. The approximate maximum speeds of various animals follows, but in different units of speed. Convert these data to m/s, and thereby arrange the animals in order of increasing maximum speed: squirrel, 19 km/h; rabbit, 30 knots; snail,

Clock	Sun.	Mon.	Tues.	Wed.	Thurs.	Fri.	Sat.
A	12:36:40	12:36:56	12:37:12	12:37:27	12:37:44	12:37:59	12:38:14
B	11:59:59	12:00:02	11:59:57	12:00:07	12:00:02	11:59:56	12:00:03
C	15:50:45	15:51:43	15:52:41	15:53:39	15:54:37	15:55:35	15:56:33
D	12:03:59	12:02:52	12:01:45	12:00:38	11:59:31	11:58:24	11:57:17
E	12:03:59	12:02:49	12:01:54	12:01:52	12:01:32	12:01:22	12:01:12

0.030 mi/h; spider, 1.8 ft/s; cheetah, 1.9 km/min; human, 1000 cm/s; fox, 1100 m/min; lion, 1900 km/day.

21. A certain spaceship has a speed of 19,200 mi/h. What is its speed in light-years per century?

22. A new car is equipped with a "real-time" dashboard display of fuel consumption. A switch permits the driver to toggle back and forth between British units and SI units. However, the British display shows mi/gal while the SI version is the inverse, L/km. What SI reading corresponds to 30.0 mi/gal?

23. Astronomical distances are so large compared to terrestrial ones that much larger units of length are used for easy comprehension of the relative distances of astronomical objects. An *astronomical unit* (AU) is equal to the average distance from Earth to the Sun, 1.50×10^8 km. A *parsec* (pc) is the distance at which 1 AU would subtend an angle of 1 second of arc. A *light-year* (ly) is the distance that light, traveling through a vacuum with a speed of 3.00×10^5 km/s, would cover in 1 year. (a) Express the distance from Earth to the Sun in parsecs and in light-years. (b) Express a light-year and a parsec in kilometers. Although the light-year is much used in popular writing, the parsec is the unit used professionally by astronomers.

24. The effective radius of a proton is about 1×10^{-15} m; the radius of the observable universe (given by the distance to the farthest observable quasar) is 2×10^{26} m (see Table 4). Identify a physically meaningful distance that is approximately halfway between these two extremes on a logarithmic scale.

25. The average distance of the Sun from Earth is 390 times the average distance of the Moon from Earth. Now consider a total eclipse of the Sun (Moon between Earth and Sun; see Fig. 8) and calculate (a) the ratio of the Sun's diameter to the Moon's diameter, and (b) the ratio of the Sun's volume to the Moon's volume. (c) The angle intercepted at the eye by the Moon is 0.52° and the distance between Earth and the Moon is 3.82×10^5 km. Calculate the diameter of the Moon.

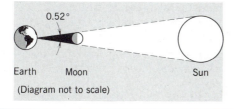

Figure 8 Problem 25.

26. The navigator of the oil tanker *Exxon Valdez* uses the satellites of the Global Positioning System (GPS/NAVSTAR) to find latitude and longitude; see Fig. 9. These are 43°36′25.3″ N and 77°31′48.2″ W. If the accuracy of these determinations is ±0.5″, what is the uncertainty in the tanker's position measured along (a) a north–south line (meridian of longitude) and (b) an east–west line (parallel of latitude)? (c) Where is the tanker?

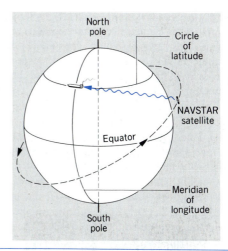

Figure 9 Problem 26.

Section 1-5 The Standard of Mass

27. Using conversions and data in the chapter, determine the number of hydrogen atoms required to obtain 1.00 kg of hydrogen.

28. One molecule of water (H_2O) contains two atoms of hydrogen and one atom of oxygen. A hydrogen atom has a mass of 1.0 u and an atom of oxygen has a mass of 16 u. (a) What is the mass in kilograms of one molecule of water? (b) How many molecules of water are in the oceans of the world? The oceans have a total mass of 1.4×10^{21} kg.

29. In continental Europe, one "pound" is half a kilogram. Which is the better buy: one Paris pound of coffee for $3.00 or one New York pound of coffee for $2.40?

30. A room has dimensions of 21 ft × 13 ft × 12 ft. What is the mass of the air it contains? The density of air at room temperature and normal atmospheric pressure is 1.21 kg/m³.

31. A typical sugar cube has an edge length of 1 cm. If you had a cubical box that contained 1 mole of sugar cubes, what would its edge length be?

32. A person on a diet loses 2.3 kg (corresponding to about 5 lb) per week. Express the mass loss rate in milligrams per second.

33. Suppose that it takes 12 h to drain a container of 5700 m³ of water. What is the mass flow rate (in kg/s) of water from the container? The density of water is 1000 kg/m³.

34. The grains of fine California beach sand have an average radius of 50 μm. What mass of sand grains would have a total surface area equal to the surface area of a cube exactly 1 m on an edge? Sand is made of silicon dioxide, 1 m³ of which has a mass of 2600 kg.

35. The standard kilogram (see Fig. 5) is in the shape of a circular cylinder with its height equal to its diameter. Show that, for a circular cylinder of fixed volume, this equality gives the smallest surface area, thus minimizing the effects of surface contamination and wear.

36. The distance between neighboring atoms, or molecules, in a solid substance can be estimated by calculating twice the

radius of a sphere with volume equal to the volume per atom of the material. Calculate the distance between neighboring atoms in (*a*) iron and (*b*) sodium. The densities of iron and sodium are 7870 kg/m³ and 1013 kg/m³, respectively; the mass of an iron atom is 9.27×10^{-26} kg, and the mass of a sodium atom is 3.82×10^{-26} kg.

Section 1-6 Precision and Significant Figures

37. For the period 1960–1983, the meter was defined to be 1,650,763.73 wavelengths of a certain orange-red light emitted by krypton atoms. Compute the distance in nanometers corresponding to one wavelength. Express your result using the proper number of significant figures.

38. (*a*) Evaluate $37.76 + 0.132$ to the correct number of significant figures. (*b*) Evaluate $16.264 - 16.26325$ to the correct number of significant figures.

39. (*a*) A rectangular metal plate has a length of 8.43 cm and a width of 5.12 cm. Calculate the area of the plate to the correct number of significant figures. (*b*) A circular metal plate has a radius of 3.7 cm. Calculate the area of the plate to the correct number of significant figures.

Section 1-7 Dimensional Analysis

40. Porous rock through which groundwater can move is called an aquifer. The volume V of water that, in time t, moves through a cross section of area A of the aquifer is given by

$$\frac{V}{t} = KA\frac{H}{L},$$

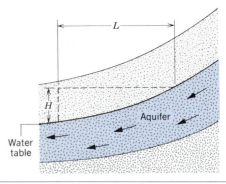

Figure 10   Problem 40.

where H is the vertical drop of the aquifer over the horizontal distance L; see Fig. 10. This relation is called Darcy's law. The quantity K is the hydraulic conductivity of the aquifer. What are the SI units of K?

41. In Sample Problem 5, the constants h, G, and c were combined to obtain a quantity with the dimensions of time. Repeat the derivation to obtain a quantity with the dimensions of length, and evaluate the result numerically. Ignore any dimensionless constants. This is the *Planck length,* the size of the observable universe at the Planck time.

42. Repeat the procedure of Problem 41 to obtain a quantity with the dimensions of mass. This gives the *Planck mass,* the mass of the observable universe at the Planck time.

CHAPTER 2

MOTION IN ONE DIMENSION

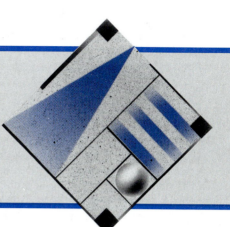

Mechanics, the oldest of the physical sciences, is the study of the motion of objects. The calculation of the path of a baseball or of a space probe sent to Mars is among its problems, as is the analysis of the tracks of elementary particles formed following collisions in our largest accelerators. When we describe motion, we are dealing with the part of mechanics called kinematics *(from the Greek word for motion, as also in cinema). When we analyze the* causes *of motion we are dealing with* dynamics *(from the Greek word for force, as in dynamite). In this chapter, we deal only with kinematics in one dimension. The next two chapters extend these results to two and three dimensions, and in Chapter 5 we begin the study of dynamics.*

2-1 PARTICLE KINEMATICS

To begin our study of kinematics, we choose a simple case: a particle moving in a straight line. We choose straight-line motion because it allows us to introduce some of the basic kinematic concepts, such as velocity or acceleration, without the mathematical complexity of vectors, which are often used to analyze two- and three-dimensional motion. Within this limitation, however, we can consider a broad range of physical situations: a falling stone, an accelerating train, a braking car, a sliding hockey puck, a crate being pulled up a ramp, a fast moving electron in an x-ray tube, and so on. The state of the motion may change (the hockey puck may be struck before it slides) and its direction may change (the stone may be thrown upward before falling downward), but the motion must be confined to a single line.

We also simplify this discussion by considering the motion only of a *particle*. That is, we treat a complex object as if it were a single mass point. This allows us to neglect all possible internal motions—for example, the rotary motion of an object (which we consider in Chapters 11–13) or the vibration of its parts (Chapter 15). For this discussion, all parts of the object move in exactly the same way. A rolling wheel does not satisfy this restriction, because a point on the rim moves in a different way than a point on the axle. (A *sliding* wheel, on the other hand, would qualify. Thus the wheel, along with other material objects, might be considered a particle for some calculations but not for others.) As long as we are concerned only with the kinematic variables, there is no reason not to consider both a speeding train and an electron on the same basis—as examples of the motion of a *particle*.

Within these limitations, we consider all possible kinds of motion. Particles can speed up, slow down, and even stop and reverse their motion. We seek a description of the motion that includes any of these possibilities.

2-2 DESCRIPTIONS OF MOTION

We describe the motion of a particle in two ways: with mathematical equations and with graphs. Either way is appropriate for the study of kinematics, and we begin by using both methods. The mathematical approach is usually better for solving problems, because it permits more precision than the graphical sketch. The graphical method is helpful because it often provides more physical insight than a set of mathematical equations.

A complete description of the motion of a particle can be obtained if we know the mathematical dependence of its position x (relative to a chosen origin of a particular frame of reference) on the time t for all times. This is just the function $x(t)$. Here are some possible kinds of motion along with the functions and graphs that describe them:

1. *No motion at all.* Here the particle occupies the position at the coordinate A at all times:

$$x(t) = A. \qquad (1)$$

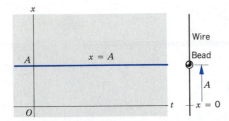

Figure 1 A bead is free to slide along a wire in one dimension; the direction of motion is arbitrary and not necessarily vertical. In this case the bead is at rest at the *x* coordinate *A*, and its "motion" is described by the horizontal straight line $x = A$.

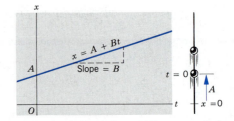

Figure 2 A bead sliding along a wire in one dimension moves with constant speed *B* in the positive *x* direction; it starts at time 0 at the *x* coordinate *A*. Its motion is described by the line $x = A + Bt$.

A graph of this "motion" is shown in Fig. 1. For the purposes of these illustrations, we imagine the particle described by the graph to be a bead sliding without friction on a long wire. In this case the bead is at rest at the location $x = A$. Notice that we plot the graph with *x* as the dependent variable (on the vertical axis) and *t* as the independent variable (on the horizontal axis).

2. *Motion at constant speed.* The rate of motion of a particle is described by its *velocity*. For motion in one dimension, the velocity can be either positive, if the particle is moving in the direction of increasing *x*, or negative, if it is moving in the opposite direction. Another measure of the rate of motion of a particle is its *speed*, which is simply the magnitude of the velocity of the particle. Speed is always positive and carries no directional information.

In the case of motion at constant speed, the graph plotting position against time is a straight line with a constant slope. From calculus, we learn that the *slope* of any function tells us about its *rate of change*. Here the rate of change of the position is the velocity, and the higher the slope of the graph, the greater the velocity. Mathematically, we have

$$x(t) = A + Bt, \qquad (2)$$

which is in the customary form of the expression for a straight line (more commonly expressed as $y = mx + b$) of slope *B*.

The graphical illustration of Fig. 2 shows the particle at the position $x = A$ at the time $t = 0$. It is moving at constant speed in the direction of increasing *x*. Its velocity is thus positive, as indicated by the positive slope.

3. *Accelerated motion.* In this case the speed is changing (acceleration being defined as the rate of change of velocity), and so the slope must change also. These graphs are therefore curves rather than straight lines. Two examples are:

$$x(t) = A + Bt + Ct^2, \qquad (3)$$

$$x(t) = A \cos \omega t. \qquad (4)$$

In the first case, assuming $C > 0$, the slope is continually increasing as the particle moves faster and faster (Fig. 3*a*). In the second case, the particle oscillates between $x = +A$

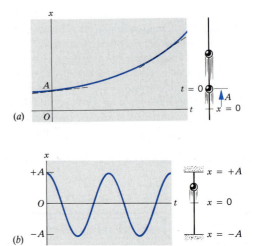

Figure 3 (*a*) A bead sliding along a wire in one dimension moves in the positive *x* direction with ever increasing speed. The speed is equal to the slope of the curve describing the particle's motion; you can see how the slope of the curve continually increases. (*b*) A bead sliding along a wire in one dimension oscillates between $x = +A$ and $x = -A$.

and $x = -A$ (Fig. 3*b*), its velocity changing from positive to negative as the slope of Fig. 3*b* changes sign.

Often the complete descriptions of motion are more complex than the simple illustrations we have shown so far. Here are some examples:

4. *Accelerating and braking car.* A car starts from rest and accelerates to a certain speed. It then moves for a time at a constant speed, after which the brakes are applied, bringing the car to rest again. Figure 4 shows the motion. No single mathematical equation describes the motion; we might use expressions of the form of Eq. 1 for the resting parts of the motion, an expression of the form of Eq. 3 for the accelerating part, one of the form of Eq. 2 for the part at constant speed, and finally another, also of the form of Eq. 3, for the braking part.

Notice that the graph has two features: $x(t)$ is continuous (there are no breaks in the graph) and the slope is

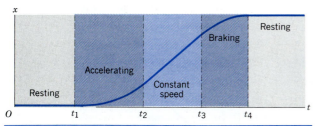

Figure 4   The curve describes a car that is at rest from $t = 0$ until $t = t_1$, at which time it begins accelerating. At $t = t_2$ it stops accelerating and begins moving at constant velocity. The brakes are applied at time $t = t_3$, and the velocity gradually decreases until it becomes 0 at time $t = t_4$.

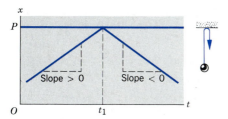

Figure 5   A hockey puck is moving across the ice at constant velocity when it collides with a rigid wall at $x = P$ at time t_1, after which it moves away from the wall with a velocity equal in magnitude but opposite in direction. The motion of the puck takes place in one dimension. For a real rebounding object, the sharp point in $x(t)$ would be slightly rounded.

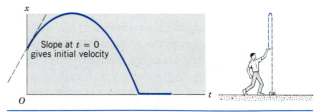

Figure 6   A ball of clay is thrown upward, rises to a certain height, and then falls to the ground. Upon striking the ground it comes to rest. The curve describes its motion. In reality, the sharp point in $x(t)$ would be slightly rounded.

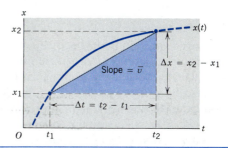

Figure 7   The average velocity in the interval Δt between t_1 and t_2 is determined by the displacement Δx during the interval; the actual shape of the $x(t)$ curve in the interval is of no consequence in determining the average velocity.

continuous (there are no sharp points). We expect that $x(t)$ must always be continuous; otherwise the car would disappear at one point and reappear at another. Sharp points in the graph, as we shall see later, mean that the speed changes *instantly* from one value to another. This is of course not a completely physical situation, but it is often a good approximation to one.

5. *Rebounding hockey puck.* A hockey puck slides across the ice at a constant speed, collides with the wall, and then rebounds in the opposite direction with the same speed. Figure 5 shows the motion, in which it is assumed that the collision instantly reverses the motion. In reality, if we examined the "point" very carefully, we would find that it is not sharp but ever so slightly rounded, resulting from the elasticity of the wall and the puck.

6. *Sticky ball of clay.* A student throws a ball of modeling clay upward; the point of release is above the student's head. The ball rises to a certain height, then falls and sticks to the floor. Figure 6 describes the motion. The slope at $t = 0$ represents the initial speed with which the clay is thrown upward. The velocity passes through zero at the top of the path (where the slope is zero), and then the clay moves downward with increasing speed. When it strikes the floor, it suddenly comes to rest and its speed becomes zero.

Remember that the graphs shown in this section are

representations of motion, not sketches of the actual paths followed by the particles. In Fig. 6, for instance, the particle moves upward and downward along the same line; it doesn't follow the curved path shown in the figure.

2-3 AVERAGE VELOCITY

If the motion of a particle is described by graphs like Figs. 1 or 2, we have no problem obtaining the velocity over any interval of time: it is constant and equal to the slope of the line. In more complicated cases, such as those of Figs. 3–6 in which the velocity changes, it is convenient to define the *mean velocity* or *average velocity* $\bar{v}$. (A bar over the symbol for *any* physical quantity indicates an average value of that quantity.)

Suppose, as indicated in Fig. 7, the particle is at point x_1 at time t_1 and then it moves to point x_2 at time t_2. The average velocity over the interval is defined to be

$$\bar{v} = \frac{x_2 - x_1}{t_2 - t_1} = \frac{\Delta x}{\Delta t}, \qquad (5)$$

where

$$\Delta x = x_2 - x_1 \qquad (6)$$

and

$$\Delta t = t_2 - t_1. \qquad (7)$$

Here Δx is the *displacement* (that is, the change in position) that occurs during the *time interval* Δt. From Fig. 7 you can see that $\bar{v}$ is simply the slope of the straight line that connects the endpoints of the interval.

The average velocity tells us about the average behavior during the time interval Δt. *The actual behavior between x_1 and x_2 is of no concern for the calculation of the average velocity.* Any details of the particular motion between x_1 and x_2 are lost when we take the average.

If we assume that our clocks are always running forward ($t_2 > t_1$), then the sign of $\bar{v}$ is determined by the sign of $\Delta x = x_2 - x_1$. If $\bar{v}$ is positive, then on the average the particle is moving so that x increases with time. (It may move backward somewhere in the interval, but it finishes with a larger x coordinate than it started with.) If $\bar{v}$ is negative, than on the average the particle moves backward. In particular, notice that according to this definition of $\bar{v}$, your average velocity is zero on any trip in which you return to your starting point, no matter how fast you may move in any particular segment, because your displacement is zero. Timing from the starting line to the finishing line, the average velocity of an Indianapolis 500 racer is zero!

Sample Problem 1   You drive your BMW down a straight road for 5.2 mi at 43 mi/h, at which point you run out of gas. You walk 1.2 mi farther, to the nearest gas station, in 27 min. What is your average velocity from the time that you started your car to the time that you arrived at the gas station?

Solution   You can find your average velocity from Eq. 5 if you know both Δx, the net distance that you covered (your displacement), and Δt, the corresponding elapsed time. These quantities are

$$\Delta x = 5.2 \text{ mi} + 1.2 \text{ mi} = 6.4 \text{ mi}$$

and

$$\Delta t = \frac{5.2 \text{ mi}}{43 \text{ mi/h}} + 27 \text{ min}$$

$$= 7.3 \text{ min} + 27 \text{ min} = 34 \text{ min} = 0.57 \text{ h}.$$

From Eq. 5 we then have

$$\bar{v} = \frac{\Delta x}{\Delta t} = \frac{6.4 \text{ mi}}{0.57 \text{ h}} = 11.2 \text{ mi/h}.$$

The $x(t)$ plot of Fig. 8 helps us to visualize the problem. Points O and P define the interval for which we want to find the average velocity, this quantity being the slope of the straight line connecting these points.

2-4   INSTANTANEOUS VELOCITY

Average velocity may be helpful in considering the overall behavior of a particle during some interval, but in describing the *details* of its motion the average velocity is

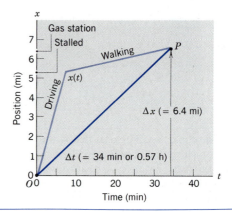

Figure 8   Sample Problem 1. The lines marked "Driving" and "Walking" show motions at different constant velocities for the two portions of the trip. The average velocity is the slope of the line OP.

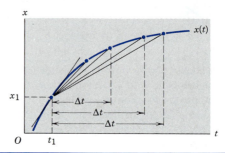

Figure 9   The interval Δt grows smaller, in this case as we keep t_1 fixed and move the other endpoint t_2 closer to t_1. In the limit, the interval goes to zero and the chord becomes a tangent.

not particularly useful. It would be more appropriate to obtain a mathematical function $v(t)$, which gives the velocity at every point in the motion. This is the *instantaneous velocity;* from now on, when we use the term "velocity" we understand it to mean instantaneous velocity.

Suppose we try to calculate the average velocity, as shown in Fig. 9, when the interval Δt gets smaller and smaller. In this limiting case of $\Delta t \to 0$, the line connecting the endpoints of the interval approaches the tangent to the $x(t)$ curve at a point, and the average velocity approaches the slope of $x(t)$, which defines the instantaneous velocity at that point:

$$v = \lim_{\Delta t \to 0} \frac{\Delta x}{\Delta t}. \qquad (8)$$

The right side of Eq. 8 is in the form of the *derivative* of $x(t)$ with respect to t, or dx/dt. Thus

$$v = \frac{dx}{dt}. \qquad (9)$$

The (instantaneous) velocity is just the rate of change of position with time.

TABLE 1 THE LIMITING PROCESS

Initial Point		Final Point		Intervals		Average Velocity
x_1 (m)	t_1 (s)	x_2 (m)	t_2 (s)	Δx (m)	Δt (s)	(m/s)
6.000	1.000	13.000	2.000	7.000	1.000	7.00
6.000	1.000	9.000	1.500	3.000	0.500	6.00
6.000	1.000	8.320	1.400	2.320	0.400	5.80
6.000	1.000	7.375	1.250	1.375	0.250	5.50
6.000	1.000	7.080	1.200	1.080	0.200	5.40
6.000	1.000	6.520	1.100	0.520	0.100	5.20
6.000	1.000	6.255	1.050	0.255	0.050	5.1
6.000	1.000	6.152	1.030	0.152	0.030	5.1
6.000	1.000	6.050	1.010	0.050	0.010	5.0

Table 1 gives an example of how the limiting process converges to the instantaneous value. The data in Table 1 were calculated using $x(t) = 3.000 + 1.000t + 2.000t^2$, with t in seconds and x in meters. We have chosen to keep the point (t_1, x_1) fixed and to move the point (t_2, x_2) gradually closer to (t_1, x_1) to simulate the limiting process. The limit seems to be approaching the value $v = 5.0$ m/s at $t_1 = 1.0$ s; differentiating the above expression for $x(t)$, we find an expression for the instantaneous velocity:

$$v(t) = \frac{dx}{dt} = \frac{d}{dt}(3.000 + 1.000t + 2.000t^2)$$

$$= 0 + 1.000 + 2(2.000t) = 1.000 + 4.000t,$$

which does indeed evaluate to 5.000 m/s for $t = 1.000$ s. Clearly the average value is converging toward the instantaneous value as the interval becomes smaller.

Thus given any $x(t)$, we can find $v(t)$ by differentiating. Graphically, we can evaluate (point by point) the slope of $x(t)$ to sketch $v(t)$. Let us now review the examples of Section 2-2, the first three of which deal with a bead sliding along a long straight wire:

1. *No motion at all.* From Eq. 1, $x(t) = A$ and so

$$v(t) = \frac{dx}{dt} = 0, \qquad (10)$$

because the derivative of any constant is zero. Figure 10 shows $x(t)$ along with $v(t)$.

2. *Motion at constant speed.* With $x(t) = A + Bt$ from Eq. 2, we find

$$v(t) = \frac{dx}{dt} = \frac{d}{dt}(A + Bt) = 0 + B. \qquad (11)$$

The (constant) instantaneous velocity is B, as shown in Fig. 11.

3. *Accelerated motion.* Using Eq. 3, $x(t) = A + Bt + Ct^2$, we have

$$v(t) = \frac{dx}{dt} = \frac{d}{dt}(A + Bt + Ct^2) = 0 + B + 2Ct. \qquad (12)$$

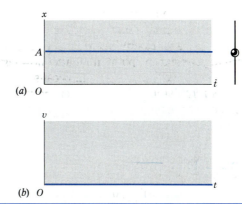

Figure 10 (a) The position and (b) the velocity of a bead on a wire at rest at $x = A$.

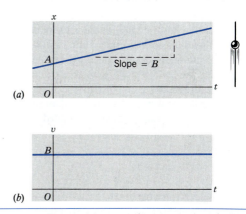

Figure 11 (a) The position and (b) the velocity of a bead sliding in one dimension along a wire with constant velocity. The velocity is equal to the slope B of the graph of $x(t)$. The graph of $v(t)$ is the horizontal line $v = B$.

The velocity changes with time; if $C > 0$, the velocity *increases* with time. Figure 12 shows $x(t)$ and $v(t)$.

4. *Accelerating and braking car.* Without writing $x(t)$, we can sketch the graph of $v(t)$ by studying Fig. 4. In the first interval, the car is at rest and $v = 0$. In the next interval,

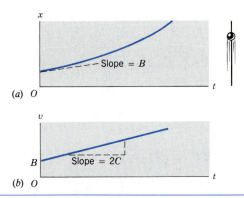

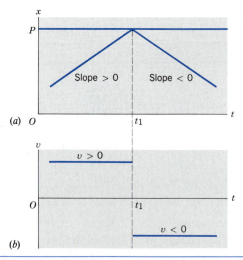

Figure 12 (*a*) The position and (*b*) the velocity of an accelerated bead sliding in one dimension along a wire. The velocity increases with time, as indicated by the increasing slope of $x(t)$ and also by the linear increase of $v(t)$.

Figure 14 (*a*) The position and (*b*) the velocity of a hockey puck rebounding from a hard surface. At $t = t_1$, the velocity "instantly" changes sign in this idealized graph, although for a real hockey puck the velocity would change over some small (but nonzero) interval and the sharp point on the $x(t)$ graph would be correspondingly rounded.

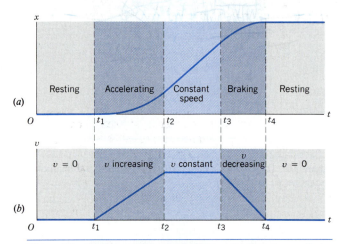

Figure 13 (*a*) The position and (*b*) the velocity of a car that starts at rest, then increases its velocity for a time, then moves for a time with constant velocity, and then decreases its velocity back to rest. The lower graph shows $v(t)$ corresponding exactly to the $x(t)$ graph above and in Fig. 4. For a real car, the changes in velocity must be smooth rather than sudden, so the sharp points in the $v(t)$ graph would be rounded.

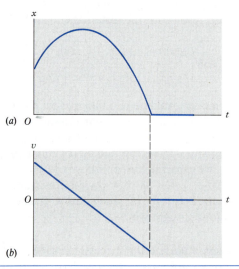

Figure 15 (*a*) The position and (*b*) the velocity of the thrown ball of clay, as in Fig. 6. In reality, the velocity cannot change instantly from a nonzero value to zero, and the sharp vertical rise in $v(t)$ when the ball strikes the floor would be more gradual.

the car is accelerating and $v(t)$ has the form of Eq. 12. In the interval at constant speed, $v = $ constant (equal to its value at the end of the accelerating interval), and therefore $C = 0$ in this interval. Finally, in the braking phase, $v(t)$ again has the form of Eq. 12 but now with $C < 0$ (negative slope). Figure 13 shows a sketch of the motion.

In reality, we cannot jump suddenly from a state of rest to a state of accelerating motion, or from an accelerating state to one of constant speed. In terms of the graph of Fig. 13, the sharp corners in the $v(t)$ plot would be slightly rounded for a real car, and the equation of motion would be more complicated than Eq. 12. For simplicity, we continue to assume the idealized behavior shown in Fig. 13.

5. *Rebounding hockey puck.* Here we have a constant velocity before the rebound and an equal but opposite (negative) velocity after the rebound. Figure 14 shows $v(t)$. Note that the "point" in the $x(t)$ graph gives a discontinuity in the $v(t)$ graph, neither of which would occur for real objects.

6. *Sticky ball of clay.* Here, as shown in Fig. 15, the clay starts with a positive initial v (we arbitrarily choose the upward direction to be positive), but its velocity is decreasing. Its motion would be described by an equation similar to Eq. 12, but with $C < 0$. At the top of its motion $v = 0$, so the $v(t)$ line must cross the axis at that point. When the ball strikes the ground, v goes instantly to zero. (Again, a "point" in the $x(t)$ graph gives a discontinuity in $v(t)$; in reality the point would be rounded and there would be no discontinuity.)

2-5 ACCELERATED MOTION

As we have seen (Figs. 12, 13, and 15), the velocity of a particle may change with time as the motion proceeds. This change in velocity with time is called *acceleration*. In analogy with Eq. 5, we can compute an average acceleration by the change in velocity $\Delta v = v_2 - v_1$ in the interval Δt:

$$\bar{a} = \frac{v_2 - v_1}{t_2 - t_1} = \frac{\Delta v}{\Delta t}. \tag{13}$$

The acceleration has units of velocity divided by time, for instance, meters per second per second, written as m/s².

As was the case with the average velocity $\bar{v}$, the average acceleration $\bar{a}$ tells us nothing about the variation of $v(t)$ with t during the interval Δt. It depends only on the net change in velocity during the interval. If $\bar{a}$ is evaluated to be a constant (possibly zero) over all such intervals, then we may conclude that we have constant acceleration. In this case, the change in velocity is the same in all intervals of the same duration. For example, the acceleration produced by the Earth's gravity is (as we discuss later in this chapter) nearly constant near the Earth's surface and has the value 9.8 m/s². The velocity of a falling object changes by 9.8 m/s every second, increasing by 9.8 m/s in the first second, then by another 9.8 m/s in the next second, and so on.

If the velocity change for successive time intervals of equal length is not the same, then we have a case of variable acceleration. For such cases it is helpful to define the instantaneous acceleration:

$$a = \lim_{\Delta t \to 0} \frac{\Delta v}{\Delta t}$$

or

$$a = \frac{dv}{dt}, \tag{14}$$

in analogy with Eq. 9 for instantaneous velocity.

Note that acceleration may be positive or negative independent of whether v is positive or negative: we can have a positive a with a negative v, for instance. The acceleration a gives the *change* in velocity; the change can be an increase or a decrease for either positive or negative velocity. For example, an elevator moving upward (which we take to be the direction of positive velocity) can accelerate

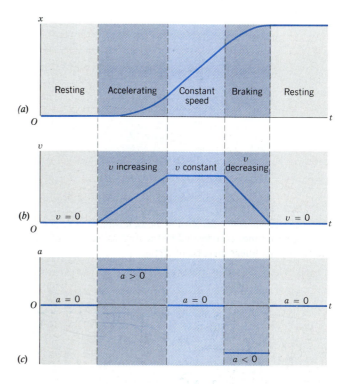

Figure 16 (a) The position, (b) the velocity, and (c) the acceleration of a car that starts at rest, accelerates for an interval, then moves with constant velocity, and then brakes with negative acceleration to rest again. In reality, we cannot instantly change the acceleration of a car from one value to another; both $a(t)$ and $v(t)$ for a real car would be smooth and continuous. Smooth curves would connect the flat $a(t)$ segments, and the sharp points in $v(t)$ would become rounded.

upward ($a > 0$) and move faster or accelerate downward ($a < 0$) and move slower (but still in the upward direction). When it is moving downward ($v < 0$), it can accelerate downward ($a < 0$) and move faster or accelerate upward ($a > 0$) and move slower. When the acceleration and velocity have opposite signs, so that the speed (the magnitude of the velocity) is decreasing, we refer to a *deceleration*.

The acceleration defined by Eq. 14 is just the slope of the $v(t)$ graph. If $v(t)$ is constant, then $a = 0$; if $v(t)$ is a straight line, then a is a constant equal to the slope of the line. If $v(t)$ is a curve, then a will be some function of t, obtained by finding the derivative of $v(t)$.

We can now include the acceleration in the graphs of Figs. 10–15. As an example, we show the case of the accelerating and braking car in Fig. 16. The remaining examples are left as exercises for the student.

Sample Problem 2 Figure 17a shows six successive "snapshots" of a particle moving along the x axis. At $t = 0$ it is at position $x = +1.00$ m to the right of the origin; at $t = 2.5$ s it has

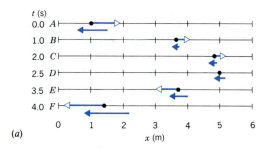

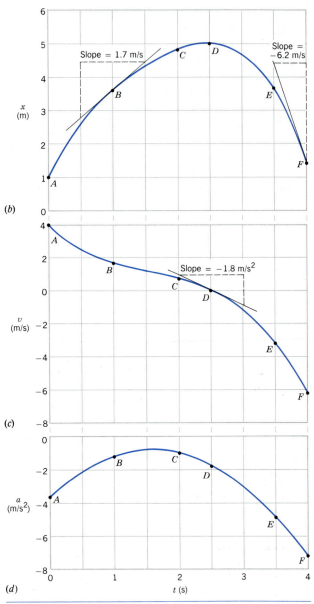

come to rest at $x = +5.00$ m; at $t = 4.0$ s it has returned to $x = 1.4$ m. Figure 17b is a plot of position x versus time t for this motion, and Figs. 17c and 17d show the corresponding velocity and acceleration of the particle. (a) Find the average velocity for the intervals AD and DF. (b) Estimate the slope of $x(t)$ at points B and F and compare with the corresponding points on the $v(t)$ curve. (c) Find the average acceleration in the intervals AD and AF. (d) Estimate the slope of $v(t)$ at point D and compare with the corresponding value of $a(t)$.

Solution   (a) From Eq. 5,

$$\bar{v}_{AD} = \frac{\Delta x_{AD}}{\Delta t_{AD}} = \frac{x_D - x_A}{t_D - t_A} = \frac{5.0 \text{ m} - 1.0 \text{ m}}{2.5 \text{ s} - 0.0 \text{ s}}$$

$$= \frac{4.0 \text{ m}}{2.5 \text{ s}} = +1.6 \text{ m/s},$$

$$\bar{v}_{DF} = \frac{\Delta x_{DF}}{\Delta t_{DF}} = \frac{x_F - x_D}{t_F - t_D} = \frac{1.4 \text{ m} - 5.0 \text{ m}}{4.0 \text{ s} - 2.5 \text{ s}}$$

$$= \frac{-3.6 \text{ m}}{1.5 \text{ s}} = -2.4 \text{ m/s}.$$

The positive sign for $\bar{v}_{AD}$ tells us that, on the average, the particle moves in the direction of increasing x (that is, to the right in Fig. 17a) during the interval AD. The negative sign for $\bar{v}_{DF}$ tells us that the particle is, on the average, moving in the direction of decreasing x (to the left in Fig. 17a) during the interval DF.

(b) From the tangents to $x(t)$ drawn at points B and F in Fig. 17b we estimate the following:

point B: slope $= \dfrac{4.5 \text{ m} - 2.8 \text{ m}}{1.5 \text{ s} - 0.5 \text{ s}} = \dfrac{1.7 \text{ m}}{1.0 \text{ s}} = +1.7 \text{ m/s},$

point F: slope $= \dfrac{1.4 \text{ m} - 4.5 \text{ m}}{4.0 \text{ s} - 3.5 \text{ s}} = \dfrac{-3.1 \text{ m}}{0.5 \text{ s}} = -6.2 \text{ m/s}.$

From $v(t)$ at points B and F in Fig. 17c we estimate $v_B = +1.7$ m/s and $v_F = -6.2$ m/s, in agreement with the slopes of $x(t)$. As expected, $v(t) = dx/dt$.

(c) From Eq. 13,

$$\bar{a}_{AD} = \frac{\Delta v_{AD}}{\Delta t_{AD}} = \frac{v_D - v_A}{t_D - t_A} = \frac{0.0 \text{ m/s} - 4.0 \text{ m/s}}{2.5 \text{ s} - 0.0 \text{ s}}$$

$$= \frac{-4.0 \text{ m/s}}{2.5 \text{ s}} = -1.6 \text{ m/s}^2,$$

$$\bar{a}_{AF} = \frac{\Delta v_{AF}}{\Delta t_{AF}} = \frac{v_F - v_A}{t_F - t_A} = \frac{-6.2 \text{ m/s} - 4.0 \text{ m/s}}{4.0 \text{ s} - 0.0 \text{ s}}$$

$$= \frac{-10.2 \text{ m/s}}{4.0 \text{ s}} = -2.6 \text{ m/s}^2.$$

(d) From the line drawn tangent to $v(t)$ at D, we estimate the following:

$$\text{slope} = \frac{-0.9 \text{ m/s} - 0.9 \text{ m/s}}{3.0 \text{ s} - 2.0 \text{ s}} = \frac{-1.8 \text{ m/s}}{1.0 \text{ s}} = -1.8 \text{ m/s}^2.$$

At point D on the $a(t)$ graph we see $a_D = -1.8$ m/s². Thus $a = dv/dt$. Examining the $v(t)$ graph of Fig. 17c, we see that its slope is negative at all times covered by the graph, and thus $a(t)$ should be negative. Figure 17d bears this out.

Figure 17   Sample Problem 2. (a) Six consecutive "snapshots" of a particle moving along the x axis. The arrow *through* the particle shows its instantaneous velocity, and the arrow *below* the particle shows its instantaneous acceleration. (b) A plot of $x(t)$ for the motion of the particle. The six points $A - F$ correspond to the six snapshots. (c) A plot of $v(t)$. (d) A plot of $a(t)$.

2-6 MOTION WITH CONSTANT ACCELERATION

It is fairly common to encounter motion with constant (or nearly constant) acceleration: the examples already cited of objects falling near the Earth's surface or braking cars are typical. In this section we derive a set of useful results for this special case. However, keep in mind that this is a special situation and the results are not applicable to cases in which a is not constant. Examples of cases with non-constant acceleration include a swinging pendulum bob, a rocket blasting off toward Earth orbit, and a raindrop falling against air resistance.

We let a represent the constant acceleration, plotted in Fig. 18a. (If a is indeed constant, the average and instantaneous accelerations are identical, and we can use the formulas previously derived for either case.) An object starts with velocity v_0 at time $t = 0$, and at a time t later it has velocity v. Equation 13 becomes, for this time interval,

$$a = \frac{\Delta v}{\Delta t} = \frac{v - v_0}{t - 0},$$

or

$$v = v_0 + at. \qquad (15)$$

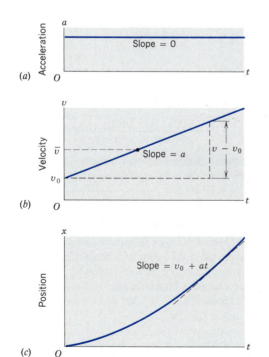

(a)

(b)

(c)

Figure 18 (a) The constant acceleration of a particle, equal to the (constant) slope of $v(t)$. (b) Its velocity $v(t)$, given at each point by the slope of the $x(t)$ curve. The average velocity $\bar{v}$, which in the case of constant acceleration is equal to the average of v and v_0, is indicated. (c) The position $x(t)$ of a particle moving with constant acceleration. The curve is drawn for initial position $x_0 = 0$.

This important result allows us to find the velocity at all later times. Equation 15 gives the velocity as a function of the time, which might be written as $v(t)$, but which we usually write simply as v. Notice that Eq. 15 is in the form of $y = mx + b$, which describes the graph of a straight line. Here a is the slope, as we have already explained, and v_0 is the intercept (the value of v at $t = 0$). This straight line is plotted in Fig. 18b.

To complete the analysis of the kinematics of constant acceleration, we must find the dependence of the position x on the time. For this we need an expression for the average velocity in the interval. If the plot of v against t is a straight line (see Fig. 18b), then the average or mean value of v occurs midway through the interval and is equal to the average or mean of the two endpoints at time 0 and time t:

$$\bar{v} = \tfrac{1}{2}(v + v_0). \qquad (16)$$

Using Eq. 15 to eliminate v, we obtain

$$\bar{v} = v_0 + \tfrac{1}{2}at. \qquad (17)$$

Now using Eq. 5, which defines average velocity, and assuming the particle moves from position x_0 at time 0 to position x at time t, the average velocity can be written

$$\bar{v} = \frac{\Delta x}{\Delta t} = \frac{x - x_0}{t - 0}. \qquad (18)$$

Combining Eqs. 17 and 18, we obtain the desired result for $x(t)$:

$$x = x_0 + v_0 t + \tfrac{1}{2}at^2. \qquad (19)$$

Given the value of a, and the *initial conditions* x_0 and v_0 (that is, the position and velocity at $t = 0$), Eq. 19 then permits us to find the position x at all subsequent times, which is the goal of our kinematic analysis. The net distance traveled from the starting point, $x - x_0$, is often called the *displacement*. For convenience we often choose the origin of coordinates so that $x_0 = 0$. Figure 18c shows the plot of x against t for this case.

Notice that there are four variables (x, v, a, t) and two initial conditions (x_0, v_0). Equations 15 and 19 are written in the customary form for analysis of kinematics as an *initial value* problem: given the physical situation (that is, the acceleration a) and the initial conditions (x_0 and v_0), we can find v and x for all t. Often, however, the problem may be cast in a different form. For example, given the acceleration a, through what distance (rather than "for what time") must the particle move for its velocity to change from v_0 to v? Here the time does not enter, and so we can treat Eqs. 15 and 19 as algebraic equations and eliminate the unwanted variable t between them:

$$v^2 = v_0^2 + 2a(x - x_0). \qquad (20)$$

Eliminating other variables or parameters, we can obtain

TABLE 2 EQUATIONS FOR MOTION WITH CONSTANT ACCELERATION[a]

Equation Number	Equation	Contains				
		x	v_0	v	a	t
15	$v = v_0 + at$	×	✓	✓	✓	✓
19	$x = x_0 + v_0 t + \frac{1}{2}at^2$	✓	✓	×	✓	✓
20	$v^2 = v_0^2 + 2a(x - x_0)$	✓	✓	✓	✓	×
21	$x = x_0 + \frac{1}{2}(v_0 + v)t$	✓	✓	✓	×	✓
22	$x = x_0 + vt - \frac{1}{2}at^2$	✓	×	✓	✓	✓

[a] Make sure that the acceleration is indeed constant before using the equations in this table.

Eqs. 21 and 22, which are displayed in Table 2 with the complete set of kinematic equations for constant acceleration.

We can verify that Eq. 19 is the correct kinematic result by differentiating, which should yield the velocity v:

$$\frac{dx}{dt} = \frac{d}{dt}(x_0 + v_0 t + \frac{1}{2}at^2) = v_0 + at = v.$$

It does indeed give the expected result.

In using the equations of Table 2 to solve a problem, you can choose the *origin of the coordinate system* at any convenient location. The four equations in Table 2 that depend on x also depend on x_0, and in fact they always depend on the difference $x - x_0$. Usually the origin is chosen to make $x_0 = 0$, so that the equations become somewhat simplified. You can also choose either *direction of the coordinate axis* to be positive. Once you have selected a particular direction to be designated as positive, then all displacements, velocities, and accelerations in that direction are positive, and those in the opposite direction are negative. The choice of the *origin* and *direction* of the coordinate axis must remain in effect throughout the solution of any particular problem.

Sample Problem 3   You brake your Porsche from a velocity of 85 km/h (about 53 mi/h, well below the speed limit, of course) to 45 km/h over a distance of 105 m. (*a*) What is the acceleration, assuming it to be constant over the interval? (*b*) How much time elapses during this interval? (*c*) If you were to continue braking with the same acceleration, how much longer would it take for you to stop and how much additional distance would you cover?

Solution   (*a*) Let us first select the positive direction to be the direction of the velocity, and choose the origin so that $x_0 = 0$ when you begin braking. We are given the initial velocity $v_0 = +85$ km/h at $t = 0$, and we know the final velocity $v = +45$ km/h at time t (which is unknown) when the displacement is $+0.105$ km. We need an equation that includes the unknown acceleration that we seek, but that does not involve the time. Equation 20 is our choice, and we solve for a:

$$a = \frac{v^2 - v_0^2}{2(x - x_0)} = \frac{(45 \text{ km/h})^2 - (85 \text{ km/h})^2}{2(0.105 \text{ km})}$$
$$= -2.48 \times 10^4 \text{ km/h}^2 = -1.91 \text{ m/s}^2.$$

The acceleration comes out to be negative, which means it is opposite to the direction we chose to be positive.

(*b*) We need an equation that does not include the acceleration, which permits us to find the time from the original data. From Table 2, we see that Eq. 21 fills the role, and we solve for t:

$$t = \frac{2(x - x_0)}{v_0 + v} = \frac{2(0.105 \text{ km})}{85 \text{ km/h} + 45 \text{ km/h}} = 1.62 \times 10^{-3} \text{ h} = 5.8 \text{ s}.$$

We have selected for this part an equation that does not include the acceleration, because otherwise an error that might have been made in solving part (*a*) would be compounded in solving part (*b*). It is good practice always to return to the original data, if possible, when solving independent parts of a problem.

(*c*) Now with a known acceleration, we seek the time t for the car to go from $v_0 = 85$ km/h to $v = 0$. Equation 15 is the choice to find t:

$$t = \frac{v - v_0}{a} = \frac{0 - 85 \text{ km/h}}{-2.48 \times 10^4 \text{ km/h}^2} = 3.43 \times 10^{-3} \text{ h} = 12.3 \text{ s}.$$

The car comes to a stop 12.3 s after you began braking, or 6.5 s ($= 12.3 \text{ s} - 5.8 \text{ s}$) after it reached the velocity of 45 km/h.

To find the distance, we can use Eq. 20:

$$x - x_0 = \frac{v^2 - v_0^2}{2a} = \frac{0 - (85 \text{ km/h})^2}{2(-2.48 \times 10^4 \text{ km/h}^2)} = 0.146 \text{ km} = 146 \text{ m}.$$

The additional distance traveled between the point at which $v = 45$ km/h and the point at which $v = 0$ is 146 m − 105 m = 41 m.

Sample Problem 4   An alpha particle (the nucleus of a helium atom) travels along the inside of a straight hollow tube 2.0 m long which forms part of a particle accelerator. (*a*) If one assumes uniform acceleration, what is the acceleration of the particle, if it enters at a speed of 1.0×10^4 m/s and leaves at 5.0×10^6 m/s? (*b*) How long is it in the tube?

Solution   (*a*) We choose an x axis parallel to the tube, its positive direction being that in which the particle is moving and its origin at the tube entrance. We are given v_0, v, and x, and we seek a. Rewriting Eq. 20, with $x_0 = 0$,

$$a = \frac{v^2 - v_0^2}{2x}$$

$$= \frac{(5.0 \times 10^6 \text{ m/s})^2 - (1.0 \times 10^4 \text{ m/s})^2}{2(2.0 \text{ m})}$$

$$= +6.3 \times 10^{12} \text{ m/s}^2.$$

(b) Here we use Eq. 21 solved for t with $x_0 = 0$, which gives

$$t = \frac{2x}{v_0 + v} = \frac{2(2.0 \text{ m})}{1.0 \times 10^4 \text{ m/s} + 5.0 \times 10^6 \text{ m/s}}$$
$$= 8.0 \times 10^{-7} \text{ s} = 0.80 \text{ } \mu s.$$

2-7 FREELY FALLING BODIES

The most common example of motion with (nearly) constant acceleration is that of a body falling toward the Earth. If we allow a body to fall in a vacuum, so that air resistance does not affect its motion, we find a remarkable fact: *all bodies, regardless of their size, shape, or composition, fall with the same acceleration at the same point near the Earth's surface.* This acceleration, denoted by the symbol g, is called the *free-fall acceleration* (or sometimes the *acceleration due to gravity*). Although the acceleration depends on the distance from the center of the Earth (as we shall show in Chapter 16), if the distance of fall is small compared with the Earth's radius (6400 km) we can regard the acceleration as constant throughout the fall.

Near the Earth's surface the magnitude of g is approximately 9.8 m/s^2, a value that we use throughout the text unless we specify otherwise. The direction of the free-fall acceleration at any point establishes what we mean by the word "down" at that point.

Although we speak of *falling* bodies, bodies in upward motion experience the same free-fall acceleration (magnitude *and* direction). That is, no matter whether the velocity of the particle is up or down, the direction of its acceleration under the influence of the Earth's gravity is always down.

The exact value of the free-fall acceleration varies with latitude and with altitude. There are also significant variations caused by differences in the local density of the Earth's crust. We discuss these variations in Chapter 16.

The equations of Table 2, which were derived for the case of constant acceleration, can be applied to free fall. For this purpose, we first make two small changes: (1) We label the direction of free fall as the y axis and take its positive direction to be upward. Later, in Chapter 4, we shall consider motion in two dimensions, and we shall want to use the x label for horizontal motion. (2) We replace the constant acceleration a in Table 2 with $-g$, since our choice of the positive y direction to be upward means that the acceleration is negative. Because we take the (downward) acceleration to be $-g$, g is a *positive* number.

With these small changes, the equations of Table 2 become

$$v = v_0 - gt, \tag{23}$$

$$y = y_0 + v_0 t - \tfrac{1}{2}gt^2, \tag{24}$$

$$v^2 = v_0^2 - 2g(y - y_0), \tag{25}$$

$$y = y_0 + \tfrac{1}{2}(v_0 + v)t, \tag{26}$$

and

$$y = y_0 + vt + \tfrac{1}{2}gt^2. \tag{27}$$

Sample Problem 5   A body is dropped from rest and falls freely. Determine the position and velocity of the body after 1.0, 2.0, 3.0, and 4.0 s have elapsed.

Solution   We choose the starting point as the origin. We know the initial speed (zero) and the acceleration, and we are given the time. To find the position, we use Eq. 24 with $y_0 = 0$ and $v_0 = 0$:

$$y = -\tfrac{1}{2}gt^2.$$

Putting $t = 1.0$ s, we obtain

$$y = -\tfrac{1}{2}(9.8 \text{ m/s}^2)(1.0 \text{ s})^2 = -4.9 \text{ m}.$$

To find the velocity, we use Eq. 23, again with $v_0 = 0$:

$$v = -gt = -(9.8 \text{ m/s}^2)(1.0 \text{ s}) = -9.8 \text{ m/s}.$$

After falling for 1.0 s, the body is 4.9 m *below* (y is negative) its starting point and is moving *downward* (v is negative) with a speed of 9.8 m/s. Continuing in this way, we can find the positions and velocities at $t = 2.0$, 3.0, and 4.0 s, which are shown in Fig. 19.

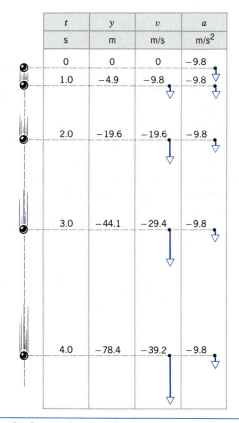

t	y	v	a
s	m	m/s	m/s^2
0	0	0	−9.8
1.0	−4.9	−9.8	−9.8
2.0	−19.6	−19.6	−9.8
3.0	−44.1	−29.4	−9.8
4.0	−78.4	−39.2	−9.8

Figure 19   Sample Problem 5. The height, velocity, and acceleration of a body in free fall are shown.

Sample Problem 6   A ball is thrown vertically upward from the ground with a speed of 25.2 m/s. (a) How long does it take to reach its highest point? (b) How high does it rise? (c) At what times will it be 27.0 m above the ground?

Solution   (a) At its highest point its velocity passes through the value zero. Given v_0 and $v(=0)$, we wish to find t and we therefore choose Eq. 23, which we solve for t:

$$t = \frac{v_0 - v}{g} = \frac{25.2 \text{ m/s} - 0}{9.8 \text{ m/s}^2} = 2.57 \text{ s.}$$

(b) Let us use only the original data for this part, to keep from compounding any error that might have been made in part (a). Equation 25, with y_0 assigned as 0, allows us to solve for y when we know the other quantities:

$$y = \frac{v_0^2 - v^2}{2g} = \frac{(25.2 \text{ m/s})^2 - 0}{2(9.8 \text{ m/s}^2)} = 32.4 \text{ m.}$$

(c) Equation 24 is useful for this case, because t is the only unknown. Since we wish to solve for t, let us rewrite Eq. 24, with $y_0 = 0$, in the usual form of a quadratic equation:

$$\tfrac{1}{2}gt^2 - v_0 t + y = 0$$

$$\tfrac{1}{2}(9.8 \text{ m/s}^2)t^2 - (25.2 \text{ m/s})t + 27.0 \text{ m} = 0.$$

Using the quadratic formula, we find the solutions to be $t = 1.52$ s and $t = 3.62$ s. At $t = 1.52$ s, the velocity of the ball is

$$v = v_0 - gt = 25.2 \text{ m/s} - (9.8 \text{ m/s}^2)(1.52 \text{ s}) = 10.3 \text{ m/s.}$$

At $t = 3.62$ s, the velocity is

$$v = v_0 - gt = 25.2 \text{ m/s} - (9.8 \text{ m/s}^2)(3.62 \text{ s}) = -10.3 \text{ m/s.}$$

The two velocities have identical magnitudes but opposite directions. You should be able to convince yourself that, in the absence of air resistance, the ball will take as long to rise to its maximum height as to fall the same distance, and that at each point it will have the same speed going up that it has coming down. Note that the answer to part (a) for the time to reach the highest point, 2.57 s, is exactly midway between the two times found in part (c). Can you explain this? Can you predict qualitatively the effect of air resistance on the times of rise and fall?

Sample Problem 7   A rocket is launched from rest from an underwater base a distance of 125 m below the surface of a body of water. It moves vertically upward with an unknown but assumed constant acceleration (the combined effect of its engines, the Earth's gravity, and the buoyancy and drag of the water), and it reaches the surface in a time of 2.15 s. When it breaks the surface its engines automatically shut off (to make it more difficult to detect) and it continues to rise. What maximum height does it reach? (Ignore any effects at the surface.)

Solution   As with any projectile in free fall, we could analyze the motion of the rocket during the portion of its motion in the air if we knew the initial velocity of that part of the motion. The plan of attack in this problem is therefore to analyze the underwater portion of the motion to find the velocity when the rocket reaches the surface, and then to treat that velocity as the initial velocity of the free-fall portion. These parts must be done sepa-

rately, because the acceleration changes at the surface of the water.

For the underwater motion, we know the displacement, the time, and the initial velocity (zero). The acceleration is not needed, but we wish to know the final velocity; Eq. 21 of Table 2 provides the proper relationship:

$$v = \frac{2(y - y_0)}{t} = \frac{2(125 \text{ m})}{2.15 \text{ s}} = 116 \text{ m/s.}$$

The velocity at the surface is 116 m/s upward. We now analyze the free-fall portion of the upward motion, taking this velocity to be the *initial* velocity. We use Eq. 25 for free fall, and as usual we find the maximum height by seeking the point at which the velocity becomes zero:

$$y - y_0 = \frac{v_0^2 - v^2}{2g} = \frac{(116 \text{ m/s})^2 - 0}{2(9.8 \text{ m/s}^2)} = 687 \text{ m.}$$

To test your understanding of this problem, you should draw graphs of $y(t)$, $v(t)$, and $a(t)$ in a fashion similar to Fig. 16. Be sure to keep in mind which variables vary continuously and smoothly and which do not in this idealized problem. How would a real rocket differ from this picture?

2-8   GALILEO AND FREE FALL   *(Optional)*

The nature of the motion of a falling object was long ago a subject of interest in natural philosophy. Aristotle had asserted that "the downward movement . . . of any body endowed with weight is quicker in proportion to its size." That is, heavier objects fall more quickly. It was not until many centuries later when Galileo Galilei (1564–1642) made the correct assertion: "if one were to remove entirely the resistance of the medium, all materials would descend with equal speed." In the later years of his life, Galileo wrote the treatise entitled *Dialogues Concerning Two New Sciences* in which he detailed his studies of motion.

Aristotle's belief that a heavier object will fall faster is a commonly held view. It appears to receive support from a well-known lecture demonstration in which a ball and a sheet of paper are dropped at the same instant, the ball reaching the floor much sooner than the paper. However, when the lecturer first crumples the paper tightly and then repeats the demonstration, both ball and paper strike the floor at essentially the same time. In the former case, it is the effect of greater resistance of the air that makes the paper fall more slowly than the ball. In the latter case, the effect of air resistance on the paper is reduced and is about the same for both bodies, so that they fall at about the same rate. Of course, a direct test can be made by dropping bodies in vacuum. Even in easily obtainable partial vacuums we can show that a feather and a ball of lead thousands of times heavier drop at rates that are practically indistinguishable. In 1971, astronaut David Scott released a feather and a geologist's hammer on the (airless) Moon, observing that—within the experimental error of his observation—they reached the lunar surface simultaneously.

In Galileo's time, however, there was no effective way to obtain a partial vacuum, nor did equipment exist to time freely falling bodies with sufficient precision to obtain reliable numerical data. (The famous story about Galileo dropping two objects from the Tower of Pisa and observing them to reach the ground at the same time is almost certainly only legend. Given the height of the tower and the objects Galileo is said to have used, the larger and heavier object would have hit the ground from one to several meters ahead of the lighter object, owing to the effects of air resistance. Thus Galileo would have appeared to prove that Aristotle was correct after all!) Nevertheless, Galileo proved his result using a ball rolling down an incline. He first showed that the kinematics of a ball rolling down an incline was the same as that of a ball in free fall. The incline merely served to reduce the accelerating effect of the Earth's gravity, thereby slowing the motion so that measurements could be made more easily. Moreover, at slow speeds air resistance is far less important.

Galileo found from his experiments that the distances covered in consecutive time intervals were proportional to the odd numbers 1, 3, 5, 7, . . . , etc. Total distances for consecutive intervals thus were proportional to 1, 1 + 3 (=4), 1 + 3 + 5 (=9), 1 + 3 + 5 + 7 (=16), and so on, that is, to the squares of the integers 1,2,3,4, and so on. But if the distance covered is proportional to the square of the elapsed time, then the gain in velocity is directly proportional to the elapsed time, a result that holds *only* in the case of constant acceleration. Finally, Galileo found that the same results held regardless of the mass of the ball, and thus, in our terminology, the free-fall acceleration is independent of the mass of the object. ■

2-9 MEASURING THE FREE-FALL ACCELERATION *(Optional)*

The measurement of g is a standard exercise in the introductory physics lab. It can be done simply by timing a freely falling object dropped from rest through a measured distance. Equation 24 then gives g directly. Even with the relatively crude equipment normally found in student labs, a precision of about 1% is possible. A better method uses a pendulum, whose driving force is the Earth's attraction for the pendulum bob. As we show in Chapter 15, the value of g can be found by measuring the period of oscillation of a pendulum of known length. By timing many swings, a precise value for the period can be found, and using typical student lab equipment, a precision of 0.1% is not difficult to obtain. This level of precision is sufficient to observe the variation in g between sea level and a high mountain (3 km or 10,000 ft), or between the equator and the poles of the Earth.

For several centuries, a pendulum method was used for precise measurements of g, and the ultimate precision was about 1 part in 10^6, sufficient to detect variations in g from one floor of a building to the next. Pendulum methods are limited to this precision by the uncertainty in the true behavior at the pivot point, which makes it difficult to determine the length to higher precision. Recently, in attempts to improve the precision of g, investigators have returned to the free-fall method of measuring g, which through the modern techniques of laser interferometry has been extended to about 1 part in 10^9. This method is sufficient to observe the change in the Earth's gravity over a vertical distance of 1 cm; equivalently, such a gravity meter can detect the gravitational change caused by the measuring scientist standing 1 m from the apparatus!

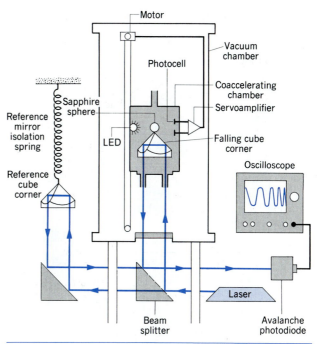

Figure 20 Diagram of free-fall apparatus. The oscilloscope records the changing pattern of cancellations and reinforcements when the laser beam reflected from the falling cube corner is recombined with the beam from the reference cube corner. A motor drives the coaccelerating chamber downward so that it falls with the cube. For a description of this apparatus and a discussion of measurements of g, see "Ballistic Methods of Measuring g" by J. E. Faller and I. Marson, *Metrologia*, v. 25 (1988), p. 49.

Achieving such precision is a remarkable tribute to careful experimental techniques. For instance, you might expect that, to eliminate the effects of air resistance in free fall, the object must be dropped in vacuum. That is certainly true, but even the best vacuums presently achievable in the laboratory are not good enough for a 10^{-9} level of precision in the measurement of g. To reduce the effects of the tiny amount of residual gas present even at high vacuum, the freely falling object is placed inside an evacuated box, which is also dropped. The small amount of residual gas is carried by the falling box, and because the gas falls with the object it offers no resistance to the free fall.

Figure 20 shows a representation of the free-fall apparatus developed by Dr. James E. Faller and his colleagues at the Joint Institute for Laboratory Astrophysics in Boulder, Colorado. The falling object is a corner reflector, which is in essence a corner of a glass cube that has a reflective coating on the three perpendicular faces. This device has the useful property that light incident on the corner from any interior direction is reflected back in exactly the opposite direction. (An array of such reflectors was placed on the Moon by the Apollo astronauts; laser beams have been reflected back to Earth from the Moon to measure the Earth–Moon distance precisely.) A laser beam is reflected from the falling object, and the incident and reflected beams are made to interfere with one another so that they continually reinforce

and then cancel as the object falls. The distance that it falls between cancellations is the wavelength of the light, and the total distance of fall can be measured, to a precision of a small fraction of the wavelength of light, by merely counting the number of cancellations. Simultaneously, the time between cancellations is measured with an atomic clock. Thus distance and time are measured simultaneously, just as you might do in your introductory physics lab. A photograph of this remarkable apparatus is shown in Fig. 21.

The construction of more precise gravity meters has important practical consequences. Mapping the Earth's gravity can assist in prospecting for oil or minerals (see Fig. 5 of Chapter 16). Changes in the Earth's crust with time can be observed through their effect on g, enabling the monitoring of plate movements and seismic activity. Small variations in gravity over the Earth's surface can affect the orbits of satellites and the trajectories of ballistic missiles. From the standpoint of basic science, precise measurements of g provide detailed tests of our understanding of the theory of gravitation, which originated with Isaac Newton more than three centuries ago. ■

Figure 21   A photograph of the free-fall apparatus of Fig. 20. The apparatus is easily portable, so that g can be measured at remote locations.

QUESTIONS

1. Can the speed of a particle ever be negative? If so, give an example; if not, explain why.

2. Each second a rabbit moves one-half the remaining distance from its nose to a head of lettuce. Does the rabbit ever get to the lettuce? What is the limiting value of the rabbit's average velocity? Draw graphs showing the rabbit's velocity and position as time increases.

3. *Average speed* can mean the magnitude of the average velocity. Another, more common, meaning given to it is that average speed is the total length of path traveled divided by the elapsed time. Are these meanings different? Give an example to support your answer.

4. A racing car, in a qualifying two-lap heat, covers the first lap with an average speed of 90 mi/h. The driver wants to speed up during the second lap so that the average speed of the two laps together will be 180 mi/h. Show that it cannot be done.

5. Bob beats Judy by 10 m in a 100-m dash. Bob, claiming to give Judy an equal chance, agrees to race her again but to start from 10 m behind the starting line. Does this really give Judy an equal chance?

6. When the velocity is constant, can the average velocity over any time interval differ from the instantaneous velocity at any instant? If so, give an example; if not, explain why.

7. Can the average velocity of a particle moving along the x axis ever be $\frac{1}{2}(v_0 + v)$ if the acceleration is not uniform? Prove your answer with the use of graphs.

8. Does the speedometer on an automobile register speed as we have defined it?

9. (a) Can an object have zero velocity and still be accelerating? (b) Can an object have a constant velocity and still have a varying speed? In each case, give an example if your answer is yes; explain why if your answer is no.

10. Can the velocity of an object reverse direction when its acceleration is constant? If so, give an example; if not, explain why.

11. Figure 30 shows Colonel John P. Stapp in his braking rocket sled; see Problem 34. (a) His body is an accelerometer, not a speedometer. Explain. (b) Can you tell the direction of the acceleration from the figure?

12. Can an object be increasing in speed as its acceleration decreases? If so, give an example; if not, explain why.

13. Of the following situations, which one is impossible? (a) A body having velocity east and acceleration east; (b) a body having velocity east and acceleration west; (c) a body having zero velocity but acceleration not zero; (d) a body having constant acceleration and variable velocity; (e) a body having constant velocity and variable acceleration.

14. What are some examples of falling objects in which it would be unreasonable to neglect air resistance?

15. Figure 22 shows a shot tower in Baltimore, Maryland. It was built in 1829 and used to manufacture lead shot pellets by pouring molten lead through a sieve at the top of the tower.

Figure 22 Question 15.

The lead pellets solidify as they fall into a tank of water at the bottom of the tower, 230 ft below. What are the advantages of manufacturing shot in this way?

16. A person standing on the edge of a cliff at some height above the ground below throws one ball straight up with initial speed v_0 and then throws another ball straight down with the same initial speed. Which ball, if either, has the larger speed when it hits the ground? Neglect air resistance.

17. What is the downward acceleration of a projectile that is released from a missile accelerating upward at 9.8 m/s²?

18. If a particle is released from rest ($v_0 = 0$) at $x_0 = 0$ at the time $t = 0$, Eq. 19 for constant acceleration says that it is at position x at two different times, namely, $+\sqrt{2x/a}$ and $-\sqrt{2x/a}$. What is the meaning of the negative root of this quadratic equation?

19. On another planet, the value of g is one-half the value on the Earth. How is the time needed for an object to fall to the ground from rest related to the time required to fall the same distance on the Earth?

20. (a) A stone is thrown upward with a certain speed on a planet where the free-fall acceleration is double that on Earth. How high does it rise compared to the height it rises on Earth? (b) If the initial speed were doubled, what change would that make?

21. Consider a ball thrown vertically up. Taking air resistance into account, would you expect the time during which the ball rises to be longer or shorter than the time during which it falls? Why?

22. Make a qualitative graph of speed v versus time t for a falling object (a) for which air resistance can be ignored and (b) for which air resistance cannot be ignored.

23. A second ball is dropped down an elevator shaft 1 s after the first ball is dropped. (a) What happens to the distance between the balls as time goes on? (b) How does the ratio v_1/v_2 of the speed of the first ball to the speed of the second ball change as time goes on? Neglect air resistance, and give qualitative answers.

24. Repeat Question 23 taking air resistance into account. Again, give qualitative answers.

25. If m is a light stone and M is a heavy one, according to Aristotle M should fall faster than m. Galileo attempted to show that Aristotle's belief was logically inconsistent by the following argument. Tie m and M together to form a double stone. Then, in falling, m should retard M, because it tends to fall more slowly, and the combination would fall faster than m but more slowly than M; but according to Aristotle the double body ($M + m$) is heavier than M and, hence, should fall faster than M. If you accept Galileo's reasoning as correct, can you conclude that M and m must fall at the same rate? What need is there for experiment in that case? If you believe Galileo's reasoning is incorrect, explain why.

26. What happens to our kinematic equations (see Table 2) under the operation of time reversal, that is, replacing t by $-t$? Explain.

27. We expect a truly general relation, such as those in Table 2, to be valid regardless of the choice of coordinate system. By demanding that general equations be dimensionally consistent we ensure that the equations are valid regardless of the choice of units. Is there any need then for units or coordinate systems?

PROBLEMS

Section 2-3 Average Velocity

1. How far does your car, moving at 55 mi/h (= 88 km/h) travel forward during the 1 s of time that you take to look at an accident on the side of the road?

2. Boston Red Sox pitcher Roger Clemens threw a fastball at a horizontal speed of 160 km/h, as verified by a radar gun. How long did it take for the ball to reach homeplate, which is 18.4 m away?

3. Figure 23 shows the relation between the age of the oldest

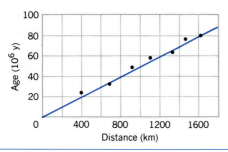

Figure 23   Problem 3.

sediment, in millions of years, and the distance, in kilometers, at which the sediment was found from a particular ocean ridge. Seafloor material is extruded from this ridge and moves away from it at approximately uniform speed. Find the speed, in centimeters per year, at which this material recedes from the ridge.

4. Carl Lewis runs the 100-m dash in about 10 s, and Bill Rodgers runs the marathon (26 mi, 385 yd) in about 2 h 10 min. (a) What are their average speeds? (b) If Carl Lewis could maintain his sprint speed during a marathon, how long would it take him to finish?

5. For many months, a well-known high-energy physicist commuted weekly between Boston, Massachusetts and Geneva, Switzerland, the two cities being separated by a distance of 4000 mi. What was the physicist's average speed during this period? Are you surprised that you do not need to know the speed of the airplane to solve this problem?

6. The legal speed limit on a thruway is changed from 55 mi/h (= 88.5 km/h) to 65 mi/h (= 104.6 km/h). How much time is thereby saved on a trip from the Buffalo entrance to the New York City exit of the New York State Thruway for someone traveling at the higher speed over this 435-mi (= 700-km) stretch of highway?

7. You drive on Interstate 10 from San Antonio to Houston, one-half the *time* at 35 mi/h (= 56.3 km/h) and the other half at 55 mi/h (= 88.5 km/h). On the way back you travel one-half the *distance* at 35 mi/h and the other half at 55 mi/h. What is your average speed (a) from San Antonio to Houston, (b) from Houston back to San Antonio, and (c) for the entire trip?

8. A high-performance jet plane, practicing radar avoidance maneuvers, is in horizontal flight 35 m above the level ground. Suddenly, the plane encounters terrain that slopes gently upward at 4.3°, an amount difficult to detect; see Fig. 24. How much time does the pilot have to make a correction if the plane is to avoid flying into the ground? The airspeed is 1300 km/h.

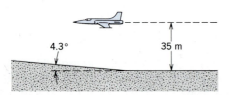

Figure 24   Problem 8.

9. The position of an object moving in a straight line is given by $x = 3t - 4t^2 + t^3$, where x is in meters and t is in seconds. (a) What is the position of the object at $t = 0, 1, 2, 3,$ and 4 s? (b) What is the object's displacement between $t = 0$ and $t = 2$ s? Between $t = 0$ and $t = 4$ s? (c) What is the average velocity for the time interval from $t = 2$ to $t = 4$ s? From $t = 0$ to $t = 3$ s?

10. A car travels up a hill at the constant speed of 40 km/h and returns down the hill at the speed of 60 km/h. Calculate the average speed for the round trip.

11. Compute your average speed in the following two cases. (a) You walk 240 ft at a speed of 4.0 ft/s and then run 240 ft at a speed of 10 ft/s along a straight track. (b) You walk for 1.0 min at a speed of 4.0 ft/s and then run for 1.0 min at 10 ft/s along a straight track.

12. Two trains, each having a speed of 34 km/h, are headed at each other on the same straight track. A bird that can fly 58 km/h flies off the front of one train when they are 102 km apart and heads directly for the other train. On reaching the other train it flies directly back to the first train, and so forth. (a) How many trips can the bird make from one train to the other before they crash? (b) What is the total distance the bird travels?

Section 2-4   Instantaneous Velocity

13. The position of a particle moving along the x axis is given in centimeters by $x = 9.75 + 1.50t^3$, where t is in seconds. Consider the time interval $t = 2$ to $t = 3$ s and calculate (a) the average velocity; (b) the instantaneous velocity at $t = 2$ s; (c) the instantaneous velocity at $t = 3$ s; (d) the instantaneous velocity at $t = 2.5$ s; and (e) the instantaneous velocity when the particle is midway between its positions at $t = 2$ and $t = 3$ s.

14. How far does the runner whose velocity–time graph is shown in Fig. 25 travel in 16 s?

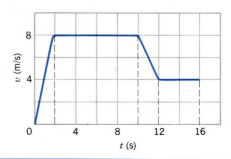

Figure 25   Problems 14 and 15.

Section 2-5   Accelerated Motion

15. What is the acceleration of the runner in Problem 14 at $t = 11$ s?

16. A particle had a velocity of 18 m/s in the $+x$ direction and 2.4 s later its velocity was 30 m/s in the opposite direction. What was the average acceleration of the particle during this 2.4-s interval?

17. An object moves in a straight line as described by the

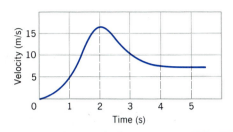

Figure 26 Problem 17.

velocity–time graph in Fig. 26. Sketch a graph that represents the acceleration of the object as a function of time.

18. The graph of x versus t in Fig. 27a is for a particle in straight-line motion. (a) State for each interval whether the velocity v is $+$, $-$, or 0, and whether the acceleration a is $+$, $-$, or 0. The intervals are *OA, AB, BC,* and *CD*. (b) From the curve, is there any interval over which the acceleration is obviously not constant? (Ignore the behavior at the endpoints of the intervals.)

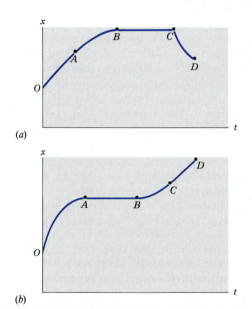

(a)

(b)

Figure 27 (a) Problem 18 and (b) Problem 19.

19. Answer the previous questions for the motion described by the graph of Fig. 27b.

20. A particle moves along the x axis with a displacement versus time as shown in Fig. 28. Roughly sketch curves of velocity versus time and acceleration versus time for this motion.

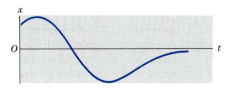

Figure 28 Problem 20.

21. For each of the following situations, sketch a graph that is a possible description of position as a function of time for a particle that moves along the x axis. At $t = 1$ s, the particle has (a) zero velocity and positive acceleration; (b) zero velocity and negative acceleration; (c) negative velocity and positive acceleration; (d) negative velocity and negative acceleration. (e) For which of these situations is the speed of the particle increasing at $t = 1$ s?

22. If the position of an object is given by $x = 2t^3$, where x is in meters and t in seconds, find (a) the average velocity and the average acceleration between $t = 1$ and $t = 2$ s and (b) the instantaneous velocities and the instantaneous accelerations at $t = 1$ and $t = 2$ s. (c) Compare the average and instantaneous quantities and in each case explain why the larger one is larger.

23. A particle moves along the x axis according to the equation $x = 50t + 10t^2$, where x is in meters and t is in seconds. Calculate (a) the average velocity of the particle during the first 3 s of its motion, (b) the instantaneous velocity of the particle at $t = 3$ s, and (c) the instantaneous acceleration of the particle at $t = 3$ s.

24. A man stands still from $t = 0$ to $t = 5$ min; from $t = 5$ to $t = 10$ min he walks briskly in a straight line at a constant speed of 2.2 m/s. What are his average velocity and average acceleration during the time intervals (a) 2 min to 8 min and (b) 3 min to 9 min?

25. A particle moving along the positive x axis has the following positions at various times:

x (m)	0.080	0.050	0.040	0.050	0.080	0.13	0.20
t (s)	0	1	2	3	4	5	6

(a) Plot displacement (not position) versus time. (b) Find the average velocity of the particle in the intervals 0 to 1 s, 0 to 2 s, 0 to 3 s, 0 to 4 s. (c) Find the slope of the curve drawn in part (a) at the points $t = 0$, 1, 2, 3, 4, and 5 s. (d) Plot the slope (units?) versus time. (e) From the curve of part (d) determine the acceleration of the particle at times $t = 2$, 3, and 4 s.

26. The position of a particle along the x axis depends on the time according to the equation

$$x = At^2 - Bt^3,$$

where x is in meters and t is in seconds. (a) What SI units must A and B have? For the following, let their numerical values in SI units be 3 and 1, respectively. (b) At what time does the particle reach its maximum positive x position? (c) What total pathlength does the particle cover in the first 4 s? (d) What is its displacement during the first 4 s? (e) What is the particle's velocity at the end of each of the first four seconds? (f) What is the particle's acceleration at the end of each of the first four seconds? (g) What is the average velocity for the time interval $t = 2$ to $t = 4$ s?

27. An electron, starting from rest, has an acceleration that increases linearly with time, that is, $a = kt$, in which $k = (1.50 \text{ m/s}^2)$/s or 1.50 m/s^3. (a) Plot a versus t during the first 10-s interval. (b) From the curve of part (a) plot the corresponding v versus t curve and estimate the electron's velocity 5 s after the motion starts. (c) From the v versus t curve of part (*b*) plot the corresponding *x* versus *t* curve and estimate how far the electron moved during the first 5 s of its motion.

28. In an arcade video game, a spot is programmed to move across the screen according to $x = 9.00t - 0.750t^3$, where *x* is the distance in centimeters measured from the left edge of the screen and *t* is the time in seconds. When the spot reaches a screen edge, at either $x = 0$ or $x = 15$ cm, it starts over. (*a*) At what time after starting is the spot instantaneously at rest? (*b*) Where does this occur? (*c*) What is its acceleration when this occurs? (*d*) In which direction does it move in the next instant after coming to rest? (*e*) When does it move off the screen?

Section 2-6 Motion with Constant Acceleration

29. A jumbo jet needs to reach a speed of 360 km/h (= 224 mi/h) on the runway for takeoff. Assuming a constant acceleration and a runway 1.8 km (= 1.1 mi) long, what minimum acceleration from rest is required?

30. A rocketship in free space moves with constant acceleration equal to 9.8 m/s². (*a*) If it starts from rest, how long will it take to acquire a speed one-tenth that of light? (*b*) How far will it travel in so doing? (The speed of light is 3.0×10^8 m/s.)

31. The head of a rattlesnake can accelerate 50 m/s² in striking a victim. If a car could do as well, how long would it take for it to reach a speed of 100 km/h from rest?

32. A muon (an elementary particle) is shot with initial speed 5.20×10^6 m/s into a region where an electric field produces an acceleration of 1.30×10^{14} m/s² directed opposite to the initial velocity. How far does the muon travel before coming to rest?

33. An electron with initial velocity $v_0 = 1.5 \times 10^5$ m/s enters a region 1.2 cm long where it is electrically accelerated (see Fig. 29). It emerges with a velocity $v = 5.8 \times 10^6$ m/s. What was its acceleration, assumed constant? (Such a process occurs in the electron gun in a cathode-ray tube, used in television receivers and video terminals.)

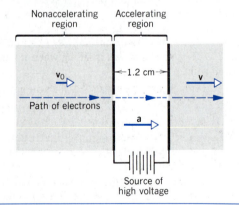

Figure 29   Problem 33.

34. A world's land speed record was set by Colonel John P. Stapp when, on March 19, 1954, he rode a rocket-propelled sled that moved down a track at 1020 km/h. He and the sled were brought to a stop in 1.4 s; see Fig. 30. What acceleration did he experience? Express your answer in terms of *g*

Figure 30   Problem 34.

(= 9.8 m/s²), the acceleration due to gravity. (Note that his body acts as an accelerometer, not a speedometer.)

35. The brakes on your automobile are capable of creating a deceleration of 17 ft/s². If you are going 85 mi/h and suddenly see a state trooper, what is the minimum time in which you can get your car under the 55-mi/h speed limit?

36. On a dry road a car with good tires may be able to brake with a deceleration of 11.0 mi/h·s (= 4.92 m/s²). (*a*) How long does such a car, initially traveling at 55 mi/h (= 24.6 m/s), take to come to rest? (*b*) How far does it travel in this time?

37. An arrow is shot straight up into the air and on its return strikes the ground at 260 ft/s, imbedding itself 9.0 in. into the ground. Find (*a*) the acceleration (assumed constant) required to stop the arrow, and (*b*) the time required for the ground to bring it to rest.

38. Suppose that you were called upon to give some advice to a lawyer concerning the physics involved in one of her cases. The question is whether a driver was exceeding a 30-mi/h speed limit before he made an emergency stop, brakes locked and wheels sliding. The length of skid marks on the road was 19.2 ft. The police officer made the assumption that the maximum deceleration of the car would not exceed the acceleration of a freely falling body (= 32 ft/s²) and did not give the driver a ticket. Was he speeding? Explain.

39. A train started from rest and moved with constant acceleration. At one time it was traveling at 33.0 m/s, and 160 m farther on it was traveling at 54.0 m/s. Calculate (*a*) the acceleration, (*b*) the time required to travel the 160 m, (*c*) the time required to attain the speed of 33.0 m/s, and (*d*) the distance moved from rest to the time the train had a speed of 33.0 m/s.

40. A car moving with constant acceleration covers the distance between two points 58.0 m apart in 6.20 s. Its speed as it passes the second point is 15.0 m/s. (*a*) What is the speed at the first point? (*b*) What is its acceleration? (*c*) At what prior distance from the first point was the car at rest?

41. A subway train accelerates from rest at one station (*a* = + 1.20 m/s²) for the first half of the distance to the next station and then decelerates to rest (*a* = − 1.20 m/s²) for the final half of the distance. The stations are 1.10 km apart. Find (*a*) the time of travel between stations and (*b*) the maximum speed of the train.

42. An elevator cab in the New York Marquis Marriott (see Fig.

Figure 31 Problem 42.

31) has a total run of 624 ft. Its maximum speed is 1000 ft/min and its (constant) acceleration is 4.00 ft/s². (*a*) How far does it move while accelerating to full speed from rest? (*b*) How long does it take to make the run, starting and ending at rest?

43. When a driver brings a car to a stop by braking as hard as possible, the stopping distance can be regarded as the sum of a "reaction distance," which is initial speed times reaction time, and "braking distance," which is the distance covered during braking. The following table gives typical values:

Initial Speed (m/s)	Reaction Distance (m)	Braking Distance (m)	Stopping Distance (m)
10	7.5	5.0	12.5
20	15	20	35
30	22.5	45	67.5

(*a*) What reaction time is the driver assumed to have? (*b*) What is the car's stopping distance if the initial speed is 25 m/s?

44. In a speed trap, two pressure-activated strips are placed 110 m apart across a highway on which the speed limit is 90 km/h. While going 120 km/h, a driver notices a police car just as he activates the first strip and slows down. What deceleration is needed so that the car's average speed is within the speed limit when the car crosses the second marker?

45. At the instant the traffic light turns green, an automobile starts with a constant acceleration of 2.2 m/s². At the same instant a truck, traveling with a constant speed of 9.5 m/s, overtakes and passes the automobile. (*a*) How far beyond the starting point will the automobile overtake the truck? (*b*) How fast will the car be traveling at that instant? (It is instructive to plot a qualitative graph of *x* versus *t* for each vehicle.)

46. The engineer of a train moving at a speed v_1 sights a freight train a distance *d* ahead of him on the same track moving in the same direction with a slower speed v_2. He puts on the brakes and gives his train a constant deceleration *a*. Show that

$$\text{if } d > \frac{(v_1 - v_2)^2}{2a} \text{ , there will be no collision;}$$

$$\text{if } d < \frac{(v_1 - v_2)^2}{2a} \text{ , there will be a collision.}$$

(It is instructive to plot a qualitative graph of *x* versus *t* for each train.)

47. An automobile traveling 35 mi/h (= 56 km/h) is 110 ft (= 34 m) from a barrier when the driver slams on the brakes. Four seconds later the car hits the barrier. (*a*) What was the automobile's constant deceleration before impact? (*b*) How fast was the car traveling at impact?

48. A sprinter, in the 100-m dash, accelerates from rest to a top speed at 2.80 m/s² and maintains the top speed to the end of the dash. (*a*) What time elapsed and (*b*) what distance did the sprinter cover during the acceleration phase if the total time taken in the dash was 12.2 s?

49. A driver's handbook states that an automobile with good brakes and going 50 mi/h can stop in a distance of 186 ft. The corresponding distance for 30 mi/h is 80 ft. Assume that the driver reaction time, during which the acceleration is zero, and the acceleration after the brakes are applied are both the same for the two speeds. Calculate (*a*) the driver reaction time and (*b*) the acceleration.

Section 2-7 Freely Falling Bodies

50. Raindrops fall to the ground from a cloud 1700 m above the Earth's surface. If they were not slowed by air resistance, how fast would the drops be moving when they struck the ground? Would it be safe to walk outside during a rainstorm?

51. The single cable supporting an unoccupied construction elevator breaks when the elevator is at rest at the top of a 120-m high building. (*a*) With what speed does the elevator strike the ground? (*b*) For how long was it falling? (*c*) What was its speed when it passed the halfway point on the way down? (*d*) For how long was it falling when it passed the halfway point?

52. At a construction site a pipe wrench strikes the ground with a speed of 24.0 m/s. (*a*) From what height was it inadvertently dropped? (*b*) For how long was it falling?

53. (*a*) With what speed must a ball be thrown vertically up in order to rise to a maximum height of 53.7 m? (*b*) For how long will it be in the air?

54. A rock is dropped from a 100-m high cliff. How long does it take to fall (*a*) the first 50.0 m and (*b*) the second 50.0 m?

55. Space explorers land on a planet in our solar system. They note that a small rock tossed at 14.6 m/s vertically upward takes 7.72 s to return to the ground. On which planet have they landed? (*Hint:* See Appendix C.)

56. A ball is thrown down vertically with an initial speed of 20.5 m/s from a height of 58.8 m. (*a*) What will be its speed just before it strikes the ground? (*b*) How long will it take for the ball to reach the ground? (*c*) What would be the answers to (*a*) and (*b*) if the ball were thrown directly up from the same height and with the same initial speed?

57. Figure 32 shows a simple device for measuring your reaction time. It consists of a strip of cardboard marked with a scale and two large dots. A friend holds the strip with his thumb and forefinger at the upper dot and you position your thumb and forefinger at the lower dot, being careful not to touch the strip. Your friend releases the strip, and you try to pinch it as soon as possible after you see it begin to fall. The mark at the place where you pinch the strip gives your reaction time. How far from the lower dot should you place the 50-, 100-, 200-, and 250-ms marks?

Figure 32   Problem 57.

58. A ball thrown straight up takes 2.25 s to reach a height of 36.8 m. (*a*) What was its initial speed? (*b*) What is its speed at this height? (*c*) How much higher will the ball go?

59. While thinking of Isaac Newton, a person standing on a bridge overlooking a highway inadvertently drops an apple over the railing just as the front end of a truck passes directly below the railing. If the vehicle is moving at 55 km/h (= 34 mi/h) and is 12 m (= 39 ft) long, how far above the truck must the railing be if the apple just misses hitting the rear end of the truck?

60. A rocket is fired vertically and ascends with a constant vertical acceleration of 20 m/s² for 1.0 min. Its fuel is then all used and it continues as a free-fall particle. (*a*) What is the maximum altitude reached? (*b*) What is the total time elapsed from takeoff until the rocket strikes the Earth? (Ignore the variation of *g* with altitude.)

61. A basketball player, about to "dunk" the ball, jumps 76 cm vertically. How much time does the player spend (*a*) in the

Figure 33   Problem 61.

top 15 cm of this jump and (*b*) in the bottom 15 cm? Does this help explain why such players seem to hang in the air at the tops of their jumps. See Fig. 33.

62. A stone is thrown vertically upward. On its way up it passes point *A* with speed *v*, and point *B*, 3.00 m higher than *A*, with speed *v*/2. Calculate (*a*) the speed *v* and (*b*) the maximum height reached by the stone above point *B*.

63. Water drips from the nozzle of a shower onto the floor 200 cm below. The drops fall at regular intervals of time, the first drop striking the floor at the instant the fourth drop begins to fall. Find the location of the individual drops when a drop strikes the floor.

64. The Zero Gravity Research Facility at the NASA Lewis Research Center includes a 145-m drop tower. This is an evacuated vertical tower through which, among other possibilities, a 1-m diameter sphere containing an experimental package can be dropped. (*a*) For how long is the experimental package in free fall? (*b*) What is its speed at the bottom of the tower? (*c*) At the bottom of the tower, the sphere experiences an average acceleration of 25*g* as its speed is reduced to zero. Through what distance does it travel in coming to rest?

65. A ball is dropped from a height of 2.2 m and rebounds to a height of 1.9 m above the floor. Assume the ball was in contact with the floor for 96 ms and determine the average acceleration (magnitude and direction) of the ball during contact with the floor.

66. A woman fell 144 ft from the top of a building, landing on the top of a metal ventilator box, which she crushed to a depth of 18 in. She survived without serious injury. What acceleration (assumed uniform) did she experience during the collision? Express your answer in terms of *g*.

67. If an object travels one-half its total path in the last second of its fall from rest, find (*a*) the time and (*b*) the height of its fall. Explain the physically unacceptable solution of the quadratic time equation.

68. Two objects begin a free fall from rest from the same height 1.00 s apart. How long after the first object begins to fall will the two objects be 10.0 m apart?

69. As Fig. 34 shows, Clara jumps from a bridge, followed closely by Jim. How long did Jim wait after Clara jumped? Assume that Jim is 170 cm tall and that the jumping-off level is at the top of the figure. Make scale measurements directly on the figure.

Figure 34 Problem 69.

70. A balloon is ascending at 12.4 m/s at a height of 81.3 m above the ground when a package is dropped. (*a*) With what speed does the package hit the ground? (*b*) How long did it take to reach the ground?

71. A parachutist after bailing out falls 52.0 m without friction. When the parachute opens, she decelerates at 2.10 m/s² and reaches the ground with a speed of 2.90 m/s. (*a*) How long is the parachutist in the air? (*b*) At what height did the fall begin?

72. A lead ball is dropped into a swimming pool from a diving board 2.6 m above the water. It hits the water with a certain velocity and then sinks to the bottom with this same constant velocity. It reaches the bottom 0.97 s after it is dropped. (*a*) How deep is the pool? (*b*) Suppose that all the water is drained from the pool. The ball is thrown from the diving board so that it again reaches the bottom in 0.97 s. What is the initial velocity of the ball?

73. At the National Physical Laboratory in England (the British equivalent of our National Institute of Standards and Technology), a measurement of the acceleration *g* was made by throwing a glass ball straight up in an evacuated tube and letting it return, as in Fig. 35. Let Δt_{L} be the time interval between the two passages across the lower level, Δt_{U} the time interval between the two passages across the upper level, and *H* the distance between the two levels. Show that

$$g = \frac{8H}{\Delta t_{\mathrm{L}}^2 - \Delta t_{\mathrm{U}}^2}.$$

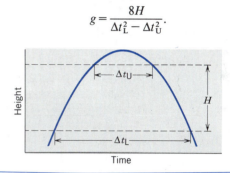

Figure 35 Problem 73.

74. A steel ball bearing is dropped from the roof of a building (the initial velocity of the ball is zero). An observer standing in front of a window 120 cm high notes that the ball takes 0.125 s to fall from the top to the bottom of the window. The ball bearing continues to fall, makes a completely elastic collision with a horizontal sidewalk, and reappears at the bottom of the window 2.0 s after passing it on the way down. How tall is the building? (The ball will have the same speed at a point going up as it had going down after a completely elastic collision.)

75. A dog sees a flowerpot sail up and then back down past a window 1.1 m high. If the total time the pot is in sight is 0.74 s, find the height above the top of the window to which the pot rises.

CHAPTER 3

VECTORS

Many of the laws of physics involve not only algebraic relationships among quantities, but geometrical relationships as well. For example, picture a spinning top that rotates rapidly about its axis, while the axis of rotation itself rotates slowly about the vertical. This geometrical relationship is complicated to represent by algebraic equations. However, if we use vectors *to represent the physical variables, a single equation is sufficient to explain the behavior. Vectors permit such economy of expression in a great variety of physical laws. Sometimes the vector form of a physical law permits us to see relationships or symmetries that would otherwise be obscured by a cumbersome algebraic equation.*

In this chapter we explore some of the properties and uses of vectors and we introduce the mathematical operations that involve vectors. In the process you will learn that familiar symbols from arithmetic, such as +, −, and ×, have different meanings when applied to vectors.

3-1 VECTORS AND SCALARS

A change of position of a particle is called a *displacement.* If a particle moves from position A to position B (Fig. 1a), we can represent its displacement by drawing a line from A to B. The direction of displacement can be shown by putting an arrowhead at B indicating that the displacement was *from A to B.* The path of the particle need not necessarily be a straight line from A to B; the arrow represents only the net effect of the motion, not the actual motion.

In Fig. 1b, for example, we plot an actual path followed by a particle from A to B. The path is not the same as the displacement AB. If we were to take snapshots of the particle when it was at A and, later, when it was at some intermediate position P, we could obtain the displacement vector AP, representing the net effect of the motion during this interval, even though we would not know the actual path taken between these points. Furthermore, a displacement such as $A'B'$ (Fig. 1a), which is parallel to AB, similarly directed, and equal in length to AB, represents the same *change* in position as AB. We make no distinction between these two displacements. A displacement is therefore characterized by a *length* and a *direction.*

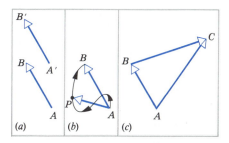

Figure 1 Displacement vectors. (*a*) Vectors AB and $A'B'$ are identical, since they have the same length and point in the same direction. (*b*) The actual *path* of the particle in moving from A to B may be the curve shown; the *displacement* is the vector AB. At the intermediate point P, the displacement is the vector AP. (*c*) After displacement AB, the particle undergoes another displacement BC. The net effect of the two displacements is the vector AC.

In a similar way, we can represent a subsequent displacement from B to C (Fig. 1c). The net effect of the two displacements is the same as a displacement from A to C. We speak then of AC as the *sum* or *resultant* of the displacements AB and BC. Notice that this sum is not an algebraic sum and that a number alone cannot uniquely specify it.

37

Quantities that behave like displacements are called *vectors*. (The word *vector* means *carrier* in Latin. Biologists use the term *vector* to mean an insect, animal, or other agent that *carries* a cause of disease from one organism to another.) Vectors, then, are quantities that have both magnitude and direction and that follow certain rules of combination, which we describe below. The displacement vector is a convenient prototype. Some other physical quantities that are represented by vectors are force, velocity, acceleration, electric field, and magnetic field. Many of the laws of physics can be expressed in compact form by using vectors, and derivations involving these laws are often greatly simplified if we do so.

Quantities that can be specified completely by a number and unit and that therefore have magnitude only are called *scalars*. Some physical quantities that are scalars are mass, length, time, density, energy, and temperature. Scalars can be manipulated by the rules of ordinary algebra.

3-2 ADDING VECTORS: GRAPHICAL METHOD

To represent a vector on a diagram we draw an arrow. We choose the length of the arrow to be proportional to the magnitude of the vector (that is, we choose a scale), and we choose the direction of the arrow to be the direction of the vector, with the arrowhead giving the sense of the direction. For example, a displacement of 42 m in a northeast direction would be represented on a scale of 1 cm per 10 m by an arrow 4.2 cm long, drawn at an angle of 45° above a line pointing east with the arrowhead at the top right extreme (Fig. 2). A vector is usually represented in printing by a boldface symbol such as **d**. In handwriting we usually put an arrow above the symbol to denote a vector quantity, such as $\vec{d}$.

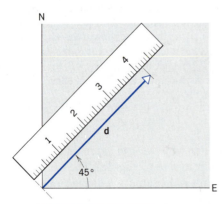

Figure 2   The vector **d** represents a displacement of magnitude 42 m (on a scale in which 10 m = 1 cm) in a direction 45° north of east.

Figure 3   The vector sum **a** + **b** = **s**. Compare with Fig. 1*c*.

Often we are interested only in the magnitude (or length) of the vector and not in its direction. The magnitude of **d** is sometimes written as |**d**|; more frequently we represent the magnitude alone by the italic letter symbol *d*. The boldface symbol is meant to signify both properties of the vector, magnitude and direction. When handwritten, the magnitude of the vector is represented by the symbol without the arrow.

Consider now Fig. 3 in which we have redrawn and relabeled the vectors of Fig. 1*c*. The relation among these vectors can be written

$$\mathbf{a} + \mathbf{b} = \mathbf{s}. \qquad (1)$$

The rules to be followed in performing this vector addition graphically are these: (1) On a diagram drawn to scale lay out the vector **a** with its proper direction in the coordinate system. (2) Draw **b** to the same scale with its tail at the head of **a**, making sure that **b** has its own proper direction (generally different from the direction of **a**). (3) Draw a line from the tail of **a** to the head of **b** to construct the vector sum **s**. If the vectors were representing displacements, then **s** would be a displacement equivalent in length and direction to the successive displacements **a** and **b**. This procedure can be generalized to obtain the sum of any number of vectors.

Since vectors differ from ordinary numbers, we expect different rules for their manipulation. The symbol "+" in Eq. 1 has a meaning different from its meaning in arithmetic or scalar algebra. It tells us to carry out a different set of operations.

By careful inspection of Fig. 4 we can deduce two important properties of vector addition:

$$\mathbf{a} + \mathbf{b} = \mathbf{b} + \mathbf{a} \quad \text{(commutative law)} \qquad (2)$$

and

$$\mathbf{d} + (\mathbf{e} + \mathbf{f}) = (\mathbf{d} + \mathbf{e}) + \mathbf{f} \quad \text{(associative law)}. \qquad (3)$$

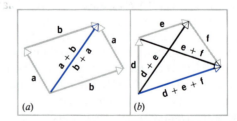

Figure 4   (*a*) The commutative law for vector addition, which states that **a** + **b** = **b** + **a**. (*b*) The associative law, which states that **d** + (**e** + **f**) = (**d** + **e**) + **f**.

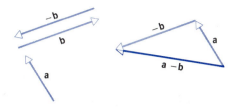

Figure 5   The vector difference **a** − **b** = **a** + (−**b**).

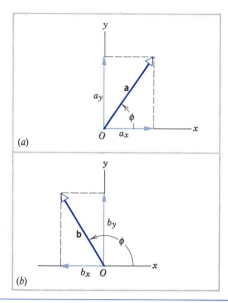

Figure 6   (*a*) The vector **a** has component a_x in the *x* direction and component a_y in the *y* direction. (*b*) The vector **b** has a negative *x* component.

These laws assert that it makes no difference in what order or in what grouping we add vectors; the sum is the same. In this respect, vector addition and scalar addition follow the same rules.

By inspection of Fig. 4*b*, you will see how the graphical method is used to find the sum of more than two vectors, in this case **d** + **e** + **f**. Each succeeding vector is placed with its tail at the head of the previous one. The vector representing the sum is then drawn from the tail of the first vector to the head of the last one.

The operation of subtraction can be included in our vector algebra by defining the negative of a vector to be another vector of equal magnitude but opposite direction. Then

$$\mathbf{a} - \mathbf{b} = \mathbf{a} + (-\mathbf{b}) \tag{4}$$

as shown in Fig. 5. Here −**b** means a vector with the same magnitude as **b** but pointing in the opposite direction. It follows from Eq. 4 that **a** − **a** = **a** + (−**a**) = 0.

Remember that, although we have used displacements to illustrate these operations, the rules apply to *all* vector quantities, such as velocities and forces.

3-3   COMPONENTS OF VECTORS

Even though we defined vector addition with the graphical method, it is not very useful for vectors in three dimensions. Often it is even inconvenient for the two-dimensional case. Another way of adding vectors is the analytical method, involving the resolution of a vector into components with respect to a particular coordinate system.

Figure 6*a* shows a vector **a** whose tail has been placed at the origin of a rectangular coordinate system. If we draw perpendicular lines from the head of **a** to the axes, the quantities a_x and a_y so formed are called the (Cartesian) *components* of the vector **a**. The process is called *resolving a vector into its components*. The vector **a** is completely and uniquely specified by these components; given a_x and a_y, we could immediately reconstruct the vector **a**.

The components of a vector can be positive, negative, or zero. Figure 6*b* shows a vector **b** that has $b_x < 0$ and $b_y > 0$.

The components a_x and a_y in Fig. 6*a* are readily found from

$$a_x = a \cos \phi \quad \text{and} \quad a_y = a \sin \phi, \tag{5}$$

where ϕ is the angle that the vector **a** makes with the positive *x* axis, measured counterclockwise from this axis. As shown in Fig. 6, the algebraic signs of the components of a vector depend on the quadrant in which the angle ϕ lies. For example, when ϕ is between 90° and 180°, as in Fig. 6*b*, the vector always has a negative *x* component and a positive *y* component. The components of a vector behave like scalar quantities because, in any particular coordinate system, only a number with an algebraic sign is needed to specify them.

Once a vector is resolved into its components, the components themselves can be used to specify the vector. Instead of the two numbers *a* (magnitude of the vector) and ϕ (direction of the vector relative to the *x* axis), we now have the two numbers a_x and a_y. We can pass back and forth between the description of a vector in terms of its components (a_x and a_y) and the equivalent description in terms of magnitude and direction (*a* and ϕ). To obtain *a* and ϕ from a_x and a_y, we note from Fig. 6*a* that

$$a = \sqrt{a_x^2 + a_y^2} \tag{6a}$$

and

$$\tan \phi = a_y/a_x. \tag{6b}$$

The quadrant in which ϕ lies is determined from the signs of a_x and a_y.

In three dimensions the process works similarly: just draw perpendicular lines from the tip of the vector to the three coordinate axes *x*, *y*, and *z*. Figure 7 shows one way

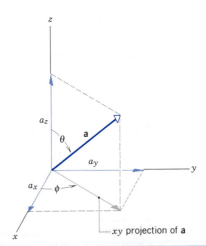

Figure 7   A vector **a** in three dimensions with components a_x, a_y, and a_z. The x and y components are conveniently found by first drawing the xy projection of **a**. The angle θ between **a** and the z axis is called the *polar angle*. The angle ϕ in the xy plane between the projection of **a** and the x axis is called the *azimuthal angle*. The azimuthal angle ϕ has the same meaning here as it does in Fig. 6.

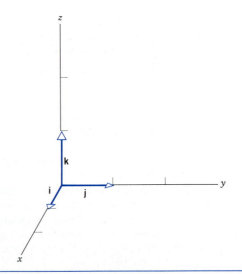

Figure 8   The unit vectors **i**, **j**, and **k** are used to specify the positive x, y, and z axes, respectively. Each vector is dimensionless and has a length of unity.

this is often drawn to make the components easier to recognize; the component (sometimes called a *projection*) of **a** in the xy plane is first drawn, and then from its tip we can find the individual components a_x and a_y. We would obtain exactly the same x and y components if we worked directly with the vector **a** instead of with its xy projection, but the drawing would not be as clear. From the geometry of Fig. 7, we can deduce the components of the vector **a** to be

$$a_x = a \sin \theta \cos \phi, \quad a_y = a \sin \theta \sin \phi, \quad \text{and}$$
$$a_z = a \cos \theta. \tag{7}$$

When resolving a vector into components it is sometimes useful to introduce a vector of unit length in a given direction. Often it is convenient to draw unit vectors along the particular coordinate axes chosen. In the rectangular coordinate system the special symbols **i**, **j**, and **k** are usually used for unit vectors in the positive x, y, and z directions, respectively (see Fig. 8). In handwritten notation, unit vectors are often indicated with a circumflex or "hat," such as $\hat{\imath}$, $\hat{\jmath}$, and $\hat{k}$.

Note that **i**, **j**, and **k** need not be located at the origin. Like all vectors, they can be translated anywhere in the coordinate space as long as their directions with respect to the coordinate axes are not changed.

In general, a vector **a** in a three-dimensional coordinate system can be written in terms of its components and the unit vectors as

$$\mathbf{a} = a_x \mathbf{i} + a_y \mathbf{j} + a_z \mathbf{k}, \tag{8a}$$

or in two dimensions as

$$\mathbf{a} = a_x \mathbf{i} + a_y \mathbf{j}. \tag{8b}$$

The vector relation Eq. 8b is equivalent to the scalar relations of Eq. 6. Each equation relates the vector (**a**, or a and ϕ) to its components (a_x and a_y). Sometimes we call quantities such as $a_x \mathbf{i}$ and $a_y \mathbf{j}$ in Eq. 8b the *vector components* of **a**. Figure 9 shows the vectors **a** and **b** of Fig. 6 drawn in terms of their vector components. Many physical problems involving vectors can be simplified by replacing a vector by its vector components. That is, the

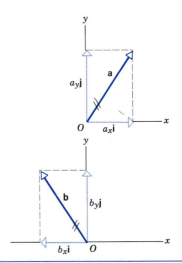

Figure 9   The *vector* components of **a** and **b**. In any physical situation that involves vectors, we get the same outcome whether we use the vector itself, such as **a**, or its two vector components, $a_x \mathbf{i}$ and $a_y \mathbf{j}$. The effect of the single vector **a** is equivalent to the net effect of the two vectors $a_x \mathbf{i}$ and $a_y \mathbf{j}$. When we have replaced a vector with its vector components, it is helpful to draw a double line through the original vector, as shown; this helps us to remember not to consider the original vector any more.

action of a quantity represented as a vector can be replaced by the actions of its vector components. When necessary, we refer explicitly to *vector components,* while the word *component* alone continues to refer to the scalar quantities a_x and a_y.

Other Coordinate Systems *(Optional)*

Many other varieties of coordinate systems may be appropriate for analyzing certain physical situations. For example, the two-dimensional xy coordinate system may be changed in either of two ways: (1) by moving the origin to another location in the xy plane, which is called a *translation* of the coordinate system, or (2) by pivoting the xy axes about the fixed origin, which is a *rotation* of the coordinate system. In both of these operations we keep the vector fixed and move the coordinate axes. Figure 10 shows the effect of these two changes. In the case shown in Fig. 10a, the components are unchanged, but in the case shown in Fig. 10b, the components do change.

When the physical situation we are analyzing has certain symmetries, it may be advantageous to choose a different coordinate system for resolving a vector into its components. For instance, we might choose the radial and tangential directions of plane polar coordinates, shown in Fig. 11. In this case, we find the components on the coordinate axes just as we did in the ordinary xyz system: we draw a perpendicular from the tip of the vector to each coordinate axis.

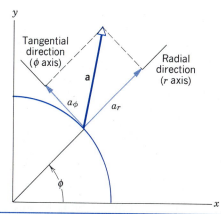

Figure 11   The vector **a** is resolved into its radial and tangential components. These components will have important applications when we consider circular motion in Chapters 4 and 11.

The three-dimensional extensions of Fig. 11 (spherical polar or cylindrical polar coordinates) in many important cases are far superior to Cartesian coordinate systems for the analysis of physical problems. For example, the gravitational force exerted by the Earth on distant objects has the symmetry of a sphere, and thus its properties are most simply described in spherical polar coordinates. The magnetic force exerted by a long straight current-carrying wire has the symmetry of a cylinder and is therefore most simply described in cylindrical polar coordinates. ∎

3-4   ADDING VECTORS: COMPONENT METHOD

Now that we have shown how to resolve vectors into their components, we can consider the addition of vectors by an analytic method.

Let **s** be the sum of the vectors **a** and **b**, or

$$\mathbf{s} = \mathbf{a} + \mathbf{b}. \tag{9}$$

If two vectors, such as **s** and **a** + **b**, are to be equal, they must have the same magnitude and must point in the same direction. This can only happen if their corresponding components are equal. We stress this important conclusion:

Two vectors are equal to each other only if their corresponding components are equal.

For the vectors of Eq. 9, we can write

$$s_x\mathbf{i} + s_y\mathbf{j} = a_x\mathbf{i} + a_y\mathbf{j} + b_x\mathbf{i} + b_y\mathbf{j}$$
$$= (a_x + b_x)\mathbf{i} + (a_y + b_y)\mathbf{j}. \tag{10}$$

Equating the x components on both sides of Eq. 10 gives

$$s_x = a_x + b_x, \tag{11a}$$

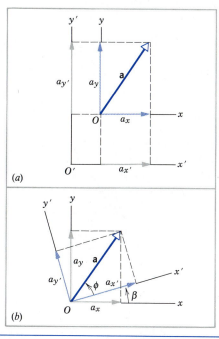

Figure 10   (a) The origin O of the coordinate system of Fig. 6a has been moved or *translated* to the new position O'. The x and y components of **a** are identical to the x' and y' components. (b) The x and y axes have been rotated through the angle β. The x' and y' components are different from the x and y components (note that the y' component is now smaller than the x' component, while in Fig. 6a the y component was greater than the x component), but *the vector* **a** *is unchanged.* By what angle should we rotate the coordinate axes to make the y' component zero?

and equating the y components gives

$$s_y = a_y + b_y. \qquad (11b)$$

These two algebraic equations, taken together, are equivalent to the single vector relation of Eq. 9.

Instead of specifying the components of **s**, we can give its length and direction:

$$s = \sqrt{s_x^2 + s_y^2} = \sqrt{(a_x + b_x)^2 + (a_y + b_y)^2} \qquad (12a)$$

and

$$\tan \phi = \frac{s_y}{s_x} = \frac{a_y + b_y}{a_x + b_x}. \qquad (12b)$$

Here is the rule for adding vectors by this method. (1) Resolve each vector into its components, keeping track of the algebraic sign of each component. (2) Add the components for each coordinate axis, taking the algebraic sign into account. (3) The sums so obtained are the components of the sum vector. Once we know the components of the sum vector, we can easily reconstruct that vector in space.

The advantage of the method of breaking up vectors into components, rather than adding directly with the use of suitable trigonometric relations, is that we always deal with right triangles and thus simplify the calculations.

In adding vectors by the component method, the choice of coordinate axes determines how simple the process will be. Sometimes the components of the vectors with respect to a particular set of axes are known at the start, so that the choice of axes is obvious. Other times a judicious choice of axes can greatly simplify the job of resolution of the vectors into components. For example, the axes can be oriented so that at least one of the vectors lies parallel to an axis; the components of that vector along the other axes will then be zero.

Sample Problem 1   An airplane travels 209 km on a straight course making an angle of 22.5° east of due north. How far north and how far east did the plane travel from its starting point?

Solution   We choose the positive x direction to be east and the positive y direction to be north. Next, we draw a displacement vector (Fig. 12) from the origin (starting point), making an angle of 22.5° with the y axis (north) inclined along the positive x direction (east). The length of the vector represents a magnitude of 209 km. If we call this vector **d**, then d_x gives the distance traveled east of the starting point and d_y gives the distance traveled north of the starting point. We have

$$\phi = 90.0° - 22.5° = 67.5°,$$

so that (see Eqs. 5)

$$d_x = d \cos \phi = (209 \text{ km}) (\cos 67.5°) = 80.0 \text{ km},$$

and

$$d_y = d \sin \phi = (209 \text{ km}) (\sin 67.5°) = 193 \text{ km}.$$

We have used Cartesian components in this sample problem, even though the Earth's surface is curved and therefore non-Cartesian. For example, a plane starting on the equator and flying northeast will eventually be due north of its starting point,

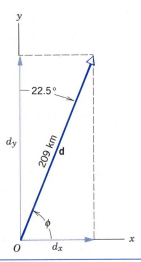

Figure 12   Sample Problem 1.

which could never occur in a flat coordinate system. Similarly, two planes starting at different points on the equator and flying due north at the same speed along parallel paths will eventually collide at the north pole. This also would be impossible in a flat coordinate system. If we restrict our calculations to distances that are small with respect to the radius of the Earth (6400 km), we can safely use Cartesian coordinates for analyzing displacements on the Earth's surface.

Sample Problem 2   An automobile travels due east on a level road for 32 km. It then turns due north at an intersection and travels 47 km before stopping. Find the resultant displacement of the car.

Solution   We choose a coordinate system fixed with respect to the Earth, with the positive x direction pointing east and the positive y direction pointing north. The two successive displacements, **a** and **b**, are then drawn as shown in Fig. 13. The resultant displacement **s** is obtained from $\mathbf{s} = \mathbf{a} + \mathbf{b}$. Since **b** has no x component and **a** has no y component, we obtain (see Eqs. 11)

$$s_x = a_x + b_x = 32 \text{ km} + 0 = 32 \text{ km},$$

$$s_y = a_y + b_y = 0 + 47 \text{ km} = 47 \text{ km}.$$

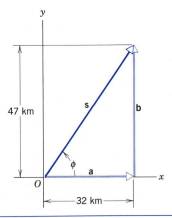

Figure 13   Sample Problem 2.

The magnitude and direction of **s** are then (see Eqs. 12)

$$s = \sqrt{s_x^2 + s_y^2} = \sqrt{(32 \text{ km})^2 + (47 \text{ km})^2} = 57 \text{ km},$$

$$\tan \phi = \frac{s_y}{s_x} = \frac{47 \text{ km}}{32 \text{ km}} = 1.47, \qquad \phi = \tan^{-1}(1.47) = 56°.$$

The resultant vector displacement **s** has a magnitude of 57 km and makes an angle of 56° north of east.

Sample Problem 3   Three coplanar vectors are expressed with respect to a certain rectangular coordinate system as

$$\mathbf{a} = 4.3\mathbf{i} - 1.7\mathbf{j},$$

$$\mathbf{b} = -2.9\mathbf{i} + 2.2\mathbf{j},$$

and

$$\mathbf{c} = -3.6\mathbf{j},$$

in which the components are given in arbitrary units. Find the vector **s** which is the sum of these vectors.

Solution   Generalizing Eqs. 11 to the case of three vectors, we have

$$s_x = a_x + b_x + c_x = 4.3 - 2.9 + 0 = 1.4,$$

and

$$s_y = a_y + b_y + c_y = -1.7 + 2.2 - 3.6 = -3.1.$$

Thus

$$\mathbf{s} = s_x\mathbf{i} + s_y\mathbf{j} = 1.4\mathbf{i} - 3.1\mathbf{j}.$$

Figure 14 shows the four vectors. From Eqs. 6 we can calculate that the magnitude of **s** is 3.4 and that the angle ϕ that **s** makes with the positive x axis, measured counterclockwise from that axis, is

$$\phi = \tan^{-1}(-3.1/1.4) = 294°.$$

Most pocket calculators return angles between $+90°$ and $-90°$ for the arctan. In this case, $-66°$ (which our calculator gives) is equivalent to 294°. However, we would obtain the same angle if we asked for $\tan^{-1}(3.1/-1.4)$, which should give an angle in the second (upper left) quadrant. Drawing a sketch similar to Fig. 14 will keep you from going too far wrong, and if necessary you can convert your calculator's value into a result in the correct quadrant by using the identity $\tan(-\phi) = \tan(180° - \phi)$.

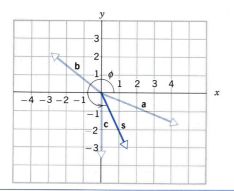

Figure 14   Sample Problem 3.

3-5 MULTIPLICATION OF VECTORS

When we add scalar quantities, the items being added must have the same dimensions, and the sum likewise has the same dimensions. The same rule applies in adding vector quantities. On the other hand, we can multiply scalar quantities of different dimensions and obtain a product of dimensions possibly different from either of the quantities being multiplied, for example, distance = velocity × time.

Like scalars, vectors of different kinds can be multiplied by one another to generate quantities of new physical dimensions. Because vectors have direction as well as magnitude, vector multiplication cannot follow exactly the same rules as the algebraic rules of scalar multiplication. We must establish new rules of multiplication for vectors.

We find it useful to define three kinds of multiplication operations for vectors: (1) multiplication of a vector by a scalar, (2) multiplication of two vectors in such a way as to yield a scalar, and (3) multiplication of two vectors in such a way as to yield another vector. There are still other possibilities, but we do not consider them here.

1.   *Multiplication of a vector by a scalar.* The multiplication of a vector by a scalar has a simple meaning: the product of a scalar c and a vector **a**, written $c\mathbf{a}$, is defined to be a new vector whose magnitude is c times the magnitude of **a**. The new vector has the same direction as **a** if c is positive and the opposite direction if c is negative, as shown in Fig. 15. To divide a vector by a scalar we simply multiply the vector by the reciprocal of the scalar. Often the scalar is not a pure number but a physical quantity with dimensions and units.

2.   *Multiplication of two vectors to yield a scalar.* The *scalar product* of two vectors **a** and **b**, written as $\mathbf{a} \cdot \mathbf{b}$, is defined to be

$$\mathbf{a} \cdot \mathbf{b} = ab \cos \phi, \tag{13}$$

where a is the magnitude of vector **a**, b is the magnitude of vector **b**, and $\cos \phi$ is the cosine of the angle ϕ between the

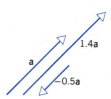

Figure 15   Multiplication of a vector **a** by a scalar c gives a vector $c\mathbf{a}$ whose magnitude is c times the magnitude of **a**. The vector $c\mathbf{a}$ has the same direction as **a** if c is positive and the opposite direction if c is negative. Examples are illustrated for $c = +1.4$ and $c = -0.5$.

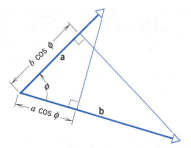

Figure 16   The scalar product $\mathbf{a} \cdot \mathbf{b}$ $(= ab \cos \phi)$ is the product of the magnitude of one vector (a, for instance) and the component of the other vector in the direction of the first ($b \cos \phi$).

two vectors (see Fig. 16). Because of the notation, $\mathbf{a} \cdot \mathbf{b}$ is also called the dot product of $\mathbf{a}$ and $\mathbf{b}$ and is spoken as "$\mathbf{a}$ dot $\mathbf{b}$." The dot product is independent of the choice of coordinate axes.

Since a and b are scalars and $\cos \phi$ is a pure number, the scalar product of two vectors is a scalar. The scalar product of two vectors can be regarded as the product of the magnitude of one vector and the component of the other vector in the direction of the first, as in Fig. 16. The scalar product can correspondingly be expressed either as $a(b \cos \phi)$ or as $b(a \cos \phi)$.

We could have defined $\mathbf{a} \cdot \mathbf{b}$ to be any operation at all, for example, $a^{\frac{1}{2}} b^{\frac{1}{4}} \tan(\phi/2)$, but this would turn out to be of no use to us in physics. With our definition of the scalar product, a number of important physical quantities can be described as the scalar product of two vectors. Some of them are mechanical work, gravitational potential energy, electrical potential, electric power, and electromagnetic energy density. Later in the text, these physical quantities will be defined in terms of scalar products of two vectors.

If two vectors are perpendicular, their dot product vanishes. Using the definition of the dot product, we can derive the following relationships for the Cartesian unit vectors $\mathbf{i}$, $\mathbf{j}$, and $\mathbf{k}$:

$$\mathbf{i} \cdot \mathbf{i} = \mathbf{j} \cdot \mathbf{j} = \mathbf{k} \cdot \mathbf{k} = 1,$$
$$\mathbf{i} \cdot \mathbf{j} = \mathbf{i} \cdot \mathbf{k} = \mathbf{j} \cdot \mathbf{k} = 0. \tag{14}$$

With these relationships, we can find (see Problem 35) an alternative form for the dot product of two vectors $\mathbf{a}$ and $\mathbf{b}$ in a three-dimensional xyz coordinate system in terms of their components:

$$\mathbf{a} \cdot \mathbf{b} = a_x b_x + a_y b_y + a_z b_z. \tag{15}$$

3. *Multiplication of two vectors to yield another vector.* The *vector product* of two vectors $\mathbf{a}$ and $\mathbf{b}$ is written as $\mathbf{a} \times \mathbf{b}$ and is another vector $\mathbf{c}$, where $\mathbf{c} = \mathbf{a} \times \mathbf{b}$. The *magnitude* of $\mathbf{c}$ is defined by

$$c = |\mathbf{a} \times \mathbf{b}| = ab \sin \phi, \tag{16}$$

where ϕ is the (smaller) angle between $\mathbf{a}$ and $\mathbf{b}$. There are two different angles between $\mathbf{a}$ and $\mathbf{b}$: ϕ as in Fig. 16 and

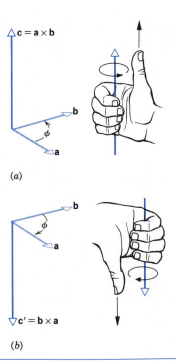

Figure 17   The right-hand rule for vector products. (*a*) Swing vector $\mathbf{a}$ into vector $\mathbf{b}$ with the fingers of your right hand. Your thumb shows the direction of $\mathbf{c}$. (*b*) Reversing the procedure shows that $(\mathbf{b} \times \mathbf{a}) = -(\mathbf{a} \times \mathbf{b})$.

$2\pi - \phi$. We always choose the smaller of these angles in vector multiplication. In Eq. 13 for the scalar product, it makes no difference which we choose, because $\cos(2\pi - \phi) = \cos \phi$. However, it does matter for Eq. 16, because $\sin(2\pi - \phi) = -\sin \phi$.

Because of the notation, $\mathbf{a} \times \mathbf{b}$ is also called the cross product of $\mathbf{a}$ and $\mathbf{b}$ and is spoken as "$\mathbf{a}$ cross $\mathbf{b}$."

The *direction* of $\mathbf{c}$, the vector product of $\mathbf{a}$ and $\mathbf{b}$, is defined to be perpendicular to the plane formed by $\mathbf{a}$ and $\mathbf{b}$. To specify the sense of the vector $\mathbf{c}$ we refer to Fig. 17. Draw the vectors $\mathbf{a}$ and $\mathbf{b}$ intersecting at their tails, and imagine an axis perpendicular to the plane of $\mathbf{a}$ and $\mathbf{b}$ through their origin. Now wrap the fingers of the *right hand* around this axis and push the vector $\mathbf{a}$ into the vector $\mathbf{b}$ through the smaller angle between them with the fingertips, keeping the thumb extended; the direction of the thumb then gives the direction of the vector product $\mathbf{a} \times \mathbf{b}$. This procedure describes a convention. The two vectors $\mathbf{a}$ and $\mathbf{b}$ form a plane, and there are two (opposite) directions for a vector $\mathbf{c}$ that is perpendicular to the plane. Our choice is based on a right-hand convention. (A left-hand convention would give the opposite direction for the vector product.)

If ϕ is $90°$, then $\mathbf{a}$, $\mathbf{b}$, and $\mathbf{c}$ $(= \mathbf{a} \times \mathbf{b})$ are all at right angles to one another and give the directions of a three-dimensional right-handed coordinate system.

Notice that $\mathbf{b} \times \mathbf{a}$ is not the same vector as $\mathbf{a} \times \mathbf{b}$, so that the order of factors in a vector product is important. This is not true for scalars because the order of factors in algebra or arithmetic does not affect the resulting product. Actually, $\mathbf{a} \times \mathbf{b} = -(\mathbf{b} \times \mathbf{a})$, as shown in Fig. 17. This can be deduced from the fact that the magnitude $ab \sin \phi$ equals the magnitude $ba \sin \phi$, but the direction of $\mathbf{a} \times \mathbf{b}$ is opposite to that of $\mathbf{b} \times \mathbf{a}$. The cross product, as we have defined it so far, is independent of the choice of coordinate axes.

The three Cartesian unit vectors $\mathbf{i}$, $\mathbf{j}$, and $\mathbf{k}$ in a right-handed coordinate system are related by the cross product $\mathbf{i} \times \mathbf{j} = \mathbf{k}$. (This in effect defines what we mean by a right-handed system. Unless stated otherwise, we always use right-handed coordinate systems.) Keeping $\mathbf{i}$, $\mathbf{j}$, and $\mathbf{k}$ in the same cyclic order we can also write $\mathbf{k} \times \mathbf{i} = \mathbf{j}$ and $\mathbf{j} \times \mathbf{k} = \mathbf{i}$. If we change the order, a minus sign enters, for example, $\mathbf{j} \times \mathbf{i} = -\mathbf{k}$. The cross product of any two like unit vectors vanishes ($\mathbf{i} \times \mathbf{i} = \mathbf{j} \times \mathbf{j} = \mathbf{k} \times \mathbf{k} = 0$), as does the cross product of any vector with itself ($\mathbf{a} \times \mathbf{a} = 0$). With these relationships for the cross products of like and of different unit vectors, you can show (see Problem 36) that

$$\mathbf{a} \times \mathbf{b} = (a_y b_z - a_z b_y)\mathbf{i} \\ + (a_z b_x - a_x b_z)\mathbf{j} + (a_x b_y - a_y b_x)\mathbf{k}. \quad (17)$$

The reason for defining the vector product in this way is that it proves to be useful in physics. We often encounter physical quantities that are vectors whose product, defined as above, is a vector quantity having important physical meaning. Some examples of physical quantities that are vector products are torque, angular momentum, the force on a moving charge in a magnetic field, and the flow of electromagnetic energy. When such quantities are discussed later, their connection with the vector product of two vectors will be pointed out.

Generalized Products of Vectors *(Optional)*
The scalar product is the simplest product of two vectors. The order of multiplication does not affect the product. The vector product is the next simplest case. Here the order of multiplication does affect the product, but only by a factor of minus one, which implies a direction reversal. Other products of vectors are useful but more involved. For example, a *tensor* can be generated by multiplying each of the three components of one vector

by the three components of another vector. Hence a tensor (of the second rank) has nine numbers associated with it, a vector three, and a scalar only one. Some physical quantities that can be represented by tensors are mechanical and electrical stress, rotational inertia, and strain. Still more complex physical quantities are possible. In this book, however, we are concerned only with scalars and vectors. ■

Sample Problem 4 A certain vector $\mathbf{a}$ in the xy plane is directed 250° counterclockwise from the positive x axis and has magnitude 7.4 units. Vector $\mathbf{b}$ has magnitude 5.0 units and is directed parallel to the z axis. Calculate (*a*) the scalar product $\mathbf{a} \cdot \mathbf{b}$ and (*b*) the vector product $\mathbf{a} \times \mathbf{b}$.

Solution (*a*) Because $\mathbf{a}$ and $\mathbf{b}$ are perpendicular to one another, the angle ϕ between them is 90° and $\cos \phi = \cos 90° = 0$. Therefore, from Eq. 13, the scalar product is

$$\mathbf{a} \cdot \mathbf{b} = ab \cos \phi = ab \cos 90° = (7.4)(5.0)(0) = 0,$$

consistent with the fact that neither vector has a component in the direction of the other.

(*b*) The *magnitude* of the vector product is, from Eq. 16,

$$|\mathbf{a} \times \mathbf{b}| = ab \sin \phi = (7.4)(5.0)\sin 90° = 37.$$

The *direction* of the vector product is perpendicular to the plane formed by $\mathbf{a}$ and $\mathbf{b}$. Therefore, as shown in Fig. 18, it lies in the xy plane (perpendicular to $\mathbf{b}$) at an angle of $250° - 90° = 160°$ from the positive x axis (perpendicular to $\mathbf{a}$ in accordance with the right-hand rule).

We can find the components of $\mathbf{a} \times \mathbf{b}$ using Eq. 17. We first need the components of $\mathbf{a}$ and $\mathbf{b}$:

$$a_x = 7.4 \cos 250° = -2.5, \quad b_x = 0, \\ a_y = 7.4 \sin 250° = -7.0, \quad b_y = 0, \\ a_z = 0; \quad b_z = 5.0.$$

Thus we have

$$\mathbf{a} \times \mathbf{b} = [(-7.0)(5.0) - (0)(0)]\mathbf{i} + [(0)(0) - (-2.5)(5.0)]\mathbf{j} \\ + [(-2.5)(0) - (-7.0)(0)]\mathbf{k} \\ = -35\mathbf{i} + 13\mathbf{j}.$$

This is consistent with the magnitude and direction shown in Fig. 18.

Figure 18 Sample Problem 4.

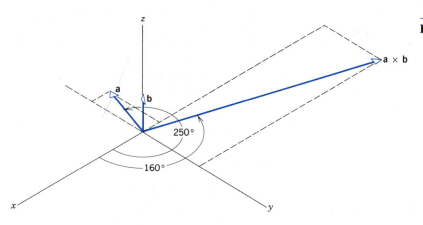

3-6 VECTOR LAWS IN PHYSICS*   *(Optional)*

Referring to Fig. 10*b*, it can be shown (see Problem 51) that the components of the displacement vector **a** in the rotated $(x'y')$ coordinate system are related to those in the original (xy) system by

$$a_{x'} = a_x \cos \beta + a_y \sin \beta \qquad (18a)$$

$$a_{y'} = -a_x \sin \beta + a_y \cos \beta, \qquad (18b)$$

where β is the angle through which the coordinate axes have been rotated.

Equations 18 are examples of *transformation* equations, which relate the components of the displacement vector in one coordinate system with its components in any rotated system. We can use these equations to formulate a more general and rigorous definition of a vector, which so far we have defined as a physical quantity that has both magnitude and direction and that obeys certain rules of combination. We can now replace that definition by a more specific one:

For any physical quantity (velocity or force, for example) to be represented by a vector, the components of that quantity must transform under rotation by the rules given in Eqs. 18.

Although Eqs. 18 hold for vectors in two-dimensional space, they can be generalized to three dimensions. The two-dimensional case, however, illustrates all the essential concepts.

As indicated by Fig. 10, a vector is unchanged or *invariant* when the coordinate axes are translated or rotated. Certain physical quantities have this same property; in the case of velocity, for example, you will measure the same value for the velocity of a passing car as will your friend in the house across the street (as long as your houses are at rest relative to each other!). Quantities that have these properties, and that obey the laws of vector arithmetic given in this chapter, are represented as vectors. Among the quantities that are represented as vectors are velocity, acceleration, force, linear momentum, angular momentum, and electric and magnetic fields. Equations relating these quantities are vector equations; examples of vector equations are $\mathbf{a} \cdot \mathbf{b} = s$ (a scalar), $\mathbf{a} + 2\mathbf{b} = 6\mathbf{c}$, $\mathbf{a} \times \mathbf{b} = \mathbf{c}$, and so on. On the other hand, many physical quantities are well described by scalars and scalar equations: temperature, pressure, mass, energy, and time. One of the characteristics of vector equations is that they not only indicate the mathematical relationship between physical quantities but also the geometrical relationship as well. Let us consider some examples of equations that we shall develop and discuss at length later in this text; here we only present the equations as examples of the basic forms.

We begin with Newton's second law, $\mathbf{F} = m\mathbf{a}$ (see Chapter 5), which gives the force **F** that must act on a particle of mass m to provide it with an acceleration **a**. On the right side we have the scalar mass times the vector acceleration, and on the left side we have the vector force. The equation looks simple, but it is rich in content. Carrying out the multiplication and equating components, we discover in reality *three* independent equations: $F_x = ma_x$, $F_y = ma_y$, and $F_z = ma_z$. Each of these equations can be solved separately in studying how the particle responds to the force. Thus the y component of the force, for instance, has abso-

lutely no effect on the x or z components of the acceleration. Equivalently, we might say that the direction of the acceleration of a system is determined by the direction of the force acting on it (because multiplying a vector by a positive scalar gives a vector in the same direction). We shall use this fact in the next chapter when we analyze the two-dimensional motion of projectiles moving under the influence of gravity.

Laws involving scalar products arise in several different contexts. Our first example will come in the definition of mechanical work W, a scalar, that is done by a force **F** acting on a system to produce a displacement **d**: $W = \mathbf{F} \cdot \mathbf{d} = Fd \cos \phi$ (see Chapter 7). In this case, the force need not necessarily be parallel to the displacement; imagine, for example, that you are pulling a sled along the ground with a rope over your shoulder. The displacement will be horizontal, but the force (which is applied along the rope) will have both horizontal and vertical components. Note that according to the geometrical relationships illustrated in Fig. 16, only the component of **F** along **d** (which is $F \cos \phi$) will contribute to the actual work done. Once again, the vector equation carries information about a geometrical relationship.

An example of a physical law involving a vector product can be found in the equation $\mathbf{F} = q\mathbf{v} \times \mathbf{B}$ (see Chapter 34), which gives the force **F** experienced by an electric charge q moving with velocity **v** through a magnetic field **B**. The geometric nature of the force, determined by the vector equation, is responsible for the bending of particle trajectories into circular orbits, as in large particle accelerators such as cyclotrons. Note that the force is always at right angles both to the velocity and to the direction of the magnetic field. Without the vector equation we would have difficulty understanding the basis for this behavior.

The physical laws that are represented by the vector equations are universal and independent of any particular choice of coordinate system. Specifically, if we were to examine the motion of a charged particle in a magnetic field from two coordinate systems, one of which was rotated with respect to the other, we would certainly find that the vectors **F** (force), **v** (velocity), and **B** (magnetic field) have different components in the rotated system (as in Fig. 10*b*), but observers in both systems would agree on the form of the physical law. That is, in the rotated system the transformed vectors must satisfy $\mathbf{F}' = q\mathbf{v}' \times \mathbf{B}'$.

This transformation property is such a sensible way for nature to behave that we often take for granted that nature *should* behave that way. Aside from any purely local effects, for example, the electric force between two electrons separated by a certain distance should *not* depend on whether the separation is measured north–south or east–west. It is not too difficult to imagine a universe that is not so well behaved; the length of a vector, for instance, might change when we translated or rotated it. Physicists and mathematicians have speculated on why our universe has these particular symmetries such as translation and rotation, and they have learned that there are fascinating relationships between symmetries of nature and certain quantities that are *conserved* (that is, their total amount is unchanged) in physical processes. For example, the invariance of physical laws under the symmetry of *time translation* (that is, if a law holds on Monday it also holds on Tuesday) leads directly to the law of conservation of energy.

Reflection Symmetry, Polar Vectors, and Axial Vectors

There is another kind of transformation that is quite different from translation and rotation. This transformation involves in-

---

* The material in this section can be skipped without loss of continuity.

verting the coordinate system, that is, $x \rightarrow -x$, $y \rightarrow -y$, and $z \rightarrow -z$. In effect, the entire system is reflected through the origin.

On the surface, you might expect that for this transformation all we need do in our equations is replace each vector component by its negative. (Scalars are unaffected by this inversion.) After all, if we invert the x axis without changing the vector **a**, then clearly $a_x \rightarrow -a_x$. Thus instead of drawing an inverted coordinate system, all we need to do is draw the vector $-\mathbf{a}$ in the original coordinate system. These expectations are quite correct for a large class of physical quantities that we represent with vectors: velocity, acceleration, force, linear momentum, electric field. Such well-behaved vectors are given the general name *polar vectors.*

Another class of vectors does not follow this kind of behavior upon inversion. For example, as illustrated in Fig. 19, it is often helpful to represent a particle moving in a circle by an *angular velocity* vector ω. The magnitude of ω tells in effect how rapidly the particle is rotating, and the direction of ω is perpendicular to the plane of the circle and determined by a right-hand rule. (If you curl the fingers of your right hand in the direction of the particle's motion, your extended thumb points in the direction of ω.)

Consider now the situation as the orbit of the particle is inverted or reflected through the origin, as in Fig. 19. The vector **r** that locates the particle P relative to the center of the circle is transformed to $\mathbf{r}' = -\mathbf{r}$, and the velocity becomes $\mathbf{v}' = -\mathbf{v}$. As the original particle moves from a to b, the reflected particle moves from a' to b'. The sense or direction of rotation (clockwise or counterclockwise) is unchanged, and so $\omega' = \omega$. Thus, unlike the polar vectors **r** and **v**, the angular velocity does *not* change sign when the coordinates are inverted. Such a vector is called an *axial vector* or a *pseudovector;* torque and magnetic field are other examples.

The vectors **r**, **v**, and ω are related through the cross product $\mathbf{v} = \omega \times \mathbf{r}$, as we shall discuss when we consider rotational motion in Chapter 11. If all three vectors changed sign upon inversion, then the relationships among the reflected vectors would be $-\mathbf{v} = (-\omega) \times (-\mathbf{r}) = \omega \times \mathbf{r}$. This is a contradiction, for $\omega \times \mathbf{r}$ cannot be both $-\mathbf{v}$ and **v** (unless **v** is zero, which is not the case here). Thus the transformation $\omega' = \omega$ is absolutely necessary, in order for the physical relationship $\mathbf{v} = \omega \times \mathbf{r}$ to have the same form $\mathbf{v}' = \omega' \times \mathbf{r}'$ in the reflected system. This is what is meant by the invariance of a physical law to a certain transformation of the coordinate system. That is, if we write an equation for a physical law in one coordinate system, transform each vector corresponding to the transformed coordinates, and substitute the transformed vectors into the physical law, the result should be an equation identical in form to the original one.

Until about 1956, it was believed that all physical laws were unchanged by inversions like that of Fig. 19 (and by translations and rotations as well). In 1956, however, it was discovered that the inversion symmetry was violated in a certain class of radioactive decays called beta decays, in which electrons are emitted from the atomic nucleus. Nuclei rotate on their axes like tiny tops, and it is possible to assign to each nucleus a vector like ω representing its rotation. In the beta-decay experiment, the direction of emission of the electrons relative to the direction of ω was studied (Fig. 20). If equal numbers of electrons were emitted parallel to ω and antiparallel to ω, then the reflected experiment would look exactly like the original and the inversion symmetry would be valid. It was discovered that nearly all the electrons

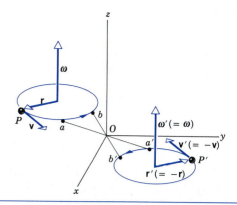

Figure 19 A particle P moving in a circle is represented by the angular velocity vector ω. If all the coordinates are reflected through the origin O, the "reflected" particle P' rotates in a circle and is represented by the angular velocity vector ω'.

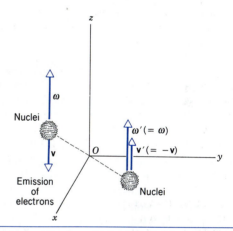

Figure 20 A collection of spinning nuclei, represented by an angular velocity vector ω, emits electrons preferentially in a direction opposite to ω. In the reflected version of the experiment, the electrons would be emitted parallel to ω'. The experiment and its reflected image look quite different from one another, showing that reflection symmetry is violated by these decays.

were emitted opposite to ω, so that in the reflected experiment more electrons would be emitted along ω (because **v** changes sign on reflection while ω does not). The experiment differs from its mirror image; the inversion symmetry and the associated conservation law called *conservation of parity* were found not to be valid in this case.*

This experiment helped to revolutionize our thinking about fundamental processes, and it provided an essential clue about the nature of the physical law that is responsible for the beta-decay process, which is one of the four basic forces. It was the forerunner of a series of experiments that have revealed other relationships between transformation properties, invariance principles, and symmetries. ∎

---

* See *The New Ambidextrous Universe,* by Martin Gardner (W. H. Freeman and Company, 1990).

QUESTIONS

1. In 1969, three Apollo astronauts left Cape Canaveral, went to the Moon and back, and splashed down at a selected landing site in the Pacific Ocean; see Fig. 21. An admiral bid them goodbye at the Cape and then sailed to the Pacific Ocean in an aircraft carrier to pick them up. Compare the displacements of the astronauts and the admiral.

Figure 21 Question 1.

2. A dog runs 100 m south, 100 m east, and 100 m north, ending up at his starting point, his displacement for the entire trip being zero. Where is his starting point? One clear answer is the North Pole but there is another solution, located near the South Pole. Describe it.

3. Can two vectors having different magnitudes be combined to give a zero resultant? Can three vectors?

4. Can a vector have zero magnitude if one of its components is not zero?

5. Can the sum of the magnitudes of two vectors ever be equal to the magnitude of the sum of these two vectors?

6. Can the magnitude of the difference between two vectors ever be greater than the magnitude of either vector? Can it be greater than the magnitude of their sum? Give examples.

7. Suppose that $d = d_1 + d_2$. Does this mean that we must have either $d \geq d_1$ or $d \geq d_2$? If not, explain why.

8. If three vectors add up to zero, they must all be in the same plane. Make this plausible.

9. Do the unit vectors $\mathbf{i}$, $\mathbf{j}$, and $\mathbf{k}$ have units?

10. Explain in what sense a vector equation contains more information than a scalar equation.

11. Name several scalar quantities. Does the value of a scalar quantity depend on the coordinate system you choose?

12. You can order events in time. For example, event b may precede event c but follow event a, giving us a time order of events a, b, c. Hence, there is a sense of time, distinguishing past, present, and future. Is time a vector therefore? If not, why not?

13. Do the commutative and associative laws apply to vector subtraction?

14. Can a scalar product be a negative quantity?

15. (a) If $\mathbf{a} \cdot \mathbf{b} = 0$, does it follow that $\mathbf{a}$ and $\mathbf{b}$ are perpendicular to one another? (b) If $\mathbf{a} \cdot \mathbf{b} = \mathbf{a} \cdot \mathbf{c}$, does it follow that $\mathbf{b} = \mathbf{c}$?

16. If $\mathbf{a} \times \mathbf{b} = 0$, must $\mathbf{a}$ and $\mathbf{b}$ be parallel to each other? Is the converse true?

17. A vector $\mathbf{a}$ lies parallel to the Earth's rotation axis, pointing from south to north. A second vector $\mathbf{b}$ points vertically upward at your location. What is the direction of the vector $\mathbf{a} \times \mathbf{b}$? At what locations on the Earth's surface is the magnitude of the vector $\mathbf{a} \times \mathbf{b}$ a maximum? A minimum?

18. Must you specify a coordinate system when you (a) add two vectors, (b) form their scalar product, (c) form their vector product, or (d) find their components?

19. (a) Show that if all the components of a vector are reversed in direction, then the vector itself is reversed in direction. (b) Show that if the components of the two vectors forming a vector product are all reversed, then the vector product is not changed. (c) Is a vector product, then, a vector?

20. We have discussed addition, subtraction, and multiplication of vectors. Why do you suppose that we do not discuss division of vectors? Is it possible to define such an operation?

21. It is conventional to use, as we did, the right-hand rule for vector algebra. What changes would be required if a left-handed convention were adopted instead?

22. (a) Convince yourself that the vector product of two polar vectors is an axial vector. (b) What is the vector product of a polar vector with an axial vector?

PROBLEMS

Section 3-2 Adding Vectors: Graphical Method

1. Consider two displacements, one of magnitude 3 m and another of magnitude 4 m. Show how the displacement vectors may be combined to get a resultant displacement of magnitude (a) 7 m, (b) 1 m, and (c) 5 m.

2. What are the properties of two vectors $\mathbf{a}$ and $\mathbf{b}$ such that (a) $\mathbf{a} + \mathbf{b} = \mathbf{c}$ and $a + b = c$; (b) $\mathbf{a} + \mathbf{b} = \mathbf{a} - \mathbf{b}$; (c) $\mathbf{a} + \mathbf{b} = \mathbf{c}$ and $a^2 + b^2 = c^2$?

3. A woman walks 250 m in the direction 35° east of north, then 170 m directly east. (*a*) Using graphical methods, find her final displacement from the starting point. (*b*) Compare the magnitude of her displacement with the distance she walked.

4. A person walks in the following pattern: 3.1 km north, then 2.4 km west, and finally 5.2 km south. (*a*) Construct the vector diagram that represents this motion. (*b*) How far and in what direction would a bird fly in a straight line to arrive at the same final point?

5. Two vectors **a** and **b** are added. Show graphically with vector diagrams that the magnitude of the resultant cannot be greater than $a + b$ or smaller than $|a - b|$, where the vertical bars signify absolute value.

6. A car is driven east for a distance of 54 km, then north for 32 km, and then in a direction 28° east of north for 27 km. Draw the vector diagram and determine the total displacement of the car from its starting point.

7. Vector **a** has a magnitude of 5.2 units and is directed east. Vector **b** has a magnitude of 4.3 units and is directed 35° west of north. By constructing vector diagrams, find the magnitudes and directions of (*a*) **a** + **b**, and (*b*) **a** − **b**.

8. A golfer takes three putts to get his ball into the hole once he is on the green. The first putt displaces the ball 12 ft north, the second 6.0 ft southeast, and the third 3.0 ft southwest. What displacement was needed to get the ball into the hole on the first putt? Draw the vector diagram.

9. A bank in downtown Boston is robbed (see the map in Fig. 22). To elude police, the thieves escape by helicopter, making three successive flights described by the following displacements: 20 mi, 45° south of east; 33 mi, 26° north of west; 16 mi, 18° east of south. At the end of the third flight they are captured. In what town are they apprehended? (Use the graphical method to add these displacements on the map.)

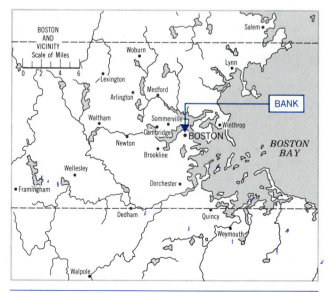

Figure 22 Problem 9.

Section 3-3 Components of Vectors

10. (*a*) What are the components of a vector **a** in the *xy* plane if its direction is 252° counterclockwise from the positive *x* axis and its magnitude is 7.34 units? (*b*) The *x* component of a certain vector is −25 units and the *y* component is +43 units. What are the magnitude of the vector and the angle between its direction and the positive *x* axis?

11. A heavy piece of machinery is raised by sliding it 13 m along a plank oriented at 22° to the horizontal, as shown in Fig. 23. (*a*) How high above its original position is it raised? (*b*) How far is it moved horizontally?

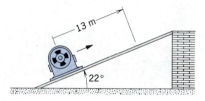

Figure 23 Problem 11.

12. The minute hand of a wall clock measures 11.3 cm from axis to tip. What is the displacement vector of its tip (*a*) from a quarter after the hour to half past, (*b*) in the next half hour, and (*c*) in the next hour?

13. A person desires to reach a point that is 3.42 km from her present location and in a direction that is 35.0° north of east. However, she must travel along streets that go either north–south or east–west. What is the minimum distance she could travel to reach her destination?

14. A ship sets out to sail to a point 124 km due north. An unexpected storm blows the ship to a point 72.6 km to the north and 31.4 km to the east of its starting point. How far, and in what direction, must it now sail to reach its original destination?

15. Rock *faults* are ruptures along which opposite faces of rock have moved past each other, parallel to the fracture surface. Earthquakes often accompany this movement. In Fig. 24 points *A* and *B* coincided before faulting. The component of the net displacement *AB* parallel to the horizontal surface fault line is called the *strike-slip (AC)*. The component of the net displacement along the steepest line of the fault plane is the *dip-slip (AD)*. (*a*) What is the net shift if the strike-slip is 22 m and the dip-slip is 17 m? (*b*) If the fault plane is inclined 52° to the horizontal, what is the net *vertical* displacement of *B* as a result of the faulting in (*a*)?

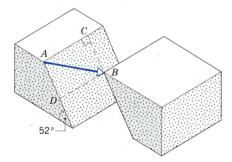

Figure 24 Problem 15.

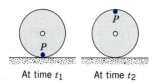

Figure 25   Problem 16.

16. A wheel with a radius of 45 cm rolls without slipping along a horizontal floor, as shown in Fig. 25. P is a dot painted on the rim of the wheel. At time t_1, P is at the point of contact between the wheel and the floor. At a later time t_2, the wheel has rolled through one-half of a revolution. What is the displacement of P during this interval?

17. A room has the dimensions 10 ft $\times$ 12 ft $\times$ 14 ft. A fly starting at one corner ends up at a diametrically opposite corner. (a) Find the displacement vector in a frame with coordinate axes parallel to the edges of the room. (b) What is the magnitude of the displacement? (c) Could the length of the path traveled by the fly be less than this distance? Greater than this distance? Equal to this distance? (d) If the fly walks rather than flies, what is the length of the shortest path it can take?

Section 3-4 Adding Vectors: Component Method

18. (a) What is the sum in unit vector notation of the two vectors $\mathbf{a} = 5\mathbf{i} + 3\mathbf{j}$ and $\mathbf{b} = -3\mathbf{i} + 2\mathbf{j}$? (b) What are the magnitude and direction of $\mathbf{a} + \mathbf{b}$?

19. Two vectors are given by $\mathbf{a} = 4\mathbf{i} - 3\mathbf{j} + \mathbf{k}$ and $\mathbf{b} = -\mathbf{i} + \mathbf{j} + 4\mathbf{k}$. Find (a) $\mathbf{a} + \mathbf{b}$, (b) $\mathbf{a} - \mathbf{b}$, and (c) a vector $\mathbf{c}$ such that $\mathbf{a} - \mathbf{b} + \mathbf{c} = 0$.

20. Given two vectors, $\mathbf{a} = 4\mathbf{i} - 3\mathbf{j}$ and $\mathbf{b} = 6\mathbf{i} + 8\mathbf{j}$, find the magnitudes and directions (with the $+x$ axis) of (a) $\mathbf{a}$, (b) $\mathbf{b}$, (c) $\mathbf{a} + \mathbf{b}$, (d) $\mathbf{b} - \mathbf{a}$, and (e) $\mathbf{a} - \mathbf{b}$.

21. (a) A man leaves his front door, walks 1400 m east, 2100 m north, and then takes a penny from his pocket and drops it from a cliff 48 m high. In a coordinate system in which the positive x, y, and z axes point east, north, and up, the origin being at the location of the penny as the man leaves his front door, write down an expression, using unit vectors, for the displacement of the penny. (b) The man returns to his front door, following a different path on the return trip. What is his resultant displacement for the round trip?

22. A particle undergoes three successive displacements in a plane, as follows: 4.13 m southwest, 5.26 m east, and 5.94 m in a direction 64.0° north of east. Choose the x axis pointing east and the y axis pointing north and find (a) the components of each displacement, (b) the components of the resultant displacement, (c) the magnitude and direction of the resultant displacement, and (d) the displacement that would be required to bring the particle back to the starting point.

23. Two vectors $\mathbf{a}$ and $\mathbf{b}$ have equal magnitudes of 12.7 units. They are oriented as shown in Fig. 26 and their vector sum is $\mathbf{r}$. Find (a) the x and y components of $\mathbf{r}$, (b) the magnitude of $\mathbf{r}$, and (c) the angle $\mathbf{r}$ makes with the $+x$ axis.

24. A radar station detects a missile approaching from the east. At first contact, the range to the missile is 12,000 ft at 40.0° above the horizon. The missile is tracked for another 123° in the east–west plane, the range at final contact being 25,800

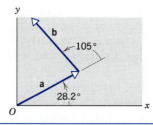

Figure 26   Problem 23.

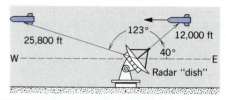

Figure 27   Problem 24.

ft; see Fig. 27. Find the displacement of the missile during the period of radar contact.

25. Two vectors of magnitudes a and b make an angle θ with each other when placed tail to tail. Prove, by taking components along two perpendicular axes, that the magnitude of their sum is

$$r = \sqrt{a^2 + b^2 + 2ab\cos\theta}.$$

26. Prove that two vectors must have equal magnitudes if their sum is perpendicular to their difference.

27. (a) Using unit vectors along three cube edges, express the diagonals (the lines from one corner to another through the center of the cube) of a cube in terms of its edges, which have length a. (b) Determine the angles made by the diagonals with the adjacent edges. (c) Determine the length of the diagonals.

28. A tourist flies from Washington, DC to Manila. (a) Describe the displacement vector. (b) What is its magnitude? The latitude and longitude of the two cities are 39° N, 77° W and 15° N, 121° E. (*Hint:* See Fig. 7 and Eqs. 7. Let the z axis be along the Earth's rotation axis, so that $\theta = 90° - $ latitude and $\phi = $ longitude. The radius of the Earth is 6370 km.)

29. Let N be an integer greater than 1; then

$$\cos 0 + \cos\frac{2\pi}{N} + \cos\frac{4\pi}{N} + \cdots + \cos(N-1)\frac{2\pi}{N} = 0;$$

that is,

$$\sum_{n=0}^{n=N-1} \cos\frac{2\pi n}{N} = 0.$$

Also

$$\sum_{n=0}^{n=N-1} \sin\frac{2\pi n}{N} = 0.$$

Prove these two statements by considering the sum of N vectors of equal length, each vector making an angle of $2\pi/N$ with that preceding.

Section 3-5 Multiplication of Vectors

30. A vector $\mathbf{d}$ has a magnitude of 2.6 m and points north. What are the magnitudes and directions of the vectors (a) $-\mathbf{d}$, (b) $\mathbf{d}/2.0$, (c) $-2.5\mathbf{d}$, and (d) $5.0\mathbf{d}$?

31. Show for any vector **a** that (*a*) $\mathbf{a} \cdot \mathbf{a} = a^2$ and (*b*) $\mathbf{a} \times \mathbf{a} = 0$.

32. A vector **a** of magnitude 12 units and another vector **b** of magnitude 5.8 units point in directions differing by 55°. Find (*a*) the scalar product of the two vectors and (*b*) the vector product.

33. Two vectors, **r** and **s**, lie in the *xy* plane. Their magnitudes are 4.5 and 7.3 units, respectively, whereas their directions are 320° and 85° measured counterclockwise from the positive *x* axis. What are the values of (*a*) $\mathbf{r} \cdot \mathbf{s}$ and (*b*) $\mathbf{r} \times \mathbf{s}$?

34. Find (*a*) "north" cross "west," (*b*) "down" dot "south," (*c*) "east" cross "up," (*d*) "west" dot "west," and (*e*) "south" cross "south." Let each vector have unit magnitude.

35. Given two vectors, $\mathbf{a} = a_x\mathbf{i} + a_y\mathbf{j} + a_z\mathbf{k}$ and $\mathbf{b} = b_x\mathbf{i} + b_y\mathbf{j} + b_z\mathbf{k}$, prove that the scalar product $\mathbf{a} \cdot \mathbf{b}$ is given in terms of the components by Eq. 15.

36. Given two vectors, $\mathbf{a} = a_x\mathbf{i} + a_y\mathbf{j} + a_z\mathbf{k}$ and $\mathbf{b} = b_x\mathbf{i} + b_y\mathbf{j} + b_z\mathbf{k}$, prove that the vector product $\mathbf{a} \times \mathbf{b}$ is given in terms of the components by Eq. 17.

37. Show that $\mathbf{a} \times \mathbf{b}$ can be expressed by a 3×3 determinant as

$$\mathbf{a} \times \mathbf{b} = \begin{vmatrix} \mathbf{i} & \mathbf{j} & \mathbf{k} \\ a_x & a_y & a_z \\ b_x & b_y & b_z \end{vmatrix}.$$

38. Use Eqs. 13 and 15 to calculate the angle between the two vectors $\mathbf{a} = 3\mathbf{i} + 3\mathbf{j} + 3\mathbf{k}$ and $\mathbf{b} = 2\mathbf{i} + \mathbf{j} + 3\mathbf{k}$.

39. Three vectors are given by $\mathbf{a} = 3\mathbf{i} + 3\mathbf{j} - 2\mathbf{k}$, $\mathbf{b} = -\mathbf{i} - 4\mathbf{j} + 2\mathbf{k}$, and $\mathbf{c} = 2\mathbf{i} + 2\mathbf{j} + \mathbf{k}$. Find (*a*) $\mathbf{a} \cdot (\mathbf{b} \times \mathbf{c})$, (*b*) $\mathbf{a} \cdot (\mathbf{b} + \mathbf{c})$, and (*c*) $\mathbf{a} \times (\mathbf{b} + \mathbf{c})$.

40. (*a*) Calculate $\mathbf{r} = \mathbf{a} - \mathbf{b} + \mathbf{c}$, where $\mathbf{a} = 5\mathbf{i} + 4\mathbf{j} - 6\mathbf{k}$, $\mathbf{b} = -2\mathbf{i} + 2\mathbf{j} + 3\mathbf{k}$, and $\mathbf{c} = 4\mathbf{i} + 3\mathbf{j} + 2\mathbf{k}$. (*b*) Calculate the angle between **r** and the $+z$ axis. (*c*) Find the angle between **a** and **b**.

41. Three vectors add to zero, as in the right triangle of Fig. 28. Calculate (*a*) $\mathbf{a} \cdot \mathbf{b}$, (*b*) $\mathbf{a} \cdot \mathbf{c}$, and (*c*) $\mathbf{b} \cdot \mathbf{c}$.

42. Three vectors add to zero, as in Fig. 28. Calculate (*a*) $\mathbf{a} \times \mathbf{b}$, (*b*) $\mathbf{a} \times \mathbf{c}$, and (*c*) $\mathbf{b} \times \mathbf{c}$.

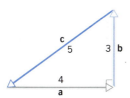

Figure 28 Problems 41 and 42.

43. Vector **a** lies in the *yz* plane 63.0° from the $+y$ axis with a positive *z* component and has magnitude 3.20 units. Vector **b** lies in the *xz* plane 48.0° from the $+x$ axis with a positive *z* component and has magnitude 1.40 units. Find (*a*) $\mathbf{a} \cdot \mathbf{b}$, (*b*) $\mathbf{a} \times \mathbf{b}$, and (*c*) the angle between **a** and **b**.

44. (*a*) We have seen that the commutative law *does not* apply to vector products; that is, $\mathbf{a} \times \mathbf{b}$ does not equal $\mathbf{b} \times \mathbf{a}$. Show that the commutative law *does* apply to scalar products; that is, $\mathbf{a} \cdot \mathbf{b} = \mathbf{b} \cdot \mathbf{a}$. (*b*) Show that the distributive law applies to both scalar products and vector products; that is, show that

$$\mathbf{a} \cdot (\mathbf{b} + \mathbf{c}) = \mathbf{a} \cdot \mathbf{b} + \mathbf{a} \cdot \mathbf{c}$$

and that

$$\mathbf{a} \times (\mathbf{b} + \mathbf{c}) = \mathbf{a} \times \mathbf{b} + \mathbf{a} \times \mathbf{c}.$$

(*c*) Does the associative law apply to vector products; that is, does $\mathbf{a} \times (\mathbf{b} \times \mathbf{c})$ equal $(\mathbf{a} \times \mathbf{b}) \times \mathbf{c}$? (*d*) Does it make any sense to talk about an associative law for scalar products?

45. Show that the area of the triangle contained between the vectors **a** and **b** in Fig. 29 is $\frac{1}{2}|\mathbf{a} \times \mathbf{b}|$, where the vertical bars signify magnitude.

46. Show that the magnitude of a vector product gives numerically the area of the parallelogram formed with the two component vectors as sides (see Fig. 29). Does this suggest how an element of area oriented in space could be represented by a vector?

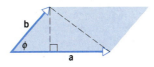

Figure 29 Problems 45 and 46.

47. Show that $\mathbf{a} \cdot (\mathbf{b} \times \mathbf{c})$ is equal in magnitude to the volume of the parallelepiped formed on the three vectors **a**, **b**, and **c** as shown in Fig. 30.

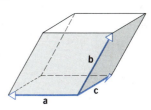

Figure 30 Problem 47.

48. Two vectors **a** and **b** have components, in arbitrary units, $a_x = 3.2$, $a_y = 1.6$; $b_x = 0.50$, $b_y = 4.5$. (*a*) Find the angle between **a** and **b**. (*b*) Find the components of a vector **c** that is perpendicular to **a**, is in the *xy* plane, and has a magnitude of 5.0 units.

49. Find the angles between the body diagonals of a cube. See Problem 27.

50. The three vectors shown in Fig. 31 have magnitudes $a = 3$, $b = 4$, $c = 10$. (*a*) Calculate the *x* and *y* components of these vectors. (*b*) Find the numbers *p* and *q* such that $\mathbf{c} = p\mathbf{a} + q\mathbf{b}$.

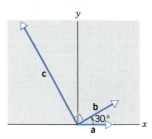

Figure 31 Problem 50.

Section 3-6 Vector Laws in Physics

51. Use Fig. 10*b* to derive Eqs. 18.

52. A vector **a** with a magnitude of 17 m is directed 56° counter-clockwise from the +x axis, as shown in Fig. 32. (a) What are the components a_x and a_y of the vector? (b) A second coordinate system is inclined by 18° with respect to the first. What are the components $a_{x'}$ and $a_{y'}$ in this "primed" coordinate system?

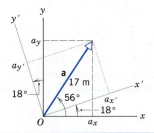

Figure 32   Problem 52.

53. Figure 33 shows two vectors **a** and **b** and two systems of coordinates which differ in that the x and x′ axes and the y and y′ axes each make an angle β with each other. Prove analytically that **a** + **b** has the same magnitude and direction no matter which system is used to carry out the analysis. (*Hint:* Use Eqs. 18.)

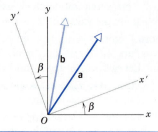

Figure 33   Problem 53.

CHAPTER 4

MOTION IN TWO AND THREE DIMENSIONS

This chapter presents a combination or synthesis of the concepts developed in Chapters 2 and 3. We continue to describe the motion of a particle in terms of its position, velocity, and acceleration, just as we did in Chapter 2. However, here we remove the restriction imposed in Chapter 2 that the particle move only in a straight line. Now we allow the particle to move throughout an ordinary three-dimensional coordinate system. Keeping track of the x, y, and z components of the motion is greatly simplified if we use a notation based on vectors. We see that the kinematic equations of Chapter 2 can be applied in the general case by merely replacing the one-dimensional variable with a corresponding vector. Two familiar examples of motion are considered as applications of the vector techniques: a projectile launched with both horizontal and vertical velocity components in the Earth's gravity, and an object moving in a circular path.

4-1 POSITION, VELOCITY, AND ACCELERATION

Figure 1 shows a particle at time t moving along a curved path in three dimensions. Its *position,* or displacement from the origin, is measured by the vector $\mathbf{r}$. The *velocity* is indicated by the vector $\mathbf{v}$ which, as we shall show below,

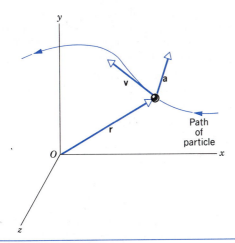

Figure 1 Position, velocity, and acceleration vectors for a particle moving along an arbitrary path. The relative lengths of the three vectors are independent of one another, as are their relative directions.

must be tangent to the path of the particle. The *acceleration* is indicated by the vector $\mathbf{a}$, whose direction, as we shall see more explicitly later, does not in general bear any unique relationship to the position of the particle or the direction of $\mathbf{v}$.

In Cartesian coordinates, the particle is located by x, y, and z, which are the components of the vector $\mathbf{r}$ that gives the position of the particle:

$$\mathbf{r} = x\mathbf{i} + y\mathbf{j} + z\mathbf{k}. \qquad (1)$$

Suppose the particle moves from a position $\mathbf{r}_1$ at time t_1 to position $\mathbf{r}_2$ at time t_2, as shown in Fig. 2a. Its displacement (change in position) in the interval $\Delta t = t_2 - t_1$ is the *vector* $\Delta \mathbf{r} = \mathbf{r}_2 - \mathbf{r}_1$, and the average velocity $\bar{\mathbf{v}}$ in the interval Δt is

$$\bar{\mathbf{v}} = \frac{\Delta \mathbf{r}}{\Delta t}. \qquad (2)$$

In Eq. 2, the vector $\Delta \mathbf{r}$ is multiplied by the scalar $1/\Delta t$ to give the vector $\bar{\mathbf{v}}$. Thus $\bar{\mathbf{v}}$ must have the same direction as $\Delta \mathbf{r}$.

Note that the three vectors, $\mathbf{r}_1$, $\Delta \mathbf{r}$, and $\mathbf{r}_2$ have the same relationship as the three vectors $\mathbf{a}$, $\mathbf{b}$, and $\mathbf{s}$ in Fig. 3 of Chapter 3. That is, using the graphical head-to-tail addition method, $\Delta \mathbf{r}$ added to $\mathbf{r}_1$ gives the resultant $\mathbf{r}_2$. Thus $\mathbf{r}_2 = \Delta \mathbf{r} + \mathbf{r}_1$, and so $\Delta \mathbf{r} = \mathbf{r}_2 - \mathbf{r}_1$.

As the interval Δt is reduced, the vector $\Delta \mathbf{r}$ approaches the actual path (as in Fig. 2b), and it becomes tangent to

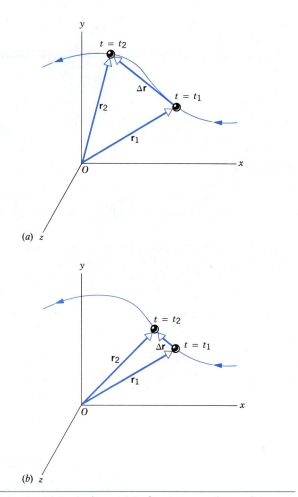

(a)

(b)

Figure 2   (a) In the interval Δt from t_1 to t_2, the particle moves from position $\mathbf{r}_1$ to position $\mathbf{r}_2$. Its displacement in that interval is $\Delta \mathbf{r} = \mathbf{r}_2 - \mathbf{r}_1$. (b) As the interval grows smaller, the displacement vector approaches the actual path of the particle.

the path in the limit $\Delta t \rightarrow 0$, in which case the average velocity approaches the instantaneous velocity $\mathbf{v}$:

$$\mathbf{v} = \lim_{\Delta t \to 0} \frac{\Delta \mathbf{r}}{\Delta t}. \qquad (3)$$

By a reasonable extension of our earlier definition of a derivative (see Eq. 8 of Chapter 2), we write the quantity on the right side of Eq. 3 as the derivative of the vector $\mathbf{r}$ with respect to time:

$$\mathbf{v} = \frac{d\mathbf{r}}{dt}. \qquad (4)$$

Like the vector $\Delta \mathbf{r}$ in the limit $\Delta t \rightarrow 0$, *the vector* $\mathbf{v}$ *is tangent to the path of the particle at every point in the motion.*

   Equation 4, like all vector equations, is equivalent to three scalar equations. To explore this, let us write $\mathbf{v}$ in terms of its components and let us substitute into Eq. 4 for $\mathbf{r}$ from Eq. 1:

$$v_x\mathbf{i} + v_y\mathbf{j} + v_z\mathbf{k} = \frac{d}{dt}(x\mathbf{i} + y\mathbf{j} + z\mathbf{k})$$

$$= \frac{dx}{dt}\mathbf{i} + \frac{dy}{dt}\mathbf{j} + \frac{dz}{dt}\mathbf{k}. \qquad (5)$$

Because two vectors can be equal to each other only if their corresponding components are equal, we see from comparing the left and right sides of Eq. 5 that

$$v_x = \frac{dx}{dt}, \qquad v_y = \frac{dy}{dt}, \qquad v_z = \frac{dz}{dt}. \qquad (6)$$

To summarize, the single vector relation of Eq. 4 is totally equivalent to the three scalar relations of Eq. 6.

   It is now straightforward to extend these concepts to acceleration, just as we did in Section 2-5. The average acceleration is

$$\bar{\mathbf{a}} = \frac{\Delta \mathbf{v}}{\Delta t}, \qquad (7)$$

and the instantaneous acceleration is obtained from the limit as the time interval vanishes:

$$\mathbf{a} = \lim_{\Delta t \to 0} \frac{\Delta \mathbf{v}}{\Delta t}. \qquad (8)$$

Once again, the quantity on the right can be expressed as a derivative with respect to time, and so

$$\mathbf{a} = \frac{d\mathbf{v}}{dt}, \qquad (9)$$

where, again equating like components,

$$a_x = \frac{dv_x}{dt}, \qquad a_y = \frac{dv_y}{dt}, \qquad a_z = \frac{dv_z}{dt}. \qquad (10)$$

Notice that vector equations serve both to simplify notation (Eq. 9, for example, represents the three relationships given as Eq. 10) as well as to separate the components (a_x, for example, has no effect on v_y or v_z).

   Also, note from Eq. 9 that, because $\mathbf{v}$ is a vector having both direction and magnitude, a change in the *direction* of the velocity can produce an acceleration, even if the magnitude of the velocity does not change. Motion at constant speed can be accelerated motion. That is, since $v^2 = v_x^2 + v_y^2 + v_z^2$, the components can change in such a way that the magnitude of $\mathbf{v}$ remains constant. The most familiar example of this case is uniform circular motion, which we discuss in Section 4-4.

Sample Problem 1   A particle moves in an xy plane in such a way that its x and y coordinates vary with time according to $x(t) = t^3 - 32t$ and $y(t) = 5t^2 + 12$. Here x and y are in units of meters when t is in units of seconds. Find the position, velocity, and acceleration of the particle when $t = 3$ s.

Solution   The position is given by Eq. 1, and inserting the expressions given for $x(t)$ and $y(t)$, we obtain

$$\mathbf{r} = x\mathbf{i} + y\mathbf{j} = (t^3 - 32t)\mathbf{i} + (5t^2 + 12)\mathbf{j}.$$

Evaluating this expression at $t = 3$ s gives

$$\mathbf{r} = -69\mathbf{i} + 57\mathbf{j},$$

where the components are in units of meters.

The velocity components are found from Eq. 6:

$$v_x = \frac{dx}{dt} = \frac{d}{dt}(t^3 - 32t) = 3t^2 - 32,$$

$$v_y = \frac{dy}{dt} = \frac{d}{dt}(5t^2 + 12) = 10t.$$

Using Eq. 5, we obtain

$$\mathbf{v} = v_x\mathbf{i} + v_y\mathbf{j} = (3t^2 - 32)\mathbf{i} + 10t\mathbf{j},$$

and at $t = 3$ s we find

$$\mathbf{v} = -5\mathbf{i} + 30\mathbf{j}$$

in units of m/s.

The components of the acceleration are

$$a_x = \frac{dv_x}{dt} = \frac{d}{dt}(3t^2 - 32) = 6t,$$

$$a_y = \frac{dv_y}{dt} = \frac{d}{dt}(10t) = 10.$$

The acceleration at $t = 3$ s is

$$\mathbf{a} = 18\mathbf{i} + 10\mathbf{j}$$

in units of m/s².

Figure 3 shows the path of the particle from $t = 0$ to $t = 4$ s. The position, velocity, and acceleration vectors at $t = 3$ s are drawn. Notice that $\mathbf{v}$ is tangent to the path at $t = 3$ s, and also

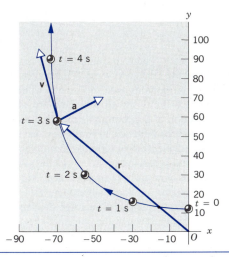

Figure 3   Sample Problem 1. The path of the moving particle is shown, and its positions at $t = 0, 1, 2, 3$, and 4 s are indicated. At $t = 3$ s, the vectors representing its position, velocity, and acceleration are shown. Note that there is no particular relationship between the directions of $\mathbf{r}$, $\mathbf{v}$, and $\mathbf{a}$.

notice that the direction of $\mathbf{a}$ has no particular relationship to the direction of either $\mathbf{r}$ or $\mathbf{v}$.

4-2   MOTION WITH CONSTANT ACCELERATION

We now consider the special case of motion with constant acceleration. As the particle moves, the acceleration $\mathbf{a}$ does not vary either in magnitude or in direction. Hence the components of $\mathbf{a}$ also do not vary. We then have a situation that can be described as the sum of three component motions occurring simultaneously with constant acceleration along each of three perpendicular directions. The particle moves, in general, along a curved path. This may be so even if one component of the acceleration, say a_x, is zero, for then the corresponding component of the velocity, say v_x, has a constant value that may *not* be zero. An example of this latter situation is the motion of a projectile that follows a curved path in a vertical plane and, neglecting the effects of air resistance, is subject to a constant acceleration $\mathbf{g}$ directed down along the vertical axis only.

We can obtain the general equations for motion with constant $\mathbf{a}$ simply by setting

$$a_x = \text{constant}, \quad a_y = \text{constant}, \quad \text{and} \quad a_z = \text{constant}.$$

The particle begins at $t = 0$ with an initial position $\mathbf{r}_0 = x_0\mathbf{i} + y_0\mathbf{j} + z_0\mathbf{k}$ and an initial velocity $\mathbf{v}_0 = v_{x0}\mathbf{i} + v_{y0}\mathbf{j} + v_{z0}\mathbf{k}$. We now proceed as we did in Section 2-6 and develop, in analogy with Eq. 15 of Chapter 2, three scalar equations: $v_x = v_{x0} + a_xt$, $v_y = v_{y0} + a_yt$, and $v_z = v_{z0} + a_zt$, which we write as the single vector equation

$$\mathbf{v} = \mathbf{v}_0 + \mathbf{a}t. \qquad (11)$$

When using this or any other vector equation, remember that it represents three independent scalar equations.

The second term on the right side of Eq. 11 involves the multiplication of a vector by a scalar. As we discussed in Section 3-5, this gives a vector of length at that points in the same direction as the original vector $\mathbf{a}$.

Continuing as we did in Section 2-6, we can develop five equations that describe motion in three dimensions with constant acceleration. These five equations are displayed in Table 1, which you should compare with the five corresponding one-dimensional equations in Table 2 of Chapter 2. With the exception of Eq. 13, which includes vectors but is a scalar equation, each equation of Table 1 represents three independent scalar equations. The x components of Eqs. 11, 12, 14, and 15 are just the corresponding equations listed in Table 2 of Chapter 2. Because Eq. 13 is a scalar equation, *it has no x (or any other) component.*

TABLE 1   VECTOR EQUATIONS FOR MOTION WITH CONSTANT ACCELERATION

Equation Number	Equation	Contains				
		$\mathbf{r}$	$\mathbf{v_0}$	$\mathbf{v}$	$\mathbf{a}$	t
11	$\mathbf{v} = \mathbf{v_0} + \mathbf{a}t$	✗	✓	✓	✓	✓
12	$\mathbf{r} = \mathbf{r_0} + \mathbf{v_0}t + \frac{1}{2}\mathbf{a}t^2$	✓	✓	✗	✓	✓
13[a]	$\mathbf{v} \cdot \mathbf{v} = \mathbf{v_0} \cdot \mathbf{v_0} + 2\mathbf{a} \cdot (\mathbf{r} - \mathbf{r_0})$	✓	✓	✓	✓	✗
14	$\mathbf{r} = \mathbf{r_0} + \frac{1}{2}(\mathbf{v_0} + \mathbf{v})t$	✓	✓	✓	✗	✓
15	$\mathbf{r} = \mathbf{r_0} + \mathbf{v}t - \frac{1}{2}\mathbf{a}t^2$	✓	✗	✓	✓	✓

[a] This equation involves the scalar or dot product of two vectors, which we introduced in Section 3-5.

Sample Problem 2   A skier is moving down a flat slope on a mountainside. The downslope (north–south) makes an angle of 10° with the horizontal. A wind blowing from the west gives the skier a lateral acceleration of 0.54 m/s² (see Fig. 4). At the northwest corner of the slope, the skier pushes off with a downhill component of velocity of 9.0 m/s and a lateral component of zero. The frictionless slope is 125 m long and 25 m wide (*a*) Where does the skier leave the slope? (*b*) What is the skier's velocity at that point? (*Hint:* The gravitational acceleration along a plane that slopes at an angle θ is $g \sin \theta$.)

Solution   (*a*) Choose the origin at the northwest corner, with the *x* axis downslope and the *y* axis lateral. The components of the acceleration are

$$a_x = g \sin 10° = 1.70 \text{ m/s}^2,$$

$$a_y = 0.54 \text{ m/s}^2.$$

Note that these components are evaluated independently. The component a_x is the downhill acceleration that would result even if there were no lateral wind, and similarly a_y is the lateral acceleration that would result from the wind even if there were no downhill slope. Handling these two components independently is the essence of vector arithmetic.

We take $t = 0$ to be the time that the skier pushes off, and we are given that $v_{x0} = 9.0$ m/s and $v_{y0} = 0$. Thus

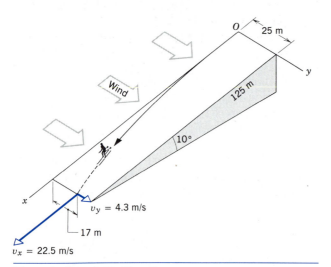

Figure 4   Sample Problem 2.

$$v_x = v_{x0} + a_x t = 9.0 \text{ m/s} + (1.70 \text{ m/s}^2)t,$$

$$v_y = v_{y0} + a_y t = 0 + (0.54 \text{ m/s}^2)t,$$

$$x = x_0 + v_{x0}t + \tfrac{1}{2}a_x t^2 = 0 + (9.0 \text{ m/s})t + (0.85 \text{ m/s}^2)t^2,$$

$$y = y_0 + v_{y0}t + \tfrac{1}{2}a_y t^2 = 0 + 0 + (0.27 \text{ m/s}^2)t^2.$$

We assume for now that the skier reaches the bottom of the slope before leaving the lateral edge. (We can check this assumption later.) We first must find the time at which this occurs (that is, when $x = 125$ m):

$$125 \text{ m} = (9.0 \text{ m/s})t + (0.85 \text{ m/s}^2)t^2.$$

Solving quadratically, we obtain $t = 7.94$ s or -18.5 s. Considering for the moment only the positive root, we evaluate the corresponding *y* coordinate:

$$y = (0.27 \text{ m/s}^2)t^2 = (0.27 \text{ m/s}^2)(7.94 \text{ s})^2 = 17.0 \text{ m}.$$

The lateral displacement of 17.0 m is indeed less than the width of the slope (25 m), as we assumed. The skier therefore leaves the bottom of the slope at a point 17.0 m from the western edge.

(*b*) The velocity components can be found directly at $t = 7.94$ s:

$$v_x = 9.0 \text{ m/s} + (1.70 \text{ m/s}^2)(7.94 \text{ s}) = 22.5 \text{ m/s},$$

$$v_y = (0.54 \text{ m/s}^2)(7.94 \text{ s}) = 4.3 \text{ m/s}.$$

Note that in solving this problem, we have chosen the *x* and *y* axes to lie in the plane of the slope, thereby reducing a three-dimensional problem to two dimensions. If we had chosen to work in a coordinate system in which the *xy* plane was that of level ground and the *z* axis was vertical, the acceleration would have had three components and the problem would have been more complicated. In solving problems, we are usually free to choose the direction of the coordinate axes and the location of the origin for our convenience, as long as we keep our choice fixed throughout the entire solution of the problem.

What about the negative root, $t = -18.5$ s? We wrote our original equations of motion starting at time 0, so positive times are those that describe the skier's subsequent motion down the slope, and negative times therefore must describe the skier's motion *before* passing through the corner of the slope that we defined as the origin. The negative solution reminds us that there might have been a previous path that the skier could have followed to pass through the origin at $t = 0$ with the correct speed. During this previous part of the motion, the skier must have passed through $x = 125$ m (presumably skiing uphill!) at 18.5 s before arriving at the northwest corner. Calculate the velocity components at $t = -18.5$ s and find out about the skier's motion at that time. What would the corresponding *y* coordinate have been at $t = -18.5$ s? Is this reasonable? What would have been the minimum x and y coordinates reached during the time between $t = -18.5$ s and $t = 0$?

The mathematical solution of a physical problem often yields an unexpected result, such as the negative time in this sample problem. If we assume in this problem that the skier's motion began at $t = 0$, the negative root is of no interest to us, but it is good practice to examine the physical meaning of such solutions when they appear.

4-3 PROJECTILE MOTION

An example of motion with constant acceleration is projectile motion. This is the two-dimensional motion of a particle thrown obliquely into the air. The ideal motion of a baseball or a golf ball is an example of projectile motion. We assume for now that we can neglect the effect of the air on this motion. In Chapter 6 we consider the (often considerable) effect of air resistance on projectile motion.

The motion of a projectile is one of constant acceleration $\mathbf{g}$, directed downward. Although there may be a horizontal component of velocity, there is no horizontal component of acceleration. If we choose a coordinate system with the positive y axis vertically upward, we may put $a_y = -g$ (as in Chapter 2, g is always a *positive* number) and $a_x = 0$. Furthermore, we assume that $\mathbf{v}_0$ is in the xy plane, so that $v_{z0} = 0$. Since a_z is also 0, the z component of Eq. 11 tells us that v_z is zero at all times, and we can therefore confine our attention to what happens in the xy plane.

Let us further choose the origin of our coordinate system to be the point at which the projectile begins its flight (see Fig. 5). Hence the origin is the point at which the ball leaves the thrower's hand, for example. This choice of origin implies that $x_0 = y_0 = 0$. The velocity at $t = 0$, the instant the projectile begins its flight, is $\mathbf{v}_0$, which makes an angle ϕ_0 with the positive x direction. The x and y components of $\mathbf{v}_0$ (see Fig. 5) are then

$$v_{x0} = v_0 \cos \phi_0 \quad \text{and} \quad v_{y0} = v_0 \sin \phi_0. \quad (16)$$

Because there is no horizontal component of acceleration, the horizontal component of the velocity is constant. In the x component of Eq. 11 we set $a_x = 0$ and $v_{x0} = v_0 \cos \phi_0$, obtaining

$$v_x = v_{x0} + a_x t = v_0 \cos \phi_0. \quad (17)$$

The horizontal velocity component retains its initial value throughout the flight.

The vertical component of the velocity changes with time due to the constant downward acceleration. In Eq. 11, we take the y components and set $a_y = -g$ and $v_{y0} = v_0 \sin \phi_0$, so that

$$v_y = v_{y0} + a_y t = v_0 \sin \phi_0 - gt. \quad (18)$$

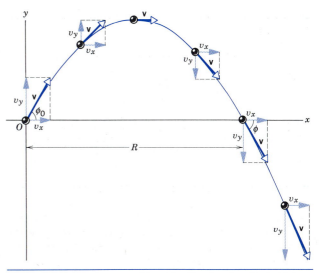

Figure 5 The trajectory of a projectile, showing the initial velocity $\mathbf{v}_0$ and its components and also the velocity $\mathbf{v}$ and its components at five later times. Note that $v_x = v_{x0}$ throughout the flight. The horizontal distance R is the range of the projectile.

The vertical velocity component is that of free fall. (Indeed, if we view the motion of Fig. 5 from a reference frame that moves to the right with a speed v_{x0}, the motion is that of an object thrown vertically upward with an initial speed $v_0 \sin \phi_0$.)

The magnitude of the resultant velocity vector at any instant is

$$v = \sqrt{v_x^2 + v_y^2}. \quad (19)$$

The angle ϕ that the velocity vector makes with the horizontal at that instant is given by

$$\tan \phi = \frac{v_y}{v_x}. \quad (20)$$

The velocity vector is tangent to the path of the particle at every point, as shown in Fig. 5.

The x coordinate of the particle's position at any time, obtained from the x component of Eq. 12 (see Table 1) with $x_0 = 0$, $a_x = 0$, and $v_{x0} = v_0 \cos \phi_0$, is

$$x = x_0 + v_{x0} t + \tfrac{1}{2} a_x t^2 = (v_0 \cos \phi_0) t. \quad (21)$$

The y coordinate, obtained from the y component of Eq. 12 with $y_0 = 0$, $a_y = -g$, and $v_{y0} = v_0 \sin \phi_0$, is

$$y = y_0 + v_{y0} t + \tfrac{1}{2} a_y t^2 = (v_0 \sin \phi_0) t - \tfrac{1}{2} g t^2. \quad (22)$$

Equations 21 and 22 give us x and y as functions of the common parameter t, the time of flight. By combining and eliminating t from them, we obtain

$$y = (\tan \phi_0) x - \frac{g}{2(v_0 \cos \phi_0)^2} x^2, \quad (23)$$

which relates y to x and is the equation of the *trajectory* of the projectile. Since v_0, ϕ_0, and g are constants, this equation has the form

$$y = bx - cx^2,$$

the equation of a parabola. Hence the trajectory of a projectile is parabolic, as shown in Fig. 5.

The *horizontal range R* of the projectile, as shown in Fig. 5, is defined as the distance along the horizontal where the projectile returns to the level from which it was launched. We can find the range by putting $y = 0$ into Eq. 23. One solution immediately arises at $x = 0$; the other gives the range:

$$R = \frac{2v_0^2}{g} \sin \phi_0 \cos \phi_0$$

$$= \frac{v_0^2}{g} \sin 2\phi_0, \tag{24}$$

using the trigonometric identity $\sin 2\theta = 2 \sin \theta \cos \theta$. Note that, for a given initial speed, we get the maximum range for $\phi_0 = 45°$, such that $\sin 2\phi_0 = 1$.

The solutions we have obtained represent an idealized view of projectile motion. We have considered one important effect, namely, gravity; but there is another factor in projectile motion that is often important—air resistance. Air resistance is an example of a velocity-dependent force; the greater the velocity, the greater the decelerating effect of air resistance. At low speed, the effect of air resistance is usually negligible, but at high speed the path of a projectile will no longer be described by a parabola, as in Eq. 23, and the range may be considerably less than that given by Eq. 24. In Chapter 6, we consider the effects of air resistance; for now we assume that the equations derived in this section adequately describe the motion of projectiles.

Figure 6 shows an example of the path of a projectile that is not severely affected by air resistance. The path certainly appears parabolic in its shape. Figure 7 shows a comparison of the motions of a projectile fired horizontally and one simultaneously dropped into free fall. Here you can see directly the predictions of Eqs. 21 and 22 when $\phi_0 = 0$. Note that (1) the horizontal motion of the first projectile does indeed follow Eq. 21: its *x* coordinate increases by equal amounts in equal intervals of time, independent of the *y* motion; and (2) the *y* motions of the two projectiles are identical: the vertical increments of the position of the two projectiles are the same, independent of the horizontal motion of one of them.

Shooting a Falling Target

In a favorite lecture demonstration an air gun is sighted at an elevated target, which is released in free fall by a trip mechanism as the "bullet" leaves the muzzle. No matter what the initial speed of the bullet, it always hits the falling target.

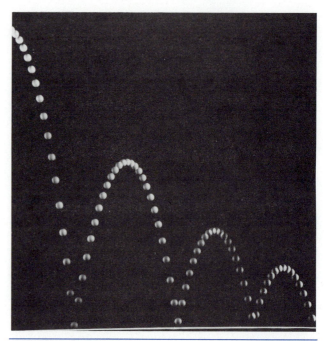

Figure 6   A strobe photo of a golf ball (which enters the photo from the left) bouncing off a hard surface. Between impacts, the ball shows the parabolic path characteristic of projectile motion. Why do you suppose the height of successive bounces is decreasing? (Chapters 8 and 10 may provide the answer.)

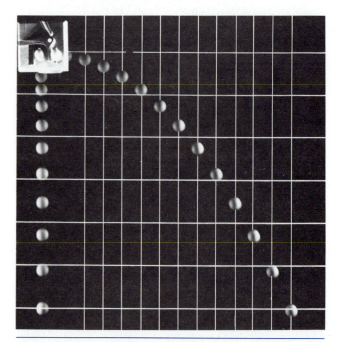

Figure 7   Ball I is released from rest at the same instant that ball II is fired to the right. Note that both balls fall at exactly the same rate; the horizontal motion of ball II does not affect its vertical rate of fall. The exposures in this strobe photo were taken at intervals of 1/30 s. Does the horizontal velocity of ball II appear to be constant?

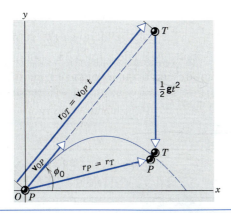

Figure 8   In the motion of a projectile, its displacement from the origin at any time t can be thought of as the sum of two vectors: $\mathbf{v}_{0P}t$, directed along $\mathbf{v}_{0P}$, and $\frac{1}{2}\mathbf{g}t^2$, directed down.

The simplest way to understand this is the following. If there were no acceleration due to gravity, the target would not fall and the bullet would move along the line of sight directly into the target (Fig. 8). The effect of gravity is to cause each body to accelerate down at the same rate from the position it would otherwise have had. Therefore, in the time t, the bullet will fall a distance $\frac{1}{2}gt^2$ from the position it would have had along the line of sight and the target will fall the same distance from its starting point. When the bullet reaches the line of fall of the target, it will be the same distance below the target's initial position as the target is and hence the collision. If the bullet moves faster than shown in the figure (v_0 larger), it will have a greater range and will cross the line of fall at a higher point; but since it gets there sooner, the target will fall a correspondingly smaller distance in the same time and collide with it. A similar argument holds for slower speeds.

For an equivalent analysis, let us use Eq. 12

$$\mathbf{r} = \mathbf{r}_0 + \mathbf{v}_0 t + \tfrac{1}{2}\mathbf{a}t^2$$

to describe the positions of the projectile and the target at any time t. For the projectile P, $\mathbf{r}_0 = 0$ and $\mathbf{a} = \mathbf{g}$, and we have

$$\mathbf{r}_P = \mathbf{v}_{0P}t + \tfrac{1}{2}\mathbf{g}t^2.$$

For the target T, $\mathbf{r}_0 = \mathbf{r}_{0T}$, $\mathbf{v}_0 = 0$, and $\mathbf{a} = \mathbf{g}$, leading to

$$\mathbf{r}_T = \mathbf{r}_{0T} + \tfrac{1}{2}\mathbf{g}t^2.$$

If there is a collision, we must have $\mathbf{r}_P = \mathbf{r}_T$. Inspection shows that this will always occur at a time t given by $\mathbf{r}_{0T} = \mathbf{v}_{0P}t$, that is, in the time t ($=r_{0T}/v_{0P}$) that it would take for an unaccelerated projectile to travel to the target position along the line of sight. Because multiplying a vector by a scalar gives another vector in the same direction, the equation $\mathbf{r}_{0T} = \mathbf{v}_{0P}t$ tells us that $\mathbf{r}_{0T}$ and $\mathbf{v}_{0P}$ must be in the same direction. That is, the gun must be aimed at the initial position of the target.

Sample Problem 3   In a contest to drop a package on a target, one contestant's plane is flying at a constant horizontal velocity of 155 km/h at an elevation of 225 m toward a point directly above the target. At what angle of sight α should the package be released to strike the target (Fig. 9)?

Solution   We choose a reference frame fixed with respect to the Earth, its origin O being the release point. The motion of the package at the moment of release is the same as that of the plane. Hence the initial package velocity $\mathbf{v}_0$ is horizontal and its magnitude is 155 km/h. The angle of projection ϕ_0 is zero.

We find the time of fall from Eq. 22. With $\phi_0 = 0$ and $y = -225$ m this gives

$$t = \sqrt{-\frac{2y}{g}} = \sqrt{-\frac{(2)(-225 \text{ m})}{9.8 \text{ m/s}^2}} = 6.78 \text{ s}.$$

Note that the time of fall does not depend on the speed of the plane for a horizontal projection. (See, however, Problem 38.)

The horizontal distance traveled by the package in this time is given by Eq. 21:

$$x = v_{x0}t = (155 \text{ km/h})(1 \text{ h}/3600 \text{ s})(6.78 \text{ s})$$
$$= 0.292 \text{ km} = 292 \text{ m},$$

so that the angle of sight (Fig. 9) should be

$$\alpha = \tan^{-1}\frac{x}{|y|} = \tan^{-1}\frac{292 \text{ m}}{225 \text{ m}} = 52°.$$

Does the motion of the package appear to be parabolic when viewed from a reference frame fixed with respect to the plane? (Can you recall having seen films of bombs dropping from a plane, taken by a camera either on that plane or on another plane flying a parallel course at the same speed?)

Sample Problem 4   A soccer player kicks a ball at an angle of 36° from the horizontal with an initial speed of 15.5 m/s. Assuming that the ball moves in a vertical plane, find (*a*) the time t_1 at which the ball reaches the highest point of its trajectory, (*b*) its maximum height, (*c*) its range and time of flight, and (*d*) its velocity when it strikes the ground.

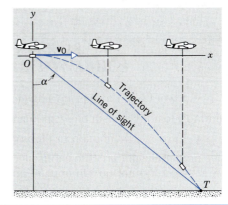

Figure 9   Sample Problem 3.

Solution   (*a*) At the highest point, the vertical component of velocity v_y is zero. Solving Eq. 18 for t, we obtain

$$t = \frac{v_0 \sin \phi_0 - v_y}{g}.$$

With

$$v_y = 0, \qquad v_0 = 15.5 \text{ m/s}, \qquad \phi_0 = 36°, \qquad g = 9.8 \text{ m/s}^2,$$

we have

$$t_1 = \frac{(15.5 \text{ m/s})(\sin 36°)}{9.8 \text{ m/s}^2} = 0.93 \text{ s}.$$

(*b*) The maximum height is reached at $t = 0.93$ s. By using Eq. 22,

$$y = (v_0 \sin \phi_0)t - \tfrac{1}{2}gt^2,$$

we have

$$y_{max} = (15.5 \text{ m/s})(\sin 36°)(0.93 \text{ s}) - \tfrac{1}{2}(9.8 \text{ m/s}^2)(0.93 \text{ s})^2$$
$$= 4.2 \text{ m}.$$

(*c*) The range R can be obtained from Eq. 24:

$$R = \frac{v_0^2}{g} \sin 2\phi_0 = \frac{(15.5 \text{ m/s})^2}{9.8 \text{ m/s}^2} \sin 72° = 23.3 \text{ m}.$$

We set $y = 0$ in Eq. 22 and find the time t_2 at which the ball returns to the ground. We obtain

$$t_2 = \frac{2v_0 \sin \phi_0}{g} = \frac{2(15.5 \text{ m/s})(\sin 36°)}{9.8 \text{ m/s}^2} = 1.86 \text{ s}.$$

Notice that $t_2 = 2t_1$, which must occur because the same time is required for the ball to go up (reach its maximum height from the ground) as is required for the ball to come down (reach the ground from its maximum height).

We can check these results for consistency with $x = x_0 + v_{x0}t$. When $t = t_2$, x should be equal to R. Thus, from Eq. 21, $R = v_{x0}t_2 = (v_0 \cos \phi_0)t_2 = 23.3$ m, as expected.

(*d*) To find the velocity of the ball when it strikes the ground, we use Eq. 17 to obtain v_x, which remains constant throughout the flight:

$$v_x = v_0 \cos \phi_0 = (15.5 \text{ m/s})(\cos 36°) = 12.5 \text{ m/s},$$

and from Eq. 18 we obtain v_y for $t = t_2$,

$$v_y = v_0 \sin \phi_0 - gt = (15.5 \text{ m/s})(\sin 36°) - (9.8 \text{ m/s}^2)(1.86 \text{ s})$$
$$= -9.1 \text{ m/s}.$$

Hence, the velocity has magnitude given by

$$v = \sqrt{v_x^2 + v_y^2} = \sqrt{(12.5 \text{ m/s})^2 + (-9.1 \text{ m/s})^2} = 15.5 \text{ m/s},$$

and direction given by

$$\tan \phi = v_y/v_x = -9.1/12.5,$$

so that $\phi = -36°$, or 36° clockwise from the x axis. Notice that $\phi = -\phi_0$, as we expect from symmetry (Fig. 5).

The final speed turned out to be equal to the initial speed. Can you explain this? Is it a coincidence?

4-4   UNIFORM CIRCULAR MOTION

In projectile motion the acceleration is constant in both magnitude and direction, but the velocity changes in both magnitude and direction. We now examine a special case in which a particle moves at constant *speed* in a circular path. As we shall see, both the velocity and acceleration are constant in magnitude, but both change their directions continuously. This situation is called *uniform circular motion*. Examples of this kind of motion may include Earth satellites and points on spinning rotors such as fans, phonograph records, and computer discs. In fact, to the extent that we can regard ourselves as particles, we participate in uniform circular motion because of the rotation of the Earth.

The situation is shown in Fig. 10*a*. Let P_1 be the position of the particle at the time t_1 and P_2 its position at the time $t_2 = t_1 + \Delta t$. The velocity at P_1 is $\mathbf{v}_1$, a vector tangent to the curve at P_1. The velocity at P_2 is $\mathbf{v}_2$, a vector tangent to the curve at P_2. Vectors $\mathbf{v}_1$ and $\mathbf{v}_2$ have the same magnitude v, because the speed is constant, but their directions are different. The length of path traversed during Δt is the arc length P_1P_2, which is equal to $r\theta$ (when θ is measured in radians) and also to $v \Delta t$. Thus we have

$$r\theta = v \Delta t. \tag{25}$$

We now redraw the vectors $\mathbf{v}_1$ and $\mathbf{v}_2$, as in Fig. 10*b*, so that they originate at a common point. We are free to do this as long as the magnitude and direction of each vector are the same as in Fig. 10*a*. Figure 10*b* enables us to see clearly the *change in velocity* as the particle moves from P_1

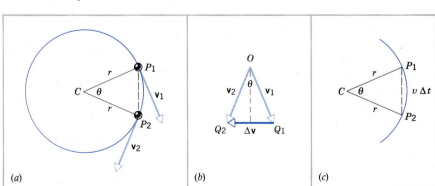

Figure 10 Uniform circular motion. (*a*) The particle travels around a circle at constant speed. Its velocity at two points P_1 and P_2 is shown. (*b*) Its change in velocity in going from P_1 to P_2 is $\Delta\mathbf{v}$. (*c*) The particle travels along the arc P_1P_2 during the time Δt.

to P_2. This change, $\mathbf{v}_2 - \mathbf{v}_1 = \Delta\mathbf{v}$, is the vector that must be added to $\mathbf{v}_1$ to get $\mathbf{v}_2$. If we represent the change in velocity in the interval P_1P_2 by drawing $\Delta\mathbf{v}$ from the midpoint of the arc P_1P_2, then $\Delta\mathbf{v}$ would point toward the center of the circle.

Now the triangle OQ_1Q_2 formed by $\mathbf{v}_1$, $\mathbf{v}_2$, and $\Delta\mathbf{v}$ is similar to the triangle CP_1P_2 (Fig. 10c) formed by the chord P_1P_2 and the radii CP_1 and CP_2. This is so because both are isosceles triangles having the same vertex angle; the angle θ between $\mathbf{v}_1$ and $\mathbf{v}_2$ is the same as the angle P_1CP_2 because $\mathbf{v}_1$ is perpendicular to CP_1 and $\mathbf{v}_2$ is perpendicular to CP_2. Drawing a bisector of the angle θ in Fig. 10b, we find

$$\tfrac{1}{2}\Delta v = v \sin \frac{\theta}{2}. \tag{26}$$

Let us now express the magnitude of the average acceleration in the interval using the results we have obtained in Eqs. 25 and 26 for Δv and Δt:

$$\bar{a} = \frac{\Delta v}{\Delta t} = \frac{2v \sin (\theta/2)}{r\theta/v} = \frac{v^2}{r}\frac{\sin (\theta/2)}{\theta/2}. \tag{27}$$

We now wish to find the instantaneous acceleration by taking the limit of this expression as $\Delta t \to 0$. When Δt is very small, the angle θ is small. In this case we can use the *small angle approximation,* $\sin x \approx x$. (This is valid *only* when the angle is in radians; for example, when $x = 5° = 0.0873$ rad, $\sin x = 0.0872$.) Thus, for small angles $\sin (\theta/2) \approx \theta/2$, and the second fraction on the right side of Eq. 27 approaches 1. Notice also that, in the first fraction on the right side of Eq. 27, neither v nor r depends on Δt and so the value of this fraction is unaffected by the limit. We therefore obtain for the magnitude of the instantaneous acceleration

$$a = \lim_{\Delta t \to 0} \frac{\Delta v}{\Delta t} = \lim_{\Delta t \to 0} \frac{v^2}{r}\frac{\sin (\theta/2)}{\theta/2} = \frac{v^2}{r} \lim_{\Delta t \to 0} \frac{\sin (\theta/2)}{\theta/2}$$

or, using the small angle approximation to replace the remaining limit by 1,

$$a = \frac{v^2}{r}. \tag{28}$$

Because the direction of the average acceleration is the same as that of $\Delta\mathbf{v}$, the direction of $\mathbf{a}$ is always radially inward toward the center of the circle or circular arc in which the particle is moving.

Figure 11 shows the instantaneous relationship between $\mathbf{v}$ and $\mathbf{a}$ at various points of the motion. The magnitude of $\mathbf{v}$ is constant, but its direction changes continuously. This gives rise to an acceleration $\mathbf{a}$, which is also constant in magnitude but continuously changing in direction. The velocity $\mathbf{v}$ is always tangent to the circle in the direction of motion; the acceleration $\mathbf{a}$ is always directed radially inward. Because of this, $\mathbf{a}$ is called a radial, or *centripetal,* acceleration. Centripetal means "seeking a center." A derivation of Eq. 28 using unit vectors is given in the next section.

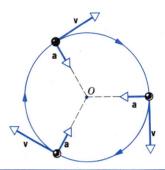

Figure 11   In uniform circular motion, the acceleration $\mathbf{a}$ is always directed toward the center of the circle and hence is always perpendicular to $\mathbf{v}$.

Both in free fall and in projectile motion $\mathbf{a}$ is constant in direction and magnitude, and we can use the equations developed for constant acceleration. We cannot use these equations for uniform circular motion because $\mathbf{a}$ varies in direction and is therefore not constant.

The units of centripetal acceleration are the same as those of an acceleration resulting from a change in the magnitude of a velocity. Dimensionally, we have

$$[a] = \frac{[v^2]}{[r]} = \frac{(L/T)^2}{L} = \frac{L}{T^2},$$

which are the usual dimensions of acceleration. The units therefore may be m/s², km/h², or similar units of dimension L/T².

The acceleration resulting from a change in direction of a velocity is just as real and just as much an acceleration in every sense as that arising from a change in magnitude of a velocity. By definition, acceleration is the time rate of change of velocity, and velocity, being a vector, can change in direction as well as magnitude. If a physical quantity is a vector, its directional aspects cannot be ignored, for their effects will prove to be every bit as important and real as those produced by changes in magnitude.

It is worth emphasizing at this point that there need not be any motion in the direction of an acceleration and that there is no fixed relation in general between the directions of $\mathbf{a}$ and $\mathbf{v}$. In Fig. 12 we give examples in which the angle between $\mathbf{v}$ and $\mathbf{a}$ varies from 0 to 180°. Only in one case, $\theta = 0°$, is the motion in the direction of $\mathbf{a}$.

Sample Problem 5   The Moon revolves about the Earth, making a complete revolution in 27.3 days. Assume that the orbit is circular and has a radius of 238,000 miles. What is the magnitude of the acceleration of the Moon toward the Earth?

Solution   We have $r = 238,000$ mi $= 3.82 \times 10^8$ m. The time for one complete revolution, called the period, is $T = 27.3$ d $= 2.36 \times 10^6$ s. The speed of the Moon (assumed constant) is therefore

$$v = \frac{2\pi r}{T} = \frac{2\pi(3.82 \times 10^8 \text{ m})}{2.36 \times 10^6 \text{ s}} = 1018 \text{ m/s}.$$

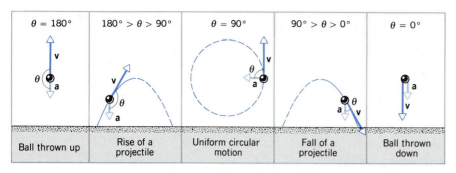

Figure 12 The geometric relationship between **v** and **a** for various motions.

The centripetal acceleration is

$$a = \frac{v^2}{r} = \frac{(1018 \text{ m/s})^2}{3.82 \times 10^8 \text{ m}}$$

$$= 0.00271 \text{ m/s}^2, \quad \text{or only } 2.76 \times 10^{-4} \, g_n.$$

Here g_n ($= 9.80665 \text{ m/s}^2$) is an internationally accepted standard value of g. It represents the approximate value of the free-fall acceleration at sea level and at a latitude of 45°. This standard value is often used as an alternative measure of acceleration. For example, the accelerations experienced by jet pilots or by patrons of amusement park rides are often expressed in this way.

Sample Problem 6 Calculate the speed of an Earth satellite, assuming that it is traveling at an altitude h of 210 km where $g = 9.2 \text{ m/s}^2$. (This value is less than 9.8 m/s^2, because g decreases with altitude above the Earth, as we discuss in Chapter 16.) The radius R of the Earth is 6370 km.

Solution Like any free object near the Earth's surface the satellite has an acceleration g toward the Earth's center. It is this acceleration, coupled with its tangential speed, that causes it to follow the circular path. Hence the centripetal acceleration is g, and from Eq. 28, $a = v^2/r$, we have, for $a = g$ and $r = R + h$,

$$g = \frac{v^2}{R + h}$$

or

$$v = \sqrt{(R + h)g} = \sqrt{(6580 \text{ km})(9.2 \text{ m/s}^2)(10^3 \text{ m/km})}$$

$$= 7780 \text{ m/s} \quad \text{or} \quad 17{,}400 \text{ mi/h}.$$

At this speed the satellite requires 1.48 h to complete one orbit.

4-5 VELOCITY AND ACCELERATION VECTORS IN CIRCULAR MOTION *(Optional)* *

As we derived in the previous section, a particle moving at constant speed along an arc of a circle experiences a centripetal acceleration. Even if its speed is not constant, it still must have a

centripetal acceleration, but it also will have a tangential acceleration that causes a change in its tangential speed. Vector methods are useful in relating the velocities and accelerations and in determining the direction of the resulting acceleration.

We begin by rederiving Eq. 28 for the centripetal acceleration at constant speed using more general vector techniques. Figure 13 shows a particle in uniform circular motion about the origin O of a reference frame. For this motion the plane polar coordinates r and ϕ are more useful than the rectangular coordinates x and y because r remains constant throughout the motion and ϕ increases in a simple linear way with time; the behavior of x and y during such motion is more complex. The two sets of coordinates are related by

$$r = \sqrt{x^2 + y^2} \quad \text{and} \quad \phi = \tan^{-1}(y/x) \quad (29)$$

or by the reciprocal relations

$$x = r \cos \phi \quad \text{and} \quad y = r \sin \phi. \quad (30)$$

In rectangular coordinate systems we use the unit vectors **i** and **j** to describe motion in the xy plane. Here we find it more convenient to introduce two new unit vectors $\mathbf{u}_r$ and $\mathbf{u}_\phi$. These, like **i** and **j**, have unit length and are dimensionless; they designate direction only.

The unit vector $\mathbf{u}_r$ at any point is in the direction of increasing **r** at that point. It is directed radially outward from the origin. The unit vector $\mathbf{u}_\phi$ at any point is in the direction of increasing ϕ at that point. It is always tangent to a circle through the point in a counterclockwise direction. As Fig. 13a shows, $\mathbf{u}_r$ and $\mathbf{u}_\phi$ are at

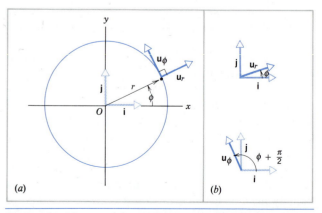

(a) *(b)*

Figure 13 (a) A particle moving counterclockwise in a circle of radius r. (b) The unit vectors $\mathbf{u}_r$ and $\mathbf{u}_\phi$ and their relation to **i** and **j**.

---

* The material in this section can be skipped or postponed to accompany the discussion of rotational motion in Chapter 11.

right angles to each other. The unit vectors $\mathbf{u}_r$ and $\mathbf{u}_\phi$ differ from the unit vectors $\mathbf{i}$ and $\mathbf{j}$ in that the directions of $\mathbf{u}_r$ and $\mathbf{u}_\phi$ vary from point to point in the plane; the unit vectors $\mathbf{u}_r$ and $\mathbf{u}_\phi$ are thus *not* constant vectors. Therefore, when we take derivatives of expressions involving the unit vectors, $\mathbf{i}$ and $\mathbf{j}$ can be treated as constants, but $\mathbf{u}_r$ and $\mathbf{u}_\phi$ cannot.

In terms of the unit vectors $\mathbf{i}$ and $\mathbf{j}$, we can write the unit vectors $\mathbf{u}_r$ and $\mathbf{u}_\phi$ as (see Fig. 13b)

$$\mathbf{u}_r = \mathbf{i}\cos\phi + \mathbf{j}\sin\phi, \tag{31}$$

$$\mathbf{u}_\phi = \mathbf{i}\cos(\phi + \pi/2) + \mathbf{j}\sin(\phi + \pi/2)$$
$$= -\mathbf{i}\sin\phi + \mathbf{j}\cos\phi. \tag{32}$$

In writing terms such as $\mathbf{i}\cos\phi$, we are multiplying a vector by a scalar, and the order of multiplication is unimportant. We could equally well express this term as $(\cos\phi)\mathbf{i}$.

If the particle moves in a circle at constant speed, it has no radial velocity component, and the velocity vector is in the direction of $\mathbf{u}_\phi$. Furthermore, the magnitude of the velocity is just the constant speed v, and we can therefore write

$$\mathbf{v} = v\mathbf{u}_\phi. \tag{33}$$

That is, $\mathbf{v}$ is tangent to the circle and of constant magnitude but changing direction.

The acceleration now follows directly:

$$\mathbf{a} = \frac{d\mathbf{v}}{dt} = \frac{d}{dt}(v\mathbf{u}_\phi) = v\frac{d\mathbf{u}_\phi}{dt}. \tag{34}$$

Note that the constant speed v passes through the differentiation. To find the derivative of the unit vector $\mathbf{u}_\phi$, we use Eq. 32:

$$\frac{d\mathbf{u}_\phi}{dt} = -\mathbf{i}\frac{d(\sin\phi)}{dt} + \mathbf{j}\frac{d(\cos\phi)}{dt}$$
$$= -\mathbf{i}\cos\phi\frac{d\phi}{dt} - \mathbf{j}\sin\phi\frac{d\phi}{dt}$$
$$= (-\mathbf{i}\cos\phi - \mathbf{j}\sin\phi)\frac{d\phi}{dt}$$
$$= -\mathbf{u}_r\frac{d\phi}{dt}. \tag{35}$$

Note that we have used Eq. 31 in the last step. Thus

$$\mathbf{a} = -\mathbf{u}_r v\frac{d\phi}{dt}. \tag{36}$$

The particle moves uniformly around the circle, and so $d\phi/dt$ is just the angular distance covered in one revolution (2π radians)

divided by the time for one revolution (the distance $2\pi r$ divided by the speed v):

$$\frac{d\phi}{dt} = \frac{2\pi}{2\pi r/v} = \frac{v}{r}. \tag{37}$$

Finally, substituting Eq. 37 into Eq. 36, we obtain

$$\mathbf{a} = -\mathbf{u}_r v\frac{v}{r}$$
$$= -\mathbf{u}_r\frac{v^2}{r}. \tag{38}$$

This equation tells us that the acceleration has the constant magnitude of v^2/r, as we obtained in Eq. 28, and that it points radially inward (that is, opposite to $\mathbf{u}_r$). As the particle travels around the circle, the directions of $\mathbf{u}_r$ and $\mathbf{a}$ change relative to the xy coordinate axes because the radial direction changes.

Tangential Acceleration in Circular Motion

We now consider the more general case of circular motion in which the speed v of the moving particle is *not* constant. We again use vector methods in plane polar coordinates.

As before, the velocity is given by Eq. 33, or

$$\mathbf{v} = v\mathbf{u}_\phi$$

except that, in this case, not only $\mathbf{u}_\phi$ but also the magnitude v varies with time. Recalling the formula for the derivative of a product, we obtain for the acceleration

$$\mathbf{a} = \frac{d\mathbf{v}}{dt} = \frac{d(v\mathbf{u}_\phi)}{dt} = v\frac{d\mathbf{u}_\phi}{dt} + \mathbf{u}_\phi\frac{dv}{dt}. \tag{39}$$

Equation 34 did not include the second term on the right-hand side of Eq. 39 because v was assumed to be a constant. The first term on the right-hand side of Eq. 39 reduces, as we derived above, to $-\mathbf{u}_r(v^2/r)$. We can now write Eq. 39 as

$$\mathbf{a} = -\mathbf{u}_r a_R + \mathbf{u}_\phi a_T, \tag{40}$$

in which $a_R = v^2/r$ and $a_T = dv/dt$. The first term, $-\mathbf{u}_r a_R$, is the vector component of $\mathbf{a}$ directed radially in toward the center of the circle and arises from a change in the *direction* of the velocity in circular motion (see Fig. 14). The vector $\mathbf{a}_R$ and its magnitude a_R are both called the *centripetal acceleration*. The second term, $\mathbf{u}_\phi a_T$, is the vector component of $\mathbf{a}$ that is tangent to the path of the particle and arises from a change in the *magnitude* of the velocity in circular motion (see Fig. 14). The vector $\mathbf{a}_T$ and its magnitude a_T are both called the *tangential acceleration*.

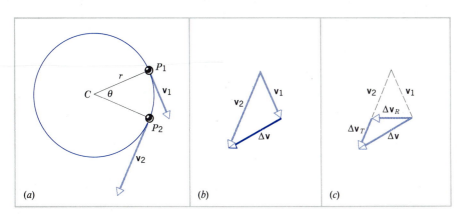

(a) (b) (c)

Figure 14   (*a*) In nonuniform circular motion the speed is variable. (*b*) The change in velocity $\Delta\mathbf{v}$ in going from P_1 to P_2. (*c*) There are two parts to $\Delta\mathbf{v}$: $\Delta\mathbf{v}_R$, caused by the change in the direction of $\mathbf{v}$, and $\Delta\mathbf{v}_T$, caused by the change in the magnitude of $\mathbf{v}$. In the limit $\Delta t \to 0$, $\Delta\mathbf{v}_R$ points toward the center C of the circle and $\Delta\mathbf{v}_T$ is tangent to the circular path.

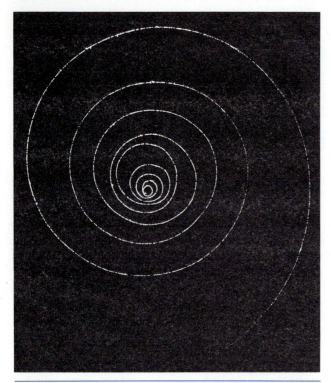

Figure 15 A track left in a liquid-hydrogen bubble chamber by an electron. There is a radial acceleration, caused by a magnetic field, which tends to produce a circular path, but because the electron is also slowing down through collisions with the hydrogen atoms, it also experiences a tangential acceleration. The resulting path is a spiral.

The magnitude of the instantaneous acceleration is

$$a = \sqrt{a_T^2 + a_R^2}. \tag{41}$$

If the speed is constant, then $a_T = dv/dt = 0$ and Eq. 40 reduces to Eq. 38. When the speed v is not constant, a_T is not zero and a_R varies from point to point. The speed v may be changing in such a way that a_T is not constant, and thus both a_T and a_R may vary from point to point.

Figure 15 shows the track left in a liquid-hydrogen-filled bubble chamber by an energetic electron that spirals inward. The electron slows down as it travels through the liquid in the chamber so that its speed v is being reduced steadily. Thus there is at every point a tangential acceleration a_T given by dv/dt. Although the electron is not traveling in a circular path, small arcs of the spiral look very much like arcs of a circle with a given radius r. The centripetal acceleration a_R at any point is thus given by v^2/r, where r is the radius of the path at the point in question; both v and r become smaller as the particle loses energy. The radial acceleration of the electron is produced by a magnetic field present in the bubble chamber and at right angles to the plane of Fig. 15 (see Chapter 34). ■

4-6 RELATIVE MOTION

Suppose you are in a car moving down a straight highway at a constant speed of 55 mi/h. The others with you in the car are moving at the same speed; even though their speed relative to the ground is 55 mi/h, their speed relative to you is zero. In the car you could carry out a normal set of physics experiments that would be unaffected by the uniform motion of the car. For example, you could toss a ball directly upward (in your reference frame), and you would observe it to fall directly downward. The ball has horizontal motion (because of the motion of the car), but you have the same horizontal motion and there is no *relative* horizontal motion.

To an observer on the ground, however, the result is different. The ball has a forward horizontal component of velocity equal to 55 mi/h and a vertical component from the motion you give it. We know that a projectile in gravity with such velocity components follows a parabolic trajectory. You and the ground-based observer would therefore use different equations to describe the motion, but you would agree on the physical laws followed by the ball; for instance, you would both deduce the same value of the free-fall acceleration.

If another car now pulls up beside you and passes at a constant speed of 57 mi/h, you observe this car (relative to your own reference frame) to move slowly ahead of you at a rate of 2 mi/h ($= 57$ mi/h $- 55$ mi/h). Take away the external clues—the scenery speeding by, the still air rushing past the moving car, the bumpiness of the road, and the noise of the engine—and consider only the two cars. You would have no way to decide which car was "really" moving. For example, the passing car could be at rest and you could be moving backward at 2 mi/h; the observed result would be the same.

In this section we consider the description of the motion of a single particle by two observers who are in uniform motion relative to one another. The two observers might be, for example, a person in a car moving at constant velocity along a long straight road and another person standing at rest on the ground. The particle they are both observing might be a ball tossed into the air or another moving car.

We call the two observers S and S'. Each has a corresponding reference frame to which is attached a Cartesian coordinate system. For convenience, we assume the observers to be located at the origins of their respective coordinate systems. We make only one restriction on this situation: *the relative velocity between S and S' must be a constant.* Here we mean constant in both magnitude and direction. Note that this restriction does not include the motion of the particle being observed by S and S'. The particle need not necessarily be moving with constant velocity, and indeed the particle may well be accelerating.

Figure 16 shows, at a particular time t, the two coordinate systems belonging to S and S'. For simplicity, we consider motion in only two dimensions, the common xy and $x'y'$ planes shown in Fig. 16. The origin of the S' system is located with respect to the origin of the S system by the vector $\mathbf{r}_{S'S}$. Note in particular the order of the subscripts we use to label the vector: the first subscript gives the system being located (in this case, the coordinate

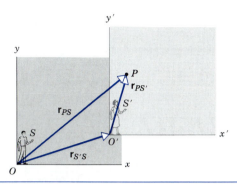

Figure 16 Observers S and S', who are moving with respect to each other, observe the same moving particle P. At the time shown, they measure the position of the particle with respect to the origins of their coordinate systems to be $\mathbf{r}_{PS}$ and $\mathbf{r}_{PS'}$, respectively. At this same instant, observer S measures the position of S' with respect to the origin O to be $\mathbf{r}_{S'S}$.

system of S') and the second subscript gives the system with respect to which we are doing the locating (in this case, the coordinate system of S). The vector $\mathbf{r}_{S'S}$ would then be read as "the position of S' with respect to S."

Figure 16 also shows a particle P in the common xy and $x'y'$ planes. Both S and S' locate the particle P with respect to their coordinate systems. According to S, the particle P is at the position indicated by the vector $\mathbf{r}_{PS}$, while according to S' the particle P is at $\mathbf{r}_{PS'}$. From Fig. 16 we can deduce the following relationship among the three vectors:

$$\mathbf{r}_{PS} = \mathbf{r}_{S'S} + \mathbf{r}_{PS'} = \mathbf{r}_{PS'} + \mathbf{r}_{S'S}, \qquad (42)$$

where we have used the commutative law of vector addition to exchange the order of the two vectors. Once again, pay careful attention to the order of the subscripts. In words, Eq. 42 tells us: "the position of P as measured by S is equal to the position of P as measured by S' plus the position of S' as measured by S."

Suppose the particle P is moving with velocity $\mathbf{v}_{PS'}$ according to S'. What velocity will S measure for the particle? To answer this question, we need only take the derivative with respect to time of Eq. 42, which gives

$$\frac{d\mathbf{r}_{PS}}{dt} = \frac{d\mathbf{r}_{PS'}}{dt} + \frac{d\mathbf{r}_{S'S}}{dt}.$$

The rate of change of each position vector gives the corresponding velocity, so that

$$\mathbf{v}_{PS} = \mathbf{v}_{PS'} + \mathbf{v}_{S'S}. \qquad (43)$$

Thus, at any instant, the velocity of P as measured by S is equal to the velocity of P as measured by S' plus the relative velocity of S' with respect to S. Although we have illustrated Eqs. 42 and 43 for motion in two dimensions, they hold equally well in three dimensions.

Equation 43 is a law of the *transformation of velocities*. It permits us to transform a measurement of velocity made by an observer in one frame of reference, say S', to

another frame of reference, say S, as long as we know the relative velocity between the two reference frames. It is a law firmly grounded both in the common sense of everyday experience and in the concepts of space and time that are essential to the classical physics of Galileo and Newton. In fact, Eq. 43 is often called the *Galilean form of the law of transformation of velocities.*

We consider here only the very important special case in which the two reference frames are moving at constant velocity with respect to one another. That is, $\mathbf{v}_{S'S}$ is constant both in magnitude and direction. The velocities $\mathbf{v}_{PS}$ and $\mathbf{v}_{PS'}$ that S and S' measure for the particle P may not be constant, and of course they will in general not be equal to one another. If, however, one of the observers, say S', measures a velocity that is constant in time, then both terms on the right-hand side of Eq. 43 are independent of time and therefore the left side of Eq. 43 must also be independent of time. Thus, if one observer concludes that the particle moves with constant velocity, then all other observers conclude the same, as long as the other observers are in frames of reference that move at constant velocity with respect to the frame of the first observer.

An even more significant result follows from differentiating Eq. 43:

$$\frac{d\mathbf{v}_{PS}}{dt} = \frac{d\mathbf{v}_{PS'}}{dt} + \frac{d\mathbf{v}_{S'S}}{dt}. \qquad (44)$$

The last term of Eq. 44 vanishes, because we assume that the relative velocity of the two reference frames is a constant. Thus

$$\frac{d\mathbf{v}_{PS}}{dt} = \frac{d\mathbf{v}_{PS'}}{dt}.$$

Replacing these two derivatives of velocity with the corresponding accelerations, we obtain

$$\mathbf{a}_{PS} = \mathbf{a}_{PS'}. \qquad (45)$$

The accelerations of P measured by the two observers are identical!

In the next chapter we shall find that the acceleration is fundamental in the dynamical behavior of an object through Newton's second law $\mathbf{F} = m\mathbf{a}$, which relates the force $\mathbf{F}$, the mass m, and the acceleration $\mathbf{a}$. Equation 45 was derived in the special circumstance that the reference frames S and S' move with a relative velocity that is constant in both magnitude and direction. Such frames, which may move relative to one another but in which all observers find the same value for the acceleration of a given moving particle, are called *inertial reference frames*. We shall see in the following chapter that they are especially important because Newton's laws of motion hold only in such frames.

Here is an example of a law of physics that can be used to test inertial reference frames. Tie a mass to one end of a string and hold the other end of the string so that the mass hangs freely. The attraction of the Earth's gravity for the mass pulls it toward the center of the Earth; the direction of the string can be used to define a vertical axis. Now try the experiment in your car moving in a straight line at a constant 55 mi/h. The result is the same: the string hangs in the same vertical direction. The car, like the ground, is an inertial reference frame. If you try the experiment again when the car is accelerating, braking, or rounding a curve, the string deviates from the vertical. These accelerated frames (even with centripetal acceleration) are noninertial frames.

Actually, the Earth is only approximately an inertial reference frame. Because of the rotation of the Earth on its axis, two observers at different latitudes have a relative tangential velocity that changes its direction with the rotation. This is a small effect and is negligible in many circumstances, although it must be taken into account for very precise work and it can have dramatic effects in large-scale circumstances. For example, the noninertial nature of the reference frame of the surface of the Earth causes the rotation of winds about a high- or low-pressure center which can produce severe and destructive storms. In Section 6-8 we discuss other effects of making observations in noninertial frames of reference.

Sample Problem 7   The compass in an airplane indicates that it is headed due east; its air speed indicator reads 215 km/h. A steady wind of 65 km/h is blowing due north. (a) What is the velocity of the plane with respect to the ground? (b) If the pilot wishes to fly due east, what must be the heading? That is, what must the compass read?

Solution   (a) The moving "particle" in this problem is the plane P. There are two reference frames, the ground (G) and the air (A). We let the ground be our S system and the air be the S' system, and by a simple change of notation, we can rewrite Eq. 43 as

$$\mathbf{v}_{PG} = \mathbf{v}_{PA} + \mathbf{v}_{AG}.$$

Figure 17a shows these vectors, which form a right triangle. The terms are, in sequence, the velocity of the plane with respect to the ground, the velocity of the plane with respect to the air, and the velocity of the air with respect to the ground (that is, the wind velocity). Note the orientation of the plane, which is consistent with a due east reading on its compass.

The magnitude of the ground velocity (the ground speed) is found from

$$v_{PG} = \sqrt{v_{PA}^2 + v_{AG}^2} = \sqrt{(215 \text{ km/h})^2 + (65 \text{ km/h})^2} = 225 \text{ km/h}.$$

The angle α in Fig. 17a follows from

$$\alpha = \tan^{-1} \frac{v_{AG}}{v_{PA}} = \tan^{-1} \frac{65 \text{ km/h}}{215 \text{ km/h}} = 16.8°.$$

Thus, with respect to the ground, the plane is flying at 225 km/h in a direction 16.8° north of east. Note that its ground speed is greater than its air speed.

(b) In this case the pilot must head into the wind so that the velocity of the plane with respect to the ground points east. The wind remains unchanged and the vector diagram representing Eq. 43 is as shown in Fig. 17b. Note that the three vectors still form a right triangle, as they did in Fig. 17a, but in this case the hypoteneuse is v_{PA} rather than v_{PG}.

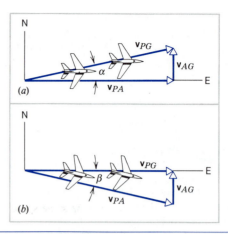

Figure 17   Sample Problem 7. (a) A plane, heading due east, is blown to the north. (b) To travel due east, the plane must head into the wind.

The pilot's ground speed is now

$$v_{PG} = \sqrt{v_{PA}^2 - v_{AG}^2} = \sqrt{(215 \text{ km/h})^2 - (65 \text{ km/h})^2} = 205 \text{ km/h}.$$

As the orientation of the plane in Fig. 17b indicates, the pilot must head into the wind by an angle β given by

$$\beta = \sin^{-1} \frac{v_{AG}}{v_{PA}} = \sin^{-1} \frac{65 \text{ km/h}}{215 \text{ km/h}} = 17.6°.$$

Note that, by heading into the wind as the pilot has done, the ground speed is now less than the air speed.

Relative Motion at High Speed   (Optional)
The above arguments about relative motion form the cornerstone of Newtonian mechanics, which we begin to discuss in Chapter 5. They place no restriction on the relative velocity of the reference frames (as long as it is constant) or on the velocity of the object being observed. Two centuries after Newton, Albert Einstein tried to imagine the result of applying Eq. 43 to a beam of light traveling at a speed of $c = 299{,}792{,}458$ m/s in vacuum. Suppose observer S' observed a beam of light traveling at c in the positive x' direction. Let S' move relative to S, again in the x' direction, with a speed $v_{S'S} = 1$ m/s. What speed will S observe for the beam of light? Newtonian mechanics would answer according to Eq. 43: $v_{PS} = 299{,}792{,}458$ m/s + 1 m/s = 299,792,459 m/s.

Einstein had studied his physics textbooks. He knew what Newtonian mechanics had to say about observers in relative motion watching beams of light. He also knew that a beam of light was not an ordinary moving object. A beam of light travels in a special way. Light is electromagnetic radiation and can be analyzed in terms of its constituent electric and magnetic fields. A moving electric field creates a magnetic field, and a moving magnetic field in turn creates an electric field. Thus the moving electric and magnetic fields of light essentially recreate themselves as the beam travels. If Eq. 43 were valid for light beams, Einstein reasoned, the observer S could emit a light beam in the x direction at speed c, and observer S' could travel in the x direction relative to S at $v_{S'S} = c$ and could catch the light beam.

Just as in the case of a car traveling beside you at the same speed as your car, to the S' observer the light beam would appear at rest. To Einstein this was a terrible contradiction: how could a light beam, which is fundamentally electromagnetic fields *in motion,* ever be observed "at rest"?

Einstein proposed what was to him an obvious solution to this dilemma: no light beam could ever be observed "at rest." Therefore, it must absolutely follow that Eq. 43 is wrong if we apply it at speeds near c. Einstein went even a step further: he asserted that both S and S' must measure precisely the same value for the speed of light, *no matter what their relative speed!* This statement seems contrary to our common sense and to the predictions of Eq. 43; if two observers are moving at a relative speed of $0.9999999c$, how can both measure the same speed of c for a beam of light emitted by one of them?

We delay until Chapter 21 the complete mathematical description of how this comes about; for now we give a brief clue in the special case that all velocities are in the x (or x') direction. Here is Einstein's result for the transformation of velocities:

$$v_{PS} = \frac{v_{PS'} + v_{S'S}}{1 + v_{PS'}v_{S'S}/c^2} . \qquad (46)$$

Notice the beauty of this result. When $v_{PS'}$ and $v_{S'S}$ are small (compared with c), the denominator of Eq. 46 is very close to 1 and Eq. 46 reduces to Eq. 43. At low speed, the Galilean velocity transformation gives acceptable results. When $v_{PS'} = c$ (S' is observing a light beam) then Eq. 46 gives $v_{PS} = c$ *no matter what the value of* $v_{S'S}$. All observers measure the same value of the speed of a light beam, no matter what their relative speeds.

Einstein's assertion, and the kinematics and mechanics that follow from it, does not require us to abandon Newtonian physics. Instead, it warns us to restrict our Newtonian calculations to speeds very small compared with c. For the moving objects we normally encounter, we are well within this restriction. Even a high-speed rocket ($v = 10^4$ m/s), one of the fastest objects of human construction, has a speed that is so far less than c (3×10^8 m/s) that we can safely use the Galilean formula without significant error. Particles such as electrons or protons, however, can easily be accelerated to speeds that are very close to c. At these high speeds, a new kind of physics, with new equations of kinematics and dynamics, must be used. This new physics is the basis of the special theory of relativity, which we discuss at greater length in Chapter 21. ■

QUESTIONS

1. Can the acceleration of a body change its direction without its velocity changing direction?

2. Let **v** and **a** represent the velocity and acceleration, respectively, of an automobile. Describe circumstances in which (*a*) **v** and **a** are parallel; (*b*) **v** and **a** are antiparallel; (*c*) **v** and **a** are perpendicular to one another; (*d*) **v** is zero but **a** is not zero; (*e*) **a** is zero but **v** is not zero.

3. In broad jumping, sometimes called long jumping, does it matter how high you jump? What factors determine the span of the jump?

4. Why doesn't the electron in the beam from an electron gun fall as much because of gravity as a water molecule in the stream from a hose? Assume hoizontal motion initially in each case.

5. At what point or points in its path does a projectile have its minimum speed? Its maximum?

6. Figure 18 shows the path followed by a NASA Learjet in a run designed to simulate low-gravity conditions for a short period of time. Make an argument to show that, if the plane follows a particular parabolic path, the passengers will experience weightlessness.

7. A shot put is thrown from above ground level. The launch angle that will produce the longest range is less than $45°$; that is, a flatter trajectory has a longer range. Explain why.

8. Consider a projectile at the top of its trajectory. (*a*) What is its speed in terms of v_0 and ϕ_0? (*b*) What is its acceleration? (*c*) How is the direction of its acceleration related to that of its velocity?

9. Trajectories are shown in Fig. 19 for three kicked footballs. Pick the trajectory for which (*a*) the time of flight is least, (*b*) the vertical velocity component at launch is greatest, (*c*) the horizontal velocity component at launch is greatest, and (*d*) the launch speed is least. Ignore air resistance.

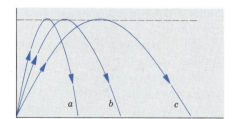

Figure 19   Question 9.

10. A rifle is bore-sighted with its barrel horizontal. Show that, at the same range, it will shoot too high when shooting either uphill or downhill. (See "A Puzzle in Elementary Ballistics," by Ole Anton Haugland, *The Physics Teacher,* April 1983, p. 246.)

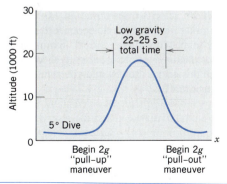

Figure 18   Question 6.

11. In his book, *Sport Science,* Peter Brancazio, with such projectiles as baseballs and golf balls in mind, writes: "Everything else being equal, a projectile will travel farther on a hot day than on a cold day, farther at high altitude than at sea level, farther in humid than in dry air." How can you explain these claims?

12. A graph of height versus time for an object thrown vertically upward is a parabola. The path of a projectile, thrown upward but not vertically upward, is also a parabola. Is this a coincidence? Justify your answer.

13. Long-range artillery pieces are not set at the "maximum range" angle of 45° but at larger elevation angles, in the range of 55° to 65°. What's wrong with 45°?

14. In projectile motion when air resistance is negligible, is it ever necessary to consider three-dimensional motion rather than two-dimensional?

15. Is it possible to be accelerating if you are traveling at constant speed? Is it possible to round a curve with zero acceleration? With constant acceleration?

16. Describe qualitatively the acceleration acting on a bead that, sliding along a frictionless wire, moves inward with constant speed along a spiral.

17. Show that, taking the Earth's rotation and revolution into account, a book resting on your table moves faster at night than it does during the daytime. In what reference frame is this statement true?

18. An aviator, pulling out of a dive, follows the arc of a circle and is said to have "pulled 3g's" in pulling out of the dive. Explain what this statement means.

19. Could the acceleration of a projectile be represented in terms of a radial and a tangential component at each point of the motion? If so, is there any advantage to this representation?

20. A tube in the shape of a rectangle with rounded corners is placed in a vertical plane, as shown in Fig. 20. You introduce two ball bearings at the upper right-hand corner. One travels by path *AB* and the other by path *CD*. Which will arrive first at the lower left-hand corner?

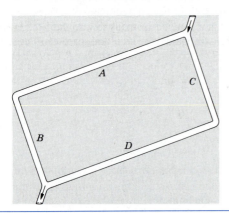

Figure 20   Question 20.

21. If the acceleration of a body is constant in a given reference frame, is it necessarily constant in all other reference frames?

22. A boy sitting in a railroad car moving at constant velocity throws a ball straight up into the air. Will the ball fall behind him? In front of him? Into his hands? What happens if the car accelerates forward or goes around a curve while the ball is in the air?

23. A woman on the rear platform of a train moving with constant velocity drops a coin while leaning over the rail. Describe the path of the coin as seen by (*a*) the woman on the train, (*b*) a person standing on the ground near the track, and (*c*) a person in a second train moving in the opposite direction to the first train on a parallel track.

24. An elevator is descending at a constant speed. A passenger drops a coin to the floor. What accelerations would (*a*) the passenger and (*b*) a person at rest with respect to the elevator shaft observe for the falling coin?

25. Water is collecting in a bucket during a steady downpour. Will the rate at which the bucket is filling change if a steady horizontal wind starts to blow?

26. A bus with a vertical windshield moves along in a rainstorm at speed v_b. The raindrops fall vertically with a terminal speed v_r. At what angle do the raindrops strike the windshield?

27. Drops are falling vertically in a steady rain. In order to go through the rain from one place to another in such a way as to encounter the least number of raindrops, should you move with the greatest possible speed, the least possible speed, or some intermediate speed? (See "An Optimal Speed for Traversing a Constant Rain," by S. A. Stern, *American Journal of Physics,* September 1983, p. 815.)

28. What's wrong with Fig. 21? The boat is sailing with the wind.

Figure 21   Question 28.

29. The Galilean velocity transformation, Eq. 43, is so instinctively familiar from everyday experience that it is sometimes claimed to be "obviously correct, requiring no proof." Many so-called refutations of relativity theory turn out to be based on this claim. How would you refute someone who made this claim?

PROBLEMS

Section 4-1 Position, Velocity, and Acceleration

1. A plane flies 410 mi east from city A to city B in 45 min and then 820 mi south from city B to city C in 1 h 30 min. (*a*) What are the magnitude and direction of the displacement vector that represents the total trip? What are (*b*) the average velocity vector and (*c*) the average speed for the trip?

2. The position of a particle moving in an xy plane is given by $\mathbf{r} = (2t^3 - 5t)\mathbf{i} + (6 - 7t^4)\mathbf{j}$. Here r is in meters and t is in seconds. Calculate (*a*) $\mathbf{r}$, (*b*) $\mathbf{v}$, and (*c*) $\mathbf{a}$ when $t = 2$ s.

3. In 3 h 24 min, a balloon drifts 8.7 km north, 9.7 km east, and 2.9 km in elevation from its release point on the ground. Find (*a*) the magnitude of its average velocity and (*b*) the angle its average velocity makes with the horizontal.

4. The velocity of a particle moving in the xy plane is given by $\mathbf{v} = (6t - 4t^2)\mathbf{i} + 8\mathbf{j}$. Here v is in meters per second and t (>0) is in seconds. (*a*) What is the acceleration when $t = 3$ s? (*b*) When (if ever) is the acceleration zero? (*c*) When (if ever) is the velocity zero? (*d*) When (if ever) does the speed equal 10 m/s?

Section 4-2 Motion with Constant Acceleration

5. In a cathode-ray tube, a beam of electrons is projected horizontally with a speed of 9.6×10^8 cm/s into the region between a pair of horizontal plates 2.3 cm long. An electric field between the plates causes a constant downward acceleration of the electrons of magnitude 9.4×10^{16} cm/s². Find (*a*) the time required for the electrons to pass through the plates, (*b*) the vertical displacement of the beam in passing through the plates, and (*c*) the horizontal and vertical components of the velocity of the beam as it emerges from the plates.

6. An iceboat sails across the surface of a frozen lake with constant acceleration produced by the wind. At a certain instant its velocity is $6.30\mathbf{i} - 8.42\mathbf{j}$ in m/s. Three seconds later the boat is instantaneously at rest. What is its acceleration during this interval?

7. A particle moves so that its position as a function of time in SI units is
$$\mathbf{r}(t) = \mathbf{i} + 4t^2\mathbf{j} + t\mathbf{k}.$$
Write expressions for (*a*) its velocity and (*b*) its acceleration as functions of time. (*c*) What is the shape of the particle's trajectory?

8. A particle leaves the origin at $t = 0$ with an initial velocity $\mathbf{v}_0 = 3.6\mathbf{i}$, in m/s. It experiences a constant acceleration $\mathbf{a} = -1.2\mathbf{i} - 1.4\mathbf{j}$, in m/s². (*a*) At what time does the particle reach its maximum x coordinate? (*b*) What is the velocity of the particle at this time? (*c*) Where is the particle at this time?

9. A particle A moves along the line $y = d$ (30 m) with a constant velocity $\mathbf{v}$ ($v = 3.0$ m/s) directed parallel to the positive x axis (Fig. 22). A second particle B starts at the origin with zero speed and constant acceleration $\mathbf{a}$ ($a = 0.40$ m/s²) at the same instant that particle A passes the y axis. What angle θ between $\mathbf{a}$ and the positive y axis would result in a collision between these two particles?

10. A ball is dropped from a height of 39.0 m. The wind is blowing horizontally and imparts a constant acceleration of

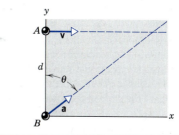

Figure 22 Problem 9.

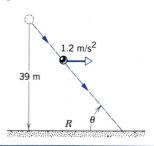

Figure 23 Problem 10.

1.20 m/s² on the ball. (*a*) Show that the path of the ball is a straight line and find the values of R and θ in Fig. 23. (*b*) How long does it take for the ball to reach the ground? (*c*) With what speed does the ball hit the ground?

Section 4-3 Projectile Motion

11. A ball rolls off the edge of a horizontal table top 4.23 ft high. It strikes the floor at a point 5.11 ft horizontally away from the edge of the table. (*a*) For how long was the ball in the air? (*b*) What was its speed at the instant it left the table?

12. Electrons, like all forms of matter, fall under the influence of gravity. If an electron is projected horizontally with a speed of 3.0×10^7 m/s (one-tenth the speed of light), how far will it fall in traversing 1 m of horizontal distance?

13. A dart is thrown horizontally toward the bull's eye, point P on the dart board, with an initial speed of 10 m/s. It hits at point Q on the rim, vertically below P, 0.19 s later; see Fig. 24. (*a*) What is the distance PQ? (*b*) How far away from the dart board did the player stand?

Figure 24 Problem 13.

14. A rifle is aimed horizontally at a target 130 ft away. The bullet hits the target 0.75 in. below the aiming point. (*a*) What is the bullet's time of flight? (*b*) What is the muzzle speed of the bullet?

15. A projectile is fired horizontally from a gun located 45.0 m above a horizontal plane with a muzzle speed of 250 m/s. (*a*) How long does the projectile remain in the air? (*b*) At what horizontal distance does it strike the ground? (*c*) What is the magnitude of the vertical component of its velocity as it strikes the ground?

16. A baseball leaves the pitcher's hand horizontally at a speed of 92.0 mi/h. The distance to the batter is 60.0 ft. (*a*) How long does it take for the ball to travel the first 30.0 ft horizontally? The second 30.0 ft? (*b*) How far does the ball fall under gravity during the first 30.0 ft of its horizontal travel? (*c*) During the second 30.0 ft? (*d*) Why are these quantities not equal? Ignore the effects of air resistance.

17. In a detective story, a body is found 15 ft out from the base of a building and beneath an open window 80 ft above. Would you guess the death to be accidental or not? Why?

18. You throw a ball from a cliff with an initial velocity of 15 m/s at an angle of 20° below the horizontal. Find (*a*) its horizontal displacement and (*b*) its vertical displacement 2.3 s later.

19. You throw a ball with a speed of 25.3 m/s at an angle of 42.0° above the horizontal directly toward a wall as shown in Fig. 25. The wall is 21.8 m from the release point of the ball. (*a*) How long is the ball in the air before it hits the wall? (*b*) How far above the release point does the ball hit the wall? (*c*) What are the horizontal and vertical components of its velocity as it hits the wall? (*d*) Has it passed the highest point on its trajectory when it hits?

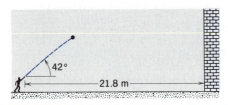

Figure 25   Problem 19.

20. Show that the maximum height reached by a projectile is $y_{max} = (v_0 \sin \phi_0)^2/2g$.

21. (*a*) Prove that for a projectile fired from the surface of level ground at an angle ϕ_0 above the horizontal, the ratio of the maximum height H to the range R is given by $H/R = \frac{1}{4} \tan \phi_0$. (*b*) Find the angle of projection at which the maximum height and the horizontal range are equal. See Fig. 26.

22. A projectile is fired from the surface of level ground at an angle ϕ_0 above the horizontal. (*a*) Show that the elevation angle θ of the highest point as seen from the launch point is related to ϕ_0 by $\tan \theta = \frac{1}{2} \tan \phi_0$. See Fig. 26. (*b*) Calculate θ for $\phi_0 = 45°$.

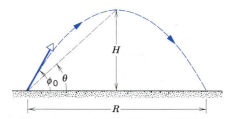

Figure 26   Problems 21 and 22.

23. A stone is projected at an initial speed of 120 ft/s directed 62° above the horizontal, at a cliff of height h, as shown in Fig. 27. The stone strikes the ground at A 5.5 s after launching. Find (*a*) the height h of the cliff, (*b*) the speed of the stone just before impact at A, and (*c*) the maximum height H reached above the ground.

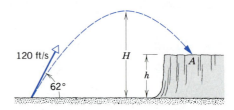

Figure 27   Problem 23.

24. In the 1968 Olympics in Mexico City, Bob Beamon shattered the record for the long jump with a jump of 8.90 m. Assume that his initial speed on takeoff was 9.50 m/s, about equal to that of a sprinter. How close did this world-class athlete come to the maximum possible range in the absence of air resistance? The value of g in Mexico City is 9.78 m/s².

25. In Sample Problem 3, find (*a*) the speed of the package as it hits the target and (*b*) the angle of impact with the vertical. (*c*) Why is the angle of impact not equal to the angle of sight?

26. (*a*) In Galileo's *Two New Sciences,* the author states that "for elevations [angles of projection] which exceed or fall short of 45° by equal amounts, the ranges are equal." Prove this statement. See Fig. 28. (*b*) For an initial speed of 30.0 m/s and a range of 20.0 m, find the two possible elevation angles of projection.

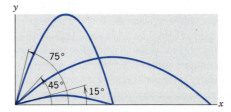

Figure 28   Problem 26.

27. A juggler manages to keep five balls in motion, throwing each sequentially up a distance of 3.0 m. (*a*) Determine the time interval between successive throws. (*b*) Give the positions of the other balls at the instant when one reaches her hand. (Neglect the time taken to transfer balls from one hand to the other.)

28. A rifle with a muzzle velocity of 1500 ft/s shoots a bullet at a target 150 ft away. How high above the target must the rifle be aimed so that the bullet will hit the target?

29. A ball rolls off the top of a stairway with a horizontal velocity of magnitude 5.0 ft/s. The steps are 8.0 in. high and 8.0 in. wide. Which step will the ball hit first?

30. A ball is thrown from the ground into the air. At a height of 9.1 m, the velocity is observed to be $\mathbf{v} = 7.6\mathbf{i} + 6.1\mathbf{j}$, in m/s (*x* axis horizontal, *y* axis vertical and up). (*a*) To what maximum height will the ball rise? (*b*) What will be the total horizontal distance traveled by the ball? (*c*) What is the velocity of the ball (magnitude and direction) the instant before it hits the ground?

31. If the pitcher's mound is 1.25 ft above the baseball field, can a pitcher release a fast ball horizontally at 92.0 mi/h and still get it into the strike zone over the plate 60.5 ft away? Assume that, for a strike, the ball must fall at least 1.30 ft but no more than 3.60 ft.

32. According to Eq. 24, the range of a projectile depends not only on v_0 and ϕ_0 but also on the value g of the gravitational acceleration, which varies from place to place. In 1936, Jesse Owens established a world's running broad jump record of 8.09 m at the Olympic Games at Berlin ($g = 9.8128$ m/s²). Assuming the same values of v_0 and ϕ_0, by how much would his record have differed if he had competed instead in 1956 at Melbourne ($g = 9.7999$ m/s²)? (In this connection see "The Earth's Gravity," by Weikko A. Heiskanen, *Scientific American,* September 1955, p. 164.)

33. During volcanic eruptions, chunks of solid rock can be blasted out of the volcano; these projectiles are called *volcanic blocks.* Figure 29 shows a cross section of Mt. Fuji, in Japan. (*a*) At what initial speed would a block have to be ejected, at 35° to the horizontal, from the vent at *A* in order to fall at the foot of the volcano at *B*? (*b*) What is the time of flight?

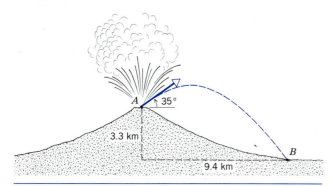

Figure 29   Problem 33.

34. A third baseman wishes to throw to first base, 127 ft distant.

His best throwing speed is 85 mi/h. (*a*) If the ball leaves his hand, 3.0 ft above the ground, in a horizontal direction, what will happen to it? (*b*) At what upward angle must the third baseman launch the ball if the first baseman is to catch it? Assume that the first baseman's glove is also 3.0 ft above the ground. (*c*) What will be the time of flight?

35. At what initial speed must the basketball player throw the ball, at 55° above the horizontal, to make the foul shot, as shown in Fig. 30? The basket rim is 18 in. in diameter. Obtain other data from Fig. 30.

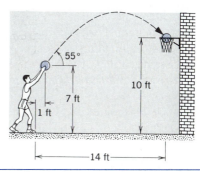

Figure 30   Problem 35.

36. A football player punts the football so that it will have a "hang time" (time of flight) of 4.50 s and land 50 yd (= 45.7 m) away. If the ball leaves the player's foot 5.0 ft (= 1.52 m) above the ground, what is its initial velocity (magnitude and direction)?

37. A certain airplane has a speed of 180 mi/h and is diving at an angle of 27° below the horizontal when a radar decoy is released. The horizontal distance between the release point and the point where the decoy strikes the ground is 2300 ft. (*a*) How long was the decoy in the air? (*b*) How high was the plane when the decoy was released? See Fig. 31.

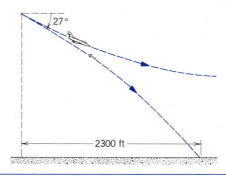

Figure 31   Problem 37.

38. A dive bomber, diving at an angle of 56.0° with the vertical, releases a bomb at an altitude of 730 m. The bomb hits the ground 5.10 s later, missing the target. (*a*) What is the speed of the bomber? (*b*) How far did the bomb travel horizontally

Figure 32   Problem 39.

during its flight? (*c*) What were the horizontal and vertical components of its velocity just before striking the ground? (*d*) At what speed and angle with the vertical did the bomb strike the ground?

39. The B-52 shown in Fig. 32 is 49 m long and is traveling at an air speed of 820 km/h (= 510 mi/h) over a bombing range. How far apart will the bomb craters be? Make any measurements you need directly from the figure. Assume that there is no wind and ignore air resistance. How would air resistance affect your answer?

40. A football is kicked off with an initial speed of 64 ft/s at a projection angle of 42° above the horizontal. A receiver on the goal line 65 yd away in the direction of the kick starts running to meet the ball at that instant. What must be his average speed if he is to catch the ball just before it hits the ground? Neglect air resistance.

41. (*a*) During a tennis match, a player serves at 23.6 m/s (as recorded by radar gun), the ball leaving the racquet, 2.37 m above the court surface, horizontally. By how much does the ball clear the net, which is 12 m away and 0.90 m high? (*b*) Suppose the player serves the ball as before except that the ball leaves the racquet at 5.0° below the horizontal. Does the ball clear the net now?

42. A batter hits a pitched ball at a height 4.0 ft above the ground so that its angle of projection is 45° and its horizontal range is 350 ft. The ball travels down the left field line where a 24-ft-high fence is located 320 ft from home plate. Will the ball clear the fence? If so, by how much?

43. The kicker on a football team can give the ball an initial speed of 25 m/s. Within what angular range must he kick the ball if he is to just score a field goal from a point 50 m in front of the goalposts whose horizontal bar is 3.44 m above the ground?

44. A cannon is arranged to fire projectiles, with initial speed v_0, directly up the face of a hill of elevation angle α, as shown in

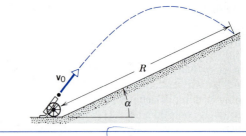

Figure 33   Problem 44.

Fig. 33. At what angle from the horizontal should the cannon be aimed to obtain the maximum possible range R up the face of the hill?

45. In a baseball game, a batter hits the ball at a height of 4.60 ft above the ground so that its angle of projection is 52.0° to the horizontal. The ball lands in the grandstand, 39.0 ft up from the bottom; see Fig. 34. The grandstand seats slope upward at 28.0°, with the bottom seats 358 ft from home plate. Calculate the speed with which the ball left the bat. (Ignore air resistance.)

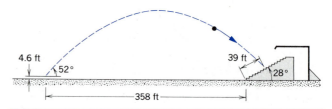

Figure 34   Problem 45.

46. Projectiles are hurled at a horizontal distance R from the edge of a cliff of height h in such a way as to land a horizontal distance x from the bottom of the cliff. If you want x to be as

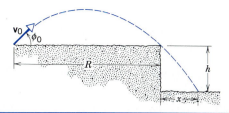

Figure 35   Problem 46.

small as possible, how would you adjust ϕ_0 and v_0, assuming that v_0 can be varied from zero to a finite maximum value v_{max} and that ϕ_0 can be varied continuously? Only one collision with the ground is allowed; see Fig. 35.

47. A radar observer on the ground is "watching" an approaching projectile. At a certain instant she has the following information: the projectile is at maximum altitude and is moving horizontally with speed v; the straight-line distance to the projectile is L; the line of sight to the projectile is at an angle θ above the horizontal. (*a*) Find the distance D between the observer and the point of impact of the projectile. D is to be expressed in terms of the observed quantities v, L, θ, and the known value of g. Assume a flat Earth; assume also that the observer lies in the plane of the projectile's trajectory. (*b*) How can you tell whether the projectile will pass over the observer's head or strike the ground before reaching her?

48. A rocket is launched from rest and moves in a straight line at $70.0°$ above the horizontal with an acceleration of 46.0 m/s². After 30.0 s of powered flight, the engines shut off and the rocket follows a parabolic path back to the Earth; see Fig. 36. (*a*) Find the time of flight from launching to impact. (*b*) What is the maximum altitude reached? (*c*) What is the distance from launch pad to impact point? (Ignore the variation of g with altitude.)

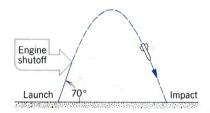

Figure 36   Problem 48.

49. An antitank gun is located on the edge of a plateau that is 60.0 m above a surrounding plain; see Fig. 37. The gun crew sights an enemy tank stationary on the plain at a horizontal distance of 2.20 km from the gun. At the same moment, the tank crew sees the gun and starts to move directly away from it with an acceleration of 0.900 m/s². If the antitank gun fires a shell with a muzzle speed of 240 m/s at an elevation

Figure 37   Problem 49.

angle of $10.0°$ above the horizontal, how long should the gun crew wait before firing if they are to hit the tank?

50. What is the maximum vertical height to which a baseball player can throw a ball if he can throw it a maximum distance of 60.0 m? Assume the ball is released at a height of 1.60 m with the same speed in both cases.

Section 4-4 Uniform Circular Motion

51. In Bohr's model of the hydrogen atom, an electron revolves around a proton in a circular orbit of radius 5.29×10^{-11} m with a speed of 2.18×10^6 m/s. What is the acceleration of the electron in this model of the hydrogen atom?

52. An astronaut is rotated in a centrifuge of radius 5.2 m. (*a*) What is the speed if the acceleration is $6.8g$? (*b*) How many revolutions per minute are required to produce this acceleration?

53. An Earth satellite moves in a circular orbit 640 km above the Earth's surface. The time for one revolution is 98.0 min. (*a*) What is the speed of the satellite? (*b*) What is the free-fall acceleration at the orbit?

54. A carnival Ferris wheel has a 15-m radius and completes five turns about its horizontal axis every minute. (*a*) What is the acceleration, magnitude and direction, of a passenger at the highest point? (*b*) What is the acceleration at the lowest point?

55. A rotating fan completes 1200 revolutions every minute. Consider a point on the tip of the blade, which has a radius of 0.15 m. (*a*) Through what distance does the point move in one revolution? (*b*) What is the speed of the point? (*c*) What is its acceleration?

56. The fast train known as the TGV Atlantique (Train Grande Vitesse) that runs south from Paris to LeMans in France has a top speed of 310 km/h. (*a*) If the train goes around a curve at this speed and the acceleration experienced by the passengers is to be limited to $0.05g$, what is the smallest radius of curvature for the track that can be tolerated? (*b*) If there is a curve with a 0.94-km radius, to what speed must the train be slowed?

57. Certain neutron stars (extremely dense stars) are believed to be rotating at about 1 rev/s. If such a star has a radius of 20 km (a typical value), (*a*) what is the speed of a point on the equator of the star and (*b*) what is the centripetal acceleration of this point?

58. A particle P travels with constant speed on a circle of radius 3.0 m and completes one revolution in 20 s (Fig. 38). The particle passes through O at $t = 0$. With respect to the origin O, find (*a*) the magnitude and direction of the vectors describing its position 5.0, 7.5, and 10 s later; (*b*) the magni-

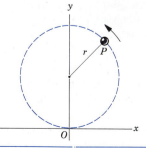

Figure 38   Problem 58.

tude and direction of the displacement in the 5.0-s interval from the fifth to the tenth second; (c) the average velocity vector in this interval; (d) the instantaneous velocity vector at the beginning and at the end of this interval; and (e) the instantaneous acceleration vector at the beginning and at the end of this interval. Measure angles counterclockwise from the x axis.

59. A particle in uniform circular motion about the origin O has a speed v. (a) Show that the time Δt required for it to pass through an angular displacement $\Delta \theta$ is given by

$$\Delta t = \frac{2\pi r}{v} \frac{\Delta \theta}{360°},$$

where $\Delta \theta$ is in degrees and r is the radius of the circle. (b) Refer to Fig. 39 and by taking x and y components of the velocities at points 1 and 2, show that $\overline{a}_x = 0$ and $\overline{a}_y = -0.9v^2/r$, for a pair of points symmetric about the y axis with $\Delta \theta = 90°$. (c) Show that if $\Delta \theta = 30°$, $\overline{a}_x = 0$ and $\overline{a}_y = -0.99v^2/r$. (d) Argue that $\overline{a}_y \rightarrow -v^2/r$ as $\Delta \theta \rightarrow 0$ and that circular symmetry requires this answer for each point on the circle.

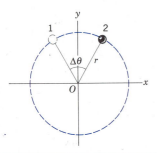

Figure 39   Problem 59.

60. A child whirls a stone in a horizontal circle 1.9 m above the ground by means of a string 1.4 m long. The string breaks, and the stone flies off horizontally, striking the ground 11 m away. What was the centripetal acceleration of the stone while in circular motion?

61. (a) Use the data of Appendix C to calculate the ratio of the centripetal accelerations of Earth and Saturn owing to their revolutions about the Sun. Assume that both planets move in circular orbits with constant speed. (b) What is the ratio of the distances of these two planets from the Sun? (c) Compare your answers in parts (a) and (b) and suggest a simple relation between centripetal acceleration and distance from the Sun. Check your hypothesis by calculating the same ratios for another pair of planets.

62. (a) What is the centripetal acceleration of an object on the Earth's equator owing to the rotation of the Earth? (b) What would the period of rotation of the Earth have to be in order that objects on the equator have a centripetal acceleration equal to 9.8 m/s²?

63. Calculate the acceleration of a person at latitude 40° owing to the rotation of the Earth.

64. A woman 1.6 m tall stands upright at latitude 50° for 24 h. (a) During this interval, how much farther does the top of her head move than the soles of her feet? (b) How much greater is the acceleration of the top of her head than the

acceleration of the soles of her feet? Consider only effects associated with the rotation of the Earth.

Section 4-5 Velocity and Acceleration Vectors in Circular Motion

65. A particle is traveling in a circular path of radius 3.64 m. At a certain instant, the particle is moving at 17.4 m/s, and its acceleration is at an angle of 22.0° from the direction to the center of the circle as seen from the particle; see Fig. 40. (a) At what rate is the speed of the particle increasing? (b) What is the magnitude of the acceleration?

Figure 40   Problem 65.

66. A particle moves in a plane according to

$$x = R \sin \omega t + \omega R t,$$

$$y = R \cos \omega t + R,$$

where ω and R are constants. This curve, called a *cycloid,* is the path traced out by a point on the rim of a wheel which rolls without slipping along the x axis. (a) Sketch the path. (b) Calculate the instantaneous velocity and acceleration when the particle is at its maximum and minimum value of y.

Section 4-6 Relative Motion

67. A person walks up a stalled 15-m-long escalator in 90 s. When standing on the same escalator, now moving, the person is carried up in 60 s. How much time would it take that person to walk up the moving escalator? Does the answer depend on the length of the escalator?

68. The airport terminal in Geneva, Switzerland has a "moving sidewalk" to speed passengers through a long corridor. Peter, who walks through the corridor but does not use the moving sidewalk, takes 150 s to do so. Paul, who simply stands on the moving sidewalk, covers the same distance in 70 s. Mary not only uses the sidewalk but walks along it. How long does Mary take? Assume that Peter and Mary walk at the same speed.

69. A transcontinental flight of 2700 mi is scheduled to take 50 min longer westward than eastward. The air speed of the jet is 600 mi/h. What assumptions about the jet stream wind velocity, presumed to be east or west, are made in preparing the schedule?

70. Snow is falling vertically at a constant speed of 7.8 m/s. (a) At what angle from the vertical and (b) with what speed do the snowflakes appear to be falling as viewed by the driver of a car traveling on a straight road with a speed of 55 km/h?

71. A train travels due south at 28 m/s (relative to the ground) in a rain that is blown to the south by the wind. The path of each raindrop makes an angle of 64° with the vertical, as measured by an observer stationary on the Earth. An observer on the train, however, sees perfectly vertical tracks of rain on the windowpane. Determine the speed of the drops relative to the Earth.

72. In a large department store, a shopper is standing on the "up" escalator, which is traveling at an angle of 42° above the horizontal and at a speed of 0.75 m/s. He passes his daughter, who is standing on the identical, adjacent "down" escalator. (See Fig. 41.) Find the velocity of the shopper relative to his daughter.

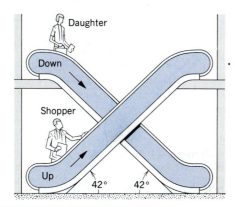

Figure 41 Problem 72.

73. A pilot is supposed to fly due east from A to B and then back again to A due west. The velocity of the plane in air is $\mathbf{v}$ and the velocity of the air with respect to the ground is $\mathbf{u}$. The distance between A and B is l and the plane's air speed is constant. (a) If $u = 0$ (still air), show that the time for the round trip is $t_0 = 2l/v$. (b) Suppose that the air velocity is due east (or west). Show that the time for a round trip is then

$$t_E = \frac{t_0}{1 - u^2/v^2} \, .$$

(c) Suppose that the air velocity is due north (or south). Show that the time for a round trip is then

$$t_N = \frac{t_0}{\sqrt{1 - u^2/v^2}} \, .$$

(d) In parts (b) and (c) one must assume that $u < v$. Why?

74. Two highways intersect, as shown in Fig. 42. At the instant

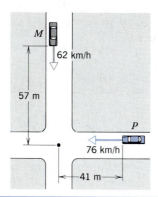

Figure 42 Problem 74.

shown, a police car P is 41 m from the intersection and moving at 76 km/h. Motorist M is 57 m from the intersection and moving at 62 km/h. At this moment, what is the velocity (magnitude and angle with the line of sight) of the motorist with respect to the police car?

75. A helicopter is flying in a straight line over a level field at a constant speed of 6.2 m/s and at a constant altitude of 9.5 m. A package is ejected horizontally from the helicopter with an initial velocity of 12 m/s relative to the helicopter, and in a direction opposite to the helicopter's motion. (a) Find the initial speed of the package relative to the ground. (b) What is the horizontal distance between the helicopter and the package at the instant the package strikes the ground? (c) What angle does the velocity vector of the package make with the ground at the instant before impact, as seen from the ground? (d) As seen from the helicopter?

76. An elevator ascends with an upward acceleration of 4.0 ft/s². At the instant its upward speed is 8.0 ft/s, a loose bolt drops from the ceiling of the elevator 9.0 ft from the floor. Calculate (a) the time of flight of the bolt from ceiling to floor and (b) the distance it has fallen relative to the elevator shaft.

77. A light plane attains an air speed of 480 km/h. The pilot sets out for a destination 810 km to the north but discovers that the plane must be headed 21° east of north to fly there directly. The plane arrives in 1.9 h. What was the vector wind velocity?

78. The New Hampshire State Police use aircraft to enforce highway speed limits. Suppose that one of the airplanes has a speed of 135 mi/h in still air. It is flying straight north so that it is at all times directly above a north–south highway. A ground observer tells the pilot by radio that a 70-mi/h wind is blowing but neglects to give the wind direction. The pilot observes that in spite of the wind the plane can travel 135 mi along the highway in 1 h. In other words, the ground speed is the same as if there were no wind. (a) What is the direction of the wind? (b) What is the heading of the plane, that is, the angle between its axis and the highway?

79. A woman can row a boat 4.0 mi/h in still water. (a) If she is crossing a river where the current is 2.0 mi/h, in what direction will her boat be headed if she wants to reach a point directly opposite from her starting point? (b) If the river is 4.0 mi wide, how long will it take her to cross the river? (c) How long will it take her to row 2.0 mi *down* the river and then back to her starting point? (d) How long will it take her to row 2.0 mi *up* the river and then back to her starting point? (e) In what direction should she head the boat if she wants to cross in the shortest possible time? What is this time?

80. A wooden boxcar is moving along a straight railroad track at a speed v_1. A sniper fires a bullet (initial speed v_2) at it from a high-powered rifle. The bullet passes through both walls of the car, its entrance and exit holes being exactly opposite to each other as viewed from within the car. From what direction, relative to the track, was the bullet fired? Assume that the bullet was not deflected upon entering the car, but that its speed decreased by 20%. Take $v_1 = 85$ km/h and $v_2 = 650$ m/s. (Are you surprised that you don't need to know the width of the boxcar?)

81. A man wants to cross a river 500 m wide. His rowing speed

(relative to the water) is 3.0 km/h. The river flows at a speed of 2.0 km/h. The man's walking speed on shore is 5.0 km/h. (*a*) Find the path (combined rowing and walking) he should take to get to the point directly opposite his starting point in the shortest time. (*b*) How long does it take?

82. A battleship steams due east at 24 km/h. A submarine 4.0 km away fires a torpedo that has a speed of 50 km/h; see Fig. 43. If the bearing of the ship as seen from the submarine is 20° east of north, (*a*) in what direction should the torpedo be fired to hit the ship, and (*b*) what will be the running time for the torpedo to reach the battleship?

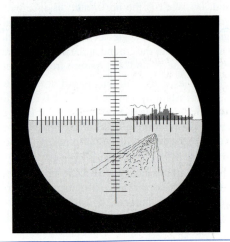

Figure 43   Problem 82.

83. An electron moves at speed $0.42c$ with respect to observer *B*. Observer *B* moves at speed $0.63c$ with respect to observer *A*, in the same direction as the electron. What does observer *A* measure for the speed of the electron?

84. Galaxy Alpha is observed to be receding from us with a speed of $0.350c$. Galaxy Beta, located in precisely the opposite direction, is found to be receding from us at this same speed. What recessional speed would an observer on galaxy Alpha find (*a*) for our galaxy and (*b*) for galaxy Beta?

Computer Projects

85. A computer can generate a table of coordinates, velocity components, and acceleration components of an object at specified times. The table can then be searched to find quantities of interest, such as the highest point on a trajectory, the time of landing, and so forth. Write a program or design a spreadsheet to compute the coordinates and velocity components of a projectile at the end of every time interval Δt from time t_1 to time t_2, assuming the projectile starts at the origin at time $t = 0$. That is, the computer should evaluate $x = v_0 t \cos \theta_0$, $y = v_0 t \sin \theta_0 - \frac{1}{2} g t^2$, $v_x = v_0 \cos \theta_0$, and $v_y = v_0 \sin \theta_0 - gt$ for $t = t_1, t_1 + \Delta t, t_1 + 2\Delta t, \ldots, t_1 + N \Delta t$. At the beginning enter values for v_0, θ_0, t_1, Δt, and N. Arrange the program so you can easily change t_1, Δt, and N on subsequent runs without reentering values for the other quantities. Test the program by solving the following problem. Compare your results with those obtained from the appropriate algebraic expressions.

A projectile is fired over level ground with $v_0 = 50$ m/s at 25° above the horizontal. (*a*) Evaluate $x(t)$, $y(t)$, $v_x(t)$, and

$v_y(t)$ at the end of every 0.1 s from $t = 0$ to $t = 4.5$ s. (*b*) Find the two values that straddle the time for which the projectile is at the highest point in its trajectory. Now rerun the program with t_1 equal to the earliest of these times and $\Delta t = 0.005$ s. Use the table to estimate the coordinates of the highest point to 2 significant figures. (*c*) Use the same technique to find the time, coordinates, and velocity components when the projectile returns to the firing height.

86. A particle moves in the xy plane subject to the acceleration $a_x = -1.7$ and $a_y = -0.45$. (In this problem, all dimensions are in centimeters and all times are in seconds.) At $t = 0$, the particle passes through the point $x = 1$, $y = 10$ moving with velocity $v_x = 10$ and $v_y = 2$. Write a computer program that tabulates the following variables describing the motion of the particle while it is in the first (upper-right) quadrant only: t, x, y, r, $\phi (= \tan^{-1} y/x)$, v_x, v_y, v, $\theta (= \tan^{-1} v_y/v_x)$. Use your table of values to answer the following questions. Give all answers to 3 significant figures. Some questions may have more than one answer. (*a*) At what time and at what location does the particle leave the first quadrant? (*b*) What is the maximum distance of the particle from the origin, and what is its speed at that time? (*c*) In what direction is the particle moving when its speed is 2.00? (*d*) Where does the particle cross the 45° line that bisects the quadrant?

87. The coordinates of an object traveling uniformly on a circle of radius R are given by $x = R \cos \omega t$ and $y = R \sin \omega t$, where ω is a constant and the angle ωt is in radians. Write a computer program or design a spreadsheet to calculate the average velocity over the time interval from t_0 to $t_0 + \Delta t$. Take $R = 1.5$ m and $\omega = 5.0$ rad/s and calculate $\bar{v}_x = [x(t_0 + \Delta t) - x(t_0)]/\Delta t$ and $\bar{v}_y = [y(t_0 + \Delta t) - y(t_0)]/\Delta t$. Arrange the program so you can easily rerun it with different values of t_0 and Δt. Loss of significance is reduced if all variables are double precision.

(*a*) Take $t_0 = 1$ s. Calculate x, y, $\bar{v}_x$, $\bar{v}_y$, and $x\bar{v}_x + y\bar{v}_y$. The last quantity is the scalar product of the position and average velocity vectors. It is zero if they are perpendicular to each other. Now repeat the calculations with $\Delta t = 0.1$, 0.01, 0.001, and 0.0001 s. Notice that the components of $\bar{v}$ approach limiting values, the components of the instantaneous velocity v, and that $\bar{v}$ is becoming more nearly perpendicular to the position vector (that is, more nearly tangent to the circle). As you can show by direct differentiation, the components of v are given by $v_x = -\omega R \sin \omega t$ and $v_y = \omega R \cos \omega t$. Evaluate these expressions to see how accurately your program estimated v. (*b*) Now revise the program so it calculates the components of the average acceleration: $\bar{a}_x = [v_x(t_0 + \Delta t) - v_x(t_0)]/\Delta t$ and $\bar{a}_y = [v_y(t_0 + \Delta t) - v_y(t_0)]/\Delta t$. Use $v_x(t) = -\omega R \sin \omega t$ and $v_y(t) = \omega R \cos \omega t$. Also compute $x\bar{a}_y + y\bar{a}_x$. This is the magnitude of the vector product of the position and average acceleration vectors. It is zero if they are parallel. Carry out the calculation for $t_0 = 1$ s and $\Delta t = 1, 0.1, 0.01, 0.001$, and 0.0001 s. Notice that $\bar{a}$ approaches a limiting value, the instantaneous acceleration a, and that it becomes more nearly parallel to the position vector. The components of a are given by $a_x = -\omega^2 R \cos \omega t$ and $a_y = -\omega^2 R \sin \omega t$. Evaluate these expressions and compare the results with the estimates generated by your program. Also verify that the results generated by your program predict $a = v^2/R$ for the magnitude of the acceleration.

CHAPTER 5

FORCE AND NEWTON'S LAWS

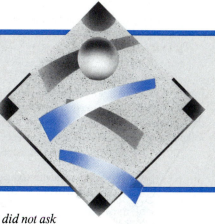

*In Chapters 2 and 4, we studied the motion of a particle. We did not ask
what "caused" the motion; we simply described it in terms of the vectors* **r**, **v**, *and* **a**.
In this chapter and the next, we discuss the causes *of motion, a field of study called* dynamics.

*The approach to dynamics we consider in this chapter and the next, which is generally
known as* classical mechanics, *was developed and successfully tested in the 17th and 18th
centuries. In our century, new theories (special and general relativity and quantum
mechanics) have indicated certain realms far from our ordinary experiences where classical
mechanics fails to give predictions that agree with experiment, but these new theories reduce
to classical mechanics in the limits of ordinary objects.*

*Without reference to special or general relativity or to quantum mechanics, we can build great
skyscrapers and study the properties of their construction materials; build airplanes that can
carry hundreds of people and fly halfway around the world; and send space probes on complex
missions to the comets, the planets, and beyond. This is the stuff of classical mechanics.*

5-1 CLASSICAL MECHANICS

We focus our attention on the motion of a particular
body. It interacts with the surrounding bodies (its *environment*) so that its velocity changes: an acceleration is produced. Table 1 shows some common accelerated motions
and the environment that is mostly responsible for the
acceleration. The central problem of classical mechanics
is this: (1) We are given a body whose characteristics
(mass, volume, electric charge, etc.) we know. (2) We
place this body, at a known initial location and with a
known initial velocity, in an environment of which we
have a complete description. (3) What is the subsequent
motion of the body?

In previous chapters, we have treated physical objects

as *particles,* that is, as bodies whose internal structures or
motions can be ignored and whose parts all move in exactly the same way. In studying the interaction of a body
with its environment, we often must consider extended
objects whose different parts may interact with the environment in different ways. For example, a worker pushes
a heavy crate along a rough surface. The worker pushes on
one vertical side of the crate, while the horizontal bottom
experiences the retarding effect of friction with the floor.
The front surface may even experience air resistance.

Later in the text we treat the mechanics of extended
bodies in detail. For the present, we continue to assume
that all parts of the body move in the same way, so that we
can treat the body as a particle. Under this assumption, it
doesn't matter where the environment acts on the body;

TABLE 1   SOME ACCELERATED MOTIONS AND THEIR CAUSES

Object	Change in Motion	Major Cause (Environment)
Apple	Falls from tree	Gravity (Earth)
Billiard ball	Bounces off another	Other ball, table, gravity (Earth)
Skier	Slides down hill	Gravity (Earth), friction (snow), air resistance
Beam of electrons (in TV set)	Focusing and deflection	Electromagnetic fields (magnets and voltage differences)
Comet Halley	Round trip through solar system	Gravity (Sun)

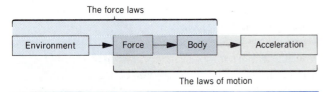

Figure 1   Our program for mechanics. The three boxes on the left suggest that force is an interaction between a body and its environment. The three boxes on the right suggest that a force acting on a body will accelerate it.

our primary concern is with the *net effect* of the environment.

This problem of classical mechanics was solved, at least for a large variety of environments, by Isaac Newton (1642–1727) when he put forward his laws of motion and formulated his law of universal gravitation. The procedure for solving this problem, in terms of our present framework of classical mechanics, is as follows: (1) We introduce the concept of *force* **F** (which we regard for now as a push or a pull), and we define it in terms of the acceleration **a** experienced by a particular standard body. (2) We develop a procedure for assigning a *mass m* to a body so that we may understand the fact that different bodies experience different accelerations in the same environment. (3) Finally, we try to find ways of calculating the forces that act on bodies from the properties of the body and of its environment; that is, we look for *force laws.* Force, which is basically a means of relating the environment to the motion of the body, appears both in the laws of motion (which tell us what acceleration a given body will experience under the action of a given force) and in the force laws (which tell us how to calculate the force that will act on a given body in a given environment). The laws of motion and the force laws, taken together, constitute the laws of mechanics, as Fig. 1 suggests.

This program of mechanics cannot be tested piecemeal. We must view it as a unit and we shall judge it to be successful if we can say "yes" to these two questions. (1) Does the program yield results that agree with experiment? (2) Are the force laws simple in form? It is the crowning glory of Newtonian mechanics that we can indeed answer each of these questions in the affirmative.

5-2   NEWTON'S FIRST LAW

For centuries the problem of motion and its causes was a central theme of natural philosophy, an early name for what we now call physics. It was not until the time of Galileo and Newton, however, that dramatic progress was made. Isaac Newton, born in England in the year of Galileo's death, is the principal architect of classical mechanics. He carried to full fruition the ideas of Galileo and others who preceded him. His three laws of motion were

first presented (in 1686) in his *Philosophiae Naturalis Principia Mathematica,* usually called the *Principia.*

Before Galileo's time most philosophers thought that some influence or "force" was needed to keep a body moving. They thought that a body was in its "natural state" when it was at rest. For a body to move in a straight line at constant speed, for example, they believed that some external agent had to continually propel it; otherwise it would "naturally" stop moving.

If we wanted to test these ideas experimentally, we would first have to find a way to free a body from all influences of its environment or from all forces. This is hard to do, but in certain cases we can make the forces very small. If we study the motion as we make the forces smaller and smaller, we shall have some idea of what the motion would be like if the external forces were truly zero.

Let us place our test body, say a block, on a rigid horizontal plane. If we let the block slide along this plane, we note that it gradually slows down and stops. This observation was used, in fact, to support the idea that motion stopped when the external force, in this case the hand initially pushing the block, was removed. We can argue against this idea, however, by reasoning as follows. Let us repeat our experiment, now using a smoother block and a smoother plane and providing a lubricant. We note that the velocity decreases more slowly than before. Let us use still smoother blocks and surfaces and better lubricants. We find that the block decreases in velocity at a slower and slower rate and travels farther each time before coming to rest. You may have experimented with an air track, on which objects can be made to float on a film of air; such a device comes close to the limit of no friction, as even a slight tap on one of the gliders can send it moving along the track at a slow and almost constant speed. We can now extrapolate and say that if all friction could be eliminated, the body would continue indefinitely in a straight line with constant speed. An external force is needed to set the body in motion, but *no external force is needed to keep a body moving with constant velocity.*

It is difficult to find a situation in which no external force acts on a body. The force of gravity will act on an object on or near the Earth, and resistive forces such as friction or air resistance oppose motion on the ground or in the air. Fortunately, we need not go to the vacuum of distant space to study motion free of external force, because, as far as the overall translational motion of a body is concerned, *there is no distinction between a body on which no external force acts and a body on which the sum or resultant of all the external forces is zero.* We usually refer to the resultant of all the forces acting on a body as the "net" force. For example, the push of our hand on the sliding block can exert a force that counteracts the force of friction on the block, and an upward force of the horizontal plane counteracts the force of gravity. The net force on the block can then be zero, and the block can move with constant velocity.

This principle was adopted by Newton as the first of his three laws of motion:

Consider a body on which no net force acts. If the body is at rest, it will remain at rest. If the body is moving with constant velocity, it will continue to do so.

Newton's first law is really a statement about reference frames. In general, the acceleration of a body depends on the reference frame relative to which it is measured. However, the laws of classical mechanics are valid only in a certain set of reference frames, namely, those from which all observers would measure the *same* acceleration for a moving body. Newton's first law helps us to identify this family of reference frames if we express it as follows:

If the net force acting on a body is zero, then it is possible to find a set of reference frames in which that body has no acceleration.

The tendency of a body to remain at rest or in uniform linear motion is called *inertia,* and Newton's first law is often called the *law of inertia.* The reference frames to which it applies are called *inertial frames,* as we discussed in Section 4-6. You will recall from that discussion that observers in different inertial reference frames (moving with constant velocity relative to one another) all measure the same value of the acceleration. Thus there is not just one frame in which the acceleration happens to be zero; there is a set of all inertial frames in which the acceleration is zero.

To test whether a particular frame of reference is an inertial frame, we place a test body at rest in the frame and ascertain that no net force acts on it. If the body does not remain at rest, the frame is not an inertial frame. Similarly, we can put the body (again subject to no net force) in motion at constant velocity; if its velocity changes, either in magnitude or direction, the frame is not an inertial frame. A frame in which these tests are everywhere passed is an inertial frame. Once we have found one inertial frame, it is easy to find many more, because a frame of reference that moves at constant velocity relative to one inertial frame is also an inertial frame.

In this book we almost always apply the laws of classical mechanics from the point of view of an observer in an inertial frame. Occasionally, we discuss problems involving observers in noninertial reference frames, such as an accelerating car, a rotating merry-go-round, or an orbiting satellite. Even though the Earth is rotating, a reference frame attached to the Earth can be considered to be approximately an inertial reference frame for most practical purposes. For large-scale applications, such as analyzing the flight of ballistic missiles or studying wind and ocean currents, the noninertial character of the rotating Earth becomes important.

Notice that there is no distinction in the first law between a body at rest and one moving with a constant velocity. Both motions are "natural" if the net force acting on the body is zero. This becomes clear when a body at rest in one inertial frame is viewed from a second inertial frame, that is, a frame moving with constant velocity with respect to the first. An observer in the first frame finds the body to be at rest; an observer in the second frame finds the same body to be moving with constant velocity. Both observers find the body to have no acceleration, that is, no change in velocity, and both may conclude from the first law that no net force acts on the body.

If there is a net interaction between the body and objects present in the environment, the effect may be to change the "natural" state of the body's motion. To investigate this, we must now examine carefully the concept of force.

5-3 FORCE

We develop our concept of force by defining it operationally. In everyday language, a force is a push or a pull. To measure such forces quantitatively, we express them in terms of the acceleration that a given standard body experiences in response to that force.

As a standard body we find it convenient to use (or rather to imagine that we use!) the standard kilogram (see Fig. 5 of Chapter 1). This body has been assigned, by definition, a mass m_0 of exactly 1 kg. Later we shall describe how masses are assigned to other bodies.

For an environment that exerts a force, we place the standard body on a horizontal table having negligible friction and we attach a spring to it. We hold the other end of the spring in our hand, as in Fig. 2a. Now we pull the spring horizontally to the right so that by trial and error we are able to give the standard body a measured constant acceleration of exactly 1 m/s². We then declare, as a matter of definition, that the spring (which is the significant body in the environment) is exerting on the standard kilogram a constant force whose magnitude we call "1 newton" (abbreviated 1 N). We note that, in imparting this

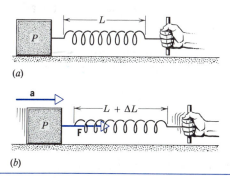

Figure 2 (*a*) A "particle" *P* (the standard kilogram) at rest on a horizontal frictionless surface. (*b*) The body is accelerated by pulling the spring to the right.

force, the spring is stretched an amount ΔL beyond its normal unextended length L, as Fig. 2*b* shows.

We can repeat the experiment, either stretching the spring more or using a stiffer spring, so that we measure an acceleration of 2 m/s^2 for the standard body. We now declare that the spring is exerting a force of 2 N on the standard body. In general, if we observe this particular standard body to have an acceleration a in a particular environment, we then say that the environment is exerting a force F on the standard 1-kg body, where F (in newtons) is numerically equal to a (in m/s^2).

Now let us see whether force, as we have defined it, is a *vector* quantity. In Fig. 2*b* we assigned a magnitude to the force F, and it is a simple matter to assign a direction to it as well, namely, the direction of the acceleration that the force produces. However, to be a vector it is not enough for a quantity to have magnitude and direction; it must also obey the laws of vector addition described in Chapter 3. We can learn only from experiment whether forces, as we defined them, do indeed obey these laws.

Let us arrange to exert a force of 4 N along the x axis and a force of 3 N along the y axis. We apply these forces first separately and then simultaneously to the standard body placed, as before, on a horizontal, frictionless surface. What will be the acceleration of the standard body? We would find by experiment that the 4-N force in the x direction produced an acceleration of 4 m/s^2 in the x direction, and that the 3-N force in the y direction produced an acceleration of 3 m/s^2 in the y direction (Fig. 3*a*). When the forces are applied simultaneously, as shown in Fig. 3*b*, we find that the acceleration is 5 m/s^2 directed along a line that makes an angle of 37° with the x axis. This is the same acceleration that would be produced if the standard body were experiencing a force of 5 N in that direction. This same result can be obtained if we first add the 4-N and 3-N forces vectorially (Fig. 3*c*) to a 5-N resultant directed at 37° from the x axis, and then apply that single 5-N net force to the body. Experiments of this kind show conclusively that forces are vectors: they have mag-

nitude and direction, *and* they add according to the vector addition law.

Note that we have two methods of analysis available, which should produce identical results: (1) Find the acceleration produced by each separate force, and add the resultant accelerations vectorially. (2) Add the forces vectorially to a single resultant, and then find the acceleration when that single net force is applied to the body.

5-4   MASS

In Section 5-3 we considered only the accelerations given to one particular body, the standard kilogram. We were able thereby to define forces quantitatively. What effect would these forces have on other bodies? Because our standard body was chosen arbitrarily in the first place, we know that for any given body the acceleration will be directly proportional to the force applied. The significant question remaining then is: What effect will the *same force* have on *different bodies*?

Everyday experience gives us a qualitative answer. The same force will produce different accelerations on different bodies. A baseball will be accelerated more by a given force than will an automobile. In order to obtain a quantitative answer to this question, we need a method to measure mass, *the property of a body which determines its resistance to a change in its motion.*

Let us attach a spring to our standard body (the standard kilogram, to which we have arbitrarily assigned a mass $m_0 = 1$ kg, exactly) and arrange to give it an acceleration a_0 of, say, 2.00 m/s^2, using the method of Fig. 2*b*. Let us measure carefully the extension ΔL of the spring associated with the force that the spring is exerting on the block.

We now attach two identical standard bodies to the spring and apply the same force as before (that is, we pull on the two bodies until the spring stretches by the same

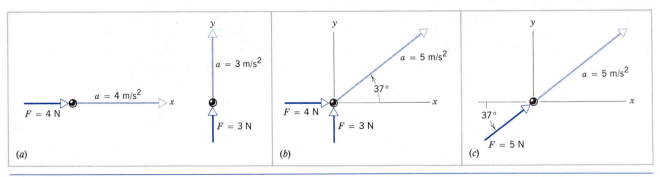

Figure 3   (*a*) A 4-N force in the x direction gives an acceleration of 4 m/s^2 in the x direction, and a 3-N force in the y direction gives an acceleration of 3 m/s^2 in the y direction. (*b*) When the forces are applied simultaneously, the resultant acceleration is 5 m/s^2 in the direction shown. (*c*) The same acceleration can be produced by a single 5-N force in the direction shown.

amount ΔL). We measure the acceleration of the two bodies, and obtain the value of 1.00 m/s^2. If we used three identical standard bodies and applied the same force, we would obtain an acceleration of 0.667 m/s^2.

From these observations, it appears that, for a given force, the greater the mass, the smaller the acceleration. More precisely, we conclude from many such experiments that *the acceleration produced by a given force is inversely proportional to the mass being accelerated.* Another way to put this is: *the mass of a body is inversely proportional to the acceleration it receives from the application of a given force.* The mass of a body can thus be regarded as *a quantitative measure of the resistance of a body to acceleration by a given force.*

This observation gives us a direct way to compare the masses of different bodies: we simply compare the accelerations we measure from the application of a given force to each body. The ratio of the masses of the two bodies is then the same as the *inverse* ratio of the accelerations given to these bodies by that force, or

$$\frac{m_1}{m_0} = \frac{a_0}{a_1} \quad \text{(same force } F \text{ acting).}$$

Here we are comparing the acceleration a_1 of the body of unknown mass m_1 with the acceleration a_0 imparted to the standard body of mass m_0.

For example, suppose as above we use a force that gives an acceleration of 2.00 m/s^2 to the standard body. We apply the same force (by stretching the spring by the same amount ΔL) to a body of unknown mass m_1, and we measure an acceleration a_1 of, say, 0.50 m/s^2. We can then solve for the unknown mass, which gives

$$m_1 = m_0 \left(\frac{a_0}{a_1}\right) = (1.00 \text{ kg}) \left(\frac{2.00 \text{ m/s}^2}{0.50 \text{ m/s}^2}\right) = 4.00 \text{ kg}.$$

The second body, which has only one-fourth the acceleration of the first body when the same force acts on it, has four times the mass of the first body. This illustrates the inverse relationship between mass and acceleration for a given force.

Let us now repeat the preceding experiment on the same two bodies using a common force F' different from that used above. This force will give the standard body an acceleration of a_0' and the unknown body an acceleration of a_1'. From our measurement we would find that the ratio of the accelerations, a_0'/a_1', is the same as in the previous experiment, or

$$\frac{m_1}{m_0} = \frac{a_0}{a_1} = \frac{a_0'}{a_1'}.$$

For example, we apply a greater force so that the extension of the spring is $1.5\Delta L$. We would then find that the standard mass m_0 is accelerated to 3.00 m/s^2 and the unknown mass m_1 is accelerated to 0.75 m/s^2. We would deduce the unknown mass to be

$$m_1 = m_0 \left(\frac{a_0'}{a_1'}\right) = (1.00 \text{ kg}) \left(\frac{3.00 \text{ m/s}^2}{0.75 \text{ m/s}^2}\right) = 4.00 \text{ kg}.$$

We obtain the same value for the unknown mass m_1, no matter what the value of the common force. The mass ratio m_1/m_0 is independent of the common force used; the mass is a fundamental property of the object, unrelated to the value of the force used to compare the unknown mass to the standard mass. In effect, this procedure allows us to measure mass by comparison with the standard kilogram.

We can extend this procedure to a direct comparison of the masses of any two bodies. For example, let us first use our previous procedure to compare a second arbitrary body with the standard body, and thus determine its mass, say m_2. We can now compare the two arbitrary bodies, m_2 and m_1, directly, obtaining accelerations a_2'' and a_1'' when the same force F'' is applied. The mass ratio, defined as usual from

$$\frac{m_2}{m_1} = \frac{a_1''}{a_2''} \quad \text{(same force acting),}$$

turns out to have the same value that we obtain by using the masses m_2 and m_1 previously determined by direct comparison with the standard.

We can show, in still another experiment of this type, that if objects of mass m_1 and m_2 are fastened together, they behave mechanically as a single object of mass $(m_1 + m_2)$. In other words, *masses add like (and are) scalar quantities.*

One practical example of the use of this technique— assigning masses by comparison of the relative accelerations produced by a given force—is in the precise measurement of the masses of atoms. The force in this case is a magnetic deflecting force and the acceleration is centripetal, but the principle is exactly the same. For a common magnetic force acting on two atoms, the ratio of their masses is equal to the inverse ratio of their accelerations. Measuring the deflection, as in the mass spectrometer shown in Fig. 6 of Chapter 1, permits precise mass ratios to be measured, and defining ^{12}C as the standard then permits precise values of masses, such as those shown in Table 6 of Chapter 1, to be obtained.

5-5 NEWTON'S SECOND LAW

We can now summarize all the previously described experiments and definitions in one equation, the fundamental equation of classical mechanics,

$$\sum \mathbf{F} = m\mathbf{a}. \tag{1}$$

In this equation $\sum \mathbf{F}$ is the (vector) *sum* of *all* the forces acting *on* the body, m is the mass of the body, and $\mathbf{a}$ is its (vector) acceleration. We shall usually refer to $\sum \mathbf{F}$ as the *resultant* force or *net* force.

Equation 1 is a statement of Newton's second law. If we write it in the form $\mathbf{a} = (\sum \mathbf{F})/m$, we can easily see that the acceleration of the body is in magnitude directly proportional to the resultant force acting on it and in direction parallel to this force. We also see that the acceleration, for a given force, is inversely proportional to the mass of the body.

Note that the first law of motion appears to be contained in the second law as a special case, for if $\Sigma \mathbf{F} = 0$, then $\mathbf{a} = 0$. In other words, if the resultant force on a body is zero, the acceleration of the body is zero and the body moves with constant velocity, as stated by the first law. However, the first law has an independent and important role in defining inertial reference frames. Without that definition, we would not be able to choose the frames of reference in which to apply the second law. We therefore need *both laws* for a complete system of mechanics.

Equation 1 is a vector equation. As in the case of all vector equations, we can write this single vector equation as three scalar equations,

$$\Sigma F_x = ma_x, \quad \Sigma F_y = ma_y, \quad \text{and} \quad \Sigma F_z = ma_z, \quad (2)$$

relating the *x*, *y*, and *z* components of the resultant force (ΣF_x, ΣF_y, and ΣF_z) to the *x*, *y*, and *z* components of acceleration (a_x, a_y, and a_z) for the mass *m*. It should be emphasized that ΣF_x is the *algebraic* sum of the *x* components of *all* the forces, ΣF_y is the *algebraic* sum of the *y* components of *all* the forces, and ΣF_z is the *algebraic* sum of the *z* components of *all* the forces acting on *m*. In taking the algebraic sum, the signs of the components (that is, the relative directions of the forces) must be taken into account.

In analyzing situations using Newton's second law, it is helpful to draw a diagram showing the body in question as a particle and showing all forces as vectors that act on the particle. Such a drawing is called a *free-body diagram* and is an essential first step both in the analysis of a problem and in the visualization of the physical situation.

Sample Problem 1 A student pushes a loaded sled whose mass *m* is 240 kg for a distance *d* of 2.3 m over the frictionless surface of a frozen lake. She exerts a constant horizontal force *F* of 130 N (= 29 lb) as she does so; see Fig. 4a. If the sled starts from rest, what is its final velocity?

Solution As Fig. 4b shows, we lay out a horizontal *x* axis, we take the direction of increasing *x* to be to the right, and we treat the sled as a particle. Figure 4b is a *partial* free-body diagram. In drawing free-body diagrams, it is important always to include *all* forces that act on the particle, but here we have omitted two vertical forces that will be discussed later in this chapter and that do not affect our solution. We assume that the force *F* exerted by the student is the only horizontal force acting on the sled. We can then find the acceleration of the sled from Newton's second law, or

$$a = \frac{F}{m} = \frac{130 \text{ N}}{240 \text{ kg}} = 0.54 \text{ m/s}^2.$$

Because the acceleration is constant, we can use Eq. 20 of Chap-

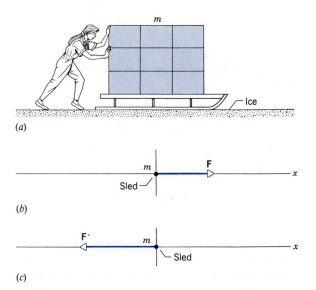

(a)

(b)

(c)

Figure 4 Sample Problems 1 and 2. (*a*) A student pushing a loaded sled over a frictionless surface. (*b*) A free-body diagram, showing the sled as a "particle" and the force acting on it. (*c*) A second free-body diagram, showing the force acting when the student pushes in the opposite direction.

ter 2 [$v^2 = v_0^2 + 2a(x - x_0)$] to find the final velocity. Putting $v_0 = 0$ and $x - x_0 = d$ and solving for *v*, we obtain

$$v = \sqrt{2ad} = \sqrt{(2)(0.54 \text{ m/s}^2)(2.3 \text{ m})} = 1.6 \text{ m/s}.$$

The force, acceleration, displacement, and final velocity of the sled are all positive, which means that they all point to the right in Fig. 4b.

Note that to continue applying the constant force, the student would have to run faster and faster to keep up with the accelerating sled. Eventually, the velocity of the sled would exceed the fastest speed at which the student could run, and thereafter the student would no longer be able to apply a force to the sled. The sled would then continue (in the absence of friction) to coast at constant velocity.

Sample Problem 2 The student in Sample Problem 1 wants to reverse the direction of the velocity of the sled in 4.5 s. With what constant force must she push on the sled to do so?

Solution Let us find the (constant) acceleration, using Eq. 15 of Chapter 2 ($v = v_0 + at$). Solving for *a* gives

$$a = \frac{v - v_0}{t} = \frac{(-1.6 \text{ m/s}) - (1.6 \text{ m/s})}{4.5 \text{ s}} = -0.71 \text{ m/s}^2.$$

This is larger in magnitude than the acceleration in Sample Problem 1 (0.54 m/s^2) so it stands to reason that the student must push harder this time. We find this (constant) force *F'* from

$$F' = ma = (240 \text{ kg})(-0.71 \text{ m/s}^2)$$
$$= -170 \text{ N} (= -38 \text{ lb}).$$

The negative sign shows that the student is pushing the sled in the direction of decreasing *x*, that is, to the left as shown in the free-body diagram of Fig. 4c.

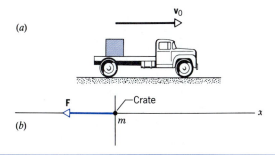

$$\mathbf{F}_{AB} = -\mathbf{F}_{BA}$$

Figure 6   Newton's third law. Body A exerts a force $\mathbf{F}_{BA}$ on body B. Body B must then exert a force $\mathbf{F}_{AB}$ on body A, and $\mathbf{F}_{AB} = -\mathbf{F}_{BA}$.

Figure 5   Sample Problem 3. (*a*) A crate on a truck that is slowing down. (*b*) The free-body diagram of the crate.

Sample Problem 3   A crate whose mass m is 360 kg rests on the bed of a truck that is moving at a speed v_0 of 120 km/h, as in Fig. 5*a*. The driver applies the brakes and slows to a speed v of 62 km/h in 17 s. What force (assumed constant) acts on the crate during this time? Assume that the crate does not slide on the truck bed.

Solution   We first find the (constant) acceleration of the crate. Solving Eq. 15 of Chapter 2 ($v = v_0 + at$) for a yields

$$a = \frac{v - v_0}{t} = \frac{(62 \text{ km/h}) - (120 \text{ km/h})}{17 \text{ s}}$$

$$= \left(-3.41 \ \frac{\text{km}}{\text{h} \cdot \text{s}}\right)\left(\frac{1 \text{ h}}{3600 \text{ s}}\right)\left(\frac{1000 \text{ m}}{1 \text{ km}}\right) = -0.95 \text{ m/s}^2.$$

Because we have taken the positive sense of the horizontal direction to the right, the acceleration vector must point to the left.

The force on the crate follows from Newton's second law:

$$F = ma$$
$$= (360 \text{ kg})(-0.95 \text{ m/s}^2) = -340 \text{ N}.$$

This force acts in the same direction as the acceleration, namely, to the left in Fig. 5*b*. The force must be supplied by an external agent, such as the straps or other mechanical means used to secure the crate to the truck bed. If the crate is not secured, then friction between the crate and the truck bed must supply the required force. If there is not enough friction to provide a force of 340 N, the crate will slide on the truck bed because, as measured by a ground-based observer, it will slow down less rapidly than the truck.

5-6   NEWTON'S THIRD LAW

Forces acting on a body result from other bodies that make up its environment. If we examine the forces acting on a second body, one that was formerly considered part of the environment, then the first body is part of the environment of the second body and is in part responsible for the forces acting on the second body. Any single force is therefore part of the mutual interaction between *two* bodies. We find by experiment that when one body exerts a force on a second body, the second body always exerts a

force on the first. Furthermore, we find these forces *always* to be equal in magnitude but opposite in direction. A single isolated force is therefore an impossibility.

Suppose this were not true. Consider two isolated bodies A and B, and suppose that body A exerts a force on body B, while no force is exerted by B on A. The total force on the combination $A + B$ is nonzero, and the combined mass must accelerate. If such a situation could occur, then we would have a limitless source of energy that could propel $A + B$ through space at no cost: sailboats could sail by passengers blowing on the sails, and spaceships could be accelerated by astronauts pushing on the walls. The impossibility of these actions is a consequence of Newton's third law.

We arbitrarily label one of the forces of the mutual interaction between two bodies as the "action" force, and the other is called the "reaction" force. Newton's third law can then be stated in traditional form:

To every action there is an equal and opposite reaction.

A more modern version of the third law concerns the mutual force exerted by two bodies on one another:

When two bodies exert mutual forces on one another, the two forces are always equal in magnitude and opposite in direction.

Formally (see Fig. 6) let body A exert a force $\mathbf{F}_{BA}$ on body B; experiment then shows that body B exerts a force $\mathbf{F}_{AB}$ on body A. (Note the order of subscripts; the force is exerted *on* the body represented by the first subscript *by* the body represented by the second.) In terms of a vector equation,

$$\mathbf{F}_{AB} = -\mathbf{F}_{BA}. \tag{3}$$

It is important to remember that the action and reaction forces always act on *different* bodies, as the differing first subscripts remind us. If they acted on the same body, there would be no net force on that body and no accelerated motion.

When a bat strikes a baseball, the bat exerts a force on the ball (the action), and the ball exerts an equal and opposite force on the bat. When a soccer player kicks the ball, the foot exerts a force on the ball (the action), and the ball exerts an opposite reaction force on the foot. When you push a stalled car, you can feel the car pushing back on you. In each case the action and reaction forces act on different bodies. If our goal were to study the dynamics of one body—the baseball, for instance—only one force of the action–reaction pair would be considered; the other is felt by a different body and would be considered only if we were studying the dynamics of that body.

The following examples illustrate applications of the third law.

1. *An orbiting satellite.* Figure 7 shows a satellite orbiting the Earth. The only force that acts on it is F_{SE}, the force exerted *on* the satellite *by* the gravitational pull of the Earth. Where is the corresponding reaction force? It is F_{ES}, the force acting on the Earth owing to the gravitational pull of the satellite.

You may think that the tiny satellite cannot exert much of a gravitational pull on the Earth but it does, exactly as Newton's third law requires. That is, considering magnitudes only, $F_{ES} = F_{SE}$. (Recall that the magnitude of any vector quantity is always positive.) The force F_{ES} causes the Earth to accelerate, but, because of the Earth's large mass, its acceleration is so small that it cannot easily be detected.

2. *A book resting on a table.* Figure 8a shows a book resting on a table. The Earth pulls downward on the book with a force F_{BE}. The book does not accelerate because

this force is canceled by an equal and opposite contact force F_{BT} exerted on the book by the table.

Even though F_{BE} and F_{BT} are equal in magnitude and oppositely directed, they do *not* form an action–reaction pair. Why not? *Because they act on the same body—the book.* They cancel each other and thus account for the fact that the book is not accelerating.

Each of these forces must then have a corresponding reaction force somewhere. Where are they? The reaction to F_{BE} is F_{EB}, the (gravitational) force with which the book attracts the Earth. We show this action–reaction pair in Fig. 8b.

Figure 8c shows the reaction force to F_{BT}. It is F_{TB}, the contact force on the table owing to the book. The action–reaction pairs involving the book in this problem, and the bodies on which they act, are

first pair: $F_{BE} = -F_{EB}$ (book and Earth)

and

second pair: $F_{BT} = -F_{TB}$ (book and table).

3. *Pushing a row of crates.* Figure 9 shows a worker W pushing two crates, each of which rests on a wheeled cart that can roll with negligible friction. The worker exerts a force F_{1W} on crate 1, which in turn pushes back on the worker with a reaction force F_{W1}. Crate 1 pushes on crate 2 with a force F_{21}, and crate 2 pushes back on crate 1 with a force F_{12}. (Note that the worker exerts no force on crate 2 directly.) To move forward, the worker must push backward against the ground. The worker exerts a force F_{GW} on the ground, and the rection force of the ground on the worker, F_{WG}, pushes the worker forward. The figure shows three action–reaction pairs:

$$F_{21} = -F_{12} \quad \text{(crate 1 and crate 2)},$$

$$F_{1W} = -F_{W1} \quad \text{(worker and crate 1)},$$

$$F_{WG} = -F_{GW} \quad \text{(worker and ground)}.$$

The acceleration of crate 2 is determined, according to Newton's second law, by the net force applied to it:

$$F_{21} = m_2 a_2.$$

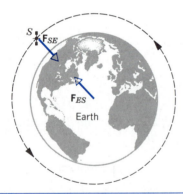

Figure 7   A satellite in Earth orbit. The forces shown are an action–reaction pair. Note that they act on different bodies.

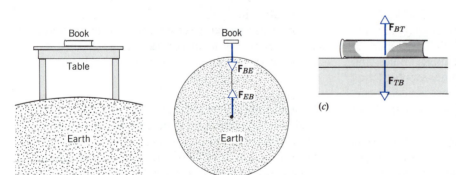

Figure 8   (*a*) A book rests on a table, which in turn rests on the Earth. (*b*) The book and the Earth exert gravitational forces on each other, forming an action–reaction pair. (*c*) The table and book exert action–reaction contact forces on each other.

(*a*)   (*b*)

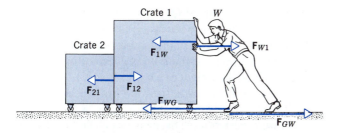

Figure 9   A worker pushes against crate 1, which in turn pushes on crate 2. The crates are on wheels that move freely, so there is no friction between the crates and the ground.

The net force on crate 1 determines its acceleration,

$$F_{1w} - F_{12} = m_1 a_1,$$

where we have written the vector sum of the forces as the difference in their magnitudes, because they act on crate 1 in opposite directions. If the two crates remain in contact, their accelerations must be equal. Letting a represent the common acceleration and adding the equations gives

$$F_{1w} = (m_1 + m_2)a.$$

This same equation would result if we considered crates 1 and 2 to be a single object of mass $m_1 + m_2$. The net external force acting on the combined object is F_{1w}. The two contact forces at the boundary between crates 1 and 2 do not appear in the equation describing the *combined* object. Nor do the internal atomic forces that bind the object together; each internal force forms an action–reaction pair acting on separate parts (individual atoms, perhaps) and such pairs sum to zero when we add together the separate parts to make the combined whole.

Note that in this example the worker is the active agent that is responsible for the motion, but it is the reaction force of the ground that makes this possible. If there were no friction between the worker's shoes and the ground, the worker could not move the system forward.

4. *Block hanging from a spring.* Figure 10a shows a block hanging at rest from a spring, the other end of which is fixed to the ceiling. The forces on the block, shown separately in Fig. 10b, are its weight **W** (acting down) and the force **F** exerted by the spring (acting up). The block is at rest under the influence of these forces, but they are *not* an action–reaction pair, because once again they act on the same body. The reaction force to the weight **W** is the gravitational force that the block exerts on the Earth, which is not shown.

The reaction force to **F** (the force exerted *on* the block *by* the spring) is the force exerted *by* the block *on* the spring. To show this force, we illustrate the forces acting on the spring in Fig. 10c. These forces include the reaction to **F**, which we show as a force **F′** (= −**F**) acting downward, the weight **w** of the spring (usually negligible), and

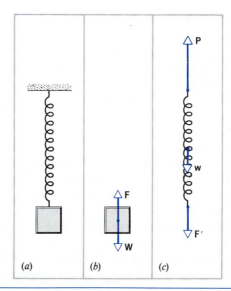

Figure 10   (a) A block hangs at rest supported by a stretched spring. (b) The forces on the block. (c) The forces on the spring.

the upward pull **P** of the ceiling. If the spring is at rest, the net force must be zero: **P** + **w** + **F′** = 0.

The reaction force to **P** acts *on* the ceiling. Since we are not showing the ceiling as an independent body in this diagram, the reaction to **P** does not appear.

5-7   UNITS OF FORCE

Like all equations, Newton's second law ($F = ma$) must be dimensionally consistent. On the right side, the dimensions are, recalling from Chapter 1 that [ ] denotes *the dimensions of,* $[m][a] = ML/T^2$, and therefore these must also be the dimensions of force:

$$[F] = ML/T^2.$$

No matter what the origin of the force—gravitational, electrical, nuclear, or whatever—and no matter how complicated the equation describing the force, these dimensions must hold for it.

In the SI system of units, mass is measured in kg and acceleration in m/s^2. To impart an acceleration of 1 m/s^2 to a mass of 1 kg requires a force of 1 kg·m/s^2. This somewhat inconvenient combination of units is given the name of newton (abbreviated N):

$$1 \text{ N} = 1 \text{ kg·m/s}^2.$$

If we measure the mass in kg and the acceleration in m/s^2, Newton's second law gives the force in N.

Two other systems of units in common use are the cgs (centimeter–gram–second) and the British systems. In the cgs system, mass is measured in grams and acceleration in cm/s^2. The force unit in this system is the *dyne*

TABLE 2   UNITS IN NEWTON'S SECOND LAW

System	Force	Mass	Acceleration
SI	newton (N)	kilogram (kg)	m/s^2
cgs	dyne	gram (g)	cm/s^2
British	pound	slug	ft/s^2

and is equivalent to the $g \cdot cm/s^2$. Since $1 \text{ kg} = 10^3 \text{ g}$ and $1 \text{ m/s}^2 = 100 \text{ cm/s}^2$, it follows that $1 \text{ N} = 10^5$ dyne. A dyne is a very small unit, roughly equal to the weight of a cubic millimeter of water. (A newton, on the other hand, is about the weight of a half-cup of water.)

In the British system, force is measured in pounds and acceleration in ft/s^2. In this system, the mass that is accelerated at 1 ft/s^2 by a force of 1 lb is called the *slug* (from the word *sluggish,* meaning slow or unresponsive).

Other variants on these basic systems are occasionally found, but these three are by far the most common. Table G summarizes these common force units; a more extensive listing can be found in Appendix G.

5-8   WEIGHT AND MASS

The *weight* of a body on the Earth is the gravitational force exerted on it by the Earth. Like all forces, weight is a vector quantity. The direction of this vector is the direction of the gravitational force, that is, toward the center of the Earth. The magnitude of the weight is expressed in force units, such as pounds or newtons.

Let us for the moment assume that the surface of the Earth provides a sufficiently good inertial frame of reference. We release a body of mass m near the Earth's surface and allow it to fall freely under the influence of gravity. Only one force acts on the body, its weight W. The acceleration of the body is the acceleration of free fall, g. We apply Newton's second law, $\mathbf{F} = m\mathbf{a}$, to this freely falling body by putting $\mathbf{W}$ for $\mathbf{F}$ and $\mathbf{g}$ for $\mathbf{a}$, which gives $\mathbf{W} = m\mathbf{g}$. Both $\mathbf{W}$ and $\mathbf{g}$ are vectors directed toward the center of the Earth. We can therefore write

$$W = mg, \qquad (4)$$

where W and g are the magnitudes of the weight and acceleration vectors.

Of course, it is not necessary for a body to be falling to determine its weight. If a body is at rest near the surface of the Earth, then Newton's second law requires that the net force on it be zero. The weight $W = mg$ acts on the body, and therefore to keep the body at rest it must experience another force numerically equal to mg but acting in a direction opposite to the weight. In Fig. 10, the spring supplies this force; the force exerted by the spring on the body must be numerically equal to mg. In Fig. 8, the table exerts an upward force $\mathbf{F}_{BT}$ on the book which keeps it in equilibrium; this upward force is in magnitude equal to the weight mg.

Because g varies from point to point on the Earth, $\mathbf{W}$, the weight of a body of mass m, is different in different localities. Thus, the weight of a body of mass 1.00 kg, in a locality where $g = 9.80 \text{ m/s}^2$, is 9.80 N; in a locality where $g = 9.78 \text{ m/s}^2$, the same body weighs 9.78 N. If these weights were determined by measuring the amount of stretch required in a spring to balance them, the difference in weight of the same 1-kg body at the two different localities would be evident in the slightly different stretch of the spring at these two localities. Hence, unlike the mass of a body, which is an *intrinsic* property of the body, the weight of a body depends on its location relative to the center of the Earth. As we discuss in the next section, spring scales may read differently but balances read the same at different locations on Earth.

Because the Earth is rotating, its surface cannot be an inertial reference frame. All reference frames on the Earth's surface are centripetally accelerated by the rotation. The free-fall acceleration that we measure in this noninertial frame has at least two components: one from the gravitational attraction of the Earth and another from its rotation. The small effect of this rotation is to change the free-fall acceleration by about 0.3% from its value at the equator, where the centripetal acceleration is largest, to its value at the poles, where the centripetal acceleration vanishes. We neglect this small noninertial contribution to the weight for now, but we shall return to it in Chapter 16. Other noninertial contributions to the force on a body are discussed in Section 6-8.

The weight of a body is zero in regions of space where the gravitational effects are nil, although properties of a body that depend on its mass, such as its resistance to being accelerated, remain unchanged from those on Earth. In a spaceship free from the influence of gravity it is a simple matter to lift slowly a large block of lead ($W = 0$), but an astronaut would still feel a painful stub of the toe when kicking the block ($m \neq 0$).

It takes the same force to accelerate a body in gravity-free space as it does to accelerate it along a horizontal frictionless surface on the Earth, for its mass is the same in each place. But it takes a greater force to hold the body up against the pull of the Earth on the Earth's surface than it does high up in space, for its weight is different in each place.

At locations where g has a specified value, mass and weight are proportional to each other. We sometimes write, for example, 1 kg "=" 2.2 lb, in which "=" means "is equivalent to." This is a numerical correspondence, not a true equation (because equations cannot equate quantities with different dimensions!). It's a little like saying 1 orange "=" x apples, where x might have one value if we were discussing cost and a very different value if we were discussing how much juice they could produce.

The relationship between mass and weight is valid only

for a specific value of g, and so it should be used with caution. Otherwise, you could find yourself in a confusing or embarrassing situation. For instance, when you some-day stop your spacecraft at a well-known fast-food restaurant on the Moon and order a quarter-pound hamburger, you would be served a sandwich with a diameter of nearly 1 foot. (The Moon's gravity is about one-sixth of the Earth's. On the Moon, 1 kg "=" 0.38 lb.) If you place the same order on the surface of the Sun, your hamburger will be barely more than 1 in. in diameter but very well cooked! (On the Sun, 1 kg "=" 62 lb.) Obviously we want to order food by quantity of matter (mass), not by weight. A 100-gram hamburger (about $\frac{1}{4}$ lb on the Earth) would be exactly the same size in all locations.

Weightlessness

Photographs of astronauts in an orbiting space vehicle (such as Fig. 11) show them floating freely in a state that is usually called "weightless." According to our definition of weight, however, they are not at all weightless; in fact, their weight is only about 10% less than it would be if they were standing on the surface of the Earth. This reduction occurs because the Earth's gravity grows weaker with increasing altitude.

The orbiting astronauts are said to be "weightless" for two reasons: (1) To an external observer, the astronauts are in free fall toward the center of the Earth. They maintain their altitude only because their tangential velocity has been chosen so that gravity provides the centripetal acceleration necessary for uniform circular motion. (2) There is no floor in contact with them and pushing up on them.

Our psychological perception of weight involves the force with which the floor pushes up on us. Floating in water, we are less aware of our weight (but we are fully aware of our *mass,* such as when we try to accelerate by swimming through the water). If we stand in an accelerat-ing elevator, we feel as if our weight increases when the elevator accelerates upward and as if our weight decreases when the elevator accelerates downward. This effect, which is considered in Sample Problem 7, is a result of the increase or decrease in the upward force exerted on us by the floor.

True weightlessness can be achieved only deep in space far from any star or planet, where the astronauts would float freely in a spacecraft drifting with its engines shut off. If the craft were to rotate about an axis, astronauts standing perpendicular to the axis on a rotating surface would feel an "artificial gravity," because the floor must push up on them to supply the centripetal force necessary to move them in a circle. The upward push of the floor is perceived as weight.

If we dive from a springboard or bounce on a trampoline, we are in free fall while we are in the air; there is no surface to push against us and we feel "weightless." Similarly, if we are in free fall near the Earth inside a chamber that is also in free fall, there is again no surface to push on us and we would again regard ourselves as "weightless." We would float freely inside the chamber. Three examples of such a situation are (1) the orbiting spacecraft discussed above, (2) an elevator cab falling after its cable breaks, and (3) an airplane flying along a chosen parabolic trajectory. Suppose the airplane climbs such that at a particular time it is moving upward with velocity v_0 in a direction at an angle ϕ_0 above the horizontal. If you were a passenger in that plane, and if at that instant the plane suddenly vanished, you would follow the parabolic trajectory of free fall given by Eq. 23 of Chapter 4. If instead at that same instant the pilot flies the plane so it follows that same trajectory, the plane is in effect in free fall and objects inside will float freely in a state of "weightlessness." In fact, such a system has been used to train astronauts to adapt to the "weightlessness" of spacecraft in orbit. In each of these three cases, your weight is only slightly changed from what it would be if you were standing on the Earth's surface, but the lack of a floor to push on you results in your perception of being "weightless."

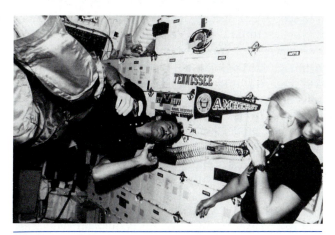

Figure 11 Astronauts in the space shuttle are in a state of free fall, in which they appear to float as if weightless.

5-9 MEASURING FORCES

In Section 5-3 we defined force by measuring the acceleration imparted to a standard body by pulling on it with a stretched spring. That may be called a dynamic method for measuring force. Although convenient for the purposes of definition, it is not always a particularly practical procedure for the measurement of forces. (Acceleration is seldom easy to measure.) Another method for measuring forces is based on measuring the change in shape or size of a body (a spring, say) on which the force is applied when the body is unaccelerated. This may be called the static method of measuring forces.

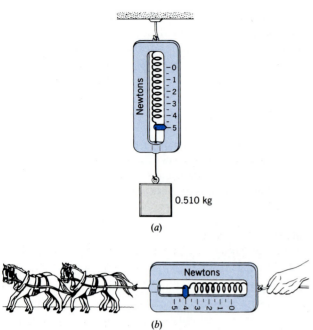

(a)

(b)

Figure 12 (a) A spring scale can be calibrated, in a region where g is known, by hanging a known mass and marking the force corresponding to the weight of the mass. In the case shown, a mass of 0.510 kg gives $F = 5.00$ N when $g = 9.8$ m/s^2. (b) The calibrated scale can then be used to measure an unknown force. This is the basis of operation of all spring scales, such as the postage meter, the produce scale in grocery stores, and the bathroom scale.

The basis of the static method is that when a body, under the action of several forces, has zero acceleration, the vector sum of all the forces acting on the body must be zero. This is, of course, just the second law of motion. A single force acting on a body would produce an acceleration; this acceleration can be made zero if we apply another force to the body equal in magnitude but oppositely directed. In practice we seek to keep the body at rest. If now we choose some force as our unit force, we are in a position to measure forces. The pull of the Earth on a standard body at a particular point can be taken as the unit force, for example.

An instrument commonly used to measure forces in this way is the spring scale (Fig. 12). It consists of a coiled spring having a pointer at one end that moves over a scale. A force exerted on the scale changes the length of the spring. If a body weighing 1.00 N ($m = 0.102$ kg, where $g = 9.80$ m/s^2) is hung from the spring, the spring stretches until the pull of the spring on the body is equal in magnitude but opposite in direction to its weight. A mark can be made on the scale next to the pointer and labeled "1.00-N force." Similarly, 2.00-N, 3.00-N, . . . weights may be hung from the spring and corresponding marks can be made on the scale next to the pointer in each case. In this way the spring is calibrated. We assume that the

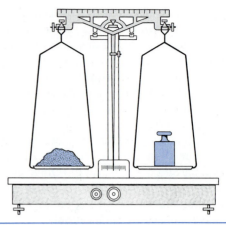

Figure 13 The equal-arm balance, which compares the weights of different masses.

force exerted on the spring is always the same when the pointer stands at the same position. The calibrated scale can now be used as in Fig. 12b to measure an unknown force, not merely the pull of the Earth on some body.

The equal-arm balance (Fig. 13) provides another static method for measuring force. The most common application involves comparing known weights with unknown weights; when the arms balance, the weights must be equal. Furthermore, because g is the same for both arms of the balance, equality of weights implies equality of masses. The equal-arm balance thus determines the relative equality of masses by weighing them. (In fact, the known weights supplied with such balances are usually marked as masses in grams.) This system works for any value of g but zero; the balance would work equally well for comparing masses on the Moon, but it would not work at all in gravity-free space or in the relative weightlessness of Earth orbits.

5-10 APPLICATIONS OF NEWTON'S LAWS

Although each problem to be solved using Newton's laws will require a unique approach, there are a few general rules that are applied in setting up the solutions to all such problems. In this section we present the rules and illustrate their application with several examples. The best way to learn the rules is to study the examples.

The basic steps in applying Newton's laws are these: (1) Clearly identify the body that will be analyzed. Sometimes there will be two or more such bodies; each is usually treated independently. (2) Identify the environment that will be exerting forces on the body—surfaces, other objects, the Earth, springs, cords, and so on. (3) Select a suitable inertial (nonaccelerating) reference frame. (4)

Pick a convenient coordinate system (in the chosen reference frame), locate the origin, and orient the axes to simplify the problem as much as possible. With suitable care, a different coordinate system can be chosen for each component of a complex problem. (5) Make a free-body diagram, showing each object as a particle and all forces acting on it. (6) Now apply Newton's second law to each component of force and acceleration.

In the following examples, we make some assumptions that simplify the problem at the cost of some physical reality. Bodies are treated as particles, so that all forces on the body are considered to act at a single point. We assume all motion to be frictionless. All strings are assumed to be massless (no force is required to accelerate the strings) and inextensible (they do not stretch, so that objects in linear motion connected by taut strings have the same speeds and accelerations). Pulleys are massless (no forces are required to rotate them) with frictionless bearings. All bodies are rigid (no deformations occur under load and forces are transmitted instantaneously through them). Despite these simplifications, the examples provide insight into the basic techniques of dynamical analysis. Later in the text, we add new techniques that will permit us to be more realistic in our analysis of physical situations. For example, in Chapter 6 we show how friction can be included in the analysis, and in Chapter 12 we show how to account for the mass of a pulley and the friction of its bearings. For now, we ignore these admittedly important effects, so that we may concentrate on the more basic methods used to solve problems.

In the following sample problem, we introduce the *tension T*, the force with which a string pulls on objects attached to it. For strings of negligible thickness, the direction of the tension must always be parallel to the string itself. (This statement does not hold for thick, solid beams, as we discuss in Chapter 14.) For strings of negligible mass, the tension is transmitted uniformly along string and is the same at each end.

Microscopically, each element of the string pulls on the element next to it (and is in turn pulled *by* that element, according to Newton's third law). In this way the force pulling on one end of the string is transmitted to the object

on the other end. Any particular element i of the string experiences a tension T acting in one direction due to element $i - 1$, and an equal tension T acting in the opposite direction due to element $i + 1$. If we were to cut the string at any point and tie a spring scale (calibrated as described in Section 5-9) to the cut ends, the spring scale would read the tension T directly.

Sample Problem 4   Figure 14*a* shows a block of mass $m = 15.0$ kg hung by three strings. What are the tensions in the three strings?

Solution   We consider the knot at the junction of the three strings to be "the body." Figure 14*b* shows the free-body diagram of the knot, which remains at rest under the action of the three forces T_A, T_B, and T_C, which are the tensions in the strings. (We assume that, like the string, the knot is massless, so its weight does not appear in the diagram.) Choosing the x and y axes as shown in Fig. 14*b*, we can resolve the forces into their x and y components. The acceleration components are zero, so we can write:

x component: $\sum F_x = T_{Ax} + T_{Bx} = ma_x = 0,$

y component: $\sum F_y = T_{Ay} + T_{By} + T_{Cy} = ma_y = 0.$

From Fig. 14*b* we see that

$$T_{Ax} = -T_A \cos 30° = -0.866 T_A,$$

$$T_{Ay} = T_A \sin 30° = 0.500 T_A,$$

$$T_{Bx} = T_B \cos 45° = 0.707 T_B,$$

$$T_{By} = T_B \sin 45° = 0.707 T_B,$$

and

$$T_{Cx} = 0,$$

$$T_{Cy} = -T_C.$$

To continue, we examine the free-body diagram of the mass m, shown in Fig. 14*c*. Only the y components enter, and again the acceleration is zero:

$$T_{Cy} - mg = ma_y = 0.$$

Because T_C has only a y component, we can write

$$T_C = mg = (15.0 \text{ kg})(9.80 \text{ m/s}^2) = 147 \text{ N}.$$

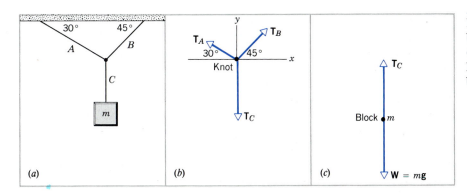

(a)   (b)   (c)

Figure 14   Sample Problem 4. (*a*) A block hangs from three strings *A*, *B*, and *C*. (*b*) The free-body diagram of the knot that joins the strings. (*c*) The free-body diagram of the block.

We can now rewrite the x and y component equations for the forces on the knot:

x component: $-0.866T_A + 0.707T_B = 0,$

y component: $0.500T_A + 0.707T_B - T_C = 0.$

Substituting the value for T_C and solving the two equations simultaneously, we find

$$T_A = 108 \text{ N},$$

$$T_B = 132 \text{ N}.$$

Check these results (as you should check all problems) to see if the vector sum of the three forces is indeed zero.

In the next sample problem, we introduce another kind of force, the *normal force N* exerted by a surface on a body. Consider the book resting on the table illustrated in Fig. 8. The Earth exerts a downward force on the book (its weight), yet the book is in equilibrium so the total force on it must be zero. The other force acting on the book is the upward normal force exerted by the table (indicated as $\mathbf{F}_{BT}$ in Fig. 8). In effect, this force keeps the book on the surface of the table. In the absence of friction, surfaces can exert only normal forces, that is, only forces perpendicular to the surface. (Note that the book also exerts a downward normal force on the table.)

If we were to place our hand on top of the book and push downward with a force P, the book would remain in equilibrium, so the normal force of the table on the book must increase accordingly, in this case being equal to the sum of the weight of the book and the force P. If P were great enough, we would exceed the ability of the table to provide the upward normal force, and the book would break through the tabletop.

Tension forces and normal forces are examples of *contact forces,* in which one body exerts a force on another by virtue of the contact between them. These forces originate with the atoms in the body, with each atom exerting a force on its neighbor. Contact forces can be maintained only if they do not exceed the interatomic forces; otherwise we break the binding between atoms, and the string or the surface splits into pieces.

Sample Problem 5   A sled of mass $m = 7.5$ kg is pulled along a frictionless horizontal surface by a cord (Fig. 15). A constant force of $P = 21.0$ N is applied to the cord. Analyze the motion if (*a*) the cord is horizontal and (*b*) the cord makes an angle of $\theta = 15°$ with the horizontal.

Solution   (*a*) The free-body diagram with the cord horizontal is shown in Fig. 15*b*. The surface exerts a force N, the normal force, on the sled. The forces are analyzed into components and Newton's second law is used as follows:

x component: $\sum F_x = P = ma_x,$

y component: $\sum F_y = N - mg = ma_y.$

If there is to be no vertical motion, the sled remains on the surface and $a_y = 0$. Thus

$$N = mg = (7.5 \text{ kg})(9.80 \text{ m/s}^2) = 74 \text{ N}.$$

The horizontal acceleration is

$$a_x = \frac{P}{m} = \frac{21.0 \text{ N}}{7.5 \text{ kg}} = 2.80 \text{ m/s}^2.$$

Note that, if the surface is truly frictionless, as we have assumed, the person cannot continue to exert this force on the sled for very long. After 30 s at this acceleration, the sled would be moving at 84 m/s or 188 mi/h!

(*b*) When the pulling force is not horizontal, the free-body diagram is shown in Fig. 15*c* and the component equations take the following forms:

x component: $\sum F_x = P \cos \theta = ma_x,$

y component: $\sum F_y = N + P \sin \theta - mg = ma_y.$

Let us assume for the moment that the sled stays on the surface; that is, $a_y = 0$. Then

$$N = mg - P \sin \theta = 74 \text{ N} - (21.0 \text{ N})(\sin 15°) = 69 \text{ N},$$

$$a_x = \frac{P \cos \theta}{m} = \frac{(21.0 \text{ N})(\cos 15°)}{7.5 \text{ kg}} = 2.70 \text{ m/s}^2.$$

A normal force is always perpendicular to the surface in contact; with the coordinates chosen as in Fig. 15*b*, N must be positive. If we increase $P \sin \theta$, N would decrease and at some point would become zero. At that point the sled would leave the surface, under the influence of the upward component of P, and we would need to analyze its vertical motion. With the values of P and θ we have used, the sled remains on the surface and $a_y = 0$.

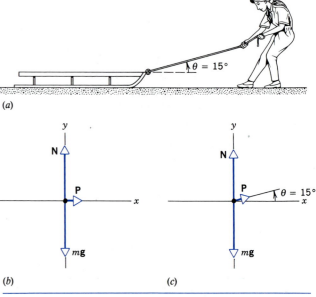

Figure 15   Sample Problem 5. (*a*) A sled is pulled along a frictionless horizontal surface. (*b*) The free-body diagram of the sled when $\theta = 0°$. (*c*) The free-body diagram of the sled when $\theta = 15°$.

Sample Problem 6   A block of mass $m = 18.0$ kg is held in place by a string on a frictionless plane inclined at an angle of 27° (see Fig. 16a). (a) Find the tension in the string and the normal force exerted on the block by the plane. (b) Analyze the subsequent motion after the string is cut.

Solution   (a) The free-body diagram of the block is shown in Fig. 16b. The block is acted on by the normal force N, its weight $W = mg$, and the tension T of the string. We choose a coordinate system with the x axis along the plane and the y axis perpendicular to it. With this choice, two of the forces (T and N) are already resolved into components, and the motion that will eventually occur along the plane has only one component as well.

In the static case there is no acceleration and the forces must sum to zero. The weight is resolved into its x ($-mg \sin \theta$) and y ($-mg \cos \theta$) components, and the force equations are as follows:

x component: $\sum F_x = T - mg \sin \theta = ma_x = 0,$

y component: $\sum F_y = N - mg \cos \theta = ma_y = 0.$

Examine these equations. Are they reasonable? What happens in the limit $\theta = 0°$? It looks like the tension would be zero. Would you expect the tension to be zero if the block were resting on a horizontal surface? What happens to the normal force when $\theta = 0°$? Is this reasonable? What happens to T and N in the limit of $\theta = 90°$? You should form the habit of asking questions like these before starting on the algebra to find the solution. If there is an error, now is the best time to find and correct it.

Solving the equations,

$T = mg \sin \theta = (18.0 \text{ kg})(9.80 \text{ m/s}^2)(\sin 27°) = 80$ N,

$N = mg \cos \theta = (18.0 \text{ kg})(9.80 \text{ m/s}^2)(\cos 27°) = 157$ N.

(b) When the string is cut, the tension disappears from the equations and the block is no longer in equilibrium. Newton's second law now gives the following:

x component: $\sum F_x = -mg \sin \theta = ma_x,$

y component: $\sum F_y = N - mg \cos \theta = ma_y.$

Cutting the string doesn't change the motion in the y direction (the block doesn't jump off the plane!), so $a_y = 0$ as before and the normal force still equals $mg \cos \theta$, or 157 N. In the x direction

$a_x = -g \sin \theta = -(9.80 \text{ m/s}^2)(\sin 27°) = -4.45 \text{ m/s}^2.$

The minus sign shows that the block accelerates in the negative x direction, that is, down the plane. Check the limits $\theta = 0°$ and $\theta = 90°$. Are they consistent with your expectations?

Sample Problem 7   A passenger of mass 72.2 kg is riding in an elevator while standing on a platform scale (Fig. 17a). What does the scale read when the elevator cab is (a) descending with constant velocity and (b) ascending with acceleration 3.20 m/s²?

Solution   Let us first develop a general result valid for any vertical acceleration a. We choose our inertial reference frame to be outside the elevator (the fixed elevator shaft of the building, for instance), because an accelerating elevator is not an inertial reference frame. Both g and a are measured by an observer in this external frame. Figure 17b shows the free-body diagram of the passenger. There is the downward force of the weight and the upward normal force exerted by the scale. The normal force is exerted *by* the scale *on* the passenger; the scale reads the downward force exerted *by* the passenger *on* the scale. By Newton's third law, these are equal in magnitude. Thus if we can find the normal force, we have the scale reading.

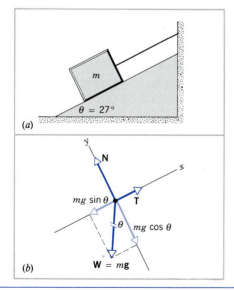

(a)

(b)

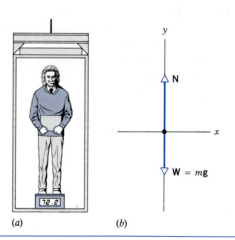

(a)   (b)

Figure 16   Sample Problem 6. (a) A mass m is supported at rest by a string on a frictionless inclined plane. (b) The free-body diagram of m. Note that the xy coordinate system is tilted so that the x axis is parallel to the plane. The weight $m\mathbf{g}$ has been resolved into its vector components.

Figure 17   Sample Problem 7. (a) A passenger riding in an elevator cab stands on a scale. (b) The free-body diagram of the passenger. The normal force N is exerted by the scale and is equal in magnitude to the scale reading. (Commercial scales, like the one shown here, are calibrated to read in kilograms, rather than newtons.)

From the free-body diagram we have

$$\sum F_y = N - mg = ma$$

or

$$N = m(g + a).$$

When $a = 0$, such that the elevator is at rest or moving with constant velocity, as in part (a), then

$$N = mg = (72.2 \text{ kg})(9.80 \text{ m/s}^2) = 708 \text{ N} (= 159 \text{ lb}).$$

When $a = 3.20$ m/s², as in part (b), we have

$$N = m(g + a) = (72.2 \text{ kg})(9.80 \text{ m/s}^2 + 3.20 \text{ m/s}^2)$$
$$= 939 \text{ N} (= 211 \text{ lb}).$$

The scale reading, which indicates the normal force with which the floor is pushing up on the passenger, increases when the elevator is accelerating upward (a is positive as we have defined the coordinate system) and decreases when it is accelerating downward. In free fall ($a = -g$) the scale reading is zero (there is no normal force).

5-11   MORE APPLICATIONS OF NEWTON'S LAWS

Here we consider some additional applications of Newton's laws. These examples involve several objects that must be analyzed separately but not quite independently, because the motion of one object is constrained by the motion of another, such as when they are attached to one another by a string of fixed length. Study these examples, and note the independent choices of coordinate systems used for the separate objects.

Sample Problem 8   Figure 18a shows a block of mass m_1 on a frictionless horizontal surface. The block is pulled by a string of negligible mass which is attached to a hanging block of mass m_2. The string passes over a pulley whose mass is negligible and whose axle rotates with negligible friction. Find the tension in the string and the acceleration of each block.

Solution   This problem differs from those we considered earlier in that two objects, rather than one, are involved. Figures 18b

and 18c show the free-body diagrams for the separate objects. It is not necessary to choose the same coordinate system for both objects; as long as we are consistent within each subsystem, it doesn't matter how the individual axes are defined.

Block 1 is acted on by a normal force N, by gravity, and by the tension in the string. Since we expect block 1 to accelerate to the right, we choose that for our positive x direction. We also expect block 1 to remain on the horizontal surface, so the y component of its acceleration is zero. The component equations of Newton's second law are then as follows:

x component: $\sum F_x = T_1 = m_1 a_{1x},$

y component: $\sum F_y = N - m_1 g = m_1 a_{1y} = 0.$

For block 2, we choose the y axis to be vertically downward, which is the direction we expect for its acceleration. No x components need be considered for block 2, and the y component of Newton's second law gives

$$\sum F_y = m_2 g - T_2 = m_2 a_{2y}.$$

We consider the string to be massless, so that the net force on it must be zero. The tensions T_1 and T_2 exerted by the string *on* the blocks result in equal reaction forces T_1 and T_2 being exerted *by* the blocks on the string. If the string were straight, the vanishing of the net force on the string would require $T_1 = T_2$. The presence of the ideal (massless and frictionless) pulley to change the direction of the tension in the string doesn't change this argument: the tension has a common magnitude all along the length of the string. We represent the common tension by the single variable T.

If the string is also inextensible (that is, it doesn't stretch), then any motion by block 1 in its x direction is exactly matched by a corresponding motion of block 2 in its y direction. In this case the accelerations of the two blocks are equal. We call this common acceleration a. We now have three equations:

$$T = m_1 a,$$
$$N = m_1 g,$$

and

$$m_2 g - T = m_2 a.$$

Solving the first and third simultaneously gives

$$a = \frac{m_2}{m_1 + m_2} g \tag{5}$$

and

$$T = \frac{m_1 m_2}{m_1 + m_2} g. \tag{6}$$

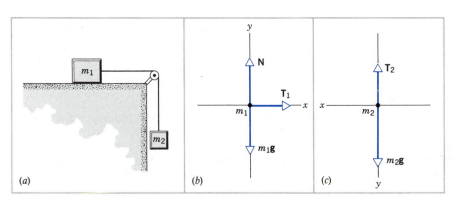

(a)   (b)   (c)

Figure 18   Sample Problem 8. (a) Block m_1 is pulled along a smooth horizontal surface by a string that passes over a pulley and is attached to block m_2. (b) The free-body diagram of block m_1. (c) The free-body diagram of block m_2.

It is helpful to consider the limiting cases of these results. What happens when m_1 is zero? We would expect the string to be slack ($T = 0$) and m_2 to be in free fall ($a = g$). The equations correctly predict these limits. When $m_2 = 0$, there is no horizontal force on block 1 and it does not accelerate; again, the equations give the correct prediction.

Note that $a < g$, as we should expect. Also, note that T is *not* equal to $m_2 g$. Only if block 2 were hanging in equilibrium ($a = 0$) would $T = m_2 g$. If block 2 accelerates downward, then $T < m_2 g$; if it accelerates upward, then $T > m_2 g$.

Do Eqs. 5 and 6 behave properly in the limit $g = 0$?

Sample Problem 9   Consider two unequal masses connected by a string that passes over an ideal pulley (whose mass is negligible and whose axle rotates with negligible friction), as shown in Fig. 19. (This arrangement is also known as an *Atwood's machine.**) Let m_2 be greater than m_1. Find the tension in the string and the acceleration of the masses.

---

* George Atwood (1745–1807) was an English mathematician who developed this device in 1784 for demonstrating the laws of accelerated motion and measuring g. By making the difference between m_1 and m_2 small, he was able to "slow down" the effect of free fall and time the motion of the falling weight with a pendulum clock, the most precise way to measure time intervals in his day.

Solution   Because we anticipate the masses to have only vertical accelerations, we choose the positive y direction to be the direction of motion for each mass. Only y components need be considered. The free-body diagrams are shown in Fig. 19b, and the equations of motion are as follows:

$$\text{block 1:} \quad \sum F_y = T_1 - m_1 g = m_1 a_1,$$
$$\text{block 2:} \quad \sum F_y = m_2 g - T_2 = m_2 a_2,$$

where a_1 and a_2 are the accelerations of m_1 and m_2, respectively. As in the previous example, if the string is massless and doesn't stretch, and if the pulley is massless and frictionless, then $T_1 = T_2 = T$ and $a_1 = a_2 = a$. (We assume that this ideal pulley doesn't change the magnitude of the tension or the acceleration from one side of the string to the other; its only function is to change their directions.) Substituting and solving the two equations simultaneously, we find

$$a = \frac{m_2 - m_1}{m_2 + m_1} g, \tag{7}$$

$$T = \frac{2 m_1 m_2}{m_1 + m_2} g. \tag{8}$$

Consider what happens in the limiting cases $m_1 = 0$, $m_2 = 0$, $g = 0$, and $m_1 = m_2$. Notice that $m_1 g < T < m_2 g$, and be sure you understand why this must be true.

Sample Problem 10   Consider the mechanical system shown in Fig. 20a, where $m_1 = 9.5$ kg, $m_2 = 2.6$ kg, and $\theta = 34°$. The system is released from rest. Describe the motion.

Solution   The free-body diagrams for blocks 1 and 2 are shown in Figs. 20b and 20c. We choose coordinate systems as shown, so that one coordinate axis is parallel to the anticipated acceleration of each body. As in the previous examples, we expect that the tension has a common value and that the vertical motion of m_2 and the motion along the plane of m_1 can be described by accelerations of the same magnitude. We assume that m_1 moves in the positive x direction (if our assumption is wrong, a will come out negative). For m_1, the component equations of Newton's second law are as follows:

$$x \text{ component:} \quad \sum F_x = T - m_1 g \sin \theta = m_1 a,$$
$$y \text{ component:} \quad \sum F_y = N - m_1 g \cos \theta = 0,$$

and for m_2,

$$\sum F_y = m_2 g - T = m_2 a.$$

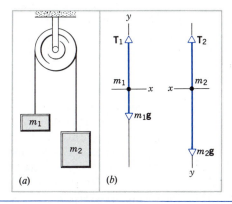

(a)   (b)

Figure 19   Sample Problem 9. (a) Diagram of Atwood's machine, consisting of two suspended masses connected by a string that passes over a pulley. (b) Free-body diagrams of m_1 and m_2.

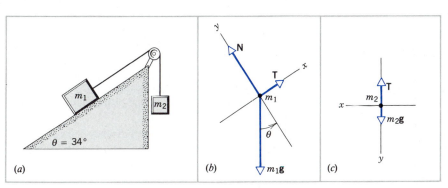

Figure 20   Sample Problem 10. (a) Block m_1 slides on a frictionless inclined plane. Block m_2 hangs from a string attached to m_1. (b) Free-body diagram of m_1. (c) Free-body diagram of m_2.

Solving simultaneously gives

$$a = \frac{m_2 - m_1 \sin \theta}{m_1 + m_2} g \qquad (9)$$

and

$$T = \frac{m_1 m_2}{m_1 + m_2} g(1 + \sin \theta). \qquad (10)$$

Note that these results duplicate Eqs. 5 and 6 of Sample Problem 8 if we put $\theta = 0$ (so that block 1 moves horizontally) and Eqs. 7 and 8 of Sample Problem 9 if we put $\theta = 90°$ (so that block 1 moves vertically).

Putting in the numbers, we have

$$a = \frac{2.6 \text{ kg} - (9.5 \text{ kg})(\sin 34°)}{9.5 \text{ kg} + 2.6 \text{ kg}} (9.80 \text{ m/s}^2) = -2.2 \text{ m/s}^2.$$

The acceleration comes out to be negative, which means our initial guess about the direction of motion was wrong. Block 1 slides down the plane, and block 2 moves upward. Because the dynamical equations do not involve forces that depend on the direction of motion, this incorrect initial guess has no effect on the equations and we can accept the final value as correct. In general, this will not be the case when we consider frictional forces that act opposite to the direction of motion.

For the tension in the string, we find

$$T = \frac{(9.5 \text{ kg})(2.6 \text{ kg})}{9.5 \text{ kg} + 2.6 \text{ kg}} (9.80 \text{ m/s}^2)(1 + \sin 34°) = 31 \text{ N}.$$

This value is greater than the weight of m_2 ($m_2 g = 26$ N), which is consistent with the acceleration of m_2 being upward.

QUESTIONS

1. Why do you fall forward when a moving bus decelerates to a stop and fall backward when it accelerates from rest? Subway standees often find it convenient to face the side of the car when the train is starting or stopping and to face the front or rear when it is running at constant speed. Why?

2. A block with mass m is supported by a cord C from the ceiling, and a similar cord D is attached to the bottom of the block (Fig. 21). Explain this: If you give a sudden jerk to D, it will break, but if you pull on D steadily, C will break.

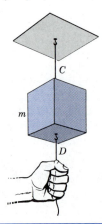

Figure 21   Question 2.

3. Criticize the statement, often made, that the mass of a body is a measure of the "quantity of matter" in it.

4. Using force, length, and time as fundamental quantities, what are the dimensions of mass?

5. Can Newton's first law be considered merely the special case $a = 0$ of the second law? If so, is the first law really needed? Discuss.

6. What's the relation — if any — between the force acting on an object and the direction in which the object is moving?

7. Suppose that a body that is acted on by exactly two forces is accelerated. Does it then follow that (a) the body cannot move with constant speed; (b) the velocity can never be zero; (c) the sum of the two forces cannot be zero; (d) the two forces must act in the same line?

8. In Fig. 22, we show four forces that are equal in magnitude. What combination of three of these, acting together on the same particle, might keep that particle in equilibrium?

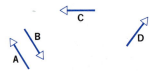

Figure 22   Question 8.

9. A horse is urged to pull a wagon. The horse refuses to try, citing Newton's third law as a defense: the pull of the horse on the wagon is equal but opposite to the pull of the wagon on the horse. "If I can never exert a greater force on the wagon than it exerts on me, how can I ever start the wagon moving?" asks the horse. How would you reply?

10. Comment on whether the following pairs of forces are examples of action–reaction: (a) The Earth attracts a brick; the brick attracts the Earth. (b) A propellered airplane pushes air toward the tail; the air pushes the plane forward. (c) A horse pulls forward on a cart, moving it; the cart pulls backward on the horse. (d) A horse pulls forward on a cart without moving it; the cart pulls back on the horse. (e) A horse pulls forward on a cart without moving it; the Earth exerts an equal and opposite force on the cart. (f) The Earth pulls down on the cart; the ground pushes up on the cart with an equal and opposite force.

11. The following statement is true; explain it. Two teams are having a tug of war; the team that pushes harder (horizontally) against the ground wins.

12. Two students try to break a rope. First they pull against each other and fail. Then they tie one end to a wall and pull together. Is this procedure better than the first? Explain your answer.

13. What is your mass in slugs? Your weight in newtons?

14. A Frenchman, filling out a form, writes "78 kg" in the space marked Poids (weight). However, weight is a force and the kilogram is a mass unit. What do Frenchmen (among others) have in mind when they use a mass unit to report their weight? Why don't they report their weight in newtons? How many newtons does this Frenchman weigh? How many pounds?

15. Comment on the following statements about mass and weight taken from examination papers. (*a*) Mass and weight are the same physical quantities expressed in different units. (*b*) Mass is a property of one object alone, whereas weight results from the interaction of two objects. (*c*) The weight of an object is proportional to its mass. (*d*) The mass of a body varies with changes in its local weight.

16. A horizontal force acts on a body that is free to move. Can it produce an acceleration if the force is less than the weight of that body?

17. Why does the acceleration of a freely falling object not depend on the weight of the object?

18. Describe several ways in which you could, even briefly, experience weightlessness.

19. Under what circumstances would your weight be zero? Does your answer depend on the choice of a reference system?

20. The "mechanical arm" on the space shuttle can handle a 2200-kg satellite when extended to 12 m; see Fig. 23. Yet, on the ground, this remote manipulator system (RMS) cannot support its own weight. In the "weightlessness" of an orbiting shuttle, why does the RMS have to be able to exert any force at all?

Figure 23 Questions 20 and 26.

21. The owner's manual of a car suggests that your seat belt should be adjusted "to fit snugly" and that the front seat head rest should *not* be adjusted so that it fits comfortably at the back of your neck but so that "the top of the head rest is level with the top of your ears." How do Newton's laws support these good recommendations?

22. You shoot an arrow into the air and you keep your eye on it as it follows a parabolic flight path to the ground. You note that the arrow turns in flight so that it is always tangent to its flight path. What makes it do that?

23. In a tug of war, three men pull on a rope to the left at *A* and three men pull to the right at *B* with forces of equal magnitude. Now a 5-lb weight is hung vertically from the center of the rope. (*a*) Can the men get the rope *AB* to be horizontal? (*b*) If not, explain. If so, determine the magnitude of the forces required at *A* and *B* to do this.

24. A bird alights on a stretched telegraph wire. Does this change the tension in the wire? If so, by an amount less than, equal to, or greater than the weight of the bird?

25. A massless rope is strung over a frictionless pulley. A monkey holds onto one end of the rope and a mirror, having the same weight as the monkey, is attached to the other end of the rope at the monkey's level. Can the monkey get away from its image seen in the mirror (*a*) by climbing up the rope, (*b*) by climbing down the rope, or (*c*) by releasing the rope?

26. In November 1984, astronauts Joe Allen and Dale Gardner salvaged a Westar-6 communications satellite from a faulty orbit and placed it into the cargo bay of the space shuttle *Discovery;* see Fig. 23. Describing the experience, Joe Allen said of the satellite, "It's not heavy; it's massive." What did he mean?

27. You are an astronaut in the lounge of an orbiting space station and you remove the cover from a long thin jar containing a single olive. Describe several ways — all taking advantage of the inertia of either the olive or the jar — to remove the olive from the jar.

28. In Fig. 24, a needle has been placed in each end of a broomstick, the tips of the needles resting on the edges of filled wine glasses. The experimenter strikes the broomstick a swift and sturdy blow with a stout rod. The broomstick breaks and falls to the floor but the wine glasses remain in place and no wine is spilled. This impressive parlor stunt was popular at the end of the last century. What is the physics behind it? (If you try it, practice first with empty soft drink cans. Come to think of it, you might ask your physics instructor to do it, as a lecture demonstration!)

Figure 24 Question 28.

29. An elevator is supported by a single cable. There is no counterweight. The elevator receives passengers at the ground floor and takes them to the top floor, where they disembark. New passengers enter and are taken down to the ground floor. During this round trip, when is the tension in the cable equal to the weight of the elevator plus passengers? Greater? Less?

30. You are on the flight deck of the orbiting space shuttle *Discovery* and someone hands you two wooden balls, outwardly identical. One, however, has a lead core but the other does not. Describe several ways of telling them apart.

31. You stand on the large platform of a spring scale and note your weight. You then take a step on this platform and note that the scale reads less than your weight at the beginning of the step and more than your weight at the end of the step. Explain.

32. Could you weigh yourself on a scale whose maximum reading is less than your weight? If so, how?

33. A weight is hung by a cord from the ceiling of an elevator. From the following conditions, choose the one in which the tension in the cord will be greatest . . . least: (*a*) elevator at rest; (*b*) elevator rising with uniform speed; (*c*) elevator descending with decreasing speed; (*d*) elevator descending with increasing speed.

34. A woman stands on a spring scale in an elevator. In which of the following cases will the scale record the minimum reading . . . the maximum reading: (*a*) elevator stationary; (*b*) elevator cable breaks, free fall; (*c*) elevator accelerating upward; (*d*) elevator accelerating downward; (*e*) elevator moving at constant velocity?

35. What conclusion might a physicist draw if unequal masses hung over a pulley inside an elevator remain balanced; that is, there is no tendency for the pulley to turn?

36. Figure 25 shows comet Kohoutek as it appeared in 1973. Like all comets, it moves around the Sun under the influence of the gravitational pull that the Sun exerts on it. The nucleus of the comet is a relatively massive core at a position

Figure 25   Questions 36 and 37.

indicated by *P*. The tail of a comet is produced by the action of the solar wind, which consists of charged particles streaming outward from the Sun. By inspection, what, if anything, can you say about the direction of the force that acts on the nucleus of the comet? What about the direction in which the nucleus is being accelerated? What about the direction in which the comet is moving?

37. In general (see Fig. 25), comets have a dust tail, consisting of dust particles pushed away from the Sun by the pressure of sunlight. Why is this tail often curved?

38. Can you think of physical phenomena involving the Earth in which the Earth cannot be treated as a particle?

PROBLEMS

Section 5-5 Newton's Second Law

1. Suppose that the Sun's gravitational force was suddenly turned off, so that Earth became a free object rather than being confined to orbit the Sun. How long would it take for Earth to reach a distance from the Sun equal to Pluto's present orbital radius? (*Hint:* You will find some of the data you need in Appendix C.)

2. A 5.5-kg block is initially at rest on a frictionless horizontal surface. It is pulled with a constant horizontal force of 3.8 N. (*a*) What is its acceleration? (*b*) How long must it be pulled before its speed is 5.2 m/s? (*c*) How far does it move in this time?

3. An electron travels in a straight line from the cathode of a vacuum tube to its anode, which is 1.5 cm away. It starts with zero speed and reaches the anode with a speed of 5.8×10^6 m/s. (*a*) Assume constant acceleration and compute the force on the electron. The electron's mass is 9.11×10^{-31} kg. This force is electrical in origin. (*b*) Calculate the gravitational force on the electron.

4. A neutron travels at a speed of 1.4×10^7 m/s. Nuclear forces are of very short range, being essentially zero outside a nucleus but very strong inside. If the neutron is captured and brought to rest by a nucleus whose diameter is 1.0×10^{-14} m, what is the minimum magnitude of the force, pre-

sumed to be constant, that acts on this neutron? The neutron's mass is 1.67×10^{-27} kg.

5. In a modified "tug-of-war" game, two people pull in opposite directions, not on a rope, but on a 25-kg sled resting on an icy road. If the participants exert forces of 90 N and 92 N, what is the acceleration of the sled?

6. A light beam from a satellite-carried laser strikes an object ejected from an accidentally launched ballistic missile; see Fig. 26. The beam exerts a force of 2.7×10^{-5} N on the target. If the "dwell time" of the beam on the target is 2.4 s, by how much is the object displaced if it is (*a*) a 280-kg

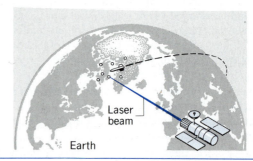

Figure 26   Problem 6.

warhead and (b) a 2.1-kg decoy? (These displacements can be measured by observing the reflected beam.)

7. A car traveling at 53 km/h hits a bridge abutment. A passenger in the car moves forward a distance of 65 cm (with respect to the road) while being brought to rest by an inflated air bag. What force (assumed constant) acts on the passenger's upper torso, which has a mass of 39 kg?

8. An electron is projected horizontally at a speed of 1.2×10^7 m/s into an electric field that exerts a constant vertical force of 4.5×10^{-16} N on it. The mass of the electron is 9.11×10^{-31} kg. Determine the vertical distance the electron is deflected during the time it has moved forward 33 mm horizontally.

9. The Sun yacht *Diana,* designed to navigate in the solar system using the pressure of sunlight, has a sail area of 3.1 km^2 and a mass of 930 kg. Near Earth's orbit, the Sun could exert a radiation force of 29 N on its sail. (a) What acceleration would such a force impart to the craft? (b) A small acceleration can produce large effects if it acts steadily for a long enough time. Starting from rest then, how far would the craft have moved after 1 day under these conditions? (c) What would then be its speed? (See "The Wind from the Sun," a fascinating science fiction account by Arthur C. Clarke of a Sun yacht race.)

10. A body with mass m is acted on by two forces $\mathbf{F}_1$ and $\mathbf{F}_2$, as shown in Fig. 27. If $m = 5.2$ kg, $F_1 = 3.7$ N, and $F_2 = 4.3$ N, find the vector acceleration of the body.

Figure 27 Problem 10.

11. An 8.5-kg object passes through the origin with a velocity of 42 m/s parallel to the x axis. It experiences a constant 19 N force in the direction of the positive y axis. Calculate (a) the velocity and (b) the position of the particle after 15 s have elapsed.

12. A certain force gives object m_1 an acceleration of 12.0 m/s^2. The same force gives object m_2 an acceleration of 3.30 m/s^2. What acceleration would the force give to an object whose mass is (a) the difference between m_1 and m_2 and (b) the sum of m_1 and m_2?

13. (a) Neglecting gravitational forces, what force would be required to accelerate a 1200-metric-ton spaceship from rest to one-tenth the speed of light in 3 days? In 2 months? (One metric ton = 1000 kg.) (b) Assuming that the engines are shut down when this speed is reached, what would be the time required to complete a 5-light-month journey for each of these two cases? (Use 1 month = 30 days.)

Section 5-6 Newton's Third Law

14. Two blocks, with masses $m_1 = 4.6$ kg and $m_2 = 3.8$ kg, are connected by a light spring on a horizontal frictionless table. At a certain instant, when m_2 has an acceleration $a_2 = 2.6$ m/s^2, (a) what is the force on m_2 and (b) what is the acceleration of m_1?

15. A 40-kg girl and an 8.4-kg sled are on the surface of a frozen lake, 15 m apart. By means of a rope the girl exerts a 5.2-N force on the sled, pulling it toward her. (a) What is the acceleration of the sled? (b) What is the acceleration of the girl? (c) How far from the girl's initial position do they meet, presuming the force to remain constant? Assume that no frictional forces act.

Section 5-8 Weight and Mass

16. What are the weight in newtons and the mass in kilograms of (a) a 5.00-lb bag of sugar, (b) a 240-lb fullback, and (c) a 1.80-ton car? (1 ton = 2000 lb.)

17. What are the mass and weight of (a) a 1420-lb snowmobile and (b) a 412-kg heat pump?

18. A space traveler whose mass is 75.0 kg leaves Earth. Compute his weight (a) on Earth, (b) on Mars, where $g = 3.72$ m/s^2, and (c) in interplanetary space. (d) What is his mass at each of these locations?

19. A certain particle has a weight of 26.0 N at a point where the acceleration due to gravity is 9.80 m/s^2. (a) What are the weight and mass of the particle at a point where the acceleration due to gravity is 4.60 m/s^2? (b) What are the weight and mass of the particle if it is moved to a point in space where the gravitational force is zero?

20. A 12,000-kg airplane is in level flight at a speed of 870 km/h. What is the upward-directed lift force exerted by the air on the airplane?

21. What is the net force acting on a 3900-lb automobile accelerating at 13 ft/s^2?

22. A 523-kg experimental rocket sled can be accelerated from rest to 1620 km/h in 1.82 s. What net force is required?

23. A jet plane starts from rest on the runway and accelerates for takeoff at 2.30 m/s^2 (=7.55 ft/s^2). It has two jet engines, each of which exerts a thrust of 1.40×10^5 N (=15.7 tons). What is the weight of the plane?

Section 5-10 Applications of Newton's Laws

24. (a) Two 10-lb weights are attached to a spring scale as shown in Fig. 28a. What is the reading of the scale? (b) A single 10-lb weight is attached to a spring scale which itself is attached to a wall, as shown in Fig. 28b. What is the reading of the scale? (Ignore the weight of the scale.)

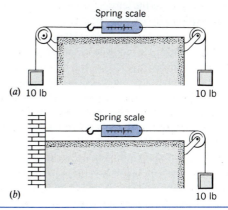

Figure 28 Problem 24.

25. A charged sphere of mass 2.8×10^{-4} kg is suspended from a string. An electric force acts horizontally on the sphere so that the string makes an angle of 33° with the vertical when at rest. Find (*a*) the magnitude of the electric force and (*b*) the tension in the string.

26. A car moving initially at a speed of 50 mi/h (≈ 80 km/h) and weighing 3000 lb ($\approx 13{,}000$ N) is brought to a stop in a distance of 200 ft (≈ 61 m). Find (*a*) the braking force and (*b*) the time required to stop. Assuming the same braking force, find (*c*) the distance and (*d*) the time required to stop if the car were going 25 mi/h (≈ 40 km/h) initially.

27. A meteor of mass 0.25 kg is falling vertically through Earth's atmosphere with an acceleration of 9.2 m/s². In addition to gravity, a vertical retarding force (due to the frictional drag of the atmosphere) acts on the meteor. What is the magnitude of this retarding force? See Fig. 29.

Figure 29 Problem 27.

28. An elevator weighing 6200 lb is pulled upward by a cable with an acceleration of 3.8 ft/s². (*a*) What is the tension in the cable? (*b*) What is the tension when the elevator is accelerating downward at 3.8 ft/s² but is still moving upward?

29. A man of mass 83 kg (weight $mg = 180$ lb) jumps down to a concrete patio from a window ledge only 0.48 m ($= 1.6$ ft) above the ground. He neglects to bend his knees on landing, so that his motion is arrested in a distance of about 2.2 cm ($= 0.87$ in.). (*a*) What is the average acceleration of the man from the time his feet first touch the patio to the time he is brought fully to rest? (*b*) With what average force does this jump jar his bone structure?

30. A block is projected up a frictionless inclined plane with a speed v_0. The angle of incline is θ. (*a*) How far up the plane does it go? (*b*) How long does it take to get there? (*c*) What is its speed when it gets back to the bottom? Find numerical answers for $\theta = 35°$ and $v_0 = 8.2$ ft/s.

31. A lamp hangs vertically from a cord in a descending eleva-

tor. The elevator has a deceleration of 2.4 m/s² ($= 7.9$ ft/s²) before coming to a stop. (*a*) If the tension in the cord is 89 N ($= 20$ lb), what is the mass of the lamp? (*b*) What is the tension in the cord when the elevator ascends with an upward acceleration of 2.4 m/s² ($= 7.9$ ft/s²)?

32. What strength fishing line is needed to stop a 19-lb salmon swimming at 9.2 ft/s in a distance of 4.5 in.?

33. A 5.1-kg block is pulled along a frictionless floor by a cord that exerts a force $P = 12$ N at an angle $\theta = 25°$ above the horizontal, as shown in Fig. 30. (*a*) What is the acceleration of the block? (*b*) The force P is slowly increased. What is the value of P just before the block is lifted off the floor? (*c*) What is the acceleration of the block just before it is lifted off the floor?

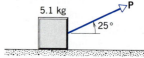

Figure 30 Problem 33.

34. How could a 100-lb object be lowered from a roof using a cord with a breaking strength of 87 lb without breaking the cord?

35. A block is released from rest at the top of a frictionless inclined plane 16 m long. It reaches the bottom 4.2 s later. A second block is projected up the plane from the bottom at the instant the first block is released in such a way that it returns to the bottom simultaneously with the first block. (*a*) Find the acceleration of each block on the incline. (*b*) What is the initial velocity of the second block? (*c*) How far up the incline does it travel? (*d*) What angle does the plane make with the horizontal?

36. A worker drags a crate across a factory floor by pulling on a rope tied to the crate. The worker exerts a force of 450 N on the rope, which is inclined at 38.0° above the horizontal. The floor exerts a horizontal resistive force of 125 N, as shown in Fig. 31. Calculate the acceleration of the crate (*a*) if its mass is 96.0 kg and (*b*) if its weight is 96.0 N.

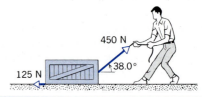

Figure 31 Problem 36.

37. An elevator and its load have a combined mass of 1600 kg. Find the tension in the supporting cable when the elevator, originally moving downward at 12.0 m/s, is brought to rest with constant acceleration in a distance of 42.0 m.

38. An object is hung from a spring balance attached to the ceiling of an elevator. The balance reads 65 N when the elevator is standing still. (*a*) What is the reading when the elevator is moving upward with a constant speed of 7.6 m/s? (*b*) What is the reading of the balance when the elevator is moving upward with a speed of 7.6 m/s and decelerating at 2.4 m/s²?

39. A plumb bob, consisting of a small weight suspended by a

string of negligible mass, hangs from the ceiling of a railroad car and acts as an accelerometer. (*a*) Show that the expression relating the horizontal acceleration *a* of the car to the angle *θ* made by the string with the vertical is given by $a = g \tan \theta$. (*b*) Find *a* when $\theta = 20°$. (*c*) Find *θ* when $a = 5.0$ ft/s².

40. A 1400-kg jet engine is fastened to the fuselage of a passenger jet by just three bolts (this is the usual practice). Assume that each bolt supports one-third of the load. (*a*) Calculate the force on each bolt as the plane waits in line for clearance to take off. (*b*) During flight, the plane encounters turbulence which suddenly imparts an upward vertical acceleration of 2.60 m/s² to the plane. Calculate the force on each bolt now. Why are only three bolts used? See Fig. 32.

Figure 32 Problem 40.

41. Workers are loading equipment into a freight elevator at the top floor of a building. However, they overload the elevator and the worn cable snaps. The mass of the loaded elevator at the time of the accident is 1600 kg. As the elevator falls, the guide rails exert a constant retarding force of 3700 N on the elevator. At what speed does the elevator hit the bottom of the shaft 72 m below?

42. A 1200-kg car is being towed up an 18° incline by means of a rope attached to the rear of a truck. The rope makes an angle of 27° with the incline. What is the greatest distance that the car can be towed in the first 7.5 s starting from rest if the rope has a breaking strength of 4.6 kN? Ignore all resistive forces on the car. See Fig. 33.

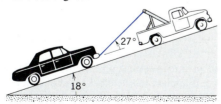

Figure 33 Problem 42.

43. A 110-kg crate is pushed at constant speed up the frictionless 34° ramp shown in Fig. 34. (*a*) What horizontal force *F* is required? (*b*) What is the force exerted by the ramp on the crate?

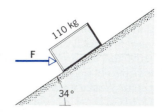

Figure 34 Problem 43.

44. A new 26-ton Navy jet (Fig. 35) requires an air speed of 280 ft/s for lift-off. Its own engine develops a thrust of 24,000 lb. The jet is to take off from an aircraft carrier with a 300-ft flight deck. What force must be exerted by the catapult of the carrier? Assume that the catapult and the jet's engine each exert a constant force over the 300-ft takeoff distance.

Figure 35 Problem 44.

45. A landing craft approaches the surface of Callisto, one of the satellites (moons) of the planet Jupiter (Fig. 36). If an up-

Figure 36 Problem 45.

ward thrust of 3260 N is supplied by the rocket engine, the craft descends with constant speed. Callisto has no atmosphere. If the upward thrust is 2200 N, the craft accelerates downward at 0.390 m/s². (*a*) What is the weight of the landing craft in the vicinity of Callisto's surface? (*b*) What is the mass of the craft? (*c*) What is the acceleration due to gravity near the surface of Callisto?

46. In earlier days, horses pulled barges down canals in the manner shown in Fig. 37. Suppose that the horse is exerting a force of 7900 N at an angle of 18° to the direction of motion of the barge, which is headed straight along the canal. The mass of the barge is 9500 kg and its acceleration is 0.12 m/s². Calculate the force exerted by the water on the barge.

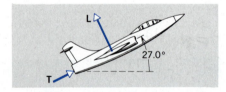

Figure 37   Problem 46.

47. A rocket and its payload have a total mass of 51,000 kg. How large is the thrust of the rocket engine when (*a*) the rocket is "hovering" over the launch pad, just after ignition, and (*b*) when the rocket is accelerating upward at 18 m/s²?

48. A jet fighter takes off at an angle of 27.0° with the horizontal, accelerating at 2.62 m/s². The weight of the plane is 79,300 N. Find (*a*) the thrust *T* of the engine on the plane and (*b*) the lift force *L* exerted by the air perpendicular to the wings; see Fig. 38. Ignore air resistance.

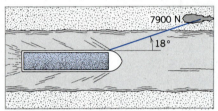

Figure 38   Problem 48.

49. A research balloon of total mass *M* is descending vertically with downward acceleration *a* (see Fig. 39). How much bal-

Figure 39   Problem 49.

last must be thrown from the car to give the balloon an *upward* acceleration *a*, presuming that the upward lift of the air on the balloon does not change?

50. A rocket with mass 3030 kg is fired from rest from the ground at an elevation angle of 58.0°. The motor exerts a thrust of 61.2 kN at a constant angle of 58.0° with the horizontal for 48.0 s and then cuts out. Ignore the mass of fuel consumed and neglect aerodynamic drag. Calculate (*a*) the altitude of the rocket at motor cut out and (*b*) the total distance from firing point to impact.

51. A block, mass *m*, slides down a frictionless incline making an angle θ with an elevator floor. Find its acceleration relative to the incline in the following cases. (*a*) Elevator descends at constant speed *v*. (*b*) Elevator ascends at constant speed *v*. (*c*) Elevator descends with acceleration *a*. (*d*) Elevator descends with deceleration *a*. (*e*) Elevator cable breaks. (*f*) In part (*c*) above, what is the force exerted on the block by the incline?

Section 5-11   More Applications of Newton's Laws

52. Refer to Fig. 18. Let $m_1 = 4.30$ kg and $m_2 = 1.80$ kg. Find (*a*) the acceleration of the two blocks and (*b*) the tension in the string.

53. A 110-kg man lowers himself to the ground from a height of 12 m by holding on to a rope passed over a frictionless pulley and attached to a 74-kg sandbag. (*a*) With what speed does the man hit the ground? (*b*) Is there anything he could do to reduce the speed with which he hits the ground?

54. An 11-kg monkey is climbing a massless rope attached to a 15-kg log over a (frictionless!) tree limb. (*a*) With what minimum acceleration must the monkey climb up the rope so that it can raise the 15-kg log off the ground? If, after the log has been raised off the ground, the monkey stops climbing and hangs on to the rope, what will now be (*b*) the monkey's acceleration and (*c*) the tension in the rope?

55. Three blocks are connected, as shown in Fig. 40, on a horizontal frictionless table and pulled to the right with a force $T_3 = 6.5$ N. If $m_1 = 1.2$ kg, $m_2 = 2.4$ kg, and $m_3 = 3.1$ kg, calculate (*a*) the acceleration of the system and (*b*) the tensions T_1 and T_2. Draw an analogy to bodies being pulled in tandem, such as an engine pulling a train of coupled cars.

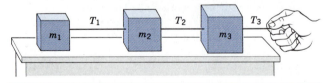

Figure 40   Problem 55.

56. Two blocks are in contact on a frictionless table. A horizontal force is applied to one block, as shown in Fig. 41. (*a*) If $m_1 = 2.3$ kg, $m_2 = 1.2$ kg, and $F = 3.2$ N, find the force of

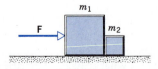

Figure 41   Problem 56.

contact between the two blocks. (*b*) Show that if the same force *F* is applied to m_2 rather than to m_1, the force of contact between the blocks is 2.1 N, which is not the same value derived in (*a*). Explain.

57. Figure 42 shows three crates with masses $m_1 = 45.2$ kg, $m_2 = 22.8$ kg, and $m_3 = 34.3$ kg on a horizontal frictionless surface. (*a*) What horizontal force *F* is needed to push the crates to the right, as one unit, with an acceleration of 1.32 m/s²? (*b*) Find the force exerted by m_2 on m_3. (*c*) By m_1 on m_2.

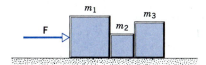

Figure 42   Problem 57.

58. A chain consisting of five links, each with mass 100 g, is lifted vertically with a constant acceleration of 2.50 m/s², as shown in Fig. 43. Find (*a*) the forces acting between adjacent links, (*b*) the force *F* exerted on the top link by the agent lifting the chain, and (*c*) the *net* force on each link.

Figure 43   Problem 58.

59. A block of mass $m_1 = 3.70$ kg on a frictionless inclined plane of angle $\theta = 28.0°$ is connected by a cord over a small frictionless, massless pulley to a second block of mass $m_2 = 1.86$ kg hanging vertically (see Fig. 44). (*a*) What is the acceleration of each block? (*b*) Find the tension in the cord.

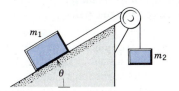

Figure 44   Problem 59.

60. A 77-kg person is parachuting and experiencing a downward acceleration of 2.5 m/s² shortly after opening the parachute. The mass of the parachute is 5.2 kg. (*a*) Find the upward force exerted on the parachute by the air. (*b*) Calculate the downward force exerted by the person on the parachute.

61. An elevator consists of the elevator cage (*A*), the counterweight (*B*), the driving mechanism (*C*), and the cable and pulleys shown in Fig. 45. The mass of the cage is 1000 kg and the mass of the counterweight is 1400 kg. Neglect friction and the mass of the cable and pulleys. The elevator acceler-

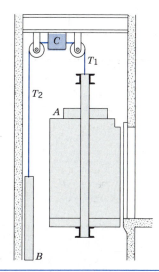

Figure 45   Problem 61.

ates upward at 2.30 m/s² and the counterweight accelerates downward at the same rate. What are the values of the tensions (*a*) T_1 and (*b*) T_2? (*c*) What force is exerted on the cable by the driving mechanism?

62. A 15,000-kg helicopter is lifting a 4500-kg car with an upward acceleration of 1.4 m/s². Calculate (*a*) the vertical force the air exerts on the helicopter blades and (*b*) the tension in the upper supporting cable; see Fig. 46.

Figure 46   Problem 62.

63. Someone exerts a force *F* directly up on the axle of the pulley shown in Fig. 47. Consider the pulley and string to be massless and the bearing frictionless. Two objects, m_1 of mass 1.2 kg and m_2 of mass 1.9 kg, are attached as shown to the opposite ends of the string, which passes over the pulley. The object m_2 is in contact with the floor. (*a*) What is the largest value the force *F* may have so that m_2 will remain at rest on the floor? (*b*) What is the tension in the string if the upward

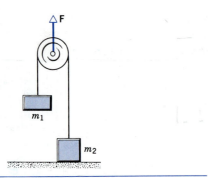

Figure 47 Problem 63.

force F is 110 N? (*c*) With the tension determined in part (*b*), what is the acceleration of m_1?

64. Two particles, each of mass m, are connected by a light string of length $2L$, as shown in Fig. 48. A steady force **F** is applied at the midpoint of the string ($x = 0$) at a right angle to the initial position of the string. Show that the acceleration of each mass in the direction at 90° to **F** is given by

$$a_x = \frac{F}{2m} \frac{x}{(L^2 - x^2)^{1/2}}$$

in which x is the perpendicular distance of one of the particles from the line of action of **F**. Discuss the situation when $x = L$.

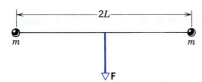

Figure 48 Problem 64.

65. A block of mass M is pulled along a horizontal frictionless surface by a rope of mass m, as shown in Fig. 49. A horizontal force **P** is applied to one end of the rope. (*a*) Show that the rope *must* sag, even if only by an imperceptible amount. Then, assuming that the sag is negligible, find (*b*) the acceleration of rope and block, (*c*) the force that the rope exerts on the block, and (*d*) the tension in the rope at its midpoint.

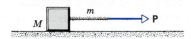

Figure 49 Problem 65.

66. Figure 50 shows a section of an alpine cable-car system. The maximum permitted mass of each car with occupants is 2800 kg. The cars, riding on a support cable, are pulled by a second cable attached to each pylon. What is the difference in tension between adjacent sections of pull cable if the cars are accelerated up the 35° incline at 0.81 m/s²?

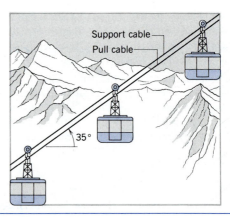

Figure 50 Problem 66.

67. The man in Fig. 51 weighs 180 lb; the platform and attached frictionless pulley weigh a total of 43 lb. Ignore the weight of the rope. With what force must the man pull up on the rope in order to lift himself and the platform upward at 1.2 ft/s²?

Figure 51 Problem 67.

68. A right triangular wedge of mass M and angle θ, supporting a small block of mass m on its side, rests on a horizontal table, as shown in Fig. 52. (*a*) What horizontal acceleration a must M have relative to the table to keep m stationary relative to the wedge, assuming frictionless contacts? (*b*) What horizontal force **F** must be applied to the system to achieve this result, assuming a frictionless table top? (*c*) Suppose no force is supplied to M and both surfaces are frictionless. Describe the resulting motion.

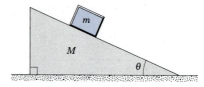

Figure 52 Problem 68.

CHAPTER 6

PARTICLE DYNAMICS

In Chapter 5 we introduced Newton's laws and gave some examples of their applications. Those examples were deliberately oversimplified, so that the use of the laws could be illustrated. In the process of oversimplification, we lost some of the physical insight. For example, a major problem in mechanics, which is of concern in the design of mechanical systems, is dealing with friction. None of the examples considered in Chapter 5 allowed for the presence of friction.

In this chapter we continue with further applications of Newton's laws. We introduce frictional forces and study their consequences. We discuss nonconstant forces and show how to solve the equations of motion for such forces. Finally, we show how using a noninertial reference frame produces effects that can be analyzed by introducing inertial forces or pseudoforces that, in contrast with the forces we discussed in Chapter 5, are not caused by specific objects in the environment.

6-1 FORCE LAWS

Before we return to applications of Newton's laws, we should briefly discuss the nature of the forces themselves. We have used the equations of motion to analyze and calculate the *effects* of forces, but they tell us nothing about the *causes* of the forces. To understand what causes a force we must have a detailed microscopic understanding of the interactions of objects with their environment. On the most fundamental level, nature appears to operate through a small number of fundamental forces. Physicists have traditionally identified four basic forces: (1) *the gravitational force,* which originates with the presence of matter (or, more in line with the general theory of relativity, matter and energy); (2) *the electromagnetic force,* which includes basic electric and magnetic interactions and is responsible for the binding of atoms and the structure of solids; (3) *the weak nuclear force,* which causes certain radioactive decay processes and certain reactions among the most fundamental particles; and (4) *the strong force,* which operates among the fundamental particles and is responsible for binding the nucleus together.

On the most microscopic scale, for example, two protons just touching at their surfaces, these forces would have the following relative strengths: strong (relative strength = 1); electromagnetic (10^{-2}); weak (10^{-7}); gravi-

tational (10^{-38}). On the fundamental scale, gravity is exceedingly weak and has negligible effects. You can get some appreciation for the weakness of gravity from some common experiments—for example, lifting a few bits of paper with an electrostatically charged comb or lifting a few nails or paper clips with a magnet. The magnetic force of a small magnet is sufficient to overcome the gravitational force exerted by the entire Earth on these objects!

The search for ever more simplification has led physicists to try to reduce the number of forces even below four. In 1967, a theory was proposed according to which the weak and electromagnetic forces could be regarded as parts of a single force, called the *electroweak* force. The combination or *unification* of these two forces is similar to the 19th-century unification of the separate electric and magnetic forces into a single electromagnetic force. Other new theories, called *grand unification theories,* have been proposed which combine the strong and electroweak forces into a single framework, and there are even "theories of everything" which attempt to include gravity as well.

One prediction of these theories is that the proton (the positively charged nuclear particle) is not stable but instead decays over a very long period, perhaps 10^{33} years (a very long time indeed, compared with the age of the universe, 10^{10} years). One way to test this theory is to watch a collection of 10^{33} protons (equivalent to a cube of water 50 feet on a side) for a year to see if one of the protons decays. Such needle-in-a-haystack experiments are necessary to test these exotic theories. We shall consider more of these speculations in Chapter 56 of the extended version of this text.

Fortunately, our analysis of mechanical systems need not invoke such theories. In fact, everything we study about ordinary mechanical systems involves only two forces: gravity and electromagnetism. The gravitational force is apparent in the Earth's attraction for objects, which gives them their weight. The much weaker gravitational attraction of one laboratory object for another is almost always negligible.

All the other forces we normally consider are ultimately electromagnetic in origin: contact forces, such as the normal force exerted when one object pushes on another and the frictional force produced when one surface rubs against another; viscous forces, such as air resistance; tensile forces, such as in a stretched rope or string; elastic forces, as in a spring; and many others. Microscopically, these forces originate with the forces between atoms. Fortunately, when we deal with ordinary mechanical systems we can ignore the microscopic basis and replace the complicated substructure with a single effective force of a specified magnitude and direction.

6-2 FRICTIONAL FORCES*

If we project a block of mass m with initial velocity $\mathbf{v}_0$ along a long horizontal table, it eventually comes to rest. This means that, while it is moving, it experiences an average acceleration $\bar{\mathbf{a}}$ that points in the direction opposite to its motion. If (in an inertial frame) we see that a body is accelerated, we always associate a force, defined from Newton's second law, with the motion. In this case we declare that the table exerts a force of *friction*, whose average value is $m\bar{\mathbf{a}}$, on the sliding block. We generally take friction to mean a contact interaction between solids. Frictionlike effects caused by liquids and gases are described by other terms (see Section 6-7).

Actually, whenever the surface of one body slides over that of another, each body exerts a frictional force on the other. The frictional force on each body is in a direction opposite to its motion relative to the other body. Frictional forces automatically oppose this relative motion and never aid it. Even when there is no relative motion, frictional forces may exist between surfaces.

Although we have ignored its effects up to now, friction is very important in our daily lives. Left to act alone it brings every rotating shaft to a halt. In an automobile,

---

* For a good general reference on friction, see the article in the *Encyclopaedia Britannica*, 14th edition.

about 20% of the engine power is used to counteract frictional forces. Friction causes wear and seizing of moving parts, and much engineering effort is devoted to reducing it. On the other hand, without friction we could not walk; we could not hold a pencil, and if we could it would not write; wheeled transport as we know it would not be possible.

We want to know how to express frictional forces in terms of the properties of the body and its environment; that is, we want to know the force law for frictional forces. In what follows we consider the sliding (not rolling) of one dry (unlubricated) surface over another. As we shall see later, friction, viewed at the microscopic level, is a very complicated phenomenon. The force laws for dry, sliding friction are empirical in character and approximate in their predictions. They do not have the elegant simplicity and accuracy that we find for the gravitational force law (Chapter 16) or for the electrostatic force law (Chapter 27). It is remarkable, however, considering the enormous diversity of surfaces one encounters, that many aspects of frictional behavior can be understood qualitatively on the basis of a few simple mechanisms.

Consider a block at rest on a horizontal table as in Fig. 1a. Attach a spring to it to measure the horizontal force $\mathbf{F}$ required to set the block in motion. We find that the block will not move even though we apply a small force (Fig. 1b). We say that our applied force is balanced by an opposite frictional force $\mathbf{f}$ exerted on the block by the table, acting along the surface of contact. As we increase the applied force (Fig. 1c,d) we find some definite force at which the block will "break away" from the surface and begin to accelerate (Fig. 1e). By reducing the force once motion has started, we find that it is possible to keep the block in uniform motion without acceleration (Fig. 1f). Figure 1g shows the results of an experiment to measure the frictional force. An increasing force F is applied starting at about $t = 2$ s, after which the frictional force increases with the applied force and the object remains at rest. At $t = 4$ s, the object suddenly begins to move and the frictional force becomes constant, independent of the applied force.

The frictional forces acting between surfaces at rest with respect to each other are called forces of *static friction*. The maximum force of static friction (corresponding to the peak at $t = 4$ s in Fig. 1g) will be the same as the smallest applied force necessary to start motion. Once motion is started, the frictional forces acting between the surfaces usually decrease so that a smaller force is necessary to maintain uniform motion (corresponding to the nearly constant force at $t > 4$ s in Fig. 1g). The forces acting between surfaces in relative motion are called forces of *kinetic friction*.

The maximum force of static friction between any pair of dry unlubricated surfaces follows these two empirical laws. (1) It is approximately independent of the area of contact over wide limits and (2) it is proportional to the

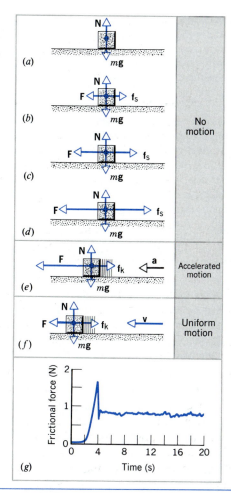

No motion

Accelerated motion

Uniform motion

Figure 1 (*a – d*) An external force **F**, applied to a resting block, is counterbalanced by an equal but opposite frictional force **f**. As **F** is increased, **f** also increases, until **f** reaches a certain maximum value. (*e*) The block then "breaks away," accelerating to the left. (*f*) If the block is to move with constant velocity, the applied force **F** must be reduced from the maximum value it had just before the block began to move. (*g*) Experimental results; here the applied force **F** is increased from zero starting at about $t = 2$ s, and the motion suddenly begins at about $t = 4$ s. For details of the experiment, see "Undergraduate Computer-Interfacing Projects," by Joseph Priest and John Snyder, *The Physics Teacher*, May 1987, p. 303.

normal force.* The normal force (sometimes called the loading force) arises from the elastic properties of the bodies in contact (see Chapter 14). Such bodies are never entirely rigid, and when a force is exerted on one body that

---

* The two laws of friction were first discovered experimentally by Leonardo da Vinci (1452 – 1519). Leonardo's statement of the two laws was remarkable, coming as it did two centuries before Newton developed the concept of force. The mathematical expressions of the laws of friction and the concept of the coefficient of friction were developed by Charles Augustin Coulomb (1736 – 1806), who is best known for his studies of electrostatics (see Chapter 27).

is prevented from moving in the direction of the force, the body pushes back to oppose being stretched or deformed. For a block resting on a horizontal table or sliding along it, the normal force is equal in magnitude to the weight of the block. Because the block has no vertical acceleration, the table must be exerting a force on the block that is directed upward and is equal in magnitude to the downward pull of the Earth on the block, that is, equal to the block's weight.

The ratio of the magnitude of the *maximum* force of static friction to the magnitude of the normal force is called the *coefficient of static friction* for the surfaces involved. If f_s represents the magnitude of the force of static friction, we can write

$$f_s \le \mu_s N, \qquad (1)$$

where μ_s is the coefficient of static friction and N is the magnitude of the normal force. The equality sign holds only when f_s has its maximum value.

The force of kinetic friction f_k between dry, unlubricated surfaces follows the same two laws as those of static friction. (1) It is approximately independent of the area of contact over wide limits and (2) it is proportional to the normal force. The force of kinetic friction is also reasonably independent of the relative speed with which the surfaces move over each other.

The ratio of the magnitude of the force of kinetic friction to the magnitude of the normal force is called the *coefficient of kinetic friction*. If f_k represents the magnitude of the force of kinetic friction, then

$$f_k = \mu_k N, \qquad (2)$$

where μ_k is the coefficient of kinetic friction.

Both μ_s and μ_k are dimensionless constants, each being the ratio of (the magnitudes of) two forces. Usually, for a given pair of surfaces $\mu_s > \mu_k$. The actual values of μ_s and μ_k depend on the nature of both the surfaces in contact. In most cases we can regard them as being constants (for a given pair of surfaces) over the range of forces and velocities we commonly encounter. Both μ_s and μ_k can exceed unity, although commonly they are less than 1. Table 1 shows some representative values of μ_s and μ_k.

TABLE 1 COEFFICIENTS OF FRICTION[a]

Surfaces	μ_s	μ_k
Wood on wood	0.25 – 0.5	0.2
Glass on glass	0.9 – 1.0	0.4
Steel on steel, clean surfaces	0.6	0.6
Steel on steel, lubricated	0.09	0.05
Rubber on dry concrete	1.0	0.8
Waxed wood ski on dry snow	0.04	0.04
Teflon on Teflon	0.04	0.04

[a] Values are approximate and are intended only as estimates. The actual coefficients of friction for any pair of surfaces depend on such conditions as the cleanliness of the surfaces, the temperature, and the humidity.

Notice that Eqs. 1 and 2 are relations between the *magnitudes only* of the normal and frictional forces. These forces are always directed perpendicularly to one another.

The Microscopic Basis of Friction *(Optional)*

On the atomic scale even the most finely polished surface is far from plane. Figure 2, for example, shows an actual profile, highly magnified, of a steel surface that would be considered to be highly polished. One can readily believe that when two bodies are placed in contact, the actual microscopic area of contact is much less than the true area of the surface; in a particular case these areas can easily be in the ratio of $1:10^4$.

The actual (microscopic) area of contact is proportional to the normal force, because the contact points deform plastically under the great stresses that develop at these points. Many contact points actually become "cold-welded" together. This phenomenon, *surface adhesion,* occurs because at the contact points the molecules on opposite sides of the surface are so close together that they exert strong intermolecular forces on each other.

When one body (a metal, say) is pulled across another, the frictional resistance is associated with the rupturing of these thousands of tiny welds, which continually reform as new chance contacts are made (see Fig. 3). Radioactive tracer experiments have shown that, in the rupturing process, small fragments of one metallic surface may be sheared off and adhere to the other surface. If the relative speed of the two surfaces is great enough, there may be local melting at certain contact areas even though the surface as a whole may feel only moderately warm. The "stick and slip" events are responsible for the noises that dry surfaces make when sliding across one another as, for example, the squealing chalk on the blackboard.*

The coefficient of friction depends on many variables, such as the nature of the materials, surface finish, surface films, tempera-

---

* See, for example, "Stick and Slip," by Ernest Rabinowicz in *Scientific American,* May 1956, p. 109.

Figure 2 A magnified section of a highly polished steel surface. The vertical scale of the surface irregularities is several thousand atomic diameters. The section has been cut at an angle so that the vertical scale is exaggerated with respect to the horizontal scale by a factor of 10.

ture, and extent of contamination. For example, if two carefully cleaned metal surfaces are placed in a highly evacuated chamber so that surface oxide films do not form, the coefficient of friction rises to enormous values and the surfaces actually become firmly "welded" together. The admission of a small amount of air to the chamber so that oxide films may form on the opposing surfaces reduces the coefficient of friction to its "normal" value.

The frictional force that opposes one body *rolling* over another is much less than that for sliding motion; this gives the advantage to the rolling wheel over the sliding sledge. This reduced friction is due in large part to the fact that, in rolling, the microscopic contact welds are "peeled" apart rather than "sheared" apart as in sliding friction. This reduces the frictional force by a large factor.

Frictional resistance in dry, sliding friction can be reduced considerably by lubrication. A mural in a grotto in Egypt dating back to 1900 B.C. shows a large stone statue being pulled on a sledge while a man in front of the sledge pours lubricating oil in its path. A still more effective technique is to introduce a layer of gas between the sliding surfaces; a laboratory air track and the gas-supported bearing are two examples. Friction can be reduced still further by suspending an object by means of magnetic forces. Magnetically levitated trains now under development have the potential for high-speed, nearly frictionless travel. ∎

Sample Problem 1 A block is at rest on an inclined plane making an angle θ with the horizontal, as in Fig. 4a. As the angle of incline is raised, it is found that slipping just begins at an angle of inclination $\theta_s = 15°$. What is the coefficient of static friction between block and incline?

Solution The forces acting on the block, considered to be a particle, are shown in Fig. 4b. The weight of the block is mg, the normal force exerted on the block by the inclined surface is N, and the force of friction exerted by the inclined surface on the block is $\mathbf{f}_s$. Notice that the resultant force exerted by the inclined surface on the block, $\mathbf{N} + \mathbf{f}_s$, is no longer perpendicular to the surface of contact, as was true for frictionless surfaces ($\mathbf{f}_s = 0$). The block is at rest, so that Newton's second law gives $\Sigma \mathbf{F} = 0$. Resolving the forces into x and y components (along the plane and normal to the plane, respectively), we obtain

x component: $\sum F_x = f_s - mg \sin \theta = 0$ or $f_s = mg \sin \theta,$

y component: $\sum F_y = N - mg \cos \theta = 0$ or $N = mg \cos \theta.$

At the angle θ_s, where slipping just begins, f_s has its maximum

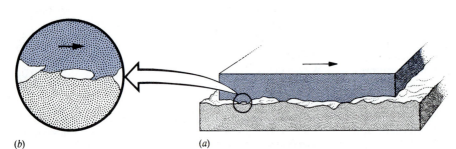

(b)   (a)

Figure 3 The mechanism of sliding friction. (*a*) The upper surface is sliding to the right over the lower surface in this enlarged view. (*b*) A detail, showing two spots where cold welding has occurred. Force is required to break these welds and maintain the motion.

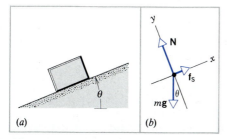

Figure 4 Sample Problem 1. (*a*) A block at rest on a rough inclined plane. (*b*) A free-body diagram of the block.

value and is equal to $\mu_s N$. Evaluating these expressions at θ_s and dividing by one another, we obtain

$$\frac{f_s}{N} = \frac{mg \sin \theta_s}{mg \cos \theta_s} = \tan \theta_s$$

or

$$\mu_s = \tan \theta_s = \tan 15° = 0.27.$$

Hence measurement of the angle of inclination at which slipping just starts provides a simple experimental method for determining the coefficient of static friction between two surfaces. Note that this determination is independent of the weight of the object.

You can use similar arguments to show that the angle of inclination θ_k required to maintain a *constant speed* for the block as it slides down the plane, once it has been started by a gentle tap, is given by

$$\mu_k = \tan \theta_k,$$

where $\theta_k < \theta_s$. With the aid of a ruler to measure the tangent of the angle of inclination, you can now determine μ_s and μ_k for a coin sliding down your textbook.

Sample Problem 2 Consider an automobile moving along a straight horizontal road with a speed v_0. If the coefficient of static friction between the tires and the road is μ_s, what is the shortest distance in which the automobile can be stopped?

Solution The forces acting on the automobile are shown in Fig. 5. The car is assumed to be moving in the positive x direction. If we assume that f_s is a constant force, we have uniformly decelerated motion.

From the relation

$$v^2 = v_0^2 + 2a(x - x_0),$$

with the initial position chosen so that $x_0 = 0$ and with a final speed $v = 0$, we obtain

$$x = -\frac{v_0^2}{2a},$$

where x is the stopping distance over which the speed changes from v_0 to 0. Because a is negative, x is positive, as we expect.

To determine a, we apply Newton's second law with components assigned according to Fig. 5*b*:

x component: $\sum F_x = -f_s = ma$ or $a = -f_s/m,$

y component: $\sum F_y = N - mg = 0$ or $N = mg,$

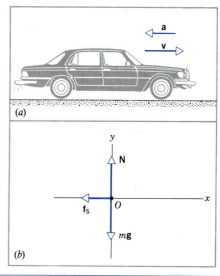

Figure 5 Sample Problem 2. (*a*) A decelerating automobile. (*b*) A free-body diagram of the decelerating automobile, considered to be a particle. For convenience, all forces are taken to act at a common point. In reality, the three forces shown are sums of the individual forces acting on each of the four tires.

so that

$$f_s = \mu_s N = \mu_s mg.$$

Substituting this quantity into the expression for a, we find

$$a = -\frac{f_s}{m} = -\mu_s g.$$

Then the distance of stopping is

$$x = -\frac{v_0^2}{2a} = \frac{v_0^2}{2\mu_s g}.$$

The greater the initial speed, the longer the distance required to come to a stop; in fact, this distance varies as the square of the initial velocity. Also, the greater the coefficient of static friction between the surfaces, the less the distance required to come to a stop.

We have used the coefficient of static friction in this problem, rather than the coefficient of kinetic friction, because we assume there is no sliding between the tires and the road. Furthermore, we have assumed that the maximum force of static friction ($f_s = \mu_s N$) operates because the problem seeks the shortest distance for stopping. With a smaller static frictional force the distance for stopping would obviously be greater. The correct braking technique required here is to keep the car just on the verge of skidding. (Cars equipped with anti-lock braking systems maintain this condition automatically.) If the surface is smooth and the brakes are fully applied, sliding may occur. In this case μ_k replaces μ_s, and the distance required to stop would increase because μ_k is smaller than μ_s.

As a specific example, if $v_0 = 60$ mi/h = 27 m/s, and $\mu_s = 0.60$ (a typical value), we obtain

$$x = \frac{v_0^2}{2\mu_s g} = \frac{(27 \text{ m/s})^2}{2(0.60)(9.8 \text{ m/s}^2)} = 62 \text{ m}.$$

Notice that this result is independent of the mass of the car.

On rear-wheel drive cars, with the engine in front, it is a common practice to "weigh down" the trunk in order to increase safety when driving on icy roads. How can this practice be consistent with our result that the stopping distance is independent of the mass of the car? (*Hint:* See Problem 2.)

Sample Problem 3   Repeat Sample Problem 10 of Chapter 5, taking into account a frictional force between block 1 and the plane. Use the values $\mu_s = 0.24$ and $\mu_k = 0.15$.

Solution   If we assume, as we learned from the solution to Sample Problem 10 of Chapter 5, that block 1 moves down the plane, then the frictional force acts up the plane. The free-body diagram of m_1 is shown in Fig. 6. The component equations of Newton's second law for m_1 are now the following:

x component: $\sum F_x = T + f - m_1 g \sin \theta = m_1 a_{1x} = -m_1 a,$

y component: $\sum F_y = N - m_1 g \cos \theta = m_1 a_{1y} = 0.$

Here we have explicitly put in that we expect a_1 to be in the negative x direction (that is, $a_{1x} = -a$). We make a similar change in the equation for m_2:

$$\sum F_y = m_2 g - T = m_2 a_{2y} = -m_2 a,$$

where we use $a_{2y} = -a$, because we expect block 2 to move in its negative y direction.

Putting $f = \mu_k N = \mu_k m_1 g \cos \theta$, we have, from the x-component equation of m_1,

$$T + \mu_k m_1 g \cos \theta - m_1 g \sin \theta = -m_1 a.$$

Solving these last two equations simultaneously for the two unknowns a and T, we obtain

$$a = -g \frac{m_2 - m_1 (\sin \theta - \mu_k \cos \theta)}{m_1 + m_2}, \qquad (3)$$

$$T = \frac{m_1 m_2 g}{m_1 + m_2} (1 + \sin \theta - \mu_k \cos \theta). \qquad (4)$$

Note that, in the limit of $\mu_k \rightarrow 0$, Eqs. 3 and 4 reduce to Eqs. 9 and 10 of Sample Problem 10 in Chapter 5 (except for the sign of a, which we have taken to be in the opposite direction in our solution to this problem).

Let us now find the numerical values of a and T:

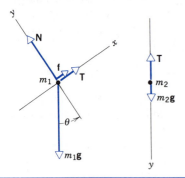

Figure 6   Sample Problem 3. The free-body diagrams of Fig. 20 of Chapter 5, in the case of friction along the plane.

$$a = (-9.80 \text{ m/s}^2) \frac{2.6 \text{ kg} - 9.5 \text{ kg} \, (\sin 34° - 0.15 \cos 34°)}{2.6 \text{ kg} + 9.5 \text{ kg}}$$

$$= 1.2 \text{ m/s}^2,$$

$$T = \frac{(9.5 \text{ kg})(2.6 \text{ kg})(9.80 \text{ m/s}^2)}{9.5 \text{ kg} + 2.6 \text{ kg}} (1 + \sin 34° - 0.15 \cos 34°)$$

$$= 29 \text{ N}.$$

The positive value of a is consistent with the way we set up our equations; the block moves down the plane, as it did in Sample Problem 10 of Chapter 5, but with less acceleration than it did in the frictionless case (2.2 m/s²).

The tension in the string is less than it was in the frictionless case (31 N). Block 1 accelerates less rapidly down the plane when there is friction, so it doesn't pull as strongly on the string attached to block 2.

One additional question that must be answered is whether the system will move at all. That is, is there enough force down the plane to exceed the static friction and start the motion? When the system is initially at rest, the tension in the string is equal to the weight of m_2, or (2.6 kg) (9.8 m/s²) = 26 N. The maximum static friction, which opposes the tendency to move down the plane, is $\mu_s N = \mu_s m_1 g \cos \theta = 19$ N. The component of the weight of m_1 acting down the plane is $m_1 g \sin \theta = 52$ N. Thus there is more than enough weight acting down the plane (52 N) to overcome the total of the tension and the static frictional force (26 N + 19 N = 45 N), and the system does indeed move. You should be able to show that if the static coefficient of friction is greater than 0.34 then there will be no motion.

6-3   THE DYNAMICS OF UNIFORM CIRCULAR MOTION

In Section 4-4 we pointed out that if a body is moving at uniform speed v in a circle or a circular arc of radius r, it experiences a centripetal acceleration $\mathbf{a}$ whose magnitude is v^2/r. The direction of $\mathbf{a}$ is always radially inward toward the center of the circle. Thus $\mathbf{a}$ is a variable vector because, even though its magnitude remains constant, its direction changes continuously as the motion progresses. You may wish to review Fig. 11 of Chapter 4, which shows the vector relationship between $\mathbf{v}$ and $\mathbf{a}$ in circular motion at constant speed.

Every accelerated body must have a net force acting on it, according to Newton's second law ($\sum \mathbf{F} = m\mathbf{a}$). Thus (assuming that we are in an inertial frame), if we see a body undergoing uniform circular motion, we can be certain that the magnitude of the net force $\sum \mathbf{F}$ acting on the body must be given by

$$\left| \sum \mathbf{F} \right| = ma = \frac{mv^2}{r}. \qquad (5)$$

The body is *not* in equilibrium because the net force is not zero. The direction of the net force $\sum \mathbf{F}$ at any instant

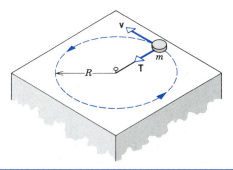

Figure 7   A disk of mass m moves with constant speed in a circular path on a horizontal frictionless surface. The only horizontal force acting on the disk is the tension **T** with which the string pulls on the disk; **T** provides the centripetal force necessary for circular motion. Vertical forces (**N** and $m\mathbf{g}$) are not shown.

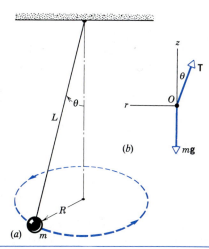

Figure 8   The conical pendulum. (*a*) A body of mass m suspended from a string of length L moves in a circle; the string describes a right circular cone of semiangle θ. (*b*) A free-body diagram of the body.

must be the direction of **a** at that instant, namely, radially inward. This force is provided by an external agent (or agents) in the environment of the accelerating mass m.

If the body in uniform circular motion is a disk moving on the end of a string in a circle on a frictionless horizontal table as in Fig. 7, the net force on the disk is provided by the tension **T** in the string. It accelerates the disk by constantly changing the direction of its velocity so that the disk moves in a circle. The direction of **T** is always toward the pin at the center, and its magnitude must equal mv^2/R.

If the string were to be cut where it joins the disk, there would be no net force exerted on the disk. The disk would then move with constant speed in a straight line along the direction of the tangent to the circle at the point at which the string was cut. The disk will *not* fly radially outward or in a curved path, but will exactly follow the straight-line direction of **v** at the instant the string breaks.

Hence, to keep the disk moving in a circle, a force must be supplied to it pulling it *inward* toward the center. Forces responsible for uniform circular motion are called *centripetal* forces because they are directed "toward the center" of the circular motion. To label a force as "centripetal," however, simply means that it always points radially inward; the name tells us nothing about the nature of the force or about the body that is exerting it. Thus, for the revolving disk of Fig. 7, the centripetal force is a tensile force provided by the string; for the Moon revolving around the Earth the centripetal force is the gravitational pull of the Earth on the Moon; for an electron circulating about an atomic nucleus the centripetal force is electrostatic. A centripetal force is not a new kind of force but simply a way of describing the behavior with time of forces that are attributable to specific bodies in the environment. Thus a force can be centripetal *and* elastic, centripetal *and* gravitational, or centripetal *and* electrostatic, among other possibilities.

Let us consider some examples of forces that act centripetally.

The Conical Pendulum

Figure 8 shows a small body of mass m revolving in a horizontal circle with constant speed v at the end of a string of length L. As the body swings around, the string sweeps over the surface of an imaginary cone. This device is called a *conical pendulum*. Let us find the time required for one complete revolution of the body.

If the string makes an angle θ with the vertical, the radius of the circular path is $R = L \sin \theta$. The forces acting on the body of mass m are its weight $m\mathbf{g}$ and the tension **T** of the string, as shown in Fig. 8*b*. Newton's second law in this case gives

$$\sum \mathbf{F} = \mathbf{T} + m\mathbf{g} = m\mathbf{a}.$$

Clearly, the net force acting on the body is nonzero, which is as it should be because a force is required to keep the body moving in a circle with constant speed.

We can resolve **T** at any instant into a radial and a vertical component

$$T_r = -T \sin \theta \quad \text{and} \quad T_z = T \cos \theta.$$

The radial component is negative if we define the radial direction to be positive outward from the axis.

Since the body has no vertical acceleration, we can write the z component of Newton's second law as

$$\sum F_z = T_z - mg = 0,$$

or

$$T \cos \theta = mg.$$

The radial acceleration is $a_r = -v^2/R$, negative because it acts radially inward (opposite to the direction of **r**, which we take to be the positive radial direction). This acceleration is supplied by T_r, the radial component of **T**, which provides the centripetal force acting on m. Hence, from the radial component of Newton's second law,

$$\sum F_r = T_r = ma_r,$$

or

$$-T \sin \theta = -mv^2/R.$$

Dividing the radial and z-component equations, we obtain

$$\frac{-T \sin \theta}{T \cos \theta} = \frac{-mv^2/R}{mg}$$

or, solving for v,

$$v = \sqrt{Rg \tan \theta},$$

which gives the constant speed of the body. If we let t represent the time for one complete revolution of the body, then

$$v = \frac{2\pi R}{t}$$

or

$$t = \frac{2\pi R}{v} = \frac{2\pi R}{\sqrt{Rg \tan \theta}} = 2\pi \sqrt{\frac{R}{g \tan \theta}}.$$

But $R = L \sin \theta$, so that

$$t = 2\pi \sqrt{\frac{L \cos \theta}{g}}. \tag{6}$$

This equation gives the relation between t, L, and θ. Note that t, called the *period* of motion, does not depend on m.

If $L = 1.2$ m and $\theta = 25°$, what is the period of the motion? We have

$$t = 2\pi \sqrt{\frac{(1.2 \text{ m})(\cos 25°)}{9.8 \text{ m/s}^2}} = 2.1 \text{ s}.$$

The Rotor

In many amusement parks* we find a device called the rotor. The rotor is a hollow cylindrical room that can be set rotating about the central vertical axis of the cylinder. A person enters the rotor, closes the door, and stands up against the wall. The rotor gradually increases its rotational speed from rest until, at a predetermined speed, the floor below the person is opened downward, revealing a deep pit. The person does not fall but remains "pinned up" against the wall of the rotor. What minimum rotational speed is necessary to prevent falling?

The forces acting on the person are shown in Fig. 9. The person's weight is mg, the force of static friction between person and rotor wall is $\mathbf{f}_s$, and **N** is the normal force

---

* See "Fear and Trembling at the Amusement Park," by John Roeder and Jearl Walker, in *Fundamentals of Physics*, 3rd ed., by David Halliday and Robert Resnick (Wiley, 1988).

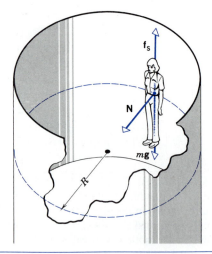

Figure 9   The rotor. Forces acting on the person are shown.

exerted by the wall on the person (which, as we shall see, provides the needed centripetal force). As we did in the previous calculation, we resolve the forces into radial and vertical components. We define a z axis to be positive upward, and if the person is not to fall there must be no acceleration in the z direction. The z component of Newton's second law gives

$$\sum F_z = f_s - mg = ma_z = 0.$$

Let the radius of the rotor be R and the person's tangential speed be v. The passenger experiences a radial acceleration $-v^2/R$, and the radial component of Newton's second law can then be written

$$\sum F_r = -N = ma_r = \frac{-mv^2}{R}.$$

Note that N provides the centripetal force in this case. If μ_s is the coefficient of static friction between person and wall necessary to prevent slipping, then $f_s = \mu_s N$ and we find

$$f_s = mg = \mu_s N = \frac{\mu_s m v^2}{R}$$

or

$$v = \sqrt{\frac{gR}{\mu_s}}. \tag{7}$$

This equation relates the coefficient of friction necessary to prevent slipping to the tangential speed of an object on the wall. Note that the result does not depend on the person's weight.

As a practical matter the coefficient of friction between the textile material of clothing and a typical rotor wall (canvas) is about 0.40. For a typical rotor the radius is 2.0 m, so that v must be about 7.0 m/s or more. The circumference of the circular path is $2\pi R = 12.6$ m, and at 7.0 m/s it takes a time of $t = 12.6$ m/(7.0 m/s) = 1.80 s to complete each revolution. The rotor must therefore turn at a rate of at least $1/1.80$ s $= 0.56$ revolution/s or about 33 rpm, the same rate of rotation as a phonograph turntable.

The Banked Curve

Let the block in Fig. 10*a* represent an automobile or railway car moving at constant speed v on a *level* roadbed around a curve having a radius of curvature R. In addition to two vertical forces, namely, the weight $m\mathbf{g}$ and a normal force $\mathbf{N}$, a horizontal force $\mathbf{P}$ must act on the car. The force $\mathbf{P}$ provides the centripetal force necessary for motion in a circle. In the case of the automobile this force is supplied by a sidewise frictional force exerted by the road on the tires; in the case of the railway car the force is supplied by the rails exerting a sidewise force on the inner rims of the car's wheels. Neither of these sidewise forces can safely be relied on to be large enough at all times, and both cause unnecessary wear. Hence, the roadbed is *banked* on curves, as shown in Fig. 10*b*. In this case, the normal force $\mathbf{N}$ has not only a vertical component, as before, but also a horizontal component that supplies the centripetal force necessary for uniform circular motion. No additional sidewise forces are needed, therefore, with a roadbed that is properly banked for vehicles of a particular speed.

The correct angle θ of banking in the absence of friction can be obtained as follows. We begin, as usual, with Newton's second law, and we refer to the free-body diagram shown in Fig. 10*b*. There is no vertical acceleration, so that the vertical component gives

$$\sum F_z = N \cos \theta - mg = ma_z = 0.$$

The radial component of the normal force is $-N \sin \theta$ and the radial acceleration is $-v^2/R$. The radial component of Newton's second law therefore gives

$$\sum F_r = -N \sin \theta = ma_r = -mv^2/R.$$

Dividing these two equations, we obtain

$$\tan \theta = v^2/Rg. \qquad (8)$$

Notice that the proper angle of banking depends on the speed of the car and the curvature of the road. It does not depend on the mass of the car; for a given banking angle, all cars will be able to travel safely. For a given curvature, the road is banked at an angle corresponding to an expected average speed. Curves are often marked by signs giving the proper speed for which the road was banked. If vehicles exceed that speed, the friction between tires and road must supply the additional centripetal force needed to travel the curve safely.

Check the banking formula for the limiting cases $v = 0$, $R \to \infty$, v large, and R small. Also, note that Eq. 8, if solved for v, gives the same result that we derived for the speed of the bob of a conical pendulum. Compare Figs. 8 and 10, noting their similarities.

6-4   EQUATIONS OF MOTION: CONSTANT AND NONCONSTANT FORCES*

Let us briefly review our progress in studying dynamics and kinematics. Our ultimate goal is to describe how a particle will move when it is acted on by a set of forces. Schematically, the analysis (in one dimension) can be represented as follows:

$$\sum F \to a \to x(t), v(t).$$

That is, Newton's laws (as described in Chapter 5) provide us with the means to calculate the acceleration of a particle from the net force acting on it. The next step is the mathematical one of finding the position and velocity (at

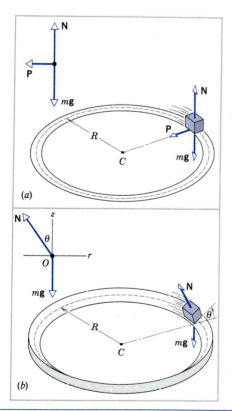

(a)

(b)

Figure 10 (*a*) A level roadbed. A free-body diagram of the moving body is shown at left. The centripetal force must be supplied by friction between tires and road. (*b*) A banked roadbed. No friction is necessary to round the curve safely.

* Sections 6-4 to 6-7 involve elementary aspects of integral calculus. The material in these sections can be skipped or postponed until the student is more familiar with integration methods.

all times t) from the acceleration and initial position and velocity.

With the exception of the previous section on circular motion, we have so far considered only constant forces (that is, forces that do not depend on the time, velocity, or position). If the force is constant, then the acceleration is constant, and for constant acceleration the solutions in one dimension for $v(t)$ and $x(t)$ are readily obtained, as we did in Section 2-6. Thus our analysis is complete for constant forces.

When the forces are not constant, we can still use Newton's laws to find the acceleration, but we certainly cannot use the constant-acceleration formulas of Section 2-6 to find $v(t)$ and $x(t)$. Instead, we must turn to methods involving integral calculus.

Before we apply our analysis to the case of nonconstant forces, let us apply integral calculus in the case of constant forces and see how the results of Section 2-6 are obtained. We assume that we have found the acceleration a (from Newton's laws) and that we wish to obtain $v(t)$ and $x(t)$. We begin with $a = dv/dt$, and so

$$dv = a\, dt. \tag{9}$$

We now integrate both sides. On the left side, the velocity is the integration variable, with limits of v_0 at time 0 and v at time t. On the right, we integrate over time between 0 and t.

$$\int_{v_0}^{v} dv = \int_{0}^{t} a\, dt. \tag{10}$$

In the case of constant acceleration, a comes out of the integral on the right and we obtain

$$v - v_0 = a \int_{0}^{t} dt \tag{11}$$

or

$$v(t) = v_0 + at \tag{12}$$

which is just Eq. 15 of Chapter 2.

Continuing, we find $x(t)$ by using $v = dx/dt$ to set up another integral:

$$dx = v\, dt = (v_0 + at)dt = v_0\, dt + at\, dt. \tag{13}$$

We integrate from position x_0 at time 0 to position x at time t:

$$\int_{x_0}^{x} dx = \int_{0}^{t} v_0\, dt + \int_{0}^{t} at\, dt, \tag{14}$$

and if a is constant we can again bring it out of the integral:

$$x - x_0 = v_0 \int_{0}^{t} dt + a \int_{0}^{t} t\, dt$$

$$= v_0 t + a(\tfrac{1}{2}t^2)$$

$$x(t) = x_0 + v_0 t + \tfrac{1}{2}at^2. \tag{15}$$

This equation is identical with Eq. 19 of Chapter 2.

If the acceleration is not constant, the integrals are more complicated to evaluate. Doing the integrals of Eqs. 10 and 14 to obtain explicit functions for $v(t)$ and $x(t)$ is called the *analytical* approach to solving the problem. An alternative is the *numerical* method, in which we can use a computer to evaluate the integrals, obtaining not the analytic functions $v(t)$ and $x(t)$ but instead the numerical values of v and x at any time t. This can be done to any desired level of precision.

Constant forces demonstrate the applications of Newton's laws, and they are certainly easier to work with than nonconstant forces. It is fortunate that practical problems often include forces that we may consider under many circumstances to be approximately constant—gravity near the Earth's surface, frictional forces, tension forces in strings, and so on. However, many physical situations are not well described by constant forces, in which cases we must use analytical or numerical techniques to carry out a solution to the problem. Here are some examples of these forces:

1. *Forces depending on the time.* In Chapter 2, we analyzed the braking of an automobile *assuming* the acceleration to be constant. In practice, this is seldom the case. Under many circumstances, especially at high speed, we usually apply the brakes slowly at first and then more strongly as the car slows. The braking force therefore depends on the time during the interval over which the car is slowing; the function $a(t)$ will depend on the details of how we apply the brakes.

Another example of a force depending on the time occurs in the case of a wave traveling through a medium. Consider a sound wave in air in which, at any given position, the wave varies sinusoidally with time. The force acting on the individual air molecules will also vary sinusoidally with time, at the same frequency as the wave. The acceleration of the particle will have the same time dependence as the force.

2. *Forces depending on the velocity.* A familiar example of a velocity-dependent force is the drag force experienced by a body moving through a fluid medium such as air or water. This frictional force increases with velocity. You may have encountered this effect when you try to walk in a swimming pool. If you walk slowly, you feel only a small resistive force, but if you try to walk quickly, the resistive forces on your legs can be quite large. The faster you try to move, the greater is the drag force.

Projectile motion is severely affected by drag forces, even though we neglected them in our analysis of falling bodies and projectiles in Chapters 2 and 4. For a given initial speed, a projectile such as a baseball has a range in air that is one-half or less of what we would expect based on the analysis of Section 4-3. A body dropped over a great distance will not follow the free-fall equations of Section 2-7, which seem to permit its velocity to increase without limit. Instead, as its speed becomes larger, so does the drag force, which tends to reduce or even prevent additional increases in speed. In fact, as we shall see in Section 6-7, the speed approaches a limit (the *terminal speed*) beyond which it does not accelerate. (For most objects, this effect occurs only at fairly high speed, following falls through distances of the order of 100 m or so. For the 1 or 2 meters of fall in our laboratory experiments, the effect is negligible and we can use the equations of Section 2-7 with confidence.)

3. *Forces depending on the position.* A familiar example of a position-dependent force is the restoring force exerted by a spring stretched a distance x from its equilibrium length: $F = -kx$. The acceleration experienced by a body of mass m attached to the spring is thus $a = F/m = -kx/m$. If we displace the body by a distance x, it experiences a force that tends to pull it back toward its equilibrium position. If we release the body, it moves toward the equilibrium position; as it does, the displacement x decreases and so does the acceleration. As it passes through the equilibrium position, its acceleration is instantaneously zero, but the acceleration increases in magnitude once again as it moves beyond $x = 0$.

Position-dependent forces are most easily analyzed using techniques of work and energy, which we discuss in Chapters 7 and 8. In the next sections, we demonstrate some methods that use Newton's laws to analyze situations involving forces that depend on time and on velocity.

6-5 TIME-DEPENDENT FORCES: ANALYTICAL METHODS

Using Newton's laws in the usual way, with some of the forces depending on the time, we obtain an acceleration $a(t)$ that depends on the time. In this case we can proceed exactly as we did in Section 6-4 to find the velocity by direct integration. Recalling that $a = dv/dt$, we write $dv = a(t)\, dt$ and integrate from time $t = 0$, when the initial velocity is v_0, to time t when the velocity is v. For simplicity we assume the motion is confined to one dimension, but the extension to three dimensions is straightforward. We proceed as follows:

$$\int_{v_0}^{v} dv = \int_0^t a(t)\, dt,$$

$$v - v_0 = \int_0^t a(t)\, dt,$$

$$v(t) = v_0 + \int_0^t a(t)\, dt. \tag{16}$$

Compare the above set of equations with Eqs. 10–12; the only difference is that a remains inside the integral.

Once we have $v(t)$, we can repeat the procedure to find $x(t)$. With $v = dx/dt$, we have $dx = v(t)\, dt$, and carrying out a similar integral from time $t = 0$, when the particle is located at x_0, to time t, when the position is x, we have the following:

$$\int_{x_0}^{x} dx = \int_0^t v(t)\, dt,$$

$$x - x_0 = \int_0^t v(t)\, dt,$$

$$x(t) = x_0 + \int_0^t v(t)\, dt. \tag{17}$$

Again, compare with Eqs. 14 and 15, and note how Eq. 17 reduces to Eq. 15 when a is constant.

Sample Problem 4   A car is moving at 105 km/h (about 65 mi/h or 29.2 m/s). The driver suddenly begins to apply the brakes, but does so with increasing force so that the deceleration increases with time according to $a(t) = ct$, where $c = -2.67$ m/s³. (*a*) How much time passes before the car comes to rest? (*b*) How far does it travel in the process?

Solution   (*a*) We need an expression for $v(t)$, so that we can find the time at which $v = 0$. Using Eq. 16 with $a(t) = ct$, we have

$$v(t) = v_0 + \int_0^t ct\, dt = v_0 + \tfrac{1}{2}ct^2.$$

Setting $v(t)$ to 0, we can solve for the time t_1 at which the car comes to rest:

$$0 = v_0 + \tfrac{1}{2}ct_1^2,$$

$$t_1 = \sqrt{\frac{-2v_0}{c}} = \sqrt{\frac{-2(29.2 \text{ m/s})}{-2.67 \text{ m/s}^3}} = 4.68 \text{ s}.$$

It takes 4.68 s to bring the car to rest.

(*b*) To find how far the car travels, we need an expression for $x(t)$, for which we must integrate $v(t)$ according to Eq. 17:

$$x(t) = x_0 + \int_0^t (v_0 + \tfrac{1}{2}ct^2)\, dt = x_0 + v_0 t + \tfrac{1}{6}ct^3.$$

With $t = t_1 = 4.68$ s, the distance traveled is (setting x_0 to 0)

$$x(t_1) = 0 + (29.2 \text{ m/s})(4.68 \text{ s}) + \tfrac{1}{6}(-2.67 \text{ m/s}^3)(4.68 \text{ s})^3$$

$$= 91.0 \text{ m}.$$

Figure 11 shows the time dependence of x, v, and a. In contrast with the case of constant acceleration, $v(t)$ is not a straight line.

With this method of braking, most of the change in velocity occurs near the end of the motion. The change in velocity in the first second after the brakes are applied is only 1.3 m/s (about 3 mi/h); in the last second, however, the change is 11.2 m/s (about 25 mi/h). (Recall that in the case of constant acceleration, the change in velocity is the same in equal time intervals.) Can you think of an advantage to braking in this manner? Are there also disadvantages?

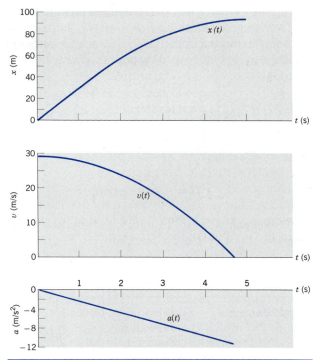

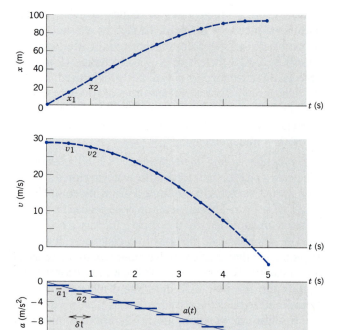

Figure 11 Sample Problem 4. The deduced position $x(t)$ and velocity $v(t)$ are shown corresponding to $a(t)$, which varies linearly with time.

Figure 12 Numerical solution (shown as dots) to Sample Problem 4; compare with analytic solution (Fig. 11 and dashed curves). For each of the 0.5-s intervals, the acceleration is assumed to be constant, and the position and velocity at the end of the interval can be calculated, giving the plotted points. Taking more (and smaller) intervals would give more points and smoother curves for $x(t)$ and $v(t)$.

6-6 TIME DEPENDENT FORCES: NUMERICAL METHODS *(Optional)*

The analytic procedure described in the previous section in principle enables us to calculate $x(t)$ and $v(t)$ from any given $a(t)$. Often, however, this method is not practical or desirable. For instance, there may be no analytic form for the integrals, or perhaps the form is so complicated that the solutions do not contribute to our physical insight into the problem. Numerical techniques offer a convenient alternative to analytical methods, and are of course particularly helpful in instances in which the analytical method cannot be used.

In the numerical method, we approximate the problem by dividing the range of times, over which we wish to solve, into a large number of small intervals. In each interval we apply the equations of constant acceleration, but the "constant" changes from one interval to the next. A convenient choice for the constant acceleration in each interval is the average acceleration in the interval.

This method works best, and gives the most precise results, if we make the intervals as small as possible; the smaller the interval, the better the (constant) average acceleration approximates the actual acceleration. On the other hand, when we decrease the size of the intervals, we must correspondingly increase their number, and we may therefore be required to do many repetitive calculations. This is just the sort of task that computers do very well, and so this method of solution can be done on a computer to any desired level of precision.

Figure 12 shows graphically how this procedure is done, in the case of the variable acceleration problem we solved in Sample Problem 4. The region between $t = 0$ and $t = 5$ s is divided into 10 small intervals each of width $\delta t = 0.5$ s. The function $a(t)$ is approximated in each interval by a different constant (the average acceleration, which in this linear case also happens to be the value of a at the midpoint of the interval). In the first interval, the average acceleration is determined from the values of a at $t = 0$ and $t = 0.5$ s:

$$\bar{a}_1 = \tfrac{1}{2}[a(0) + a(0.5 \text{ s})] = \tfrac{1}{2}[0 + (-2.67 \text{ m/s}^3)(0.5 \text{ s})]$$
$$= -0.67 \text{ m/s}^2.$$

The change in velocity in the first interval, δv_1, is approximately

$$\delta v_1 = \bar{a}_1 \, \delta t = (-0.67 \text{ m/s}^2)(0.5 \text{ s}) = -0.34 \text{ m/s},$$

and the velocity at $t = 0.5$ s is therefore

$$v_1 = v_0 + \delta v_1 = 29.2 \text{ m/s} - 0.34 \text{ m/s} = 28.9 \text{ m/s}.$$

To find the displacement during the first interval, we first find the average velocity during that interval:

$$\bar{v}_1 = \tfrac{1}{2}(v_0 + v_1) = \tfrac{1}{2}(29.2 \text{ m/s} + 28.9 \text{ m/s}) = 29.1 \text{ m/s},$$

and the displacement δx_1 in that interval is approximately

$$\delta x_1 = \bar{v}_1 \, \delta t = (29.1 \text{ m/s})(0.5 \text{ s}) = 14.6 \text{ m}.$$

If we assign the starting point as $x_0 = 0$, then the position at the end of the first interval is

$$x_1 = x_0 + \delta x_1 = 0 + 14.6 \text{ m} = 14.6 \text{ m}.$$

The values of v_1 and x_1 are plotted at $t = 0.5$ s in Fig. 12.

We now move to the second interval and repeat the procedure. Here the average acceleration is

$$\bar{a}_2 = \tfrac{1}{2}[a(0.5\ \text{s}) + a(1.0\ \text{s})]$$
$$= \tfrac{1}{2}[(-2.67\ \text{m/s}^3)(0.5\ \text{s}) + (-2.67\ \text{m/s}^3)(1.0\ \text{s})]$$
$$= -2.00\ \text{m/s}^2.$$

Continuing as we did for the first interval, in the second interval

$$\delta v_2 = \bar{a}_2\ \delta t = (-2.00\ \text{m/s}^2)(0.5\ \text{s}) = -1.0\ \text{m/s},$$

$$v_2 = v_1 + \delta v_2 = 28.9\ \text{m/s} - 1.0\ \text{m/s} = 27.9\ \text{m/s},$$

and

$$\bar{v}_2 = \tfrac{1}{2}(v_1 + v_2) = \tfrac{1}{2}(28.9\ \text{m/s} + 27.9\ \text{m/s}) = 28.4\ \text{m/s},$$

$$\delta x_2 = \bar{v}_2\ \delta t = (28.4\ \text{m/s})(0.5\ \text{s}) = 14.2\ \text{m},$$

$$x_2 = x_1 + \delta x_2 = 14.6\ \text{m} + 14.2\ \text{m} = 28.8\ \text{m}.$$

The values of v_2 and x_2 give the velocity and position at the end of the second interval, and they are plotted at $t = 1.0$ s in Fig. 12.

Continuing through all 10 intervals, we find the remaining points plotted in Fig. 12.

Comparing Figs. 11 and 12, you can see how well the numerical solution agrees with the analytic one, even for so few as 10 intervals. A computer could easily do this calculation for 100 or 1000 intervals, so that the points plotted for x and v would appear very nearly as smooth curves.

Interpolating between the velocity endpoints of the last interval, we see that the car stops at about 4.7 s, just as we found in the analytic solution. Estimating the distance traveled from Fig. 12, we find about 91 m, again in agreement with the analytic value.

The negative value found for v at the end of the tenth interval is of course not meaningful in this problem: the original dynamical situation does not permit negative values, because applying the brakes will not cause a car to move backward. It is convenient for us to continue the numerical calculation up to that point, to help us analyze the last interval.

In Appendix I, you will find a computer program (written in the BASIC language), which can do this calculation. By making minor changes in the program, you can do this type of calculation for any form of $a(t)$.   ■

6-7   DRAG FORCES AND THE MOTION OF PROJECTILES

Raindrops fall from clouds whose height h above the ground is about 2 km. Using our equation for freely falling bodies (Eq. 25 of Chapter 2), we expect the raindrop to strike the ground with a speed of $v = \sqrt{2gh} \approx 200$ m/s, or about 440 mi/h. Impact with a projectile, even a raindrop, at that speed would be lethal; since raindrops move at much slower speeds, we have obviously made an error in the calculation.

The error occurs when we neglect the effect of the frictional force exerted by the air on the falling raindrop. This frictional force is an example of a drag force, experienced by any object that moves through a fluid medium. Drag forces have important effects on a variety of objects, such as baseballs, which deviate considerably from the ideal drag-free trajectory, and downhill skiers, who try to streamline their bodies and skiing position to reduce the drag. Drag forces must be taken into account in the design of aircraft and seacraft. From the standpoint of falling bodies, from raindrops to skydivers, drag forces prevent the velocity from increasing without limit and they impose a maximum or *terminal* speed that can be attained by a falling body.

One particular characteristic of drag forces is that they depend on velocity: the faster the object moves, the greater is the drag force. We therefore must use integral methods to analyze the kinematics.

When the force, and therefore the acceleration, is a function of the velocity, the methods of Section 6-5 for time-dependent forces must be modified somewhat. We begin, as we did in Eq. 16, with $a = dv/dt$, but now a is a function of the velocity, $a(v)$:

$$a(v) = \frac{dv}{dt}$$

$$\frac{dv}{a(v)} = dt.$$

This can now be integrated directly:

$$\int_{v_0}^{v} \frac{dv}{a(v)} = \int_{0}^{t} dt = t. \tag{18}$$

The left side of Eq. 18 gives some function of v, and thus Eq. 18 is in effect t as a function of v or $t(v)$, instead of $v(t)$. Often we are able to "invert" this result to find $v(t)$, which is generally more useful for calculations.

Sample Problem 5   Assume that an object of mass m falling in air experiences a drag force D that increases *linearly* with velocity,

$$D = bv,$$

and always acts in a direction opposite to the velocity. The constant b depends on the properties of the object (its size and shape, for instance) and on the properties of the fluid (especially its density). Find the velocity as a function of the time, $v(t)$, for an object of mass m dropped from rest.

Solution   Figure 13 shows the free-body diagram, which changes with time because D varies with v. When the object is released D is zero (because v is zero), and D increases with v. At a certain point in the motion $D = mg$ and the object has no net force acting on it and therefore no acceleration, as in Fig. 13c. From this point the velocity remains constant. Our mathematical solution should show this property.

Newton's second law for this problem is

$$\sum \mathbf{F} = \mathbf{D} + m\mathbf{g} = m\mathbf{a}.$$

We choose the y axis downward, so that the vertical component is

$$\sum F_y = mg - bv = ma,$$

or

$$a = g - \frac{b}{m}v.$$

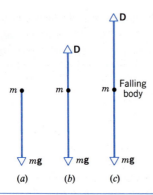

Figure 13   Forces acting on a body falling in air. (*a*) At the instant of release, $v = 0$ and there is no drag force. (*b*) The drag force increases as the body gains speed. (*c*) Eventually the drag force equals the weight; for all later times it remains equal to the weight and the body falls at its constant terminal speed.

You can see from this expression that as v increases, we eventually reach a point where the right side becomes zero, when $bv/m = g$. At this point $a = 0$, and it remains zero through the rest of the motion, so that the velocity stays constant from that point on. This is the terminal velocity, $v_\mathrm{T} = mg/b$.

To find $v(t)$, we use Eq. 18 with $v_0 = 0$:

$$\int_0^v \frac{dv}{g - (b/m)v} = t.$$

The integral can be written

$$-\frac{m}{b} \int_0^v \frac{-b\,dv}{mg - bv}$$

which is of the form $\int du/u = \ln u$, where $u = mg - bv$. Thus

$$-\frac{m}{b} \int_0^v \frac{-b\,dv}{mg - bv} = -\frac{m}{b} \ln (mg - bv)\Big|_0^v$$

$$= -\frac{m}{b} \ln (mg - bv) + \frac{m}{b} \ln (mg)$$

$$= -\frac{m}{b} \ln \left(\frac{mg - bv}{mg}\right) = t.$$

This expression is a perfectly acceptable relationship between v and t, but it is somewhat easier to use and interpret if we invert it to find $v(t)$:

$$\ln \left(\frac{mg - bv}{mg}\right) = -\frac{bt}{m}$$

$$\frac{mg - bv}{mg} = e^{-bt/m},$$

and finally, solving for v,

$$v(t) = \frac{mg}{b}(1 - e^{-bt/m}). \tag{19}$$

When t is small (near the beginning of the projectile's fall), we can approximate the exponential by using $e^x \approx 1 + x$ for small x ($x \ll 1$). Thus

$$v(t) \approx \frac{mg}{b}\left[1 - \left(1 - \frac{bt}{m}\right)\right] = gt \quad \text{(small } t\text{)}.$$

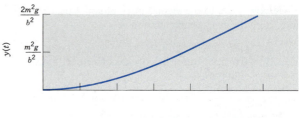
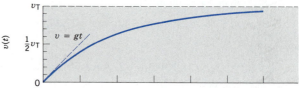
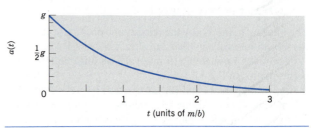

Figure 14   Sample Problem 5. Position, velocity, and acceleration for a falling body subject to a drag force. Note that the acceleration starts at g and falls to zero; the velocity starts at zero and approaches v_T.

Early in the motion, before the drag force has increased significantly, the object is very nearly in free fall with acceleration g.

For large t, the exponential approaches zero ($e^{-x} \to 0$ as $x \to \infty$). The speed then approaches the terminal speed v_T:

$$v_\mathrm{T} = \frac{mg}{b}. \tag{20}$$

Once we have a general expression for $v(t)$, we can differentiate it to find $a(t)$ and also integrate it to find $y(t)$. Doing these calculations and checking the results at small t and large t are left as exercises for the student (see Problem 66). Figure 14 illustrates the time dependence of a, v, and y.

This example shows one way of analyzing the drag force. Another approach assumes D to be proportional to v^2 instead of v. Similar methods are used to find the solutions in that case, but the mathematics is somewhat more complicated. A terminal speed is also obtained in that case, although its mathematical expression is different from the one derived here.

Table 2 shows typical measured values of terminal speeds of different objects in air.

Projectile Motion with Air Resistance   (*Optional*)

Drag calculations are also important for two-dimensional projectile motion. A baseball, for example, leaves the bat with a speed of the order of 100 mi/h or 45 m/s. This is already greater than its terminal speed in air when dropped from rest (Table 2). The drag force $D = bv$ can be estimated from our solution to Sample Problem 5. From Eq. 20 we see that the constant b is the weight

TABLE 2   SOME TERMINAL SPEEDS IN AIR

Object	Terminal Speed (m/s)	95% Distance[a] (m)
16-lb shot	145	2500
Skydiver (typical)	60	430
Baseball	42	210
Tennis ball	31	115
Basketball	20	47
Ping-Pong ball	9	10
Raindrop (radius = 1.5 mm)	7	6
Parachutist (typical)	5	3

[a] This is the distance through which the body must fall from rest to reach 95% of its terminal speed.

Source: Adapted from Peter J. Brancazio, *Sport Science,* Simon & Schuster Inc., New York, © 1984.

mg of the baseball (about 1.4 N, corresponding to a mass of 0.14 kg) divided by its terminal velocity, 42 m/s. Thus $b = 0.033$ N/(m/s). If the ball travels at 45 m/s, it experiences a drag force bv of about 1.5 N, which is greater than its weight and therefore has a substantial effect on its motion.

Figure 15 shows the free-body diagram at a particular point in the baseball's trajectory. Like all frictional forces, $\mathbf{D}$ is in a direction opposite to $\mathbf{v}$, and we assume no wind is blowing. If we take $\mathbf{D} = -b\mathbf{v}$, we can use Newton's laws to find an analytic solution

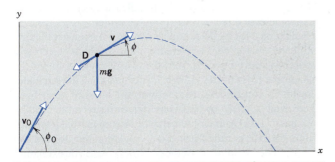

Figure 15   A projectile in motion. It is launched with velocity v_0 at an angle ϕ_0 with the horizontal. At a certain time later its velocity is v at the angle ϕ. The weight and the drag force (which always points in a direction opposite to v) are shown at that time.

for the trajectory, an example of which is illustrated in Fig. 16. When air resistance is taken into account, the range is reduced from 179 m to 72 m and the maximum height from 78 m to 48 m. Note also that the trajectory is no longer symmetric about the maximum; the descending motion is much steeper than the ascending motion. For $\phi_0 = 60°$, the projectile strikes the ground at an angle of $-79°$, while in the absence of drag it would strike the ground at an angle equal to $-\phi_0$.

The drag force depends on the velocity of the projectile in still air. If a wind is blowing, the calculation must be changed accordingly, and the result will differ.

For other (and more realistic) choices for the drag force $\mathbf{D}$, the calculation must be done numerically.* ■

6-8   NONINERTIAL FRAMES AND PSEUDOFORCES   *(Optional)*

In our treatment of classical mechanics thus far, we have assumed that measurements and observations were made from an inertial reference frame. This is one of the set of reference frames defined by Newton's first law, namely, that set of frames in which a body will not be accelerated ($\mathbf{a} = 0$) if there are no identifiable force-producing bodies in its environment ($\Sigma \mathbf{F} = 0$). The choice of a reference frame is always ours to make, so that if we choose to select only inertial frames, we do not restrict in any way our ability to apply classical mechanics to natural phenomena.

Nevertheless, we can, if we find it convenient, apply classical mechanics from the point of view of an observer in a *noninertial frame,* that is, a frame attached to a body that is accelerating as viewed from an inertial frame. The frames defined by an accelerating car or a rotating merry-go-round are examples of noninertial frames.

To apply classical mechanics in noninertial frames we must introduce additional forces known as *pseudoforces* (sometimes

---

* You can find more information about this calculation in "Trajectory of a Fly Ball," by Peter J. Brancazio, *The Physics Teacher,* January 1985, p. 20, and in his book *SportScience* (Simon & Schuster Inc., 1984), which contains many fascinating applications of the principles of physics to sports. See also "Physics and Sports: the Aerodynamics of Projectiles," by Peter Brancazio, in *Fundamentals of Physics,* 3rd ed., by David Halliday and Robert Resnick (Wiley, 1988).

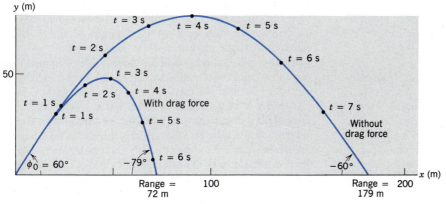

Figure 16   Projectile motion with and without a drag force, calculated for $v_0 = 45$ m/s and $\phi_0 = 60°$.

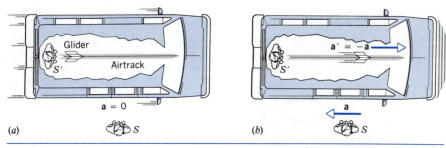

Figure 17   (*a*) Ground-based observer *S* watches observer *S'* traveling in a van at constant velocity. Both observers are in inertial reference frames. (*b*) The van brakes with constant acceleration *a* according to observer *S*. Observer *S'*, now in a noninertial reference frame, sees the glider move forward on its airtrack with constant acceleration $\mathbf{a'} = -\mathbf{a}$. Observer *S'* accounts for this motion in terms of a pseudoforce.

called inertial forces). Unlike the forces we have examined thus far, we cannot associate pseudoforces with any particular object in the environment of the body on which they act, and we cannot classify them into any of the categories listed in Section 6-1. Moreover, if we view the body from an inertial frame, the pseudoforces disappear. Pseudoforces are simply devices that permit us to apply classical mechanics in the normal way to events if we insist on viewing the events from a noninertial reference frame.

As an example, consider an observer *S'* riding in a van that is moving at constant velocity. The van contains a long airtrack with a frictionless 0.25-kg glider resting at one end (Fig. 17*a*). The driver of the van applies the brakes, and the van begins to decelerate. An observer *S* on the ground measures the constant acceleration of the van to be -2.8 m/s². The observer *S'* riding in the van is therefore in a noninertial frame of reference when the van begins to decelerate. *S'* observes the glider to move down the track with an acceleration of $+2.8$ m/s². How might each observer use Newton's second law to account for the motion of the glider?

For ground observer *S*, who is in an inertial reference frame, the analysis is straightforward. The glider, which had been moving forward at constant velocity before the van started to brake, simply continues to do so. According to *S*, the glider has no acceleration and therefore no horizontal force need be acting on it.

Observer *S'*, however, sees the glider accelerate and can find no object in the environment of the glider that exerted a force on it to provide its observed forward acceleration. To preserve the applicability of Newton's second law, *S'* must assume that a pseudoforce acts on the glider. According to *S'*, this force **F'** must equal *ma'*, where **a'** ($= -\mathbf{a}$) is the acceleration of the glider measured by *S'*. The magnitude of this pseudoforce is

$$F' = ma' = (0.25 \text{ kg})(2.8 \text{ m/s}^2) = 0.70 \text{ N},$$

and its direction is the same as **a'**, that is, toward the front of the van. This force, which is very real from the point of view of *S'*, is not apparent to ground-based observer *S*, who has no need to introduce it to account for the motion of the glider.

One indication that pseudoforces are non-Newtonian is that they violate Newton's third law. To apply Newton's third law, *S'* must find a reaction force exerted *by* the glider *on* some other body. No such reaction force can be found, and so Newton's third law is violated.

Pseudoforces are very real to those that experience them. Imagine yourself riding in a car that is rounding a curve to the left. To a ground observer, the car is experiencing a centripetal acceleration and therefore constitutes a noninertial reference frame. If the car has smooth vinyl seats, you will find yourself sliding across the seat to the right. To the ground observer, who is in an inertial frame, this is quite natural; your body is simply trying to obey Newton's first law and move in a straight line, and it is the car that is sliding to the left under you. From your point of view in the noninertial reference frame of the car, you must ascribe your sliding motion to a pseudoforce pulling you to the right. This type of pseudoforce is called a *centrifugal force*, meaning a force directed *away from* the center.

Riding on a merry-go-round, you are again in an accelerated and therefore noninertial reference frame in which objects will apparently move outward from the axis of rotation under the influence of the centrifugal force. If you hold a ball in your hand, it seems to you to be in equilibrium, the outward centrifugal force being balanced by the inward force exerted on the ball by your hand. To a ground observer, who is in an inertial reference frame, the ball is moving in a circle, accelerating toward the center under the influence of the *centripetal* force you exert on it with your hand. To the ground observer, there is no centrifugal force because the ball is not in equilibrium: it is accelerating radially inward.

Pseudoforces can be used as the basis of practical devices. Consider the centrifuge, one of the most useful of laboratory instruments. As a mixture of substances moves rapidly in a circle, the more massive substances experience a larger centrifugal force mv^2/r and move farther away from the axis of rotation. The centrifuge thus uses a pseudoforce to separate substances by mass, just as the mass spectrometer (Sections 1-5 and 5-4) uses an electromagnetic force to separate atoms by mass.

Another pseudoforce is called a *Coriolis* force. Suppose that you roll a ball inward with constant speed along a radial line painted on the floor of a rotating merry-go-round. At the instant you release it at the radius *r*, it has just the right tangential velocity (the same as yours) to be in circular motion. As it moves inward it would take a smaller tangential speed to maintain its circular motion at the same rate as its immediate surroundings. Because it has no way to lose tangential speed (we assume little friction between the ball and the floor), it moves a bit ahead of the painted line representing a uniform rotational speed. That is,

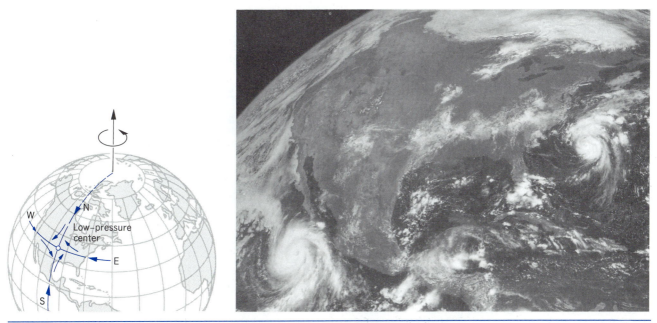

Figure 18   A low-pressure center on the rotating Earth. As the air flows inward, it appears to rotate counterclockwise to noninertial observers in the northern hemisphere of the rotating Earth. A hurricane (photo) is such a low-pressure center.

in your rotating noninertial reference frame you would suggest that a sideways pseudoforce—a Coriolis force—causes the ball to veer steadily away from the line as it rolls inward. To a ground observer in an inertial frame, there is no Coriolis force: the ball moves in a straight line at a speed determined by the components of its velocity at the instant of release.

Perhaps the most familiar example of the effects of the Coriolis force is in the motion of the atmosphere around centers of low or high pressure. Figure 18 shows a diagram of a low-pressure center in the northern hemisphere. Because the pressure is lower than the surroundings, air flows radially inward in all directions. As the Earth rotates (making it a noninertial frame), the effect is similar to that of the ball on the merry-go-round: air rushing inward from the south moves a bit ahead of an imaginary line drawn on the rotating Earth, while air from the north (like a ball rolled outward on the merry-go-round) lags a bit behind the line. The total effect is that the air rotates in a counterclockwise direction around the low-pressure center. This Coriolis effect is thus responsible for the circulation of winds in a cyclone or hurricane. In the southern hemisphere the effects are reversed.

It is necessary to correct for the Coriolis effect of the rotating Earth in the motion of long-range artillery shells. For a typical shell of range 10 km, the Coriolis effect may cause a deflection as large as 20 m. Such corrections are built into the computer programs used to control the aiming and firing of long-range weapons. Things can go wrong, however, as the British Navy discovered in a World War I battle near the Falkland Islands. Its fire control manuals were written for the northern hemisphere, and the Falklands are in the southern hemisphere where the Coriolis correction would be in the opposite direction. The British shells were landing about 100 m from their targets, because the correction for the Coriolis effect was being made in the wrong direction!

In mechanical problems, then, we have two choices: (1) select an *inertial* reference frame and consider only "real" forces, that is, forces that we can associate with definite bodies in the environment, or (2) select a *noninertial* reference frame and consider not only the "real" forces but suitably defined pseudoforces. Although we usually choose the first alternative, we sometimes choose the second; both are completely equivalent and the choice is a matter of convenience.   ■

6-9   LIMITATIONS OF NEWTON'S LAWS   *(Optional)*

In the first six chapters, we have described a system for analyzing mechanical behavior with a seemingly vast range of applications. With little more than the equations of Newton's laws, you can design great skyscrapers and suspension bridges, or even plan the trajectory of an interplanetary spacecraft (Fig. 19). Newtonian mechanics, which provided these computational tools, was the first truly revolutionary development in theoretical physics.

Here is an example of our faith in Newton's laws. Galaxies and clusters of galaxies are often observed to rotate, and by observation we can deduce the speed of rotation. From this we can calculate the amount of matter that must be present in the galaxy or cluster for gravity to supply the centripetal force corresponding to the observed rotation. Yet the amount of matter that we actually observe with telescopes is far less than we expect. Therefore, it has been proposed that there is additional "dark matter" that we cannot see with telescopes but that must be present to provide the needed gravitational force. There is as yet no convincing candidate for the type or nature of this dark matter, and so other explanations have been proposed for the apparent inconsistency between the amount of matter actually observed in the galaxies and the amount we think is needed to

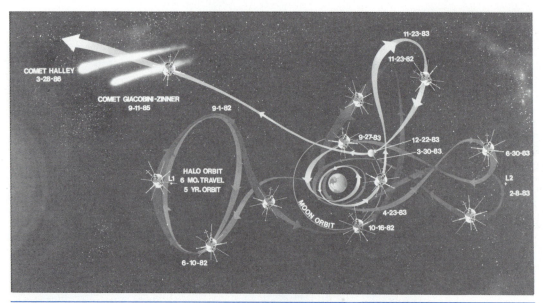

Figure 19   A triumph of Newtonian mechanics. Launched in 1978, the spacecraft International Planetary Explorer spent 4 years orbiting about point L_1, monitoring the solar wind. It then explored Earth's magnetotail from a nightside orbit; then it encountered the tail of Comet Giacobini–Zinner in 1985 and passed upstream of Comet Halley in 1986. Now on an interplanetary cruise, it will return to Earth's vicinity in the year 2012. Its journey so far has involved 37 rocket burns and 5 lunar flybys.

satisfy Newton's laws. One proposed explanation is that our calculations are incorrect because Newton's laws do not hold in the conditions that we find on the very large scale, that is, when the accelerations are very small (below a few times 10^{-10} m/s^2). In particular, it has been proposed that for these very small accelerations, the force is proportional to a^2 instead of a.

Figure 20 shows the results of a recently reported experiment testing this supposition. If force depended on the acceleration to some power other than 1, the data would not fall on a straight line. From this extremely precise experiment, we conclude that down to accelerations of about 10^{-10} m/s^2, force is proportional to acceleration and Newton's second law holds.

In our own century, we have experienced three other revolu-

tionary developments: Einstein's special theory of relativity (1905), his general theory of relativity (1915), and quantum mechanics (in about 1925). Special relativity teaches us that we cannot extrapolate the use of Newton's laws to particles moving at speeds comparable to the speed of light. General relativity shows that we cannot use Newton's laws in the vicinity of a very strong gravitational force. Quantum mechanics teaches us that we cannot extrapolate Newton's laws to objects as small as atoms.

Special relativity, which involves a distinctly non-Newtonian view of space and time, can be applied under all circumstances, at both high speeds and low speeds. In the limit of low speeds, it can be shown that the dynamics of special relativity reduces directly to Newton's laws. Similarly, general relativity can be applied to weak as well as strong gravitational forces, but its equations reduce to Newton's laws for weak forces. Quantum mechanics can be applied to individual atoms, where a certain randomness of behavior is predicted, or to ordinary objects containing a huge number of atoms, in which case the randomness averages out to give Newton's laws once again.

Within the past decade, another apparently revolutionary development has emerged. This new development concerns mechanical systems whose behavior is described as *chaotic*. One of the hallmarks of Newton's laws is their ability to predict the future behavior of a system, if we know the forces that act and the initial motion. For example, from the initial position and velocity of a space probe that experiences known gravitational forces from the Sun and the planets, we can calculate its exact trajectory. On the other hand, consider a twig floating in a turbulent stream. Even though it is acted on at all times by forces governed by Newtonian mechanics, its path downstream is totally unpredictable. If two twigs are released side-by-side in the stream, they may be found very far apart downstream. One particular theme of chaotic dynamics is that tiny changes in the initial conditions

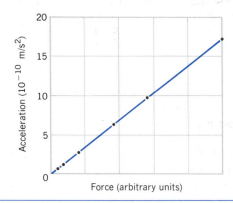

Figure 20   Results of a recent experiment to test whether Newton's second law holds for small accelerations below 10^{-9} m/s^2. The straight line shows that acceleration is proportional to the applied force down to 10^{-10} m/s^2, and so Newton's law remains valid even at such small accelerations.

Computer Projects

68. Section 6–6 describes a numerical technique to integrate Newton's second law and to obtain a table giving the position and velocity of an object for a sequence of times. Divide the period from some initial time t_0 to some final time t_f into N small intervals Δt. If x_b, v_b, and F_b are the coordinate, velocity, and force at the beginning of an interval, then $x_e = x_b + v_b \Delta t$ and $v_e = v_b + (F_b/m)\Delta t$ give estimates of the coordinate and velocity at the end. These values are then used as the coordinate and velocity at the beginning of the next interval. The smaller Δt, the better the estimate, but Δt cannot be taken too small or significant figures will be lost during the calculation. The force may be a function of position, velocity, and time. The explicit function is determined by the physical situation, and once it is known you use values of x_b, v_b, and t_b to evaluate F_b. Write a computer program or design a spreadsheet to carry out the integration. You will input x_0, v_0, t_0, Δt, and N. Here's an example to try.

Starting from rest a person pushes a 95-kg crate across a rough floor with a force given by $F = 200e^{-0.15t}$, where F is in newtons and t is in seconds. The force decreases exponentially because the person tires. As long as the crate is moving a constant frictional force of 80 N opposes the motion. (*a*) How long after starting does the crate stop? (*b*) How far does it go? Obtain 2 significant figure accuracy.

For purposes of integration divide the time between $t = 0$ and $t = 15$ s into 1500 intervals, each of 0.01 s duration. You do not want to display or print the coordinate and velocity at the end of every interval. For the first run, display results at the end of every 100 intervals. On subsequent runs you may wish to display results for shorter intervals over a limited range. Once the table of results is generated, search it for two values of the velocity that straddle $v_e = 0$. If the values of x are the same to 2 significant figures you are finished. If they are not, repeat the calculation with a smaller display interval or perhaps a smaller integration interval.

69. A 150-g ball is thrown straight upward from the edge of cliff with an initial speed of 25 m/s. On the way down it misses the cliff edge and continues to fall to the ground 300 m below. In addition to the force of gravity it is subjected to a force of air resistance given by $F_D = -0.0150v$, where F_D is in newtons and v is in m/s. (*a*) How long is the ball in flight? (*b*) What is its speed just before it hits the ground? (*c*) What is the ratio of this speed to its terminal speed?

Use a computer program or spreadsheet to integrate Newton's second law (see Section 6–6 and the previous problem for hints). Use an integration interval of 0.001 s and display the coordinate and velocity for every 0.1 s from $t = 0$ to $t = 12$ s. This should give an accuracy of 2 significant figures.

70. A 2.5-kg projectile is launched over level ground with an initial speed of 150 m/s, at an angle of 40° above the horizontal. In addition to the force of gravity it is subjected to a force of air resistance $\mathbf{F}_D = -0.30\mathbf{v}$, where $\mathbf{F}_D$ is in newtons and $\mathbf{v}$ is in m/s. Numerically integrate Newton's second law from $t = 0$ (the time of launching) to $t = 20$ s. Use an integration interval of 0.001 s but display results for every 0.5 s. You will need to consider both the x and y coordinates and velocity components. Use $a_x = -(b/m)v_x$ and $a_y = -g - (b/m)v_y$, where b is the drag coefficient. See the previous computer projects. (*a*) Plot the trajectory y vs. x from launch to the time the projectile hits the ground. Notice that the trajectory is not symmetric about the highest point as it would be if air resistance were absent. Use your graph or list of values to estimate: (*b*) the time the projectile reaches the highest point on its trajectory and the coordinates of the highest point; (*c*) the time it lands, its range, and its velocity just before landing. (*d*) Compare these quantities with the values they would have if there were no air resistance. How does air resistance influence the height of the highest point? How does it influence the range? How does it influence the impact speed?

71. Air resistance may significantly influence the launch angle for which a projectile has maximum range. To see the influence consider a 2.5-kg projectile launched over level ground with an initial speed of 150 m/s and suppose the force of the air is given by $\mathbf{F}_D = -0.30\mathbf{v}$, where $\mathbf{F}_D$ is in newtons and $\mathbf{v}$ is in m/s. For each of the launch angles 25°, 30°, 35°, and 40° numerically integrate Newton's second law with an integration interval of 0.001 s. Display results for every 0.5 s from $t = 0$ (the time of launch) to $t = 25$ s. See the previous computer projects. Use the results to estimate the range. For which of these launch angles is the range the greatest?

72. The velocity of a projectile subject to air resistance approaches a terminal velocity. Suppose the net force is given by $-mg\mathbf{j} - b\mathbf{v}$, where b is the drag coefficient and the y axis is chosen to be positive in the upward direction. At terminal velocity $\mathbf{v}_T$ the net force vanishes, so $\mathbf{v}_T = -(mg/b)\mathbf{j}$. Notice it has no horizontal component. The projectile eventually falls straight down.

You can use a computer program or spreadsheet to "watch" a projectile approach terminal velocity. Consider a 2.5-kg projectile launched with an initial speed of 150 m/s, at an angle of 40° above the horizontal. Take the drag coefficient to be $b = 0.50$ kg/s. Numerically integrate Newton's second law and display results for every 0.5 s from $t = 0$ (the time of launch) to the time the y component of the velocity is 90% of v_T. Plot $v_x(t)$ and $v_y(t)$ on the same graph. Notice that v_x approaches 0 as v_y approaches v_T.

73. When the effect of the air on a projectile is taken into account the coordinates are given by

$$x(t) = (v_{0x}/b)(1 - e^{-bt})$$

$$y(t) = (1/b^2)(g + bv_{0y})(1 - e^{-bt}) - (g/b)t,$$

where the positive y direction is chosen to be upward and the origin is at the launch point. The drag coefficient b tells the strength of the interaction between the air and the projectile. Differentiate the expressions for the coordinates to show that the velocity components are given by $v_x = v_{0x}e^{-bt}$ and $v_y = (1/b)(g + bv_{0y})e^{-bt} - g/b$ and that the acceleration components are given by $a_x = -bv_{0x}e^{-bt}$ and $a_y = -(g + bv_{0y})e^{-bt}$. Write a computer program or design a spreadsheet to compute the coordinates, velocity components, and acceleration components at the end of every time interval of duration Δt from time t_1 to time t_2.

Now use the program to investigate the influence of the air on a projectile fired over level ground with an initial speed of 50 m/s, at a firing angle of 25° above the horizontal. (*a*) Take $b = 0.10\ \text{s}^{-1}$ and use the program to find the coordinates of the highest point, the velocity, and acceleration when the projectile is there. Start by using the program to evaluate the coordinates, velocity, and acceleration at the end of every 0.1 s from $t = 0$ to $t = 4.5$ s. To obtain 2 significant figure accuracy you may want to narrow the interval on subsequent runs. Once you have obtained an answer notice that the highest point is reached in less time than when air resistance is absent, that the highest point is lower and closer to the firing point, and that the velocity is less. (*b*) To see if the trend continues, repeat the calculation with $b = 0.20$ s^{-1}. (*c*) How is the range of a projectile affected by the air?

Take $b = 0.10$ s^{-1} and use the program to find the range (the value of x when $y = 0$). Repeat with $b = 0.20$ s^{-1}. (*d*) How is the velocity just before landing affected by the air? Use the program with $b = 0.10$ s^{-1}, then 0.20 s^{-1}. Recall that in the absence of drag each velocity component has the same value as at launch. (*e*) Notice that the equations predict $a_x = -bv_x$ and $a_y = -g - bv_y$. Use these relationships to explain why $a_y = -g$ at the highest point, why a_x is not zero anywhere, and why a_y just before landing decreases in magnitude if b is increased.

CHAPTER 7

WORK
AND
ENERGY

*A fundamental problem of particle dynamics is to find how a particle
will move, given the forces that act on it. By "how a particle will move" we
mean how its position varies with time. In the previous two chapters we solved this problem
for the special case of a constant force, in which case the formulas for constant acceleration
can be used to find **r**(t), completing the solution of the problem.*

*The problem is more difficult, however, when the force acting on a particle and thus its
acceleration are not constant. We can solve such problems by integration methods, as
illustrated in Sections 6-5 and 6-7, respectively, for forces depending on time and velocity.
In this chapter, we extend the analysis to forces that depend on the position of the particle,
such as the gravitational force exerted by the Earth on any nearby object and the force
exerted by a stretched spring on a body to which it is attached. This analysis leads us to the
concepts of* work *and* kinetic energy *and to the development of the* work–energy theorem,
*which is the central feature of this chapter. In Chapter 8 we consider a broader view of
energy, embodied in the law of conservation of energy, a concept that has played a major
role in the development of physics.*

7-1 WORK DONE BY A CONSTANT FORCE

Consider a particle acted on by a constant force **F**, and assume the simplest case in which the motion takes place in a straight line in the direction of the force. In such a situation we define the *work W done by the force on the particle* as the product of the magnitude of the force *F* and the magnitude of the displacement *s* through which the force acts. We write this as

$$W = Fs. \qquad (1)$$

In a more general case, the constant force acting on a particle may not act in the direction in which the particle moves. In this case we define the work done by the force on the particle as the product of the component of the force along the line of motion and the magnitude of the displacement *s*. In Fig. 1, a particle experiences a constant force **F** that makes an angle ϕ with the direction of the displacement **s** of the particle. The work *W* done by **F** during this displacement is, according to our definition,

$$W = (F \cos \phi)s. \qquad (2)$$

Of course, other forces may also act on the particle. Equation 2 refers only to the work done on the particle by

one particular force **F**. The work done on the particle by the other forces must be calculated separately. To find the total work done on the particle, we add the values of the work done by all the separate forces. (Alternatively, as we discuss in Section 7-4, we can first find the net force on the particle and then calculate the work that would be done by a single force equal to the net force. The two methods of finding the work done on a particle are equivalent, and they always yield the same result for the work done on the particle.)

When ϕ is zero, the work done by **F** is simply *Fs*, in agreement with Eq. 1. Thus, when a horizontal force moves a body horizontally, or when a vertical force lifts a body vertically, the work done by the force is the product of the magnitude of the force and the distance moved.

Particle
(initial position)

s

(final position)

Figure 1 A force **F** acts on a particle as it undergoes a displacement **s**. The component of **F** that does work on the particle is $F \cos \phi$. The work done by the force **F** on the particle is $Fs \cos \phi$, which we can also write as **F·s**.

Figure 2   The weightlifter is exerting a great force on the weights, but at the instant shown he is doing no work because he is holding them in place. There is a force but no displacement. Of course, he probably has already done some work to have lifted them off the floor to that height.

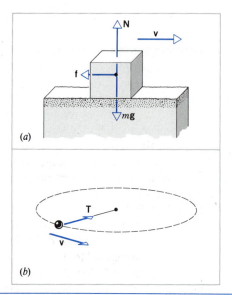

Figure 3   Work is not necessarily done by all the forces applied to a body, even if the body is in motion. In (*a*), the weight and the normal force do no work, because they are perpendicular to the displacement (which is in the direction of the velocity **v**). Work is done by the frictional force. In (*b*), which shows a body attached to a cord and revolving in a horizontal circle, the tension **T** in the cord does no work on the body, because it has no component in the direction of the displacement.

When ϕ is 90°, the force has no component in the direction of motion. That force then does no work on the body. For instance, a weightlifter (Fig. 2) does work in lifting the weights off the ground, but he does no work in holding them up (because there is no displacement). If he were to carry the weights above his head while walking, he would again (according to our definition of work) do no work on them, assuming there to be no vertical displacement, because the vertical force he exerts would be perpendicular to the horizontal displacement. Figure 3 shows other examples of forces applied to a body that do no work on the body.

Notice that we can write Eq. 2 either as $(F \cos \phi)s$ or $F(s \cos \phi)$. This suggests that the work can be calculated in two different ways, which give the same result: either we multiply the magnitude of the displacement by the component of the force in the direction of the displacement, or we multiply the magnitude of the force by the component of the displacement in the direction of the force. Each way reminds us of an important part of the definition of work: there must be a component of **s** in the direction of **F**, and there must be a component of **F** in the direction of **s** (Fig. 4).

Work is a *scalar,* although the two quantities involved in its definition, force and displacement, are vectors. In Section 3-5 we defined the *scalar product* of two vectors as the scalar quantity that we find when we multiply the magnitude of one vector by the component of a second vector along the direction of the first. Equation 2 shows that work is calculated in exactly this way, so work must be expressible as a scalar product. Comparing Eq. 2 with Eq. 13 of Chapter 3, we find that we can express work as

$$W = \mathbf{F} \cdot \mathbf{s}, \qquad (3)$$

where the dot indicates a scalar (or dot) product.

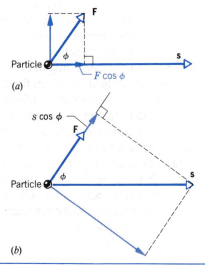

Figure 4   (*a*) The work *W* interpreted as $W = (s)(F \cos \phi)$. (*b*) The work *W* interpreted as $W = (F)(s \cos \phi)$.

Work can be either positive or negative. If a force has a component opposite to the direction of the motion, the work done by that force is negative. This corresponds to an obtuse angle between the force and displacement vectors. For example, when you lower an object to the floor, the work done on the object by the upward force of your hand holding the object is negative. In this case ϕ is 180°, for **F** points up and **s** points down. (The gravitational force in this case does positive work as the object moves down.)

Although the force **F** is an invariant, independent in both magnitude and direction of our choice of inertial frames, the displacement **s** is *not*. Depending on the inertial frame from which the measurement is made, an observer could measure essentially any magnitude and direction for the displacement **s**. Thus observers in different inertial frames, who will agree on the forces that act on a body, will disagree in their evaluation of the work done by the forces acting on the body. Different observers might find the work to be positive, negative, or even zero. We explore this point later in Section 7-6.

Work as we have defined it (Eq. 3) proves to be a very useful concept in physics. Our special definition of the word "work" does not correspond to the colloquial usage of the term. This may be confusing. A person holding a heavy weight at rest in the air may be working hard in the physiological sense, but from the point of view of physics that person is not doing any work on the weight. We say this because the applied force causes no displacement of the weight.

If, on the other hand, we consider the weightlifter to be a system of particles (which we treat in Chapter 9), we find that microscopically work is indeed being done. A muscle is not a solid support and cannot sustain a load in a static manner. The individual muscle fibers repeatedly relax and contract, and if we analyze the situation in this manner we would find that work is done in each contraction. That is why the weightlifter becomes tired in supporting the weight. In this chapter we do not consider this "internal" work. The word *work* is used only in the strict sense of Eq. 3, so that it does indeed vanish in the case of no displacement of the particle on which the force acts.

The unit of work is determined from the work done by a unit force in moving a body a unit distance in the direction of the force. The SI unit of work is 1 *newton-meter,* called 1 *joule* (abbreviation J). In the British system the unit of work is the foot-pound. In cgs systems the unit of work is 1 dyne-centimeter, called 1 erg. Using the relations between the newton, dyne, and pound, and between the meter, centimeter, and foot, we obtain 1 joule $= 10^7$ ergs $= 0.7376$ ft·lb.

A convenient unit of work when dealing with atomic or subatomic particles is the *electron-volt* (abbreviation eV), where 1 eV $= 1.60 \times 10^{-19}$ J. The work required to remove an outer electron from an atom has a typical magnitude of several eV. The work required to remove a proton or a neutron from a nucleus has a typical magnitude of several MeV (10^6 eV).

Sample Problem 1 A block of mass $m = 11.7$ kg is to be pushed a distance of $s = 4.65$ m along an incline so that it is raised a distance of $h = 2.86$ m in the process (Fig. 5a). Assuming frictionless surfaces, calculate how much work you would do

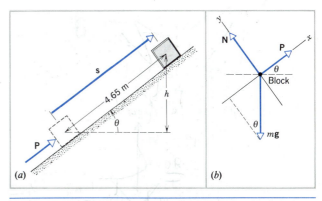

Figure 5 Sample Problem 1. (*a*) A force **P** moves a block up a plane through a displacement **s**. (*b*) A free-body diagram for the block.

if you applied a force parallel to the incline to push the block up at constant speed.

Solution A free-body diagram of the block is given in Fig. 5*b*. We must first find *P*, the magnitude of the force pushing the block up the incline. Because the motion is not accelerated (we are given that the speed is constant), the net force parallel to the plane must be zero. If we choose our *x* axis parallel to the plane, with its positive direction up the plane, we have, from Newton's second law,

$$x \text{ component:} \quad P - mg \sin \theta = 0,$$

or

$$P = mg \sin \theta = (11.7 \text{ kg})(9.80 \text{ m/s}^2)\left(\frac{2.86 \text{ m}}{4.65 \text{ m}}\right) = 70.5 \text{ N}.$$

Then the work done by **P**, from Eq. 3 with $\phi = 0°$, is

$$W = \mathbf{P} \cdot \mathbf{s} = Ps \cos 0° = Ps = (70.5 \text{ N})(4.65 \text{ m}) = 328 \text{ J}.$$

Note that the angle ϕ ($=0°$) used in this expression is the angle between the applied force and the displacement of the block, both of which are parallel to the incline. The angle ϕ must not be confused with the angle θ of the incline.

If you were to raise the block vertically at constant speed without using the incline, the work you do would be the vertical force, which is equal to *mg*, times the vertical distance *h*, or

$$W = mgh = (11.7 \text{ kg})(9.80 \text{ m/s}^2)(2.86 \text{ m}) = 328 \text{ J},$$

the same as before. The only difference is that the incline permits a smaller force ($P = 70.5$ N) to raise the block than would be required without the incline ($mg = 115$ N). On the other hand, you must push the block a greater distance (4.65 m) up the incline than you would if you raised it directly (2.86 m).

Sample Problem 2 A child pulls a 5.6-kg sled a distance of $s = 12$ m along a horizontal surface at a constant speed. What work does the child do on the sled if the coefficient of kinetic friction μ_k is 0.20 and the cord makes an angle of $\phi = 45°$ with the horizontal?

Solution The situation is shown in Fig. 6*a* and the forces acting on the sled are shown in the free-body diagram of Fig. 6*b*. **P** is the

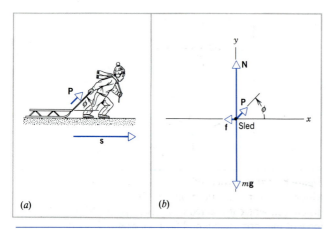

Figure 6 Sample Problem 2. (a) A child displaces a sled an amount **s** by pulling with a force **P** on a rope that makes an angle ϕ with the horizontal. (b) A free-body diagram for the sled.

child's pull, $m\mathbf{g}$ the sled's weight, $\mathbf{f}$ the frictional force, and **N** the normal force exerted by the surface on the sled. The work done by the child on the sled is

$$W = \mathbf{P} \cdot \mathbf{s} = Ps \cos \phi.$$

To evaluate this we first must determine P, whose value has not been given. To obtain P we refer to the free-body diagram of Fig. 6b.

The sled is unaccelerated, so that from the second law of motion we obtain the following:

$$x \text{ component:} \qquad P \cos \phi - f = 0,$$

$$y \text{ component:} \qquad P \sin \phi + N - mg = 0.$$

We know that f and N are related by

$$f = \mu_k N.$$

These three equations contain three unknown quantities: $P, f,$ and N. To find P we eliminate f and N from these equations and solve the remaining equation for P. You should verify that

$$P = \frac{\mu_k mg}{\cos \phi + \mu_k \sin \phi}.$$

With $\mu_k = 0.20$, $mg = (5.6 \text{ kg})(9.8 \text{ m/s}^2) = 55 \text{ N}$, and $\phi = 45°$ we obtain

$$P = \frac{(0.20)(55 \text{ N})}{\cos 45° + (0.20)(\sin 45°)} = 13 \text{ N}.$$

Then with $s = 12$ m, the work done by the child on the sled is

$$W = Ps \cos \phi = (13 \text{ N})(12 \text{ m})(\cos 45°) = 110 \text{ J}.$$

The vertical component of the pull **P** does no work on the sled. Note, however, that it reduces the normal force between the sled and the surface ($N = mg - P \sin \phi$) and thereby reduces the magnitude of the force of friction ($f = \mu_k N$).

Would the child do more work, less work, or the same amount of work on the sled if P were applied horizontally instead of at 45° from the horizontal? Do any of the other forces acting on the sled do work on it?

7-2 WORK DONE BY A VARIABLE FORCE: ONE-DIMENSIONAL CASE

We now consider the work done by a force that is not constant. Let the force act only in one direction, which we take to be the x direction, and let it vary in magnitude with x according to the function $F(x)$. Suppose a body that moves in the x direction is acted on by this force. What is the work done by this variable force when the body moves from an initial position x_i to a final position x_f?

In Fig. 7 we plot F versus x. Let us divide the total displacement into a number N of small intervals of equal width δx (Fig. 7a). Consider the first interval, in which there is a small displacement δx from x_i to $x_i + \delta x$. During this small displacement the force $F(x)$ has a nearly constant value F_1, and the small amount of work δW_1 it does in that interval is approximately

$$\delta W_1 = F_1 \, \delta x. \qquad (4)$$

Likewise, during the second interval, there is a small displacement from $x_i + \delta x$ to $x_i + 2\delta x$, and the force $F(x)$

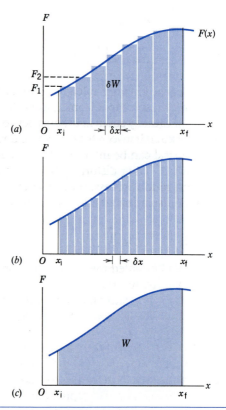

Figure 7 (a) The area under the curve of the variable one-dimensional force $F(x)$ is approximated by dividing the region between the limits x_i and x_f into a number of intervals of width δx. The sum of the areas of the rectangular strips is approximately equal to the area under the curve. (b) A better approximation is obtained using a larger number of narrower strips. (c) In the limit $\delta x \to 0$, the actual area is obtained.

has a nearly constant value F_2. The work done by the force in the second interval is approximately $\delta W_2 = F_2 \, \delta x$. The total work W done by $F(x)$ in displacing the body from x_i to x_f is approximately the sum of a large number of terms like that of Eq. 4, in which F has a different value for each term. Hence

$$W = \delta W_1 + \delta W_2 + \delta W_3 + \cdots$$
$$= F_1 \, \delta x + F_2 \, \delta x + F_3 \, \delta x + \cdots$$

or

$$W = \sum_{n=1}^{N} F_n \, \delta x, \qquad (5)$$

where the Greek letter sigma (Σ) stands for the sum over all N intervals from x_i to x_f.

To make a better approximation we can divide the total displacement from x_i to x_f into a larger number of intervals, as in Fig. 7b, so that δx is smaller and the value of F_n in each interval is more typical of the force within the interval. It is clear that we can obtain better and better approximations by taking δx smaller and smaller so as to have a larger and larger number of intervals. We can obtain an exact result for the work done by F if we let δx go to zero and the number of intervals N go to infinity. Hence the exact result is

$$W = \lim_{\delta x \to 0} \sum_{n=1}^{N} F_n \, \delta x. \qquad (6)$$

The relation

$$\lim_{\delta x \to 0} \sum F_n \, \delta x = \int_{x_i}^{x_f} F(x) \, dx,$$

as you may have learned in your calculus course, defines the integral of F with respect to x from x_i to x_f. Numerically, this quantity is exactly equal to the area between the force curve and the x axis between the limits x_i and x_f (Fig. 7c). Hence, an integral can be interpreted graphically as an area. The symbol $\int$ is a distorted S (for sum) and symbolizes the integration process. We can write the total work done by F in displacing a body from x_i to x_f as

$$W = \int_{x_i}^{x_f} F(x) \, dx. \qquad (7)$$

Because we have eliminated the vector notation from this one-dimensional equation, we must take care explicitly to put in the sign of F, positive if F is in the direction of increasing x and negative if F is in the direction of decreasing x.

As an example of a variable force, we consider a spring that acts on a particle of mass m (Fig. 8). The particle moves in the horizontal direction, which we take to be the x direction, with the origin ($x = 0$) representing the position of the particle when the spring is relaxed (Fig. 8a). An external force F_{ext} acts on the particle in a direction opposite to the spring force. We assume that the external force is always approximately equal to the spring force, so that the particle is nearly in equilibrium at all times ($a = 0$).

Let the particle be displaced a distance x from its original position at $x = 0$ (Fig. 8b). As the agent exerts a force

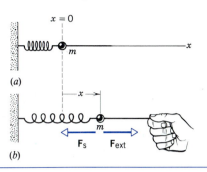

Figure 8 (a) A particle of mass m is attached to a spring, which is in its relaxed position. (b) The particle is displaced a distance x, where it is acted on by two forces, the restoring force of the spring and a pull from an external agent.

F_{ext} on the particle, the spring exerts an opposing force F_s. This force is given to a good approximation by

$$F_s = -kx, \qquad (8)$$

where k is a positive constant called the *force constant* of the spring. The constant k is a measure of the force necessary to produce a given stretching of the spring; stiffer springs have larger values of k. Equation 8 is the *force law* for springs and is known as *Hooke's law*. The minus sign in Eq. 8 reminds us that the direction of the force exerted by the spring is always opposite to the direction of the displacement of the particle. When the spring is stretched, $x > 0$ and F_s is negative; when the spring is compressed, $x < 0$ and F_s is positive. The force exerted by the spring is a *restoring force*: it always tends to *restore* the particle to its position at $x = 0$. Most real springs will obey Eq. 8 reasonably well if we do not stretch them beyond a limited range.

Let us first consider the work done *on* the particle *by* the spring when the particle moves from initial position x_i to final position x_f. We use Eq. 7 with the force F_s:

$$W_s = \int_{x_i}^{x_f} F_s(x) \, dx = \int_{x_i}^{x_f} (-kx) \, dx$$
$$= \tfrac{1}{2} k x_i^2 - \tfrac{1}{2} k x_f^2. \qquad (9)$$

The sign of the work done by the spring on the particle is positive if $x_i^2 > x_f^2$ (that is, if the magnitude of the initial displacement of the particle is greater than that of its final displacement). Note that the spring does *positive* work when it acts to restore the particle to its position at $x = 0$. If the magnitude of the initial displacement is smaller than that of the final displacement, the spring does *negative* work on the particle.

If we are interested in knowing the work done by the spring on the particle when the particle moves from its original position at $x = 0$ through a displacement x, we let $x_i = 0$ and $x_f = x$ and obtain

$$W_s = \int_0^x (-kx) \, dx = -\tfrac{1}{2} k x^2. \qquad (10)$$

Note that the work done by the spring in compression through a displacement x is the same as that done in extension through x, because the displacement x is squared in Eq. 10; either sign for x gives a positive value for x^2 and a negative value for W_s.

How much work does the *external agent* do when the particle moves from $x_i = 0$ to $x_f = x$? To keep the particle in equilibrium, the external force F_{ext} must be equal in magnitude to the spring force but opposite in sign, so $F_{ext} = +kx$. Repeating the calculation as in Eq. 10 for the work done by the external agent gives

$$W_{ext} = +\tfrac{1}{2}kx^2. \tag{11}$$

Note that this is exactly the negative of Eq. 10.

We can also find W_s and W_{ext} by computing the area between the appropriate force–displacement curve and the x axis from $x = 0$ to any arbitrary value x. In Fig. 9 the two straight sloping lines passing through the origin are plots of the external force against displacement ($F_{ext} = +kx$) and of the spring force against displacement ($F_s = -kx$). The right-hand side of the plot ($x > 0$) corresponds to stretching the spring and the left-hand side ($x < 0$) to compressing it.

In *stretching* the spring, the work done by the external force is positive and is represented by the upper triangle on the right of Fig. 9 labeled W_{ext}. The base of this triangle is $+x$ and its altitude is $+kx$; its area is therefore

$$\tfrac{1}{2}(+x)(+kx) = +\tfrac{1}{2}kx^2$$

in agreement with Eq. 11. When the spring is stretched, the work done by the spring force is negative and is represented by the lower triangle labeled W_s on the right side of Fig. 9; this triangle can be shown by a similar geometrical argument to have an area of $-\tfrac{1}{2}kx^2$, in agreement with Eq. 10.

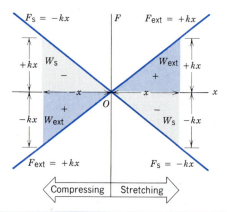

Figure 9   The work W_s done by the spring force is represented by the negative areas (shown with gray shading), and the work W_{ext} done by the external force, which is in equilibrium with the spring force, is represented by the positive areas (shown with colored shading). Whether the spring is stretched ($x > 0$) or compressed ($x < 0$), W_s is negative and W_{ext} is positive.

In *compressing* the spring, as the left side of Fig. 9 shows, the work W_{ext} done by the external agent is still positive, and the work W_s done by the spring is still negative, just as we expect from the signs of the forces and the displacement.

Sample Problem 3   A spring hangs vertically in equilibrium. A block of mass $m = 6.40$ kg is attached to the spring, but the block is held in place so that at first the spring does not stretch. Now the hand holding the block is slowly lowered, allowing the block to descend at constant speed until equilibrium is reached, at which point the hand is removed. A measurement shows that the spring has been stretched by a distance $s = 0.124$ m over its previous equilibrium length. Find the work done on the block in this process by (*a*) gravity, (*b*) the spring, and (*c*) the hand.

Solution   We are not given the force constant of the spring, but we can find it because we know that at the stretched position the block is in equilibrium between the upward spring force and the downward force of gravity:

$$\sum F = mg - ks = 0.$$

We have chosen the downward direction to be positive here. Solving for k, we find

$$k = mg/s = (6.40 \text{ kg})(9.80 \text{ m/s}^2)/(0.124 \text{ m}) = 506 \text{ N/m}.$$

To find the work done by gravity, W_g, we note that gravity is a constant force, and the force and the displacement are parallel, so we can use Eq. 1:

$$W_g = Fs = mgs = (6.40 \text{ kg})(9.80 \text{ m/s}^2)(0.124 \text{ m}) = +7.78 \text{ J}.$$

This is positive, because the force and displacement are in the same direction. To find the work W_s done by the spring, we use Eq. 10 with $x = s$:

$$W_s = -\tfrac{1}{2}ks^2 = -\tfrac{1}{2}(506 \text{ N/m})(0.124 \text{ m})^2 = -3.89 \text{ J}.$$

This is negative, because the force and displacement are in opposite directions.

One way to find the work W_h done by the hand is to find the force exerted by the hand as the block is lowered. If the block is in equilibrium during the entire process, then the upward force F_h exerted by the hand can be found from Newton's second law with $a = 0$:

$$\sum F = -kx - F_h + mg = 0,$$

or

$$F_h = mg - kx.$$

The work can be found from an integral of the form of Eq. 7, with a negative sign introduced to indicate that the force is opposite to the displacement:

$$W_h = -\int_0^s F_h \, dx = -\int_0^s (mg - kx) \, dx = -mgs + \tfrac{1}{2}ks^2$$

$$= -mgs + \frac{1}{2}\left(\frac{mg}{s}\right)s^2 = -\tfrac{1}{2}mgs = -3.89 \text{ J}.$$

A simpler way to obtain this result is to recognize that if the block (which we treat as a particle) is lowered slowly and uniformly, then the net force is zero, and the total work done by all the forces acting on the particle must therefore be zero:

$$W_{net} = W_s + W_g + W_h = 0,$$

$$W_h = -W_s - W_g = -(-3.89 \text{ J}) - 7.78 \text{ J} = -3.89 \text{ J}.$$

Note that the work done by the hand is equal to the work done by the spring.

7-3   WORK DONE BY A VARIABLE FORCE: TWO-DIMENSIONAL CASE *(Optional)*

The force **F** acting on a particle may vary in direction as well as in magnitude, and the particle may move along a curved path. To compute the work in this general case we divide the path into a large number of small displacements δs, each pointing tangent to the path in the direction of motion. Figure 10 shows two selected displacements for a particular situation; it also shows the force **F** and the angle ϕ between **F** and δs at each location. We can find the amount of work δW done on the particle during a displacement δs from

$$\delta W = \mathbf{F} \cdot \delta \mathbf{s} = F \cos \phi \, \delta s. \tag{12}$$

Here **F** is the force at the point where we take δs. The work done by the variable force **F** on the particle as the particle moves from i to f in Fig. 10 is found approximately by adding up (summing) the elements of work done over each of the line segments that make up the path from i to f. If the line segments δs become infinitesimally small, they may be replaced by differentials ds and the sum over the line segments may be replaced by an integral, as in Eq. 7. The work is then found from

$$W = \int_i^f \mathbf{F} \cdot d\mathbf{s} = \int_i^f F \cos \phi \, ds. \tag{13}$$

We cannot evaluate this integral until we are able to say how F and ϕ in Eq. 13 vary from point to point along the path; both are functions of the x and y coordinates of the particle in Fig. 10.

We can obtain an expression equivalent to Eq. 13 by writing **F**

and $d\mathbf{s}$ in terms of their components. Thus $\mathbf{F} = F_x \mathbf{i} + F_y \mathbf{j}$ and $d\mathbf{s} = dx \mathbf{i} + dy \mathbf{j}$, so that $\mathbf{F} \cdot d\mathbf{s} = F_x \, dx + F_y \, dy$. In this evaluation recall that $\mathbf{i} \cdot \mathbf{i} = \mathbf{j} \cdot \mathbf{j} = 1$ and $\mathbf{i} \cdot \mathbf{j} = \mathbf{j} \cdot \mathbf{i} = 0$ (see Eq. 14, Chapter 3). Substituting this result into Eq. 13, we obtain

$$W = \int_i^f (F_x \, dx + F_y \, dy). \tag{14}$$

Integrals such as those in Eq. 13 and 14 are called *line integrals;* to evaluate them we must know how $F \cos \phi$ or F_x and F_y vary as the particle moves along a particular line (or curve). The extension of Eq. 14 to three dimensions is straightforward.

Sample Problem 4   A small object of mass m is suspended from a string of length L. The object is pulled sideways by a force P that is always horizontal, until the string finally makes an angle ϕ_m with the vertical (Fig. 11a). The displacement is accomplished so slowly that we may regard the system as being in equilibrium during the process. Find the work done by all the forces that act on the object.

Solution   The motion is along an arc of radius L, and the displacement $d\mathbf{s}$ is always along the arc. At an intermediate point in the motion, the cord makes an angle ϕ with the vertical, and from the free-body diagram of Fig. 11b we see by applying Newton's second law that

$$x \text{ component:} \qquad P - T \sin \phi = 0,$$

$$y \text{ component:} \qquad T \cos \phi - mg = 0.$$

Combining these two equations to eliminate T, we find

$$P = mg \tan \phi.$$

Since P acts only in the x direction, we can use Eq. 14 with $F_x = P$ and $F_y = 0$ to find the work done by P. Thus

$$W_P = \int P \, dx = \int_0^{\phi_m} mg \tan \phi \, dx.$$

To carry out the integral over ϕ, we must have a single integration variable; we choose to define x in terms of ϕ. At an arbitrary intermediate position, when the horizontal coordinate is x, we

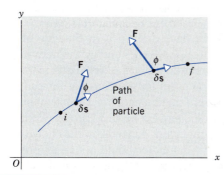

Figure 10   A particle moves from point i to point f along the path shown. During its motion it is acted on by a force **F** that varies in both magnitude and direction. As $\delta s \to 0$, we replace the interval by $d\mathbf{s}$, which is in the direction of the instantaneous velocity and therefore tangent to the path.

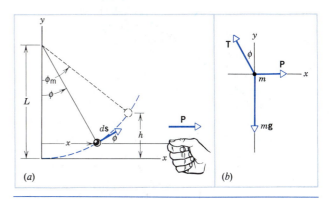

Figure 11   Sample Problem 4. (*a*) A particle is suspended from a string of length L and is pulled aside by a horizontal force P. The maximum angle reached is ϕ_m. (*b*) A free-body diagram for the particle.

see that $x = L \sin \phi$ and thus $dx = L \cos \phi \, d\phi$. Substituting for dx, we can now carry out the integration:

$$W_P = \int_0^{\phi_m} mg \tan \phi \, (L \cos \phi \, d\phi)$$

$$= mgL \int_0^{\phi_m} \sin \phi \, d\phi = mgL(-\cos \phi)\Big|_0^{\phi_m}$$

$$= mgL(1 - \cos \phi_m).$$

From Fig. 11a, we can see that $h = L(1 - \cos \phi_m)$, and thus

$$W_P = mgh.$$

The work W_g done by the (constant) gravitational force mg can be evaluated using a similar technique based on Eq. 14 (taking $F_x = 0$, $F_y = -mg$) to give $W_g = -mgh$ (see Problem 16). The minus sign enters because the direction of the vertical displacement is opposite to the direction of the gravitational force. The work W_T done by the tension in the string is zero, because **T** is perpendicular to the displacement ds at every point of the motion. Now you can see that the total work is zero: $W_{net} = W_P + W_g + W_T = mgh - mgh + 0 = 0$, consistent with the net force on the particle being zero at all times during its motion.

Note that in this problem the (positive) work done by the horizontal force **P** in effect cancels the (negative) work done by the vertical force mg. This can occur because work is a *scalar:* it has no direction or components. The motion of the particle depends on the *total* work done on it, which is the scalar sum of the values of the work associated with each of the individual forces. ∎

7-4 KINETIC ENERGY AND THE WORK–ENERGY THEOREM

In this section we consider the effect of work on the motion of a particle. An unbalanced force applied to a particle will certainly change the particle's state of motion. Newton's second law provides us with one way to analyze this change of motion. We now consider a different approach that ultimately gives the same result as Newton's laws but is often simpler to apply. It also leads us into one of the many important *conservation laws* that play such an important role in our interpretation of physical processes.

In this discussion, we consider not the work done on a particle by a single force, but the net work W_{net} done by all the forces that act on the particle. There are two ways to find the net work. The first is to find the net force, that is, the vector sum of all the forces that act on the particle,

$$\mathbf{F}_{net} = \mathbf{F}_1 + \mathbf{F}_2 + \mathbf{F}_3 + \cdots, \tag{15}$$

and then treat this net force as a single force in calculating the work according to Eq. 7 in one dimension or Eq. 13 in more than one dimension. In the second approach, we calculate the work done by each of the forces that act on the particle,

$$W_1 = \int \mathbf{F}_1 \cdot d\mathbf{s}, \qquad W_2 = \int \mathbf{F}_2 \cdot d\mathbf{s},$$

$$W_3 = \int \mathbf{F}_3 \cdot d\mathbf{s}, \cdots,$$

and then, since work is a scalar, we can add the work done by each of the individual forces to find the net work:

$$W_{net} = W_1 + W_2 + W_3 + \cdots. \tag{16}$$

The two methods give equal results, and the choice between them is merely a matter of convenience.

We know that a net unbalanced force applied to a particle will change its state of motion by accelerating it, let us say from initial velocity v_i to final velocity v_f. What is the effect of the work done on the particle by this net unbalanced force?

We first look at the answer to this question in the case of the constant force in one dimension. Under the influence of this force, the particle moves from x_i to x_f, and it accelerates uniformly from v_i to v_f. The work done is

$$W_{net} = F_{net}(x_f - x_i) = ma(x_f - x_i).$$

Because the acceleration a is constant, we can use Eq. 20 of Chapter 2, written $v_f^2 = v_i^2 + 2a(x_f - x_i)$, to obtain

$$W_{net} = \tfrac{1}{2}mv_f^2 - \tfrac{1}{2}mv_i^2. \tag{17}$$

That is, the result of the net work on the particle has been to bring about a change in the value of the quantity $\tfrac{1}{2}mv^2$ from point i to point f. This quantity is called the *kinetic energy K* of the particle, with the definition

$$K = \tfrac{1}{2}mv^2. \tag{18}$$

In terms of the kinetic energy K, we can rewrite Eq. 17 as

$$W_{net} = K_f - K_i = \Delta K. \tag{19}$$

Equation 19 is the mathematical representation of an important result called the *work–energy theorem*, which in words can be stated as follows:

The net work done by the forces acting on a particle is equal to the change in the kinetic energy of the particle.

Although we have derived it in the case of a constant resultant force, the work–energy theorem holds in general for nonconstant forces as well. Later in this section we give a general proof for nonconstant forces.

Like work, kinetic energy is a scalar quantity; unlike work, kinetic energy is never negative. We have already mentioned that work depends on the choice of reference frame, and it therefore should not be surprising that kinetic energy does also. Of course, we already know that observers in different inertial frames will differ in their measurements of velocity, and they will therefore differ in assigning kinetic energies to particles. Although the observers disagree on the numbers to be assigned to work and to kinetic energy, they nevertheless find the same relation to hold between these quantities, namely, $W_{net} = \Delta K$.

For Eq. 19 to be dimensionally consistent, kinetic energy must have the same units as work, namely, joules, ergs, foot-pounds, electron-volts, and so on.

When the magnitude of the velocity of a particle is constant, there is no change in kinetic energy, and therefore the resultant force does no work. In uniform circular motion, for example, the resultant force acts toward the center of the circle and is always at right angles to the direction of motion. Such a force does no work on the particle: it changes the direction of the velocity of the particle but not its magnitude. Only when the resultant force has a component in the direction of motion does it do work on the particle and change its kinetic energy.

The work–energy theorem does *not* represent a new, independent law of classical mechanics. We have simply *defined* work (Eq. 7, for instance) and kinetic energy (Eq. 18) and *derived* the relation between them from Newton's second law. The work–energy theorem is useful, however, for solving problems in which the net work done on a particle by external forces is easily computed and in which we are interested in finding the particle's speed at certain positions. Of even more significance is the work–energy theorem as a starting point for a broad generalization of the concept of energy and how energy can be stored or shared among the parts of a complex system. The principle of conservation of energy is the subject of the next chapter.

General Proof of the Work–Energy Theorem

The following calculation gives a proof of Eq. 19 in the case of nonconstant forces in one dimension. The equivalent calculation in two or three dimensions is left as an exercise (see Problem 34). We let F_{net} represent the net force acting on the particle. The net work done by all the external forces that act on the particle is just $W_{net} = \int F_{net}\, dx$. With a bit of mathematical manipulation we can accomplish a change of integration variable and put this in a more useful form:

$$F_{net} = ma = m\frac{dv}{dt} = m\frac{dv}{dx}\frac{dx}{dt} = m\frac{dv}{dx}\,v = mv\frac{dv}{dx}.$$

Thus

$$W_{net} = \int F_{net}\, dx = \int mv\frac{dv}{dx}\, dx = \int mv\, dv.$$

The variable of integration is now the velocity v. Let us integrate from initial velocity v_i to final velocity v_f:

$$W_{net} = \int_{v_i}^{v_f} mv\, dv = m\int_{v_i}^{v_f} v\, dv = \tfrac{1}{2}m(v_f^2 - v_i^2)$$
$$= \tfrac{1}{2}mv_f^2 - \tfrac{1}{2}mv_i^2.$$

This is identical with Eq. 19 and shows that the work–energy theorem holds even for nonconstant forces.

Sample Problem 5   One method of determining the kinetic energy of neutrons in a beam, such as from a nuclear reactor, is to measure how long it takes a particle in the beam to pass two fixed points a known distance apart. This technique is known as the *time-of-flight* method. Suppose a neutron travels a distance of $d = 6.2$ m in a time of $t = 160$ μs. What is its kinetic energy? The mass of a neutron is 1.67×10^{-27} kg.

Solution   We find the speed from

$$v = \frac{d}{t} = \frac{6.2 \text{ m}}{160 \times 10^{-6}\text{ s}} = 3.88 \times 10^4 \text{ m/s}.$$

From Eq. 18, the kinetic energy is

$$K = \tfrac{1}{2}mv^2 = \tfrac{1}{2}(1.67 \times 10^{-27}\text{ kg})(3.88 \times 10^4\text{ m/s})^2$$
$$= 1.26 \times 10^{-18}\text{ J} = 7.9\text{ eV}.$$

In nuclear reactors, neutrons are produced in nuclear fission with typical kinetic energies of a few MeV. Negative work has been done on the neutrons in this example by an external agent (called a moderator), thereby reducing their kinetic energies by a considerable factor from a few MeV to a few eV.

Sample Problem 6   A body of mass $m = 4.5$ g is dropped from rest at a height $h = 10.5$ m above the Earth's surface. What will its speed be just before it strikes the ground?

Solution   We assume that the body can be treated as a particle. We could solve this problem using a method based on Newton's laws, such as we considered in Chapter 5. We choose instead to solve it here using the work–energy theorem. The gain in kinetic energy is equal to the work done by the resultant force, which here is the force of gravity. This force is constant and directed along the line of motion, so that the work done by gravity is

$$W = \mathbf{F} \cdot \mathbf{s} = mgh.$$

Initially, the body has a speed $v_0 = 0$ and finally a speed v. The gain in kinetic energy of the body is

$$\Delta K = \tfrac{1}{2}mv^2 - \tfrac{1}{2}mv_0^2 = \tfrac{1}{2}mv^2 - 0.$$

According to the work–energy theorem, $W = \Delta K$ and so

$$mgh = \tfrac{1}{2}mv^2.$$

The speed of the body is then

$$v = \sqrt{2gh} = \sqrt{2(9.80 \text{ m/s}^2)(10.5 \text{ m})} = 14.3 \text{ m/s}.$$

Note that this result is independent of the mass of the object, as we have previously deduced using Newton's laws.

Sample Problem 7   A block of mass $m = 3.63$ kg slides on a horizontal frictionless table with a speed of $v = 1.22$ m/s. It is brought to rest in compressing a spring in its path. By how much is the spring compressed if its force constant k is 135 N/m?

Solution   The change in kinetic energy of the block is

$$\Delta K = K_f - K_i = 0 - \tfrac{1}{2}mv^2.$$

The work W done by the spring on the block when the spring is compressed from its relaxed length through a distance d is, according to Eq. 10,

$$W = -\tfrac{1}{2}kd^2.$$

Using the work–energy theorem, $W = \Delta K$, we obtain

$$-\tfrac{1}{2}kd^2 = -\tfrac{1}{2}mv^2$$

or

$$d = v\sqrt{\frac{m}{k}} = (1.22 \text{ m/s})\sqrt{\frac{3.63 \text{ kg}}{135 \text{ N/m}}} = 0.200 \text{ m}.$$

Limitation of the Work–Energy Theorem

We derived the work–energy theorem, Eq. 19, directly from Newton's second law, which, in the form in which we have stated it, applies *only to particles*. Hence the work–energy theorem, as we have presented it so far, likewise applies only to particles. We can apply this important theorem to real objects only if those objects behave like particles. Previously, we considered an object to behave like a particle if all parts of the object move in exactly the same way. In the use of the work–energy theorem, we can treat an extended object as a particle if the only kind of energy it has is directed kinetic energy.

Consider, for example, a test car that is crashed head-on into a heavy, rigid concrete barrier. The directed kinetic energy of the car certainly decreases as the car hits the barrier, crumples up, and comes to rest. However, there are forms of energy other than directed kinetic energy that enter into this situation. There is internal energy associated with the bending and crumpling of the body of the car; some of this internal energy may appear, for instance, as an increase in the temperature of the car, and some may be transferred to the surroundings as heat. Note that, even though the barrier may exert a large force on the car during the crash, the force does no work because *the point of application of the force on the car does not move*. (Recall our original definition of work—given by Eq. 1 and illustrated in Fig. 1—the force must act through some distance to do work.) Thus in this case $\Delta K \neq 0$, but $W = 0$; clearly, Eq. 19 does not hold. The car does *not* behave like a particle: every part of it does *not* move in exactly the same way.

For similar reasons, from the work–energy standpoint, we cannot treat a sliding block acted on by a frictional force as a particle (even though we *can* continue to treat it as a particle, as we did in Chapter 6, when analyzing its behavior using Newton's laws). The frictional force, which we represented as a constant force **f**, is in reality quite complicated, involving the making and breaking of many microscopic welds (see Section 6-2), which deform the surfaces and result in changes in internal energy of the surfaces (which may in part be revealed as an increase in the temperature of the surfaces). Because of the difficulty of accounting for these other forms of energy, and because the objects do not behave as particles, it is generally not correct to apply the particle form of the work–energy theorem to objects subject to frictional forces.

In these examples, we must view the crashing car and the sliding block not as particles but as systems containing large numbers of particles. Although it would be correct to apply the work–energy theorem to each individual particle in the system, it would be hopelessly complicated to do so. In Chapter 9, we begin to develop a simpler method for dealing with complex systems of particles, and we show how to extend the work–energy theorem so that we may apply it in such cases.

7-5   POWER

In designing a mechanical system, it is often necessary to consider not only how much work must be done but also how rapidly the work is to be done. The same amount of work is done in raising a given body through a given height whether it takes 1 second or 1 year to do so. However, the *rate at which work is done* is very different in the two cases.

We define *power* as the rate at which work is done. (Here we consider only *mechanical* power, which results from mechanical work. A more general view of power as energy delivered per unit time permits us to broaden the concept of power to include electrical power, solar power, and so on.) The average power $\overline{P}$ delivered by an agent that exerts a particular force on a body is the total work done by that force on the body divided by the total time interval, or

$$\overline{P} = \frac{W}{t}. \tag{20}$$

The instantaneous power P delivered by an agent is

$$P = \frac{dW}{dt}, \tag{21}$$

where dW is the small amount of work done in the infinitesimal time interval dt. If the power is constant in time, then $P = \overline{P}$ and

$$W = Pt. \tag{22}$$

The SI unit of power is the joule per second, which is called 1 *watt* (abbreviation W). This unit is named in honor of James Watt (1736–1819), who made major improvements to the steam engines of his day and pointed the way toward today's more efficient engines. In the British system, the unit of power is 1 ft·lb/s, although a more common practical unit, the *horsepower* (hp), is generally used to describe the power of such devices as electric motors or automobile engines. One horsepower is defined to be 550 ft·lb/s, which is equivalent to about 746 W.

Work can also be expressed in units of power × time. This is the origin of the term *kilowatt-hour,* which the electric company uses to measure how much work (in the form of electrical energy) it has delivered to your house. One kilowatt-hour is the work done in 1 hour by an agent working at a constant rate of 1 kW.

We can also express the power delivered to a body in terms of the velocity of the body and the force that acts on it. In general, we can rewrite Eq. 21 as

$$P = \frac{dW}{dt} = \frac{\mathbf{F} \cdot d\mathbf{s}}{dt} = \mathbf{F} \cdot \frac{d\mathbf{s}}{dt}$$

which becomes, after substituting the velocity **v** for $d\mathbf{s}/dt$,

$$P = \mathbf{F} \cdot \mathbf{v}. \tag{23}$$

If **F** and **v** are parallel to one another, this can be written

$$P = Fv. \tag{24}$$

Note that the power can be negative if **F** and **v** are antiparallel. Delivering negative power to a body means doing negative work on it: the force exerted on the body by the external agent is in a direction opposite to its displacement $d\mathbf{s}$ and therefore opposite to **v**.

Sample Problem 8 An elevator has an empty weight of 5160 N (1160 lb). It is designed to carry a maximum load of 20 passengers from the ground floor to the 25th floor of a building in a time of 18 seconds. Assuming the average weight of a passenger to be 710 N (160 lb) and the distance between floors to be 3.5 m (11 ft), what is the minimum constant power needed for the elevator motor? (Assume that all the work that lifts the elevator comes from the motor and that the elevator has no counterweight.)

Solution The minimum total force that must be exerted is the total weight of the elevator and passengers, $F = 5160$ N + $20(710)$ N = 19,400 N. The work that must be done is

$$W = Fs = (19,400 \text{ N})(25 \times 3.5 \text{ m}) = 1.7 \times 10^6 \text{ J}.$$

The minimum power is therefore

$$P = \frac{W}{t} = \frac{1.7 \times 10^6 \text{ J}}{18 \text{ s}} = 94 \text{ kW}.$$

This is the same as 126 hp, roughly the power delivered by the engine of an automobile. Of course, frictional losses and other inefficiencies will increase the power that the motor must provide to lift the elevator.

In practice, an elevator usually has a counterweight that falls as the elevator cab rises. The motor delivers positive power to the rising cab and negative power to the falling counterweight. Thus the *net* power that the motor must provide is greatly reduced.

7-6 REFERENCE FRAMES (Optional)

Newton's laws hold only in inertial reference frames (see Section 6-8), and if they hold in one particular inertial frame then they hold in all reference frames that move at constant velocity relative to that frame. Certain physical quantities, if observed in different inertial frames, always give the same measured result. In Newtonian mechanics, these *invariant* quantities include force, mass, acceleration, and time. Other quantities, such as displacement or velocity, are not invariant when measured from different inertial frames. For example, we discussed in Section 4-6 how to relate velocities measured from two reference frames in relative motion at constant velocity.

Two observers in different inertial frames will measure the same acceleration for a particle, and so they must deduce the same value for the change in its velocity, Δv, but they will in general *not* measure the same change in its kinetic energy. Observers in relative motion will also measure different values for the displacement of a particle, so that (although they measure the same values for the forces acting on the particle, force being an invariant) they will deduce different values for the work done on the particle. In this section we clarify these statements with a specific numerical example that demonstrates the validity of the work–energy theorem from the points of view of observers in different inertial frames.

Consider the following example. A worker on a flatbed railroad car is pushing a crate. The train is moving at a constant speed of 15.0 m/s. The crate has a mass of 12 kg, and in being pushed forward over a distance of 2.4 m its velocity is increased (relative to the car) at constant acceleration from rest to 1.5 m/s. Figure 12a shows the starting and finishing positions according

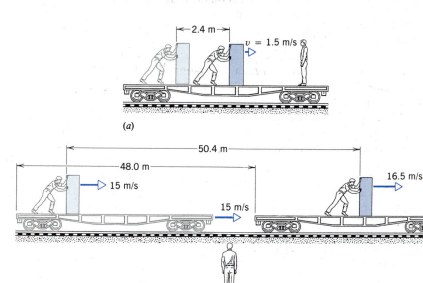

Figure 12 A worker on a flatbed railroad car pushing a crate forward, as viewed by (a) an observer on the train and (b) an observer on the ground.

to an observer who is riding on the train. This observer determines the change in kinetic energy to be

$$\Delta K = K_{\rm f} - K_{\rm i} = \tfrac{1}{2}(12 \text{ kg})(1.5 \text{ m/s})^2 - 0 = 13.5 \text{ J}.$$

The assumed constant acceleration of the crate can be found from Eq. 20 of Chapter 2, which gives

$$a = \frac{v_{\rm f}^2 - v_{\rm i}^2}{2(x_{\rm f} - x_{\rm i})} = \frac{(1.5 \text{ m/s})^2 - 0}{2(2.4 \text{ m})} = 0.469 \text{ m/s}^2.$$

This acceleration results from a correspondingly constant net force given by $F = ma = (12 \text{ kg})(0.469 \text{ m/s}^2) = 5.63 \text{ N}$. As the crate moves through a displacement Δx of 2.4 m, the work done on the crate by this force is

$$W = F\,\Delta x = (5.63 \text{ N})(2.4 \text{ m}) = 13.5 \text{ J}.$$

The observer on the train thus happily concludes that $W = \Delta K$ and that the work–energy theorem is satisfied.

How does an observer on the ground interpret a similar measurement? (We use primed coordinates to represent the measurements of the ground-based observer.) When the crate is at rest on the railroad car, it is moving forward at $v_{\rm i}' = 15.0$ m/s according to the ground-based observer. After the crate is pushed, the ground-based observer concludes its speed to be $v_{\rm f}' = 15.0$ m/s + 1.5 m/s = 16.5 m/s, and thus deduces the change in kinetic energy to be

$$\Delta K' = K_{\rm f}' - K_{\rm i}' = \tfrac{1}{2}mv_{\rm f}'^2 - \tfrac{1}{2}mv_{\rm i}'^2$$
$$= \tfrac{1}{2}(12 \text{ kg})(16.5 \text{ m/s})^2 - \tfrac{1}{2}(12 \text{ kg})(15.0 \text{ m/s})^2$$
$$= 284 \text{ J}.$$

This is very different from what the observer on the train measures for the change in kinetic energy ($\Delta K = 13.5$ J).

Before we begin to doubt the applicability of the work–energy theorem, let us go quickly to a calculation of the work done on the crate according to the ground-based observer. The total displacement of the crate also depends on the observer's reference frame, as Fig. 12b shows. To the ground-based observer, the force is exerted not over a distance of 2.4 m but over a larger distance of 50.4 m, because at a speed of 15.0 m/s the train travels 48.0 m in the 3.2 s ($= \Delta v/a$ or $\Delta v'/a'$, both acceleration and time being invariant in Newtonian mechanics) it takes to move the crate; the total displacement $\Delta x'$ of the crate in this time is 48.0 m + 2.4 m = 50.4 m. Force, on the other hand, is an invariant; for the ground-based observer, $F' = F = 5.63$ N. The ground-based observer concludes that the work is

$$W' = F'\,\Delta x' = (5.63 \text{ N})(50.4 \text{ m}) = 284 \text{ J}.$$

The work–energy theorem also holds for the ground-based observer! Even though the two observers use differing numerical values for displacements and velocities and disagree on the numerical values they assign to work and to kinetic energy, each nevertheless concludes that there is the same numerical equality between work and change in kinetic energy.

An *invariant* law of physics is one that has the same *form* in all inertial reference frames. A good example is the work–energy theorem, as we have seen. In the inertial reference frame of observer S, who measures work W and kinetic energy change ΔK for a particular process, the work–energy theorem is $W = \Delta K$. Observer S', who is in motion at constant velocity relative to S, measures work W' and change in kinetic energy $\Delta K'$ for the same process; while it will in general be true that $W \neq W'$ and

$\Delta K \neq \Delta K'$, observer S' determines that $W' = \Delta K'$. For yet another inertial observer S'', $W'' = \Delta K''$. To every observer in an inertial frame, the work–energy theorem has the same form. Invariance principles often give us a clue about the working of the natural world; they signal that a particular relationship is not an accident of one observer's preferred position but is instead an effect of some deep underlying symmetry of nature.

Sample Problem 9   Two identical airplanes, each of mass m, are flying together just above still water at a constant speed v measured relative to the water. Plane 1 lands on an aircraft carrier, which is at rest in the water. A hook on the plane engages a wire on the landing deck, and the wire exerts a springlike force that brings the plane to rest. An observer on the carrier measures a distance d between the point where the plane first engaged the wire and the point where it finally came to rest. Neglecting the other forces on the plane (friction, for example), discuss the validity of the work–energy theorem from the standpoint of (*a*) an observer on the carrier and (*b*) the pilot of the plane 2, which continues to fly at the original speed and in the original direction.

Solution   (*a*) The observer on the carrier measures a change in kinetic energy of

$$\Delta K = K_{\rm f} - K_{\rm i} = 0 - \tfrac{1}{2}mv^2 = -\tfrac{1}{2}mv^2.$$

It is reasonable that this observer, who sees the plane come to rest, measures a *negative* change in kinetic energy.

The external springlike force is exerted on the plane by the wire; if we assume an effective force constant k, this force does work on the plane that can be expressed as

$$W_s = -\tfrac{1}{2}kd^2.$$

This work is clearly negative for this observer, because the springlike force and the displacement are in opposite directions as the plane lands. If we neglect the other forces (including friction), we can certainly treat the plane as a particle, and for this observer the work–energy theorem would be valid. In particular, in this frame of reference, both W and ΔK are negative.

(*b*) According to the pilot of plane 2, the initial kinetic energy of plane 1 is zero: the planes are flying side-by-side, and there is no relative motion between them. When plane 1 finally comes to rest on the carrier, its velocity relative to plane 2 is $-v$, and its change in kinetic energy is therefore

$$\Delta K = K_{\rm f} - K_{\rm i} = \tfrac{1}{2}mv^2 - 0 = +\tfrac{1}{2}mv^2.$$

It may seem surprising that the pilot of plane 2 should note an *increase* in the kinetic energy of plane 1. However, in the reference frame of plane 2, plane 1 was initially at rest and finally in motion at speed v.

The wire does *positive* work in this reference frame. To the pilot of plane 2, the springlike force is in the direction opposite that which plane 2 is facing, and the displacement of plane 1 under the influence of that force is in the same direction as the force. In the reference frame of plane 2, the work–energy theorem applies, and both W and ΔK are positive.

We conclude from this example that both the sign of the work done by a given force *and* the sign of the change in kinetic energy of a given particle can depend on the reference frame of the observer. Despite this difference in interpretation, however, both observers will agree on the validity of the work–energy theorem. (See also Question 22, which considers the interpretation according to the pilot of plane 1.) ∎

7-7   KINETIC ENERGY AT HIGH SPEED* (Optional)

In the previous section, we used the Galilean–Newtonian formula for transforming velocities from one reference frame to another: $v = v' + u$. We derived this formula in Section 4-6 and discussed its failure at high speeds, where we must use the correct relationship from special relativity, Eq. 46 of Chapter 4. If you are now suspecting that the $\frac{1}{2}mv^2$ formula for kinetic energy also fails at high speed, you would be correct.

The general formula for kinetic energy, applicable at any speed, is

$$K = mc^2 \left[\frac{1}{\sqrt{1 - v^2/c^2}} - 1 \right]. \quad (25)$$

(We derive this result in Chapter 21.) Does this mean that $\frac{1}{2}mv^2$ is wrong? It certainly is at high speed, but it is not too difficult to show that Eq. 25 does indeed reduce to $\frac{1}{2}mv^2$ at low speed. For this we need the binomial expansion of expressions of the form $(1 + x)^p$:

$$(1 + x)^p = 1 + px + \frac{p(p-1)}{2!} x^2 + \frac{p(p-1)(p-2)}{3!} x^3 + \cdots,$$

where $n!$ (read as "n factorial") means the product of all the integers from 1 to n. Thus $3! = 1 \times 2 \times 3 = 6$.

The binomial expansion is a useful result, but it is of particular value when x is small compared with 1. For example, suppose that x is about 0.01. Then the second term of the expansion, px, is (if p is not too large) much smaller than the first term, the third term is even smaller than the second, and so on. As the terms get smaller and smaller, we may decide that it is important for a certain calculation to keep only a few terms and neglect the rest.

In Eq. 25 for the kinetic energy, the brackets contain the factor $(1 - v^2/c^2)^{-1/2}$. This factor can be expanded with the binomial formula, with $x = -v^2/c^2$ and $p = -\frac{1}{2}$. Let's try keeping three terms in the expansion:

$$\left(1 - \frac{v^2}{c^2}\right)^{-1/2} \approx 1 + \left(-\frac{1}{2}\right)\left(-\frac{v^2}{c^2}\right) + \frac{(-\frac{1}{2})(-\frac{3}{2})}{2}\left(-\frac{v^2}{c^2}\right)^2$$

$$= 1 + \frac{1}{2}\frac{v^2}{c^2} + \frac{3}{8}\frac{v^4}{c^4}. \quad (26)$$

Now we substitute Eq. 26 back into Eq. 25:

$$K \approx mc^2 \left[1 + \frac{1}{2}\frac{v^2}{c^2} + \frac{3}{8}\frac{v^4}{c^4} - 1 \right] = \frac{1}{2}mv^2 \left[1 + \frac{3}{4}\frac{v^2}{c^2} \right]. \quad (27)$$

You can see that the fractional error we make by using $\frac{1}{2}mv^2$ is about $\frac{3}{4}(v^2/c^2)$. Even at a speed of 1% of the speed of light, this error is less than 1 part in 10^4. At our ordinary laboratory speeds, which are seldom beyond 10^{-6} of the speed of light, the error in using $\frac{1}{2}mv^2$ is far smaller than the precision of our ability to measure energies, and $\frac{1}{2}mv^2$ is an excellent approximation.

---

* This section can be skipped or delayed until relativity is discussed in Chapter 21.

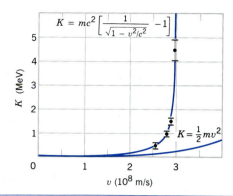

Figure 13   A comparison of the classical and relativistic formulas for the kinetic energy of electrons. At low speeds, the two formulas give identical results, but at speeds near the speed of light the data clearly show that the relativistic formula is correct.

If Eq. 25 is always correct, at high as well as low speeds, why not use it always and forget about $\frac{1}{2}mv^2$? Here we run up against a practical problem. Try using Eq. 25 when $v = 300$ m/s, quite a respectable speed by most standards (roughly the speed of sound in air), but far below the speed of light ($v/c = 10^{-6}$ and $v^2/c^2 = 10^{-12}$). Using your pocket calculator to evaluate the term in brackets in Eq. 25, you will probably find a result of zero. The reason is that your calculator uses only 8 or 9 digits and therefore will obtain "exactly" 1 when it tries to evaluate $1 - 10^{-12}$. In practice, we use $\frac{1}{2}mv^2$ because it is accurate enough and computationally easier when v is below about 1% of the speed of light, and we reserve Eq. 25 for higher speeds.

Figure 13 shows the results of an experimental test of Eq. 25. This test was made by accelerating electrons to a known kinetic energy and then measuring their speed by timing them over a known distance. Obviously the data do favor the result from relativity theory at high speed. Notice also how the two curves are indistinguishable at low speed.

Sample Problem 10   The Tevatron accelerator at the Fermi National Accelerator Laboratory accelerates protons to a kinetic energy of about 1 TeV (= 10^{12} eV, where 1 eV is 1.6×10^{-19} J). What is the speed of a 1-TeV proton? The mass of a proton is 1.67×10^{-27} kg.

Solution   In SI units, the kinetic energy of a 1-TeV proton is

$$K = 1 \text{ TeV} = 1.6 \times 10^{-7} \text{ J}.$$

Thus, using Eq. 25,

$$1.6 \times 10^{-7} \text{ J}$$

$$= (1.67 \times 10^{-27} \text{ kg})(3.00 \times 10^8 \text{ m/s})^2 \left[\frac{1}{\sqrt{1 - v^2/c^2}} - 1 \right].$$

Solving, we find

$$v/c = 0.99999956,$$

so that v, while less than c, is exceedingly close to it, differing from c by only 132 m/s. ∎

QUESTIONS

1. Can you think of other words like work whose colloquial meanings are often different from their scientific meanings?

2. Explain why you become physically tired when you push against a wall, fail to move it, and therefore do no work on the wall.

3. Suppose that three constant forces act on a particle as it moves from one position to another. Prove that the work done on the particle by the resultant of these three forces is equal to the sum of the work done by each of the three forces calculated separately.

4. The inclined plane (Sample Problem 1) is a simple "machine" that enables us to do work with the application of a smaller force than is otherwise necessary. The same statement applies to a wedge, a lever, a screw, a gear wheel, and a pulley combination (Problem 9). But far from saving us work, such machines in practice require that we do a little more work with them than without them. Why is this so? Why do we use such machines?

5. In a tug of war, one team is slowly giving way to the other. What work is being done and by whom?

6. Why can you much more easily ride a bicycle for a mile on level ground than run the same distance? In each case, you transport your own weight for a mile and in the first you must also transport the bicycle and, moreover, do so in a shorter time! (See *The Physics Teacher,* March 1981, p. 194.)

7. Suppose that the Earth revolves around the Sun in a perfectly circular orbit. Does the Sun do any work on the Earth?

8. You slowly lift a bowling ball from the floor and put it on a table. Two forces act on the ball: its weight, $m\mathbf{g}$, and your upward force, $-m\mathbf{g}$. These two forces cancel each other so that it would seem that no work is done. On the other hand, you know that you have done some work. What is wrong?

9. You cut a spring in half. What is the relation of the force constant k for the original spring to that for either of the half-springs?

10. Springs A and B are identical except that A is stiffer than B; that is, $k_A > k_B$. On which spring is more work expended if they are stretched (a) by the same amount and (b) by the same force?

11. Does kinetic energy depend on the direction of the motion involved? Can it be negative? Does its value depend on the reference frame of the observer?

12. In picking up a book from the floor and putting it on a table, you do work. However, the kinetic energy of the book does not change. Is there a violation of the work–energy theorem here? Explain why or why not.

13. Does the work–energy theorem hold if friction acts on an object? Explain your answer.

14. The work done by the net force on a particle is equal to the change in kinetic energy. Can it happen that the work done by one of the component forces alone will be greater than the change in kinetic energy? If so, give examples.

15. Why can a car so easily pass a loaded truck when going uphill? The truck is heavier, of course, but its engine is more powerful in proportion (or is it?). What considerations enter into choosing the design power of a truck engine and of a car engine?

16. Does the power needed to raise a box onto a platform depend on how fast it is raised?

17. You lift some library books from a lower to a higher shelf in time Δt. Does the work that you do depend on (a) the mass of the books, (b) the weight of the books, (c) the height of the upper shelf above the floor, (d) the time Δt, and (e) whether you lift the books sideways or directly upward?

18. The world record for the pole vault is about 5.5 m. Could the record be raised to, say, 8 m by using a pole long enough? If not, why not? How high might an athlete get?

19. We hear a lot about the "energy crisis." Would it be more accurate to speak of a "power crisis"?

20. Does the work done by the net force acting on a particle depend on the (inertial) reference frame of the observer? Does the change in kinetic energy so depend? If so, give examples.

21. A man rowing a boat upstream is at rest with respect to the shore. (a) Is he doing any work? (b) If he stops rowing and moves down with the stream, is any work being done on him?

22. Consider the work–energy theorem from the frame of reference of the pilot of plane 1 in Sample Problem 9. Does the theorem fail in this case? Explain.

23. We say that a 1-keV electron is a "classical" particle, a 1-MeV electron is a "relativistic" particle, and a 1-GeV electron is an "extremely relativistic" particle. What exactly do these terms mean?

PROBLEMS

Section 7-1 Work Done by a Constant Force

1. To push a 52-kg crate across a floor, a worker applies a force of 190 N, directed 22° below the horizontal. As the crate moves 3.3 m, how much work is done on the crate by (a) the worker, (b) the force of gravity, and (c) the normal force of the floor on the crate?

2. A 106-kg object is initially moving in a straight line with a speed of 51.3 m/s. (a) If it is brought to a stop with a deceleration of 1.97 m/s², what force is required, what distance does the object travel, and how much work is done by the force? (b) Answer the same questions if the object's deceleration is 4.82 m/s².

3. To push a 25-kg crate up a 27° incline, a worker exerts a force of 120 N, parallel to the incline. As the crate slides 3.6 m, how much work is done on the crate by (a) the worker, (b) the force of gravity, and (c) the normal force of the incline?

4. Electric fields can be used to pull electrons out of metals. To remove an electron from tungsten, the electric field must do 4.5 eV of work. Suppose that the distance over which the electric field acts is 3.4 nm. Calculate the minimum force that the field must exert on the electron being removed.

5. A cord is used to lower vertically a block of mass M a distance d at a constant downward acceleration of $g/4$. (*a*) Find the work done by the cord on the block. (*b*) Find the work done by the force of gravity.

6. A worker pushed a 58.7-lb block ($m = 26.6$ kg) a distance of 31.3 ft ($= 9.54$ m) along a level floor at constant speed with a force directed 32.0° below the horizontal. The coefficient of kinetic friction is 0.21. How much work did the worker do on the block?

7. A 52.3-kg trunk is pushed 5.95 m at constant speed up a 28.0° incline by a constant horizontal force. The coefficient of kinetic friction between the trunk and the incline is 0.19. Calculate the work done by (*a*) the applied force and (*b*) the force of gravity.

8. A 47.2-kg block of ice slides down an incline 1.62 m long and 0.902 m high. A worker pushes up on the ice parallel to the incline so that it slides down at constant speed. The coefficient of kinetic friction between the ice and the incline is 0.110. Find (*a*) the force exerted by the worker, (*b*) the work done by the worker on the block of ice, and (*c*) the work done by gravity on the ice.

9. Figure 14 shows an arrangement of pulleys designed to facilitate the lifting of a heavy load L. Assume that friction can be ignored everywhere and that the pulleys to which the load is attached weigh a total of 20.0 lb. An 840-lb load is to be raised 12.0 ft. (*a*) What is the minimum applied force F that can lift the load? (*b*) How much work must be done against gravity in lifting the 840-lb load 12.0 ft? (*c*) Through what distance must the applied force be exerted to lift the load 12.0 ft? (*d*) How much work must be done by the applied force F to accomplish this task?

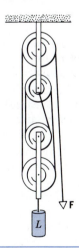

Figure 14   Problem 9.

Section 7-2 Work Done by a Variable Force:
One-Dimensional Case

10. A 5.0-kg block moves in a straight line on a horizontal frictionless surface under the influence of a force that varies with position as shown in Fig. 15. How much work is done

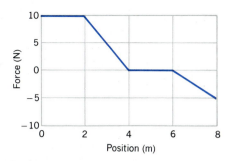

Figure 15   Problem 10.

by the force as the block moves from the origin to $x = 8.0$ m?

11. A 10-kg object moves along the x axis. Its acceleration as a function of its position is shown in Fig. 16. What is the net work performed on the object as it moves from $x = 0$ to $x = 8.0$ m?

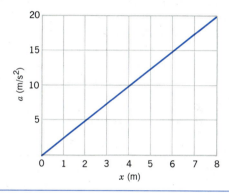

Figure 16   Problem 11.

12. A spring has a force constant of 15.0 N/cm. (*a*) How much work is required to extend the spring 7.60 mm from its relaxed position? (*b*) How much work is needed to extend the spring an additional 7.60 mm?

13. The force exerted on an object is $F = F_0(x/x_0 - 1)$. Find the work done in moving the object from $x = 0$ to $x = 3x_0$ (*a*) by plotting $F(x)$ and finding the area under the curve, and (*b*) by evaluating the integral analytically.

14. (*a*) Estimate the work done by the force shown on the graph (Fig. 17) in displacing a particle from $x = 1$ m to $x = 3$ m. Refine your method to see how close you can come to the exact answer of 6 J. (*b*) The curve is given analytically by $F = A/x^2$, where $A = 9$ N·m². Show how to calculate the work by the rules of integration.

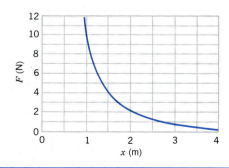

Figure 17   Problem 14.

15. Figure 18 shows a spring with a pointer attached, hanging next to a scale graduated in millimeters. Three different weights are hung from the spring, in turn, as shown. (*a*) If all weight is removed from the spring, which mark on the scale will the pointer indicate? (*b*) Find the weight *W*.

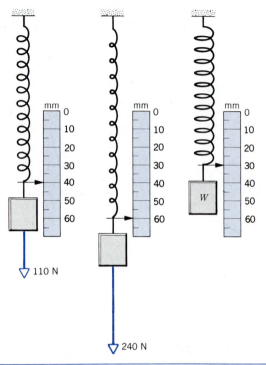

Figure 18   Problem 15.

Section 7-3 Work Done by a Variable Force: Two-Dimensional Case

16. By integrating along the arc, show that the work done by gravity in Sample Problem 4 is equal to $-mgh$.

17. An object of mass 0.675 kg on a frictionless table is attached to a string that passes through a hole in the table at the center of the horizontal circle in which the object moves with constant speed. If the radius of the circle is 0.500 m and the speed is 10.0 m/s, compute the tension in the string. (*b*) It is found that drawing an additional 0.200 m of the string down through the hole, thereby reducing the radius of the circle to 0.300 m, has the effect of multiplying the original tension in the string by 4.63. Compute the total work done by the string on the revolving object during the reduction of the radius.

Section 7-4 Kinetic Energy and the Work–Energy Theorem

18. Calculate the kinetic energies of the following objects moving at the given speeds: (*a*) a 110-kg football linebacker running at 8.1 m/s; (*b*) a 4.2-g bullet at 950 m/s; (*c*) the aircraft carrier *Nimitz*, 91,400 tons at 32.0 knots.

19. A conduction electron in copper near the absolute zero of temperature has a kinetic energy of 4.2 eV. What is the speed of the electron?

20. A proton (nucleus of the hydrogen atom) is being accelerated in a linear accelerator. In each stage of such an accelera-

tor the proton is accelerated along a straight line at 3.60×10^{15} m/s². If a proton enters such a stage moving initially with a speed of 2.40×10^7 m/s and the stage is 3.50 cm long, compute (*a*) its speed at the end of the stage and (*b*) the gain in kinetic energy resulting from the acceleration. The mass of the proton is 1.67×10^{-27} kg. Express the energy in electron-volts.

21. A single force acts on a particle in rectilinear motion. A plot of velocity versus time for the particle is shown in Fig. 19. Find the sign (positive or negative) of the work done *by* the force *on* the particle in each of the intervals *AB*, *BC*, *CD*, and *DE*.

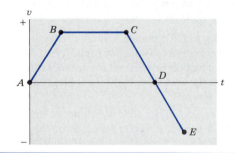

Figure 19   Problem 21.

22. To travel to the Moon, a 2.9×10^5-kg Saturn V rocket with an Apollo spacecraft attached must achieve an escape velocity of 11.2 km/s (=25,000 mi/h) near the surface of the Earth. How much energy must the fuel contain? Would the system actually need as much or would it need more or less? Why?

23. From what height would a 2800-lb automobile have to fall to gain the kinetic energy equivalent to what it would have when going 55 mi/h? Does the answer depend on the weight of the car?

24. An 1100-kg car is traveling at 46 km/h on a level road. The brakes are applied long enough to remove 51 kJ of kinetic energy. (*a*) What is the final speed of the car? (*b*) How much more kinetic energy must be removed by the brakes to stop the car?

25. An outfielder throws a baseball with an initial speed of 120 ft/s (=36.6 m/s). Just before an infielder catches the ball at the same level, its speed is reduced to 110 ft/s (=33.5 m/s). How much energy has been lost due to air drag? The weight of a baseball is 9.0 oz (*m* = 255 g).

26. The Earth circles the Sun once a year. How much work would have to be done on the Earth to bring it to rest relative to the Sun? See Appendix C for numerical data and ignore the rotation of the Earth about its own axis.

27. A running man has half the kinetic energy that a boy of half his mass has. The man speeds up by 1.00 m/s and then has the same kinetic energy as the boy. What were the original speeds of man and boy?

28. A 0.550-kg projectile is launched from the edge of a cliff with an initial kinetic energy of 1550 J and at its highest point is 140 m above the launch point. (*a*) What is the horizontal component of its velocity? (*b*) What was the vertical component of its velocity just after launch? (*c*) At one instant during its flight the vertical component of its velocity is found to

be 65.0 m/s. At that time, how far is it above or below the launch point?

29. A comet having a mass of 8.38×10^{11} kg strikes the Earth at a relative speed of 30 km/s. (*a*) Compute the kinetic energy of the comet in "megatons of TNT"; the detonation of 1 million tons of TNT releases 4.2×10^{15} J of energy. (*b*) The diameter of the crater blasted by a large explosion is proportional to the one-third power of the explosive energy released, with 1 megaton of TNT producing a crater about 1 km in diameter. What is the diameter of the crater produced by the impact of the comet? (In the past, atmospheric effects produced by impacts of comets may have been the cause of mass extinctions of many species of animals and plants; it is thought by many that dinosaurs became extinct by this mechanism.)

30. A 125-g frisbee is thrown from a height of 1.06 m above the ground with a speed of 12.3 m/s. When it has reached a height of 2.32 m, its speed is 9.57 m/s. (*a*) How much work was done on the frisbee by gravity? (*b*) How much kinetic energy was lost due to air drag? Ignore the spin of the frisbee.

31. A ball loses 15.0% of its kinetic energy when it bounces back from a concrete walk. With what speed must you throw it vertically down from a height of 12.4 m to have it bounce back to that same height? Ignore air resistance.

32. A rubber ball dropped from a height of exactly 6 ft bounces (hits the floor) several times, losing 10% of its kinetic energy each bounce. After how many bounces will the ball subsequently not rise above 3 ft?

33. A 263-g block is dropped onto a vertical spring with force constant $k = 2.52$ N/cm (Fig. 20). The block sticks to the spring, and the spring compresses 11.8 cm before coming momentarily to rest. While the spring is being compressed, how much work is done (*a*) by the force of gravity and (*b*) by the spring? (*c*) What was the speed of the block just before it hit the spring? (*d*) If this initial speed of the block is doubled, what is the maximum compression of the spring? Ignore friction.

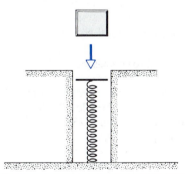

Figure 20 Problem 33.

34. Show that Eq. 19 holds true for the case of motion in two or three dimensions by extending the proof for one-dimensional motion.

Section 7-5 Power

35. A 57-kg woman runs up a flight of stairs having a rise of 4.5 m in 3.5 s. What average power must she supply?

36. In a 100-person ski lift, a machine raises passengers averaging 667 N in weight a height of 152 m in 55.0 s, at constant speed. Find the power output of the motor, assuming no frictional losses.

37. A swimmer moves through the water at a speed of 0.22 m/s. The drag force opposing this motion is 110 N. How much power is developed by the swimmer?

38. Starting a race, a 68.2-kg sprinter runs the first 7.04 m in 1.60 s, starting from rest and accelerating uniformly. (*a*) What is the sprinter's speed at the end of the 1.60 s? (*b*) What is the sprinter's kinetic energy? (*c*) What average power does the sprinter generate during the 1.60-s interval?

39. A horse pulls a cart with a force of 42.0 lb at an angle of 27.0° with the horizontal and moves along at a speed of 6.20 mi/h. (*a*) How much work does the horse do in 12.0 min? (*b*) Find the power output of the horse, in hp of course.

40. An auto manufacturer reports that the maximum power delivered by the engine of a car of mass 1230 kg is 92.4 kW. Find the minimum time in which the car could accelerate from rest to 29.1 m/s (= 65 mi/h). In a test the time to do this is found to be 12.3 s. Account for the difference in these times.

41. The hydrogen-filled airship *Hindenburg* (see Fig. 21) could cruise at 77 knots with the engines providing 4800 hp. Calculate the air drag force in newtons on the airship at this speed.

Figure 21 Problem 41.

42. The luxury liner *Queen Elizabeth 2* (see Fig. 22) is powered by a new diesel-electric powerplant, which replaced the original steam engines. The maximum power output is 92 MW

Figure 22 Problem 42.

at a cruising speed of 32.5 knots. What force is exerted by the propellers on the water at this highest attainable speed?

43. How much power, in horsepower, must be developed by the engine of a 1600-kg car moving at 26 m/s (= 94 km/h) on a level road if the forces of resistance total 720 N?

44. Each minute, 73,800 m³ of water passes over a waterfall 96.3 m high. Assuming that 58.0% of the kinetic energy gained by the water in falling is converted to electrical energy by a hydroelectric generator, calculate the power output of the generator. (The density of water is 1000 kg/m³.)

45. Suppose that your car averages 30 mi/gal of gasoline. (a) How far could you travel on 1 kW·h of energy consumed? (b) If you are driving at 55 mi/h, at what rate are you expending energy? The heat of combustion of gasoline is 140 MJ/gal.

46. The motor on a water pump is rated at 6.6 hp. From how far down a well can water be pumped up at the rate of 220 gal/min?

47. A 1380-kg block of granite is dragged up an incline at a constant speed of 1.34 m/s by a steam winch (Fig. 23). The coefficient of kinetic friction between the block and the incline is 0.41. How much power must be supplied by the winch?

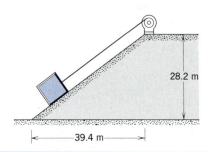

28.2 m

39.4 m

Figure 23   Problem 47.

48. A 3700-lb automobile (m = 1680 kg) starts from rest on a level road and gains a speed of 45 mi/h (= 72 km/h) in 33 s. (a) What is the kinetic energy of the auto at the end of the 33 s? (b) What is the average net power delivered to the car during the 33-s interval? (c) What is the instantaneous power at the end of the 33-s interval assuming that the acceleration was constant?

49. An object of mass m accelerates uniformly from rest to a speed v_f in time t_f. (a) Show that the work done on the object as a function of time t, in terms of v_f and t_f, is

$$W = \tfrac{1}{2}m \frac{v_f^2}{t_f^2} t^2.$$

(b) As a function of time t, what is the instantaneous power delivered to the object?

50. A force acts on a 2.80-kg particle in such a way that the position of the particle as a function of time is given by $x = 3t - 4t^2 + t^3$, where x is in meters and t is in seconds. (a) Find the work done by the force during the first 4.0 s. (b) At what instantaneous rate is the force doing work on the particle at the instant t = 3.0 s?

51. A fully loaded freight elevator has a total mass of 1220 kg. It is required to travel downward 54.5 m in 43.0 s. The counterweight has a mass of 1380 kg. Find the power output, in hp, of the elevator motor. Ignore the work required to start and stop the elevator; that is, assume that it travels at constant speed.

52. Show that the speed v reached by a car of mass m that is driven with constant power P is given by

$$v = \left(\frac{3xP}{m}\right)^{1/3},$$

where x is the distance traveled from rest.

53. (a) Show that the power output of an airplane cruising at constant speed v in level flight is proportional to v^3. Assume that the aerodynamic drag force is given by $D = bv^2$. (b) By what factor must the engines' power be increased to increase the air speed by 25.0%?

54. What power is developed by a grinding machine whose wheel has a radius of 20.7 cm and runs at 2.53 rev/s when the tool to be sharpened is held against the wheel with a force of 180 N? The coefficient of friction between the tool and the wheel is 0.32.

55. An escalator joins one floor with another one 8.20 m above. The escalator is 13.3 m long and moves along its length at 62.0 cm/s. (a) What power must its motor deliver if it is required to carry a maximum of 100 persons per minute, of average mass 75.0 kg? (b) An 83.5-kg man walks up the escalator in 9.50 s. How much work does the motor do on him? (c) If this man turned around at the middle and walked down the escalator so as to stay at the same level in space, would the motor do work on him? If so, what power does it deliver for this purpose? (d) Is there any (other?) way the man could walk along the escalator without consuming power from the motor?

56. At full power, a 1.5-MW railroad locomotive accelerates a train from a speed of 10 to 25 m/s in 6.0 min. (a) Neglecting friction, calculate the mass of the train. (b) Find the speed of the train as a function of time in seconds during the interval. (c) Find the force accelerating the train as a function of time during the interval. (d) Find the distance moved by the train during the interval.

57. The resistance to motion of an automobile depends on road friction, which is almost independent of its speed v, and on aerodynamic drag, which is proportional to v^2. For a particular 12,000-N car, the total resistant force F is given by $F = 300 + 1.8v^2$, where F is in newtons and v is in meters per second. Calculate the power required from the motor to accelerate the car at 0.92 m/s² when the speed is 80 km/h.

58. A governor consists of two 200-g spheres attached by light, rigid 10.0-cm rods to a vertical rotating axle. The rods are hinged so that the spheres swing out from the axle as they rotate with it. However, when the angle θ is 45.0°, the spheres encounter the wall of the cylinder in which the governor is rotating; see Fig. 24. (a) What is the minimum rate of rotation, in revolutions per minute, required for the spheres to touch the wall? (b) If the coefficient of kinetic friction between the spheres and the wall is 0.35, what power is dissipated as a result of the spheres rubbing against the wall when the mechanism is rotating at 300 rev/min?

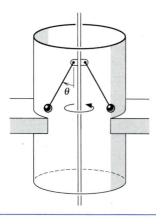

Figure 24 Problem 58.

Section 7-6 Reference Frames

59. Consider two observers, one whose frame is attached to the ground and another whose frame is attached, say, to a train moving with uniform velocity **u** with respect to the ground. Each observes that a particle, initially at rest with respect to the train, is accelerated by a constant force applied to it for time t in the forward direction. (a) Show that for each observer the work done by the force is equal to the gain in kinetic energy of the particle, but that one observer mea- sures these quantities to be $\frac{1}{2}ma^2t^2$, whereas the other observer measures them to be $\frac{1}{2}ma^2t^2 + maut$. Here a is the common acceleration of the particle of mass m. (b) Explain the differences in work done by the same force in terms of the different distances through which the observers measure the force to act during the time t. Explain the different final kinetic energies measured by each observer in terms of the work the particle could do in being brought to rest relative to each observer's frame.

Section 7-7 Kinetic Energy at High Speed

60. Calculate the kinetic energy of a proton traveling at a speed of 2.94×10^8 m/s. Give your answer in both joules and MeV.

61. An electron is moving at a speed such that it could circumnavigate the Earth at the equator in 1.0 s. (a) What is its speed in terms of the speed of light? (b) What is its kinetic energy in electron-volts? (c) What percent error do you make if you use the classical formula to compute the kinetic energy?

62. An electron has a speed of $0.999c$. (a) What is its kinetic energy? (b) If its speed is increased by 0.05%, by what percentage will its kinetic energy increase?

63. The work–energy theorem applies to particles at all speeds. How much work must be done to increase the speed of an electron from rest (a) to $0.50c$, (b) to $0.99c$, and (c) to $0.999c$?

CHAPTER 8

CONSERVATION
OF
ENERGY

*In Chapter 7 we discussed the work–energy theorem, according to
which the total work done by the forces acting on a particle is equal to the
change in the particle's kinetic energy. In this chapter, we see that the work done on a
system (which may be more complex than the simple particles we considered previously) by
a certain class of forces depends only on the initial and final states of the system and not at
all on the path followed between the states. Such forces are called* conservative forces *and
are also distinguished by their ability to store energy merely from the configuration of the
system. The stored energy is called* potential energy. *Other forces, called* nonconservative
forces, *cannot store energy in this way.*

The central theme of this chapter is conservation of energy, *one of the main guiding
principles of physics. We show that in the storage, conversion, or transfer of energy in
mechanical systems, the total energy remains constant. We begin with simple frictionless
mechanical systems, in which only kinetic and potential energies play a role. Later we
include systems in which friction and other dissipative forces may occur. Further expansions
permit other kinds of energy, including heat and electromagnetic, to be incorporated into
this same framework, making the principle of conservation of energy one of the most widely
applicable and general laws of physics.*

8-1 CONSERVATIVE FORCES

To illustrate the behavior of conservative systems, we consider the one-dimensional motion of a particle acted on by three separate forces: the spring force, $F = -kx$; the gravitational force, $F = mg$; and the frictional force, $F = \mu N$.

1. *The spring force.* Figure 1 shows a block of mass m attached to a spring of force constant k; the block slides without friction across a horizontal surface. Initially (Fig. 1a) an external agent has compressed the spring so that the block is displaced to $x = +d$ from its position at $x = 0$ when the spring is relaxed. The external agent is suddenly removed at $t = 0$, and the spring begins to do work on the block. As the block moves from $x = +d$ to $x = 0$, the spring does work $+\frac{1}{2}kd^2$ according to Eq. 9 of Chapter 7. According to the work–energy theorem, this work appears as kinetic energy of the block.

As the block passes through $x = 0$ (see Fig. 1b), the sign of the spring force reverses, and the spring now acts to slow down the block, doing negative work on it. When the

block has been brought momentarily to rest, as in Fig. 1c, the amount of this negative work done by the spring force between $x = 0$ and $x = -d$ is $-\frac{1}{2}kd^2$. Similarly, from $x = -d$ to $x = 0$, the spring force does work $+\frac{1}{2}kd^2$, and from $x = 0$ back to $x = +d$, it does work $-\frac{1}{2}kd^2$. The block is now back in its original position (compare Figs. 1a and 1e), and we see from adding the four separate contributions that the total work done on the block by the spring force in the complete cycle is zero.

2. *The force of gravity.* Figure 2 shows an example of a system consisting of a ball acted on by the Earth's gravity. The ball is projected upward by an external agent that gives it an initial speed v_0 and thus an initial kinetic energy $\frac{1}{2}mv_0^2$. As the ball rises, the Earth does work on it and eventually brings it momentarily to rest at $y = h$. The work done by the Earth as the ball rises from $y = 0$ to $y = h$ is $-mgh$ (the constant force mg times the distance h, negative because the force and displacement are in opposite directions as the ball rises). The work–energy theorem relates the change in kinetic energy, $-\frac{1}{2}mv_0^2$, to the net work done by the only force (gravity), $-mgh$. As the ball falls from $y = h$ to $y = 0$, the force of gravity does

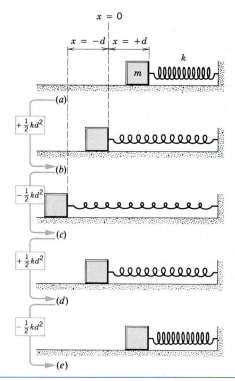

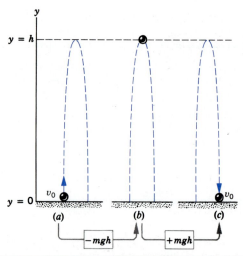

Figure 1   A block moves under the action of a spring force from (a) $x = +d$ to (b) $x = 0$, moving left, to (c) $x = -d$, to (d) $x = 0$, moving right, and (e) back to $x = +d$. The work done by the spring force between each pair of successive positions is shown at the left. Note that the total work done by the spring force on the block is zero for the round trip.

Figure 2   A ball is thrown upward against the Earth's gravity. In (a) it is just leaving its starting point, in (b) it has reached the top of its trajectory, and in (c) it has returned to its original height. The work done by the Earth's gravity between the pairs of successive positions is shown at the bottom. Note that the total work done by the force of gravity on the ball is zero for the round trip.

work $+mgh$; therefore by analogy with the upward trip, the kinetic energy must increase from 0 to $+\frac{1}{2}mv_0^2$. The total work done by the force of gravity during the round trip is zero.

3.   *The frictional force.* For our third example, we consider a disk of mass m on the end of a string of length R.

The disk is given an initial speed v_0, and the string constrains it to move in a circle of radius R over a horizontal surface that exerts a frictional force on the disk (see Fig. 3). The only force that does work on the disk is the frictional force exerted by the surface on the bottom of the disk. This force *always* acts in a direction opposite to the velocity of the disk, so that the work done by the frictional force on the disk is always negative. After the disk has returned to its starting point, the work done by the frictional force for this round trip is definitely not zero; the total work for the round trip is a negative quantity. The disk returns to its starting point with a smaller kinetic energy after the round trip.

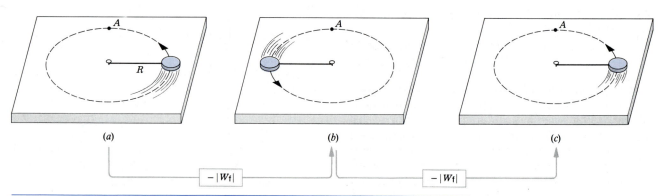

Figure 3   A disk moves with friction in a circle on a horizontal surface. The positions shown represent (a) an arbitrary starting point, (b) one-half revolution later, and (c) another half revolution later. The work done by friction between successive positions is indicated at the bottom. Note that the total work done by the frictional force on the disk is *not* zero for the round trip, but instead has the negative value $-2|W_f|$.

Notice the differences between these three examples. In the first two (the spring force and gravity), the object returned to its starting point after a round trip with no net work done on it (and thus with no change in kinetic energy). In the third example, there was net work done on the object by the frictional force during the round trip, and there was a loss in kinetic energy. This basic difference in behavior between the two kinds of forces leads us to our first way of distinguishing conservative forces.

If a body moves under the action of a force that does no total work during any round trip, then the force is conservative; otherwise it is nonconservative.

The elastic restoring force (spring force) and gravity are two examples of conservative forces, and friction is an example of a nonconservative force.*

A second way of distinguishing conservative from nonconservative forces concerns the work done in taking the body through different paths leading to the same final position. As an example, let us calculate the work done by the spring force as the block of Fig. 1 moves from $x = +d$ to $x = -d/2$ along two different paths (see Fig. 4): path 1, directly; path 2, moving first from $x = +d$ to $x = -d$, and then from $x = -d$ to $x = -d/2$. Letting W_1 and W_2 represent the work done by the spring along paths 1 and 2, we have

$$W_1 = \int_{+d}^{-d/2} (-kx)\,dx = -\tfrac{1}{2}kx^2 \Big|_{+d}^{-d/2}$$

$$= -\tfrac{1}{2}k[(-d/2)^2 - d^2] = \tfrac{3}{8}kd^2$$

and

$$W_2 = \int_{+d}^{-d} (-kx)\,dx + \int_{-d}^{-d/2} (-kx)\,dx$$

$$= -\tfrac{1}{2}kx^2 \Big|_{+d}^{-d} - \tfrac{1}{2}kx^2 \Big|_{-d}^{-d/2}$$

$$= 0 - \tfrac{1}{2}k[(-d/2)^2 - (-d)^2] = \tfrac{3}{8}kd^2.$$

Thus $W_1 = W_2$, and the work is the same for the two different paths.

On the other hand, let us consider the behavior of the nonconservative frictional force for the system illustrated

---

* When an object moves subject to the frictional force, microscopic welds are repeatedly formed and broken, as described in Section 6-2. When the object retraces its path, the changes in the surface are not restored, and therefore the frictional force, viewed macroscopically, is certainly nonconservative. However, the surface interatomic forces responsible for friction are electrostatic forces, which *are* conservative (see Chapter 30). If in retracing the path we restored all the disturbed atoms to their original locations, we would find the frictional force to be microscopically conservative. Such a process is highly improbable (in fact, more atoms would be displaced by new welds when the path is retraced), and so the frictional force is macroscopically nonconservative.

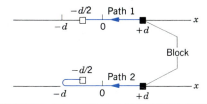

Figure 4 The block (represented here by a square) in the system of Fig. 1 is moved from $x = +d$ to $x = -d/2$ along two different paths.

in Fig. 3 as the disk moves through two different paths to the position shown as point A in Fig. 3. If we compare the work done by friction when the disk moves from the starting position to point A through one-quarter of a revolution with the work done by friction when it moves through $1\tfrac{1}{4}$ revolutions (reaching exactly the same final location), we find that the (negative) work done by friction will be five times larger in magnitude for the second path. Thus, in the case of the frictional force, the work depends on the path taken between the initial and final locations.

This leads to our second way of distinguishing conservative forces.

If the work done by a force in moving a body from an initial location to a final location is independent of the path taken between the two points, then the force is conservative; otherwise it is nonconservative.

With the help of Fig. 5, we can show that the two criteria we have developed for identifying conservative forces are precisely equivalent. In Fig. 5a, a particle moves in the round trip from a to b and back again. If only a conservative force **F** acts on the particle, the total work done on the particle by that force during the cycle must be zero. That is,

$$W_{ab,1} + W_{ba,2} = 0$$

or

$$\int_{a \atop \text{path 1}}^{b} \mathbf{F} \cdot d\mathbf{s} + \int_{b \atop \text{path 2}}^{a} \mathbf{F} \cdot d\mathbf{s} = 0, \qquad (1)$$

where $W_{ab,1}$ means "the work done by the force when the

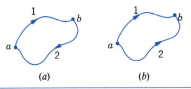

Figure 5 (a) A particle, acted on by a conservative force, moves in a round trip starting at point a, passing through point b, and returning to point a. (b) A particle starts from point a and travels to point b following either of two possible paths.

particle moves from a to b along path 1" and $W_{ba,2}$ means "the work done by the force when the particle moves from b to a along path 2." Equation 1 is the mathematical statement equivalent to the *first* criterion for a conservative force.

Reversing the direction in which we travel any particular path interchanges the limits of integration and changes the sign of the displacement; that is, the work in going from a to b is related to the work in going from b to a:

$$\int_a^b \mathbf{F} \cdot d\mathbf{s} = - \int_b^a \mathbf{F} \cdot d\mathbf{s} \quad \text{(any particular path)}$$

or, in the case of path 2,

$$W_{ab,2} = - W_{ba,2}. \tag{2}$$

Combining Eqs. 1 and 2 gives

$$W_{ab,1} = W_{ab,2}$$

or

$$\int_{\substack{a \\ \text{path 1}}}^b \mathbf{F} \cdot d\mathbf{s} = \int_{\substack{a \\ \text{path 2}}}^b \mathbf{F} \cdot d\mathbf{s}. \tag{3}$$

This is the mathematical representation of the *second* definition of a conservative force: the work done by the force is the same for any arbitrary path between a and b. Thus the first definition leads directly to the second and (by a similar argument) the second leads to the first, so that the two definitions are equivalent.

8-2 POTENTIAL ENERGY

We obtain a fresh insight into the analysis of systems with conservative forces if we introduce a new concept, *potential energy*. As we shall see, potential energy can be defined only for conservative forces such as the spring force or the force of gravity; it does not exist for nonconservative forces such as friction.

Potential energy, represented by the symbol U, is the *energy of configuration* of a system. It is the energy stored in a system because of the relative position or orientation of the parts of a system (for example, the compressing of the block–spring system or the separation of the ball–Earth system).

Consider a system in which only one force acts, and let that force be conservative. When we change the configuration of the system, such as by moving one of its parts, work W is done by the conservative force. We define the *change* in potential energy ΔU corresponding to a particular change in configuration to be

$$\Delta U = - W. \tag{4}$$

The change in potential energy in the process is the negative of the work done by the conservative force.

When the block–spring system of Fig. 1 changes its configuration from that of Fig. 1d (in which the spring is in its relaxed state) to that of Fig. 1e (in which the block is momentarily at rest), the work done by the spring force on the block is $W = -\frac{1}{2}kd^2$. The change in potential energy of the system is therefore $\Delta U = -W = +\frac{1}{2}kd^2$. From the work–energy theorem, however, the change in the kinetic energy of the block is $\Delta K = W = -\frac{1}{2}kd^2$. For the block–spring system, we therefore have the following result:

$$\Delta U + \Delta K = 0. \tag{5}$$

Equation 5, which we obtained for the block–spring system, is in fact a general result that follows directly from Eq. 4 and the work–energy theorem, $W = \Delta K$. It states that, in a system in which only conservative forces act, any change in the potential energy must be balanced by an equal and opposite change in the kinetic energy.

As an example, let us release the block (see Fig. 1a again) from $x = +d$ when the spring is compressed. The spring pushes against the block and accelerates it. The displacement from equilibrium decreases, the spring does positive work on the block, and the change in potential energy is therefore negative by Eq. 4. As the potential energy decreases, the kinetic energy increases.

We can also write Eq. 5 as

$$\Delta(U + K) = 0. \tag{6}$$

The change in the total $U + K$ is zero in these processes. If there is no change in the sum $U + K$, then the value of the sum must be a constant during the motion. We call this constant E, the *mechanical energy* of the conservative system:

$$U + K = E. \tag{7}$$

Equation 7 is the mathematical representation of the law of *conservation of mechanical energy*. In any isolated system of objects interacting only through conservative forces, such as the block and spring, energy can be transferred back and forth from kinetic to potential, but the total change is zero; the sum of the kinetic plus potential energy remains constant. Figure 6 shows a representation of the distribution of kinetic and potential energy in the system consisting of block and spring as the system oscillates freely.

Suppose that more than one conservative force acts on an object, for example, the block in Fig. 7, which is acted on by two forces, $\mathbf{F}_{\text{spring}}$ and $\mathbf{F}_{\text{grav}}$, each of which does work on the block. The work-energy theorem, which was used in deriving Eq. 5, always refers to the net work done by all the forces that act on the object, which in this case is $W_{\text{spring}} + W_{\text{grav}}$. Using Eq. 4 ($\Delta U = -W$), we can associate a potential energy change with the work done by each force, so that Eq. 5 becomes

$$\Delta U_{\text{spring}} + \Delta U_{\text{grav}} + \Delta K = 0$$

and Eq. 7 becomes

$$U_{\text{spring}} + U_{\text{grav}} + K = E. \tag{8}$$

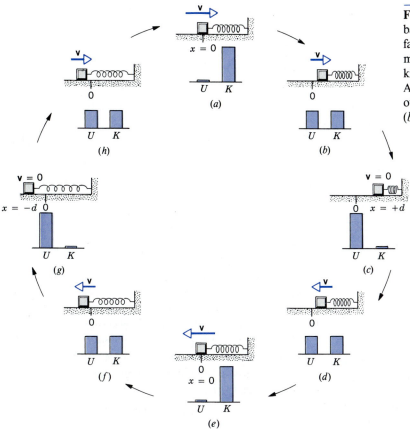

Figure 6 A block attached to a spring oscillates back and forth on a horizontal frictionless surface. The mechanical energy E of the system remains constant but is shared differently between kinetic and potential energy as the system moves. At certain times (a, e) the energy is all kinetic, at others (c, g) it is all potential, and at still others (b, d, f, h) it is shared equally between the two forms.

The potential energy should be considered to be a property of the entire system, rather than of any particular part of a system. It is not the ball of Fig. 2, for example, that has potential energy; it is the system consisting of Earth + ball. When the ball rises through a vertical height h, the potential energy of the *system* increases by mgh, and the kinetic energy of the *system* decreases by this same amount. As the ball falls freely back to ground through this same height h, the potential energy of the *system* decreases by mgh, and the kinetic energy of the *system* increases by the same amount.

Because the ball is much less massive than the Earth, virtually all of the increase in the kinetic energy of the system of Earth + ball is given to the ball. It is for this reason that we sometimes refer to the potential energy of the ball rather than (more precisely) of the Earth + ball system. In other systems, in which the masses are more nearly equal, both objects might acquire measurable kinetic energy as a result of the change in the potential energy. The method for calculating how the kinetic energy is shared between the two objects is discussed in Chapter 9.

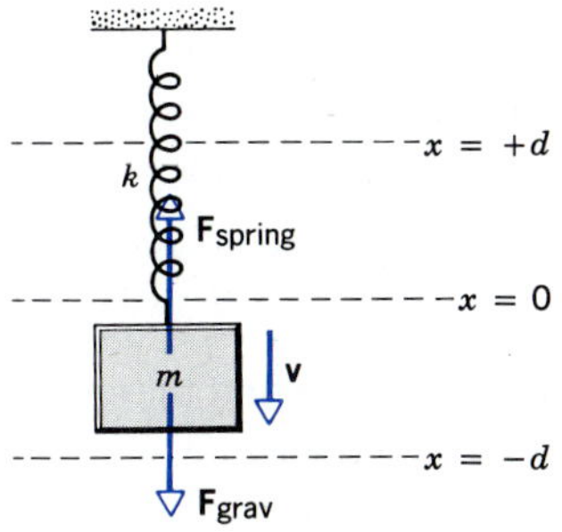

8-3 ONE-DIMENSIONAL CONSERVATIVE SYSTEMS

We can use Eq. 4 to obtain the change in potential energy for a particle in one-dimensional motion in a system in which it is acted on by a single conservative force $F(x)$:

$$\Delta U = -W = -\int_{x_0}^{x} F(x)\,dx. \qquad (9)$$

The particle moves from the initial coordinate x_0 to the

Figure 7 A block of mass m, suspended from a spring, oscillates vertically between $x = +d$ and $x = -d$. The motion of the block is governed by two conservative forces, the spring force $\mathbf{F}_{\text{spring}}$ and the force of the Earth's gravity $\mathbf{F}_{\text{grav}}$.

final coordinate x. Because the potential energy depends only on position, the change ΔU between x_0 and x is $\Delta U = U(x) - U(x_0)$, and we obtain

$$U(x) - U(x_0) = -\int_{x_0}^{x} F(x)\,dx. \qquad (10)$$

If we consider the point x_0 to be an arbitrary reference point, we can then obtain the potential energy function $U(x)$. We are free to choose any convenient value for the potential energy at the reference point, $U(x_0)$, because only *changes* in potential energy are significant. For a particular choice of $U(x_0)$, the resulting function $U(x)$ could then be used to calculate the potential energy at particular points in the motion, for instance, x_1 and x_2. A different choice of $U(x_0)$ will change the values of $U(x_1)$ and $U(x_2)$ by the same constant, but the difference in potential energy, $U(x_2) - U(x_1)$, is unchanged. The analysis of the dynamical behavior is thus independent of the choice of $U(x_0)$.

In effect, the choice of the reference point for $U(x)$ is similar to the choice of a reference frame for kinetic energy. As we discussed in Section 7-6, observers in relative motion may differ on the particular values of kinetic energy they measure. Observers in different reference frames will differ in their values of U, K, and the mechanical energy E, but they all will agree on the constancy of E and on the conservation of mechanical energy.

In moving from x_0 to x, the particle's velocity will change from v_0 to v, and according to the work–energy theorem the work done by the force F is

$$W = \Delta K = \tfrac{1}{2}mv^2 - \tfrac{1}{2}mv_0^2. \qquad (11)$$

Combining Eqs. 9, 10, and 11, we have

$$\tfrac{1}{2}mv^2 + U(x) = \tfrac{1}{2}mv_0^2 + U(x_0) \qquad (12)$$
$$= E.$$

The quantity on the right in Eq. 12 depends only on the initial position x_0 and the initial speed v_0, which have definite values; it is therefore constant during the motion. This is the constant *mechanical energy* E. Notice that force and acceleration do not appear in this equation, only position and speed. Equation 12 is another form of the law of *conservation of mechanical energy* for conservative forces.

Instead of starting with Newton's laws, we can simplify the solution of problems involving only conservative forces by starting with Eq. 12. This relation is derived from Newton's laws, of course, but it is one step closer to the solution (the so-called first integral of the motion). We often solve problems without analyzing the forces or writing down Newton's laws by looking instead for something in the motion that is constant; here the mechanical energy is constant and we can write down Eq. 12 as the first step.

In one dimension we can write the relation between force and potential energy (Eq. 9) as

$$F(x) = -\frac{dU(x)}{dx}. \qquad (13)$$

To show this, substitute this expression for $F(x)$ into Eq. 9 and observe that you get an identity. Equation 13 gives us another way of looking at potential energy. *The potential energy is a function of position whose negative derivative gives the force.* The force F is exerted *by the system* whose potential energy is U.

We now illustrate the calculation of potential energy with the two examples of conservative forces we considered in Section 8-1, the block–spring system and the ball–Earth system.

The Spring Force

We choose the reference position x_0 of the block in the block–spring system of Fig. 1 to be that in which the spring is in its relaxed state ($x_0 = 0$), and we declare the potential energy of the system to be zero when the block is at that location [$U(x_0) = 0$]. The potential energy of the block–spring system can be found by substituting these values into Eq. 10 and evaluating the integral for the spring force, $F(x) = -kx$:

$$U(x) - 0 = -\int_{0}^{x} (-kx)\,dx$$

or

$$U(x) = \tfrac{1}{2}kx^2. \qquad (14)$$

Whenever the block is displaced a distance x from its reference position, the potential energy of the system is $\tfrac{1}{2}kx^2$. The same result is obtained whether x is positive or negative; that is, whether the spring is stretched or compressed by a given amount x, the stored energy is the same.

Differentiating Eq. 14, we see that Eq. 13 is satisfied:

$$-\frac{dU}{dx} = -\frac{d}{dx}\left(\tfrac{1}{2}kx^2\right) = -kx = F.$$

Suppose we stretch the block–spring system until the block is a distance x_m from its reference position; the potential energy is $\tfrac{1}{2}kx_m^2$. If we release the spring from rest in this configuration, the mechanical energy E is equal to $\tfrac{1}{2}kx_m^2$, since there is no kinetic energy at the instant of release. Equation 12 can be written, in this case,

$$\tfrac{1}{2}mv^2 + \tfrac{1}{2}kx^2 = E$$
$$= \tfrac{1}{2}kx_m^2. \qquad (15)$$

This expression allows us to find the speed at any particular value of the displacement:

$$v = \sqrt{\frac{k}{m}\left(x_m^2 - x^2\right)}. \qquad (16)$$

As we expect, when $x = \pm x_m$, Eq. 16 predicts that the

speed is zero. When the block passes through the reference point ($x = x_0 = 0$), the speed v_0 is

$$v_0 = \sqrt{\frac{k}{m}} \, x_m. \tag{17}$$

The mechanical energy can be expressed in terms of either the speed v_0 at the reference position ($E = \frac{1}{2}mv_0^2$) or the maximum displacement x_m from the reference position ($E = \frac{1}{2}kx_m^2$).

The Force of Gravity

For the ball–Earth system, we represent the vertical coordinate by y rather than x. We choose the reference point $y_0 = 0$ at the surface of the Earth, and we define $U(y_0) = 0$ at that point. We can now evaluate the potential energy $U(y)$ of the system from Eq. 10 with $F(y) = -mg$:

$$U(y) - 0 = -\int_0^y -mg \, dy$$

$$U(y) = mgy. \tag{18}$$

Note that Eq. 13 is satisfied for this potential energy: $-dU/dy = -mg = F$.

The initial velocity of the ball at the reference point is v_0, and Eq. 12 gives

$$\tfrac{1}{2}mv^2 + mgy = \tfrac{1}{2}mv_0^2. \tag{19}$$

This equation, which is equivalent to Eq. 25 of Chapter 2, permits us to find the speed at any height y.

This example illustrates the slightly different language of the energy and force approaches to the analysis of dynamics. The force approach analyzes this system as follows: "The ball begins with initial velocity v_0. The Earth exerts a force $-mg$, and the resulting acceleration is $-g$. This downward acceleration causes the velocity to decrease until the velocity passes through zero at a height h. The ball then begins to move downward under the influence of the Earth's gravity and reaches the ground with velocity $-v_0$."

The energy approach is as follows: "The ball begins with kinetic energy $\frac{1}{2}mv_0^2$. As it rises, the potential energy of the ball–Earth system increases, and so the kinetic energy must decrease in order to keep the mechanical energy E constant. At the highest point in the motion, all the kinetic energy has been converted to gravitational potential energy. The falling ball reverses the process, with the potential energy converting back into kinetic energy and becoming fully converted again when the falling ball reaches the ground." These two approaches of course give the same result. Often we find the energy approach to be more useful and to provide more insight. There are also cases in which it is easier to work with energy, which is a scalar, than with force, which is a vector.

Sample Problem 1 An elevator cab of mass m ($= 920$ kg) moves from street level to the top of the World Trade Center in New York, a height $h = 412$ m above the ground. What is the change in the gravitational potential energy of the cab?

Solution Strictly, we are talking about the change in potential energy of the cab–Earth system. From Eq. 18

$$\Delta U = mg \, \Delta y = mgh = (920 \text{ kg})(9.8 \text{ m/s}^2)(412 \text{ m})$$
$$= 3.7 \times 10^6 \text{ J} = 3.7 \text{ MJ}.$$

This is almost exactly 1 kW·h; the equivalent quantity of electrical energy costs a few cents at commercial rates.

Sample Problem 2 The spring of a spring gun is compressed a distance d of 3.2 cm from its relaxed state, and a ball of mass m ($= 12$ g) is put in the barrel. With what speed will the ball leave the barrel once the gun is fired? The force constant k of the spring is 7.5 N/cm. Assume no friction and a horizontal gun barrel.

Solution We can apply Eq. 12 directly, with the initial position of the spring $x_0 = d$ and initial velocity of the ball $v_0 = 0$. In the final state the spring is relaxed ($x = 0$) and the ball moves with velocity v. Thus

$$\tfrac{1}{2}mv^2 + 0 = 0 + \tfrac{1}{2}kd^2.$$

Solving for v yields

$$v = d\sqrt{\frac{k}{m}} = (0.032 \text{ m})\sqrt{\frac{750 \text{ N/m}}{12 \times 10^{-3} \text{ kg}}} = 8.0 \text{ m/s}.$$

Sample Problem 3 A roller coaster (Fig. 8) slowly lifts a car filled with passengers to a height of $y = 25$ m, from which it accelerates downhill. Neglecting friction in the system, with what speed will the car reach the bottom?

Figure 8 A device for converting gravitational potential energy into kinetic energy.

Solution   At first glance, this problem looks hopeless, for there is no information given about the shape of the path followed by the car. However, in the absence of friction, the track does no work on the car, and the only force that does work on the coasting car is gravity. The mechanical energy E_t, when the car is at the top of the track, is

$$E_t = U_t + K_t = mgy + 0,$$

where we have taken $y = 0$ at the bottom of the track. When the car reaches the bottom, the mechanical energy E_b is

$$E_b = U_b + K_b = 0 + \tfrac{1}{2}mv^2,$$

with the reference for U chosen so that $U = 0$ at $y = 0$. Conservation of energy means $E_t = E_b$, and thus

$$mgy = \tfrac{1}{2}mv^2.$$

Solving for v, we obtain

$$v = \sqrt{2gy} = \sqrt{(2)(9.8 \text{ m/s}^2)(25 \text{ m})} = 22 \text{ m/s}.$$

This is the same speed with which an object dropped vertically from a height of 25 m would hit the ground. The track does not change the speed of the "falling" car; it merely changes the car's direction. Notice that the result is independent of the mass of the car or of its occupants.

As the roller coaster car travels, its speed increases and decreases as it passes through the valleys and peaks of the track. As long as no peak is higher than the starting point, there is enough mechanical energy in the system to overcome any of the intermediate hills of potential energy and carry the system through to the finish.

You can readily appreciate the advantages of the energy technique from this problem. To use Newton's laws would require knowing the exact shape of the track, and then we would need to find the force components and the acceleration at every point. This could be quite a difficult procedure. On the other hand, the solution using Newton's laws would provide more information than the solution using the energy method, for instance, the time it takes the car to reach the bottom.

8-4   ONE-DIMENSIONAL CONSERVATIVE SYSTEMS: THE COMPLETE SOLUTION

Our goal in the analysis of a mechanical system is often to describe the motion of a particle as a function of the time. In Chapters 5 and 6, we showed how to solve this problem by applying Newton's laws; we refer to this procedure as the *dynamical* method. An alternative and sometimes more useful procedure is the *energy* method, which we discuss in this section.

Equation 12 gives the relation between coordinate and speed for one-dimensional motion when the force depends on position only. (In one dimension, forces that depend only on position are *always* conservative; this is not necessarily true in two or three dimensions, as we discuss in Section 8-5.) The force and the acceleration

have been eliminated in arriving at Eq. 12. To complete the analysis we must eliminate the speed and determine position as a function of time.

We can do this in a formal way, as follows. From Eq. 12 we have

$$U(x) + \tfrac{1}{2}mv^2 = E.$$

Solving for v, we obtain

$$v = \pm\sqrt{\frac{2}{m}\left[E - U(x)\right]}. \tag{20}$$

Here $U(x)$ is the potential energy associated with the force that acts in the system, while E is the (constant) mechanical energy that is supplied to the system. For a given value of E, Eq. 20 tells us that the motion is restricted to regions of the x axis where $E \geq U(x)$. That is, we cannot have an imaginary speed or a negative kinetic energy, so $[E - U(x)]$ must be zero or greater. Furthermore, we can obtain a good qualitative description of the possible types of motion by plotting $U(x)$ versus x. This description depends on the fact that the speed is proportional to the square root of the difference between E and U.

For example, consider the potential energy function shown in Fig. 9a. (While this function looks like the profile of a roller coaster, it represents the potential energy of a conservative system in which the motion is confined to only one dimension. A roller coaster confined to a track moves in two or three dimensions.) Since we must have $E \geq U(x)$ for real motion, the lowest mechanical energy possible is E_0. At this value of the energy $E = E_0 = U$, and the kinetic energy must be zero. The particle must be at rest at the point x_0. If the system were given a slightly larger energy E_1, the particle could move only between x_1 and x_2. As it moves from x_0 its speed decreases on approaching either x_1 or x_2. At x_1 or x_2 the particle stops and

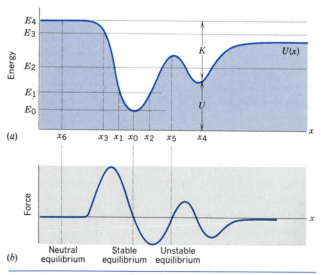

Figure 9   (a) A potential energy function $U(x)$. (b) The force corresponding to that potential energy.

reverses its direction. These points x_1 and x_2 are therefore called *turning points* of the motion. At an energy E_2 there are four turning points, and the particle can oscillate in either one of the two potential valleys. At the energy E_3 there is only one turning point of the motion, at x_3. If the particle is initially moving in the negative x direction, it will stop at x_3 and then move in the positive x direction. At energies above E_4 there are no turning points, and the particle will not reverse direction. Its speed will change according to the value of the potential energy at each point; as shown at the point x_4, the kinetic energy at any point is always the difference between the mechanical energy (E_4, for instance, as shown in Fig. 9a) and the potential energy $U(x)$ evaluated at that point.

At a point where $U(x)$ has a minimum value, such as at $x = x_0$, the slope of the curve is zero, and therefore the force is zero; that is, $F(x_0) = -(dU/dx)_{x=x_0} = 0$. A particle at rest at this point will remain at rest. Furthermore, if the particle is displaced slightly in either direction, the force, $F(x) = -dU/dx$, will tend to return it, and it will oscillate about the equilibrium point. This equilibrium point is therefore called a point of *stable equilibrium.* Figure 9b shows the force $F(x)$ corresponding to the potential energy $U(x)$. If the particle moves slightly to the left of x_0 (that is, to smaller x), the force is positive and the particle is pushed toward larger x (that is, back toward x_0). If the particle moves to the right of x_0, it experiences a negative force that again moves it back toward x_0.

At a point where $U(x)$ has a maximum value, such as at $x = x_5$, the slope of the curve is zero so that the force is again zero; that is, $F(x_5) = -(dU/dx)_{x=x_5} = 0$. A particle at rest at this point will remain at rest. However, if the particle is displaced even the slightest distance from this point, the force $F(x)$ will tend to push it farther from the equilibrium position. Such an equilibrium point is therefore called a point of *unstable equilibrium.* At the point in Fig. 9b corresponding to x_5, moving away from x_5 to the right (toward larger x) results in a positive force that pushes the particle toward even larger x.

In an interval in which $U(x)$ is constant, such as near $x = x_6$, the slope of the curve is zero, and so the force is zero; that is, $F(x_6) = -(dU/dx)_{x=x_6} = 0$. Such an interval is called one of *neutral equilibrium,* since a particle can be displaced slightly without experiencing either a repelling or a restoring force.

From this it is clear that if we know the potential energy function for the region of x in which the body moves, we know a great deal about the motion of the body.

Sample Problem 4   The potential energy function for the force between two atoms in a diatomic molecule can be expressed approximately as follows:

$$U(x) = \frac{a}{x^{12}} - \frac{b}{x^6},$$

where a and b are positive constants and x is the distance between atoms. Find (a) the equilibrium separation between the atoms, (b) the force between the atoms, and (c) the minimum energy necessary to break the molecule apart (that is, to separate the atoms from the equilibrium position to $x = \infty$).

Solution   (a) In Fig. 10a we show $U(x)$ as a function of x. Equilibrium occurs at the coordinate x_m, where $U(x)$ is a minimum, which is found from

$$\left(\frac{dU}{dx}\right)_{x=x_m} = 0.$$

That is,

$$\frac{-12a}{x_m^{13}} + \frac{6b}{x_m^7} = 0$$

or

$$x_m = \left(\frac{2a}{b}\right)^{1/6}.$$

(b) From Eq. 13, we can find the force corresponding to this potential energy:

$$F(x) = -\frac{dU}{dx} = -\frac{d}{dx}\left(\frac{a}{x^{12}} - \frac{b}{x^6}\right) = \frac{12a}{x^{13}} - \frac{6b}{x^7}.$$

We plot the force as a function of the separation between the atoms in Fig. 10b. When the force is positive (from $x = 0$ to $x = x_m$), the atoms are repelled from one another (the force is directed toward increasing x). When the force is negative (from $x = x_m$ to $x = \infty$), the atoms are attracted to one another (the force is directed toward decreasing x). At $x = x_m$ the force is zero; this is the equilibrium point and is a point of stable equilibrium.

(c) The minimum energy needed to break up the molecule into separate atoms is called the *dissociation energy*, E_d. From the potential energy plotted in Fig. 10a, we see that we can

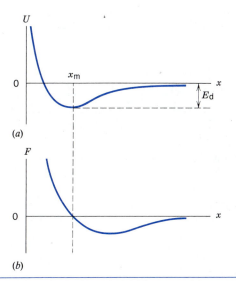

(a)

(b)

Figure 10   Sample Problem 4. (a) The potential energy and (b) the force between two atoms in a diatomic molecule as a function of the distance x separating the atoms. Note that the potential energy is taken as zero when the atoms are infinitely separated.

separate the atoms to $x = \infty$, where $U = 0$, whenever $E \geq 0$. The *minimum* energy needed corresponds to $E = 0$, which means that the atoms will be infinitely separated ($U = 0$) and at rest ($K = 0$) in their final state. In the molecule's equilibrium state, however, its energy is all potential so that (see Fig. 10a) $E = U(x_m)$, a negative quantity. The energy that we must add to the molecule in its equilibrium state to raise its energy from this negative value to zero is what we have called its dissociation energy E_d. Thus

$$U(x_m) + E_d = 0,$$

or

$$E_d = -U(x_m) = -\frac{a}{x_m^{12}} + \frac{b}{x_m^6}.$$

Inserting the value for x_m, we find

$$E_d = \frac{b^2}{4a},$$

which is a positive quantity, as it must be. This energy could be supplied by doing external work on the molecule, perhaps using electric forces, or else by increasing the kinetic energy of one atom of the molecule relative to the other.

Analytical Solution for $x(t)$ (Optional)

The complete description of the one-dimensional motion of a particle is contained in the function $x(t)$, which specifies the position x of the particle at any time t. We can obtain $x(t)$ starting with Eq. 20, which we write as

$$\frac{dx}{dt} = \pm \sqrt{\frac{2}{m}[E - U(x)]},$$

or

$$\frac{dx}{\pm\sqrt{\frac{2}{m}[E - U(x)]}} = dt. \qquad (21)$$

Integrating both sides of Eq. 21 from the initial position ($x = x_0$ when $t = t_0$) to an arbitrary final position x at time t, we obtain

$$\int_{x_0}^{x} \frac{dx}{\pm\sqrt{\frac{2}{m}[E - U(x)]}} = \int_{t_0}^{t} dt = t - t_0. \qquad (22)$$

After carrying out the integration on the left side of Eq. 22, we can in principle solve the resulting equation for $x(t)$.

In applying this equation, the sign taken for the square root depends on whether **v** points in the positive or in the negative x direction. When **v** changes direction during the motion, it may be necessary to carry out the integration separately for each part of the motion.

In some cases, we can carry out the procedure represented by Eq. 22 to obtain an analytical solution for $x(t)$. In other cases, it may be more convenient to find a numerical solution using a computer, which we illustrate later in this section. Here we carry out the analytical solution in the case of a particle of mass m moving in one dimension and acted on by a spring of force constant k, for which $U(x) = \frac{1}{2}kx^2$. Let us assume that at $t = 0$ the particle is located at $x = x_0$ and is moving with velocity $v = 0$. The mechanical energy E is therefore $\frac{1}{2}kx_0^2$ according to Eq. 12. In this case Eq. 22 becomes

$$\sqrt{\frac{m}{k}} \int_{x_0}^{x} \frac{dx}{\pm\sqrt{x_0^2 - x^2}} = t.$$

This integral is a standard form that can be found in integral tables:

$$\int \frac{dx}{\sqrt{a^2 - x^2}} = -\cos^{-1}\left(\frac{x}{a}\right).$$

In our case, we have

$$\int_{x_0}^{x} \frac{dx}{\sqrt{x_0^2 - x^2}} = -\cos^{-1}\left(\frac{x}{x_0}\right)\Big|_{x_0}^{x} = \pm\sqrt{\frac{k}{m}}\,t,$$

and after some manipulation we can write this result as

$$x(t) = x_0 \cos\sqrt{\frac{k}{m}}\,t.$$

The one-dimensional motion of a particle acted on by a spring force is sinusoidal. We know from experience that the motion is oscillatory (that is, it repeats over the same path); this result shows that the oscillation is sinusoidal. We consider oscillatory motion in more general terms in Chapter 15, where we obtain this result for $x(t)$ from Newton's laws instead of from the energy method.

Numerical Solution

As we did in the case of forces that depend on the time (Section 6-6) or on the velocity (Section 6-7), we can obtain a numerical solution for the motion resulting from forces that depend on the position. The numerical technique we discuss is based on Newton's laws rather than on energy methods.

Suppose we have a force $F(x)$ acting on a particle of mass m. At $t = 0$, the particle is located at x_0 and is moving with velocity v_0. Our goal is to find the description of the resulting motion, that is, $x(t)$ and $v(t)$ at all times t.

We divide the motion into a series of small intervals of time δt. Each interval is so small that we can take the acceleration to be approximately constant over the interval. (Over a small enough interval, x does not change very much; thus, $F(x)$ is nearly constant, and so is $a = F/m$.)

In the first interval, which lasts from $t = 0$ to $t = \delta t$, the acceleration has its initial value $a_1 = F(x_0)/m$. (Subscripts here indicate the number of the time interval, and the variable corresponds to the value at the *end* of the interval. Thus v_2 means the velocity at the end of the second interval.)

We can now easily adapt the kinematical equations of constant acceleration to the motion within each interval. Equation 15 of Chapter 2 gives the velocity at the end of the first interval:

$$v_1 = v_0 + a_1\,\delta t,$$

and Eq. 19 of Chapter 2 gives the position at the end of the first interval:

$$x_1 = x_0 + v_0\,\delta t + \tfrac{1}{2}a_1(\delta t)^2.$$

We use this new position x_1 to find the (approximately constant) acceleration during the second interval, $a_2 = F(x_1)/m$, and then we apply the equations of constant acceleration to the second interval, obtaining

$$v_2 = v_1 + a_2\,\delta t$$

and

$$x_2 = x_1 + v_1\,\delta t + \tfrac{1}{2}a_2(\delta t)^2.$$

We can continue this procedure for as many intervals as we like. The smaller we take the interval δt, the more precise will be the result of the calculation.

As an example, we consider the spring force, $F(x) = -kx$ with $k = 9.6$ N/m, acting on a particle of mass $m = 2.5$ kg. Let the particle start at $t = 0$ at position $x_0 = 0.5$ m and velocity $v_0 = 0$. Figure 11 shows the results of the numerical calculation for $x(t)$ and $v(t)$, done using 400 intervals of 0.01 s each.

A computer program that carries out the numerical calculation is presented in Appendix I. Using this program, you can analyze the one-dimensional motion resulting from the action of any force that depends on the position of the particle, even for those forces for which the integral of Eq. 10 does not yield an analytical form for the potential energy or for which the integral in Eq. 22 cannot be evaluated in analytical form.

The results shown in Fig. 11 look very familiar: they appear to be sine and cosine curves. In fact, we have previously used Eq. 22 to obtain the analytical solution in this system, which we showed to be a cosine function. The numerical approach verifies this result. ■

8-5   TWO- AND THREE-DIMENSIONAL CONSERVATIVE SYSTEMS *(Optional)*

Thus far we have discussed potential energy and energy conservation for one-dimensional systems in which the force was directed along the line of motion. We can easily generalize the discussion to three-dimensional motion and obtain an expression for conservation of mechanical energy.

Consider a system in which a particle moves over a path and is acted on by a force arising from other parts of the system. If the work done by the force **F** depends only on the endpoints of the motion and is independent of the path taken between these points, the force is conservative. We define the potential energy U by analogy with the one-dimensional system and find that it is a function of three spatial coordinates, that is, $U = U(x, y, z)$. The generalization of Eq. 9 to motion in three dimensions is

$$\Delta U = -\int_{x_0}^{x} F_x \, dx - \int_{y_0}^{y} F_y \, dy - \int_{z_0}^{z} F_z \, dz \qquad (23)$$

or, written more compactly in vector notation,

$$\Delta U = -\int_{\mathbf{r}_0}^{\mathbf{r}} \mathbf{F}(\mathbf{r}) \cdot d\mathbf{r} \qquad (24)$$

in which ΔU is the change in potential energy for the system as the particle moves from the point (x_0, y_0, z_0), described by the position vector $\mathbf{r}_0$, to the point (x, y, z), described by the position vector $\mathbf{r}$. F_x, F_y, and F_z are the components of the conservative force $\mathbf{F}(\mathbf{r}) = \mathbf{F}(x, y, z)$.

The generalization of Eq. 12 to three-dimensional motion is

$$\tfrac{1}{2}mv^2 + U(x, y, z) = \tfrac{1}{2}mv_0^2 + U(x_0, y_0, z_0), \qquad (25)$$

which can be written in vector notation as

$$\tfrac{1}{2}m\mathbf{v} \cdot \mathbf{v} + U(\mathbf{r}) = \tfrac{1}{2}m\mathbf{v}_0 \cdot \mathbf{v}_0 + U(\mathbf{r}_0), \qquad (26)$$

in which $\mathbf{v} \cdot \mathbf{v} = v_x^2 + v_y^2 + v_z^2 = v^2$ and $\mathbf{v}_0 \cdot \mathbf{v}_0 = v_{0x}^2 + v_{0y}^2 + v_{0z}^2 = v_0^2$. In terms of the mechanical energy E, Eq. 25 can be written

$$\tfrac{1}{2}mv^2 + U(x, y, z) = E.$$

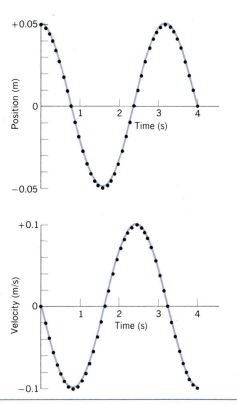

Figure 11   Numerical solution for the motion of a particle acted on by a spring force $F = -kx$. The dots represent values obtained directly from the computer solution. For clarity, only one of every 10 computer points is shown as a dot. The curves are sketched through the dots and certainly resemble sine and cosine curves, which are the results of the analytical solution.

Finally, the generalization of Eq. 13 to three dimensions is*

$$\mathbf{F}(\mathbf{r}) = -\mathbf{i}\,\frac{\partial U}{\partial x} - \mathbf{j}\,\frac{\partial U}{\partial y} - \mathbf{k}\,\frac{\partial U}{\partial z}. \qquad (27)$$

If we substitute this expression for **F** into Eq. 24, we again obtain an identity, which demonstrates that Eqs. 24 and 27 are equivalent. In vector language the conservative force **F** is said to be the negative of the *gradient* of the potential energy $U(x, y, z)$. You can show that all these expressions reduce to the corresponding one-dimensional equations for motion along the x axis. In Eqs. 24 and 27, **F** represents the force exerted *by the system* whose potential energy is U.

Sample Problem 5   In a certain system of particles confined to the xy plane, the force has the form $\mathbf{F}(x, y) = F_x\mathbf{i} + F_y\mathbf{j} = -ky\mathbf{i} - kx\mathbf{j}$, where k is a positive constant. (A particle located at an arbitrary point (x, y) is pushed toward the diagonal line $y = -x$ by this force. You can verify this by drawing the line $y = -x$

---

* The *partial derivative* $\partial/\partial x$ means that we take the derivative of $U(x, y, z)$ with respect to x as if y and z were constants. Similarly, $\partial/\partial y$ and $\partial/\partial z$ indicate that we differentiate with respect to one variable and hold all other variables constant.

and sketching the force components F_x and F_y at various points in the xy plane.) (a) Show that the work done by this force when a particle moves from the origin $(0, 0)$ to the point (a, b) is independent of path along the three paths shown in Fig. 12. (b) Assuming this force to be conservative, find the corresponding potential energy $U(x, y)$ of this system. Take the reference point to be $x_0 = 0$, $y_0 = 0$ and assume $U(0, 0) = 0$.

Solution   (a) The work done along path 1 can be found by breaking the path into two parts: path 1a from $x = 0$ to $x = a$ along the x axis, and path 1b vertically from point $(a, 0)$ to point (a, b). The work along path 1a is

$$W_{1a} = \int \mathbf{F} \cdot d\mathbf{s} = \int F_x \, dx + \int F_y \, dy$$

$$= \int (-ky) \, dx + \int (-kx) \, dy.$$

Along path 1a, $y = 0$ and $dy = 0$. Hence both of the above integrals vanish and $W_{1a} = 0$. Along path 1b, $d\mathbf{s} = dy\mathbf{j}$ and $x = a$, so

$$W_{1b} = \int \mathbf{F} \cdot d\mathbf{s} = \int_{y=0}^{y=b} (-kx) \, dy = (-ka) \int_0^b dy = -kab.$$

The total work along path 1 is therefore

$$W_1 = W_{1a} + W_{1b} = -kab.$$

Along path 2 we proceed in similar fashion:

$$W_{2a} = \int \mathbf{F} \cdot d\mathbf{s} = \int_{y=0}^{y=b} (-kx) \, dy = 0$$

$$W_{2b} = \int \mathbf{F} \cdot d\mathbf{s} = \int_{x=0}^{x=a} (-ky) \, dx = (-kb) \int_0^a dx = -kab.$$

Along path 3, $d\mathbf{s} = dx\mathbf{i} + dy\mathbf{j}$, and

$$W_3 = \int \mathbf{F} \cdot d\mathbf{s} = \int (-ky \, dx - kx \, dy).$$

Let the variable r run along the straight line from $(0, 0)$ to (a, b). With $y = r \sin \phi$, then $dy = dr \sin \phi$ (because ϕ is constant along the line). Also, $x = r \cos \phi$ and $dx = dr \cos \phi$. We treat r as our integration variable, with values in the range from 0 at the origin to $d = (a^2 + b^2)^{1/2}$ at the point (a, b). The integral for W_3 then becomes

$$W_3 = \int_0^d [-k(r \sin \phi)(dr \cos \phi) - k(r \cos \phi)(dr \sin \phi)]$$

$$= -2k \sin \phi \cos \phi \int_0^d r \, dr = -kd^2 \sin \phi \cos \phi.$$

With $\sin \phi = b/d$ and $\cos \phi = a/d$, this becomes $W_3 = -kab$. Thus $W_1 = W_2 = W_3$. This doesn't prove $\mathbf{F}$ is conservative (we would need to evaluate *all* such paths to make that conclusion), but it certainly leads us to suspect that $\mathbf{F}$ might be conservative.

(b) The potential energy can be found from Eq. 24, which we have in effect already evaluated in finding the work done along path 3. The only difference is that we must integrate to the arbitrary point (x, y) instead of to (a, b). We simply relabel point (a, b) as point (x, y) and thus

$$\Delta U = U(x, y) - U(0, 0) = -W = kxy,$$

where we have taken $U(0, 0) = 0$. You should be able to show

that we can apply Eq. 27 to this potential energy function and obtain the force $\mathbf{F}(x, y)$.

If we change the force slightly to $\mathbf{F} = -k_1 y\mathbf{i} - k_2 x\mathbf{j}$, then the methods of part (a) show that this force is not conservative when $k_1 \neq k_2$. (See Problem 46.) Even when $k_1 = -k_2$, the force is still nonconservative. Such a force has important applications to the magnetic focusing of electrically charged particles, but it cannot be represented by a potential energy function, because it is not conservative. ∎

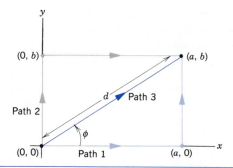

Figure 12   Sample Problem 5. Three different paths are used to evaluate the work done in moving a particle from the origin $(0, 0)$ to the point (a, b).

8-6   CONSERVATION OF ENERGY IN A SYSTEM OF PARTICLES

When an object interacts with one or more objects in its environment, we are free to define our *system* to be as many or as few of the objects as we choose. For any definition of the system, conservation of energy holds as long as we are careful about keeping track of energies within the system and energy transfers between the system and its surroundings.

Figure 13 shows an arbitrary system, around which we have drawn an imaginary closed curve called the *system boundary*. The system inside the boundary has an energy that may include many possible forms, some of which are

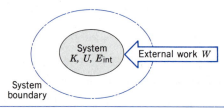

Figure 13   A system enclosed within the boundary has kinetic energy K, potential energy U (representing only the interactions among components within the system), and internal energy E_{int}. The environment can exchange energy with the system through the performance of external work W.

indicated: kinetic energy K, potential energy U, and *internal energy* E_{int}. Here U refers to the potential energy that results from the interaction of the parts of the system among themselves; interactions with the environment are represented not in terms of changes of potential energies but in terms of (external) work W. Later in this section we make a precise definition of internal energy in terms of the microscopic potential and kinetic energies of the molecules of which the components of the system are made. Examples of changes in internal energy are changes in the arrangement of the molecules of a system (such as the microscopic welds formed in sliding friction) and changes in the speed of the molecules of a system (observed as a change in its temperature; temperature is discussed in Chapter 22 and is related to internal energy in Chapter 23).

The energy of the system within the boundary can be changed when external work W is done on the system by its environment, as represented in Fig. 13. (Internal work, done within the boundary by one part of the system acting on another, does not change the total energy, although it may convert energy from one form to another, such as from potential to kinetic.) We can therefore write conservation of energy for the system as

$$\Delta U + \Delta K + \Delta E_{\text{int}} = W, \tag{28}$$

where W represents the total external work done by all the forces through which the environment acts on the system.

Figure 13 also reminds us of the important *sign convention* we have chosen for external work. Positive work done on the system by the environment tends to *increase* the energy of the system. Negative work done on the system by the environment (which is equivalent to positive work done on the environment by the system) tends to *decrease* the energy of the system.

Let us illustrate these principles by considering the block–spring system of Fig. 1, now assuming a frictional force to be present between the block and the table on which it slides. We first define our system to be the block itself (Fig. 14a). The figure shows two transfers of energy through the system boundary: the positive conservative work W_s done on the block by the spring and the negative nonconservative work W_f done on the block by the frictional force exerted by the table. For this system, conservation of energy can be written as

$$\Delta K + \Delta E_{\text{int}} = W_s + W_f. \tag{29}$$

Here $\Delta U = 0$, because the system within the boundary experiences no change in potential energy. The spring is not part of the system, so the spring potential energy is not considered; instead, we account for the spring as a part of the environment through the conservative work W_s it does on the system. Note the directions of the arrows indicating the energy transfers in Fig. 14a; Eq. 29 indicates that positive work done by the spring (which we assume to be compressed from its relaxed length) tends to

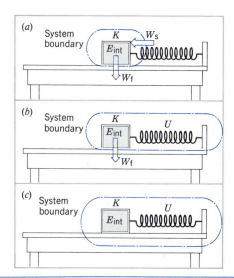

Figure 14 A block acted on by a spring slides on a table that exerts a frictional force. (a) The system consists only of the block; the spring force and friction do work on the system, changing its energy. (b) The system now consists of the block and spring, and it has both kinetic and potential energy. (c) The system now includes the table. The frictional force is now an internal force and contributes to the internal energy of the system.

increase the energy of the block, and negative work done by the horizontal surface tends to decrease the energy of the block.

Now let us consider the system to consist of the block and the spring (Fig. 14b). The system now has potential energy (that associated with the spring force). The frictional force is the only external force that does work on the system. For this definition of the system, we write conservation of energy as

$$\Delta U + \Delta K + \Delta E_{\text{int}} = W_f. \tag{30}$$

The energy of the system is now $U + K + E_{\text{int}}$; transfers of energy between the spring and the block do not change the energy of the system in this case. The spring force is an *internal force* that can transfer energy within the system from one form to another ($U \leftrightarrow K$), but it cannot change the total energy of the system. Negative (frictional) work by the horizontal surface can decrease the energy of the system.

Finally, let us define the system to include the table (Fig. 14c). Now there is no external force, conservative or nonconservative, that is responsible for energy transfers that penetrate the system boundary. With this definition of the system, the external work is zero and thus

$$\Delta U + \Delta K + \Delta E_{\text{int}} = 0. \tag{31}$$

The frictional force is now an internal force, along with the spring force. Energy can be transferred within the system from the mechanical energy $U + K$ of the block + spring to the internal energy of the block + table, but the total energy (mechanical + internal) remains constant. Suppose, for example, that we release the block from rest with the spring compressed. The block slides across the table and eventually comes to rest. In this case $\Delta K = 0$ (because $K_f = K_i = 0$), and so $\Delta E_{int} = -\Delta U$. The loss in potential energy that was originally stored in the system becomes a gain in the internal energy of the system. From this analysis, we cannot determine the separate changes in internal energy of the block and the table, only the total change for the system as a whole.

The frictional force is an example of a nonconservative, dissipative force. In a closed mechanical system such as that illustrated here, mechanical energy is transformed into internal energy by the frictional force. Mechanical energy is *not* conserved in this case, the loss in mechanical energy being compensated by an equivalent gain in the internal energy. (Not all nonconservative forces are dissipative. Some nonconservative forces, such as the magnetic force, can *increase* the mechanical energy of a system. Even the frictional force can, under certain conditions, result in an increase in the mechanical energy of a system. Can you think of an example in which this can occur?)

Notice that in the above examples we have written the macroscopic potential energy of the spring as an explicit term. We could have regarded the energy stored in the spring as a part of the internal energy of the system. However, for convenience, we choose to separate macroscopic terms that can easily be accounted for, leaving in E_{int} the remaining microscopic terms that are not included in U. That is, the rearrangement of the molecules of the spring is included in U, while the rearrangement of the molecules of the block and the table is included in E_{int}. This somewhat arbitrary classification is made for convenience in discussing the energy of this particular system.

Equation 28 represents our first step in progressing from a law of conservation of *mechanical* energy to a generalized law of conservation of energy. In words, this generalized law can be stated as follows.

Energy can be transformed from one kind to another in an isolated system, but it cannot be created or destroyed; the total energy of the system remains constant.

By "isolated" it is meant that no external work, conservative or nonconservative, is done on the system. This statement of the conservation of energy is a generalization from our experience, so far not contradicted by any laboratory experiment or observation of nature.

Occasionally in the history of physics this law seemed to fail, but its apparent failure stimulated the search for some other form of energy that could be included in an even more general law that would explain the observation. For example, in the 1920s many experimental studies were made of nuclear beta decay, a form of radioactive decay in which electrons are emitted from an atomic nucleus.

These experiments showed that the electrons were emitted with less energy than was expected, based on knowledge of the total energy of the nucleus before and after the decay. Many suggestions were made to account for this "missing" energy. According to one proposal, as the electrons made their way outward from the nucleus, they collided with the ordinary atomic electrons and gave up some of their energy in these collisions. If this were true, this process should cause an increase in the internal energy of the system consisting of the emitted electrons plus the decaying atoms. Such an increase in internal energy should be revealed as an increase in the temperature of the radioactive sample. Precise experiments failed to reveal any temperature increase, and the hypothesis was rejected. In 1930, the correct hypothesis was put forward by the Swiss physicist Wolfgang Pauli. Pauli proposed that, in addition to the electron, a *second* particle was emitted in beta decay and carried away the "missing" energy. This particle, called the *neutrino,* proved to be very elusive; even though Pauli's hypothesis was soon confirmed by indirect methods, it was not until 25 years later that the neutrino was directly observed. This prediction of the existence of the neutrino, based on faith in the conservation of energy, had a dramatic effect on the development of the physics of elementary particles in the following decades. The neutrino is one of the most fundamental of elementary particles, and the study of its properties and its interactions with other particles has advanced our understanding of the underlying structure of the material world.

Sample Problem 6 A Chicago Cubs fan drops a baseball (of mass $m = 0.143$ kg) from the top of the Sears Tower at a height h of 443 m (=1450 ft). The ball reaches a terminal speed v of 42 m/s (see Section 6-7). Find the change in the internal energy of the ball and the surrounding air during the fall to the surface of the Earth.

Solution Let us regard the system as the baseball, the air through which it falls, and the Earth. No external force acts on this system; the gravitational pull of the Earth on the ball and the drag force of the air on the ball are internal forces in the system as we have defined it. The change in potential energy of the system is

$$\Delta U = U_f - U_i = 0 - mgh$$
$$= -(0.143 \text{ kg})(9.80 \text{ m/s}^2)(443 \text{ m}) = -621 \text{ J}.$$

The change in kinetic energy during the fall is

$$\Delta K = K_f - K_i = \tfrac{1}{2}mv^2 - 0 = \tfrac{1}{2}(0.143 \text{ kg})(42 \text{ m/s})^2 = 126 \text{ J}.$$

(We are neglecting the motion of the Earth under the gravitational attraction of the ball.) According to Eq. 28, we can write conservation of energy as $\Delta U + \Delta K + \Delta E_{int} = 0$, because there is no external work done on the system. Solving for the internal energy, we obtain

$$\Delta E_{int} = -\Delta U - \Delta K = -(-621 \text{ J}) - 126 \text{ J} = 495 \text{ J}.$$

This internal energy increase might be observed as a temperature rise of the ball and the surrounding air, or perhaps as kinetic energy of the air left in the wake of the falling ball. Using Eq. 28 alone, we cannot allocate the energy among these forms. To do so, we must isolate the ball or the air as our system and calculate the work done by the external forces that act. This procedure, which requires knowledge of the drag force between the ball and the air as well as the details of the ball's motion, is too complex for us to solve here.

In this problem we have assumed that the increase in internal energy stays within the system as we have defined it. In practice, temperature differences between the ball or the air and their environment will result in another kind of energy transfer called *heat,* which we discuss in Chapter 25.

Sample Problem 7   A 4.5-kg block is thrust up a 30° incline with an initial speed v of 5.0 m/s. It is found to travel a distance $d = 1.5$ m up the plane as its speed gradually decreases to zero. (*a*) How much mechanical energy does the block lose in this process due to friction? (*b*) The block then slides from rest back down the plane. Assuming friction to produce the same loss in mechanical energy during the downward journey, what is the speed of the block as it passes through its initial location?

Solution   (*a*) As we did in Sample Problem 6, we ignore the energy changes of the Earth in our calculation and consider the changes in kinetic energy of the block alone. The change in potential energy is

$$\Delta U = U_f - U_i = mgh - 0$$
$$= (4.5 \text{ kg})(9.8 \text{ m/s}^2)(1.5 \text{ m})(\sin 30°) = 33 \text{ J}.$$

The change in kinetic energy between the bottom and the top of the plane is

$$\Delta K = K_f - K_i = 0 - \tfrac{1}{2}mv^2 = -\tfrac{1}{2}(4.5 \text{ kg})(5.0 \text{ m/s})^2 = -56 \text{ J}.$$

The change in mechanical energy is

$$\Delta E = \Delta U + \Delta K = 33 \text{ J} - 56 \text{ J} = -23 \text{ J}.$$

Note that, according to Eq. 28, this loss in mechanical energy can be written as $-\Delta E_{int} + W_f$. Here ΔE_{int} is a positive quantity representing the increase in the internal energy of the block (not the block + plane), and W_f is the (negative) external work done on the block by the frictional force of the plane. Without additional information, we cannot calculate these quantities separately.

(*b*) Now we let $\Delta K'$ represent the change in kinetic energy between the first and second passages through the bottom of the plane. The corresponding change in potential energy $\Delta U'$ is zero. Equation 28 gives

$$\Delta K' = -\Delta U' + (-\Delta E'_{int} + W'_f).$$

The quantity in parentheses has the value $2(-23 \text{ J}) = -46 \text{ J}$, because we are given that the loss in mechanical energy during the downward journey equals that of the upward journey. Thus $\Delta K' = K_f - K_i = -46 \text{ J}$, and the kinetic energy at the bottom of the plane is

$$K_f = 56 \text{ J} - 46 \text{ J} = 10 \text{ J}.$$

The corresponding speed is

$$v_f = \sqrt{\frac{2K_f}{m}} = \sqrt{\frac{2(10 \text{ J})}{4.5 \text{ kg}}} = 2.1 \text{ m/s}.$$

Microscopic Basis of Internal Energy   *(Optional)*
Let us consider an object such as the sliding block discussed above or the falling baseball of Sample Problem 6. The work–energy theorem applied to a particular particle (say, an atom) within the composite system can be written $\Delta K_i = W_i$, where the index i indicates one of the N particles of the object. Here W_i means the total work due to all the forces on that particle. We can apply the work–energy theorem separately to every particle of the system and then add the resulting N equations to obtain

$$\Sigma \, \Delta K_i = \Sigma \, W_i, \tag{32}$$

where the index i ranges from 1 to N. On the right side of Eq. 32, we split the total work done on the object into two parts, such that $\Sigma \, W_i = W_{int} + W_{ext}$. The term W_{int} includes the work done by the forces that the atoms or molecules of the system exert on one another, and the term W_{ext} includes the work done by all external forces. On the left side of Eq. 32, we split the total kinetic energy into two parts: one part, indicated by K, represents the overall directed motion of the object; the second part, indicated by K_{int}, represents the total of all the random internal motions of the atoms or molecules of the object. (The procedure for making this division will be explained in Chapter 9 when we consider center-of-mass motion; for now we simply assume that such a division is possible.) We thus rewrite Eq. 32 as

$$\Delta K + \Delta K_{int} = W_{int} + W_{ext}. \tag{33}$$

We assume that at the microscopic level all forces are conservative, and thus the total internal work can be replaced by a corresponding total interatomic or intermolecular potential energy, such that $W_{int} = -\Delta U - \Delta U_{int}$. We could have written this simply as $-\Delta U_{int}$, but for convenience we may want to group some microscopic potential energies into an easily calculable macroscopic term represented by U, such as the spring potential energy discussed above. Making this substitution and rearranging terms, we obtain

$$\Delta U + \Delta K + (\Delta U_{int} + \Delta K_{int}) = W_{ext}. \tag{34}$$

With $\Delta E_{int} = \Delta U_{int} + \Delta K_{int}$, we obtain Eq. 28. Thus the internal energy term follows directly from applying the work–energy theorem microscopically to an object. ■

8-7   MASS AND ENERGY*   *(Optional)*

A common type of radioactivity easily observed in the laboratory is *positron emission,* a process by which an atomic nucleus emits a positron, which is a particle of the same mass as the electron but of opposite (positive) electric charge. When positrons encounter electrons in ordinary matter, we observe the process called *electron–positron annihilation.* In this process the electron and the positron both disappear and in their place we find only electromagnetic radiation. Symbolically, we can represent this process as

$$e^+ + e^- \rightarrow \text{radiation,}$$

where e^+ and e^- stand for the positron and electron, respectively. Figure 15 illustrates the reverse process in which gamma radiation is converted into an electron and a positron; this process is known as *pair production.*

---

* This section can be skipped or delayed until relativity is discussed in Chapter 21.

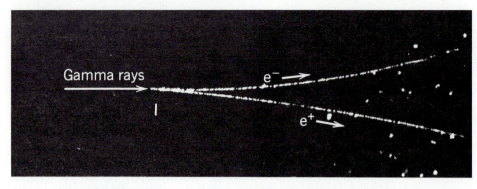

Figure 15 Gamma radiation has its energy converted into a positron and an electron. The two particles leave visible tracks in the bubble chamber in which they were created. The tracks are curved because a strong magnetic field exerts a force that is always perpendicular to the velocity of the particles but is in opposite directions for particles of opposite electric charges.

Consider a system (Fig. 16a) consisting of a positron and an electron of negligibly small kinetic energies and separated by a large enough distance that the potential energy (due to the electrostatic force between them) can also be neglected. Eventually, the positron and electron meet and annihilate, and the resulting radiation escapes through the system boundary (Fig. 16b). By suitable measurements in the environment, we can determine the energy of the radiation that leaves the system, and we find that, for each annihilation event, the radiation carries 1.022 MeV of energy out of the system. When this radiation is absorbed by atoms in the environment, work in the amount of 1.022 MeV is done on the environment through the electromagnetic forces associated with the radiation. Because Eq. 28 is written in terms of the work done *on* a system *by* its environment, we consider in this case that the environment does negative work W in the amount of -1.022 MeV on the system.

Applying Eq. 28 to this system leads to an apparent violation of the conservation of energy; the right-hand side of Eq. 28 equals the negative value W, but the corresponding change in an energy on the left-hand side necessary to maintain the equality is not obvious. We might, for example, propose a decrease in the internal energy that is numerically equal to W, but it is certainly not apparent what sort of internal energy present in the original system is missing from the final system.

The resolution to this dilemma can be found in Albert Einstein's famous equation relating mass and energy, which he proposed in 1905, long before any such experiments as electron–positron annihilation were done:

$$E_0 = mc^2, \qquad (35)$$

where c represents the speed of light.* This equation tells us that mass is a form of energy and that a particle of mass m has associated with it a *rest energy E_0* given by mc^2. This rest energy can be regarded as the internal energy of a body at rest. Thus the electron and positron have internal energy merely because of their masses. For either particle, we can compute the rest energy as

$$E_0 = mc^2 = \frac{(9.11 \times 10^{-31}\ \text{kg})(3.00 \times 10^8\ \text{m/s})^2}{1.60 \times 10^{-13}\ \text{J/MeV}} = 0.511\ \text{MeV}.$$

---

* Although physicists agree about the results of relativistic calculations, there is not universal agreement on the interpretation of Eq. 35. See "The Concept of Mass," by Lev B. Okun, *Physics Today*, June 1989, p. 31, which summarizes the views held by many physicists and adopted for use in this book.

The total internal energy (rest energy) of the two initial particles is then $2(0.511\ \text{MeV}) = 1.022\ \text{MeV}$, and thus the change in the rest energy of the system is -1.022 MeV. *The negative work done on the system of Fig. 16 is balanced by an equivalent loss in the rest energy of the system.* By taking proper account of the rest energies of the particles, we find that energy is conserved.

Equation 35 also tells us that whenever we add energy ΔE to a material object that remains at rest, we increase its mass by an amount $\Delta m = \Delta E/c^2$. If we compress a spring and increase its potential energy by an amount ΔU, then its mass increases by $\Delta U/c^2$. If we raise the temperature of an object, increasing its internal energy by ΔE_{int} in the process, we increase its mass by $\Delta E_{\text{int}}/c^2$. These mass changes are very small and normally beyond our ability to measure in the case of ordinary objects (because c^2 is a very large number), but in the case of decays and reactions of nuclei and subnuclear particles, the relative mass change can be large enough to be measurable.

Within the system boundary of Fig. 13, changes in potential energy U and internal energy E_{int} can thus be associated with changes in the rest energy E_0 of the system. In this case, we can write Eq. 28 as

$$\Delta E_0 + \Delta K = W. \qquad (36)$$

Here we take W to represent the energy (in the form of work) exchanged between the system and its environment. Notice that the left side of Eq. 36 includes only two terms: the rest energy (which includes all types of energy of a system at rest) and the motional (kinetic) energy. Applied to the case of electron–positron annihilation (in which $\Delta K = 0$), Eq. 36 shows directly that the (negative) external work associated with the radiation originates from a decrease in rest energy of the original system.

Examining the situation of Fig. 16b at a time after the radiation has been emitted but before it has been absorbed by the

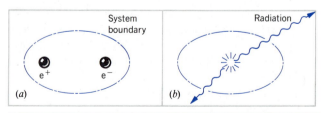

Figure 16 (a) A system consisting of a positron e^+ and an electron e^-. (b) After the positron and electron annihilate, the resulting radiation emerges through the system boundary.

environment, we find that Eq. 35 leads us to another conclusion. For energy to be conserved at that intermediate time, we must assign to the radiation a mass $\Delta m = \Delta E_0/c^2$. Thus Eq. 35 asserts that *energy has mass.*

We therefore conclude that the conservation of energy is equivalent to the conservation of mass. As Einstein wrote: "Prerelativity physics contains two conservation laws of fundamental importance, namely, the law of conservation of energy and the law of conservation of mass; these two appear there as completely independent of each other. Through relativity theory they melt together into *one* principle."

We can apply Eq. 36 to other isolated systems consisting of particles and radiation. Let us consider a star such as the Sun as our system. The Sun radiates an energy of 4×10^{26} J every second. As we did in the case of electron–positron annihilation, we regard this radiant energy as a decrease in the rest energy of the system, and the corresponding change in the mass is

$$\Delta m = \frac{\Delta E_0}{c^2} = \frac{-4 \times 10^{26} \text{ J}}{(3 \times 10^8 \text{ m/s})^2} = -4 \times 10^9 \text{ kg}$$

in every second. This decrease in mass is quite significant by ordinary standards but quite small compared with the total mass of the Sun (2×10^{30} kg). In one year, the Sun's mass decreases by a fraction of only 6×10^{-14}.

Let us instead draw our system boundary around the 1987 supernova (Fig. 17), the first in nearly 400 years to be visible with the naked eye.* A supernova is a star that has used up its supply of thermonuclear fuel and explodes spectacularly. In a time of

---

* See "The Great Supernova of 1987," by Stan Woosley and Tom Weaver, *Scientific American,* August 1989, p. 32.

Figure 17   The 1987 supernova, at the center, easily outshines all the other stars in this photograph.

about 10 seconds, the 1987 supernova is believed to have converted about 10% of its rest energy, roughly equivalent to the mass of two Suns, into radiation and other forms of energy. The change in rest energy corresponding to two solar masses would be

$$\Delta E_0 = \Delta mc^2 = -2(2 \times 10^{30} \text{ kg})(3 \times 10^8 \text{ m/s})^2 = -4 \times 10^{47} \text{ J}.$$

The energy radiated during this 10-second period, corresponding to a power of 4×10^{46} W, is about equal to that of the combined total of all the other stars and galaxies in the rest of the visible universe!

Sample Problem 8   Two 35-g putty balls are thrown toward each other, each with a speed of 1.7 m/s. The balls strike each other head-on and stick together. By how much does the mass of the combined ball differ from the sum of the masses of the two original balls?

Solution   We treat the two putty balls as an isolated system and apply Eq. 36. There is a (negative) change in kinetic energy of this system, with a final value of zero after the collision and a total value K_i for the two balls before the collision. No external work is involved here, so we have

$$\Delta K + \Delta E_0 = (0 - K_i) + \Delta E_0 = 0,$$

or

$$\Delta E_0 = K_i = 2(\tfrac{1}{2}mv^2) = (0.035 \text{ kg})(1.7 \text{ m/s})^2 = 0.101 \text{ J}.$$

This increase in rest energy might be in the form of internal energy, perhaps resulting in an increase in the temperature of the combined system. The corresponding increase in mass is

$$\Delta m = \frac{\Delta E_0}{c^2} = \frac{0.101 \text{ J}}{(3.00 \times 10^8 \text{ m/s})^2} = 1.1 \times 10^{-18} \text{ kg}.$$

Such a tiny increase in mass is hopelessly beyond our ability to measure.

Sample Problem 9   In a 1989 experiment at the Stanford Linear Collider, Z^0 particles were produced when a beam of electrons collided head-on with a beam of positrons of the same kinetic energy. Find the kinetic energy of the two beams needed to produce the Z^0, which has a rest energy of 91.2 GeV (1 GeV = 10^9 eV).

Solution   As in the collision between the putty balls considered in Sample Problem 8, let us assume that there is no external work (that is, no radiation) to be accounted for before or after the collision. The change in rest energy between the initial state (an electron and a positron of rest energy 0.511 MeV each) and the final state (the Z^0) is

$$\Delta E_0 = 91.2 \text{ GeV} - 2(0.511 \text{ MeV}) = 91.2 \text{ GeV},$$

the total rest energy of the electron and positron (1.022 MeV = 0.001022 GeV) being quite negligible here. From Eq. 36, we obtain

$$\Delta K = -\Delta E_0 = -91.2 \text{ GeV} = K_f - K_i.$$

If we assume that the Z^0 is produced at rest, then $K_f = 0$ and the energies of the positron and electron must each be

0.5(91.2 GeV) = 45.6 GeV. In contrast with the previous sample problem, the relative change in rest energy (or in mass) within the system is substantial in this case, the final mass being about 100,000 times the initial mass.* ∎

8-8 QUANTIZATION OF ENERGY *(Optional)*

In the previous section, we discussed how conservation of energy is consistent with relativity, according to which we broaden our concept of energy to include the rest energy of a system. Here we consider the conservation of energy in a different limiting case, in which we approach the *quantum limit* of systems on the atomic or nuclear scale.

If we give the block–spring system some initial energy and release it, the system will oscillate back and forth. If friction is present, the motion gradually dies away. The loss in energy of the system due to external work by the frictional force appears to be smooth and continuous.

On the other hand, let us consider an oscillator consisting of a diatomic molecule: two atoms coupled by a springlike force. If we provide this system with some energy and allow it to oscillate, we find it emits radiation and eventually dissipates as much energy as it can. However, there is an important difference between this atomic oscillator and the block–spring system: *on the atomic scale, the changes in motion occur not continuously but in discontinuous discrete jumps.* Conservation of energy applies on this microscopic scale: the energy difference between the initial and final states is equal to the energy ΔE carried by the radiation, or

$$|\Delta E| = E_i - E_f. \qquad (37)$$

Note that $E_i > E_f$ if the system gives up energy.

Radiations emitted on the atomic scale are discrete: only certain energy changes are possible, in contrast to the classical situation in which the energy change can be treated as a continuous variable. As we discuss in Chapter 49 of the extended version of this text, the permitted energy jumps are related to the frequency ν of the oscillator according to

$$E_i - E_f = h\nu, \qquad (38)$$

where h is a constant called the *Planck constant* and having the value

$$h = 6.63 \times 10^{-34} \text{ J} \cdot \text{s}$$
$$= 4.14 \times 10^{-15} \text{ eV} \cdot \text{s}.$$

Figure 18 shows a schematic view of a process in which a system (perhaps an atom or nucleus) jumps from an initial energy E_i to a final energy E_f, with the emission of radiation of energy $h\nu$. This discrete bundle of energy is called a *quantum*, and the energy states are said to be *quantized*, which means that they have definite and discrete values.

Figure 19 shows an example of some quantized energy states of a sodium atom. The atom may exist in any one of these energy states, but it is not permitted to have an energy intermediate between these allowed values. This structure is responsible for

---

* See "The Stanford Linear Collider," by John R. Rees, *Scientific American*, October 1989, p. 58.

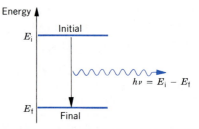

Figure 18 A system in the initial state emits radiation of energy $h\nu$ leading to the final state.

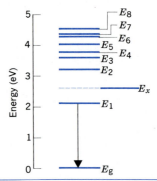

Figure 19 Some of the energy levels of a sodium atom, corresponding to the various quantum states in which the atom may exist. The lowest state, indicated by E_g, is called the ground state. The atom emits characteristic yellow sodium light when it changes from the state of energy E_1 to the ground state, as indicated by the vertical arrow. The atom may exist only in the states indicated; it is not permitted, for example, to have the energy E_x shown between E_2 and E_1.

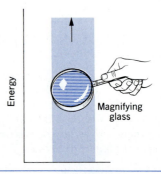

Figure 20 The energy levels of a pendulum are also quantized, but the levels are so close together that they cannot be distinguished, even under the closest scrutiny. No "magnifying glass" could ever reveal the quantized structure of a pendulum.

the discrete radiations emitted by atoms; for example, sodium emits light of a characteristic yellow color (as seen in sodium-vapor street lamps) when the atom jumps from the energy state E_1 (called the first excited state) to the state E_g (called the ground state).

Figure 20 shows the "quantized" structure of a classical oscillator, such as a pendulum. The states may be discrete, but they are so close together that jumps between individual states can be regarded as a continuous process. Suppose the frequency of the pendulum is one swing per second ($v = 1/s$). According to Eq. 38 the "energy quantum" has the value

$$hv = (6.63 \times 10^{-34} \text{ J} \cdot \text{s})(1 \text{ s}^{-1}) = 6.63 \times 10^{-34} \text{ J}.$$

This tiny quantity is far below our ability to determine energies in a macroscopic object such as a pendulum, and thus this discrete structure cannot be observed. In a pendulum, for example, it corresponds to changing the amplitude of the swing by a distance of the order of 10^{-32} m, or about $1/10^{22}$ of the diameter of an atom! We are perfectly safe in ignoring the quantum behavior of ordinary objects.

Conservation of energy on the microscopic scale can be tested by observing the radiations emitted by atoms or nuclei in making discrete jumps between levels, either in the emission of radiation (as in Fig. 18) or in the reverse process in which an atom originally in the ground state (the lower energy state) *absorbs* a quantum of radiation and makes an upward jump to the higher state. Such experiments involving emission and absorption can be made to extraordinary precision, of the order of 1 part in 10^{15} of the energy difference between the states. Every experiment of this sort has been consistent with conservation of energy on the microscopic scale. ∎

QUESTIONS

1. What happens to the potential energy that an elevator loses in coming down from the top of a building to a stop at the ground floor?

2. Mountain roads rarely go straight up the slope but wind up gradually. Explain why.

3. Air bags greatly reduce the chance of injury in a car accident. Explain how they do so, in terms of energy transfers.

4. Pole vaulting was transformed when the wooden pole was replaced by the fiberglass pole. Explain why.

5. You drop an object and observe that it bounces to one and one-half times its original height. What conclusions can you draw?

6. A ball dropped to Earth cannot rebound higher than its release point. However, spray from the bottom of a waterfall can sometimes rise higher than the top of the falls. Why is this?

7. An earthquake can release enough energy to devastate a city. Where does this energy reside an instant before the earthquake takes place?

8. Figure 21 shows a circular glass tube fastened to a vertical wall. The tube is filled with water except for an air bubble that is temporarily at rest at the bottom of the tube. Discuss the subsequent motion of the bubble in terms of energy transfers. Do so both neglecting viscous and frictional forces and also taking them fully into account.

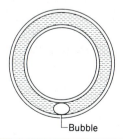

Figure 21 Question 8.

9. In Sample Problem 3 (see Fig. 8) we concluded that the speed of the roller coaster at the bottom does not depend at all on the shape of the track. Would this still be true if friction were present?

10. Taking into account how the potential energy of a system of two identical molecules is related to the separation of their centers, explain why a liquid that is spread out in a thin layer has more potential energy than the same mass of liquid in the shape of a sphere.

11. A swinging pendulum eventually comes to rest. Is this a violation of the law of conservation of mechanical energy?

12. A scientific article ("The Energetic Cost of Moving About," by V. A. Tucker, *American Scientist*, July–August 1975, p. 413) asserts that walking and running are extremely inefficient forms of locomotion and that much greater efficiency is achieved by birds, fish, and bicyclists. Can you suggest an explanation?

13. An automobile is moving along a highway. The driver jams on the brakes and the car skids to a halt. In what form does the lost kinetic energy of the car appear?

14. In the above question, assume that the driver operates the brakes in such a way that there is no skidding or sliding. In this case, in what form does the lost kinetic energy of the car appear?

15. An automobile accelerates from rest to a speed v, under conditions such that no slipping of the driving wheels occurs. From where does the mechanical energy of the car come? In particular, is it true that it is provided by the (static) frictional force exerted by the road on the car?

16. In the case of work done against friction, the internal energy change is independent of the velocity (or inertial reference frame) of the observer. That is, different observers would assign the same quantity of mechanical energy transformed into internal energy due to friction. How can this be explained, considering that such observers measure different quantities of total work done and different changes in kinetic energy in general?

17. Give physical examples of unstable equilibrium, of neutral equilibrium, and of stable equilibrium.

18. In an article "Energy and the Automobile," which appeared in the October 1980 issue of *The Physics Teacher* (p. 494), the author (Gene Waring) states: "It is interesting to note that *all* the fuel input energy is eventually transformed to thermal energy and strung out along the car's path." Analyze the various mechanisms by which this might come about. Consider, for example, road friction, air resistance,

braking, the car radio, the headlamps, the battery, internal engine and drive train losses, the horn, and so on. Assume a straight and level roadway.

19. Trace back to the Sun as many of our present energy sources as you can. Can you think of any that cannot be so traced?

20. Explain, using work and energy ideas, how a child pumps a swing up to large amplitudes from a rest position. (See "How to Make a Swing Go," by R. V. Hesheth, *Physics Education,* July 1975, p. 367.)

21. Two disks are connected by a stiff spring. Can you press the upper disk down enough so that when it is released it will spring back and raise the lower disk off the table (see Fig. 22)? Can mechanical energy be conserved in such a case?

Figure 22   Question 21.

22. Discuss the words "energy conservation" as used (*a*) in this chapter and (*b*) in connection with an "energy crisis" (for example, turning off lights). How do these two usages differ?

23. The electric power for a small town is provided by a hydroelectric plant at a nearby river. If you turn off a light bulb in this closed-energy system, conservation of energy requires that an equal amount of energy, perhaps in another form, appears somewhere else in the system. Where and in what form does this energy appear?

24. A spring is compressed by tying its ends together tightly. It is then placed in acid and dissolves. What happens to its stored potential energy?

25. The expression $E_0 = mc^2$ tells us that perfectly ordinary objects such as coins or pebbles contain enormous amounts of energy. Why did these large stores of energy go unnoticed for so long?

26. "Nuclear explosions—weight for weight—release about a million times more energy than do chemical explosions because nuclear explosions are based on Einstein's $E_0 = mc^2$ relation." What do you think of this statement?

27. How can mass and energy be "equivalent" in view of the fact that they are totally different physical quantities, defined in different ways and measured in different units?

28. A hot metallic sphere cools off as it rests on the pan of a scale. If the scale were sensitive enough, would it indicate a change in mass?

29. Are there quantized quantities in classical (that is, non-quantum) physics? If so, give examples.

PROBLEMS

Section 8-3 One-Dimensional Conservative Systems

1. To disable ballistic missiles during the early boost-phase of their flight, an "electromagnetic rail gun," to be carried in low-orbit Earth satellites, is being developed. The gun might fire a 2.38-kg maneuverable projectile at 10.0 km/s. The kinetic energy carried by the projectile is sufficient on impact to disable a missile even if it carries no explosive. (A weapon of this kind is a "kinetic energy" weapon.) The projectile is accelerated to muzzle speed by electromagnetic forces. Suppose instead that we wish to fire the projectile using a spring (a "spring" weapon). What must the force constant be in order to achieve the desired speed after compressing the spring 1.47 m?

2. It is claimed that as much as 900 kg of water per day can evaporate from large trees. Evaporation takes place from the leaves. To get there the water must be raised from the roots of the tree. (*a*) Assuming the average rise of water to be 9.20 m from the ground, how much energy must be supplied? (*b*) What is the average power if the evaporation is assumed to occur during 12 h of the day?

3. The summit of Mount Everest is 8850 m above sea level. (*a*) How much energy would a 90-kg climber expend against gravity in climbing to the summit from sea level? (*b*) How many Mars bars, at 300 kcal per bar, would supply an energy equivalent to this? Your answer should suggest that work done against gravity is a very small part of the energy expended in climbing a mountain.

4. A 220-lb man jumps out a window into a fire net 36 ft below. The net stretches 4.4 ft before bringing him to rest and tossing him back into the air. What is the potential energy of the stretched net if no energy is dissipated by nonconservative forces?

5. A very small ice cube is released from the edge of a hemispherical frictionless bowl whose radius is 23.6 cm; see Fig. 23. How fast is the cube moving at the bottom of the bowl?

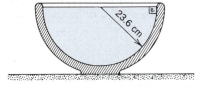

Figure 23   Problem 5.

6. A volcanic ash flow is moving across horizontal ground when it encounters a 10° upslope. It is observed to travel 920 m on the upslope before coming to rest. The volcanic ash contains trapped gas, so the force of friction with the ground is very small and can be ignored. At what speed was the ash flow moving just before encountering the upslope?

7. A projectile with a mass of 2.40 kg is fired from a cliff 125 m high with an initial velocity of 150 m/s, directed 41.0°

above the horizontal. What are (*a*) the kinetic energy of the projectile just after firing and (*b*) its potential energy? (*c*) Find the speed of the projectile just before it strikes the ground. Which answers depend on the mass of the projectile? Ignore air drag.

8. A ball of mass *m* is attached to the end of a very light rod of length *L*. The other end of the rod is pivoted so that the ball can move in a vertical circle. The rod is pulled aside to the horizontal and given a downward push as shown in Fig. 24 so that the rod swings down and just reaches the vertically upward position. What initial speed was imparted to the ball?

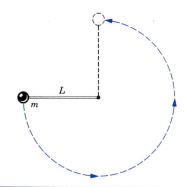

Figure 24   Problems 8 and 38.

9. A 112-g ball is thrown from a window with an initial velocity of 8.16 m/s at an angle of 34.0° above the horizontal. Using conservation of energy, determine (*a*) the kinetic energy of the ball at the top of its flight and (*b*) its speed when it is 2.87 m below the window. Ignore air drag.

10. A frictionless roller-coaster car starts at point *A* in Fig. 25 with speed v_0. What will be the speed of the car (*a*) at point *B*, (*b*) at point *C*, and (*c*) at point *D*? Assume that the car can be considered a particle and that it always remains on the track.

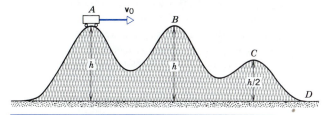

Figure 25   Problem 10.

11. A runaway truck with failed brakes is barreling downgrade at 80 mi/h. Fortunately, there is an emergency escape ramp at the bottom of the hill. The inclination of the ramp is 15°; see Fig. 26. What must be its minimum length *L* to make certain of bringing the truck to rest, at least momentarily?

Figure 26   Problem 11.

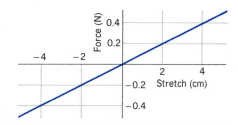

Figure 27   Problem 12.

12. Figure 27 shows the force in newtons as a function of stretch or compression in centimeters for the spring in a cork gun. The spring is compressed by 5.50 cm and used to propel a cork of mass 3.80 g from the gun. (*a*) What is the speed of the cork if it is released as the spring passes through its relaxed position? (*b*) Suppose now that the cork sticks to the spring, causing the spring to extend 1.50 cm beyond its unstretched length before separation occurs. What is the speed of the cork at the time of release in this case?

13. A thin rod whose length is *L* = 2.13 m and whose mass is negligible is pivoted at one end so that it can rotate in a vertical circle. The rod is pulled aside through an angle $\theta =$ 35.0° and then released, as shown in Fig. 28. How fast is the lead ball at the end of the rod moving at its lowest point?

Figure 28   Problem 13.

14. Figure 29 shows a 7.94-kg stone resting on a spring. The spring is compressed 10.2 cm by the stone. (*a*) Calculate the force constant of the spring. (*b*) The stone is pushed down an additional 28.6 cm and released. How much potential energy is stored in the spring just before the stone is released? (*c*) How high above this new (lowest) position will the stone rise?

Figure 29   Problem 14.

15. Approximately 3.3×10^5 m³ of water drops 50 m over Niagara Falls every minute. (*a*) What would be the power output of an electric generating plant that could convert 48% of the water's potential energy to electrical energy? (*b*) If the utility company sold this energy at an industrial rate of 1.2 cent/kW·h, what would be their annual income from this source? One cubic meter (1 m³) of water has a mass of 1000 kg.

16. The area of the continental United States is about 8 × 10⁶ km², and the average elevation of its land surface is about 500 m. The average yearly rainfall is 75 cm. Two-thirds of this rainwater returns to the atmosphere by evaporation, but the rest eventually flows into the oceans. If all this water could be used to generate electricity in hydroelectric power plants, what average power output could be produced?

17. An object falls from rest from a height *h*. Determine the kinetic energy and the potential energy of the object as a function (*a*) of time and (*b*) of height. Graph the expressions and show that their sum—the total energy—is constant in each case.

18. In the 1984 Olympic Games, the West German high jumper Ulrike Meyfarth set a women's Olympic record for this event with a jump of 2.02 m; see Fig. 30. Other things being equal, how high might she have jumped on the Moon, where the surface gravity is only 1.67 m/s²? (*Hint:* The height that "counts" is the vertical distance that her center of gravity rose after her feet left the ground. Assume that, at the instant her feet lost contact, her center of gravity was 110 cm above ground level. Assume also that, as she clears the bar, her center of gravity is at the same height as the bar.)

Figure 30   Problem 18.

19. A 1.93-kg block is placed against a compressed spring on a frictionless 27.0° incline (see Fig. 31). The spring, whose force constant is 20.8 N/cm, is compressed 18.7 cm, after

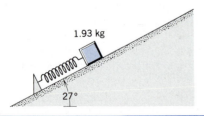

Figure 31   Problem 19.

which the block is released. How far up the incline will the block go before coming to rest? Measure the final position of the block with respect to its position just before being released.

20. An ideal massless spring can be compressed 2.33 cm by a force of 268 N. A block whose mass is *m* = 3.18 kg is released from rest at the top of the incline as shown in Fig. 32, the angle of the incline being 32.0°. The block comes to rest momentarily after it has compressed this spring by 5.48 cm. (*a*) How far has the block moved down the incline at this moment? (*b*) What is the speed of the block just as it touches the spring?

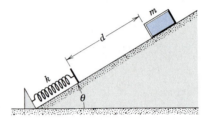

Figure 32   Problems 20 and 35.

21. The spring of a spring gun has a force constant of 4.15 lb/in. When the gun is inclined at an angle of 36.0°, a 2.80-oz ball is projected to a height of 6.33 ft above the muzzle of the gun. (*a*) What was the muzzle speed of the ball? (*b*) By how much must the spring have been compressed initially?

22. A pendulum is made by tying a 1.33-kg stone to a string 3.82 m long. The stone is projected perpendicular to the string, away from the ground, with the string at an angle of 58.0° with the vertical. It is observed to have a speed of 8.12 m/s when it passes its lowest point. (*a*) What was the speed of the stone when projected? (*b*) What is the largest angle with the vertical that the string will reach during the stone's motion? (*c*) Using the lowest point of the swing as the zero of gravitational potential energy, calculate the total mechanical energy of the system.

23. A chain is held on a frictionless table with one-fourth of its length hanging over the edge, as shown in Fig. 33. If the chain has a length *L* and a mass *m*, how much work is required to pull the hanging part back on the table?

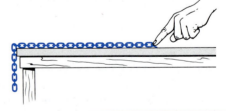

Figure 33   Problem 23.

24. One end of a vertical spring is fastened to the ceiling. A weight is attached to the other end and slowly lowered to its equilibrium position. Show that the loss of gravitational potential energy of the weight equals one-half the gain in

spring potential energy. (Why are these two quantities not equal?)

25. A 2.14-kg block is dropped from a height of 43.6 cm onto a spring of force constant $k = 18.6$ N/cm, as shown in Fig. 34. Find the maximum distance the spring will be compressed.

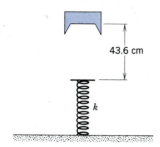

Figure 34 Problem 25.

26. Two children are playing a game in which they try to hit a small box on the floor with a marble fired from a spring-loaded gun that is mounted on a table. The target box is 2.20 m horizontally from the edge of the table; see Fig. 35. Bobby compresses the spring 1.10 cm, but the marble falls 27.0 cm short. How far should Rhoda compress the spring to score a hit?

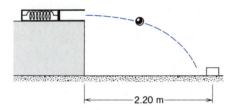

Figure 35 Problem 26.

27. A small block of mass m slides along the frictionless loop-the-loop track shown in Fig. 36. (*a*) The block is released from rest at point P. What is the net force acting on it at point Q? (*b*) At what height above the bottom of the loop should the block be released so that it is on the verge of losing contact with the track at the top of the loop?

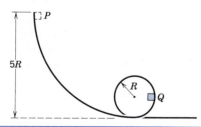

Figure 36 Problem 27.

28. Tarzan, who weighs 180 lb, swings from a cliff at the end of a convenient 50-ft vine; see Fig. 37. From the top of the cliff to the bottom of the swing, Tarzan would fall by 8.5 ft. The vine has a breaking strength of 250 lb. Will the vine break?

Figure 37 Problem 28.

29. The magnitude of the gravitational force of attraction between a particle of mass m_1 and one of mass m_2 is given by

$$F(x) = G\frac{m_1 m_2}{x^2},$$

where G is a constant and x is the distance between the particles. (*a*) What is the potential energy function $U(x)$? Assume that $U(x) \to 0$ as $x \to \infty$. (*b*) How much work is required to increase the separation of the particles from $x = x_1$ to $x = x_1 + d$?

30. A 1.18-kg object is acted on by a net conservative force given exactly by $F = -3x - 5x^2$, where F is in newtons if x is in meters. (*a*) Find the potential energy of the object at $x = 2.26$ m. Assume that $U(0) = 0$. (*b*) The object has a speed of 4.13 m/s in the negative x direction when it is at $x = 4.91$ m. Find its speed as it passes $x = 1.77$ m.

31. A certain spring is found *not* to conform to Hooke's law. The force (in newtons) it exerts when stretched a distance x (in meters) is found to have magnitude $52.8x + 38.4x^2$ in the direction opposing the stretch. (*a*) Compute the work required to stretch the spring from $x = 0.522$ m to $x = 1.34$ m. (*b*) With one end of the spring fixed, a particle of mass 2.17 kg is attached to the other end of the spring when it is extended by an amount $x = 1.34$ m. If the particle is then released from rest, compute its speed at the instant the spring has returned to the configuration in which the extension is $x = 0.522$ m. (*c*) Is the force exerted by the spring conservative or nonconservative? Explain.

32. The string in Fig. 38 has a length $L = 120$ cm, and the distance d to the fixed peg is 75.0 cm. When the ball is released from rest in the position shown, it will swing along the dot-

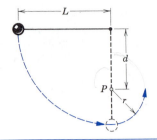

Figure 38 Problems 32 and 33.

ted arc. How fast will it be going (a) when it reaches the lowest point in its swing and (b) when it reaches its highest point, after the string catches on the peg?

33. In Fig. 38 show that, if the pendulum bob is to swing completely around the fixed peg, then $d > 3L/5$. (*Hint:* The bob must be moving at the top of its swing; otherwise the string will collapse.)

34. A block of mass m at the end of a string is whirled around in a vertical circle of radius R. Find the critical speed below which the string would become slack at the highest point.

35. A 3.22-kg block starts at rest and slides a distance d down a frictionless 28.0° incline where it runs into a spring of negligible mass; see Fig. 32. The block slides an additional 21.4 cm before it is brought to rest momentarily by compressing the spring, whose force constant is 427 N/m. (a) What is the value of d? (b) The speed of the block continues to increase for a certain interval after the block makes contact with the spring. What additional distance does the block slide before it reaches its maximum speed and begins to slow down?

36. A boy is seated on the top of a hemispherical mound of ice (Fig. 39). He is given a very small push and starts sliding down the ice. Show that he leaves the ice at a point whose height is $2R/3$ if the ice is frictionless. (*Hint:* The normal force vanishes as he leaves the ice.)

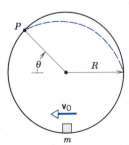

Figure 39 Problem 36.

37. The particle m in Fig. 40 is moving in a vertical circle of radius R inside a track. There is no friction. When m is at its lowest position, its speed is v_0. (a) What is the minimum

Figure 40 Problem 37.

value v_m of v_0 for which m will go completely around the circle without losing contact with the track? (b) Suppose v_0 is $0.775v_m$. The particle will move up the track to some point at P at which it will lose contact with the track and travel along a path shown roughly by the dashed line. Find the angular position θ of point P.

38. Suppose that the rod in Fig. 24 is replaced with a very elastic string, made of rubber, say, and that the string is unextended at length L when the ball is released. (a) Explain why you would expect the ball to reach a low point greater than a distance L below the point of suspension. (b) Show, using dynamic and energy considerations, that if ΔL is small compared to L, the string will stretch by an amount $\Delta L = 3mg/k$, where k is the assumed force constant of the string. Note that the larger k is, the smaller ΔL is, and the better the approximation $\Delta L \ll L$. (c) Show, under these circumstances, that the speed of the ball at the bottom is $v = \sqrt{2g(L - 3mg/2k)}$, *less* than it would be for an inelastic string ($k = \infty$). Give a physical explanation for this result using energy considerations.

Section 8-4 One-Dimensional Conservative Systems: The Complete Solution

39. A particle moves along the x axis through a region in which its potential energy $U(x)$ varies as in Fig. 41. (a) Make a quantitative plot of the force $F(x)$ that acts on the particle, using the same x axis scale as in Fig. 41. (b) The particle has a (constant) mechanical energy E of 4.0 J. Sketch a plot of its kinetic energy $K(x)$ directly on Fig. 41.

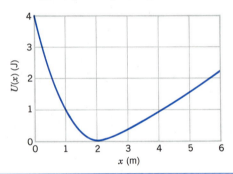

Figure 41 Problem 39.

40. A particle of mass 2.0 kg moves along the x axis through a region in which its potential energy $U(x)$ varies as shown in Fig. 42. When the particle is at $x = 2.0$ m, its velocity is -2.0 m/s. (a) Calculate the force acting on the particle at

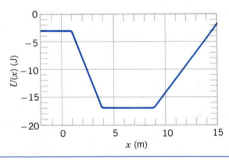

Figure 42 Problem 40.

this position. (*b*) Between what limits does the motion take place? (*c*) How fast is it moving when it is at $x = 7.0$ m?

41. Figure 43*a* shows an atom of mass m at a distance r from a resting atom of mass M, where $m \ll M$. Figure 43*b* shows the potential energy function $U(r)$ for various positions of the lighter atom. Describe the motion of this atom if (*a*) the total mechanical energy is greater than zero, as at E_1, and (*b*) if it is less than zero, as at E_2. For $E_1 = 1.0 \times 10^{-19}$ J and $r = 0.30$ nm, find (*c*) the potential energy, (*d*) the kinetic energy, and (*e*) the force (magnitude and direction) acting on the moving atom.

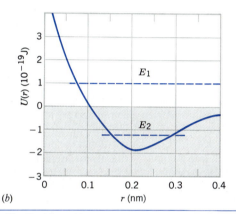

(*b*)

Figure 43 Problem 41.

42. An alpha particle (helium nucleus) inside a large nucleus is bound by a potential energy like that shown in Fig. 44. (*a*) Construct a function of x, which has this general shape, with a minimum value U_0 at $x = 0$ and a maximum value U_1 at $x = x_1$ and $x = -x_1$. (*b*) Determine the force between the alpha particle and the nucleus as a function of x. (*c*) Describe the possible motions.

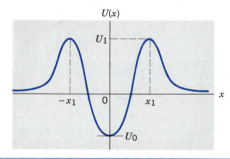

Figure 44 Problem 42.

Section 8-5 Two- and Three-Dimensional Conservative Systems

43. Show that for the same initial speed v_0, the speed v of a projectile will be the same at all points at the same elevation, regardless of the angle of projection. Ignore air drag.

44. The potential energy corresponding to a certain two-dimensional force is given by $U(x, y) = \frac{1}{2}k(x^2 + y^2)$. (*a*) Derive F_x

and F_y and describe the vector force at each point in terms of its coordinates x and y. (*b*) Derive F_r and F_θ and describe the vector force at each point in terms of the polar coordinates r and θ of the point. (*c*) Can you think of a physical model of such a force?

45. The so-called Yukawa potential energy

$$U(r) = -\frac{r_0}{r} U_0 e^{-r/r_0}$$

gives a fairly accurate description of the interaction between nucleons (i.e., neutrons and protons, the constituents of the nucleus). The constant r_0 is about 1.5×10^{-15} m and the constant U_0 is about 50 MeV. (*a*) Find the corresponding expression for the force of attraction. (*b*) To show the short range of this force, compute the ratio of the force at $r = 2r_0$, $4r_0$, and $10r_0$ to the force at $r = r_0$.

46. By integrating along the same three paths as Sample Problem 5, show that the force $\mathbf{F} = -k_1 y \mathbf{i} - k_2 x \mathbf{j}$ is nonconservative when $k_1 \neq k_2$.

Section 8-6 Conservation of Energy in a System of Particles

47. A 25.3-kg bear slides, from rest, 12.2 m down a lodgepole pine tree, moving with a speed of 5.56 m/s at the bottom. (*a*) What is the initial potential energy of the bear? (*b*) Find the kinetic energy of the bear at the bottom. (*c*) What is the change in the mechanical energy of the bear, associated with the action of frictional forces?

48. When a space shuttle (mass 79,000 kg) returns to Earth from orbit, it enters the atmosphere at a speed of 18,000 mi/h, which is gradually reduced to a touchdown speed of 190 knots (= 220 mi/h). What is its kinetic energy (*a*) at atmospheric entry and (*b*) at touchdown? See Fig. 45. (*c*) What happens to the "missing" energy?

Figure 45 Problem 48.

49. A 68-kg skydiver falls at a constant terminal speed of 59 m/s. At what rate is the internal energy of the skydiver and surrounding air increasing?

50. A river descends 15 m in passing through rapids. The speed of the water is 3.2 m/s upon entering the rapids and is 13 m/s as it leaves. What percentage of the potential energy lost by the water in traversing the rapids appears as kinetic energy of water downstream? What happens to the rest of the energy?

51. During a rockslide, a 524-kg rock slides from rest down a hillslope that is 488 m long and 292 m high. The speed of the rock as it reaches the bottom of the hill is 62.6 m/s. How much mechanical energy does the rock lose in the slide due to friction?

52. A projectile whose mass is 9.4 kg is fired vertically upward. On its upward flight, 68 kJ of mechanical energy is dissipated because of air drag. How much higher would it have gone if the air drag had been made negligible (for example, by streamlining the projectile)?

53. A 4.26-kg block starts up a 33.0° incline at 7.81 m/s. How far will it slide if it loses 34.6 J of mechanical energy due to friction?

54. A stone of weight w is thrown vertically upward into the air with an initial speed v_0. Suppose that the air drag force f dissipates an amount fy of mechanical energy as the stone travels a distance y. (a) Show that the maximum height reached by the stone is

$$h = \frac{v_0^2}{2g(1 + f/w)}.$$

(b) Show that the speed of the stone upon impact with the ground is

$$v = v_0 \left(\frac{w - f}{w + f} \right)^{1/2}.$$

55. A 1.34-kg block sliding on a horizontal surface collides with a spring of force constant 1.93 N/cm. The block compresses the spring 4.16 cm from the unextended position. Friction between the block and the surface dissipates 117 mJ of mechanical energy as the block is brought to rest. Find the speed of the block at the instant of collision with the spring.

56. A small object of mass $m = 234$ g slides along a track with elevated ends and a central flat part, as shown in Fig. 46. The flat part has a length $L = 2.16$ m. The curved portions of the track are frictionless. In traversing the flat part, the object loses 688 mJ of mechanical energy, due to friction. The object is released at point A, which is a height $h = 1.05$ m above the flat part of the track. Where does the object finally come to rest?

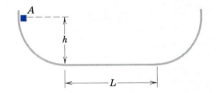

Figure 46   Problem 56.

57. Two snow-covered peaks are at elevations of 862 m and 741 m above the valley between them. A ski run extends from the top of the higher peak to the top of the lower one; see Fig. 47. (a) A skier starts from rest on the higher peak. At what speed would he arrive at the lower peak if he just coasted without using the poles? Assume icy conditions, so

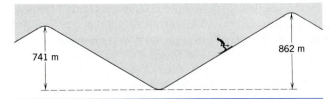

Figure 47   Problem 57.

that there is no friction. (b) After a snowfall, a 54.4-kg skier making the same run also without using the poles only just makes it to the lower peak. By how much does the internal energy of her skis and the snow over which she traveled increase?

58. The magnitude of the force of attraction between the positively charged proton and the negatively charged electron in the hydrogen atom is given by

$$F = k \frac{e^2}{r^2},$$

where e is the charge of the electron, k is a constant, and r is the separation between electron and proton. Assume that the proton is fixed. Imagine that the electron is initially moving in a circle of radius r_1 about the proton and jumps suddenly into a circular orbit of smaller radius r_2; see Fig. 48. (a) Calculate the change in kinetic energy of the electron, using Newton's second law. (b) Using the relation between force and potential energy, calculate the change in potential energy of the atom. (c) By how much has the total energy of the atom changed in this process? (This energy is often given off in the form of radiation.)

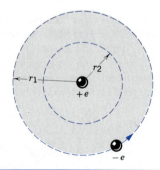

Figure 48   Problem 58.

59. The cable of a 4000-lb elevator in Fig. 49 snaps when the elevator is at rest at the first floor so that the bottom is a distance $d = 12.0$ ft above a cushioning spring whose force constant is $k = 10,000$ lb/ft. A safety device clamps the guide rails removing 1000 ft·lb of mechanical energy for each 1.00 ft that the elevator moves. (a) Find the speed of the elevator just before it hits the spring. (b) Find the distance that the spring is compressed. (c) Find the distance that the elevator will bounce back up the shaft. (d) Calculate approximately the total distance that the elevator will move before coming to rest. Why is the answer not exact?

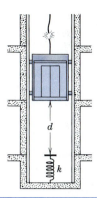

Figure 49 Problem 59.

60. While a 1700-kg automobile is moving at a constant speed of 15 m/s, the motor supplies 16 kW of power to overcome friction, wind resistance, and so on. (*a*) What is the effective retarding force associated with all the frictional forces combined? (*b*) What power must the motor supply if the car is to move up an 8.0% grade (8.0 m vertically for each 100 m horizontally) at 15 m/s? (*c*) At what downgrade, expressed in percentage terms, would the car coast at 15 m/s?

Section 8-7 Mass and Energy

61. (*a*) How much energy in joules is equivalent to a mass of 120 g? (*b*) For how many years would this supply the energy needs of a one-family home consuming energy at the average rate of 1.30 kW?

62. The magnitude *M* of an earthquake on the Richter scale is related to the released energy *E* in joules by the equation

$$\log E = 1.44M + 5.24.$$

(*a*) The 1989 San Francisco area earthquake (see Fig. 50) was of magnitude 7.1. How much energy was released? (*b*) What is the reduction in mass equivalent to this energy release?

63. A nuclear power plant in Oregon supplies 1030 MW of useful power steadily for a year. In addition, 2100 MW of power is discharged as thermal energy to the Columbia river. Compute the change in mass of the nuclear fuel after 1 year of operation.

64. The United States generated about 2.31×10^{12} kW·h of electrical energy in 1983. Suppose the energy was produced in nuclear power plants. Find the reduction in the mass of the fuel that would accompany the production of this amount of energy.

65. An aspirin tablet has a mass of 320 mg. For how many miles would the energy equivalent of this mass, in the form of gasoline, power a car? Assume 30.0 mi/gal and a heat of combustion of gasoline of 130 MJ/gal. Express your answer in terms of the equatorial circumference of the Earth.

66. A spaceship is powered by matter–antimatter annihilation. How much matter and antimatter must annihilate to accelerate the 1820-ton spaceship from rest to one-tenth the speed of light? Use the nonrelativistic formula for kinetic energy.

67. The Sun radiates energy at the rate of 4×10^{26} W. How many "tons of sunlight" does the Earth intercept in 1 day?

68. The *binding energy* of the nucleus of an atom is the difference between the total of the rest energies of its constituent protons and neutrons and the rest energy of the nucleus itself. The nucleus of an atom of gold contains 79 protons and 118 neutrons and has a mass of 196.9232 u. Calculate the binding energy of the nucleus. (A proton has a mass of 1.00728 u, a neutron a mass of 1.00867 u; the rest energy of one atomic mass unit is 931.5 MeV.)

Section 8-8 Quantization of Energy

69. By how much must the energy of an atom change in order to emit light of frequency 5.34×10^{14} s^{-1}?

70. (*a*) A hydrogen atom has an energy of -3.4 eV. If its energy changes to -13.6 eV, what is the frequency of the light? (*b*) Is the light emitted or absorbed?

Figure 50 Problem 62.

Computer Projects

71. Suppose the force acting on a particle is given by $\mathbf{F} = 8xy^3\mathbf{i} + 12x^2y^2\mathbf{j}$. This force is conservative and the potential energy function associated with it is $U = -4x^2y^3$. You can use this function to demonstrate some of the important properties of a conservative force. First, the potential energy of the particle depends only on its coordinates. On a piece of graph paper lay out a coordinate system, with x and y each running from -5 m to $+5$ m. Now use a computer program or spreadsheet to generate values of the potential energy for all integer values of x and y (in meters) between these limits and write the values at appropriate places on your chart. Use the chart to answer the following questions: (a) What work does the force do as the particle moves from $x = -5$ m, $y = -5$ m to the origin? (b) What work does the force do as the particle moves from the origin to $x = +5$ m, $y = +3$ m? (c) What work does the force do as the particle moves from $x = -5$ m, $y = -5$ m to $x = +5$ m, $y = +3$ m? Your answer should be the sum of your answers to parts (a) and (b). (d) The particle starts at the origin with a kinetic energy of 900 J and gets to $x = +5$ m, $y = +2$ m. If this is the only force acting on it what is its kinetic energy when it gets to the second point? (e) The particle starts at the origin with a kinetic energy of 900 J and gets to $x = +5$ m, $y = -2$ m. If this is the only force acting on it what is its kinetic energy when it gets to the second point? (f) The particle starts at the origin with a kinetic energy of 600 J and moves along the line $x = -y$ toward $x = +5$ m, $y = -5$ m. Another force is required to keep it on the path, but assume the second force is always perpendicular to the path. Where does the particle stop?

72. A robot pushes a 20-kg crate at constant velocity across a floor from $x = 0$ to $x = 5.0$ m. Owing to the varying condition of the floor's surface, the robot must push with a variable horizontal force to keep the crate moving at constant velocity. A good representation of this variable force is found to be $F(x) = 0.30mg\sqrt{x}e^{-0.20x}$, where x is in meters and F is in newtons. Evaluate the work done by the robot between $x = 0$ and $x = 5$ m.

The work done is given by $W = \int_0^5 F\, dx$. The integral cannot be evaluated analytically but it can be estimated numerically by means of a computer. Divide the integration region into N intervals, each of width Δx, and let F_i be the value of the force at the center of interval i. Then $\int_0^5 F\, dx \approx \Delta x \sum_{i=1}^{N} F_i$. The smaller you make Δx the better the estimate, but you cannot make it so small that significance is lost when the sum is evaluated. (You may wish to use Simpson's rule, which gives a better estimate. See a calculus text for details.)

Write a computer program or design a spreadsheet to calculate the work done by the force. You should be able to input values of x_0, x_f, and N. The sum can be handled as a loop. Each time around, the force at the center of an interval is evaluated and added to the running sum. For the first run take $N = 20$, then make several more runs, doubling N each time. Stop when two successive results agree to 3 significant figures.

73. The conservative force $\mathbf{F}$, with components $F_x = y(1 - x)e^{-x}$, $F_y = xe^{-x}$, and $F_z = 0$, acts on a particle. (a) Suppose the particle moves from the origin along the x axis to $x = 2.0$ m and then along a line parallel to the y axis to $x = 2.0$ m, $y = 2.0$ m. The work done by the force can be calculated analytically with ease. Do it. Now suppose the object moves from the origin along the y axis to $y = 2.0$ m and then along a line parallel to the x axis to $x = 2.0$ m, $y = 2.0$ m. Again, calculate the work done but this time use numerical integration. See the previous problem for details. Finally, use a numerical integration program to calculate the work done as the object moves along the line $x = y$ from the origin to the $x = 2.0$ m, $y = 2.0$ m. Since the force is conservative, you should get the same answer (within the accuracy of the calculation) for each path. (b) The force $\mathbf{F}$, with components $F_x = y^2(1 - x)e^{-x}$, $F_y = xe^{-x}$, and $F_z = 0$, is not conservative. Calculate the work it does as the particle moves from the origin to $x = 2.0$ m, $y = 2.0$ m along each of the paths described in part (a). Notice that you do not obtain the same answer for different paths.

CHAPTER 9

SYSTEMS OF PARTICLES

Thus far we have treated objects as if they were point particles, having mass but no size. This is really not such a terrible restriction, because all points of an object in simple translational motion move in identical fashion, and it makes no difference whether we treat the object as a particle or as an extended body. For many objects in motion, however, this restriction is not valid. When an object rotates as it moves, for instance, or when its parts vibrate relative to one another, it would not be valid to treat the entire object as a single particle. Even in these more complicated cases, there is one point of the object whose motion under the influence of external forces can be analyzed as that of a simple particle. This point is called the center of mass. *In this chapter we describe how to find the center of mass of objects, and we show that simple rules (Newton's laws again) for the motion of the center of mass of a complex system lead us to the second of the great conservation laws that we shall encounter, the* conservation of linear momentum.

9-1 TWO-PARTICLE SYSTEMS

In Chapters 7 and 8, we used energy concepts to study the motion of a body acted on by a spring force. Let us now look at a problem only slightly more complicated—the one-dimensional motion of two bodies connected by a spring. For simplicity, we shall for the time being assume that, other than the spring force, no net external force acts on the bodies. That is, we assume they are free to slide without friction on a level horizontal surface. As a practical example of such a system, we might consider the motion on an airtrack of two gliders connected by a spring.

When the spring is extended or compressed from its relaxed length, it exerts a force on both bodies, which we can treat individually as particles. The forces on the two particles have equal magnitudes. (We can think of the spring as merely a physical representation of forces the two bodies might exert directly on one another as, for example, two atoms in a molecule. In that case, Newton's third law requires that the forces on the two particles be equal and opposite. The presence of the spring, assumed to be massless, doesn't change this requirement.)

We cannot independently analyze the motions of the two bodies using Newton's laws, because the motion of one depends on the motion of the other. For example, if one body is very much more massive than the other, its

displacement is relatively small, and the displacement of the less massive body is roughly equal to the change in length of the spring. On the other hand, if the two bodies have equal masses, they have displacements that are each equal in magnitude to half the extension of the spring.

Figure 1 illustrates an example of the type of motion we wish to analyze. In this special case, the spring (of force constant k) is given an initial extension, and the two bodies are released from rest. Let the initial extension of the spring be d_i, so that the initial energy is $E_i = U_i + K_i = \frac{1}{2}kd_i^2 + 0$. At any particular instant of time, when the extension of the spring is d, the energy is

$$E = U + K = \tfrac{1}{2}kd^2 + \tfrac{1}{2}m_1v_1^2 + \tfrac{1}{2}m_2v_2^2, \qquad (1)$$

representing the potential energy of the spring and the kinetic energy of the two bodies. Conservation of energy requires that the energy E at any time must equal the initial energy E_i, which gives

$$\tfrac{1}{2}kd_i^2 = \tfrac{1}{2}kd^2 + \tfrac{1}{2}m_1v_1^2 + \tfrac{1}{2}m_2v_2^2. \qquad (2)$$

As Fig. 1 shows, the positions of the two bodies are related by

$$x_2 = x_1 + L + d, \qquad (3)$$

where L is the relaxed length of the spring. Equations 2 and 3 do not provide sufficient information to solve for x_1 and x_2 as functions of the time, and thus we are unable to

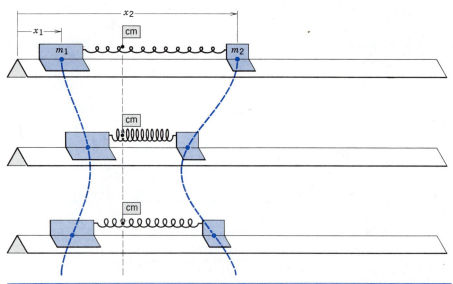

Figure 1   Two gliders connected by a stretched spring are released from rest on an air-track. The resulting motion is not simple, except for the point marked with the flag, which remains at rest. The snapshots are separated by equal time intervals. In the case shown, $m_1 = 2m_2$.

complete the solution to this problem without additional information.

The additional information we need results from the analysis of a particular point of the system of Fig. 1. This point, called the *center of mass* (cm) of the system, is marked with the flag in Fig. 1. In this special case, the center of mass does not move at all.

Let us see how using the center of mass helps us to complete the solution to this problem. The position of the center of mass is defined, for the special case of two particles in one dimension, as

$$x_{cm} = \frac{1}{M}(m_1 x_1 + m_2 x_2), \tag{4}$$

where x_1 and x_2 are the respective x coordinates of the two particles. Here M is the *total mass* of the system:

$$M = m_1 + m_2.$$

The center of mass of a system of two bodies is a point in space defined by Eq. 4 in one dimension. It need not necessarily be a part of either body.

The velocity of the center of mass, v_{cm}, is found by taking the time derivative of Eq. 4:

$$v_{cm} = \frac{dx_{cm}}{dt} = \frac{1}{M}\frac{d}{dt}(m_1 x_1 + m_2 x_2)$$

$$= \frac{1}{M}\left(m_1 \frac{dx_1}{dt} + m_2 \frac{dx_2}{dt}\right)$$

$$= \frac{1}{M}(m_1 v_1 + m_2 v_2), \tag{5}$$

which is the velocity of the flag in Fig. 1. The acceleration

of the center of mass is found by differentiating again. The result is

$$a_{cm} = \frac{dv_{cm}}{dt} = \frac{1}{M}\frac{d}{dt}(m_1 v_1 + m_2 v_2)$$

$$= \frac{1}{M}\left(m_1 \frac{dv_1}{dt} + m_2 \frac{dv_2}{dt}\right)$$

$$= \frac{1}{M}(m_1 a_1 + m_2 a_2), \tag{6}$$

where a_1 and a_2 are the respective accelerations of m_1 and m_2.

We continue by applying Newton's laws separately to m_1 and m_2. Let the force exerted on m_1 by m_2 be $\mathbf{F}_{12}$, and let the force exerted on m_2 by m_1 be $\mathbf{F}_{21}$. Newton's second law applied separately to m_1 and m_2 gives $\mathbf{F}_{12} = m_1 \mathbf{a}_1$ and $\mathbf{F}_{21} = m_2 \mathbf{a}_2$. (In our example, it is the spring that exerts the forces on m_1 and m_2. However, we lose no generality by assuming that the bodies exert direct forces on one another, as long as the spring is considered to be massless.) Newton's *third* law requires that $\mathbf{F}_{12} = -\mathbf{F}_{21}$. Substituting into Eq. 6 gives

$$a_{cm} = \frac{1}{M}(F_{12} + F_{21}) = 0.$$

In this special case, in which no net external force acts on the system, the center of mass has no acceleration and thus moves with constant velocity (which happens to be zero in Fig. 1). Combining Eqs. 2 and 3, and using Eqs. 4 and 5 to eliminate either x_1 and v_1 or x_2 and v_2, we could then complete the solution. (See Problem 1.)

Figure 2 illustrates the slightly more general case in which the spring is given an initial extension and the two

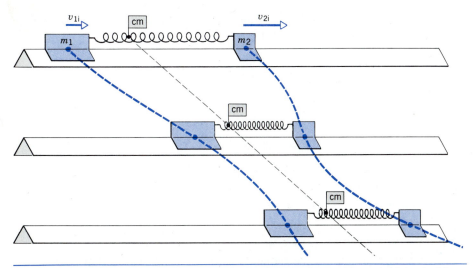

Figure 2 Two gliders connected by a stretched spring are given arbitrary initial velocities. The two gliders move in a complex pattern, while the center of mass, marked with the flag, moves at constant velocity. The snapshots are separated by equal time intervals.

bodies are given initial velocities v_{1i} and v_{2i}. Here you can see that the center of mass moves with constant velocity, even though the motion of the system as a whole is quite complex.

Equations 4–6 are even more general than this particular experiment would suggest. In order to take the most general situation, we let there now be an external force $\mathbf{F}_{\text{ext,1}}$ on m_1 in addition to the internal force $\mathbf{F}_{12}$ on m_1 caused by m_2. (For example, the airtrack might be inclined, so that gravity would act, or the experiment could be done on a surface with friction.) Newton's second law applied to m_1 is

$$\mathbf{F}_{\text{ext,1}} + \mathbf{F}_{12} = m_1 \mathbf{a}_1. \tag{7}$$

Similarly, we assume both an external force $\mathbf{F}_{\text{ext,2}}$ and an internal force $\mathbf{F}_{21}$ act on m_2, and we have

$$\mathbf{F}_{\text{ext,2}} + \mathbf{F}_{21} = m_2 \mathbf{a}_2. \tag{8}$$

Adding Eqs. 7 and 8 gives

$$\mathbf{F}_{\text{ext,1}} + \mathbf{F}_{\text{ext,2}} + \mathbf{F}_{12} + \mathbf{F}_{21} = m_1 \mathbf{a}_1 + m_2 \mathbf{a}_2. \tag{9}$$

The first two terms in this equation give the net external force $\sum \mathbf{F}_{\text{ext}}$ acting on the system (which we previously assumed to be zero in our discussion). The sum of the next two terms, $\mathbf{F}_{12} + \mathbf{F}_{21}$, vanishes because of Newton's third law, which demands that $\mathbf{F}_{21} = -\mathbf{F}_{12}$. The right side of Eq. 9 can be written $M\mathbf{a}_{\text{cm}}$, using Eq. 6. Thus we have the general result

$$\sum \mathbf{F}_{\text{ext}} = M\mathbf{a}_{\text{cm}}. \tag{10}$$

This looks very much like Newton's second law again, as it would be applied to a particle of the same mass M as our system, moving with the same velocity v_{cm} at the location x_{cm}.

Summarizing our results in the case of a two-particle,

one-dimensional system, we see that the whole system can be regarded for certain purposes as moving with velocity v_{cm} and having its total mass M concentrated at the location x_{cm}. Furthermore, in the absence of net external force, $a_{\text{cm}} = 0$ and the center of mass moves with constant velocity. We now develop more general expressions for these concepts.

9-2 MANY-PARTICLE SYSTEMS

In this section we generalize the results of the previous section to systems in three dimensions that contain more than two particles.

We consider a system consisting of N particles of masses $m_1, m_2, \ldots, m_N$. The total mass is

$$M = m_1 + m_2 + \cdots + m_N = \sum m_n.$$

Each particle in the system can be represented by its mass m_n (where $n = 1, 2, \ldots, N$), its location at the coordinate $\mathbf{r}_n$ (whose components are x_n, y_n, and z_n), its velocity $\mathbf{v}_n$ (whose components are v_{nx}, v_{ny}, and v_{nz}), and its acceleration $\mathbf{a}_n$. Each particle is acted on by a force $\mathbf{F}_n$, which in general differs from one particle to another. This force may arise partly from the other $N - 1$ particles and partly from an external agent.

The center of mass of the system can be defined by a logical extension of Eq. 4:

$$x_{\text{cm}} = \frac{1}{M}(m_1 x_1 + m_2 x_2 + \cdots + m_N x_N)$$

$$= \frac{1}{M}\sum m_n x_n, \tag{11a}$$

$$y_{cm} = \frac{1}{M}(m_1 y_1 + m_2 y_2 + \cdots + m_N y_N)$$

$$= \frac{1}{M} \sum m_n y_n, \tag{11b}$$

$$z_{cm} = \frac{1}{M}(m_1 z_1 + m_2 z_2 + \cdots + m_N z_N)$$

$$= \frac{1}{M} \sum m_n z_n. \tag{11c}$$

In the more compact vector notation, these three equations can be written as a single expression giving the position of the center of mass:

$$\mathbf{r}_{cm} = \frac{1}{M}(m_1 \mathbf{r}_1 + m_2 \mathbf{r}_2 + \cdots + m_N \mathbf{r}_N)$$

$$= \frac{1}{M} \sum m_n \mathbf{r}_n. \tag{12}$$

Taking the derivative of this expression, we find the velocity of the center of mass:

$$\mathbf{v}_{cm} = \frac{d\mathbf{r}_{cm}}{dt} = \frac{1}{M}\left(m_1 \frac{d\mathbf{r}_1}{dt} + m_2 \frac{d\mathbf{r}_2}{dt} + \cdots + m_N \frac{d\mathbf{r}_N}{dt}\right)$$

or

$$\mathbf{v}_{cm} = \frac{1}{M}(m_1 \mathbf{v}_1 + m_2 \mathbf{v}_2 + \cdots + m_N \mathbf{v}_N)$$

$$= \frac{1}{M} \sum m_n \mathbf{v}_n. \tag{13}$$

Differentiating once again, we find the acceleration of the center of mass:

$$\mathbf{a}_{cm} = \frac{d\mathbf{v}_{cm}}{dt} = \frac{1}{M}(m_1 \mathbf{a}_1 + m_2 \mathbf{a}_2 + \cdots + m_N \mathbf{a}_N)$$

$$= \frac{1}{M} \sum m_n \mathbf{a}_n. \tag{14}$$

We can rewrite Eq. 14 as

$$M\mathbf{a}_{cm} = m_1 \mathbf{a}_1 + m_2 \mathbf{a}_2 + \cdots + m_N \mathbf{a}_N$$

or

$$M\mathbf{a}_{cm} = \mathbf{F}_1 + \mathbf{F}_2 + \cdots + \mathbf{F}_N, \tag{15}$$

where the last result follows from applying Newton's second law, $\mathbf{F}_n = m_n \mathbf{a}_n$, to each individual particle. The total force acting on a system of particles is thus equal to the total mass of the system times the acceleration of the center of mass. Equation 15 is just Newton's second law for the system of N particles treated as a single particle of mass M located at the center of mass, moving with velocity $\mathbf{v}_{cm}$ and experiencing acceleration $\mathbf{a}_{cm}$.

It is helpful to simplify Eq. 15 even a bit more. Among the forces acting on the particles there are *internal forces*, which arise from the interactions with other particles that are part of the system, and *external forces*, which originate beyond the system under consideration. Any given particle m_n may experience force exerted on it by particle m_k, which we write as $\mathbf{F}_{nk}$. This particular force is one among the many that make up $\mathbf{F}_n$, the total force on m_n. Similarly, the total force on particle m_k includes a term $\mathbf{F}_{kn}$ due

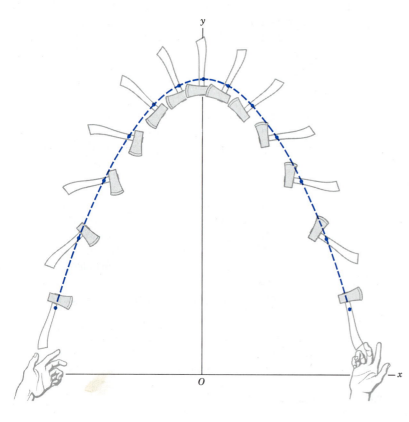

Figure 3   An axe is tossed between two performers and rotates as it travels. The parabolic path of the center of mass (represented by the dot on the axe) is indicated by the dashed line. A particle tossed in the same way would follow that same path. No other point on the axe moves in such a simple way.

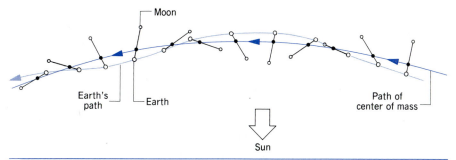

Figure 4   The center of mass of the Earth–Moon system follows a nearly circular orbit about the Sun, while the Earth and Moon rotate about their common center of mass, just like the axe of Fig. 3. This effect, which causes a slight "wobble" in the orbit of the Earth, is greatly exaggerated in the figure. The center of mass of the Earth–Moon system actually lies within the Earth, so the Earth always overlaps the orbital path of the center of mass.

to the interaction with particle m_n. By Newton's third law, $\mathbf{F}_{nk} = -\mathbf{F}_{kn}$, and thus these two particular forces cancel when we carry out the sum of all the forces in Eq. 15. In fact, all such internal forces are part of action–reaction pairs and cancel. (In Chapter 5 we cautioned that the action and reaction forces must apply to different particles and thus cannot oppose one another. We are not violating that caution here, because we are applying the action to one particle and the reaction to another. The distinction here is that we are adding to get the net force on the *two* particles, in which case the action and reaction components, which still apply to different particles, do indeed cancel.)

All that remains in Eq. 15 is the total of all the *external* forces, and Eq. 15 reduces to

$$\sum \mathbf{F}_{\text{ext}} = M\mathbf{a}_{\text{cm}}, \qquad (16)$$

which can be written in terms of its components as

and
$$\sum F_{\text{ext},x} = Ma_{\text{cm},x}, \quad \sum F_{\text{ext},y} = Ma_{\text{cm},y},$$
$$\sum F_{\text{ext},z} = Ma_{\text{cm},z}.$$

We can summarize this important result as follows:

The overall translational motion of a system of particles can be analyzed using Newton's laws as if all the mass were concentrated at the center of mass and the total external force were applied at that point.

A corollary follows immediately in the case $\sum \mathbf{F}_{\text{ext}} = 0$:

If the net external force on a system of particles is zero, then the center of mass of the system moves with constant velocity.

This explains the observation we made in Section 9-1 in studying the problem of the two masses connected by a spring.

These are general results that apply equally well to collections of individual particles as they do to particles

joined together by internal forces, as in a solid object. The object itself may be executing any sort of complicated motion, but the center of mass moves according to Eq. 16. Figure 3 shows a complex object moving under the influence of gravity. As it travels, it also rotates. Its center of mass, however, follows a simple parabolic path. As far as the external force (gravity) is concerned, the system behaves as if it were a particle of mass M located at the center of mass. A complicated problem is therefore reduced to two relatively simple problems—the parabolic path of the center of mass and a rotation about the center of mass.

For another example, consider the Earth–Moon system moving under the Sun's gravity (the external force). Figure 4 shows that the center of mass of the system follows a stable orbit around the Sun; this is the path that would be followed by a particle of mass $m_{\text{Earth}} + m_{\text{Moon}}$. The Earth and Moon also rotate about their center of mass, resulting in a slight oscillation of the Earth about the path of the stable orbit. Using the data in Appendix C, you should be able to show that the center of mass of the Earth–Moon system is about 4600 km from the center of the Earth and thus lies within the Earth.

Figure 5 shows the motion of a ballistic missile that breaks apart into three multiple re-entry vehicles

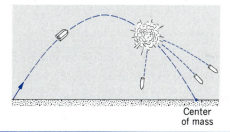

Figure 5   A missile containing three warheads follows a parabolic path. An explosion releases the three warheads, which travel so that their center of mass follows the original parabolic path. For simplicity the "bus" that carries the three warheads is not shown.

(MRVs). In effect, an explosion releases the three separate vehicles, but since the explosion produces only internal forces it does not affect the motion of the center of mass. The center of mass continues to follow the ballistic path as if the explosion had not occurred, until one or more of the vehicles experiences an external force, such as from atmospheric drag or impact with a target.

Sample Problem 1   Figure 6*a* shows a system of three initially resting particles of masses $m_1 = 4.1$ kg, $m_2 = 8.2$ kg, and $m_3 = 4.1$ kg. The particles are acted on by different external forces, which have magnitudes $F_1 = 6$ N, $F_2 = 12$ N, and $F_3 = 14$ N. The directions of the forces are shown in the figure. Where is the center of mass of this system, and what is the acceleration of the center of mass?

Solution   The position of the center of mass is marked by a dot in the figure. As Fig. 6*b* suggests, we treat this point as if it held a real particle, assigning to it a mass M equal to the system mass of $m_1 + m_2 + m_3 = 16.4$ kg and assuming that all external forces are applied at that point. We find the center of mass from Eqs. 11*a* and 11*b*:

$$x_{cm} = \frac{1}{M}(m_1 x_1 + m_2 x_2 + m_3 x_3)$$

$$= \frac{1}{16.4 \text{ kg}}[(4.1 \text{ kg})(-2 \text{ cm}) + (8.2 \text{ kg})(4 \text{ cm})$$

$$+ (4.1 \text{ kg})(1 \text{ cm})] = 1.8 \text{ cm},$$

$$y_{cm} = \frac{1}{M}(m_1 y_1 + m_2 y_2 + m_3 y_3)$$

$$= \frac{1}{16.4 \text{ kg}}[(4.1 \text{ kg})(3 \text{ cm}) + (8.2 \text{ kg})(2 \text{ cm})$$

$$+ (4.1 \text{ kg})(-2 \text{ cm})] = 1.3 \text{ cm}.$$

Note the quite acceptable use of mixed units here.

The x component of the net external force acting on the center of mass is (see Fig. 6*b*)

$$F_{ext, x} = F_{1x} + F_{2x} + F_{3x}$$
$$= -6 \text{ N} + (12 \text{ N})(\cos 45°) + 14 \text{ N} = 16.5 \text{ N},$$

and the y component is

$$F_{ext, y} = F_{1y} + F_{2y} + F_{3y}$$
$$= 0 + (12 \text{ N})(\sin 45°) + 0 = 8.5 \text{ N}.$$

The net external force thus has a magnitude of

$$F_{ext} = \sqrt{(F_{ext, x})^2 + (F_{ext, y})^2} = \sqrt{(16.5 \text{ N})^2 + (8.5 \text{ N})^2} = 18.6 \text{ N}$$

and makes an angle with the x axis given by

$$\phi = \tan^{-1}\frac{F_{ext, y}}{F_{ext, x}} = \tan^{-1}\frac{8.5 \text{ N}}{16.5 \text{ N}} = 27°.$$

This is also the direction of the acceleration vector. From Eq. 16, the magnitude of the acceleration of the center of mass is given by

$$a_{cm} = \frac{F_{ext}}{M} = \frac{18.6 \text{ N}}{16.4 \text{ kg}} = 1.1 \text{ m/s}^2.$$

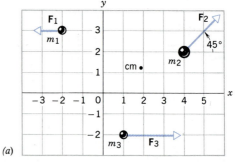

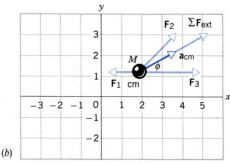

Figure 6   Sample Problem 1. (*a*) Three particles, placed at rest at the positions shown, are acted on by the forces shown. The center of mass of the system is marked. (*b*) The translational motion of the entire system can be represented by the motion of a particle with the total mass M located at the center of mass and acted on by the three external forces. The resultant force and the acceleration of the center of mass are shown.

All three particles of Fig. 6*a*, and also their center of mass, move with (different) constant accelerations. If the particles start from rest, each will move, with ever-increasing speed, along a straight line in the direction of the force acting on it.

Sample Problem 2   In the system illustrated in Fig. 7*a*, find the common magnitude of the accelerations of the two blocks. We have already solved this problem, as Sample Problem 8 of Chapter 5, by applying Newton's laws separately to each block. Solve the problem in this case by considering the motion of the center of mass of the two-particle system.

Solution   Figure 7*b* shows a free-body diagram for the two-particle system. We first find the center of mass by applying Eqs. 11*a* and 11*b* to the system shown in Fig. 7*b*:

$$x_{cm} = -\frac{m_1}{M}(L - y) \quad \text{and} \quad y_{cm} = \frac{m_2}{M}y,$$

in which L is the length of the cord and y is the vertical coordinate of m_2.

Differentiating with respect to time, we can find the velocity components of the center of mass:

$$v_{cm, x} = \frac{m_1}{M}v \quad \text{and} \quad v_{cm, y} = \frac{m_2}{M}v,$$

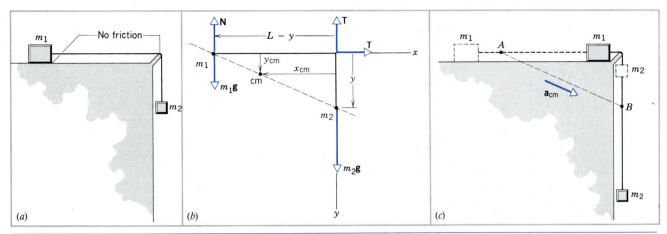

Figure 7   Sample Problem 2. (*a*) Two masses are connected by a string of length *L* that passes over a frictionless support. (*b*) A partial free-body diagram shows the relevant *external* forces on the system. The frictionless support exerts an external force on the string whose components are each equal to the tension *T* in the string (which is an *internal* force and so is not shown). (*c*) The center of mass moves from point *A*, when m_2 is at its highest position, to point *B*, when m_1 reaches the support. As m_2 falls, m_1 moves to the right, and so the center of mass must move to the right. The horizontal force *T* is the only possible external force that can result in the horizontal motion of the center of mass. Gravity, of course, provides the external force that is responsible for the downward motion of the center of mass.

where $v\ (= dy/dt)$ is the common magnitude of the velocities of the two blocks. Differentiating once again, we can find the components of the acceleration:

$$a_{cm,x} = \frac{m_1}{M}\,a \quad \text{and} \quad a_{cm,y} = \frac{m_2}{M}\,a,$$

in which $a\ (= dv/dt)$ is the common magnitude of the accelerations of the two blocks.

We now apply Newton's laws. In Fig. 7*b*, the external force exerted by the frictionless support on the connecting cord is broken down into its *x* and *y* components, each of magnitude *T* (the tension in the cord). Using Eq. 16 gives

x component: $\qquad\qquad T = M a_{cm,x},$

y component: $\quad m_1 g - N + m_2 g - T = M a_{cm,y}.$

Substituting for $a_{cm,x}$ and $a_{cm,y}$, we can then eliminate *T* from these two equations and obtain, with $m_1 g = N$,

$$a = g\,\frac{m_2}{M},$$

in agreement with the result obtained previously in Chapter 5.

Notice that in this sample problem, we must consider the external force exerted by the frictionless support on the system, which does not enter when we consider the forces on bodies 1 and 2 separately.

If the system is released from rest when m_2 is at its highest position, the resulting motion of the center of mass is along the straight line shown in Fig. 7*c*. The direction of $\mathbf{a}_{cm}$ can be found by taking the vector sum of the five forces that act on the system shown in Fig. 7*b*.

9-3   CENTER OF MASS OF SOLID OBJECTS

It is far too tedious to find the center of mass of a solid object by using Eq. 12 and summing over every atom in the system. Instead we divide the object into tiny elements of mass δm_n. As these elements become infinitesimally small, the sums of Eqs. 11 and 12 transform into integrals:

$$x_{cm} = \frac{1}{M}\lim_{\delta m \to 0}\sum x_n\,\delta m_n = \frac{1}{M}\int x\,dm, \quad (17a)$$

$$y_{cm} = \frac{1}{M}\lim_{\delta m \to 0}\sum y_n\,\delta m_n = \frac{1}{M}\int y\,dm, \quad (17b)$$

$$z_{cm} = \frac{1}{M}\lim_{\delta m \to 0}\sum z_n\,\delta m_n = \frac{1}{M}\int z\,dm. \quad (17c)$$

In vector form, these equations can be written

$$\mathbf{r}_{cm} = \frac{1}{M}\int \mathbf{r}\,dm. \qquad (18)$$

In many cases it is possible to use arguments based on geometry or symmetry to simplify the calculation of the center of mass of solid objects. If an object has spherical symmetry, the center of mass must lie at the geometrical center of the sphere. (It is not necessary that its density be constant; a baseball, for example, has spherical symmetry even though it is composed of layers of different materials.

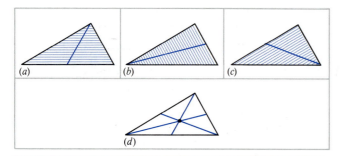

Figure 8   In (*a*), (*b*), and (*c*), the triangle is divided into thin slats, parallel to each of the three sides. The center of mass must lie along the symmetrical dividing lines shown. (*d*) The dot, the only point common to all three lines, is the position of the center of mass.

Its center of mass is at its geometric center. When we refer to spherical symmetry, we mean that the density may vary with r but it must have the same variation in every direction.) If a solid has cylindrical symmetry (that is, if its mass is distributed symmetrically about an axis), then the center of mass must lie on the axis. If its mass is distributed symmetrically about a plane, then the center of mass must be in the plane.

Often we encounter solid, irregular objects that can be divided into several parts. We can find the center of mass of each part, and then by treating each part as a particle located at its own center of mass we can find the center of mass of the combination.

As an example, consider the triangular plate shown in Fig. 8. We divide the plate into a large number of thin strips parallel to the base of the triangle, as in Fig. 8*a*. The center of mass of each strip must lie at its geometrical center, and therefore the center of mass of the plate must lie somewhere along the line connecting the centers of the strips. (Replace each strip with a point mass located at the center of mass of the strip. The row of point masses forms in effect a one-dimensional object whose center of mass must surely lie along its length.) Repeating this procedure for strips drawn parallel to the other two sides (Figs. 8*b* and 8*c*), we obtain two additional lines which each must also include the center of mass of the plate. Superimposing all three lines, as in Fig. 8*d*, we find they have only one point in common, which must therefore be the center of mass.

Sample Problem 3   Figure 9*a* shows a circular metal plate of radius $2R$ from which a disk of radius R has been removed. Let us call it object X. Its center of mass is shown as a dot on the x axis. Locate this point.

Solution   Figure 9*b* shows object X, its hole filled with a disk of radius R, which we call object D. Let us label as object C the large uniform composite disk so formed. From symmetry, the center

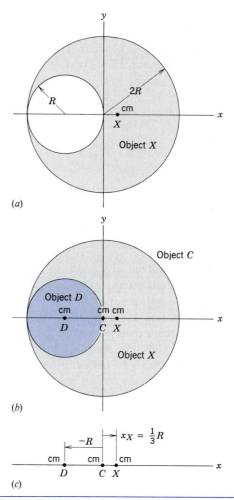

Figure 9   Sample Problem 3. (*a*) Object X is a metal disk of radius $2R$ with a hole of radius R cut in it. (*b*) Object D is a metal disk that fills the hole in object X; its center of mass is at $x_D = -R$. Object C is the composite disk made up of objects X and D; its center of mass is at the origin. (*c*) The centers of mass of the three objects.

of mass of object C is at the origin of the coordinate system, as shown.

In finding the center of mass of a composite object, we can assume that the masses of its components are concentrated at their individual centers of mass. Thus object C can be treated as equivalent to two mass points, representing objects X and D. Figure 9*c* shows the positions of the centers of mass of these three objects.

The position of the center of mass of object C is given from Eq. 11*a* as

$$x_C = \frac{m_D x_D + m_X x_X}{m_D + m_X},$$

in which x_D and x_X are the positions of the centers of mass of objects D and X, respectively. Noting that $x_C = 0$ and solving for x_X, we obtain

$$x_X = -\frac{x_D m_D}{m_X}.$$

The ratio m_D/m_X must be the same as the ratio of the areas of objects D and X (assuming the plate is of uniform density and thickness). That is,

$$\frac{m_D}{m_X} = \frac{\text{area of } D}{\text{area of } X} = \frac{\text{area of } D}{\text{area of } C - \text{area of } D}$$

$$= \frac{\pi R^2}{\pi (2R)^2 - \pi R^2} = \frac{1}{3}.$$

With $x_D = -R$, we obtain

$$x_X = \tfrac{1}{3} R.$$

Sample Problem 4   A thin strip of material is bent into the shape of a semicircle of radius R (Fig. 10). Find its center of mass.

Solution   In this case, using an angular coordinate simplifies the integration to be performed. Furthermore, from the symmetry of the object, we conclude that the center of mass must lie on the y axis (that is, $x_{\text{cm}} = 0$). We therefore use Eq. 17b to find y_{cm}. Consider the small element of mass dm shown in Fig. 10b. It subtends an angle $d\phi$, and since the total mass M of the strip subtends an angle π (a full circle would subtend an angle 2π), the mass dm must be the same fraction of M as $d\phi$ is of π. That is, $dm/M = d\phi/\pi$, or $dm = (M/\pi)d\phi$. The element dm is located at the coordinate $y = R \sin \phi$. In this case we can write Eq. 17b as

$$y_{\text{cm}} = \frac{1}{M} \int y \, dm = \frac{1}{M} \int_0^\pi (R \sin \phi) \frac{M}{\pi} \, d\phi$$

$$= \frac{R}{\pi} \int_0^\pi \sin \phi \, d\phi = \frac{2R}{\pi} = 0.637R.$$

The center of mass is roughly two-thirds of a radius along the y axis. Note that, as this case illustrates, the center of mass does not need to be within the volume or the material of an object.

Sample Problem 5   A ball of mass m and radius R is placed inside a spherical shell of the same mass m and inner radius $2R$. The combination is at rest on a table top as shown in Fig. 11a. The ball is released, rolls back and forth inside, and finally comes to rest at the bottom, as in Fig. 11c. What will be the displacement d of the shell during this process?

Solution   The only external forces acting on the ball–shell system are the downward force of gravity and the normal force exerted vertically upward by the table. Neither force has a horizontal component so that $\Sigma F_{\text{ext}, x} = 0$. From Eq. 16 the acceleration component $a_{\text{cm}, x}$ of the center of mass must also be zero. Thus the horizontal position of the center of mass of the system must remain fixed, and the shell must move in such a way as to make sure that this happens.

We can represent both ball and shell by single particles of mass m, located at their respective centers. Figure 11b shows the system before the ball is released and Fig. 11d after the ball has come

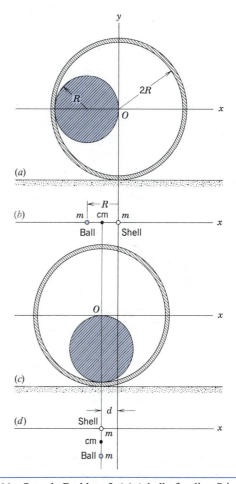

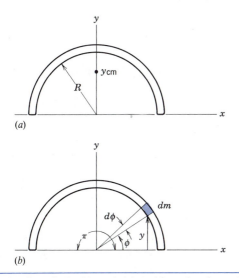

Figure 10   Sample Problem 4. (*a*) A thin strip of metal bent into the shape of a semicircle. (*b*) An element of the strip of mass dm located at the coordinate ϕ.

Figure 11   Sample Problem 5. (*a*) A ball of radius R is released from this initial position and is free to roll inside a spherical shell of radius $2R$. (*b*) The centers of mass of the ball, the shell, and their combination. (*c*) The final state after the ball has come to rest. The shell has moved so that the center of mass of the system remains in place. (*d*) The centers of mass of the ball, the shell, and their combination.

to rest at the bottom of the shell. We choose our origin to coincide with the initial position of the center of the shell. Figure 11*b* shows that, with respect to this origin, the center of mass of the ball–shell system is located a distance $\frac{1}{4}R$ to the left, halfway between the two particles. Figure 11*d* shows that the displacement of the shell is given by

$$d = \tfrac{1}{2}R.$$

The shell must move to the left through this distance as the ball comes to rest.

The ball is brought to rest by the frictional force that acts between it and the shell. Why does this frictional force not affect the final location of the center of mass?

9-4   LINEAR MOMENTUM OF A PARTICLE

The momentum of a single particle is a vector **p** defined as the product of its mass m and its velocity **v**:

$$\mathbf{p} = m\mathbf{v}. \tag{19}$$

Momentum, being the product of a scalar by a vector, is itself a vector. Because it is proportional to **v**, the momentum **p** of a particle depends on the reference frame of the observer; we must always specify this frame.

Newton, in his famous *Principia,* expressed the second law of motion in terms of momentum (which he called "quantity of motion"). Expressed in modern terminology Newton's second law reads:

The rate of change of momentum of a body is equal to the resultant force acting on the body and is in the direction of that force.

In symbolic form this becomes

$$\sum \mathbf{F} = \frac{d\mathbf{p}}{dt}. \tag{20}$$

Here $\sum \mathbf{F}$ represents the resultant force acting on the particle.

For a single particle of constant mass, this form of the second law is equivalent to the form $\mathbf{F} = m\mathbf{a}$ that we have used up to now. That is, if m is constant, then

$$\mathbf{F} = \frac{d\mathbf{p}}{dt} = \frac{d}{dt}(m\mathbf{v}) = m\frac{d\mathbf{v}}{dt} = m\mathbf{a}.$$

The relations $\mathbf{F} = m\mathbf{a}$ and $\mathbf{F} = d\mathbf{p}/dt$ for single particles are completely equivalent in classical mechanics.

A convenient relationship between momentum and kinetic energy is found by combining $K = \frac{1}{2}mv^2$ and $p = mv$, which gives

$$K = \frac{p^2}{2m}. \tag{21}$$

Momentum at High Speeds   *(Optional)*

At particle speeds close to the speed of light (a region in which relativity theory must be used in place of Newtonian mechanics), Newton's second law in the form $\mathbf{F} = m\mathbf{a}$ is no longer valid. However, it turns out that Newton's second law in the form $\mathbf{F} = d\mathbf{p}/dt$ is still a valid law if the momentum **p** for a single particle is defined not as $m\mathbf{v}$ but as

$$\mathbf{p} = \frac{m\mathbf{v}}{\sqrt{1 - v^2/c^2}}, \tag{22}$$

in which c is the speed of light. At ordinary speeds ($v \ll c$), Eq. 22 reduces to Eq. 19.

For relativistic particles, the basic relationship between momentum and kinetic energy can be shown to be

$$K = \sqrt{(pc)^2 + (mc^2)^2} - mc^2. \tag{23}$$

We shall derive this result in Chapter 21. Figure 12 shows a comparison between the classical (Eq. 21) and relativistic (Eq. 23) results for particles of a range of velocities. Obviously the classical result fails at high speed. As expected (see Problem 27), Eq. 23 reduces to Eq. 21 at ordinary speeds.

No matter in what form we write the kinetic energy, it has dimensions of mass times velocity squared, which is the same as momentum times velocity. We can therefore write, using our notation of Section 1-7 to indicate dimensions,

$$[p] = \frac{[K]}{[v]}.$$

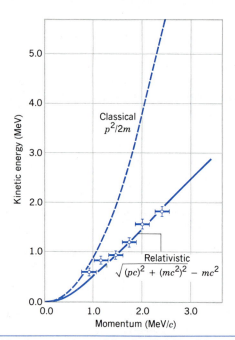

Figure 12   A comparison of the classical (Eq. 21) and relativistic (Eq. 23) relationships between momentum and kinetic energy for electrons emitted in certain radioactive decay processes. The circles represent the experimental measurements; the horizontal and vertical bars passing through the circles represent the range of uncertainty of these measurements. The data obviously favor the relativistic relationship. Notice that at low velocity (small energy and momentum) the two relationships are indistinguishable.

It is often convenient to express momentum in units of energy divided by velocity, and convenient choices in working with particles are eV/c, MeV/c, and so on. This allows us to express the quantity pc in energy units such as MeV, which makes it much more convenient in working with expressions like Eq. 23. For an electron with a momentum given as 1.5 MeV/c, for example, the term pc in Eq. 23 is 1.5 MeV and the kinetic energy of the electron can easily be calculated from that equation to be 1.1 MeV.

In the region of *very* high particle speeds, the particle momentum p can be so great that the term pc in Eq. 23 becomes much larger than the term mc^2, and that equation then reduces to $K = pc$ to a good approximation. Expressing momentum in units of energy divided by c is especially useful in this region. For example, an electron whose momentum is given as 500 MeV/c has a kinetic energy very close to 500 MeV. (Note that this approximation is a very poor one for the 1.5-MeV electron considered above.) ∎

9-5 LINEAR MOMENTUM OF A SYSTEM OF PARTICLES

Suppose that instead of a single particle we have a system of N particles, with masses $m_1, m_2, \ldots, m_N$. We assume that no mass enters or leaves the system, so that the total mass M (= Σm_n) of the system remains constant with time. The particles may interact with each other, and external forces may act on them as well. Each particle has a certain velocity and momentum in the particular reference frame being used. The system as a whole has a total momentum **P**, which is defined to be simply the vector sum of the momenta of the individual particles in that same frame, or

$$\begin{aligned}\mathbf{P} &= \mathbf{p}_1 + \mathbf{p}_2 + \cdots + \mathbf{p}_N \\ &= m_1\mathbf{v}_1 + m_2\mathbf{v}_2 + \cdots + m_N\mathbf{v}_N.\end{aligned} \quad (24)$$

If we compare this relation with Eq. 13, we see at once that

$$\mathbf{P} = M\mathbf{v}_{cm}, \quad (25)$$

which is an equivalent definition for the momentum of a system of particles:

The total linear momentum of a system of particles is equal to the product of the total mass of the system and the velocity of its center of mass.

If we differentiate Eq. 25 with respect to time we obtain, for an assumed constant mass M,

$$\frac{d\mathbf{P}}{dt} = M\frac{d\mathbf{v}_{cm}}{dt} = M\mathbf{a}_{cm}. \quad (26)$$

Comparison of Eq. 26 with Eq. 16, $\Sigma \mathbf{F}_{ext} = M\mathbf{a}_{cm}$, allows us to write Newton's second law for a system of particles in the form:

$$\sum \mathbf{F}_{ext} = \frac{d\mathbf{P}}{dt}. \quad (27)$$

Equation 27 states that, in a system of particles, the net external force equals the rate of change of the linear momentum of the system. This equation is the generalization of the single-particle equation $\Sigma \mathbf{F} = d\mathbf{p}/dt$ (Eq. 20) to a system of many particles, when no mass enters or leaves the system. Equation 27 reduces to Eq. 20 for the special case of a single particle, since only external forces can act on a one-particle system. In Section 9-8 we consider modifications of Eq. 27 for systems of variable mass.

9-6 CONSERVATION OF LINEAR MOMENTUM

Suppose that the sum of the external forces acting on a system is zero. Then, from Eq. 27,

$$\frac{d\mathbf{P}}{dt} = 0 \quad \text{or} \quad \mathbf{P} = \text{a constant.}$$

When the net external force acting on a system is zero, the total vector momentum of the system remains constant.

This simple but quite general result is called the law of *conservation of linear momentum.* Like the law of conservation of energy, the law of conservation of linear momentum applies to a vast range of physical situations and has no known exceptions.

Conservation laws (such as those of energy and linear momentum, which we have already encountered, and those of angular momentum and electric charge, which we shall encounter later in the text) are of theoretical and practical importance in physics because they are simple and universal. The laws of conservation of energy and of linear momentum, for example, go beyond the limitations of classical mechanics and remain valid in both the relativistic and quantum realms.

Conservation laws all have the following form. While the system is changing there is one aspect of the system that remains unchanged. Different observers, each in a different reference frame, would all agree, if they watched the same changing system, that the conservation laws applied to the system. For the conservation of linear momentum, for example, observers in different inertial reference frames would assign different values of **P** to the linear momentum of the system, but each would agree (assuming $\Sigma \mathbf{F}_{ext} = 0$) that the value of **P** remained unchanged as the particles that make up the system move about. The force **F** is an invariant with respect to Galilean transformations (all inertial observers agree on its measurement). If $\Sigma \mathbf{F}_{ext} = 0$ in *any* inertial frame, then *all* inertial observers will also find $\Sigma \mathbf{F}_{ext} = 0$ and will conclude that momentum is conserved.

The total momentum of a system can be changed only

by external forces acting on the system. The internal forces, being equal and opposite, produce equal and opposite changes in momentum, which cancel each other. For a system of particles on which no net external force acts,

$$\mathbf{p}_1 + \mathbf{p}_2 + \cdots + \mathbf{p}_N = \text{a constant.} \quad (28)$$

The momenta of the individual particles may change, but their sum remains constant if there is no external force.

Momentum is a vector quantity. Equation 28 is therefore equivalent to three scalar equations, one for each coordinate direction. Hence the conservation of linear momentum supplies us with three conditions on the motion of a system to which it applies. The conservation of energy, on the other hand, supplies us with only one condition on the motion of a system to which it applies, because energy is a scalar.

If our system of particles consists of only a single particle, then Eq. 28 reduces to a statement that when no net force acts on it the momentum of the particle is a constant, which (for a single particle) is equivalent to stating that its velocity is a constant. This is simply a restatement of Newton's first law.

Sample Problem 6   A stream of bullets whose mass m is each 3.8 g is fired horizontally with a speed v of 1100 m/s into a large wooden block of mass M $(= 12$ kg) that is initially at rest on a horizontal table; see Fig. 13. If the block is free to slide without friction across the table, what speed will it acquire after it has absorbed 8 bullets?

Solution   Equation 28 ($P = $ constant) is valid only for closed systems, in which no particles leave or enter. Thus our system must include both the block and the 8 bullets, taken together. In Fig. 13, we have identified this system by drawing a closed curve around it.

For the moment we consider only the horizontal direction. No external horizontal force acts on the system of block + bullets. The forces that act when the bullets strike the block are internal forces and do not contribute to $\mathbf{F}_{\text{ext}}$, which has no horizontal component.

Because no (horizontal) external forces act, we can apply the law of conservation of momentum (Eq. 28). The initial (horizontal) momentum, measured while the bullets are still in flight and the block is at rest, is

$$P_{\text{i}} = N(mv),$$

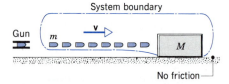

Figure 13   Sample Problem 6. A gun fires a stream of bullets toward a block of wood. We analyze the system that we define to be the block plus the bullets in flight.

in which mv is the momentum of an individual bullet and $N = 8$. The final momentum, measured when all the bullets are in the block and the block is sliding over the table with speed V, is

$$P_{\text{f}} = (M + Nm)V.$$

Conservation of momentum requires that

$$P_{\text{i}} = P_{\text{f}}$$

or

$$N(mv) = (M + Nm)V.$$

Solving for V yields

$$V = \frac{Nm}{M + Nm}\,v = \frac{(8)(3.8 \times 10^{-3}\ \text{kg})}{12\ \text{kg} + (8)(3.8 \times 10^{-3}\ \text{kg})}(1100\ \text{m/s})$$
$$= 2.8\ \text{m/s}.$$

With the choice of system that we made, we did not have to consider the forces exerted when the bullets hit the block. Those forces are all internal.

In the vertical direction, the external forces are the weight of the bullets, the weight of the block, and the normal force on the block. While the bullets are in flight, they acquire a small vertical momentum component as a result of the action of gravity. When the bullets strike the block, the block must exert on each bullet a force with both horizontal and vertical components. Along with the vertical force on the bullet, which is necessary to change its vertical momentum to zero, there must (according to Newton's third law) be a corresponding increase in the normal force exerted on the block by the horizontal surface. This increase is not only from the weight of the imbedded bullet; it has an additional contribution arising from the rate of change of the vertical momentum of the bullet. When all the bullets have come to rest relative to the block, the normal force will equal the combined weights of block and imbedded bullets.

For simplicity in solving this problem, we have assumed that the bullets are fired so rapidly that all 8 are in flight before the first bullet strikes the block. Can you solve this problem without making this assumption?

Suppose the system boundary is enlarged so that it includes the gun, which is fixed to the Earth. Does the horizontal momentum of this system change before and after the firing? Is there a horizontal external force?

Sample Problem 7   As Fig. 14 shows, a cannon whose mass M is 1300 kg fires a 72-kg ball in a horizontal direction with a muzzle speed v of 55 m/s. The cannon is mounted so that it can recoil freely. (a) What is the velocity V of the recoiling cannon with respect to the Earth? (b) What is the initial velocity v_{E} of the ball with respect to the Earth?

Solution   (a) We choose the cannon plus the ball as our system. By doing so, the forces associated with the firing of the cannon are internal to the system, and we do not have to deal with them. The external forces acting on the system have no horizontal components. Thus the horizontal component of the total linear momentum of the system must remain unchanged as the cannon is fired.

We choose a reference frame fixed with respect to the Earth, and we assume that all velocities are positive if they point to the right in Fig. 14.

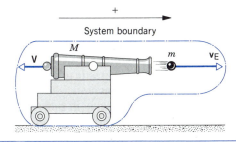

Figure 14 Sample Problem 7. A cannon of mass M fires a ball of mass m. The velocities of the ball and the recoiling cannon are shown in a reference frame fixed with respect to the Earth. Velocities are taken as positive to the right.

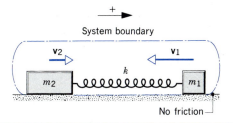

Figure 15 Sample Problem 8. Two blocks, resting on a frictionless surface and connected by a spring, have been pulled apart and then released from rest. The initial total momentum is zero, and so it must remain zero at all subsequent times. Velocities are taken as positive to the right.

Before the cannon is fired, the system has an initial momentum P_i of zero. After the cannon has fired, the ball has a horizontal velocity v with respect to the recoiling cannon, v being the ball's muzzle speed. In the reference frame of the Earth, however, the horizontal velocity of the ball is $v + V$. Thus, the total linear momentum of the system after firing is

$$P_f = MV + m(v + V),$$

in which the first term on the right is the momentum of the recoiling cannon and the second term that of the speeding ball.

Conservation of linear momentum in the horizontal direction requires that $P_i = P_f$, or

$$0 = MV + m(v + V).$$

Solving for V yields

$$V = -\frac{mv}{M + m} = -\frac{(72 \text{ kg})(55 \text{ m/s})}{1300 \text{ kg} + 72 \text{ kg}} = -2.9 \text{ m/s}.$$

The minus sign tells us that the cannon recoils to the left in Fig. 14, as we expect it should.

(*b*) The velocity of the ball with respect to the (recoiling) cannon is its muzzle speed v. With respect to the Earth, the velocity of the ball is

$$\begin{aligned} v_E &= v + V \\ &= 55 \text{ m/s} + (-2.9 \text{ m/s}) = 52 \text{ m/s}. \end{aligned}$$

Because of the recoil, the ball is moving a little slower with respect to the Earth than it otherwise would. Note the importance in this problem of choosing the system (cannon + ball) wisely and being absolutely clear about the reference frame (Earth or recoiling cannon) to which the various measurements are referred.

Sample Problem 8   Figure 15 shows two blocks, connected by a spring and free to slide on a frictionless horizontal surface. The blocks, whose masses are m_1 and m_2, are pulled apart and then released from rest. What fraction of the total kinetic energy of the system will each block have at any later time?

Solution   We take the two blocks and the spring (assumed massless) as our system and the horizontal surface on which they slide as our reference frame. We assume that velocities are positive if they point to the right in Fig. 15.

The initial momentum P_i of the system before the blocks are

released is zero. The final momentum, at any time after the blocks are released, is

$$P_f = m_1 v_1 + m_2 v_2,$$

in which v_1 and v_2 are the velocities of the blocks. Conservation of momentum requires that $P_i = P_f$, or

$$0 = m_1 v_1 + m_2 v_2 .$$

Thus we have

$$\frac{v_1}{v_2} = -\frac{m_2}{m_1}, \qquad (29)$$

the minus sign telling us that the two velocities always have opposite directions. This holds at every instant after release, no matter what the individual speeds of the blocks.

The kinetic energies of the blocks are $K_1 = \frac{1}{2}m_1 v_1^2$ and $K_2 = \frac{1}{2}m_2 v_2^2$. The fraction we seek, for the block of mass m_1, is

$$f_1 = \frac{K_1}{K_1 + K_2} = \frac{\frac{1}{2}m_1 v_1^2}{\frac{1}{2}m_1 v_1^2 + \frac{1}{2}m_2 v_2^2}.$$

Substituting $v_2 = -v_1 (m_1/m_2)$ leads, after a little algebra, to

$$f_1 = \frac{m_2}{m_1 + m_2}.$$

Similarly, for the block of mass m_2,

$$f_2 = \frac{m_1}{m_1 + m_2}.$$

Thus, although the kinetic energy of the oscillating system varies with time, the division of this energy between the two blocks is a constant, independent of time, the least massive block receiving the largest share of the available kinetic energy. If, for example, $m_2 = 10m_1$, then

$$f_1 = \frac{10m_1}{m_1 + 10m_1} = 0.91 \quad \text{and} \quad f_2 = \frac{m_1}{m_1 + 10m_1} = 0.09.$$

In this case, the lighter block (m_1) gets 91% of the available kinetic energy and the heavier block (m_2) gets the remaining 9%. In the limit $m_2 \gg m_1$, the lighter block gets essentially all the kinetic energy.

The expressions for f_1 and f_2 apply equally well to a stone falling in the gravitational field of the Earth. Let m_2 represent the mass of the Earth and m_1 the mass of the stone. In the reference frame of their center of mass, the stone takes nearly all the kinetic energy ($f_1 \approx 1$) and the Earth takes very little ($f_2 \approx 0$). The magnitudes of the linear momenta of the stone and Earth

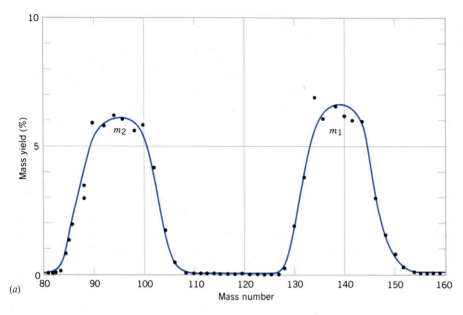

(a)

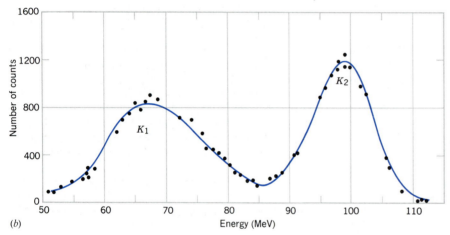

(b)

Figure 16 (*a*) The mass distribution of the fragments emitted in nuclear fission. The vertical scale gives the fraction of fissions that result in a fragment with the mass number given by the horizontal scale. (*b*) The energy distribution for fragments emitted in fission.

are equal, but the small velocity of the Earth is compensated by its enormous mass. This argument justifies neglecting the kinetic energy of the Earth when we used the conservation of energy in Chapter 8 to analyze objects falling in the Earth's gravity.

Another practical example of this effect occurs in the case of nuclear fission, in which a heavy nucleus such as ^{235}U splits into two lighter fragments. The fragments are driven apart by their mutual electrical repulsion from an initial position in which they are very close together and nearly at rest. From Eq. 29, we expect the ratio of the kinetic energies to be

$$\frac{K_1}{K_2} = \frac{\frac{1}{2}m_1 v_1^2}{\frac{1}{2}m_2 v_2^2} = \left(\frac{m_1}{m_2}\right)\left(\frac{v_1}{v_2}\right)^2 = \frac{m_2}{m_1}.$$

That is, the heavier fragment gets the smaller kinetic energy.

Fission is a statistical process, in which there is a distribution of possible masses of the fragments and a corresponding distribution in the fragment kinetic energies. Figure 16*a* shows the mass distribution and Fig. 16*b* shows the kinetic energy distribution. Note that fission into fragments of equal mass is very rare; one fragment usually has a mass number of about 138 and the other about 94. A typical mass ratio m_2/m_1 is thus about $94/138 = 0.68$. A typical kinetic energy ratio K_1/K_2 is about

67 MeV/99 MeV = 0.68, equal to the typical mass ratio, as expected. Thus the sharing of kinetic energy between the fission fragments is done according to the restriction that momentum is conserved.

9-7 WORK AND ENERGY IN A SYSTEM OF PARTICLES *(Optional)*

Figure 17 shows a skater pushing away from a railing, gaining kinetic energy in the process. If you ask the skater where this kinetic energy comes from, he will probably tell you that, judging by his muscular exertions, the required energy must come from his own store of internal energy. Let us try to verify this claim by applying conservation of energy to the system consisting of the skater alone.

From Eq. 28 of Chapter 8 we have

$$\Delta U + \Delta K_{\text{cm}} + \Delta E_{\text{int}} = W. \qquad (30)$$

In deriving Eq. 33 of Chapter 8, we divided the kinetic energy of a system into two terms: ΔK_{int}, which represented the internal

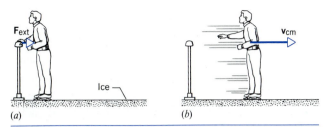

Figure 17   (*a*) A skater pushing away from a railing. The railing exerts a force $\mathbf{F}_{\text{ext}}$ on the skater. (*b*) After pushing off, the skater is moving with velocity $\mathbf{v}_{\text{cm}}$.

motions of the particles in a system, and ΔK, which represented the "overall" motion of the system. Here we indicate explicitly that this "overall" motion is in fact the center-of-mass motion, and the corresponding kinetic energy is

$$K_{\text{cm}} = \tfrac{1}{2}Mv_{\text{cm}}^2, \tag{31}$$

which is the kinetic energy the system of total mass M would have if it moved as a particle with speed v_{cm}. The internal kinetic energy is included in Eq. 30 as a part of ΔE_{int}. (See Problem 49 for a derivation of this division of the kinetic energy.)

There are no changes in potential energy for the skater (the ice is horizontal), so we put $\Delta U = 0$. Furthermore, the railing does no work on the skater because *the point of application of the force does not move*. Recall the discussion accompanying Fig. 13 of Chapter 8. When external work is done on a system, energy is transferred through the system boundary. No energy is transferred from the railing to the skater, so the railing does no external work on the skater. Thus $W = 0$, and Eq. 30 reduces to

$$\Delta K_{\text{cm}} = -\Delta E_{\text{int}}. \tag{32}$$

Since ΔK_{cm} is a positive quantity (the skater gains kinetic energy in pushing away from the railing), ΔE_{int} must be a negative quantity. This confirms the skater's claim: the kinetic energy he acquires in pushing away from the railing comes from his store of internal energy and not from any external source.

The energy analysis is useful, but we may wish to go further and analyze the system in terms of forces and accelerations. Let us see what we can learn about the skater by applying Newton's second law. The railing exerts a force $\mathbf{F}_{\text{ext}}$ on the skater (which we continue to regard as our system). To push away from the railing, the skater must necessarily straighten his arm. As a result, different parts of his body may have different displacements, velocities, and accelerations while he is pushing. Thus the skater must be treated not as a single particle but as a system of particles. In this case, using Eq. 16, we can find the acceleration of the skater's center of mass if we know the external force exerted on him by the railing:

$$F_{\text{ext}} = Ma_{\text{cm}}. \tag{33}$$

For a single particle, we found the work–energy theorem ($W = \Delta K$) to be a useful result. It is clear that we *cannot* apply this theorem to the skater, because the skater does not move as a single particle. As we have already deduced, $W = 0$, but $\Delta K \neq 0$. Thus the single-particle form of the work–energy theorem is not valid. Let us try to find a relationship that is applicable to a system of particles.

Let a net external force $\mathbf{F}_{\text{ext}}$ act on a system of particles. We consider the general case in which the point of application of this

force may move or (as in the case of the skater of Fig. 17) may *not* move in our chosen inertial reference frame. We assume that all forces and motions are in the x direction. Because we are dealing with a system of particles, we focus our attention not on the motion of the point of application of the external force but on the motion of the center of mass of the system.

Let the center of mass of the system move a distance dx_{cm} along the x axis. Multiplying each side of Eq. 33 by dx_{cm} gives

$$F_{\text{ext}}\, dx_{\text{cm}} = Ma_{\text{cm}}\, dx_{\text{cm}} = M\frac{dv_{\text{cm}}}{dt}\, v_{\text{cm}}\, dt,$$

where we have replaced a_{cm} by dv_{cm}/dt and dx_{cm} by $v_{\text{cm}}\, dt$. This gives

$$F_{\text{ext}}\, dx_{\text{cm}} = Mv_{\text{cm}}\, dv_{\text{cm}}. \tag{34}$$

Suppose the center of mass moves from x_i to x_f as this force acts. Integrating Eq. 34 between these limits gives

$$\int_{x_i}^{x_f} F_{\text{ext}}\, dx_{\text{cm}} = \int_{v_{\text{cm,i}}}^{v_{\text{cm,f}}} Mv_{\text{cm}}\, dv_{\text{cm}} = \tfrac{1}{2}Mv_{\text{cm,f}}^2 - \tfrac{1}{2}Mv_{\text{cm,i}}^2. \tag{35}$$

The right side of Eq. 35 can be written using Eq. 31 as $K_{\text{cm,f}} - K_{\text{cm,i}} = \Delta K_{\text{cm}}$. This represents the change in kinetic energy that would be experienced by a particle of mass M whose velocity changed from $v_{\text{cm,i}}$ to $v_{\text{cm,f}}$.

The left side of Eq. 35 looks somewhat like the definition of work, and in fact the integral has the dimension of work. It is, however, not work in the sense that we have defined it, because dx_{cm} is not the displacement of the point of application of the external force. (In our original definition of work $W = \int F\, dx$ in Chapter 7, dx was the displacement of the point of application of F.) Note again that the displacement of the point of application of the external force is zero in Fig. 17, and so $W = 0$ in that case, but the left side of Eq. 35 is not zero.*

In many cases of interest to us, the external force is constant and can be taken out of the integral in Eq. 35. The remaining integral gives the net displacement s_{cm} of the center of mass of the system. In this case we can rewrite Eq. 35 as

$$F_{\text{ext}}s_{\text{cm}} = \Delta K_{\text{cm}}. \tag{36}$$

Equation 35 resembles the work–energy theorem for a particle, and in fact it reduces to that result if our system consists only of a single particle (or of a body that can be treated as a particle). There is, however, an important difference between Eq. 35 and the work–energy theorem for a particle. The work–energy theorem for a single particle is also a statement about conservation of energy in the motion of a particle, because translational energy is the only kind of energy a particle can have. Equation 35, on the other hand, is in no sense an expression of energy conservation, because a system of particles can have energy in other forms, including internal, potential, and rotational, among others. For a system of particles, Eq. 35 and the conser-

---

* Some authors use the terms *pseudowork* or *center-of-mass work* to describe the left side of Eq. 35. This equation is sometimes known as the *center-of-mass equation*. We prefer *not* to introduce a term closely related to work to describe a quantity that is unrelated to the accepted meaning of work. For a comprehensive summary of work and energy in a system of particles, see "Developing the Energy Concepts in Introductory Physics," by A. B. Arons, *The Physics Teacher,* October 1989, p. 506.

vation of energy (Eq. 30) can be applied as separate and independent relationships.

As an example of the application of these principles, let us consider the result of pushing on a meter stick (initially at rest) that is free to slide without friction on a horizontal surface. We supply a constant force of magnitude F_{ext}, which can be applied anywhere on the stick. If we apply the force at the 50-cm mark (Fig. 18a), the stick moves like a particle with an acceleration $a_{cm} = F_{ext}/m$; every point on the stick moves with this acceleration. The displacement s of the point at which we apply the force is equal to the displacement s_{cm} of the center of mass. In this case, we do work of magnitude $F_{ext}s$ when the entire stick (moving as a particle) is displaced by s_{cm}. The particle form of the work–energy theorem can be used to find the resulting speed v of every point of the stick. Now consider the case when the force is applied at the 25-cm mark (Fig. 18b). If you try this experiment yourself, you will find that the stick does *not* move like a particle. As we shall discuss in Chapter 12, we can divide this complex motion into two parts—translational motion as a particle and rotation about the center of mass. The point at which we apply the force moves through a distance *greater than* s_{cm} as you can see from Fig. 18b. The work that we do on the stick is therefore *greater than* $F_{ext}s_{cm}$. To analyze this motion we must use both Eqs. 30 and 36. The product $F_{ext}s_{cm}$ gives, using Eq. 36, the change in the *translational* kinetic energy of the stick. The product $F_{ext}s$, where s is the distance moved by the 25-cm point at

which the force is applied, gives the work W that appears in Eq. 30, which is an expression of conservation of energy. As we shall discuss in Chapter 12, we can assign part of the total kinetic energy K to translational motion and part to rotational motion.

Sample Problem 9   A 72-kg skater pushes away from a railing as shown in Fig. 17, exerting a constant force $F = 55$ N on the railing as he does so. His center of mass moves through a distance $s_{cm} = 32$ cm until he loses contact with the railing. (a) What is the speed of the center of mass of the skater as he breaks away from the railing? (b) What is the change in the stored internal energy of the skater during this process?

Solution   (a) Once again we take the skater as our system. From Newton's third law, the railing exerts on the skater a force of 55 N to the right in Fig. 17. This force is the only external force that we need consider. From Eq. 36, we have

$$F_{ext}s_{cm} = \tfrac{1}{2}Mv_{cm}^2 - 0,$$

or

$$v_{cm} = \sqrt{\frac{2F_{ext}s_{cm}}{M}} = \sqrt{\frac{2(55 \text{ N})(0.32 \text{ m})}{72 \text{ kg}}} = 0.70 \text{ m/s}.$$

(b) Now we apply the law of conservation of energy, which, under the conditions that apply in this problem, takes the form of Eq. 32, or

$$\Delta E_{int} = -\Delta K_{cm} = -\tfrac{1}{2}Mv_{cm}^2 = -\tfrac{1}{2}(72 \text{ kg})(0.70 \text{ m/s})^2$$
$$= -17.6 \text{ J}.$$

This amount of internal energy could be replenished by digesting about $\tfrac{1}{4}$ teaspoon of diet soda.

Sample Problem 10   In this case, our skater pushes himself away from his partner, who is standing with her back against a wall, as in Fig. 19a. Both have their arms bent initially. Each pushes against the other as they straighten their arms, until finally they lose contact (Fig. 19b). The partner exerts a constant force $F_{ext} = 55$ N through a distance of $s = 32$ cm; this is the distance her hands actually move as she straightens her arms. At the instant contact is broken, the center of mass of the skater has moved through a total distance of $s_{cm} = 58$ cm as a result of the

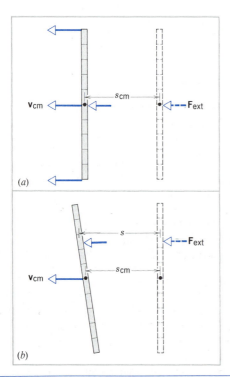

(a)

(b)

Figure 18   (a) A meter stick is pushed across a frictionless horizontal surface by a force $\mathbf{F}_{ext}$. The force is applied at the 50-cm mark. Here the stick moves as a particle. (b) The force is now applied at the 25-cm mark. The stick now rotates as well as translates and no longer moves as a particle. The force is applied through a displacement s that is greater than the displacement s_{cm} of the center of mass.

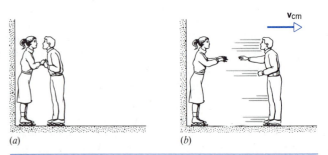

(a)

(b)

Figure 19   Sample Problem 10. (a) A skater and his partner are preparing to exert forces on one another by extending their arms. The partner has her back against a wall and so does not move. (b) After the arms have been extended, the skater is moving with speed v_{cm}.

extension of *both* pairs of arms. (*a*) What is the speed of the center of mass of the skater after contact is broken? (*b*) What is the change in the stored internal energy of the skater during this process?

Solution   (*a*) We take the skater as our system. Note that in this case there is external work done on the system, so there is a transfer of energy through the system boundary. From Eq. 36, we have

$$F_{ext}s_{cm} = \Delta K_{cm} = \tfrac{1}{2}Mv_{cm}^2 - 0,$$

or

$$v_{cm} = \sqrt{\frac{2F_{ext}s_{cm}}{M}} = \sqrt{\frac{2(55\ \text{N})(0.58\ \text{m})}{72\ \text{kg}}} = 0.94\ \text{m/s}.$$

(*b*) From the equation of conservation of energy for a system, we have

$$\Delta K_{cm} + \Delta E_{int} = W,$$

where $W(= F_{ext}s)$ is the external work done on the skater by his partner. Solving for the internal energy change ΔE_{int} and substituting the result $\Delta K_{cm} = F_{ext}s_{cm}$ from part (*a*), we obtain

$$\Delta E_{int} = W - \Delta K_{cm} = F_{ext}s - F_{ext}s_{cm}$$
$$= (55\ \text{N})(0.32\ \text{m}) - (55\ \text{N})(0.58\ \text{m})$$
$$= +17.6\ \text{J} - 31.9\ \text{J} = -14.3\ \text{J}.$$

Thus to attain his final kinetic energy, the skater must supply 14.3 J of energy from his internal resources. His partner supplies 17.6 J by doing work on the skater, which, of course, comes from *her* internal store. If the partner were not present and the skater had attained the same kinetic energy by pushing directly on the wall, he would need to supply the full 31.9 J of kinetic energy from his internal energy source.

Sample Problem 11   A 5.2-kg block is projected over a horizontal surface with an initial horizontal velocity of 0.65 m/s. The coefficient of kinetic friction between the block and the surface is 0.12. (*a*) What happens to the initial kinetic energy of the block? (*b*) How far does the block travel in coming to rest?

Solution   (*a*) In applying energy conservation, the most useful system to consider is the block plus the portion of the horizontal surface over which it slides. In using Eq. 30, we have $\Delta U = 0$, because no change of potential energy occurs on the horizontal surface. Furthermore, $W = 0$, because no external force acts on the system. (We have defined the system so that friction is an *internal* force.) Thus Eq. 30 becomes

$$\Delta E_{int} = -\Delta K_{cm},$$

in which ΔK_{cm} is negative, corresponding to a loss in kinetic energy. Substituting values, we have

$$\Delta E_{int} = -(0 - \tfrac{1}{2}Mv_{cm}^2) = +\tfrac{1}{2}(5.2\ \text{kg})(0.65\ \text{m/s})^2$$
$$= +1.1\ \text{J}.$$

This increase in internal energy of the system reveals itself as a small increase in the temperature of the block and of the horizontal surface. It is difficult to calculate how this energy is shared between the block and the surface; it is largely to avoid this difficulty that we have chosen to analyze the combined system of the block plus the surface, rather than the block alone.

(*b*) In this case we choose the block alone as our system. We

cannot treat the block as a particle, because energy transfers (specifically, internal energy) other than translational kinetic energy are involved. Applying Eq. 36, we have

$$F_{ext}s_{cm} = \Delta K_{cm},$$

where F_{ext} is the external frictional force ($= -\mu Mg$, taking the direction of motion to be positive) that acts on the block and s_{cm} is the displacement of the center of mass of the block. Thus we have

$$(-\mu Mg)(s_{cm}) = 0 - \tfrac{1}{2}Mv_{cm}^2$$

or

$$s_{cm} = \frac{v_{cm}^2}{2\mu g} = \frac{(0.65\ \text{m/s})^2}{2(0.12)(9.8\ \text{m/s}^2)} = 0.18\ \text{m}. \quad\blacksquare$$

9-8 SYSTEMS OF VARIABLE MASS (Optional)

Imagine that the cart holding the cannon of Fig. 14 also holds a large stack of cannonballs. As the cannon is repeatedly fired, the cart (which we assume to move without friction) recoils to the left, and with each recoil its speed increases. With the system boundary drawn as in Fig. 14, we know that the total horizontal momentum must be zero and that there is no net horizontal force on the system. If, however, we consider a system including only the cannon plus cart, then the previous statement is no longer true. The momentum of the cannon increases each time it is fired, and it is appropriate for us to use the familiar language of Newtonian physics to account for the change in momentum through the action of a suitable force. In this case, the force that accelerates the cannon is a reaction force: the cannon, by virtue of its exploding charge, pushes on the cannonballs to eject them, and the reaction force (the cannonballs pushing back on the cannon) moves the cannon to the left.

As the cannon is repeatedly fired, the total mass on the cart decreases by the quantity of cannonballs that have been ejected. The methods of Sample Problem 7 cannot easily be used to solve this problem, because the mass M of the recoiling object is different every time the cannon fires.

We refer to the system S consisting of cannon plus cart in this example as a "variable-mass" system. Of course, the larger system S' consisting of cannon plus fired cannonballs is a constant-mass system *and* a constant-momentum system (in the absence of external force). The smaller system S, however, does not have constant mass. Moreover, the ejected cannonballs carry momentum, and there is a net outflow of momentum from S which is responsible for its acceleration.

The above example gives a reasonably good mental image of how a rocket works. Fuel is burned and ejected at high speed; the combustion products correspond to the cannonballs. The rocket (less the consumed fuel) experiences an acceleration that depends on the rate at which fuel is consumed and the speed with which it is ejected.

The goal in analyzing systems similar to the rocket is *not* to consider the kinematics of the entire system S'. Instead, we focus our attention on one particular subsystem S, and we ask how S moves as the mass within the entire system S' is redistributed so that the mass within subsystem S changes. The total mass within S' remains constant, but the particular subsystem S we consider can change its state of motion as it gains or loses mass (and momentum).

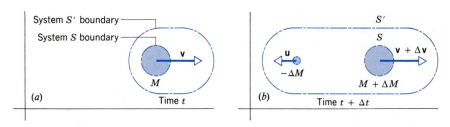

Figure 20   (*a*) A system *S'* at time *t* consists of a mass *M* moving with velocity **v**. (*b*) At a time Δt later, the original mass *M* has ejected some mass $-\Delta M$. The remaining mass $M + \Delta M$, which we call the subsystem *S*, now moves with velocity $\mathbf{v} + \Delta\mathbf{v}$.

Figure 20 shows a schematic view of a generalized system. At time *t*, the subsystem *S* has a mass *M* and moves with velocity **v** in the particular frame of reference from which we are observing. At a time $t + \Delta t$, the mass of *S* has changed by an amount ΔM (a negative quantity, in the case of ejected mass) to $M + \Delta M$, while the mass of the remainder of the full system *S'* has changed by a corresponding amount $-\Delta M$. The system *S* now moves with a velocity $\mathbf{v} + \Delta\mathbf{v}$, and the ejected matter moves with velocity **u**, both measured from our frame of reference.

To make the situation as general as possible, we also allow for an external force F_{ext} that may act on the entire system. This is not the force that propels the rocket (which is an internal force for the system *S'*), but instead is the force due to some external agent, perhaps gravity or atmospheric drag. The total momentum of the entire system *S'* is **P**, and Newton's second law can be written

$$\mathbf{F}_{ext} = \frac{d\mathbf{P}}{dt}.$$ (37)

In the time interval Δt, the change in momentum $\Delta\mathbf{P}$ is

$$\Delta\mathbf{P} = \mathbf{P}_f - \mathbf{P}_i$$ (38)

where $\mathbf{P}_f$, the final momentum of the system *S'* at time $t + \Delta t$, and $\mathbf{P}_i$, the initial momentum of *S'* at time *t*, are given by

$$\mathbf{P}_i = M\mathbf{v},$$ (39*a*)

$$\mathbf{P}_f = (M + \Delta M)(\mathbf{v} + \Delta\mathbf{v}) + (-\Delta M)\mathbf{u}.$$ (39*b*)

The change in momentum of *S'* is thus

$$\Delta\mathbf{P} = \mathbf{P}_f - \mathbf{P}_i = (M + \Delta M)(\mathbf{v} + \Delta\mathbf{v}) + (-\Delta M)\mathbf{u} - M\mathbf{v}.$$ (40)

Rewriting the derivative in Eq. 37 as a limit and substituting this expression for $\Delta\mathbf{P}$, we obtain

$$\mathbf{F}_{ext} = \lim_{\Delta t \to 0} \frac{\Delta\mathbf{P}}{\Delta t}$$

$$= \lim_{\Delta t \to 0} \frac{(M + \Delta M)(\mathbf{v} + \Delta\mathbf{v}) + (-\Delta M)\mathbf{u} - M\mathbf{v}}{\Delta t}$$

$$= \lim_{\Delta t \to 0} \left[M\frac{\Delta\mathbf{v}}{\Delta t} + (\mathbf{v} - \mathbf{u})\frac{\Delta M}{\Delta t} + \Delta\mathbf{v}\frac{\Delta M}{\Delta t} \right]$$

$$= M\frac{d\mathbf{v}}{dt} + (\mathbf{v} - \mathbf{u})\frac{dM}{dt}.$$ (41)

Note that, in taking the limit, the last term in the square brackets vanishes, because $\Delta\mathbf{v} \to 0$ as $\Delta t \to 0$. In Eq. 41, *M* is the mass of the subsystem *S* at time *t*, and $d\mathbf{v}/dt$ is its acceleration as it ejects mass at velocity **u** (in our frame of reference) and at a rate $|dM/dt|$.

We can also express Eq. 41 in a slightly more general form:

$$\mathbf{F}_{ext} = \frac{d}{dt}(M\mathbf{v}) - \mathbf{u}\frac{dM}{dt}.$$ (42)

Equation 42 does not look at all like $\mathbf{F}_{ext} = M\mathbf{a}$ or $\mathbf{F}_{ext} = d(M\mathbf{v})/dt$, which we have used previously to analyze the motion of particles or systems of constant mass. We can reduce Eq. 42 to the particle form of Newton's second law in only two very special cases: (1) when $dM/dt = 0$ so that *M* is a constant, in which case we are back to discussing systems of constant mass, or (2) when $\mathbf{u} = 0$, in which case we are viewing the variable-mass system from a very special reference frame in which the ejected matter is at rest.

In general, when we apply $\mathbf{F}_{ext} = d\mathbf{P}/dt$ to a system *S* that gains or loses mass, we must take into account the change in momentum of the mass that is gained or lost.* That is, as suggested by Eq. 42 and Fig. 20, we must consider the larger system *S'*, which includes the system *S* and the additional mass. This approach to the dynamics of variable-mass systems highlights the importance of the law of conservation of momentum, and it gives us a relatively simple recipe for treating complicated systems.

Equation 41 has been derived in a special form that can easily be adapted to the analysis of the motion of a rocket. The quantity $\mathbf{u} - \mathbf{v}$ is $\mathbf{v}_{rel}$, the velocity of the ejected gases *relative to the rocket*. This is a reasonable quantity to introduce, because the speed of the ejected gases is a fundamental design characteristic of the rocket engine and should not be expressed in a form that depends on any frame of reference other than the rocket itself. In terms of $\mathbf{v}_{rel}$, we can write Eq. 41 in the following form:

$$M\frac{d\mathbf{v}}{dt} = \mathbf{F}_{ext} + \mathbf{v}_{rel}\frac{dM}{dt}.$$ (43)

The last term in Eq. 43 gives the rate at which momentum is being transferred into (or perhaps out of) the subsystem *S*. It can be interpreted as a force exerted on *S* by the mass that enters or leaves *S*. In the case of a rocket, this term is called the *thrust;* to make the thrust as large as possible, rocket designers attempt to make both $\mathbf{v}_{rel}$ (the exhaust velocity) and $|dM/dt|$ (the rate at which mass is ejected) as large as possible.

---

* See "Force, Momentum Change, and Motion," by Martin S. Tiersten, *American Journal of Physics,* January 1969, p. 82, for an excellent general reference on systems of fixed and variable mass.

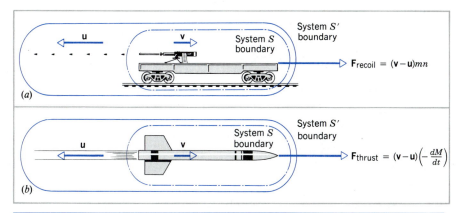

Figure 21 (*a*) A machine gun fires a stream of bullets at a rate of *n* per unit time. The total momentum of the system S' remains constant, but the subsystem S experiences a recoil force that changes its momentum. Its change in momentum in a time dt is exactly equal to the opposite momentum $mn\,\mathbf{u}\,dt$ carried by the bullets. (*b*) A rocket ejects a stream of combustion products. The total momentum of the system S' remains constant, but the subsystem S experiences a thrust that changes its momentum. Its change in momentum in a time dt is exactly equal to the opposite momentum $\mathbf{u}\,dM$ carried by the ejected gas.

The Rocket Equation

Consider a rocket in distant space, where it is subject to no external force. Let us assume for simplicity that the motion is confined to one dimension, with dv/dt defining the positive direction as the rocket accelerates and $\mathbf{v}_{\text{rel}}$ therefore pointing in the negative direction. Equation 43 can be written in this case as

$$M\frac{dv}{dt} = -v_{\text{rel}}\frac{dM}{dt}, \qquad (44)$$

where v_{rel} is the magnitude of the exhaust velocity. Note that dM/dt is negative, so that the right side of Eq. 44 is positive, as is the left side.

Equation 44 is the fundamental equation governing the behavior of a rocket. During a stable burn of the engine, the thrust (the right side of Eq. 44) is a constant (but the resulting acceleration of the rocket dv/dt is not a constant because the mass M changes as the fuel burns).

Let us consider the velocity change of the rocket corresponding to the burning of a specific quantity of fuel m_{b}. The initial velocity is v_i, and the final velocity after the burn is v_f. We rewrite Eq. 44 as

$$dv = -v_{\text{rel}}\frac{dM}{M}. \qquad (45)$$

The total mass M of the rocket is a variable. The original mass of rocket plus fuel is M_0, and at any time t, the remaining mass M of the rocket plus the mass m_{b} of the fuel burned up to that time must sum to M_0; thus $M = M_0 - m_{\text{b}}$. We integrate Eq. 45 between the limits v_i, when the mass of the rocket is M_0, and v_f, when the mass is $M_0 - m_{\text{b}}$, obtaining

$$\int_{v_i}^{v_f} dv = -v_{\text{rel}}\int_{M_0}^{M_0-m_{\text{b}}} \frac{dM}{M}$$

$$v_f - v_i = -v_{\text{rel}}\ln M\,\Big|_{M_0}^{M_0-m_{\text{b}}}$$

$$= -v_{\text{rel}}\,[\ln\,(M_0 - m_{\text{b}}) - \ln M_0]$$

$$= -v_{\text{rel}}\ln\left(\frac{M_0 - m_{\text{b}}}{M_0}\right). \qquad (46)$$

Equation 46 gives the change in velocity of the rocket following the burning of a quantity m_{b} of fuel.

Assuming that the rocket starts from rest ($v_i = 0$) with an initial mass M_0 and reaches a final velocity v_f at burnout when its mass is $M_f = M_0 - m_{\text{b}}$, we can write Eq. 46 as

$$\frac{M_f}{M_0} = e^{-v_f/v_{\text{rel}}}. \qquad (47)$$

The analogy between the rocket and the recoiling gun is apparent from Fig. 21. In each case momentum is conserved for the entire system, consisting of the ejected mass (bullets or fuel) plus the object that ejects the mass. When we focus our attention on the gun or the rocket within the larger system, we see that its mass changes and that there is a force that drives it, a recoil in the case of the gun and a thrust in the case of the rocket. If we view the system from a reference frame at the center of mass, then as time passes there is more ejected mass, and it has traveled further to the left in Fig. 21, meaning that the object must travel to the right to keep the center of mass fixed.

Sample Problem 12 A rocket has a mass of 13,600 kg when fueled on the launching pad. It is fired vertically upward and, at burnout, has consumed and ejected 9100 kg of fuel. Gases are exhausted at the rate of 146 kg/s with a speed of 1520 m/s, relative to the rocket, both quantities being assumed to be constant while the fuel is burning. (*a*) What is the thrust? (*b*) If we could neglect all external forces, including gravity and air resistance, what would be the speed of the rocket at burnout?

Solution (*a*) The thrust F is the last term in Eq. 43, or

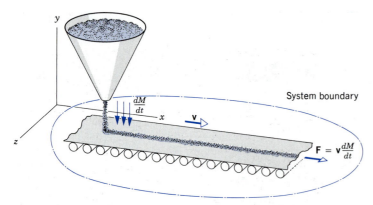

System boundary

$F = v\dfrac{dM}{dt}$

Figure 22   Sample Problem 13. Sand drops from a hopper at a rate dM/dt onto a conveyor belt moving with constant velocity **v** in the reference frame of the laboratory. The force required to keep the belt moving at constant velocity is **v** dM/dt. The hopper is at rest in the reference frame of the laboratory.

$$F = v_{rel}\left|\frac{dM}{dt}\right| = (1520 \text{ m/s})(146 \text{ kg/s}) = 2.22 \times 10^5 \text{ N}.$$

Note that initially, when the fuel tanks are full, the net upward force acting on the rocket (neglecting air resistance) is the thrust minus the initial weight Mg, or 88,600 N. Just before burnout the net upward force is the thrust minus the final weight, or 1.78×10^5 N.

(*b*) From Eq. 46, we can find the speed at burnout:

$$v_f = -v_{rel} \ln\left(\frac{M_0 - m_b}{M_0}\right)$$

$$= -(1520 \text{ m/s}) \ln\left(\frac{13,600 \text{ kg} - 9100 \text{ kg}}{13,600 \text{ kg}}\right) = 1680 \text{ m/s}.$$

If the external forces of gravity and air resistance were taken into account, the final speed would be smaller.

Sample Problem 13   Sand drops from a stationary hopper at a rate dM/dt onto a conveyor belt moving with velocity v in the reference frame of the laboratory, as in Fig. 22. What power is required to keep the belt moving at v?

Solution   Figure 20 describes this situation, with system *S* representing the belt plus the accumulated sand and ΔM representing the additional sand dropping onto the belt. The system *S'* includes the belt and all of the supply of sand in the hopper. We take as our object (system *S*) the belt (including the sand) of varying mass M, and using Eq. 41 we must put $d\mathbf{v}/dt = 0$, because the velocity of the belt is constant, and $\mathbf{u} = 0$, because the falling sand has no horizontal velocity in our frame of reference. We obtain

$$\mathbf{F}_{ext} = \mathbf{v}\frac{dM}{dt}.$$

In this example, dM/dt is positive because the system is gaining mass with time. Hence, as expected, the necessary external force must point in the direction in which the belt moves. Note that the mass of the belt itself does not enter the problem, because we have assumed the belt to move at constant speed.

The power supplied by the external force is

$$P_{ext} = \mathbf{F}_{ext}\cdot\mathbf{v} = \mathbf{v}\cdot\mathbf{F}_{ext} = \mathbf{v}\cdot\left(\mathbf{v}\frac{dM}{dt}\right) = v^2\left(\frac{dM}{dt}\right).$$

Since v is a constant, we can write this as

$$P_{ext} = \frac{d(Mv^2)}{dt} = 2\frac{d}{dt}\left(\tfrac{1}{2}Mv^2\right) = 2\frac{dK}{dt}.$$

This tells us that the external power required to keep the belt moving is twice the rate at which the kinetic energy of the system is increasing; note that we need not consider the kinetic energy of the belt itself because its speed is constant, and so its kinetic energy does not change.

It is clear that mechanical energy is not conserved in this case. Only one-half of the work done by the motor driving the belt appears as mechanical energy of the system. Where is the other half of the work going? To answer this question, let us apply conservation of energy, Eq. 30, to a small element of mass dM that drops onto the belt. We assume it drops from a height sufficiently small that its change in potential energy can be neglected. In the interval of time dt that it takes for dM to begin moving with the speed of the belt, the work done by the external source is $dW = P_{ext}\, dt = v^2\, dM$. The change in kinetic energy of this element of mass is $+\tfrac{1}{2}(dM)v^2$. Applying Eq. 30 then gives

$$\Delta E_{int} = v^2\, dM - \tfrac{1}{2}(dM)v^2 = \tfrac{1}{2}(dM)v^2.$$

The internal energy of the system increases by the same amount as the kinetic energy. Thus half of the input power goes into the kinetic energy of the moving sand, while the other half ends up as internal energy of the sand and the belt (resulting perhaps from friction between the sand and the belt occurring after the sand drops to the belt but before it is moving with the speed of the belt).

This sample problem gives an example of a force exerted with a change in mass, the velocity being constant. It is also possible for the velocity of a variable-mass system to decrease as a result of mass added to the system, in effect reversing the operation of a rocket.* ■

---

* See, for example, "The Falling Raindrop: Variations on a Theme of Newton," by K. S. Krane, *American Journal of Physics*, February 1981, p. 113.

QUESTIONS

1. Does the center of mass of a solid object necessarily lie within the object? If not, give examples.

2. Figure 23 shows (*a*) an isosceles triangular prism and (*b*) a right circular cone whose diameter is the same length as the base of the triangle. The center of mass of the triangle is one-third of the way up from the base but that of the cone is only one-fourth of the way up. Can you explain the difference?

Figure 23 Question 2.

3. How is the center of mass concept related to the concept of geographic center of the country? To the population center of the country? What can you conclude from the fact that the geographic center differs from the population center?

4. Where is the center of mass of the Earth's atmosphere?

5. An amateur sculptor decides to portray a bird (Fig. 24). Luckily, the final model is actually able to stand upright. The model is formed of a single thick sheet of metal of uniform thickness. Of the points shown, which is most likely to be the center of mass?

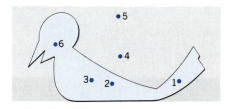

Figure 24 Question 5.

6. Someone claims that when a skillful high jumper clears the bar, the jumper's center of mass actually goes *under* the bar. Is this possible?

7. A ballet dancer doing a *grand jete* (great leap; see Fig. 25) seems to float horizontally in the central portion of her leap. Show how the dancer can maneuver her legs in flight so that, although her center of mass does indeed follow the expected parabolic trajectory, the top of her head moves more or less horizontally. (See "The Physics of Dance," by Kenneth Laws, *Physics Today,* February 1985, p. 24.)

8. A light object and a heavy object have equal kinetic energies of translation. Which one has the larger momentum?

9. A bird is in a wire cage hanging from a spring balance. Is the reading of the balance when the bird is flying about greater than, less than, or the same as that when the bird sits in the cage?

10. Can a sailboat be propelled by air blown at the sails from a fan attached to the boat? Explain your answer.

Figure 25 Question 7.

11. Can a body have energy without having momentum? Explain. Can a body have momentum without having energy? Explain.

12. A canoeist in a still pond can reach shore by jerking sharply on the rope attached to the bow of the canoe. How do you explain this? (It really can be done.)

13. How might a person sitting at rest on a frictionless horizontal surface get altogether off it?

14. A man stands still on a large sheet of slick ice; in his hand he holds a lighted firecracker. He throws the firecracker at an angle (that is, not vertically) into the air. Describe briefly, but as exactly as you can, the motion of the center of mass of the firecracker and the motion of the center of mass of the system consisting of man and firecracker. It will be most convenient to describe each motion during each of the following periods: (*a*) after he throws the firecracker, but before it explodes; (*b*) between the explosion and the first piece of firecracker hitting the ice; (*c*) between the first fragment hitting the ice and the last fragment landing; and (*d*) during the time when all fragments have landed but none has reached the edge of the ice.

15. Justify the following statement. "The law of conservation of linear momentum, as applied to a single particle, is equivalent to Newton's first law of motion."

16. You throw an ice cube with velocity **v** into a hot gravity-free, evacuated space. The cube gradually melts to liquid water and then boils to water vapor. (*a*) Is it a system of particles all the time? (*b*) If so, is it the same system of particles? (*c*) Does the motion of the center of mass undergo any abrupt changes? (*d*) Does the total linear momentum change?

17. A particle with mass $m = 0$ (a neutrino, possibly) carries momentum. How can this be in view of Eq. 22, in which we see that the momentum is directly proportional to the mass?

18. If only an external force can change the state of motion of the center of mass of a body, how does it happen that the internal force of the brakes can bring a car to rest?

19. We say that a car is not accelerated by internal forces but rather by external forces exerted on it by the road. Why then do cars need engines?

20. Can the work done by internal forces decrease the kinetic energy of a body? . . . increase it?

21. (*a*) If you do work on a system, does the system necessarily acquire kinetic energy? (*b*) If a system acquires kinetic energy, does it necessarily mean that some external agent did work on it? Give examples. (By "kinetic energy" here we mean kinetic energy associated with the motion of the center of mass.)

22. In Sample Problem 9, we saw an example (a skater) in which kinetic energy appeared but no external work was done. Consider the opposite case. A screwdriver is held tightly against a rotating grinding wheel. Here external work is done but the kinetic energy of the screwdriver does not change. Explain this apparent contradiction.

23. Can you think of variable-mass systems other than the examples given in the text?

24. As stated in the text, one cannot use the equation $F_{ext} = d(Mv)/dt$ for a system of variable mass. To show this (*a*) put the equation in the form $(\mathbf{F}_{ext} - M \, d\mathbf{v}/dt)/(dM/dt) = \mathbf{v}$ and (*b*) show that one side of this equation has the same value in all inertial frames, whereas the other side does not. Hence the equation cannot be generally valid. (*c*) Show that Eq. 42 leads to no such contradiction.

25. In 1920 a prominent newspaper editorialized as follows about the pioneering rocket experiments of Robert H. Goddard, dismissing the notion that a rocket could operate in a vacuum: "That Professor Goddard, with his 'chair' in Clark College and the countenancing of the Smithsonian Institution, does not know the relation of action to reaction, and of the need to have something better than a vacuum against which to react—to say that would be absurd. Of course, he seems only to lack the knowledge ladled out daily in high schools." What is wrong with this argument?

26. The final velocity of the final stage of a multistage rocket is much greater than the final velocity of a single-stage rocket of the same total mass and fuel supply. Explain this fact.

27. Can a rocket reach a speed greater than the speed of the exhaust gases that propel it? Explain why or why not.

28. Are there any possible methods of propulsion in outer space other than rockets? If so, what are they and why are they not used?

29. Equation 46 suggests that the speed of a rocket can increase without limit if enough fuel is burned. Is this reasonable? What is the limit of applicability of Eq. 46? Where in our derivation of Eq. 46 did we introduce this limit? (See "The Equation of Motion for Relativistic Particles and Systems with Variable Rest Mass," by Kalman B. Pomeranz, *American Journal of Physics*, December 1964, p. 955.)

PROBLEMS

Section 9-1 Two-Particle Systems

1. (*a*) Solve Eq. 4 for x_1 and Eq. 5 for v_1, and substitute both results along with Eq. 3 into Eq. 2 to obtain

$$m_1^2 k d_i^2 = ku^2 + \mu \left(\frac{du}{dt}\right)^2,$$

where $u = Mx_2 - Mx_{cm} - m_1 L$ and $\mu = m_1 m_2/M$. (*b*) Show that this result can be solved for $u(t)$ using techniques presented in Section 8-4 to give

$$u(t) = m_1 d_i \cos \omega t,$$

where $\omega = \sqrt{k/\mu}$. (*c*) Solve for $x_1(t)$, $x_2(t)$, $v_1(t)$, and $v_2(t)$. This problem demonstrates that the center-of-mass equations permit us to solve for the motions of m_1 and m_2 in the situation shown in Fig. 1.

Section 9-2 Many-Particle Systems

2. Where is the center of mass of the three particles shown in Fig. 26?

3. How far is the center of mass of the Earth–Moon system from the center of the Earth? (From Appendix C, obtain the masses of the Earth and Moon and the distance between the centers of the Earth and Moon. It is interesting to compare the answer to the Earth's radius.)

4. Show that the ratio of the distances x_1 and x_2 of two particles from their center of mass is the inverse ratio of their masses: that is, $x_1/x_2 = m_2/m_1$.

5. A Chrysler with a mass of 2210 kg is moving along a straight

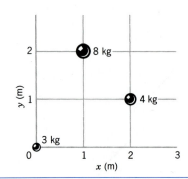

Figure 26   Problem 2.

stretch of road at 105 km/h. It is followed by a Ford with mass 2080 kg moving at 43.5 km/h. How fast is the center of mass of the two cars moving?

6. Two skaters, one with mass 65 kg and the other with mass 42 kg, stand on an ice rink holding a pole with a length of 9.7 m and a mass that is negligible. Starting from the ends of the pole, the skaters pull themselves along the pole until they meet. How far will the 42-kg skater move?

7. A man of mass m clings to a rope ladder suspended below a balloon of mass M; see Fig. 27. The balloon is stationary with respect to the ground. (*a*) If the man begins to climb the ladder at a speed v (with respect to the ladder), in what direction and with what speed (with respect to the Earth)

Figure 27 Problem 7.

will the balloon move? (*b*) What is the state of motion after the man stops climbing?

8. Two particles *P* and *Q* are initially at rest 1.64 m apart. *P* has a mass of 1.43 kg and *Q* a mass of 4.29 kg. *P* and *Q* attract each other with a constant force of 1.79×10^{-2} N. No external forces act on the system. (*a*) Describe the motion of the center of mass. (*b*) At what distance from *P*'s original position do the particles collide?

9. A cannon and a supply of cannonballs are inside a sealed railroad car of length *L*, as in Fig. 28. The cannon fires to the right; the car recoils to the left. The cannonballs remain in the car after hitting the far wall. (*a*) After all the cannonballs have been fired, what is the greatest distance the car can have moved from its original position? (*b*) What is the speed of the car after all the cannonballs have been fired?

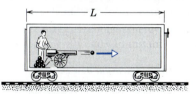

Figure 28 Problem 9.

10. In the ammonia (NH_3) molecule, the three hydrogen (H) atoms form an equilateral triangle, the distance between centers of the atoms being 16.28×10^{-11} m, so that the center of the triangle is 9.40×10^{-11} m from each hydrogen atom. The nitrogen (N) atom is at the apex of a pyramid, the three hydrogens constituting the base (see Fig. 29). The nitrogen–hydrogen distance is 10.14×10^{-11} m and the ni-

trogen/hydrogen atomic mass ratio is 13.9. Locate the center of mass relative to the nitrogen atom.

11. Two bodies, each made up of weights from a set, are connected by a light cord that passes over a light, frictionless pulley with a diameter of 56.0 mm. The two bodies are at the same level. Each originally has a mass of 850 g. (*a*) Locate their center of mass. (*b*) Thirty-four grams are transferred from one body to the other, but the bodies are prevented from moving. Locate the center of mass. (*c*) The two bodies are now released. Describe the motion of the center of mass and determine its acceleration.

12. A shell is fired from a gun with a muzzle velocity of 466 m/s, at an angle of 57.4° with the horizontal. At the top of the trajectory, the shell explodes into two fragments of equal mass. One fragment, whose speed immediately after the explosion is zero, falls vertically. How far from the gun does the other fragment land, assuming level terrain?

13. A uniform flexible chain of length *L*, with weight per unit length λ, passes over a small, frictionless peg; see Fig. 30. It is released from a rest position with a length of chain *x* hanging from one side and a length *L* − *x* from the other side. Find the acceleration *a* as a function of *x*.

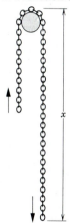

Figure 30 Problem 13.

14. A dog, weighing 10.8 lb, is standing on a flatboat so that he is 21.4 ft from the shore. He walks 8.50 ft on the boat toward shore and then halts. The boat weighs 46.4 lb, and one can assume there is no friction between it and the water. How far is he from the shore at the end of this time? (*Hint:* The center of mass of boat + dog does not move. Why?) The shoreline is also to the left in Fig. 31.

Figure 31 Problem 14.

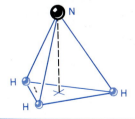

Figure 29 Problem 10.

15. Richard, mass 78.4 kg, and Judy, who is lighter, are enjoying Lake George at dusk in a 31.6-kg canoe. When the canoe is at rest in the placid water, they change seats, which are 2.93 m apart and symmetrically located with respect to the canoe's center. Richard notices that the canoe moved 41.2 cm relative to a submerged log and calculates Judy's mass. What is it?

16. An 84.4-kg man is standing at the rear of a 425-kg iceboat that is moving at 4.16 m/s across ice that may be considered to be frictionless. He decides to walk to the front of the 18.2-m-long boat and does so at a speed of 2.08 m/s with respect to the boat. How far did the boat move across the ice while he was walking?

Section 9-3 Center of Mass of Solid Objects

17. Three thin rods each of length L are arranged in an inverted U, as shown in Fig. 32. The two rods on the arms of the U each have mass M; the third rod has mass $3M$. Where is the center of mass of the assembly?

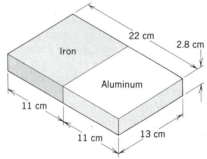

Figure 32   Problem 17.

18. Figure 33 shows a composite slab with dimensions 22.0 cm $\times$ 13.0 cm $\times$ 2.80 cm. Half the slab is made of aluminum (density = 2.70 g/cm³) and half of iron (density = 7.85 g/cm³), as shown. Where is the center of mass of the slab?

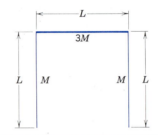

Figure 33   Problem 18.

19. A box, open at the top, in the form of a cube of edge length 40 cm, is constructed from thin metal plate. Find the coordinates of the center of mass of the box with respect to the coordinate system shown in Fig. 34.

20. A cylindrical storage tank is initially filled with aviation gasoline. The tank is then drained through a valve on the bottom. See Fig. 35. (a) As the gasoline is withdrawn, describe qualitatively the motion of the center of mass of the tank and its remaining contents. (b) What is the depth x to

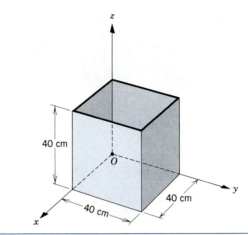

Figure 34   Problem 19.

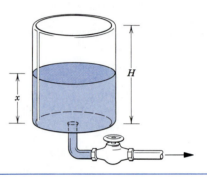

Figure 35   Problem 20.

which the tank is filled when the center of mass of the tank and its remaining contents reaches its lowest point? Express your answer in terms of H, the height of the tank; M, its mass; and m, the mass of gasoline it can hold.

21. Find the center of mass of a homogenous semicircular plate. Let R be the radius of the circle.

Section 9-4 Linear Momentum of a Particle

22. How fast must an 816-kg Volkswagen travel (a) to have the same momentum as a 2650-kg Cadillac going 16.0 km/h and (b) to have the same kinetic energy? (c) Make the same calculations using a 9080-kg truck instead of a Cadillac.

23. A 2000-kg truck traveling north at 40.0 km/h turns east and accelerates to 50.0 km/h. (a) What is the change in kinetic energy of the truck? (b) What is the magnitude and direction of the change of the truck's momentum?

24. A 4.88-kg object with a speed of 31.4 m/s strikes a steel plate at an angle of 42.0° and rebounds at the same speed and angle (Fig. 36). What is the change (magnitude and direction) of the linear momentum of the object?

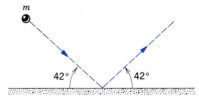

Figure 36   Problem 24.

25. A 52.4-g ball is thrown from the ground into the air, with an initial speed of 16.3 m/s at an angle of 27.4° above the horizontal. (*a*) What are the values of the kinetic energy of the ball initially and just before it hits the ground? (*b*) Find the corresponding values of the momentum (magnitude and direction) and the change in momentum. (*c*) Show that the change in momentum is equal to the weight of the ball multiplied by the time of flight, and thereby find the time of flight.

26. A particle with mass m has linear momentum p equal to mc. What is its speed in terms of c, the speed of light?

27. Show that Eq. 23 reduces to Eq. 21 at speeds $v \ll c$. *Hint:* Show that Eq. 23 can be written

$$K = mc^2 \left(\frac{1}{\sqrt{1 - v^2/c^2}} - 1 \right).$$

28. An electron has a speed of $0.990c$. (*a*) Find its linear momentum, in kg·m/s. (*b*) Express this momentum in the units MeV/c.

Section 9-6 Conservation of Linear Momentum

29. A 195-lb man standing on a surface of negligible friction kicks forward a 0.158-lb stone lying at his feet so that it acquires a speed of 12.7 ft/s. What velocity does the man acquire as a result?

30. A 75.2-kg man is riding on a 38.6-kg cart traveling at a speed of 2.33 m/s. He jumps off in such a way as to land on the ground with zero horizontal speed. Find the resulting change in the speed of the cart.

31. A railroad flatcar of weight W can roll without friction along a straight horizontal track. Initially, a man of weight w is standing on the car, which is moving to the right with speed v_0. What is the change in velocity of the car if the man runs to the left (Fig. 37) so that his speed relative to the car is v_{rel} just before he jumps off at the left end?

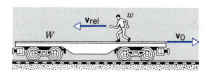

Figure 37   Problem 31.

32. A rocket sled with a mass of 2870 kg moves at 252 m/s on a set of rails. At a certain point, a scoop on the sled dips into a trough of water located between the tracks and scoops water into an empty tank on the sled. Determine the speed of the sled after 917 kg of water have been scooped up.

33. Each minute, a special game warden's machine gun fires 220 12.6-g rubber bullets with a muzzle velocity of 975 m/s. How many bullets must be fired at an 84.7-kg animal charging toward the warden at 3.87 m/s in order to stop the animal in its tracks? (Assume that the bullets travel horizontally and drop to the ground after striking the target.)

34. A space vehicle is traveling at 3860 km/h with respect to the Earth when the exhausted rocket motor is disengaged and sent backward with a speed of 125 km/h with respect to the command module. The mass of the motor is four times the mass of the module. What is the speed of the command module after the separation?

35. The last stage of a rocket is traveling at a speed of 7600 m/s. This last stage is made up of two parts which are clamped together, namely, a rocket case with a mass of 290.0 kg and a payload capsule with a mass of 150.0 kg. When the clamp is released, a compressed spring causes the two parts to separate with a relative speed of 910.0 m/s. (*a*) What are the speeds of the two parts after they have separated? Assume that all velocities are along the same line. (*b*) Find the total kinetic energy of the two parts before and after they separate and account for the difference, if any.

36. A vessel at rest explodes, breaking into three pieces. Two pieces, one with twice the mass of the other, fly off perpendicular to one another with the same speed of 31.4 m/s. The third piece has three times the mass of the lightest piece. Find the magnitude and direction of its velocity immediately after the explosion. (Specify the direction by giving the angle from the line of travel of the least massive piece.)

37. A radioactive nucleus, initially at rest, decays by emitting an electron and a neutrino at right angles to one another. The momentum of the electron is 1.2×10^{-22} kg·m/s and that of the neutrino is 6.4×10^{-23} kg·m/s. (*a*) Find the direction and magnitude of the momentum of the recoiling nucleus. (*b*) The mass of the residual nucleus is 5.8×10^{-26} kg. What is its kinetic energy of recoil? A *neutrino* is one of the fundamental particles of nature.

38. A 1930-kg railroad flatcar, which can move on the tracks with virtually no friction, is sitting motionless next to a station platform. A 108-kg football player is running along the platform parallel to the tracks at 9.74 m/s. He jumps onto the back of the flatcar. (*a*) What is the speed of the flatcar after he is aboard and at rest on the flatcar? (*b*) Now he starts to walk, at 0.520 m/s relative to the flatcar, to the front of the car. What is the speed of the flatcar as he walks?

39. A 3.54-g bullet is fired horizontally at two blocks resting on a frictionless tabletop, as shown in Fig. 38*a*. The bullet passes through the first block, with mass 1.22 kg, and embeds itself in the second, with mass 1.78 kg. Speeds of 0.630 m/s and 1.48 m/s, respectively, are thereby imparted to the blocks, as shown in Fig. 38*b*. Neglecting the mass removed from the first block by the bullet, find (*a*) the speed of the bullet immediately after emerging from the first block and (*b*) the original speed of the bullet.

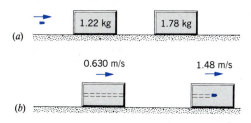

Figure 38   Problem 39.

40. A body of mass 8.0 kg is traveling at 2.0 m/s under the influence of no external force. At a certain instant an internal explosion occurs, splitting the body into two chunks of 4.0 kg mass each; 16 J of translational kinetic energy are imparted to the two-chunk system by the explosion. Neither chunk leaves the line of the original motion. Determine the speed and direction of motion of each of the chunks after the explosion.

41. Assume that the car in Problem 31 is initially at rest. It holds *n* people each of weight *w*. If each person in succession runs with a relative velocity v_{rel} and jumps off the end, do they impart to the car a greater velocity than if they all run and jump at the same time?

42. A 1400-kg cannon, which fires a 70.0-kg shell with a muzzle speed of 556 m/s, is set at an elevation angle of 39.0° above the horizontal. The cannon is mounted on frictionless rails, so that it recoils freely. (*a*) What is the speed of the shell with respect to the Earth? (*b*) At what angle with the ground is the shell projected? (*Hint:* The horizontal component of the momentum of the system remains unchanged as the gun is fired.)

43. A block of mass *m* rests on a wedge of mass *M* which, in turn, rests on a horizontal table, as shown in Fig. 39. All surfaces are frictionless. If the system starts at rest with point *P* of the block a distance *h* above the table, find the velocity of the wedge the instant point *P* touches the table.

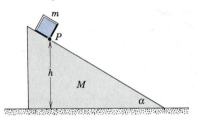

Figure 39 Problem 43.

Section 9-7 Work and Energy in a System of Particles

44. An automobile with passengers has weight 3680 lb (= 16,400 N) and is moving at 70.0 mi/h (= 113 km/h) when the driver brakes to a stop. The road exerts a force of 1850 lb (= 8230 N) on the wheels and there is no skidding. Calculate the stopping distance.

45. You crouch from a standing position, lowering your center of mass 18.0 cm in the process. Then you jump vertically into the air. The force that the floor exerts on you while you are jumping is three times your weight. What is your upward speed as you pass through your standing position leaving the floor?

46. A 55.0-kg woman leaps vertically into the air from a crouching position in which her center of mass is 40.0 cm above the ground. As her feet leave the floor her center of mass is 90.0 cm above the ground and rises to 120 cm at the top of her leap. (*a*) What upward force, assumed constant, does the ground exert on her? (*b*) What maximum speed does she attain?

47. A 116-kg ice hockey player skates at 3.24 m/s toward a railing at the edge of the ice and stops himself by grasping the railing with his outstretched arms. During this stopping process his center of mass moves 34.0 cm toward the rail. (*a*) Find the average force he must exert on the rail. (*b*) How much internal energy does he lose?

48. The National Transportation Safety Board is testing the crash-worthiness of a new car. The 2340-kg vehicle is driven at 12.6 km/h into an abutment. During impact, the center of mass of the car moves forward 64.0 cm; the abutment is compressed by 8.30 cm. Ignore friction between the car and the road. (*a*) Find the force, assumed constant, exerted by the abutment on the car. (*b*) By how much does the internal energy of the car increase?

49. Let the total energy of a system of *N* particles be measured in an arbitrary frame of reference, such that $K = \Sigma \frac{1}{2} m_i v_i^2$. In the center-of-mass reference frame, the velocities are $\mathbf{v}'_i = \mathbf{v}_i - \mathbf{v}_{cm}$, where $\mathbf{v}_{cm}$ is the velocity of the center of mass relative to the original frame of reference. Keeping in mind that $v_i^2 = \mathbf{v}_i \cdot \mathbf{v}_i$, show that the kinetic energy can be written

$$K = K_{int} + K_{cm}$$

where $K_{int} = \Sigma \frac{1}{2} m_i v_i'^2$ and $K_{cm} = \frac{1}{2} M v_{cm}^2$. This demonstrates that the kinetic energy of a system of particles can be divided into an internal term and a center-of-mass term, as we asserted in Section 9-7. The internal kinetic energy is measured in a frame of reference in which the center of mass is at rest; for example, the random motions of the molecules of gas in a container at rest are responsible for its internal translational kinetic energy.

Section 9-8 Systems of Variable Mass

50. A rocket at rest in space, where there is virtually no gravity, has a mass of 2.55×10^5 kg, of which 1.81×10^5 kg is fuel. The engine consumes fuel at the rate of 480 kg/s, and the exhaust speed is 3.27 km/s. The engine is fired for 250 s. (*a*) Find the thrust of the rocket engine. (*b*) What is the mass of the rocket after the engine burn? (*c*) What is the final speed attained?

51. Consider a rocket at rest in empty space. What must be its *mass ratio* (ratio of initial to final mass) in order that, after firing its engine, the rocket's speed is (*a*) equal to the exhaust speed and (*b*) equal to twice the exhaust speed?

52. During a lunar mission, it is necessary to make a midcourse correction of 22.6 m/s in the speed of the spacecraft, which is moving at 388 m/s. The exhaust speed of the rocket engine is 1230 m/s. What fraction of the initial mass of the spacecraft must be discarded as exhaust?

53. A rocket of total mass 1.11×10^5 kg, of which 8.70×10^4 kg is fuel, is to be launched vertically. The fuel will be burned at the constant rate of 820 kg/s. Relative to the rocket, what is the minimum exhaust speed that allows liftoff at launch?

54. A 5.4-kg toboggan carrying 35 kg of sand slides from rest down an icy slope 93 m long, inclined 26° below the horizontal. The sand leaks from the back of the toboggan at the rate of 2.3 kg/s. How long does it take the toboggan to reach the bottom of the slope?

55. To keep a conveyor belt moving when it transports luggage requires a greater driving force than for an empty belt. What additional driving force is needed if the belt moves at a constant speed of 1.5 m/s and the rate at which luggage is placed on one end of the belt and removed at the other end is 20 kg/s? Assume that the luggage is dropped vertically onto the belt; persons removing luggage grab hold of it and bring it to rest relative to themselves before lifting it off the belt.

56. A freight car, open at the top, weighing 9.75 metric tons, is coasting along a level track with negligible friction at 1.36 m/s when it begins to rain hard. The raindrops fall vertically with respect to the ground. What is the speed of the car when it has collected 0.50 metric tons of rain? What assumptions, if any, must you make to get your answer?

57. A 5860-kg rocket is set for vertical firing. The exhaust speed is 1.17 km/s. How much gas must be ejected each second to supply the thrust needed (*a*) to overcome the weight of the rocket and (*b*) to give the rocket an initial upward acceleration of 18.3 m/s²? Note that, in contrast to the situation described in Sample Problem 12, gravity is present here as an external force.

58. Two long barges are floating in the same direction in still water, one with a speed of 9.65 km/h and the other with a speed of 21.2 km/h. While they are passing each other, coal is shoveled from the slower to the faster one at a rate of 925 kg/min; see Fig. 40. How much additional force must be provided by the driving engines of each barge if neither is to change speed? Assume that the shoveling is always perfectly sideways and that the frictional forces between the barges and the water do not depend on the weight of the barges.

59. A jet airplane is traveling 184 m/s (= 604 ft/s). The engine takes in 68.2 m³ (= 2410 ft³) of air making a mass of 70.2 kg (= 4.81 slugs) each second. The air is used to burn 2.92 kg (= 0.200 slug) of fuel each second. The energy is used to compress the products of combustion and to eject them at the rear of the engine at 497 m/s (= 1630 ft/s) relative to the plane. Find (*a*) the thrust of the jet engine and (*b*) the delivered power (horsepower).

60. A flexible inextensible string of length L is threaded into a smooth tube, into which it snugly fits. The tube contains a

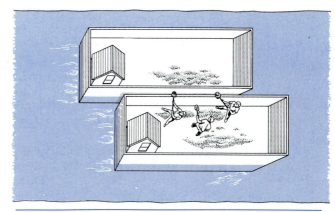

Figure 40 Problem 58.

right-angled bend and is positioned in the vertical plane so that one arm is vertical and the other horizontal. Initially, at $t = 0$, a length y_0 of the string is hanging down in the vertical arm. The string is released and slides through the tube, so that at any subsequent time t later, it is moving with a speed dy/dt, where $y(t)$ is the length of the string that is then hanging vertically. (*a*) Show that in terms of the variable-mass problem $\mathbf{v}_{rel} = 0$, so that the equation of motion has the form $m\,d\mathbf{v}/dt = \mathbf{F}_{ext}$. (*b*) Show that the specific equation of motion is $d^2y/dt^2 = gy$. (*c*) Show that conservation of mechanical energy leads to $(dy/dt)^2 - gy^2 = $ a constant, and that this is consistent with (*b*). (*d*) Show that $y = (y_0/2)(e^{\sqrt{g/L}\,t} + e^{-\sqrt{g/L}\,t})$ is a solution to the equation of motion [by substitution into (*b*)] and discuss the solution.

CHAPTER 10

COLLISIONS

One of the principal applications of the conservation of linear momentum is in the analysis of collisions between objects. No matter what the size of the objects, ranging from elementary particles to galaxies, and no matter what the forces involved, from the strongest (the nuclear force) to the weakest (gravity), the law of conservation of linear momentum holds and enables us to study these processes.

In this chapter, we show how collision processes can be analyzed by using the laws of conservation of energy and momentum, and we give examples from the realm of subatomic physics to show how fundamental information about the physical world is obtained from studying the outcomes of various kinds of collisions.

10-1 WHAT IS A COLLISION?

In a collision a relatively large force acts on each colliding particle for a relatively short time. The basic idea of a collision is that the motion of the colliding particles (or of at least one of them) changes rather abruptly and that we can make a relatively clean separation of times that are "before the collision" and those that are "after the collision."

When a bat strikes a baseball, for example, the beginning and the end of the collision can be determined fairly precisely. The bat is in contact with the ball for an interval that is quite short compared with the time during which we are watching the ball. During the collision the bat exerts a large force on the ball (Fig. 1). This force varies with time in a complex way that we can measure only with difficulty. Both the ball and the bat are deformed during the collision. Forces that act for a time that is short compared with the time of observation of the system are called *impulsive* forces.

When an alpha particle (⁴He nucleus) collides with another nucleus (Fig. 2), the force acting between them may be the well-known repulsive electrostatic force associated with the charges on the particles. The particles may not actually touch, but we still may speak of a collision because a relatively strong force, acting for a time that is short compared with the time that the alpha particle is under observation, has a substantial effect on the motion of the alpha particle.

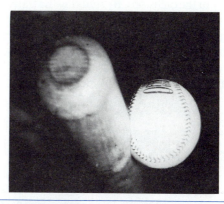

Figure 1 A high-speed photograph of a bat striking a baseball. Note the deformation of the ball, indicating the large impulsive force exerted by the bat.

Figure 2 An alpha particle collides with a helium nucleus in a cloud chamber. Most of the incident particles (coming from the left) pass through without colliding.

Figure 3 Two galaxies colliding.

(a)

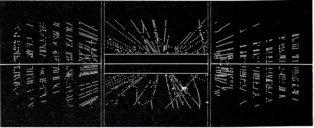

(b)

We can even speak about a collision between two galaxies (Fig. 3), if we are prepared to observe them over a time scale of the order of millions or billions of years. (But a more feasible alternative is to shorten this long time span by computer modeling!)

Collisions between elementary particles provide the principal source of information about their internal structure. When two particles collide at high energy, often the products of the collision are very different from the original particles (Fig. 4). Sometimes these collisions produce hundreds of product particles, whose total mass may be far greater than the masses of the colliding particles (kinetic energy of the incident particles being converted into rest energy in the collision). By studying the trajectories of the outgoing particles and applying the fundamental conservation laws, we can reconstruct the original event.

On a different scale, those who study traffic accidents also try to reconstruct collisions. From the paths and impact patterns of the colliding vehicles (Fig. 5), it is often

Figure 4 (*a*) The massive detector UA1 used at the proton–proton collider at CERN, the particle physics research facility near Geneva, Switzerland. (*b*) A computer reconstruction of the paths of the particles produced in one proton–proton collision. Such reconstructions were used in 1983 to confirm the existence of the particles called W and Z, which verified a theory that treats the electromagnetic force and the weak nuclear force as different aspects of a single more basic force.

Figure 5 A collision between two automobiles. Much of the incident kinetic energy is converted into energy of deformation of the two automobiles. Momentum conservation is used by accident reconstruction experts to calculate the velocities before the collision.

possible to deduce such important details as the speed and direction of motion of the two vehicles before the collision.

Another kind of collision is one that takes place between a space probe and a planet, called the "slingshot effect," in which the speed and direction of the space probe can be altered in a "close encounter" with a (moving) planet. The probe does not actually touch the planet, but it does come strongly under its gravitational influence for a time that is very short compared to the duration of the space probe's journey. Thus we are justified in calling such encounters "collisions." (See Problem 4.)

10-2 IMPULSE AND MOMENTUM

In studying collisions in this chapter, our aim is to learn what we can about the final motions of colliding particles from the principles of conservation of momentum and energy, given the initial motions of the colliding particles and assuming that we know nothing about the forces acting during the collision.

Let us assume that Fig. 6 shows the magnitude of the net force exerted on a body during a collision. The collision begins at time t_i and ends at time t_f, the force being zero before and after collision. From Newton's second law in the form $\mathbf{F} = d\mathbf{p}/dt$, we can write the change in momentum $d\mathbf{p}$ of a particle in a time dt during which a force $\mathbf{F}$ acts on it as

$$d\mathbf{p} = \mathbf{F} \, dt.$$

We can find the change in momentum of the body during a collision by integrating over the time of collision, that is,

between the initial conditions (momentum $\mathbf{p}_i$ at time t_i) and the final conditions (momentum $\mathbf{p}_f$ at time t_f):

$$\int_{p_i}^{p_f} d\mathbf{p} = \int_{t_i}^{t_f} \mathbf{F} \, dt. \qquad (1)$$

The left side of Eq. 1 is just the change in momentum, $\mathbf{p}_f - \mathbf{p}_i$. The right side, which depends on both the strength of the force and its duration, is called the *impulse* $\mathbf{J}$ of the force:

$$\mathbf{J} = \int_{t_i}^{t_f} \mathbf{F} \, dt, \qquad (2)$$

and from Eq. 1 it follows that

$$\mathbf{J} = \mathbf{p}_f - \mathbf{p}_i. \qquad (3)$$

Equation 3 is the mathematical statement of the *impulse–momentum theorem*:

The impulse of the net force acting on a particle during a given time interval is equal to the change in momentum of the particle during that interval.

Both impulse and momentum are vectors and they have the same units and dimensions.

Although we use Eq. 3 in this chapter only in situations involving impulsive forces (that is, those of short duration compared with the time of observation), no such limitation is built into that equation. Equation 3 is just as general as Newton's second law, from which it was derived. We could use Eq. 3, for example, to find the momentum acquired by a body falling in the Earth's gravity.

The impulse–momentum theorem is very similar to the work–energy theorem, which we derived in Chapter 7. Both apply to single particles and both are derived directly from Newton's second law. Work involves an integral of the net force over position, while impulse involves the integral of the net force over time. The work–energy theorem is a scalar equation dealing with the change in kinetic energy of the particle, while the impulse–momentum theorem is a vector equation dealing with the change in momentum of the particle.

The impulsive force whose magnitude is represented in Fig. 6 is assumed to have a constant direction. The magnitude of the impulse of this force is represented by the area under the $F(t)$ curve. We can represent that same area by the rectangle in Fig. 6 of width Δt and height $\overline{F}$, where $\overline{F}$ is the magnitude of the *average force* that acts during the interval Δt. Thus

$$J = \overline{F} \, \Delta t. \qquad (4)$$

In a collision such as the ball and bat of Fig. 1, it is difficult to measure $F(t)$ directly, but we can estimate Δt (perhaps a few milliseconds) and obtain a reasonable value for $\overline{F}$ based on the impulse computed according to Eq. 3 from the change in momentum of the ball (see Sample Problem 1).

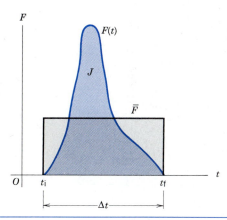

Figure 6   An impulsive force $F(t)$ varies in an arbitrary way with time during a collision that lasts from t_i to t_f. The area under the $F(t)$ curve is the impulse J, and the rectangle bounded by the average force $\overline{F}$ has an equal area.

10-3 CONSERVATION OF MOMENTUM DURING COLLISIONS

Consider now a collision between two particles, such as those of masses m_1 and m_2, shown in Fig. 7. During the brief collision these particles exert large forces on one another. At any instant $\mathbf{F}_{12}$ is the force exerted on particle 1 by particle 2 and $\mathbf{F}_{21}$ is the force exerted on particle 2 by particle 1. By Newton's third law these forces are equal in magnitude but oppositely directed.

The change in momentum of particle 1 resulting from the collision is

$$\Delta \mathbf{p}_1 = \int_{t_i}^{t_f} \mathbf{F}_{12} \, dt = \overline{\mathbf{F}}_{12} \, \Delta t \qquad (5)$$

in which $\overline{\mathbf{F}}_{12}$ is the average value of the force $\mathbf{F}_{12}$ during the time interval of the collision $\Delta t = t_f - t_i$.

The change in momentum of particle 2 resulting from the collision is

$$\Delta \mathbf{p}_2 = \int_{t_i}^{t_f} \mathbf{F}_{21} \, dt = \overline{\mathbf{F}}_{21} \, \Delta t \qquad (6)$$

in which $\overline{\mathbf{F}}_{21}$ is the average value of the force $\mathbf{F}_{21}$ during the time interval of the collision $\Delta t = t_f - t_i$.

If no other forces act on the particles, then $\Delta \mathbf{p}_1$ and $\Delta \mathbf{p}_2$ give the total change in momentum for each particle. But we have seen that at each instant $\mathbf{F}_{12} = -\mathbf{F}_{21}$, so that $\overline{\mathbf{F}}_{12} = -\overline{\mathbf{F}}_{21}$, and therefore

$$\Delta \mathbf{p}_1 = -\Delta \mathbf{p}_2. \qquad (7)$$

If we consider the two particles as an isolated system, the total momentum of the system is

$$\mathbf{P} = \mathbf{p}_1 + \mathbf{p}_2, \qquad (8)$$

and the total change in momentum of the system as a result of the collision is zero; that is,

$$\Delta \mathbf{P} = \Delta \mathbf{p}_1 + \Delta \mathbf{p}_2 = 0. \qquad (9)$$

Hence, *if there are no external forces the total momentum of the two-particle system is not changed by the collision.* This is simply the law of conservation of linear momentum (see Section 9-6) applied to this two-particle system. The impulsive forces acting during the collision are internal forces that have no effect on the total momentum of the system.

We have defined a collision as an interaction that occurs in a time Δt that is negligible compared to the time during which we are observing the system. We can also characterize a collision as an event in which the external forces that may act on the system are negligible compared to the impulsive collision forces (see Sample Problem 1). When a bat strikes a baseball, a golf club strikes a golf ball, or one billiard ball strikes another, external forces act on the system. Gravity or friction may exert forces on these bodies, for example; these external forces may not be the same on each colliding body nor are they necessarily canceled by other external forces. Even so, it is quite safe to neglect these external forces during the collision and to assume momentum conservation provided, as is almost always true, that the external forces are negligible compared to the impulsive forces of collision. As a result, the change in momentum of a particle during a collision arising from an external force is negligible compared to the change in momentum of that particle arising from the impulsive collisional force (Fig. 8).

For example, when a bat strikes a baseball, the collision lasts only a few milliseconds. Because the change in momentum of the ball is large and the time of collision is small, it follows from

$$\Delta \mathbf{p} = \overline{\mathbf{F}} \, \Delta t$$

that the average impulsive force $\overline{\mathbf{F}}$ is relatively large. Compared to this force, the external force of gravity is negligible. During the collision we can safely ignore this external force in determining the change in motion of the ball; the shorter the duration of the collision the more likely this is to be true.

In practice, therefore, we can apply the law of momentum conservation during collisions if the time of collision is small enough. We can then say that the momentum of a system of particles just before the particles collide is equal to the momentum of the system just after the particles collide.

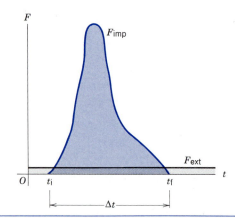

Figure 8   The impulsive force F_{imp} that acts during a collision is generally much stronger that any external force F_{ext} (shown here as constant) that may also act.

Figure 7   Two particles of masses m_1 and m_2 collide and experience equal and opposite forces.

Sample Problem 1   A baseball (which has an official weight of about 5 oz) is moving horizontally at a speed of 93 mi/h (about 150 km/h) when it is struck by the bat (see Fig. 1). It leaves the bat in a direction at an angle $\phi = 35°$ above its incident path and with a speed of 180 km/h. (*a*) Find the impulse of the force exerted on the ball. (*b*) Assuming the collision lasts for 1.5 ms (= 0.0015 s), what is the average force? (*c*) Find the change in the momentum of the bat.

Solution   (*a*) Figure 9*a* shows the initial momentum $\mathbf{p}_i$ and the final momentum $\mathbf{p}_f$ of the baseball. The mass corresponding to 5 oz is 0.14 kg, and the final speed of the ball in more appropriate units is 50 m/s. The components of the final momentum are given by

$$p_{fx} = mv_f \cos \phi = (0.14 \text{ kg})(50 \text{ m/s})(\cos 35°) = 5.7 \text{ kg·m/s},$$

$$p_{fy} = mv_f \sin \phi = (0.14 \text{ kg})(50 \text{ m/s})(\sin 35°) = 4.0 \text{ kg·m/s}.$$

In this coordinate system, the initial momentum has only an x component, whose (negative) value is

$$p_{ix} = mv_i = (0.14 \text{ kg})(-42 \text{ m/s}) = -5.9 \text{ kg·m/s}.$$

The impulse can now be found:

$$J_x = p_{fx} - p_{ix} = 5.7 \text{ kg·m/s} - (-5.9 \text{ kg·m/s}) = 11.6 \text{ kg·m/s},$$

$$J_y = p_{fy} - p_{iy} = 4.0 \text{ kg·m/s} - 0 = 4.0 \text{ kg·m/s}.$$

In other terms, the impulse has magnitude

$$J = \sqrt{J_x^2 + J_y^2} = \sqrt{(11.6 \text{ kg·m/s})^2 + (4.0 \text{ kg·m/s})^2}$$
$$= 12.3 \text{ kg·m/s}$$

and acts in a direction determined by

$$\theta = \tan^{-1}(J_y/J_x) = \tan^{-1}[(4.0 \text{ kg·m/s})/(11.6 \text{ kg·m/s})] = 19°$$

above the horizontal. Figure 9*b* shows the impulse vector $\mathbf{J}$ and verifies graphically that, as the definition of Eq. 3 requires,

$$\mathbf{J} = \mathbf{p}_f - \mathbf{p}_i = \mathbf{p}_f + (-\mathbf{p}_i).$$

(*b*) With $\mathbf{J} = \overline{\mathbf{F}} \Delta t$, we have $\overline{\mathbf{F}} = \mathbf{J}/\Delta t$. Thus $\overline{\mathbf{F}}$ has magnitude

$$\overline{F} = (12.3 \text{ kg·m/s})/0.0015 \text{ s} = 8200 \text{ N},$$

which is nearly 1 ton. This force acts in the same direction as $\mathbf{J}$,

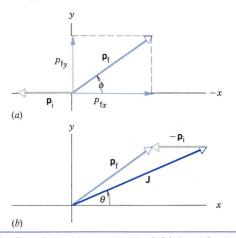

Figure 9   Sample Problem 1. (*a*) The initial and final momenta of the baseball. (*b*) The difference $\mathbf{p}_f - \mathbf{p}_i$ is equal to the impulse $\mathbf{J}$.

that is, 19° above the horizontal. Note that this is the *average* force; the *maximum* force is considerably greater, as Fig. 6 shows. Also, note that $\overline{F} (= 8200 \text{ N}) \gg mg (= 1.4 \text{ N})$. Thus we are quite safe in assuming that the impulsive force greatly exceeds the external force (gravity, in this case) and therefore is very nearly equal to the net force that acts during the collision.

(*c*) The change in momentum of the bat is, by Eq. 7, equal and opposite to that of the ball. Thus, for the bat,

$$\Delta p_x = -11.6 \text{ kg·m/s},$$

$$\Delta p_y = -4.0 \text{ kg·m/s}.$$

Is this a large change or a small one? Try to estimate the momentum of the bat in motion to answer this question.

10-4   COLLISIONS IN ONE DIMENSION

In this section we consider the effect of a collision between two objects. Usually we know the initial velocities of the two objects before the collision, and our goal is to apply conservation laws or laws of motion to find the velocities after the collision.

We can always calculate the motions of objects after they collide from their previous motions if we know the forces that act during the collision, and if we can solve the equations of motion. In most collisions, however, we do not know these forces. The law of conservation of momentum must hold during any collision in which only internal forces act, and *it can be applied even if we do not know the forces.* Although we may not know the details of the interaction, we can use the conservation of momentum and the conservation of energy in many cases to predict the results of the collision.

Linear momentum is *always* conserved in collisions. Total energy is also conserved: the initial total energy of the colliding particles is equal to the final total energy of the products. This energy may include not only kinetic energy but other forms as well, such as internal energy, energy of deformation, rotational energy, radiant energy, and so on.

In one special category of collision, called an *elastic* collision, we neglect all these other forms of energy and consider only mechanical energy $U + K$. Furthermore, we assume that in an impulsive collision, the internal forces act for a short time and therefore over a short distance; we observe the particles only at much greater relative separation, so that the effects of their internal potential energy can be neglected. In an elastic collision, translational kinetic energy is the only form of energy that we must account for, and conservation of mechanical energy is therefore equivalent to conservation of kinetic energy: *the initial kinetic energy K_i equals the final kinetic energy K_f in an elastic collision.*

In another category of collision, which is called *inelastic,* energy appears in other forms, and the initial and final kinetic energies are not equal. In some cases $K_i > K_f$, as, for example, when initial kinetic energy is converted into internal energy of the products, while in other cases $K_i < K_f$, such as when internal energy stored in the colliding particles is released. Mechanical energy $U + K$ is *not* conserved in an inelastic collision, but *total* energy is (see Section 8-6). When the colliding bodies are simple, such as atoms or molecules, we can often account directly for the difference between K_i and K_f in terms of the known discrete internal energy states of the system. In more complex systems, such as colliding automobiles, we regard the difference simply as "lost" or "gained" kinetic energy.

All collisions between real objects are to some extent inelastic. When the objects are very rigid, such as billiard balls, we can often treat the collision as approximately elastic. In this case the energy that is changed from kinetic to other forms (such as the sound wave you hear when the balls collide) is negligible compared with the kinetic energy. Note that the classification of a collision as elastic or inelastic is independent of the reference frame from which the collision is viewed.

When the two bodies stick together after collision, the collision is said to be *completely inelastic.* For example, the collision between a bullet and a block of wood into which it is fired is completely inelastic when the bullet remains embedded in the block. The term "completely inelastic" does not necessarily mean that all the initial kinetic energy is lost; as we shall see, it means rather that the loss is as large as it can be, consistent with momentum conservation.

Even if the forces of collision are not known, we can find the motions of the particles after collision from the motions before collision, provided the collision is completely inelastic, or, if the collision is elastic, provided the collision takes place in one dimension. For a one-dimensional collision the relative motion after collision is along the same line as the relative motion before collision. We restrict ourselves to one-dimensional motion for the present.

Elastic Collisions

Consider first an elastic one-dimensional collision. We imagine two objects (perhaps gliders on an airtrack) moving initially along the line joining their centers, then colliding head-on and moving along the same straight line after collision (see Fig. 10). These bodies exert forces on each other during the collision that are along the initial line of motion, so that the final motion is also along this same line.

The masses of the colliding particles are m_1 and m_2, the velocity components being v_{1i} and v_{2i} before collision and v_{1f} and v_{2f} after collision. [In our notation, the numerical subscripts 1 and 2 specify the particle, while the subscripts i and f refer, respectively, to initial values (before the collision) and final values (after the collision).] We take the positive direction of the momentum and velocity to be to the right in Fig. 10. We assume, unless we specify otherwise, that the speeds of the colliding particles are low enough so that we need not use the relativistic expressions for momentum and kinetic energy. Then from conservation of momentum we obtain

$$m_1 v_{1i} + m_2 v_{2i} = m_1 v_{1f} + m_2 v_{2f}. \qquad (10)$$

Because we are considering an elastic collision, the kinetic energy is conserved by definition, and we obtain, from $K_i = K_f$,

$$\tfrac{1}{2} m_1 v_{1i}^2 + \tfrac{1}{2} m_2 v_{2i}^2 = \tfrac{1}{2} m_1 v_{1f}^2 + \tfrac{1}{2} m_2 v_{2f}^2. \qquad (11)$$

If we know the masses and the initial velocities, we can calculate the two (unknown) final velocities v_{1f} and v_{2f} from these two equations.

The momentum equation can be written

$$m_1(v_{1i} - v_{1f}) = m_2(v_{2f} - v_{2i}), \qquad (12)$$

and the energy equation can be written

$$m_1(v_{1i}^2 - v_{1f}^2) = m_2(v_{2f}^2 - v_{2i}^2). \qquad (13)$$

Dividing Eq. 13 by Eq. 12, and assuming $v_{2f} \neq v_{2i}$ and $v_{1f} \neq v_{1i}$ (see Question 15), we obtain

$$v_{1i} + v_{1f} = v_{2f} + v_{2i}$$

and, after rearrangement,

$$v_{1i} - v_{2i} = -(v_{1f} - v_{2f}). \qquad (14)$$

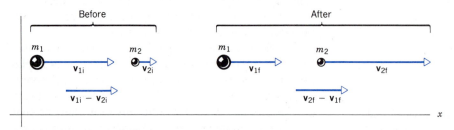

Figure 10   Two particles before and after an elastic collision. Note that the relative velocities before and after are equal.

This tells us that in an elastic one-dimensional collision, the relative velocity of approach before collision is equal and opposite to the relative velocity of separation after collision, no matter what the masses of the colliding particles may be.

To find the velocity components v_{1f} and v_{2f} after collision from the velocity components v_{1i} and v_{2i} before collision, we combine Eqs. 12 and 14 to eliminate v_{2f} and solve for v_{1f}:

$$v_{1f} = \left(\frac{m_1 - m_2}{m_1 + m_2}\right)v_{1i} + \left(\frac{2m_2}{m_1 + m_2}\right)v_{2i}. \quad (15)$$

Similarly, we eliminate v_{1f} and solve for v_{2f}:

$$v_{2f} = \left(\frac{2m_1}{m_1 + m_2}\right)v_{1i} + \left(\frac{m_2 - m_1}{m_1 + m_2}\right)v_{2i}. \quad (16)$$

Equations 15 and 16, which hold in all inertial reference frames, are general results that allow us to find the final velocities in any one-dimensional elastic collision. Often we simplify these equations by choosing a frame in which the target particle (say, m_2) is initially at rest, thus allowing us to put $v_{2i} = 0$ in Eqs. 15 and 16. We now consider certain special cases that are of interest.

1. *Equal masses.* When the colliding particles have equal masses ($m_1 = m_2$), Eqs. 15 and 16 become simply

$$v_{1f} = v_{2i} \quad \text{and} \quad v_{2f} = v_{1i}. \quad (17)$$

That is, the particles exchange velocities: the final velocity of one particle is equal to the initial velocity of the other.

2. *Target particle at rest.* Another case of interest is that in which particle m_2 is initially at rest. Then $v_{2i} = 0$ and

$$v_{1f} = \left(\frac{m_1 - m_2}{m_1 + m_2}\right)v_{1i} \quad \text{and} \quad v_{2f} = \left(\frac{2m_1}{m_1 + m_2}\right)v_{1i}. \quad (18)$$

Combining this special case with the previous one (that is, a collision between equal mass particles in which one is initially at rest), we see that the first particle is "stopped cold" and the second one "takes off" with the velocity the first one originally had. It is often possible to observe this effect in collisions of nonrotating billiard balls.

3. *Massive target.* If $m_2 \gg m_1$, then Eqs. 15 and 16 reduce to

$$v_{1f} \approx -v_{1i} + 2v_{2i} \quad \text{and} \quad v_{2f} \approx v_{2i}. \quad (19)$$

When the massive particle is moving slowly or at rest, then

$$v_{1f} \approx -v_{1i} \quad \text{and} \quad v_{2f} \approx 0. \quad (20)$$

That is, when a light projectile collides with a very much more massive one at rest, the velocity of the light particle is approximately reversed, and the massive particle remains approximately at rest. For example, a ball dropped from a height h rebounds from the Earth after the collision with reversed velocity and, if the collision were perfectly elastic and there were no air resistance, it would reach the same height h. Similarly, an electron rebounds

from a (relatively massive) atom in a head-on collision with its motion reversed, while the target atom is essentially unaffected by the collision.

4. *Massive projectile.* When $m_1 \gg m_2$, Eqs. 15 and 16 become

$$v_{1f} \approx v_{1i} \quad \text{and} \quad v_{2f} \approx 2v_{1i} - v_{2i}. \quad (21)$$

If the light target particle is initially at rest (or moving much slower than m_1), then after the collision the target particle moves at twice the speed of m_1. The motion of m_1 is nearly unaffected by the collision with the much lighter target.

In alpha-particle scattering (Fig. 2), the incident alpha particle (whose mass is about 8000 times the electron mass) is essentially unaffected by collisions with the electrons of the target atoms (as indicated by the many straight-line paths in Fig. 2). The alpha particle is deflected only in the rare encounters with the massive nucleus of a target atom.

Inelastic Collisions

We now consider inelastic collisions, in which by definition kinetic energy is not conserved, although of course conservation of momentum *always* holds. Conservation of *total* energy holds as well, but the inclusion of forms of energy other than kinetic adds more terms to Eq. 11, and unless we can exactly specify the energy transfers (for instance, how much internal energy is converted into kinetic), we no longer have a solvable system of equations.

In one special inelastic case, that of the completely inelastic collision, the final outcome can be found from the initial values alone. In this case, the particles stick together and move with a common velocity v_f after the collision. Thus there is only one unknown, and the momentum equation alone (Eq. 10) is sufficient. Replacing both v_{1f} and v_{2f} in that equation by the common velocity v_f leads to

$$v_f = \left(\frac{m_1}{m_1 + m_2}\right)v_{1i} + \left(\frac{m_2}{m_1 + m_2}\right)v_{2i}. \quad (22)$$

When m_2 is initially at rest, this reduces to

$$v_f = \left(\frac{m_1}{m_1 + m_2}\right)v_{1i}. \quad (23)$$

The velocity of m_1 is "scaled down" by the mass ratio $m_1/(m_1 + m_2)$. The larger m_1 is, the faster the combination moves; the smaller m_1, the slower.

Equation 22 can be applied equally well in reverse. That is, a particle of mass M moving with velocity v_f splits into two particles, one of mass m_1 moving with velocity v_{1i} and another of mass m_2 ($= M - m_1$) moving in the opposite direction with velocity v_{2i}. This result applies even though the explosion might impart considerable kinetic energy to the two particles. In the special case that $v_f = 0$ (the initial particle is at rest), we have $v_{1i}/v_{2i} = -m_2/m_1$.

The more massive particle has the smaller velocity, as we would expect in order to make the total momentum zero, and the two particles move in opposite directions. Applications of this principle to spontaneous decay processes are considered in Section 10-7.

Sample Problem 2   (a) By what fraction is the kinetic energy of a neutron (mass m_1) decreased in a head-on elastic collision with an atomic nucleus (mass m_2) initially at rest? (b) Find the fractional decrease in the kinetic energy of a neutron when it collides in this way with a lead nucleus, a carbon nucleus, and a hydrogen nucleus. The ratio of nuclear mass to neutron mass ($= m_2/m_1$) is 206 for lead, 12 for carbon, and 1 for hydrogen.

Solution   (a) The initial kinetic energy K_i of the neutron, assumed to be nonrelativistic, is $\frac{1}{2}m_1v_{1i}^2$. Its final kinetic energy K_f is $\frac{1}{2}m_1v_{1f}^2$. The fractional decrease in kinetic energy is

$$\frac{K_i - K_f}{K_i} = \frac{v_{1i}^2 - v_{1f}^2}{v_{1i}^2} = 1 - \frac{v_{1f}^2}{v_{1i}^2}.$$

But, for such a collision (see Eq. 18),

$$v_{1f} = \left(\frac{m_1 - m_2}{m_1 + m_2}\right)v_{1i},$$

so that

$$\frac{K_i - K_f}{K_i} = 1 - \left(\frac{m_1 - m_2}{m_1 + m_2}\right)^2 = \frac{4m_1m_2}{(m_1 + m_2)^2}.$$

(b) For lead, with $m_2 = 206m_1$,

$$\frac{K_i - K_f}{K_i} = \frac{4m_1(206m_1)}{(207m_1)^2} = 0.02 = 2\%.$$

For carbon, with $m_2 = 12m_1$,

$$\frac{K_i - K_f}{K_i} = \frac{4m_1(12m_1)}{(13m_1)^2} = 0.28 = 28\%.$$

For hydrogen, with $m_2 = m_1$,

$$\frac{K_i - K_f}{K_i} = \frac{4m_1(m_1)}{(2m_1)^2} = 1 = 100\%.$$

These results explain why a material such as paraffin, with its high content of hydrogen, is far more effective in slowing down neutrons than is a heavy material such as lead. Note, however, that collisions are not always "head-on" as we have assumed throughout this section. Although a neutron would not lose *all* its energy in a more typical glancing collision with a resting hydrogen, it will still lose energy much more efficiently in hydrogenous materials than it will in carbon or lead.

The fission of uranium in a reactor gives neutrons with relatively large kinetic energies, in the MeV range. To produce a chain reaction, these neutrons must be used to initiate other fission events, but the probability for a neutron to cause fission decreases rapidly with increasing neutron kinetic energy. It is therefore necessary to slow down or *moderate* the neutrons until their energy is in the eV range, where the fission probability is nearly three orders of magnitude greater. The above calculation, while somewhat oversimplified, shows that a hydrogen-rich material, such as water or paraffin, would be a good choice as a moderator.

Sample Problem 3   A ballistic pendulum (Fig. 11) is a device that was used to measure the speeds of bullets before electronic timing devices were available. It consists of a large block of wood of mass M, hanging from two long pairs of cords. A bullet of mass m is fired into the block and comes quickly to rest relative to the block. The block + bullet combination swings upward, its center of mass rising a vertical distance h before the pendulum comes momentarily to rest at the end of its arc. Take the mass of the block to be $M = 5.4$ kg and the mass of the bullet to be $m = 9.5$ g. (a) What is the initial speed of the bullet if the block rises to a height of $h = 6.3$ cm? (b) What is the initial kinetic energy of the bullet? How much of this energy remains as mechanical energy of the swinging pendulum?

Solution   (a) As the bullet collides with the block, we have, from the conservation of momentum in the horizontal direction,

$$mv = (M + m)V,$$

where v is the velocity of the bullet before impact and V is the velocity of the combination after the impact. Although mechanical energy is certainly *not* conserved *during* the collision of the bullet with the block, it *is* conserved in the swinging pendulum *after* the impact. The kinetic energy of the system when the block is at the bottom of its arc must then equal the potential energy of the system when the block is at the top, or

$$\tfrac{1}{2}(M + m)V^2 = (M + m)gh.$$

Eliminating V between these two equations leads to

$$v = \left(\frac{M + m}{m}\right)\sqrt{2gh}$$

$$= \left(\frac{5.4\ \text{kg} + 0.0095\ \text{kg}}{0.0095\ \text{kg}}\right)\sqrt{(2)(9.8\ \text{m/s}^2)(0.063\ \text{m})} = 630\ \text{m/s}.$$

We can look at the ballistic pendulum as a kind of transformer, exchanging the high speed of a light object (the bullet) for the low—and thus more easily measureable—speed of a massive object (the block).

(b) The kinetic energy of the bullet is

$$K_b = \tfrac{1}{2}mv^2 = \tfrac{1}{2}(0.0095\ \text{kg})(630\ \text{m/s})^2 = 1900\ \text{J}.$$

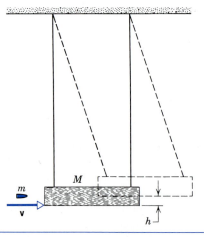

Figure 11   Sample Problem 3. A ballistic pendulum is used to measure the speed of a bullet.

The mechanical energy of the swinging pendulum is equal to its potential energy when the block is at the top of its swing, or

$$E = (M + m)gh = (5.4 \text{ kg} + 0.0095 \text{ kg})(9.8 \text{ m/s}^2)(0.063 \text{ m})$$
$$= 3.3 \text{ J}.$$

Thus only 3.3/1900 or 0.2% of the original kinetic energy of the bullet is transferred to mechanical energy of the pendulum. The rest is stored inside the pendulum block as internal energy or transferred to the environment, for example, as heat or sound waves.

10-5 TWO-DIMENSIONAL COLLISIONS

If two particles collide in a manner other than head-on, the particles may move in directions that do not coincide with the original directions of motion. Figure 12 shows the geometry of such a collision. We have selected our coordinate system so that $\mathbf{p}_1$ has only an x component, thereby simplifying the calculation somewhat. The target particle m_2 is assumed to be at rest. The distance b between the line of motion of the incident particle and a parallel line through m_2 is called the *impact parameter*. A head-on collision corresponds to $b = 0$, and larger values of b give a more glancing collision. The figure might represent the paths of two nuclei that collide through their mutual electrostatic force of repulsion; the force depends inversely on the square of the distance between the nuclei, which need not be in actual contact for the collision. At large enough distances the force becomes small, and the

particles move in straight lines essentially unaffected by the force.

No matter what the force that acts between the particles, momentum must be conserved. The force between the particles is an internal force, which cannot change the total momentum of the two-particle system. Furthermore, since momentum is a vector, we know that the x components and y components will give us two independent scalar equations. For the x components, the initial momentum is $m_1 v_{1i}$ along the x axis, and the total final momentum is the sum of the x components of the final momenta of the two particles:

$$p_{ix} = p_{fx}$$
$$m_1 v_{1i} = m_1 v_{1f} \cos \phi_1 + m_2 v_{2f} \cos \phi_2. \qquad (24)$$

Here we account for the directions of $\mathbf{v}_{1f}$ and $\mathbf{v}_{2f}$ through the angles ϕ_1 and ϕ_2, respectively; thus v_{1f} and v_{2f} in Eq. 24 stand for the *magnitudes* of the velocities and are always positive. This contrasts with Eqs. 15 and 16 or Eq. 22, which dealt with velocity *components* that could be positive or negative.

The initial y momentum is zero (thanks to our simplifying choice of coordinate axes), and the final y momentum is the difference between that of the two particles (we have chosen ϕ_1 and ϕ_2 on opposite sides of the x axis, so that the sum of the y components becomes an algebraic difference):

$$p_{iy} = p_{fy}$$
$$0 = m_1 v_{1f} \sin \phi_1 - m_2 v_{2f} \sin \phi_2. \qquad (25)$$

If the collision is elastic, the usual result for conservation

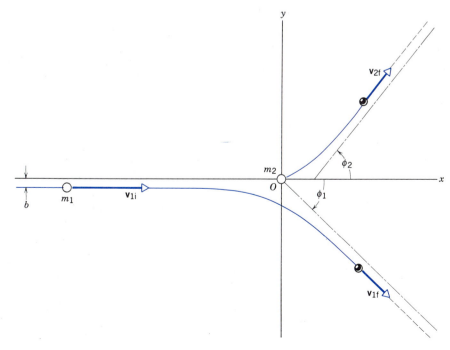

Figure 12 Two particles undergoing a collision. The open circles indicate their positions before the collision and the filled circles after collision. Initially, particle 2 is at rest. The impact parameter b is the distance by which the collision misses being head-on.

of energy holds. Equating the initial and final kinetic energies gives

$$\tfrac{1}{2}m_1 v_{1i}^2 = \tfrac{1}{2}m_1 v_{1f}^2 + \tfrac{1}{2}m_2 v_{2f}^2. \tag{26}$$

Given the initial conditions (m_1, m_2, and v_{1i}), there are four unknowns in Eqs. 24–26 (v_{1f}, v_{2f}, ϕ_1, ϕ_2) but only three equations relating them. There is no unique solution to such an underdetermined system of equations; in fact, there is an infinite number of solutions. To obtain a single solution, we must provide an additional constraint or restriction. For example, we can choose to observe particle 1 at a particular angle ϕ_1, as in Sample Problem 4. Once we make that choice, the three equations can then be solved for the three remaining unknowns.

Sample Problem 4 A gas molecule having a speed of 322 m/s collides elastically with another molecule of the same mass which is initially at rest. After the collision the first molecule moves at an angle of 30° to its initial direction. Find the speed of each molecule after collision and the angle made with the incident direction by the recoiling target molecule.

Solution This example corresponds exactly to Eqs. 24–26, with $m_1 = m_2$, $v_{1i} = 322$ m/s, and $\phi_1 = 30°$. Setting m_1 equal to m_2, we have the relations

$$v_{1i} = v_{1f} \cos \phi_1 + v_{2f} \cos \phi_2, \tag{27}$$

$$v_{1f} \sin \phi_1 = v_{2f} \sin \phi_2, \tag{28}$$

and

$$v_{1i}^2 = v_{1f}^2 + v_{2f}^2. \tag{29}$$

We must solve for v_{1f}, v_{2f}, and ϕ_2. To do this we first eliminate ϕ_2 by squaring Eq. 27 (rewriting it as $v_{1i} - v_{1f} \cos \phi_1 = v_{2f} \cos \phi_2$), and we then add this to the square of Eq. 28. Noting that $\sin^2 \phi_2 + \cos^2 \phi_2 = 1$, we obtain

$$v_{1i}^2 + v_{1f}^2 - 2v_{1i}v_{1f} \cos \phi_1 = v_{2f}^2.$$

Combining this with Eq. 29, we obtain (provided that $v_{1f} \neq 0$)

$$v_{1f} = v_{1i} \cos \phi_1 = (322 \text{ m/s})(\cos 30°) = 279 \text{ m/s}.$$

From Eq. 29

$$v_{2f}^2 = v_{1i}^2 - v_{1f}^2 = (322 \text{ m/s})^2 - (279 \text{ m/s})^2,$$

or

$$v_{2f} = 161 \text{ m/s}.$$

Finally, from Eq. 28

$$\sin \phi_2 = \frac{v_{1f}}{v_{2f}} \sin \phi_1$$

$$= \frac{279 \text{ m/s}}{161 \text{ m/s}} \sin 30° = 0.866$$

or

$$\phi_2 = 60°.$$

The two molecules move apart at right angles ($\phi_1 + \phi_2 = 90°$ in Fig. 12).

You should be able to show that in an elastic collision between particles of equal mass, one of which is initially at rest, the recoiling particles always move off at right angles to one another. Figure 13 shows a series of four successive elastic collisions between protons caused when a high-energy proton enters a bubble

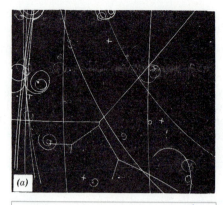

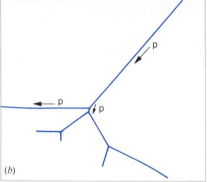

Figure 13 (*a*) Four proton–proton collisions in a bubble chamber. (*b*) A schematic representation of the paths of the protons. The original proton enters from the upper right. The tracks do not all lie in the plane of the photograph, and stereoscopic viewing shows that the angle between the incident and outgoing proton in each collision is 90°, as expected. The other tracks in the photo are caused by mesons (slight curvature) and electrons (tight spirals).

chamber filled with liquid hydrogen, which supplies the target protons. The tracks of the particles are made visible by the trail of bubbles left in their wake. Since the interacting particles are of equal mass and the collisions are elastic, the particles recoil at right angles to each other; this is apparent when the tracks of Fig. 13 are viewed stereoscopically. See Fig. 2 for another example.

Inelastic Collisions in Two Dimensions

If the collision is inelastic, Eq. 26 no longer applies. Often we can replace it with an equivalent expression that accounts for the energy converted to or from kinetic energy and that therefore yields a relationship between the initial and final kinetic energies.

A *completely* inelastic collision in two dimensions must start with both bodies in motion. (Why?) We again let the motion of one body define the x axis, and we arrange the collision so that the two bodies meet and stick together at the origin. The final object then moves in the direction ϕ_f with velocity $\mathbf{v}_f$ (Fig. 14). Conservation of momentum for the x and y components gives the following:

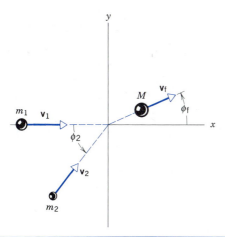

Figure 14 A completely inelastic collision in two dimensions. Particles with masses m_1 and m_2 collide to form the combined particle of mass M.

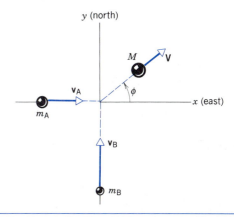

Figure 15 Sample Problem 5. Two skaters, Alfred (A) and Barbara (B), have a completely inelastic collision. Afterward, they move together in a direction given by the angle ϕ.

x component: $m_1v_1 + m_2v_2 \cos\phi_2 = Mv_f \cos\phi_f,$ (30)

y component: $m_2v_2 \sin\phi_2 = Mv_f \sin\phi_f.$ (31)

Here $M = m_1 + m_2$ is the total mass of the combination after the collision. Because the combination moves with a common velocity (magnitude and direction), the four unknowns of the elastic case are reduced to two: v_f and ϕ_f. The two equations (Eqs. 30 and 31) are sufficient for a unique solution.

Sample Problem 5 Two skaters collide and embrace in a completely inelastic collision. That is, they stick together after impact, as Fig. 15 suggests. Alfred, whose mass m_A is 83 kg, is originally moving east with a speed $v_A = 6.4$ km/h. Barbara, whose mass m_B is 55 kg, is originally moving north with a speed $v_B = 8.8$ km/h. (a) What is the velocity **V** of the couple after impact? (b) What is the fractional change in the kinetic energy of the skaters because of the collision?

Solution (a) Momentum is conserved during the collision. We can write, for the two momentum components:

x component: $m_A v_A = MV \cos\phi,$ (32)

y component: $m_B v_B = MV \sin\phi,$ (33)

in which $M = m_A + m_B$. Dividing Eq. 33 by Eq. 32 yields

$$\tan\phi = \frac{m_B v_B}{m_A v_A} = \frac{(55 \text{ kg})(8.8 \text{ km/h})}{(83 \text{ kg})(6.4 \text{ km/h})} = 0.911.$$

Thus

$$\phi = \tan^{-1} 0.911 = 42.3°.$$

From Eq. 33 we then have

$$V = \frac{m_B v_B}{M \sin\phi} = \frac{(55 \text{ kg})(8.8 \text{ km/h})}{(83 \text{ kg} + 55 \text{ kg})(\sin 42.3°)}$$
$$= 5.21 \text{ km/h}.$$

(b) The initial kinetic energy is

$$K_i = \tfrac{1}{2}m_A v_A^2 + \tfrac{1}{2}m_B v_B^2$$
$$= \tfrac{1}{2}(83 \text{ kg})(6.4 \text{ km/h})^2 + \tfrac{1}{2}(55 \text{ kg})(8.8 \text{ km/h})^2$$
$$= 3830 \text{ kg} \cdot \text{km}^2/\text{h}^2.$$

The final kinetic energy is

$$K_f = \tfrac{1}{2}MV^2$$
$$= \tfrac{1}{2}(83 \text{ kg} + 55 \text{ kg})(5.21 \text{ km/h})^2$$
$$= 1870 \text{ kg} \cdot \text{km}^2/\text{h}^2.$$

The fraction we seek is then

$$f = \frac{K_f - K_i}{K_i} = \frac{1870 \text{ kg} \cdot \text{km}^2/\text{h}^2 - 3830 \text{ kg} \cdot \text{km}^2/\text{h}^2}{3830 \text{ kg} \cdot \text{km}^2/\text{h}^2} = -0.51.$$

Thus 51% of the initial kinetic energy is lost in the collision. It must be dissipated in some form or other as internal energy of the skating couple.

10-6 CENTER-OF-MASS REFERENCE FRAME

When collision experiments are carried out in practice, the measurements are naturally made in a reference frame fixed in the laboratory (the *laboratory frame*). Very often such experiments involve a projectile fired at a target that is at rest in the laboratory. In many experiments in particle physics, on the other hand, two particles of the same mass and speed (two protons, perhaps, or two electrons) are fired directly at each other. Regardless of how the experiment is carried out, the analysis is often easier and the physical understanding is clarified if we view the collision from a reference frame attached to the center of mass of the colliding particles (the *center-of-mass frame*).

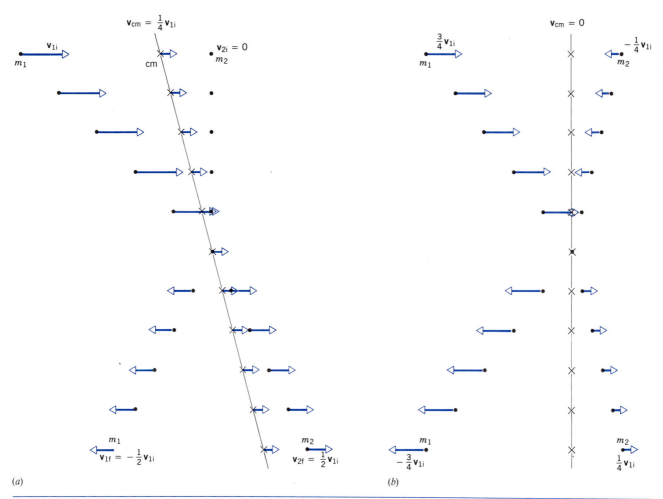

(a) (b)

Figure 16 A series of "snapshots" of two particles of masses m_1 and $m_2 = 3m_1$ colliding elastically in one dimension. The center of mass of the two particles is indicated by $\times$. (a) Laboratory reference frame. (b) Center-of-mass reference frame.

For example, consider the simple case of a one-dimensional (head-on) elastic collision between two identical particles. If one particle (the target) is fixed in the laboratory, the other particle (the projectile, initially moving with speed v) is brought to rest, and the original target particle moves forward with speed v. In the center-of-mass frame, however, the two particles approach each other before the collision, each with speed $\frac{1}{2}v$, and they recede from each other with the same speed after the collision. The distinction between projectile and target no longer exists, and the description of the event is completely symmetrical in this frame of reference.

Figure 16a shows a series of "snapshots" of an elastic collision between an initially moving particle of mass m_1 and a stationary particle of mass $m_2 = 3m_1$. Because only internal forces act in the collision, the motion of the center of mass, as shown in Fig. 16a, is unaffected by the collision. The center of mass of m_1 and m_2, calculated from Eq. 4 of Chapter 9, moves with the same constant velocity v_{cm} both before and after the collision.

The velocity of the center of mass is found from Eq. 5 of Chapter 9,

$$v_{cm} = \left(\frac{m_1}{m_1 + m_2}\right)v_{1i}, \tag{34}$$

with $v_{2i} = 0$. Let us now diagram the same collision from a frame of reference moving, relative to the laboratory, at the velocity v_{cm}. This is the center-of-mass reference frame. The velocities of m_1 and m_2 in this frame can be obtained from Eq. 43 of Section 4-6 for velocity transformations between reference frames:

$$v = v' + u, \tag{35}$$

where v is the velocity measured in the laboratory frame, v' is the velocity measured in the reference frame that is moving with respect to the laboratory, and u is the velocity of the moving frame relative to the laboratory frame. In our case, the moving frame is the center-of-mass frame and $u = v_{cm}$. We can then find the initial velocities of m_1 and m_2 in the moving frame:

$$v'_{1i} = v_{1i} - v_{cm} = v_{1i} - \left(\frac{m_1}{m_1 + m_2}\right)v_{1i}$$

$$= \left(\frac{m_2}{m_1 + m_2}\right)v_{1i}, \tag{36}$$

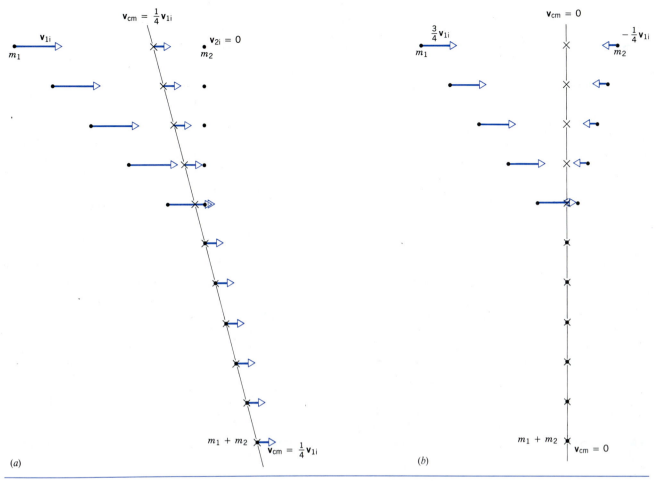

Figure 17   A series of "snapshots" of two particles of masses m_1 and $m_2 = 3m_1$ colliding completely inelastically in one dimension. (*a*) Laboratory reference frame. (*b*) Center-of-mass reference frame.

$$v'_{2i} = v_{2i} - v_{cm} = 0 - \left(\frac{m_1}{m_1 + m_2}\right)v_{1i}$$

$$= -\left(\frac{m_1}{m_1 + m_2}\right)v_{1i}. \tag{37}$$

The final velocities in the laboratory frame are given in Eq. 18, and these can also be transformed into the center-of-mass frame:

$$v'_{1f} = v_{1f} - v_{cm} = \left(\frac{m_1 - m_2}{m_1 + m_2}\right)v_{1i} - \left(\frac{m_1}{m_1 + m_2}\right)v_{1i}$$

$$= -\left(\frac{m_2}{m_1 + m_2}\right)v_{1i}, \tag{38}$$

$$v'_{2f} = v_{2f} - v_{cm} = \left(\frac{2m_1}{m_1 + m_2}\right)v_{1i} - \left(\frac{m_1}{m_1 + m_2}\right)v_{1i}$$

$$= \left(\frac{m_1}{m_1 + m_2}\right)v_{1i}. \tag{39}$$

Note the symmetry of these results. In the center-of-mass frame, the velocities of m_1 and m_2 each simply reverse directions in the collision, with the velocity of m_1 going from $+m_2 v_{1i}/(m_1 + m_2)$ to $-m_2 v_{1i}/(m_1 + m_2)$,

and the velocity of m_2 going from $-m_1 v_{1i}/(m_1 + m_2)$ to $+m_1 v_{1i}/(m_1 + m_2)$. A series of snapshots of the collision in the center-of-mass reference frame is shown in Fig. 16*b*. In this special reference frame, each particle performs a motion similar to a ball rebounding from a hard surface; the other particle is there just to provide the impulse for the reversal of the motion. It is also obvious in this reference frame that the total kinetic energy remains constant in the collision. (In fact, it remains constant for each particle separately.) Viewing the collision from this perspective, we gain fresh insight into the meaning of an "elastic" collision.

Now let us look at the completely inelastic one-dimensional collision in the center-of-mass frame. Again we assume m_1 to be incident on m_2 ($= 3m_1$) at rest in the laboratory. After the collision, there is a composite particle of mass $M = m_1 + m_2$. The center-of-mass velocity is again given by Eq. 34. The sequence of snapshots in Fig. 17*a* shows the collision in the laboratory frame; once again the center of mass moves with the same velocity before and after the collision.

The initial velocities of m_1 and m_2 transform in exactly the same way as in the previous case and are given by Eqs. 36 and 37. The final center-of-mass velocity of M after the collision can be found from transforming the general result for v_f in the laboratory frame, Eq. 23:

$$v_f' = v_f - v_{cm} = \left(\frac{m_1}{m_1 + m_2} \right) v_{1i} - \left(\frac{m_1}{m_1 + m_2} \right) v_{1i} = 0.$$

Of course this result should come as no surprise and should have been expected. The combined mass M is always at the center of mass, since it contains all the mass in the system after the collision. In the laboratory frame, M must move at the center-of-mass velocity, and comparing Eqs. 23 and 34 you see that indeed it does. In a reference frame in which the center of mass is at rest, M must also be at rest.

In the center-of-mass frame (Fig. 17*b*) there is again a symmetry: before the collision, m_1 and m_2 approach with equal and opposite momenta. After they collide and combine, the momentum must be zero.

There is another interesting property of the completely inelastic collision in the center-of-mass reference frame. In the laboratory frame, the kinetic energy lost (that is, converted to internal energy, energy of deformation, and so on) is always less than 100%; for instance, in a collision between two particles of equal mass, one of which is initially at rest, the loss in kinetic energy is 50%. In the center-of-mass frame, the loss is always 100%, regardless of the values of m_1 and m_2. When the purpose in colliding the particles is to convert kinetic energy into another form, it is advantageous not only to analyze but actually to *perform* the experiment in the center-of-mass frame.

In studies of the properties of the fundamental particles of nature, the goal is often to smash particles together at high energy to produce new and exotic varieties of particles of larger masses; in this case, kinetic energy is transformed in the collision into rest energy mc^2 of the new particles. The energy available to form new particles is just the "lost" kinetic energy in the inelastic collision; in the realm of these high-energy collisions, where we must use equations of relativistic kinematics, we learn that the initial laboratory kinetic energy required to produce new particles increases as the square of the rest energy of the particle we hope to produce. That is, to produce a particle with 10 times the rest energy requires 100 times the kinetic energy and therefore an accelerator that is 100 times as large and as costly. If, however, we could arrange to do the collision in the center-of-mass frame, then particles with 10 times the rest energy can be produced with only 10 (rather than 100) times the kinetic energy, since the collisions are 100% efficient in converting kinetic energy.

The current generation of particle accelerators includes many examples of such *colliding beam* devices. At the Stanford Linear Accelerator Center (SLAC) in California, beams of electrons and positrons (antielectrons) are made to collide at 50 GeV each (Fig. 18), and at the Fermi National Accelerator Laboratory in Illinois, beams of

Figure 18 The 2-mile-long electron accelerator of the Stanford Linear Accelerator Center. Electrons and positrons are accelerated along the straight section. The dashed lines show the underground paths of the electrons and positrons as they are made to collide in a laboratory near the bottom of the photo. See "The Stanford Linear Collider," by John R. Rees, *Scientific American,* October 1989, p. 58.

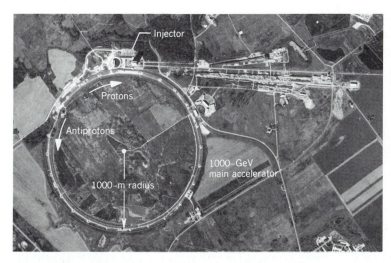

Figure 19 The Fermi National Accelerator Laboratory. Protons and antiprotons are injected from the small ring near the top of the photo into the large ring (of radius 1 km). They circulate in opposite directions and collide once in each circuit.

protons and antiprotons at 1000 GeV (1 TeV) circulate in the same ring in opposite directions and intersect once in each circuit (Fig. 19). Of course, the reaction is the same no matter from what frame of reference it is viewed, but the interpretation will differ.

So far we have studied only one-dimensional collisions in the center-of-mass frame. The two-dimensional elastic collision also takes on a more symmetric structure when viewed from the center-of-mass reference frame. We again consider m_2 to be at rest initially in the laboratory. We shall not go through the algebra in this case, which is a bit more complicated than the one-dimensional case; instead, we show a graphical description of the center-of-mass collision in Fig. 20. As in the one-dimensional case, each particle's velocity is merely reflected in the collision. The only difference here is that the two particles move after the collision along a line that is in general different from the axis of the initial velocities. The symmetry requires that the angles made by the final velocities with the incident velocities be the same for both particles; when we transform back to the laboratory frame, they will become the generally unequal angles ϕ_1 and ϕ_2 of Fig. 12.

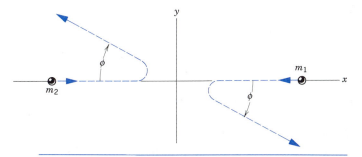

Figure 20 A two-dimensional elastic collision viewed from the center-of-mass reference frame. In this frame the particles must move in opposite directions, so each is deflected through the same angle ϕ.

10-7 SPONTANEOUS DECAY PROCESSES *(Optional)*

Of the more than 2000 species of atomic nuclei that have been identified so far, the majority are unstable and—sooner or later—get rid of all or part of their excess energy by breaking up into two or more parts. The average lifetime for such *radioactive decay* processes ranges from billions of years (for ^{238}U, for example) to very small fractions of a second. All these decays occur *spontaneously*. That is, in a given sample of radioactive material containing a large number (perhaps 10^{20}) of nuclei, we can estimate accurately how many of them will decay during any given time interval, but there is no way to predict which particular nuclei will decay.

Atoms, such as those that form the gas that fills a fluorescent lamp, can also exist in states of excess energy and can settle down to a stable configuration by emitting (again spontaneously, for an isolated atom) a quantum of radiation. Elementary particles, formed in proton–proton collisions in a high-energy accelerator, can also decay spontaneously, transforming themselves into other particles (see Fig. 4). The spontaneous decays of some of these particles occur so rapidly (10^{-20} s for the J/ψ particle, for example) that the only evidence we have for their existence is the observation of the decay products under conditions favoring the formation of the particle.

In this section, we discuss spontaneous decays of the form

$$A \rightarrow B + C,$$

in which A is the decaying particle and B and C the decay products. We usually observe such decays in the laboratory from a frame of reference in which A is at rest. The decay $A \rightarrow B + C$ is therefore simply the inverse of the totally inelastic collision $B + C \rightarrow A$ viewed from the frame of reference of the center of mass, as in Fig. 17b. In fact, we get a good mental picture of the decay process by making time go backward in Fig. 17b: read it from bottom to top and reverse the direction of the velocity vectors.

In a completely inelastic collision, kinetic energy of the colliding particles is "lost" in the collision. Of course, total energy must be conserved so the "lost" kinetic energy must appear in the composite system in another form, in which we observe it as a contribution to the rest energy of the composite system, as we discussed in Section 8-7. In a decay process, the reverse occurs:

rest energy of A is converted into kinetic energy of B and C. We can therefore write conservation of energy in the decay process as

$$E_A = E_B + E_C$$

$$m_A c^2 + K_A = (m_B c^2 + K_B) + (m_C c^2 + K_C), \quad (40)$$

with each particle's total energy E given as the sum of its rest energy mc^2 and its kinetic energy K. We have written Eq. 40 in the most general case by including the possibility that A is moving with kinetic energy K_A when it decays; we usually consider the case when $K_A = 0$.

We can rewrite Eq. 40 by grouping the rest energy terms on one side and the kinetic energy terms on the other:

$$m_A c^2 - m_B c^2 - m_C c^2 = K_B + K_C - K_A. \quad (41)$$

We define Q, the energy released in the decay, to be the difference between the initial rest energy $m_i c^2$ and the final rest energy $m_f c^2$:

$$Q = m_i c^2 - m_f c^2, \quad (42)$$

which in our case becomes

$$Q = (m_A - m_B - m_C) c^2 \quad (43)$$

or, using Eq. 41,

$$Q = K_B + K_C - K_A. \quad (44)$$

That is, Q is equal to the net gain in kinetic energy of the decay products. If A decays from rest, then Q is just the total kinetic energy of the decay products.

The decay process must conserve linear momentum. If A is at rest, the total initial momentum is zero, and thus the final momentum must be zero:

$$p_i = p_f$$

$$0 = p_B - p_C. \quad (45)$$

Equations 44 (with $K_A = 0$) and 45 provide two equations in two unknowns that can be solved for the energies or momenta of the decay products B and C. The results, for the case in which neither B nor C has a rest energy of zero, are

$$K_B = Q \frac{m_C}{m_B + m_C}, \quad (46)$$

$$K_C = Q \frac{m_B}{m_B + m_C}. \quad (47)$$

In many decay processes that are studied in the laboratory, one of the product particles, let us say B, has a much smaller rest

energy than the other, so that $m_B \ll m_C$. For example, B might be an electron (rest energy = 0.511 MeV) or an alpha particle (rest energy = 3727 MeV), while C might be a heavy atom or nucleus (rest energy = 10^5 MeV, typically). Often it is this lighter particle that is the one observed in the experiment. In this case, as Eqs. 46 and 47 show, $K_B \approx Q$ and $K_C \ll K_B$. Note that even though the kinetic energies of the two particles are quite different, the magnitudes of their (oppositely directed) momenta remain exactly equal, as Eq. 45 requires. In this case, we often refer to the *recoil momentum* or *recoil (kinetic) energy* of C, as if C were a heavy gun recoiling after firing a light bullet B, as in Sample Problem 7 of Chapter 9.

Sample Problem 6   Consider the emission of alpha particles, which are nuclei of helium atoms, in the decay of the naturally occurring radioactive element radium (^{226}Ra) to the gaseous element radon (^{222}Rn):

$$^{226}\text{Ra} \rightarrow {}^{222}\text{Rn} + \alpha.$$

If the ^{226}Ra decays from rest, find the kinetic energies of the products.

Solution   The atomic masses are:

$$^{226}\text{Ra}: \quad 226.025403 \text{ u}; \quad {}^{222}\text{Rn}: \quad 222.017571 \text{ u};$$
$$\alpha: \quad 4.002603 \text{ u}.$$

We can calculate Q from Eq. 43, using the value of $c^2 = 932$ MeV/u:

$$Q = [m(^{226}\text{Ra}) - m(^{222}\text{Rn}) - m(\alpha)]c^2$$
$$= (226.025403 \text{ u} - 222.017571 \text{ u}$$
$$- 4.002603 \text{ u})(932 \text{ MeV/u})$$
$$= 4.87 \text{ MeV}.$$

The kinetic energies can then be found from Eqs. 46 and 47:

$$K_{\text{Rn}} = (4.87 \text{ MeV}) \frac{4.002603 \text{ u}}{222.017571 \text{ u} + 4.002603 \text{ u}} = 0.09 \text{ MeV},$$

$$K_\alpha = (4.87 \text{ MeV}) \frac{222.017571 \text{ u}}{222.017571 \text{ u} + 4.002603 \text{ u}} = 4.78 \text{ MeV}.$$

Note that, as Eq. 44 (with $K_A = 0$) requires, the two kinetic energies add to Q; note also that the lighter alpha particle takes most (but not all) of the energy, about 98% in this case. ∎

QUESTIONS

1. Explain how conservation of momentum applies to a handball bouncing off a wall.

2. Can the impulse of a force be zero, even if the force is not zero? Explain why or why not.

3. Figure 21 shows a popular carnival device, in which the contestant tries to see how high a weighted marker can be raised by hitting a target with a sledge hammer. What physical quantity does the device measure? Is it the average force, the maximum force, the work done, the impulse, the energy transferred, the momentum transferred, or something else? Discuss your answer.

4. Although the acceleration of a baseball after it has been hit does not depend on who hit it, something about the baseball's flight must depend on the batter. What is it?

5. Explain how an airbag in an automobile may help to protect a passenger from serious injury in case of a collision.

6. It is said that, during a 30-mi/h collision, a 10-lb child can exert a 300-lb force against a parent's grip. How can such a large force come about?

7. Comment on these statements: In a car collision, the force the car exerts on being stopped can be determined either from its momentum or its kinetic energy. In one case the

Figure 21 Question 3.

time of stopping and in the other the *distance* of stopping also need be known.

8. Steel is more elastic than rubber. Explain what this means.

9. Discuss the possibility that, if only we could take into account internal motions of atoms in objects, *all* collisions are elastic.

10. If (only) two particles collide, are we ever forced to resort to a three-dimensional description to describe the event? Explain.

11. We have seen that the conservation of momentum may apply whether kinetic energy is conserved or not. What about the reverse; that is, does the conservation of kinetic energy imply the conservation of momentum in classical physics? (See "Connection Between Conservation of Energy and Conservation of Momentum," by Carl G. Adler, *American Journal of Physics,* May 1976, p. 483.)

12. The following statement was taken from an exam paper: "The collision between two helium atoms is perfectly elastic, so that momentum is conserved." What do you think of this statement?

13. You are driving along a highway at 50 mi/h, followed by another car moving at the same speed. You slow to 40 mi/h but the other driver does not and there is a collision. What are the initial velocities of the colliding cars as seen from the reference frame of (*a*) yourself, (*b*) the other driver, and (*c*) a state trooper, who is in a patrol car parked by the roadside? (*d*) A judge asks whether you bumped into the other driver or the other driver bumped into you. As a physicist, how would you answer?

14. C. R. Daish has written that, for professional golfers, the initial speed of the ball off the clubhead is about 140 mi/h. He also says: (*a*) "if the Empire State Building could be swung at the ball at the same speed as the clubhead, the initial ball velocity would only be increased by about 2%" and (*b*) that, once the golfer has started his or her downswing, camera clicking, sneezing, and so on can have no effect on the motion of the ball. Can you give qualitative arguments to support these two statements?

15. It is obvious from inspection of Eqs. 12 and 13 that a valid solution to the problem of finding the final velocities of two particles in a one-dimensional elastic collision is $v_{1f} = v_{1i}$ and $v_{2f} = v_{2i}$. What does this mean physically?

16. Two clay balls of equal mass and speed strike each other head-on, stick together, and come to rest. Kinetic energy is certainly not conserved. What happened to it? How is momentum conserved?

17. A football player, momentarily at rest on the field, catches a football as he is tackled by a running player on the other team. This is certainly a collision (inelastic!) and momentum must be conserved. In the reference frame of the football field, there is momentum before the collision but there seems to be none after the collision. Is linear momentum really conserved? If so, explain how. If not, explain why.

18. Consider a one-dimensional elastic collision between a moving object *A* and an object *B* initially at rest. How would you choose the mass of *B*, in comparison to the mass of *A*, in order that *B* should recoil with (*a*) the greatest speed, (*b*) the greatest momentum, and (*c*) the greatest kinetic energy?

19. Two identical cubical blocks, moving in the same direction with a common speed *v*, strike a third such block initially at rest on a horizontal frictionless surface. What is the motion of the blocks after the collision? Does it matter whether or not the two initially moving blocks were in contact? Does it matter whether these two blocks were glued together? Assume that the collisions are (*a*) completely inelastic or (*b*) elastic.

20. How would you design a recoilless gun?

21. In a two-body collision *in the center-of-mass reference frame* the momenta of the particles are equal and opposite to one another both before and after the collision. Is the line of relative motion necessarily the same after collision as before? Under what conditions would the magnitudes of the velocities of the bodies increase? decrease? remain the same as a result of the collision?

22. An hourglass is being weighed on a sensitive balance, first when sand is dropping in a steady stream from the upper to the lower part and then again after the upper part is empty. Are the two weights the same or not? Explain your answer.

23. Give a plausible explanation for the breaking of wooden boards or of bricks by a karate punch. (See "Karate Strikes," by Jearl D. Walker, *American Journal of Physics,* October 1975, p. 845.)

24. An evacuated box is at rest on a frictionless table. You punch a small hole in one face so that air can enter. (See Fig. 22.) How will the box move? What argument did you use to arrive at your answer?

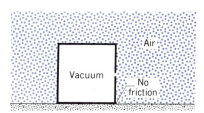

Figure 22 Question 24.

25. In commenting on the fact that kinetic energy is not conserved in a totally inelastic collision, a student observed that kinetic energy is not conserved in an explosion and that a totally inelastic collision is merely the reverse of an explosion. Is this a useful or valid observation?

26. Under what conditions, if any are needed, is it correct to say that the decay $A \rightarrow B + C$ is simply the reverse of the totally inelastic collision $B + C \rightarrow A$?

PROBLEMS

Section 10-3 Conservation of Momentum During Collisions

1. The bumper of a new car is being tested. The 2300-kg vehicle, moving at 15 m/s, is allowed to collide with a bridge abutment, being brought to rest in a time of 0.54 s. Find the average force that acted on the car during impact.

2. A ball of mass m and speed v strikes a wall perpendicularly and rebounds with undiminished speed. (a) If the time of collision is Δt, what is the average force exerted by the ball on the wall? (b) Evaluate this average force numerically for a rubber ball with mass 140 g moving at 7.8 m/s; the duration of the collision is 3.9 ms.

3. A stream of water impinges on a stationary "dished" turbine blade, as shown in Fig. 23. The speed of the water is u, both before and after it strikes the curved surface of the blade, and the mass of water striking the blade per unit time is constant at the value μ. Find the force exerted by the water on the blade.

Figure 23 Problem 3.

4. Spacecraft *Voyager 2* (mass m and speed v relative to the Sun) approaches the planet Jupiter (mass M and speed V relative to the Sun) as shown in Fig. 24. The spacecraft rounds the planet and departs in the opposite direction. What is its speed, relative to the Sun, after this "slingshot" encounter? Assume $v = 12$ km/s and $V = 13$ km/s (the orbital speed of Jupiter). The mass of Jupiter is very much greater than the mass of the spacecraft; $M \gg m$. (See "The Slingshot Effect: Explanation and Analogies," by Albert A. Bartlett and Charles W. Hord, *The Physics Teacher,* November 1985, p. 466.)

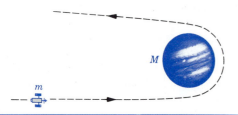

Figure 24 Problem 4.

5. A golfer hits a golf ball, imparting to it an initial velocity of magnitude 52.2 m/s directed 30° above the horizontal. Assuming that the mass of the ball is 46.0 g and the club and ball are in contact for 1.20 ms, find (a) the impulse imparted to the ball, (b) the impulse imparted to the club, (c) the average force exerted on the ball by the club, and (d) the work done on the ball.

6. A 1420-kg car moving at 5.28 m/s is initially traveling north. After completing a 90° right-hand turn in 4.60 s, the inattentive operator drives into a tree, which stops the car in 350 ms. What is the magnitude of the impulse delivered to the car (a) during the turn and (b) during the collision? What average force acts on the car (c) during the turn and (d) during the collision?

7. A 150-g (weight = 5.30 oz) baseball pitched at a speed of 41.6 m/s (= 136 ft/s) is hit straight back to the pitcher at a speed of 61.5 m/s (= 202 ft/s). The bat was in contact with the ball for 4.70 ms. Find the average force exerted by the bat on the ball.

8. A force that averages 984 N is applied to a 420-g steel ball moving at 13.8 m/s by a collision lasting 27.0 ms. If the force is in a direction opposite to the initial velocity of the ball, find the final speed of the ball.

9. A 325-g ball with a speed v of 6.22 m/s strikes a wall at an angle θ of 33.0° and then rebounds with the same speed and angle (Fig. 25). It is in contact with the wall for 10.4 ms. (a) What impulse was experienced by the ball? (b) What was the average force exerted by the ball on the wall?

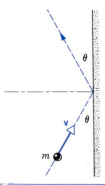

Figure 25 Problem 9.

10. Figure 26 shows an approximate representation of force versus time during the collision of a 58-g tennis ball with a wall. The initial velocity of the ball is 32 m/s perpendicular to the wall; it rebounds with the same speed, also perpendic-

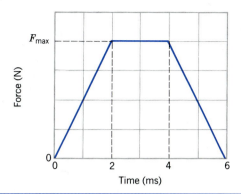

Figure 26 Problem 10.

ular to the wall. What is the value of F_{max}, the maximum value of the contact force during the collision?

11. A 2500-kg unmanned space probe is moving in a straight line at a constant speed of 300 m/s. A rocket engine on the space probe executes a burn in which a thrust of 3000 N acts for 65.0 s. (*a*) What is the change in momentum (magnitude only) of the probe if the thrust is backward, forward, or sideways? (*b*) What is the change in kinetic energy under the same three conditions? Assume that the mass of the ejected fuel is negligible compared to the mass of the space probe.

12. A force exerts an impulse J on an object of mass m, changing its speed from v to u. The force and the object's motion are along the same straight line. Show that the work done by the force is $\frac{1}{2}J(u + v)$.

13. Two parts of a spacecraft are separated by detonating the explosive bolts that hold them together. The masses of the parts are 1200 kg and 1800 kg; the magnitude of the impulse delivered to each part is 300 N·s. What is the relative speed of recession of the two parts?

14. A croquet ball with a mass 0.50 kg is struck by a mallet, receiving the impulse shown in the graph (Fig. 27). What is the ball's velocity just after the force has become zero?

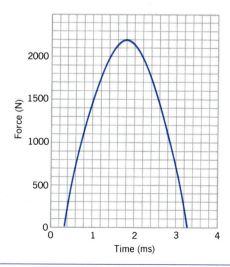

Figure 27 Problem 14.

15. It is well known that bullets and other missiles fired at Superman simply bounce off his chest as in Fig. 28. Suppose that a gangster sprays Superman's chest with 3.0-g bullets at the rate of 100 bullets/min, the speed of each bullet being 500 m/s. Suppose too that the bullets rebound straight back with no loss in speed. Show that the average force exerted by the stream of bullets on Superman's chest is only 5.0 N (= 18 oz).

Figure 28 Problem 15.

16. A karate expert breaks a pine board, 2.2 cm thick, with a hand chop. Strobe photography shows that the hand, whose mass may be taken as 540 g, strikes the top of the board with a speed of 9.5 m/s and comes to rest 2.8 cm below this level. (*a*) What is the time duration of the chop (assuming a constant force)? (*b*) What average force is applied?

17. A pellet gun fires ten 2.14-g pellets per second with a speed of 483 m/s. The pellets are stopped by a rigid wall. (*a*) Find the momentum of each pellet. (*b*) Find the kinetic energy of each pellet. (*c*) Calculate the average force exerted by the stream of pellets on the wall. (*d*) If each pellet is in contact with the wall for 1.25 ms, what is the average force exerted on the wall by each pellet while in contact? Why is this so different from (*c*)?

18. During a violent thunderstorm, hail the size of marbles (diameter = 1.0 cm) falls at a speed of 25 m/s. There are estimated to be 120 hailstones per cubic meter of air. Ignore the bounce of the hail on impact. (*a*) What is the mass of each hailstone? (*b*) What force is exerted by hail on a 10 m × 20 m flat roof during the storm? Assume that, as for ice, 1.0 cm³ of hail has a mass of 0.92 g.

19. Suppose that the blades on a helicopter push vertically down the cylindrical column of air they sweep out as they rotate. The total mass of the helicopter is 1820 kg and the length of the blades is 4.88 m. Find the minimum power needed to keep the helicopter airborne. Assume that the density of air is 1.23 kg/m³.

20. A very flexible uniform chain of mass M and length L is suspended from one end so that it hangs vertically, the lower end just touching the surface of a table. The upper end is suddenly released so that the chain falls onto the table and coils up in a small heap, each link coming to rest the instant it strikes the table; see Fig. 29. Find the force exerted by the table on the chain at any instant, in terms of the weight of chain already on the table at that moment.

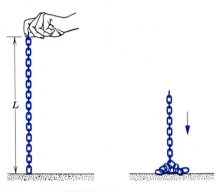

Figure 29 Problem 20.

Section 10-4 Collisions in One Dimension

21. The blocks in Fig. 30 slide without friction. (*a*) What is the velocity **v** of the 1.6-kg block after the collision? (*b*) Is the collision elastic?

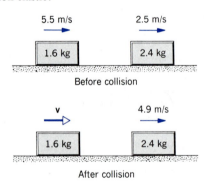

Figure 30 Problems 21 and 22.

22. Refer to Fig. 30. Suppose the initial velocity of the 2.4-kg block is reversed; it is headed directly toward the 1.6-kg block. (*a*) What would be the velocity **v** of the 1.6-kg block after the collision? (*b*) Would this collision be elastic?

23. A hovering fly is approached by an enraged elephant charging at 2.1 m/s. Assuming that the collision is elastic, at what speed does the fly rebound? Note that the projectile (the elephant) is much more massive than the target (the fly).

24. Two titanium spheres approach each other head-on with the same speed and collide elastically. After the collision, one of the spheres, whose mass is 300 g, remains at rest. What is the mass of the other sphere?

25. A bullet of mass 4.54 g is fired horizontally into a 2.41 kg wooden block at rest on a horizontal surface. The coefficient of kinetic friction between block and surface is 0.210. The

bullet comes to rest in the block, which moves 1.83 m. (*a*) What is the speed of the block immediately after the bullet comes to rest within it? (*b*) What is the speed of the bullet?

26. A cart with mass 342 g moving on a frictionless linear air-track at an initial speed of 1.24 m/s strikes a second cart of unknown mass at rest. The collision between the carts is elastic. After the collision, the first cart continues in its original direction at 0.636 m/s. (*a*) What is the mass of the second cart? (*b*) What is its speed after impact?

27. Meteor Crater in Arizona (see Fig. 31) is thought to have been formed by the impact of a meteorite with the Earth some 20,000 years ago. The mass of the meteorite is estimated to be 5×10^{10} kg and its speed to have been 7.2 km/s. What speed would such a meteorite impart to the Earth in a head-on collision?

Figure 31 Problem 27.

28. A 5.18-g bullet moving at 672 m/s strikes a 715-g wooden block at rest on a frictionless surface. The bullet emerges with its speed reduced to 428 m/s. Find the resulting speed of the block.

29. An object of 2.0-kg mass makes an elastic collision with another object at rest and continues to move in the original direction but with one-fourth of its original speed. What is the mass of the struck object?

30. In a breech-loading automatic firearm of early vintage the reloading mechanism at the rear of the bore is activated when the breech-block, which recoils after the bullet is fired, compresses a spring by a predetermined distance d. (*a*) Show that the speed of the bullet of mass m must be at least $d\sqrt{kM}/m$ on firing, for automatic loading, where k is the force constant of the spring and M is the mass of the breech-block. (*b*) In what sense, if any, can this process be regarded as a collision?

31. The head of a golf club moving at 45.0 m/s strikes a golf ball (mass = 46.0 g) resting on a tee. The effective mass of the clubhead is 220 g. (*a*) With what speed does the ball leave the tee? (*b*) With what speed would it leave the tee if you doubled the mass of the clubhead? If you tripled it? What conclusions can you draw about the use of heavy clubs? Assume that the collisions are perfectly elastic and that the golfer can bring the heavier clubs up to the same speed at impact. See Question 14.

32. A steel ball of mass 0.514 kg is fastened to a cord 68.7 cm long and is released when the cord is horizontal. At the

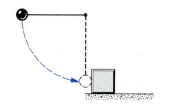

Figure 32   Problem 32.

bottom of its path, the ball strikes a 2.63-kg steel block initially at rest on a frictionless surface (Fig. 32). The collision is elastic. Find (*a*) the speed of the ball and (*b*) the speed of the block, both just after the collision. (*c*) Suppose now that, on collision, one-half the mechanical kinetic energy is converted to internal energy and sound energy. Find the final speeds.

33. Two cars *A* and *B* slide on an icy road as they attempt to stop at a traffic light. The mass of *A* is 1100 kg and the mass of *B* is 1400 kg. The coefficient of kinetic friction between the locked wheels of both cars and the road is 0.130. Car *A* succeeds in coming to rest at the light, but car *B* cannot stop and rear-ends car *A*. After the collision, *A* comes to rest 8.20 m ahead of the impact point and *B* 6.10 m ahead: see Fig. 33. Both drivers had their brakes locked throughout the incident. (*a*) From the distances each car moved after the collision, find the speed of each car immediately after impact. (*b*) Use conservation of momentum to find the speed at which car *B* struck car *A*. On what grounds can the use of momentum conservation be criticized here?

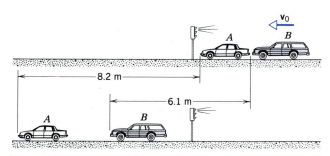

Figure 33   Problem 33.

34. A 2.9-ton weight falling through a distance of 6.5 ft drives a 0.50-ton pile 1.5 in. into the ground. Assuming that the weight–pile collision is completely inelastic, find the average force of resistance exerted by the ground.

35. A 35.0-ton railroad freight car collides with a stationary caboose car. They couple together and 27.0% of the initial kinetic energy is dissipated as heat, sound, vibrations, and so on. Find the weight of the caboose.

36. The bumper of a 1220-kg car is designed so that it can just absorb all the energy when the car runs head-on into a solid stone wall at 5.20 km/h. The car is involved in a collision in which it runs at 75.5 km/h into the rear of a 934-kg car ahead moving at 62.0 km/h in the same direction. The 934-kg car is accelerated to 71.3 km/h as a result of the collision. (*a*) What is the speed of the 1220-kg car immediately after impact? (*b*) What is the ratio of the kinetic energy absorbed

in the collision to that which can be absorbed by the bumper of the 1220-kg car?

37. A railroad freight car weighing 31.8 tons and traveling at 5.20 ft/s overtakes one weighing 24.2 tons and traveling at 2.90 ft/s in the same direction. (*a*) Find the speeds of the cars after collision and the loss of kinetic energy during collision if the cars couple together. (*b*) If instead, as is very unlikely, the collision is elastic, find the speeds of the cars after collision.

38. A platform scale is calibrated to indicate the mass in kilograms of an object placed on it. Particles fall from a height of 3.5 m and collide with the balance pan of the scale. The collisions are elastic; the particles rebound upward with the same speed they had before hitting the pan. Each particle has a mass of 110 g and collisions occur at the rate of 42 s^{-1}. Find the scale reading.

39. A box is put on a scale that is adjusted to read zero when the box is empty. A stream of marbles is then poured into the box from a height *h* above its bottom at a rate of *R* (marbles per second). Each marble has a mass *m*. The collisions between the marbles and the box are completely inelastic. Find the scale reading of weight at time *t* after the marbles begin to fill the box. Determine a numerical answer when $R = 115$ s^{-1}, $h = 9.62$ m, $m = 4.60$ g, and $t = 6.50$ s.

40. A ball of mass *m* is projected with speed v_i into the barrel of a spring gun of mass *M* initially at rest on a frictionless surface; see Fig. 34. The ball sticks in the barrel at the point of maximum compression of the spring. No energy is lost in friction. (*a*) What is the speed of the spring gun after the ball comes to rest in the barrel? (*b*) What fraction of the initial kinetic energy of the ball is stored in the spring?

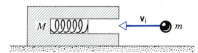

Figure 34   Problem 40.

41. A block of mass $m_1 = 1.88$ kg slides along a frictionless table with a speed of 10.3 m/s. Directly in front of it, and moving in the same direction, is a block of mass $m_2 = 4.92$ kg moving at 3.27 m/s. A massless spring with a force constant $k = 11.2$ N/cm is attached to the backside of m_2, as shown in Fig. 35. When the blocks collide, what is the maximum compression of the spring? (*Hint:* At the moment of maximum compression of the spring, the two blocks move as one; find the velocity by noting that the collision is completely inelastic to this point.)

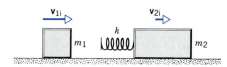

Figure 35   Problem 41.

42. Two 22.7-kg ice sleds are placed a short distance apart, one directly behind the other, as shown in Fig. 36. A 3.63-kg cat,

Figure 36 Problem 42.

standing on one sled, jumps across to the other and immediately back to the first. Both jumps are made at a speed of 3.05 m/s relative to the sled the cat is standing on when the jump is made. Find the final speeds of the two sleds.

43. An electron, mass m, collides head-on with an atom, mass M, initially at rest. As a result of the collision, a characteristic amount of energy E is stored internally in the atom. What is the minimum initial speed v_0 that the electron must have? (*Hint:* Conservation principles lead to a quadratic equation for the final electron speed v and a quadratic equation for the final atom speed V. The minimum value v_0 follows from the requirement that the radical in the solutions for v and V be real.)

44. The two spheres on the right of Fig. 37 are slightly separated and initially at rest; the left sphere is incident with speed v_0. Assuming head-on elastic collisions, (*a*) if $M \leq m$, show that there are two collisions and find all final velocities; (*b*) if $M > m$, show that there are three collisions and find all final velocities.

Figure 37 Problems 44 and 45.

45. Consider a situation such as that in the previous problem (Fig. 37) but in which the collisions now may be either all elastic, all inelastic, or some elastic and some inelastic; also, the masses are now m, m', and M. Show that to transfer the maximum kinetic energy from m to M, the intermediate body should have a mass $m' = \sqrt{mM}$, that is, the geometric mean of the adjacent masses. (It is interesting to note that this same relationship exists between masses of successive layers of air in the exponential horn in acoustics. See "Energy Transfer in One-Dimensional Collisions of Many Objects," by John B. Hart and Robert B. Herrmann, *American Journal of Physics,* January 1968, p. 46.)

Section 10-5 Two-Dimensional Collisions

46. Two vehicles A and B are traveling west and south, respectively, toward the same intersection where they collide and lock together. Before the collision, A (weight 2720 lb) is moving with a speed of 38.5 mi/h and B (weight 3640 lb) has a speed of 58.0 mi/h. Find the magnitude and direction of the velocity of the (interlocked) vehicles immediately after the collision.

47. Two objects, A and B, collide. A has mass 2.0 kg, and B has mass 3.0 kg. The velocities before the collision are $\mathbf{v}_{iA} = 15\mathbf{i} + 30\mathbf{j}$ and $\mathbf{v}_{iB} = -10\mathbf{i} + 5.0\mathbf{j}$. After the collision, $\mathbf{v}_{fA} = -6.0\mathbf{i} + 30\mathbf{j}$. All speeds are given in meters per second. (*a*) What is the final velocity of B? (*b*) How much kinetic energy was gained or lost in the collision?

48. An alpha particle collides with an oxygen nucleus, initially at rest. The alpha particle is scattered at an angle of 64.0° above its initial direction of motion and the oxygen nucleus recoils at an angle of 51.0° below this initial direction. The final speed of the nucleus is 1.20×10^5 m/s. What is the final speed of the alpha particle? (The mass of an alpha particle is 4.00 u and the mass of an oxygen nucleus is 16.0 u.)

49. Show that a slow neutron (called a *thermal* neutron) that is scattered through 90° in an elastic collision with a deuteron, that is initially at rest, loses two-thirds of its initial kinetic energy to the deuteron. (The mass of a neutron is 1.01 u; the mass of a deuteron is 2.01 u.)

50. After a totally inelastic collision, two objects of the same mass and initial speed are found to move away together at half their initial speed. Find the angle between the initial velocities of the objects.

51. A proton (atomic mass 1.01 u) with a speed of 518 m/s collides elastically with another proton at rest. The original proton is scattered 64.0° from its initial direction. (*a*) What is the direction of the velocity of the target proton after the collision? (*b*) What are the speeds of the two protons after the collision?

52. Two balls A and B, having different but unknown masses, collide. A is initially at rest and B has a speed v. After collision, B has a speed $v/2$ and moves at right angles to its original motion. (*a*) Find the direction in which ball A moves after the collision. (*b*) Can you determine the speed of A from the information given? Explain.

53. In a game of billiards, the cue ball is given an initial speed V and strikes the pack of 15 stationary balls. All 16 balls then engage in numerous ball–ball and ball–cushion collisions. Some time later, it is observed that (by some accident) all 16 balls have the same speed v. Assuming that all collisions are elastic and ignoring the rotational aspect of the balls' motion, calculate v in terms of V.

54. Two pendulums each of length L are initially situated as in Fig. 38. The first pendulum is released from height d and strikes the second. Assume that the collision is completely inelastic and neglect the mass of the strings and any frictional effects. How high does the center of mass rise after the collision?

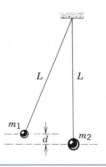

Figure 38 Problem 54.

55. A barge with mass 1.50×10^5 kg is proceeding downriver at 6.20 m/s in heavy fog when it collides broadside with a barge heading directly across the river; see Fig. 39. The second barge has mass 2.78×10^5 kg and was moving at 4.30 m/s. Immediately after impact, the second barge finds its course

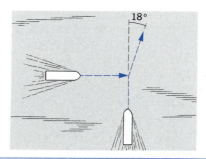

Figure 39 Problem 55.

deflected by 18.0° in the downriver direction and its speed increased to 5.10 m/s. The river current was practically zero at the time of the accident. (a) What is the speed and direction of motion of the first barge immediately after the collision? (b) How much kinetic energy was lost in the collision?

56. A ball with an initial speed of 10.0 m/s collides elastically with two identical balls whose centers are on a line perpendicular to the initial velocity and that are initially in contact with each other (Fig. 40). The first ball is aimed directly at the contact point and all the balls are frictionless. Find the velocities of all three balls after the collision. (*Hint:* With friction absent, each impulse is directed along the line of centers of the balls, normal to the colliding surfaces.)

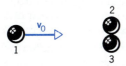

Figure 40 Problem 56.

57. In a game of pool, the cue ball strikes another ball initially at rest. After the collision, the cue ball moves at 3.50 m/s along a line making an angle of 65.0° with its original direction of motion. The second ball acquires a speed of 6.75 m/s. Using momentum conservation, find (a) the angle between the direction of motion of the second ball and the original direction of motion of the cue ball and (b) the original speed of the cue ball.

58. In 1932 James Chadwick, in England, demonstrated the existence and properties of the neutron (one of the fundamental particles making up the atom) with the device shown in Fig. 41. In an evacuated chamber, a sample of radioactive polonium decays to yield alpha particles (helium nuclei). These nuclei impinge on a block of beryllium, inducing a

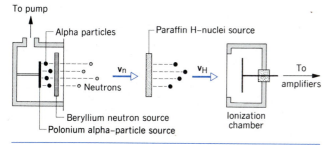

Figure 41 Problem 58.

process whereby neutrons are emitted. (In the reaction He and Be combine to form stable carbon + neutrons.) The neutrons strike a film of paraffin (CH_4), releasing hydrogen nuclei that are detected in an ionization chamber. In other words, an elastic collision occurs in which the momentum of the neutron is partially transferred to the hydrogen nucleus. (a) Find an expression for the maximum speed v_H that the hydrogen nucleus (mass m_H) can achieve. Let the incoming neutrons have mass m_n and speed v_n. (*Hint:* Will more energy be transferred in a head-on collision or in a glancing collision?) (b) One of Chadwick's goals was to find the mass of his new particle. Inspection of expression (a), which contains this parameter, however, shows that *two* unknowns are present, v_n and m_n (v_H is known; it can be measured with the ionization chamber). To eliminate the unknown v_n, he substituted a paracyanogen (CN) block for the paraffin. The neutrons then underwent elastic collisions with nitrogen nuclei instead of hydrogen nuclei. Of course, expression (a) still holds if v_N is written for v_H and m_N for m_H. Therefore if v_H and v_N are measured in separate experiments, v_N can be eliminated between the two expressions for hydrogen and nitrogen to yield a value for m_n. Chadwick's values were

$$v_H = 3.3 \times 10^9 \text{ cm/s},$$

$$v_N = 0.47 \times 10^9 \text{ cm/s}.$$

What is his value for m_n? How does this compare with the established value $m_n = 1.00867$ u? (Take $m_H = 1.0$ u, $m_N = 14$ u.)

59. Show that, in the case of an elastic collision between a particle of mass m_1 with a particle of mass m_2 initially at rest, (a) the maximum angle θ_m through which m_1 can be deflected by the collision is given by $\cos^2 \theta_m = 1 - m_2^2/m_1^2$, so that $0 \leqq \theta_m \leqq \pi/2$, when $m_1 > m_2$; (b) $\theta_1 + \theta_2 = \pi/2$, when $m_1 = m_2$; (c) θ_1 can take on all values between 0 and π, when $m_1 < m_2$.

Section 10-6 Center-of-Mass Reference Frame

60. (a) Show that in a one-dimensional elastic collision the speed of the center of mass of two particles, m_1 moving with initial speed v_{1i} and m_2 moving with initial speed v_{2i} is

$$v_{cm} = \left(\frac{m_1}{m_1 + m_2}\right)v_{1i} + \left(\frac{m_2}{m_1 + m_2}\right)v_{2i}.$$

(b) Use Eqs. 15 and 16 for v_{1f} and v_{2f}, the particles' speeds after collision, to derive the same result for v_{cm} *after* the collision.

61. In the laboratory, a particle of mass 3.16 kg moving at 15.6 m/s to the left collides head-on with a particle of mass 2.84 kg moving at 12.2 m/s to the right. Find the velocity of the center of mass of the system of two particles after the collision.

62. A particle of mass m_1 moving with speed v_{1i} collides head-on with m_2, initially at rest, in a completely inelastic collision. (a) What is the kinetic energy of the system before collision? (b) What is the kinetic energy of the system after collision? (c) What fraction of the original kinetic energy was lost? (d) Let $\mathbf{v}_{cm}$ be the velocity of the center of mass of the system. View the collision from a primed reference frame moving with the center of mass so that $v'_{1i} = v_{1i} - v_{cm}$, $v'_{2i} = -v_{cm}$.

Repeat parts (*a*), (*b*), and (*c*), as seen by an observer in this reference frame. Is the kinetic energy lost the same in each case? Explain.

Section 10-7 Spontaneous Decay Processes

63. A particle called Σ^- (sigma minus), at rest in a certain reference frame, decays spontaneously into two other particles according to

$$\Sigma^- \rightarrow \pi^- + \text{n}.$$

The masses are

$$m_\Sigma = 2340.5 m_e,$$

$$m_\pi = 273.2 m_e,$$

$$m_n = 1838.65 m_e,$$

where m_e is the electron mass. (*a*) Find the total kinetic energy of the decay products. (*b*) How much kinetic energy does each decay product get?

64. A particle of mass m at rest spontaneously decays into two particles of masses m_1 and m_2 with respective speeds v_1 and v_2. Show that $m > m_1 + m_2$.

65. A certain nucleus, at rest, spontaneously disintegrates into three particles. Two of them are detected; their masses and velocities are as shown in Fig. 42. (*a*) What is the momentum of the third particle, which is known to have a mass of 11.7×10^{-27} kg? (*b*) How much kinetic energy in MeV appears in the disintegration process?

66. A pion at rest spontaneously decays according to the scheme

$$\pi \rightarrow \mu + \nu,$$

in which μ represents a muon (rest energy = 105.7 MeV)

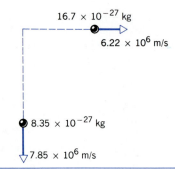

Figure 42   Problem 65.

and ν a neutrino (rest energy = zero). The measured kinetic energy of the muon is 4.100 MeV. (*a*) Find the momentum of the neutrino, in units MeV/c. (*b*) Calculate the rest energy of the pion.

Computer Project

67. Write a program describing the head-on elastic collision between two particles, of mass m_1 and m_2 and initial velocities v_{1i} and v_{2i}. The program should accept numerical values of these four quantities as input data and should yield as outputs numerical values of the final velocities, v_{1f} and v_{2f}, and v_{cm}, the velocity of the center of mass. Use your program to explore as many special cases as you can think of, such as $m_1 = m_2$, $m_1 \gg m_2$, $m_1 \ll m_2$, $v_{1i} \gg v_{2i}$, $v_{1i} = v_{2i}$, $v_{1i} = -v_{2i}$.

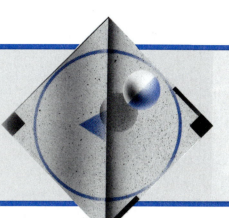

CHAPTER 11

ROTATIONAL KINEMATICS

Up to this point, we have studied only the translational motion of objects. We considered both rigid bodies (in which all parts of the body are fixed with respect to each other) and nonrigid systems (parts of which can move with respect to each other).

The most general motion of a rigid body includes rotational as well as translational motions. In this chapter, we begin to consider this general motion. We start by describing the rotation with appropriate variables and relating them to one another; this is rotational kinematics and is the subject of this chapter. Relating rotational motion to the interaction of an object with its environment (rotational dynamics) is discussed in the next two chapters.

11-1 ROTATIONAL MOTION

Figure 1 shows a fixed exercise bicycle. The axle of the spinning front wheel is fixed in space; let it define the z axis of our coordinate system. An arbitrary point P on the wheel is a perpendicular distance r from point A on the z axis. The line AB is drawn through P from A. The motion of point P traces out the arc of a circle as the wheel turns. It does not necessarily do so at constant speed, because the rider might be changing the rate at which she is pedaling.

The motion of the wheel is an example of *pure rotation of a rigid body,* which we define as follows:

A rigid body moves in pure rotation if every point of the body (such as P in Fig. 1) moves in a circular path. The centers of these circles must lie on a common straight line called the axis of rotation (the z axis of Fig. 1).

We can also characterize the motion of the wheel by the reference line AB in Fig. 1. As the wheel rotates, the line AB moves through a certain angle in the xy plane. Another way to define a pure rotation is the following:

A rigid body moves in pure rotation if a reference line perpendicular to the axis (such as AB in Fig. 1) moves through the same angle in a given time interval

(a)

Figure 1 (a) The wheel of a fixed exercise bicycle is an example of the pure rotation of a rigid body. (b) The coordinates used to describe the rotation of the wheel. The axis of rotation, which is perpendicular to the plane of the figure, is the z axis. An arbitrary point P at a distance r from the axis A moves in a circle of radius r.

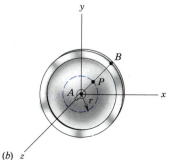

(b) z

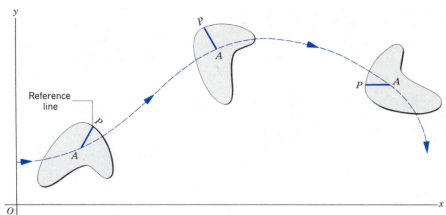

Figure 2 An arbitrary rigid body in both rotational and translational motion. In this special two-dimensional case, the translational motion is confined to the xy plane. The dashed line shows the path in the xy plane corresponding to the translational motion of the axis of rotation, which is parallel to the z axis through point A. The rotational motion is indicated by the line AP.

as any other reference line perpendicular to the axis in the body.

In the case of an ordinary bicycle wheel, the line AB might represent one of the (assumed radial) spokes of the wheel. The above definition thus means that, for a wheel in pure rotation, if one spoke turns through a certain angle $\Delta\phi$ in a time interval Δt, then any other spoke must also turn through $\Delta\phi$ during that same interval.

The general motion of a rigid object will include both rotational and translational components, as, for example, in the case of a wheel on a *moving* bicycle. A point P on such a wheel moves in a circle according to an observer in the same reference frame as the wheel (the bicycle rider, for instance), but another observer fixed to the ground would describe the motion differently. In even more complex cases, such as a wobbling football in flight, we may have a combination of translational motion, rotational motion about an axis, *and* a variation in the direction of the axis. In general, the three-dimensional description of a rigid body requires six coordinates: three to locate the center of mass, two angles (such as latitude and longitude) to orient the axis of rotation, and one angle to describe rotations about the axis. Figure 2 shows a two-dimensional arbitrary rigid body undergoing both rotational and translational motion. In this case only three coordinates are needed: two for the center of mass and one for the angular coordinate of a reference line in the body.

In this chapter, only pure rotational motion is considered. (In the next chapter, the more complicated case of combined rotation and translation is discussed.) We consider only rigid objects, in which there is no relative motion of the parts as the object rotates; a liquid in a spinning container, for instance, is excluded.

11-2   THE ROTATIONAL VARIABLES

Figure 3a shows a body of arbitrary shape rotating about the z axis. We can tell exactly where the entire rotating

body is in our reference frame if we know the location of any single point P of the body in this frame. Thus, for the kinematics of this problem, we need consider only the (two-dimensional) motion of a point in a circle of radius r equal to the perpendicular distance from P to the point A on the z axis. Figure 3b shows a slice through the body parallel to the xy plane that includes the point P.

The angle ϕ in Fig. 3b is the angular position of the reference line AP with respect to the x' axis. *We arbitrarily choose the positive sense of the rotation to be counterclockwise,* so that (in Fig. 3b) ϕ increases for counterclockwise rotation and decreases for clockwise rotation, according to an observer who is farther along the positive z axis than the rotating object.

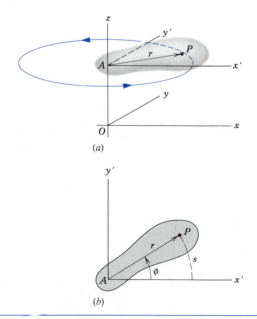

Figure 3 (*a*) An arbitrary rigid body rotating about the z axis. (*b*) A cross-sectional slice through the body. The x' and y' axes are parallel to the x and y axes, respectively, but pass through point A. The reference line AP, which connects a point P of the body to the axis, is instantaneously located at an angle ϕ with respect to the x' axis. The point P moves through an arc length s as the line AP rotates through the angle ϕ.

It is convenient to measure ϕ in radians rather than in degrees. By definition ϕ is given in radians (rad) by the relation

$$\phi = s/r, \tag{1}$$

in which s is the arc length shown in Fig. 3b.

The radian, being the ratio of two lengths, is a pure number and has no dimensions. It may therefore be inserted into or dropped from the units attached to physical quantities, as convenient.

Because the circumference of a circle of radius r is $2\pi r$, it follows from Eq. 1 that a particle that moves in an arc length of one circumference must trace out an angle of 2π rad. Thus

$$1 \text{ revolution} = 2\pi \text{ radians} = 360°,$$

or

$$1 \text{ radian} = 57.3° = 0.159 \text{ revolution.}$$

Let the body of Fig. 3b rotate counterclockwise. At time t_1 the angular position of the line AP is ϕ_1, and at a later time t_2 its angular position is ϕ_2. This is shown in Fig. 4, which gives the positions of P and of the reference line at these times; the outline of the body itself has been omitted for simplicity.

The *angular displacement* of P will be $\phi_2 - \phi_1 = \Delta\phi$ during the time interval $t_2 - t_1 = \Delta t$. We define the *average angular velocity* $\overline{\omega}$ of particle P in this time interval as

$$\overline{\omega} = \frac{\phi_2 - \phi_1}{t_2 - t_1} = \frac{\Delta\phi}{\Delta t}. \tag{2}$$

The *instantaneous angular velocity* ω is the limit approached by this ratio as Δt approaches zero:

$$\omega = \lim_{\Delta t \to 0} \frac{\Delta\phi}{\Delta t}$$

or

$$\omega = \frac{d\phi}{dt}. \tag{3}$$

For a rigid body in pure rotation, all lines fixed in it that are perpendicular to the axis of rotation rotate through the

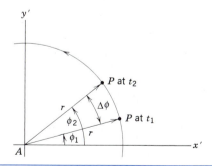

Figure 4   The reference line AP of Fig. 3b is at the angular coordinate ϕ_1 at time t_1 and at the angular coordinate ϕ_2 at time t_2. In the time interval $\Delta t = t_2 - t_1$, the net angular displacement is $\Delta\phi = \phi_2 - \phi_1$.

same angle in the same time, so that the angular velocity ω about this axis is the same for every point of the body. Thus ω is characteristic of the body as a whole. Angular velocity has the dimensions of an inverse time (T^{-1}); its units may be radians/second (rad/s) or revolutions/second (rev/s).

If the angular velocity of P is not constant, then the point has an angular acceleration. Let ω_1 and ω_2 be the instantaneous angular velocities at the times t_1 and t_2, respectively; then the *average angular acceleration* $\overline{\alpha}$ of the point P is defined as

$$\overline{\alpha} = \frac{\omega_2 - \omega_1}{t_2 - t_1} = \frac{\Delta\omega}{\Delta t}. \tag{4}$$

The *instantaneous angular acceleration* is the limit of this ratio as Δt approaches zero:

$$\alpha = \lim_{\Delta t \to 0} \frac{\Delta\omega}{\Delta t}$$

or

$$\alpha = \frac{d\omega}{dt}. \tag{5}$$

Because ω is the same for all points in the rigid body, it follows from Eq. 5 that α must be the same for each point, and thus α, like ω, is a characteristic of the body as a whole. Angular acceleration has the dimensions of an inverse time squared (T^{-2}); its units may be radians/second² (rad/s²) or revolutions/second² (rev/s²).

Instead of the rotation of a rigid body, we could have considered the motion of a single particle in a circular path. That is, P in Fig. 4 can represent a particle of mass m, constrained to move in a circle of radius r (perhaps held by a rigid massless rod of length r pivoted on the z axis). All the results derived in this section are valid whether we regard P as a mathematical point or as a physical particle; we could, for example, refer to the angular velocity or angular acceleration of the *particle P* as it rotates about the z axis. Later, we shall find it useful to regard the rotating rigid body of Fig. 3 as a collection of particles, each of which is rotating about the axis with the same angular velocity and angular acceleration.

The rotation of a particle (or a rigid body) *about a fixed axis* has a formal correspondence to the translational motion of a particle (or a rigid body) *along a fixed direction*. The kinematical variables are ϕ, ω, and α in the first case and x, v, and a in the second. These quantities correspond in pairs: ϕ to x, ω to v, and α to a. Note that the angular quantities differ dimensionally from the corresponding linear quantities by a length factor. Note, too, that all six quantities may be treated as scalars in this special case. For example, a particle at any instant can be moving in one direction or the other along its straight-line path, corresponding to a positive or a negative value for v; similarly, a particle at any instant can be rotating in one direction or another about its fixed axis, corresponding to a positive or a negative value for ω.

When, in translational motion, we remove the restriction that the motion be along a straight line and consider the general case of motion in three dimensions along a curved path, the scalar variables x, v, and a must be replaced by the kinematic vectors **r**, **v**, and **a**. In Section 11-4, we see to what extent the rotational kinematic variables reveal themselves as vectors when we remove the restriction of a fixed axis of rotation.

11-3 ROTATION WITH CONSTANT ANGULAR ACCELERATION

For translational motion of a particle or a rigid body along a fixed direction, such as the x axis, we have seen (in Chapter 2) that the simplest type of motion is that in which the acceleration a is zero. The next simplest type corresponds to $a = $ a constant (other than zero); for this motion we derived the equations of Table 2 of Chapter 2, which connect the kinematic variables x, v, a, and t in all possible combinations.

For the rotational motion of a particle or a rigid body around a fixed axis, the simplest type of motion is that in which the angular acceleration α is zero (such as uniform circular motion). The next simplest type of motion, in which $\alpha = $ a constant (other than zero), corresponds exactly to linear motion with $a = $ a constant (other than zero). As before, we can derive five equations linking the four kinematic variables ϕ, ω, α, and t in all possible combinations. These angular equations can be derived, using methods we used to derive the linear equations, or they can simply be written down, by substituting corresponding angular quantities for the linear quantities in the linear equations.

As an example, we derive the expression linking ω, α, and t. We begin by rewriting Eq. 5 as

$$d\omega = \alpha \, dt.$$

We now integrate on the left from ω_0 (the angular velocity at time $t = 0$) to ω (the angular velocity at time t), and on the right from time 0 to time t:

$$\int_{\omega_0}^{\omega} d\omega = \int_0^t \alpha \, dt = \alpha \int_0^t dt,$$

where the last step can be taken *only* when the angular acceleration α is constant. Carrying out the integration, we obtain

$$\omega - \omega_0 = \alpha t,$$

or

$$\omega = \omega_0 + \alpha t. \tag{6}$$

This is the rotational analogue of Eq. 15 of Chapter 2, $v = v_0 + at$. Note that we could obtain the rotational expression by substituting ω for v and α for a in the translational expression.

By means of such derivations, we can find five basic expressions of rotational kinematics with constant angular acceleration, which are listed in Table 1 along with their translational counterparts. Equation 7 can be derived by writing Eq. 3 as $d\phi = \omega \, dt$ and integrating. Equations 8, 9, and 10 can be derived by eliminating, respectively, t, α, and ω_0 from Eqs. 6 and 7 (which can be considered the two basic equations, because they are derived from the definitions of angular acceleration and angular velocity). You should check all equations dimensionally before verifying them. Both sets of equations hold for particles as well as for rigid bodies.

The positive sense of the angular quantities ω and α is determined by the direction in which ϕ is increasing. From Eq. 3 we see that ω is positive if ϕ is increasing with time (that is, the object is rotating in a counterclockwise direction). Similarly, from Eq. 5, we see that α is positive if ω is increasing with time, even if ω is negative and becoming less negative. These are similar to the corresponding sign conventions for the linear quantities.

Sample Problem 1 Starting from rest at time $t = 0$, a grindstone has a constant angular acceleration α of 3.2 rad/s². At $t = 0$ the reference line AB in Fig. 5 is horizontal. Find (a) the angular displacement of the line AB (and hence of the grindstone) and (b) the angular speed of the grindstone 2.7 s later.

TABLE 1 MOTION WITH CONSTANT LINEAR OR ANGULAR ACCELERATION

Equation Number (Chapter 2)	Translational Motion (Fixed Direction)	Rotational Motion (Fixed Axis)	Equation Number (This Chapter)
(15)	$v = v_0 + at$	$\omega = \omega_0 + \alpha t$	(6)
(19)	$x = x_0 + v_0 t + \frac{1}{2}at^2$	$\phi = \phi_0 + \omega_0 t + \frac{1}{2}\alpha t^2$	(7)
(20)	$v^2 = v_0^2 + 2a(x - x_0)$	$\omega^2 = \omega_0^2 + 2\alpha(\phi - \phi_0)$	(8)
(21)	$x = x_0 + \dfrac{v_0 + v}{2} t$	$\phi = \phi_0 + \dfrac{\omega_0 + \omega}{2} t$	(9)
(22)	$x = x_0 + vt - \frac{1}{2}at^2$	$\phi = \phi_0 + \omega t - \frac{1}{2}\alpha t^2$	(10)

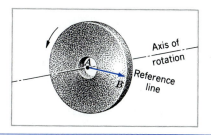

Figure 5   Sample Problem 1. The reference line AB is horizontal at $t = 0$ and rotates with the grindstone.

Solution   (a) α and t are given; we wish to find ϕ. Hence we use Eq. 7 (see Table 1):

$$\phi = \phi_0 + \omega_0 t + \tfrac{1}{2}\alpha t^2.$$

At $t = 0$, we have $\phi_0 = 0$, $\omega = \omega_0 = 0$, and $\alpha = 3.2$ rad/s². Therefore, after 2.7 s,

$$\phi = 0 + (0)(2.7 \text{ s}) + \tfrac{1}{2}(3.2 \text{ rad/s}^2)(2.7 \text{ s})^2$$
$$= 11.7 \text{ rad} = 1.9 \text{ rev}.$$

(b) α and t are given; we wish to find ω. Hence we use Eq. 6:

$$\omega = \omega_0 + \alpha t = 0 + (3.2 \text{ rad/s}^2)(2.7 \text{ s})$$
$$= 8.6 \text{ rad/s} = 1.4 \text{ rev/s}.$$

Sample Problem 2   Suppose the power driving the grinding wheel of Sample Problem 1 is turned off when the wheel is spinning with an angular speed of 8.6 rad/s. A small frictional force on the shaft causes a constant angular deceleration, and the wheel eventually comes to rest in a time of 192 s. Find (a) the angular acceleration and (b) the total angle turned through during the slowing down.

Solution   (a) Given $\omega_0 = 8.6$ rad/s and $t = 192$ s, we seek α and therefore use Eq. 6 ($\omega = \omega_0 + \alpha t$), or

$$\alpha = \frac{\omega - \omega_0}{t} = \frac{0 - 8.6 \text{ rad/s}}{192 \text{ s}} = -0.045 \text{ rad/s}^2.$$

(b) From Eq. 9 of Table 1 we have

$$\phi = \phi_0 + \frac{\omega_0 + \omega}{2} t = 0 + \frac{8.6 \text{ rad/s} + 0}{2} (192 \text{ s})$$
$$= 826 \text{ rad} = 131 \text{ rev}.$$

11-4   ROTATIONAL QUANTITIES AS VECTORS

The linear displacement, velocity, and acceleration are vectors. The corresponding angular quantities *may* be vectors also, for in addition to a magnitude we must also specify a direction for them, namely, the direction of the axis of rotation in space. Because we considered rotation only about a fixed axis, we were able to treat ϕ, ω, and α as scalar quantities. If the direction of the axis changes, however, we can no longer avoid the question, "Are rotational quantities vectors?"

We learned in Section 3-2 that, to be represented as a vector, a physical quantity must not only have both magnitude and direction, it must also obey the laws of vector addition. Only through experiment can we learn whether a physical quantity obeys these laws.

Let us discuss first the angular displacement ϕ. The magnitude of the angular displacement of a body is the angle through which the body turns. Angular displacements, however, are not vectors because they do not add like vectors. For example, give two successive rotations ϕ_1 and ϕ_2 to a book which initially lies in a horizontal plane (Fig. 6). Let rotation ϕ_1 be a 90° clockwise turn about a vertical axis through the center of the book as we view it from above. Let ϕ_2 be a 90° clockwise turn about a north–south axis through the center of the book as we view it looking north. In one case, apply operation ϕ_1 first and then ϕ_2. In the other case, apply operation ϕ_2 first and then ϕ_1. You should try this for yourself. Now, if angular displacements are vector quantities, they must add like vectors. In particular, they must obey the associative law of vector addition, $\mathbf{A} + \mathbf{B} = \mathbf{B} + \mathbf{A}$, which tells us that the order in which we add vectors does not affect their sum. As indicated by Fig. 6a, this law fails for finite angular displacements, and thus $\phi_1 + \phi_2 \neq \phi_2 + \phi_1$, where ϕ_1 and ϕ_2 stand for the operations shown in Fig. 6a. Hence finite angular displacements cannot be represented as vector quantities.

As the two angular displacements are made smaller, the result of the operation $\phi_1 + \phi_2$ approaches that of the operation $\phi_2 + \phi_1$ (Fig. 6b,c). If the angular displacements are made infinitesimal, the order of addition no longer affects the result; thus $d\phi_1 + d\phi_2 = d\phi_2 + d\phi_1$. Hence *infinitesimal angular displacements can be represented as vectors.*

Quantities defined in terms of infinitesimal angular displacements may also be vectors. For example, the angular velocity is $\omega = d\phi/dt$. Since $d\phi$ is a vector and dt is a scalar, the quotient ω is a vector. *Angular velocity can therefore be represented as a vector.* In Fig. 7a, for example, we represent the angular velocity ω of the rotating rigid body by an arrow drawn along the axis of rotation; in Fig. 7b we represent the rotation of a particle P about a fixed axis in just the same way. The length of the arrow is made proportional to the magnitude of the angular velocity. The sense of the rotation determines the direction in which the arrow points along the axis. By convention, if the fingers of the *right hand* curl around the axis in the direction of rotation of the body, the extended thumb points along the direction of the angular velocity vector. For the wheel of Fig. 1, therefore, the angular velocity vector points perpendicularly into the page (in the nega-

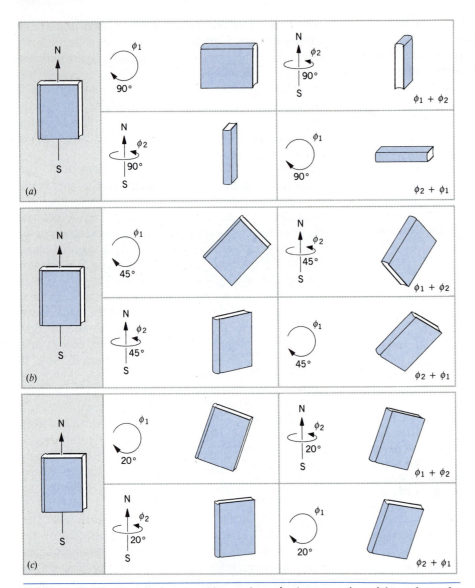

Figure 6   (*a*) The book is given two 90° rotations, ϕ_1 about an axis at right angles to the page and ϕ_2 about an axis in the page designated north–south. As shown, the final orientation depends on the order in which we perform these rotations. Thus the result of operation $\phi_1 + \phi_2$ differs from that of $\phi_2 + \phi_1$. (*b*) For 45° rotations, the difference between the final orientations is smaller than it was in the case of the 90° rotation. (*c*) For 20° rotations, the final orientations are nearly identical. As the angles ϕ_1 and ϕ_2 become smaller, the final orientations become more alike.

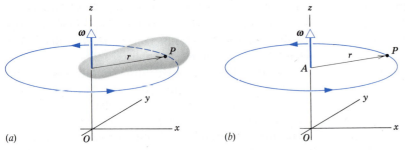

Figure 7   The angular velocity vector of (*a*) a rotating rigid body and (*b*) a rotating particle, both taken about a fixed axis.

tive *z* direction) if the rider is pedaling forward. In Fig. 3*b*, ω is perpendicular to the page pointing up out of the page, corresponding to the counterclockwise rotation. The angular velocity of the turntable of a phonograph (which rotates clockwise when viewed from above) is a vector pointing down. Note that nothing moves in the direction of the angular velocity vector. The vector represents the angular velocity of the rotational motion taking place in a plane perpendicular to it.

Angular acceleration is also a vector quantity. This follows from the definition $\alpha = d\omega/dt$, in which $d\omega$ is a vector and dt a scalar. Later we shall encounter other rotational quantities that are vectors, such as torque and angular momentum. The use of the right-hand rule to define the direction of the vectors $d\phi$, ω, and α leads to a consistent vector formalism for all rotational quantities.

Sample Problem 3   A disk spins on a horizontal shaft mounted in bearings, with an angular speed ω_1 of 84 rad/s as in Fig. 8*a*. The entire disk and shaft assembly are placed on a turntable rotating about a vertical axis at $\omega_2 = 43$ rad/s, counterclockwise as we view it from above. Describe the rotation of the disk as seen by an observer in the room.

Solution   The disk is subject to two angular velocities simultaneously; we can describe its resultant motion by the vector sum of these vectors. The angular velocity ω_1 associated with the shaft rotation has a magnitude of 84 rad/s and occurs about an axis that is not fixed but, as seen by an observer in the room, rotates in a horizontal plane at 43 rad/s. The angular velocity ω_2 associated with the turntable is fixed vertically and has a magnitude of 43 rad/s.

The resultant angular velocity ω of the disk is the vector sum of ω_1 and ω_2. The magnitude of ω is

$$\omega = \sqrt{\omega_1^2 + \omega_2^2} = \sqrt{(84 \text{ rad/s})^2 + (43 \text{ rad/s})^2}$$
$$= 94 \text{ rad/s}.$$

The direction of ω is not fixed in our observer's reference frame but rotates at the same angular rate as the turntable. The vector

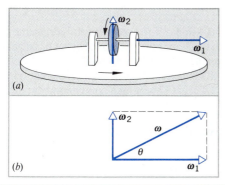

(a)

(b)

Figure 8   Sample Problem 3. (*a*) A spinning disk on a rotating turntable. (*b*) The addition of the angular velocity vectors.

ω does not lie in the horizontal plane but points above it by an angle θ (see Fig. 8*b*), where

$$\theta = \tan^{-1}\frac{\omega_2}{\omega_1} = \tan^{-1}\frac{43 \text{ rad/s}}{84 \text{ rad/s}} = 27°.$$

We can describe the motion of the disk as a simple rotation about this new axis (whose direction in our observer's reference frame is changing with time as described above) at an angular rate of 94 rad/s. How would the situation change if the direction of rotation of the disk, or of the turntable, or of both, were reversed?

11-5   RELATIONSHIPS BETWEEN LINEAR AND ANGULAR VARIABLES: SCALAR FORM

In Sections 4-4 and 4-5 we discussed the linear velocity and acceleration of a particle moving in a circle. When a rigid body rotates about a fixed axis, every particle in the body moves in a circle. Hence we can describe the motion of such a particle either in linear variables or in angular variables. The relationship between the linear and angular variables enables us to pass back and forth from one description to another and is very useful.

Consider a particle at *P* in the rigid body, a perpendicular distance *r* from the axis through *A*, as in Fig. 7. This particle moves in a circle of radius *r*. The angular position ϕ of the reference line *AP* is measured with respect to the *x* or *x′* axis, as in Fig. 3*b*. The particle moves through a distance *s* along the arc when the body rotates through an angle ϕ, such that

$$s = \phi r, \qquad (11)$$

where ϕ is in radians.

Differentiating both sides of this equation with respect to the time, and noting that *r* is constant, we obtain

$$\frac{ds}{dt} = \frac{d\phi}{dt} r.$$

But ds/dt is the (tangential) linear speed of the particle at *P* and $d\phi/dt$ is the angular speed ω of the rotating body so that

$$v = \omega r. \qquad (12)$$

This is a relation between the *magnitudes* of the tangential linear velocity and the angular velocity; the linear speed of a particle in circular motion is the product of the angular speed and the distance *r* of the particle from the axis of rotation.

Differentiating Eq. 12 with respect to the time, we have

$$\frac{dv}{dt} = \frac{d\omega}{dt} r.$$

But dv/dt is the magnitude of the *tangential* component

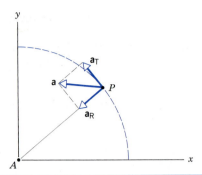

Figure 9   The radial and tangential components of the acceleration of a particle at point P of the rigid body rotating about the z axis.

a_T of the acceleration of the particle (see Section 4-5), and $d\omega/dt$ is the magnitude of the angular acceleration of the rotating body, so that

$$a_T = \alpha r. \qquad (13)$$

Hence the magnitude of the tangential component of the linear acceleration of a particle in circular motion is the product of the magnitude of the angular acceleration and the distance r of the particle from the axis of rotation.

We have seen in Section 4-4 that the *radial* component a_R of the acceleration is v^2/r for a particle moving in a circle. This can be expressed in terms of angular speed by use of Eq. 12. We have

$$a_R = \frac{v^2}{r} = \omega^2 r. \qquad (14)$$

The resultant acceleration **a** of point P is shown in Fig. 9.

Equations 11 through 14 enable us to describe the motion of one point on a rigid body rotating about a fixed axis *either* in angular variables *or* in linear variables. We might ask why we need the angular variables when we are already familiar with the equivalent linear variables. The answer is that the angular description offers a distinct advantage over the linear description when various points on the same rotating body must be considered. On a rotating body, points that are at different distances from the axis do not have the same *linear* displacement, speed, or acceleration, but *all* points on a rigid body rotating about a fixed axis do have the same *angular* displacement, speed, or acceleration at any instant. By the use of angular variables we can describe the motion of the entire body in a simple way.

Figure 10 shows an interesting example of the relation between linear and angular variables. When a tall chimney is toppled by an explosive charge at its base, it will often break as it falls, the rupture starting on the downward side of the falling chimney.

Before rupture, the chimney is a rigid body, rotating about an axis near its base with a certain angular acceleration α. According to Eq. 13, the top of the chimney has a tengential acceleration a_T given by αL, where L is the

Figure 10   A falling chimney often is not strong enough to provide the tangential acceleration at large radius that is necessary if the entire object is to rotate like a rigid body with constant angular acceleration. See "More on the Falling Chimney," by A. A. Bartlett, *The Physics Teacher,* September 1976, p. 351, for an account of this phenomenon.

length of the chimney. The vertical component of a_T can easily exceed g, the acceleration of free fall. That is, the top of the chimney is falling downward with a vertical acceleration greater than that of a freely falling brick.

This can happen only as long as the chimney remains a single rigid body. Put simply, the bottom part of the chimney, acting through the mortar that holds the bricks together, must "pull down" on the top part of the chimney to cause it to fall so fast. This shearing force is often more than the mortar can tolerate, and the chimney breaks. The chimney has now become two rigid bodies, its top part being in free fall and reaching the ground later than it would if the chimney had not broken.

Sample Problem 4   If the radius of the grindstone of Sample Problem 1 is 0.24 m, calculate (*a*) the linear or tangential speed of a point on the rim, (*b*) the tangential acceleration of a point on the rim, and (*c*) the radial acceleration of a point on the rim, at the end of 2.7 s. (*d*) Repeat for a point halfway in from the rim, that is, at $r = 0.12$ m.

Solution   We have $\alpha = 3.2$ rad/s^2, $\omega = 8.6$ rad/s after 2.7 s, and $r = 0.24$ m. Then,

(*a*) $v = \omega r = (8.6 \text{ rad/s})(0.24 \text{ m}) = 2.1$ m/s;

(*b*) $a_T = \alpha r = (3.2 \text{ rad/s}^2)(0.24 \text{ m}) = 0.77$ m/s^2;

(*c*) $a_R = \omega^2 r = (8.6 \text{ rad/s})^2(0.24 \text{ m}) = 18$ m/s^2.

(*d*) The *angular* variables are the same for this point at

$r = 0.12$ m as for a point on the rim. That is, once again $\alpha = 3.2$ rad/s^2 and $\omega = 8.6$ rad/s. Using Eqs. 12–14 with $r = 0.12$ m, we obtain for this point

$$v = 1.0 \text{ m/s}, \qquad a_T = 0.38 \text{ m/s}^2, \qquad a_R = 8.9 \text{ m/s}^2.$$

These are each half of their respective values for the point on the rim. The linear variables scale in proportion to the radius from the axis of rotation.

Note once again that, in equations that involve *only* angular variables, such as those listed in Table 1, you may express the angular quantities in any angular unit (degress, radians, revolutions), as long as you do so consistently. However, in equations in which angular and linear quantities are mixed, such as Eqs. 11, 12, 13, and 14, you *must* express the angular quantities in radians, as we have done in this sample problem. We must do so because Eqs. 12, 13, and 14 were based on Eq. 11, which in effect defines radian measure.

11-6   RELATIONSHIPS BETWEEN LINEAR AND ANGULAR VARIABLES: VECTOR FORM   *(Optional)*

In the previous section, we expressed both the linear and angular variables in scalar form. We now turn to vector methods, making an analysis similar to that of Section 4-5 but now using angular variables. We continue to work with the rigid body rotating about the fixed axis, as in Fig. 3.

Figure 11a shows a particle P in the rotating rigid body of Fig. 3. (Here, for convenience, we have eliminated even the outline of the body, showing only the particle at P and the circle it follows as the body rotates.) The body may have any arbitrary angular acceleration, even a nonconstant one. As in Fig. 7b, we represent the angular velocity of the particle at P at the chosen time t by the *vector* $\boldsymbol{\omega}$, which is parallel to the z axis and which has been drawn in Fig. 11a at the origin for convenience.

The particle P is located in the three-dimensional coordinate system of Fig. 11a by the position vector $\mathbf{r}$ *drawn from the origin*. This represents a change in notation from our previous usage, in which r represented the *perpendicular* distance from P to the z

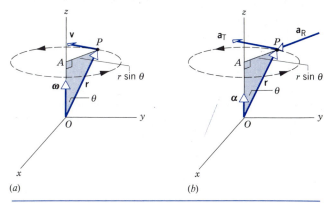

axis (see Fig. 3). In Fig. 11, that perpendicular distance is now $r \sin \theta$.

From Eq. 12 we can find the magnitude of the velocity of the particle P (bearing in mind that r in Eq. 12 is now $r \sin \theta$):

$$v = \omega r \sin \theta. \qquad (15)$$

Here ω is the magnitude of the vector $\boldsymbol{\omega}$ in Fig. 11, r is the magnitude of the vector $\mathbf{r}$, and θ is the angle between these two vectors.

Equation 15 has the same form as Eq. 16 of Chapter 3 for the magnitude of the cross or vector product between two vectors: If $\mathbf{a}$ and $\mathbf{b}$ are any two vectors, their vector product $\mathbf{c}$ has magnitude $c = ab \sin \theta$, where θ is the angle between $\mathbf{a}$ and $\mathbf{b}$. Comparison with Eq. 15 suggests that we might be able to write the velocity as a vector product:

$$\mathbf{v} = \boldsymbol{\omega} \times \mathbf{r}. \qquad (16)$$

Equation 16 certainly gives the correct magnitude for $\mathbf{v}$, because as we have seen the definition of the magnitude of the vector product will yield Eq. 15. Let us see whether Eq. 16 also gives the correct direction for $\mathbf{v}$.

According to the definition of the vector product in Section 3-5, if $\mathbf{c} = \mathbf{a} \times \mathbf{b}$, then the vector $\mathbf{c}$ lies along a line that is at right angles to the plane formed by $\mathbf{a}$ and $\mathbf{b}$. The direction of $\mathbf{c}$ is determined by the right-hand rule, in which we swing the *first* vector (the order *is* important) $\mathbf{a}$ into the second vector $\mathbf{b}$ with the fingers of the right hand moving through the smaller angle between $\mathbf{a}$ and $\mathbf{b}$; the extended thumb of the right hand then gives the direction of $\mathbf{c}$. (See Fig. 17 of Chapter 3.)

Applying this rule carefully to the vectors $\boldsymbol{\omega}$ and $\mathbf{r}$ of Fig. 11a, you will find that Eq. 16 does indeed give the correct direction for the velocity $\mathbf{v}$ at the time t, that is, tangent to the path of the particle at P. (In applying the right-hand rule, it is quite permissible to slide a vector to any position in the coordinate space, as long as you do not change its direction. Thus you can temporarily place $\mathbf{v}$ at the origin if that helps you apply the right-hand rule to find the direction of $\mathbf{v}$.)

Let us now turn to the acceleration. We can find the (linear) acceleration by differentiating the velocity given by Eq. 16. That is,

$$\mathbf{a} = \frac{d\mathbf{v}}{dt} = \frac{d}{dt}(\boldsymbol{\omega} \times \mathbf{r}) = \frac{d\boldsymbol{\omega}}{dt} \times \mathbf{r} + \boldsymbol{\omega} \times \frac{d\mathbf{r}}{dt}. \qquad (17)$$

Note that, in taking the derivative of a vector product, we follow the same rule that we use in taking the derivative of an ordinary algebraic product, except that we must take care that the order of the vectors in the terms remains the same. (That is, in both of the two terms on the right of Eq. 17, $\boldsymbol{\omega}$ comes *before* $\mathbf{r}$, as it does in the original product.) Keeping the proper order is important because, as we learned in Fig. 17b of Chapter 3, $\mathbf{a} \times \mathbf{b} = -\mathbf{b} \times \mathbf{a}$.

In Eq. 17, replacing $d\boldsymbol{\omega}/dt$ with the angular acceleration $\boldsymbol{\alpha}$ (see Eq. 5), and also replacing $d\mathbf{r}/dt$ with the velocity $\mathbf{v}$, we obtain

$$\mathbf{a} = \boldsymbol{\alpha} \times \mathbf{r} + \boldsymbol{\omega} \times \mathbf{v}. \qquad (18)$$

Let us discuss each of the two terms on the right side of Eq. 18 in turn.

As long as the axis of rotation remains fixed, $d\boldsymbol{\omega}$ will also point along the z axis. Since the direction of $\boldsymbol{\alpha}$ must be the same as the direction of $d\boldsymbol{\omega}$, the angular acceleration $\boldsymbol{\alpha}$ also lies along the z axis, as shown in Fig. 11b. Since $\boldsymbol{\alpha}$ is parallel to $\boldsymbol{\omega}$, it follows that $\boldsymbol{\alpha} \times \mathbf{r}$ must be parallel to $\boldsymbol{\omega} \times \mathbf{r}$ and thus (see Eq. 16) to $\mathbf{v}$. There-

Figure 11   (a) A particle at P in the rotating rigid body of Fig. 3 is located at $\mathbf{r}$ with respect to the origin O. The particle has angular velocity $\boldsymbol{\omega}$ (directed along the z axis) and tangential velocity $\mathbf{v}$. (b) The particle at P has angular acceleration $\boldsymbol{\alpha}$ along the z axis. The particle also has tangential acceleration $\mathbf{a}_T$ and radial acceleration $\mathbf{a}_R$.

fore $\boldsymbol{\alpha} \times \mathbf{r}$, like $\mathbf{v}$ in Fig. 11a, is tangent to the circular path of the particle at P, and it is the *tangential* component $\mathbf{a}_T$ of the acceleration $\mathbf{a}$ of the particle. The magnitude of the tangential acceleration was given by Eq. 13, which we write as $a_T = \alpha r \sin \theta$ after replacing r by $r \sin \theta$.

Now let us consider the second term on the right side of Eq. 18. If we imagine the vector $\boldsymbol{\omega}$ in Fig. 11a moved to P (remaining parallel to the z axis in the process), we see immediately by using the right-hand rule that the second term of Eq. 18, $\boldsymbol{\omega} \times \mathbf{v}$, points radially inward at P. This second term has magnitude ωv (because the angle between the vectors $\boldsymbol{\omega}$ and $\mathbf{v}$ is 90°), which becomes $\omega^2 r \sin \theta$ using Eq. 15. Comparing with Eq. 14, we see

that this is just the radial component $\mathbf{a}_R$ of the acceleration vector.

We can therefore write Eq. 18 as

$$\mathbf{a} = \mathbf{a}_T + \mathbf{a}_R, \tag{19}$$

in which

$$\mathbf{a}_T = \boldsymbol{\alpha} \times \mathbf{r} \tag{20}$$

and

$$\mathbf{a}_R = \boldsymbol{\omega} \times \mathbf{v}. \tag{21}$$

Equations 16 and 18 are then the vector relationships between the linear and angular variables for the rotation of a rigid body about a fixed axis. ■

QUESTIONS

1. In Section 11-1 we stated that, in general, six variables are required to locate a rigid body with respect to a particular reference frame. How many variables are required to locate the body of Fig. 2 with respect to the xy frame shown in that figure? If this number is not six, account for the difference.

2. In what sense is the radian a "natural" measure of angle and the degree an "arbitrary" measure of that same quantity? Hence what advantages are there in using radians rather than degrees?

3. Could the angular quantities ϕ, ω, and α be expressed in terms of degrees instead of radians in the rotational equations of Table 1?

4. Does the vector representing the angular velocity of a wheel rotating about a fixed axis necessarily have to lie along that axis? Could it be pictured as merely parallel to the axis, but located anywhere? Recall that we are free to slide a displacement vector along its own direction or translate it sideways without changing its value.

5. Experiment rotating a book after the fashion of Fig. 6a, but this time use angular displacements of 180° rather than 90°. What do you conclude about the final positions of the book? Does this change your mind about whether (finite) angular displacements can be treated as vectors?

6. The rotation of the Sun can be monitored by tracking sunspots, magnetic storms on the Sun that appear dark against the otherwise bright solar disk. Figure 12a shows the initial positions of five spots and Fig. 12b the positions of these same spots one solar rotation later. What can we conclude about the physical nature of the Sun from these observations?

7. Why is it suitable to express α in rev/s^2 in Eq. 7 in Table 1 ($\phi = \phi_0 + \omega_0 t + \frac{1}{2}\alpha t^2$) but not in Eq. 13 ($a_T = \alpha r$)?

8. A rigid body is free to rotate about a fixed axis. Can the body have nonzero angular acceleration even if the angular velocity of the body is (perhaps instantaneously) zero? What is the linear equivalent of this question? Give physical examples to illustrate both the angular and linear situations.

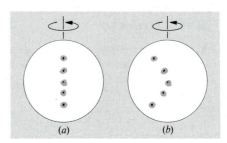

Figure 12   Question 6.

9. A golfer swings a golf club, making a long drive from the tee. Do all points on the club have the same angular velocity ω at any instant while the club is in motion?

10. When we say that a point on the equator of the Earth has an angular speed of 2π rad/day, what reference frame do we have in mind?

11. Taking the rotation and the revolution of the Earth into account, does a tree move faster during the day or during the night? With respect to what reference frame is your answer given? (The Earth's rotation and revolution are in the same direction.)

12. A wheel is rotating about its axle. Consider a point on the rim. When the wheel rotates with constant angular velocity, does the point have a radial acceleration? A tangential acceleration? When the wheel rotates with constant angular acceleration, does the point have a radial acceleration? A tangential acceleration? Do the magnitudes of these accelerations change with time?

13. Suppose that you were asked to determine the distance traveled by a needle in playing a vinyl phonograph record. What information do you need? Discuss from the point of view of reference frames (*a*) fixed in the room, (*b*) fixed on the rotating record, and (*c*) fixed on the arm of the record player.

14. What is the relation between the angular velocities of a pair of coupled gears of different radii?

15. The planet Venus (see Fig. 13) moves in a circular orbit around the Sun, completing one revolution every 225 days. Venus also rotates about a polar axis, completing one rotation every 243 days. The sense (direction) of the rotational motion is opposite, but parallel, to that of the orbital motion. (a) Describe a vector that represents the rotation of Venus about its axis. (b) Describe the vector that represents the angular velocity of Venus about the Sun. (c) Describe the resultant angular velocity, obtained by adding the orbital and rotational angular velocities.

Figure 13 Question 15.

PROBLEMS

Section 11-2 The Rotational Variables

1. Show that 1 rev/min = 0.105 rad/s.

2. The angle turned through by the flywheel of a generator during a time interval t is given by

$$\phi = at + bt^3 - ct^4,$$

where a, b, and c are constants. What is the expression for its (a) angular speed and (b) angular acceleration?

3. Our Sun is 2.3×10^4 ly (light-years) from the center of our Milky Way galaxy and is moving in a circle around this center at a speed of 250 km/s. (a) How long does it take the Sun to make one revolution about the galactic center? (b) How many revolutions has the Sun completed since it was formed about 4.5×10^9 years ago?

4. A wheel rotates with an angular acceleration α given by

$$\alpha = 4at^3 - 3bt^2,$$

where t is the time and a and b are constants. If the wheel has an initial angular speed ω_0, write the equations for (a) the angular speed and (b) the angle turned through as functions of time.

5. What is the angular speed of (a) the second hand, (b) the minute hand, and (c) the hour hand of a watch?

6. A good baseball pitcher can throw a baseball toward home plate at 85 mi/h with a spin of 1800 rev/min. How many revolutions does the baseball make on its way to home plate? For simplicity, assume that the 60-ft trajectory is a straight line.

7. A diver makes 2.5 complete revolutions on the way from a 10-m platform to the water below. Assuming zero initial vertical velocity, calculate the average angular velocity for this dive.

8. The angular position of a point on the rim of a rotating wheel is described by $\phi = 4.0t - 3.0t^2 + t^3$, where ϕ is in radians if t is given in seconds. (a) What is the angular velocity at $t = 2.0$ s and at $t = 4.0$ s? (b) What is the average angular acceleration for the time interval that begins at $t = 2.0$ s and ends at $t = 4.0$ s? (c) What is the instantaneous angular acceleration at the beginning and end of this time interval?

9. A wheel has eight spokes and a radius of 30 cm. It is mounted on a fixed axle and is spinning at 2.5 rev/s. You want to shoot a 24-cm arrow parallel to this axle and through the wheel without hitting any of the spokes. Assume that the arrow and the spokes are very thin; see Fig. 14. (a) What minimum speed must the arrow have? (b) Does it matter where between the axle and the rim of the wheel you aim? If so, where is the best location?

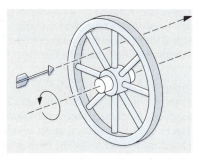

Figure 14 Problem 9.

10. A wheel with 16 spokes rotating in the clockwise direction is photographed on film. The film is passed through a projector at the rate of 24 frames/s, which is the proper rate for the projector. On the screen, however, the wheel appears to rotate counterclockwise at 4.0 rev/min. Find the smallest possible angular speed at which the wheel was rotating.

11. A solar day is the time interval between two successive appearances of the Sun overhead at a given longitude, that is, the time for one complete rotation of Earth relative to the Sun. A sidereal day is the time for one complete rotation of Earth relative to the fixed stars, that is, the time interval between two successive overhead observations of a fixed direction in the heavens called the vernal equinox. (a) Show that there is exactly one less (mean) solar day in a year than there are (mean) sidereal days in a year. (b) If the (mean) solar day is exactly 24 hours, how long is a (mean) sidereal day?

Section 11-3   Rotation with Constant Angular Acceleration

12. A phonograph turntable rotating at 78 rev/min slows down and stops in 32 s after the motor is turned off. (a) Find its (uniform) angular acceleration in rev/min². (b) How many revolutions did it make in this time?

13. The angular speed of an automobile engine is increased from 1170 rev/min to 2880 rev/min in 12.6 s. (a) Find the angular acceleration in rev/min². (b) How many revolutions does the engine make during this time?

14. As part of a maintenance inspection, the compressor of a jet engine is made to spin according to the graph shown in Fig. 15. How many revolutions does the compressor make during the test?

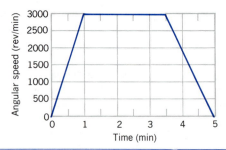

Figure 15   Problem 14.

15. The flywheel of an engine is rotating at 25.2 rad/s. When the engine is turned off, the flywheel decelerates at a constant rate and comes to rest after 19.7 s. Calculate (a) the angular acceleration (in rad/s²) of the flywheel, (b) the angle (in rad) through which the flywheel rotates in coming to rest, and (c) the number of revolutions made by the flywheel in coming to rest.

16. While waiting to board a helicopter, you notice that the rotor's motion changed from 315 rev/min to 225 rev/min in 1.00 min. (a) Find the angular acceleration during the interval. (b) Assuming that this acceleration remains constant, calculate how long it will take for the rotor to stop. (c) How many revolutions will the rotor make after your second observation?

17. A certain wheel turns through 90 rev in 15 s, its angular speed at the end of the period being 10 rev/s. (a) What was the angular speed of the wheel at the beginning of the 15-s interval, assuming constant angular acceleration? (b) How much time had elapsed between the time the wheel was at rest and the beginning of the 15-s interval?

18. A pulley wheel 8.14 cm in diameter has a 5.63-m-long cord wrapped around its periphery. Starting from rest, the wheel is given an angular acceleration of 1.47 rad/s². (a) Through what angle must the wheel turn for the cord to unwind? (b) How long does it take?

19. A flywheel completes 42.3 rev as it slows from an angular speed of 1.44 rad/s to a complete stop. (a) Assuming uniform acceleration, what is the time required for it to come to rest? (b) What is the angular acceleration? (c) How much time is required for it to complete the first one-half of the 42.3 rev?

20. Starting from rest at $t = 0$, a wheel undergoes a constant angular acceleration. When $t = 2.33$ s, the angular velocity of the wheel is 4.96 rad/s. The acceleration continues until $t = 23.0$ s, when it abruptly ceases. Through what angle does the wheel rotate in the interval $t = 0$ to $t = 46.0$ s?

21. A pulsar is a rapidly rotating neutron star from which we receive radio pulses with precise synchronization, there being one pulse for each rotation of the star. The period T of rotation is found by measuring the time between pulses. At present, the pulsar in the central region of the Crab nebula (see Fig. 16) has a period of rotation of $T = 0.033$ s, and this is observed to be increasing at the rate of 1.26×10^{-5} s/y. (a) Show that the angular velocity ω of the star is related to the period of rotation by $\omega = 2\pi/T$. (b) What is the value of the angular acceleration in rad/s²? (c) If its angular acceleration is constant, when will the pulsar stop rotating? (d) The pulsar originated in a supernova explosion in the year A.D. 1054. What was the period of rotation of the pulsar when it was born? (Assume constant angular acceleration.)

Figure 16   Problem 21.

Section 11-4   Rotational Quantities as Vectors

22. A planet P revolves around the Sun in a circular orbit, with the Sun at the center, which is coplanar with and concentric to the circular orbit of Earth E around the Sun. P and E revolve in the same direction. The times required for the revolution of P and E around the Sun are T_P and T_E. Let

T_S be the time required for P to make one revolution around the Sun relative to E: show that $1/T_S = 1/T_E - 1/T_P$. Assume $T_P > T_E$.

Section 11-5 Relationships Between Linear and Angular Variables: Scalar Form

23. A phonograph record on a turntable rotates at $33\frac{1}{3}$ rev/min. (*a*) What is the angular speed in rad/s? What is the linear speed of a point on the record at the needle at (*b*) the beginning and (*c*) the end of the recording? The distances of the needle from the turntable axis are 5.90 in. and 2.90 in., respectively, at these two positions. (*d*) Find the acceleration at each of these two positions.

24. What is the angular speed of a car rounding a circular turn of radius 110 m at 52.4 km/h?

25. A point on the rim of a 0.75-m diameter grinding wheel changes speed uniformly from 12 m/s to 25 m/s in 6.2 s. What is the angular acceleration of the grinding wheel during this interval?

26. What are (*a*) the angular speed, (*b*) the radial acceleration, and (*c*) the tangential acceleration of a spaceship negotiating a circular turn of radius 3220 km at a constant speed of 28,700 km/h?

27. An astronaut is being tested in a centrifuge. The centrifuge has a radius of 10.4 m and, in starting, rotates according to $\theta = 0.326t^2$, where t in seconds gives θ in radians. When $t = 5.60$ s, what are the astronaut's (*a*) angular speed, (*b*) tangential speed, (*c*) tangential acceleration, and (*d*) radial acceleration?

28. Earth's orbit about the Sun is almost a circle. (*a*) What is the angular speed of Earth (regarded as a particle) about the Sun? (*b*) What is its linear speed in its orbit? (*c*) What is the acceleration of Earth with respect to the Sun?

29. A threaded rod with 12.0 turns/cm and diameter 1.18 cm is mounted horizontally. A bar with a threaded hole to match the rod is screwed onto the rod; see Fig. 17. The bar spins at 237 rev/min. How long will it take for the bar to move 1.50 cm along the rod?

30. (*a*) What is the angular speed about the polar axis of a point on the Earth's surface at a latitude of 40°N? (*b*) What is the linear speed? (*c*) What are the values for a point at the equator?

31. The flywheel of a steam engine runs with a constant angular speed of 156 rev/min. When steam is shut off, the friction of

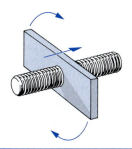

Figure 17 Problem 29.

the bearings and of the air brings the wheel to rest in 2.20 h. (*a*) What is the constant angular acceleration of the wheel, in rev/min²? (*b*) How many rotations will the wheel make before coming to rest? (*c*) What is the tangential linear acceleration of a particle 52.4 cm from the axis of rotation when the flywheel is turning at 72.5 rev/min? (*d*) What is the magnitude of the total linear acceleration of the particle in part (*c*)?

32. A gyroscope flywheel of radius 2.83 cm is accelerated from rest at 14.2 rad/s² until its angular speed is 2760 rev/min. (*a*) What is the tangential acceleration of a point on the rim of the flywheel? (*b*) What is the radial acceleration of this point when the flywheel is spinning at full speed? (*c*) Through what distance does a point on the rim move during the acceleration?

33. If an airplane propeller of radius 5.0 ft (=1.5 m) rotates at 2000 rev/min and the airplane is propelled at a ground speed of 300 mi/h (=480 km/h), what is the speed of a point on the tip of the propeller, as seen by (*a*) the pilot and (*b*) an observer on the ground? Assume that the plane's velocity is parallel to the propeller's axis of rotation.

34. An early method of measuring the speed of light makes use of a rotating toothed wheel. A beam of light passes through a slot at the outside edge of the wheel, as in Fig. 18, travels to a distant mirror, and returns to the wheel just in time to pass through the next slot in the wheel. One such toothed wheel has a radius of 5.0 cm and 500 teeth at its edge. Measurements taken when the mirror was a distance $L = 500$ m from the wheel indicated a speed of light of 3.0×10^5 km/s. (*a*) What was the (constant) angular speed of the wheel? (*b*) What was the linear speed of a point on its edge?

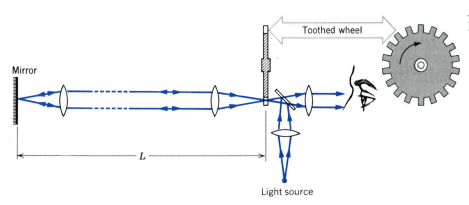

Figure 18 Problem 34.

35. Wheel *A* of radius $r_A = 10.0$ cm is coupled by a belt *B* to wheel *C* of radius $r_C = 25.0$ cm, as shown in Fig. 19. Wheel *A* increases its angular speed from rest at a uniform rate of 1.60 rad/s². Determine the time for wheel *C* to reach a rotational speed of 100 rev/min, assuming the belt does not slip. (*Hint:* If the belt does not slip, the linear speeds at the rims of the two wheels must be equal.)

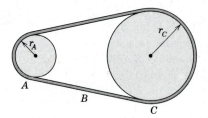

Figure 19 Problem 35.

36. The blades of a windmill start from rest and rotate with an angular acceleration of 0.236 rad/s². How much time elapses before a point on a blade experiences the same value for the magnitudes of the centripetal acceleration and tangential acceleration?

37. A rigid body, starting at rest, rotates about a fixed axis with constant angular acceleration α. Consider a particle a distance *r* from the axis. Express (*a*) the radial acceleration and (*b*) the tangential acceleration of this particle in terms of α, *r*, and the time *t*. (*c*) If the resultant acceleration of the particle at some instant makes an angle of 57.0° with the tangential acceleration, what total angle has the body turned through to that instant?

38. The disc of a Compact Disc/Digital Audio system has an inner and outer radius for its recorded material (the Tchaikovsky and Mendelssohn violin concertos) of 2.50 cm and 5.80 cm. At playback, the disc is scanned at a constant linear speed of 130 cm/s, starting from its inner edge and moving outward. (*a*) If the initial angular speed of the disc is 50.0 rad/s, what is its final angular speed? (*b*) The spiral scan lines are 1.60 μm apart; what is the total length of the scan? (*c*) What is the playing time?

39. An automobile traveling at 97 km/h has wheels of diameter 76 cm. (*a*) Find the angular speed of the wheels about the axle. (*b*) The car is brought to a stop uniformly in 30 turns of the wheels. Calculate the angular acceleration. (*c*) How far does the car advance during this braking period?

40. A speedometer on the front wheel of a bicycle gives a reading that is directly proportional to the angular speed of the wheel. Suppose that such a speedometer is calibrated for a wheel of diameter 72 cm but is mistakenly used on a wheel of diameter 62 cm. Would the linear speed reading be wrong? If so, in what sense and by what fraction of the true speed?

Section 11-6 *Relationships Between Linear and Angular Variables: Vector Form*

41. An object moves in the *xy* plane such that $x = R \cos \omega t$ and $y = R \sin \omega t$. Here *x* and *y* are the coordinates of the object, *t* is the time, and *R* and ω are constants. (*a*) Eliminate *t* between these equations to find the equation of the curve in which the object moves. What is this curve? What is the meaning of the constant ω? (*b*) Differentiate the equations for *x* and *y* with respect to the time to find the *x* and *y* components of the velocity of the body, v_x and v_y. Combine v_x and v_y to find the magnitude and direction of **v**. Describe the motion of the object. (*c*) Differentiate v_x and v_y with respect to the time to obtain the magnitude and direction of the resultant acceleration.

42. A rigid object rotating about the *z* axis is slowing down at 2.66 rad/s². Consider a particle located at $\mathbf{r} = 1.83\mathbf{j} + 1.26\mathbf{k}$ (in meters). At the instant that $\omega = 14.3\mathbf{k}$ (in rad/s), find (*a*) the velocity of the particle and (*b*) its acceleration. (*c*) What is the radius of the circular path of the particle?

CHAPTER 12

ROTATIONAL DYNAMICS

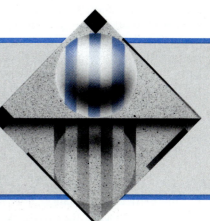

In Chapter 11 we considered rotational kinematics and pointed out that it contained no basic new features, the rotational parameters ϕ, ω, and α being related to corresponding translational parameters x, v, and a for the particles that make up the rotating system. In this chapter, following the pattern of our study of translational motion, we consider the causes of rotation, a subject called rotational *dynamics. Rotating systems are made up of particles, and we have already learned how to apply the laws of classical mechanics to the motion of particles. For this reason rotational dynamics, like kinematics, should contain no features that are fundamentally new. As in Chapter 11, however, it is very useful to recast the concepts of translational motion into a new form, especially chosen for its convenience in describing rotating systems.*

12-1 ROTATIONAL DYNAMICS: AN OVERVIEW

In Chapter 5 we posed the fundamental problem of dynamics: when external forces are applied to a body of mass m, what is the resulting motion? We showed how the solution to this problem can be found using Newton's second law, which we state in words as

$$\text{force} = \text{mass} \times \text{acceleration}.$$

In this chapter, we seek the dynamical relationship that enables us to analyze the similar problem in rotational dynamics: when a force is applied at a certain location to a rigid body free to rotate about a particular axis, what is the resulting motion? The location at which the force is applied must be significant, because we know from experience that a given force applied to a body at one location may produce a different rotation than the same force applied at another location. The quantity in rotational dynamics that takes into account both the magnitude of the force and the location and direction in which it is applied is called *torque;* we can think of a torque as a twist in the same sense that we regard a force as a push or a pull.

We also know from experience that the effort required to put a body into rotation depends on how the body's mass is distributed: it is easier for a given force to rotate a body whose mass is close to the axis of rotation than one whose mass is far from the axis. The inertial quantity that

takes into account the distribution of a body's mass is called the *rotational inertia.** Unlike mass, rotational inertia is not an intrinsic property of a body; the rotational inertia depends on the axis of rotation about which the body rotates.

Given the analogies between the translational quantities (force and mass) and the rotational quantities (torque and rotational inertia), we are led to guess at a rotational analogue of Newton's second law in the form

$$\text{torque} = \text{rotational inertia} \times \text{angular acceleration}.$$

This is in fact the correct result, as we demonstrate in Section 12-5.

Like force and acceleration, torque and angular acceleration are vector quantities. However, *in this chapter we consider only cases in which the rotational axis can be regarded as fixed in direction.* This restriction is similar to considering only one-dimensional motion in the case of translational dynamics. Although torque is a vector quantity, as we show in Section 12-4, we can use the scalar form of the dynamical equations in which all vector quantities refer to the components along the rotational axis. (In Chapter 13, we discuss situations in which we *must* consider the vector nature of the rotational quantities.)

There are two approaches that can be taken to derive

---

* Also known as the *moment of inertia.*

the equations of rotational dynamics. In the first approach, the force that acts on each particle of the body is considered, and the torques acting on each particle are summed to find the total torque acting on the body. To carry through this method, we must know how the external forces are transmitted from their points of application to the location of every particle.*

The second approach, which we adopt, is based on conservation of energy, in particular the work–energy theorem discussed in Chapter 8,

$$W = \Delta K.$$

For this particular calculation, W represents the net work done on the object by the external forces that change the rotational motion, and ΔK represents the change in rotational kinetic energy, which in this case we assume to be the only form of energy the body can have.

We begin with a discussion of rotational kinetic energy and rotational inertia in the next two sections. The subsequent discussion of torque leads to the equations of rotational dynamics.

12-2   KINETIC ENERGY OF ROTATION AND ROTATIONAL INERTIA

Figure 1 shows a rigid body rotating about a fixed vertical axis. We consider the body as a collection of particles, and we analyze the rotation of an individual particle as we did in Chapter 11. A particle of mass m at a distance r from the axis of rotation moves in a circle of radius r with an angular speed ω about this axis and has a tangential linear speed $v = \omega r$. The kinetic energy of the particle therefore is $\frac{1}{2}mv^2 = \frac{1}{2}mr^2\omega^2$.

The total kinetic energy K of the rotating body is the sum of the kinetic energies of all the particles of which the body is composed, which can be written

$$K = \tfrac{1}{2}m_1 r_1^2 \omega^2 + \tfrac{1}{2}m_2 r_2^2 \omega^2 + \cdots = \tfrac{1}{2}\left(\sum m_i r_i^2\right)\omega^2. \quad (1)$$

Here we have assumed that the body is rigid, so that all particles have the same angular velocity ω; hence the common factor ω^2 can be removed from each term in the sum of Eq. 1. The quantity $\sum m_i r_i^2$ is the sum of the products of the mass of each particle by the square of its perpendicular distance from the axis of rotation. It is called the *rotational inertia* of the body with respect to the particular axis of rotation and is represented by the symbol I:

$$I = \sum m_i r_i^2. \quad (2)$$

---

* For a critical discussion of this method, see "Rotational Motion and the Law of the Lever," by Hans C. Ohanian, *American Journal of Physics,* February 1991, p. 182.

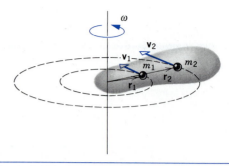

Figure 1   A rigid body rotates about a fixed axis. Each particle in the body has the same angular velocity ω, but the tangential velocity v varies with the distance r of the particle from the axis of rotation. Here m_1 and m_2 have the same angular velocity ω, but $v_2 > v_1$ because $r_2 > r_1$.

Note that *the rotational inertia of a body depends on the axis about which it is rotating* as well as on the manner in which its mass is distributed. Rotational inertia has the dimensions ML^2 and is usually expressed in $kg \cdot m^2$.

Combining Eqs. 1 and 2, we can write the kinetic energy of the rotating rigid body as

$$K = \tfrac{1}{2}I\omega^2. \quad (3)$$

This is analogous to the expression for the translational kinetic energy of a body, $K = \frac{1}{2}Mv^2$. We have already seen that the angular speed ω is analogous to the linear speed v. Now we see that the rotational inertia I is analogous to the mass M (which we can regard as translational inertia).

In Eq. 3, as in all equations that mix angular quantities with nonangular quantities, the angular variable (ω in this case) *must be expressed in radian measure.*

The rotational kinetic energy given by Eq. 3 is not a new kind of energy; it is simply the sum of the ordinary translational kinetic energies of all the particles of the body. Even though the entire body may not be in translational motion, each particle of it has a tangential velocity, and thus each particle has a kinetic energy. The instantaneous direction of each particle's velocity changes as the body rotates, but kinetic energy depends on v^2 and is a scalar, so there is no direction associated with it. It is therefore quite proper to add the kinetic energies of all the particles of the rotating body. The rotational kinetic energy $\frac{1}{2}I\omega^2$ is merely a convenient way of expressing the total kinetic energy of all the particles in the rigid body.

Figure 2 shows a simple demonstration you can do to convince yourself that it takes a greater effort (applied at a particular location) to rotate a body of large rotational inertia than to give the same rotation to a body of small rotational inertia. Turning the stick about an axis along its length (Fig. 2a) requires relatively little effort; relative to the long axis, all the particles of the stick have small values of r, and the rotational inertia is small. When you try to rotate about an axis perpendicular to the long axis (Fig. 2b), this is no longer true. The mass has of course not changed, but more mass is located far from this axis; in

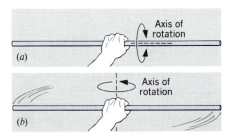

Figure 2   To rotate a long stick about the axis along its length, as in (*a*), takes less effort than it does to rotate it about an axis perpendicular to its length, as in (*b*). In (*a*), the particles of the stick lie closer to the axis of rotation than they do in (*b*), and so the stick has a smaller rotational inertia in (*a*).

accordance with Eq. 2, which indicates that the mass contributes to *I* as the *square* of its distance from the axis, this distant mass gives a much larger contribution to *I* than the mass close to the axis. You can feel that it takes a greater effort to rotate the stick about this axis.

Sample Problem 1   Three particles of masses m_1 (2.3 kg), m_2 (3.2 kg), and m_3 (1.5 kg) are at the vertices of a 3-4-5 right triangle, as shown in Fig. 3. (*a*) Find the rotational inertia about axes perpendicular to the *xy* plane and passing through each of the three particles. (*b*) Find the rotational inertia about an axis perpendicular to the *xy* plane and passing through the center of mass.

Solution   (*a*) Consider first the axis through m_1. For point masses, m_1 lies on the axis, so $r_1 = 0$ and m_1 does not contribute to the rotational inertia. The distances from this axis to m_2 and m_3 are $r_2 = 3.0$ m and $r_3 = 4.0$ m. The rotational inertia about the axis through m_1 is then

$$I_1 = \sum m_i r_i^2 = (2.3 \text{ kg})(0 \text{ m})^2 + (3.2 \text{ kg})(3.0 \text{ m})^2$$
$$+ (1.5 \text{ kg})(4.0 \text{ m})^2$$
$$= 52.8 \text{ kg} \cdot \text{m}^2.$$

Similarly for the axis through m_2, we have

$$I_2 = \sum m_i r_i^2 = (2.3 \text{ kg})(3.0 \text{ m})^2 + (3.2 \text{ kg})(0 \text{ m})^2$$
$$+ (1.5 \text{ kg})(5.0 \text{ m})^2$$
$$= 58.2 \text{ kg} \cdot \text{m}^2.$$

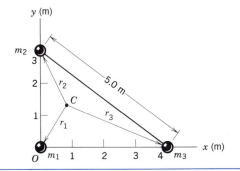

Figure 3   Sample Problem 1. Point *C* marks the center of mass of the system consisting of the three particles.

For the axis through m_3,

$$I_3 = \sum m_i r_i^2 = (2.3 \text{ kg})(4.0 \text{ m})^2 + (3.2 \text{ kg})(5.0 \text{ m})^2$$
$$+ (1.5 \text{ kg})(0 \text{ m})^2$$
$$= 116.8 \text{ kg} \cdot \text{m}^2.$$

About which axis will rotations require the greatest effort? The least?

(*b*) First we must locate the center of mass:

$$x_{cm} = \frac{\sum m_i x_i}{\sum m_i}$$
$$= \frac{(2.3 \text{ kg})(0 \text{ m}) + (3.2 \text{ kg})(0 \text{ m}) + (1.5 \text{ kg})(4.0 \text{ m})}{2.3 \text{ kg} + 3.2 \text{ kg} + 1.5 \text{ kg}} = 0.86 \text{ m},$$

$$y_{cm} = \frac{\sum m_i y_i}{\sum m_i}$$
$$= \frac{(2.3 \text{ kg})(0 \text{ m}) + (3.2 \text{ kg})(3.0 \text{ m}) + (1.5 \text{ kg})(0 \text{ m})}{2.3 \text{ kg} + 3.2 \text{ kg} + 1.5 \text{ kg}} = 1.37 \text{ m}.$$

The squared distances from the center of mass to each of the particles are

$$r_1^2 = x_{cm}^2 + y_{cm}^2 = (0.86 \text{ m})^2 + (1.37 \text{ m})^2 = 2.62 \text{ m}^2,$$

$$r_2^2 = x_{cm}^2 + (y_2 - y_{cm})^2 = (0.86 \text{ m})^2 + (3.0 \text{ m} - 1.37 \text{ m})^2$$
$$= 3.40 \text{ m}^2,$$

$$r_3^2 = (x_3 - x_{cm})^2 + y_{cm}^2 = (4.0 \text{ m} - 0.86 \text{ m})^2 + (1.37 \text{ m})^2$$
$$= 11.74 \text{ m}^2.$$

The rotational inertia then follows directly:

$$I_{cm} = \sum m_i r_i^2 = (2.3 \text{ kg})(2.62 \text{ m}^2) + (3.2 \text{ kg})(3.40 \text{ m}^2)$$
$$+ (1.5 \text{ kg})(11.74 \text{ m}^2)$$
$$= 34.5 \text{ kg} \cdot \text{m}^2.$$

Note that the rotational inertia about the center of mass is the smallest of those we have calculated. This is a general result, which we shall prove later. It is easier to rotate a body about an axis through the center of mass than about any other parallel axis.

The result of the previous sample problem leads us to an important general result, the *parallel-axis theorem*:

The rotational inertia of any body about an arbitrary axis equals the rotational inertia about a parallel axis through the center of mass plus the total mass times the squared distance between the two axes.

Mathematically, the parallel-axis theorem has the following form:

$$I = I_{cm} + Mh^2, \tag{4}$$

where *I* is the rotational inertia about the arbitrary axis, I_{cm} is the rotational inertia about the parallel axis through the center of mass, *M* is the total mass of the object, and *h* is the perpendicular distance between the axes. Note that the two axes must be parallel.

Before we prove the parallel-axis theorem, let us show how it could have been used to obtain the results of the previous sample problem. We start with the rotational inertia about the center of mass, which was found in part (b): $I_{cm} = 34.5$ kg·m². The distance h between the axis through the center of mass and the axis through m_1 is just r_1, which was computed in part (b). Thus

$$I_1 = I_{cm} + Mh^2$$
$$= 34.5 \text{ kg·m}^2 + (2.3 \text{ kg} + 3.2 \text{ kg} + 1.5 \text{ kg})(2.62 \text{ m}^2)$$
$$= 52.8 \text{ kg·m}^2,$$

in agreement with the result of part (a). You should check that I_2 and I_3 are similarly verified.

The parallel-axis theorem has an important corollary: since the term Mh^2 is always positive, I_{cm} is always the smallest rotational inertia of any group of parallel axes. (It may not be the absolute smallest rotational inertia of the object; an axis pointing in a different direction may yield a smaller value.) Thus for rotations in a given plane and at a given angular speed, choosing an axis through the center of mass costs the least energy (because $K = \frac{1}{2}I\omega^2$).

Proof of Parallel-Axis Theorem

Figure 4 shows a thin slab in the xy plane, which can be regarded as a collection of particles. We wish to calculate the rotational inertia of this object about the z axis, which passes through the origin O in Fig. 4 at right angles to the plane of that figure. We represent each particle in the slab by its mass m_i, its coordinates x_i and y_i with respect to the origin O, and its coordinates x_i' and y_i' with respect to the center of mass C. The rotational inertia about an axis through O is

$$I = \sum m_i r_i^2 = \sum m_i(x_i^2 + y_i^2).$$

Relative to O, the center of mass has coordinates x_{cm} and y_{cm}, and from the geometry of Fig. 4 you can see that the relationships between the coordinates x_i, y_i and x_i', y_i' are $x_i = x_i' + x_{cm}$ and $y_i = y_i' + y_{cm}$. Substituting these transformations, we have

$$I = \sum m_i[(x_i' + x_{cm})^2 + (y_i' + y_{cm})^2]$$
$$= \sum m_i(x_i'^2 + 2x_i' x_{cm} + x_{cm}^2 + y_i'^2 + 2y_i' y_{cm} + y_{cm}^2).$$

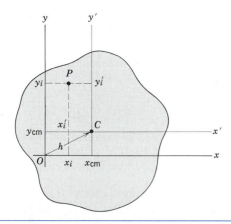

Figure 4 A thin slab in the xy plane is to be rotated about the z axis, which is perpendicular to the page at the origin O. Point C labels the center of mass of the slab. A particle P is located at coordinates x_i, y_i relative to the origin O and at coordinates x_i', y_i' relative to the center of mass C.

Regrouping the terms, we can write this as

$$I = \sum m_i(x_i'^2 + y_i'^2) + 2x_{cm} \sum m_i x_i' + 2y_{cm} \sum m_i y_i'$$
$$+ (x_{cm}^2 + y_{cm}^2) \sum m_i.$$

The first summation above is just $I_{cm} = \sum m_i r_i'^2$. The next two terms look like the formulas used to calculate the coordinates of a center of mass (Eq. 11 of Chapter 9), but (as Fig. 4 shows) they are calculated *in* the center-of-mass system. For instance, $\sum m_i x_i' = Mx_{cm}' = 0$ because $x_{cm}' = 0$, and similarly $\sum m_i y_i' = My_{cm}' = 0$: in the center-of-mass coordinate system, the center of mass is by definition at the origin, and so these terms vanish. In the last term, we let h represent the distance between the origin O and the center of mass C, so that $h^2 = x_{cm}^2 + y_{cm}^2$; also, $\sum m_i = M$, the total mass. Thus

$$I = I_{cm} + Mh^2,$$

which proves the parallel-axis theorem. ■

Sample Problem 2 The object shown in Fig. 5 consists of two particles, of masses m_1 and m_2, connected by a light rigid rod of length L. (a) Neglecting the mass of the rod, find the rotational inertia I of this system for rotations of this object about an axis perpendicular to the rod and a distance x from m_1. (b) Show that I is a minimum when $x = x_{cm}$.

Solution (a) From Eq. 2, we obtain

$$I = m_1 x^2 + m_2(L - x)^2.$$

(b) We find the minimum value of I by setting dI/dx equal to 0:

$$\frac{dI}{dx} = 2m_1 x + 2m_2(L - x)(-1) = 0.$$

Solving, we find the value of x at which this minimum occurs:

$$x = \frac{m_2 L}{m_1 + m_2}.$$

This is identical to the expression for the center of mass of the object, and thus the rotational inertia does reach its minimum value at $x = x_{cm}$. This is consistent with the parallel-axis theorem, which requires that I_{cm} be the smallest rotational inertia among parallel axes.

Points at which the first derivative of a function equals zero may not all be minima of the function. Can you show, using the *second* derivative, that we have indeed found a minimum of I?

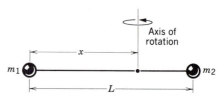

Figure 5 Sample Problem 2. The object is to be rotated about an axis perpendicular to the connecting rod and a distance x from m_1.

12-3 ROTATIONAL INERTIA OF SOLID BODIES

If we regard a body as made up of a number of discrete particles, we can calculate its rotational inertia about any axis from Eq. 2, in which the sum is taken over all the particles. If, however, we regard it as a continuous distribution of matter, we can imagine it divided into a large number of small mass elements δm_i. Each δm_i is located at a particular perpendicular distance r_i from the axis of rotation. By considering each δm_i as approximately a point mass, we can calculate the rotational inertia according to Eq. 2:

$$I = \sum r_i^2 \, \delta m_i. \qquad (5)$$

We shall soon take this to the limit of infinitesimally small δm_i, so that the sum becomes an integral. For now, let us illustrate the transition to integral calculus by using Eq. 5 to approximate the rotational inertia of a uniform solid rod rotated about an axis perpendicular to the rod at its midpoint. Figure 6*a* illustrates the situation. The rod has length L and mass M. Let us imagine that the rod is divided into 10 pieces, each of length $L/10$ and mass $M/10$. The pieces are numbered from $i = 1$ to $i = 10$, so that the ith piece is a distance r_i from the axis; for this calculation, we take r_i to be measured from the axis to the center of the piece. Thus the pieces on each end have $r_1 = r_{10} = 0.45L$, the pieces next to the ends have $r_2 = r_9 = 0.35L$, and the pieces nearest the axis have $r_5 = r_6 = 0.05L$. We now carry out the sum over the 10 pieces according to Eq. 5:

$$I = r_1^2 \, \delta m_1 + r_2^2 \, \delta m_2 + \cdots + r_{10}^2 \, \delta m_{10}$$
$$= (0.1M)(0.45L)^2 + (0.1M)(0.35L)^2 + (0.1M)(0.25L)^2$$
$$+ (0.1M)(0.15L)^2 + (0.1M)(0.05L)^2 + \cdots,$$

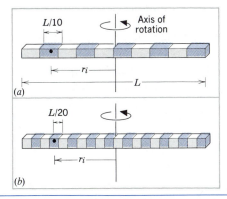

Figure 6 (*a*) The rotational inertia of a solid rod of length L, rotated about an axis through its center and perpendicular to its length, can be approximately computed by dividing the rod into 10 equal pieces, each of length $L/10$. Each piece is treated as a point mass a distance r_i from the axis. (*b*) A more accurate approximation to the rotational inertia of the rod is obtained by dividing it into 20 pieces.

where in the second equation the five terms listed correspond to half of the rod and $\cdots$ means we have five identical terms from the other half. Evaluating the numerical factors, we obtain the result

$$I = 0.825ML^2 = \frac{1}{12.12}ML^2 \quad (10 \text{ pieces}).$$

Our reason for writing the result in this form will soon be apparent.

Suppose now we divide the rod into 20 pieces, each of length $L/20$ and mass $M/20$ (Fig. 6*b*). Repeating the above calculation, we obtain the result

$$I = 0.831ML^2 = \frac{1}{12.03}ML^2 \quad (20 \text{ pieces}).$$

As we increase the number of pieces, does the result approach a limiting value that we can regard as the rotational inertia? In Problem 12, you are asked to derive the result for any arbitrary number N of pieces:

$$I = \frac{1}{12}ML^2 \left(\frac{N^2 - 1}{N^2} \right) \quad (N \text{ pieces}). \qquad (6)$$

Clearly this approaches a limit of $ML^2/12$ as $N \to \infty$, and we can assign this as the value of the rotational inertia of the rod. Note that the numerical coefficients for $N = 10$ ($\frac{1}{12.12}$) and $N = 20$ ($\frac{1}{12.03}$) show the approach to the $N \to \infty$ limit ($\frac{1}{12}$).

The above algebraic method works easily in a few cases, and it is helpful in forming an image in our minds of how integral calculus divides a solid object into infinitesimal pieces and sums over the pieces. For calculations involving most solids, the algebraic method is cumbersome, and it is far easier to use calculus techniques directly. Let us take the limit of Eq. 5 as the number of pieces becomes very large or, equivalently, as their masses δm become very small:

$$I = \lim_{\delta m_i \to 0} \sum r_i^2 \, \delta m_i,$$

and in the usual way the sum becomes an integral in the limit:

$$I = \int r^2 \, dm. \qquad (7)$$

The integration is carried out over the entire volume of the object, but often certain geometrical simplifications can reduce the integral to more manageable terms.

As an example, let us return to the rod rotated about an axis through its center. Figure 7 shows the problem drawn for the integral approach. We choose an *arbitrary* element of mass dm located a distance x from the axis. (We use x as the variable of integration.) The mass of this element is equal to its density (mass per unit volume) ρ times the volume element dV. The volume element is equal to the area times its thickness dx:

$$dV = A \, dx$$
$$dm = \rho \, dV = \rho A \, dx.$$

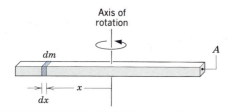

Figure 7   The rotational inertia of a solid rod is computed by integrating along its length. An element of mass *dm* is located at a perpendicular distance *x* from the axis of rotation.

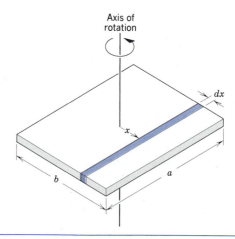

Figure 8   A solid rectangular plate of sides *a* and *b* is rotated about an axis through its center and perpendicular to its surface. To compute the rotational inertia, we consider the plate to be divided into strips. The shaded strip can be considered a rod, whose rotational inertia about the central axis can be found using the parallel-axis theorem.

We assume the rod has uniform cross-sectional area A and uniform density ρ, the latter being equal to the total mass M divided by the total volume AL: $\rho = M/V = M/AL$. Evaluating Eq. 7, we obtain

$$I = \int r^2 \, dm = \int x^2 \frac{M}{AL} A \, dx = \frac{M}{L} \int x^2 \, dx.$$

With $x = 0$ at the midpoint of the rod, the limits of integration are from $x = -L/2$ to $x = +L/2$. The rotational inertia is then

$$I = \frac{M}{L} \int_{-L/2}^{+L/2} x^2 \, dx = \frac{M}{L} \frac{x^3}{3} \Big|_{-L/2}^{+L/2}$$

$$I = \tfrac{1}{12}ML^2. \tag{8}$$

This result is identical with the one deduced from the algebraic method, Eq. 6, in the limit $N \rightarrow \infty$.

If we wish to rotate the rod about an axis through one end perpendicular to its length, we can use the parallel-axis theorem (Eq. 4). We have already found I_{cm}, and the distance h between the parallel axes is just half the length, so

$$I = \tfrac{1}{12}ML^2 + M(L/2)^2 = \tfrac{1}{3}ML^2.$$

Often we can calculate the rotational inertia of a solid body by decomposing it into elements of known rotational inertia. For example, suppose we have a uniform solid rectangular plate of length a and width b, as shown in Fig. 8. We wish to calculate the rotational inertia about an axis perpendicular to the plate and through its center.

The plate can be divided into a series of strips, each of which is to be regarded as a rod. Consider the strip of mass dm, length a, and width dx shown in Fig. 8. The mass dm of the strip is related to the total mass M as the surface area of the strip ($a \, dx$) is related to the total surface area ab:

$$\frac{dm}{M} = \frac{a \, dx}{ab} = \frac{dx}{b}$$

$$dm = \frac{M}{b} \, dx.$$

The rotational inertia dI of the strip about the axis is, by the parallel-axis theorem, related to the rotational inertia

of the strip (regarded as a rod) about its center of mass, given by Eq. 8 as $dI_{\mathrm{cm}} = \tfrac{1}{12}dm \, a^2$:

$$dI = dI_{\mathrm{cm}} + dm \, h^2$$
$$= \tfrac{1}{12}dm \, a^2 + dm \, x^2.$$

Substituting for dm yields

$$dI = \frac{Ma^2}{12b} \, dx + \frac{M}{b} x^2 \, dx,$$

and I follows from the integral

$$I = \int dI = \frac{Ma^2}{12b} \int dx + \frac{M}{b} \int x^2 \, dx.$$

The limits of integration on x are from $-b/2$ to $+b/2$. Carrying out the integrations, we obtain

$$I = \tfrac{1}{12}M(a^2 + b^2). \tag{9}$$

Note that this result is independent of the thickness of the plate: we would get the same result for a stack of plates of total mass M or, equivalently, for a solid rectangular block of the same surface dimensions. Note also that our result depends on the diagonal of the plate rather than on a and b separately. Can you explain this?

Working in this way, we can evaluate the rotational inertia of almost any regular solid object. Figure 9 shows some common objects and their rotational inertias. Although it is relatively straightforward to use two- and three-dimensional integrals to calculate these rotational inertias, it is often possible, as we did in the above calculation, to decompose a complex solid into simpler solids of known rotational inertia. Problem 14 at the end of this chapter describes such a calculation for a solid sphere.

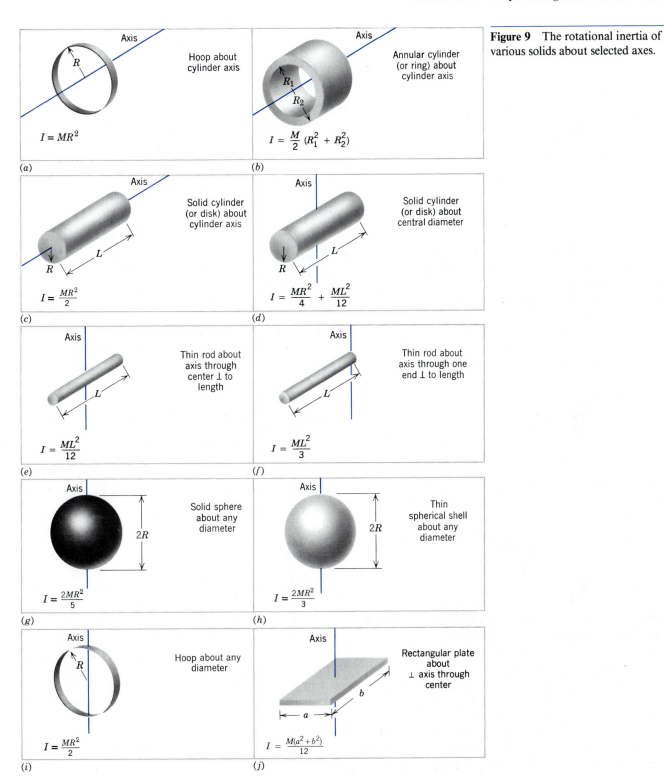

Figure 9 The rotational inertia of various solids about selected axes.

12-4 TORQUE ACTING ON A PARTICLE

Experience with a heavy door teaches us that a given force can produce various angular accelerations depending on where the force is applied to the door and how it is di-

rected (see Fig. 10). A force (such as F_1) applied at the edge and directed along the door cannot produce an angular acceleration, nor can a force (such as F_2) applied to the hinge line along the door; but a force (such as F_3) applied at right angles to the door at its outer edge produces the largest angular acceleration.

The rotational analogue of force is called *torque*. We

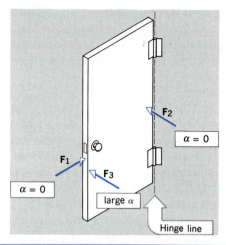

Figure 10   Applying a given force **F** to a door produces an angular acceleration α that varies with the point at which **F** is applied and with its direction relative to the hinge line. Force F_1 is applied along a line that would pass through the hinge line, and it produces no angular acceleration (the door doesn't move). Force F_2 is applied at the hinge line; it likewise produces no angular acceleration. Force F_3 is applied at a point far from the hinge line and in a direction perpendicular to the line connecting the point of application of F_3 with the hinge line; this force produces the largest possible angular acceleration.

now define it for the special case of a single particle observed from an inertial reference frame. Later, we extend the torque concept to systems of particles (including rigid bodies) and show that torque is intimately associated with angular acceleration.

Let a force **F** act on a single particle at a point P whose position with respect to the origin O of the inertial reference frame is given by the vector **r** (Fig. 11). Since two vectors determine a plane, we have chosen the xy plane to contain the vectors **r** and **F**. The torque τ acting on the particle *with respect to the origin O* is defined in terms of the vector (cross) product of **r** and **F** as

$$\tau = \mathbf{r} \times \mathbf{F}. \qquad (10)$$

Torque is a vector quantity. Its magnitude is given by

$$\tau = rF \sin \theta, \qquad (11)$$

where θ is the angle between **r** and **F**; its direction is normal to the plane formed by **r** and **F** (that is, parallel to the z axis when **r** and **F** lie in the xy plane), given by the right-hand rule for the vector product of two vectors: if you swing **r** into **F** (when they are drawn tail to tail) through the smaller angle between them with the curled fingers of your right hand, then the direction of your extended thumb gives the direction of τ. (You may wish to review the definition of the vector cross product in Section 3-5.)

We have drawn the torque vector in Fig. 11 so that it passes through the origin, but we need not have done so. If

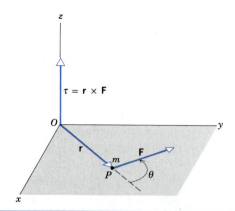

Figure 11   A force **F** acts on a particle of mass m located at position **r** in the xy plane. This force exerts a torque $\tau = \mathbf{r} \times \mathbf{F}$ on the particle with respect to the origin O. The torque vector points in the direction of increasing z; it could be drawn anywhere we choose, as long as it is parallel to the z axis.

r and **F** lie in the xy plane, as we have assumed, then Eq. 10 requires only that the cross product τ be *parallel* to the z axis, not necessarily *along* the z axis. We could locate the vector τ at any point in the coordinate space of Fig. 11 without changing the validity of Eq. 10, as long as τ remains parallel to the z axis.

Torque has the dimensions of force times distance; in terms of our fundamental dimensions M, L, and T, it has the dimensions ML^2T^{-2}. These are the same as the dimensions of work. However, torque and work are very different physical quantities. Torque is a vector, and work is a scalar, for example. The unit of torque may be the newton-meter ($N \cdot m$) or pound-foot ($lb \cdot ft$), among other possibilities. (Even though $1 \, N \cdot m = 1 \, J$, we do *not* express torque in units of J.)

Note from Eq. 10 that the torque produced by a force depends not only on the magnitude and on the direction of the force but also on the point of application of the force relative to the origin, that is, on the vector **r**. In particular, when particle P in Fig. 11 is at the origin, **r** is zero and the torque τ about the origin is zero. The torque about a point O' midway between O and P is a vector (which can be drawn through O') parallel to the vector τ shown in Fig. 11 but half its length.

We can also write the magnitude of τ (Eq. 11) either as

$$\tau = (r \sin \theta)F = Fr_{\perp}, \qquad (12a)$$

or as

$$\tau = r(F \sin \theta) = rF_{\perp}, \qquad (12b)$$

in which, as Fig. 12 shows, $r_{\perp} (= r \sin \theta)$ is the component of **r** at right angles to the line along which **F** acts (called the *line of action* of **F**), and $F_{\perp} (= F \sin \theta)$ is the component of **F** at right angles to **r**. Torque is often called the *moment of force*, and $r_{\perp}$ in Eq. 12a is called the *moment arm*. Equation 12b shows that only the component of **F** perpendicular to **r** contributes to the torque. In particular, when θ

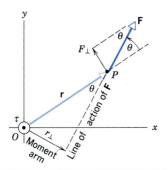

Figure 12 The xy plane, containing the force **F** and position vector **r** of Fig. 11. The magnitude of $\boldsymbol{\tau}$ is given by $Fr_\perp$ (Eq. 12a) or by $rF_\perp$ (Eq. 12b). The direction of $\boldsymbol{\tau}$ (out of the page) is indicated in the figure by $\odot$ (suggesting the tip of an arrow).

equals 0° or 180°, there is no perpendicular component ($F_\perp = F \sin \theta = 0$); the line of action of the force passes through the origin, and the moment arm $r_\perp$ about the origin is also zero. In this case both Eqs. 12a and Eq. 12b show that the torque τ is zero.

Sample Problem 3 A pendulum consists of a body of mass $m = 0.17$ kg on the end of a rigid rod of length $L = 1.25$ m and negligible mass (Fig. 13). (a) What is the magnitude of the torque due to gravity about the pivot point O at the instant the pendulum is displaced as shown through an angle of $\theta = 10°$ from the vertical? (b) What is the direction of the torque about O at that instant? Does its direction depend on whether the pendulum is displaced to the left or right of vertical?

Solution (a) We can use Eq. 11 directly to find the magnitude of the torque, with $r = L$ and $F = mg$:

$$\tau = Lmg \sin \theta = (1.25 \text{ m})(0.17 \text{ kg})(9.8 \text{ m/s}^2)(\sin 10°)$$
$$= 0.36 \text{ N} \cdot \text{m}.$$

(b) With the displacement as shown in Fig. 13, the torque

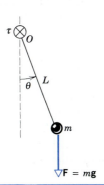

Figure 13 Sample Problem 3. A pendulum, consisting of a body of mass m on the end of a massless rigid rod of length L. Gravity exerts a torque into the page at O, indicated there by the symbol $\otimes$ (suggesting the tail of an arrow).

about the pivot is into the plane of the paper. You should be able to convince yourself that, if the pendulum is displaced on the opposite side of the vertical, the torque has the opposite direction. As we discuss in the next section, the effect of a torque is to produce a parallel angular acceleration. In the first instance, the angular acceleration into the paper tends to move the pendulum toward its equilibrium position. When the pendulum is displaced on the opposite side of the vertical, the torque out of the paper tends once again to restore the pendulum to its equilibrium position. Check these conclusions using the right-hand rule to relate the sense of the rotation to the direction of the angular acceleration vector (assumed parallel to the torque).

12-5 ROTATIONAL DYNAMICS OF A RIGID BODY

In translational motion, techniques involving work and energy gave us a different and sometimes more illuminating way of approaching problems. In this section we consider the use of work and energy in rotational motion.

Let an arbitrary rigid body be constrained to pivot about the z axis. An external force **F** acting in an arbitrary direction is applied to the body at some point P in the xy plane. Figure 14 shows the force **F** and the point P, the rest of the body being omitted for clarity. We consider the work dW done by this force as the body rotates through an angle $d\phi$.

The point P, which is at a distance r from the axis of rotation, moves through a distance $ds = r \, d\phi$ as the body rotates through the angle $d\phi$. The work dW can therefore be expressed as

$$dW = \mathbf{F} \cdot d\mathbf{s}, \tag{13}$$

where $d\mathbf{s}$ is a vector of magnitude ds in the direction of motion of P.

The z component of **F** does not contribute to the dot

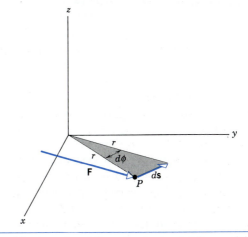

Figure 14 An external force **F** acts at point P on a rigid body (not shown) constrained to rotate about the z axis. The body rotates through an angle $d\phi$.

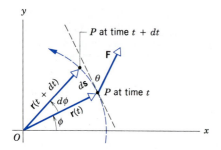

Figure 15   In a time dt, point P in a rigid body moves a distance ds along the arc of a circle of radius r. The rigid body (which is not shown) and the vector $\mathbf{r}$ that locates point P in the body each rotate through an angle $d\phi$ during this interval.

product in Eq. 13, because $d\mathbf{s}$ has no z component. (Recall from Eq. 15 of Chapter 3 that the dot product of $\mathbf{F}$ and $d\mathbf{s}$ can be written $F_x\,dx + F_y\,dy + F_z\,dz$. In our case, $dz = 0$, so the dot product does not depend on the component F_z.) *In cases in which the direction of the axis of rotation is fixed, we need consider only force components that lie in the plane perpendicular to the axis.*

Figure 15 shows the motion of point P during an infinitesimal time interval dt. An external force, now assumed to lie entirely in the xy plane, acts on the body at point P. The work dW done by this force during this infinitesimal rotation is

$$dW = \mathbf{F} \cdot d\mathbf{s} = F \cos \theta\, ds = (F \cos \theta)(r\, d\phi).$$

The term $F \cos \theta$ is the component of $\mathbf{F}$ in the direction of $d\mathbf{s}$; it is therefore perpendicular to $\mathbf{r}$ and can be represented as $F_\perp$. According to Eq. 12b, $F_\perp r$ is the magnitude of the instantaneous torque exerted by $\mathbf{F}$ on the rigid body about the axis perpendicular to the page through O, so that the above equation becomes

$$dW = F_\perp r\, d\phi = \tau\, d\phi. \tag{14}$$

This expression for the work done in rotation (about a fixed axis) is equivalent to the expression $dW = F\,dx$ for the work done in translation (along a straight line).

Suppose now several forces, $\mathbf{F}_1, \mathbf{F}_2, \ldots$, are applied to different points of the body in the plane normal to its axis of rotation (the xy plane of Fig. 15). The net work done by these forces on the body in a rotation $d\phi$ is

$$
\begin{aligned}
dW_{\text{net}} &= (F_1 \cos \theta_1)\, r_1\, d\phi + (F_2 \cos \theta_2)\, r_2\, d\phi + \cdots \\
&= (\tau_1 + \tau_2 + \cdots)d\phi,
\end{aligned}
$$

in which $(F_i \cos \theta_i)r_i$ gives the component of the torque τ_i about O. Note that, as must be true for a rigid body in pure rotation, we assume that $d\phi$ is the angular displacement of any point of the body during the time interval dt, no matter where the point may be located on the body. We can write this as

$$dW_{\text{net}} = \left(\sum \tau_{\text{ext}}\right) d\phi = \left(\sum \tau_{\text{ext}}\right) \omega\, dt, \tag{15}$$

where in the last result we have used $d\phi = \omega\, dt$ from Eq. 3 of Chapter 11. Here $\sum \tau_{\text{ext}}$ represents the total external torque acting on the body, which is computed by considering each individual external torque as positive if, acting alone, it would tend to rotate the body counterclockwise (thereby increasing ϕ) and negative if it would tend to rotate the body clockwise.

During the time interval dt, the kinetic energy of the body changes by an amount dK as a result of the action of the external forces. We assume that rotational kinetic energy is the only form of energy the body can possess. Using Eq. 3, $K = \frac{1}{2}I\omega^2$, we find

$$dK = d(\tfrac{1}{2}I\omega^2) = I\omega\, d\omega = I\omega\alpha\, dt, \tag{16}$$

using $d\omega = \alpha\, dt$ from Eq. 5 of Chapter 11.

During the interval dt, the work–energy theorem gives

$$dW = dK, \tag{17}$$

and substituting Eqs. 15 and 16 gives

$$\left(\sum \tau_{\text{ext}}\right) \omega\, dt = I\omega\alpha\, dt,$$

or, canceling the common factors of $\omega\, dt$,

$$\sum \tau_{\text{ext}} = I\alpha. \tag{18}$$

Equation 18 is the rotational analogue of Newton's second law in the scalar form of $\sum F_{\text{ext}} = ma$, corresponding to motion in one dimension. In the rotational case, it should be noted once again that we are considering only rotation about a fixed axis. Furthermore, note that a positive torque tends to produce positive angular acceleration; that is, the same right-hand rule used to obtain the sign of α can be used to determine the sign of τ. As we did in the case of translational dynamics, we drop the "ext" subscript on τ for convenience.

To obtain the rate at which work is done in rotational motion (about a fixed axis), we divide Eq. 14 by the infinitesimal time interval dt during which the body is displaced through $d\phi$ and obtain

$$P = \tau\omega, \tag{19}$$

which gives the instantaneous mechanical power P. Equation 19 is the rotational analogue of $P = Fv$ for translational motion.

The analogy between translational motion in a fixed direction and rotational motion about a fixed axis is emphasized in Table 1, which compares the corresponding equations in the two cases. The rotation of a rigid body about a fixed axis is not the most general kind of rotary motion; often the axis is not fixed in an inertial reference frame. We shall consider this more general case, along with the angular momentum listed in the last line of Table 1, in Chapter 13.

Given the dynamical relationship between torque and angular acceleration, we can now reconsider the effect of an arbitrary torque on a rigid body constrained to rotate about the z axis. Let us once again allow the force to have

TABLE 1   COMPARISON OF LINEAR AND ROTATIONAL DYNAMICAL EQUATIONS

Linear Motion		*Rotation About a Fixed Axis*	
Displacement	x	Angular displacement	ϕ
Velocity	$v = dx/dt$	Angular velocity	$\omega = d\phi/dt$
Acceleration	$a = dv/dt$	Angular acceleration	$\alpha = d\omega/dt$
Mass (translational inertia)	M	Rotational inertia	I
Force	$F = Ma$	Torque	$\tau = I\alpha$
Work	$W = \int F\, dx$	Work	$W = \int \tau\, d\phi$
Kinetic energy	$K = \frac{1}{2}Mv^2$	Kinetic energy	$K = \frac{1}{2}I\omega^2$
Power	$P = Fv$	Power	$P = \tau\omega$
Linear momentum	$p = Mv$	Angular momentum[a]	$L = I\omega$

[a] Angular momentum is discussed in Chapter 13.

an arbitrary direction, as shown in Fig. 16. The torque due to that force, given by $\boldsymbol{\tau} = \mathbf{r} \times \mathbf{F}$, is in a direction perpendicular to the plane formed by $\mathbf{r}$ and $\mathbf{F}$. We can resolve $\boldsymbol{\tau}$ into its x, y, and z components as shown in Fig. 16. Each component of the torque tends to produce rotation about its corresponding axis. However, we have assumed that the body is fixed in such a way that rotation about only the z axis is possible. The x and y components of the torque produce no motion. In this case, the bearings serve to constrain the system to rotate about only the z axis, and they must therefore provide torques that cancel the x and y components of the torque from the applied force. This indicates what is meant by a body constrained to move about a fixed axis: only torque components parallel to that axis are effective in rotating the body, and torque components perpendicular to the axis are assumed to be balanced by other parts of the system. The bearings *must* provide torques with x and y components to keep the direction of the axis of rotation fixed; the bearings *may* also provide a torque in the z direction, such as in the case of nonideal bearings that exert frictional forces on the axle of the wheel.

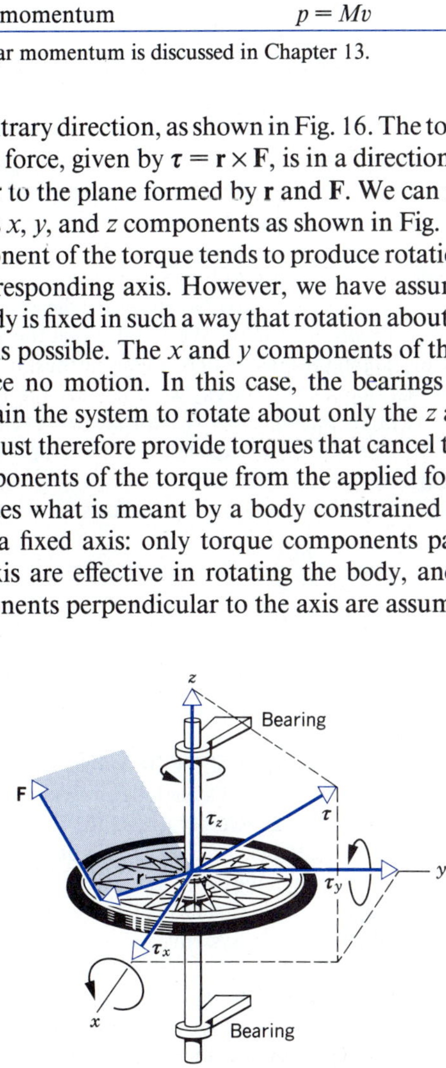

Sample Problem 4   A playground merry-go-round is pushed by a parent who exerts a force $\mathbf{F}$ of magnitude 115 N at a point P on the rim a distance of $r = 1.75$ m from the axis of rotation (Fig. 17). The force is exerted in a direction at an angle 32° below the horizontal, and the horizontal component of the force is in a direction 15° inward from the tangent at P. Find the magnitude of the component of the torque that accelerates the merry-go-round.

Solution   Only the horizontal component of $\mathbf{F}$ produces a vertical torque. Let us find $F_\perp$, the component of $\mathbf{F}$ along the horizontal line perpendicular to $\mathbf{r}$. The horizontal component of $\mathbf{F}$ is

$$F_h = F \cos 32° = 97.5 \text{ N}.$$

Figure 16   A rigid body, in this case a wheel, is free to rotate about the z axis. An arbitrary force $\mathbf{F}$, shown acting at a point on the rim, can produce torque components along the three coordinate axes. Only the z component is successful in rotating the wheel. The x and y components of the torque would tend to tip the axis of rotation away from the z axis. This tendency must be opposed by equal and opposite torques (not shown) exerted by the bearings, which hold the axis in a fixed direction.

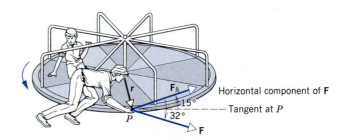

Figure 17   Sample Problem 4. A parent pushes a playground merry-go-round. The parent is leaning down, so the force has a downward component. Furthermore, because the parent is outside the rim, the force is directed slightly inward. The horizontal component of the force, F_h, is in the plane of the rotating platform and makes an angle of 15° with the tangent at P, the point at which the force is applied.

The component of F_h perpendicular to **r** is

$$F_\perp = F_h \cos 15° = 94.2 \text{ N.}$$

The (vertical) torque along the axis of rotation is thus

$$\tau = rF_\perp = (1.75 \text{ m})(94.2 \text{ N}) = 165 \text{ N·m.}$$

The component of F_h parallel to r $(= F_h \sin 15°)$ produces no torque at all about the axis of rotation, and the vertical component of F $(= F \sin 32°)$ produces a torque perpendicular to the axis that would tend to tip the rotating platform out of the horizontal plane (because the parent is pushing *down* on the platform) if that torque were not opposed by an equal and opposite torque from the bearings.

The parent must exert a torque that exceeds any frictional torque exerted by the bearings. When the parent stops pushing, this frictional torque, now acting alone, slows the merry-go-round and eventually brings it to rest.

Sample Problem 5   Figure 18*a* shows a uniform disk of mass $M = 2.5$ kg and radius $R = 20$ cm mounted on a fixed (frictionless) horizontal axle. A block of mass $m = 1.2$ kg hangs from a light cord that is wrapped around the rim of the disk. Find the acceleration of the falling block, the tension in the cord, and the angular acceleration of the disk.

Solution   Figure 18*b* shows a free-body diagram for the block. The block accelerates downward so that its weight mg must exceed the tension T in the cord. We take the downward direction as positive and, from Newton's second law, we have

$$\sum F = mg - T = ma.$$

Figure 18*c* shows a partial free-body diagram for the disk. The only torque τ acting on the disk, taken about its axis of rotation, is TR, and the rotational inertia of the disk is $\frac{1}{2}MR^2$. (Two other forces also act on the disk, its weight and the upward force exerted on the disk by its support. Both of these forces act at the axis of the disk, however, so that they exert no torque on the disk.) Applying Newton's second law in angular form (Eq. 18), and taking both τ and α to be positive for counterclockwise rotation, we obtain

$$\sum \tau = TR = \frac{1}{2}MR^2\left(\frac{a}{R}\right).$$

This reduces to

$$T = \frac{1}{2}Ma.$$

We are able to replace α by a/R because the cord does not slip, and thus the linear acceleration of the block is equal to the linear acceleration of the rim of the disk. Combining the two dynamical equations, one for the block and the other for the disk, we eliminate T to obtain

$$a = g\frac{2m}{M + 2m} = (9.8 \text{ m/s}^2)\frac{(2)(1.2 \text{ kg})}{2.5 \text{ kg} + (2)(1.2 \text{ kg})} = 4.8 \text{ m/s}^2,$$

and by eliminating a from the same two equations, we find

$$T = mg\frac{M}{M + 2m} = (1.2 \text{ kg})(9.8 \text{ m/s}^2)\frac{2.5 \text{ kg}}{2.5 \text{ kg} + (2)(1.2 \text{ kg})}$$
$$= 6.0 \text{ N.}$$

As expected, the acceleration of the falling block is less than g, and the tension in the cord $(= 6.0 \text{ N})$ is less than the weight of the hanging block $(= mg = 11.8 \text{ N})$. We see also that the acceleration of the block and the tension depend on the mass of the disk but not on its radius. As a check, we note that the formulas derived above predict $a = g$ and $T = 0$ for the case of a massless disk $(M = 0)$. This is what we expect; the block simply falls as a free body, trailing the cord behind it.

The angular acceleration of the disk follows from

$$\alpha = \frac{a}{R} = \frac{4.8 \text{ m/s}^2}{0.20 \text{ m}} = 24 \text{ rad/s}^2 = 3.8 \text{ rev/s}^2.$$

Sample Problem 6   Reconsider Sample Problem 5 from the standpoint of work and energy.

Solution   Suppose the system of Fig. 18 is released from rest. We examine it later when the block has fallen a distance L; at this point the block moves with velocity v, and the disk is turning with angular velocity ω. If the cord does not slip on the disk, then $v = \omega R$; furthermore, when the block falls a distance L, the disk must rotate through an angle ϕ such that $L = R\phi$.

We consider three different systems:

1.   *System = block + disk.* Gravity (the only external force) does external work mgL on the system as the block moves down a distance L. The net external work is then

$$W_\text{net} = mgL.$$

There is no frictional work done at the (frictionless) axle or between the cord and the disk (where there is no relative motion).

The change in kinetic energy is just the final kinetic energy, since the system was released from rest:

$$\Delta K = K_f - K_i = K_f = \frac{1}{2}I\omega^2 + \frac{1}{2}mv^2.$$

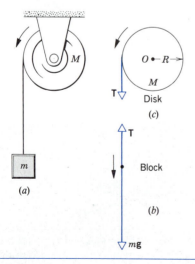

Figure 18   Sample Problem 5. (*a*) A falling block causes the disk to rotate. (*b*) A free-body diagram for the block. (*c*) A partial free-body diagram for the disk. The directions taken as positive are shown by the arrows.

The work–energy theorem gives

$$W_{net} = \Delta K$$

$$mgL = \tfrac{1}{2}I\omega^2 + \tfrac{1}{2}mv^2 = \tfrac{1}{2}I\left(\frac{v}{R}\right)^2 + \tfrac{1}{2}mv^2$$

$$v^2 = 2\left[\frac{2mg}{M+2m}\right]L.$$

The quantity in brackets is the acceleration we found in Sample Problem 5, so this result is clearly just the familiar equation $v^2 = v_0^2 + 2ax$ for linear motion with constant acceleration.

2. *System = block.* Here external work is done on the system by gravity and by the tension in the cord:

$$W_{net} = mgL - TL.$$

The change in kinetic energy of the system (the block) is just $\tfrac{1}{2}mv^2$, and the work–energy theorem gives

$$mgL - TL = \tfrac{1}{2}mv^2.$$

Substituting the previous result for v^2, we can show that this gives the tension found in the solution to Sample Problem 5.

3. *System = disk.* In this case only T does external work, and using Eq. 15 for a rotation through a total angle ϕ by a constant torque TR, we obtain

$$W_{net} = TR\phi = TL,$$

and the change in kinetic energy of the system (the disk) is

$$\Delta K = \tfrac{1}{2}I\omega^2 = \tfrac{1}{2}(\tfrac{1}{2}MR^2)v^2/R^2 = \tfrac{1}{4}Mv^2.$$

Again applying the work–energy theorem, we obtain

$$TL = \tfrac{1}{4}Mv^2.$$

You can easily show that this result is consistent with the previous results for v^2 and T.

In each case, by carefully defining the system and classifying the external forces and torques, we obtain identical results from the approaches based on Newton's laws and energy considerations.

12-6   COMBINED ROTATIONAL AND TRANSLATIONAL MOTION

The general motion of a system of particles includes both translation and rotation. So far we have considered only pure rotation in which the axis of rotation is fixed in the chosen inertial reference frame. We now generalize somewhat and allow the system also to have translational motion. When the center of mass moves with a translational speed v_{cm}, an observer viewing the system from an inertial reference frame moving at that speed will see the center of mass standing still. To this observer, the basic equation of rotational dynamics (Eq. 18, $\Sigma \tau = I\alpha$) will still apply provided (1) the axis of rotation passes through the center of mass, and (2) the axis always has the same direction in space (that is, as the system moves, its axis at one instant is parallel to the axis at any other instant). In this section we consider this special case of combined rotational and translational motion.

Perhaps the most familiar example of this motion, with the rotational axis remaining fixed in direction as the body undergoes translational motion, is the rolling wheel. Figure 19 compares the translational motion of the center of mass of a rolling wheel with the more complex motion of a point on the rim, which must be described by a combination of translational and rotational displacements.

Let us first show that the kinetic energy of an arbitrary body in this special case can be written as the sum of independent translational and rotational terms. Figure 20 shows an arbitrary body of mass M. The center of mass C is located instantaneously at the position $\mathbf{r}_{cm}$ relative to the origin of the chosen inertial reference frame. A particle P of mass m_i is located at the position $\mathbf{r}_i$ relative to the origin and at the position $\mathbf{r}'_i$ relative to the center of mass of the body. The translational motion is restricted to the xy plane; that is, the vector $\mathbf{v}_i$ describing the motion of m_i

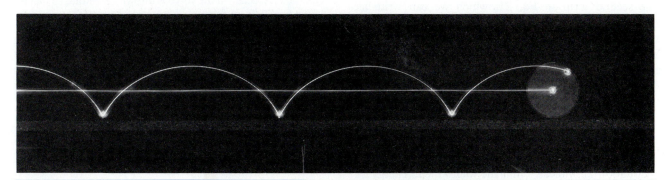

Figure 19   A time-exposure photo of a rolling disk. Small lights have been attached to the disk, one at its center and another at its edge. The latter traces out a curve called a *cycloid.*

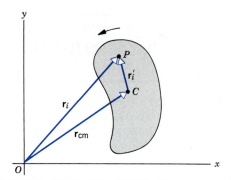

Figure 20 The center of mass C of a body undergoing both rotational and translational motion is located instantaneously at the position $\mathbf{r}_{cm}$. An arbitrary particle P of the body is located at $\mathbf{r}_i$ relative to the origin O and at $\mathbf{r}_i'$ relative to the center of mass C.

has only x and y components. The body also rotates with instantaneous angular velocity ω about an axis passing through the center of mass. Relative to O, the kinetic energy of the particle of mass m_i is $\frac{1}{2}m_iv_i^2$, and the total kinetic energy of the body is found from the sum over all such particles:

$$K = \sum \tfrac{1}{2}m_iv_i^2. \qquad (20)$$

From Fig. 20, we see that $\mathbf{r}_i = \mathbf{r}_{cm} + \mathbf{r}_i'$. Differentiating, we find the corresponding relationship between the velocities: $\mathbf{v}_i = \mathbf{v}_{cm} + \mathbf{v}_i'$, where $\mathbf{v}_i$ is the velocity of the particle in the xy system, $\mathbf{v}_{cm}$ is the velocity of the center of mass, and $\mathbf{v}_i'$ is the velocity of the particle relative to the center of mass. Observed from the reference frame of the center of mass, the motion is pure rotation about an axis through the center of mass; thus $\mathbf{v}_i'$ has magnitude $\omega r_i'$.

The quantity v_i^2 that appears in Eq. 20 can be written $\mathbf{v}_i \cdot \mathbf{v}_i$. Substituting the velocity transformation expression, $\mathbf{v}_i = \mathbf{v}_{cm} + \mathbf{v}_i'$, we find

$$K = \sum \tfrac{1}{2}m_iv_i^2 = \sum \tfrac{1}{2}m_i\mathbf{v}_i \cdot \mathbf{v}_i = \sum \tfrac{1}{2}m_i(\mathbf{v}_{cm} + \mathbf{v}_i') \cdot (\mathbf{v}_{cm} + \mathbf{v}_i')$$
$$= \sum \tfrac{1}{2}m_i(v_{cm}^2 + 2\mathbf{v}_{cm} \cdot \mathbf{v}_i' + v_i'^2). \qquad (21)$$

The second term in Eq. 21, which we can write as $\mathbf{v}_{cm} \cdot (\sum m_i\mathbf{v}_i')$, includes as a factor the total momentum of all the particles in the center-of-mass frame ($\sum \mathbf{p}_i' = \sum m_i\mathbf{v}_i' = M\mathbf{v}_{cm}'$), which equals zero because $\mathbf{v}_{cm}' = 0$ in the center-of-mass frame. Thus we have, substituting $v_i' = r_i'\omega$ in the last term of Eq. 21,

$$K = \sum \tfrac{1}{2}m_iv_{cm}^2 + \sum \tfrac{1}{2}m_iv_i'^2$$
$$= \tfrac{1}{2}Mv_{cm}^2 + \sum \tfrac{1}{2}m_ir_i'^2\omega^2$$
$$= \tfrac{1}{2}Mv_{cm}^2 + \tfrac{1}{2}I_{cm}\omega^2. \qquad (22)$$

Equation 22 indicates that the total kinetic energy of the moving object consists of two terms, one associated with the pure translation of the center of mass of the object at velocity $\mathbf{v}_{cm}$, and the other associated with pure

rotation about an axis through the center of mass. The two terms are quite independent: the rotation would be present even in the absence of translation (for example, as observed from a frame of reference moving at $\mathbf{v}_{cm}$). The velocities $\mathbf{v}_{cm}$ and ω are, in this general case, independent of one another: we can provide any amount of rotational kinetic energy and any amount of translational kinetic energy. For example, in the launch of a satellite from the space shuttle (see Fig. 34 of this chapter and Fig. 15 of Chapter 13), the satellite is set spinning about its axis for stability (as we discuss in Chapter 13) and independently given the translational velocity necessary to boost it into orbit.

Sample Problem 7 A yo-yo (Fig. 21) of mass $M = 0.023$ kg, consisting of two disks of radius $R = 2.6$ cm connected by a shaft of radius $R_0 = 0.3$ cm, is spinning at the end of the string of length $L = 0.84$ m with angular velocity ω_0. What angular velocity is needed for the yo-yo to climb up the string? Assume the string to be of negligible thickness.

Solution At the start of the climb, the energy is purely rotational kinetic, but at the end it is partly rotational kinetic, partly translational kinetic, and partly gravitational potential. Conservation of energy then gives

$$\tfrac{1}{2}I\omega_0^2 = \tfrac{1}{2}I\omega^2 + \tfrac{1}{2}Mv^2 + MgL,$$

where ω and v are the final angular and linear velocities. We cannot solve this problem simply for a real yo-yo, but we can solve for the ideal yo-yo with a string of negligible thickness by finding the condition necessary for the yo-yo just to reach the hand (arriving with $v = \omega = 0$):

$$\tfrac{1}{2}I\omega_0^2 = MgL.$$

Using the rotational inertia of a disk ($I = \frac{1}{2}MR^2$) and neglecting the contribution of the shaft to the rotational inertia, we solve for ω_0 and find

$$\omega_0 = \sqrt{\frac{4gL}{R^2}} = \sqrt{\frac{4(9.8 \text{ m/s}^2)(0.84 \text{ m})}{(0.026 \text{ m})^2}} = 221 \text{ rad/s} = 35 \text{ rev/s}.$$

This considerable rotational speed is just a lower limit. Values of the rotational angular velocity in excess of 100 rev/s are quite common, especially if the yo-yo is thrown downward so that its initial translational energy is converted to rotational energy. In

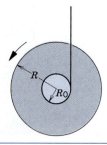

Figure 21 Sample Problem 7. A yo-yo, shown in cross section. The string, of assumed negligible thickness, is wound around an axle of radius R_0.

the case of such a large initial angular velocity, it would reach the hand with a considerable linear velocity. In fact, a popular trick is to release the string from the finger at the last instant, allowing the yo-yo's vertical speed to carry it several meters upward.

The interplay of translational kinetic, rotational kinetic, and gravitational potential energy is responsible for the behavior of the yo-yo and the many tricks that can be done with it.*

Rolling Without Slipping

We now consider a particular special case of combined translational and rotational motion, in which the object rolls across a surface in such a way that there is no relative motion between the object and the surface at the instantaneous point of contact. This special case is known as *rolling without slipping.* Figure 22 shows a photograph of a rolling bicycle wheel. You can see how the spokes near the bottom, which is instantaneously at rest, are in sharper focus than those near the top, which appear blurred. The top of the wheel is clearly moving faster than the bottom! It is of course friction between the wheel and the surface that is responsible for rolling without slipping, but in this special case the frictional force does no work and dissipates no energy, because there is no relative motion between the wheel and the surface at the point of contact. Even though there is motion in the problem, the frictional force is that of *static* friction.

Not all cases of rolling on a frictional surface will result in rolling without slipping. For example, imagine a car trying to start on an icy street. At first perhaps the wheels spin in place; in this case we have pure rotation with no translation, and there is a great amount of frictional work done, as indicated by the ice that is melted by the internal energy increase resulting from the frictional work. If sand is placed on the ice, the wheels still spin rapidly, but the car begins to inch forward. Now there is still some slipping

* See "The Yo-Yo: A Toy Flywheel," by Wolfgang Burger, *American Scientist,* March–April 1984, p. 137.

Figure 22 A photo of a rolling bicycle wheel. Note that the spokes near the top of the wheel are more blurred than those near the bottom. This is because the top has a greater linear velocity.

between the tire and the ice, so that the frictional force is still doing work, but we now have some translational motion. It is only when the tires stop slipping on the ice, so that there is no relative motion at the point of contact between tire and ice, that we have the condition of rolling without slipping and no frictional work.

Figure 23 shows one way to view rolling without slipping, as a superposition of rotational and translational motion. Figure 23a shows the translational motion, in which the center of mass C moves with speed v_{cm}, and Fig. 23b shows the rotational motion at angular speed ω. When the two motions are superposed, the bottom B of

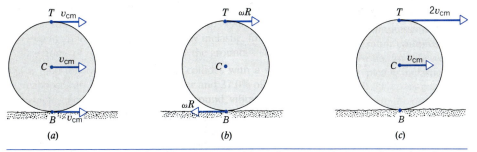

Figure 23 Rolling can be viewed as a superposition of pure translation and rotation about the center of mass. (*a*) The translational motion, in which all points move with the same linear velocity. (*b*) The rotational motion, in which all points move with the same angular velocity about the central axis. (*c*) The superposition of (*a*) and (*b*), in which the velocities at T, C, and B have been obtained by vector addition of the translational and rotational components.

the wheel will have speed $v_{cm} - \omega R$; if this is to be zero, so that the point of contact is at rest, then we must have $v_{cm} = \omega R$. Superimposing the resulting translational and rotational motions, we obtain Fig. 23c. Note that the linear speed at the top of the wheel T is exactly twice that at the center.

For pure rotation the tangential speed has magnitude $v = \omega R$. Thus, for the special case of rolling without slipping, the rotational and translational motions must be related by

$$v_{cm} = \omega R. \tag{23}$$

This result applies *only* in the case of rolling without slipping. In the general case of combined translational and rotational motion, the tangential speed $v (= \omega R)$ does not equal v_{cm}.

The kinetic energy of combined rotational and translational motion, Eq. 22, no longer has two independent terms in the case of rolling without slipping. We can regard the kinetic energy as being completely determined by either the translational speed or the rotational speed, and we obtain the corresponding expressions by substituting Eq. 23 into Eq. 22:

$$K = \tfrac{1}{2}Mv_{cm}^2 + \tfrac{1}{2}I_{cm}v_{cm}^2/R^2, \tag{24a}$$

$$K = \tfrac{1}{2}M\omega^2 R^2 + \tfrac{1}{2}I_{cm}\omega^2. \tag{24b}$$

In either case, only one parameter (v_{cm} or ω) is sufficient to determine the kinetic energy.

Another View of Rolling Without Slipping

There is another instructive way to analyze rolling without slipping: we consider the point of contact at B to be an instantaneous axis of rotation, as illustrated in Fig. 24. At each instant there is a new point of contact B and therefore a new axis of rotation, but instantaneously the motion consists of a pure rotation about B. The kinetic energy is

$$K = \tfrac{1}{2}I_B\omega_B^2, \tag{25}$$

where I_B is the rotational inertia and ω_B is the angular speed, both taken about the rotational axis through B.

From the parallel-axis theorem, $I_B = I_{cm} + MR^2$, where R is the distance between B and the center of mass. We know that the center of mass moves with speed v_{cm}, and thus the rotational motion about B must give the center of mass the proper tangential speed: $v_{cm} = R\omega_B$. The kinetic energy can now be written, using Eq. 25,

$$K = \tfrac{1}{2}(I_{cm} + MR^2)\left(\frac{v_{cm}}{R}\right)^2$$

$$= \tfrac{1}{2}I_{cm}\omega_B^2 + \tfrac{1}{2}Mv_{cm}^2. \tag{26}$$

The angular speed of the center of mass relative to B must be the same as the angular speed of B relative to the center of mass; thus $\omega_B = \omega$, and the kinetic energy becomes identical to that given by Eqs. 24. In this case the derivation has been done by assuming that the rotational and translational parts of the kinetic energy are not independent.

Sample Problem 8 A solid cylinder of mass M and radius R rolls without slipping down an inclined plane of length L and height h (Fig. 25). Find the speed of its center of mass when the cylinder reaches the bottom.

Solution We use conservation of energy to solve this problem. The cylinder is initially at rest. At the bottom of the incline, the change in potential energy is $\Delta U = -Mgh$. If the cylinder starts from rest, its kinetic energy changes by the amount given in Eq. 24a. For a cylinder rotating about its axis, $I_{cm} = \tfrac{1}{2}MR^2$. If there are no other exchanges of energy, then $\Delta E = 0$ gives $\Delta K = -\Delta U$, or

$$\tfrac{1}{2}Mv_{cm}^2 + \tfrac{1}{2}(\tfrac{1}{2}MR^2)\left(\frac{v_{cm}}{R}\right)^2 = Mgh.$$

Solving for v_{cm}, we obtain

$$v_{cm} = \sqrt{\tfrac{4}{3}gh}.$$

The speed of the center of mass would have been $v_{cm} = \sqrt{2gh}$ if the cylinder were sliding (without rolling) down a frictionless incline. The speed of the rolling cylinder is therefore less than the speed of the sliding cylinder because, for the rolling cylinder, part of the initial potential energy has been transformed into rotational kinetic energy, leaving less energy available for the

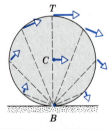

Figure 24 A rolling body can be considered to be rotating about an instantaneous axis at the point of contact B. The vectors show the instantaneous linear velocities of selected points.

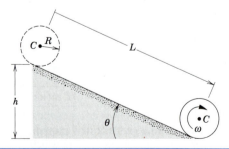

Figure 25 Sample Problem 8. A cylinder rolls without slipping down an incline.

translational part of the kinetic energy. Although the rolling cylinder arrives later at the bottom of the incline than an identical sliding cylinder started at the same time down a frictionless, but otherwise identical, incline, both arrive at the bottom with the same amount of energy; the rolling cylinder happens to be rotating as it moves, whereas the sliding one does not rotate as it moves.

The above sample problem was solved using energy techniques. We can also solve problems of this type using dynamical methods based on forces and torques. In doing so, it is convenient to use the rotational form of Newton's second law, $\Sigma \tau = I\alpha$, about an axis through the center of mass. We state again the two special conditions that permit us to apply this result when the rotational axis is not fixed in space: (1) the axis passes through the center of mass of the rotating object, and (2) the axis does not change its direction in space as the object moves. This problem satisfies both conditions.

The free-body diagram for this problem is shown in Fig. 26. $M\mathbf{g}$ is the weight of the cylinder acting vertically down through the center of mass, $\mathbf{N}$ is the normal force exerted by the incline on the cylinder, and $\mathbf{f}$ is the force of static friction acting upward along the incline at the point of contact.

Using Newton's second law for the translational motion, we obtain, for motion perpendicular to the incline,

$$N - Mg \cos \theta = 0,$$

and, for motion along the incline,

$$Mg \sin \theta - f = Ma_{cm}.$$

We now consider the rotational motion. Neither $\mathbf{N}$ nor $M\mathbf{g}$ have torques about the center of mass C because their lines of action pass through C, and they have zero moment arms. The force of friction has a moment arm R about C, so that $\tau = fR$ and thus

$$fR = I_{cm}\alpha$$

$$f = I_{cm}\alpha/R.$$

The rotational inertia about the center of mass is $I_{cm} = \frac{1}{2}MR^2$. For rolling without slipping, $v_{cm} = \omega R$; differentiating, it follows that $a_{cm} = \alpha R$, and the previous equation becomes

$$f = (\tfrac{1}{2}MR^2)(a_{cm}/R^2) = \tfrac{1}{2}Ma_{cm}.$$

Substituting this into the second translational equation, we find

$$a_{cm} = \tfrac{2}{3}g \sin \theta.$$

That is, the acceleration of the center of mass for the rolling cylinder ($\tfrac{2}{3}g \sin \theta$) is less than its acceleration would be if the cylinder were sliding down the incline ($g \sin \theta$). This result holds at any instant, regardless of the position of the cylinder along the incline.

Because a_{cm} is constant, we can find the speed of the center of mass, starting from rest. From Eq. 20 of Chapter 2, $v^2 = v_0^2 + 2ax$, or

$$v_{cm}^2 = 2a_{cm}L,$$

so that

$$v_{cm}^2 = 2(\tfrac{2}{3}g \sin \theta)L = \tfrac{4}{3}g\left(\frac{h}{L}\right)L = \tfrac{4}{3}gh$$

or

$$v_{cm} = \sqrt{\tfrac{4}{3}gh}.$$

This result is the same as that obtained before by the energy method. The energy method is certainly simpler and more direct. However, if we are interested in knowing the values of the forces, such as $\mathbf{N}$ and $\mathbf{f}$, we must use a dynamical method.

This method determines the force of static friction needed for rolling:

$$f = Ma_{cm}/2 = (M/2)(\tfrac{2}{3}g \sin \theta) = \tfrac{1}{3}Mg \sin \theta.$$

What would happen if the force of static friction between the surfaces were less than this value?

Sample Problem 9   A sphere, a cylinder, and a hoop start from rest and roll down the same incline. Which body gets to the bottom first?

Solution   We solve this problem by comparing the accelerations of the centers of mass of the three objects. The one with the greatest acceleration will reach the bottom first.

From the previous calculation, we have the following general dynamical equation for motion along the plane:

$$Mg \sin \theta - f = Ma_{cm},$$

where

$$f = I_{cm}\alpha/R = I_{cm}a_{cm}/R^2.$$

Substituting for f and solving for a_{cm}, we find

$$a_{cm} = \frac{g \sin \theta}{1 + I_{cm}/MR^2}. \qquad (27)$$

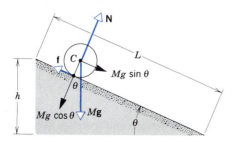

Figure 26   The forces acting on the rolling cylinder of Sample Problem 8.

We can evaluate this expression for each of the objects:

Sphere: $\dfrac{I_{cm}}{MR^2} = \tfrac{2}{5}, \quad a_{cm} = \dfrac{g\sin\theta}{1+\tfrac{2}{5}} = \tfrac{5}{7}g\sin\theta$
$$= 0.714g\sin\theta,$$

Cylinder: $\dfrac{I_{cm}}{MR^2} = \tfrac{1}{2}, \quad a_{cm} = \dfrac{g\sin\theta}{1+\tfrac{1}{2}} = \tfrac{2}{3}g\sin\theta$
$$= 0.667g\sin\theta,$$

Hoop: $\dfrac{I_{cm}}{MR^2} = 1, \quad a_{cm} = \dfrac{g\sin\theta}{1+1} = \tfrac{1}{2}g\sin\theta$
$$= 0.500g\sin\theta.$$

Clearly the sphere has the largest acceleration and reaches the bottom first, followed by the cylinder and then the hoop. The sphere is the most "compact" object and can accept rotation with the least cost in kinetic energy, since its rotational inertia is the smallest of the three. Each body has a kinetic energy equal to *Mgh* at the bottom of the incline; for the sphere, more of the kinetic energy is of the translational type and less is of the rotational type.

You should also be able to solve this problem using energy methods by finding which object has the largest v_{cm} at the bottom of the incline.

Note that our final result for the acceleration of each object depends neither on the mass nor on the radius of the object. The three objects can be of quite different sizes, but the sphere will always reach the bottom first. Furthermore, all rolling spheres will have the same acceleration, no matter what their relative sizes or masses; a marble and a bowling ball will reach the bottom at the same time and with the same speed.

Sample Problem 10 A uniform solid cylinder of radius R ($=12$ cm) and mass M ($=3.2$ kg) is given an initial (clockwise) angular velocity ω_0 of 15 rev/s and then lowered on to a flat horizontal surface. The coefficient of kinetic friction between the surface and the cylinder is $\mu_k = 0.21$. Initially, the cylinder slips as it moves along the surface, but after a time t pure rolling without slipping begins. (*a*) What is the velocity v_{cm} of the center of mass at the time t? (*b*) What is the value of t?

Solution (*a*) Figure 27 shows the forces that act on the cylinder. Since all the forces are constant while slipping occurs, the acceleration a_{cm} of the center of mass in the *x* direction is constant. Thus for the translational motion we can write

$$\sum F_x = Ma_{cm} = M\left(\frac{v_f - v_i}{t - 0}\right).$$

Here, $v_i = 0$ and $v_f = v_{cm}$, the velocity at t when pure rolling

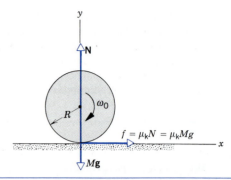

Figure 27 Sample Problem 10. A cylinder initially spinning with angular velocity ω_0 makes contact with a horizontal surface that exerts a frictional force **f** on the cylinder.

begins. Also, the only horizontal force is that of friction, given by $\mu_k Mg$, so that

$$\mu_k Mg = Mv_{cm}/t. \qquad (28)$$

The angular acceleration α about an axis through the center of mass is also constant (why?), so that for the rotational motion we can write

$$\sum \tau = I_{cm}\alpha = I_{cm}\left(\frac{\omega_f - \omega_i}{t - 0}\right).$$

Here, choosing counterclockwise rotations to be positive, $\omega_f = -v_{cm}/R$, the angular velocity at time t, and $\omega_i = -\omega_0$. Only the force f produces a torque about the center of mass; the resultant torque is $\mu_k MgR$, a positive quantity. Using $\sum\tau = I\alpha$, we obtain

$$\mu_k MgR = (\tfrac{1}{2}MR^2)\left(\frac{-v_{cm}/R - (-\omega_0)}{t}\right). \qquad (29)$$

Eliminating t from Eqs. 28 and 29 and solving for v_{cm}, we obtain

$$v_{cm} = \tfrac{1}{3}\omega_0 R = \tfrac{1}{3}(15\text{ rev/s})(2\pi\text{ rad/rev})(0.12\text{ m}) = 3.8\text{ m/s}.$$

Note that v_{cm} does not depend on the values of M, g, or μ_k. What, however, would occur if any of these quantities were zero?

(*b*) By eliminating v_{cm} between Eqs. 28 and 29, we can solve for t and find

$$t = \frac{\omega_0 R}{3\mu_k g} = 1.8\text{ s}.$$

As an exercise, you should check these results using energy methods. Find the change in rotational kinetic energy and compare it with the work done by the frictional torque. Note that, because rotation *with* slipping occurs between time 0 and time t, frictional work is done during that period.

QUESTIONS

1. Can the mass of an object be considered as concentrated at its center of mass for purposes of computing its rotational inertia? If yes, explain why. If no, offer a counterexample.

2. About what axis is the rotational inertia of your body the least? About what axis through your center of mass is your rotational inertia the greatest?

3. About what axis would a uniform cube have its minimum rotational inertia?

4. If two circular disks of the same weight and thickness are made from metals having different densities, which disk, if either, will have the larger rotational inertia about its symmetry axis?

5. The rotational inertia of a body of rather complicated shape is to be determined. The shape makes a mathematical calculation from $\int r^2 \, dm$ exceedingly difficult. Suggest ways in which the rotational inertia about a particular axis could be measured experimentally.

6. Five solids are shown in cross section in Fig. 28. The cross sections have equal heights and equal maximum widths. The solids have equal masses. Which one has the largest rotational inertia about a perpendicular axis through the center of mass? Which has the smallest?

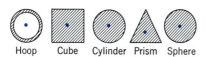

Hoop   Cube   Cylinder   Prism   Sphere

Figure 28 Question 6.

7. Does Eq. 9 still hold if the slab is not "thin," that is, if its thickness is comparable to (or even greater than) *a* or *b*?

8. Figure 29*a* shows a meter stick, half of which is wood and half of which is steel, that is pivoted at the wooden end at *O*. A force is applied to the steel end at *a*. In Fig. 29*b*, the stick is pivoted at the steel end at *O'* and the same force is applied at the wooden end at *a'*. Does one get the same angular acceleration in each case? If not, in which case is the angular acceleration greater?

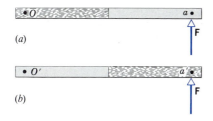

(*a*)

(*b*)

Figure 29 Question 8.

9. In cutting down a tree, a logger makes a cut on the side facing the direction in which the tree is to fall. Explain why. Would it be safe to stand directly behind the tree on the opposite side of the fall?

10. You can distinguish between a raw egg and a hardboiled one by spinning each one on a table. Explain how. Also, if you stop a spinning raw egg with your fingers and release it very quickly, it will resume spinning. Why?

11. Comment on each of the following assertions about skiing. (*a*) In downhill racing, one wants skis that do not turn easily. (*b*) In slalom racing, one wants skis that turn easily. (*c*) Therefore, the rotational inertia of downhill skis should be larger than that of slalom skis. (*d*) Considering that there is low friction between skis and snow, how does a skier exert torques to turn or stop a turn? (See "The Physics of Ski Turns," by J. I. Shonie and D. L. Mordick, *The Physics Teacher,* December 1972, p. 491.)

12. Consider a straight stick standing on end on (frictionless) ice. What would be the path of its center of mass if it falls?

13. For storing wind energy or solar energy, flywheels have been suggested. The amount of energy that can be stored in a

flywheel depends on the density and tensile strength of the material making up the flywheel and for a given weight one wants the lowest density strong material available. Can you make this plausible? (See "Flywheels," by R. F. Post and S. F. Post, *Scientific American,* December 1973, p. 17.)

14. Explain why a wheel rolling on a flat horizontal surface cannot be slowed down by static friction. Assuming no slipping, what does slow the wheel down?

15. Describe qualitatively what happens to the system of Fig. 18 if the disk is given an initial clockwise angular velocity before being released. What changes, if any, occur in the linear acceleration of the block, or the angular acceleration of the disk? See Sample Problem 5.

16. Explain why the wheel is such an important invention.

17. Apart from appearance, why do sports cars have wire wheels?

18. A cannonball and a marble roll from rest down an incline. Which gets to the bottom first?

19. A cylindrical can filled with beef and an identical can filled with apple juice both roll down an incline. Compare their angular and linear accelerations. Explain the difference.

20. A solid wooden cylinder rolls down two different inclined planes of the same height but with different angles of inclination. Will it reach the bottom with the same speed in each case? Will it take longer to roll down one incline than the other? Explain your answers.

21. A solid brass cylinder and a solid wooden cylinder have the same radius and mass, the wooden cylinder being longer. You release them together at the top of an incline. Which will beat the other to the bottom? Suppose that the cylinders are now made to be the same length (and radius) and that the masses are made to be equal by boring a hole along the axis of the brass cylinder. Which cylinder will win the race now? Explain your answers. Assume that the cylinders roll without slipping.

22. Ruth and Roger are cycling along a path at the same speed. The wheels of Ruth's bike are a little larger in diameter than the wheels of Roger's bike. How do the angular speeds of their wheels compare? What about the speeds of the top portions of their wheels?

23. A cylindrical drum, pushed along by a board from an initial position shown in Fig. 30, rolls forward on the ground a distance *L/2*, equal to half the length of the board. There is no slipping at any contact. Where is the board then? How far has the man walked?

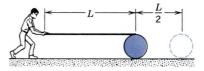

Figure 30 Question 23.

24. Two heavy disks are connected by a short rod of much smaller radius. The system is placed on a ramp so that the disks hang over the sides as in Fig. 31. The system rolls down the ramp without slipping. (*a*) Near the bottom of the ramp the disks touch the horizontal table and the system takes off

Figure 31   Question 24.

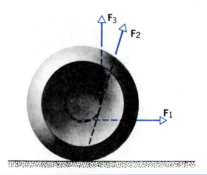

Figure 32   Question 26.

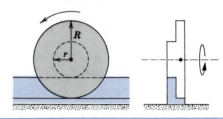

Figure 33   Question 27.

with greatly increased translational speed. Explain why. (*b*) If this system raced a hoop (of any radius) down the ramp, which would reach the bottom first?

25. A yo-yo falls to the bottom of its cord and then climbs back up. Does it reverse its direction of rotation at the bottom? Explain your answer.

26. A yo-yo is resting on a horizontal table and is free to roll (see Fig. 32). If the string is pulled by a horizontal force such as F_1, which way will the yo-yo roll? What happens when the force F_2 is applied (its line of action passes through the point of contact of the yo-yo and table)? If the string is pulled vertically with the force F_3, what happens?

27. A solid flanged wheel consists of two joined concentric disks, the larger of radius R and the smaller of radius r. The wheel is to roll along the two-level rail, as shown in Fig. 33. However, in making one rotation, the center of the wheel moves a distance $2\pi r$ according to the smaller disk and $2\pi R$

according to the larger disk. Explain the apparent discrepancy.

28. State Newton's three laws of motion in words suitable for rotating bodies.

PROBLEMS

Section 12-2 Kinetic Energy of Rotation and Rotational Inertia

1. The masses and coordinates of four particles are as follows: 50 g, $x = 2.0$ cm, $y = 2.0$ cm; 25 g, $x = 0$, $y = 4.0$ cm; 25 g, $x = -3.0$ cm, $y = -3.0$ cm; 30 g, $x = -2.0$ cm, $y = 4.0$ cm. Calculate the rotational inertia of this collection with respect to the (*a*) x, (*b*) y, and (*c*) z axes.

2. A molecule has a rotational inertia of 14,000 u·pm² and is spinning at an angular speed of 4.30×10^{12} rad/s. (*a*) Express the rotational inertia in kg·m². (*b*) Calculate the rotational kinetic energy in eV.

3. The oxygen molecule has a total mass of 5.30×10^{-26} kg and a rotational inertia of 1.94×10^{-46} kg·m² about an axis through the center perpendicular to the line joining the atoms. Suppose that such a molecule in a gas has a mean speed of 500 m/s and that its rotational kinetic energy is two-thirds of its translational kinetic energy. Find its average angular velocity.

Section 12-3 Rotational Inertia of Solid Bodies

4. A communications satellite is a uniform cylinder with mass 1220 kg, diameter 1.18 m, and length 1.72 m. Prior to launching from the shuttle cargo bay, it is set spinning at

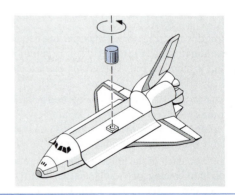

Figure 34   Problem 4.

1.46 rev/s about the cylinder axis; see Fig. 34. Calculate the satellite's rotational kinetic energy.

5. Each of three helicopter rotor blades shown in Fig. 35 is 5.20 m long and has a mass of 240 kg. The rotor is rotating at 350 rev/min. (*a*) What is the rotational inertia of the rotor assembly about the axis of rotation? (Each blade can be

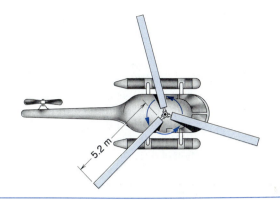

Figure 35 Problem 5.

considered a thin rod.) (*b*) What is the kinetic energy of rotation?

6. Figure 36 shows a uniform block of mass *M* and edge lengths *a*, *b*, and *c*. Calculate its rotational inertia about an axis through one corner and perpendicular to the large face of the block. (*Hint:* See Fig. 9.)

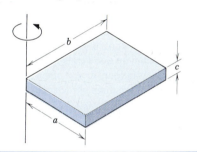

Figure 36 Problem 6.

7. Calculate the rotational inertia of a meter stick, with mass 0.56 kg, about an axis perpendicular to the stick and located at the 20-cm mark.

8. Two particles, each with mass *m*, are fastened to each other and to a rotation axis by two rods, each with length *L* and mass *M*, as shown in Fig. 37. The combination rotates around the rotation axis with angular velocity ω. Obtain algebraic expressions for (*a*) the rotational inertia of the combination about *O* and (*b*) the kinetic energy of rotation about *O*.

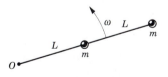

Figure 37 Problem 8.

9. (*a*) Show that the sum of the rotational inertias of a plane laminar body about any two perpendicular axes in the plane of the body is equal to the rotational inertia of the body about an axis through their point of intersection perpendicular to the plane. (*b*) Apply this to a circular disk to find its rotational inertia about a diameter as axis.

10. Delivery trucks that operate by making use of energy stored in a rotating flywheel have been used in Europe. The trucks are charged by using an electric motor to get the flywheel up to its top speed of 624 rad/s. One such flywheel is a solid, homogeneous cylinder with a mass of 512 kg and a radius of 97.6 cm. (*a*) What is the kinetic energy of the flywheel after charging? (*b*) If the truck operates with an average power requirement of 8.13 kW, for how many minutes can it operate between chargings?

11. (*a*) Show that a solid cylinder of mass *M* and radius *R* is equivalent to a thin hoop of mass *M* and radius $R/\sqrt{2}$, for rotation about a central axis. (*b*) The radial distance from a given axis at which the mass of a body could be concentrated without altering the rotational inertia of the body about that axis is called the *radius of gyration*. Let *k* represent the radius of gyration and show that

$$k = \sqrt{I/M}.$$

This gives the radius of the "equivalent hoop" in the general case.

12. Figure 38 shows the solid rod considered in Section 12-3 (see also Fig. 6) divided into an arbitrary number *N* of pieces. (*a*) What is the mass m_i of each piece? (*b*) Show that the distance of each piece from the axis of rotation can be written $r_i = (i - 1)L/N + (\frac{1}{2})L/N = (i - \frac{1}{2})L/N$. (*c*) Use Eq. 5 to evaluate the rotational inertia of this rod, and show that it reduces to Eq. 6. You may need the following sums:

$$\sum_{i=1}^{n} 1 = n, \qquad \sum_{i=1}^{n} i = n(n + 1)/2,$$

$$\sum_{i=1}^{n} i^2 = n(n + 1)(2n + 1)/6.$$

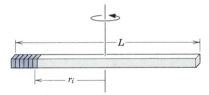

Figure 38 Problem 12.

13. In this problem we seek to compute the rotational inertia of a disk of mass *M* and radius *R* about an axis through its center and perpendicular to its surface. Consider a mass element *dm* in the shape of a ring of radius *r* and width *dr* (see Fig. 39). (*a*) What is the mass *dm* of this element, expressed as a fraction of the total mass *M* of the disk? (*b*) What is the rotational inertia *dI* of this element? (*c*) Integrate the result of part (*b*) to find the rotational inertia of the entire disk.

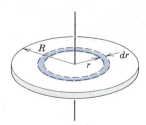

Figure 39 Problem 13.

14. In this problem, we use the result of the previous problem for the rotational inertia of a disk to compute the rotational inertia of a uniform solid sphere of mass M and radius R about an axis through its center. Consider an element dm of the sphere in the form of a disk of thickness dz at a height z above the center (see Fig. 40). (*a*) Expressed as a fraction of the total mass M, what is the mass dm of the element? (*b*) Considering the element as a disk, what is its rotational inertia dI? (*c*) Integrate the result of (*b*) over the entire sphere to find the rotational inertia of the sphere.

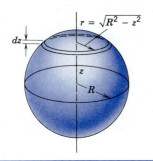

Figure 40   Problem 14.

Section 12-4 Torque Acting on a Particle

15. Figure 41 shows the lines of action and the points of application of two forces about the origin O. Imagine these forces to be acting on a rigid body pivoted at O, all vectors being in the plane of the figure. (*a*) Find an expression for the magnitude of the resultant torque on the body. (*b*) If $r_1 = 1.30$ m, $r_2 = 2.15$ m, $F_1 = 4.20$ N, $F_2 = 4.90$ N, $\theta_1 = 75.0°$, and $\theta_2 = 58.0°$, what are the magnitude and direction of the resultant torque?

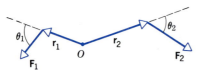

Figure 41   Problem 15.

16. Redraw Fig. 12 under the following transformations: (*a*) $\mathbf{F} \rightarrow -\mathbf{F}$, (*b*) $\mathbf{r} \rightarrow -\mathbf{r}$, and (*c*) $\mathbf{F} \rightarrow -\mathbf{F}$ and $\mathbf{r} \rightarrow -\mathbf{r}$, in each case showing the new direction of the torque. Check for consistency with the right-hand rule.

17. The object shown in Fig. 42 is pivoted at O. Three forces act on it in the directions shown on the figure: $F_A = 10$ N at point A, 8.0 m from O; $F_B = 16$ N at point B, 4.0 m from O; and $F_C = 19$ N at point C, 3.0 m from O. What are the magnitude and direction of the resultant torque about O?

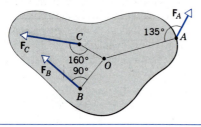

Figure 42   Problem 17.

18. (*a*) Given that $\mathbf{r} = \mathbf{i}x + \mathbf{j}y + \mathbf{k}z$ and $\mathbf{F} = \mathbf{i}F_x + \mathbf{j}F_y + \mathbf{k}F_z$, find the torque $\boldsymbol{\tau} = \mathbf{r} \times \mathbf{F}$. (*b*) Show that if $\mathbf{r}$ and $\mathbf{F}$ lie in a given plane, then $\boldsymbol{\tau}$ has no component in that plane.

Section 12-5 Rotational Dynamics of a Rigid Body

19. A cylinder having a mass of 1.92 kg rotates about its axis of symmetry. Forces are applied as shown in Fig. 43: $F_1 = 5.88$ N, $F_2 = 4.13$ N, and $F_3 = 2.12$ N. Also, $R_1 = 4.93$ cm and $R_2 = 11.8$ cm. Find the magnitude and direction of the angular acceleration of the cylinder.

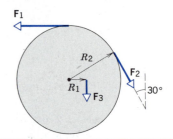

Figure 43   Problem 19.

20. A thin spherical shell has a radius of 1.88 m. An applied torque of 960 N·m imparts an angular acceleration equal to 6.23 rad/s² about an axis through the center of the shell. Calculate (*a*) the rotational inertia of the shell about the axis of rotation and (*b*) the mass of the shell.

21. In the act of jumping off a diving board, a diver changed his angular velocity from zero to 6.20 rad/s in 220 ms. The diver's rotational inertia is 12.0 kg·m². (*a*) Find the angular acceleration during the jump. (*b*) What external torque acted on the diver during the jump?

22. An automobile engine develops 133 hp (= 99.18 kW) when rotating at 1820 rev/min. How much torque does it deliver?

23. A 31.4-kg wheel with radius 1.21 m is rotating at 283 rev/min. It must be brought to a stop in 14.8 s. Find the required average power. Assume the wheel to be a thin hoop.

24. If $R = 12.3$ cm, $M = 396$ g, and $m = 48.7$ g in Fig. 18*a*, find the speed of the block after it has descended 54.0 cm starting from rest. Solve the problem using energy-conservation principles.

25. Assume the Earth to be a sphere of uniform density. (*a*) Calculate its rotational kinetic energy. (*b*) Suppose that this energy could be harnessed for our use. For how long could the Earth supply 1.00 kW of power to each of the 4.20×10^9 persons on the Earth?

26. Figure 44 shows the massive shield door at a neutron test facility at Lawrence Livermore Laboratory; this is the world's heaviest hinged door. The door has a mass of 44,000 kg, a rotational inertia about its hinge line of 8.7×10^4 kg·m², and a width of 2.4 m. What steady force, applied at its outer edge at right angles to the door, can move it from rest through an angle of 90° in 30 s?

27. A pulley having a rotational inertia of 1.14×10^{-3} kg·m² and a radius of 9.88 cm is acted on by a force, applied tangentially at its rim, that varies in time as $F = 0.496t + 0.305t^2$, where F is in newtons and t is in seconds. If the pulley was initially at rest, find its angular speed after 3.60 s.

Figure 44 Problem 26.

28. Figure 45 shows two blocks each of mass m suspended from the ends of a rigid weightless rod of length $L_1 + L_2$, with $L_1 = 20.0$ cm and $L_2 = 80.0$ cm. The rod is held in the horizontal position shown in the figure and then released. Calculate the linear accelerations of the two blocks as they start to move.

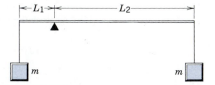

Figure 45 Problem 28.

29. Two identical blocks, each of mass M, are connected by a light string over a frictionless pulley of radius R and rotational inertia I (Fig. 46). The string does not slip on the pulley, and it is not known whether or not there is friction between the plane and the sliding block. When this system is

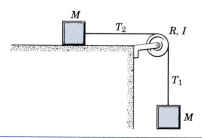

Figure 46 Problem 29.

released, it is found that the pulley turns through an angle θ in time t and the acceleration of the blocks is constant. (*a*) What is the angular acceleration of the pulley? (*b*) What is the acceleration of the two blocks? (*c*) What are the tensions in the upper and lower sections of the string? All answers are to be expressed in terms of M, I, R, θ, g, and t.

30. A wheel of mass M and radius of gyration k (see Problem 11) spins on a fixed horizontal axle passing through its hub. Assume that the hub rubs the axle of radius a at only the topmost point, the coefficient of kinetic friction being μ_k. The wheel is given an initial angular velocity ω_0. Assume uniform deceleration and find (*a*) the elapsed time and (*b*) the number of revolutions before the wheel comes to a stop.

31. In an Atwood's machine one block has a mass of 512 g and the other a mass of 463 g. The pulley, which is mounted in horizontal frictionless bearings, has a radius of 4.90 cm. When released from rest, the heavier block is observed to fall 76.5 cm in 5.11 s. Calculate the rotational inertia of the pulley.

32. A wheel in the form of a uniform disk of radius 23.0 cm and mass 1.40 kg is turning at 840 rev/min in frictionless bearings. To stop the wheel, a brake pad is pressed against the rim of the wheel with a radially directed force of 130 N. The wheel makes 2.80 revolutions in coming to a stop. Find the coefficient of friction between the brake pad and the rim of the wheel.

33. A stick 1.27 m long is held vertically with one end on the floor and is then allowed to fall. Find the speed of the other end when it hits the floor, assuming that the end on the floor does not slip.

34. A uniform spherical shell rotates about a vertical axis on frictionless bearings (Fig. 47). A light cord passes around the equator of the shell, over a pulley, and is attached to a small object that is otherwise free to fall under the influence of gravity. What is the speed of the object after it has fallen a distance h from rest?

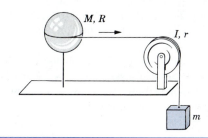

Figure 47 Problem 34.

35. A uniform steel rod of length 1.20 m and mass 6.40 kg has attached to each end a small ball of mass 1.06 kg. The rod is constrained to rotate in a horizontal plane about a vertical axis through its midpoint. At a certain instant, it is observed to be rotating with an angular speed of 39.0 rev/s. Because of axle friction, it comes to rest 32.0 s later. Compute, assuming a constant frictional torque, (*a*) the angular acceleration, (*b*) the retarding torque exerted by axle friction, (*c*) the energy dissipated by the axle friction, and (*d*) the number of revolutions executed during the 32.0 s. (*e*) Now suppose that the frictional torque is known not to be constant.

Which, if any, of the quantities (a), (b), (c), or (d) can still be computed without requiring any additional information? If such exists, give its value.

36. A rigid body is made of three identical thin rods fastened together in the form of a letter H (Fig. 48). The body is free to rotate about a horizontal axis that passes through one of the legs of the H. The body is allowed to fall from rest from a position in which the plane of the H is horizontal. What is the angular speed of the body when the plane of the H is vertical?

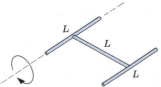

Figure 48 Problem 36.

37. A helicopter rotor blade is 7.80 m long and has a mass of 110 kg. (a) What force is exerted on the bolt attaching the blade to the rotor axle when the rotor is turning at 320 rev/min? (*Hint:* For this calculation the blade can be considered to be a point mass at the center of mass. Why?) (b) Calculate the torque that must be applied to the rotor to bring it to full speed from rest in 6.70 s. Ignore air resistance. (The blade cannot be considered to be a point mass for this calculation. Why not? Assume the distribution of a uniform rod.)

38. A tall chimney cracks near its base and falls over. Express (a) the radial and (b) the tangential linear acceleration of the top of the chimney as a function of the angle θ made by the chimney with the vertical. (c) Can the resultant linear acceleration exceed g? (d) The chimney cracks during the fall. Explain how this can happen. (See "More on the Falling Chimney," by Albert A. Bartlett, *The Physics Teacher,* September 1976, p. 351.)

39. The length of the day is increasing at the rate of about 1 ms/century. This is primarily due to frictional forces generated by movement of water in the world's shallow seas as a response to the tidal forces exerted by the Sun and Moon. (a) At what rate is the Earth losing rotational kinetic energy? (b) What is its angular acceleration? (c) What tangential force, exerted at latitudes 60°N and 60°S, is applied by the seas on the near-coastal seabed?

40. A uniform disk of radius R and mass M is spinning with angular speed ω_0. It is placed on a flat horizontal surface; the coefficient of kinetic friction between disk and surface is μ_k. (a) Find the frictional torque on the disk. (b) How long will it take for the disk to come to rest?

41. A car is fitted with an energy-conserving flywheel, which in operation is geared to the driveshaft so that it rotates at 237 rev/s when the car is traveling at 86.5 km/h. The total mass of the car is 822 kg, the flywheel weighs 194 N, and it is a uniform disk 1.08 m in diameter. The car descends a 1500-m long, 5.00° slope, from rest, with the flywheel engaged and no power supplied from the motor. Neglecting friction and the rotational inertia of the wheels, find (a) the speed of the car at the bottom of the slope, (b) the angular acceleration of the flywheel at the bottom of the slope, and (c) the

power being absorbed by the rotation of the flywheel at the bottom of the slope.

Section 12-6 Combined Rotational and Translational Motion

42. A solid sphere of radius 4.72 cm rolls up an inclined plane of inclination angle 34.0°. At the bottom of the incline the center of mass of the sphere has a translational speed of 5.18 m/s. (a) How far does the sphere travel up the plane? (b) How long does it take to return to the bottom? (c) How many rotations does the sphere make during the round trip?

43. A hoop rolling down an inclined plane of inclination angle θ keeps pace with a block sliding down the same plane. Show that the coefficient of kinetic friction between block and plane is given by $\mu_k = \frac{1}{2} \tan \theta$.

44. A hoop of radius 3.16 m has a mass of 137 kg. It rolls along a horizontal floor so that its center of mass has a speed of 0.153 m/s. How much work must be done on the hoop to stop it?

45. An automobile traveling 78.3 km/h has tires of 77.0 cm diameter. (a) What is the angular speed of the tires about the axle? (b) If the car is brought to a stop uniformly in 28.6 turns of the tires (no skidding), what is the angular acceleration of the wheels? (c) How far does the car advance during this braking period?

46. A 1040-kg car has four 11.3-kg wheels. What fraction of the total kinetic energy of the car is due to rotation of the wheels about their axles? Assume that the wheels have the same rotational inertia as disks of the same mass and size. Explain why you do not need to know the radius of the wheels.

47. A yo-yo (see Sample Problem 7) has a rotational inertia of 950 g·cm² and a mass of 120 g. Its axle radius is 3.20 mm and its string is 134 cm long. The yo-yo rolls from rest down to the end of the string. (a) What is its acceleration? (b) How long does it take to reach the end of the string? (c) If the yo-yo "sleeps" at the bottom of the string in pure rotary motion, what is its angular speed, in rev/s? (d) Repeat (c), but this time assume that the yo-yo was thrown down with an initial speed of 1.30 m/s.

48. A uniform sphere rolls down an incline. (a) What must be the incline angle if the linear acceleration of the center of the sphere is to be 0.133g? (b) For this angle, what would be the acceleration of a frictionless block sliding down the incline?

49. A homogeneous sphere starts from rest at the upper end of the track shown in Fig. 49 and rolls without slipping until it rolls off the right-hand end. If H = 60 m and h = 20 m and the track is horizontal at the right-hand end, determine the distance to the right of point A at which the ball strikes the horizontal base line.

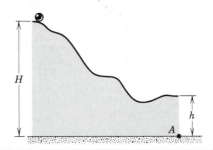

Figure 49 Problem 49.

50. A small solid marble of mass m and radius r rolls without slipping along the loop-the-loop track shown in Fig. 50, having been released from rest somewhere on the straight section of track. (*a*) From what minimum height above the bottom of the track must the marble be released in order that it just stay on the track at the top of the loop? (The radius of the loop-the-loop is R; assume $R \gg r$.) (*b*) If the marble is released from height $6R$ above the bottom of the track, what is the horizontal component of the force acting on it at point Q?

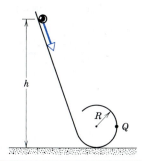

Figure 50   Problem 50.

51. A solid cylinder of length L and radius R has a weight W. Two cords are wrapped around the cylinder, one near each end, and the cord ends are attached to hooks on the ceiling. The cylinder is held horizontally with the two cords exactly vertical and is then released (Fig. 51). Find (*a*) the tension in each cord as they unwind and (*b*) the linear acceleration of the cylinder as it falls.

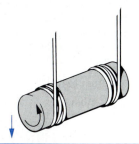

Figure 51   Problem 51.

52. A length L of flexible tape is tightly wound. It is then allowed to unwind as it rolls down a steep incline that makes an angle θ with the horizontal, the upper end of the tape being tacked down (Fig. 52). Show that the tape unwinds completely in a time $T = \sqrt{3L/g \sin \theta}$.

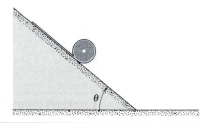

Figure 52   Problem 52.

53. Show that a cylinder will slip on an inclined plane of inclination angle θ if the coefficient of static friction between plane and cylinder is less than $\frac{1}{3} \tan \theta$.

54. A body is rolling horizontally without slipping with speed v. It then rolls up a hill to a maximum height h. If $h = 3v^2/4g$, what might the body be?

55. A uniform disk, of mass M and radius R, lies on one side initially at rest on a frictionless horizontal surface. A constant force $\mathbf{F}$ is then applied tangentially at its perimeter by means of a string wrapped around its edge. Describe the subsequent (rotational and translational) motion of the disk.

56. An apparatus for testing the skid resistance of automobile tires is constructed as shown in Fig. 53. The tire is initially motionless and is held in a light framework that is freely pivoted at points A and B. The rotational inertia of the wheel about its axis is 0.750 kg·m², its mass is 15.0 kg, and its radius is 30.0 cm. The tire is placed on the surface of a conveyor belt that is moving with a surface velocity of 12.0 m/s, such that AB is horizontal. (*a*) If the coefficient of kinetic friction between the tire and the conveyor belt is 0.600, what time will be required for the wheel to achieve its final angular velocity? (*b*) What will be the length of the skid mark on the conveyor surface?

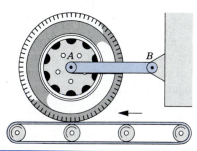

Figure 53   Problem 56.

57. A solid cylinder of radius 10.4 cm and mass 11.8 kg starts from rest and rolls without slipping a distance of 6.12 m down a house roof that is inclined at $27.0°$. (*a*) What is the angular speed of the cylinder about its center as it leaves the house roof? (*b*) The outside wall of the house is 5.16 m high. How far from the wall does the cylinder hit the level ground? See Fig. 54.

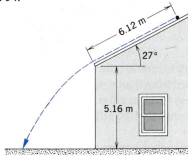

Figure 54   Problem 57.

58. A solid cylinder of mass 23.4 kg and radius 7.60 cm has a light thin tape wound around it. The tape passes over a light,

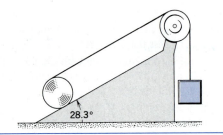

Figure 55   Problem 58.

frictionless pulley to an object of mass 4.48 kg, hanging vertically (see Fig. 55). The plane on which the cylinder moves is inclined 28.3° to the horizontal. Find (*a*) the linear acceleration of the cylinder down the incline and (*b*) the tension in the tape, assuming no slipping.

59. A student throws a stick of length L up into the air. At the moment it leaves her hand the speed of the stick's closest end is zero. The stick completes N turns just as it is caught by the student at the initial release point. Show that the height h that the center of mass rose is $h = \pi NL/4$.

60. A billiard ball is struck by a cue as in Fig. 56. The line of action of the applied impulse is horizontal and passes through the center of the ball. The initial velocity v_0 of the ball, its radius R, its mass M, and the coefficient of friction μ_k between the ball and the table are all known. How far will the ball move before it ceases to slip on the table?

Figure 56   Problem 60.

CHAPTER 13

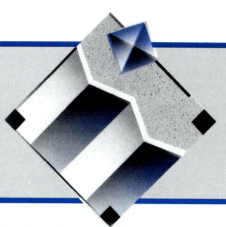

ANGULAR MOMENTUM

In Chapter 12 we discussed the dynamics of the rotational motion of a rigid body about an axis that is fixed in an inertial reference frame. We saw that the scalar relation $\Sigma \tau = I\alpha$, in which only external torque components along the axis of rotation were considered, was sufficient to solve dynamical problems in this special case.

In this chapter we continue this analysis and extend it to situations in which the axis of rotation may not *be fixed in an inertial reference frame. To solve these dynamical problems we develop and use a vector relation for rotational motion, which is analogous to the vector form of Newton's second law, $\mathbf{F} = d\mathbf{P}/dt$. We also introduce* angular momentum *and show its importance as a dynamical property of rotations.*

Finally, we show that, for systems on which no net external torque acts, the important law of conservation of angular momentum *can be applied.*

13-1 ANGULAR MOMENTUM OF A PARTICLE

We have found *linear momentum* to be useful in dealing with the translational motion of single particles or of systems of particles, including rigid bodies. For example, linear momentum is conserved in collisions. For a single particle the linear momentum is $\mathbf{p} = m\mathbf{v}$ (Eq. 19 of Chapter 9); for a system of particles it is $\mathbf{P} = M\mathbf{v}_{\mathrm{cm}}$ (Eq. 25 of Chapter 9), in which M is the total system mass and $\mathbf{v}_{\mathrm{cm}}$ is the velocity of the center of mass. In rotational motion, the analogue of linear momentum is called *angular momentum,* which we define below for the special case of a single particle. Later, we broaden the definition to include systems of particles, and we show that angular momentum is as useful a concept in rotational motion as linear momentum is in translational motion.

Consider a particle of mass m and linear momentum $\mathbf{p}$ at a position $\mathbf{r}$ relative to the origin O of an inertial reference frame; for convenience (see Fig. 1) we have chosen the plane defined by the vectors $\mathbf{p}$ and $\mathbf{r}$ to be the xy plane. We define the *angular momentum* $\mathbf{l}$ of the particle *with respect to the origin O* to be

$$\mathbf{l} = \mathbf{r} \times \mathbf{p}. \qquad (1)$$

Note that we must specify the origin O in order to define

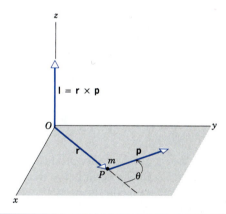

Figure 1 A particle of mass m, located at point P by the position vector $\mathbf{r}$, has a linear momentum $\mathbf{p} = m\mathbf{v}$. (For simplicity both $\mathbf{r}$ and $\mathbf{p}$ are assumed to lie in the xy plane.) Relative to the origin O, the particle has an angular momentum of $\mathbf{l} = \mathbf{r} \times \mathbf{p}$, which is parallel to the z axis in this case.

the position vector $\mathbf{r}$ in the definition of angular momentum.

Angular momentum is a vector. Its magnitude is given by

$$l = rp \sin \theta, \qquad (2)$$

where θ is the smaller angle between $\mathbf{r}$ and $\mathbf{p}$; its direction is normal to the plane formed by **r** and **p**. The sense is given by the right-hand rule: swing **r** into **p**, through the smaller angle between them, with the curled fingers of the right hand; the extended right thumb then points in the direction of **l** (parallel to the *z* axis in Fig. 1).

We can also write the magnitude of **l** either as

$$l = (r \sin \theta) p = p r_\perp, \tag{3a}$$

or as

$$l = r(p \sin \theta) = r p_\perp, \tag{3b}$$

in which $r_\perp (= r \sin \theta)$ is the component of **r** at right angles to the line of action of **p**, and $p_\perp (= p \sin \theta)$ is the component of **p** at right angles to **r**. Equation 3b shows that only the component of **p** perpendicular to **r** contributes to the angular momentum. When the angle θ between **r** and **p** is 0° or 180°, there is no perpendicular component ($p_\perp = p \sin \theta = 0$); then the line of action of **p** passes through the origin, and $r_\perp$ is also zero. In this case both Eqs. 3a and 3b show that the angular momentum *l* is zero.

We now derive an important relation between torque and angular momentum for a single particle. First, we differentiate Eq. 1 and obtain

$$\frac{d\mathbf{l}}{dt} = \frac{d}{dt}(\mathbf{r} \times \mathbf{p}). \tag{4}$$

The derivative of a vector product is taken in the same way as the derivative of an ordinary product, except that we must not change the order of the terms. We have

$$\frac{d\mathbf{l}}{dt} = \frac{d\mathbf{r}}{dt} \times \mathbf{p} + \mathbf{r} \times \frac{d\mathbf{p}}{dt}.$$

But $d\mathbf{r}/dt$ is the instantaneous velocity **v** of the particle, and **p** equals $m\mathbf{v}$. Making these substitutions into the first product on the right, we obtain

$$\frac{d\mathbf{l}}{dt} = (\mathbf{v} \times m\mathbf{v}) + \mathbf{r} \times \frac{d\mathbf{p}}{dt}. \tag{5}$$

Now $\mathbf{v} \times m\mathbf{v} = 0$, because the vector product of two parallel vectors is zero. Replacing $d\mathbf{p}/dt$ in the second product by the net force $\Sigma \mathbf{F}$ acting on the particle, we have

$$\frac{d\mathbf{l}}{dt} = \mathbf{r} \times \sum \mathbf{F}.$$

The right side of this equation is just the net torque $\Sigma \boldsymbol{\tau}$. We therefore obtain

$$\sum \boldsymbol{\tau} = \frac{d\mathbf{l}}{dt}, \tag{6}$$

which states that *the net torque acting on a particle is equal to the time rate of change of its angular momentum.* Both the torque $\boldsymbol{\tau}$ and the angular momentum **l** in this equation must be defined with respect to the same origin. Equation 6 is the rotational analogue of Eq. 20 of Chapter 9, $\Sigma \mathbf{F} = d\mathbf{p}/dt$, which states that the net *force* acting on a particle is equal to the time rate of change of its *linear* momentum.

Equation 6, like all vector equations, is equivalent to three scalar equations, namely,

$$\sum \tau_x = \frac{dl_x}{dt}, \quad \sum \tau_y = \frac{dl_y}{dt}, \quad \sum \tau_z = \frac{dl_z}{dt}. \tag{7}$$

Hence, the *x* component of the net external torque is given by the change with time of the *x* component of the angular momentum. Similar results hold for the *y* and *z* directions.

Sample Problem 1 A particle of mass *m* is released from rest at point *P* in Fig. 2, falling parallel to the (vertical) *y* axis. (*a*) Find the torque acting on *m* at any time *t*, with respect to origin *O*. (*b*) Find the angular momentum of *m* at any time *t*, with respect to this same origin. (*c*) Show that the relation $\Sigma \boldsymbol{\tau} = d\mathbf{l}/dt$ (Eq. 6) yields a correct result when applied to this familiar problem.

Solution (*a*) The torque is given by $\boldsymbol{\tau} = \mathbf{r} \times \mathbf{F}$, and its magnitude is

$$\tau = rF \sin \theta.$$

In this example $r \sin \theta = b$ and $F = mg$, so that

$$\tau = mgb = \text{a constant}.$$

Note that the torque is simply the product of the force *mg* times the moment arm *b*. The right-hand rule shows that $\boldsymbol{\tau}$ is directed perpendicularly into the figure.

(*b*) The angular momentum is given by Eq. 1, $\mathbf{l} = \mathbf{r} \times \mathbf{p}$. Its magnitude is, from Eq. 2,

$$l = rp \sin \theta.$$

In this example $r \sin \theta = b$ and $p = mv = m(gt)$, so that

$$l = mgbt.$$

The right-hand rule shows that **l** is directed perpendicularly into the figure, which means that **l** and $\boldsymbol{\tau}$ are parallel vectors. The

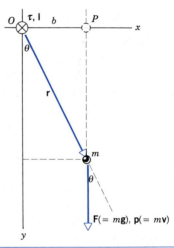

Figure 2 Sample Problem 1. A particle of mass *m* drops vertically from point *P*. The torque $\boldsymbol{\tau}$ and the angular momentum **l** with respect to the origin *O* are directed perpendicularly into the figure, as indicated by the symbol $\otimes$ at point *O*.

vector **l** changes with time in magnitude only, its direction always remaining the same in this case.

(*c*) Writing Eq. 6 in terms of magnitudes, we have

$$\tau = \frac{dl}{dt}.$$

Substituting the expression for τ and l from (*a*) and (*b*) above gives

$$mgb = \frac{d}{dt}(mgbt) = mgb,$$

which is an identity. Thus the relation $\tau = dl/dt$ yields correct results in this simple case. Indeed, if we cancel the constant b out of the first two terms above and if we substitute for gt the equivalent quantity v, we have

$$mg = \frac{d}{dt}(mv).$$

Since $mg = F$ and $mv = p$, this is the familiar result $F = dp/dt$. Thus, as we indicated earlier, relations such as $\tau = dl/dt$, though often vastly useful, are not new basic postulates of classical mechanics but are rather the reformulation of the Newtonian laws in the case of rotational motion.

Note that the values of τ and l depend on our choice of origin, that is on b. In particular, if $b = 0$, then $r = 0$ and $l = 0$.

13-2 SYSTEMS OF PARTICLES

So far we have discussed only single particles. To calculate the total angular momentum **L** of a *system of particles* about a given point, we must add vectorially the angular momenta of all the individual particles about this point. For a system containing N particles we then have

$$\mathbf{L} = \mathbf{l}_1 + \mathbf{l}_2 + \cdots + \mathbf{l}_N = \sum_{n=1}^{N} \mathbf{l}_n$$

in which the (vector) sum is taken over all particles in the system.

As time goes on, the total angular momentum **L** of the system about a fixed reference point (which we choose, as in our basic definition of **l** in Eq. 1, to be the origin of an inertial reference frame) may change. That is,

$$\frac{d\mathbf{L}}{dt} = \frac{d\mathbf{l}_1}{dt} + \frac{d\mathbf{l}_2}{dt} + \cdots = \sum_{n=1}^{N} \frac{d\mathbf{l}_n}{dt}.$$

For each particle, $d\mathbf{l}_n/dt = \boldsymbol{\tau}_n$, and making this substitution we obtain

$$\frac{d\mathbf{L}}{dt} = \sum \boldsymbol{\tau}_n.$$

That is, the time rate of change of the *total* angular momentum of a system of particles equals the net torque acting on the system.

Among the torques acting on the system will be (1) torques exerted on the particles of the system by internal forces between the particles and (2) torques exerted on the particles of the system by external forces. If Newton's third law holds in its so-called strong form, that is, if the forces between any two particles not only are equal and opposite but are also directed along the line joining the two particles, then the total internal torque is zero because the torque resulting from each internal action–reaction force pair is zero.

Hence the first source, the torque from internal forces, contributes nothing to the change in **L**. Only the second source (the torque from external forces) remains, and we can write

$$\sum \boldsymbol{\tau}_{\text{ext}} = \frac{d\mathbf{L}}{dt}, \qquad (8)$$

where $\sum \boldsymbol{\tau}_{\text{ext}}$ is the sum of the *external* torques acting on the system. In words, *the net external torque acting on a system of particles is equal to the time rate of change of the total angular momentum of the system.* The torque and the angular momentum must be calculated with respect to the same origin of an inertial reference frame. In situations in which no confusion is likely to arise, we drop the subscript on $\boldsymbol{\tau}_{\text{ext}}$ for convenience.

Equation 8 is the generalization of Eq. 6 to many particles. It holds whether the particles that make up the system are in motion relative to each other or whether they have fixed spatial relationships, as in a rigid body.

Equation 8 is the rotational analogue of Eq. 27 of Chapter 9, $\sum \mathbf{F}_{\text{ext}} = d\mathbf{P}/dt$, which tells us that for a system of particles (rigid body or not) the net external force acting on the system is equal to the time rate of change of its total linear momentum.

Let us extend further the analogy between the way a force changes linear momentum and the way a torque changes angular momentum. Suppose a force **F** acts on a particle moving with linear momentum **p**. We can resolve **F** into two components, as shown in Fig. 3: one component ($\mathbf{F}_{\parallel}$) is parallel to the (instantaneous) direction of **p**

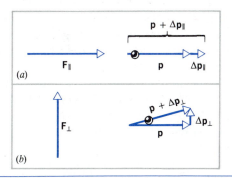

Figure 3 (*a*) When a force component $\mathbf{F}_{\parallel}$ acts parallel to the linear momentum **p** of a particle, the linear momentum changes by $\Delta\mathbf{p}_{\parallel}$, which is parallel to **p**. (*b*) When a force component $\mathbf{F}_{\perp}$ acts perpendicular to the linear momentum **p** of a particle, the linear momentum changes by $\Delta\mathbf{p}_{\perp}$, which is perpendicular to **p**. The particle now moves in the direction of the vector sum $\mathbf{p} + \Delta\mathbf{p}_{\perp}$.

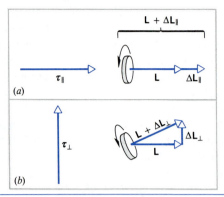

(a)

(b)

Figure 4   (*a*) When a torque component $\tau_\parallel$ acts parallel to the angular momentum **L** of a system, the angular momentum changes by $\Delta\mathbf{L}_\parallel$, which is parallel to **L**. (*b*) When a torque component $\tau_\perp$ acts perpendicular to the angular momentum **L** of a system, the angular momentum changes by $\Delta\mathbf{L}_\perp$, which is perpendicular to **L**. The axis of rotation now points in the direction corresponding to the vector sum $\mathbf{L} + \Delta\mathbf{L}_\perp$.

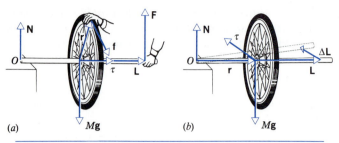

(a)

(b)

Figure 5   (*a*) A tangential force **f** on the rim of the wheel gives a torque τ (about the center of the wheel) along the axis of rotation, increasing the magnitude of the angular velocity of the wheel but leaving its direction unchanged. (*b*) When the end of the axle is released, the gravitational torque about the point *O* points into the paper, that is, perpendicular to the rotational axis, as in Fig. 4*b*. This torque changes the direction of the rotational axis, and the shaft of the wheel moves in the horizontal plane toward the position shown by the dashed line.

and another ($\mathbf{F}_\perp$) is perpendicular to **p**. In a small interval of time Δt, the force produces a change in momentum $\Delta\mathbf{p}$ determined according to $\mathbf{F} = \Delta\mathbf{p}/\Delta t$. Thus $\Delta\mathbf{p}$ is parallel to **F**. The component $\mathbf{F}_\parallel$ gives a change in momentum $\Delta\mathbf{p}_\parallel$ parallel to **p**, which adds to **p** and changes its magnitude but not its direction (see Fig. 3*a*). The perpendicular component $\mathbf{F}_\perp$, on the other hand, gives an increment $\Delta\mathbf{p}_\perp$ that changes the direction of **p** but, when $\Delta\mathbf{p}_\perp$ is small compared with **p**, leaves the magnitude of **p** unchanged (see Fig. 3*b*). An example of the latter is a particle moving in a circle at constant speed subject only to a centripetal force, which is always perpendicular to the tangential velocity.

The same analysis holds for the action of a torque, as shown in Fig. 4. In this case $\tau = \Delta\mathbf{L}/\Delta t$, and $\Delta\mathbf{L}$ must be parallel to τ. We once again resolve τ into two components, $\tau_\parallel$ parallel to **L** and $\tau_\perp$ perpendicular to **L**. The component of τ parallel to **L** changes the angular momentum in magnitude but not in direction (Fig. 4*a*). The component of τ perpendicular to **L** gives an increment $\Delta\mathbf{L}_\perp$ perpendicular to **L**, which changes the direction of **L** but not its magnitude (Fig. 4*b*). This latter condition is responsible for the motion of tops and gyroscopes, as we discuss in Section 13-5. Comparing Figs. 3 and 4, you can see the similarities between translational and rotational dynamics.

A further comparison between linear and rotational phenomena is that *no work is done* if (1) the force acts at right angles to the linear momentum (Fig. 3*b*), or (2) the torque acts at right angles to the angular momentum (Fig. 4*b*). In each case, the external agent causes no change in the kinetic energy, and the motion continues at the same linear or rotational speed.

An example of the application of Eq. 8 for rotational

dynamics is shown in Fig. 5. In Fig. 5*a*, one end of the axle of a spinning bicycle wheel rests freely on a post, and the other end is supported by a student's hand. The student pushes tangentially on the wheel at its rim, in order to make it spin faster. Taken about the center of the wheel, the torque exerted by the student is parallel to the angular momentum of the wheel, both vectors (τ and **L**) pointing toward the student. The result of this torque is an increase in the angular momentum of the wheel. In Fig. 5*b*, the student has released one support of the axle. Now we consider the torques about the remaining point of support. There are two forces acting, a normal force at the point of support, which gives no torque about that point, and the wheel's weight acting downward at the center of mass. The torque about point *O* due to the weight is perpendicular to **L**, and its effect is therefore to change the direction of **L**, as in Fig. 4*b*. However, since the direction of **L** is also the direction of the axle,* the effect of the (downward) force of gravity is to turn the axle to the side. The wheel will pivot sideways about the point of support. Try it! (If you don't have a freely mounted bicycle wheel handy, a toy gyroscope works as well.)

As we have derived it, Eq. 8 holds when τ and **L** are measured with respect to the origin of an inertial reference frame. We may well ask whether it still holds if we measure these two vectors with respect to an arbitrary point (a particular particle, say) in the moving system. In general, such a point would move in a complicated way as the body or system of particles translated, tumbled, and changed its configuration, and Eq. 8 would not apply to such a reference point. However, if the reference point is chosen to be the center of mass of the system, even though this point may be accelerating in our inertial reference

---

* This holds only if the axis of rotation is also an axis of symmetry of the body; see Section 13-3.

frame, then Eq. 8 does hold. (See Problem 8.) This is another remarkable property of the center of mass. Thus we can separate the general motion of a system of particles into the translational motion of its center of mass (Eq. 27 of Chapter 9) and the rotational motion about its center of mass (Eq. 8).

13-3   ANGULAR MOMENTUM AND ANGULAR VELOCITY

To introduce cases in which it is absolutely necessary to consider the vector nature of angular velocity, torque, and angular momentum, we first consider a simple example of a rotating particle that illustrates an instance in which the angular velocity and angular momentum are not parallel.

Figure 6a shows a single particle of mass m attached to a rigid, massless shaft by a rigid, massless arm of length r' perpendicular to the shaft. The particle moves in a circle of radius r', and we assume it does so at constant speed v. We imagine this experiment to be done in a region of negligible gravity, so that we need not consider the force of gravity acting on the particle. The only force that acts on the particle is the centripetal force exerted by the arm connecting the particle to the shaft.

The shaft is confined to the z axis by two thin ideal (frictionless) bearings. We let the lower bearing define the origin O of our coordinate system. The upper bearing, as we shall see, is necessary to prevent the shaft from wobbling about the z axis, which occurs when the angular velocity is not parallel to the angular momentum.

The angular velocity $\boldsymbol{\omega}$ of the particle points upward along (or, equivalently, parallel to) the z axis, as shown in Fig. 6b. This is consistent with the vector relationship $\mathbf{v} = \boldsymbol{\omega} \times \mathbf{r}$ (Eq. 16 of Chapter 11). No matter where the origin is chosen along the z axis, the angular velocity vector will be parallel to the axis. Its magnitude is similarly independent of the location of the origin, being given (from the cross product) by $v/(r \sin \theta) = v/r'$.

The angular momentum $\mathbf{l}$ of the particle with respect to the origin O of the reference frame is given by Eq. 1, or

$$\mathbf{l} = \mathbf{r} \times \mathbf{p},$$

where $\mathbf{r}$ and $\mathbf{p}\,(= m\mathbf{v})$ are shown in Fig. 6b. The vector $\mathbf{l}$ is perpendicular to the plane formed by $\mathbf{r}$ and $\mathbf{p}$, which means that $\mathbf{l}$ is not parallel to $\boldsymbol{\omega}$. Note (see Fig. 6c) that $\mathbf{l}$ has a (vector) component $\mathbf{l}_z$ that is parallel to $\boldsymbol{\omega}$, but it has another (vector) component $\mathbf{l}_\perp$ that is perpendicular to $\boldsymbol{\omega}$. Here is a case in which our analogy between linear and circular motion is not valid: $\mathbf{p}$ is always parallel to $\mathbf{v}$, but $\mathbf{l}$ is *not* always parallel to $\boldsymbol{\omega}$. If we choose our origin to lie in the plane of the circulating particle, then $\mathbf{l}$ *is* parallel to $\boldsymbol{\omega}$; otherwise, it is not.

Let us now consider the relationship between $\mathbf{l}_z$ and $\boldsymbol{\omega}$

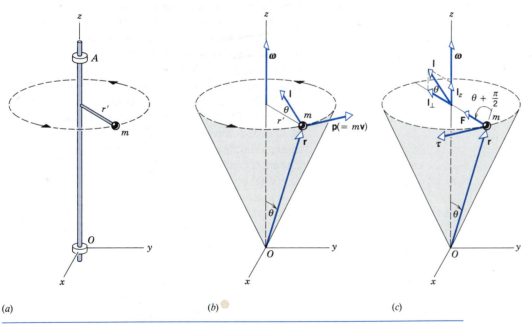

(a)   (b)   (c)

Figure 6   (a) A particle of mass m is attached through an arm of length r' to a shaft fixed by two bearings (at O and A) to rotate about the z axis. (b) The particle rotates with tangential speed v in a circle of radius r' about the z axis (the rods and bearings being omitted to simplify the drawing). The angular momentum $\mathbf{l} = \mathbf{r} \times \mathbf{p}$ about the origin O is shown. (c) For the particle to move in a circle, there must be a centripetal force $\mathbf{F}$ acting as shown, resulting in a torque $\boldsymbol{\tau}$ about O. For convenience, the angular momentum vector $\mathbf{l}$ and its components along and perpendicular to z are shown at the center of the circle.

for the rotating particle. From Fig. 6c, in which we have translated l to the center of the circle, we obtain

$$l_z = l \sin \theta = r(mv) \sin \theta = r(mr'\omega) \sin \theta,$$

using $v = r'\omega$. Substituting r' (the radius of the circle in which the particle moves) for the product $r \sin \theta$ gives

$$l_z = mr'^2\omega. \qquad (9)$$

Now mr'^2 is the rotational inertia I of the particle with respect to the z axis. Thus

$$l_z = I\omega. \qquad (10)$$

Note that the vector relation $l = I\omega$ (which is analogous to the linear relation $p = mv$) is *not* correct in this case, because l and ω do not point in the same direction.

Under what circumstances will the angular momentum and angular velocity point in the same direction? To illustrate, let us add another particle of the same mass m to the system, as shown in Fig. 7, by attaching another arm to the central shaft of Fig. 6a in the same location as the first arm but pointing in the opposite direction. The component $l_\perp$ due to this second particle will be equal and opposite to that of the first particle, and the two $l_\perp$ vectors sum to zero. The two l_z vectors point in the same direction, however, and add. Thus for this two-particle system, the total angular momentum L is parallel to ω.

We can now extend our system to a rigid body, made up of many particles. If the body is symmetric about the axis of rotation, by which we mean that for every mass element in the body there must be an identical mass element diametrically opposite the first element and at the same distance from the axis of rotation, then the body can be regarded as made up of sets of particle pairs of the kind we

have been discussing. Since L and ω are parallel for all such pairs, they are also parallel for rigid bodies that possess this kind of symmetry, which is called *axial* symmetry.

For such symmetrical rigid bodies L and ω are parallel and we can write in vector form

$$L = I\omega. \qquad (11)$$

Do not forget, however, that if L stands for the *total* angular momentum, then Eq. 11 applies *only* to bodies that have symmetry about the rotational axis. If L stands for the vector component of angular momentum along the rotational axis (that is, for L_z), then Eq. 11 holds for *any* rigid body, symmetrical or not, that is rotating about a fixed axis.

For symmetrical bodies (such as the two-particle system of Fig. 7), the upper bearing in Fig. 6a may be removed, and the shaft will remain parallel to the z axis. You can verify this by noting how easy it is to spin a symmetrical object such as a small top or a grinding wheel on a shaft held only between the thumb and forefinger of one hand. Any small asymmetry in the object requires the second bearing to keep the shaft in a fixed direction; the bearing must exert a torque on the shaft, which wobbles as the object rotates, as we discuss at the end of this section. This is particularly serious for objects that rotate at high speeds, such as turbine rotors. Although designed to be symmetrical, such rotors, because of small errors of blade placement, for example, may be slightly asymmetrical. They may be restored to symmetry by the addition or removal of metal at appropriate places; this is done by spinning the wheel in a special device such that the wobble can be measured quantitatively and the appropriate corrective measure computed and indicated automatically. In a similar manner, lead weights are placed at strategic points on automobile tire rims to reduce wobble at high speeds. In "balancing" a wheel of your car, your mechanic is really just determining that the angular momentum and angular velocity vectors of the wheel are parallel, thereby reducing the strain on the wheel bearings.

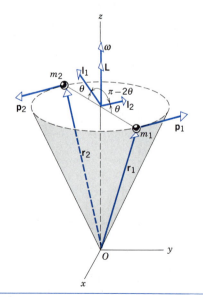

Figure 7   Two particles of mass m rotating as in Fig. 6 but at opposite ends of a diameter. The total angular momentum L of the two particles is in this case parallel to the angular velocity ω.

Sample Problem 2   Which is greater, the angular momentum of the Earth associated with its rotation on its axis or the angular momentum of the Earth associated with its orbital motion around the Sun?

Solution   For rotation on its axis, we treat the Earth as a uniform sphere ($I = \frac{2}{5}MR_E^2$). The angular velocity is $\omega = 2\pi/T$, where T is the rotational period (24 h = 8.64×10^4 s). The rotational angular momentum is then

$$L_{\text{rot}} = I\omega = \frac{2}{5}MR_E^2 \frac{2\pi}{T}$$

$$= \frac{2}{5}(5.98 \times 10^{24} \text{ kg})(6.37 \times 10^6 \text{ m})^2 \frac{2\pi}{8.64 \times 10^4 \text{ s}}$$

$$= 7.05 \times 10^{33} \text{ kg} \cdot \text{m}^2/\text{s}.$$

To calculate the orbital angular momentum, we need the rotational inertia of the Earth about an axis through the Sun. For this we can treat the Earth as a "particle," with angular momentum $L = R_{orb}p$, where R_{orb} is the radius of the orbit and p is the linear momentum of the Earth. The angular velocity is again given by $\omega = 2\pi/T$, where now T is the orbital period (1 y = 3.16×10^7 s). The orbital angular momentum is

$$L_{orb} = R_{orb}p = R_{orb}Mv = R_{orb}M(\omega R_{orb}) = MR_{orb}^2 \frac{2\pi}{T}$$

$$= (5.98 \times 10^{24} \text{ kg})(1.50 \times 10^{11} \text{ m})^2 \frac{2\pi}{3.16 \times 10^7 \text{ s}}$$

$$= 2.67 \times 10^{40} \text{ kg} \cdot \text{m}^2/\text{s}.$$

The orbital angular momentum is thus far greater than the rotational angular momentum.

The orbital angular momentum vector points at right angles to the plane of the Earth's orbit (Fig. 8), while the rotational angular momentum is inclined at an angle of 23.5° to the normal to the plane. Neglecting the very slow precession of the rotational axis, the two vectors remain constant in both magnitude and direction as the Earth moves in its orbit.

Sample Problem 3   In Sample Problem 5 of Chapter 12, find the acceleration of the falling block by direct application of Eq. 8 ($\tau = dL/dt$).

Solution   The system shown in Fig. 9, consisting of the disk of mass M and the block of mass m, is acted on by two external forces, the downward pull of gravity mg acting on m and the upward force exerted by the bearings of the shaft of the disk, which we take as our origin. (The tension in the cord is an internal force and does not act from the outside on the system of disk + block.) Only the first of these external forces exerts a torque about the origin, and its magnitude is $(mg)R$.

The angular momentum of the system about the origin O at any instant is

$$L = I\omega + (mv)R,$$

in which $I\omega$ is the angular momentum of the (symmetrical) disk and $(mv)R$ is the angular momentum (= linear momentum ×

moment arm) of the falling block about the origin. Both these contributions to L point in the same direction, namely, perpendicularly out of the plane of Fig. 9.

Applying $\tau = dL/dt$ (in scalar form) yields

$$(mg)R = \frac{d}{dt}(I\omega + mvR)$$

$$= I\left(\frac{d\omega}{dt}\right) + mR\left(\frac{dv}{dt}\right)$$

$$= I\alpha + mRa.$$

Since $a = \alpha R$ and $I = \frac{1}{2}MR^2$, this reduces to

$$mgR = (\tfrac{1}{2}MR^2)(a/R) + mRa$$

or

$$a = \frac{2mg}{M + 2m}.$$

This result is identical with the result of Sample Problem 5 of Chapter 12.

The Torque on a Particle Moving in a Circular Path (Optional)

The perhaps unexpected result that $\mathbf{l}$ and $\boldsymbol{\omega}$ are not parallel in the simple case shown in Fig. 6 may cause some concern. However, this result is consistent with the general relationship $\tau = d\mathbf{l}/dt$ for the torque acting on a single particle. The vector $\mathbf{l}$ is changing with time as the particle moves; the change is entirely in direction and not in magnitude. As the particle revolves, $\mathbf{l}_z$ remains constant in both magnitude and direction, but $\mathbf{l}_\perp$ changes its direction. This change in $\mathbf{l}_\perp$ must arise from the application of a torque. What is the source of this torque?

For the particle to move in a circle, it must be acted on by a centripetal force, as in Fig. 6c, provided by the supporting arm that connects the particle to the shaft. (We have neglected other external forces, such as gravity.) The only torque about O is provided by $\mathbf{F}$ and is given by

$$\boldsymbol{\tau} = \mathbf{r} \times \mathbf{F}.$$

The torque $\boldsymbol{\tau}$ is tangent to the circle (perpendicular to the plane

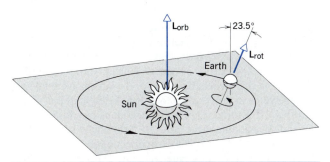

Figure 8   Sample Problem 2. The Earth rotates in an (assumed circular) orbit around the Sun, and it also rotates about its axis. The two angular momentum vectors are not parallel, because the Earth's rotational axis is inclined at an angle of 23.5° with respect to the normal to the plane of the orbit. The lengths of the vectors are not drawn to scale; L_{orb} should be greater than L_{rot} by a factor of about 4×10^6.

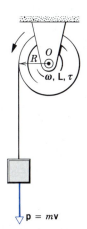

Figure 9   Sample Problem 3. The angular velocity, angular momentum, and net torque all point out of the page, as indicated by the symbol ⊙ at O.

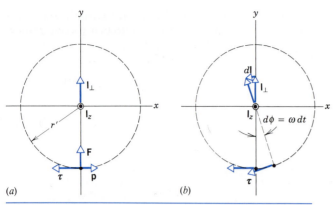

Figure 10 (a) A two-dimensional view of the plane of the rotating particle of Fig. 6. The z component of the angular momentum points out of the paper. (b) When the particle rotates through an angle $d\phi$, the vector component $\mathbf{l}_\perp$ in the plane changes by $d\mathbf{l}$. Note that $d\mathbf{l}$ is parallel to $\boldsymbol{\tau}$.

formed by $\mathbf{r}$ and $\mathbf{F}$) and in the direction shown in Fig. 6c, as you may verify from the right-hand rule.

Let us show that this torque satisfies the rotational form of Newton's second law, $\boldsymbol{\tau} = d\mathbf{l}/dt$. Figure 10a shows a two-dimensional view of the rotating particle, looking down along the z axis toward the xy plane. As the particle moves through the small angle $d\phi = \omega\,dt$ (Fig. 10b), the vector $\mathbf{l}_\perp$ changes by the small increment $d\mathbf{l}$. You can see from Fig. 10b that $d\mathbf{l}$ will always be parallel to $\boldsymbol{\tau}$, and so the directions of $d\mathbf{l}$ and $\boldsymbol{\tau}$ are consistent with $\boldsymbol{\tau} = d\mathbf{l}/dt$. We can also show that the magnitudes agree. The torque about O is, referring again to Fig. 6c,

$$\tau = rF \sin(\tfrac{1}{2}\pi + \theta) = rF \cos \theta.$$

In this case, $\mathbf{F}$ is the centripetal force and has magnitude $F = mv^2/r' = m\omega^2 r'$, where r' is the radius of the circular path ($r' = r \sin \theta$). Thus

$$\tau = m\omega^2 r^2 \sin \theta \cos \theta. \tag{12}$$

From Fig. 10b, $dl = l_\perp\, d\phi = l_\perp \omega\, dt$, from which we obtain

$$\frac{dl}{dt} = \omega l_\perp.$$

With $l = mvr$, then $l_\perp = mvr \cos \theta$. The tangential velocity v is $\omega r' = \omega r \sin \theta$, so

$$l_\perp = m\omega r^2 \sin \theta \cos \theta$$

and

$$\frac{dl}{dt} = \omega l_\perp = m\omega^2 r^2 \sin \theta \cos \theta. \tag{13}$$

Comparing Eqs. 12 and 13, we see that $\tau = dl/dt$, as expected.

Symmetrical Versus Asymmetrical Bodies

How does the situation differ for symmetrical and asymmetrical rotating bodies? Suppose the rod connecting the two particles in the symmetrical body of Fig. 7 were inclined at an arbitrary angle β with respect to the central shaft. Figure 11 shows the connecting rod, the shaft, and the two bearings (assumed frictionless) that holds the shaft along the z axis. The shaft rotates at a constant angular velocity ω about this axis, the vector $\boldsymbol{\omega}$ thus pointing along this axis. Experience tells us that such a system is

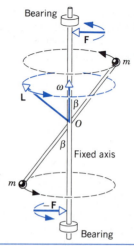

Figure 11 A rotating two-particle system, similar to Fig. 7, but with the axis of rotation making an angle β with the connecting rod. The angular momentum vector $\mathbf{L}$ rotates with the system, as do the forces $\mathbf{F}$ and $-\mathbf{F}$ exerted by the bearings.

"unbalanced" or "lop-sided," and if the connecting rod were not rigidly fastened to the vertical shaft near O, it would tend to move until the angle β became 90°, in which position the system would then be symmetrical about the shaft.

At the instant shown in Fig. 11, the upper particle is moving into the page at right angles to it, and the lower particle is moving out of the page at right angles to it. The linear momentum vectors of the two particles are therefore equal but opposite, and so are their position vectors with respect to O. Hence, by application of the right-hand rule in $\mathbf{r} \times \mathbf{p}$, we find that $\mathbf{l}$ is the same for each particle and that their sum, the total angular momentum vector $\mathbf{L}$ of the system, is as shown in the figure, at right angles to the connecting rod and in the plane of the page. Hence $\mathbf{L}$ and $\boldsymbol{\omega}$ are not parallel at this instant. As the system rotates, the angular momentum vector, while constant in magnitude, rotates around the fixed axis of rotation.

The rotation of $\mathbf{L}$ about the fixed axis of Fig. 11 is perfectly consistent with the fundamental relation $\boldsymbol{\tau} = d\mathbf{L}/dt$. The external torque on the entire system arises from the unbalanced sideways forces exerted by the bearings on the shaft and transmitted by the shaft to the connecting rod. At the instant shown in the figure, the upper particle would tend to move outward to the right. The shaft would be pulled to the right against the upper bearing, which in turn exerts a force $\mathbf{F}$ on the shaft that points to the left. Similarly, the lower particle tends to move outward to the left. The shaft would be pulled to the left against the lower bearing, which in turn exerts a force $-\mathbf{F}$ on the shaft that points to the right. The torque $\boldsymbol{\tau}$ about O as a result of these forces points perpendicularly out of the page, at right angles to the plane formed by $\mathbf{L}$ and $\boldsymbol{\omega}$, and in the right direction to account for the rotary motion of $\mathbf{L}$. (Compare with Fig. 10b, in which $\boldsymbol{\tau}$ was parallel to $d\mathbf{l}$ but perpendicular to $\mathbf{l}$.) Note that because $\boldsymbol{\tau}$ is perpendicular to $\boldsymbol{\omega}$, it does no work and hence does not change the kinetic energy of the rotating system. In the absence of friction, the system will spin forever. Friction in the bearings would give rise to a torque directed along the shaft (parallel to $\boldsymbol{\omega}$), which *would* do work on the system and change its kinetic energy.

The forces **F** and −**F** lie in the plane of Fig. 11 at the instant shown. As the system rotates, these forces, and therefore the torque τ, rotate with it, so that τ always remains at right angles to the plane formed by ω and **L**. The rotating forces **F** and −**F** cause a wobble in the upper and lower bearings. The bearings and their supports must be made strong enough to provide these forces. For a symmetrical rotating body there is no bearing wobble, and the shaft rotates smoothly. ∎

13-4 CONSERVATION OF ANGULAR MOMENTUM

In Eq. 8, we found that the time rate of change of the total angular momentum of a system of particles about a point fixed in an inertial reference frame (or about the center of mass) is equal to the net *external* torque acting on the system, that is,

$$\sum \tau_{\text{ext}} = \frac{d\mathbf{L}}{dt}. \tag{8}$$

If no net external torque acts on the system, then the angular momentum of the system does not change with time:

$$\frac{d\mathbf{L}}{dt} = 0 \quad \text{or} \quad \mathbf{L} = \text{a constant.} \tag{14}$$

Equation 14 is the mathematical statement of the principle of *conservation of angular momentum.*

When the net external torque acting on a system is zero, the total vector angular momentum of the system remains constant.

This is the third of the major conservation laws we have discussed. Along with conservation of energy and linear momentum, conservation of angular momentum is a general result valid for a wide range of systems. It holds true in both the relativistic limit and in the quantum limit. No exceptions have ever been found.

Like conservation of linear momentum in a system on which no net external *force* acts, conservation of angular momentum applies to the total angular momentum of a system of particles on which no net external *torque* acts. The angular momentum of individual particles in a system may change (just as the linear momentum of each particle in a collision may change), but the total remains constant.

Angular momentum is (like linear momentum) a *vector* quantity so that Eq. 14 is equivalent to three scalar equations, one for each coordinate direction through the reference point. Conservation of angular momentum therefore supplies us with three conditions on the motion of a system to which it applies. Any component of the angular momentum will be constant if the corresponding component of the torque is zero; it might be the case that

only one of the three components of torque is zero, which means that only one component of the angular momentum will be constant, the other components changing as determined by the corresponding torque components.

For a system consisting of a rigid body rotating about an axis (the z axis, say) that is fixed in an inertial reference frame, we have

$$L_z = I\omega, \tag{15}$$

where L_z is the (scalar) component of the angular momentum along the rotation axis and I is the rotational inertia for this same axis. It is possible for the rotational inertia I of a rotating body to change (from I_i to I_f) by rearrangement of its parts. If no net external torque acts, then L_z must remain constant and, if I does change, there must be a compensating change in ω from ω_i to ω_f. The principle of conservation of angular momentum in this case is expressed as

$$I_i\omega_i = I_f\omega_f = \text{a constant.} \tag{16}$$

Equation 16 holds not only for rotation about a fixed axis but also for rotation about an axis through the center of mass of a system that moves so that the axis always remains parallel to itself (see the discussion at the beginning of Section 12-6).

Conservation of angular momentum is a principle that regulates a wide variety of physical processes, from the subatomic world (see Section 13-6) to the motion of acrobats, divers, and ballet dancers, to the contraction of stars that have run out of fuel, and to the condensation of galaxies. The following examples show some of these applications.

The Spinning Skater

A spinning ice skater pulls her arms close to her body to spin faster and extends them to spin slower. When she does this, she is applying Eq. 16. Another application of this principle is illustrated in Fig. 12, which shows a student sitting on a stool that can rotate freely about a vertical axis. Let the student extend his arms holding the weights, and we will set him into rotation at an angular velocity ω_i. His angular momentum vector **L** lies along the vertical axis in the figure.

The system consisting of student + stool + weights is an isolated system on which no external vertical torque acts. The vertical component of angular momentum must therefore be conserved.

When the student pulls his arms (and the weights) closer to his body, the rotational inertia of his system is reduced from its initial value I_i to a smaller value I_f, because the weights are now closer to the axis of rotation. His final angular velocity, from Eq. 16, is $\omega_f = \omega_i (I_i/I_f)$, which is greater than his initial angular velocity (because $I_f < I_i$), and the student rotates faster. To slow down, he need only extend his arms again.

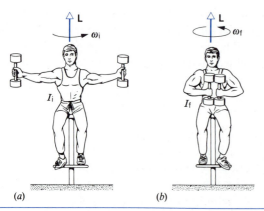

(a)   (b)

Figure 12   (a) In this configuration, the system (student + weights) has a larger rotational inertia and a smaller angular velocity. (b) Here the student has pulled the weights inward, giving a smaller rotational inertia and hence a larger angular velocity. The angular momentum **L** has the same value in both situations.

Does the kinetic energy of the system change? If so, what is the source of the work that changes the kinetic energy?

The Springboard Diver*

Figure 13a shows a diver leaving the springboard. As she jumps, she pushes herself slightly forward so that she acquires a small rotational speed, just enough to carry her head-first into the water as her body rotates through one-half revolution during the arc.

While she is in the air, no external torques act on her to change her angular momentum about her center of mass. (The only external force, gravity, acts *through* her center of mass and thus produces no torque about that point. We neglect air resistance, which could produce a net torque and change her angular momentum.) When she pulls her body into the *tuck position,* she lowers her rotational inertia, and therefore according to Eq. 16 her angular velocity must increase. The increased angular velocity enables her to complete $1\frac{1}{2}$ revolutions where she had previously completed only one-half (Fig. 13b). At the end of the dive, she pulls back out into the *layout position* and slows her angular speed as she enters the water.

The Rotating Bicycle Wheel

Figure 14a shows a student seated on a stool that is free to rotate about a vertical axis. The student holds a bicycle wheel that has been set spinning. When the student in-

---

* See "The Mechanics of Swimming and Diving," by R. L. Page, *The Physics Teacher,* February 1976, p. 72; "The Physics of Somersaulting and Twisting," by Cliff Frohlich, *Scientific American,* March 1980, p. 155.

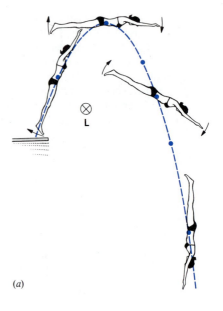

(a)

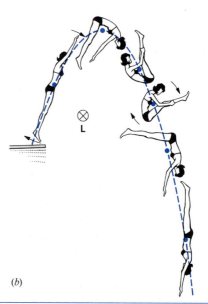

(b)

Figure 13   (a) A diver leaves the springboard in such a way that the springboard imparts to her an angular momentum **L**. She rotates about her center of mass (indicated by the dot) by one-half revolution as the center of mass follows the parabolic trajectory. (b) By entering the tuck position, she reduces her rotational inertia and thus increases her angular velocity, enabling her to make $1\frac{1}{2}$ revolutions. The external forces and torques on her are the same in (a) and (b), as indicated by the constant value of the angular momentum **L**.

verts the spinning wheel, the stool begins to rotate (Fig. 14b).

No net vertical torque acts on the system consisting of student + stool + wheel, and therefore the vertical component of the total angular momentum of the system

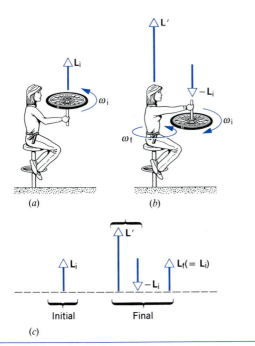

Figure 14 (*a*) A student holds a rotating bicycle wheel. The total angular momentum of the system is L_i. (*b*) When the wheel is inverted, the student begins to rotate. (*c*) The total final angular momentum must be equal to the initial angular momentum.

must remain constant. Initially, the wheel is rotating with angular momentum L_i upward, which is the total for the system. When the wheel is turned over, the vertical component of the wheel's angular momentum is now $-L_i$, but the vertical component of the *total* angular momentum must remain constant at $+L_i$. The student + stool must therefore acquire an angular momentum of $L' = +2L_i$, so that the final angular momentum of $+2L_i - L_i$ remains equal to the initial angular momentum. If I_s is the rotational inertia of the student + stool, the rotational speed will be $\omega_f = 2L_i/I_s$.

We can also consider this situation from the standpoint of two separate systems, one being the wheel and the other being the student + stool. Neither of these systems is now isolated: the student's hand forms the connection between them. When the student attempts to invert the wheel, she must apply a torque to change the wheel's angular momentum. The force she exerts on the wheel to produce that torque is returned by the wheel as a reaction force on her, by Newton's third law. This external force on the system of student + stool causes that system to rotate. In this view the student exerts an external torque on the wheel to change its angular momentum, while the wheel exerts a torque on the student to change her angular momentum. If we consider the complete system consisting of student + stool + wheel, as we did above, this torque is an internal torque which did not enter into our calcula-

tions. Whether we consider the torque as internal or external depends on how we define our system.

The Stability of Spinning Objects

Consider again Fig. 3*b*. An object moving with linear momentum $\mathbf{p} = M\mathbf{v}$ has a *directional stability*; a deflecting force provides the impulse corresponding to a sideways momentum increment $\Delta \mathbf{p}_\perp$, and as a result the direction of motion is changed by an angle $\theta = \tan^{-1}(\Delta p_\perp/p)$. The larger the momentum p, the smaller is the angle θ. The same deflecting force is less effective in diverting an object with large linear momentum than it is in diverting an object with small linear momentum.

Angular momentum provides an object with *orientational stability* in much the same way. A spinning object has a certain angular momentum $\mathbf{L}$. A torque τ perpendicular to $\mathbf{L}$ changes the direction of $\mathbf{L}$, and therefore the direction of the axis of rotation, by an angle $\theta = \tan^{-1}(\Delta L_\perp/L)$. Once again, the larger the angular momentum L, the less successful a given torque will be in changing the direction of the axis of the spinning object.

When we give an object rotational angular momentum about a symmetry axis, we in effect stabilize its orientation and make it more difficult for external forces to change its orientation. There are many common examples of this effect. A riderless bicycle given a slight push is able to remain upright for a far longer distance than we might expect. In this case it is the angular momentum of the spinning wheels that gives the stability. Minor bumps and curves of the roadway, which might otherwise topple or deflect a nonrotating object balanced on so narrow a base as a bicycle tire, have less effect in this case because of the tendency of the angular momentum of the wheels to fix their orientation.*

A football is thrown for a long forward pass such that it rotates about an axis that is roughly parallel to its translational velocity. This stabilizes the orientation of the football and keeps it from tumbling, which makes it possible to throw more accurately and catch more effectively. It also keeps the smallest profile of the football in the forward direction, thereby minimizing air resistance and increasing the range.

It is important to stabilize the orientation of a satellite, particularly if it is using its thrusters to move to a specific orbital position (Fig. 15). The orientation might be changed, for example, by friction from the thin residual atmosphere at orbital altitudes, by the solar wind (a beam of charged particles from the Sun), or by impacts from tiny meteoroids. To reduce the effects of such encounters, the craft is made to spin about an axis, thereby stabilizing its orientation.

---

* See "The Stability of the Bicycle," by David E. H. Jones, *Physics Today*, April 1970, p. 34.

Figure 15 Deployment of the Morelos-D satellite, a communications satellite for Mexico, from the bay of the space shuttle on November 17, 1985. The satellite is made to spin about its central axis (the vertical axis in this photo) to stabilize its orientation in space as it makes its way upward to its geosynchronous orbit.

Collapsing Stars

Most stars rotate, as our Sun does. It turns once on its axis every month or so. (The Sun is a ball of gas and doesn't rotate quite like a rigid body; the regions near the poles have a rotational period of about 37 days, while the equator rotates once every 26 days.) The Sun is kept from collapsing by *radiation pressure,* in essence the effect of impulsive collisions of the emerging radiation with the atoms of the Sun. When the Sun's nuclear fuel is used up, the radiation pressure will vanish, and the Sun will begin to collapse, its density correspondingly increasing. At some point the density will become so great that the atoms simply cannot be crowded any closer together, and the collapse will be halted.

In stars about 1.4 times as massive as the Sun, however,

the gravitational force is so strong that the atoms cannot prevent further collapse. The atoms are in effect crushed by gravity, and the collapse continues until the nuclei are touching one another. The star has in effect become one giant atomic nucleus; it is called a *neutron star.* The radius of a neutron star of about 1.5 solar masses is 11 km.

Suppose the star began its collapse like our Sun, rotating once every month. The forces during the collapse are clearly internal forces, which cannot change the angular momentum. The final angular velocity is therefore related to the initial angular velocity by Eq. 16: $\omega_f = \omega_i (I_i/I_f)$. The ratio of the rotational inertias will be the same as the ratio of the squares of the radii: $I_i/I_f = r_i^2/r_f^2$. If the initial radius were about the same as the Sun's (about 7×10^5 km), then

$$I_i/I_f = r_i^2/r_f^2 = (7 \times 10^5 \text{ km})^2/(11 \text{ km})^2 = 4 \times 10^9.$$

That is, its rotational speed goes from once per month to 4×10^9 per month, or more than 1000 revolutions per second!

Neutron stars can be observed from Earth, because (again like the Sun) they have magnetic fields that trap electrons, and the electrons are accelerated to very high tangential speeds as the star rotates. Such accelerated electrons emit radiation, which we see on Earth somewhat like a searchlight beacon as the star rotates. These sharp pulses of radiation earned rotating neutron stars the name *pulsars.* A sample of the radiation observed from a pulsar is shown in Fig. 16.

Conservation of angular momentum applies to a wide variety of astrophysical phenomena. The rotation of our galaxy, for example, is a result of the much slower initial rotation of the gas cloud from which the galaxy condensed; the rotation of the Sun and the orbits of the planets were determined by the original rotation of the material that formed our solar system.

Sample Problem 4 A 120-kg astronaut, carrying out a "space walk," is tethered to a spaceship by a fully extended cord 180 m long. An unintended operation of the propellant pack causes the astronaut to acquire a small tangential velocity of 2.5 m/s. To return to the spacecraft, the astronaut begins pulling along the tether at a slow and constant rate. With what force must the

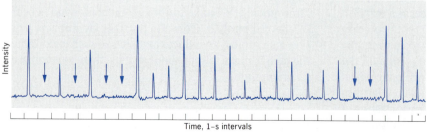

Time, 1–s intervals

Figure 16 Electromagnetic pulses received on Earth from a rapidly rotating neutron star. The vertical arrows suggest pulses too weak to be detected. The interval between pulses is remarkably constant, being equal to 1.187,911,164 s.

astronaut pull at distances of (*a*) 50 m and (*b*) 5 m from the spacecraft? What will be the astronaut's tangential speed at these points?

Solution   No external torques act on the astronaut, so that conservation of angular momentum holds. That is, the astronaut's initial angular momentum relative to the spaceship as origin $(Mv_i r_i)$ when beginning to pull on the tether must equal the angular momentum (Mvr) at any point in the motion. Thus

$$Mvr = Mv_i r_i$$

or

$$v = \frac{v_i r_i}{r}.$$

The centripetal force at any stage is given by

$$F = \frac{Mv^2}{r} = \frac{Mv_i^2 r_i^2}{r^3}.$$

Initially, the required centripetal force is

$$F = \frac{(120 \text{ kg})(2.5 \text{ m/s})^2}{180 \text{ m}} = 4.2 \text{ N} \quad \text{(about 1 lb)}.$$

(*a*) When the astronaut is 50 m from the spacecraft, the tangential speed is

$$v = \frac{(2.5 \text{ m/s})(180 \text{ m})}{50 \text{ m}} = 9.0 \text{ m/s},$$

and the centripetal force is

$$F = \frac{(120 \text{ kg})(2.5 \text{ m/s})^2(180 \text{ m})^2}{(50 \text{ m})^3} = 194 \text{ N} \quad \text{(about 44 lb)}.$$

(*b*) At 5 m from the ship, the speed goes up by a factor of 10 to 90 m/s, while the force increases by a factor of 10^3 to 1.94×10^5 N, or about 22 tons! It is clear that the astronaut cannot exert such a large force to return to the spacecraft. Even if the astronaut were being pulled toward the ship by a winch from within the spacecraft, the tether could not withstand such a large tension; at some point it would break and the astronaut would go shooting into space with whatever the tangential speed was at the time the tether broke. Moral: Space-walking astronauts should avoid acquiring tangential velocity. What could the astronaut do to move safely back to the ship?

Sample Problem 5   A turntable consisting of a disk of mass 125 g and radius 7.2 cm is spinning with an angular velocity of 0.84 rev/s about a vertical axis (Fig. 17*a*). An identical, initially nonrotating disk is suddenly dropped onto the first. The friction between the two disks causes them eventually to rotate at the same speed. A third identical nonrotating disk is then dropped onto the combination, and eventually all three are rotating together (Fig. 17*b*). (*a*) What is the rotational angular velocity of the combination? (*b*) How much rotational kinetic energy is lost to friction? (*c*) A motor driving the first disk must restore the angular velocity of the combination to its original value within one revolution. How much constant torque must the motor exert?

Solution   (*a*) This problem is the rotational analogue of the completely inelastic collision. There is no net vertical external

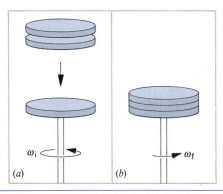

Figure 17   Sample Problem 5. (*a*) A disk is spinning with initial angular velocity ω_i. (*b*) Two identical disks, neither of which is initially rotating, are dropped onto the first, and the entire system then rotates with angular velocity ω_f.

torque, so the vertical component of angular momentum is constant. The frictional force between the disks is an internal force, which cannot change the angular momentum. Thus Eq. 16 applies, and we can write

$$I_i \omega_i = I_f \omega_f$$

$$\omega_f = \omega_i (I_i / I_f).$$

Without doing any detailed calculations, we know that the rotational inertia of three identical disks about their common axis will be three times the rotational inertia of a single disk. Thus $I_i / I_f = \frac{1}{3}$ and

$$\omega_f = (0.84 \text{ rev/s})(\tfrac{1}{3}) = 0.28 \text{ rev/s}.$$

(*b*) The rotational inertia of a disk about its axis is $\frac{1}{2}MR^2$, so for each disk,

$$I = \tfrac{1}{2}(0.125 \text{ kg})(0.072 \text{ m})^2 = 3.24 \times 10^{-4} \text{ kg} \cdot \text{m}^2.$$

The initial rotational kinetic energy is

$$K_i = \tfrac{1}{2}I\omega_i^2$$
$$= \tfrac{1}{2}(3.24 \times 10^{-4} \text{ kg} \cdot \text{m}^2)(2\pi \text{ rad/rev} \times 0.84 \text{ rev/s})^2$$
$$= 4.51 \times 10^{-3} \text{ J}.$$

We can take a shortcut in figuring the final kinetic energy, because we know the final rotational inertia goes up by a factor of 3, while the final angular velocity goes down by a factor of $\frac{1}{3}$. Because the kinetic energy depends on the square of the angular velocity, we have

$$K_f = K_i \times 3 \times (\tfrac{1}{3})^2 = (\tfrac{1}{3})(4.51 \times 10^{-3} \text{ J})$$
$$= 1.50 \times 10^{-3} \text{ J}.$$

The change in kinetic energy is

$$\Delta K = K_f - K_i = (1.50 \times 10^{-3} \text{ J}) - (4.51 \times 10^{-3} \text{ J})$$
$$= -3.01 \times 10^{-3} \text{ J}.$$

The minus sign indicates that kinetic energy is lost.

(*c*) To restore the initial angular velocity, the motor will have to increase ω from 0.28 rev/s back to 0.84 rev/s, that is, by a factor of 3. That means the kinetic energy must increase by a factor of $3^2 = 9$, from 1.50×10^{-3} J to 13.5×10^{-3} J. The change in kinetic energy, which is equal to the work done by the motor, is

$$\Delta K = 13.5 \times 10^{-3} \text{ J} - 1.50 \times 10^{-3} \text{ J} = 12.0 \times 10^{-3} \text{ J}.$$

For rotational motion, the work is given by $W = \tau\phi$, in which ϕ ($= 2\pi$ rad in this case) is the angular displacement of the rotating body through which the torque must be maintained. Thus

$$\tau = \frac{W}{\phi} = \frac{\Delta K}{\phi} = \frac{12.0 \times 10^{-3} \text{ J}}{2\pi \text{ rad}}$$
$$= 1.91 \times 10^{-3} \text{ N} \cdot \text{m}.$$

13-5 THE SPINNING TOP*

A spinning top provides us with what is perhaps the most familiar example of the phenomenon shown in Fig. 4b, in which a lateral torque changes the direction but not the magnitude of an angular momentum. Figure 18a shows a top spinning about its axis. The bottom point of the top is assumed to be fixed at the origin O of our inertial reference frame. We know from experience that the axis of this rapidly spinning top will move slowly about the vertical axis. This motion is called *precession*, and it arises from the configuration illustrated in Fig. 4b, with gravity supplying the external torque.

Figure 18b shows a simplified diagram, with the top replaced by a particle of mass M located at the top's center of mass. The gravitational force Mg gives a torque about O of magnitude

$$\tau = Mgr \sin \theta. \qquad (17)$$

The torque, which is perpendicular to the axis of the top

---

* See "The Amateur Scientist: The Physics of Spinning Tops, Including Some Far-Out Ones," by Jearl Walker, *Scientific American*, March 1981, p. 185.

and therefore perpendicular to **L** (Fig. 18c), can change the direction of **L** but not its magnitude. The change in **L** in a time Δt is given by

$$\Delta \mathbf{L} = \boldsymbol{\tau} \, \Delta t \qquad (18)$$

and is in the same direction as $\boldsymbol{\tau}$, that is, perpendicular to **L**. The effect of $\boldsymbol{\tau}$ is therefore to change **L** to $\mathbf{L} + \Delta \mathbf{L}$, a vector of the same length as **L** but pointing in a different direction. (We assume that the top rotates so rapidly that **L** is large, and thus $\mathbf{L} \gg \Delta \mathbf{L}$.)

If the top has axial symmetry, then the angular momentum will be along the axis of rotation of the top. As **L** changes direction, the axis changes direction too. The tip of the **L** vector and the axis of the top trace out a circle about the z axis, as shown in Fig. 18a. This motion is the precession of the top.

In a time Δt, the axis rotates through an angle $\Delta\phi$ (see Fig. 18d), and thus the angular speed of precession ω_P is

$$\omega_P = \frac{\Delta\phi}{\Delta t}. \qquad (19)$$

From Fig. 18d, we see that

$$\Delta\phi = \frac{\Delta L}{L \sin \theta} = \frac{\tau \, \Delta t}{L \sin \theta}. \qquad (20)$$

Thus

$$\omega_P = \frac{\Delta\phi}{\Delta t} = \frac{\tau}{L \sin \theta} = \frac{Mgr \sin \theta}{L \sin \theta} = \frac{Mgr}{L}. \qquad (21)$$

The precessional speed is inversely proportional to the angular momentum; the faster the top is spinning, the slower it will precess.

The precessional motion takes place about the z axis, and thus the vector $\boldsymbol{\omega}_P$ is in the z direction. You should be able to show that the following vector equation gives the proper relationship among the magnitudes *and* directions of the dynamical variables in this calculation:

$$\boldsymbol{\tau} = \boldsymbol{\omega}_P \times \mathbf{L}. \qquad (22)$$

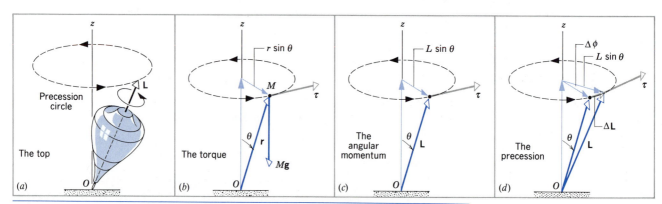

Figure 18 (a) A spinning top precesses about a vertical axis. (b) The weight of the top exerts a torque about the point of contact with the floor. (c) The torque is perpendicular to the angular momentum vector. (d) The torque changes the direction of the angular momentum vector, causing precession.

Can you write a similar vector equation for the corresponding case of a particle moving in a circle at constant speed under the influence of a centripetal force?

13-6   QUANTIZATION OF ANGULAR MOMENTUM   *(Optional)*

In Section 8-8 we discussed the quantization of energy, which restricted the emission or absorption of energy to take place only in discrete bundles or quanta. In the microscopic world of atomic or subatomic systems, we cannot change the energy by an arbitrary amount, but only by amounts of a predetermined size. These quanta are so tiny that we are unable to observe this discrete structure in the energy changes of systems of ordinary size.

Angular momentum is similarly quantized. We shall develop this concept at greater length, providing experimental evidence and theoretical support, in Chapter 51 of the extended version of this text when we discuss the structure of atoms. For now, we merely introduce some of the general ideas and show how they relate to the properties associated with angular momentum that we have developed in this chapter.

The quantized changes in the rotational motion of a system are restricted to taking place in units given by integer multiples of a fundamental constant:

$$\Delta L = n(h/2\pi) \qquad (n = 1, 2, 3, \dots). \qquad (23)$$

Here h is the Planck constant, with a value of 6.63×10^{-34} J·s. This basic unit is an extremely small quantity of angular momentum. A phonograph record, for example, spins relatively slowly but has an angular momentum of the order of 10^{32} units of $h/2\pi$. As we fine-tune the speeds of our turntables, we are certainly not able to observe these discrete single jumps on a scale of 1 part in 10^{32}!

Equation 23 for the quantization of angular momentum applies to the motion of electrons in atoms in their orbits about the nucleus. This system has *orbital angular momentum,* which must remain constant during the orbit, because the force between the electron and the nucleus is an internal force in the system and therefore cannot change its angular momentum. External forces, such as electric or magnetic fields, can cause the electron to jump to another orbit, where its angular momentum might have a different value, but the change in L must be an integral multiple of $h/2\pi$, as required by Eq. 23. The orbital angular momentum thus serves as a convenient and useful label of electronic orbits in atoms.

Experiments done in the 1920s indicated that electrons in atoms have another kind of angular momentum, which could not be accounted for by the orbital motion. This new kind of angular momentum, which is known as *intrinsic angular momentum,* is a characteristic property of the particle itself and not a result of its particular state of motion. A useful (but strictly incorrect) way of visualizing intrinsic angular momentum is in terms of the particle spinning on its axis; for this reason, intrinsic angular momentum is often called "spin" and is indicated by the symbol **s**.

The electron has an intrinsic angular momentum of $\frac{1}{2}(h/2\pi)$. This means that, relative to any z axis we may choose to define, the z component of angular momentum must be either

$$s_z = +\tfrac{1}{2}(h/2\pi) \quad \text{or} \quad s_z = -\tfrac{1}{2}(h/2\pi).$$

Note that the difference between these two possibilities, which might correspond to a *change* in the direction of the intrinsic angular momentum of an electron, is $h/2\pi$, consistent with Eq. 23.

Usually the intrinsic angular momentum is expressed by giving the *spin quantum number,* which is the intrinsic angular momentum in units of $h/2\pi$; the electron thus has a spin quantum number of $\frac{1}{2}$. The proton and neutron also have spin quantum numbers of $\frac{1}{2}$. The photon (the quantized bundle of electromagnetic radiation) has a spin quantum number of 1. All elementary particles can be characterized by their spin quantum number, which is considered a fundamental property of the particle along with its mass and electric charge.

An important application of the principle of conservation of quantized angular momentum occurs in the effect known as *nuclear magnetic resonance.* Consider the proton (the nucleus of the hydrogen atom), with its spin quantum number of $\frac{1}{2}$. A representation of the intrinsic angular momentum of the proton in a particular orientation is shown in Fig. 19a. The z component of the angular momentum is $s_z = +\frac{1}{2}(h/2\pi)$. If we expose protons to radiation of the proper energy, the absorption of an electromagnetic photon (spin = 1, angular momentum = $h/2\pi$) can change the z component of the angular momentum of the proton by one unit, from $+\frac{1}{2}(h/2\pi)$ to $-\frac{1}{2}(h/2\pi)$, as in Fig. 19b. The addition of the z components of **s** and **L** in Fig. 19c shows how the initial proton spin s_z and the photon angular momentum L_z add to give the (inverted) final proton spin s_z'. Figure 19c is yet another example of conservation of angular momentum, with the initial angular momentum $(\mathbf{s} + \mathbf{L})$ equal to the final angular momentum $(\mathbf{s}')$ in the absence of external torque.

In nuclear magnetic resonance (NMR), a static magnetic field

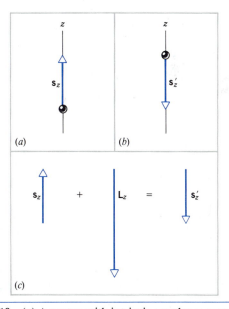

Figure 19   (a) A proton with intrinsic angular momentum (spin) **s** has a component s_z along the z axis. (b) After absorbing a photon, the z component of the spin is reversed. (c) The z component of the initial angular momentum, equal to $+\frac{1}{2}$ unit, adds to the z component of the angular momentum of the photon, equal to -1 unit, giving a resultant of $-\frac{1}{2}$ unit.

in the z direction is used to line up the proton spins with the z axis, as in Fig. 19a. A separate time-varying electromagnetic field at radio frequencies provides photons of just the right energy to be absorbed and causes the proton spins to be inverted.

Because the human body is mostly water, which is rich in hydrogen, the absorption of this electromagnetic radiation provides a way of forming an image of the body's internal organs (Fig. 20). The electromagnetic radiation in the form of radio waves is thought to pose little hazard to the body; x rays, which are also used to form images, have a much greater potential for causing damage to the body. Such *magnetic resonance imaging* may largely replace x-ray photographs as a diagnostic technique. ■

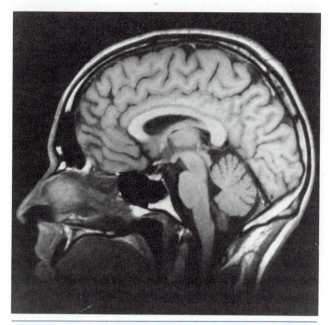

Figure 20 Magnetic resonance imaging (MRI) view of a patient's skull.

13-7 ROTATIONAL DYNAMICS: A REVIEW

In Chapters 11–13, we have presented an overview of the subjects of rotational dynamics and kinematics. A complete treatment of the subject is beyond the scope of this text, but there are many physical situations that can be analyzed using the results we have derived. It is important to keep in mind that some of our results can be applied *only* in certain special situations. To help you in this regard, we have collected some fundamental equations of rotational dynamics in Table 1.

TABLE 1 SUMMARY OF EQUATIONS OF ROTATIONAL DYNAMICS

Equation	Remarks
	I. Defining Equations
$\boldsymbol{\tau} = \mathbf{r} \times \mathbf{F}$	Torque on a particle about a point O due to a force $\mathbf{F}$.
$\boldsymbol{\tau}_{\text{ext}} = \sum \boldsymbol{\tau}_n$	Resultant external torque of a system of particles acted on by several individual torques $\boldsymbol{\tau}_n$ about the point O.
$\mathbf{l} = \mathbf{r} \times \mathbf{p}$	Angular momentum of a particle about a point O.
$\mathbf{L} = \sum \mathbf{l}_n$	Resultant angular momentum of a system of particles about a point O.
	II. General Relations
$\boldsymbol{\tau} = \dfrac{d\mathbf{l}}{dt}$	The law of motion for a single particle acted on by a torque $\boldsymbol{\tau}$. Both $\boldsymbol{\tau}$ and $\mathbf{l}$ are measured with respect to a point O of an inertial reference frame. This expression is the rotational analogue of the expression $\mathbf{F} = d\mathbf{p}/dt$ for translational motion.
$\sum \boldsymbol{\tau}_{\text{ext}} = \dfrac{d\mathbf{L}}{dt}$	The law of motion for a system of particles acted on by a resultant external torque. It holds only if $\boldsymbol{\tau}_{\text{ext}}$ and $\mathbf{L}$ are measured with respect to (1) any point O fixed in an inertial reference frame, or (2) the center of mass of the system. This expression is the rotational analogue of $\sum \mathbf{F}_{\text{ext}} = d\mathbf{P}/dt$.
	III. Special Case
The following results hold in the case of a rigid body rotating about an axis that is fixed in an inertial reference frame.	
$\tau = I\alpha$	α must lie along the axis; I must also refer to the axis, and τ is the scalar component of $\boldsymbol{\tau}_{\text{ext}}$ along the same axis. This is the rotational analogue of $F = Ma$.
$L = I\omega$	ω must lie along the axis; I must also refer to the axis, and L must be the scalar component of the total angular momentum along this axis. This is the rotational analogue of $P = Mv$.

QUESTIONS

1. We have encountered many vector quantities so far, including position, displacement, velocity, acceleration, force, momentum, and angular momentum. Which of these are defined independent of the choice of the origin in the reference frame?

2. A famous physicist (R. W. Wood), who was fond of practical jokes, mounted a rapidly spinning flywheel in a suitcase which he gave to a porter with instructions to follow him. What happens when the porter is led quickly around a corner? Explain in terms of $\tau = d\mathbf{L}/dt$.

3. A cylinder rotates with angular speed ω about an axis through one end, as in Fig. 21. Choose an appropriate origin and show qualitatively the vectors $\mathbf{L}$ and $\boldsymbol{\omega}$. Are these vectors parallel? Do symmetry considerations enter here?

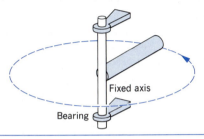

Fixed axis

Bearing

Figure 21 Question 3.

4. Assume that a uniform rod rests in a vertical position on a surface of negligible friction. The rod is then given a horizontal blow at its lower end. Describe the motion of the center of mass of the rod and of its upper endpoint.

5. If the apparatus in Fig. 5 is anchored to the floor of a large spaceship that is drifting in a region free from gravity, in what way, if any, would this affect performance of the experiment?

6. A rear-wheel-drive car accelerates quickly from rest. The driver observes that the car "noses up." Why does it do that? Would a front-wheel-drive car behave differently?

7. An arrow turns in flight so as to be tangent to its flight path at all times. However, a football (thrown with considerable spin about its long axis) does not do this. Why this difference in behavior?

8. A passer throws a spiraling football to a receiver. Is its angular momentum constant, or nearly so? Distinguish between the cases in which the football wobbles and when it doesn't.

9. Can you suggest a simple theory to explain the stability of a moving bicycle? You must explain why it is much more difficult to balance yourself on a bicycle that's at rest than on one that is rolling. (See "The Stability of the Bicycle," by David E. H. Jones, *Physics Today*, April 1970, p. 34.)

10. Why does a long bar help a tightrope walker to keep his or her balance?

11. You are walking along a narrow rail and you start to lose your balance. If you start falling to the right, which way do you turn your body to regain balance? Explain.

12. The mounting bolts that fasten the engines of jet planes to the structural framework of the plane are designed to snap apart if the (rapidly rotating) engine suddenly seizes up because of some mishap. Why are such "structural fuses" used?

13. A disgruntled hockey player throws a hockey stick along the ice. It rotates about its center of mass as it slides along and is eventually brought to rest by the action of friction. Its motion of rotation stops at the same moment that its center of mass comes to rest, not before and not after. Explain why.

14. When the angular speed ω of an object increases, its angular momentum may or may not also increase. Give an example in which it does and one in which it does not.

15. A student stands on a table rotating with an angular speed ω while holding two equal dumbbells at arm's length. Without moving anything else, the two dumbbells are dropped. What change, if any, is there in the student's angular speed? Is angular momentum conserved? Explain your answers.

16. A helicopter flies off, its propellers rotating. Why doesn't the body of the helicopter rotate in the opposite direction?

17. A single-engine airplane must be "trimmed" to fly level. (Trimming consists of raising one aileron and lowering the opposite one.) Why is this necessary? Is this necessary on a twin-engine plane under normal circumstances?

18. The propeller of an aircraft rotates clockwise as seen from the rear. When the pilot pulls upward out of a steep dive, she finds it necessary to apply left rudder at the bottom of the dive if she is to maintain her heading. Explain.

19. Many great rivers flow toward the equator. What effect does the sediment they carry to the sea have on the rotation of the Earth?

20. If the entire population of the world moved to Antarctica, would it affect the length of the day? If so, in what way?

21. A circular turntable rotates at constant angular speed about a vertical axis. There is no friction and no driving torque. A circular pan rests on the turntable and rotates with it; see Fig. 22. The bottom of the pan is covered with a layer of ice of uniform thickness, which is, of course, also rotating with the pan. The ice melts but none of the water escapes from the pan. Is the angular speed now greater than, the same as, or less than the original speed? Give reasons for your answer.

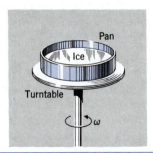

Pan

Ice

Turntable

ω

Figure 22 Question 21.

22. Figure 23a shows an acrobat propelled upward by a trampoline with zero angular momentum. Can the acrobat, by ma-

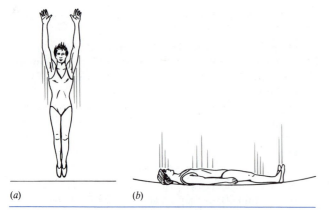

(a)   (b)

Figure 23   Question 22.

neuvering his body, manage to land on his back as in Fig. 23b? Interestingly, 38% of questioned diving coaches and 34% of a sample of physicists gave the wrong answer. What do you think? (See "Do Springboard Divers Violate Angular Momentum Conservation?," by Cliff Frohlich, *American Journal of Physics,* July 1979, p. 583, for a full discussion.)

23. Explain, in terms of angular momentum and rotational inertia, exactly how one "pumps up" a swing in the sitting position. (See "How to Get the Playground Swing Going: A First Lesson in the Mechanics of Rotation," by Jearl Walker, *Scientific American,* March 1989, p. 106.)

24. Can you "pump" a swing so that it turns in a complete circle, moving completely around its support? Assume (if you wish) that the seat of the swing is connected to its support by a rigid rod rather than a rope or chain. Explain your answer.

25. A circular turntable is rotating freely about a vertical axis. There is no friction at the axis of rotation. (a) A bug, initially at the center of the turntable, walks out to the rim and stops. How will the angular momentum of the system (turntable plus bug) change? How will the angular velocity of the turntable change? (b) If the bug falls off the edge of the turntable (without jumping), how will the angular velocity of the turntable change?

26. A massive spinning wheel can be used for a stabilizing effect on a ship. If mounted with its axis of rotation at right angles to the ship deck, what is its effect when the ship tends to roll from side to side?

27. If the top of Fig. 18 were not spinning, it would tip over. If its spin angular momentum is large compared to the change caused by the applied torque, the top precesses. What happens in between when the top spins slowly?

28. A Tippy-Top, having a section of a spherical surface of large radius on one end and a stem for spinning it on the opposite end, will rest on its spherical surface with no spin but slips over when spun, so as to stand on its stem. Explain. (See "The Tippy-Top," by George D. Freier, *The Physics Teacher,* January 1967, p. 36.) If you can't find a Tippy-Top, use a hard-boiled egg; the "standing-on-end" behavior of the spinning egg is most easily followed if you put an ink mark on the "pointed" end of the egg.

PROBLEMS

Section 13-1 Angular Momentum of a Particle

1. If we are given r, p, and θ, we can calculate the angular momentum of a particle from Eq. 2. Sometimes, however, we are given the components (x, y, z) of $\mathbf{r}$ and (v_x, v_y, v_z) of $\mathbf{v}$ instead. (a) Show that the components of $\mathbf{l}$ along the x, y, and z axes are then given by

$$l_x = m(yv_z - zv_y),$$
$$l_y = m(zv_x - xv_z),$$
$$l_z = m(xv_y - yv_x).$$

(b) Show that if the particle moves only in the xy plane, the resultant angular momentum vector has only a z component. (*Hint:* See Eq. 17 in Chapter 3.)

2. A particle P with mass 2.13 kg has position $\mathbf{r}$ and velocity $\mathbf{v}$ as shown in Fig. 24. It is acted on by the force $\mathbf{F}$. All three vectors lie in a common plane. Presume that $r = 2.91$ m, $v = 4.18$ m/s, and $F = 1.88$ N. Compute (a) the angular momentum of the particle and (b) the torque, about the origin, acting on the particle. What are the directions of these two vectors?

3. Show that the angular momentum about any point of a single particle moving with constant velocity remains constant throughout the motion.

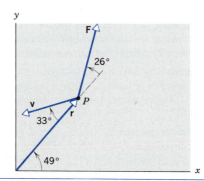

Figure 24   Problem 2.

4. (a) Use the data given in the appendices to compute the total angular momentum of all the planets owing to their revolution about the Sun. (b) What fraction of this is associated with the planet Jupiter?

5. Two particles, each of mass m and speed v, travel in opposite directions along parallel lines separated by a distance d. Find an expression for the total angular momentum of the system about any origin.

6. Calculate the angular momentum, about the Earth's center, of an 84.3-kg person on the equator of the rotating Earth.

Section 13-2 Systems of Particles

7. The total angular momentum of a system of particles relative to the origin O of an inertial reference frame is given by $\mathbf{L} = \Sigma\, \mathbf{r}_i \times \mathbf{p}_i$, where $\mathbf{r}_i$ and $\mathbf{p}_i$ are measured with respect to O. (a) Use the relations $\mathbf{r}_i = \mathbf{r}_{cm} + \mathbf{r}_i'$ and $\mathbf{p}_i = m_i\mathbf{v}_{cm} + \mathbf{p}_i'$ to express $\mathbf{L}$ in terms of the positions $\mathbf{r}_i'$ and momenta $\mathbf{p}_i'$ relative to the center of mass C; see Fig. 25. (b) Use the definition of center of mass and the definition of angular momentum $\mathbf{L}'$ with respect to the center of mass to obtain $\mathbf{L} = \mathbf{L}' + \mathbf{r}_{cm} \times M\mathbf{v}_{cm}$. (c) Show how this result can be interpreted as regarding the total angular momentum to be the sum of spin angular momentum (angular momentum relative to the center of mass) and orbital angular momentum (angular momentum of the motion of the center of mass C with respect to O if all the system's mass were concentrated at C).

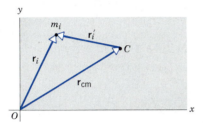

Figure 25   Problems 7 and 8.

8. Let $\mathbf{r}_{cm}$ be the position vector of the center of mass C of a system of particles with respect to the origin O of an inertial reference frame, and let $\mathbf{r}_i'$ be the position vector of the ith particle, of mass m_i, with respect to the center of mass C. Hence $\mathbf{r}_i = \mathbf{r}_{cm} + \mathbf{r}_i'$ (see Fig. 25). Now define the total angular momentum of the system of particles relative to the center of mass C to be $\mathbf{L}' = \Sigma \mathbf{r}_i' \times \mathbf{p}_i'$, where $\mathbf{p}_i' = m_i\, d\mathbf{r}_i'/dt$. (a) Show that $\mathbf{p}_i' = m_i\, d\mathbf{r}_i/dt - m_i d\mathbf{r}_{cm}/dt = \mathbf{p}_i - m_i\mathbf{v}_{cm}$. (b) Show next that $d\mathbf{L}'/dt = \Sigma \mathbf{r}_i' \times d\mathbf{p}_i'/dt$. (c) Combine the results of (a) and (b) and, using the definition of center of mass and Newton's third law, show that $\tau_{ext}' = d\mathbf{L}'/dt$, where τ_{ext}' is the sum of all the external torques acting on the system about its center of mass.

Section 13-3 Angular Momentum and Angular Velocity

9. The time integral of a torque is called the *angular impulse*. (a) Starting from $\tau = d\mathbf{L}/dt$, show that the resultant angular impulse equals the change in angular momentum. This is the rotational analogue of the linear impulse—momentum relation. (b) For rotation around a fixed axis, show that

$$\int \tau\, dt = \overline{F}r(\Delta t) = I(\omega_f - \omega_i),$$

where r is the moment arm of the force, $\overline{F}$ is the average value of the force during the time it acts on the object, and ω_i and ω_f are the angular velocities of the object just before and just after the force acts.

10. A sanding disk with rotational inertia 1.22×10^{-3} kg·m² is attached to an electric drill whose motor delivers a torque of 15.8 N·m. Find (a) the angular momentum and (b) the angular speed of the disk 33.0 ms after the motor is turned on.

11. A wheel of radius 24.7 cm, moving initially at 43.3 m/s, rolls to a stop in 225 m. Calculate (a) its linear acceleration

and (b) its angular acceleration. (c) The wheel's rotational inertia is 0.155 kg·m²; calculate the torque exerted by rolling friction on the wheel.

12. Two wheels, A and B, are connected by a belt as in Fig. 26. The radius of B is three times the radius of A. What would be the ratio of the rotational inertias I_A/I_B if (a) both wheels have the same angular momenta and (b) both wheels have the same rotational kinetic energy? Assume that the belt does not slip.

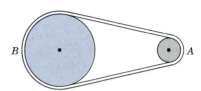

Figure 26   Problem 12.

13. Show that $\mathbf{L} = I\boldsymbol{\omega}$ for the two-particle system of Fig. 7.

14. Using data in the appendices, find the angular momentum of the Earth's spin about its own axis of rotation. Assume that the Earth is a uniform sphere.

15. The angular momentum of a flywheel having a rotational inertia of 0.142 kg·m² decreases from 3.07 to 0.788 kg·m²/s in 1.53 s. (a) Find the average torque acting on the flywheel during this period. (b) Assuming a uniform angular acceleration, through what angle will the flywheel have turned? (c) How much work was done on the wheel? (d) How much average power was supplied by the flywheel?

16. Figure 27 shows a symmetrical rigid body rotating about a fixed axis. The origin of coordinates is fixed for convenience at the center of mass. Prove, by summing over the contributions made to the angular momentum by all the mass elements m_i into which the body is divided, that $\mathbf{L} = I\boldsymbol{\omega}$, where $\mathbf{L}$ is the *total* angular momentum.

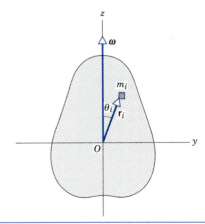

Figure 27   Problem 16.

17. A stick has a mass of 4.42 kg and a length of 1.23 m. It is initially at rest on a frictionless horizontal surface and is struck perpendicularly by a puck imparting a horizontal impulsive force of impulse 12.8 N·s at a distance of 46.4 cm from the center. Determine the subsequent motion of the stick.

18. A cylinder rolls down an inclined plane of angle θ. Show, by

direct application of Eq. 8 ($\Sigma\tau_{ext} = d\mathbf{L}/dt$), that the acceleration of its center of mass is $\frac{2}{3}g \sin \theta$. Compare this method with that used in Sample Problem 8 of Chapter 12.

19. To get a billiard ball to roll without sliding from the start, the cue must hit the ball not at the center (that is, a height above the table equal to the ball's radius R) but exactly at a height $2R/5$ above the center. Prove this result. [See Arnold Sommerfeld, *Mechanics, Volume I of Lectures on Theoretical Physics,* Academic Press, Orlando (1964 paperback edition), pp. 158–161, for a supplement on the mechanics of billiards.]

20. The axis of the cylinder in Fig. 28 is fixed. The cylinder is initially at rest. The block of mass M is initially moving to the right without friction with speed v_1. It passes over the cylinder to the dashed position. When it first makes contact with the cylinder, it slips on the cylinder, but the friction is large enough so that slipping ceases before M loses contact with the cylinder. The cylinder has a radius R and a rotational inertia I. Find the final speed v_2 in terms of v_1, M, I, and R. This can be done most easily by using the relation between impulse and change in momentum.

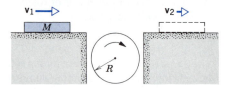

Figure 28   Problem 20.

21. A stick of length L and mass M lies on a frictionless horizontal table on which it is free to move in any way. A hockey puck of mass m, moving as shown in Fig. 29 with speed v, collides elastically with the stick. (a) What quantities are conserved in the collision? (b) What must be the mass m of the puck so that it remains at rest immediately after the collision?

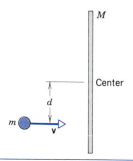

Figure 29   Problem 21.

22. Two cylinders having radii R_1 and R_2 and rotational inertias I_1 and I_2, respectively, are supported by axes perpendicular to the plane of Fig. 30. The large cylinder is initially rotating with angular velocity ω_0. The small cylinder is moved to the right until it touches the large cylinder and is caused to rotate by the frictional force between the two. Eventually, slipping ceases, and the two cylinders rotate at constant rates in opposite directions. Find the final angular velocity ω_2 of the small cylinder in terms of I_1, I_2, R_1, R_2, and ω_0. (*Hint:* Neither angular momentum nor kinetic energy is con-

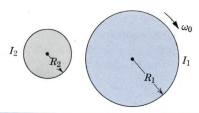

Figure 30   Problem 22.

served. Apply the angular impulse equation to each cylinder. See Problem 9.)

23. A billiard ball, initially at rest, is given a sharp impulse by a cue. The cue is held horizontally a distance h above the centerline as in Fig. 31. The ball leaves the cue with a speed v_0 and, because of its "forward english," eventually acquires a final speed of $9v_0/7$. Show that $h = 4R/5$, where R is the radius of the ball.

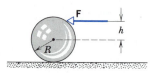

Figure 31   Problem 23.

24. In Problem 23, imagine $\mathbf{F}$ to be applied below the centerline. (a) Show that it is impossible, with this "reverse english," to reduce the forward speed to zero, without rolling having set in, unless $h = R$. (b) Show that it is impossible to give the ball a backward velocity unless $\mathbf{F}$ has a downward vertical component.

25. A bowler throws a bowling ball of radius $R = 11.0$ cm down the lane with initial speed $v_0 = 8.50$ m/s. The ball is thrown in such a way that it skids for a certain distance before it starts to roll. It is not rotating at all when it first hits the lane, its motion being pure translation. The coefficient of kinetic friction between the ball and the lane is 0.210. (a) For what length of time does the ball skid? (*Hint:* As the ball skids, its speed v decreases and its angular speed ω increases; skidding ceases when $v = R\omega$.) (b) How far down the lane does it skid? (c) How many revolutions does it make before it starts to roll? (d) How fast is it moving when it starts to roll?

Section 13-4 Conservation of Angular Momentum

26. Astronomical observations show that from 1870 to 1900 the length of the day increased by about 6.0×10^{-3} s. (a) What corresponding fractional change in the Earth's angular velocity resulted? (b) Suppose that the cause of this change was a shift of molten material in the Earth's core. What resulting fractional change in the Earth's rotational inertia could account for the answer to part (a)?

27. Suppose that the Sun runs out of nuclear fuel and suddenly collapses to form a so-called white dwarf star, with a diameter equal to that of the Earth. Assuming no mass loss, what would then be the new rotation period of the Sun, which currently is about 25 days? Assume that the Sun and the white dwarf are uniform spheres.

28. A man stands on a frictionless platform that is rotating with an angular speed of 1.22 rev/s; his arms are outstretched and he holds a weight in each hand. With his hands in this posi-

tion the total rotational inertia of the man, the weights, and the platform is 6.13 kg·m². If by moving the weights the man decreases the rotational inertia to 1.97 kg·m², (a) what is the resulting angular speed of the platform and (b) what is the ratio of the new kinetic energy to the original kinetic energy?

29. In a lecture demonstration, a toy train track is mounted on a large wheel that is free to turn with negligible friction about a vertical axis; see Fig. 32. A toy train of mass m is placed on the track and, with the system initially at rest, the electrical power is turned on. The train reaches a steady speed v with respect to the track. What is the angular velocity ω of the wheel, if its mass is M and its radius R? (Neglect the mass of the spokes of the wheel.)

Figure 32 Problem 29.

30. The rotor of an electric motor has a rotational inertia $I_m = 2.47 \times 10^{-3}$ kg·m² about its central axis. The motor is mounted parallel to the axis of a space probe having a rotational inertia $I_p = 12.6$ kg·m² about its axis. Calculate the number of revolutions required to turn the probe through $25.0°$ about its axis.

31. A wheel with rotational inertia 1.27 kg·m² is rotating with an angular speed of 824 rev/min on a shaft whose rotational inertia is negligible. A second wheel, initially at rest and with rotational inertia 4.85 kg·m², is suddenly coupled to the same shaft. (a) What is the angular speed of the resultant combination of the shaft and two wheels? (b) What fraction of the original kinetic energy is lost?

32. With center and spokes of negligible mass, a certain bicycle wheel has a thin rim of radius 36.3 cm and mass 3.66 kg; it can turn on its axle with negligible friction. A man holds the wheel above his head with the axis vertical while he stands on a turntable free to rotate without friction; the wheel rotates clockwise, as seen from above, with an angular speed of 57.7 rad/s, and the turntable is initially at rest. The rotational inertia of wheel-plus-man-plus-turntable about the common axis of rotation is 2.88 kg·m². (a) The man's hand suddenly stops the rotation of the wheel (relative to the turntable). Determine the resulting angular velocity (magnitude and direction) of the system. (b) The experiment is repeated with noticeable friction introduced into the axle of the wheel, which, starting from the same initial angular speed (57.7 rad/s), gradually comes to rest (relative to the turntable) while the man holds the wheel as described above. (The turntable is still free to rotate without friction.) Describe what happens to the system, giving as much quantitative information as the data permit.

33. A girl of mass 50.6 kg stands on the edge of a frictionless merry-go-round of mass 827 kg and radius 3.72 m that is

not moving. She throws a 1.13-kg rock in a horizontal direction that is tangent to the outer edge of the merry-go-round. The speed of the rock, relative to the ground, is 7.82 m/s. Calculate (a) the angular speed of the merry-go-round and (b) the linear speed of the girl after the rock is thrown. Assume that the merry-go-round is a uniform disk.

34. In a playground there is a small merry-go-round of radius 1.22 m and mass 176 kg. The radius of gyration (see Problem 11 of Chapter 12) is 91.6 cm. A child of mass 44.3 kg runs at a speed of 2.92 m/s tangent to the rim of the merry-go-round when it is at rest and then jumps on. Neglect friction between the bearings and the shaft of the merry-go-round and find the angular speed of the merry-go-round and child.

35. A uniform flat disk of mass M and radius R rotates about a horizontal axis through its center with angular speed ω_0. (a) What is its kinetic energy? Its angular momentum? (b) A chip of mass m breaks off the edge of the disk at an instant such that the chip rises vertically above the point at which it broke off (Fig. 33). How high above the point does it rise before starting to fall? (c) What is the final angular speed of the broken disk?

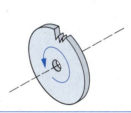

Figure 33 Problem 35.

36. A cockroach, mass m, runs counterclockwise around the rim of a lazy Susan (a circular dish mounted on a vertical axle) of radius R and rotational inertia I with frictionless bearings. The cockroach's speed (relative to the Earth) is v, whereas the lazy Susan turns clockwise with angular speed ω. The cockroach finds a bread crumb on the rim and, of course, stops. (a) Find the angular speed of the lazy Susan after the cockroach stops. (b) How much kinetic energy is lost, if any?

37. A particle is projected horizontally along the interior of a frictionless hemispherical bowl of radius r, which is kept at rest (Fig. 34). We wish to find the initial speed v_0 required for the particle to just reach the top of the bowl. Find v_0 as a function of θ_0, the initial angular position of the particle. (*Hint:* Use conservation principles.)

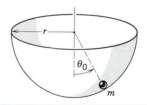

Figure 34 Problem 37.

38. On a large horizontal frictionless circular track, radius R, lie two small balls of masses m and M, free to slide on the track.

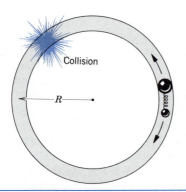

Figure 35   Problem 38.

Between the two balls is squeezed a spring, which, however, is not attached to the balls. The two balls are held together by a string. (*a*) If the string breaks, the compressed spring (assumed massless) shoots off the two balls in opposite directions; the spring itself is left behind. The balls collide when they meet again on the track (Fig. 35). Where does this collision take place? Express the answer in terms of the angle, in rad, through which ball *M* travels. (*b*) The potential energy initially stored in the spring was U_0. Find the time it takes after the string breaks for the collision to take place. (*c*) Assuming the collision to be perfectly elastic and head-on, where will the balls collide again after the first collision?

39. Two skaters, each of mass 51.2 kg, approach each other along parallel paths separated by 2.92 m. They have equal and opposite velocities of 1.38 m/s. The first skater carries a long light pole 2.92 m long, and the second skater grabs the end of it as he passes; see Fig. 36. Assume frictionless ice. (*a*) Describe quantitatively the motion of the skaters after they are connected by the pole. (*b*) By pulling on the pole, the skaters reduce their separation to 0.940 m. Find their angular speed then. (*c*) Calculate the kinetic energy of the system in parts (*a*) and (*b*). From where does the change come?

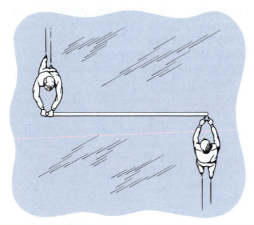

Figure 36   Problem 39.

40. If the polar ice caps of the Earth were to melt and the water returned to the oceans, the oceans would be made deeper by about 30 m. What effect would this have on the Earth's rotation? Make an estimate of the resulting change in the length of the day. (Concern has been expressed that warm-

ing of the atmosphere resulting from industrial pollution could cause the ice caps to melt.)

41. The Earth was formed about 4.5 billion years ago, probably as a sphere of roughly uniform density. Shortly thereafter, heat from the decay of radioactive elements caused much of the Earth to melt. This allowed the heavier material to sink toward the center of the Earth, forming the core. Today, we can picture the Earth as made up of a core of radius 3570 km and density 10.3 g/cm³ surrounded by a mantle of density 4.50 g/cm³ extending to the surface of the Earth (radius 6370 km). We ignore the crust of the Earth. Calculate the fractional change in the length of the day due to the formation of the core.

Section 13-5 The Spinning Top

42. A top is spinning at 28.6 rev/s about an axis making an angle of 34.0° with the vertical. Its mass is 492 g and its rotational inertia is 5.12×10^{-4} kg·m². The center of mass is 3.88 cm from the pivot point. The spin is clockwise as seen from above. Find the magnitude (in rev/s) and direction of the angular velocity of precession.

43. A gyroscope consists of a rotating disk with a 48.7-cm radius suitably mounted at the midpoint of a 12.2-cm long axle so that it can spin and precess freely. Its spin rate is 975 rev/min. The mass of the disk is 1.14 kg and the mass of the axle is 130 g. Find the time required for one precession if the axle is supported at one end and is horizontal.

Section 13-6 Quantization of Angular Momentum

44. In 1913, Niels Bohr postulated that the angular momentum of any mechanical rotating system with rotational inertia *I* is quantized. That is,

$$L = I\omega = n(h/2\pi),$$

where *L* is the angular momentum and *n* is any positive integer or zero. (*a*) Show that this postulate restricts the kinetic energy the rotating system can have to a set of discrete values—that is, the energy is quantized; find an expression for the energy. (*b*) Consider the *rigid rotator,* consisting of a particle of mass *m* constrained to rotate in a circle of radius *R*. With what angular speeds could the particle rotate if the postulate were correct? What values of kinetic energy could it have? (*c*) Draw an energy-level diagram like that of Fig. 37, indicating how the spacing between the energy levels varies as *n* increases. Certain low-energy diatomic molecules behave like a rigid rotator.

Figure 37   Problem 44.

45. (*a*) Assume that the electron moves in a circular orbit about the proton in a hydrogen atom. The centripetal force on the electron is supplied by an electric force $e^2/4\pi\epsilon_0 r^2$, where e is the magnitude of the charge of an electron and of a proton, r is the orbit radius, and ϵ_0 is a constant. Show that the radius of the orbit is

$$r = \frac{e^2}{4\pi\epsilon_0 mv^2}$$

where m is the mass of the electron and v is its speed. (*b*) Assume now that the angular momentum of the electron about the proton can only have values that are integral multiples n of $h/2\pi$, where h is *Planck's constant.* Show that the only electron orbits possible are those with a radius

$$r = \frac{nh}{2\pi mv} .$$

(*c*) Combine these results to eliminate v and show that the only orbits consistent with both requirements have radii

$$r = \frac{n^2\epsilon_0 h^2}{\pi me^2} .$$

Hence the allowed radii are proportional to the square of the integers $n = 1, 2, 3$, etc. When $n = 1$, r is smallest and has the value 0.529×10^{-10} m.

Computer Project

46. Consider two flywheels that are mounted on the same axle but are free to turn independently of each other. Wheel 1, which is initially turning at 100 rad/s, has a rotational inertia of 2.5 kg·m². Wheel 2, which is initially at rest, has a rotational inertia of 1.5 kg·m². By sliding one wheel along the axle the wheels are brought into contact with each other, face against face. They exert torques on each other and eventually reach the same angular speed. (*a*) Assume no torques act except those of each wheel on the other and use the principle of angular momentum conservation to find the final angular speed of the wheels.

If the torque of each wheel on the other can be calculated, you can use a computer to follow the wheels as they come to the same speed. Assume the torque of wheel 2 on wheel 1 is given by $\tau_1 = -0.20(\omega_1 - \omega_2)$, where ω_1 is the angular ve-

locity of wheel 1 and ω_2 is the angular velocity of wheel 2. Here τ_1 is in N·m and the angular velocities are in rad/s. The torque of wheel 1 on wheel 2 is $\tau_2 = +0.20(\omega_1 - \omega_2)$. The torques continue to act until the wheels have the same angular speed, then they vanish. While they are interacting, wheel 1 obeys $\tau_1 = I_1\alpha_1$ and wheel 2 obeys $\tau_2 = I_2\alpha_2$. These equations are mathematically similar to Newton's second law equations and can be integrated numerically in the manner described in Section 6-6 and in the computer projects at the end of Chapter 6. Consider time intervals of duration Δt and suppose wheel 1 has angular position θ_{1b} and angular velocity ω_{1b} at the beginning of an interval. Then its angular position and angular speed at the end of the interval can be approximated by $\theta_{1e} = \theta_{1b} + \omega_{1b}\Delta t$ and $\omega_{1e} = \omega_{1b} + \alpha_{1b}\Delta t = \omega_{1b} + (\tau_{1b}/I_1)\Delta t$, where τ_{1b} is the torque at the beginning of the interval. Similar equations hold for wheel 2. The smaller Δt the better the approximation.

(*b*) Write a computer program or design a spreadsheet to calculate the angular speed of the wheels at the end of each second from $t = 0$ to $t = 25$ s. Use an integration interval of 0.001 s. Plot the angular velocities as functions of time on the same graph, then use the graph or the list of values to find the final angular velocities and compare the result with the value you found in part (*a*).

(*c*) To see the influence of an external torque suppose the torque acting on wheel 1 is given by $\tau_1 = -4.0 - 0.20(\omega_1 - \omega_2)$ and the torque acting on wheel 2 is $\tau_2 = +0.20(\omega_1 - \omega_2)$, where the torques are in N·m and the angular velocities are in rad/s. This represents an external torque of -4.0 N·m. Use your computer program to find the angular speeds of the wheels and the total angular momentum for every 1 s from $t = 0$ to $t = 25$ s. Again, use an integration interval of 0.001 s. Plot the angular velocities as functions of time. Since $\tau_{ext} = dL_{total}/dt$, the external torque should produce a change in the total angular momentum of $\Delta L = \tau_{ext}\Delta t = -4.0 \times 25 = -100$ kg·m/s over the first 25 s. Do your results agree? Which wheel suffers the change (compared to the case of zero external torque) or is the change shared?

(*d*) The final angular velocity does not depend on the details of the torque each wheel exerts on the other. What does depend on the torques?

CHAPTER 14

EQUILIBRIUM OF RIGID BODIES

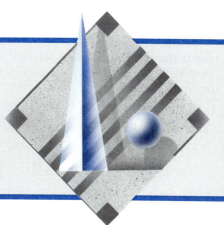

The towers supporting a suspension bridge must be strong enough so that they do not collapse under the weight of the bridge and its traffic load; the landing gear of an aircraft must not collapse if the pilot makes a poor landing; a chair must not collapse or tip over when we sit in it. In all such problems the designer is concerned that these presumed rigid structures do indeed remain rigid under the forces, and the associated torques, that act on them.

In such problems we must ask two questions: (1) What forces and torques act on the presumed rigid body? (2) Considering its design and the materials used, will the body remain rigid under the action of these forces and torques? In this chapter we consider the first question in detail. To answer the second question we must deal in great detail with the properties of materials. A full treatment of this subject is beyond the scope of this text; a brief discussion is given in the last section of this chapter.

14-1 CONDITIONS OF EQUILIBRIUM

A rigid body, such as a chair, a bridge, or a building, is said to be in *mechanical equilibrium* if, as viewed from an inertial reference frame, *both the linear momentum* **P** *and the angular momentum* **L** *of the rigid body have a constant value.* Equivalently, we could say that both the linear acceleration $\mathbf{a}_{cm}$ of its center of mass and the angular acceleration α about any axis fixed in the reference frame are zero.

This definition of mechanical equilibrium does not require that the body be at rest; that is, **P** and **L** do not necessarily have the constant value of zero. If they *are* zero (or, equivalently, if the velocity of the center of mass $\mathbf{v}_{cm}$ and the angular velocity ω about any axis in the frame are both zero), then we have the situation of *static equilibrium*.

In this chapter we seek the restrictions that must be imposed on the forces and torques that act on the body to bring about a condition of equilibrium. We concentrate on cases of static equilibrium, even though, as we shall see, the same restrictions apply whether or not the equilibrium is static.

The translational motion of the center of mass of a rigid body is governed by Eq. 27 of Chapter 9,

$$\sum \mathbf{F}_{\text{ext}} = \frac{d\mathbf{P}}{dt},$$

in which $\sum \mathbf{F}_{\text{ext}}$ is the sum of all the external forces that act on the body. If **P** has any constant value, including zero, we must have $d\mathbf{P}/dt = 0$. Thus the first condition of equilibrium is that *the vector sum of all the external forces acting on the body must be zero,* or

$$\sum \mathbf{F}_{\text{ext}} = 0. \tag{1}$$

This vector equation is equivalent to three scalar equations:

$$\sum F_x = 0, \qquad \sum F_y = 0, \qquad \sum F_z = 0, \tag{2}$$

where for convenience we have dropped the "ext" subscript on F_{ext}. Equations 1 and 2 state that the sum of the components of the external forces along each of any three mutually perpendicular directions is zero.

The rotational motion of a rigid body is governed by Eq. 8 of Chapter 13, or

$$\sum \tau_{\text{ext}} = \frac{d\mathbf{L}}{dt},$$

in which $\sum \boldsymbol{\tau}_{\text{ext}}$ is the sum of all the external torques that act on the body. If the angular momentum **L** has any constant value, including zero, we must have $d\mathbf{L}/dt = 0$. Thus the second condition for equilibrium is that *the vector sum of all the external torques acting on the body must be zero,* or

$$\sum \boldsymbol{\tau}_{\text{ext}} = 0. \qquad (3)$$

This vector equation can be written as three scalar equations (again dropping the "ext" subscript):

$$\sum \tau_x = 0, \qquad \sum \tau_y = 0, \qquad \sum \tau_z = 0, \qquad (4)$$

which state that, at equilibrium, the sum of the components of the torques acting on a body, along each of three mutually perpendicular directions, is zero.

The second condition for equilibrium is independent of the choice of the origin and coordinate axes used for calculating the components of the torques. If the net torque is zero, then its components are zero for *any* choice of the *x, y,* and *z* axes. Furthermore, for a body in equilibrium, the choice of origin for calculating torques is unimportant and can be made for convenience; if $\tau = 0$ about a particular origin *O*, then it is zero about any other point in the reference frame for a body in equilibrium.

Let us prove this last statement. Suppose that *N* external forces are applied to the object. With respect to the origin *O*, force $\mathbf{F}_1$ is applied at the point located at $\mathbf{r}_1$, force $\mathbf{F}_2$ at $\mathbf{r}_2$, and so on. The net torque about *O* is therefore

$$\boldsymbol{\tau}_O = \boldsymbol{\tau}_1 + \boldsymbol{\tau}_2 + \cdots + \boldsymbol{\tau}_N$$
$$= \mathbf{r}_1 \times \mathbf{F}_1 + \mathbf{r}_2 \times \mathbf{F}_2 + \cdots + \mathbf{r}_N \times \mathbf{F}_N. \qquad (5)$$

Suppose a point *P* is located at displacement $\mathbf{r}_P$ with respect to *O* (Fig. 1). The point of application of $\mathbf{F}_1$, with respect to *P*, is $(\mathbf{r}_1 - \mathbf{r}_P)$. The torque about *P* is

$$\boldsymbol{\tau}_P = (\mathbf{r}_1 - \mathbf{r}_P) \times \mathbf{F}_1 + (\mathbf{r}_2 - \mathbf{r}_P) \times \mathbf{F}_2$$
$$+ \cdots + (\mathbf{r}_N - \mathbf{r}_P) \times \mathbf{F}_N$$
$$= [\mathbf{r}_1 \times \mathbf{F}_1 + \mathbf{r}_2 \times \mathbf{F}_2 + \cdots + \mathbf{r}_N \times \mathbf{F}_N]$$
$$- [\mathbf{r}_P \times \mathbf{F}_1 + \mathbf{r}_P \times \mathbf{F}_2 + \cdots + \mathbf{r}_P \times \mathbf{F}_N].$$

The first group of terms in the brackets gives $\boldsymbol{\tau}_O$ according to Eq. 5. We can rewrite the second group by removing the constant factor of $\mathbf{r}_P$:

$$\boldsymbol{\tau}_P = \boldsymbol{\tau}_O - [\mathbf{r}_P \times (\mathbf{F}_1 + \mathbf{F}_2 + \cdots + \mathbf{F}_N)]$$
$$= \boldsymbol{\tau}_O - \left[\mathbf{r}_P \times \left(\sum \mathbf{F}_{\text{ext}} \right) \right]$$
$$= \boldsymbol{\tau}_O,$$

where we make the last step because $\sum \mathbf{F}_{\text{ext}} = 0$ for a body in translational equilibrium. Thus the torque about any two points has the same value when the body is in translational equilibrium.

Often we deal with problems in which all the forces lie in a plane. In this case the six conditions of Eqs. 2 and 4 are reduced to three. We resolve the forces into two components:

$$\sum F_x = 0, \qquad \sum F_y = 0, \qquad (6)$$

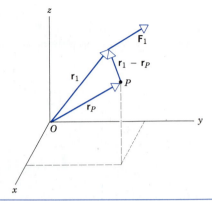

Figure 1   The force $\mathbf{F}_1$ is one of *N* external forces that act on a rigid body (not shown). The vector $\mathbf{r}_1$ locates the point of application of $\mathbf{F}_1$ relative to *O* and is used in calculating the torque of $\mathbf{F}_1$ about *O*. The vector $\mathbf{r}_1 - \mathbf{r}_P$ is used in calculating the torque of $\mathbf{F}_1$ about *P*.

and, if we calculate torques about a point that also lies in the *xy* plane, all torques must be in the direction perpendicular to the *xy* plane. In this case we have

$$\sum \tau_z = 0. \qquad (7)$$

We limit ourselves mostly to planar problems to simplify the calculations; this condition does not impose any fundamental restriction on the application of the general principles of equilibrium.

14-2   CENTER OF GRAVITY

One of the forces encountered in rigid-body dynamics is the force of gravity, which is responsible for the weight of a body. Previously (and without justification), we represented that force on a body of mass *M* by a single vector *M***g** acting at the center of mass of the body. Here we justify that step and discuss the conditions under which it is valid.

The weight of an extended body is actually the resultant of a great many forces, each one due to gravity acting on one of the particles of the body. That is, we can replace the vector sum of the gravitational forces on all the particles of a body with a single force, the weight. Furthermore, the net resultant of the corresponding gravitational torques on all the particles can be replaced by the torque due to that single force if we imagine it to act at a point of the body called the *center of gravity.*

If the gravitational acceleration **g** has the same value at all points of the body, which is true in all practical cases of interest, then two simplifications occur: (1) the weight is equal to *M***g**, and (2) the center of gravity coincides with the center of mass. Let us prove these two results.

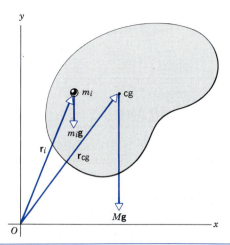

Figure 2   Each particle in a body, such as the one with mass m_i, experiences a gravitational force such as $m_i\mathbf{g}$. The entire weight of a body, though distributed throughout its volume as the sum of the gravitational forces on all such particles, may be replaced by a single force of magnitude Mg acting at the center of gravity. If the gravitational field is uniform (that is, the same for all particles), the center of gravity coincides with the center of mass, and so $\mathbf{r}_{cm}$ is the same as $\mathbf{r}_{cg}$.

Imagine the body of mass M to be divided into a large number of particles. The gravitational force exerted by the Earth on the ith particle of mass m_i is $m_i\mathbf{g}$. This force is directed down toward the center of the Earth. The net force on the entire object due to gravity is the sum over all the individual particles, or

$$\sum \mathbf{F} = \sum m_i\mathbf{g}. \qquad (8)$$

Because we have assumed that $\mathbf{g}$ has the same value for every particle of the body, we can factor $\mathbf{g}$ out of the summation of Eq. 8, which gives

$$\sum \mathbf{F} = \mathbf{g} \sum m_i = M\mathbf{g}. \qquad (9)$$

This proves the first of the assertions we made above, that we can replace the resultant force of gravity acting on the entire body by the single force $M\mathbf{g}$.

Let us now apply the torque condition, Eq. 3, taking the torques about the arbitrary point O, as shown in Fig. 2. The vector $\mathbf{r}_i$ locates the particle of mass m_i relative to this origin. The net torque about this point due to gravity acting on all the particles is

$$\sum \tau = \sum (\mathbf{r}_i \times m_i\mathbf{g}) = \sum (m_i\mathbf{r}_i \times \mathbf{g}), \qquad (10)$$

where the last step is taken by moving the scalar m_i within the sum. Once again we use the constancy of $\mathbf{g}$ to remove it from the summation, being careful not to change the order of the vectors $\mathbf{r}_i$ and $\mathbf{g}$ so that the sign of the cross product does not change. According to Eq. 12 of Chapter 9, the remaining summation, $\sum m_i\mathbf{r}_i$, is just $M\mathbf{r}_{cm}$, where $\mathbf{r}_{cm}$ is the vector that locates the center of mass of the body

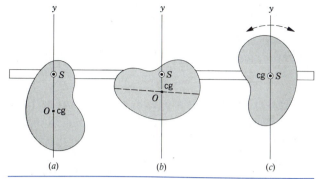

Figure 3   A body suspended from an arbitrary point S, as in (*a*) and (*b*), will be in stable equilibrium only if its center of gravity (cg) hangs vertically below its suspension point S. The dashed line in (*b*) represents the vertical line in (*a*), showing that the center of gravity can be located by suspending the body successively from two different points. (*c*) If a body is suspended at its center of gravity, it is in equilibrium no matter what its orientation.

relative to the origin O. Taking these two steps, we can write Eq. 10 as

$$\sum \tau = \left(\sum m_i\mathbf{r}_i \right) \times \mathbf{g} = M\mathbf{r}_{cm} \times \mathbf{g} = \mathbf{r}_{cm} \times M\mathbf{g}. \qquad (11)$$

The resultant torque on the body thus equals the torque that would be produced by the single force $M\mathbf{g}$ acting at the center of mass of the body, and thus the center of gravity (cg) coincides with the center of mass, which proves the second assertion we made above. A useful corollary of Eq. 11 is that *the torque due to gravity about the center of mass of a body is zero.*

Under what conditions will a body in the Earth's gravity be in equilibrium? Equations 9 and 11 show that, if we apply a single upward force $\mathbf{F}'$ of magnitude Mg at the center of mass, then both the net force and the net torque will be zero, and our conditions for equilibrium will be satisfied. However, it is also true that the body will be in equilibrium if the upward force $\mathbf{F}'$ is applied at any point on a vertical line through the center of mass. The net torque is zero in this case because $M\mathbf{g}$ and $\mathbf{F}' (= -M\mathbf{g})$ have the same line of action. We can therefore balance an object by applying a vertical force F' not only *at* the center of mass, but at any point directly *below* or *above* the center of mass.

We can use this property to find the center of mass of an extended object. Consider a body of arbitrary shape suspended from a point S (Fig. 3). The point of support, which exerts an upward force $\mathbf{F}' = -M\mathbf{g}$, must be on a vertical line with the center of mass. If we draw a vertical line through S, then we know that the center of mass must lie somewhere on the line. We can repeat the procedure with a new choice of point S, as in Fig. 3*b*, and we find a second line that must contain the center of mass. The center of mass must therefore lie at the intersection of the two lines.

If we suspend the object from the center of mass, as in Fig. 3c, the body will be in equilibrium no matter what its orientation. We can turn it any way we wish, and it remains in equilibrium. This illustrates the corollary of Eq. 11: the torque due to gravity is zero about the center of mass.

In this section, we have used "center of mass" and "center of gravity" interchangeably. The center of mass is defined for any body and can be calculated, according to methods described in Chapter 9, from the size and shape of the body. The center of gravity, on the other hand, is defined only for bodies in a gravitational field. To calculate the center of gravity, we must know not only the geometrical details of the body, but also the variation of **g** over the body. If **g** is not constant over the body, then the center of gravity and the center of mass do not coincide, and **g** cannot be removed from the sums in Eqs. 8 and 10. Consider the uniform stick shown in Fig. 4, whose axis is inclined at some nonzero angle from the horizontal. The center of mass C lies at the geometrical center of the stick. If the axis of the stick were horizontal, the center of gravity P would coincide with the center of mass; that is, a single upward force $\mathbf{F}'$ (equal in magnitude to Mg) at C would keep the stick in equilibrium. When the axis is not horizontal, this is no longer the case. Because g decreases slightly with distance from the Earth, particle N at the lower end of the stick experiences a greater gravitational attraction than an identical particle 1 at the upper end. To compensate for the resulting tendency of the stick to rotate clockwise about C, the center of gravity P (the point of application of the upward equilibrium force) must be located slightly below C. As the angle with the horizontal changes, the location of P will change. Furthermore, if we move the stick to a place where g has a different value, the relationship between P and C for a given angle of inclination will be different. Thus the center of gravity may in general depend on the orientation of the object, as well as on the local gravitational field. For a meter stick inclined at an angle of 45° near the Earth's surface, the distance between the center of mass and the center of gravity is about 18 nm, far smaller than the precision at which we normally work in problems of equilibrium and therefore

entirely negligible. In equilibrium problems, we can safely assume that the center of gravity and the center of mass coincide.

14-3   EXAMPLES OF EQUILIBRIUM

In applying the conditions for equilibrium (zero resultant force and zero resultant torque about any point), we can clarify and simplify the procedure as follows.

First, we draw an imaginary boundary around the system under consideration. This helps us to see clearly just what body or system of bodies to which we are applying the laws of equilibrium. This process is called *isolating the system*.

Second, we draw vectors representing the magnitude, direction, and point of application of all *external* forces. An external force is one that acts from outside the boundary that was drawn earlier. Examples of external forces often encountered are gravitational forces and forces exerted by strings, wires, rods, and beams that cross the boundary. Note that only external forces acting on the system need be considered; all internal forces cancel one another in pairs.

There are some cases in which the direction of a force may not be obvious. To determine the direction of a certain force, make an imaginary cut through the member exerting the force at the point where it crosses the boundary. If the ends of this cut tend to pull apart, the force acts outward. If you are in doubt, choose the direction arbitrarily. A negative value for a force in the solution means that the force acts in the direction opposite to that assumed.

Third, we choose a convenient coordinate system along whose axes we resolve the external forces before applying the first condition of equilibrium (Eqs. 1 or 2). The goal here is to simplify the calculations. The preferable coordinate system is usually one that minimizes the number of forces that must be resolved into components.

Fourth, we choose a convenient coordinate system along whose axes we resolve the external torques before applying the second condition of equilibrium (Eqs. 3 or 4). The goal again is to simplify calculations, and we may use different coordinate systems in applying the two conditions for static equilibrium if this proves to be convenient. For example, calculating torques about a point through which several forces act eliminates those forces from the torque equation.

The torque components resulting from all external forces must be zero about *any* axis for equilibrium. Internal torques will cancel in pairs and need not be considered. We follow the same convention as in previous chapters for the algebraic sign of the torque about a particular axis: we take a torque to be positive if by itself it would produce a counterclockwise rotation about the axis.

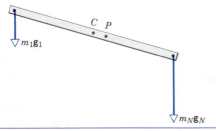

Figure 4   A uniform stick in a nonuniform gravitational field. The center of gravity is at P, which does not coincide with the center of mass C.

Sample Problem 1 A uniform beam of length L whose mass m is 1.8 kg rests with its ends on two digital scales, as in Fig. 5a. A block whose mass M is 2.7 kg rests on the beam, its center one-fourth of the way from the beam's left end. What do the scales read?

Solution We choose as our system the beam and the block, taken together. Figure 5b is a free-body diagram for this system, showing all the external forces that act on the system. The weight of the beam, $m\mathbf{g}$, acts downward at its center of mass, which is at its geometric center, since the beam is uniform. Similarly, $M\mathbf{g}$, the weight of the block, acts downward at its center of mass. The scales push upward at the ends of the beam with forces $\mathbf{F}_l$ and $\mathbf{F}_r$. The magnitudes of these latter two forces are the scale readings that we seek.

Our system is in static equilibrium so that the balance of forces equation (Eq. 6) and the balance of torques equation (Eq. 7) apply. We solve this problem in two equivalent ways.

1. *First solution.* The forces have no x components, and so the condition that $\Sigma F_x = 0$ provides no information. For the y components, we have

$$\sum F_y = F_l + F_r - Mg - mg = 0. \qquad (12)$$

There are two unknown forces (F_l and F_r) but we cannot find them separately because we have (so far) only one equation. Fortunately, we have another equation at hand, namely, Eq. 7, the balance of torques equation.

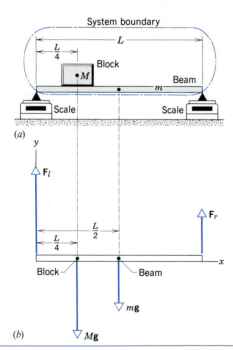

System boundary

L

$\frac{L}{4}$

Block

$\cdot M$

Beam

m

Scale   Scale

(a)

y

F_l

F_r

$\frac{L}{2}$

$\frac{L}{4}$

Block   Beam

x

$m\mathbf{g}$

(b) $M\mathbf{g}$

Figure 5 Sample Problem 1. (a) A beam of mass m supports a block of mass M. The digital scales display the vertical forces exerted on the two ends of the beam. (b) A free-body diagram showing the forces that act on the system, consisting of beam + block.

We can apply Eq. 7 to any axis at right angles to the plane of Fig. 5. Let us choose an axis through the left end of the beam, so that the unknown F_l will disappear from the torque equation. We then have, from Eq. 7,

$$\sum \tau_z = (F_l)(0) + (F_r)(L) - (mg)(L/2) - (Mg)(L/4) = 0, \qquad (13)$$

or

$$F_r = (g/4)(M + 2m)$$
$$= (\tfrac{1}{4})(9.8 \text{ m/s}^2)[2.7 \text{ kg} + 2(1.8 \text{ kg})] = 15 \text{ N}.$$

Note how our choice of axis eliminates the force F_l from the torque equation and allows us to solve directly for the other force. If we had chosen to take torques about any arbitrary point, we would have obtained an equation involving F_l and F_r which could be solved simultaneously with Eq. 12. Our choice of axis helps to simplify the algebra somewhat, but of course it in no way changes the ultimate solution.

If we substitute the value of F_r into Eq. 12 and solve for F_l, we find

$$F_l = (M + m)g - F_r$$
$$= (2.7 \text{ kg} + 1.8 \text{ kg})(9.8 \text{ m/s}^2) - 15 \text{ N} = 29 \text{ N}.$$

Note that the height of the center of mass of the block does not enter the solution to this problem. Is this physically reasonable?

2. *Second solution.* As a check, let us solve this problem in a different way, applying the balance of torques equation about two different axes. By choosing an axis through the left end of the beam, as we did above, we find the solution $F_r = 15$ N.

For a second axis passing through the right end of the beam, Eq. 7 yields

$$\sum \tau_z = (F_r)(0) - (F_l)(L) + (mg)(L/2) + (Mg)(3L/4) = 0. \qquad (14)$$

Solving for F_l, we find

$$F_l = (g/4)(3M + 2m)$$
$$= (\tfrac{1}{4})(9.8 \text{ m/s}^2)[3(2.7 \text{ kg}) + 2(1.8 \text{ kg})] = 29 \text{ N},$$

in agreement with our earlier result. Note that the length of the beam does not enter this problem explicitly.

The solution for the two unknowns in this problem (F_l and F_r) requires two independent equations. In this second method, our two equations (Eqs. 13 and 14) came from the two torque equations; the force equation (Eq. 12) gives no independent information. In fact, as you can show, subtracting the two torque equations gives the force equation.

Sample Problem 2 A bowler holds a bowling ball whose mass M is 7.2 kg in the palm of the hand. As Fig. 6a shows, the upper arm is vertical and the lower arm is horizontal. What forces must the biceps muscle and the bony structure of the upper arm exert on the lower arm? The forearm and hand together have a mass m of 1.8 kg, and the needed dimensions are $d = 4.0$ cm, $D = 15$ cm, and $L = 33$ cm.

Solution Our system is the lower arm and the bowling ball, taken together. Figure 6b shows a free-body diagram. The unknown forces are $\mathbf{T}$, the force exerted by the biceps muscle, and $\mathbf{F}$, the force exerted by the upper arm on the lower arm. As in Sample Problem 1, all the forces are vertical.

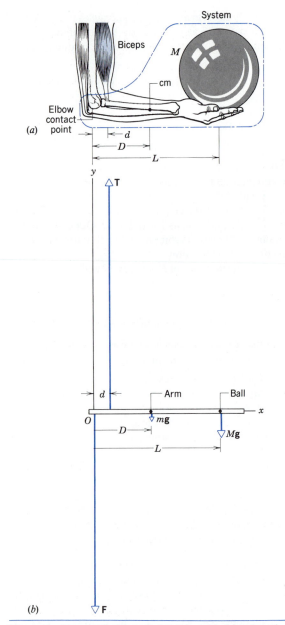

(a)

(b)

Figure 6   Sample Problem 2. (*a*) A hand holds a bowling ball. The system boundary is marked. (*b*) A free-body diagram, showing the forces that act. The vectors are to scale, showing the powerful forces exerted by the biceps muscle and by the upper arm at the elbow joint (point *O*).

From Eq. 6, $\Sigma F_y = 0$, we find

$$\sum F_y = T - F - mg - Mg = 0. \qquad (15)$$

Applying Eq. 7 about an axis through *O* and taking counterclockwise rotations as positive, we obtain

$$\sum \tau_z = (T)(d) + (F)(0) - (mg)(D) - (Mg)(L) = 0. \qquad (16)$$

By choosing our axis to pass through point *O*, we have eliminated the variable *F* from this equation. Equation 16, solved for *T*, yields

$$T = g\,\frac{mD + ML}{d}$$

$$= (9.8 \text{ m/s}^2)\,\frac{(1.8 \text{ kg})(15 \text{ cm}) + (7.2 \text{ kg})(33 \text{ cm})}{4.0 \text{ cm}}$$

$$= 648 \text{ N} = 146 \text{ lb}.$$

Thus the biceps muscle must pull up on the forearm with a force that is about nine times larger than the weight of the bowling ball.

If we solve Eq. 15 for *F* and substitute into it the value of *T* given above, we find

$$F = T - g(M + m)$$

$$= 648 \text{ N} - (9.8 \text{ m/s}^2)(7.2 \text{ kg} + 1.8 \text{ kg})$$

$$= 560 \text{ N} = 126 \text{ lb}.$$

The force *F* is also large, being about eight times the weight of the bowling ball.

Sample Problem 3   A ladder whose length *L* is 12 m and whose mass *m* is 45 kg rests against a wall. Its upper end is a distance *h* of 9.3 m above the ground, as in Fig. 7*a*. The center of mass of the ladder is one-third of the way up the ladder. A firefighter whose mass *M* is 72 kg climbs halfway up the ladder. Assume that the wall, but not the ground, is frictionless. What forces are exerted on the ladder by the wall and by the ground?

Solution   Figure 7*b* shows a free-body diagram. The wall exerts a horizontal force $\mathbf{F}_w$ on the ladder; it can exert no vertical force because the wall–ladder contact is assumed to be frictionless. The ground exerts a force on the ladder with a horizontal component *f* due to friction and a vertical component *N*, the normal force. We choose coordinate axes as shown, with the origin *O* at the point where the ladder meets the ground. The distance *a* from the wall to the foot of the ladder is readily found from

$$a = \sqrt{L^2 - h^2} = \sqrt{(12 \text{ m})^2 - (9.3 \text{ m})^2} = 7.6 \text{ m}.$$

From Eq. 6, the balance of forces equation, we have, respectively,

$$\sum F_x = F_w - f = 0 \qquad (17)$$

and

$$\sum F_y = N - Mg - mg = 0. \qquad (18)$$

Equation 18 yields

$$N = g(M + m)$$

$$= (9.8 \text{ m/s}^2)(72 \text{ kg} + 45 \text{ kg}) = 1150 \text{ N}.$$

From Eq. 7, the balance of torques equation, we have, taking an axis through *O*, the point of contact of the ladder with the ground,

$$\sum \tau_z = -(F_w)(h) + (Mg)(a/2) + (mg)(a/3) = 0. \qquad (19)$$

This wise choice of location for the axis eliminated two variables, *f* and *N*, from the balance of torques equation. We find, solving Eq. 19 for F_w,

$$F_w = \frac{ga(M/2 + m/3)}{h}$$

$$= \frac{(9.8 \text{ m/s}^2)(7.6 \text{ m})[(72 \text{ kg})/2 + (45 \text{ kg})/3]}{9.3 \text{ m}} = 410 \text{ N}.$$

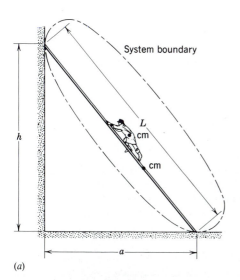

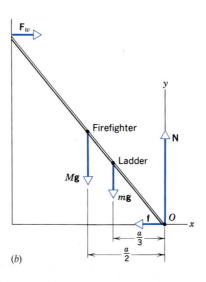

(a)   (b)

Figure 7   Sample Problems 3 and 4. (a) A firefighter climbs halfway up a ladder that is leaning against a frictionless wall. (b) A free-body diagram, showing (to scale) all the forces that act.

From Eq. 17 we have at once

$$f = F_w = 410 \text{ N}.$$

Sample Problem 4   In Sample Problem 3, the coefficient of static friction μ_s between the ladder and the ground is 0.54. How far up the ladder can the firefighter go before the ladder starts to slip?

Solution   In Sample Problem 3, we found that, when the firefighter is halfway up the ladder, the normal force N is 1150 N. The *maximum* force of static friction is $f_{max} = \mu_s N = (0.54)(1150 \text{ N}) = 620 \text{ N}$. The actual frictional force we found in that problem was $f = 410 \text{ N}$, which is less than f_{max}. As the firefighter continues to climb, f will increase, and slippage will occur when the firefighter has climbed a distance d along the ladder such that $f = f_{max}$. We wish to find the distance d.

The forces that act have the same labels as in Fig. 7. Applying Eq. 7 about an axis through the point of contact of the ladder with the ground, we have

$$\sum \tau_z = -(F_w)(h) + (mg)(a/3) + (Mg)(da/L) = 0,$$

where da/L is the horizontal distance between O and the line of action of the firefighter's weight Mg. Solving for F_w, we find

$$F_w = \frac{ga}{h}\left(M\frac{d}{L} + \frac{m}{3}\right). \tag{20}$$

Equation 20 shows us that as the firefighter climbs the ladder (that is, as d increases), the force F_w exerted by the wall must increase to maintain equilibrium. To find d at the slippage point, we must first find F_w.

Equation 6 for the balance of forces in the x direction gives

$$\sum F_x = F_w - f = 0.$$

At the point of slipping, we then have

$$F_w = f = f_{max} = \mu_s N. \tag{21}$$

From Eq. 6 for the balance of forces in the y direction, we have

$$\sum F_y = N - Mg - mg = 0,$$

or

$$N = g(M + m). \tag{22}$$

We find, combining Eqs. 21 and 22,

$$F_w = \mu_s g(M + m). \tag{23}$$

If, finally, we combine Eqs. 20 and 23 and solve for d, we have

$$d = L\left[\frac{\mu_s h}{a}\frac{(M+m)}{M} - \frac{m}{3M}\right]$$

$$= (12 \text{ m})\left[\frac{(0.54)(9.3 \text{ m})}{7.6 \text{ m}}\frac{(72 \text{ kg} + 45 \text{ kg})}{72 \text{ kg}} - \frac{45 \text{ kg}}{(3)(72 \text{ kg})}\right]$$

$$= 10.4 \text{ m}.$$

The firefighter can climb 87% of the way up the ladder before it starts to slip.

What is the minimum coefficient of friction that permits the firefighter to climb the entire ladder ($d = L$)? What is the minimum coefficient of friction necessary to keep the ladder from slipping before the firefighter starts to climb?

Sample Problem 5   A uniform beam of length $L = 3.3$ m and mass $m = 8.5$ kg is hinged at a wall as in Fig. 8a. A wire connected to the wall a distance $d = 2.1$ m above the hinge is connected to the other end of the beam, the length of the wire being such that the beam makes an angle of $\theta = 30°$ with the horizontal. A body of mass $M = 56$ kg is suspended from the upper end of the beam. Find the tension in the wire and the force exerted by the hinge on the beam.

Solution   Figure 8b shows all the external forces that act on the beam, which we have chosen as our system. Because two of the forces are directed vertically downward, we choose our axes to be horizontal and vertical. The tension in the wire and the force exerted by the hinge on the beam are represented by their horizontal and vertical components.

From Eq. 6 for translational equilibrium, we obtain

$$\sum F_x = F_h - T_h = 0, \tag{24}$$

and

$$\sum F_y = F_v + T_v - mg - Mg = 0. \tag{25}$$

To apply the condition for rotational equilibrium, we choose

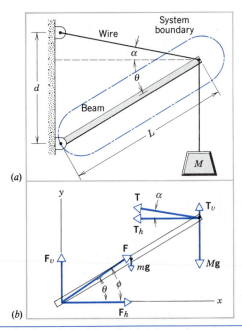

(a)

(b)

Figure 8 Sample Problem 5. (*a*) A beam is supported by a hinge on a wall at its lower end and by a wire to the wall at its upper end. An object of mass M hangs from the upper end of the beam. (*b*) A free-body diagram, showing the forces acting on the beam. A force F is exerted by the hinge and a force T is supplied by the tension in the wire.

an axis through the upper end of the beam. (Why?) From Eq. 7, we then have

$$\sum \tau_z = -F_v(L \cos\theta) + F_h(L \sin\theta) + mg\left(\frac{L}{2}\cos\theta\right) = 0,$$

or

$$F_v = F_h \tan\theta + \frac{mg}{2}. \qquad (26)$$

If we substitute the numerical values, Eqs. 24–26 become

$$F_h = T_h,$$

$$F_v + T_v = 632 \text{ N},$$

and

$$F_v = (0.577)F_h + 41.7 \text{ N}.$$

Inspection shows that we have four unknowns, namely, F_v, F_h, T_v, and T_h, but only three equations relating them. We need another relation among these quantities if we are to solve this problem. This final relation follows from the fact that T_v and T_h must add to give a resultant vector **T** directed along the wire. The (flexible) wire cannot support a force transverse to its long dimension. [Note that this is not true for the (rigid) beam.] Hence our fourth equation is

$$T_v = T_h \tan\alpha, \qquad (27)$$

where $\tan\alpha = (d - L\sin\theta)/(L\cos\theta) = 0.157$, corresponding to $\alpha = 8.9°$. Thus our fourth equation becomes

$$T_v = 0.157 T_h.$$

Combining the four equations we find, after doing the necessary algebra,

$$F_v = 506 \text{ N}, \qquad F_h = 804 \text{ N}, \qquad T_v = 126 \text{ N}, \qquad T_h = 804 \text{ N}.$$

The tension in the wire will then be

$$T = \sqrt{T_h^2 + T_v^2} = 814 \text{ N},$$

and the force exerted by the hinge on the beam is

$$F = \sqrt{F_h^2 + F_v^2} = 950 \text{ N}.$$

Note that both T and F are considerably larger than the combined weights of the beam and the suspended body (632 N).

The vector **F** makes an angle with the horizontal of

$$\phi = \tan^{-1}\frac{F_v}{F_h} = 32.2°.$$

Thus the resultant force vector acting on the beam at the hinge does not point along the beam direction.

In the preceding examples we have been careful to limit the number of unknown forces to the number of independent equations relating the forces. When all the forces act in a plane, we can have only three independent equations of equilibrium, one for rotational equilibrium about any axis normal to the plane and two others for translational equilibrium in the plane. However, we often have more than three unknown forces. For example, in Sample Problems 3 and 4, if we drop the artificial assumption of a frictionless wall, we have four unknown scalar quantities, namely, the horizontal and vertical components of the force acting on the ladder at the wall and the horizontal and vertical components of the force acting on the ladder at the ground. Because we have only three scalar equations, these forces cannot be determined. For any value assigned to one unknown force, the other three forces can be determined. But if we have no basis for assigning any particular value to an unknown force, an infinite number of solutions is mathematically possible. We must therefore be able to find another independent relation between the unknown forces if we hope to solve the problem uniquely. (In Sample Problem 5, this last equation came from a physical property of one of the elements of the system.) Taking torques about a second axis does not give a fourth independent equation; you can show that such an equation is a linear combination of the first torque equation and the two force equations, and so it contains no new information.

Another simple example of an undetermined structure occurs when we wish to determine the forces exerted by the ground on each of the four tires of an automobile when it is at rest on a horizontal surface. If we assume that these forces are normal to the ground, we have four unknown scalar quantities. All other forces, such as the weight of the car and passengers, act normal to the ground. Therefore, we have only three independent equations giving the equilibrium conditions, one for translational equilibrium in the single direction of all the forces and two for rotational equilibrium about the two axes perpendicular to each other in a horizontal plane. Again the solution of the problem is mathematically indeterminate. A four-legged table with all its legs in contact with the floor is a similar example.

Of course, since there is actually a unique solution to this physical problem, we must find a physical basis for the additional independent relation between the forces that enables us to solve the problem. The difficulty is removed when we realize that structures are never perfectly rigid, as we have tacitly assumed throughout. All structures are actually somewhat deformed. For example, the automobile tires and the ground are deformed, as are the ladder and wall. The laws of elasticity and the elastic properties of the structure determine the nature of the deformation and provide the necessary additional relation between the four forces. A complete analysis therefore requires not only the laws of rigid body mechanics but also the laws of elasticity. We briefly consider these matters in Section 14-5.

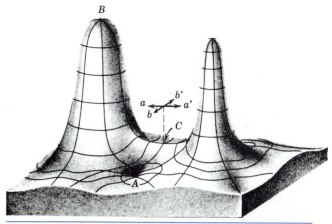

Figure 9   A gravitational potential energy surface. A particle experiencing the corresponding gravitational force would behave in similar fashion to a particle sliding without friction on a real solid surface of this shape. A particle placed at *A, B,* or *C* will be at equilibrium. Point *A* represents *stable* equilibrium, because a particle slightly displaced from *A* will tend to return there. Point *B* represents *unstable* equilibrium, because a particle slightly displaced from *B* tends to increase its displacement. At point *C,* a particle displaced along the *aa′* axis will tend to return to *C,* but if it is displaced along the *bb′* axis it will tend to increase its displacement. Point *C* is called a *saddle point,* because the surface has roughly the shape of a saddle in this region. Neutral equilibrium, which is not illustrated, would be represented by a plane horizontal surface.

14-4   STABLE, UNSTABLE, AND NEUTRAL EQUILIBRIUM OF RIGID BODIES IN A GRAVITATIONAL FIELD

In Chapter 8, we saw that the gravitational force is a conservative force. For conservative forces we can define a potential energy function $U(x, y, z)$, where U is related to **F** by

$$F_x = -\frac{\partial U}{\partial x}, \qquad F_y = -\frac{\partial U}{\partial y}, \qquad F_z = -\frac{\partial U}{\partial z}.$$

At points where $\partial U/\partial x$ is zero, a particle subject to this conservative force will be in translational equilibrium in the x direction, for then F_x equals zero. Likewise, at points were $\partial U/\partial y$ or $\partial U/\partial z$ are zero, a particle will be in translational equilibrium in the y and z directions, respectively. The derivative of U at a point, and the corresponding component of the force on a particle, will be zero when U has an extreme value (maximum or minimum) at that point or when U is constant with respect to the variable coordinate. Thus the particle can be in equilibrium when U is a maximum, a minimum, or a constant. Let us consider in turn each of these three possibilities.

When U is a *minimum* (point *A* of Fig. 9), the particle is in *stable* equilibrium; any displacement from this position will result in a restoring force tending to return the particle to the equilibrium position. Equivalently, we can say that if a body is in stable equilibrium, work must be done on it by an external agent to change its position. This results in an increase in its potential energy.

When U is a *maximum* (point *B* of Fig. 9), the particle is

in *unstable* equilibrium; any displacement from this position will result in a force tending to push the particle farther from the equilibrium position. In this case no work must be done on the particle by an external agent to change its position; the work done in displacing the body is supplied by the conservative force, resulting in a decrease in potential energy.

When U is *constant,* the particle is in *neutral* equilibrium. In this case a particle can be displaced slightly without experiencing either a repelling or restoring force.

All these remarks apply to particles, that is, to translational motion. Suppose now we treat a rigid body. We must consider rotational equilibrium as well as translational equilibrium. The problem of a rigid body in a gravitational field is particularly simple, however, because *all the gravitational forces on the particles of the rigid body can be considered to act at one point, both for translational and rotational purposes.* We can replace this entire rigid body, for purposes of equilibrium under gravitational forces, by a single particle having the equivalent mass at the center of gravity.

For example, consider a cube at rest on one face on a horizontal table. The center of gravity is shown at the center of the central cross section of the cube in Fig. 10*a.* Let us supply a force to the cube so as to rotate it, without its slipping, about an axis along an edge. Note that the center of gravity is raised and that work is done on the

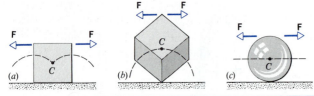

Figure 10 Equilibrium of an extended body. (*a*) A cube resting on one face is in *stable equilibrium,* since its center of gravity *C* is raised if the cube is tipped by a horizontal force **F**. (*b*) A cube balanced on one corner is in *unstable equilibrium,* since *C* falls if the cube is tipped by **F**. (*c*) A sphere is in *neutral equilibrium* with respect to a horizontal force, since *C* neither rises nor falls when **F** is applied. Compare these criteria for the equilibrium of an extended body with those of a particle illustrated in Fig. 9.

cube, which increases its potential energy. If the force is removed, the cube tends to return to its original position. This initial position is therefore one of *stable* equilibrium. In terms of a particle of equivalent mass at the center of gravity, this process is described by the dashed line that indicates the path taken by the center of gravity during this motion. The particle is seen to have a minimum potential energy in the position of stable equilibrium, as required. We can conclude that the rigid body will be in stable equilibrium if the application of any force can raise the center of gravity of the body but not lower it.

If the cube is rotated until it balances on a corner, as in Fig. 10*b*, then once again the cube is in equilibrium. This equilibrium position is seen to be unstable. The application of even the slightest horizontal force will cause the cube to fall away from this position with a decrease of potential energy. The particle of equivalent mass at the center of gravity follows the dashed path shown. At the position of unstable equilibrium this particle has a maximum potential energy, as required. We can conclude that the rigid body will be in unstable equilibrium if the application of any horizontal force tends to lower the center of gravity of the body.

A cube balanced on an edge can be considered in unstable equilibrium if a horizontal force is applied perpendicular to the edge, but it is in stable equilibrium with respect to a horizontal force parallel to the edge. Thus a particle may be in stable equilibrium with respect to one coordinate and unstable equilibrium with respect to another. Such a condition is called a *saddle point* and corresponds to point *C* of Fig. 9.

The neutral equilibrium of a rigid body is illustrated by a sphere on a horizontal table (Fig. 10*c*). If the sphere is subjected to a horizontal force, the center of gravity is neither raised nor lowered but moves along the horizontal dashed line. The potential energy of the sphere is constant during the displacement, as is that of the particle of equivalent mass at the center of gravity. The system has no tendency to move in any direction when the applied force

is removed. A rigid body will be in neutral equilibrium if the application of any horizontal force neither raises nor lowers the center of gravity of the body.

Under what circumstances would a *suspended* rigid body be in stable equilibrium? When would a *suspended* rigid body be in unstable equilibrium, and when would it be in neutral equilibrium?

14-5 ELASTICITY

A three-legged table is a structure that can be analyzed using the techniques of this chapter. All three legs are in contact with the floor, which exerts a vertical normal force on each leg. By using one force equation for equilibrium (the weight, acting at the center of gravity, must equal the sum of the three normal forces) and two torque equations (taking torques about two perpendicular axes in the horizontal plane of the floor), we can find the three unknown normal forces from the three equations.

A four-legged table, however, gives us four unknowns and cannot be analyzed by these techniques without more information about the relationship between the normal forces. Suppose, for instance, that the legs are of slightly different lengths. As we put a very heavy weight on the table, we can compress the legs by differing amounts so that the four legs will all be in contact with the floor. From the compression of the legs, we can find the missing relationship between the forces that permits us to solve the problem (see Sample Problem 8).

The rigidity of so-called rigid bodies is in reality an illusion. Solids are composed of atoms that are not in rigid contact. Atoms do not have hard surfaces that can pack closely together; their electron clouds can be shaped or deformed by external forces. In a solid, the atoms are bound together by forces that behave very much like spring forces. Figure 11 shows a representation of a por-

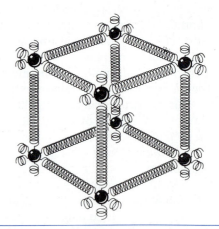

Figure 11 The atoms of a solid are distributed on a repetitive three-dimensional lattice. The interatomic forces are represented here as springs.

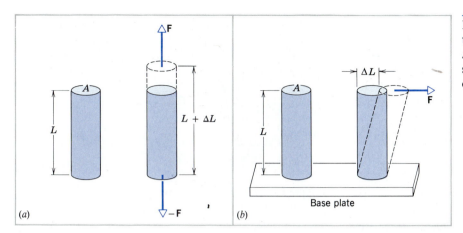

Figure 12   (*a*) A cylinder, subject to tensile stress, is stretched by an amount ΔL. (*b*) A cylinder, subject to shearing stress, deforms like a deck of playing cards.

tion of a solid *lattice,* which is a regular arrangement of atoms such as one might find in a crystal. Each atom is in equilibrium under the influence of the six springs that surround it; the effective spring constants are very large, so that it takes a large force to change the separation. This is responsible for our impression of rigidity. In other solids, the atoms may be arranged in long rows rather than in cubic lattices; these materials are not terribly rigid, as, for example, rubber. When we stretch such a material, we are applying enough force to change the atomic spacings.

All real "rigid" bodies are to some extent *elastic,* which means that we can change their dimensions slightly by pulling, pushing, twisting, or compressing them. To get a feeling for the orders of magnitude involved, consider a steel rod, 1 m long and 1 cm in diameter. If you hang a subcompact car from the end of such a rod, the rod will stretch, but only by about 0.5 mm, or 0.05%. Furthermore, the rod will return to its original length when the car is removed.

If you hang two cars from the rod, the rod will be permanently stretched and will not recover its original length when you remove the load. If you hang three cars from the rod, the rod will break. Just before rupture, the elongation of the rod will be less than 0.2%. Although deformations like this seem small, they are important in engineering practice.

Figure 12 shows two ways that a solid might change its dimensions when forces act on it. In Fig. 12*a*, a cylinder is stretched. In Fig. 12*b*, a cylinder is deformed by so-called shearing forces, such as might deform a pack of cards or a book. (A third way is uniform compression, which results from the application of forces uniformly in all directions. Uniform compression will be considered in Chapter 17.) The three modes have in common that there is a *stress,* related to the applied forces, and there is a *strain,* or a deformation of some kind.

The stress and the strain take different forms in the cases of Fig. 12, but—over the range of useful engineering practice—they are proportional to each other. The con-

stant of proportionality is called a *modulus of elasticity.* Thus

$$\text{stress} = \text{modulus} \times \text{strain}. \qquad (28)$$

Figure 13 shows the relation between stress and strain for a steel test cylinder such as that of Fig. 14. For a substantial portion of the range of applied stresses, the stress–strain curve is linear and Eq. 28 applies, with a constant modulus (corresponding to the linear portion of Fig. 13). As the stress continues to increase, the stress–strain relationship may become nonlinear, but the material remains elastic: that is, if the stress is removed, the specimen returns to its original dimensions.

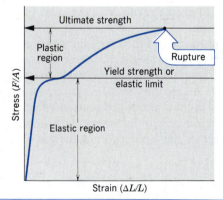

Figure 13   A stress–strain curve for a steel test specimen, such as that of Fig. 14. The specimen deforms permanently when the stress is equal to the *yield strength* of the material. It ruptures when the stress is equal to the *ultimate strength* of the material.

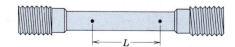

Figure 14   A test specimen, used to determine a stress–strain curve such as that of Fig. 13.

If the stress is increased beyond the *yield strength* or *elastic limit* of the material, the specimen becomes permanently changed and does not recover its original dimensions when the stress is removed; this kind of behavior is called *plasticity.* Beyond yielding—inevitably—comes rupture, which occurs at a stress called the *ultimate strength.*

Tension and Compression

For simple stretching or compressing, the stress is defined as F/A, the force divided by the area over which it acts, and the strain, or deformation, as the dimensionless quantity $\Delta L/L$, the fractional change in length of the specimen. If specimen is a long rod, note that not only the entire rod but also any section of it experiences the same strain when a given stress is applied. Because the strain is dimensionless, the modulus in Eq. 28 has the same dimensions as the stress, namely, force per unit area.

The modulus for tensile and compressive stresses is called *Young's modulus,* represented, in engineering practice, by the symbol E. Equation 28 becomes

$$\frac{F}{A} = E \frac{\Delta L}{L}$$

or

$$\Delta L = \frac{FL}{EA}. \qquad (29)$$

The strain in a specimen, $\Delta L/L$, can often be measured conveniently with a *strain gage;* see Fig. 15. These simple and useful devices, which can be attached directly to operating machinery with adhesives, are based on the principle that the electrical resistance of wires made of certain materials is a function of the strain in the wire.

Although the modulus may be almost the same for both tension and compression, the *ultimate strength* may well be different for the two cases. Concrete, for example, is very strong in compression but is so weak in tension that it is almost never used in this way in engineering practice. Table 1 shows the Young's modulus values and other

Figure 15   A strain gage, of overall dimensions 9.8 mm by 4.6 mm. The gage is fastened with adhesive to the object whose strain is to be measured. The electrical resistance of the gage varies with the strain, permitting strains up to about 3% to be measured.

elastic properties for some materials of engineering interest.

Shearing

In the case of shearing, the stress is also a force per unit area but the force vector lies in the plane of the area rather than at right angles to it. The strain is again the dimensionless ratio $\Delta L/L$, the quantities being defined as shown in Fig. 12b. The modulus, which is given the symbol G in engineering practice, is called the *shear modulus.* Equation 29 applies to shearing stresses, with the modulus E replaced by the modulus G.

Shearing stresses play a critical role in shafts that rotate under load, in bone fractures caused by twisting, and in springs.

Sample Problem 6   A structural steel rod has a radius R of 9.5 mm and a length L of 81 cm. A force F of 6.2×10^4 N (about 7 tons) stretches it axially. (*a*) What is the stress in the rod? (*b*) What is the elongation of the rod under this load?

TABLE 1   SOME ELASTIC PROPERTIES OF SELECTED MATERIALS OF ENGINEERING INTEREST

Material	Density (kg/m³)	Young's Modulus (10^9 N/m²)	Ultimate Strength (10^6 N/m²)	Yield Strength (10^6 N/m²)
Steel[a]	7860	200	400	250
Aluminum	2710	70	110	95
Glass	2190	65	50[b]	—
Concrete[c]	2320	30	40[b]	—
Wood[d]	525	13	50[b]	—
Bone	1900	9[b]	170[b]	—
Polystyrene	1050	3	48	—

[a] Structural steel (ASTM-A36).
[b] In compression.
[c] High strength.
[d] Douglas fir.

Solution (*a*) The stress is defined from

$$\text{stress} = \frac{F}{A} = \frac{F}{\pi R^2} = \frac{6.2 \times 10^4 \text{ N}}{(\pi)(9.5 \times 10^{-3} \text{ m})^2}$$

$$= 2.2 \times 10^8 \text{ N/m}^2.$$

The yield strength for structural steel is 2.5×10^8 N/m², so that this rod is dangerously close to its yield strength.

(*b*) From Eq. 29, using the result we have just calculated, we obtain

$$\Delta L = \frac{(F/A)L}{E} = \frac{(2.2 \times 10^8 \text{ N/m}^2)(0.81 \text{ m})}{2.0 \times 10^{11} \text{ N/m}^2}$$

$$= 8.9 \times 10^{-4} \text{ m} = 0.89 \text{ mm}.$$

Thus the strain $\Delta L/L$ is $(8.9 \times 10^{-4}$ m$)/(0.81$ m$)$, which is 1.1×10^{-3} or 0.11%.

Sample Problem 7 The femur, which is the principal bone of the thigh, has a minimum diameter in an adult male of about 2.8 cm, corresponding to a cross section A of 6×10^{-4} m². At what compressive load would it break?

Solution From Table 1 we see that the ultimate strength S_u for bone in compression is 170×10^6 N/m². The compressive force is then

$$F = S_u A = (170 \times 10^6 \text{ N/m}^2)(6 \times 10^{-4} \text{ m}^2)$$

$$= 1.0 \times 10^5 \text{ N}.$$

This is 23,000 lb or 11 tons. Although this is a large force, it can be encountered during, for example, an unskillful parachute landing on hard ground. The force need not be sustained; a few milliseconds will do it.

We are now prepared to understand how the elastic properties of materials can help to determine their equilibrium conditions, as suggested by the following sample problem.

Sample Problem 8 A four-legged table has three legs of length $D = 1.00$ m; the fourth leg is longer by a small distance $d = 0.50$ mm, so that the table wobbles slightly. A heavy steel cylin-der whose mass M is 290 kg is placed upright on the table so that all four legs compress and the table no longer wobbles. Each leg is a wooden cylinder whose cross-sectional area A is 1.0 cm². Young's modulus E for the wood is 1.3×10^{10} N/m². Assume that the tabletop remains level and that the legs do not buckle. With what force does the floor push upward on each leg?

Solution We take the tabletop as our system. If the tabletop remains level, each of the three short legs must be compressed by the same amount ΔL_3, by the same force F_3. The single long leg must be compressed by a larger amount ΔL_1, by a force F_1, and we must have

$$\Delta L_3 + d = \Delta L_1.$$

From Eq. 29 ($\Delta L = FL/EA$), we can write this relation as

$$F_3 D + dAE = F_1 (D + d) \approx F_1 D, \qquad (30)$$

where we neglect d in comparison with D in the last term. From Eq. 6 for the balance of forces in the vertical direction, we have

$$\sum F_y = 3F_3 + F_1 - Mg = 0. \qquad (31)$$

If we solve Eqs. 30 and 31 for the unknown forces, we find

$$F_3 = \frac{Mg}{4} - \frac{dAE}{4D}$$

$$= \frac{(290 \text{ kg})(9.8 \text{ m/s}^2)}{4}$$

$$- \frac{(5.0 \times 10^{-4} \text{ m})(10^{-4} \text{ m}^2)(1.3 \times 10^{10} \text{ N/m}^2)}{(4)(1.00 \text{ m})}$$

$$= 711 \text{ N} - 163 \text{ N} = 548 \text{ N}.$$

Similarly,

$$F_1 = \frac{Mg}{4} + \frac{3dAE}{4D}$$

$$= 711 \text{ N} + 489 \text{ N} = 1200 \text{ N}.$$

You can show that, to reach their equilibrium configuration, the three short legs were each compressed by 0.42 mm and the single long leg by 0.92 mm, the difference being 0.50 mm, as expected.

The cylinder must be placed closer to the longer leg than any of the shorter legs if the tabletop is to be horizontal. The equilibrium condition on the torques can be used to find its position, if we know the dimensions of the tabletop and the placement of its legs.

QUESTIONS

1. Are Eqs. 1 and 3 both necessary and sufficient conditions for mechanical equilibrium? For static equilibrium?

2. Is a baseball in equilibrium at the instant it comes to rest at the top of a vertical pop fly?

3. In a simple pendulum, is the bob in equilibrium at any point of its swing? If so, where?

4. A wheel rotating at constant angular velocity ω about a fixed axis is in mechanical equilibrium because no net external force or torque acts on it. However, the particles that make up the wheel undergo a centripetal acceleration **a** directed toward the axis. Since **a** $\neq 0$, how can the wheel be said to be in equilibrium?

5. Give several examples of an object that is not in equilibrium even though the resultant of all the forces acting on it is zero.

6. Do the center of mass and the center of gravity coincide for a building? For a lake? Under what conditions does the differ-

ence between the center of mass and the center of gravity become significant? Give an example.

7. If a rigid body is thrown into the air without spinning, it does not spin during its flight, provided air resistance can be ignored. What does this simple result imply about the location of the center of gravity?

8. The Olympic gymnast Mary Lou Retton did some amazing things on the uneven parallel bars. A friend tells you that careful analysis of films of her exploits shows that, no matter what she does, her center of mass is above her point(s) of support at all times, as required by the laws of physics. Comment on your friend's statement.

9. Which is more likely to break in use: a hammock stretched tightly between two trees or one that sags quite a bit? Explain your answer.

10. A ladder is at rest with its upper end against a wall and its lower end on the ground. Is it more likely to slip when someone stands on it at the bottom or at the top? Explain.

11. A book rests on a table. The table pushes up on the book with a force just equal to the weight of the book. Speaking loosely, just how does the table "know" what upward force it must provide? What is the mechanism by which this force comes into play? (See "The Smart Table," by Earl Zwicker, *The Physics Teacher,* December 1981, p. 633.)

12. Stand facing the edge of an open door, one foot on each side of the door. You will find that you are not able to stand on your toes. Why?

13. Sit in a straight-backed chair and try to stand up without leaning forward. Why can't you do it?

14. Long balancing poles help a tightrope walker to maintain balance. How?

15. A composite block made up of wood and metal rests on a tabletop. In which orientation of the two shown in Fig. 16 can you tip it over with the least force?

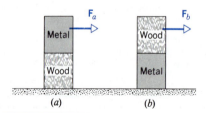

Figure 16 Question 15.

16. In Sample Problem 5, why isn't it necessary to consider friction at the hinge?

17. A picture hangs from a wall by two wires. What orientation should the wires have to be under minimum tension? Explain how equilibrium is possible with any number of orientations and tensions, even though the picture has a definite mass.

18. Show how to use a spring balance to weigh objects well beyond the maximum reading of the balance.

19. Explain, using forces and torques, how a tree can maintain equilibrium in a high wind.

20. A virus in a rotating liquid-filled centrifuge tube is in uniform circular motion (that is, in *accelerated* motion) as viewed by an observer in the laboratory. An observer rotating with the centrifuge, however, would declare the virus to be *unaccelerated.* Explain how the virus can be in equilibrium for this second observer but not for the first.

21. A uniform block, in the shape of a rectangular parallelepiped of sides in the ratio 1 : 2 : 3, lies on a horizontal surface. In what position, that is, on which of its three different faces, can it be said to be most stable, if any?

22. Is there such a thing as a truly rigid body? If so, give an example. If not, explain why.

23. You are sitting in the driver's seat of a parked automobile. You are told that the forces exerted upward by the ground on each of the four tires are different. Discuss the factors that enter into a consideration of whether this statement is true or not.

24. In Sample Problem 3, if the wall were not frictionless, would the empirical laws of friction supply us with the extra condition needed to determine the extra (vertical) force exerted by the wall on the ladder?

25. As the test cylinder in Fig. 14 stretches under the applied stress it gets longer. What change, if any, do you expect in the diameter of the cylinder?

26. Is Young's modulus for rubber higher or lower than Young's modulus for steel? By this criterion, is rubber more elastic than steel?

27. A horizontal beam supported at both ends is loaded in the middle. Show that the upper part of the beam is under compression whereas the lower part is under tension.

28. Why are reinforcing rods used in concrete structures? (Compare the tensile strength of concrete to its compressive strength.)

PROBLEMS

Section 14-1 Conditions of Equilibrium

1. An eight-member family, whose weights in pounds are indicated, is balanced on a see-saw, as shown in Fig. 17. What is the number of the person who causes the largest torque, about the pivot point, directed (a) out of the page and (b) into the page?

2. A rigid square object of negligible weight is acted on by three forces that pull on its corners as shown, to scale, in Fig. 18. (a) Is the first condition of equilibrium satisfied? (b) Is the second condition of equilibrium satisfied? (c) If either of the

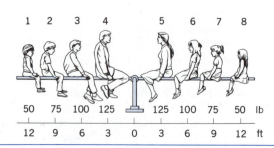

Figure 17 Problem 1.

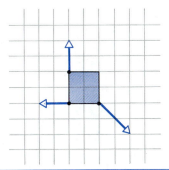

Figure 18 Problem 2.

preceding answers is no, could a fourth force restore the equilibrium of the object? If so, specify the magnitude, direction, and point of application of the needed force.

3. Prove that when only three forces act on an object in equilibrium, they must be coplanar and their lines of action must either meet at a point or be parallel.

Section 14-3 Examples of Equilibrium

4. A certain nut is known to require forces of 46 N exerted on it from both sides to crack it. What forces F will be required when it is placed in the nutcracker shown in Fig. 19?

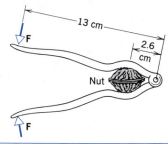

Figure 19 Problem 4.

5. The leaning Tower of Pisa (see Fig. 20) is 55 m high and 7.0 m in diameter. The top of the tower is displaced 4.5 m from the vertical. Treating the tower as a uniform, circular

Figure 20 Problem 5.

cylinder, (*a*) what additional displacement, measured at the top, will bring the tower to the verge of toppling? (*b*) What angle with the vertical will the tower make at that moment? (The current rate of movement of the top is 1 mm/year.)

6. A cube stays at rest on a horizontal table when a small horizontal force is applied perpendicular to and at the center of an upper edge. The force is now steadily increased. Does the cube slide or topple first? The coefficient of static friction between the surfaces is equal to 0.46.

7. A crate in the form of a 1.12-m cube contains a piece of machinery whose design is such that the center of gravity of the crate and its contents is located 0.28 m above its geometrical center. The crate rests on a ramp that makes an angle θ with the horizontal. As θ is increased from zero, an angle will be reached at which the crate will either start to slide down the ramp or tip over. Which event will occur if the coefficient of static friction is (*a*) 0.60? (*b*) 0.70? In each case give the angle at which the event occurs.

8. A flexible chain of weight W hangs between two fixed points, A and B, at the same level, as shown in Fig. 21. Find (*a*) the force exerted by the chain on each endpoint and (*b*) the tension in the chain at the lowest point.

Figure 21 Problem 8.

9. In Fig. 22 a man is trying to get his car out of the mud on the shoulder of a road. He ties one end of a rope tightly around the front bumper and the other end around a telephone pole 62 ft away. He then pushes sideways on the rope at its midpoint with a force $F = 120$ lb, displacing the center of the rope 1.5 ft from its previous position, and the car almost moves. Find the force exerted by the rope on the car. (The rope stretches somewhat under the tension.)

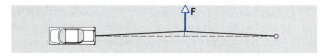

Figure 22 Problem 9.

10. A uniform sphere of weight w and radius r is being held by a rope attached to a frictionless wall a distance L above the center of the sphere, as in Fig. 23. Find (*a*) the tension in the rope and (*b*) the force exerted on the sphere by the wall.

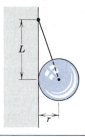

Figure 23 Problem 10.

11. A parked automobile of mass 1360 kg has a wheel base (distance between front and rear axles) of 305 cm. Its center of gravity is located 178 cm behind the front axle. Determine (*a*) the upward force exerted by the level ground on each of the front wheels (assumed the same) and (*b*) the upward force exerted by the level ground on each of the rear wheels (assumed the same).

12. A 160-lb person is walking across a level bridge and stops three-fourths of the way from one end. The bridge is uniform and weighs 600 lb. What are the values of the vertical forces exerted on each end of the bridge by its supports?

13. A diver of weight 582 N stands at the end of a uniform 4.48-m diving board of weight 142 N. The board is attached by two pedestals 1.55 m apart, as shown in Fig. 24. Find the tension (or compression) in each of the two pedestals.

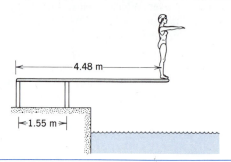

Figure 24   Problem 13.

14. A meter stick balances on a knife edge at the 50.0-cm mark. When two nickels are stacked over the 12.0-cm mark, the loaded stick is found to balance at the 45.5-cm mark. A nickel has a mass of 5.00 g. Find the mass of the meter stick.

15. A beam is carried by three workers, one at one end and the other two supporting the beam between them on a crosspiece so placed that the load is equally divided among the three. Find where the crosspiece is placed. Neglect the mass of the crosspiece.

16. A 74.6-kg window cleaner uses a 10.3-kg ladder that is 5.12 m long. He places one end 2.45 m from a wall and rests the upper end against a cracked window and climbs the ladder. He climbs 3.10 m up the ladder when the window breaks. Neglecting friction between the ladder and the window and assuming that the base of the ladder did not slip, find (*a*) the force exerted on the window by the ladder just before the window breaks and (*b*) the magnitude and direction of the force exerted on the ladder by the ground just before the window breaks.

17. Figure 25 shows the anatomical structures in the lower leg and foot that are involved when the heel is raised off the floor so that the foot effectively contacts the floor at only one point, shown as *P* in the figure. Calculate the forces that must be exerted on the foot by the calf muscle and by the lower-leg bones when a 65-kg person stands tip-toe on one foot. Compare these forces to the person's weight. Assume that $a = 5.0$ cm and $b = 15$ cm.

18. Two identical uniform frictionless spheres, each of weight *W*, rest as shown in Fig. 26 at the bottom of a fixed, rectangular container. The line of centers of the spheres makes an

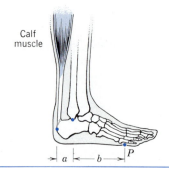

Figure 25   Problem 17.

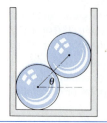

Figure 26   Problem 18.

angle θ with the horizontal. Find the forces exerted on the spheres (*a*) by the container bottom, (*b*) by the container sides, and (*c*) by one another.

19. What minimum force *F* applied horizontally at the axle of the wheel in Fig. 27 is necessary to raise the wheel over an obstacle of height *h*? Take *r* as the radius of the wheel and *W* as its weight.

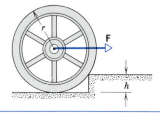

Figure 27   Problem 19.

20. A balance is made up of a rigid rod free to rotate about a point not at the center of the rod. It is balanced by unequal weights placed in the pans at each end of the rod. When an object of unknown mass *m* is placed in the left-hand pan, it is balanced by an object of mass m_1 placed in the right-hand pan, and similarly when the object of mass *m* is placed in the right-hand pan, it is balanced by an object of mass m_2 in the left-hand pan. Show that

$$m = \sqrt{m_1 m_2}.$$

21. A uniform sphere of weight *w* lies at rest wedged between two inclined planes of inclination angles θ_1 and θ_2 (Fig. 28). (*a*) Assume that no friction is involved and determine the forces (directions and magnitudes) that the planes exert on the sphere. (*b*) What change would it make, in principle, if friction were taken into account?

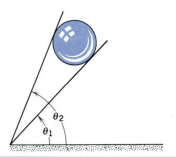

Figure 28   Problem 21.

22. A 15.4-kg object is being lifted by the pulley system shown in Fig. 29. The upper arm is vertical, whereas the forearm makes an angle of 27.0° with the horizontal. What forces are being exerted on the forearm by (*a*) the triceps muscle and (*b*) the upper-arm bone (the humerus)? The forearm and hand together have a mass of 2.13 kg with a center of mass 14.7 cm (measured along the arm) from the point where the two bones are in contact. The triceps muscle pulls vertically upward at a point 2.50 cm behind the contact point.

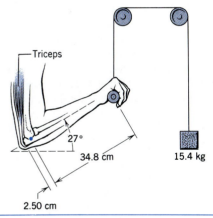

Triceps

27°

34.8 cm

15.4 kg

2.50 cm

Figure 29   Problem 22.

23. A 52.3-kg uniform square sign, 1.93 m on a side, is hung from a 2.88-m rod of negligible mass. A cable is attached to the end of the rod and to a point on the wall 4.12 m above the point where the rod is fixed to the wall, as shown in Fig.

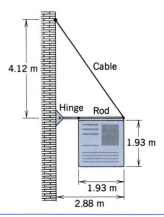

4.12 m

Cable

Hinge   Rod

1.93 m

1.93 m

2.88 m

Figure 30   Problem 23.

30. (*a*) Find the tension in the cable. (*b*) Calculate the horizontal and vertical components of the force exerted by the wall on the rod.

24. A trap door in a ceiling is 3.0 ft (= 0.91 m) square, weighs 25 lb (mass = 11 kg), and is hinged along one side with a catch at the opposite side. If the center of gravity of the door is 4.0 in. (= 10 cm) from the door's center and closer to the hinged side, what forces must (*a*) the catch and (*b*) the hinge sustain?

25. One end of a uniform beam weighing 52.7 lb and 3.12 ft long is attached to a wall with a hinge. The other end is supported by a wire making equal angles of 27.0° with the beam and wall (see Fig. 31). (*a*) Find the tension in the wire. (*b*) Compute the horizontal and vertical components of the force on the hinge.

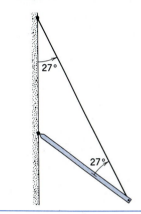

27°

27°

Figure 31   Problem 25.

26. A door 2.12 m high and 0.907 m wide has a mass of 26.8 kg. A hinge 0.294 m from the top and another 0.294 m from the bottom each support half the door's weight. Assume that the center of gravity is at the geometric center of the door and determine the horizontal and vertical force components exerted on each hinge by the door.

27. The system shown in Fig. 32 is in equilibrium. The object hanging from the end of the strut *S* weighs 513 lb and the strut itself weighs 107 lb. Find (*a*) the tension in the cable *C* and (*b*) the horizontal and vertical force components exerted on the strut by the pivot *P*.

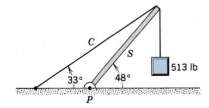

C

S

513 lb

33°   48°

P

Figure 32   Problem 27.

28. A nonuniform bar of weight W is suspended at rest in a horizontal position by two light cords as shown in Fig. 33; the angle one cord makes with the vertical is θ; the other makes the angle ϕ with the vertical. The length of the bar is

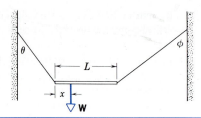

Figure 33   Problem 28.

L. Find the distance x from the left-hand end to the center of gravity.

29. A thin horizontal bar AB of negligible weight and length L is pinned to a vertical wall at A and supported at B by a thin wire BC that makes an angle θ with the horizontal. A weight W can be moved anywhere along the bar as defined by the distance x from the wall (Fig. 34). (*a*) Find the tension T in the thin wire as a function of x. Find (*b*) the horizontal and (*c*) the vertical components of the force exerted on the bar by the pin at A.

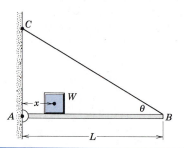

Figure 34   Problems 29 and 30.

30. In Fig. 34, the length L of the bar is 2.76 m and its weight w is 194 N. Also, $W = 315$ N and $\theta = 32.0°$. The wire can withstand a maximum tension of 520 N. (*a*) What is the maximum distance x possible before the wire breaks? (*b*) With W placed at this maximum x, what are the horizontal and vertical components of the force exerted on the bar by the pin?

31. Two uniform beams are attached to a wall with hinges and then loosely bolted together as in Fig. 35. Find the horizontal and vertical components of (*a*) the force on each hinge and (*b*) the force exerted by the bolt on each beam.

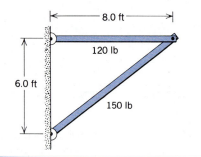

Figure 35   Problem 31.

32. A 274-N plank, of length $L = 6.23$ m, rests on the ground and on a frictionless roller at the top of a wall of height

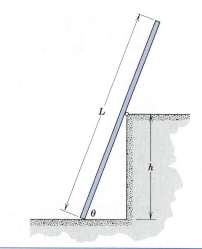

Figure 36   Problem 32.

$h = 2.87$ m (see Fig. 36). The center of gravity of the plank is at its center. The plank remains in equilibrium for any value of $\theta \geq 68.0°$ but slips if $\theta < 68.0°$. Find the coefficient of static friction between the plank and the ground.

33. In the stepladder shown in Fig. 37 AC and CE are 8.0 ft long and hinged at C. BD is a tie rod 2.5 ft long, halfway up. A man weighing 192 lb climbs 6.0 ft along the ladder. Assuming that the floor is frictionless and neglecting the weight of the ladder, find (*a*) the tension in the tie rod and (*b*) the forces exerted on the ladder by the floor. (*Hint:* It will help to isolate parts of the ladder in applying the equilibrium conditions.)

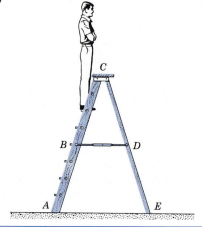

Figure 37   Problem 33.

34. By means of a turnbuckle G, a tension $\mathbf{T}$ is produced in bar AB of the square frame $ABCD$ in Fig. 38. Determine the

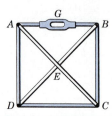

Figure 38   Problem 34.

forces produced in the other bars. The diagonals *AC* and *BD* pass each other freely at *E*. Symmetry considerations can lead to considerable simplification in this and similar problems.

35. A cubical box is filled with sand and weighs 892 N. It is desired to "roll" the box by pushing horizontally on one of the upper edges. (*a*) What minimum force is required? (*b*) What minimum coefficient of static friction is required? (*c*) Is there a more efficient way to roll the box? If so, find the smallest possible force that would be required to be applied directly to the box.

36. A car on a horizontal road makes an emergency stop by applying the brakes so that all four wheels lock and skid along the road. The coefficient of kinetic friction between tires and road is 0.40. The separation between front and rear axles is 4.20 m, and the center of mass of the car is located 1.80 m behind the front axle and 0.750 m above the road; see Fig. 39. The car weighs 11.0 kN, with occupant. Calculate (*a*) the braking deceleration of the car, (*b*) the normal force on each front and rear wheel, and (*c*) the braking force on each front and rear wheel. (*Hint:* Although the car is not in translational equilibrium, it *is* in rotational equilibrium.)

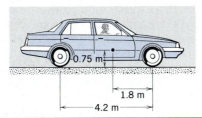

Figure 39   Problem 36.

37. A well-known problem is the following (see, for example, *Scientific American*, November 1964, p. 128): Uniform bricks are placed one upon another in such a manner as to have the maximum offset. This is accomplished by having the center of gravity of the top brick directly above the edge of the brick below it, the center of gravity of the two top bricks combined directly above the edge of the third brick from the top, and so on. (*a*) Justify this criterion for maximum offset; find the largest equilibrium offsets for four bricks. (*b*) Show that, if the process is continued downward, one can obtain as large an offset as one wants. (Martin Gardner, in the article referred to above, states: "With 52 playing cards, the first placed so that its end is flush with a table edge, the maximum overhang is a little more than $2\frac{1}{4}$ card lengths . . . ") (*c*) Suppose now, instead, one piles up uniform bricks so that the end of one brick is offset from the one below it by a constant fraction, $1/n$, of a brick length *L*. How many bricks, *N*, can one use in this process before the pile will fall over? Check the plausibility of your answer for $n = 1$, $n = 2$, $n = \infty$.

38. A homogeneous sphere of radius *r* and weight *W* slides along the floor under the action of a constant horizontal force **P** applied to a string, as shown in Fig. 40. (*a*) Show that if μ is the coefficient of kinetic friction between sphere and floor, the height *h* is given by $h = r(1 - \mu W/P)$. (*b*) Show that the sphere is not in translational equilibrium under these cir-

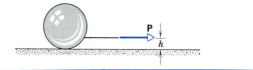

Figure 40   Problem 38.

cumstances. Is there any point about which the sphere is in rotational equilibrium? (*c*) Can one get the sphere to be in both rotational *and* translational equilibrium by a different choice of *h*? By a different direction for **P**? Explain.

Section 14-4 Stable, Unstable, and Neutral Equilibrium of Rigid Bodies in a Gravitational Field

39. A bowl having a radius of curvature *r* rests on a horizontal table. Show that the bowl will be in stable equilibrium about the center point at its bottom only if the center of mass of the material piled up in the bowl is not as high as *r* above the center of the bowl.

40. A cube of uniform density and edge *a* is balanced on a cylindrical surface of radius *r* as shown in Fig. 41. Show that the criterion for stable equilibrium of the cube, assuming that friction is sufficient to prevent slipping, is $r > a/2$.

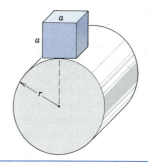

Figure 41   Problem 40.

Section 14-5 Elasticity

41. Figure 42 shows the stress–strain curve for quartzite. Calculate Young's modulus for this material.

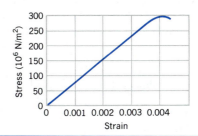

Figure 42   Problem 41.

42. After a fall, a 95-kg rock climber finds himself dangling from the end of a rope 15 m long and 9.6 mm in diameter. The rope stretches by 2.8 cm. Calculate Young's modulus of the rope.

43. A mine elevator is supported by a single steel cable 2.52 cm in diameter. The total mass of the elevator cage plus occupants is 873 kg. By how much does the cable stretch when

the elevator is suspended 42.6 m below the elevator motor? (Neglect the mass of the cable.)

44. A horizontal aluminum pole 48.0 cm in diameter projects 5.30 cm from a wall. A 120-kg object is suspended from the end of the pole. The shear modulus of aluminum is 3.00×10^{10} N/m^2. (*a*) Calculate the shear stress on the pole. (*b*) Find the vertical deflection of the end of the pole.

45. Calculate the force F needed to punch a 1.46-cm diameter hole in a steel plate 1.27 cm thick; see Fig. 43. The ultimate shear strength of steel is 345 MN/m^2.

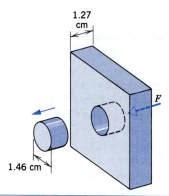

Figure 43 Problem 45.

46. A uniform bar of mass 4.7 kg and length 1.3 m is suspended at the ends by two vertical wires. One wire is of steel and has a diameter of 1.2 mm; the other wire is of aluminum and has a diameter of 0.84 mm. Before the bar was attached, the wires were of the same 1.7-m length. Find the angle θ between the bar and the horizontal; see Fig. 44. (Ignore the change in the diameters of the wires; the bar and wires are in the same plane.)

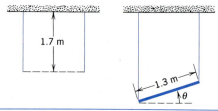

Figure 44 Problem 46.

47. A rotor blade 5.27 m long is composed of material of density 4.55 g/cm^3 and ultimate tensile strength 446 MN/m^2. Calculate the greatest possible rotation speed. Assume that the blade rotates about an axis perpendicular to and through one end of the blade.

48. A 152-m long tunnel 7.18 m high and 5.77 m wide (with a flat roof) is to be constructed 61.5 m beneath the ground. The tunnel roof is to be supported entirely by square steel columns, each with a cross-sectional area of 962 cm^2. The density of the ground material is 2.83 g/cm^3. (*a*) Calculate the weight that the columns must support. (*b*) How many

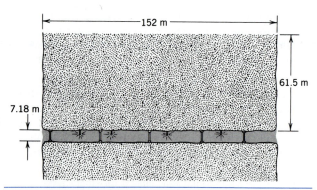

Figure 45 Problem 48.

columns are needed to provide a safety factor of 2 against rupture? See Fig. 45.

49. A rectangular slab of rock rests on a 26.0° incline; see Fig. 46. The slab has dimensions 43.3 m long, 2.50 m thick, and 12.2 m wide. Its density is 3.17 g/cm^3. The coefficient of static friction between the slab and the underlying rock is 0.390. (*a*) Calculate the component of the slab's weight acting parallel to the incline. (*b*) Calculate the static force of friction. (*c*) Comparing (*a*) and (*b*), convince yourself that the slab is in danger of sliding. Only cohesion between slab and incline prevents this. It is desired to stabilize the slab with rock bolts driven perpendicular to the incline so that, ignoring cohesion, the slab is stable. Each rock bolt has an area of 6.38 cm^2 and shear strength 362 MN/m^2. Find the minimum number of bolts needed. (The bolts are not tightened and therefore do not affect the normal force.)

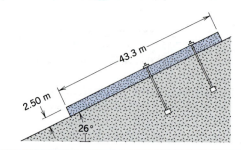

Figure 46 Problem 49.

50. Consider a metal bar of length L, cross-sectional area A, equilibrium atomic separation x, and Young's modulus E. When a tension force F is applied to the bar, it causes an extension ΔL. Calculate the atomic force constant k by deriving expressions for (*a*) the number of chains of atoms in any cross section, (*b*) the number of atoms in a single chain of length L, (*c*) the microscopic extension Δx between atoms, and (*d*) the tensile force f between atoms. (*e*) Write $f = k \Delta x$ and show that $k = Ex$. (*f*) Calculate the value of k for a typical metal for which $E = 1.2$ GN/m^2 and $x = 0.16$ nm.

CHAPTER 15

OSCILLATIONS

Each day we encounter many kinds of oscillatory motion. Common examples include the swinging pendulum of a clock, a person bouncing on a trampoline, and a vibrating guitar string. Examples on the microscopic scale are vibrating atoms in the quartz crystal of a wristwatch and vibrating molecules of air that transmit sound waves. The above cases are mechanical oscillations. We are also familiar with electromagnetic oscillations, such as electrons surging back and forth in circuits that are responsible for transmitting and receiving radio or TV signals.

One common feature of all these systems, despite the differences in their characteristics and in the laws that govern their behavior, is the mathematical formulation used to describe their oscillations. In all cases, the oscillating quantity, whether it be the displacement of a particle or the magnitude of an electric field, can be described in terms of sine or cosine functions, which are the periodic functions most familiar to us.

In this chapter we concentrate on mechanical oscillations and their description. Later in this book, we deal with various kinds of waves and with electromagnetic oscillations, which use the same mathematical description.

15-1 OSCILLATING SYSTEMS

Imagine an oscillating system, such as the pendulum of a clock or a mass on a spring. What must be the properties of the force that produces such oscillations?

If you displace a pendulum in one direction from its equilibrium position, the force (which is due to gravity) pushes it back toward equilibrium. If you displace it in the other direction, the force still acts toward the equilibrium position. *No matter what the direction of the displacement, the force always acts in a direction to restore the system to its equilibrium position.* Such a force is called a *restoring force.* (The equilibrium position is the kind we called *stable* in Chapter 14; the system tends to return to equilibrium when slightly displaced.)

Let us consider a simple example. Suppose we have a particle that is free to move only in the x direction, and let the particle experience a force of constant magnitude F_m that acts in the $+x$ direction when $x < 0$ and in the $-x$ direction when $x > 0$, as shown in Fig. 1a. The force, which is shown in Fig. 1b, is similar to the piecewise constant forces we considered in Chapter 2.

A particle of mass m at coordinate $x = +x_m$ experiences a force whose x component is $-F_m$, and the corre-

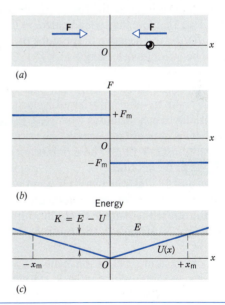

Figure 1 (a) A particle is acted on by a constant force **F** that is always directed toward the origin. (b) A plot of this piecewise constant force, equal to $+F_m$ when $x < 0$ and to $-F_m$ when $x > 0$. Any real force of this type must be represented by a continuous function, even though it may be very steep as it goes through $x = 0$. (c) The potential energy corresponding to this force. If the system has total mechanical energy E, then at any location the difference $E - U$ gives the kinetic energy.

sponding x component of the acceleration of the particle is $-a_m = -F_m/m$. The particle moves toward its equilibrium position at $x = 0$ and reaches that position with velocity $v = -v_m$. When it passes through the origin to negative x, the force becomes $+F_m$, and the acceleration is $+a_m$. The particle slows and comes to rest for an instant at $x = -x_m$ before reversing its motion through the origin and returning eventually to $x = +x_m$. In the absence of friction and other dissipative forces, the cycle repeats endlessly.

Figure 2 shows the resulting motion, plotted in the style of the examples we considered in Chapter 2. The position $x(t)$ consists of a sequence of smoothly joined segments of parabolas, as is always the case for motion at constant acceleration. The particle oscillates back and forth between $x = +x_m$ and $x = -x_m$. The magnitude of the maximum displacement from equilibrium (x_m in this case) is called the *amplitude* of the motion. The time necessary for one complete cycle (a complete repetition of the motion) is called the *period T,* as indicated in Fig. 2a. The number of cycles per unit time is called the *frequency* v. The frequency and the period are reciprocals of one another:

$$v = 1/T. \qquad (1)$$

Period is measured in time units (seconds, for instance), while frequency is measured in the SI unit of hertz (Hz),* where 1 Hz = 1 cycle/s. Thus, for example, an oscillation with a period of $T = 5$ s has a frequency $v = 0.2$ Hz.

So far we have used a dynamical description of the oscillation. Often a description in terms of energy is useful. Figure 1c shows the potential energy corresponding to the force of Fig. 1b. Note that, as indicated by the expression $F = -dU/dx$, the negative of the slope of $U(x)$ gives the force. The mechanical energy $E = K + U$ remains constant for an isolated system. At every point, the difference $E - U$ gives the kinetic energy K at that point. If we extended the graph to sufficiently large displacements, we would eventually reach locations where $E = U$ and thus $K = 0$. At these points, as Fig. 2 shows, the velocity is zero and the position is $x = \pm x_m$. These points are called the *turning points* of the motion.

Figures 1b and 1c illustrate two equivalent ways of describing the conditions for oscillation: the force must always act to restore the particle to equilibrium, and the potential energy must have a minimum at the equilibrium position.

The case of constant acceleration is always pleasant to work with, because the mathematics is simple, but it is seldom an accurate description of nature. Figure 3a shows an example of a more realistic force that can produce oscillatory motion. Such a force is responsible for the

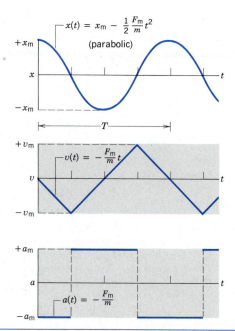

Figure 2   The position, velocity, and acceleration of the particle of Fig. 1 are plotted as functions of the time. The acceleration consists of alternating horizontal segments with values $+F_m/m$ and $-F_m/m$, the velocity consists of alternating linear segments with slopes $+F_m/m$ and $-F_m/m$, and the position consists of smoothly joined sections of parabolas. Because the force $F(x)$ is in reality a continuous function, $a(t)$ is also continuous, the horizontal segments having steep connections. Moreover, the sharp corners of $v(t)$ are rounded. The curves shown, however, are excellent approximations if the force changes from $+F_m$ to $-F_m$ over a very short interval.

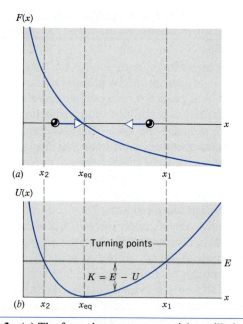

Figure 3   (a) The force that acts on a particle oscillating between the limits x_1 and x_2. Note that the force always tends to push the particle toward its equilibrium position, as in Fig. 1. Such a force might act on an atom in a molecule. (b) The potential energy corresponding to this force.

---

* The frequency unit is named after Heinrich Hertz (1857–1894), whose research provided the experimental confirmation of electromagnetic waves.

binding of molecules containing two atoms. The force increases rapidly as we try to push one atom close to the other; this repulsive component keeps the molecule from collapsing. As we try to pull the atoms to larger spacings, the force tends to oppose our efforts; this force may be an electrostatic force between two opposite electric charges, but often it is more complex and involves the spatial distribution of electronic orbits in atoms.

Figure 3b shows the corresponding potential energy function $U(x)$. Note that, as was the case in Fig. 1, the force changes sign at the equilibrium position, and the potential energy has a minimum at that position. Note that in this case the turning points (labeled x_1 and x_2 in Fig. 3) are *not* symmetrically located about the equilibrium position. If we were to stretch the molecule a bit beyond its equilibrium configuration and release it (which often occurs when a molecule absorbs infrared radiation), it would execute periodic motion about equilibrium, although the mathematical description would be more complex than that of Fig. 2. The study of these oscillations is an important technique for learning about molecular structure, as we discuss in Section 15-10.

15-2 THE SIMPLE HARMONIC OSCILLATOR

The motion of a particle in a complex system, such as an atom in the vibrating molecule discussed in the previous section, is easier to analyze if we consider the motion to be a superposition of *harmonic* oscillations, which can be described in terms of sine and cosine functions.

Consider an oscillating system consisting of a particle subject to a force

$$F(x) = -kx, \qquad (2)$$

in which k is a constant and x is the displacement of the particle from its equilibrium position. Such an oscillating system is called a *simple harmonic oscillator,* and its motion is called *simple harmonic motion.* The potential energy corresponding to this force is

$$U(x) = \tfrac{1}{2}kx^2. \qquad (3)$$

The force and potential energy are of course related by $F(x) = -dU/dx$. As indicated by Eq. 2 and plotted in Fig. 4a, the force acting on the particle is directly proportional to the displacement but is opposite to it in direction. Equation 3 shows that the potential energy varies as the square of the displacement, as illustrated by the parabolic curve in Fig. 4b.

You will recognize Eqs. 2 and 3 as the expressions for the force and potential energy of an "ideal" spring of force constant k, compressed or extended by a distance x; see Section 8-3. Hence, *a body of mass m attached to an ideal spring of force constant k and free to move over a friction-*

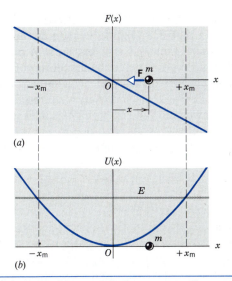

(a)

(b)

Figure 4   (a) The force and (b) the corresponding potential energy of a simple harmonic oscillator. Note the similarities and differences with Fig. 3.

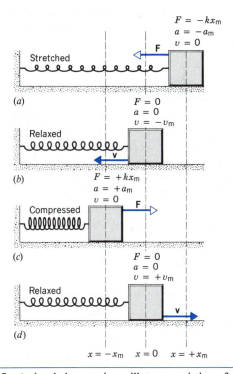

Figure 5   A simple harmonic oscillator, consisting of a spring acting on a body that slides on a frictionless horizontal surface. In (a), the spring is stretched so that the body has its maximum displacement from equilibrium. In (c) the spring is fully compressed. In (b) and (d), the body is passing through equilibrium with its maximum speed, and the spring is relaxed.

less horizontal surface is an example of a simple harmonic oscillator (see Fig. 5). Note that there is a position (the equilibrium position; see Fig. 5b) in which the spring exerts no force on the body. If the body is displaced to the right (as in Fig. 5a), the force exerted by the spring on the body points to the left. If the body is displaced to the left (as in Fig. 5c), the force points to the right. In each case the force is a *restoring* force. (It is in this case a *linear* restoring force, that is, proportional to the first power of x.)

Let us apply Newton's second law, $F = ma$, to the motion of Fig. 5. For F we substitute $-kx$ and for the acceleration a we put in d^2x/dt^2 ($= dv/dt$). This gives us

$$-kx = m\frac{d^2x}{dt^2}$$

or

$$\frac{d^2x}{dt^2} + \frac{k}{m}x = 0. \qquad (4)$$

Equation 4 is called the *equation of motion* of the simple harmonic oscillator. Its solution, which we describe in the next section, is a function $x(t)$ that describes the position of the oscillator as a function of the time, in analogy with Fig. 2a, which represents the variation of position with time of a different oscillator.

The simple harmonic oscillator problem is important for two reasons. First, many problems involving mechanical vibrations at small amplitudes reduce to that of the simple harmonic oscillator, or to a combination of such oscillators. This is equivalent to saying that if we consider a small enough portion of a restoring force curve near the equilibrium position, Fig. 3a, for instance, it becomes arbitrarily close to a straight line, which, as Fig. 4a shows, is characteristic of simple harmonic motion. Or, in other words, the potential energy curve of Fig. 3b is very nearly parabolic near the equilibrium position.

Second, as we have indicated, equations like Eq. 4 occur in many physical problems in acoustics, in optics, in mechanics, in electrical circuits, and even in atomic physics. The simple harmonic oscillator exhibits features common to many physical systems.

15-3 SIMPLE HARMONIC MOTION

Let us now solve the equation of motion of the simple harmonic oscilllator,

$$\frac{d^2x}{dt^2} + \frac{k}{m}x = 0. \qquad (4)$$

We derived Eq. 4 for a spring force $F = -kx$ (where the force constant k is a measure of the stiffness of the spring) acting on a particle of mass m. We shall see later that other oscillating systems are governed by similar equations of motion, in which the constant k is related to other physical features of the system. We can use the oscillating mass–spring system as our prototype.

Equation 4 gives a relation between a function of the time $x(t)$ and its second time derivative d^2x/dt^2. Our goal

is to find a function $x(t)$ that satisfies this relation. We begin by rewriting Eq. 4 as

$$\frac{d^2x}{dt^2} = -\left(\frac{k}{m}\right)x. \qquad (5)$$

Equation 5 requires that $x(t)$ be a function whose second derivative is the negative of the function itself, except for a constant factor k/m. We know from calculus that the sine and cosine functions have this property. For example,

$$\frac{d}{dt}\cos \omega t = -\omega \sin \omega t$$

and

$$\frac{d^2}{dt^2}\cos \omega t = \frac{d}{dt}(-\omega \sin \omega t) = -\omega^2 \cos \omega t.$$

The second derivative of a cosine (or of a sine) gives us back the original function multiplied by a negative factor $-\omega^2$. This property is not affected if we multiply the cosine function by any constant. We choose the constant to be x_m, so that the maximum value of x (the amplitude of the motion) will be x_m.

We write a tentative solution to Eq. 5 as

$$x = x_m \cos(\omega t + \phi). \qquad (6)$$

Here, since

$$x_m \cos(\omega t + \phi) = x_m \cos\phi \cos\omega t - x_m \sin\phi \sin\omega t$$
$$= a \cos\omega t + b \sin\omega t,$$

the constant ϕ allows for any combination of sine and cosine solutions.

With the (as yet) unknown constants x_m, ω, and ϕ, we have written as general a solution to Eq. 5 as we can. To determine these constants such that Eq. 6 is actually the solution of Eq. 5, we differentiate Eq. 6 twice with respect to the time. We have

$$\frac{dx}{dt} = -\omega x_m \sin(\omega t + \phi)$$

and

$$\frac{d^2x}{dt^2} = -\omega^2 x_m \cos(\omega t + \phi).$$

Putting this into Eq. 5, we obtain

$$-\omega^2 x_m \cos(\omega t + \phi) = -\frac{k}{m}x_m \cos(\omega t + \phi).$$

Therefore, if we choose the constant ω such that

$$\omega^2 = \frac{k}{m}, \qquad (7)$$

then Eq. 6 is in fact a solution of the equation of motion of a simple harmonic oscillator.

The constants x_m and ϕ are still undetermined and therefore still completely arbitrary. This means that *any* choice of x_m and ϕ whatsoever will satisfy Eq. 5, so that a large variety of motions (all of which have the same ω) is possible for the oscillator. We shall see later that x_m and ϕ

are determined for a particular harmonic motion by how the motion starts.

Let us find the physical significance of the constant ω. If we increase the time t in Eq. 6 by $2\pi/\omega$, the function becomes

$$x = x_m \cos \left[\omega(t + 2\pi/\omega) + \phi\right]$$
$$= x_m \cos (\omega t + 2\pi + \phi)$$
$$= x_m \cos (\omega t + \phi).$$

That is, the function merely repeats itself after a time $2\pi/\omega$. Therefore $2\pi/\omega$ is the period of the motion T. Since $\omega^2 = k/m$, we have

$$T = \frac{2\pi}{\omega} = 2\pi \sqrt{\frac{m}{k}}. \tag{8}$$

Hence all motions given by Eq. 5 have the same period of oscillation, which is determined only by the mass m of the oscillating particle and the force constant k of the spring. The frequency ν of the oscillator is the number of complete vibrations per unit time and is given by

$$\nu = \frac{1}{T} = \frac{1}{2\pi} \sqrt{\frac{k}{m}}. \tag{9}$$

Hence

$$\omega = 2\pi\nu = \frac{2\pi}{T}. \tag{10}$$

The quantity ω is called the *angular frequency;* it differs from the frequency ν by a factor 2π. It has the dimension of reciprocal time (the same as angular speed), and its unit is the radian/second. In Section 15-6 we give a geometric meaning to this angular frequency.

The constant x_m has a simple physical meaning. The cosine function takes on values from -1 to $+1$. The displacement x from the central equilibrium position $x = 0$ therefore has a maximum value of x_m; see Eq. 6. We call x_m the *amplitude* of the motion. Because x_m is not fixed by Eq. 4, motions of various amplitudes are possible, but all have the same frequency and period. *The frequency of a simple harmonic motion is independent of the amplitude of the motion.*

The quantity $(\omega t + \phi)$ is called the *phase* of the motion. The constant ϕ is called the *phase constant.* Two motions may have the same amplitude and frequency but differ in phase. If $\phi = -\pi/2 = -90°$, for example,

$$x = x_m \cos (\omega t + \phi) = x_m \cos (\omega t - 90°)$$
$$= x_m \sin \omega t$$

so that the displacement is zero at the time $t = 0$. If $\phi = 0$, on the other hand, the displacement $x = x_m \cos \omega t$ has its maximum value $x = x_m$ at the time $t = 0$. Other initial displacements correspond to other phase constants. See Sample Problem 3 for an example of the method of finding x_m and ϕ from the initial displacement and velocity.

The amplitude x_m and the phase constant ϕ of the oscillation are determined by the initial position and

speed of the particle. These two initial conditions will specify x_m and ϕ exactly (except that ϕ may be increased or decreased by any multiple of 2π without changing the motion). Once the motion has started, however, the particle will continue to oscillate with a constant amplitude and phase constant at a fixed frequency, unless other forces disturb the system.

In Fig. 6 we plot the displacement x versus the time t for several simple harmonic motions described by Eq. 6. Three comparisons are made. In Fig. 6a, the two curves have the same amplitude and frequency but differ in phase by $\phi = \pi/4$ or 45°. In Fig. 6b, the two curves have the same frequency and phase constant but differ in amplitude by a factor of 2. In Fig. 6c, the curves have the same amplitude and phase constant but differ in frequency by a factor of $\frac{1}{2}$ or in period by a factor of 2. Study these curves carefully to become familiar with the terminology used in simple harmonic motion.

Another distinctive feature of simple harmonic motion is the relation between the displacement, the velocity, and the acceleration of the oscillating particle. Let us compare

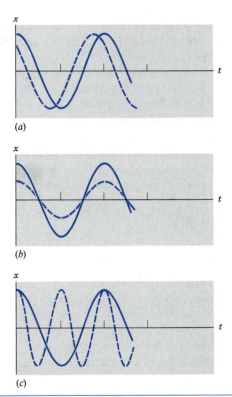

(a)

(b)

(c)

Figure 6 (a) Comparison of the motions of two simple harmonic oscillators of the same amplitude and frequency but differing in phase constant by 45°. If the motion is represented by Eq. 6, then the solid curve has $\phi = 0°$ and the dashed curve has $\phi = 45°$. (b) Two simple harmonic motions of the same phase constant and frequency but differing in amplitude by a factor of 2. (c) Two simple harmonic motions of the same amplitude and phase constant (0°) but differing in frequency by a factor of 2. The solid curve has twice the *period,* and therefore half the *frequency,* of the dashed curve.

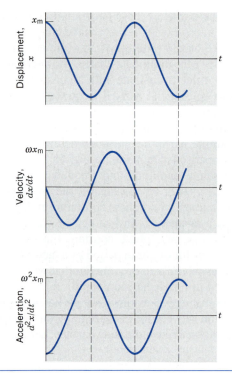

Figure 7   The displacement, velocity, and acceleration of a simple harmonic oscillator, according to Eqs. 11.

these quantities. In Fig. 7 we plot separately the displacement x versus the time t, the velocity $v = dx/dt$ versus the time t, and the acceleration $a = dv/dt = d^2x/dt^2$ versus the time t. The equations of these curves are

$$x = x_m \cos (\omega t + \phi),$$

$$v = \frac{dx}{dt} = -\omega x_m \sin (\omega t + \phi), \qquad (11)$$

$$a = \frac{dv}{dt} = -\omega^2 x_m \cos (\omega t + \phi).$$

For the case plotted we have taken $\phi = 0$. The units and scale of displacement, velocity, and acceleration are omitted for simplicity of comparison. The displacement, velocity, and acceleration all oscillate harmonically. Notice that the maximum displacement (amplitude) is x_m, the maximum speed (velocity amplitude) is ωx_m, and the maximum acceleration (acceleration amplitude) is $\omega^2 x_m$. When the displacement is a maximum in either direction, the speed is zero because the velocity must now change its direction. The acceleration at this instant, like the restoring force, has a maximum magnitude but is directed opposite to the displacement. When the displacement is zero, the speed of the particle is a maximum and the acceleration is zero, corresponding to a zero restoring force. The speed increases as the particle moves toward the equilibrium position and then decreases as it moves out to the maximum displacement. Compare Fig. 7 with Fig. 2, and note their similarities and differences.

Sample Problem 1   A certain spring hangs vertically. When a body of mass $M = 1.65$ kg is suspended from it, its length increases by 7.33 cm. The spring is then mounted horizontally, and a block of mass $m = 2.43$ kg is attached to the spring. The block is free to slide along a frictionless horizontal surface, as in Fig. 5. (*a*) What is the force constant k of the spring? (*b*) How much horizontal force is required to stretch the spring by a distance of 11.6 cm? (*c*) When the block is displaced a distance of 11.6 cm and released, with what period will it oscillate?

Solution   (*a*) The force constant k is determined from the force Mg necessary to stretch the spring by the measured distance of 7.33 cm. When the suspended body is in equilibrium, the spring force kx balances the weight Mg:

$$kx = Mg$$

$$k = Mg/x = (1.65 \text{ kg})(9.80 \text{ m/s}^2)/(0.0733 \text{ m})$$

$$= 221 \text{ N/m}.$$

(*b*) The magnitude of the force needed to stretch the spring by 11.6 cm is determined from Hooke's law (Eq. 2) using the force constant we found in part (*a*):

$$F = kx = (221 \text{ N/m})(0.116 \text{ m}) = 25.6 \text{ N}.$$

(*c*) The period is independent of the amplitude and depends only on the values of the mass of the block and the force constant. From Eq. 8,

$$T = 2\pi \sqrt{\frac{m}{k}} = 2\pi \sqrt{\frac{2.43 \text{ kg}}{221 \text{ N/m}}} = 0.6589 \text{ s} = 659 \text{ ms}.$$

(We display the value of T to four significant figures, more than are justified by the input data, because we shall need this result in the solution of Sample Problem 3. To avoid rounding errors in intermediate steps, it is standard practice to carry excess significant figures in this way. The final result, of course, must be properly rounded.)

15-4   ENERGY CONSIDERATIONS IN SIMPLE HARMONIC MOTION

For harmonic motion, including simple harmonic motion, in which no dissipative forces act, the total mechanical energy $E (= K + U)$ is conserved (remains constant). We can now study this in more detail for the special case of simple harmonic motion, for which the displacement is given by

$$x = x_m \cos (\omega t + \phi).$$

The potential energy U at any instant is given by

$$U = \tfrac{1}{2}kx^2 = \tfrac{1}{2}kx_m^2 \cos^2(\omega t + \phi). \qquad (12)$$

The potential energy thus oscillates with time and has a maximum value of $\tfrac{1}{2}kx_m^2$. During the motion, the potential energy varies between zero and this maximum value, as the curves in Figs. 8*a* and 8*b* show.

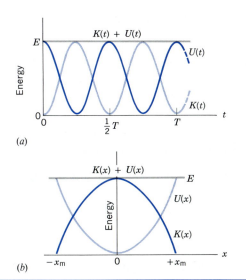

(a)

(b)

Figure 8 The potential energy U, kinetic energy K, and total mechanical energy E of a particle undergoing simple harmonic motion are shown as functions of (a) the time and (b) the displacement. Note that in (a) the kinetic and potential energies each reach their maxima twice during each period of the motion. See also Fig. 6 of Chapter 8.

The kinetic energy K at any instant is $\frac{1}{2}mv^2$. Using Eq. 11 for $v(t)$ and Eq. 7 for ω^2, we obtain

$$K = \frac{1}{2}mv^2$$
$$= \frac{1}{2}m\omega^2 x_m^2 \sin^2(\omega t + \phi)$$
$$= \frac{1}{2}kx_m^2 \sin^2(\omega t + \phi). \tag{13}$$

The kinetic energy, like the potential energy, oscillates with time and has a maximum value of $\frac{1}{2}kx_m^2$. During the motion, the kinetic energy varies between zero and this maximum value, as shown by the curves in Figs. 8a and 8b. Notice that the kinetic and potential energies vary with twice the frequency (half the period) of the displacement and velocity. Can you explain this?

The total mechanical energy is the sum of the kinetic energy and the potential energy. Using Eqs. 12 and 13, we obtain

$$E = K + U = \frac{1}{2}kx_m^2 \sin^2(\omega t + \phi) + \frac{1}{2}kx_m^2 \cos^2(\omega t + \phi)$$
$$= \frac{1}{2}kx_m^2 . \tag{14}$$

We see that the total mechanical energy is constant, as we expect, and has the value $\frac{1}{2}kx_m^2$. At the maximum displacement the kinetic energy is zero, but the potential energy has the value $\frac{1}{2}kx_m^2$. At the equilibrium position the potential energy is zero, but the kinetic energy has the value $\frac{1}{2}kx_m^2$. At other positions the kinetic and potential energies each contribute terms whose sum is always $\frac{1}{2}kx_m^2$. This constant total energy E is shown in Figs. 8a and 8b. The total energy of a particle executing simple harmonic motion is proportional to the square of the amplitude of the motion. It can be shown (see Problem 38) that the average kinetic energy for the motion during one period is exactly equal to the average potential energy and that each

of these average quantities is half the total energy, or $\frac{1}{4}kx_m^2$.

Equation 14 can be written quite generally as

$$K + U = \frac{1}{2}mv^2 + \frac{1}{2}kx^2 = \frac{1}{2}kx_m^2 . \tag{15}$$

From this relation we obtain $v^2 = (k/m)(x_m^2 - x^2)$ or

$$v = \frac{dx}{dt} = \pm\sqrt{\frac{k}{m}(x_m^2 - x^2)}. \tag{16}$$

This relation shows clearly that the speed is a maximum at the equilibrium position ($x = 0$) and is zero at the extreme displacements ($x = \pm x_m$). In fact, we can start from the conservation of energy, Eq. 15 (in which $\frac{1}{2}kx_m^2 = E$), and by integration of Eq. 16 obtain the displacement as a function of time. The result is identical with Eq. 6, which we deduced from the equation of motion, Eq. 4. (See Problem 32.)

Sample Problem 2 The block–spring combination of Sample Problem 1 is stretched in the positive x direction a distance of 11.6 cm from equilibrium and released. (a) What is the total energy stored in the system? (b) What is the maximum velocity of the block? (c) What is the maximum acceleration? (d) If the block is released at $t = 0$, what are its position, velocity, and acceleration at $t = 0.215$ s?

Solution (a) The amplitude of the motion is given as $x_m = 0.116$ m. The total energy is given by Eq. 14:

$$E = \frac{1}{2}kx_m^2 = \frac{1}{2}(221 \text{ N/m})(0.116 \text{ m})^2 = 1.49 \text{ J}.$$

(b) The maximum kinetic energy is numerically equal to the total energy; when $U = 0$, $K = K_{max} = E$. The maximum velocity is then

$$v_{max} = \sqrt{\frac{2K_{max}}{m}} = \sqrt{\frac{2(1.49 \text{ J})}{2.43 \text{ kg}}} = 1.11 \text{ m/s}.$$

(c) The maximum acceleration occurs just at the instant of release, when the force is greatest:

$$a_{max} = \frac{F_{max}}{m} = \frac{kx_m}{m} = \frac{(221 \text{ N/m})(0.116 \text{ m})}{2.43 \text{ kg}} = 10.6 \text{ m/s}^2.$$

(d) From the period found in Sample Problem 1, we can obtain the angular frequency:

$$\omega = \frac{2\pi}{T} = \frac{2\pi}{0.6589 \text{ s}} = 9.536 \text{ radians/s}.$$

Since the block has its maximum displacement of $x_m = 0.116$ m at $t = 0$, its motion can be described by a cosine function:

$$x(t) = x_m \cos \omega t,$$

a result that follows by putting $\phi = 0$ in Eq. 6. At $t = 0.215$ s, we find

$$x = (0.116 \text{ m}) \cos (9.536 \text{ radians/s})(0.215 \text{ s}) = -0.0535 \text{ m}.$$

Note that the angle ωt, whose cosine we must find, is expressed in radians. The velocity is given by Eq. 11, which, with $\phi = 0$, becomes $v(t) = -\omega x_m \sin \omega t$. At 0.215 s, we obtain

$$v = -(9.536 \text{ radians/s})(0.116 \text{ m}) \sin (9.536 \text{ radians/s})(0.215 \text{ s})$$
$$= -0.981 \text{ m/s}.$$

To find the acceleration, we again use Eq. 11 and note that, at all times, $a = -\omega^2 x$:

$$a = -(9.536 \text{ radians/s})^2 (-0.0535 \text{ m}) = +4.87 \text{ m/s}^2.$$

Let us examine our results to see if they are reasonable. The time $t = 0.215$ s is between $T/4 = 0.165$ s and $T/2 = 0.330$ s. If the block begins at $x = +0.116$ m, then at $T/4$ it will pass through equilibrium, and it is certainly reasonable that at $t = 0.215$ s it is at a negative x coordinate, as we found. Since it is at that time moving toward $x = -x_m$, its velocity must be negative, as we found. However, it has already passed through the point of most negative velocity, and it is slowing as it approaches $x = -x_m$; therefore the acceleration should be positive. We can check the value of the acceleration from $a = kx/m$. We can also check the relationship between v and x using Eq. 16.

Sample Problem 3   The block of the block–spring system of Sample Problem 1 is pushed from equilibrium by an external force in the positive x direction. At $t = 0$, when the displacement of the block is $x = +0.0624$ m and its velocity is $v = +0.847$ m/s, the external force is removed and the block begins to oscillate. Write an equation for $x(t)$ during the oscillation.

Solution   Since we have the same mass (2.43 kg) and force constant (221 N/m), the angular frequency is still 9.536 radians/s, as we found in Sample Problem 2. The most general equation for $x(t)$ is given by Eq. 6,

$$x(t) = x_m \cos(\omega t + \phi),$$

and we must find x_m and ϕ to complete the solution. To find x_m, let us compute the total energy, which at $t = 0$ has both kinetic and potential terms:

$$E = K + U = \tfrac{1}{2} mv^2 + \tfrac{1}{2} kx^2$$
$$= \tfrac{1}{2}(2.43 \text{ kg})(0.847 \text{ m/s})^2 + \tfrac{1}{2}(221 \text{ N/m})(0.0624 \text{ m})^2$$
$$= 0.872 \text{ J} + 0.430 \text{ J} = 1.302 \text{ J}.$$

Setting this equal to $\tfrac{1}{2} kx_m^2$, as Eq. 15 requires, we have

$$x_m = \sqrt{\frac{2E}{k}} = \sqrt{\frac{2(1.302 \text{ J})}{221 \text{ N/m}}} = 0.1085 \text{ m}.$$

To find the phase constant, we use the information given for $t = 0$:

$$x(0) = x_m \cos \phi$$
$$\cos \phi = \frac{x(0)}{x_m} = \frac{+0.0624 \text{ m}}{0.1085 \text{ m}} = +0.5751.$$

In the range of 0 to 2π, there are two values of ϕ whose cosine is $+0.5751$; the possible values are $\phi = 54.9°$ or $\phi = 305.1°$. Either one will satisfy the condition that $x(0)$ have the proper value, but only one will give the correct initial velocity:

$$v(0) = -\omega x_m \sin \phi = -(9.536 \text{ rad/s})(0.1085 \text{ m}) \sin \phi$$
$$= -(1.035 \text{ m/s}) \sin \phi$$
$$= -0.847 \text{ m/s} \text{ for } \phi = 54.9°$$
$$= +0.847 \text{ m/s} \text{ for } \phi = 305.1°.$$

Obviously the second choice is the one we want, and we therefore take $\phi = 305.1° = 5.33$ radians. We can now write

$$x(t) = 0.109 \cos(9.54t + 5.33),$$

where x is in meters and t in seconds.

See Problem 31 for a derivation of the general relationships that permit x_m and ϕ to be calculated from $x(0)$ and $v(0)$.

15-5   APPLICATIONS OF SIMPLE HARMONIC MOTION

A few physical systems that move with simple harmonic motion are considered here. Others are found throughout the text.*

The Torsional Oscillator

Figure 9 shows a disk suspended by a wire or shaft attached to the center of mass of the disk. The wire is securely fixed to a solid support or clamp and to the disk. With the disk in equilibrium, a radial line is drawn from its center to a point P on its rim, as shown. If the disk is rotated in a horizontal plane so that the reference line OP moves to the position OQ, the wire will be twisted. The twisted wire will exert a restoring torque on the disk tending to return the reference line to its equilibrium position. For small twists the restoring torque is found to be proportional to the angular displacement (Hooke's law), so that

$$\tau = -\kappa \theta. \tag{17}$$

---

* See "A Repertoire of S.H.M.," by Eli Maor, *The Physics Teacher,* October 1972, p. 377, for a full discussion of 16 physical systems that exhibit simple harmonic motion.

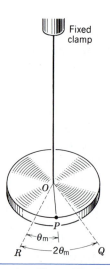

Figure 9   The torsional oscillator. The line drawn from O to P oscillates between OQ and OR, sweeping out an angle $2\theta_m$, where θ_m is the angular amplitude of the motion.

Here κ (the Greek letter kappa) is a constant that depends on the properties of the wire and is called the *torsional constant*. The minus sign shows that the torque is directed opposite to the angular displacement θ. Equation 17 is the condition for *angular simple harmonic motion.*

The equation of motion for such a system is based on the angular form of Newton's second law,

$$\tau = I\alpha = I\frac{d^2\theta}{dt^2}, \tag{18}$$

so that, using Eq. 17, we obtain

$$-\kappa\theta = I\frac{d^2\theta}{dt^2}$$

or

$$\frac{d^2\theta}{dt^2} = -\left(\frac{\kappa}{I}\right)\theta. \tag{19}$$

Notice the similarity between Eq. 19 for angular simple harmonic motion and Eq. 5 for linear simple harmonic motion. In fact, the equations are mathematically identical. Just as in Chapter 11, we can simply substitute angular displacement θ for linear displacement x, rotational inertia I for mass m, and torsional constant κ for force constant k. By making these substitutions, we find the solution of Eq. 19 to be a simple harmonic oscillation in the angle coordinate θ; namely,

$$\theta = \theta_m \cos(\omega t + \phi). \tag{20}$$

Here θ_m is the maximum angular displacement, that is, the amplitude of the angular oscillation. Note that ω here means angular frequency, not angular velocity. In Eq. 20, $\omega \neq d\theta/dt$.

In Fig. 9 the disk oscillates about the equilibrium position $\theta = 0$, the total angular range being $2\theta_m$ (from OQ to OR). The period of the oscillation by analogy with Eq. 8 is

$$T = 2\pi\sqrt{\frac{I}{\kappa}}. \tag{21}$$

If κ is known and T is measured, the rotational inertia I about the axis of rotation of any oscillating rigid body can be determined. If I is known and T is measured, the torsional constant κ of any sample of wire can be determined.

A torsional oscillator like that of Fig. 9 is also called a *torsional pendulum*. The Cavendish balance, used to measure the gravitational force constant G (see Chapter 16), is a torsional pendulum. Like the simple pendulum (discussed below) the torsional pendulum is often used for timekeeping, a common example being the balance wheel of a mechanical watch, in which the restoring torque is supplied by a spiral hairspring.

The Simple Pendulum

A simple pendulum is an idealized body consisting of a particle suspended by a light inextensible cord. When pulled to one side of its equilibrium position and released, the pendulum swings in a vertical plane under the influence of gravity. The motion is periodic and oscillatory. We wish to determine the period of the motion.

Figure 10 shows a pendulum of length L and particle mass m. At the instant shown, the cord makes an angle θ with the vertical. The forces acting on m are the weight $m\mathbf{g}$ and the tension $\mathbf{T}$ in the cord. The motion will be along an arc of the circle with radius L, and so we choose axes tangent to the circle and along the radius. The weight $m\mathbf{g}$ is resolved into a radial component of magnitude $mg\cos\theta$ and a tangential component of magnitude $mg\sin\theta$. The radial components of the forces supply the necessary centripetal acceleration to keep the particle moving on a circular arc. The tangential component is the restoring force acting on m tending to return it to the equilibrium position. Hence the restoring force is

$$F = -mg\sin\theta, \tag{22}$$

the minus sign indicating that F is opposite to the direction of increasing θ.

Notice that the restoring force is not proportional to the angular displacement θ but to $\sin\theta$ instead. The resulting motion is therefore not simple harmonic. However, if the angle θ is small, $\sin\theta$ is very nearly equal to θ in radians. For example, if $\theta = 5°$ ($= 0.0873$ rad), then $\sin\theta = 0.0872$, which differs from θ by only about 0.1%. The displacement along the arc is $x = L\theta$, and for small angles this is nearly straight-line motion. Hence, assuming

$$\sin\theta \approx \theta,$$

we obtain

$$F = -mg\theta = -mg\frac{x}{L} = -\left(\frac{mg}{L}\right)x. \tag{23}$$

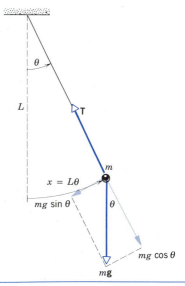

Figure 10   The simple pendulum. The forces acting on the pendulum are the tension $\mathbf{T}$ and the gravitational force $m\mathbf{g}$, which is resolved into its radial and tangential components.

For *small displacements,* the restoring force is proportional to the displacement and is oppositely directed. This is exactly the criterion for simple harmonic motion, and in fact Eq. 23 has the same form as Eq. 2, $F = -kx$, with the constant mg/L representing the constant k. (Check that the dimensions of k and mg/L are the same.) The period of a simple pendulum when its amplitude is small is then found by putting $k = mg/L$ into Eq. 8:

$$T = 2\pi \sqrt{\frac{m}{k}} = 2\pi \sqrt{\frac{m}{mg/L}}$$

or

$$T = 2\pi \sqrt{\frac{L}{g}}. \tag{24}$$

Notice that the period is independent of the mass of the suspended particle.

When the amplitude of the oscillation is not small, the general equation for the period can be shown* to be

$$T = 2\pi \sqrt{\frac{L}{g}} \left(1 + \frac{1}{2^2} \sin^2 \frac{\theta_m}{2} + \frac{1}{2^2}\frac{3^2}{4^2} \sin^4 \frac{\theta_m}{2} + \cdots \right). \tag{25}$$

Here θ_m is the maximum angular displacement. Note that T increases with increasing amplitude. Succeeding terms in the infinite series become smaller and smaller, and the period can be computed to any desired degree of accuracy by taking enough terms. When $\theta_m = 15°$, the true period differs from that given by Eq. 24 by less than 0.5%.

For the past three centuries, the pendulum has been our most reliable timekeeper, succeeded only in the last decades by clocks based on atomic or electronic oscillations. For a pendulum clock to be an accurate timekeeper, the amplitude of the swing must be kept constant despite the frictional losses that affect all mechanical systems. Even so small a change in amplitude as from 5° to 4° would cause a pendulum clock to run fast by 0.25 minute per day, an unacceptable amount even for household timekeeping. To keep the amplitude constant in a pendulum clock, energy is automatically supplied in small increments from a weight or a spring by an escapement mechanism to compensate for frictional losses. The pendulum clock with escapement was invented by Christiaan Huygens (1629–1695).

The simple pendulum also provides a convenient method for measuring the value of g, the acceleration due to gravity. We can easily determine L and T using student laboratory equipment to a precision of less than 0.1%, and thus Eq. 24 permits us to determine g to about that precision. With better apparatus, this can be extended to about 0.0001%.

---

* See K. R. Symon, *Mechanics,* 3rd edition (Addison-Wesley, 1971), Section 5.3.

The Physical Pendulum

Any rigid body mounted so that it can swing in a vertical plane about some axis passing through it is called a *physical pendulum.* This is a generalization of the simple pendulum, in which a weightless cord holds a single particle. Actually all real pendulums are physical pendulums.

In Fig. 11 a body of irregular shape is pivoted about a horizontal frictionless axis through P and displaced from the equilibrium position by an angle θ. The equilibrium position is that in which the center of mass C of the body lies vertically below P. The distance from the pivot to the center of mass is d, the rotational inertia of the body about an axis through the pivot is I, and the mass of the body is M. The restoring torque for an angular displacement θ is

$$\tau = -Mgd \sin \theta \tag{26}$$

and is due to the tangential component of the weight. Since τ is proportional to $\sin \theta$, rather than θ, the condition for angular simple harmonic motion does not, in general, hold here. For small angular displacements, however, the relation $\sin \theta \approx \theta$ is, as before, an excellent approximation, so that for small amplitudes,

$$\tau = -Mgd\theta. \tag{27}$$

This is in the form of Eq. 17, and the period follows directly from Eq. 21 with the substitution $\kappa = Mgd$, which gives

$$T = 2\pi \sqrt{\frac{I}{Mgd}}. \tag{28}$$

Equation 28 can be solved for the rotational inertia I, giving

$$I = \frac{T^2 Mgd}{4\pi^2}. \tag{29}$$

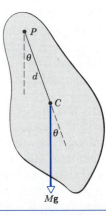

Figure 11 A physical pendulum. The center of mass is at C, and the pivot is at point P. The pendulum is displaced by an angle θ from its equilibrium position, which occurs when C hangs directly below P. The weight $M\mathbf{g}$ provides the restoring torque.

The quantities on the right are all directly measurable. Hence the rotational inertia about an axis of rotation (other than through the center of mass) of a body of any shape can be determined by suspending the body as a physical pendulum from that axis.

The physical pendulum includes the simple pendulum as a special case. Locating the pivot far from the object, using a weightless cord of length L, we would have $I = ML^2$ and $d = L$, so

$$T = 2\pi \sqrt{\frac{I}{Mgd}} = 2\pi \sqrt{\frac{ML^2}{MgL}} = 2\pi \sqrt{\frac{L}{g}},$$

which is the period of a simple pendulum.

If the mass of a physical pendulum were concentrated at the properly chosen distance L from the pivot, the resulting simple pendulum will have the same period as the original physical pendulum if

$$T = 2\pi \sqrt{\frac{L}{g}} = 2\pi \sqrt{\frac{I}{Mgd}}$$

or

$$L = \frac{I}{Md}. \tag{30}$$

Hence, as far as its period of oscillation is concerned, the mass of a physical pendulum may be considered to be concentrated at a point O whose distance from the pivot is $L = I/Md$. This point is called the *center of oscillation* of the physical pendulum. Notice that it depends on the location of the pivot for any given body. Furthermore, if we pivot the original physical pendulum from point O, it will have the same period as it does when pivoted from point P.

Sample Problem 4   A thin uniform rod of mass $M = 0.112$ kg and length $L = 0.096$ m is suspended by a wire that passes through its center and is perpendicular to its length. The wire is twisted and the rod set oscillating. The period is found to be 2.14 s. When a flat body in the shape of an equilateral triangle is suspended similarly through its center of mass, the period is found to be 5.83 s. Find the rotational inertia of the triangle about this axis.

Solution   The rotational inertia of a rod, rotated about a central axis perpendicular to its length, is $ML^2/12$. Hence

$$I_{\text{rod}} = \frac{(0.112 \text{ kg})(0.096 \text{ m})^2}{12} = 8.60 \times 10^{-5} \text{ kg} \cdot \text{m}^2.$$

From Eq. 21,

$$\frac{T_{\text{rod}}}{T_{\text{triangle}}} = \left(\frac{I_{\text{rod}}}{I_{\text{triangle}}}\right)^{1/2} \quad \text{or} \quad I_{\text{triangle}} = I_{\text{rod}} \left(\frac{T_{\text{triangle}}}{T_{\text{rod}}}\right)^2,$$

so that

$$I_{\text{triangle}} = (8.60 \times 10^{-5} \text{ kg} \cdot \text{m}^2) \left(\frac{5.83 \text{ s}}{2.14 \text{ s}}\right)^2$$
$$= 6.38 \times 10^{-4} \text{ kg} \cdot \text{m}^2.$$

Does the amplitude of either oscillation affect the period in these cases?

Sample Problem 5   A uniform disk is pivoted at its rim (Fig. 12). Find its period for small oscillations and the length of the equivalent simple pendulum.

Solution   The rotational inertia of a disk about an axis through its center is $\frac{1}{2}MR^2$, where R is the radius and M is the mass of the disk. The rotational inertia about the pivot at the rim is, using the parallel axis theorem,

$$I = \tfrac{1}{2}MR^2 + MR^2 = \tfrac{3}{2}MR^2.$$

The period of this physical pendulum, found from Eq. 28 with $d = R$, is then

$$T = 2\pi \sqrt{\frac{I}{MgR}} = 2\pi \sqrt{\frac{3}{2}\frac{MR^2}{MgR}} = 2\pi \sqrt{\frac{3}{2}\frac{R}{g}},$$

independent of the mass of the disk.

The simple pendulum having the same period has a length

$$L = \frac{I}{MR} = \tfrac{3}{2}R$$

or three-fourths the diameter of the disk. The center of oscillation of the disk pivoted at P is therefore at O, a distance $\frac{3}{2}R$ below the point of support. Is any particular mass required of the equivalent simple pendulum?

If we pivot the disk at a point midway between the rim and the center, as at O, we find that $I = \frac{1}{2}MR^2 + M(\frac{1}{2}R)^2 = \frac{3}{4}MR^2$ and $d = \frac{1}{2}R$. The period T is

$$T = 2\pi \sqrt{\frac{I}{Mgd}} = 2\pi \sqrt{\frac{\frac{3}{4}MR^2}{Mg(R/2)}} = 2\pi \sqrt{\frac{3}{2}\frac{R}{g}}$$

just as before. This illustrates the equality of the periods of the physical pendulum when pivoted about O and P.

If the disk were pivoted at the center, what would be its period of oscillation?

Sample Problem 6   The center of oscillation of a physical pendulum has another interesting property. If an impulsive force (assumed horizontal and in the plane of oscillation) acts at the center of oscillation, no reaction is felt at the point of support.

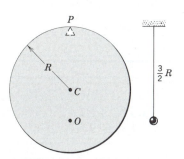

Figure 12   Sample Problem 5. A disk pivoted at its rim oscillates as a physical pendulum. To the right is shown a simple pendulum with the same period. Point O is the center of oscillation.

Prove this for an impulsive force **F** acting toward the left at point *O* in Fig. 12. Assume the pendulum to be initially at rest.

Solution This is a case of combined translation and rotation about the center of mass (see Section 12-6). The translation effect, acting alone, would make *P* (along with the entire disk) in Fig. 12 move to the left with an acceleration

$$a_{\text{left}} = F/M.$$

The rotational effect, acting alone, would produce a clockwise angular acceleration about *C* of

$$\alpha = \tau/I$$
$$= (F)(\tfrac{1}{2}R)/(\tfrac{1}{2}MR^2)$$
$$= F/MR.$$

Because of this angular acceleration *P* would move to the right with an acceleration

$$a_{\text{right}} = \alpha R$$
$$= (F/MR)(R) = F/M.$$

Thus $a_{\text{left}} = a_{\text{right}}$ and there is no movement at point *P*.

When considered from this point of view the center of oscillation is often called the *center of percussion*. Baseball players know that unless the ball hits the bat at just the right spot (center of percussion) the impact will sting their hands. The "sting" has a different direction depending on whether the ball strikes on one side or the other of this spot. The "sweet spot" on a tennis racket has a similar explanation; hitting the ball on the "sweet spot" eliminates any reaction force on the hand.*

Sample Problem 7 The period of a disk of radius 10.2 cm executing small oscillations about a pivot at its rim is measured to be 0.784 s. Find the value of *g*, the acceleration due to gravity at that location.

Solution From Sample Problem 5, we have

$$T = 2\pi \sqrt{\frac{3R}{2g}},$$

and solving for *g*, we obtain

$$g = \frac{6\pi^2 R}{T^2}.$$

With $T = 0.784$ s and $R = 0.102$ m, we find

$$g = \frac{6\pi^2(0.102 \text{ m})}{(0.784 \text{ s})^2} = 9.82 \text{ m/s}^2.$$

15-6 SIMPLE HARMONIC MOTION AND UNIFORM CIRCULAR MOTION

In 1610, Galileo used his newly constructed telescope to observe the moons of Jupiter. As he watched night after

---

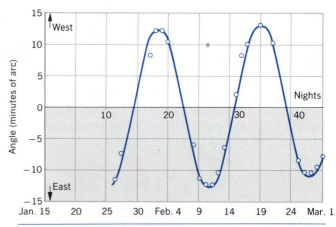

Figure 13 The angular position as a function of time of Jupiter's moon Callisto, as measured from Earth. The circles are based on Galileo's 1610 measurements. The curve is a best fit and strongly suggests simple harmonic motion. Nearly 400 years after Galileo, the motions of Jupiter's moons continue to delight the amateur astronomer. Each month the magazine *Sky and Telescope* publishes a chart showing their motions, in terms of a sinusoidally varying angular coordinate similar to this figure.

night, he measured the position of each moon relative to the planet. He observed the moons to travel back and forth in motion that we would call simple harmonic. Figure 13 shows Galileo's original data, plotted to show the sidewise displacement of one moon (Callisto) as a function of the time. The sinusoidal dependence characteristic of simple harmonic motion is apparent.

Actually, Callisto does not oscillate back and forth; it moves in a very nearly circular orbit about the planet, and what Galileo observed was uniform circular motion in a plane viewed edge on. Since this corresponds exactly with the displacement versus time relationship of simple harmonic motion, we are led to the following conclusion:

Simple harmonic motion can be described as the projection of uniform circular motion along a diameter of the circle.

Let us examine in more detail the mathematical basis for this conclusion. Figure 14 shows a particle *P* in uniform circular motion; its angular velocity is ω and the radius of the circle is *R*. At time 0 (Fig. 14*a*) the radius *OP* makes an angle ϕ with the *x* axis. At a time *t* later (Fig. 14*b*), the radius *OP* makes an angle $\omega t + \phi$ with the *x* axis, and the projection of *OP* along the *x* axis (or, equivalently, the *x* component of the radius vector corresponding to *OP*) is

$$x(t) = R \cos (\omega t + \phi). \tag{31}$$

This is of course identical to Eq. 6 for the displacement of the simple harmonic oscillator, with x_m corresponding to *R*. If we let *P'* represent the projection of *P* on the *x* axis, then *P'* executes simple harmonic motion along the *x* axis.

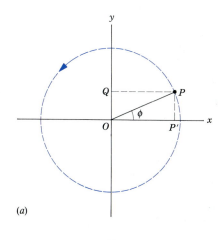

(a)

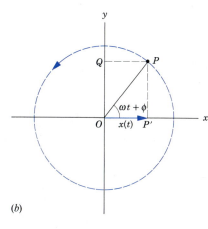

(b)

Figure 14 (a) A point P moves at a constant speed in a circle of radius R. The reference line makes an angle ϕ with the x axis at $t = 0$. The projection P' on the x axis executes simple harmonic motion. (b) After a time t, point P has rotated through an additional angle ωt. (c) The velocity of P and its x component, which represents the velocity of P' in simple harmonic motion. (d) The acceleration of P and its x component.

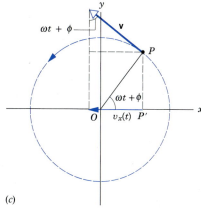

(c)

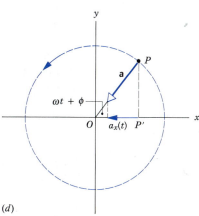

(d)

In uniform circular motion, the magnitude of the constant tangential speed is ωR. Figure 14c shows the vector representing the instantaneous velocity $\mathbf{v}$ at time t. The x component of $\mathbf{v}$, which gives the velocity of P' along the x direction, is

$$v_x(t) = -\omega R \sin (\omega t + \phi). \qquad (32)$$

The centripetal acceleration in circular motion is $\omega^2 R$, and as shown in Fig. 14d, the x component of the acceleration of P is

$$a_x(t) = -\omega^2 R \cos (\omega t + \phi). \qquad (33)$$

Equations 32 and 33 are identical with Eqs. 11 for simple harmonic motion, again with x_m replaced by R. Thus displacement, velocity, and acceleration are identical in simple harmonic motion and in the projection of circular motion.

Reversing the above argument, we can state that Eq. 31 for the displacement of a simple harmonic oscillator is sufficient to describe the x component of a vector whose tip traces a circular path at constant speed. If we can also describe the y component, then we have a complete description of the vector. Figures 14a and 14b show the y projection OQ at times 0 and t. The y component can be described by

$$y(t) = R \sin (\omega t + \phi). \qquad (34)$$

Note that the projection of uniform circular motion along the y direction also gives simple harmonic motion, as would projection along *any* direction. Notice also that, at all times, $x^2 + y^2 = R^2$ as we expect for circular motion. You should be able to find expressions for the y components of the velocity and acceleration and show that, as expected, $v_x^2 + v_y^2 = (\omega R)^2$ and $a_x^2 + a_y^2 = (\omega^2 R)^2$.

Using the trigonometric identity $\sin \theta = \cos (\theta - \pi/2)$ we can rewrite Eq. 34 as

$$y(t) = R \cos (\omega t + \phi - \pi/2). \qquad (35)$$

Thus circular motion can be regarded as the combination of two simple harmonic motions at right angles, with identical amplitudes and frequencies but differing in phase by 90°. In the next section, we see how other more complicated motions can be analyzed as combinations of simple harmonic motions with appropriately chosen amplitudes, frequencies, and phases.

Sample Problem 8 Consider a body executing a horizontal simple harmonic motion. The equation of that motion is

$$x = 0.35 \cos (8.3t),$$

where x is in meters and t in seconds. This motion can also be

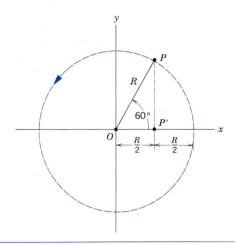

Figure 15   Sample Problem 8. The radius OP moves from $\phi = 0$ at $t = 0$ to $\omega t = 60°$ at time t. The projection P' moves correspondingly from $x = R$ to $x = R/2$.

represented as the projection of uniform circular motion along a horizontal diameter. (*a*) Give the properties of the corresponding uniform circular motion. (*b*) From the motion of the reference point determine the time required for the body to come halfway in toward the center of motion from its initial position.

Solution   (*a*) The x component of the circular motion is given by

$$x = R \cos(\omega t + \phi).$$

Therefore the reference circle must have a radius $R = 0.35$ m, the initial phase or phase constant must be $\phi = 0$, and the angular velocity must be $\omega = 8.3$ rad/s, in order to obtain the equation $x = 0.35 \cos(8.3t)$ for the horizontal projection.

(*b*) As the body moves halfway in, the reference point moves through an angle of $\omega t = \pi/3 = 60°$ (Fig. 15). The angular velocity is constant at 8.3 rad/s so that the time required to move through 60° is

$$t = \frac{60°}{\omega} = \frac{\pi/3 \text{ rad}}{8.3 \text{ rad/s}} = 0.13 \text{ s}.$$

The time may also be computed directly from the equation of motion. With

$$x = 0.35 \cos(8.3t) \quad \text{and} \quad x = \tfrac{1}{2}R = \tfrac{1}{2}(0.35),$$

we obtain

$$\tfrac{1}{2} = \cos(8.3t) \quad \text{or} \quad 8.3t = \cos^{-1}(\tfrac{1}{2}) = \pi/3 \text{ rad}.$$

Therefore

$$t = \frac{\pi/3 \text{ rad}}{8.3 \text{ rad/s}} = 0.13 \text{ s}.$$

15-7   COMBINATIONS OF HARMONIC MOTIONS

Often two simple harmonic motions at right angles are combined. The resulting motion is the sum of two independent oscillations. Consider first the case in which the frequencies of the vibrations are the same, such as

$$x = x_m \cos(\omega t + \phi_x) \quad \text{and} \quad y = y_m \cos(\omega t + \phi_y). \quad (36)$$

The x and y motions may have different amplitudes and different phase constants.

If the phase constants are the same, the resulting motion is a straight line. This can be shown analytically by taking the ratio between the expressions for x and y in Eq. 36 when $\phi_x = \phi_y$, which gives

$$y = (y_m/x_m)x.$$

This is the equation of a straight line, whose slope is y_m/x_m. In Figs. 16*a* and 16*b* we show the resultant motion for two cases, $y_m/x_m = 1$ and $y_m/x_m = 2$. In these cases both the x and y displacements reach a maximum at the same time and reach a minimum at the same time. They are *in phase*. The point P, whose x and y coordinates are given by Eqs. 36, moves back and forth along the line as t varies.

If the phase constants are different, the resulting motion will not be a straight line. For example, if the phase constants differ by $\pi/2$, the maximum x displacement occurs when the y displacement is zero and *vice versa*. When the amplitudes are equal, the resulting motion is circular; when the amplitudes are unequal, the resulting motion is elliptical. Two cases, $y_m/x_m = 1$ and $y_m/x_m = 2$, are shown in Figs. 16*c* and 16*d*, for $\phi_x = \phi_y + \pi/2$. The cases $y_m/x_m = 1$ and $y_m/x_m = 2$, for $\phi_x = \phi_y - \pi/4$, are shown in Figs. 16*e* and 16*f*.

All possible combinations of two simple harmonic motions at right angles having the same frequency correspond to elliptical paths, the circle and straight line being special cases of an ellipse. This can be shown analytically by combining Eqs. 36 and eliminating the time t; you can show that the resulting equation is that of an ellipse. The shape of the ellipse depends only on the ratio of the amplitudes, y_m/x_m, and the difference in phase between the two oscillations, $\phi_x - \phi_y$. The actual motion can be either clockwise or counterclockwise, depending on which component leads in phase.

If two oscillations of *different frequencies* are combined at right angles, the resulting motion is more complicated. The motion is not even periodic unless the two component frequencies ω_x and ω_y are the ratio of two integers (see Problem 61). The mathematical analysis of such motions is often difficult, but the patterns can be displayed graphically on an oscilloscope screen, in which a beam of electrons can simultaneously be deflected in the vertical and horizontal directions by sinusoidal electronic signals whose frequencies, amplitudes, and relative phase can be varied. Figure 17 is an example of the complex and lovely patterns that result.

In this section, we have considered only combinations of simple harmonic motions in different directions (at right angles to one another). Combinations of simple harmonic motions in the *same direction,* with the same fre-

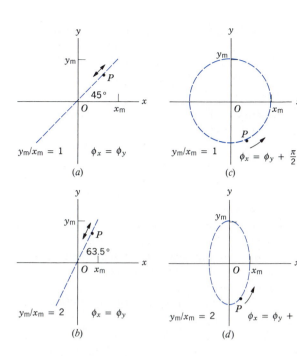

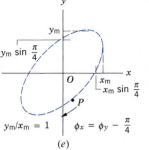

Figure 16 Combinations of simple harmonic motions along two perpendicular directions. Each figure shows the motion of point P when the amplitudes and phases of the motions have the indicated relationships. The x and y motions have equal frequencies.

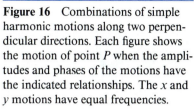

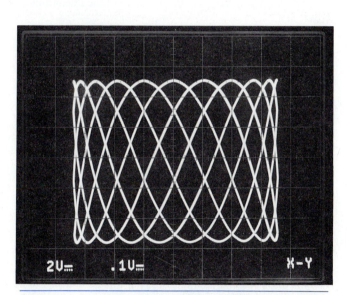

Figure 17 A Lissajous figure, which is produced on an oscilloscope screen when the horizontal and vertical deflections are sinusoidal signals whose frequencies have integer ratios. In the case shown, the ratio of the frequencies is 1/20.

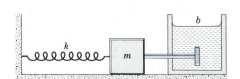

Figure 18 A representation of a damped harmonic oscillator. We consider the oscillating body (of mass m) to be attached to a (massless) vane immersed in a fluid, in which it experiences a viscous damping force $-bv$. We do not consider sliding friction at the horizontal surface.

quency but w... ..., are of special interes... ...terference of light, sound, ...ation, which will be discussed late... ...scillations of different frequencies in the s... direction may also be combined. The treatment of this motion is particularly important in the case of sound vibrations and will be discussed in Chapter 20.

15-8 DAMPED HARMONIC MOTION (Optional)

Up to this point we have assumed that no frictional forces act on the oscillator. If this assumption held strictly, a pendulum or a mass on a spring would oscillate indefinitely. Actually, the amplitude of the oscillation gradually decreases to zero as a result of friction. The motion is said to be *damped* by friction and is called *damped harmonic motion*. Often the friction arises from air resistance or internal forces. In most cases of interest the frictional force is proportional to the velocity of the body but directed opposite to it. An example of a damped oscillator is shown in Fig. 18.

The net force on the oscillating body is the sum of the restoring force $-kx$ and the damping force, which we assume to be in the form of $-bv$ as in the case of the drag force considered in Section 6-7. Here b is a positive constant, which depends on properties of the fluid, such as its density, and on the shape and dimensions of the immersed object. From Newton's second law in the form $\Sigma F = ma$, we obtain

$$-kx - b\frac{dx}{dt} = m\frac{d^2x}{dt^2}$$

or

$$m\frac{d^2x}{dt^2} + b\frac{dx}{dt} + kx = 0. \tag{37}$$

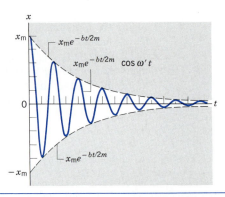

Figure 19 Damped harmonic motion. The displacement x is plotted against the time t with the phase constant ϕ taken to be 0. The motion is oscillatory, but the amplitude decreases exponentially with time.

A solution of this equation (given here without proof; see Problem 63 for verification)* is

$$x = x_m e^{-bt/2m} \cos(\omega' t + \phi), \tag{38}$$

where

$$\omega' = \sqrt{\frac{k}{m} - \left(\frac{b}{2m}\right)^2}. \tag{39}$$

This form of the solution to Eq. 37 is valid for damping constants b that are small enough that the quantity in the radical in Eq. 39 is positive. In Fig. 19 we plot the displacement x as a function of the time t in this case.

There are two notable features of this solution. First, the frequency is smaller (and the period longer) when friction is present. Friction slows down the motion, as might be expected. If no friction were present, b would equal zero and ω' would equal $\sqrt{k/m}$, which is the angular frequency ω of undamped motion. When friction is present, ω' is slightly less than ω, as shown by Eq. 39. In the case shown in Fig. 19, representing strong damping in which the amplitude decreases by a factor of 10 in 5 cycles, ω' differs from ω by only 0.3%.

Second, the amplitude of the motion, represented in Eq. 38 by the factor $x_m e^{-bt/2m}$ and in Fig. 19 by the dashed curves, decreases exponentially to zero. The time interval τ during which the amplitude drops to $1/e$ of its initial value is called the *mean lifetime* of the oscillation. The exponential factor in Eq. 38 will have the value e^{-1} when $t = \tau = 2m/b$. Once again, if there were no friction present, b would equal zero and the amplitude would have the constant value x_m as time went on; the lifetime would be infinite.

Equations 38 and 39 are valid only for $b \leq 2\sqrt{km}$. If b has its largest possible value in this range ($b = 2\sqrt{km}$), then $\omega' = 0$, and the displacement approaches zero exponentially with no oscillation. The lifetime τ has its smallest value, which can be shown to equal ω^{-1}, the inverse of the angular frequency of the undamped oscillation. This condition, called *critical damping,* is often the

---

* For a more complete discussion of the derivation and interpretation of the equations of the damped oscillator, see K. R. Symon, *Mechanics,* 3rd edition (Addison-Wesley, 1971), Section 2.9.

goal of mechanical engineers in designing a system in which the oscillations disappear in the shortest possible time.

In damped harmonic motion the energy of the oscillator is gradually dissipated by friction and falls to zero in time. In the case of small damping, when Eq. 38 is valid, we can approximate the instantaneous value of the energy by Eq. 14, replacing the (constant) amplitude x_m by the instantaneous value of the amplitude, $x_m e^{-bt/2m}$. Thus

$$E(t) = \tfrac{1}{2}k(x_m e^{-bt/2m})^2$$
$$= \tfrac{1}{2}kx_m^2 e^{-bt/m}. \tag{40}$$

Sample Problem 9 In a damped oscillator, such as that of Fig. 18, let $m = 250$ g, $k = 85$ N/m, and $b = 0.070$ kg/s. In how many periods of oscillation will the mechanical energy of the oscillator drop to one-half of its initial value?

Solution For small damping, $\omega' \approx \omega$ and the period is

$$T = 2\pi\sqrt{\frac{m}{k}} = 2\pi\sqrt{\frac{0.25 \text{ kg}}{85 \text{ N/m}}} = 0.34 \text{ s}.$$

At $t = 0$, the initial mechanical energy is $\tfrac{1}{2}kx_m^2$. According to Eq. 40, the energy will have half this value at a time t determined from

$$\tfrac{1}{2}(\tfrac{1}{2}kx_m^2) = \tfrac{1}{2}kx_m^2 e^{-bt/m}.$$

Solving for t, we obtain

$$t = \frac{m \ln 2}{b} = \frac{(0.25 \text{ kg})(\ln 2)}{0.070 \text{ kg/s}} = 2.5 \text{ s}.$$

The time t is about $7.5T$; thus about 7.5 cycles of the oscillation are required for the mechanical energy to drop by half.

Total energy must of course be conserved. Where does this energy go? ∎

15-9 FORCED OSCILLATIONS AND RESONANCE *(Optional)*

Thus far we have discussed only the natural oscillations of a body, that is, the oscillations that occur, for example, when the body is displaced and then released. For a mass attached to a spring the *natural* frequency is

$$\omega = 2\pi v = \sqrt{\frac{k}{m}}$$

in the absence of friction and

$$\omega' = 2\pi v' = \sqrt{\frac{k}{m} - \left(\frac{b}{2m}\right)^2}$$

in the presence of a small frictional force bv.

A different situation arises, however, when the body is subject to a sinusoidal external force. As examples, a bridge vibrates under the influence of marching soldiers, the housing of a motor vibrates owing to periodic impulses from an irregularity in the shaft, and our eardrums vibrate when exposed to the periodic force of a sound wave. The oscillations that result are called *forced* oscillations. These forced oscillations have the frequency of the external force and not the natural frequency of the body.

However, the response of the body depends on the relation between the forcing and the natural frequencies. A succession of small impulses applied at the proper frequency can produce an oscillation of large amplitude. A child using a swing learns to pump at proper time intervals to make the swing move with a large amplitude. The problem of forced oscillations is a very general one. Its solution is useful in acoustic systems, alternating current circuits, and atomic physics as well as in mechanics.

The equation of motion of a forced oscillator follows from the second law of motion. In addition to the restoring force $-kx$ and the damping force $-bv$, we have also the applied oscillating external force. For simplicity let this external force be given by $F_m \cos \omega''t$. Here F_m is the maximum value of the external force and $\omega'' (= 2\pi\nu'')$ is its angular frequency. We can imagine such a force applied directly to the oscillating mass of Fig. 18, for example, by replacing the fixed wall on the left with a movable support attached to the shaft of a motor. The motor moves the support at the angular frequency ω''.

From Newton's second law, we obtain

$$-kx - b\frac{dx}{dt} + F_m \cos \omega''t = m\frac{d^2x}{dt^2}$$

or

$$m\frac{d^2x}{dt^2} + b\frac{dx}{dt} + kx = F_m \cos \omega''t. \qquad (41)$$

The solution of this equation (given without proof)* is

$$x = \frac{F_m}{G} \sin (\omega''t - \phi), \qquad (42)$$

where

$$G = \sqrt{m^2(\omega''^2 - \omega^2)^2 + b^2\omega''^2}, \qquad (43)$$

and

$$\phi = \cos^{-1} \frac{b\omega''}{G}. \qquad (44)$$

Let us consider the resulting motion in a qualitative way.

Note (Eq. 42) that the system vibrates with the angular frequency ω'' of the driving force, rather than with its natural frequency ω, and that the amplitude of the motion is constant. Damping is present, which would normally cause a loss in amplitude, but the source of the driving force provides the energy necessary to keep the amplitude constant. In effect, the oscillator carries energy from the driving source to the damping medium, where the energy is dissipated.

The simplest case is that in which there is no damping, which means that $b = 0$ in Eq. 43. The factor G, which has the value $|m(\omega''^2 - \omega^2)|$ for $b = 0$, is large when the angular frequency ω'' of the driving force is very different from the natural undamped angular frequency ω of the system. This means that the amplitude of the resultant motion, F_m/G, is small. As the driving frequency approaches the natural frequency, that is, as $\omega'' \to \omega$, we see that $G \to 0$ and the amplitude $F_m/G \to \infty$. Actually, some damping is always present so that the amplitude of oscillation, although it may become large, remains finite in practice.

---

*See K. R. Symon, *Mechanics,* 3rd edition (Addison-Wesley, 1971), Section 2.10. Equation 42 is a steady-state solution that applies after some time has elapsed. When the motion first begins, it is a superposition of this solution and short-lived transient terms that decay rapidly. We examine the motion after these terms have become negligible.

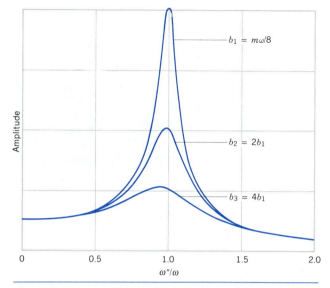

Figure 20   The amplitude F_m/G of a forced oscillator as the angular frequency ω'' of the driving force is varied. The three curves correspond to different levels of damping, the smallest damping giving the sharpest resonance curve.

For damped oscillators (for which $b \neq 0$ in Eq. 43), there is a characteristic value of the driving frequency ω'' at which the amplitude of oscillation is a maximum. This condition is called *resonance* and the value of ω'' at which resonance occurs is called the *resonant angular frequency.* (Resonance, defined here to occur at the frequency at which the forced oscillations have their maximum amplitude, may be defined in other ways as, for example, the frequency at which maximum power is transferred from the driving unit to the oscillating system or at which the speed of the oscillating mass is a maximum. The definitions are not equivalent; we shall discuss the matter further when we deal with forced electrical oscillations; see Problem 68.) The smaller the damping in a given system, the closer is the resonant angular frequency to the natural undamped angular frequency ω. Often the damping is small enough so that the resonant angular frequency can be taken to equal the natural undamped angular frequency ω with small error.

In Fig. 20 we have drawn three curves giving the amplitude of the forced vibrations as a function of the ratio of the driving frequency ω'' to the undamped natural angular frequency ω. Each of the curves corresponds to a different value of the damping constant b. When the damping is small, the resonance curve is sharp and the amplitude reaches a maximum when $\omega'' = \omega$. As the damping increases, the resonance curve becomes smaller and wider, and the resonance is slightly displaced from $\omega'' = \omega$.

All mechanical structures—such as buildings, bridges, and airplanes—have one or more natural resonant frequencies. It can be disastrous to subject the structure to an external driving force at one of those frequencies. The image of a soprano shattering a wine glass is an example of the result.

Another example of resonance occurred in the Tacoma Narrows Bridge in Washington State in 1940. The wind blowing through the Tacoma Narrows broke up into vortices, in effect providing puffs of wind that shook the bridge at a frequency that

Figure 21   The Tacoma Narrows Bridge on Puget Sound, Washington. Completed and opened to traffic in July 1940, it immediately showed gentle rolling oscillations due to resonance. Later the bridge developed violent torsional oscillations shown at left. Eventually the main span broke up, sending the bridge roadway crashing into the water below, as shown at right.

matched one of its natural vibrational frequencies. The result was a gentle vertical rolling motion, somewhat like a roller coaster, which earned the bridge the nickname "Galloping Gertie." About 5 months after the bridge opened, the gentle rolling oscillations became violent torsional oscillations, which soon caused the collapse of the bridge (Fig. 21). These oscillations were not a consequence of resonance but of nonlinear effects due to particularly strong wind gusts. Such complex effects cannot be analyzed in terms of the forced linear oscillator we have discussed here. ∎

15-10   TWO-BODY OSCILLATIONS (Optional)

On the microscopic level (molecules, atoms, nuclei), there are many examples of oscillations that are approximately simple harmonic. One example is the diatomic molecule, in which two atoms are bound together with a force of the form illustrated in Fig. 3. Near the equilibrium position, the potential energy can be approximated as a parabola of the form $U(x) = \frac{1}{2}k(x - x_{eq})^2$, and if displaced a small distance from x_{eq}, the molecule will oscillate about the equilibrium position. For our purposes, we can imagine the molecule to be represented by two particles of masses m_1 and m_2 connected by a spring of force constant k, as shown in Fig. 22. In this section, we examine the motion of this system.

One way to describe the motion of the system is in terms of the separate motions of the two particles, which are located relative to the origin O by the two coordinates x_1 and x_2, as shown in Fig. 22a. As we see below, this leads to a different and often more useful description, which is given in terms of the *relative* separation and velocity of the two particles. In effect, we replace the two coordinates x_1 and x_2 with two other coordinates: the relative separation $x_1 - x_2$ and the location x_{cm} of the center of mass. In the absence of external forces, the center of mass moves at constant velocity, and its motion is of no real interest in studying the oscillation of the system, so we can analyze the system in terms of the relative coordinate alone.

The relative separation $x_1 - x_2$ gives the length of the spring at any time. Suppose its unstretched length is L; then $x = (x_1 - x_2) - L$ is the change in length of the spring, and $F = kx$ is the magnitude of the force exerted on *each particle* by the spring. As shown in Fig. 22a, if the spring exerts a force $-\mathbf{F}$ on m_1, then it exerts a force $+\mathbf{F}$ on m_2.

Let us apply Newton's second law separately to the two particles, taking force components along the x axis:

$$m_1 \frac{d^2x_1}{dt^2} = -kx,$$

$$m_2 \frac{d^2x_2}{dt^2} = +kx.$$

We now multiply the first of these equations by m_2 and the second by m_1, and then subtract. The result is

$$m_1 m_2 \frac{d^2x_1}{dt^2} - m_1 m_2 \frac{d^2x_2}{dt^2} = -m_2 kx - m_1 kx,$$

which we can write as

$$\frac{m_1 m_2}{m_1 + m_2} \frac{d^2}{dt^2}(x_1 - x_2) = -kx. \tag{45}$$

The quantity $m_1 m_2/(m_1 + m_2)$ has the dimension of mass and is known as the *reduced mass* m:

$$m = \frac{m_1 m_2}{m_1 + m_2}. \tag{46}$$

Because the unstretched length L of the spring is a constant, the derivatives of $(x_1 - x_2)$ are the same as the derivatives of x:

$$\frac{d}{dt}(x_1 - x_2) = \frac{d}{dt}(x + L) = \frac{dx}{dt},$$

and so Eq. 45 becomes

$$\frac{d^2x}{dt^2} + \frac{k}{m}x = 0.$$

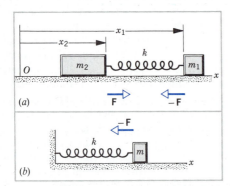

Figure 22   (a) Two oscillating bodies of masses m_1 and m_2 connected by a spring. (b) The relative motion can be represented by the oscillation of a single body having the reduced mass m.

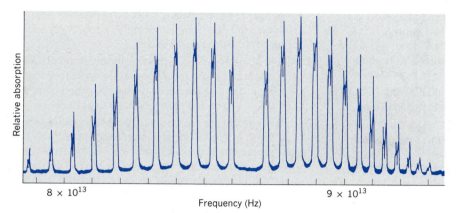

Figure 23 The absorption spectrum of infrared radiation by molecular HCl. Each peak corresponds to a change in the vibrational motion of the molecules. The closely spaced pairs of peaks are due to the two isotopes of Cl.

This is identical in form to Eq. 4 for the single oscillating mass, thus demonstrating that, from the standpoint of oscillations, the system of Fig. 22*a* can be replaced by a single particle, as represented in Fig. 22*b*, with a mass equal to the reduced mass of the system. In particular, the frequency of oscillation of the system of Fig. 22 is given by Eq. 9, using the reduced mass.

If we wish to examine the detailed motion of the system, we can simply write down the solution for $x(t)$, $v(t)$, and $a(t)$ given by Eqs. 11, keeping in mind that x represents the relative coordinate of the two particles, and thus v and a represent their *relative* velocity $v_1 - v_2$ and acceleration $a_1 - a_2$, respectively.

Note that the reduced mass m is always smaller than either mass. If one of the masses is very much smaller than the other, then m is roughly equal to the smaller mass. If the masses are equal, then m is half as large as either mass.

Sample Problem 10   Naturally occurring chlorine consists of two isotopes: ^{35}Cl, of relative abundance 76% and atomic mass 34.968853 u, and ^{37}Cl, of relative abundance 24% and atomic mass 36.965903 u. (*a*) What is the reduced mass of a molecule of HCl when it contains ^{35}Cl and when it contains ^{37}Cl? (*b*) The vibrational frequency of a molecule of HCl is 8.5×10^{13} Hz. Assuming HCl to behave like a simple two-body oscillator, find the effective force constant k.

Solution   (*a*) The reduced mass for $H^{35}Cl$ is found from Eq. 46, using the H mass of 1.007825 u:

$$m = \frac{m_1 m_2}{m_1 + m_2} = \frac{(1.007825 \text{ u})(34.968853 \text{ u})}{1.007825 \text{ u} + 34.968853 \text{ u}} = 0.979593 \text{ u}.$$

For $H^{37}Cl$ we have similarly

$$m = \frac{(1.007825 \text{ u})(36.965903 \text{ u})}{1.007825 \text{ u} + 36.965903 \text{ u}} = 0.981077 \text{ u}.$$

(*b*) Solving Eq. 9 for the force constant, we obtain

$$k = 4\pi^2 v^2 m = 4\pi^2 (8.5 \times 10^{13} \text{ Hz})^2 (0.98 \text{ u})(1.66 \times 10^{-27} \text{ kg/u})$$
$$= 464 \text{ N/m}.$$

This is of the same order of magnitude as the force constant of ordinary springs (for example, see Sample Problem 1). Can you explain how the force constant for one molecule can be the same as that of a spring?

Molecules can absorb or emit electromagnetic radiation and change their state of vibrational motion in the process. In fact, observing the radiation that is absorbed or emitted is one of the ways we learn about the structure of molecules. Figure 23 shows an example of the infrared absorption spectrum of HCl. Each peak corresponds to a change in the vibrational state of the HCl when it absorbs radiation at that frequency. The two components to each peak are due to the two isotopes of Cl; their different masses result in slightly different reduced masses for molecules of $H^{35}Cl$ and $H^{37}Cl$, as we found in part (*a*), and therefore in slightly different vibrational frequencies. ∎

QUESTIONS

1. Give some examples of motions that are approximately simple harmonic. Why are motions that are exactly simple harmonic rare?

2. A typical screen-door spring is tension-stressed in its normal state; that is, adjacent turns cling to each other and resist separation. Does such a spring obey Hooke's law?

3. Is Hooke's law obeyed, even approximately, by a diving board? A trampoline? A coiled spring made of lead wire?

4. What would happen to the motion of an oscillating system if the sign of the force term, $-kx$ in Eq. 2, were changed?

5. A spring has a force constant k, and an object of mass m is suspended from it. The spring is cut in half and the same object is suspended from one of the halves. How are the frequencies of oscillation, before and after the spring is cut, related?

6. An unstressed spring has a force constant k. It is stretched by a weight hung from it to an equilibrium length well within the elastic limit. Does the spring have the same force constant k for displacements from this new equilibrium position?

7. Suppose we have a block of unknown mass and a spring of unknown force constant. Show how we can predict the period of oscillation of this block–spring system simply by measuring the extension of the spring produced by attaching the block to it.

8. Any real spring has mass. If this mass is taken into account, explain qualitatively how this will affect the period of oscillation of a spring–block system.

9. Can one have an oscillator that even for small amplitudes is not simple harmonic? That is, can one have a nonlinear restoring force in an oscillator even at arbitrarily small amplitudes?

10. How are each of the following properties of a simple harmonic oscillator affected by doubling the amplitude: period, force constant, total mechanical energy, maximum velocity, maximum acceleration?

11. What changes could you make in a harmonic oscillator that would double the maximum speed of the oscillating object?

12. A person stands on a bathroom-type scale, which rests on a platform suspended by a large spring. The whole system executes simple harmonic motion in a vertical direction. Describe the variation in scale reading during a period of motion.

13. Could we ever construct a true simple pendulum? Explain your answer.

14. Could standards of mass, length, and time be based on properties of a pendulum? Explain.

15. Considering the elastic and the inertial aspects involved, explain the fact that whereas when an object of mass m oscillates vertically on a spring the period depends on m but is independent of g, the reverse is true for a simple pendulum.

16. Predict by qualitative arguments whether a pendulum oscillating with large amplitude will have a period longer or shorter than the period for oscillations with small amplitude. (Consider extreme cases.)

17. As the amplitude θ_m in Eq. 25 approaches 180°, what value do you expect the period to approach? Explain in physical terms.

18. What happens to the frequency of a swing as its oscillations die down from large amplitude to small?

19. How is the period of a pendulum affected when its point of suspension is (a) moved horizontally in the plane of oscillation with acceleration a; (b) moved vertically upward with acceleration a; (c) moved vertically downward with acceleration $a < g$; with acceleration $a > g$? Which case, if any, applies to a pendulum mounted on a cart rolling down an inclined plane?

20. Why was an axis through the center of mass excluded in using Eq. 29 to determine I? Does this equation apply to such an axis? How can you determine I for such an axis using physical pendulum methods?

21. A hollow sphere is filled with water through a small hole in it. It is hung by a long thread and, as the water flows out of the hole at the bottom, one finds that the period of oscillation first increases and then decreases. Explain.

22. (a) The effect of the mass, m, of the cord attached to the bob, of mass M, of a pendulum is to increase the period over that for a simple pendulum in which $m = 0$. Make this plausible. (b) Although the effect of the mass of the cord on the pendulum is to increase its period, a cord of length L swinging without anything on the end ($M = 0$) has a period less than that of a simple pendulum of length L. Make that plausible.

23. If taken to the Moon, will there be any change in the frequency of oscillation of a torsional pendulum? A simple pendulum? A spring–block oscillator? A physical pendulum?

24. How can a pendulum be used so as to trace out a sinusoidal curve?

25. What component simple harmonic motions would give a figure 8 as the resultant motion?

26. Is there a connection between the F versus x relation at the molecular level and the macroscopic relation between F and x in a spring? Explain your answer.

27. (a) Under what circumstances would the reduced mass of a two-body system be equal to the mass of one of the bodies? Explain. (b) What is the reduced mass if the bodies have equal mass? (c) Do cases (a) and (b) give the extreme values of the reduced mass?

28. Why is the tub of a washing machine often mounted on springs?

29. Why are damping devices often used on machinery? Give an example.

30. Give some examples of common phenomena in which resonance plays an important role.

31. The lunar ocean tide is much more important than the solar ocean tide. The opposite is true for tides in the Earth's atmosphere, however. Explain this, using resonance ideas, given the fact that the atmosphere has a natural period of oscillation of nearly 12 hours.

32. In Fig. 20, what value does the amplitude of the forced oscillations approach as the driving frequency ω'' approaches (a) zero and (b) infinity?

33. Buildings of different heights sustain different amounts of damage in an earthquake. Explain why.

34. A singer, holding a note of the right frequency, can shatter a glass if the glassware is of high quality. This cannot be done if the glassware quality is low. Explain why.

PROBLEMS

Section 15-3 Simple Harmonic Motion

1. A 3.94-kg block extends a spring 15.7 cm from its unstretched position. The block is removed and a 0.520-kg object is hung from the same spring. Find the period of its oscillation.

2. An oscillator consists of a block of mass 512 g connected to a spring. When set into oscillation with amplitude 34.7 cm, it is observed to repeat its motion every 0.484 s. Find (a) the period, (b) the frequency, (c) the angular frequency, (d) the

force constant, (*e*) the maximum speed, and (*f*) the maximum force exerted on the block.

3. The vibration frequencies of atoms in solids at normal temperatures are of the order of 10.0 THz. Imagine the atoms to be connected to one another by "springs." Suppose that a single silver atom vibrates with this frequency and that all the other atoms are at rest. Compute the effective force constant. One mole of silver has a mass of 108 g and contains 6.02×10^{23} atoms.

4. A loudspeaker produces a musical sound by the oscillation of a diaphragm. If the amplitude of oscillation is limited to 1.20×10^{-3} mm, what frequencies will result in the acceleration of the diaphragm exceeding *g*?

5. A 5.22-kg object is attached to the bottom of a vertical spring and set vibrating. The maximum speed of the object is 15.3 cm/s and the period is 645 ms. Find (*a*) the force constant of the spring, (*b*) the amplitude of the motion, and (*c*) the frequency of oscillation.

6. In an electric shaver, the blade moves back and forth over a distance of 2.00 mm. The motion is simple harmonic, with frequency 120 Hz. Find (*a*) the amplitude, (*b*) the maximum blade speed, and (*c*) the maximum blade acceleration.

7. An automobile can be considered to be mounted on four springs as far as vertical oscillations are concerned. The springs of a certain car of mass 1460 kg are adjusted so that the vibrations have a frequency of 2.95 Hz. (*a*) Find the force constant of each of the four springs (assumed identical). (*b*) What will be the vibration frequency if five persons, averaging 73.2 kg each, ride in the car?

8. A body oscillates with simple harmonic motion according to the equation

$$x = (6.12 \text{ m}) \cos [(8.38 \text{ rad/s})t + 1.92 \text{ rad}].$$

Find (*a*) the displacement, (*b*) the velocity, and (*c*) the acceleration at the time $t = 1.90$ s. Find also (*d*) the frequency and (*e*) the period of the motion.

9. The scale of a spring balance reading from 0 to 50.0 lb is 4.00 in. long. A package suspended from the balance is found to oscillate vertically with a frequency of 2.00 Hz. How much does the package weigh?

10. The piston in the cylinder head of a locomotive has a stroke of 76.5 cm. What is the maximum speed of the piston if the drive wheels make 193 rev/min and the piston moves with simple harmonic motion?

11. Figure 24 shows an astronaut on a Body Mass Measurement Device (BMMD). Designed for use on orbiting space vehicles, its purpose is to allow astronauts to measure their mass in the weightless conditions in Earth orbit. The BMMD is a spring-mounted chair; an astronaut measures his or her period of oscillation in the chair; the mass follows from the formula for the period of an oscillating block–spring system. (*a*) If *M* is the mass of the astronaut and *m* the effective mass of that part of the BMMD that also oscillates, show that

$$M = (k/4\pi^2)T^2 - m,$$

where *T* is the period of oscillation and *k* is the force constant. (*b*) The force constant is $k = 605.6$ N/m for the BMMD on Skylab Mission Two; the period of oscillation of the empty chair is 0.90149 s. Calculate the effective mass of the chair. (*c*) With an astronaut in the chair, the period of

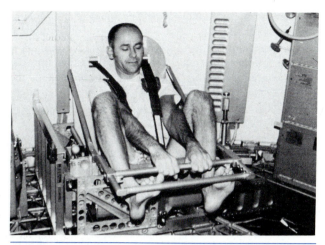

Figure 24   Problem 11.

oscillation becomes 2.08832 s. Calculate the mass of the astronaut.

12. A 2.14-kg object hangs from a spring. A 325-g body hung below the object stretches the spring 1.80 cm farther. The 325-g body is removed and the object is set into oscillation. Find the period of the motion.

13. At a certain harbor, the tides cause the ocean surface to rise and fall in simple harmonic motion, with a period of 12.5 h. How long does it take for the water to fall from its maximum height to one-half its maximum height above its average (equilibrium) level?

14. Two blocks ($m = 1.22$ kg and $M = 8.73$ kg) and a spring ($k = 344$ N/m) are arranged on a horizontal, frictionless surface as shown in Fig. 25. The coefficient of static friction between the blocks is 0.42. Find the maximum possible amplitude of the simple harmonic motion if no slippage is to occur between the blocks.

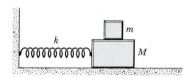

Figure 25   Problem 14.

15. A block is on a horizontal surface (a shake table) that is moving horizontally with a simple harmonic motion of frequency 2.35 Hz. The coefficient of static friction between block and plane is 0.630. How great can the amplitude be if the block does not slip along the surface?

16. A block is on a piston that is moving vertically with simple harmonic motion. (*a*) At what amplitude of motion will the block and the piston separate if the period of the piston's motion is 1.18 s? (*b*) If the piston has an amplitude of 5.12 cm in its motion, find the maximum frequency for which the block and piston will be in contact continuously.

17. The force of interaction between two atoms in certain diatomic molecules can be represented by $F = -a/r^2 + b/r^3$, in which *a* and *b* are positive constants and *r* is the separation distance of the atoms. Make a graph of *F* versus *r*. Then (*a*) show that the separation at equilibrium is *b/a*; (*b*) show

that for small oscillations about this equilibrium separation the force constant is a^4/b^3; (c) find the period of this motion.

18. An oscillator consists of a block attached to a spring ($k = 456$ N/m). At some time t, the position (measured from the equilibrium location), velocity, and acceleration of the block are $x = 0.112$ m, $v = -13.6$ m/s, $a = -123$ m/s². Calculate (a) the frequency, (b) the mass of the block, and (c) the amplitude of oscillation.

19. Two particles oscillate in simple harmonic motion along a common straight line segment of length L. Each particle has a period of 1.50 s but they differ in phase by 30.0°. (a) How far apart are they (in terms of L) 0.500 s after the lagging particle leaves one end of the path? (b) Are they moving in the same direction, toward each other, or away from each other at this time?

20. Two particles execute simple harmonic motion of the same amplitude and frequency along the same straight line. They pass one another when going in opposite directions each time their displacement is half their amplitude. Find the phase difference between them.

21. Two springs are attached to a block of mass m, free to slide on a frictionless horizontal surface, as shown in Fig. 26. Show that the frequency of oscillation of the block is

$$v = \frac{1}{2\pi} \sqrt{\frac{k_1 + k_2}{m}} = \sqrt{v_1^2 + v_2^2} \,,$$

where v_1 and v_2 are the frequencies at which the block would oscillate if connected only to spring 1 or spring 2. (The electrical analog of this system is a series combination of two capacitors.)

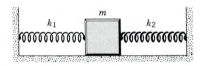

Figure 26 Problem 21.

22. Two springs are joined and connected to a block of mass m as shown in Fig. 27. The surfaces are frictionless. If the springs separately have force constants k_1 and k_2, show that the frequency of oscillation of the block is

$$v = \frac{1}{2\pi} \sqrt{\frac{k_1 k_2}{(k_1 + k_2)m}} = \frac{v_1 v_2}{\sqrt{v_1^2 + v_2^2}} \,,$$

where v_1 and v_2 are the frequencies at which the block would oscillate if connected only to spring 1 or spring 2. (The electrical analog of this system is a parallel combination of two capacitors.)

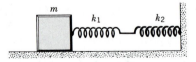

Figure 27 Problem 22.

23. Three 10,000-kg ore cars are held at rest on a 26.0° incline on a mine railway using a cable that is parallel to the incline (Fig. 28). The cable is observed to stretch 14.2 cm just before a coupling breaks, detaching one of the cars. Find (a) the

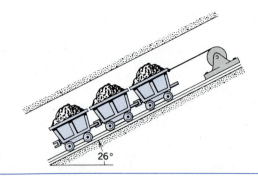

Figure 28 Problem 23.

frequency of the resulting oscillations of the remaining two cars and (b) the amplitude of the oscillations.

24. A massless spring of force constant 3.60 N/cm is cut into halves. (a) What is the force constant of each half? (b) The two halves, suspended separately, support a block of mass M (see Fig. 29). The system vibrates at a frequency of 2.87 Hz. Find the value of the mass M.

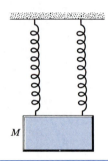

Figure 29 Problem 24.

25. If the mass of a spring m_s is not negligible but is small compared to the mass m of the object suspended from it, the period of motion is $T = 2\pi \sqrt{(m + m_s/3)/k}$. Derive this result. (*Hint:* The condition $m_s \ll m$ is equivalent to the assumption that the spring stretches proportionally along its length.) (See H. L. Armstrong, *American Journal of Physics,* Vol. 37, p. 447, 1969, for a complete solution of the general case.)

Section 15-4 Energy Considerations in Simple Harmonic Motion

26. An oscillating block–spring system has a mechanical energy of 1.18 J, an amplitude of 9.84 cm, and a maximum speed of 1.22 m/s. Find (a) the force constant of the spring, (b) the mass of the block, and (c) the frequency of oscillation.

27. A (hypothetical) large slingshot is stretched 1.53 m to launch a 130-g projectile with speed sufficient to escape from the Earth (11.2 km/s). (a) What is the force constant of the device, if all the potential energy is converted to kinetic energy? (b) Assume that an average person can exert a force of 220 N. How many people are required to stretch the slingshot?

28. (a) When the displacement is one-half the amplitude x_m, what fraction of the total energy is kinetic and what fraction is potential in simple harmonic motion? (b) At what displacement is the energy half kinetic and half potential?

29. A 12.3-kg particle is undergoing simple harmonic motion with an amplitude of 1.86 mm. The maximum acceleration experienced by the particle is 7.93 km/s². (*a*) Find the period of the motion. (*b*) What is the maximum speed of the particle? (*c*) Calculate the total mechanical energy of this simple harmonic oscillator.

30. A 5.13-kg object moves on a horizontal frictionless surface under the influence of a spring with force constant 9.88 N/cm. The object is displaced 53.5 cm and given an initial velocity of 11.2 m/s back toward the equilibrium position. Find (*a*) the frequency of the motion, (*b*) the initial potential energy of the system, (*c*) the initial kinetic energy, and (*d*) the amplitude of the motion.

31. Show that the general relationships between the two initial values of position $x(0)$ and velocity $v(0)$, and the amplitude x_m and phase angle ϕ of Eq. 6, are

$$x_m = \sqrt{[x(0)]^2 + [v(0)/\omega]^2}, \qquad \tan \phi = -v(0)/\omega x(0).$$

32. Solve Eq. 16, which expresses conservation of energy, for dt and integrate the result. Assume that $x = x_m$ at $t = 0$, and show that Eq. 6 (with $\phi = 0$), the displacement as a function of time, is obtained.

33. An object of mass 1.26 kg attached to a spring of force constant 5.38 N/cm is set into oscillation by extending the spring 26.3 cm and giving the object a velocity of 3.72 m/s toward the equilibrium position of the spring. Using the results obtained in Problem 31, calculate (*a*) the amplitude and (*b*) the phase angle of the resulting simple harmonic motion.

34. A block of mass M, at rest on a horizontal, frictionless table, is attached to a rigid support by a spring of force constant k. A bullet of mass m and speed v strikes the block as shown in Fig. 30. The bullet remains embedded in the block. Determine the amplitude of the resulting simple harmonic motion, in terms of m, M, v, and k.

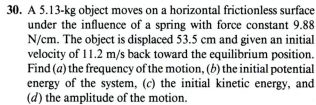

Figure 30 Problem 34.

35. Consider a massless spring of force constant k in a uniform gravitational field. Attach an object of mass m to the spring. (*a*) Show that if $x = 0$ marks the slack position of the spring, the static equilibrium position is given by $x = mg/k$ (see Fig. 31). (*b*) Show that the equation of motion of the mass–spring system is

$$m \frac{d^2x}{dt^2} + kx = mg$$

and that the solution for the displacement as a function of time is $x = x_m \cos(\omega t + \phi) + mg/k$, where $\omega = \sqrt{k/m}$ as before. (*c*) Show therefore that the system has the same ω, v, a, v, and T in a uniform gravitational field as in the absence of such a field, with the one change that the equilibrium position has been displaced by mg/k. (*d*) Now consider the energy of the system, $\frac{1}{2}mv^2 + \frac{1}{2}kx^2 + mg(h - x) = $ constant, and show that time differentiation leads to the equation of motion of part (*b*). (*e*) Show that when the object

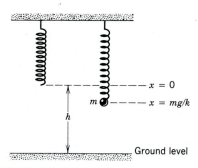

Figure 31 Problem 35.

falls from $x = 0$ to the static equilibrium position, $x = mg/k$, the loss in gravitational potential energy goes half into a gain in elastic potential energy and half into a gain in kinetic energy. (*f*) Finally, consider the system in motion about the static equilibrium position. Compute separately the change in gravitational potential energy and in elastic potential energy when the object moves *up* through a displacement x_m, and when the object moves *down* through a displacement x_m. Show that the *total* change in potential energy is the same in each case, namely, $\frac{1}{2}kx_m^2$. In view of the results (*c*) and (*f*), one can simply ignore the uniform gravitational field in the analysis merely by shifting the reference position from $x = 0$ to $x_0 = x - mg/k = 0$. The new potential energy curve $[U(x_0) = \frac{1}{2}kx_0^2 + \text{constant}]$ has the same parabolic shape as the potential energy curve in the absence of a gravitational field $[U(x) = \frac{1}{2}kx^2]$.

36. A 4.00-kg block is suspended from a spring with a force constant of 5.00 N/cm. A 50.0-g bullet is fired into the block from below with a speed of 150 m/s and comes to rest in the block. (*a*) Find the amplitude of the resulting simple harmonic motion. (*b*) What fraction of the original kinetic energy of the bullet appears as mechanical energy in the oscillator?

37. A solid cylinder is attached to a horizontal massless spring so that it can *roll without slipping* along a horizontal surface, as in Fig. 32. The force constant k of the spring is 2.94 N/cm. If the system is released from rest at a position in which the spring is stretched by 23.9 cm, find (*a*) the translational kinetic energy and (*b*) the rotational kinetic energy of the cylinder as it passes through the equilibrium position. (*c*) Show that under these conditions the center of mass of the cylinder executes simple harmonic motion with a period

$$T = 2\pi \sqrt{3M/2k},$$

where M is the mass of the cylinder.

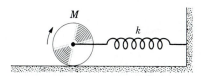

Figure 32 Problem 37.

38. (*a*) Prove that in simple harmonic motion the average potential energy equals the average kinetic energy when the average is taken with respect to time over one period of the motion, and that each average equals $\frac{1}{4}kx_m^2$. (*b*) Prove that when the average is taken with respect to position over one cycle, the average potential energy equals $\frac{1}{6}kx_m^2$ and the average kinetic energy equals $\frac{1}{3}kx_m^2$. (*c*) Explain physically why the two results above (*a* and *b*) are different.

Section 15-5 Applications of Simple Harmonic Motion

39. Find the length of a simple pendulum whose period is 1.00 s at a location where $g = 9.82$ m/s^2.

40. A simple pendulum of length 1.53 m makes 72.0 complete oscillations in 180 s at a certain location. Find the acceleration due to gravity at this point.

41. A 2500-kg demolition ball swings from the end of a crane, as shown in Fig. 33. The length of the swinging segment of cable is 17.3 m. Find the period of swing, assuming that the system can be treated as a simple pendulum.

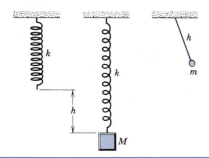

Figure 33   Problem 41.

42. There is an interesting relation between the block–spring system and the simple pendulum. Suppose that you hang an object of mass M on the end of a spring, and when the object is in equilibrium the spring is stretched a distance h. Show that the frequency of this block–spring system is the same as that of a simple pendulum of mass m and length h, even if $m \neq M$; see Fig. 34.

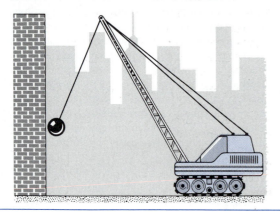

Figure 34   Problem 42.

43. A circular hoop of radius 65.3 cm and mass 2.16 kg is suspended on a horizontal nail. (*a*) Find its frequency of oscillation for small displacements from equilibrium. (*b*) What is the length of the equivalent simple pendulum?

44. An engineer wants to find the rotational inertia of an odd-shaped object of mass 11.3 kg about an axis through its center of mass. The object is supported with a wire through its center of mass and along the desired axis. The wire has a

torsional constant $\kappa = 0.513$ N·m. The engineer observes that this torsional pendulum oscillates through 20.0 complete cycles in 48.7 s. What value of rotational inertia is calculated?

45. A physical pendulum consists of a uniform solid disk of mass $M = 563$ g and radius $R = 14.4$ cm supported in a vertical plane by a pivot located a distance $d = 10.2$ cm from the center of the disk, as shown in Fig. 35. The disk is displaced by a small angle and released. Find the period of the resulting simple harmonic motion.

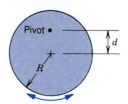

Figure 35   Problem 45.

46. A 95.2-kg solid sphere with a 14.8-cm radius is suspended by a vertical wire attached to the ceiling of a room. A torque of 0.192 N·m is required to twist the sphere through an angle of 0.850 rad. Find the period of oscillation when the sphere is released from this position.

47. A physical pendulum consists of a meter stick that is pivoted at a small hole drilled through the stick a distance x from the 50.0-cm mark. The period of oscillation is observed to be 2.50 s. Find the distance x.

48. A pendulum consists of a uniform disk with radius 10.3 cm and mass 488 g attached to a 52.4-cm long uniform rod with mass 272 g; see Fig. 36. (*a*) Calculate the rotational inertia of the pendulum about the pivot. (*b*) What is the distance between the pivot and the center of mass of the pendulum? (*c*) Calculate the small-angle period of oscillation.

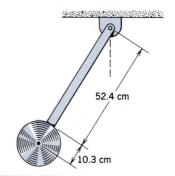

Figure 36   Problem 48.

49. A pendulum is formed by pivoting a long thin rod of length L and mass m about a point on the rod which is a distance d above the center of the rod. (*a*) Find the small-amplitude period of this pendulum in terms of d, L, m, and g. (*b*) Show that the period has a *minimum* value when $d = L/\sqrt{12} = 0.289L$.

50. A wheel is free to rotate about its fixed axle. A spring is attached to one of its spokes a distance r from the axle, as shown in Fig. 37. Assuming that the wheel is a hoop of mass M and radius R, obtain the angular frequency of small oscil-

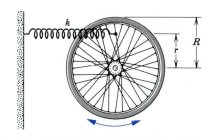

Figure 37   Problem 50.

lations of this system in terms of M, R, r, and the force constant k. Discuss the special cases $r = R$ and $r = 0$.

51. A meter stick swinging from one end oscillates with a frequency ν_0. What would be the frequency, in terms of ν_0, if the bottom third of the stick were cut off?

52. A particle is released from rest at a point P inside a frictionless hemispherical bowl of radius R. (a) Show that when P is near the bottom of the bowl the particle undergoes simple harmonic motion. (b) Find the length of the equivalent simple pendulum.

53. A physical pendulum has two possible pivot points; one has a fixed position and the other is adjustable along the length of the pendulum, as shown in Fig. 38. The period of the pendulum when suspended from the fixed pivot is T. The pendulum is then reversed and suspended from the adjustable pivot. The position of this pivot is moved until, by trial and error, the pendulum has the same period as before, namely, T. Show that the free-fall acceleration g is given by

$$g = \frac{4\pi^2 L}{T^2},$$

in which L is the distance between the two pivot points. Note that g can be measured in this way without needing to know the rotational inertia of the pendulum or any of its other dimensions except L.

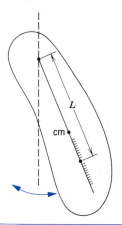

Figure 38   Problem 53.

54. A 2.50-kg disk, 42.0 cm in diameter, is supported by a light rod, 76.0 cm long, which is pivoted at its end, as shown in Fig. 39. (a) The light, torsional spring is initially not connected. What is the period of oscillation? (b) The torsional spring is now connected so that, in equilibrium, the rod hangs vertically. What should be the torsional constant of

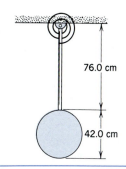

Figure 39   Problem 54.

the spring so that the new period of oscillation is 500 ms shorter than before?

55. A simple pendulum of length L and mass m is suspended in a car that is traveling with a constant speed v around a circle of radius R. If the pendulum undergoes small oscillations in a radial direction about its equilibrium position, what will its frequency of oscillation be?

56. Figure 40 shows a physical pendulum constructed from equal-length sections of identical pipe. The inner radius of the pipe is 10.2 cm and the thickness is 6.40 mm. (a) Calculate the period of oscillation about the pivot shown. (b) Suppose that a new physical pendulum is constructed by rotating the bottom section 90° about a vertical axis through its center. Show that the new period of oscillation about the same pivot is about 2% less than the period of the original pendulum.

Figure 40   Problem 56.

Section 15-7 Combinations of Harmonic Motions

57. Sketch the path of a particle that moves in the xy plane according to $x = x_\mathrm{m} \cos(\omega t - \pi/2)$, $y = 2x_\mathrm{m} \cos(\omega t)$.

58. The diagram shown in Fig. 41 is the result of combining the two simple harmonic motions $x = x_\mathrm{m} \cos \omega_x t$ and $y =$

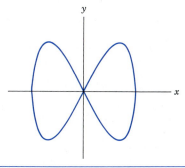

Figure 41   Problem 58.

$y_m \cos(\omega_y t + \phi_y)$. (*a*) What is the value of x_m/y_m? (*b*) What is the value of ω_x/ω_y? (*c*) What is the value of ϕ_y?

59. Electrons in an oscilloscope are deflected by two mutually perpendicular electric fields in such a way that at any time t the displacement is given by

$$x = A \cos \omega t, \qquad y = A \cos(\omega t + \phi_y).$$

Describe the path of the electrons and determine their equation when (*a*) $\phi_y = 0°$, (*b*) $\phi_y = 30°$, and (*c*) $\phi_y = 90°$.

60. A particle of mass m moves in a fixed plane along the trajectory $\mathbf{r} = \mathbf{i} A \cos \omega t + \mathbf{j} A \cos 3\omega t$. (*a*) Sketch the trajectory of the particle. (*b*) Find the force acting on the particle. Also find (*c*) its potential energy and (*d*) its total energy as functions of time. (*e*) Is the motion periodic? If so, find the period.

61. When oscillations at right angles are combined, the frequencies for the motion of the particle in the x and y directions need not be equal, so that in the general case Eqs. 36 become

$$x = x_m \cos(\omega_x t + \phi_x) \quad \text{and} \quad y = y_m \cos(\omega_y t + \phi_y).$$

The path of the particle is no longer an ellipse but is called a *Lissajous curve*, after Jules Antoine Lissajous who first demonstrated such curves in 1857. (*a*) If ω_x/ω_y is a rational number, so that the angular frequencies ω_x and ω_y are "commensurable," then the curve is closed and the motion repeats itself at regular intervals of time. Assume $x_m = y_m$ and $\phi_x = \phi_y$ and draw the Lissajous curve for $\omega_x/\omega_y = \frac{1}{2}, \frac{1}{3}$, and $\frac{2}{3}$. (*b*) Let ω_x/ω_y be a rational number, either $\frac{1}{2}, \frac{1}{3}$, or $\frac{2}{3}$, say, and show that the shape of the Lissajous curve depends on the phase difference $\phi_x - \phi_y$. Draw curves for $\phi_x - \phi_y = 0, \pi/4$, and $\pi/2$ rad. (*c*) If ω_x/ω_y is not a rational number, then the curve is "open." Convince yourself that after a long time the curve will have passed through every point lying in the rectangle bounded by $x = \pm x_m$ and $y = \pm y_m$, the particle never passing twice through a given point with the same velocity. For definiteness, assume $\phi_x = 0$ throughout.

Section 15-8 Damped Harmonic Motion

62. For the system shown in Fig. 18, the block has a mass of 1.52 kg and the force constant is 8.13 N/m. The frictional force is given by $-b(dx/dt)$, where $b = 227$ g/s. Suppose that the block is pulled aside a distance 12.5 cm and released. (*a*) Calculate the time interval required for the amplitude to fall to one-third of its initial value. (*b*) How many oscillations are made by the block in this time?

63. Verify, by taking derivatives, that Eq. 38 is a solution of Eq. 37 for the damped oscillator, provided that the frequency ω' is given by Eq. 39.

64. A damped harmonic oscillator involves a block ($m = 1.91$ kg), a spring ($k = 12.6$ N/m), and a damping force $F = -bv$. Initially, it oscillates with an amplitude of 26.2 cm; because of the damping, the amplitude falls to three-fourths of this initial value after four complete cycles. (*a*) What is the value of b? (*b*) How much energy has been "lost" during these four cycles?

65. Assume that you are examining the characteristics of a suspension system of a 2000-kg automobile. The suspension "sags" 10 cm when the weight of the entire automobile is placed on it. In addition, the amplitude of oscillation decreases by 50% during one complete oscillation. Estimate the values of k and b for the spring and shock absorber system of each wheel. Assume each wheel supports 500 kg.

Section 15-9 Forced Oscillations and Resonance

66. Consider the forced oscillations of a damped block–spring system. Show that at resonance (*a*) the amplitude of oscillation is $x_m = F_m/b\omega$, and (*b*) the maximum speed of the oscillating block is $v_{\max} = F_m/b$.

67. A 2200-lb car carrying four 180-lb people is traveling over a rough "washboard" dirt road. The corrugations in the road are 13 ft apart. The car is observed to bounce with maximum amplitude when its speed is 10 mi/h. The car now stops and the four people get out. By how much does the car body rise on its suspension owing to this decrease in weight?

68. Starting from Eq. 42, find the velocity $v (= dx/dt)$ in forced oscillatory motion. Show that the velocity amplitude is $v_m = F_m/[(m\omega'' - k/\omega'')^2 + b^2]^{1/2}$. The equations of Section 15-9 are identical in form with those representing an electrical circuit containing a resistance R, an inductance L, and a capacitance C in series with an alternating emf $V = V_m \cos \omega'' t$. Hence b, m, k, and F_m are analogous to R, L, $1/C$, and V_m, respectively, and x and v are analogous to electric charge q and current i, respectively. In the electrical case the current amplitude i_m, analogous to the velocity amplitude v_m above, is used to describe the quality of the resonance.

Section 15-10 Two-Body Oscillations

69. Suppose that the spring in Fig. 22*a* has a force constant $k = 252$ N/m. Let $m_1 = 1.13$ kg and $m_2 = 3.24$ kg. Calculate the period of oscillation of the two-body system.

70. (*a*) Show that when $m_2 \to \infty$ in Eq. 46, $m \to m_1$. (*b*) Show that the effect of a noninfinite wall ($m_2 < \infty$) on the oscillations of a body of mass m_1 at the end of a spring attached to the wall is to reduce the period, or increase the frequency, of oscillation compared to (*a*). (*c*) Show that when $m_2 = m_1$ the effect is as though the spring were cut in half, each body oscillating independently about the center of mass at the middle.

71. (*a*) Calculate the reduced mass of each of the following diatomic molecules: O_2, HCl, and CO. Express your answers in unified atomic mass units, the mass of a hydrogen atom being 1.00 u. (*b*) An HCl molecule is known to vibrate at a fundamental frequency of $\nu = 8.7 \times 10^{13}$ Hz. Find the effective "force constant" k for the coupling forces between the atoms. In terms of your experience with ordinary springs, would you say that this "molecular spring" is relatively stiff or not?

72. Show that the kinetic energy of the two-body oscillator of Fig. 22*a* is given by $K = \frac{1}{2}mv^2$, where m is the reduced mass and $v (= v_1 - v_2)$ is the relative velocity. It may help to note that linear momentum is conserved while the system oscillates.

Computer Projects

73. Write a computer program or design a spreadsheet to calculate the amplitude and phase of a simple harmonic motion when you supply the force constant k, the mass m, the initial coordinate x_0, and the initial velocity v_0. Write the coordinate as $x(t) = x_m \cos(\omega t + \phi)$ and use $\omega = \sqrt{k/m}$, $x_m = \sqrt{x_0^2 + v_0^2/\omega^2}$, and $\phi = \tan^{-1}(-v_0/\omega x_0)$. Be sure to check that the value calculated for ϕ is correct by verifying that $\cos\phi$ has the same sign as x_0 and $\sin\phi$ has the same sign as v_0. If they do not, add $180°$ (or π rad) to the calculated value. Also be careful of division by zero. If $x_0 = 0$ automatically set $\phi = +90°$ or $-90°$ without attempting to calculate $v_0/\omega x_0$. Which angle you choose, of course, depends on the sign of v_0. Write the program so that once it has finished a calculation it returns to the beginning and requests data for the next problem. Here are some oscillations to try. All involve a mass of 250 g on a spring with a force constant of 200 N/m. (a) $x_0 = 2.8$ cm, $v_0 = 0$. (b) $x_0 = -2.8$ cm, $v_0 = 0$. (c) $x_0 = 0$, $v_0 = 56$ cm/s. (d) $x_0 = 0$, $v_0 = -56$ cm/s. (e) $x_0 = 2.8$ cm, $v_0 = 56$ cm/s. (f) $x_0 = 2.8$ cm, $v_0 = -56$ cm/s. (g) $x_0 = -2.8$ cm, $v_0 = 56$ cm/s. (h) $x_0 = -2.8$ cm, $v_0 = -56$ cm/s.

74. You can use a computer to study damped oscillations. Consider a mass m on the end of a spring with force constant k, subject to a drag force that is proportional to its velocity. Newton's second law yields $m\,d^2x/dt^2 = -kx - bv$. Write a computer program or design a spreadsheet to compute the coordinate x, the velocity v, and the total mechanical energy E at the end of every time interval of duration Δt from $t = 0$ to $t = t_f$. See Section 6-6 and the computer projects at the end of Chapter 6. Use the program to solve the following problems. In each case take $m = 2.0$ kg, $k = 350$ N/m, $x_0 = 0.070$ m, $v_0 = 0$, $t_f = 1.0$ s. Use an integration interval of 0.001 s. (a) Take $b = 2.8$ kg/s and on separate graphs plot $x(t)$ and $E(t)$. Notice the decrease in amplitude as time goes on. The decrease is intimately associated with a loss in energy via the drag force. Notice that the energy graph has small oscillations and that there are short regions where the energy is nearly constant. Where in the oscillatory motion do these regions occur? Give a physical explanation for their occurrence. Does the drag force change the period of the oscillation? Use the graph to estimate the time between successive maxima and compare the result with $2\pi\sqrt{m/k}$. (b) If the drag force is increased sufficiently, no oscillations occur and the motion is said to be *overdamped*. Take $b = 110$ kg/s and use your program to plot $x(t)$ and $E(t)$.

75. If a sinusoidal force is applied to an object on the end of a spring, Newton's second law becomes $m\,d^2x/dt^2 = -kx - bv + F_m \cos\omega''t$. Write a computer program or design a spreadsheet to compute the coordinate x, velocity v, and total mechanical energy E of the oscillator at the end of every time interval of duration Δt from $t = 0$ to $t = t_f$. See Section 6-6 and the computer projects at the end of Chapter 6. For the following problems take $m = 2.0$ kg, $k = 350$ N/m, $x_0 = 0.070$ m, $v_0 = 0$, $t_f = 2.0$ s. Use an integration interval of 0.001 s. (a) Neglect damping by setting $b = 0$ and take $F_m = 18$ N and $\omega'' = 35$ rad/s. Use your program to plot $x(t)$ and $E(t)$ on separate graphs. Notice that the applied force causes slight deviations from a sinusoidal shape. Also notice that the applied force transfers energy to the oscillator during some portions of the motion and removes it during others. As a result, the amplitude changes slightly with time. Use the graph to estimate the average amplitude. Also estimate the period and use its value to calculate the angular frequency. Is it closer to 35 rad/s or to $\omega = \sqrt{k/m}$? (b) Again set $b = 0$ and take $F_m = 18$ N but now take $\omega'' = 15$ rad/s, much closer to $\sqrt{k/m}$. Plot $x(t)$ and $E(t)$ and note the increasing amplitude and energy. The applied force puts energy into the oscillator over much longer periods than it takes energy out. (c) Now take damping into account by setting $b = 15$ kg/s. Again let $F_m = 18$ N and $\omega'' = 35$ rad/s, far from $\sqrt{k/m}$. Plot $x(t)$ and $E(t)$. Use your graph to find the angular frequency near $t = 0$ and near $t = 2$ s. You might measure half a period and double your result. Notice that at first the motion is close to the natural motion, the motion in the absence of an applied force. By about 1 s the natural motion has been damped considerably and the subsequent motion is that which is applied to the mass by the external force. Estimate the amplitude near $t = 2$ s. (d) Repeat part (c) but take $\omega'' = 15$ rad/s. Estimate the amplitude. (e) For which of the situations considered is the amplitude near $t = 2$ s the greatest? the least?

CHAPTER 16

GRAVITATION

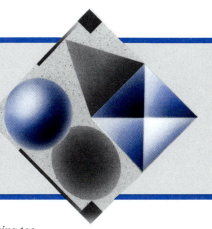

Up to now we have discussed the effects of forces, without being too specific about what determines their magnitude and direction. In this chapter we study the details of one particularly important force, gravitation. In 1665 Newton deduced that the force that governs the fall of apples near the Earth is the same as that which holds the Moon in its orbit. This was the first step toward developing a law of gravitation that could be applied to any pair of bodies in the universe.

After introducing Newton's law of universal gravitation, we discuss its consequences and its experimental tests. We show that the Earth's gravity can be understood as a particular case of this universal law, and that the motions of the planets can be similarly explained. We conclude with a look at modern gravitational theory, namely Einstein's general theory of relativity, which gives correct results when the gravitational force is strong (where Newton's theory fails) and agrees with Newton's theory when the gravitational force is weak.

As you study this chapter, you should note that many of the basic concepts of dynamics discussed in previous chapters find an application here. We apply basic laws for forces, potential energy, the conservation of energy and angular momentum, harmonic motion, and properties of extended bodies. We also introduce new concepts, including the notion of fields, which will have application in later chapters.

16-1 GRAVITATION FROM THE ANCIENTS TO KEPLER

From at least the time of the ancient Greeks, two problems were the subjects of searching inquiry: (1) the tendency of objects such as stones to fall to Earth when released, and (2) the motions of the planets, including the Sun and the Moon, which were classified with the planets in those times. In early days these problems were thought of as completely separate. One of Newton's great achievements is that he saw them clearly as aspects of a single problem and subject to the same laws.

The earliest serious attempts to explain the kinematics of the solar system were made by the ancient Greeks. Ptolemy (Claudius Ptolemaeus, 2nd century A.D.) developed a geocentric (Earth-centered) scheme for the solar system in which, as the name implies, the Earth remains stationary at the center while the planets, including the Sun and the Moon, revolve around it. This should not be a surprising deduction. The Earth seems to us to be a substantial body. Shakespeare referred to it as "this goodly frame, the Earth. . . . " Even today, in navigational astronomy we use a geocentric reference frame, and in ordinary conversation we use terms such as "sunrise," which implies such a frame.

Because simple circular orbits cannot account for the complicated motions of the planets, Ptolemy had to use the concept of epicycles, in which a planet moves around a circle whose center moves around another circle centered on the Earth (see Fig. 1*a*). He also had to resort to several other geometrical arrangements, each of which preserved the supposed sanctity of the circle as a central feature of planetary motions. We now know that it is not a circle that is fundamental but an ellipse, with the Sun at one focus, as we shall discuss.

In the 16th century Nicolaus Copernicus (1473–1543) proposed a heliocentric (Sun-centered) scheme, in which the Earth (along with the other planets) moves about the Sun (see Fig. 1*b*). Even though the Copernican scheme seems much simpler than that of Ptolemy, it was not immediately accepted. Copernicus still believed in the

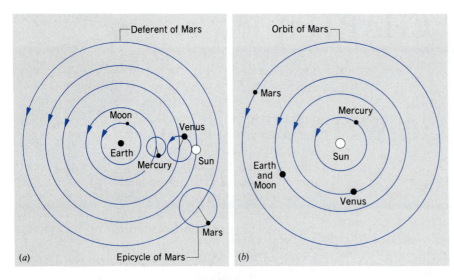

Figure 1 (*a*) The Ptolemaic view of the solar system. The Earth is at the center, and the Sun and planets move around it. The planets move in small circles (epicycles), whose centers travel along large circles (deferents). (*b*) The Copernican view of the solar system. The Sun is at the center, and the planets move around it.

sanctity of circles, and his use of epicycles and other arrangements (which are not shown in Fig. 1*b*) was about as great as that of Ptolemy. However, by putting the Sun at the center of things, Copernicus provided the correct reference frame from which our modern view of the solar system could develop.

To resolve the conflict between the Copernican and Ptolemaic schemes, more accurate observational data were needed. Such data were compiled by Tycho Brahe* (1546–1601), who was the last great astronomer to make observations without the use of a telescope. His data on planetary motions were analyzed and interpreted by Johannes Kepler (1571–1630), who had been Brahe's assistant. Kepler found important regularities in the motion of the planets, which led him to develop three laws (discussed in Section 16-8) that govern the motion of the planets.

Kepler's laws showed the great simplicity with which planetary motions could be described when the Sun was taken as the central body, if we give up the notion of perfect circles on which both the Ptolemaic and Copernican systems were based. However, Kepler's laws were empirical; they simply described the observed motions of the planets without any basis in terms of forces.† It was therefore a great triumph when Newton was later able to derive Kepler's laws from his laws of motion and his law of gravitation, which specified the force that acts between each planet and the Sun.

In this way Newton was able to account for the motion of the planets in the solar system and of bodies falling near the surface of the Earth with one common concept. He

thereby unified into one theory the previously separate sciences of terrestrial mechanics and celestial mechanics. The real scientific significance of Copernicus' work lies in the fact that the heliocentric theory opened the way for this synthesis. Subsequently, on the assumption that the Earth rotates and revolves about the Sun, it became possible to explain such diverse phenomena as the daily and the annual apparent motion of the stars, the flattening of the Earth from a spherical shape, the behavior of the trade-winds, and many other observations that could not have been explained so easily in a geocentric theory.

The historical development of gravitational theory can be viewed as a model example of the way the method of scientific inquiry leads to insight. Copernicus provided the appropriate reference frame for viewing the problem, and Brahe supplied systematic and precise experimental data. Kepler used the data to propose some empirical laws, and Newton proposed a universal force law from which Kepler's laws could be derived. Finally, Einstein was led to a new theory which explained certain small discrepancies in the Newtonian theory.

16-2 NEWTON AND THE LAW OF UNIVERSAL GRAVITATION

In 1665 the 23-year-old Newton left Cambridge University for Lincolnshire when the college was dismissed because of the plague. About 50 years later he wrote: "In the same year (1665) I began to think of gravity extending to the orb of the Moon . . . and having thereby compared the force requisite to keep the Moon in her orb with the force of gravity at the surface of the Earth, and found them to answer pretty nearly."

Newton's young friend William Stukeley wrote of having tea with Newton under some apple trees when New-

* See "Copernicus and Tycho," by Owen Gingerich, *Scientific American,* December 1973, p. 86.

† See "How Did Kepler Discover His First Two Laws," by Curtis Wilson, *Scientific American,* March 1972, p. 92.

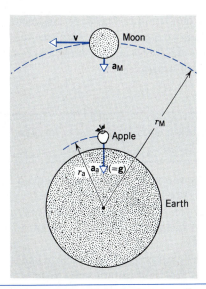

Figure 2   Both the Moon and the apple are accelerated toward the center of the Earth. The difference in their motions arises because the Moon has enough tangential speed v to maintain a circular orbit.

ton said that the setting was the same as when he got the idea of gravitation. "It was occasion'd by the fall of an apple, as he sat in a contemplative mood . . . and thus by degrees he began to apply this property of gravitation to the motion of the Earth and the heavenly bodys . . . " (see Fig. 2).

We can compute the acceleration of the Moon toward the Earth from its period of revolution and the radius of its orbit. We obtain 0.0027 m/s² (see Sample Problem 5, Chapter 4). This value is about a factor of 3600 smaller than g, the free-fall acceleration at the surface of the Earth. Newton, guided by Kepler's third law (see Problem 58), sought to account for this difference by assuming that the acceleration of a falling body is inversely proportional to the square of its distance from the Earth.

The question of what we mean by "distance from the Earth" immediately arises. Newton eventually came to regard every particle of the Earth as contributing to the gravitational attraction it had on other bodies. He made the daring assumption that the mass of the Earth could be treated as if it were all concentrated at its center. (See Section 16-5.)

We can treat the Earth as a particle with respect to the Sun, for example. It is not obvious, however, that we can treat the Earth as a particle with respect to an apple located only a couple of meters above its surface. If we do make this assumption, a falling body near the Earth's surface is a distance of one Earth radius (6400 km) from the effective center of attraction of the Earth. The Moon is about 380,000 km away. The inverse square of the ratio of these distances is $(6400/380,000)^2 = 1/3600$, in agreement with the ratio of the accelerations of the Moon and

the apple. In Newton's words quoted above, it does indeed "answer pretty nearly."

There are three overlapping realms in which we can discuss gravitation. (1) The gravitational attraction between two bowling balls, for example, although measurable by sensitive techniques, is too weak to fall within our ordinary sense perceptions. (2) The attraction of ourselves and objects around us by the Earth is a controlling feature of our lives from which we can escape only by extreme measures. The designers of our space program have the gravitational force constantly in mind. (3) On the scale of the solar system and of the interaction of stars and galaxies, gravitation is by far the dominant force. It is remarkable that all three situations can be described by the same force law.

This force law, Newton's law of universal gravitation, can be stated as follows:

Every particle in the universe attracts every other particle with a force directly proportional to the product of their masses and inversely proportional to the square of the distance between them. The direction of this force is along the line joining the particles.

Thus the magnitude of the gravitational force F that two particles of masses m_1 and m_2 separated by a distance r exert on each other is

$$F = G \frac{m_1 m_2}{r^2}. \qquad (1)$$

Here G, called the gravitational constant, is a universal constant that has the same value for all pairs of particles.

It is important to note that the gravitational forces between two particles are an action–reaction pair. The first particle exerts a force on the second particle that is directed toward the first particle along the line joining them. Likewise, the second particle exerts a force on the first particle that is directed toward the second particle along the line joining them. These forces are equal in magnitude but oppositely directed.

The universal constant G must not be confused with the g that is the acceleration of a body arising from the Earth's gravity. The constant G has the dimensions L^3/MT^2 and is a scalar, while g is the magnitude of a vector, has the dimensions L/T^2, and is neither universal nor constant.

Notice that Newton's law of universal gravitation is not a defining equation for any of the physical quantities (force, mass, or length) contained in it. According to our program for classical mechanics in Chapter 5, force is defined from Newton's second law, $\mathbf{F} = m\mathbf{a}$. The force $\mathbf{F}$ on a particle is assumed to be related in a simple way to the measurable properties of a particle and its environment. The law of universal gravitation is such a simple law. Once G is determined from experiment for any pair of bodies, that value of G can be used in the law of gravitation to determine the gravitational force between any other pair of bodies.

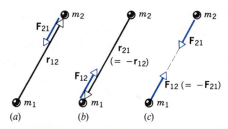

Figure 3   (*a*) The force $\mathbf{F}_{21}$ exerted on m_2 (by m_1) is directed opposite to the displacement, $\mathbf{r}_{12}$, of m_2 from m_1. (*b*) The force $\mathbf{F}_{12}$ exerted on m_1 (by m_2) is directed opposite to the displacement, $\mathbf{r}_{21}$, of m_1 from m_2. (*c*) $\mathbf{F}_{12} = -\mathbf{F}_{21}$, the forces being an action–reaction pair.

Notice also that Eq. 1 expresses the force between *particles*. If we want to determine the force between extended bodies, as, for example, the Earth and the Moon, we must regard each body as composed of particles. Then the interaction between all particles must be computed. Integral calculus makes such a calculation possible. Newton's motive in developing the calculus arose in part from a desire to solve such problems. Although it is in general incorrect to assume that all the mass of a body can be concentrated at its center of mass for gravitational purposes, this assumption *is* correct for spherically symmetric bodies. We often use this result, which we prove in Section 16-5.

Experiment strongly suggests that the gravitational force between two particles is independent of the presence of other bodies and of the properties of the medium in which the particles are immersed. The gravitational force between two bowling balls remains unchanged whether the balls are in free space, are under water, or are separated by a brick wall. The "gravity screens" of science fiction have no basis in fact.

The law of universal gravitation is a *vector law,* which can be expressed as follows. Let the displacement vector $\mathbf{r}_{12}$ point from the particle of mass m_1 to the particle of mass m_2, as Fig. 3*a* shows. The gravitational force $\mathbf{F}_{21}$, exerted on m_2 by m_1, is given in direction and magnitude by the vector relation

$$\mathbf{F}_{21} = -G\frac{m_1 m_2}{r_{12}^3}\mathbf{r}_{12} = -G\frac{m_1 m_2}{r_{12}^2}\frac{\mathbf{r}_{12}}{r_{12}}, \qquad (2a)$$

in which r_{12} is the magnitude of $\mathbf{r}_{12}$. The minus sign in Eq. 2*a* shows that $\mathbf{F}_{21}$ points in a direction opposite to $\mathbf{r}_{12}$; that is, the gravitational force is attractive, m_2 experiencing a force directed toward m_1. The displacement vector divided by its own magnitude, $\mathbf{r}_{12}/r_{12}$, is simply a unit vector $\mathbf{u}_r$ in the direction of the displacement, so the last part of Eq. 2*a* shows the inverse-square nature of the force.

The force exerted on m_1 by m_2 (see Fig. 3*b*) is similarly

$$\mathbf{F}_{12} = -G\frac{m_2 m_1}{r_{21}^3}\mathbf{r}_{21} = -G\frac{m_2 m_1}{r_{21}^2}\frac{\mathbf{r}_{21}}{r_{21}}. \qquad (2b)$$

Note in Eqs. 2*a* and 2*b* that $\mathbf{r}_{21} = -\mathbf{r}_{12}$ (see Figs. 3*a* and 3*b*) so that, as we expect, $\mathbf{F}_{12} = -\mathbf{F}_{21}$ (see Fig. 3*c*); that is, the gravitational forces acting on the two bodies form an action–reaction pair.

16-3   THE GRAVITATIONAL CONSTANT G

Determining the value of G would seem to be a simple task. All we need to do is to measure the gravitational force F between two known masses m_1 and m_2 separated by a known distance r. We can then calculate G from Eq. 1.

A large-scale system such as the Earth and the Moon or the Earth and the Sun cannot serve to determine G. The distances are large enough that the objects can be regarded as approximately point masses, but the values of the masses are not determined independently. In fact, the masses of these bodies, as we shall soon discuss, are determined using the value of G.

Instead, we must turn to a small-scale measurement, in which we use two laboratory objects of known mass and measure the force between them. The force is very weak, and the masses must be placed close together to make the force as large as possible. When we do this, we can usually no longer regard the masses as point particles, and Eq. 1 may not be applicable. There is, however, one special case in which we can use Eq. 1 for large objects. As we prove in Section 16-5, for spherical mass distributions we can regard the object as a point mass concentrated at its center. This is *not* an approximation; it is an exact relationship.

The first laboratory determination of G from the force between spherical masses at close distance was done by Henry Cavendish in 1798. He used a method based on the torsion balance, illustrated in Fig. 4. Two small balls, each of mass m, are attached to the ends of a light rod. This rigid "dumbbell" is suspended, with its axis horizontal, by a fine vertical fiber. Two large balls each of mass M are placed near the ends of the dumbbell on opposite sides. When the large masses are in the positions A, they attract the small masses according to the law of gravitation, and a torque is exerted on the dumbbell, rotating it counterclockwise as viewed from above. The rod reaches an equilibrium position under the opposing actions of the gravitational torque exerted by the masses M and the restoring torque exerted by the fiber. When the large masses are in the positions B, the dumbbell rotates clockwise to a new equilibrium position. The angle 2θ, through which the fiber is twisted when the balls are moved from one position (AA) to the other (BB), is measured by observing the deflection of a beam of light reflected from the small mirror attached to the fiber. From the value of θ and the torsional constant of the fiber (determined by measuring its period of oscillation—see Section 15-5), the torque

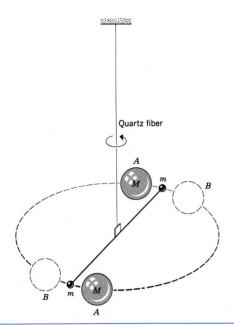

Quartz fiber

Figure 4 A schematic view of the apparatus used in 1798 by Henry Cavendish to measure the gravitational constant G. The large spheres of mass M, shown in location AA, can also be moved to location BB.

can be determined and the gravitational force can be obtained. Knowing the values of the masses m and M and the separation of their centers, we can calculate G. (See Sample Problem 1.)

Cavendish's original experiment gave a value for G of 6.75×10^{-11} N·m²/kg². In the nearly 200 years since the time of Cavendish, the same basic technique using the torsion balance has been used to repeat this measurement many times, leading to the presently accepted value of G,

$$G = 6.67259 \times 10^{-11} \text{ N·m}^2/\text{kg}^2,$$

with an uncertainty of $\pm 0.00085 \times 10^{-11}$ N·m²/kg² or about $\pm 0.013\%$. Compared with the results of measuring other physical constants, this precision is not impressive; for example, the speed of light was measured to a precision of about $10^{-8}\%$ before its value was set as a standard. It is difficult to improve substantially on the precision of the measured value of G because of its small magnitude and the correspondingly small value of the force between the two objects in our laboratory experiments. If we use two lead spheres of diameter 10 cm (and mass 6 kg), the maximum gravitational force between them when they are as close as possible is about 2×10^{-7} N, corresponding roughly to the weight of a piece of paper of area 1 mm².

This difficulty of measuring G is unfortunate, because gravitation has such an essential role in theories of the origin and structure of the universe. For example, we would like to know if G really is a constant. Does it change

with time? Does it depend on the chemical or physical state of the masses? Does it depend on their temperature? Despite many experimental searches, no such variations in G have so far been unambiguously confirmed, but measurements continue to be refined and improved, and the experimental tests continue.*

The large gravitational force exerted by the Earth on all bodies near its surface is due to the large mass of the Earth. In fact, the mass of the Earth can be determined from the law of universal gravitation and the value of G calculated from the Cavendish experiment. For this reason Cavendish is said to have been the first person to "weigh" the Earth. (In fact, the title of the paper written by Cavendish describing his experiments referred not to measuring G but instead to determining the density of the Earth from its weight and volume.) Consider the Earth, of mass M_E, and an object on its surface of mass m. The force of attraction is given both by

$$F = mg \quad \text{and} \quad F = \frac{GmM_E}{R_E^2}.$$

Here R_E is the radius of the Earth, which is the separation of the two bodies, and g is the free-fall acceleration at the Earth's surface. Combining these equations we obtain

$$M_E = \frac{gR_E^2}{G} = \frac{(9.80 \text{ m/s}^2)(6.37 \times 10^6 \text{ m})^2}{6.67 \times 10^{-11} \text{ N·m}^2/\text{kg}^2}$$

$$= 5.97 \times 10^{24} \text{ kg}.$$

Dividing the mass of the Earth by its volume, we obtain the average density of the Earth to be 5.5 g/cm³, or about 5.5 times the density of water. The average density of the rocks on the Earth's surface is much less than this value. We conclude that the interior of the Earth contains material of density greater than 5.5 g/cm³. The Cavendish experiment has given us information about the Earth's core! (See Problem 26.)

Sample Problem 1   In the Cavendish apparatus illustrated in Fig. 4, suppose $M = 12.7$ kg and $m = 9.85$ g. The length L of the rod connecting the two small spheres is 52.4 cm. The rod and the fiber form a torsion pendulum whose rotational inertia I about the central axis is 1.25×10^{-3} kg·m² and whose period of oscillation T is 769 s. The angle 2θ between the two equilibrium positions of the rod is 0.516° when the distance R between the centers of the large and small spheres is 10.8 cm. What is the value of the gravitational constant resulting from these data?

Solution   Let us first find κ, the torsional constant of the fiber.

---

* For a list of references to measurements of G, see "The Newtonian Gravitational Constant," by George T. Gillies, *Metrologia*, Vol. 24, p. 1, 1987. A discussion of these experiments and others testing the inverse-square law can be found in "Experiments on Gravitation," by Alan Cook, *Reports on Progress in Physics*, Vol. 51, p. 707, 1988.

The period of torsional oscillation is given by Eq. 21 of Chapter 15,

$$T = 2\pi \sqrt{\frac{I}{\kappa}}.$$

Solving for κ yields

$$\kappa = \frac{4\pi^2 I}{T^2} = \frac{(4\pi^2)(1.25 \times 10^{-3}\ \text{kg}\cdot\text{m}^2)}{(769\ \text{s})^2} = 8.34 \times 10^{-8}\ \text{N}\cdot\text{m}.$$

The rod is in equilibrium under the influence of two opposing torques resulting from the actions of the fiber and of the large spheres. The magnitude of the torque exerted by the fiber is related to the angular displacement θ according to Eq. 17 of Chapter 15,

$$\tau = \kappa\theta = (8.34 \times 10^{-8}\ \text{N}\cdot\text{m})\left(\frac{0.516°}{2} \times \frac{2\pi\ \text{rad}}{360°}\right)$$

$$= 3.75 \times 10^{-10}\ \text{N}\cdot\text{m}.$$

This torque is balanced by the total torque due to the gravitational force exerted by each large sphere on the nearby small sphere. The force F on each small sphere is equal to GMm/R^2, and the moment arm is one-half the length L of the rod. The total gravitational torque is then

$$\tau = (2F)(L/2) = FL = \frac{GMmL}{R^2}.$$

Solving for G yields

$$G = \frac{\tau R^2}{MmL} = \frac{(3.75 \times 10^{-10}\ \text{N}\cdot\text{m})(0.108\ \text{m})^2}{(12.7\ \text{kg})(0.00985\ \text{kg})(0.524\ \text{m})}$$

$$= 6.67 \times 10^{-11}\ \text{N}\cdot\text{m}^2/\text{kg}^2.$$

Sample Problem 2 Calculate the gravitational forces (*a*) between two 7.3-kg bowling balls separated by 0.65 m between their centers and (*b*) between the Earth and the Moon.

Solution (*a*) Using Eq. 1, we have

$$F = \frac{Gm_1 m_2}{r^2} = \frac{(6.67 \times 10^{-11}\ \text{N}\cdot\text{m}^2/\text{kg}^2)(7.3\ \text{kg})(7.3\ \text{kg})}{(0.65\ \text{m})^2}$$

$$= 8.4 \times 10^{-9}\ \text{N}.$$

(*b*) Using data for the Earth and the Moon from Appendix C, we find

$$F = \frac{(6.67 \times 10^{-11}\ \text{N}\cdot\text{m}^2/\text{kg}^2)(5.98 \times 10^{24}\ \text{kg})(7.36 \times 10^{22}\ \text{kg})}{(3.82 \times 10^8\ \text{m})^2}$$

$$= 2.01 \times 10^{20}\ \text{N}.$$

16-4 GRAVITY NEAR THE EARTH'S SURFACE

Let us assume, for the time being, that the Earth is spherical and that its density depends only on the radial distance from its center. The magnitude of the gravitational force

TABLE 1 VARIATION OF g_0 WITH ALTITUDE

Altitude (km)	g_0 (m/s²)
0	9.83
5	9.81
10	9.80
50	9.68
100	9.53
400 [a]	8.70
35,700 [b]	0.225
380,000 [c]	0.0027

[a] A typical space shuttle altitude.
[b] The altitude of communication satellites.
[c] The distance to the Moon.

acting on a particle of mass m, located at an external point a distance r from the Earth's center, can then be written, from Eq. 1, as

$$F = G\frac{M_E m}{r^2},$$

in which M_E is the mass of the Earth. This gravitational force can also be written, from Newton's second law, as

$$F = mg_0.$$

Here g_0 is the free-fall acceleration due only to the gravitational pull of the Earth. Combining the two equations above gives

$$g_0 = \frac{GM_E}{r^2}. \tag{3}$$

Table 1 shows some values of g_0 at various altitudes above the surface of the Earth, calculated from this equation. Note that, contrary to the impression that gravity drops to zero in an orbiting satellite, we find $g_0 = 8.7$ m/s² at typical space shuttle altitudes.

The real Earth differs from our model Earth in three ways.

1. *The Earth's crust is not uniform.* There are local density variations everywhere. The precise measurement of local variations in the free-fall acceleration gives information that is useful, for example, for oil prospecting. Figure 5 shows a gravity survey over an underground salt dome. The contours connect points with the same free-fall acceleration, plotted as deviations from a convenient reference value. The unit, named to honor Galileo, is the milligal, where 1 gal = 10^3 mgal = 1 cm/s².

2. *The Earth is not a sphere.* The Earth is approximately an ellipsoid, flattened at the poles and bulging at the equator. The Earth's equatorial radius is greater than its polar radius by 21 km. Thus a point at the poles is closer to the dense core of the Earth than is a point on the equator. We would expect that the free-fall acceleration would increase as one proceeds, at sea level, from the equator toward the

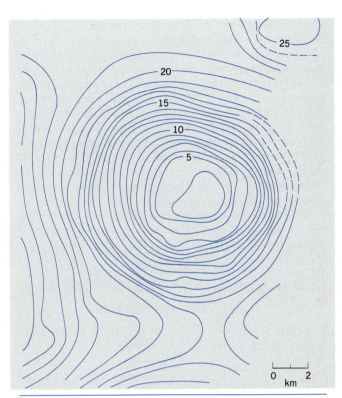

Figure 5 A surface gravity survey over an underground salt dome in Denmark. The lines connect points with the same value of g. The difference between the value of g on a contour and the value at the center is in units of milligal, equivalent to 10^{-5} m/s² or about 10^{-6} g. It is clear that something buried here is exerting a force centered in this region. Oil is often found in such formations.

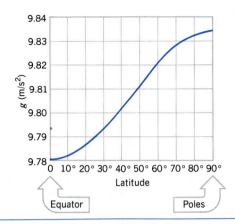

Figure 6 The variation of g with latitude at sea level. About 65% of the effect is due to the rotation of the Earth, with the remaining 35% coming from the Earth's slightly flattened shape.

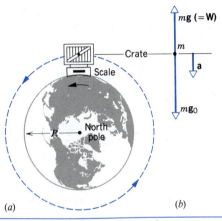

Figure 7 (a) A crate on the rotating Earth, resting on a platform scale at the equator. The view is along the Earth's rotational axis, looking down on the north pole. (b) A free-body diagram of the crate. The crate is in uniform circular motion and is thus accelerated toward the center of the Earth.

poles. Figure 6 shows that this is indeed what happens. The measured values of g in this figure include both the equatorial bulge effect and effects resulting from the rotation of the Earth.

3. *The Earth is rotating.* Figure 7a shows the rotating Earth from a position in space above the north pole. A crate of mass m rests on a platform scale at the equator. This crate is in uniform circular motion because of the Earth's rotation and is accelerated toward the center of the Earth. The resultant force acting on it must then point in that direction.

Figure 7b is a free-body diagram for the crate. The Earth exerts a downward gravitational pull of magnitude mg_0. The scale platform pushes up on the crate with a force mg, the weight of the crate. These two forces do not quite balance, and we have, from Newton's second law,

$$F = mg_0 - mg = ma$$

or

$$g_0 - g = a,$$

in which a is the centripetal acceleration of the crate. For a we can write $\omega^2 R_E$, where ω is the Earth's angular rota-

tion rate and R_E is its radius. Making this substitution leads to

$$g_0 - g = \omega^2 R_E = \left(\frac{2\pi}{T}\right)^2 R_E, \tag{4}$$

in which $T = 24$ h, the Earth's period of rotation. Substituting numerical values in Eq. 4 yields

$$g_0 - g = 0.034 \text{ m/s}^2.$$

We see that g, the measured free-fall acceleration on the equator of the rotating Earth, is less than g_0, the expected result if the Earth were not rotating, by only 0.034/9.8 or 0.35%. The effect decreases as one goes to higher latitudes and vanishes at the poles.

Sample Problem 3   (*a*) A neutron star is a collapsed star of extremely high density. The blinking pulsar in the Crab nebula is the best known of many examples. Consider a neutron star with a mass M equal to the mass of the Sun, 1.99×10^{30} kg, and a radius R of 12 km. What is the free-fall acceleration at its surface? Ignore rotational effects. (*b*) The asteroid Ceres has a mass of 1.2×10^{21} kg and a radius of 470 km. What is the free-fall acceleration at its surface?

Solution   (*a*) From Eq. 3 we have

$$g_0 = \frac{GM}{R^2} = \frac{(6.67 \times 10^{-11}\ \text{N} \cdot \text{m}^2/\text{kg}^2)(1.99 \times 10^{30}\ \text{kg})}{(12{,}000\ \text{m})^2}$$

$$= 9.2 \times 10^{11}\ \text{m/s}^2.$$

Even though pulsars rotate extremely rapidly, rotational effects have only a small influence on the value of g, because of the small size of pulsars.

(*b*) In the case of the asteroid Ceres, we have

$$g_0 = \frac{GM}{R^2} = \frac{(6.67 \times 10^{-11}\ \text{N} \cdot \text{m}^2/\text{kg}^2)(1.2 \times 10^{21}\ \text{kg})}{(4.7 \times 10^5\ \text{m})^2}$$

$$= 0.36\ \text{m/s}^2.$$

There is quite a contrast between the gravitational forces on the surfaces of these two bodies!

16-5   GRAVITATIONAL EFFECT OF A SPHERICAL DISTRIBUTION OF MATTER   *(Optional)*

We now prove a result we have already used: *a spherically symmetric body attracts particles outside it as if its mass were concentrated at its center.* We begin by considering a uniformly dense spherical shell of mass M whose thickness t is small compared to its radius R (Fig. 8). We seek the gravitational force it exerts on an external particle P of mass m.

We assume that each particle of the shell exerts on P a force that is proportional to the mass of the particle, inversely propor-

tional to the square of the distance between that particle of the shell and P, and directed along the line joining them. We must then obtain the resultant force on P, attributable to all parts of the spherical shell.

A small part of the shell at A attracts m with a force $\mathbf{F}_A$. A small part of equal mass at B, equally far from m but diametrically opposite A, attracts m with a force $\mathbf{F}_B$. The resultant of these two forces on m is $\mathbf{F}_A + \mathbf{F}_B$. Each of these forces has a component $F \cos \alpha$ along the symmetry axis and a component $F \sin \alpha$ perpendicular to the axis. The perpendicular components of $\mathbf{F}_A$ and $\mathbf{F}_B$ cancel, as they do for all such pairs of opposite points. To find the resultant force on P for all points on the shell, we need consider only the components parallel to the axis.

Let us take as our element of mass of the shell a circular strip dM. Its radius is $R \sin \theta$, its length is $2\pi(R \sin \theta)$, its width is $R\ d\theta$, and its thickness is t. Hence it has a volume

$$dV = 2\pi t R^2 \sin \theta\ d\theta.$$

Let the density of the shell be ρ, so that the mass within the strip is

$$dM = \rho\ dV = 2\pi t \rho R^2 \sin \theta\ d\theta.$$

Every particle in the ring, such as one of mass dm_A at A, attracts P with a force that has an axial component

$$dF_A = G\ \frac{m\ dm_A}{x^2} \cos \alpha.$$

Adding the contributions for all the particles in the ring gives

$$dF_A + dF_B + \cdots = \frac{Gm}{x^2} (\cos \alpha)(dm_A + dm_B + \cdots)$$

or

$$dF = \frac{Gm\ dM}{x^2} \cos \alpha,$$

where dM is the total mass of the ring and dF is the total force on m exerted by the ring.

Substituting for dM, we obtain

$$dF = 2\pi G t \rho m R^2\ \frac{\sin \theta\ d\theta}{x^2} \cos \alpha. \tag{5}$$

The variables x, α, and θ are related. From the figure we see that

$$\cos \alpha = \frac{r - R \cos \theta}{x}. \tag{6}$$

Using the law of cosines, $x^2 = r^2 + R^2 - 2rR \cos \theta$, we obtain

$$R \cos \theta = \frac{r^2 + R^2 - x^2}{2r}. \tag{7}$$

Differentiating Eq. 7 gives

$$\sin \theta\ d\theta = \frac{x}{rR}\ dx. \tag{8}$$

We now put Eq. 7 into Eq. 6 and then put Eqs. 6 and 8 into Eq. 5. As a result we eliminate θ and α and obtain

$$dF = \frac{\pi G t \rho m R}{r^2} \left(\frac{r^2 - R^2}{x^2} + 1 \right) dx. \tag{9}$$

This is the force exerted by the circular strip dM on the particle m.

We must now consider every element of mass in the shell by summing over all the circular strips in the entire shell. This

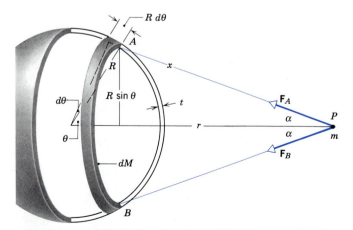

Figure 8   Gravitational attraction of a section of a spherical shell of matter on a particle of mass m at P.

involves an integration over the shell with respect to the variable x, which ranges from a minimum value of $r - R$ to a maximum value $r + R$. The needed integral is

$$\int_{r-R}^{r+R} \left(\frac{r^2 - R^2}{x^2} + 1 \right) dx = \left[\frac{-(r^2 - R^2)}{x} + x \right]\Bigg|_{r-R}^{r+R} = 4R,$$

which gives for the force, using Eq. 9,

$$F = \int_{r-R}^{r+R} dF = \frac{\pi G t \rho m R}{r^2} (4R) = G \frac{Mm}{r^2}, \qquad (10)$$

where

$$M = 4\pi R^2 t \rho$$

is the total mass of the shell. Equation 10 is exactly the same result we would obtain for the force between *particles* of mass M and m separated by a distance r. We have therefore proved the following important general result:

A uniformly dense spherical shell attracts an external point mass as if all the mass of the shell were concentrated at its center.

A solid sphere can be regarded as composed of a large number of concentric shells. If each spherical shell has a uniform density, even though different shells may have different densities, the same result applies to the solid sphere. Hence a body such as the Earth, the Moon, or the Sun, to the extent that they are such spheres, may be regarded gravitationally as point particles to bodies outside them.

Keep in mind that our proof applies only to spheres and only when the density is uniform over the sphere or a function of radius alone.

Force on an Interior Particle

We now prove another important result: the force exerted by a spherical shell on a particle *inside* it is *zero*. Figure 9 shows the particle at point P inside the shell. Notice that r is now smaller than R. The integration over x, now with the limits $R - r$ to $r + R$, gives

$$\int_{R-r}^{r+R} \left(\frac{r^2 - R^2}{x^2} + 1 \right) dx = \left[\frac{-(r^2 - R^2)}{x} + x \right]\Bigg|_{R-r}^{r+R} = 0,$$

and so $F = 0$. Thus we obtain another general result:

A uniform spherical shell of matter exerts no gravitational force on a particle located inside it.

This last result, although not obvious, is plausible because the mass elements of the shell to the left and to the right of m in Fig. 9 now exert forces of opposite directions on m. There is more mass on the left that pulls m to the left, but the smaller mass on the right is closer to m; the two effects exactly cancel only if the force varies precisely as an inverse square of the separation distance of two particles. (See Problem 29.) Important consequences of this result will be discussed in the chapters on electricity. There we shall see that the electrical force between charged particles also depends inversely on the square of the distance between them.

The above result for a particle inside a spherical shell implies that the gravitational force exerted by the Earth on a particle decreases as the particle goes deeper into the Earth, assuming a constant density for the Earth. As the particle goes deeper, more

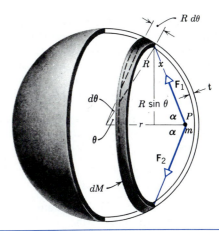

Figure 9 Gravitational attraction of a section of a spherical shell of matter on a particle of mass m at a point P inside the shell.

of the Earth's mass is in shells that are external to the location of the particle, and the net force on the particle from those shells is zero. The gravitational force becomes zero at the center of the Earth. Hence g would be a maximum at the Earth's surface and decrease both outward and inward from that point if the Earth had constant density. Can you imagine a spherically symmetric distribution of the Earth's mass which would not give this result? (See Problem 26.)

Sample Problem 4 Suppose a tunnel could be dug through the Earth from one side to the other along a diameter, as shown in Fig. 10. (*a*) Show that the motion of a particle dropped into the tunnel is simple harmonic motion. Neglect all frictional forces and assume that the Earth has a uniform density. (*b*) If mail were delivered through this chute, how much time would elapse between deposit at one end and delivery at the other end?

Solution (*a*) The gravitational attraction of the Earth for the particle at a distance r from the center of the Earth arises entirely from that portion of matter of the Earth in shells internal to the

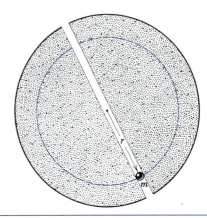

Figure 10 Sample Problem 4. A particle moves in a tunnel through the Earth.

position of the particle. The external shells exert no force on the particle. Let us assume that the Earth's density is uniform with the value ρ. Then the mass M' inside a sphere of radius r and volume V' is

$$M' = \rho V' = \rho \frac{4\pi r^3}{3}.$$

This mass can be treated as though it were concentrated at the center of the Earth for gravitational purposes. Hence the radial component of the force on the particle of mass m is

$$F = -\frac{GM'm}{r^2}.$$

The minus sign indicates that the force is attractive and thus directed toward the center of the Earth.

Substituting for M', we obtain

$$F = -G \frac{\rho 4\pi r^3 m}{3r^2} = -\left(G\rho \frac{4\pi m}{3}\right) r = -kr.$$

Here $G\rho 4\pi m/3$ is a constant, which we have called k. The force is therefore proportional to the displacement r but oppositely directed. This is exactly the criterion for simple harmonic motion.

(b) The period of this simple harmonic motion is

$$T = 2\pi \sqrt{\frac{m}{k}} = 2\pi \sqrt{\frac{3m}{G\rho 4\pi m}} = \sqrt{\frac{3\pi}{G\rho}}.$$

With $\rho = 5.51 \times 10^3$ kg/m^3, we have

$$T = \sqrt{\frac{3\pi}{G\rho}} = \sqrt{\frac{3\pi}{(6.67 \times 10^{-11} \text{ N} \cdot \text{m}^2/\text{kg}^2)(5.51 \times 10^3 \text{ kg/m}^3)}}$$
$$= 5060 \text{ s} = 84.4 \text{ min}.$$

The time for delivery is one-half period, or about 42 min. This time is *independent of the mass of the mail*. It can be shown that the same period results if the tunnel is dug along any chord instead of along a diameter.

The Earth's density is not really uniform. What would be the effect on this problem if we took ρ to be some function of r, rather than a constant?

Testing the Inverse-Square Law

As we discuss in Section 16-8, Kepler's laws give direct evidence for a $1/r^2$ gravitational force. We can therefore regard the $1/r^2$ law to be well tested at distances of the order of the size of the solar system (10^{13} m). Small exceptions in the motion of the inner planets are explained by Einstein's general theory of relativity, which supersedes Newton's law when the gravitational force is intense but which reduces to Newton's law when the force is weaker; see Section 16-10.

We would therefore like to test the $1/r^2$ law at laboratory distances. Because the force is so weak, it is difficult to make such a test by repeating the Cavendish experiment with different separations between the masses. A more precise method makes use of the vanishing of the gravitational force on a test particle inside a spherical shell. If we could isolate a test particle, say on one arm of a torsion balance, and then surround it with a spherical shell, any slight rotation of the balance as the test particle moves within the shell would indicate a deviation from the $1/r^2$ law. The rotation could be detected by a suitable mechanism attached to the other arm of the balance.

Unfortunately, surrounding a test mass with a spherical shell

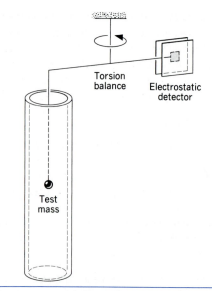

Figure 11 A test mass inside a long cylinder. For a $1/r^2$ force, the gravitational attraction between the test mass and the cylinder should vanish (neglecting effects of the ends). A torsion balance allows changes in the force on the test mass to be measured at different locations inside the cylinder.

and moving it inside present great technical difficulties; as an alternative a long cylinder is used instead. From a calculation similar to the one we used for the spherical shell, it can be shown that the gravitational force exerted by a long cylindrical shell on a test mass inside the cylinder vanishes if the cylinder is infinitely long; for a cylinder of finite length a small but easily calculable correction must be applied.

Figure 11 shows the geometry for a typical experiment. As the test mass is moved in a horizontal plane, variations in the gravitational force between the cylinder and the test mass would be detectable with the torsion balance. If the gravitational force between particles had a variation other than $1/r^2$, the force on the test mass would not vanish and would vary as the test mass moves in the horizontal plane.

Such experiments show that the force is indeed of the form $1/r^2$ at laboratory distances (centimeters to meters). One way of expressing the results of these experiments is to assume the force to be of the form $1/r^{2+\delta}$ where $\delta = 0$ in the Newtonian theory, and then to show that the experiment places a small upper limit on δ. The present upper limit on δ is about 10^{-4}; at the best precision obtainable from laboratory studies, there appears to be no deviation from the $1/r^2$ form of the law of gravitation. By comparison, experiments testing the $1/r^2$ force between electric charges (see Section 29-6) give an upper limit of about 10^{-16} on δ in that case. ∎

16-6 GRAVITATIONAL POTENTIAL ENERGY

In Chapter 8 we discussed the gravitational potential energy of a particle (mass m) and the Earth (mass M). We considered only the special case in which the particle remains close to the Earth so that we could assume the gravitational force acting on the particle to be of constant magnitude mg. In this section we remove that restriction and consider particle–Earth separations that may be appreciably greater than the Earth's radius.

Equation 4 of Chapter 8, which we may write as

$$\Delta U = U_b - U_a = -W_{ab},$$

defines the change ΔU in the potential energy of any system, in which a conservative force (gravity, say) acts, as the system changes from configuration a to configuration b. W_{ab} is the work done by that conservative force as the system changes.

The potential energy of the system in any arbitrary configuration b is

$$U_b = -W_{ab} + U_a. \qquad (11)$$

To give a value to U_b we choose configuration a to be an agreed-upon reference configuration, and we assign to U_a an arbitrary reference value, usually zero. For instance, in Chapter 8 we considered the potential energy of a particle subject to the force of gravity mg near the Earth's surface. At a height y, the potential energy is $U(y) = mgy$, where the reference value $U = 0$ is taken at $y = 0$.

We now consider the more general case of two particles of masses m and M separated by a distance r. Initially, the particles are separated by r_a, and the separation changes to r_b. To find the corresponding change in potential energy ΔU we must, according to Eq. 11, evaluate W_{ab}. Figure 12 shows the geometrical arrangement. We let M be at the origin of coordinates, and we move m toward M. Note that $\mathbf{r}$ and $d\mathbf{s}$ (the displacement vector) are in opposite directions, so that $d\mathbf{s} = -d\mathbf{r}$. The work done by $\mathbf{F}$ when the particle moves from a to b is

$$W_{ab} = \int_a^b \mathbf{F} \cdot d\mathbf{s} = -\int_a^b F\,dr$$
$$= -\int_{r_a}^{r_b} \frac{GmM}{r^2}\,dr = -GmM \int_{r_a}^{r_b} \frac{dr}{r^2}$$
$$= -GmM\left(-\frac{1}{r}\right)\Bigg|_{r_a}^{r_b} = GmM\left(\frac{1}{r_b} - \frac{1}{r_a}\right). \quad (12)$$

Figure 12 A particle M exerts a gravitational force $\mathbf{F}$ on a particle of mass m located at $\mathbf{r}$. The particle of mass m is displaced a small distance $d\mathbf{s}$.

Thus

$$\Delta U = -W_{ab} = GmM\left(\frac{1}{r_a} - \frac{1}{r_b}\right). \quad (13)$$

We choose our reference configuration to be an infinite separation of the particles ($r_a \to \infty$), and we define $U(\infty)$ to be zero. At an arbitrary separation r, the potential energy is

$$U(r) = -W_{\infty r} + 0 \qquad (14)$$

or

$$U(r) = -\frac{GMm}{r}. \qquad (15)$$

The minus sign indicates that the potential energy is negative at any finite distance; that is, the potential energy is zero at infinity and decreases as the separation distance decreases. This corresponds to the fact that the gravitational force exerted on m by M is attractive. As the particle moves in from infinity, the work $W_{\infty r}$ done by this force on the particle is positive, which means, based on Eq. 14, that $U(r)$ is negative.

Equation 15 holds for any path followed by the particle in moving from infinity to radius r. We can show this by breaking up any arbitrary path into steplike portions, which are drawn alternately along the radius and perpendicular to it (Fig. 13). No work is done along perpendicular segments, such as AB, because along them the force is perpendicular to the displacement. The total work done along all the radial parts of the path, one of which is BC, equals the work done in going directly along a radial path, such as AE. *The work done by the gravitational force as the particle moves between any two points is therefore independent of the actual path connecting these points.* Hence the gravitational force is a *conservative* force.

Equation 15 shows that the potential energy is a property of the *system* consisting of the two particles M and m, rather than of either body alone. The potential energy changes whether M or m is displaced; each is acted on by the gravitational force of the other. Nor does it make any sense to assign part of the potential energy to M and part of it to m. Often, however, we do speak of the potential energy of a body m (planet or stone, say) acted on by the gravitational force of a much more massive body M (Sun or Earth, respectively). The justification for speaking as though the potential energy belongs to the planet or to the stone alone is this: When the potential energy of a system

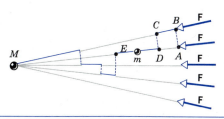

Figure 13 Work done in taking a particle from A to E is independent of the path.

of two bodies changes into kinetic energy, the lighter body gets most of the kinetic energy. The Sun is so much more massive than a planet that the Sun receives hardly any of the kinetic energy; the same is true for the Earth in the Earth–stone system.

We can reverse the previous calculation and derive the gravitational force from the potential energy. For spherically symmetric potential energy functions, the relation $F = -dU/dr$ gives the radial component of the force; see Eq. 13 of Chapter 8. With the potential energy of Eq. 15, we obtain

$$F = -\frac{dU}{dr} = -\frac{d}{dr}\left(-\frac{GMm}{r}\right) = -\frac{GMm}{r^2}. \quad (16)$$

The minus sign here shows that the force is attractive, directed inward along a radius.

We can show that the potential energy defined according to Eq. 13 leads to the familiar mgy for a small difference in elevation y near the surface of the Earth. Let us evaluate Eq. 13 for the difference in potential energy between the location at a height y above the surface (that is, $r_b = R_E + y$, where R_E is the radius of the Earth) and the surface ($r_a = R_E$):

$$\Delta U = U(R_E + y) - U(R_E) = GM_E m\left(\frac{1}{R_E} - \frac{1}{R_E + y}\right)$$
$$= \frac{GM_E m}{R_E}\left(1 - \frac{1}{1 + y/R_E}\right).$$

When $y \ll R_E$, which would be the case for small displacements of bodies near the Earth's surface, we can use the binomial expansion to approximate the last term as $(1 + x)^{-1} = 1 - x + \cdots \approx 1 - x$, which gives

$$\Delta U \approx \frac{GM_E m}{R_E}\left[1 - \left(1 - \frac{y}{R_E}\right)\right] = \frac{GM_E my}{R_E^2} = mgy,$$

using Eq. 3 to replace GM_E/R_E^2 with g. This shows that Eq. 13 for the difference in gravitational potential energy is consistent with our previous use of mgy for situations near the Earth's surface. In fact, we can use the approximation $\Delta U = mgy$ for the difference in potential energy between two elevations at any distance R from the center of the Earth, as long as $y \ll R$ and we use the value of g (see Table 1) appropriate for that R.

Sample Problem 5 What is the gravitational potential energy of the Moon–Earth system, relative to the potential energy at infinite separation?

Solution The masses of the Earth and the Moon are 5.98×10^{24} kg and 7.36×10^{22} kg, respectively, and their mean separation distance d is 3.82×10^8 m. From Eq. 15 then,

$$U = -\frac{GMm}{d}$$
$$= -\frac{(6.67 \times 10^{-11} \text{ N} \cdot \text{m}^2/\text{kg}^2)(5.98 \times 10^{24} \text{ kg})(7.36 \times 10^{22} \text{ kg})}{3.82 \times 10^8 \text{ m}}$$
$$= -7.68 \times 10^{28} \text{ J}.$$

An energy of this magnitude is about equal to world industrial energy production, at its present rate, for about 10^8 years.

Sample Problem 6 What minimum initial speed must a projectile have at the Earth's surface if it is to escape from the Earth? Ignore effects caused by atmospheric friction and the Earth's rotation.

Solution A projectile fired upward will usually slow down, come momentarily to rest, and return to Earth. For a certain initial speed, however, it will move upward forever, coming to rest only at infinity.

Consider such a projectile, of mass m, leaving the Earth's surface with this critical initial speed v. It has a kinetic energy K given by $\frac{1}{2}mv^2$ and a potential energy U given by Eq. 15 or

$$U(R_E) = -\frac{GM_E m}{R_E},$$

in which M_E is the mass of the Earth and R_E its radius.

When the projectile has reached infinity, it has no kinetic energy—recall that we seek the *minimum* speed for escape— and no potential energy—recall that this is our zero-potential-energy configuration. Its total energy at infinity is therefore zero. From the conservation of energy, its total energy at the surface must also be zero, or

$$K + U = 0.$$

This leads to

$$\frac{1}{2}mv^2 + \left(\frac{-GM_E m}{R_E}\right) = 0,$$

or

$$v = \sqrt{\frac{2GM_E}{R_E}}. \quad (17)$$

Substituting values into Eq. 17 gives

$$v = \sqrt{\frac{2GM_E}{R_E}} = \sqrt{\frac{2(6.67 \times 10^{-11} \text{ N} \cdot \text{m}^2/\text{kg}^2)(5.98 \times 10^{24} \text{ kg})}{6.37 \times 10^6 \text{ m}}}$$
$$= 1.12 \times 10^4 \text{ m/s} = 11.2 \text{ km/s} = 25,000 \text{ mi/h}.$$

The escape speed does not depend on the direction in which the projectile is fired. The Earth's rotation—which we have ignored so far—does play a role, however. Firing eastward has an advantage in that the Earth's tangential surface speed, which is 0.46 km/s at Cape Canaveral, can be subtracted from the value calculated from Eq. 17. Table 2 shows escape speeds for the Earth and some other bodies.

TABLE 2 SOME ESCAPE SPEEDS

Body	Mass (kg)	Radius (m)	Escape Speed (km/s)
Ceres[a]	1.17×10^{21}	3.8×10^5	0.64
Moon	7.36×10^{22}	1.74×10^6	2.38
Earth	5.98×10^{24}	6.37×10^6	11.2
Jupiter	1.90×10^{27}	7.15×10^7	59.5
Sun	1.99×10^{30}	6.96×10^8	618
Sirius B[b]	2×10^{30}	1×10^7	5200
Neutron star	2×10^{30}	1×10^4	2×10^5

[a] The most massive of the asteroids.
[b] A white dwarf, the companion of the bright star Sirius.

Potential Energy of Many-Particle Systems

Let us now consider another interpretation of $U(r)$. Suppose we balance the gravitational force by an *external force* applied by some external agent, and let us arrange it so that, at all times, this external force is equal and opposite to the gravitational force for each particle. (For example, we hold each particle in our hand and move it in equilibrium.) The work done by the *external* force as the particles move from an infinite separation to separation r is not $W_{\infty r}$ but $-W_{\infty r}$; this follows because the displacements are the same but the forces are equal and opposite. Thus we may interpret Eq. 14 as follows:

The potential energy of a system of particles is equal to the work that must be done by an external agent to assemble the system, starting from the standard reference configuration.

Thus if you lift a stone of mass m a distance of y above the Earth's surface, you are the external agent (separating Earth and stone) and the work you do in "assembling the system" is $+mgy$, which is also the potential energy. Similarly, the work done *by the external agent* as a body of mass m moves from infinity to a distance r from the Earth is *negative* because the agent must exert a restraining force on the body; this is in agreement with Eq. 14.

These considerations also hold for systems that contain more than two particles. Consider three bodies of masses m_1, m_2, and m_3. Let them initially be at rest infinitely far from one another. The problem is to compute the work done by an external agent to bring them into the positions shown in Fig. 14. We first bring m_1 in from infinity to its final position. No work is done by gravity or the external agent because the separation between the three particles remains infinite. Let us then bring m_2 in toward m_1 from an infinite separation to the separation r_{12}. The work done by the external agent in opposing the gravitational force exerted by m_1 on m_2 is $-Gm_1m_2/r_{12}$. Now let us bring m_3 in from infinity to the separation r_{13} from m_1 and r_{23} from m_2. The work done by the external agent opposing the gravitational force exerted by m_1 on m_3 is $-Gm_1m_3/r_{13}$, and that opposing the gravitational force exerted by m_2 on m_3 is $-Gm_2m_3/r_{23}$. The total potential energy of this system is equal to the total work done by the external agent in assembling the system, or

$$U = -\left(\frac{Gm_1m_2}{r_{12}} + \frac{Gm_1m_3}{r_{13}} + \frac{Gm_2m_3}{r_{23}} \right). \quad (18)$$

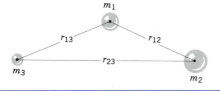

Figure 14   Three masses brought together from infinity.

Notice that no vector calculations are needed in this procedure.

No matter how we assemble the system, that is, regardless of the order in which the particles are moved or the paths they take, we always find this same amount of work required to bring the bodies into the configuration of Fig. 14 from an initial infinite separation. The potential energy must therefore be associated with the system rather than with any one or two bodies. If we wanted to separate the system into three isolated masses once again, we would have to supply an amount of energy

$$E = +\left(\frac{Gm_1m_2}{r_{12}} + \frac{Gm_1m_3}{r_{13}} + \frac{Gm_2m_3}{r_{23}} \right).$$

This energy is regarded as the *binding energy* holding the particles together in the configuration shown. In Sample Problem 5, for example, we found that the potential energy of the Earth–Moon system was -7.68×10^{28} J, and so the binding energy of the Earth–Moon system is 7.68×10^{28} J. This is the amount of energy that an external agent must provide to move the Earth and Moon from their present separation to an infinite separation, in an isolated system consisting only of the Earth and Moon (ignoring for this calculation the important effect of the Sun).

These concepts occur again in connection with forces of electric or magnetic origin, or, in fact, of nuclear origin. Their application is rather broad in physics. An advantage of the energy method over the dynamical method is that the energy method uses scalar quantities and scalar operations rather than vector quantities and vector operations. When the actual forces are not known, as is often the case in nuclear physics, the energy method is essential.

16-7   THE GRAVITATIONAL FIELD AND POTENTIAL   *(Optional)*

A basic fact of gravitation is that two particles exert forces on one another. We can think of this as a direct interaction between the two particles, if we wish. This point of view is called *action-at-a-distance*, the particles interacting even though they are not in contact. Another point of view is the *field* concept, which regards a particle as modifying the space around it in some way and setting up a *gravitational field*. This field, the strength of which depends on the mass of the particle, then acts on any other particle, exerting the force of gravitational attraction on it. The field therefore plays an intermediate role in our thinking about the forces between particles.

According to this view we have two separate parts to our problem. First, we must determine the gravitational field established by a given distribution of particles. Second, we must calculate the gravitational force that this field exerts on another particle placed in it.

We use this same approach later in the text when we study electromagnetism, in which case particles with electric charge set up an *electric* field, and the force on another charged particle is determined by the strength of the electric field at the location of the particle.

Let us consider the Earth as an isolated particle and ignore all rotational and other nongravitational effects. We use a small test body of mass m_0 as a probe of the gravitational field. If this body is placed in the vicinity of the Earth, it will experience a force having a definite direction and magnitude at each point in space. The direction is radially in toward the center of the Earth, and the magnitude is $m_0 g$. We can associate with each point near the Earth a vector $\mathbf{g}$, which is the acceleration that a body would experience if it were released at this point. We define the *gravitational field strength* at a point as the *gravitational force per unit mass* at that point or, in terms of our test mass,

$$\mathbf{g} = \frac{\mathbf{F}}{m_0} . \tag{19}$$

By moving the test mass to various positions, we can make a map showing the gravitational field at any point in space. We can then find the force on a particle at any point in that field by multiplying the mass m of the particle by the value of the gravitational field $\mathbf{g}$ at that point: $\mathbf{F} = m\mathbf{g}$.

The gravitational field is an example of a *vector field,* each point in this field having a vector associated with it. There are also *scalar fields,* such as the temperature field in a heat-conducting solid. The gravitational field arising from a fixed distribution of matter is also an example of a *static field,* because the value of the field at a given point does not change with time.

The field concept is particularly useful for understanding electromagnetic forces between moving electric charges. It has distinct advantages, both conceptually and in practice, over the action-at-a-distance concept. The field concept is particularly superior in the analysis of electromagnetic waves (for example, light or radio waves); action-at-a-distance suggests that forces can be transmitted instantly over any distance, while in theories based on fields the forces propagate at a finite speed (at most the speed of light). Gravitational waves (see Section 16-10), which have been predicted but not yet directly observed, would be similarly difficult to understand in the action-at-a-distance theory. The field concept, which was not used in Newton's day, was developed much later by Faraday for electromagnetism and only then applied to gravitation. Subsequently, this point of view was adopted for gravitation in the general theory of relativity. All present theories dealing with the ultimate nature of matter and the interactions between the fundamental particles are field theories of one kind or another.

We can also describe the gravitational field of a body by a scalar function called the *potential.* (The potential is not the same as the potential energy, although they are closely related.) We again measure the strength of the field using a test particle of mass m_0. Let us begin with the test particle at an infinite separation from the body (where the field is zero) and move the test particle toward the body until the separation is r, where the potential energy is $U(r)$. We then define the *gravitational potential V* at that point as

$$V(r) = \frac{U(r)}{m_0} . \tag{20}$$

That is, the potential is the same as the *potential energy per unit test mass.* Note that the potential is a *scalar,* being defined as the ratio of the scalars U and m.

For example, the potential energy of m_0 in the field of a spherically symmetric body of mass M is given by Eq. 15 as $U(r) = -GMm_0/r$. The gravitational potential may then be found using Eq. 20:

$$V(r) = \frac{U(r)}{m_0} = -\frac{GM}{r} . \tag{21}$$

Note that the potential $V(r)$ is independent of the value of the test mass m_0; similarly, the gravitational field $\mathbf{g}$, defined according to Eq. 19, is independent of m_0.

Just as we can find the radial component of the force $\mathbf{F}$ from $U(r)$ according to $F = -dU/dr$, we can find the radial component of the field $\mathbf{g}$ from $V(r)$ according to $g = -dV/dr$. We can therefore regard field and potential as alternative ways of analyzing gravitation; in a similar way, force and potential energy can be regarded as alternative ways of describing the dynamics of a system. ∎

16-8 THE MOTIONS OF PLANETS AND SATELLITES

Using Newton's laws of motion and law of universal gravitation, we can understand and analyze the behavior of all the bodies in the solar system: the orbits of the planets and comets about the Sun and of natural and artificial satellites about their planets. We make two assumptions that simplify the analysis: (1) we consider the gravitational force only between the orbiting body (the Earth, for instance) and the central body (the Sun), ignoring the perturbing effect of the gravitational force of other bodies (such as other planets); (2) we assume that the central body is so much more massive than the orbiting body that we can ignore its motion under their mutual interaction. In reality, both objects orbit about their common center of mass, but if one object is very much more massive than the other, the center of mass is approximately at the center of the more massive body. Exceptions to this second assumption will be noted.

The empirical basis for understanding the motions of the planets is Kepler's three laws, and we now show how these laws are related to the analytical results of Newton's laws.

1. *The Law of Orbits: All planets move in elliptical orbits having the Sun at one focus.* Newton was the first to realize that there is a direct mathematical relationship between inverse-square ($1/r^2$) forces and elliptical orbits. Figure 15 shows a typical elliptical orbit. The origin of coordinates is at the central body, and the orbiting body is located at polar coordinates r and θ. The orbit is described by two parameters: the *semimajor axis a* and the *eccentricity e.* The distance from the center of the ellipse to either focus is *ea.* A circular orbit is a special case of an elliptical orbit with $e = 0$, in which case the two foci merge to a single point at the center of the circle. For the planets in the solar system, the eccentricities are small and the orbits are nearly circular, as shown in Appendix C.

The maximum distance R_a of the orbiting body from the central body is indicated by the prefix *apo-* (or sometimes *ap-*), as in *aphelion* (the maximum distance from

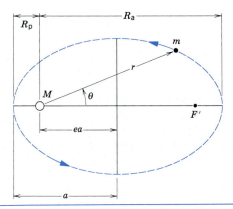

Figure 15 A planet of mass m moving in an elliptical orbit around the Sun. The Sun, of mass M, is at one focus of the ellipse. F' marks the other or "empty" focus. The semimajor axis a of the ellipse, the perihelion distance R_p, and the aphelion distance R_a are also shown. The distance ea locates the focal points, e being the eccentricity of the orbit.

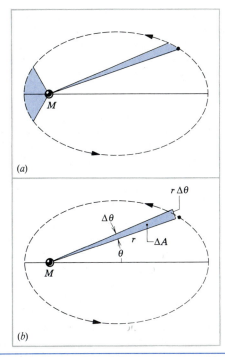

Figure 16 (a) The equal shaded areas are covered in equal times by a line connecting the planet to the Sun, demonstrating the law of areas. (b) The area ΔA is covered in a time Δt, during which the line sweeps through an angle $\Delta\theta$.

the Sun) or *apogee* (the maximum distance from Earth). Similarly, the closest distance R_p is indicated by the prefix *peri-,* as in *perihelion* or *perigee.* As you can see from Fig. 15, $R_a = a(1 + e)$ and $R_p = a(1 - e)$. For circular orbits, $R_a = R_p = a$.

2. *The Law of Areas: A line joining any planet to the Sun sweeps out equal areas in equal times.* Figure 16a illustrates this law; in effect it says that the orbiting body moves more rapidly when it is close to the central body than it does when it is far away. We now show that the law of areas is identical with the law of conservation of angular momentum.

Consider the small area increment ΔA covered in a time interval Δt, as shown in Fig. 16b. The area of this approximately triangular wedge is one-half its base, $r\,\Delta\theta$, times its height r. The rate at which this area is swept out is $\Delta A/\Delta t = \frac{1}{2}(r\,\Delta\theta)(r)/\Delta t$. In the instantaneous limit this becomes

$$\frac{dA}{dt} = \lim_{\Delta t \to 0} \frac{\Delta A}{\Delta t} = \lim_{\Delta t \to 0} \tfrac{1}{2}r^2\frac{\Delta\theta}{\Delta t} = \tfrac{1}{2}r^2\omega.$$

The instantaneous angular momentum of the orbiting body is $L = mr^2\omega$, and so

$$\frac{dA}{dt} = \frac{L}{2m}. \tag{22}$$

To the extent that we can regard the two bodies as an isolated system, L is a constant, and therefore dA/dt is a constant. The speeding up of a comet as it passes close to the Sun is therefore just a demonstration of the conservation of angular momentum.

It should be noted that conservation of angular momentum is valid for *any* central force, that is, for any force that acts along a line joining two particles and that depends only on the magnitude of the separation between

two particles. Note also that, in verifying the law of areas, we have not used the inverse-square force law; the validity of the law of areas tells us nothing about how F varies with r.

3. *The Law of Periods: The square of the period of any planet about the Sun is proportional to the cube of the planet's mean distance from the Sun.* Let us prove this result for circular orbits. The gravitational force provides the necessary centripetal force for circular motion:

$$\frac{GMm}{r^2} = m\omega^2 r. \tag{23}$$

Replacing ω by $2\pi/T$, we obtain

$$T^2 = \left(\frac{4\pi^2}{GM}\right) r^3. \tag{24}$$

A similar result is obtained for elliptical orbits, with the radius r replaced by the semimajor axis a.

The relationship between T^2 and a^3 should be determined by the quantity $4\pi^2/GM$. For all planets orbiting the Sun, the ratio T^2/a^3 should be a constant; Table 3 shows that this is indeed the case. If we can measure T and a for an orbiting body, we can determine the mass of the central body. This procedure is independent of the mass of the orbiting body, and so it gives no information about its mass.

TABLE 3   KEPLER'S LAW OF PERIODS FOR THE SOLAR SYSTEM

Planet	Semimajor Axis $a\,(10^{10}\,\text{m})$	Period $T\,(\text{y})$	T^2/a^3 $(10^{-34}\,\text{y}^2/\text{m}^3)$
Mercury	5.79	0.241	2.99
Venus	10.8	0.615	3.00
Earth	15.0	1.00	2.96
Mars	22.8	1.88	2.98
Jupiter	77.8	11.9	3.01
Saturn	143	29.5	2.98
Uranus	287	84.0	2.98
Neptune	450	165	2.99
Pluto	590	248	2.99

Figure 17   Halley's comet, photographed during its 1986 approach to the Sun.

Sample Problem 7   (*a*) Compute the mass of the Sun from the period and radius of the Earth's orbit. (*b*) Compute the mass of Jupiter from the period (1.77 d) and radius (4.22 × 10⁵ km) of its second closest moon, Io.

Solution   (*a*) From Eq. 24, we have

$$M = \frac{4\pi^2 r^3}{GT^2} = \frac{4\pi^2(1.50 \times 10^{11}\,\text{m})^3}{(6.67 \times 10^{-11}\,\text{N}\cdot\text{m}^2/\text{kg}^2)(3.15 \times 10^7\,\text{s})^2}$$

$$= 2.01 \times 10^{30}\,\text{kg}.$$

(*b*)

$$M = \frac{4\pi^2(4.22 \times 10^8\,\text{m})^3}{(6.67 \times 10^{-11}\,\text{N}\cdot\text{m}^2/\text{kg}^2)(1.53 \times 10^5\,\text{s})^2}$$

$$= 1.90 \times 10^{27}\,\text{kg}.$$

Note that the mass of Jupiter cannot be obtained from the parameters of its orbit about the Sun; to determine the mass of an object from Kepler's third law, we need to know the period and semimajor axis of objects that orbit about it as the central body.

Sample Problem 8   A satellite orbits at a height of $h = 230$ km above the Earth's surface. What is the period of the satellite?

Solution   Again using Eq. 24, with $r = R_E + h = 6370$ km + 230 km = 6600 km, we obtain

$$T = \left(\frac{4\pi^2 r^3}{GM_E}\right)^{1/2} = \left(\frac{4\pi^2(6.60 \times 10^6\,\text{m})^3}{(6.67 \times 10^{-11}\,\text{N}\cdot\text{m}^2/\text{kg}^2)(5.98 \times 10^{24}\,\text{kg})}\right)^{1/2}$$

$$= 5330\,\text{s} = 88.9\,\text{min}.$$

This period is not very dependent on h when h is much less than R_E; orbits of low-lying satellites of Earth have periods of about 90 min.

Sample Problem 9   It is desired to place a communications satellite into orbit so that it remains fixed above a given spot on the equator of the rotating Earth. What is the height above the Earth of such an orbit?

Solution   For the satellite to remain above a given point on the Earth's surface, it must rotate with the same angular velocity as

the point. The period of the satellite must therefore be 24 h or 86,400 s. The radius of the orbit must then be

$$r = \left(\frac{GT^2 M_E}{4\pi^2}\right)^{1/3}$$

$$= \left(\frac{(6.67 \times 10^{-11}\,\text{N}\cdot\text{m}^2/\text{kg}^2)(86,400\,\text{s})^2(5.98 \times 10^{24}\,\text{kg})}{4\pi^2}\right)^{1/3}$$

$$= 4.22 \times 10^7\,\text{m},$$

and its height above the Earth's surface is

$$h = r - R_E = 4.22 \times 10^7\,\text{m} - 6.37 \times 10^6\,\text{m}$$

$$= 3.58 \times 10^7\,\text{m} = 22{,}300\,\text{mi}.$$

This orbit is called the Clarke Geosynchronous Orbit after Arthur C. Clarke, who first proposed the idea in 1948. Clarke is also well known as the author of many works of science fiction, including *2001 — A Space Odyssey.*

Sample Problem 10   Halley's comet (Fig. 17) has a period of 76 years. In 1986, its closest approach to the Sun (perihelion) was 8.9 × 10¹⁰ m (between the orbits of Mercury and Venus). Find its aphelion, or farthest distance from the Sun, and the eccentricity of its orbit.

Solution   From Eq. 24 we find the semimajor axis:

$$a = \left(\frac{GT^2 M}{4\pi^2}\right)^{1/3}$$

$$= \left(\frac{(6.67 \times 10^{-11}\,\text{N}\cdot\text{m}^2/\text{kg}^2)(2.4 \times 10^9\,\text{s})^2(2.0 \times 10^{30}\,\text{kg})}{4\pi^2}\right)^{1/3}$$

$$= 2.7 \times 10^{12}\,\text{m}.$$

From Fig. 15, we have $R_a = a - ae$ and $R_p = a + ae$, so

$$R_a + R_p = 2a$$

$$R_p = 2a - R_a = 2(2.7 \times 10^{12}\,\text{m}) - 8.9 \times 10^{10}\,\text{m}$$

$$= 5.3 \times 10^{12}\,\text{m},$$

between the orbits of Neptune and Pluto. The eccentricity is

$$e = \frac{R_a - R_p}{2a} = \frac{6.3 \times 10^{12}\,\text{m} - 8.9 \times 10^{10}\,\text{m}}{2(2.7 \times 10^{12}\,\text{m})} = 0.96.$$

Such a large eccentricity (1.0 is the maximum possible) corresponds to a long, thin ellipse.

Motion About the Center of Mass

Figure 18 shows, for the case of circular orbits, two objects moving about their common center of mass. If we consider the motion of the smaller body, then Eq. 23 becomes

$$\frac{GMm}{(r+R)^2} = m\omega^2 r,$$

and the revised result for the law of periods is

$$T^2 = \left(\frac{4\pi^2}{GM}\right) r^3 \left(1 + \frac{R}{r}\right)^2. \tag{25}$$

The difference between Eqs. 24 and 25 is the factor $(1 + R/r)^2$. In the case of the Earth and the Sun,

$$\frac{R}{r} = \frac{m}{M} = \frac{5.98 \times 10^{24} \text{ kg}}{1.99 \times 10^{30} \text{ kg}} = 3.01 \times 10^{-6},$$

and the error made in neglecting the center-of-mass factor in applying the law of periods is less than 0.001%. Figure 19, on the other hand, shows a diagram of a binary star system. Here the two objects have comparable masses, and the correction for the center of mass is significant.

Energy Considerations in Planetary and Satellite Motion

Consider again the motion of a body of mass m (planet or satellite, say) about a massive body of mass M (Sun or Earth, say). We consider M to be at rest in an inertial reference frame with the body m moving about it in a circular orbit. The potential energy of the system is

$$U(r) = -\frac{GMm}{r},$$

where r is the radius of the circular orbit. The kinetic energy of the system is

$$K = \tfrac{1}{2} m\omega^2 r^2,$$

the Sun being at rest. From Eq. 23 we obtain

$$\omega^2 r^2 = \frac{GM}{r},$$

so that

$$K = \frac{1}{2}\frac{GMm}{r}. \tag{26}$$

The total energy is

$$E = K + U = \frac{1}{2}\frac{GMm}{r} - \frac{GMm}{r} = -\frac{GMm}{2r}. \tag{27}$$

This energy is constant and is negative. The kinetic energy can never be negative, but from Eq. 26 we see that it must go to zero as the separation goes to infinity. The potential energy is always negative except for its zero value at infinite separation. The meaning of the total negative energy then is that the system is a closed one, the planet m always being bound to the attracting solar center M and never escaping from it (Fig. 20).

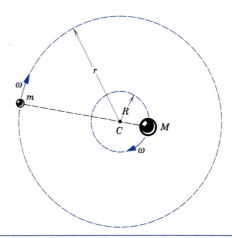

Figure 18   Two bodies moving in circular orbits under the influence of each other's gravitational attraction. They both have the same angular velocity ω. Point C is their center of mass.

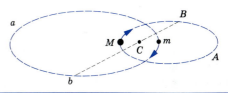

Figure 19   A binary star system, in which each star orbits about the center of mass C. Points A and B show the locations of the star of mass M when star m is at the respective locations a and b.

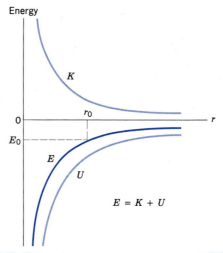

Figure 20   Kinetic energy K, potential energy U, and total energy $E = K + U$ of a body in circular planetary motion. A planet with total energy $E_0 < 0$ will remain in an orbit with radius r_0. The greater the distance from the Sun, the greater (that is, less negative) its total energy E. To escape from the center of force and still have kinetic energy at infinity, the planet would need positive total energy.

It can be shown* that Eq. 27 is also valid for elliptical orbits, if we replace r by the semimajor axis a. The total energy is still negative, and it is also constant, corresponding to the fact that gravitational forces are conservative. Hence both the total energy and the total angular momentum are constant in planetary motion. These quantities are often called *constants of the motion.*

Because the total energy does not depend on the eccentricity of the orbit, all orbits with the same semimajor axis a have the same total energy. Figure 21 shows several different orbits that have the same energy.

If we supply the proper amount of kinetic energy, we can arrange for the total energy to be zero or positive, in which case the orbits are no longer elliptical. The orbits are parabolic for $E = 0$ and hyperbolic for $E > 0$. This case often occurs in the scattering of particles from a nucleus, where the electrostatic force also varies as $1/r^2$. The spacecraft *Pioneer 10* was given enough initial kinetic energy to allow it to escape from the solar system; launched on March 3, 1972, it passed the orbit of Pluto, the outermost planet, on June 14, 1983, outward bound on a hyperbolic path.

Equation 27 shows that we cannot change the speed of an orbiting satellite without also changing the radius of its orbit. For example, suppose two satellites are following one another in the same circular orbit. If the trailing satellite tries to catch the leading one by accelerating forward, thereby increasing the kinetic energy, the total energy becomes less negative and the radius increases. Docking two spacecraft is not just a simple exercise in edging one craft forward! In fact, as the following sample problem shows, the proper procedure to use in overtaking an orbiting spacecraft often involves slowing down rather than speeding up.

Sample Problem 11   Two identical spacecraft, each with a mass of 3250 kg, are in the same circular orbit at a height of 270 km above the Earth's surface. Spacecraft A leads spacecraft B by 105 s; that is, A arrives at any fixed point 105 s before B. At a particular point P, the pilot of B fires a short rocket burst in the forward direction, reducing the speed of B by 0.95%. Find the orbital parameters (energy, period, semimajor axis) of B before and after the "burn," and find the order of the two ships when they next return to point P.

Solution   For $h = 270$ km, $r = R_E + h = 6370$ km $+ 270$ km $= 6640$ km. Thus before firing the rockets, $a = 6640$ km and

$$E = -\frac{GmM_E}{2a}$$
$$= -\frac{(6.67 \times 10^{-11} \text{ N} \cdot \text{m}^2/\text{kg}^2)(3250 \text{ kg})(5.98 \times 10^{24} \text{ kg})}{2(6.64 \times 10^6 \text{ m})}$$
$$= -9.76 \times 10^{10} \text{ J},$$

---

* See, for example, *Newtonian Mechanics,* by A. P. French (Norton, 1971), pp. 585–591.

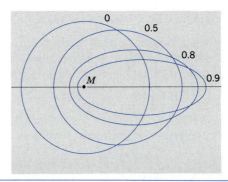

Figure 21   All four orbits have the same semimajor axis a and thus correspond to the same total energy E. Their eccentricities are marked.

$$T = \left(\frac{4\pi^2 a^3}{GM_E}\right)^{1/2}$$
$$= \left(\frac{4\pi^2 (6.64 \times 10^6 \text{ m})^3}{(6.67 \times 10^{-11} \text{ N} \cdot \text{m}^2/\text{kg}^2)(5.98 \times 10^{24} \text{ kg})}\right)^{1/2}$$
$$= 5.38 \times 10^3 \text{ s.}$$

Equations 26 and 27 show that (for a circular orbit only!) the kinetic energy is numerically equal to the negative of the total energy, so $K = +9.76 \times 10^{10}$ J and

$$v = \sqrt{\frac{2K}{m}} = \sqrt{\frac{2(9.76 \times 10^{10} \text{ J})}{3250 \text{ kg}}} = 7.75 \times 10^3 \text{ m/s.}$$

After the burn, the velocity decreases by the given amount of 0.95% to $v' = (1 - 0.0095)v = 7.68 \times 10^3$ m/s, and the new kinetic energy of B is

$$K' = \tfrac{1}{2}(3250 \text{ kg})(7.68 \times 10^3 \text{ m/s})^2 = 9.58 \times 10^{10} \text{ J.}$$

The potential energy of B at point P immediately after the short burn is unchanged, equal to the initial value $E - K$ or $2E$, according to Eq. 27. The total energy E' of B after the burn must then be

$$E' = K' + U' = 9.58 \times 10^{10} \text{ J} + 2(-9.76 \times 10^{10} \text{ J})$$
$$= -9.94 \times 10^{10} \text{ J,}$$

and the new semimajor axis is

$$a' = -\frac{GmM_E}{2E'}$$
$$= -\frac{(6.67 \times 10^{-11} \text{ N} \cdot \text{m}^2/\text{kg}^2)(3250 \text{ kg})(5.98 \times 10^{24} \text{ kg})}{2(-9.94 \times 10^{10} \text{ J})}$$
$$= 6.52 \times 10^6 \text{ m.}$$

The corresponding period is

$$T' = \left(\frac{4\pi^2 a'^3}{GM_E}\right)^{1/2}$$
$$= \left(\frac{4\pi^2 (6.52 \times 10^6 \text{ m})^3}{(6.67 \times 10^{-11} \text{ N} \cdot \text{m}^2/\text{kg}^2)(5.98 \times 10^{24} \text{ kg})}\right)^{1/2}$$
$$= 5.24 \times 10^3 \text{ s.}$$

The difference in the periods is 140 s. That is, if A originally passes through point P at $t = 0$ and B passes through (and fires its rockets) at $t = 105$ s, then A returns to P at $t = 5380$ s (determined by the period T), and B returns to P at 5240 s after its

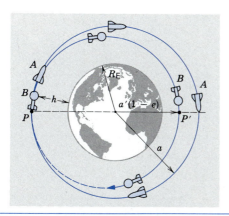

Figure 22   Sample Problem 11. The orbits of spacecraft *A* and *B* are shown. Note that *B* catches *A* by moving to a noncircular orbit at lower height above the Earth. The relative size of the Earth and the orbital heights is not to scale.

initial passage, or at $t = 105$ s $+ 5240$ s $= 5345$ s. Thus *B* is now 35 s *ahead of A* at point *P*. Now *B* can fire a second rocket burst identical in strength and duration to the first but in the reverse direction. This returns *B* to the original circular orbit, now 35 s ahead of *A*. Figure 22 shows the relationship between *A* and *B* during the first orbit after the burn.

See Problem 71 to help understand how *B* can reduce its speed at *P* and still get ahead of *A*.

16-9   UNIVERSAL GRAVITATION

So far we have discussed applications of Newton's law of gravitation on the scale of the solar system, where its predictions have been well tested. Let us now turn to gravitational effects on much larger scales.

Figure 23 shows photographs of galaxies whose spiral structures are very similar to that of our own Milky Way galaxy. Perhaps 10^{11} stars are bound together by their mutual gravitational forces in such a structure. The diameter of a typical spiral galaxy might be 50 kpc.* The Andromeda galaxy, a prominent galactic neighbor, is about 0.7 Mpc or 2×10^6 light-years away.

The spiral structure is a common one for galaxies. A bright central region is surrounded by a flat disk with several spiral arms. The entire structure rotates about an axis perpendicular to the plane of the disk.

The individual stars are gravitationally bound to the galaxy by a force directed toward its center. We can use Kepler's laws, just as we did for the solar system, to ana-

---

* One astronomical unit (AU) is equal to the average distance between the Earth and the Sun. One parsec (pc) is defined as the distance at which 1 AU would subtend an angle of 1″ and has the numerical value 3.084×10^{13} km or 3.26 light-years. Galactic sizes are typically measured in kiloparsec (kpc) and their separations in megaparsec (Mpc). See Chapter 1, Problem 23.

Figure 23   Typical spiral galaxies similar to our Milky Way, viewed from two different perspectives, one normal to the plane and one along the plane.

lyze the gravitational force. The tangential velocity of the Sun about the galactic center has been measured to be about 220 km/s. (Compare this with the Earth's tangential speed about the Sun, 30 km/s.) Our distance from the center of the galaxy is 8.5 kpc; we are in one of the spiral arms about two-thirds of the way out from the center of the galaxy. From these figures we can calculate the Sun's angular velocity about the galactic center to be $\omega = v/r = 8.4 \times 10^{-16}$ rad/s. At this rate, a complete rotation takes 240 million years, and thus far during its lifetime of about 5000 million years the Sun has already made perhaps 20 revolutions.

The galaxy does not rotate like a rigid body; instead, its rotation is more like that of the solar system. Assuming we can apply Kepler's laws to this system, we can find the relationship between tangential speed and radius. For this purpose it is useful to rewrite Kepler's third law, Eq. 24, replacing *T* by $2\pi r/v$:

$$T^2 = \left(\frac{2\pi r}{v}\right)^2 = \left(\frac{4\pi^2}{GM}\right) r^3$$

or

$$v = \sqrt{\frac{GM}{r}}. \qquad (28)$$

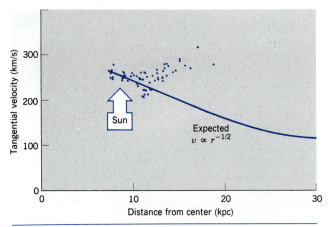

Figure 24   Tangential velocities of stars in our galaxy, deduced from measuring the Doppler shifts of their light. The solid line shows the dependence of v on r given by Eq. 28 and calculated from Kepler's third law, assuming the stars to be attracted only by the large central mass of the galaxy. The discrepancy between the measured points and the curve suggests that there is unseen matter attracting the stars in the outer region of the galaxy.

Here M refers to the mass contained in the region within the radius r. Based on the Sun's tangential speed, we would estimate that a mass equivalent to 10^{11} solar masses lies within the Sun's orbit.

If we assume that nearly all the mass of our galaxy lies in this interior region, then, on the basis of Eq. 28 with M taken as constant, we would expect that v for stars beyond the Sun will decrease as r increases. To the extent that Kepler's third law is valid and the additional mass beyond the Sun is negligible, v should decrease as $r^{-1/2}$.

Instead, we observe v to be constant or perhaps to increase slightly, even out to the very edge of the visible region of our galaxy (Fig. 24). Other spiral galaxies show the same effect. (These observations are based on the Doppler shift of the light emitted by the galaxy; the motion of the galaxy relative to us causes a change in the wavelength or frequency of its light compared with its value for no relative motion; see Sections 21-7 and 42-3. In a distant rotating galaxy, the parts whose rotational motion is toward us have Doppler shifts opposite to those of the parts moving away from us, and the rotational velocity at different distances from the center can be determined directly.) The velocity remains roughly constant to the limit of the visible part of the galaxies.

We can account for this effect in Eq. 28 if M, which represents the mass contained within a spherical region of radius r, increases at least linearly with r. This must remain true even to the outermost limit of the galaxy.

If we assume that the mass in a galaxy is mostly in the form of stars (planets add very little mass), then this increase of M with r is inconsistent with observations (Fig. 23), which clearly show the light (and therefore presum-

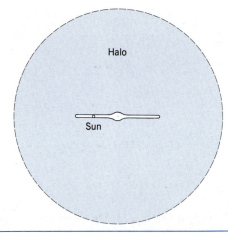

Figure 25   A representation of the suggested "halo" of dark matter in our galaxy. This matter is needed to explain the discrepancy illustrated in Fig. 24, but no direct evidence for it has yet been found.

Figure 26   A cluster of galaxies (called the Coma cluster). These galaxies are bound into a cluster by their gravitational attractions for one another.

ably the number of stars) concentrated near the center and diminishing as we go to larger r. There must therefore be a considerable quantity of *dark matter* in the galaxy, which may take the shape of a nearly spherical "halo," as shown in Fig. 25. The actual form of this dark matter is unknown; proposals range from burnt-out stars to Jupiter-sized objects to free elementary particles, but as yet there is no direct experimental evidence for the existence of any of these.

Galaxies are observed for form clusters (Fig. 26) of perhaps 100, bound by gravitational forces. The size of a typical cluster is of the order of 1 Mpc, that is, 100 times the size of a typical galaxy. As in the case of the galaxies themselves, there is a "missing mass" problem in the clus-

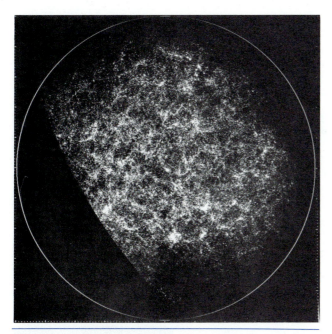

Figure 27   Each dot represents a galaxy. The map, with its clumps and filaments, suggests the existence of superclusters. Three-dimensional projections confirm this interpretation.

ters: the calculated amount of matter necessary for the clusters to form a gravitationally bound system is 10–100 times the total amount of visible matter in the galaxies that comprise the clusters. It has therefore been speculated that there may be dark matter not only in the galactic halos but also in the space between the galaxies; a similar halo permeating the clusters is needed to supply the gravitational force to make a bound system.

There are also gravitationally bound superclusters— clusters of clusters on a scale beyond 10 Mpc. A map of the clusters (Fig. 27) shows that they tend to concentrate on surfaces leaving great empty voids that contain little matter. The explanation for these cosmic "bubbles" is not yet clear.

The chain of reasoning that leads from the Cavendish balance to the superclusters is a linear one, proceeding from a laboratory measurement of G through a set of dynamical laws well tested in our solar system, to an extrapolation that the laws of Newton and our measured G hold throughout the universe. Although there are still a number of open questions, there is as yet no firm evidence that this reasoning is incorrect or that gravitation is anything but universal.

16-10   THE GENERAL THEORY OF RELATIVITY   *(Optional)*

Newton's law of universal gravitation has had remarkable success in its applications. It provides us with the means to calculate with great precision the motions of bodies in gravitational fields.

For example, we can send space probes to the planets and control their trajectories to within a few meters.

In 1916, Albert Einstein offered a different approach to understanding gravitation. In his general theory of relativity, he proposed that, in contrast to the Newtonian view, it is not possible to separate a coordinate system from the matter it contains. Gravitating matter, according to Einstein, varies the geometry of its surroundings and thus controls the behavior of nearby bodies.

General relativity is both conceptually and mathematically difficult and far beyond the level of this text. Instead, we give some of the background of the theory, discuss some of its implications, and summarize some of the major experimental tests that distinguish Einstein's theory from Newton's. It should be noted that, in weak gravitational fields, Einstein's theory reduces to Newton's, so that everything we have done so far in this chapter remains correct. It is only in strong gravitational fields, such as close to the Sun, that the differences become important.

Inertial Mass and Gravitational Mass

In Chapter 5 we discussed a procedure for assigning mass to an object, by comparing its response to a given force (that is, its acceleration) to that of a standard mass. This comparison is made on the basis of Newton's second law, and the mass that appears in $F = ma$ is called *inertial mass.* We can also use a procedure based on Newton's law of gravitation to measure the mass of an object. Let us measure the force on a standard kilogram in the Earth's gravitational field (that is, its weight), and let us then determine the force on our unknown mass in the same manner. According to Eq. 1, the ratio between those forces should be the same as the ratio of the masses, and we thus have a second method of determining mass. In this case we are measuring the *gravitational* mass.

It seems reasonable to ask whether these masses are in fact the same. Is inertial mass equal to gravitational mass? There is nothing in Newton's framework of dynamics that requires them to be equal. Their equality must be regarded in Newton's theory as an amazing coincidence, but it arises in a natural way in general relativity.

Newton was the first to test the equality of inertial and gravitational mass, using a pendulum made in the form of an empty box. He filled the box with different quantities of material and measured the period of the resulting pendulum. If we repeat the derivation of Section 15-5 for the period of the simple pendulum, but now taking care to account separately for gravitational and inertial mass, the result is

$$T = 2\pi \sqrt{\frac{m_i L}{m_g g}} \qquad (29)$$

where the m_i in the numerator refers to the inertial mass of the pendulum bob, and the m_g in the denominator refers to its gravitational mass. Of course, this reduces to the familiar result when $m_i = m_g$. Newton used identical weights of different substances and was careful to keep the physical circumstances (the amplitude, for instance) identical in all trials. He concluded that inertial and gravitational mass were the same to about one part in 10^3.

A considerable improvement in the experiment was made by Eötvös in 1909. He used a torsion balance with different materials on the two ends, and he compared for each material the gravitational mass (its weight) and the inertial mass (determined from the inertial centrifugal force owing to the Earth's rotation). Any difference in inertial and gravitational mass for the two materials would be observed as a rotation of the torsion balance. Eötvös concluded that inertial and gravitational mass were equal to within one part in 10^9. Later experiments by Dicke in 1964 and Braginsky in 1972 extended the limits to one part in 10^{11} to 10^{12} using a similar torsion balance technique but referring it to the Sun's gravitational attraction and to the inertial centrifugal force produced by the Earth's orbit about the Sun. These exceedingly precise experiments suggest that there is no difference between inertial and gravitational mass, and they force us to re-examine our laws of dynamics to account for this apparently accidental equality.*

The Principle of Equivalence

Here is how the idea occurred to Einstein: "I was sitting in a chair in the patent office in Bern when all of a sudden a thought occurred to me: If a person falls freely he will not feel his own weight. I was startled. This simple thought made a deep impression on me. It impelled me toward a theory of gravitation."

Figure 28*a* shows a person in an isolated chamber in free fall in the Earth's gravity, and Fig. 28*b* shows a person floating freely in interstellar space where the gravitational fields are negligibly weak. No measuring instruments that operate completely inside the chamber are able to distinguish between the two cases.

Einstein went one step further, as shown in Fig. 29. Consider now the person in the chamber at rest on the Earth (Fig. 29*a*). A ball is observed to accelerate toward the floor at 9.8 m/s². A simple pendulum of a specified length has a certain period of oscillation. A mass hung from a spring stretches the spring by a certain amount. The floor exerts a certain normal force on bodies resting on it.

Now suppose the chamber is part of a rocket in interstellar space, and further suppose that the engines are fired to give the rocket an acceleration of exactly 9.8 m/s² (see Fig. 29*b*). Our traveler now releases a ball and observes it to move relative to the floor with that acceleration. The pendulum oscillates normally, the mass stretches the spring by the proper amount, and the floor exerts its correct normal force. In short, there is no experiment that can be done inside the chamber that will distinguish between Fig. 29*a*—the condition of rest in an inertial frame in a gravitational field **g**—and Fig. 29*b*—acceleration **a** = −**g** relative to an inertial frame in space of negligible gravity. This is the *principle of equivalence.*

The equality of inertial and gravitational mass follows directly from the principle of equivalence. Let an object rest on a spring scale on the floor of the chamber. When the chamber accelerates in the rocket, the floor must exert an upward force $m_i a$ to accelerate the object; here m_i is the inertial mass, and the spring balance reads the reaction force (also $m_i a$) exerted by the object. When the chamber is at rest in a gravitational field, on the other hand, the scale reads the weight $m_g g$ (which depends on the gravitational mass). We have arranged our experiments so that $a = g$, and if the scale readings are to be identical (as demanded by the principle of equivalence) then the inertial and gravitational masses must be equal.

---

* See "Searching for the Secrets of Gravity," by John Boslough, *National Geographic,* May 1989, p. 563.

(a)   (b)

Figure 28 The effects of freely falling in the Earth's gravity (*a*) are identical to those of freely floating in interstellar space (*b*). No experiment done within the chamber could tell the difference.

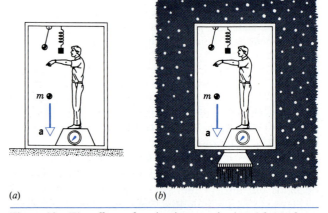

(a)   (b)

Figure 29 The effects of resting in a gravitational field of strength *g* (*a*) are identical to those of accelerating at $a = g$ in interstellar space (*b*). No experiment done within the chamber could tell the difference. This illustrates Einstein's principle of equivalence.

The General Theory of Relativity

General relativity is essentially a theory of geometry. It provides a procedure for constructing a coordinate system whose very shape depends on the presence of matter and energy. In Einstein's theory, matter bends or curves space; our familiar rectangular coordinate system is no longer strictly valid in the presence of matter. The effect of one gravitating mass on another is then merely the movement of the second mass in the distorted geometry established by the first.

This approach is similar to the concept of fields discussed earlier in this chapter. In field theory, one mass establishes a gravitational field, and the second mass then interacts directly with the field (rather than directly with the first mass, as in the action-at-a-distance approach).

Figure 30 shows a two-dimensional analogy for the bending or

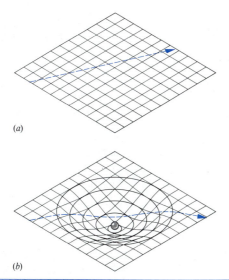

(a)

(b)

Figure 30 An analogy showing the bending or curving of space that results from the presence of gravitating mass, according to the general theory of relativity. Mass distorts the coordinate grid and changes the geometry itself.

curving of space. Imagine a rubber sheet with a coordinate grid laid out on it. All motion is confined to the coordinate system on the sheet. Now imagine a ball bearing stretching the sheet. The shortest distance between two points is no longer a straight line; in fact, in such a geometry we must redefine just what we mean by the term "straight line."

The relationship between matter and geometry in general relativity has been summarized as follows: "Geometry tells matter how to move, and matter tells geometry how to curve." The formulas of general relativity give the curvature for any given distribution of matter and energy, and the subsequent motion of beams of light or particles then follows directly.

Tests of General Relativity

Many experimental tests have been done to study the very small deviations between Newton's gravitational theory and Einstein's. The differences between the two are apparent only in strong gravitational fields, and in most cases we must therefore do measurements close to the Sun, which provides the strongest nearby gravitational field. There are four major tests of the theory:

1. *Precession of the perihelion of Mercury.* According to general relativity, the orbit of a planet is not quite a closed ellipse; the axis of the ellipse rotates a bit upon each orbit (Fig. 31). For Mercury, which is closest to the Sun and which should therefore show the largest effect, the predicted rotation is 42.98 seconds of arc per century. This is an incredibly small rotation, but one that can be measured with great accuracy; the current measured value is 43.11 ± 0.21 arc seconds per century, in excellent agreement with the predictions of general relativity. (It is interesting to note that this deviation was first noted in 1859 and was a serious problem for Newtonian gravitational theory before Einstein provided the correct explanation.)

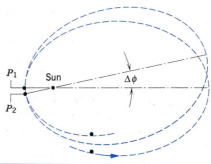

Figure 31 The precession of the perihelion of Mercury. With each orbit about the Sun, the position of perihelion rotates through the small angle $\Delta\phi$. The drawing is greatly exaggerated; the actual angle is about 0.1″ per orbit.

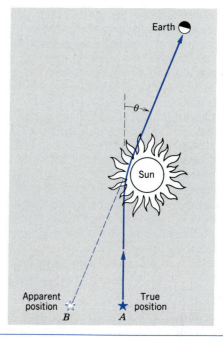

Figure 32 The deflection of starlight that passes near the Sun. The bending of space by the Sun's mass forces the light to travel in a path similar to that shown in Fig. 30*b*. As a result, we view a star from Earth at the apparent position *B* rather than at the true position *A* where we would see it if the Sun were not present.

2. *Bending of starlight near the Sun.* Here the rubber sheet analogy of Fig. 30 gives a good picture of what is happening. When the light from a distant star travels to Earth after first grazing the edge of the Sun, its path is bent as it follows the most direct route through the curved space (Fig. 32). The apparent position of the star viewed from Earth is shifted somewhat from its true position. For stars near the Sun to be visible, the observation must be made during a solar eclipse. Several measurements have been done, the earliest coming in 1919 just after Einstein proposed his theory. Here again theory and experiment are in excellent agreement despite the small effect; the prediction for the angle of deflection is 1.75 arc seconds, and experimental observations of stars and quasars confirm this value to within about 1%.

3.   *Delay of radar echoes.* When a planet such as Venus is behind the Sun as viewed from Earth, a radar signal sent from Earth to Venus and reflected back is delayed somewhat, because it must pass through the distorted spacetime near the Sun (Fig. 33). Again the rubber sheet analogy gives a means of understanding the effect. From that viewpoint, the delay is not associated with the bending of the path but with the "stretching" of the space near the Sun. The expected delay is of the order of one part in 10^4, and it has been confirmed to a few percent. The limit on the precision is imposed by the uncertainties in the surface of the planet; we don't know whether the signals are being reflected by mountains or valleys. A great improvement was made by the use of the *Viking* landers on Mars in the late 1970s, which gave results consistent with general relativity to within 0.1%.

4.   *Gravitational radiation.* Just as accelerated electric charges emit electromagnetic radiation that travels with the speed of light, so accelerated masses emit gravitational waves that also travel with the speed of light. Many experimental groups have built antennas to search for this gravitational radiation, but so far none has yet produced an unambiguous observation. Indirect but nevertheless very strong evidence for the emission of gravitational radiation comes from a binary pulsar. Pulsars emit very sharp pulses of electromagnetic radiation, which vary little in time and can be measured with great precision. (See Fig. 16 of Chapter 13.) One such pulsar, called PSR 1913 + 16, orbits with a companion star as part of a binary system; the orbital period is very short, about 7.75 h. The sharp radio spikes provide a direct way of timing the orbit with great precision, and soon after its discovery in 1974 it was known that its orbital period was decreasing by about 64 ns per orbit. The system appeared to be losing rotational kinetic energy, and the only reasonable explanation for the loss is energy radiated away as gravitational radiation. The loss in energy agrees with Einstein's theory to within about 3%.

These precise experimental tests have confirmed the predictions of general relativity in spectacular fashion. While there are

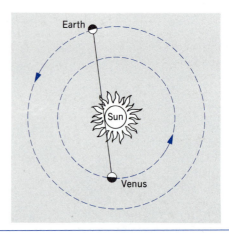

Figure 33   The time necessary for an electromagnetic wave to travel from Earth to Venus is slowed by the distortion of the geometry caused by the Sun's mass, as in Fig. 30*b*. This delay can be measured by observing radar signals reflected back to Earth from Venus.

other non-Newtonian gravitational theories, only general relativity has survived the test of experiment. Like special relativity, general relativity offers new notions about space and time, and several tests of even more exotic features of the theory remain to be done. Even though the distinction between Newtonian and Einsteinian gravitation has little effect on our daily lives, the fundamental implications for our understanding of this most basic aspect of nature demand that we continue to extend these measurements to the limit.* ∎

---

* For an elementary and highly readable account of these measurements, see *Was Einstein Right?*, by Clifford M. Will (Basic Books, 1986).

QUESTIONS

1. Modern observational astronomy and navigation procedures make use of the geocentric (or Ptolemaic) point of view (by using the rotating "celestial sphere"). Is this wrong? If not, what criterion determines the system (the Copernican or Ptolemaic) we use? When would we use the heliocentric (or Copernican) system?

2. Two planets are never seen at midnight. Which ones and why not? Can this be considered as evidence in favor of the heliocentric and against the geocentric theory?

3. If the force of gravity acts on all bodies in proportion to their masses, why doesn't a heavy body fall correspondingly faster than a light body?

4. How does the weight of a space probe vary en route from the Earth to the Moon? Would its mass change?

5. Our analysis of the Cavendish experiment (see Fig. 4 and Sample Problem 1) considered the attraction of each large sphere only for the small sphere closest to it. Each large sphere also attracts the small sphere on the opposite end of the rod. What is the effect of this attraction on the calculation done in Sample Problem 1? Estimate the error that neglecting this attraction might make in the calculated value of *G*.

6. Is the mutual gravitational force exerted by a pair of objects affected by the nature of the intervening medium? By the temperatures of the objects? By the orientation of the objects? How could you check these effects by experiment?

7. Because the Earth bulges near the equator, the source of the Mississippi River (at about 50° N latitude), although high

above sea level, is about 5 km closer to the center of the Earth than is its mouth (at about 30° N latitude). How can the river flow "uphill" as it flows south?

8. Would we have more sugar to the pound at the pole or at the equator? What about sugar to the kilogram?

9. How could you determine the mass of the Moon?

10. One clock is based on an oscillating spring, the other on a pendulum. Both are taken to Mars. Will they keep the same time there that they kept on Earth? Will they agree with each other? Explain. Mars has a mass about one-tenth that of the Earth and a radius about one-half as great.

11. At the Earth's surface, an object resting on a horizontal, frictionless surface is given a horizontal blow by a hammer. The object is then taken to the Moon, supported in the same manner, and given an equal blow by the same hammer. To the best of our knowledge, what would be the speed imparted to the object on the Moon when compared with the speed resulting from the blow on Earth (neglecting any atmospheric effects)?

12. Use qualitative arguments to explain why the following four periods are equal (all are 84 min, assuming a uniform Earth density): (*a*) time of revolution of a satellite just above the Earth's surface; (*b*) period of oscillation of mail in a tunnel through the Earth; (*c*) period of a simple pendulum having a length equal to the Earth's radius in a uniform field 9.8 m/s²; (*d*) period of an infinite simple pendulum in the Earth's real gravitational field.

13. The gravitational force exerted by the Sun on the Moon is about twice as great as the gravitational force exerted by the Earth on the Moon. Why then doesn't the Moon escape from the Earth?

14. Explain why the following reasoning is wrong. "The Sun attracts all bodies on the Earth. At midnight, when the Sun is directly below, it pulls on an object in the same direction as the pull of the Earth on that object; at noon, when the Sun is directly above, it pulls on an object in a direction opposite to the pull of the Earth. Hence, all objects should be heavier at midnight (or night) than they are at noon (or day)."

15. The gravitational attraction of the Sun and the Moon on the Earth produces tides. The Sun's tidal effect is about half as great as the Moon's. The direct pull of the Sun on the Earth, however, is about 175 times that of the Moon. Why is it then that the Moon causes the larger tides?

16. Particularly large tides, called *spring tides,* occur at full moon and at new moon, when the configurations of the Sun, Earth, and Moon are as shown in Fig. 34. From the figure you might conclude (incorrectly!) that the tidal effects of the Sun and of the Moon tend to add at new moon but to cancel at full moon. Instead, they add at both these configurations. Explain why.

17. If lunar tides slow down the rotation of the Earth (owing to friction), the angular momentum of the Earth decreases. What happens to the motion of the Moon as a consequence of the conservation of angular momentum? Does the Sun (and solar tides) play a role here? (See "Tides and the Earth–Moon System," by Peter Goldreich, *Scientific American*, April 1972, p. 42.)

18. From Kepler's second law and observations of the Sun's motion as seen from the Earth, how can we deduce that the Earth is closer to the Sun during winter in the northern hemisphere than during summer? Why isn't it colder in summer than in winter?

19. In Sample Problem 4, the tunnel transit time was derived on the assumption of an Earth of uniform density. Would this time be larger or smaller if the actual density distribution of the Earth, with its dense inner core, were taken into account? Explain your answer.

20. Why can we learn more about the shape of the Earth by studying the motion of an artificial satellite than by studying the motion of the Moon?

21. A satellite in Earth orbit experiences a small drag force as it starts to enter the Earth's atmosphere. What happens to its speed? (Be careful!)

22. Would you expect the total energy of the solar system to be constant? The total angular momentum? Explain your answers.

23. Does a rocket always need the escape speed of 11.2 km/s to escape from the Earth? If not, what then does "escape speed" really mean?

24. Objects at rest on the Earth's surface move in circular paths with a period of 24 h. Are they "in orbit" in the sense that an Earth satellite is in orbit? Why not? What would the length of the "day" have to be to put such objects in true orbit?

25. Neglecting air friction and technical difficulties, can a satellite be put into an orbit by being fired from a huge cannon at the Earth's surface? Explain your answer.

26. What advantage does Florida have over California for launching (nonpolar) U.S. satellites?

27. Can a satellite coast in a stable orbit in a plane not passing through the Earth's center? Explain your answer.

28. As measured by an observer on Earth, would there be any difference in the periods of two satellites, each in a circular orbit near the Earth in an equatorial plane, but one moving eastward and the other westward?

29. After *Sputnik I* was put into orbit, it was said that it would not return to Earth but would burn up in its *descent*. Considering the fact that it did not burn up in its *ascent*, how is this possible?

30. An artificial satellite is in a circular orbit about the Earth. How will its orbit change if one of its rockets is momentarily fired (*a*) toward the Earth, (*b*) away from the Earth, (*c*) in a forward direction, (*d*) in a backward direction, and (*e*) at right angles to the plane of the orbit?

31. Inside a spaceship, what difficulties would you encounter in walking, in jumping, and in drinking?

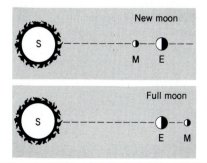

Figure 34   Question 16.

32. We have all seen TV transmissions from orbiting shuttles and watched objects floating around in effective zero gravity. Suppose that an astronaut, braced against the shuttle frame, kicks a floating bowling ball. Will a stubbed toe result? Explain your answer.

33. If a planet of given density were made larger by accreting material from space, its force of attraction for an object on its surface would increase because of the planet's greater mass but would decrease because of the greater distance from the object to the center of the planet. Which effect dominates?

34. The gravitational field associated with the Earth is zero both at infinity and at the center of the Earth. Is the gravitational potential also zero at each place? Is it indeed the same at each place? *Can* it be zero at either place? *Need* it be zero at either place?

35. The orbits of satellites around the Earth are elliptical (or circular) and yet we claimed in Chapter 4 that projectiles launched from the Earth follow parabolic trajectories. Which is correct?

36. Artificial Earth satellites can locate the mean sea level with great precision. Above oil-bearing rock, however, the mean sea level can be as much as 1 m higher than that above non-oil-bearing rock (which is usually denser). Explain this.

37. (a) In order for two observers at any two positions on the Earth's equator to maintain radio communication by using satellites in the geosynchronous orbit, there must be at least three such satellites. Explain. (b) Find the maximum angular separation of any two of these satellites.

38. A stone is dropped along the center of a deep vertical mine shaft. Assume no air resistance but consider the Earth's rotation. Will the stone continue along the center of the shaft? If not, describe its motion.

39. Why is there virtually no atmosphere on the Moon?

40. Does the law of universal gravitation require the planets of the solar system to have the actual orbits observed? Would planets of another star, similar to our Sun, have the same orbits? Suggest factors that might have determined the special orbits observed.

41. Does it matter which way a rocket is pointed for it to escape from Earth? Assume, of course, that it is pointed above the horizon and neglect air resistance.

42. For a flight to Mars, a rocket is fired in the direction the Earth is moving in its orbit. For a flight to Venus, it is fired backward along that orbit. Explain why.

43. Saturn is about six times farther from the Sun than Mars. Which planet has (a) the greater period of revolution, (b) the greater orbital speed, and (c) the greater angular speed?

44. See Fig. 35. What is being plotted? Put numbers with units on each axis.

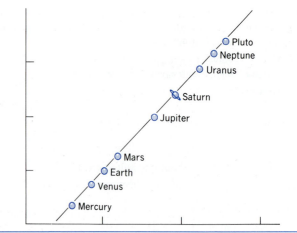

Figure 35   Question 44.

45. How can the captain of a spaceship, coasting toward a previously unknown planet, infer the value of *g* at the surface of the planet?

46. An iron cube is placed near an iron sphere at a location remote from the Earth's gravity. What can you say about the location of the center of gravity of the cube? Of the sphere? In general, does the location of the center of gravity of an object depend on the nature of the gravitational field in which the object is placed?

47. How could you determine whether two objects have (a) the same gravitational mass, (b) the same inertial mass, and (c) the same weight?

48. You are a passenger on the *S.S. Arthur C. Clarke,* the first interstellar spaceship. The *Clarke* rotates about a central axis to simulate Earth gravity. If you are in an enclosed cabin, how could you tell that you are not on Earth?

49. Can one regard gravity as a "fictitious" force arising from the acceleration of one's reference frame relative to an inertial reference frame, rather than a "real" force?

50. The "action-at-a-distance" view of the gravitational force implies that the action is spontaneous. Actually, present physical theory assumes that gravitation propagates with a finite speed and this is taken into account in the modification of classical physics represented by general relativity theory. (See "Gravitational Waves—A Progress Report," by Jonothan L. Logan, *Physics Today,* March 1973, p. 44, for a discussion of the ideas and attempts at experimental verification.) What would happen to classical deductions if it were assumed that the action is not instantaneous? (See also "Infinite Speed of Propagation of Gravitation in Newtonian Physics," by I. J. Good, *American Journal of Physics,* July 1975, p. 640.)

PROBLEMS

Section 16-3   The Gravitational Constant G

1. In the Cavendish balance of Sample Problem 1, calculate the gravitational force exerted by one of the large spheres on the other large sphere.

2. The Sun and Earth each exert a gravitational force on the Moon. Calculate the ratio F_{Sun}/F_{Earth} of these two forces. (The average Sun–Moon distance is equal to the Sun–Earth distance.)

3. How far from the Earth must a space probe be along a line toward the Sun so that the Sun's gravitational pull balances the Earth's?

Section 16-4 Gravity Near the Earth's Surface

4. One of the *Echo* satellites consisted of an inflated aluminum balloon 30 m in diameter and of mass 20 kg. A meteor having a mass of 7.0 kg passes within 3.0 m of the surface of the satellite. If the effect of all bodies other than the meteor and satellite are ignored, what gravitational force does the meteor experience at closest approach to the satellite?

5. If a pendulum has a period of 1.00 s at the equator, what would be its period at the south pole? See Fig. 6.

6. You weigh 120 lb at the sidewalk level outside the World Trade Center in New York City. Suppose that you ride from this level to the top of one of its 1350-ft towers. How much less would you weigh there because you are slightly farther away from the center of the Earth?

7. At what altitude above the Earth's surface is the free-fall acceleration equal to 7.35 m/s² (three-quarters of its value at the surface)?

8. Show that on a hypothetical planet having half the diameter of the Earth but twice its density, the acceleration of free fall is the same as on Earth.

9. A typical neutron star may have a mass equal to that of the Sun but a radius of only 10.0 km. (*a*) What is the gravitational acceleration at the surface of such a star? (*b*) How fast would an object be moving if it fell from rest through a distance of 1.20 m on such a star?

10. (*a*) Calculate g_0 on the surface of the Moon from values of the mass and radius of the Moon found in Appendix C. (*b*) What is the period of a "seconds pendulum" (period = 2.00 s on the Earth) on the surface of the Moon? (*c*) What will an object weigh on the Moon's surface if it weighs 100 N on the Earth's surface? (*d*) How many Earth radii must this same object be from the surface of the Earth if it is to weigh the same as it does on the surface of the Moon?

11. The fact that *g* varies from place to place over the Earth's surface drew attention when Jean Richer in 1672 took a pendulum clock from Paris to Cayenne, French Guiana, and found that it lost 2.5 min/day. If *g* = 9.81 m/s² in Paris, calculate *g* in Cayenne.

12. (*a*) If *g* is to be determined by dropping an object through a distance of (exactly) 10 m, how accurately must the time be measured to obtain a result good to 0.1%? Calculate a percent error and an absolute error, in milliseconds. (*b*) How accurately (in seconds) would you have to measure the time for 100 oscillations of a 10-m long pendulum to achieve the same percent error in the measurement of *g*?

13. Consider an inertial reference frame whose origin is fixed at the center of mass of the system Earth + falling object. (*a*) Show that the acceleration toward the center of mass of *either* body is independent of the mass of that body. (*b*) Show that the mutual, or relative, acceleration of the two bodies depends on the sum of the masses of the two bodies. Comment on the meaning, then, of the statement that a body falls toward the Earth with an acceleration that is independent of its mass.

14. Two objects, each of mass *m*, hang from strings of different lengths on a balance at the surface of the Earth, as shown in

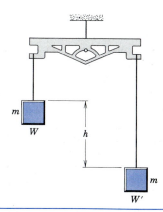

Figure 36   Problem 14.

Fig. 36. If the strings have negligible mass and differ in length by *h*, (*a*) show that the error in weighing, associated with the fact that *W'* is closer to the Earth than *W*, is *W'* − *W* = 8π*Gρmh*/3 in which *ρ* is the mean density of the Earth (5.5 g/cm³). (*b*) Find the difference in length that will give an error of one part in a million.

15. (*a*) Write an expression for the force exerted by the Moon, mass *M*, on a particle of water, mass *m*, on the Earth at *A*, directly under the Moon, as shown in Fig. 37. The radius of the Earth is *R*, and the center-to-center Earth–Moon distance is *r*. (*b*) Suppose that the particle of water was at the center of the Earth. What force would the Moon exert on it there? (*c*) Show that the difference in these forces is given by

$$F_T = \frac{2GMmR}{r^3}$$

and represents the *tidal force*, the force on water relative to the Earth. What is the direction of the tidal force? (*d*) Repeat for a particle of water at *B*, on the far side of the Earth from the Moon. What is the direction of this tidal force? (*e*) Explain why there are two tidal bulges in the oceans (and solid Earth), one pointing toward the Moon and the other away from it.

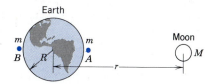

Figure 37   Problem 15.

16. An object is suspended on a spring balance in a ship sailing along the equator with a speed *v*. Show that the scale reading will be very close to $W_0(1 \pm 2\omega v/g)$, where *ω* is the angular speed of the Earth and W_0 is the scale reading when the ship is at rest. Explain the plus or minus.

17. The fastest possible rate of rotation of a planet is that for which the gravitational force on material at the equator barely provides the centripetal force needed for the rotation. (Why?) (*a*) Show then that the corresponding shortest period of rotation is given by

$$T = \sqrt{\frac{3\pi}{G\rho}},$$

where ρ is the density of the planet, assumed to be homogeneous. (b) Evaluate the rotation period assuming a density of 3.0 g/cm³, typical of many planets, satellites, and asteroids. No such object is found to be spinning with a period shorter than found by this analysis.

18. Sensitive meters that measure the local free-fall acceleration g can be used to detect the presence of deposits of near-surface rocks of density significantly greater or less than that of the surroundings. Cavities such as caverns and abandoned mine shafts can also be located. (a) Show that the vertical component of g a distance x from a point directly above the center of a spherical cavern (see Fig. 38) is less than what would be expected, assuming a uniform distribution of rock of density ρ, by the amount

$$\Delta g = \frac{4\pi}{3} R^3 G \rho \, \frac{d}{(d^2 + x^2)^{3/2}},$$

where R is the radius of the cavern and d is the depth of its center. (b) These values of Δg, called *anomalies,* are usually very small and expressed in milligals, where 1 gal = 1 cm/s². Oil prospectors doing a gravity survey find Δg varying from 10.0 milligals to a maximum of 14.0 milligals over a 150-m distance. Assuming that the larger anomaly was recorded directly over the center of a spherical cavern known to be in the region, find its radius and the depth to the roof of the cavern at that point. Nearby rocks have a density of 2.80 g/cm³. (c) Suppose that the cavern, instead of being empty, is completely flooded with water. What do the gravity readings in (b) now indicate for its radius and depth?

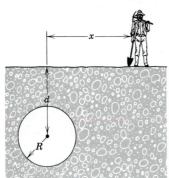

Figure 38 Problem 18.

19. A pendulum whose upper end is attached so as to allow the pendulum to swing freely in any direction can be used to repeat an experiment first shown publicly by Foucault in Paris in 1851. If the pendulum is set oscillating, the plane of oscillation slowly rotates with respect to a line drawn on the floor, even though the tension in the wire supporting the bob and the gravitational pull of the Earth on the bob lie in a vertical plane. (a) Show that this is a result of the fact that the Earth is not an inertial reference frame. (b) Show that for a Foucault pendulum at a latitude θ, the period of rotation of the plane, in hours, is 24/sin θ. (c) Explain in simple terms the result at $\theta = 90°$ (the poles) and $\theta = 0°$ (the equator).

Section 16-5 Gravitational Effect of a Spherical Distribution of Matter

20. Two concentric shells of uniform density having masses M_1 and M_2 are situated as shown in Fig. 39. Find the force on a

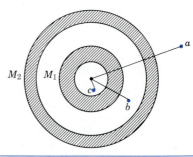

Figure 39 Problem 20.

particle of mass m when the particle is located at (a) $r = a$, (b) $r = b$, and (c) $r = c$. The distance r is measured from the center of the shells.

21. With what speed would mail pass through the center of the Earth if it were delivered by the chute of Sample Problem 4?

22. Show that, at the bottom of a vertical mine shaft dug to depth D, the measured value of g will be

$$g = g_s \left(1 - \frac{D}{R} \right),$$

g_s being the surface value. Assume that the Earth is a uniform sphere of radius R.

23. The following problem is from the 1946 "Olympic" examination of Moscow State University (see Fig. 40): A spherical hollow is made in a lead sphere of radius R, such that its surface touches the outside surface of the lead sphere and passes through its center. The mass of the sphere before hollowing was M. With what force, according to the law of universal gravitation, will the hollowed lead sphere attract a small sphere of mass m, which lies at a distance d from the center of the lead sphere on the straight line connecting the centers of the spheres and of the hollow?

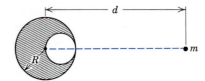

Figure 40 Problem 23.

24. (a) Show that in a chute through the Earth along a chord line, rather than along a diameter, the motion of an object will be simple harmonic; assume a uniform Earth density. (b) Find the period. (c) Will the object attain the same maximum speed along a chord as it does along a diameter?

25. Figure 41 shows, not to scale, a cross section through the interior of the Earth. Rather than being uniform throughout, the Earth is divided into three zones: an outer *crust,* a *mantle,* and an inner *core.* The dimensions of these zones and the mass contained within them are shown on the figure. The Earth has total mass 5.98×10^{24} kg and radius 6370 km. Ignore rotation and assume that the Earth is spherical. (a) Calculate g at the surface. (b) Suppose that a bore hole is driven to the crust–mantle interface (the *Moho*); what would be the value of g at the bottom of the hole? (c) Suppose that the Earth were a uniform sphere with the same total mass and size. What would be the value of g at a

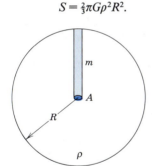

Figure 41 Problems 25 and 26.

depth of 25 km? Use the result of Problem 22. Precise measurements of g are sensitive probes of the interior structure of the Earth, although results can be clouded by local density variations and lack of a precise knowledge of the value of G.

26. Use the model of the Earth shown in Fig. 41 to examine the variation of g with depth in the interior of the Earth. (a) Find g at the core-mantle interface. How does g vary from this interface to the center of the Earth? (b) Show that g has a local minimum within the mantle; find the distance from the Earth's center where this occurs and the associated value of g. (c) Make a sketch showing the variation of g within the Earth.

27. (a) Figure 42a shows a planetary object of uniform density ρ and radius R. Show that the compressive stress S near the center is given by

$$S = \tfrac{2}{3}\pi G\rho^2 R^2.$$

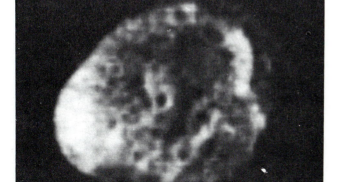

(a)

(b)

Figure 42 Problem 27.

(*Hint:* Construct a narrow column of cross-sectional area A extending from the center to the surface. The weight of the material in the column is mg_{av}, where m is the mass of material in the column and g_{av} is the value of g midway between center and surface.) (b) In our solar system, objects (for example, asteroids, small satellites, comets) with "diameters" less than 600 km can be very irregular in shape (see Fig. 42b, which shows Hyperion, a small satellite of Saturn), whereas those with larger diameters are spherical. Only if the rocks have sufficient strength to resist gravity can an object maintain a nonspherical shape. Calculate the ultimate compressive strength of the rocks making up asteroids. Assume a density of 4000 kg/m³. (c) What is the largest possible size of a nonspherical self-gravitating satellite made of concrete (see Table 1, Chapter 14); assume $\rho = 3000$ kg/m³.

28. A particle of mass m is located a distance y from an infinitely long thin rod of linear mass density λ. Show that the gravitational force between the rod and the particle is $F = 2Gm\lambda/y$, directed perpendicular to the rod. (*Hint:* Let the perpendicular from the particle to the rod define the origin. Consider two mass increments $dm = \lambda\,dx$ located at $\pm x$ along the rod. Calculate the total force dF (magnitude and direction) exerted on the particle by these two mass increments. Then integrate over x from zero to infinity.)

29. Consider a particle at a point P anywhere inside a spherical shell of matter. Assume the shell is of uniform thickness and density. Construct a narrow double cone with apex at P intercepting areas dA_1 and dA_2 on the shell (Fig. 43). (a) Show that the resultant gravitational force exerted on the particle at P by the intercepted mass elements is zero. (b) Show then that the resultant gravitational force of the entire shell on an internal particle is zero. (This method was devised by Newton.)

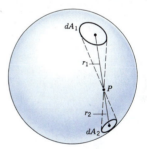

Figure 43 Problem 29.

Section 16-6 Gravitational Potential Energy

30. It is conjectured that a "burned-out" star could collapse to a "gravitational radius," defined as the radius for which the work needed to remove an object of mass m from the star's surface to infinity equals the rest energy mc^2 of the object. Show that the gravitational radius of the Sun is GM_S/c^2 and determine its value in terms of the Sun's present radius. (For a review of this phenomenon see "Black Holes: New Horizons in Gravitational Theory," by Philip C. Peters, *American Scientist,* Sept.–Oct. 1974, p. 575.)

31. A spaceship is idling at the fringes of our galaxy, 80,000 light-years from the galactic center. What minimum speed

must it have if it is to escape entirely from the gravitational attraction of the galaxy? The mass of the galaxy is 1.4×10^{11} times that of our Sun. Assume, for simplicity, that the matter forming the galaxy is distributed with spherical symmetry.

32. Show that the velocity of escape from the Sun at the Earth's distance from the Sun is $\sqrt{2}$ times the speed of the Earth in its orbit, assumed to be a circle. (This is a specific case of a general result for circular orbits: $v_{esc} = \sqrt{2}v_{orb}$.)

33. A rocket is accelerated to a speed of $v = 2\sqrt{gR_E}$ near the Earth's surface and then coasts upward. (a) Show that it will escape from the Earth. (b) Show that very far from the Earth its speed is $V = \sqrt{2gR_E}$.

34. The Sun, mass 2.0×10^{30} kg, is revolving about the center of the Milky Way galaxy, which is 2.2×10^{20} m away. It completes one revolution every 2.5×10^8 years. Estimate the number of stars in the Milky Way. (*Hint:* Assume for simplicity that the stars are distributed with spherical symmetry about the galactic center and that our Sun is essentially at the galactic edge.)

35. A sphere of matter, of mass M and radius a, has a concentric cavity of radius b, as shown in cross section in Fig. 44. (a) Sketch the gravitational force F exerted by the sphere on a particle of mass m, located a distance r from the center of the sphere, as a function of r in the range $0 \le r \le \infty$. Consider points $r = 0$, b, a, and ∞ in particular. (b) Sketch the corresponding curve for the potential energy $U(r)$ of the system. (c) From these graphs, how would you obtain graphs of the gravitational field strength and the gravitational potential due to the sphere?

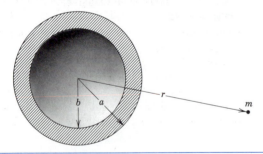

Figure 44   Problem 35.

36. A projectile is fired vertically from the Earth's surface with an initial speed of 9.42 km/s. Neglecting atmospheric friction, how far above the Earth's surface will it go?

37. Spheres of masses 2.53 kg and 7.16 kg are fixed a distance 1.56 m apart (center to center). A 212-g sphere is positioned 42.0 cm from the center of the 7.16-kg sphere, along the line of centers. How much work must be done by an external agent to move the 212-g sphere along the line of centers and place it 42.0 cm from the center of the 2.53-kg sphere?

38. A rocket burns out at an altitude h above the Earth's surface. Its speed v_0 at burnout exceeds the escape speed v_{esc} appropriate to the burnout altitude. Show that the speed v of the rocket very far from the Earth is given by

$$v = (v_0^2 - v_{esc}^2)^{1/2}.$$

39. (a) Calculate the escape speed on Europa, a satellite of the planet Jupiter. The radius of Europa is 1569 km and its

free-fall acceleration at the surface is 1.30 m/s². (b) How high will a particle rise if it leaves the surface of the satellite with a vertical velocity of 1.01 km/s? (c) With what speed will an object hit the satellite if it is dropped from a height of 1000 km? (d) Calculate the mass of Europa.

40. In a particular double star system, two stars of mass 3.22×10^{30} kg each revolve about their common center of mass, 1.12×10^{11} m away. (a) Calculate their common period of revolution, in years. (b) Suppose that a meteoroid (small solid particle in space) passes through this center of mass moving at right angles to the orbital plane of the stars. What must its speed be if it is to escape from the gravitational field of the double star?

41. Two neutron stars are separated by a center-to-center distance of 93.4 km. They each have a mass of 1.56×10^{30} kg and a radius of 12.6 km. They are initially at rest with respect to one another. (a) How fast are they moving when their separation has decreased to one-half of its initial value? (b) How fast are they moving just before they collide? Ignore relativistic effects.

42. Several planets (the gas giants Jupiter, Saturn, Uranus, and Neptune) possess nearly circular surrounding rings, perhaps composed of material that failed to form a satellite. In addition, many galaxies contain ringlike structures. Consider a homogeneous ring of mass M and radius R. (a) Find an expression for the gravitational force exerted by the ring on a particle of mass m located a distance x from the center of the ring along its axis. See Fig. 45. (b) Suppose that the particle falls from rest as a result of the attraction of the ring of matter. Find an expression for the speed with which it passes through the center of the ring.

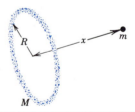

Figure 45   Problem 42.

43. Two particles of mass m and M are initially at rest an infinite distance apart. Show that at any instant their relative velocity of approach attributable to gravitational attraction is $\sqrt{2G(M + m)/d}$, where d is their separation at that instant.

Section 16-8 The Motions of Planets and Satellites

44. The mean distance of Mars from the Sun is 1.52 times that of the Earth from the Sun. From this, calculate the number of years required for Mars to make one revolution about the Sun; compare your answer with the value given in Appendix C.

45. The planet Mars has a satellite, Phobos, which travels in an orbit of radius 9400 km with a period of 7 h 39 min. Calculate the mass of Mars from this information. (The mass of Phobos is negligible compared with that of Mars.)

46. Determine the mass of the Earth from the period T and the radius r of the Moon's orbit about the Earth: $T = 27.3$ days and $r = 3.82 \times 10^5$ km.

47. A satellite is placed in a circular orbit with a radius equal to one-half the radius of the Moon's orbit. What is its period of revolution in lunar months? (A lunar month is the period of revolution of the Moon.)

48. Spy satellites have been placed in the geosynchronous orbit above the Earth's equator. What is the greatest latitude L from which the satellites are visible from the Earth's surface? See Fig. 46.

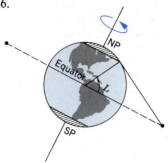

Figure 46   Problem 48.

49. A reconnaissance spacecraft circles the Moon at very low altitude. Calculate (a) its speed and (b) its period of revolution. Take needed data for the Moon from Appendix C.

50. Consider two satellites A and B of equal mass m, moving in the same circular orbit of radius r around the Earth E but in opposite senses of revolution and therefore on a collision course (see Fig. 47). (a) In terms of G, M_E, m, and r, find the total mechanical energy of the two-satellite-plus-Earth system before collision. (b) If the collision is completely inelastic so that wreckage remains as one piece of tangled material, find the total mechanical energy immediately after collision. (c) Describe the subsequent motion of the wreckage.

Figure 47   Problem 50.

51. The Sun's center is at one focus of the Earth's orbit. How far is it from the other focus? Express your answer in terms of the radius of the Sun $R_S = 6.96 \times 10^8$ m. The eccentricity of the Earth's orbit is 0.0167 and the semimajor axis is 1.50×10^{11} m.

52. Use conservation of energy and Eq. 27 for the total energy to show that the speed v of an object in an elliptical orbit satisfies the relation

$$v^2 = GM \left(\frac{2}{r} - \frac{1}{a} \right).$$

53. A comet moving in an orbit of eccentricity 0.880 has a speed of 3.72 km/s when it is most distant from the Sun. Find its speed when it is closest to the Sun.

54. A satellite, moving in an elliptical orbit, is 2360 km above the Earth's surface at its farthest point and 1180 km above at

its closest point. Calculate (a) the semimajor axis, (b) the eccentricity of the orbit, and (c) the period of revolution.

55. Consider an artificial satellite in a circular orbit about the Earth. State how the following properties of the satellite vary with the radius r of its orbit: (a) period, (b) kinetic energy, (c) angular momentum, and (d) speed.

56. (a) Express the universal gravitational constant G that appears in Newton's law of gravity in terms of the astronomical unit AU as a length unit, the solar mass M_S as a mass unit, and the year as a time unit. (1 AU = 1.496×10^{11} m, 1 M_S = 1.99×10^{30} kg, 1 y = 3.156×10^7 s.) (b) What form does Kepler's third law (Eq. 24) take in these units?

57. In the year 1610, Galileo made a telescope, turned it on Jupiter, and discovered four prominent moons. Their mean orbit radii a and periods T are

Name	a (10^8 m)	T (days)
Io	4.22	1.77
Europa	6.71	3.55
Ganymede	10.7	7.16
Callisto	18.8	16.7

(a) Plot $\log a$ (y axis) against $\log T$ (x axis) and show that you get a straight line. (b) Measure its slope and compare it with the value that you expect from Kepler's third law. (c) Find the mass of Jupiter from the intercept of this line with the y axis. (*Note:* You may also use log–log graph paper.)

58. Show how, guided by Kepler's third law, Newton could deduce that the force holding the Moon in its orbit, assumed circular, must vary as the inverse square of the distance from the center of the Earth.

59. Most asteroids revolve around the Sun between Mars and Jupiter. However, several "Apollo asteroids" with diameters of about 30 km move in orbits that cross the orbit of the Earth. The orbit of one of these is shown in Fig. 48. By taking measurements directly from the figure, deduce the asteroid's period of revolution in years. (All these asteroids are expected eventually to collide with the Earth.)

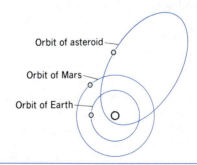

Figure 48   Problem 59.

60. A satellite in an elliptical orbit of eccentricity e has a speed v_a at apogee, v_p at perigee, and v_0 at the ends of the minor axis of its orbit. Show that (a) $v_p/v_a = (1 + e)/(1 - e)$, and (b) $v_0/v_a = (v_p/v_a)^{1/2}$.

61. A certain triple-star system consists of two stars, each of mass m, revolving about a central star, mass M, in the same circular orbit. The two stars stay at opposite ends of a diame-

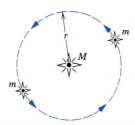

Figure 49   Problem 61.

ter of the circular orbit; see Fig. 49. Derive an expression for the period of revolution of the stars; the radius of the orbit is r.

62. A pair of stars revolves about their common center of mass. One of the stars has a mass M that is twice the mass m of the other; that is, $M = 2m$. Their centers are a distance d apart, d being large compared to the size of either star. (*a*) Derive an expression for the period of revolution of the stars about their common center of mass in terms of d, m, and G. (*b*) Compare the angular momenta of the two stars about their common center of mass by calculating the ratio L_m/L_M. (*c*) Compare the kinetic energies of the two stars by calculating the ratio K_m/K_M.

63. A satellite travels initially in an approximately circular orbit 640 km above the surface of the Earth; its mass is 220 kg. (*a*) Determine its speed. (*b*) Determine its period of revolution. (*c*) For various reasons the satellite loses mechanical energy at the (average) rate of 1.40×10^5 J per orbital revolution. Adopting the reasonable approximation that the trajectory is a "circle of slowly diminishing radius," determine the distance from the surface of the Earth, the speed, and the period of the satellite at the end of its 1500th orbital revolution. (*d*) What is the magnitude of the average retarding force? (*e*) Is angular momentum conserved?

64. A binary star system consists of two stars, each with the same mass as the Sun, revolving about their center of mass. The distance between them is the same as the distance between the Earth and the Sun. What is the period of revolution in years?

65. (*a*) Does it take more energy to get a satellite up to 1600 km above the Earth than to put it in orbit once it is there? (*b*) What about 3200 km? (*c*) What about 4800 km? Take the Earth's radius to be 6400 km.

66. One possibility for damaging a satellite in Earth orbit is to launch a swarm of pellets in such a way that they move in the same orbit as the satellite but in the opposite direction. Consider a satellite in a circular orbit whose altitude above the Earth's surface is 500 km. An on-board sensor detects a 10.0-g pellet approaching and determines that a head-on collision is inevitable. (*a*) What is the kinetic energy of the approaching pellet in the reference frame of the satellite? (*b*) How does this compare with the kinetic energy of a slug from a modern army rifle? Such a slug has a mass of 4.00 g and a muzzle velocity of 950 m/s.

67. The asteroid Eros, one of the many minor planets that orbit the Sun in the region between Mars and Jupiter, has a radius of 7.0 km and a mass of 5.0×10^{15} kg. (*a*) If you were standing on Eros, could you lift a 2000-kg pickup truck? (*b*) Could you run fast enough to put yourself into orbit? Ignore effects

due to the rotation of the asteroid. (*Note:* The Olympic records for the 400-m run correspond to speeds of 9.1 m/s for men and 8.2 m/s for women.)

68. The orbit of the Earth about the Sun is *almost* circular. The closest and farthest distances are 1.47×10^8 km and 1.52×10^8 km, respectively. Determine the maximum variations in (*a*) potential energy, (*b*) kinetic energy, (*c*) total energy, and (*d*) orbital speed that result from the changing Earth–Sun distance in the course of 1 year. (*Hint:* Use conservation of energy and angular momentum.)

69. Assume that a geosynchronous communications satellite is in orbit at the longitude of Chicago. You are in Chicago and want to pick up its signals. In what direction should you point the axis of your parabolic antenna? The latitude of Chicago is 47.5° N.

70. What minimum initial speed (measured with respect to the Earth) must be imparted to an object resting on the Earth's surface if it is to escape not only from the gravitational field of the Earth but also from that of the Sun? Ignore the Earth's rotation but not its orbital motion around the Sun. (*Hint:* Note that for minimum speed the object must be projected in the direction of the Earth's orbital motion. Treat the problem in two steps, escape from the Sun following that from the Earth. The Earth's orbital speed, v_0, connects the two reference frames involved.)

71. Using the data of Sample Problem 11, calculate (*a*) the speed of spacecraft B as it passes through point P', and (*b*) the average speed of spacecraft B in the orbit after the burn. Approximate the path of B as a circle. Compare these results with the corresponding quantities of spacecraft A.

72. A weather satellite is in a geosynchronous orbit, hovering over Nairobi, which lies very close to the equator. If its orbit radius is increased by 1.00 km, at what rate and in what direction would its reference spot, which was formerly stationary, move across the Earth's surface?

73. Three identical stars of mass M are located at the vertices of an equilateral triangle with side L. At what speed must they move if they all revolve under the influence of one another's gravity in a circular orbit circumscribing, while still preserving, the equilateral triangle?

74. How long will it take a comet, moving in a parabolic path, to move from its point of closest approach to the Sun at A (see Fig. 50) through an angle of 90°, measured at the Sun, to B? Let the distance of closest approach to the Sun be equal to the radius of the Earth's orbit, assumed circular.

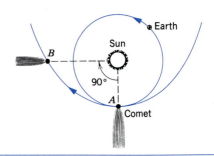

Figure 50   Problem 74.

75. Imagine a planet of mass M with a small moon of mass m and radius a orbiting it and keeping the same face toward it.

If the moon now approaches the planet, there will be a critical distance from the planet's center at which loose material lying on the moon's surface will be lifted off. Show that this distance is given by $r_c = a(3M/m)^{1/3}$. This critical distance is called *Roche's limit*.

76. (*a*) Show that the two-body problem of Section 16-8 can be simplified to a one-body problem by use of the reduced mass concept of Section 15-10. That is, show that if we use $\mu = mM/(m + M)$ instead of m, where μ is the reduced mass, we may solve for the motion of m relative to M exactly as though M were the origin of our inertial reference frame. (*b*) Show that the assumption made in Section 16-8 that R is negligibly small compared to r is equivalent to assuming that the reduced mass μ is equal to m. (*c*) Compare μ for the Earth–Sun system with the Earth's mass; compare μ for the Moon–Earth system with the Moon's mass. (*d*) If we were to use the reduced mass μ of the two-body system instead of m, how would this affect the equations of Section 16-8?

CHAPTER 17

FLUID STATICS

Most matter can conveniently be described as being in one of three phases—solid, liquid, or gas. Solids and liquids (also called condensed matter) have a certain set of properties in common; for example, they are relatively incompressible, and their densities stay relatively constant as we vary the temperature (keeping other properties, such as pressure, constant). Gases, on the other hand, are easily compressible, and their density changes substantially with temperature if we hold the pressure constant.

From a different perspective, we can usually group gases and liquids together under the common designation fluids. The word "fluid" comes from a Latin word meaning "to flow." Fluids will flow, for example, to take the shape of any container that holds them; solids do not share this property. In a solid, the atoms remain relatively fixed in their arrangement; in a fluid, the atoms can move relative to one another.

In this chapter we consider the properties of fluids at rest and the laws that govern them. In the next chapter we discuss the dynamical properties of fluids in motion.

17-1 FLUIDS AND SOLIDS

In our everyday experience we have a clear idea of the distinction between fluids and solids but, as we often find in science, everyday experiences are gathered over a very limited range of circumstances, and extrapolating too far can lead to incorrect conclusions. For example, from everyday experience we can propose the following distinction: a solid holds its shape but a fluid flows to take the shape of its container. Some substances cannot easily be classified. Glass, for example, should be classified as a fluid; even though it appears to hold its shape, glass does flow over a long period of time. Glass windows that have been in place for many years are measurably thicker at the bottom than at the top.

Another somewhat intermediate form is the *plastic* substance—one that can be molded or shaped. Consider modeling clay, for instance. It holds its shape relatively well, and we would be reluctant to classify it as a fluid, but by applying *pressure* to it we can force it to take the shape of its container. Other substances, which we might identify as solids in ordinary experience, can also be made to flow at high enough pressure.

Of course, we are familiar with changing the state of matter by changing its temperature, which might melt or vaporize it. We are somewhat less familiar with changing the state of matter by changing the pressure on it, in part because the range of pressures required is generally beyond our normal experience. Aluminum, for instance, can be drawn into wires by pulling it through a small hole and can be *extruded* into various shapes by forcing it through a die at high pressure. The deeply folded rock layers that are often seen in highway cuts are evidence that "solid rock" also flows readily at sufficiently high pressure.

There is still another phase of matter which cannot be easily classified as solid, liquid, or gas. A *plasma* is a gas in which the atoms are ionized, so that they form an overall electrically neutral mixture containing equal numbers of positively charged ions and negatively charged electrons. The resulting strong electrical interactions with the environment and among the atoms make the behavior of the plasma quite different from that of an ordinary gas. The gas inside a fluorescent light fixture turns into a plasma when the light is turned on. On a grander scale, the Sun and the other stars are balls of plasma, and thus much of the matter in the Universe exists in this form. Creating and confining plasmas of sufficient size in the laboratory are the major hurdles facing researchers who are seeking ways of using controlled fusion reactions to generate electrical power.

Microscopically, how do these forms of matter differ from one another? Solids are able to support a variety of stresses, as we discussed in Chapter 14. These stresses include tension, compression, and shearing, among others. Solids can support and transmit such stresses because there are relatively strong forces between the molecules of a solid and because they have a *long-range order;* that is, their molecules are arranged in an orderly way, like bricks in a wall, so that we cannot easily displace an atom in one location without displacing many other atoms as well.

In liquids, the intermolecular distances are generally larger than in solids; hence the intermolecular forces, which vary strongly with distance, tend to be weaker in liquids than in solids. Many liquids are, like solids, relatively incompressible, and so liquids can often support and transmit compressional stresses; as we discuss later in this chapter, hydraulic systems depend on this property of fluids. To a limited extent, liquids can also support tensile stresses, as we discuss in Section 17-6. However, liquids cannot support shearing stresses, because layers of the liquid can easily slide over one another.

In gases, the molecules interact only weakly, and therefore gases are unable to transmit static tensile or shearing stresses, and gases generally are far more compressible than solids or liquids. In a plasma, however, there are long-range electromagnetic forces among the particles. Thus even though a plasma resembles a gaseous state of matter, it is more like a liquid in its ability to transmit stresses.

We developed a set of mechanical laws that allowed us to analyze the dynamics of individual particles. We developed another similar set of laws that permitted us to analyze the dynamics of collections of particles in rigid solids. It is important to note that we did this even without a theory to explain the forces between the particles of which the solid is composed. Even in the case of solids that cannot be regarded as perfectly rigid, we have a theory of elasticity (see Chapter 14).

The mechanics of fluids takes a similar approach. Like the mechanics of rigid bodies, fluid mechanics takes its start from Newton's laws. For fluids, as for solids, it is convenient to develop a special formulation for these laws.

17-2   PRESSURE AND DENSITY

Pressure

We can apply a force to a solid at an arbitrary angle with its surface. In Section 14-5 we considered the effect on a solid of a shearing stress, in which the force lies in the plane of an element of surface area. The ability to flow makes a fluid unable to sustain a shearing stress, and under static conditions the only force component that

need be considered is one that acts *normal* or *perpendicular* to the surface of the fluid. No matter what the shape of a fluid, forces between the interior and exterior are everywhere at right angles to the fluid boundary.

The magnitude of the normal force per unit surface area is called the *pressure.* Pressure is a scalar quantity; it has no directional properties. When you swim underwater, for example, the water presses on your body from all directions. Even though the pressure is produced by a force that has directional properties and is a vector, the pressure itself is a scalar.

Microscopically, the pressure exerted by a fluid on a surface in contact with it is caused by collisions of molecules of the fluid with the surface. As a result of a collision, the component of a molecule's momentum perpendicular to the surface is reversed. The surface must exert an impulsive force on the molecule, and by Newton's third law the molecules exert an equal force perpendicular to the surface. The net result of the reaction force exerted by many molecules on the surface gives rise to the pressure on the surface. We develop this picture more quantitatively in the case of gases in Chapter 23.

A fluid under pressure exerts an outward force on any surface in contact with it. Consider a closed surface containing a fluid, as in Fig. 1. The fluid within the surface pushes out against the environment. A small element of surface area can be represented by the vector $\Delta \mathbf{A}$, whose magnitude is numerically equal to the element of area and whose direction is along the outward normal to the surface. The force $\Delta \mathbf{F}$ exerted by the fluid against this surface depends on the pressure p according to

$$\Delta \mathbf{F} = p\,\Delta \mathbf{A}. \qquad (1)$$

Since the vectors representing the force and the area are parallel, we can write the pressure in terms of the scalar relationship

$$p = \frac{\Delta F}{\Delta A}. \qquad (2)$$

We take the element ΔA small enough that the pressure p defined according to Eq. 2 is independent of the size of the

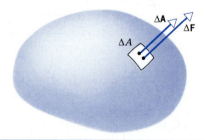

Figure 1   An element of surface ΔA can be represented by a vector $\Delta \mathbf{A}$ of length equal to the magnitude of the area of the element and of direction perpendicular to the element. The fluid enclosed by the surface exerts a force $\Delta \mathbf{F}$ against the element. The force is perpendicular to the element and therefore parallel to $\Delta \mathbf{A}$.

element. In general, the pressure may vary from point to point along the surface.

Pressure has dimensions of force divided by area, and a common unit for pressure is N/m². This unit is given the SI designation *pascal* (abbreviation Pa; 1 Pa = 1 N/m²). A wide variety of other units can be found. Tire pressure gauges usually read in lb/in.² in the United States. The standard pressure exerted by the atmosphere of the Earth at sea level is designated as 1 atmosphere (atm; 1 atm = 14.7 lb/in.² = 1.01325 × 10⁵ Pa, exactly). Because the pascal is a small unit (1 Pa ≈ 10⁻⁵ atm), weather forecasters often use the unit of the bar (1 bar = 10⁵ Pa, or approximately 1 atm) to express atmospheric pressure. Another common unit is based on the pressure exerted at its base by a vertical column of mercury of a specified height; a column 760 mm in height at a temperature of 0°C at a location where $g = 9.80665$ m/s² exerts a pressure equal to that of the atmosphere, and thus we have the equivalence 760 mm Hg = 1 atm. The height of such a column in inches is 29.9 in.; ordinary barometers (and TV weather forecasters) give the atmospheric pressure in inches of mercury. Pressure readings in the laboratory are often expressed in the unit of the *torr*, which is the pressure exerted by a column of mercury 1 mm high under the specified conditions.

Table 1 gives some representative pressures in pascal units. The term "overpressure" indicates a pressure value in excess of normal atmospheric pressure. Note that in the laboratory we can produce pressures that range over 22 orders of magnitude. In Appendix G you will find the conversion factors necessary to convert pressure measurements from one set of units to another.

Density

The density ρ of a small element of any material is the mass Δm of the element divided by its volume ΔV:

$$\rho = \frac{\Delta m}{\Delta V}. \qquad (3)$$

The density at a point is the limiting value of this ratio as the volume element becomes small. Density has no directional properties and is a scalar.

If the density of an object has the same value at all points, the density of the object is equal to the mass of the entire object divided by its volume:

$$\rho = \frac{m}{V}. \qquad (4)$$

The density of a material in general depends on environmental factors, including the pressure and temperature. For liquids and solids, the variation in density is very small over wide ranges of variation of pressure and temperature, and for many applications we can regard the density as a constant. Table 2 gives some representative densities, which vary by about 21 orders of magnitude in the laboratory and by nearly 40 orders of magnitude from the densest objects in the Universe (a hypothetical black hole) to the near vacuum of space itself.

In analogy with the stress versus strain discussion of Section 14-5, a change Δp in the pressure applied to any material is a stress. The corresponding strain is a change in volume, which we write as $\Delta V/V$. The ratio between the stress and strain is called the *bulk modulus B:*

$$B = -\frac{\Delta p}{\Delta V/V}. \qquad (5)$$

The minus sign is inserted in this definition to make B a positive quantity, because Δp and ΔV have opposite signs. That is, an *increase* in pressure ($\Delta p > 0$) causes a *decrease* in volume ($\Delta V < 0$). Note that B has the same dimension as pressure, because $\Delta V/V$ is a dimensionless quantity.

If the bulk modulus of a material is large, then (according to Eq. 5) a large pressure change Δp produces only a

TABLE 1 SOME PRESSURES

System	Pressure (Pa)
Center of the Sun	2 × 10¹⁶
Center of the Earth	4 × 10¹¹
Highest sustained laboratory pressure	1.5 × 10¹⁰
Deepest ocean trench (bottom)	1.1 × 10⁸
Spiked heels on a dance floor	2 × 10⁷
Automobile tire (overpressure)	2 × 10⁵
Atmosphere at sea level	1.0 × 10⁵
Normal blood pressure[a]	1.6 × 10⁴
Loudest tolerable sound[b]	30
Faintest detectable sound[b]	3 × 10⁻⁵
Best laboratory vacuum	10⁻¹²

[a] The systolic overpressure, corresponding to 120 mm Hg on the physician's pressure gauge.
[b] Overpressure at the eardrum, at 1000 Hz.

TABLE 2 SOME DENSITIES

Material or Object	Density (kg/m³)
Interstellar space	10⁻²⁰
Best laboratory vacuum	10⁻¹⁷
Air: 20°C and 1 atm	1.21
20°C and 50 atm	60.5
Styrofoam	1 × 10²
Ice	0.917 × 10³
Water: 20°C and 1 atm	0.998 × 10³
20°C and 50 atm	1.000 × 10³
Seawater: 20°C and 1 atm	1.024 × 10³
Whole blood	1.060 × 10³
Iron	7.8 × 10³
Mercury	13.6 × 10³
The Earth: average	5.5 × 10³
core	9.5 × 10³
crust	2.8 × 10³
The Sun: average	1.4 × 10³
core	1.6 × 10⁵
White dwarf star (core)	10¹⁰
Uranium nucleus	3 × 10¹⁷
Neutron star (core)	10¹⁸
Black hole (1 solar mass)	10¹⁹

small change in its volume. In this case, we can regard the material as being nearly incompressible. The bulk modulus of water, for example, is 2.2×10^9 N/m². At the pressure at the bottom of the Pacific Ocean (4.0×10^7 N/m², about 400 atm), the relative change in volume caused by pressure alone is only 1.8%. Solids usually have higher bulk moduli than liquids, because of the tighter coupling of the atoms in solids. A given pressure thus produces a smaller change in volume of a solid than a liquid. Under ordinary circumstances, we can therefore regard both solids and liquids as incompressible.

If B is small, the volume can be changed by a modest change in pressure, and the material is said to be compressible. Typical gases have bulk moduli of about 10^5 N/m². A small pressure change of 0.1 atm can change the volume of a gas by 10%. Gases are thus easily compressible.

17-3 VARIATION OF PRESSURE IN A FLUID AT REST

If a fluid is in equilibrium, every portion of the fluid is in equilibrium. That is, both the net force and the net torque on every element of the fluid must be zero. Consider a small element of fluid volume submerged within the body of the fluid. Let this element have the shape of a thin disk and be a distance y above some reference level, as shown in Fig. 2a. The thickness of the disk is dy and each face has area A. The mass of this element is $dm = \rho\, dV = \rho A\, dy$, and its weight is $(dm)g = \rho g A\, dy$. The forces exerted on the element by the surrounding fluid are perpendicular to its surface at each point (Fig. 2b).

The resultant horizontal force is zero, because the element has no horizontal acceleration. The horizontal forces are due only to the pressure of the fluid, and by symmetry the pressure must be the same at all points within a horizontal plane at y.

The fluid element is also unaccelerated in the vertical direction, so the resultant vertical force on it must be zero.

A free-body diagram of the fluid element is shown in Fig. 2c. The vertical forces are due not only to the pressure of the surrounding fluid on its faces but also to the weight of the element. If we let p be the pressure on the lower face and $p + dp$ the pressure on its upper face, the upward force is pA, and the downward forces are $(p + dp)A$ and the weight of the element $(dm)g = \rho g A\, dy$. Hence, for vertical equilibrium,

$$\sum F_y = pA - (p + dp)A - \rho g A\, dy = 0,$$

from which we obtain

$$\frac{dp}{dy} = -\rho g. \tag{6}$$

This equation tells us how the pressure varies with elevation above some reference level in a fluid in static equilibrium. As the elevation increases (dy positive), the pressure decreases (dp negative). The cause of this pressure variation is the weight per unit cross-sectional area of the layers of fluid lying between the points whose pressure difference is being measured.

The quantity ρg is often called the *weight density* of the fluid; it is the weight per unit volume of the fluid. For water, for example, the weight density is 9800 N/m³ = 62.4 lb/ft³.

If p_1 is the pressure at elevation y_1, and p_2 the pressure at elevation y_2 above some reference level, integration of Eq. 6 gives

$$\int_{p_1}^{p_2} dp = -\int_{y_1}^{y_2} \rho g\, dy$$

or

$$p_2 - p_1 = -\int_{y_1}^{y_2} \rho g\, dy. \tag{7}$$

For liquids, which are nearly incompressible, ρ is practically constant, and differences in level are rarely so great that any change in g need be considered. Hence, taking ρ and g as constants, we obtain

$$p_2 - p_1 = -\rho g(y_2 - y_1) \tag{8}$$

for a homogeneous liquid.

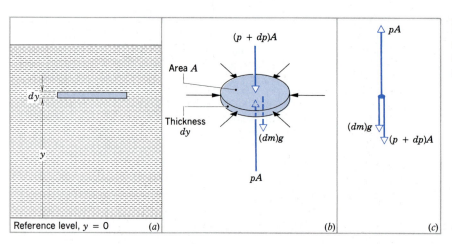

Figure 2 (a) A small volume element of the fluid at rest. (b) The forces on the element. (c) A free-body diagram of the element.

If a liquid has a free surface, this is the natural level from which to measure distances (Fig. 3). Let y_2 be the elevation of the surface, at which point the pressure p_2 acting on the fluid is usually that exerted by the Earth's atmosphere p_0. We take y_1 to be at any level in the fluid, and we represent the pressure there as p. Then

$$p_0 - p = -\rho g(y_2 - y_1).$$

But $y_2 - y_1$ is the depth h below the surface at which the pressure is p (see Fig. 3), so that

$$p = p_0 + \rho g h. \tag{9}$$

This shows clearly that the pressure in a liquid increases with depth but is the same at all points at the same depth. The second term on the right of Eq. 9 gives the contribution to the pressure at a point in the liquid due to the weight of the fluid of height h above that point.

Equation 8 gives the relation between the pressures at any two points in a fluid, regardless of the shape of the containing vessel. For no matter what the shape of the containing vessel, two points in the fluid can be connected

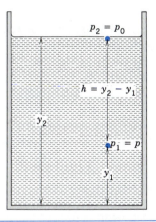

Figure 3   A container holds a quantity of a liquid whose top surface is open to the atmosphere. The pressure at any point in the liquid depends on the depth h.

by a path made up of vertical and horizontal steps. For example, consider points A and B in the homogeneous liquid contained in the U-tube of Fig. 4a. Along the zigzag path from A to B there is a difference in pressure $\rho g y'$ for each vertical segment of length y', whereas along each horizontal segment there is no change in pressure. Hence the difference in pressure $p_B - p_A$ is ρg times the algebraic sum of the vertical segments from A to B, or $\rho g(y_2 - y_1)$.

If the U-tube contains different immiscible liquids, say, a dense liquid in the right tube and a less dense one in the left tube, as shown in Fig. 4b, the pressure can be different at the same level (points A and B) on different sides. The liquid below the line CC is in equilibrium; thus the force exerted by the left column above C must equal the force exerted by the right column above C. The pressure at C is the same on both sides, but the pressure falls less from C to A than from C to B, because the liquid on the left is less dense than the liquid on the right. Thus the pressure at A is greater than at B.

Variation of Pressure in the Atmosphere

For gases ρ is comparatively small, and the difference in pressure at two nearby points is usually negligible (see Eq. 8). Thus in a reasonably small vessel containing a gas, the pressure can be taken as the same everywhere. However, this is not the case if $y_2 - y_1$ is very great. The pressure of the air varies greatly as we ascend to great heights in the atmosphere. Moreover, the density ρ varies with altitude, and ρ must be known as a function of y before we can integrate Eq. 7.

We can get a reasonable idea of the variation of pressure with altitude in the Earth's atmosphere if we assume that the density ρ is proportional to the pressure. This would be very nearly true (according to the ideal gas law, which we discuss in Chapter 23) if the temperature of the air remained the same at all altitudes. Using this assumption, and also assuming that the variation of g with altitude is negligible, we can find the pressure p at any altitude y above sea level.

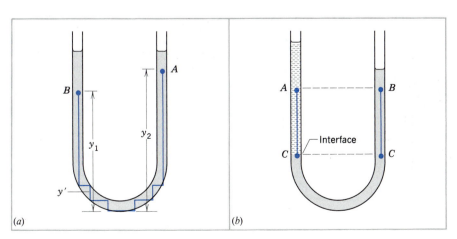

Figure 4   (a) The difference in pressure between two points A and B in a homogeneous liquid depends only on their difference in elevation $y_2 - y_1$. (b) Two points A and B at the same elevation can be at different pressures if the densities there differ.

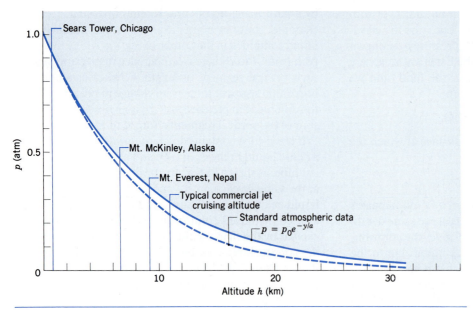

Figure 5 Comparison of standard atmospheric pressure data (dashed line) with predictions of Eq. 13 (solid line). The two curves differ because our calculation neglected the variation of the density with temperature as the altitude increases.

From Eq. 6 we have

$$\frac{dp}{dy} = -\rho g.$$

Since ρ is proportional to p, we have

$$\frac{\rho}{\rho_0} = \frac{p}{p_0}, \qquad (10)$$

where ρ_0 and p_0 are the values of density and pressure at sea level. Then

$$\frac{dp}{dy} = -g\rho_0 \frac{p}{p_0},$$

so that

$$\frac{dp}{p} = -\frac{g\rho_0}{p_0} dy. \qquad (11)$$

Integrating Eq. 11 from the pressure p_0 at the altitude $y = 0$ (sea level) to the pressure p at the altitude y, we obtain

$$\int_{p_0}^{p} \frac{dp}{p} = -\int_{0}^{y} \frac{g\rho_0}{p_0} dy,$$

which gives

$$\ln \frac{p}{p_0} = -\frac{g\rho_0}{p_0} y$$

or

$$p = p_0 e^{-(g\rho_0/p_0)y}. \qquad (12)$$

Using the values $g = 9.80$ m/s², $\rho_0 = 1.21$ kg/m³ (at 20°C), and $p_0 = 1.01 \times 10^5$ Pa, we obtain

$$\frac{g\rho_0}{p_0} = 1.17 \times 10^{-4} \text{ m}^{-1} = 0.117 \text{ km}^{-1}.$$

Hence

$$p = p_0 e^{-y/a} \qquad (13)$$

where $1/a = g\rho_0/p_0 = 0.117$ km^{-1} or $a = 8.55$ km. The constant a gives the change in altitude over which the pressure drops by a factor of e. Put another way, the atmospheric pressure drops by a factor of 10 when the altitude changes by $a \ln 10 = 2.30a = 20$ km. At an altitude of $h = 20$ km above sea level, the atmospheric pressure would thus be 0.1 atm; at $h = 40$ km above sea level, it would be 0.01 atm. Figure 5 shows a comparison between the pressure variation with altitude predicted by Eq. 13 and that measured for the atmosphere.

For gases at uniform temperature the density ρ of any layer is proportional to the pressure p at that layer. Liquids, however, are almost incompressible, so the lower layers are not noticeably compressed by the weight of the upper layers superimposed on them, and the density ρ is practically constant at all levels. The variation of pressure with distance above the bottom of the fluid for a gas is different from that for a liquid, as indicated by Eq. 9 for a liquid and Eq. 13 for a gas.

Sample Problem 1 A U-tube, in which both ends are open to the atmosphere, is partly filled with water. Oil, which does not mix with water, is poured into one side until it stands a distance $d = 12.3$ mm above the water level on the other side, which has meanwhile risen a distance $a = 67.5$ mm from its original level (Fig. 6). Find the density of the oil.

Solution In Fig. 6 points C are at the same pressure. (If this were not true, then the U-shaped fluid element below the CC level would experience a net unbalanced force and would accelerate, violating the static assumption we make in this problem.) The pressure drop from C to the surface on the water side is $\rho_w g 2a$, where $2a$ is the height of the water column above C.

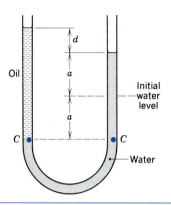

Figure 6 Sample Problem 1. A U-tube is filled partly with water and partly with oil of unknown density.

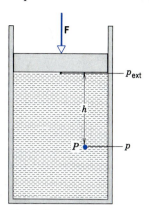

Figure 7 A fluid in a cylinder fitted with a movable piston. The pressure at any point P is due not only to the weight of the fluid above the level of P but also to the force exerted by the piston.

The pressure drop on the other side from C to the surface is $\rho g(2a + d)$, where ρ is the unknown density of the oil. Equating the pressures at point C on each side, we obtain

$$p_0 + \rho_w g2a = p_0 + \rho g(2a + d)$$

and so

$$\rho = \rho_w \frac{2a}{(2a + d)}$$

$$= (1.000 \times 10^3 \text{ kg/m}^3) \frac{2(67.5 \text{ mm})}{2(67.5 \text{ mm}) + 12.3 \text{ mm}}$$

$$= 916 \text{ kg/m}^3.$$

The ratio of the density of a substance to the density of water is called the *relative density* (or the *specific gravity*) of that substance. In this case the specific gravity of the oil is 0.916.

Note that in solving this problem, we have assumed that the pressure is continuous across the interface between the oil and the water at point C on the left side of the tube. If this were not so and the pressures were different, then the force exerted by the fluid on one side of the interface would differ from that of the fluid on the other side, and the interface would accelerate under the influence of the unbalanced force. Since we are assuming a static situation, there can be no motion and the pressures must therefore be the same. When we first pour the oil into the tube, however, there may be a difference in pressure and an unbalanced force that would cause the system to move until it reached the static situation shown in Fig. 6.

17-4 PASCAL'S PRINCIPLE AND ARCHIMEDES' PRINCIPLE

When you squeeze a tube of toothpaste, the toothpaste flows out of the open top of the tube. This demonstrates the action of *Pascal's principle.* When pressure is applied anywhere on the tube, it is felt everywhere in the tube and forces the toothpaste out of the top. Here is the statement of Pascal's principle, which was first stated by Blaise Pascal in 1652:

Pressure applied to an enclosed fluid is transmitted undiminished to every portion of the fluid and to the walls of the containing vessel.

That is, if you increase the external pressure on a fluid at one location by an amount Δp, the same increase in pressure is experienced everywhere in the fluid.

Pascal's principle is the basis for the operation of all hydraulic force-transmitting mechanisms, such as might be found in earth-moving machinery or the brake system of your car. It enables us to amplify a relatively small applied force to raise a much greater weight (as in the automobile lift or the dentist's chair) and to transmit forces over long distances to relatively inaccessible locations (as in the control mechanisms for the wing flaps used in aircraft).

We shall prove Pascal's principle for an incompressible liquid. Figure 7 shows the liquid in a cylinder that is fitted with a piston. An external force is applied to the piston, for instance, by the weight of some objects stacked on it. The external force results in an external pressure p_{ext} being applied to the liquid immediately beneath the piston. If the liquid has a density ρ, then from Eq. 9 we can write the pressure at an arbitrary point P a distance h below the surface:

$$p = p_{ext} + \rho gh. \qquad (14)$$

Suppose now the external pressure is increased by an amount Δp_{ext}, perhaps by adding some more weight to the piston. How does the pressure p in the fluid change as a result of this change in the external pressure? We assume the liquid to be incompressible, so that the density ρ remains constant. The change in external pressure results in a change in pressure in the fluid that follows from Eq. 14:

$$\Delta p = \Delta p_{ext} + \Delta(\rho gh). \qquad (15)$$

Since the liquid is incompressible, the density is constant, and the second term on the right of Eq. 15 equals zero. In this case, we obtain

$$\Delta p = \Delta p_{\text{ext}} . \tag{16}$$

The change in pressure at any point in the fluid is simply equal to the change in the externally applied pressure. This result confirms Pascal's principle and shows that it follows directly from our previous consideration of the static pressure in a fluid. It is therefore not an independent principle but a direct consequence of our formulation of fluid statics.

Although we derived the above result for incompressible liquids, Pascal's principle is true for all real (compressible) fluids, gases as well as liquids. The change in external pressure causes a change in density that propagates as a wave with the speed of sound in the fluid, but once the disturbance has died out and equilibrium has been established, Pascal's principle is found to remain valid.

The Hydraulic Lever

Figure 8 shows an arrangement that is often used to lift a heavy object such as an automobile. An external force F_i is exerted on a piston of area A_i. The object to be lifted exerts a force Mg on the larger piston of area A_o. In equilibrium, the magnitude of the upward force F_o exerted by the fluid on the larger piston must equal that of the downward force Mg of the weight of the object (neglecting the weight of the piston itself). We wish to find the relationship between the applied force F_i and the "output force" F_o that the system can exert on the larger piston.

The pressure on the fluid at the smaller piston, due to our externally applied force, is $p_i = F_i/A_i$. According to Pascal's principle, this "input" pressure must be equal to the "output" pressure $p_o = F_o/A_o$, which is exerted *by* the fluid *on* the larger piston. Thus $p_i = p_o$, and so

$$\frac{F_i}{A_i} = \frac{F_o}{A_o} ,$$

or

$$F_i = F_o \frac{A_i}{A_o} = Mg \frac{A_i}{A_o} . \tag{17}$$

The ratio A_i/A_o is generally much smaller than 1, and thus

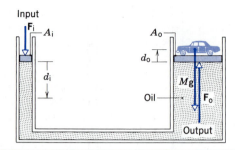

Figure 8   The hydraulic lever. A force $\mathbf{F}_i$ applied to the smaller piston can give a much larger force $\mathbf{F}_o$ on the larger piston, which can lift a weight Mg.

the applied force can be much smaller than the weight Mg that is lifted.

The downward movement of the smaller piston through a distance d_i displaces a volume of fluid $V = d_i A_i$. If the fluid is incompressible, then this volume must be equal to the volume displaced by the upward motion of the larger piston:

$$V = d_i A_i = d_o A_o ,$$

or

$$d_o = d_i \frac{A_i}{A_o} . \tag{18}$$

If A_i/A_o is a small number, then the distance moved by the larger piston is much smaller than the distance the applied force moves the smaller piston. The price we pay for gaining the ability to lift a large load is losing the ability to move it very far.

By combining Eqs. 17 and 18, we see that $F_i d_i = F_o d_o$, which shows that the work done by the external force on the smaller piston is equal to the work done by the fluid on the larger piston. Thus (ignoring friction and other dissipative forces), there is no net gain (or loss) in energy in using this hydraulic system.

Sample Problem 2   Figure 9 shows a schematic view of a hydraulic jack used to lift an automobile. The hydraulic fluid is oil (density = 812 kg/m³). A hand pump is used, in which a force of

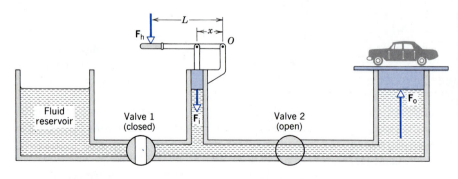

Figure 9   Sample Problem 2. A hydraulic pump is used to lift a car. For the downstroke, valve 1 is closed and valve 2 is open. During the upstroke, valve 1 is opened and valve 2 is closed, permitting additional fluid to be drawn into the hydraulic chamber.

magnitude F_i is applied to the smaller piston (of diameter 2.2 cm) when the hand applies a force of magnitude F_h to the end of the pump handle. The combined mass of the car to be lifted and the lifting platform is $M = 1980$ kg, and the large piston has a diameter of 16.4 cm. The length L of the pump handle is 36 cm, and the distance x from the pivot to the piston is 9.4 cm. (*a*) What is the applied force F_h needed to lift the car? (*b*) For each downward stroke of the pump, in which the hand moves a vertical distance of 28 cm, how far is the car raised?

Solution   (*a*) From Eq. 17,

$$F_i = Mg\frac{A_i}{A_o} = (1980 \text{ kg})(9.8 \text{ m/s}^2)\frac{\pi(1.1 \text{ cm})^2}{\pi(8.2 \text{ cm})^2} = 349 \text{ N}.$$

Taking torques on the pump handle about the pivot point O, neglecting the masses of the pump handle and the small piston, and assuming the pump handle moves with a negligibly small angular acceleration, we obtain

$$\sum \tau = F_h L - F_i x = 0,$$

where we have used Newton's third law to relate the force $\mathbf{F}_i$ exerted *by* the pump handle *on* the piston to the force $-\mathbf{F}_i$ exerted *by* the piston *on* the pump handle. Solving for F_h, we find that

$$F_h = F_i \frac{x}{L} = (349 \text{ N})\frac{9.4 \text{ cm}}{36 \text{ cm}} = 91 \text{ N}.$$

Such a force, about 20 lb, can easily be applied by hand.

(*b*) When the hand moves through a vertical distance h, the smaller piston will move through the distance

$$d_i = h\frac{x}{L} = (28 \text{ cm})\frac{9.4 \text{ cm}}{36 \text{ cm}} = 7.3 \text{ cm}.$$

Equation 18 then gives the distance moved by the larger piston:

$$d_o = d_i \frac{A_i}{A_o} = (7.3 \text{ cm})\frac{\pi(1.1 \text{ cm})^2}{\pi(8.2 \text{ cm})^2} = 0.13 \text{ cm} = 1.3 \text{ mm}.$$

Raising the car by only such a tiny distance is the price we pay for exerting such a small force to lift the car. Of course, to make a useful device we must be able to lift the car by a larger distance, which is accomplished through many strokes of the pump. To keep the car from moving downward during the upward stroke

of the pump, the valve arrangement shown in Fig. 9 is used. During the downstroke, the valves are in the positions shown in Fig. 9, and the car is raised by the distance d_o. During the return stroke valve 2 is closed, trapping fluid in the right side of the chamber and keeping the car at a fixed height, and valve 1 is opened, so the return stroke draws additional fluid from the reservoir into the left side of the chamber. For the next down-stroke, the valves return to the positions shown in the figure, and the car is raised by another increment d_o. In effect, the volume of hydraulic fluid drawn into the left side of the chamber during the upstroke is pumped into the right side of the chamber during the downstroke. When the process is completed, the car can be lowered by opening both valves and allowing fluid to drain directly into the reservoir.

How does the operation of the hydraulic jack change as the car is raised and the height of the fluid in the right-hand column increases? Make a numerical estimate.

Archimedes' Principle

Figure 10*a* shows a volume of water contained in a thin plastic sack placed underwater. The water in the sack is in static equilibrium. Therefore its weight must be balanced by an upward force of equal magnitude. This upward force is the vector sum of all the inward forces exerted by the fluid that surrounds the sack. The arrows in Fig. 10*a* represent the forces exerted on the volume of liquid as a result of the pressure of the surrounding fluid. Notice that the upward forces on the bottom of the sack are greater than the downward forces on the top, because the pressure increases with depth. The net upward force resulting from this pressure difference is called the *buoyant force* or *buoyancy*.

The pressure exerted on a submerged object by the surrounding fluid certainly cannot depend on the material of which the object is made. We could therefore replace the sack of water by a piece of wood of exactly the same size and shape, and the buoyant force would be unchanged. The upward force is still equal to the weight of

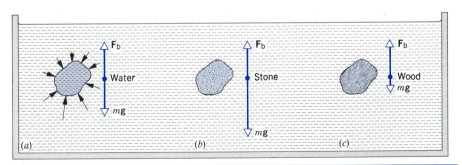

Figure 10   (*a*) A thin plastic sack full of water in equilibrium underwater. The water surrounding the sack exerts pressure forces on its surface, the resultant being an upward buoyant force $\mathbf{F}_b$ acting on the sack. (*b*) For a stone of the same volume, the buoyant force is the same, but the weight exceeds the buoyant force so the stone is not in equilibrium. (*c*) For a piece of wood of the same volume, the weight is less than the buoyant force.

the original volume of water. This leads us to Archimedes' principle:

A body wholly or partially immersed in a fluid is buoyed up by a force equal in magnitude to the weight of the fluid displaced by the body.

An object of density greater than water (Fig. 10*b*) displaces a volume of water whose weight is less than the weight of the object. The object therefore sinks in the water, because the buoyant force is less than the weight of the object. If we try to lift the object while it is underwater, we find that it takes less force than the object's normal weight, the difference being the buoyant force. Submerged objects appear to weigh less than they normally do. Astronauts prepare for their voyages by practicing tasks in huge underwater tanks, to simulate somewhat the weightless condition of space.

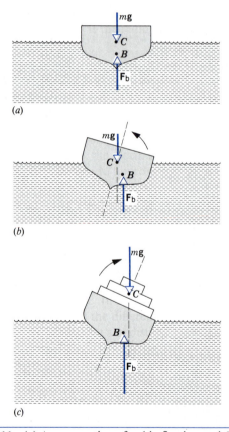

(a)

(b)

(c)

Figure 11   (*a*) A cross section of a ship floating upright. The buoyant force $\mathbf{F}_b$ acts at the center of buoyancy *B*, and the weight acts at the center of gravity *C*. The ship is in equilibrium under the action of these forces. (*b*) When the ship tips, the center of buoyancy may no longer lie on the same vertical line as the center of gravity, and a net torque may act on the ship. Here the torque about *C* acts to restore the ship to the upright position. (*c*) Here the center of gravity lies higher, so that the torque about *C* due to the buoyant force tends to tip the ship even further.

An object of density less than water (Fig. 10*c*) experiences a net upward force when completely submerged, because the weight of the water displaced is greater than the weight of the object. The object therefore rises until it breaks the surface, and it continues to rise until the only part of it still submerged is that volume necessary to displace water whose weight is equal to the total weight of the object. In that situation the object floats in equilibrium.

The buoyant force can be regarded as acting at the center of gravity of the fluid displaced by the submerged part of a floating object. This point is known as the *center of buoyancy.* The weight acts at the center of gravity of the entire object. These two points are in general not the same (Fig. 11*a*). If the two points lie on the same vertical line, then the object can float in equilibrium: both the net force and the net torque are zero. If the floating object is tipped slightly from its equilibrium position, then in general the shape of the displaced fluid changes, and the center of buoyancy shifts its position with respect to the center of gravity of the floating object. Thus a torque acts on the floating object that might tip the object back to its equilibrium position (Fig. 11*b*), or it might act in the other direction to tip it completely over (Fig. 11*c*).

Sample Problem 3   What fraction of the total volume of an iceberg is exposed?

Solution   The weight of the iceberg is

$$W_i = \rho_i V_i g,$$

where V_i is the volume of the iceberg. The weight of the volume V_w of seawater displaced (or, equivalently, the volume of the submerged part of the iceberg) is the buoyant force

$$F_b = \rho_w V_w g.$$

But F_b equals W_i, because the iceberg is in equilibrium, so that

$$\rho_w V_w g = \rho_i V_i g,$$

and, using densities from Table 2,

$$\frac{V_w}{V_i} = \frac{\rho_i}{\rho_w} = \frac{917 \text{ kg/m}^3}{1024 \text{ kg/m}^3} = 0.896 = 89.6\%.$$

The volume of water displaced V_w is the volume of the submerged portion of the iceberg, so that 10.4% of the iceberg is exposed.

17-5   MEASUREMENT OF PRESSURE

The pressure exerted by a fluid can be measured using either static or dynamic techniques. The dynamic methods are based on the speed of flow of a moving fluid and are discussed in Chapter 18. In this section, we discuss static methods for measuring pressure.

Most pressure gauges use atmospheric pressure as a reference level and measure the difference between the actual pressure and atmospheric pressure, called the *gauge pressure*. The actual pressure at a point in a fluid is called the *absolute pressure*, which is then the atmospheric pressure plus the gauge pressure. Gauge pressure is given either above or below atmospheric pressure and may thus be positive or negative; absolute pressure is always positive.

The mercury barometer is a long glass tube that has been filled with mercury and then inverted into a dish of mercury, as in Fig. 12. The space above the mercury column is in effect a vacuum containing only mercury vapor, whose pressure p_2 is so small at ordinary temperatures that it can be neglected. The pressure p_1 on the surface of the dish of mercury is the unknown pressure p we wish to measure. From Eq. 8, we obtain

$$p_2 - p_1 = 0 - p = -\rho g(y_2 - y_1) = -\rho gh,$$

or

$$p = \rho gh.$$

Measuring the height of the column above the surface of the dish then gives the pressure.

The mercury barometer is most often used for measuring the pressure of the atmosphere, p_0. The mercury column in the barometer will have a height of about 760 mm at sea level, varying with the atmospheric pressure. The pressure of 1 atmosphere (1 atm) is equivalent to that exerted by a column of mercury of height 760 mm at 0°C under standard gravity ($g = 9.80665$ m/s²). The density of mercury at this temperature is 1.35955×10^4 kg/m³. Hence 1 atmosphere is equivalent to

$$1 \text{ atm} = (1.35955 \times 10^4 \text{ kg/m}^3)(9.80665 \text{ m/s}^2)(0.76 \text{ m})$$
$$= 1.013 \times 10^5 \text{ N/m}^2 \quad (\equiv 1.013 \times 10^5 \text{ Pa}).$$

The pressure of the atmosphere at any point is numerically equal to the weight of a column of air of unit cross-sectional area extending from that point to the top of the atmosphere. Since normal atmospheric pressure can be expressed as 14.7 lb/in.², we know that the vertical column of air that extends from each square inch of the Earth's surface to the top of the atmosphere has a weight of 14.7 pounds. As we have already derived in Section 17-3, the atmospheric pressure decreases with altitude. There are also variations in atmospheric pressure at a given location from day to day because the atmosphere is not static.

Barometer readings are sometimes expressed in torr, where 1 torr is the pressure exerted by a column of mercury 1 mm high at a place where $g = 9.80665$ m/s² and at a temperature (0°C) at which mercury has a density of 1.35955×10^4 kg/m³. Thus

$$1 \text{ torr} = (1.35955 \times 10^4 \text{ kg/m}^3)(9.80665 \text{ m/s}^2)(0.001 \text{ m})$$
$$= 133.326 \text{ Pa}.$$

The open-tube manometer (Fig. 13) measures gauge pressure. It consists of a U-shaped tube containing a liquid, one end of the tube being open to the atmosphere and the other end being connected to the system (tank) whose pressure p we want to measure. From Eq. 9 we obtain

$$p - p_0 = \rho gh.$$

Thus the gauge pressure, $p - p_0$, is proportional to the difference in height of the liquid columns in the U-tube. If the vessel contains gas under high pressure, a dense liquid

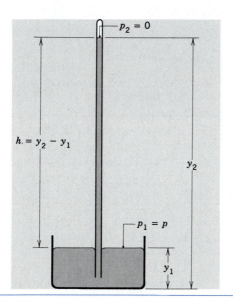

Figure 12 The mercury barometer. The mercury in the dish is in equilibrium under the influence of atmospheric pressure and the weight of the mercury in the vertical column.

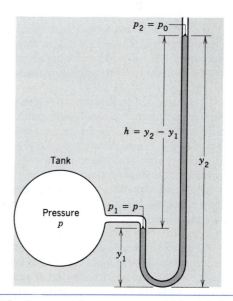

Figure 13 An open-tube manometer, which might be used to measure the pressure of a fluid in a tank.

like mercury is used in the tube; water can be used when low gas pressures are involved.

Sample Problem 4   The mercury column in a barometer has a measured height h of 740.35 mm. The temperature is $-5.0°C$, at which temperature the density of mercury is 1.3608×10^4 kg/m³. The free-fall acceleration g at the site of the barometer is 9.7835 m/s². What is the atmospheric pressure?

Solution   From Eq. 8 we have

$$p_0 = \rho g h$$
$$= (1.3608 \times 10^4 \text{ kg/m}^3)(9.7835 \text{ m/s}^2)(0.74035 \text{ m})$$
$$= 9.8566 \times 10^4 \text{ Pa} = 739.29 \text{ torr.}$$

Note that the value of the pressure in torr (739.29 torr) is numerically close to the value of the height h of the mercury column expressed in mm (740.35 mm). These two quantities will be numerically equal only if the barometer is located at a place where g has its standard value and where the mercury temperature is 0°C.

Another way to express the result of this sample problem would be as 0.98566 bar or 985.66 millibar, where 1 bar = 10^5 Pa.

Historical Notes   *(Optional)*

The mercury barometer was invented by the Italian Evangelista Torricelli (1608–1647), after whom the unit torr was named. Torricelli described his experiments with the mercury barometer in letters in 1644 to his friend Michelangelo Ricci in Rome. He wrote that the aim of his investigation was "not simply to produce a vacuum, but to make an instrument which shows the mutations of the air, now heavier and dense, and now lighter and thin." On hearing of the Italian experiments, Blaise Pascal, in France, reasoned that if the mercury column were held up simply by the pressure of the air, the column ought to be shorter at a high altitude. He tried it on a church steeple in Paris, but desiring more decisive results, he wrote to his brother-in-law to try the experiment on the Puy de Dôme, a high mountain in Auvergne. The measured difference in the height of the mercury was 8 cm, a result "which ravished us with admiration and astonishment." Pascal himself constructed a barometer using red wine and a glass tube 14 m long.

The chief significance of these experiments at the time was the realization it brought that an evacuated space could be created. Aristotle believed that a vacuum could not exist, and as late a writer as Descartes held the same view. For 2000 years philosophers spoke of the "horror" that nature had for empty space — the *horror vacui*. Nature was said to prevent the formation of a vacuum by laying hold of anything nearby and with it instantly filling up any evacuated space. Hence the mercury or wine should fill up the inverted tube because "nature abhorred a vacuum." The experiments of Torricelli and Pascal showed that there were limitations to nature's ability to prevent a vacuum. They created a sensation at the time. The goal of producing a vacuum became more of a practical reality through the development of pumps by Otto von Guericke in Germany around 1650 and by Robert Boyle in England around 1660. Even though these pumps were relatively crude, they did provide a tool for

experimentation. With a pump and a glass jar, an experimental space could be provided in which to study how the properties of heat, light, sound, and later electricity and magnetism are affected by an increasingly rarefied atmosphere. Although even today we cannot completely remove every trace of gas from a closed vessel, these 17th-century experimenters freed science from the false principle of *horror vacui* and stimulated efforts to create highly evacuated systems.

Within several decades in the 17th century no fewer than six important instruments were developed. They are the barometer, air pump, pendulum clock, telescope, microscope, and thermometer. All excited great wonder and curiosity. ∎

17-6   SURFACE TENSION   *(Optional)*

Leaves and insects can be observed to float on the *surface* of a body of water (Fig. 14a). They are *not* partially submerged and thus *not* buoyed up because of Archimedes' principle. In this case the object is completely on the surface and none of it is submerged.

The object is kept afloat by the *surface tension* of the liquid. You can demonstrate the surface tension of water by carefully floating a steel needle or a razor blade. There is of course no way for steel to float by Archimedes' principle, since its density is greater than that of water. If you submerge the needle or the

(a)

(b)

Figure 14   (a) A razor blade floating on the surface of water, supported only by surface tension. (b) The surface is distorted by the floating object, which is kept afloat by the vertical components of the surface force $\mathbf{F}_s$.

razor blade, it will sink as Archimedes' principle predicts. Only when it is entirely on the surface can it float. You can add to the water a chemical, called a surface-active agent or surfactant, which reduces the surface tension (by reducing the cohesive force between molecules) and makes it more difficult to float the object. Detergents are common surfactants. If you carefully introduce detergent into the water on which a razor blade is floating, the surface tension suddenly decreases and the razor blade sinks to the bottom.

A floating object, such as that shown in Fig. 14*a*, depresses the surface layer of the fluid slightly (Fig. 14*b*), which stretches the surface layer and thus tends to increase its potential energy. Somewhat like a trampoline, the stretched surface layer exerts a restoring force, the vertical component of which can maintain equilibrium with the weight of the object. We shall soon see, however, that this analogy of the behavior of the surface layer is not strictly correct.

Figure 15 shows a way of measuring the surface tension of a liquid. A thin wire is bent into the shape of three sides of a rectangle and fitted with a sliding wire as the fourth side. If a film of the liquid covers the vertical loop (established perhaps by dipping the loop into a container of the liquid), the surface tension will tend to draw the sliding wire downward. We apply an external upward force *P* necessary to maintain the sliding wire in equilibrium. This upward force must balance the total downward force on the sliding wire, equal to its weight plus the force *F* due to the surface tension.

By experiment we find that the force *F* depends on the length *d* of the slide wire but does not depend at all on the height *h* of the rectangle. Although it is tempting to regard the surface layer as a sort of elastic sheet stretched over the liquid, this observation shows that such a picture is incorrect. Imagine the film of Fig. 15 to be cut into a large number *N* of narrow vertical strips of length *h* and width $\Delta d = d/N$. If the film behaved like an elastic sheet, each strip would behave like a spring, and so the total force would depend both on the number of springlike strips (and hence on *d*) and on the length *h* of each strip. Because the surface tension depends only on *d* and not on *h*, the analogy of the elastic sheet is not correct.

The surface tension γ is defined as *the surface force F per unit length L over which it acts,* or

$$\gamma = \frac{F}{L}. \qquad (19)$$

Note that the surface tension γ is not a force but a force per unit length. Our previous use of the term *tension* has always indicated a force, but here the usage is somewhat different.

For the film of Fig. 15, the force acts over a length *L* of 2*d*, because there are *two* surface layers each of length *d*. The surface tension in the experimental arrangement shown in Fig. 15 would therefore be

$$\gamma = \frac{F}{2d}.$$

For water at room temperature, the value of the surface tension is $\gamma = 0.073$ N/m. Adding soap reduces the surface tension to 0.025 N/m. Organic liquids and aqueous solutions typically have surface tensions in this range. The surface tension of liquid metals is typically an order of magnitude larger than that of water. Liquid mercury at room temperature, for example, has a surface tension of 0.487 N/m. (This higher surface tension in metals occurs because the forces between molecules are typically an order of magnitude larger in metals than in water. For this same reason, the boiling points of metals are typically much higher than that of water.)

We can also analyze surface tension from the standpoint of energy. If we move the sliding wire of Fig. 15 through a displacement Δx, the work done by the force of surface tension equals $F \Delta x$ in magnitude and is positive or negative according to whether Δx is in the direction of the surface force or in the opposite direction. The surface force meets our definition of a conservative force from Chapter 8, and we can therefore associate a change in potential energy ΔU with the action of the surface force, so that

$$\Delta U = F \Delta x = \gamma L \Delta x, \qquad (20)$$

where *L* is the length of the surface layer. The product $L \Delta x$ is just the change in area ΔA of the surface that occurs when we stretch it. We can therefore write the surface tension as

$$\gamma = \frac{\Delta U}{\Delta A}. \qquad (21)$$

This provides us with another interpretation of the surface tension in terms of *the surface potential energy per unit area of surface.*

Surface tension causes suspended droplets of a liquid to acquire a spherical shape (Fig. 16). For a drop of a given mass or volume, the surface energy (equal to γ times the surface area) is least when the area is smallest, and a sphere has the smallest surface-to-volume ratio of any geometric shape. If no other forces act on the drop, it will naturally assume a spherical shape. In equilibrium, the surface tension gives a net inward force on an element of surface, which is balanced by an equal outward force due to the pressure of the liquid within the drop. In a soap bubble (which has two surfaces and therefore twice the surface tension of a liquid drop of equal size), the gauge pressure of the gas confined in the bubble provides the outward force needed for equilibrium.

Like the molecules in a liquid drop, the protons and neutrons

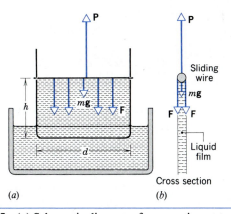

(a)   (b)

Figure 15 (*a*) Schematic diagram of an experiment to measure the surface tension of a liquid. A film of the liquid is supported in the vertical rectangular area, the top border of which is a sliding wire. An external force balances the weight of the sliding wire plus the total downward force *F* of the surface tension. (*b*) A cross-sectional sketch of the film, showing that the surface tension acts on two surfaces.

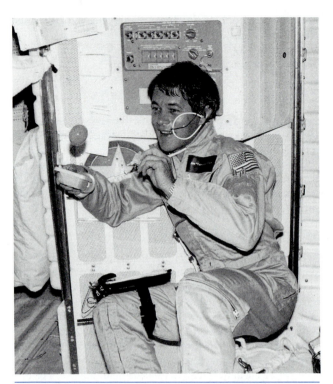

Figure 16 Freely floating droplets of liquid naturally assume a spherical shape. Here astronaut Dr. Joseph P. Allen, in Earth orbit on Space Shuttle *Columbia,* watches a ball of orange juice he created using a beverage dispenser.

in a nucleus experience short-range forces exerted by their neighbors. The nucleus experiences a surface tension that is similar to that of a liquid drop. In the case of the nucleus, the outward force originates with the electrostatic repulsion of the charged protons. For many nuclei, the equilibrium shape is determined by the balance between the surface and electrostatic forces, and it should therefore not be surprising that for these

nuclei the preferred shape is spherical. The calculation of the binding energy of nuclei must include a term corresponding to the surface energy, which typically accounts for 30% of the total binding energy.

Considering the nucleus to behave like a charged liquid drop provides great insight in understanding many properties of the nucleus, especially nuclear fission, in which the nucleus splits into two parts of comparable size. Such a procedure is called *modeling,* in which we try to understand a complex system, whose properties often cannot be calculated or understood directly, on the basis of a simpler physical system of relatively similar behavior, whose properties can be calculated and then tested against experiment. The *liquid drop model of the nucleus* has played an important role in our understanding of atomic nuclei, as we discuss in Chapters 54 and 55 of the extended text.

Sample Problem 5   In the experiment shown in Fig. 15*a*, it is found that the movable wire is in equilibrium when the upward force *P* is 3.45×10^{-3} N. The wire has a length *d* of 4.85 cm and a linear mass density μ of 1.75×10^{-3} kg/m. Find the surface tension of the liquid.

Solution   From the equilibrium condition of Fig. 15*b*, we have

$$\sum F_y = P - F - mg = 0,$$

or

$$F = P - mg.$$

With $F = 2d\gamma$ and $m = \mu d$, we obtain

$$2d\gamma = P - \mu dg$$

or

$$\begin{aligned}
\gamma &= \frac{P - \mu dg}{2d} \\
&= \frac{3.45 \times 10^{-3}\,\text{N} - (1.75 \times 10^{-3}\,\text{kg/m})(0.0485\,\text{m})(9.80\,\text{m/s}^2)}{2(0.0485\,\text{m})} \\
&= 0.027\ \text{N/m}. \qquad \blacksquare
\end{aligned}$$

QUESTIONS

1. Explain how it can be that pressure is a scalar quantity when forces, which are vectors, can be produced by the action of pressures.

2. Make an estimate of the average density of your body. Explain a way in which you could get an accurate value using ideas in this chapter.

3. In Chapter 20, we shall learn that an overpressure of only 20 Pa corresponds to the threshold of pain for intense sound. Yet a diver 2 m below the surface of water experiences a much greater pressure than this (how much?) and feels no pain. Why this difference?

4. Persons confined to bed are less likely to develop sores on their bodies if they use a water bed rather than an ordinary mattress. Explain.

5. Explain why one could lie on a bed of nails without pain.

6. Explain the statement "water seeks its own level."

7. Water is poured to the same level in each of the vessels shown, all having the same base area (Fig. 17). If the pressure is the same at the bottom of each vessel, the force experienced by the base of each vessel is the same. Why then do the three vessels have different weights when put on a

Figure 17 Question 7.

scale? This apparently contradictory result is commonly known as the *hydrostatic paradox.*

8. Does Archimedes' principle hold in a vessel in free fall or in a satellite moving in a circular orbit?

9. A spherical bob made of cork floats half submerged in a pot of tea at rest on the Earth. Will the cork float or sink aboard a spaceship (*a*) coasting in free space and (*b*) on the surface of Mars?

10. How does a suction cup work?

11. Is the buoyant force acting on a submerged submarine the same at all depths?

12. Explain how a submarine rises, falls, and maintains a fixed depth. Do fish use the same principles? (See "The Buoyancy of Marine Animals," by Eric Denton, *Scientific American,* July 1960, p. 118, and "Submarine Physics," by G. P. Harnwell, *American Journal of Physics,* March 1948, p. 127.)

13. A block of wood floats in a pail of water in an elevator. When the elevator starts from rest and accelerates down, does the block float higher above the water surface?

14. Two identical buckets are filled to the rim with water, but one has a block of wood floating in the water. Which bucket, if either, is heavier?

15. Estimate with some care the buoyant force exerted by the atmosphere on you.

16. According to Sample Problem 3, 89.6% of an iceberg is submerged. Yet occasionally icebergs turn over, with possibly disastrous results to nearby shipping. How can this happen considering that so much of their mass is below sea level?

17. Can you sink an iron ship by siphoning seawater into it?

18. Scuba divers are warned not to hold their breath when swimming upward. Why?

19. A beaker is exactly full of liquid water at its freezing point and has an ice cube floating in it, also at its freezing point. As the cube melts, what happens to the water level in these three cases: (*a*) the cube is solid ice; (*b*) the cube contains some grains of sand; and (*c*) the cube contains some bubbles?

20. Although parachutes are supposed to brake your fall, they are often designed with a hole at the top. Explain why.

21. A ball floats on the surface of water in a container exposed to the atmosphere. Will the ball remain immersed at its former depth or will it sink or rise somewhat if (*a*) the container is covered and the air is removed or (*b*) the container is covered and the air is compressed?

22. Explain why an inflated balloon will only rise to a definite height once it starts to rise, whereas a submarine will always sink to the very bottom of the ocean once it starts to sink, if no changes are made.

23. Why does a balloon weigh the same when empty as when filled with air at atmospheric pressure? Would the weights be the same if measured in a vacuum?

24. Liquid containers tend to leak when taken aloft in an airplane. Why? Does it matter whether or not they are right-side up? Does it matter whether or not they are initially completely full?

25. During World War II, a damaged freighter that was barely able to float in the North Sea steamed up the Thames es-

tuary toward the London docks. It sank before it could arrive. Why?

26. Is it true that a floating object will be in stable equilibrium only if its center of buoyancy lies above its center of gravity? Illustrate with examples.

27. Logs dropped upright into a pond do not remain upright, but float "flat" in the water. Explain.

28. Why will a sinking ship often turn over as it becomes immersed in water?

29. A barge filled with scrap iron is in a canal lock. If the iron is thrown overboard, what happens to the water level in the lock? What if it is thrown onto the land beside the canal?

30. A bucket of water is suspended from a spring balance. Does the balance reading change when a piece of iron suspended from a string is immersed in the water? When a piece of cork is put in the water?

31. If enough iron is added to one end of a uniform wooden stick or log, it will float vertically, rather than horizontally (see Question 27). Explain.

32. Although there are practical difficulties, it is possible in principle to float an ocean liner in a few barrels of water. How would you go about doing this?

33. An open bucket of water is on a frictionless plane inclined at an angle α to the horizontal. Find the equilibrium inclination to the horizontal of the free surface of the water when (*a*) the bucket is held at rest; (*b*) the bucket is allowed to slide down at constant speed ($a = 0$, v = constant); and (*c*) the bucket slides down without restraint (a = constant). If the plane is curved so that $a \neq$ constant, what will happen?

34. In a barometer, how important is it that the inner diameter of the barometer be uniform? That the barometer tube be absolutely vertical?

35. An open-tube manometer has one tube twice the diameter of the other. Explain how this would affect the operation of the manometer. Does it matter which end is connected to the chamber whose pressure is to be measured?

36. We have considered liquids under compression. Can liquids be put under tension? If so, will they tear under sufficient tension as do solids? (See "The Tensile Strength of Liquids," by Robert E. Apfel, *Scientific American,* December 1972, p. 58.)

37. Explain why two glass plates with a thin film of water between them are difficult to separate by a direct pull but can easily be separated by sliding.

38. Give a molecular explanation of why surface tension decreases with increasing temperature.

39. Soap films are much more stable than films of water. Why? (Consider how surface tension reacts to stretching.)

40. Explain why a soap film collapses if a small hole appears in it.

41. Explain these observations: (*a*) water forms globules on a greasy plate but not on a clean one; (*b*) small bubbles on the surface of water cluster together.

42. If soap reduces the surface tension of water, why do we blow soap bubbles instead of water bubbles?

43. Some water beetles can walk on water. Estimate the maximum weight such an insect can have and still be supported in this way.

44. What is the source of the energy that allows a fluid in a capillary (e.g., a thin, hollow, glass tube) to rise?

45. What does it mean to say that certain liquids can exert a small negative pressure?

PROBLEMS

Section 17-2 Pressure and Density

1. Find the pressure increase in the fluid in a syringe when a nurse applies a force of 42.3 N to the syringe's piston of diameter 1.12 cm.

2. Three liquids that will not mix are poured into a cylindrical container. The amounts and densities of the liquids are 0.50 L, 2.6 g/cm³; 0.25 L, 1.0 g/cm³; and 0.40 L, 0.80 g/cm³ (L = liter). Find the total force on the bottom of the container. (Ignore the contribution due to the atmosphere.) Does it matter if the fluids mix?

3. An office window is 3.43 m by 2.08 m. As a result of the passage of a storm, the outside air pressure drops to 0.962 atm, but inside the pressure is held at 1.00 atm. What net force pushes out on the window?

4. A solid copper cube has an edge length of 85.5 cm. How much pressure must be applied to the cube to reduce the edge length to 85.0 cm? The bulk modulus of the copper is 140 GPa.

5. An airtight box having a lid with an area of 12 in.² is partially evacuated. If a force of 108 lb is required to pull the lid off the box, and the outside atmospheric pressure is 15 lb/in.², what was the pressure in the box?

6. In 1654 Otto von Guericke, Bürgermeister of Magdeburg and inventor of the air pump, gave a demonstration before the Imperial Diet in which two teams of horses could not pull apart two evacuated brass hemispheres. (a) Show that the force F required to pull apart the hemispheres is $F = \pi R^2 \Delta p$, where R is the (outside) radius of the hemispheres and Δp is the difference in pressure outside and inside the sphere (Fig. 18). (b) Taking R equal to 0.305 m and the inside pressure as 0.100 atm, what force would the team of horses have had to exert to pull apart the hemispheres? (c) Why were two teams of horses used? Would not one team prove the point just as well?

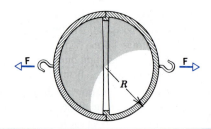

Figure 18 Problem 6.

Section 17-3 Variation of Pressure in a Fluid at Rest

7. The human lungs can operate against a pressure differential of less than 0.050 atm. How far below the water level can a diver, breathing through a snorkel (long tube), swim?

8. Calculate the hydrostatic difference in blood pressure in a person of height 1.83 m between the brain and the foot.

9. Find the total pressure, in pascal, 118 m below the surface of

the ocean. The density of seawater is 1.024 g/cm³ and the atmospheric pressure at sea level is 1.013×10^5 Pa.

10. The sewer outlets of a house constructed on a slope are 8.16 m below street level. If the sewer is 2.08 m below street level, find the minimum pressure differential that must be created by the sewage pump to transfer waste of average density 926 kg/m³.

11. Figure 19 displays the phase diagram of carbon, showing the ranges of temperature and pressure in which carbon will crystallize either as diamond or graphite. What is the minimum depth at which diamonds can form if the local temperature is 1000°C and the subsurface rocks have density 3.1 g/cm³? Assume that, as in a fluid, the pressure is due to the weight of material lying above.

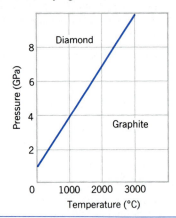

Figure 19 Problem 11.

12. According to the constant temperature model of the Earth's atmosphere, (a) what is the pressure (in atm) at an altitude of 5.00 km, and (b) at what altitude is the pressure equal to 0.500 atm? Compare your answers with Fig. 5.

13. A simple U-tube contains mercury. When 11.2 cm of water is poured into the right arm, how high does the mercury rise in the left arm from its initial level?

14. Water stands at a depth D behind the vertical upstream face of a dam, as shown in Fig. 20. Let W be the width of the dam. (a) Find the resultant horizontal force exerted on the

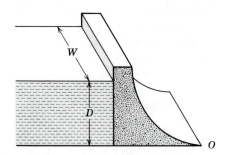

Figure 20 Problem 14.

dam by the gauge pressure of the water and (*b*) the net torque due to the gauge pressure of the water exerted about a line through *O* parallel to the width of the dam. (*c*) Where is the line of action of the equivalent resultant force?

15. A swimming pool has the dimensions 80 ft × 30 ft × 8.0 ft. (*a*) When it is filled with water, what is the force (due to the water alone) on the bottom? On the ends? On the sides? (*b*) If you are concerned with whether or not the concrete walls will collapse, is it appropriate to take the atmospheric pressure into account?

16. What would be the height of the atmosphere if the air density (*a*) were constant and (*b*) decreased linearly to zero with height? Assume a sea-level density of 1.21 kg/m³.

17. Crew members attempt to escape from a damaged submarine 112 m below the surface. How much force must they apply to a pop-out hatch, which is 1.22 m by 0.590 m, to push it out?

18. A cylindrical barrel has a narrow tube fixed to the top, as shown with dimensions in Fig. 21. The vessel is filled with water to the top of the tube. Calculate the ratio of the hydrostatic force exerted on the bottom of the barrel to the weight of the water contained inside. Why is the ratio not equal to one? (Ignore the presence of the atmosphere.)

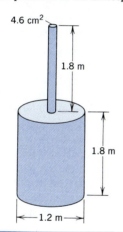

4.6 cm²

1.8 m

1.8 m

1.2 m

Figure 21 Problem 18.

19. In analyzing certain geological features of the Earth, it is often appropriate to assume that the pressure at some horizontal *level of compensation,* deep in the Earth, is the same over a large region and is equal to that exerted by the weight of the overlying material. That is, the pressure on the level of compensation is given by the hydrostatic (fluid) pressure formula. This requires, for example, that mountains have low-density *roots;* see Fig. 22. Consider a mountain 6.00 km high. The continental rocks have a density of 2.90 g/cm³; beneath the continent is the mantle, with a density of 3.30 g/cm³. Calculate the depth *D* of the root. (*Hint:* Set the pressure at points *a* and *b* equal; the depth *y* of the level of compensation will cancel out.)

20. (*a*) Show that the density ρ of water at a depth *y* in the ocean is related to the surface density ρ_s by

$$\rho \approx \rho_s[1 + (\rho_s g/B)y],$$

where *B* = 2.2 GPa is the bulk modulus of water. Ignore temperature variations. (*b*) By what fraction does the density at a depth of 4200 m exceed the surface density?

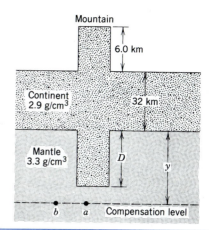

Mountain

6.0 km

Continent 2.9 g/cm³

32 km

Mantle 3.3 g/cm³

D

y

b   *a*   Compensation level

Figure 22 Problem 19.

21. A test tube 12.0 cm long is filled with water and set spinning in a horizontal plane in a centrifuge at 655 rev/s. Calculate the hydrostatic pressure on the outer base of the tube. The inner end of the tube is 5.30 cm from the axis of rotation.

22. The surface of contact of two fluids of different densities that are at rest and do not mix is horizontal. Prove this general result (*a*) from the fact that the potential energy of a system must be a minimum in stable equilibrium; (*b*) from the fact that at any two points in a horizontal plane in either fluid the pressures are equal.

23. Two identical cylindrical vessels with their bases at the same level each contain a liquid of density ρ. The area of either base is *A*, but in one vessel the liquid height is h_1 and in the other h_2. Find the work done by gravity in equalizing the levels when the two vessels are connected.

24. A U-tube is filled with a single homogeneous liquid. The liquid is temporarily depressed in one side by a piston. The piston is removed and the level of the liquid in each side oscillates. Show that the period of oscillation is $\pi\sqrt{2L/g}$, where *L* is the total length of the liquid in the tube.

25. (*a*) Show that Eq. 13, the variation of pressure with altitude in the atmosphere (temperature assumed to be uniform), can be written in terms of density ρ as

$$\rho = \rho_0 e^{-y/a},$$

where ρ_0 is the density at the ground ($y = 0$). (*b*) Assume that the drag force *D* due to the air on an object moving at speed *v* is given by $D = CA\rho v^2$, where *C* is a constant, *A* is the frontal cross-sectional area of the object, and ρ is the local air density. Find the altitude at which the drag force on a rocket is a maximum if the rocket is launched vertically and moves with constant upward acceleration a_r.

26. (*a*) Consider a container of fluid subject to a *vertical upward* acceleration *a*. Show that the pressure variation with depth in the fluid is given by

$$p = \rho h(g + a),$$

where *h* is the depth and ρ is the density. (*b*) Show also that if the fluid as a whole undergoes a *vertical downward* acceleration *a*, the pressure at a depth *h* is given by

$$p = \rho h(g - a).$$

(*c*) What is the state of affairs in free fall?

27. (*a*) Consider the horizontal acceleration of a mass of liquid in an open tank. Acceleration of this kind causes the liquid surface to drop at the front of the tank and to rise at the rear. Show that the liquid surface slopes at an angle θ with the horizontal, where $\tan \theta = a/g$, *a* being the horizontal acceleration. (*b*) How does the pressure vary with *h*, the *vertical* depth below the surface?

28. The tension in a string holding a solid block below the surface of a liquid (of density greater than the solid) is T_0 when the containing vessel (Fig. 23) is at rest. Show that the tension *T*, when the vessel has an upward vertical acceleration *a*, is given by $T_0(1 + a/g)$.

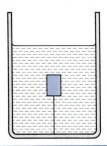

Figure 23 Problem 28.

29. (*a*) A fluid is rotating at constant angular velocity ω about the central vertical axis of a cylindrical container. Show that the variation of pressure in the radial direction is given by

$$\frac{dp}{dr} = \rho \omega^2 r.$$

(*b*) Take $p = p_c$ at the axis of rotation ($r = 0$) and show that the pressure *p* at any point *r* is

$$p = p_c + \tfrac{1}{2}\rho \omega^2 r^2.$$

(*c*) Show that the liquid surface is of paraboloidal form (Fig. 24); that is, a vertical cross section of the surface is the curve $y = \omega^2 r^2/2g$. (*d*) Show that the variation of pressure with depth is $p = \rho g h$.

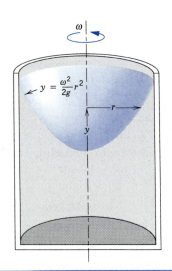

Figure 24 Problem 29.

Section 17-4 Pascal's Principle and Archimedes' Principle

30. (*a*) If the small piston of a hydraulic lever has a diameter of 3.72 cm, and the large piston one of 51.3 cm, what weight on the small piston will support 18.6 kN (e.g., a car) on the large piston? (*b*) Through what distance must the small piston move to raise the car 1.65 m?

31. A boat floating in fresh water displaces 35.6 kN of water. (*a*) What weight of water would this boat displace if it were floating in salt water of density 1024 kg/m³? (*b*) Would the volume of water displaced change? If so, by how much?

32. A block of wood floats in water with 0.646 of its volume submerged. In oil it has 0.918 of its volume submerged. Find the density of (*a*) the wood and (*b*) the oil.

33. A tin can has a total volume of 1200 cm³ and a mass of 130 g. How many grams of lead shot could it carry without sinking in water? The density of lead is 11.4 g/cm³.

34. About one-third of the body of a physicist swimming in the Dead Sea will be above the water line. Assuming that the human body density is 0.98 g/cm³, find the density of the water in the Dead Sea. Why is it so much greater than 1.0 g/cm³?

35. Assume the density of brass weights to be 8.0 g/cm³ and that of air to be 0.0012 g/cm³. What fractional error arises from neglecting the buoyancy of air in weighing an object of density 3.4 g/cm³ on a beam balance?

36. An iron casting containing a number of cavities weighs 6130 N in air and 3970 N in water. What is the volume of the cavities in the casting? The density of iron is 7870 kg/m³.

37. A cubic object of dimensions $L = 0.608$ m on a side and weight $W = 4450$ N in a vacuum is suspended by a wire in an open tank of liquid of density $\rho = 944$ kg/m³, as in Fig. 25. (*a*) Find the total downward force exerted by the liquid and the atmosphere on the top of the object. (*b*) Find the total upward force on the bottom of the object. (*c*) Find the tension in the wire. (*d*) Calculate the buoyant force on the object using Archimedes' principle. What relation exists among all these quantities?

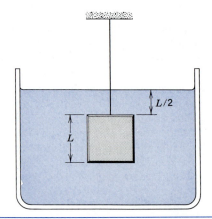

Figure 25 Problem 37.

38. A fish maintains its depth in seawater by adjusting the air content of porous bone or air sacs to make its average den-

sity the same as that of the water. Suppose that with its air sacs collapsed a fish has a density of 1.08 g/cm³. To what fraction of its expanded body volume must the fish inflate the air sacs to reduce its average density to that of the water? Assume that the air density is 0.00121 g/cm³.

39. It has been proposed to move natural gas from the North Sea gas fields in huge dirigibles, using the gas itself to provide lift. Calculate the force required to tether such an airship to the ground for off-loading when it arrives fully loaded with 1.17×10^6 m³ of gas at a density of 0.796 kg/m³. The density of the air is 1.21 kg/m³. (The weight of the airship is negligible by comparison.)

40. The Goodyear blimp *Columbia* (see Fig. 26) is cruising slowly at low altitude, filled as usual with helium gas. Its maximum useful payload, including crew and cargo, is 1280 kg. How much more payload could the *Columbia* carry if you replaced the helium with hydrogen? Why not do it? The volume of the helium-filled interior space is 5000 m³. The density of helium gas is 0.160 kg/m³ and the density of hydrogen is 0.0810 kg/m³.

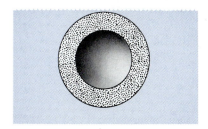

Figure 26   Problem 40.

41. A hollow spherical iron shell floats almost completely submerged in water; see Fig. 27. The outer diameter is 58.7 cm and the density of iron is 7.87 g/cm³. Find the inner diameter of the shell.

Figure 27   Problem 41.

42. A block of wood has a mass of 3.67 kg and a density of 594 kg/m³. It is to be loaded with lead so that it will float in water with 0.883 of its volume immersed. What mass of lead

is needed (*a*) if the lead is on top of the wood and (*b*) if the lead is attached below the wood? The density of lead is 1.14×10^4 kg/m³.

43. Three children each of weight 82.4 lb make a log raft by lashing together logs of diameter 1.05 ft and length 5.80 ft. How many logs will be needed to keep them afloat? Take the density of the wood to be 47.3 lb/ft³.

44. (*a*) What is the minimum area of a block of ice 0.305 m thick floating on water that will hold up an automobile of mass 1120 kg? (*b*) Does it matter where the car is placed on the block of ice? The density of ice is 917 kg/m³.

45. An object floating in mercury has one-fourth of its volume submerged. If enough water is added to cover the object, what fraction of its volume will remain immersed in mercury?

46. A cylindrical wooden log is loaded with lead at one end so that it floats upright in water as in Fig. 28. The length of the submerged portion is $L = 2.56$ m. The log is set into vertical oscillation. (*a*) Show that the oscillation is simple harmonic. (*b*) Find the period of the oscillation. Neglect the fact that the water has a damping effect on the motion.

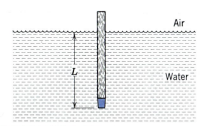

Figure 28   Problem 46.

47. A car has a total mass of 1820 kg. The volume of air space in the passenger compartment is 4.87 m³. The volume of the motor and front wheels is 0.750 m³, and the volume of the rear wheels, gas tank, and luggage is 0.810 m³. Water cannot enter these areas. The car is parked on a hill; the handbrake cable snaps and the car rolls down the hill into a lake; see Fig. 29. (*a*) At first, no water enters the passenger compartment. How much of the car, in cubic meters, is below the water surface with the car floating as shown? (*b*) As water slowly enters, the car sinks. How many cubic meters of water are in the car as it disappears below the water surface? (The car remains horizontal, owing to a heavy load in the trunk.)

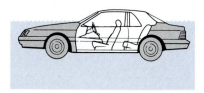

Figure 29   Problem 47.

48. You place a glass beaker, partially filled with water, in a sink (Fig. 30). It has a mass of 390 g and an interior volume of 500 cm³. You now start to fill the sink with water and you find, by experiment, that if the beaker is less than half full, it will float; but if it is more than half full, it remains on the

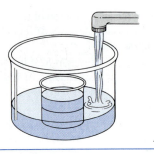

Figure 30   Problem 48.

bottom of the sink as the water rises to its rim. What is the density of the material of which the beaker is made?

Section 17-5 Measurement of Pressure

49. Estimate the density of the red wine that Pascal used in his 14-m-long barometer. Assume that the wine filled the tube.

50. The pressure at the surface of the planet Venus is 90 atm (i.e., 90 times the pressure at the surface of the Earth). How long would a mercury barometer have to be to measure this pressure? Assume that the mercury is maintained at 0°C.

Section 17-6 Surface Tension

51. How much energy is stored in the surface of a soap bubble 2.1 cm in radius if its surface tension is 4.5×10^{-2} N/m?

52. A thin film of water of thickness 80.0 μm is sandwiched between two glass plates and forms a circular patch of radius 12.0 cm. Calculate the normal force needed to separate the plates if the surface tension of water is 0.072 N/m.

53. Using a soap solution for which the surface tension is 0.025 N/m, a child blows a soap bubble of radius 1.40 cm. How much energy is expended in stretching the soap surface?

54. The surface tension of liquid ^{4}He is 0.35 mN/m and the liquid density is 145 kg/m³. Estimate (*a*) the number of atoms/m² in the surface and (*b*) the energy per bond, in eV,

in the liquid at this temperature. The mass of a helium atom is 6.64×10^{-27} kg. Picture each atom as a cube and assume that each atom interacts only with its four nearest neighbors.

55. Show that the pressure difference between the inside and the outside of a bubble of radius r is $4\gamma/r$, where γ is the surface tension of the liquid from which the bubble is blown.

56. A solid glass rod of radius $r = 1.3$ cm is placed inside and coaxial with a glass cylinder of internal radius $R = 1.7$ cm. Their bottom ends are aligned and placed in contact with, and perpendicular to, the surface of an open tank of water (see Fig. 31). To what height y will the water rise in the region between the rod and the cylinder? Assume that the angle of contact is 0° and use 72.8 mN/m for the surface tension of water.

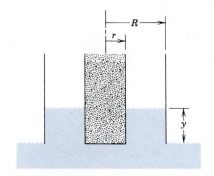

Figure 31   Problem 56.

57. A soap bubble in air has a radius of 3.20 cm. It is then blown up to a radius of 5.80 cm. Use 26.0 mN/m for the (constant) surface tension of the bubble. (*a*) What is the initial pressure difference across the bubble film? (*b*) Find the pressure difference across the film at the larger size. (*c*) How much work was done on the atmosphere in blowing up the bubble? (*d*) How much work was done in stretching the bubble surface?

CHAPTER 18

FLUID DYNAMICS

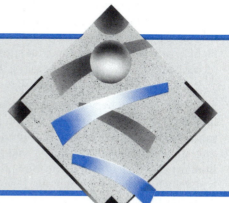

We now turn from fluid statics to the dynamics of fluids in motion. In principle, there is little difference between the dynamics of fluid motion and the dynamics of particle and rigid-body motions, which we have studied in previous chapters. We use familiar concepts to analyze fluid dynamics, including Newton's laws of motion and the conservation of mass and of energy. In this chapter we apply these principles to fluids, which are described using variables such as pressure and density that we introduced in Chapter 17.

We begin with a simplified model of fluid flow, in which we ignore dissipative forces. This approach is similar to our previous study of particle dynamics, in which we at first ignored dissipative (frictional) forces. An advantage of this approach is that it permits an analysis in terms of conservation of mechanical energy, as we did in Chapter 8 in the case of particles. Later in this chapter we give a brief description of the interesting and unusual results that occur in real fluids when dissipative forces, called viscous forces, *are taken into account.*

18-1 GENERAL CONCEPTS OF FLUID FLOW

One way of describing the motion of a fluid is to divide the fluid into infinitesimal volume elements, which we may call *fluid particles,* and to follow the motion of each particle. If we knew the forces acting on each fluid particle, we could then solve for the coordinates and velocities of each particle as functions of the time. This procedure, which is a direct generalization of particle mechanics, was first developed by Joseph Louis Lagrange (1736–1813). Because the number of fluid particles is generally very large, using this method is a formidable task.

There is a different treatment, developed by Leonhard Euler (1707–1783), that is more convenient for most purposes. In it we give up the attempt to specify the history of each fluid particle and instead specify the density and the velocity of the fluid at each point in space at each instant of time. This is the method we shall use. We describe the motion of the fluid by specifying the density $\rho(x,y,z,t)$ and the velocity $\mathbf{v}(x,y,z,t)$ at the point x,y,z at the time t. We thus focus our attention on what is happening at a particular point in space at a particular time, rather than on what is happening to a particular fluid particle. Any quantity used in describing the state of the

fluid, for example, the pressure p, will have a definite value at each point in space and at each instant of time. Although this description of fluid motion focuses attention on a point in space rather than on a fluid particle, we cannot avoid following the fluid particles themselves, at least for short time intervals dt. After all, the laws of mechanics apply to particles and not to points in space.

We first consider some general characteristics of fluid flow.

1. *Fluid flow can be steady or nonsteady.* We describe the flow in terms of the values of such variables as pressure, density, and flow velocity at every point of the fluid. If these variables are constant in time, the flow is said to be *steady.* The values of these variables will generally change from one point to another, but they do not change with time at any particular point. This condition can often be achieved at low flow speeds; a gently flowing stream is an example. In nonsteady flow, as in a tidal bore, the velocities $\mathbf{v}$ are functions of the time. In the case of *turbulent* flow, such as rapids or a waterfall, the velocities vary erratically from point to point as well as from time to time.

2. *Fluid flow can be compressible or incompressible.* If the density ρ of a fluid is a constant, independent of x, y, z, and t, its flow is called *incompressible flow.* Liquids can usually be considered as flowing incompressibly. But even for a highly compressible gas the variation in density may be insignificant, and for practical purposes we can consider its flow to be incompressible. For example, in flight at speeds much lower than the speed of sound in air (described by subsonic aerodynamics), the flow of the air over the wings is nearly incompressible.

3. *Fluid flow can be viscous or nonviscous.* Viscosity in fluid motion is the analogue of friction in the motion of solids. When a fluid flows such that no energy is dissipated through viscous forces, the flow is said to be *nonviscous*. In many cases, such as in lubrication problems, viscosity is extremely important; motor oils, for example, are rated according to their viscosity and its variation with temperature. In other cases, viscosity may be relatively unimportant, and by neglecting it we can use a simpler description in terms of nonviscous flow.

4. *Fluid flow can be rotational or irrotational.* If an element of the moving fluid does not rotate about an axis through the center of mass of the element, the flow is said to be *irrotational*. We can imagine a small paddle wheel immersed in the moving fluid (Fig. 1). If the wheel moves without rotating, the motion is irrotational; otherwise it is rotational. Note that a particular element of fluid can move in a circular path and still experience irrotational flow; an analogy is the motion of the hanging cars in a Ferris wheel — even though the wheel rotates, the people in the cars do not rotate about their center of mass. The vortex formed when water flows around a bathtub drain is an example of this kind of irrotational flow.

To simplify the mathematical description of fluid motion, we confine our discussion of fluid dynamics for the most part to steady, incompressible, nonviscous, irrotational flow. We run the danger, however, of making so many simplifying assumptions that we are no longer talking about a real fluid. Furthermore, it is sometimes difficult to decide whether a given property of a fluid — its viscosity, say — can be neglected in a particular situation. In spite of all this, the restricted analysis that we are going to give has wide application in practice, as we shall see.

18-2 STREAMLINES AND THE EQUATION OF CONTINUITY

In steady flow the velocity **v** at a given point is constant in time. Consider the point P (Fig. 2) within the fluid. Since **v** at P does not change in time in steady flow, every fluid particle arriving at P will pass on with the same speed in the same direction. The motion of every particle passing through P thus follows the same path, called a *streamline*. Every fluid particle that passes through P will later pass through points further along the streamline, such as Q and R in Fig. 2. Moreover, every fluid particle that passes through R must have previously passed through P and Q.

The magnitude of the velocity vector of the fluid particle will, in general, change as it moves along the streamline. The direction of the velocity vector at any point along the streamline is always tangent to the streamline.

No two streamlines can cross one another, for if they did, an oncoming fluid particle could go either one way or the other, and the flow could not be steady. In steady flow the pattern of streamlines does not change with time.

In principle we can draw a streamline through every point in the fluid. Assuming steady flow, we select a finite number of streamlines to form a bundle, like the streamline pattern of Fig. 3. This tubular region is called a *tube of flow*. The boundary of such a tube consists of streamlines to which the velocity of the fluid particles is always tangent. Thus no fluid can cross the boundaries of a tube of flow, and the tube behaves somewhat like a pipe of the same shape. The fluid that enters at one end must leave at the other.

Let us consider in detail the flow of fluid through a tube of flow shown in Fig. 4. Fluid enters at P where the cross-

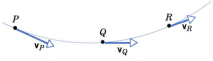

Figure 2   In steady flow, a fluid particle passing through P traces out a streamline, later passing through downstream points Q and R. Any other particle passing through P must follow this same path.

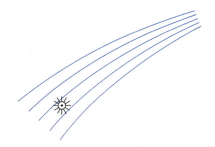

Figure 1   A small free-floating paddle wheel in a flowing liquid. If the wheel rotates, we call the flow *rotational*; if not, it is *irrotational*.

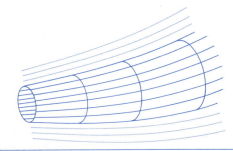

Figure 3   A bundle of streamlines forms a tube of flow.

sectional area is A_1 and leaves at Q where the area is A_2. Let the speed be v_1 for fluid particles at P and v_2 for fluid particles at Q. In the time interval Δt a fluid element travels approximately the distance $v\,\Delta t$. Then the fluid that crosses A_1 in the time interval Δt has a volume of approximately $A_1 v_1\,\Delta t$. If its density at that location is ρ_1, then the mass of fluid Δm_1 crossing A_1 is approximately

$$\Delta m_1 = \rho_1 A_1 v_1\,\Delta t.$$

The *mass flux*, defined as the mass of fluid per unit time passing through any cross section, is thus approximately $\Delta m_1/\Delta t = \rho_1 A_1 v_1$ at P. We must take Δt small enough so that in this time interval neither v nor A varies appreciably over the distance the fluid travels. In the limit as $\Delta t \to 0$, we obtain the precise result:

$$\text{mass flux at } P = \rho_1 A_1 v_1,$$

and, from a similar analysis,

$$\text{mass flux at } Q = \rho_2 A_2 v_2,$$

where ρ_2, A_2, and v_2 represent, respectively, the density, cross-sectional area, and flow velocity at Q.

We have assumed that fluid enters the tube only at P and leaves only at Q. That is, between P and Q there are no other "sources" where fluid can enter the tube nor "sinks" where it can leave. Furthermore, the flow is steady, so the density of fluid between P and Q doesn't change with time (even though it may change from place to place). Thus the mass flux at P must equal that at Q:

$$\rho_1 A_1 v_1 = \rho_2 A_2 v_2, \tag{1}$$

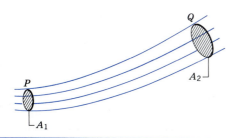

Figure 4 A tube of flow, which has cross-sectional area A_1 at P and area A_2 at Q.

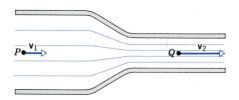

Figure 5 As the area of a horizontal tube narrows, the flow velocity must increase. If no other force acts on the fluid, the pressure at P must be greater than the pressure at Q, so that a force acts in the direction PQ to provide the necessary acceleration.

or, in more general terms referring to any location in the tube of flow,

$$\rho A v = \text{constant}. \tag{2}$$

This result expresses the *law of conservation of mass* in fluid dynamics.

If the fluid is incompressible, as we shall assume from now on, then $\rho_1 = \rho_2$, and Eq. 1 takes on the simpler form

$$A_1 v_1 = A_2 v_2, \tag{3}$$

or, defining R to be the *volume flow rate* (or *volume flux*) Av,

$$R = Av = \text{constant}. \tag{4}$$

The SI units of R are m^3/s. Note that Eq. 3 predicts that in steady incompressible flow the speed of flow varies inversely with the cross-sectional area, being larger in narrower parts of the tube.

The constancy of the volume flux along a tube of flow gives an important graphical interpretation to the streamlines, as shown in Fig. 5. In a narrow part of the tube, the streamlines must crowd closer together than in a wide part. Hence, as the distance between streamlines decreases, the fluid speed must increase. Therefore we conclude that widely spaced streamlines indicate regions of relatively low speed, and closely spaced streamlines indicate regions of relatively high speed.

We can obtain another interesting result by applying Newton's second law of motion to the flow of fluid between P and Q (Fig. 5). A fluid particle at P with speed v_1 must be accelerated in the forward direction in acquiring the higher forward speed v_2 at Q. This acceleration can come about only from a force exerted in the direction PQ, and (if there is no other external force, for instance, gravity) the force must arise from a change in pressure within the fluid. To provide this force, the pressure must be greater at P than at Q. Therefore, in the absence of other sources of acceleration, regions of higher fluid velocity must be associated with lower fluid pressure. We make this preliminary conclusion about fluid dynamics more rigorous in the next section.

Were you ever part of a "human fluid," in which a large crowd of people was pushing through a narrow doorway? Toward the back of the crowd, the cross-sectional area is large, the pressure is great, but the speed of advance is rather small. Once through the door, the crowd moves more rapidly—the flow velocity becomes greater. This "fluid" is compressible and viscous, and the flow may be rotational as well as turbulent!

Sample Problem 1 Figure 6 shows how the stream of water emerging from a faucet "necks down" as it falls. The cross-sectional area A_0 is 1.2 cm^2 and that of A is 0.35 cm^2. The two levels are separated by a vertical distance h (=45 mm). At what rate does water flow from the tap?

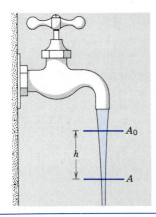

Figure 6 Sample Problem 1. As water falls from a tap, its speed increases. Because the flow rate must be the same at all cross sections, the stream must become narrower as it falls. (Effects associated with surface tension are neglected.)

Solution From the equality of the volume flux (Eq. 3) we have

$$A_0 v_0 = A v,$$

where v_0 and v are the water velocities at the corresponding levels. From Eq. 20 of Chapter 2 we can also write, because each water element is falling freely under gravity,

$$v^2 = v_0^2 + 2gh.$$

Eliminating v between these two equations and solving for v_0, we obtain

$$v_0 = \sqrt{\frac{2ghA^2}{A_0^2 - A^2}} = \sqrt{\frac{(2)(9.8 \text{ m/s}^2)(0.045 \text{ m})(0.35 \text{ cm}^2)^2}{(1.2 \text{ cm}^2)^2 - (0.35 \text{ cm}^2)^2}}$$

$$= 0.286 \text{ m/s} = 28.6 \text{ cm/s}.$$

The volume flow rate R is then

$$R = A_0 v_0 = (1.2 \text{ cm}^2)(28.6 \text{ cm/s})$$

$$= 34 \text{ cm}^3/\text{s}.$$

At this rate, it would take about 3 s to fill a 100-mL beaker.

The Equation of Continuity *(Optional)*

Equations 2 and 4 are examples of mathematical relationships known as *equations of continuity.* An equation of continuity is in effect a conservation law for matter. It says that if there are no sources (places where new matter is introduced) or sinks (places where flowing matter can escape), then the mass dm contained in any volume dV must remain constant. We derived these equations for very special conditions; a more general equation of continuity is

$$\frac{\partial(\rho v_x)}{\partial x} + \frac{\partial(\rho v_y)}{\partial y} + \frac{\partial(\rho v_z)}{\partial z} + \frac{\partial \rho}{\partial t} = 0. \qquad (5)$$

This equation is based on considering an arbitrary volume of *space* (not of fluid) $dV = dx\, dy\, dz$. The volume element remains fixed in space while the fluid flows through it. The first three terms, when multiplied by dV, give the net outflow of fluid mass from the volume, in terms of its velocity components v_x, v_y, and v_z. The fourth term, again multiplied by dV, gives the rate at

which mass is changing within the volume element. When these terms add to zero, it means that any net outflow (or inflow) of mass must be compensated by an equivalent change in the mass within the element. This is again just a statement of conservation of mass. If the volume element dV contains sources or sinks, these would be included in Eq. 5 by making the right-hand side equal to the rate at which matter is entering or leaving dV. (The derivatives that appear in Eq. 5 are partial derivatives, because the density and the velocity components may be functions of more than one variable.)

Equations of continuity are common in physics and play a fundamental role not only in fluid mechanics but in any subject in which a flow is involved. In electromagnetism, for example, we consider not the flow of mass but the flow of electric charge. The components of **v** in Eq. 5 are replaced with corresponding components of the electric current, while the mass density is replaced with the charge density. The interpretation of the equation is unchanged, except that it then refers to the conservation of electric charge rather than mass. (See Chapters 27 and 32.) ∎

18-3 BERNOULLI'S EQUATION*

Bernoulli's equation, which is a fundamental relation in fluid mechanics, is not a new principle but is derivable from the basic laws of Newtonian mechanics. We find it convenient to derive it from the work–energy theorem (see Section 7-4), for it is essentially a statement of the work–energy theorem for fluid flow.

Consider the steady, incompressible, nonviscous, and irrotational flow of a fluid through the pipeline or tube of flow in Fig. 7. The portion of pipe shown in the figure has a uniform cross section A_1 at the left. It is horizontal there at an elevation y_1 above some reference level. It gradually widens and rises and at the right has a uniform cross section A_2. It is horizontal there at an elevation y_2. Let us concentrate on the portion of fluid represented by both the gray and blue shadings and call this fluid the "system." Consider then the motion of the system from the position shown in Fig. 7a to that in Fig. 7b. At all points in the narrow part of the pipe the pressure is p_1 and the speed v_1; at all points in the wide portion the pressure is p_2 and the speed v_2.

The work–energy theorem (see Eq. 19 of Chapter 7) states: the work done by the resultant force acting on a system is equal to the change in kinetic energy of the system. In Fig. 7 the forces that do work on the system, assuming that we can neglect viscous forces, are the pressure forces $p_1 A_1$ and $p_2 A_2$ that act on the left- and right-hand ends of the system, respectively, and the force of gravity. As fluid flows through the pipe the net effect, as a

---

* There are eight Bernoullis listed in the *Encyclopaedia Britannica* (eleventh edition). We refer here to Daniel Bernoulli (1700–1782), perhaps the most renowned member of this famous family.

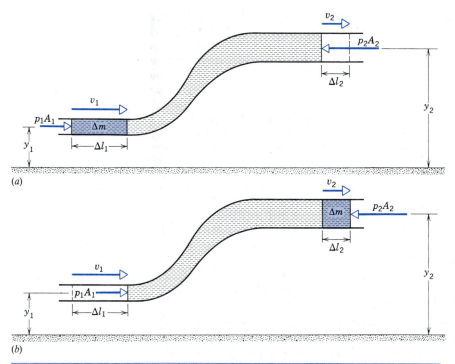

Figure 7   Fluid flows through a pipe at a steady rate. During the interval from (*a*) to (*b*), the net effect of the flow is the transfer of the element of fluid indicated by the blue shading from the input end of the tube to the output end.

comparison of Figs. 7*a* and 7*b* shows, it to raise the fluid represented by the area shaded in blue in Fig. 7*a* to the position shown in Fig. 7*b*. The amount of fluid represented by the gray shading is unchanged by the flow.

We can find the work W done on the system by the resultant force as follows:

1. The work done on the system by the pressure force p_1A_1 is $p_1A_1 \Delta l_1$.

2. The work done on the system by the pressure force p_2A_2 is $-p_2A_2 \Delta l_2$. Note that it is negative, because the force acts in a direction opposite to the horizontal displacement.

3. The work done on the system by gravity is associated with lifting the fluid element shaded in blue from height y_1 to height y_2 and is $-\Delta m\, g(y_2 - y_1)$ in which Δm is the mass of fluid in either area with blue shading. This contribution is also negative because the gravitational force acts in a direction opposite to the vertical displacement.

The net work W done on the system by all the forces is found by adding these three terms, or

$$W = p_1A_1 \Delta l_1 - p_2A_2 \Delta l_2 - \Delta m\, g(y_2 - y_1).$$

Now $A_1 \Delta l_1 (= A_2 \Delta l_2)$ is the volume ΔV of the fluid element shaded in blue, which we can write as $\Delta m/\rho$, in which ρ is the (constant) fluid density. Recall that the two fluid elements have the same mass, so that in setting

$A_1 \Delta l_1 = A_2 \Delta l_2$ we have assumed the fluid to be incompressible. With this assumption we have

$$W = (p_1 - p_2)(\Delta m/\rho) - \Delta m\, g(y_2 - y_1). \quad (6)$$

The change in kinetic energy of the fluid element is

$$\Delta K = \tfrac{1}{2}\Delta m\, v_2^2 - \tfrac{1}{2}\Delta m\, v_1^2.$$

From the work–energy theorem, $W = \Delta K$, we then have

$$(p_1 - p_2)(\Delta m/\rho) - \Delta m\, g(y_2 - y_1)$$
$$= \tfrac{1}{2}\Delta m\, v_2^2 - \tfrac{1}{2}\Delta m\, v_1^2, \quad (7)$$

which, after canceling the common factor of Δm, can be rearranged to read

$$p_1 + \tfrac{1}{2}\rho v_1^2 + \rho g y_1 = p_2 + \tfrac{1}{2}\rho v_2^2 + \rho g y_2. \quad (8)$$

Since the subscripts 1 and 2 refer to any two locations along the pipeline, we can drop the subscripts and write

$$p + \tfrac{1}{2}\rho v^2 + \rho g y = \text{constant.} \quad (9)$$

Equation 9 is called *Bernoulli's equation* for steady, incompressible, nonviscous, and irrotational flow. It was first presented by Daniel Bernoulli in his *Hydrodynamica* in 1738.

Bernoulli's equation is strictly applicable only to steady flow, the quantities involved being evaluated along a streamline. In our figure the streamline used is along the axis of the pipeline. If the flow is irrotational, however, it can be shown (see Problem 33 for a special case) that the constant in Eq. 9 is the same for *all* streamlines.

Just as the statics of a particle is a special case of particle dynamics, so fluid statics is a special case of fluid dynamics. It should come as no surprise therefore that the law of pressure change with height in a fluid at rest is included in Bernoulli's equation as a special case. If the fluid is at rest, then $v_1 = v_2 = 0$ and Eq. 8 becomes

$$p_1 + \rho g y_1 = p_2 + \rho g y_2$$

or

$$p_2 - p_1 = -\rho g(y_2 - y_1),$$

which is the same as Eq. 8 of Chapter 17.

Another basic result follows from Eq. 8 when $y_1 = y_2$ (that is, the tube is horizontal, so that gravitational effects need not be considered). Thus

$$p_1 + \tfrac{1}{2}\rho v_1^2 = p_2 + \tfrac{1}{2}\rho v_2^2. \qquad (10)$$

Where the speed is large, the pressure must be small, and conversely. This is the mathematical statement confirming the conclusion of the discussion following Eq. 4. In Eq. 9 all terms have the dimension of a pressure (which you should verify). The pressure $p + \rho g y$, which would be present even if there were no flow ($v = 0$), is called the *static pressure;* the term $\tfrac{1}{2}\rho v^2$ is called the *dynamic pressure.*

Bernoulli's equation is in effect a statement about the conservation of mechanical energy in a system. In analogy to our treatment of conservation of energy in Chapter 8, we can rewrite Eq. 7 as

$$\Delta K + \Delta U = W,$$

where the three terms, respectively, refer to the changes in kinetic and potential energy and the work done by the pressure force, all quantities being taken as per unit volume of the fluid. If the fluid is compressible, it can acquire internal energy by mechanical means; for example, the pressure force can push the molecules closer together, thereby increasing their internal potential energy. Thus for a compressible fluid, we should include another term ΔE_{int} corresponding to the change in internal energy per unit volume, and the statement of conservation of energy becomes

$$\Delta K + \Delta U + \Delta E_{\text{int}} = W.$$

This result is identical with Eq. 28 of Chapter 8. If in addition the flow is viscous, the friction-like forces do work that may appear as an increase in the internal energy of the fluid.

In practice, we can modify Bernoulli's equation as necessary to account for the conversion of mechanical energy of the fluid into internal energy. If the flow can be regarded as approximately incompressible and nonviscous, these corrections are negligible.

Sample Problem 2   A storage tower of height $h = 32$ m and diameter $D = 3.0$ m supplies water to a house (Fig. 8). A horizontal pipe at the base of the tower has a diameter $d = 2.54$ cm

($= 1$ in., typical of the supply pipes for many homes in the United States). To satisfy the needs of the home, the supply pipe must be able to deliver water at a rate $R = 0.0025$ m³/s (about $\tfrac{2}{3}$ of a gallon per second). (*a*) If water were flowing at the maximum rate, what would be the pressure in the horizontal pipe? (*b*) A smaller pipe, of diameter $d' = 1.27$ cm ($= 0.5$ in.) supplies the second floor of the house, a distance of 7.2 m above the ground level. What are the flow speed and water pressure in this pipe? Neglect the viscosity of the water.

Solution   (*a*) We apply Bernoulli's equation along the streamline *ABC* shown in Fig. 8. At points *A* and *B* we have

$$p_A + \tfrac{1}{2}\rho v_A^2 + \rho g y_A = p_B + \tfrac{1}{2}\rho v_B^2 + \rho g y_B.$$

At *A*, the pressure is that of the atmosphere, p_0. With $y_A = h$ and $y_B = 0$, we obtain, for the unknown pressure,

$$p_B = p_0 + \rho g h + \tfrac{1}{2}\rho(v_A^2 - v_B^2).$$

We can find v_A and v_B from the equality of the volume flux (Eq. 4), which gives

$$v_A A_A = v_B A_B = R,$$

where R is the constant volume flow rate. Thus

$$v_A = \frac{R}{A_A} = \frac{0.0025 \text{ m}^3/\text{s}}{\pi(1.5 \text{ m})^2} = 3.5 \times 10^{-4} \text{ m/s},$$

$$v_B = \frac{R}{A_B} = \frac{0.0025 \text{ m}^3/\text{s}}{\pi(0.0127 \text{ m})^2} = 4.9 \text{ m/s}.$$

Note that the term $\tfrac{1}{2}\rho v_A^2$ in the expression for p_B is negligible compared with the term $\tfrac{1}{2}\rho v_B^2$. That is, the flow velocity at the top of the tank is quite small, owing to its large cross-sectional area.

We can now solve for the pressure in the pipe:

$$
\begin{aligned}
p_B &= p_0 + \rho g h - \tfrac{1}{2}\rho v_B^2 \\
&= 1.01 \times 10^5 \text{ Pa} + (1.0 \times 10^3 \text{ kg/m}^3)(9.8 \text{ m/s}^2)(32 \text{ m}) \\
&\quad - \tfrac{1}{2}(1.0 \times 10^3 \text{ kg/m}^3)(4.9 \text{ m/s})^2 \\
&= 1.01 \times 10^5 \text{ Pa} + 3.14 \times 10^5 \text{ Pa} - 0.12 \times 10^5 \text{ Pa} \\
&= 4.03 \times 10^5 \text{ Pa}.
\end{aligned}
$$

If the water in the horizontal pipe were not flowing (that is, if the valve were closed), the *static* pressure at *B* would include only the first two terms above, which give 4.15×10^5 Pa. The pressure when the water is flowing is reduced from this static value by the amount of the dynamic pressure.

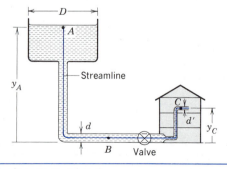

Figure 8   Sample Problem 2.

(b) If the narrower pipe to the second floor is to have the same flow rate R, the velocity at C must be

$$v_C = \frac{R}{A_C} = \frac{0.0025 \text{ m}^3/\text{s}}{\pi(0.0064 \text{ m})^2} = 19.7 \text{ m/s},$$

or four times the value at B. Bernoulli's equation gives

$$p_A + \tfrac{1}{2}\rho v_A^2 + \rho g y_A = p_C + \tfrac{1}{2}\rho v_C^2 + \rho g y_C$$

or

$$
\begin{aligned}
p_C &= p_0 + \tfrac{1}{2}\rho(v_A^2 - v_C^2) + \rho g(y_A - y_C) \\
&= 1.01 \times 10^5 \text{ Pa} - \tfrac{1}{2}(1.0 \times 10^3 \text{ kg/m}^3)(19.7 \text{ m/s})^2 \\
&\quad + (1.0 \times 10^3 \text{ kg/m}^3)(9.8 \text{ m/s}^2)(32 \text{ m} - 7.2 \text{ m}) \\
&= 1.01 \times 10^5 \text{ Pa} - 1.95 \times 10^5 \text{ Pa} + 2.43 \times 10^5 \text{ Pa} \\
&= 1.49 \times 10^5 \text{ Pa}.
\end{aligned}
$$

Because of the larger flow velocity through the smaller pipe, the dynamic contribution to the pressure is much larger at C than it is at B. Both the static and dynamic effects tend to reduce the pressure at this location relative to B.

18-4   APPLICATIONS OF BERNOULLI'S EQUATION AND THE EQUATION OF CONTINUITY

In this section, we consider a number of applications of Bernoulli's equation, which illustrate its use and demonstrate the range of its applicability.

The Venturi Meter

This device (Fig. 9) is a gauge to measure the flow speed of a fluid in a pipe. A fluid of density ρ flows through a pipe of cross-sectional area A. At the throat the area is reduced to a, and a manometer tube is attached, as shown. Let the manometer liquid, such as mercury, have a density ρ'. By applying Bernoulli's equation and the equality of the vol-

ume flux at points 1 and 2, you can show (see Problem 31) that the speed of flow at point 1 is

$$v = a\sqrt{\frac{2(\rho' - \rho)gh}{\rho(A^2 - a^2)}}. \tag{11}$$

The Pitot Tube

This device (Fig. 10) is used to measure the flow speed of a gas. Consider the gas—say, air—flowing with density ρ and velocity $\mathbf{v}_a$ parallel to the planes of the openings at a. The pressure in the left arm of the manometer, which is connected to these openings, is then the static pressure in the gas stream, p_a. The opening of the right arm of the manometer is at right angles to the stream. The velocity is reduced to zero at b, and the gas is stagnant at that point. Applying Bernoulli's equation to points a and b, we obtain

$$p_a + \tfrac{1}{2}\rho v_a^2 = p_b.$$

Substituting the manometer reading $\rho' gh$ for the pressure difference $p_b - p_a$, we can solve for v_a to obtain

$$v_a = \sqrt{\frac{2gh\rho'}{\rho}}. \tag{12}$$

This device can be calibrated to read v_a directly. The air-speed indicator found on airplane wingtips is based on this principle.

Dynamic Lift

Dynamic lift is the force that acts on a body, such as an airplane wing, a hydrofoil, or a helicopter rotor, by virtue of its motion through a fluid. It is not the same as static lift, which is the buoyant force that acts on a balloon or an iceberg in accord with Archimedes' principle (Section 17-4).

Familiar examples of dynamic lift occur in the flight of a baseball, tennis ball, or golf ball. The dynamic lift, which originates with the rotation of the ball in flight, can cause the ball to curve or to rise or fall relative to a para-

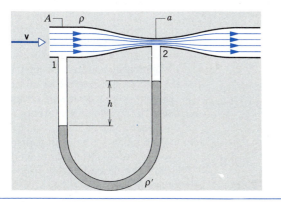

Figure 9   A Venturi meter, used to measure the speed of flow of a fluid in a pipe.

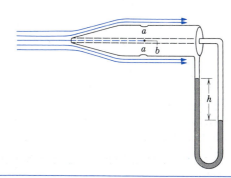

Figure 10   A Pitot tube, which is used to measure the flow speed of a gas.

bolic trajectory. Because the fluid (in this case air) is somewhat viscous, there is friction as the ball travels, and the ball tends to carry with it a thin layer of fluid called the *boundary layer.* Viewed from the rest frame of a nonrotating ball, the fluid speed drops from its value beyond the boundary layer (equal to the flight speed of the ball) to zero at the surface of the ball.

Figure 11*a* shows, in the rest frame of the ball, streamlines for the steady flow of air rushing past a nonrotating ball, at speeds low enough so that turbulence does not occur. Figure 11*b* shows streamlines for the air carried around by a rapidly rotating ball. Without viscosity and the boundary layer, the spinning ball could not carry air around in this way and this circulation (as it is called) would not exist. Golf balls are systematically roughened by means of dimples to increase this circulation and the dynamic lift that results from it. Baseballs are sometimes artificially (and illegally) roughened by pitchers for the same reason.

Figure 11*c* shows the effect of combining the circulation (resulting from the rotation of the ball) and the steady flow (resulting from the translation of the ball through the air). For the case shown, the two velocities add above the ball and subtract below. From the spacing of the resultant streamlines, we see that the velocity of air below the ball is less than that above the ball. From Bernoulli's equation, the pressure of air below the ball must then be greater than that above, so the ball experiences a dynamic lift force.

A pitched baseball curves for essentially the same reason. For example, if Fig. 11 represents a top view of the spinning ball as it travels toward the batter, the "lift" acts in a sidewise direction to move the ball horizontally toward or away from the batter, as in the case of a curveball. If Fig. 11 represents a side view, the ball is thrown with backspin, as in the case of a fastball. The lift acts upward, causing the ball to rise relative to its parabolic trajectory.

The lift acting on an airplane wing has a similar explanation. Figure 12 shows the streamlines about an airfoil (or wing cross section) attached to an aircraft. Let us choose the aircraft as our frame of reference, as in a wind tunnel

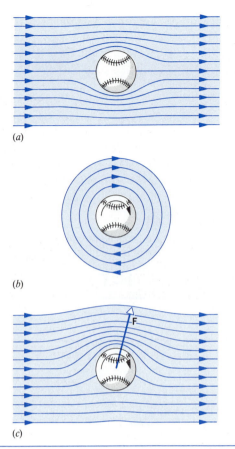

(a)

(b)

(c)

Figure 11 (*a*) Streamline flow around a nonrotating ball. (*b*) The circulation of air around a rotating ball, which results from the boundary layer. The thickness of the boundary layer has been greatly exaggerated. (*c*) The combined effects of both motions. From Bernoulli's equation, we see that a dynamic lift acts upward on the ball. The fluid exerts on the ball a net force **F** having a component transverse to the fluid flow (lift) and a component parallel to the fluid flow (drag).

experiment, and let us assume that the air is moving past the wing from left to right. Note the similarities between Figs. 12 and 11*c*. (In fact, the explanation of the lift on an airplane wing involves a circulation similar to Fig. 11*b*.)

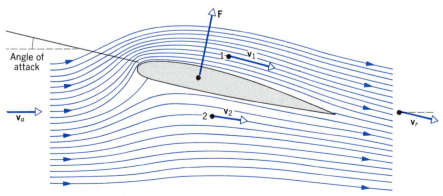

Figure 12 Streamlines around an airfoil or airplane wing. The velocity v_a of the approaching air is horizontal, while the air receding from the airfoil has a velocity v_r with a downward component. The airfoil has thus exerted a downward force on the air, and by Newton's third law the air must therefore have exerted an upward force on the airfoil. This upward force is represented by the lift **F**.

The *angle of attack* of the wing causes air to be deflected downward. From Newton's third law the reaction of this downward force of the wing on the air is an upward force **F**, the lift, exerted by the air on the wing.

Although Bernoulli's equation alone cannot be used to predict the streamline pattern around an airfoil, that equation can be used to verify that a lift force is exerted on the airfoil. Above the wing (point 1) the streamlines are closer together than they are below the wing (point 2). Thus $v_1 > v_2$ and, from Bernoulli's principle, $p_1 < p_2$, which must be true if there is to be a lift.*

Thrust on a Rocket

As our final example let us compute the thrust on a rocket produced by the escape of its exhaust gases. Consider a chamber (Fig. 13) of cross-sectional area A filled with a gas of density ρ at a pressure p. Let there be a small orifice of cross-sectional area A_0 at the bottom of the chamber. We wish to find the speed v_0 with which the gas escapes through the orifice.

Let us write Bernoulli's equation (Eq. 8) as

$$p - p_0 = \rho g(y_0 - y) + \tfrac{1}{2}\rho(v_0^2 - v^2),$$

where p_0 represents atmospheric pressure just outside the orifice. For a gas the density is so small that we can neglect the variation in pressure with height in a chamber, which gives

$$p - p_0 = \tfrac{1}{2}\rho(v_0^2 - v^2)$$

or

$$v_0^2 = \frac{2(p - p_0)}{\rho} + v^2, \tag{13}$$

where v is the speed of the flowing gas inside the chamber and v_0 is the speed of the gas through the orifice. Although a gas is compressible and the flow may become turbulent, we can treat the flow as steady and incompressible for pressure and exhaust speeds that are not too high.

Now let us assume continuity of mass flow (in a rocket engine this is achieved when the mass of escaping gas equals the mass of gas created by burning the fuel), so that (for an assumed constant density)

$$Av = A_0 v_0.$$

---

* For more information on how airplanes fly, see "The Science of Flight," by Peter P. Wegener, *American Scientist,* May–June 1986, p. 268. Also see "Bernoulli's Law and Aerodynamic Lifting Force," by Klaus Weltner, *The Physics Teacher,* February 1990, p. 84. Dynamic lift is discussed in "Physics and Sports: The Aerodynamics of Projectiles," by Peter J. Brancazio, in *Fundamentals of Physics,* 3rd ed., by David Halliday and Robert Resnick (Wiley, 1988). Dynamic lift can also be used to provide a horizontal force that propels a ship; see "The Flettner Ship," by Albert Einstein, in *Essays in Science* (Philosophical Library, 1955), p. 92.

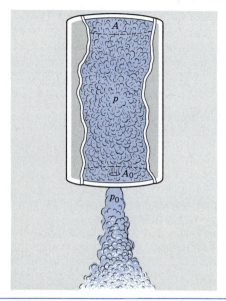

Figure 13   Fluid streaming out of a chamber, which might represent the exhaust chamber of a rocket.

If the orifice is very small so that $A_0 \ll A$, then $v_0 \gg v$, and we can neglect v^2 compared to v_0^2 in Eq. 13. Hence the exhaust speed is

$$v_0 = \sqrt{\frac{2(p - p_0)}{\rho}}. \tag{14}$$

If our chamber is the exhaust chamber of a rocket, the thrust on the rocket (Section 9-8) is $v_0\, dM/dt$. But the mass of gas flowing out in time dt is $dM = \rho A_0 v_0\, dt$, so that

$$v_0 \frac{dM}{dt} = v_0 \rho A_0 v_0 = \rho A_0 v_0^2,$$

and using Eq. 14 the thrust is

$$v_0 \frac{dM}{dt} = 2A_0(p - p_0). \tag{15}$$

18-5 FIELDS OF FLOW *(Optional)*

In Section 16-7 we discussed how to represent the situation near masses by use of a gravitational field. Each point in the field can be regarded as having a vector associated with it, namely, **g**, the gravitational force per unit mass at that point. We can make a graphical representation of the field by drawing lines in the direction of the field whose spacing is proportional to the strength of the field. We use a similar approach in our study of the electric field in Chapter 28.

Likewise, in fluid dynamics we can summarize the situation within a moving fluid by means of a field of flow. In general, the field of flow is a *vector* field. We associate a vector quantity with each point in space, namely, the flow velocity **v** at that point. For a steady flow the field of flow is stationary. Of course, even in this case a particular fluid particle may still have a variable velocity as it moves from point to point in the field. The field gives some of

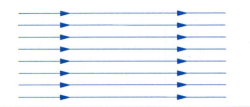

Figure 14 Streamlines (horizontal lines) for a homogeneous nonviscous field of flow.

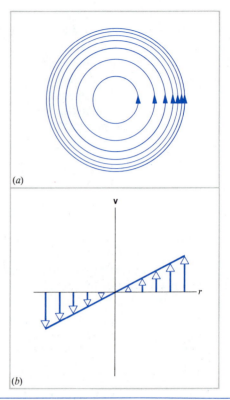

Figure 15 (*a*) A uniform rotational field of flow. (*b*) Variation of fluid velocity from the center.

the properties of the flow, and we can use it to deduce the behavior of the moving particles.*

A flowing fluid mass can always be divided into tubes of flow. When the flow is steady, the tubes remain unchanged in shape, and the fluid that is at one instant in a tube remains inside the same tube thereafter. We have seen that the flow velocity inside a tube of flow is parallel to the tube and has a magnitude inversely proportional to the area of the cross section (Eq. 2). Let us assign cross sections to the tubes so that the constant of proportionality is the same for all of them; if possible we take this constant to be unity. That is, the volume flux is the same for all tubes, namely, unit flux. Then the magnitude of the flow velocity can be determined from the areas of the cross sections of the tubes of flow. There is another procedure equivalent to this which consists of setting up a unit area perpendicular to the direction of flow and drawing through it just as many streamlines as the number of units of magnitude of the velocity at that point.

Let us consider some examples of fields of flow. For drawing purposes we consider only two-dimensional examples. In these the flow velocity is the same at all points on a line perpendicular to the plane at any point.

In Fig. 14 we have drawn a *homogeneous field of flow,* such as might occur in the steady, nonviscous flow of a liquid through a pipe with smooth interior walls. Here all the streamlines are parallel, and the flow velocity **v** is the same at all points.

In Fig. 15 we show the field for *uniform rotational flow,* such as might be produced by rotating a bucket of water on a turntable (see Problem 29, Chapter 17). Here v is proportional to r, because the angular velocity ω is constant. In Fig. 16 we draw the field of flow of a *vortex,* such as might be obtained by pulling the plug in a bathtub full of water. In this case v is proportional to $1/r$, because the angular momentum $L = mvr$ is constant, and the flow is irrotational (see Problem 36). Note that both uniform rotation and vortex motion are represented by circular streamlines but are entirely different kinds of flow. Obviously, the shapes of the streamlines give only limited information; their spacing is needed too.

Figure 17 represents the field of flow for a source. All streamlines are directed radially outward. The source is a line through the center perpendicular to the paper. The strength of a source is specified by giving the mass per unit time it emits. The field of

---

* If the flow is irrotational as well as steady, we call it *potential flow.* Then the flow velocity **v** can be related to a velocity potential ψ, just as in gravitation **g** can be related to the gravitational potential V (see Section 16-7). Hence a vector field for potential flow is analogous to a conservative force field.

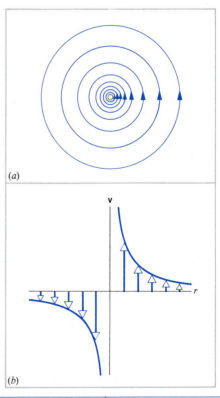

Figure 16 (*a*) Field of flow of a vortex. (*b*) Variation of fluid velocity from the center.

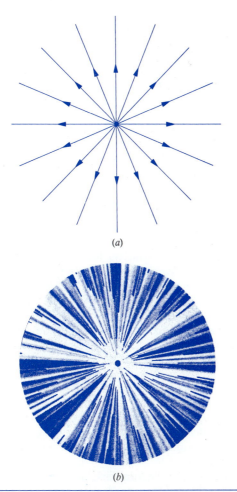

(a)

(b)

Figure 17   (a) Flow from a linear source. (b) Map of fluid flow from a linear source. The map is made by allowing water to flow between a horizontal layer of plate glass and a horizontal layer of plaster. The water comes through a hole in the center and flows out toward the edges. The direction of flow is made visible by sprinkling the plaster with crystals of potassium permangenate, which dissolve in the water and color it purple. The fluid flow map was made and photographed by Professor A. D. Moore at the University of Michigan and is taken from *Introduction to Electric Fields,* by W. E. Rogers (McGraw-Hill, 1954).

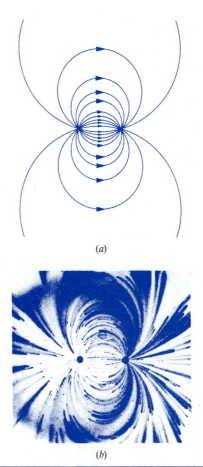

(a)

(b)

Figure 18   (a) Linear dipole flow. The source is on the left and the sink is on the right. (b) A fluid flow map of a linear dipole, made as described for Fig. 17.

flow around a linear sink is the same as that of a source except for the sign of the flow, which is directed radially inward.

For a linear source and linear sink which have the same strengths and are slightly separated, we obtain the combined field called linear dipole flow, shown in Fig. 18.

As we shall see later the electrostatic field, the magnetic field, and the field of flow for an electric current are also vector fields. In this connection, the homogeneous field (Fig. 14) corresponds to the electric field of a plane capacitor, the source field or sink field (Fig. 17) corresponds to the electric field of a cylindrical capacitor or straight wire of positive or negative charge, respectively, and the linear dipole field (Fig. 18) corresponds to the

electric field of two oppositely charged wires. In all these the field of flow is potential flow, and the electric fields are conservative.

The homogeneous field of Fig. 14 also represents the magnetic field inside a solenoid. The vortex field of Fig. 16 represents the magnetic field around a straight current-carrying wire. This last is an example of a field that is rotational (about the vortex axis).

Because of these analogies between fluid and electromagnetic fields, we can often determine a field of flow, which is difficult to calculate by present mathematical methods, by experimental measurements on appropriate electrical devices.

As we have seen throughout this chapter, the basic field ideas and conservation principles find application in many areas of physics. We shall encounter them many times again.   ■

18-6 VISCOSITY, TURBULENCE, AND CHAOTIC FLOW   *(Optional)*

Viscosity in fluid flow is similar to friction in the motion of solid bodies. When we slide one solid body over another, we must supply an external force **F** to oppose the frictional force **f** if we wish to keep the body in motion at constant velocity. In the case of fluid motion, we can consider a fluid between two parallel

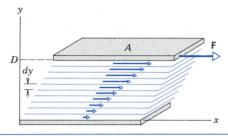

Figure 19   A viscous fluid fills the space between two flat plates separated by a distance D. The lower plate is at rest and the upper plate is pulled to the right with a constant force **F**. The velocity of each layer of fluid decreases uniformly from the upper plate to the lower plate.

plates, illustrated in Fig. 19. A force **F** is applied to the upper plate, so that it is in motion at constant velocity **v** relative to the lower plate, which we assume to be at rest. The force **F** opposes the viscous drag on the upper plate to keep its velocity constant.

The fluid can be imagined to be divided into layers parallel to the plates. Viscosity acts not only between the fluid and the upper plate, but between each layer of fluid and the adjacent layers. The speed of each layer differs by an amount dv from the one below it. Fluid flow in which the speed varies layer-by-layer is called *laminar* flow. For this discussion, we assume that the top fluid layer has the same speed v as the top plate and the bottom fluid layer has the same speed as the bottom plate, namely, zero.

In analogy with the shearing stress applied to a solid (Section 14-5), we can define the shearing stress on the fluid to be F/A, where A is the area of the layer of fluid. A solid can respond to this shearing stress with a change in its shape (the shearing strain, which is a lateral displacement across each layer), but a fluid responds through motion, that is, through a change in speed dv across each layer of thickness dy. The ratio between stress and strain in the fluid is called the *coefficient of viscosity η* (Greek letter eta) of the fluid:

$$\eta = \frac{F/A}{dv/dy}. \qquad (16)$$

Under our assumption that the top layer moves at speed v and the bottom layer at $v = 0$, the *velocity gradient dv/dy* is simply v/D, where D is the spacing between the plates. Thus

$$\eta = \frac{F/A}{v/D} = \frac{FD}{vA}. \qquad (17)$$

TABLE 1   VISCOSITIES OF SELECTED FLUIDS

Fluid	η (N·s/m²)
Glycerine (20°C)	1.5
Motor oil[a] (0°C)	0.11
Motor oil[a] (20°C)	0.03
Blood (37°C)	4.0×10^{-3}
Water (20°C)	1.0×10^{-3}
Water (90°C)	0.32×10^{-3}
Gasoline (20°C)	2.9×10^{-4}
Air (20°C)	1.8×10^{-5}
CO_2 (20°C)	1.5×10^{-5}

[a] Medium weight (S.A.E. 30).

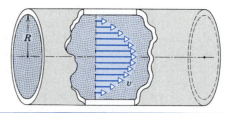

Figure 20   Fluid flows through a cylindrical pipe of radius R. The variation in the velocity from the wall to the center is shown.

The SI unit of viscosity is the N·s/m². The equivalent cgs unit is the dyne·s/cm², which is called the *poise*. (The unit is named for the French physician Jean-Louis-Marie Poiseuille, who first investigated the flow of viscous fluids through tubes, as an aid in understanding the circulation of blood.) Comparing the units shows that 1 poise = 0.1 N·s/m². Table 1 shows some typical values of the viscosities of fluids.

A practical application of viscosity occurs in the fluid flow through cylindrical pipes. The flow is again laminar, but in this case the layers of fluid are thin-walled cylinders of varying radii. The flow velocity varies with the radius; its maximum value occurs on the axis and its minimum value, which we assume to be zero, at the walls (Fig. 20). Note that the flow illustrated in Fig. 20 is rotational, even though the fluid elements travel in straight lines. If we were to place a small paddlewheel anywhere in the flow except along the central streamline, it would be set into rotation because of the variation in velocity of the fluid particles striking its vanes.

In the case of a cylindrical pipe, as shown in Fig. 20, the variation of the velocity with location across the pipe is not linear. Assuming once again that the layer next to the walls is at rest, the velocity in the cylindrical shell of radius r can be shown to be (see Problem 41)

$$v = v_0 \left(1 - \frac{r^2}{R^2}\right), \qquad (18)$$

where v_0 is the velocity at the center of the pipe. In terms of the pressure difference Δp across the length L of the pipe, the central velocity is

$$v_0 = \frac{\Delta p \, R^2}{4\eta L}. \qquad (19)$$

By considering the flow through each thin cylindrical shell, we can show (see Problem 42) that the total mass flux dm/dt (fluid mass flowing through the pipe per unit time) is

$$\frac{dm}{dt} = \frac{\rho \pi R^4 \, \Delta p}{8\eta L}. \qquad (20)$$

This result is known as *Poiseuille's law*. Knowing the coefficient of viscosity of the fluid, we can then determine the pressure difference that must be provided by an external agent (a pump, perhaps) to sustain a given mass flux through the pipe. Equivalently, if we force fluid through a pipe with a known pressure difference, measuring the mass flux permits us to determine the coefficient of viscosity of the fluid.

Viscosity in liquids originates with the intermolecular cohesive forces. As the temperature increases, the coefficient of viscosity of a liquid decreases, because the increasing kinetic energy of the molecules weakens the effect of the intermolecular forces. In gases, on the other hand, the viscosity increases with increasing temperature, because the molecules themselves can migrate between the layers. At higher temperatures, there is more molecular motion and therefore more mixing. However, note that in a pipe there are always more slow molecules near the walls than there are fast molecules near the central axis, so more mixing always means more slow molecules moving toward the axis and impeding the motion of the faster-moving molecules. (The effect is similar to that of slow-moving traffic merging into the fast lane of a highway.)

Sample Problem 3 Castor oil, which has a density of 0.96×10^3 kg/m³ at room temperature, is forced through a pipe of circular cross section by a pump that maintains a gauge pressure of 950 Pa. The pipe has a diameter of 2.6 cm and a length of 65 cm. The castor oil emerging from the free end of the pipe at atmospheric pressure is collected. After 90 s, a total of 1.23 kg has been collected. What is the coefficient of viscosity of the castor oil at this temperature?

Solution The mass flux is

$$\frac{dm}{dt} = \frac{1.23 \text{ kg}}{90 \text{ s}} = 0.0137 \text{ kg/s}.$$

The coefficient of viscosity can now be found directly from Eq. 20 if we first solve for η, which gives

$$\eta = \frac{\rho \pi R^4 \, \Delta p}{8(dm/dt)L} = \frac{(0.96 \times 10^3 \text{ kg/m}^3)\pi(0.013 \text{ m})^4(950 \text{ Pa})}{8(0.0137 \text{ kg/s})(0.65 \text{ m})}$$

$$= 1.15 \text{ N} \cdot \text{s/m}^2.$$

Heavy oils typically have viscosities in this range.

Turbulence

After rising a short distance, the smooth column of smoke from a cigarette breaks up into an irregular and seemingly random pattern (Fig. 21). In a similar fashion, a stream of fluid flowing past an obstacle breaks up into eddies and vortices (Fig. 22), which give the flow irregular velocity components transverse to the flow direction. An example of this case is the flapping of a flag in a breeze—if the flow of air were laminar, the flag would occupy a fixed position along streamlines, but the flagpole breaks the flow into an irregular pattern similar to Fig. 22, which causes the transverse flapping motion of the flag. These are examples of *turbulent* fluid flow. Other examples include the wakes left in water by moving ships and in air by moving cars and airplanes. The sounds produced by whistling and by wind instruments result from the turbulent flow of air.

In a viscous fluid, the flow at low speed can be described as laminar, which suggests layers sliding smoothly over one another. When the flow speed is sufficiently large, the motion becomes disordered and irregular; this is turbulent flow. An analogy from mechanics is a block that is pushed across a rough surface. If the frictional force is small, the block will slide across the surface if the applied force F is at least as great as the frictional force f. If the frictional force were greater, the applied force

F must also be greater, eventually becoming great enough that it tips the block over. The tipping of the block is analogous to the transition from laminar to turbulent flow.

We can determine the critical speed at which the flow becomes turbulent through a dimensional analysis. We let v_c represent the critical speed, which we take to be an average over the pipe, because, as Fig. 20 suggests, the speed varies over the cross section of the pipe. We expect this critical speed to depend on the viscosity η and density ρ of the fluid and the diameter D of the

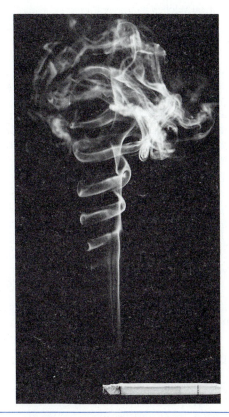

Figure 21 Rising smoke is at first in laminar flow, but the flow soon becomes turbulent.

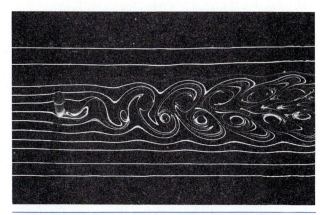

Figure 22 Fluid flowing left to right past a cylindrical obstacle clearly goes from laminar to turbulent. Note the eddies and vortices that form downstream from the obstacle.

pipe. Using our standard technique of dimensional analysis, we proceed as follows:

$$v_c \propto \eta^a \rho^b D^c$$

$$[v_c] = [\eta^a][\rho^b][D^c]$$

$$LT^{-1} = (ML^{-1}T^{-1})^a(ML^{-3})^b(L)^c,$$

where the dimensions of viscosity have been obtained from its units of N·s/m². Solving, we obtain

$$a = 1, \qquad b = -1, \qquad c = -1.$$

Thus the critical velocity can be written

$$v_c \propto \frac{\eta}{\rho D},$$

or, introducing a constant of proportionality R,

$$v_c = R \frac{\eta}{\rho D}. \qquad (21)$$

The dimensionless constant R is called the *Reynolds number.* Solving Eq. 21 for R, we can write the Reynolds number for *any* flow velocity v as

$$R = \frac{\rho D v}{\eta}. \qquad (22)$$

In this interpretation, the Reynolds number can be used to characterize *any* flow, and we can determine by experiment the value of the Reynolds number at which the flow becomes turbulent.

For cylindrical pipes, the Reynolds number corresponding to the critical velocity is about 2000. Thus for water flowing through a pipe of diameter 2 cm (a typical household garden hose, for example), the critical speed is

$$v_c = 2000 \frac{1 \times 10^{-3} \, N \cdot s/m^2}{(10^3 \, kg/m^3)(0.02 \, m)} = 0.1 \, m/s = 10 \, cm/s.$$

This is quite a low speed, which suggests that the flow of water is turbulent in ordinary household plumbing. (The flow speed from a typical household tap is about 1 m/s.)

Note from Eq. 21 that the critical flow speed increases with the viscosity. That is, the greater the viscous friction exerted by the surrounding fluid, the more likely the flow will be steady.

Chaotic Flow

The geometry of Fig. 19 is not particularly convenient for measuring viscosity. Figure 23 shows a more convenient arrangement. The space between the coaxial cylinders is filled with the fluid whose viscosity is to be determined. The inner cylinder is made to rotate, while the outer cylinder is held fixed. From the force necessary to keep the inner cylinder rotating at constant speed, the viscosity of the fluid can be determined.

For small rotational speeds, the flow in Fig. 23 will be steady and laminar. As the rotational speed of the inner cylinder is increased, the flow eventually becomes turbulent. We can observe that the transition from laminar to turbulent flow takes place in an orderly fashion. Figure 24 shows two intermediate stages. The fluid first forms toroidal vortices (somewhat like a stack of doughnuts) and then shows a pattern of waves of definite frequency that becomes superimposed on the vortices. As the rotational speed continues to increase, waves appear with new frequencies. We can imagine the turbulent flow to be the exten-

sion of this motion to include so many frequency components that the motion appears to become completely disordered and confused (somewhat like electronic noise). There may be an underlying periodic structure, but it is too complex to follow.

Chaos theory (see Section 6-9) takes a different approach in explaining the onset of turbulence. The turbulent motion result-

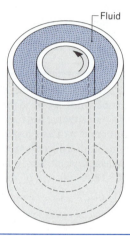

Figure 23 Experimental apparatus to measure fluid viscosities. The fluid is placed between the two cylinders, the outer cylinder being fixed and the inner cylinder rotating with angular velocity ω. The torque needed to turn the inner cylinder at this angular velocity is determined by the viscosity of the fluid.

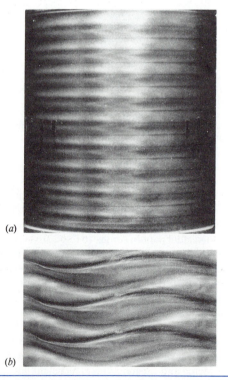

(a)

(b)

Figure 24 When the fluid speed in the apparatus of Fig. 23 exceeds the critical velocity, the flow becomes unstable and breaks up into (a) toroidal vortices and then (b) waves superimposed on the vortices.

ing from chaos theory is truly *nonperiodic,* not simply the combination of a large number of periodic motions. There is a critical distinction between these two cases. If the transition from laminar to turbulent flow takes place through a succession of orderly periodic motions, then two particles of fluid that in the laminar flow are moving similarly will remain in closely related states of motion throughout the transition into turbulent flow. However, if the intermediate condition can be described as chaotic, then the motion loses its predictability, and the two particles can be found in the turbulent flow in very different states of motion. Chaos theory, which is applicable to a wide variety of physical systems, provides an alternative theoretical basis for understanding complex systems such as the turbulent motion of fluids. ■

QUESTIONS

1. Briefly describe what is meant by each of the following and illustrate with an example: (*a*) steady fluid flow; (*b*) nonsteady fluid flow; (*c*) rotational fluid flow; (*d*) irrotational fluid flow; (*e*) compressible fluid flow; (*f*) incompressible fluid flow; (*g*) viscous fluid flow; (*h*) nonviscous fluid flow.

2. Explain the pressure variations in your blood as it circulates through your body.

3. Explain how a physician can measure your blood pressure.

4. In steady flow, the velocity vector at any point is constant. Can there then be accelerated motion of the fluid particles? Explain.

5. Describe the forces acting on an element of fluid as it flows through a pipe of nonuniform cross section.

6. In a lecture demonstration, a Ping-Pong ball is kept in midair by a vertical jet of air. Is the equilibrium stable, unstable, or neutral? Explain.

7. The height of the liquid in the standpipes of Fig. 25 indicates that the pressure drops along the channel, even though the channel has a uniform cross section and the flowing liquid is incompressible. Explain.

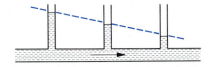

Figure 25   Question 7.

8. Explain why a taller chimney creates a better draft for taking the smoke out of a fireplace. Why doesn't the smoke pour into the room containing the fireplace?

9. (*a*) Explain how a baseball pitcher can make the baseball curve to his right or left. Can we justify applying Bernoulli's equation to such a spinning baseball? (See "Bernoulli and Newton in Fluid Mechanics," by Norman F. Smith, *The Physics Teacher,* November 1972, p. 451.) (*b*) Why is it easier to throw a curve with a tennis ball than with a baseball?

10. Not only a ball with a rough surface but also a smooth ball can be made to curve when thrown, but these balls will curve in *opposite* directions. Why? (See "Effect of Spin and Speed on the Curve of a Baseball and the Magnus Effect for Smooth Spheres," by Lyman J. Briggs, *American Journal of Physics,* November 1959, p. 589.)

11. Two rowboats moving parallel to one another in the same direction are pulled toward one another. Two automobiles moving parallel are also pulled together. Explain such phenomena on the basis of Bernoulli's equation.

12. In building "skyscrapers," what forces produced by the movement of air must be counteracted? How is this done? (See "The Wind Bracing of Buildings," by Carl W. Condit, *Scientific American,* February 1974, p. 92.)

13. Explain the action of a parachute in retarding free fall using Bernoulli's equation.

14. Why does a stream of water from a faucet become narrower as it falls?

15. Can you explain why water flows in a continuous stream down a vertical pipe, whereas it breaks into drops when falling freely?

16. How does the flush toilet work? Really. (See *Flushed with Pride: The Story of Thomas Crapper,* by W. Reyburn, Prentice-Hall, Englewood Cliffs, N.J., 1969.)

17. Sometimes people remove letters from envelopes by cutting a sliver from a narrow end, holding it firmly, and blowing toward it. Explain, using Bernoulli's equation, why this procedure is successful.

18. On takeoff would it be better for an airplane to move into the wind or with the wind? On landing?

19. Explain how the difference in pressure between the lower and upper surfaces of an airplane wing depend on the altitude of the moving plane.

20. The accumulation of ice on an airplane wing may greatly reduce its lift. Explain. (The *weight* of the ice is not the issue here.)

21. How is an airplane able to fly upside down?

22. "The characteristic banana-like shape of most returning boomerangs has hardly anything to do with their ability to return. . . . The essential thing is the cross section of the arms, which should be more convex on one side than on the other, like the wing profile of an airplane." (From "The Aerodynamics of Boomerangs," by Felix Hess, *Scientific American,* November 1968, p. 124.) Explain.

23. What powers the flight of soaring birds? (See "The Soaring Flight of Birds," by C. D. Cone, Jr., *Scientific American,* April 1962, p. 130.)

24. Why does the factor "2" appear in Eq. 15, rather than "1"? One might naively expect that the thrust would simply be the pressure difference times the area, that is, $A_0(p - p_0)$.

25. Explain why the destructive effect of a tornado is greater near the center of the disturbance than near the edge.

26. When a stopper is pulled from a filled basin, the water drains out while circulating like a small whirlpool. The angular velocity of a fluid element about a vertical axis through the orifice appears to be greatest near the orifice. Explain.

27. Is it true that in bathtubs in the northern hemisphere the water drains out with a counterclockwise rotation and in those in the southern hemisphere with a clockwise rotation? If so, explain and predict what would happen at the equator. (See "Bath-Tub Vortex," by Ascher H. Shapiro, *Nature,* December 15, 1962, p. 1080.)

28. Explain why you cannot remove the filter paper from the funnel of Fig. 26 by blowing into the narrow end.

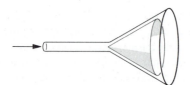

Figure 26   Question 28.

29. According to Bernoulli's equation, an increase in velocity should be associated with a decrease in pressure. Yet, when you put your hand outside the window of a moving car, increasing the speed at which the air flows by, you sense an *increase* in pressure. Why is this not a violation of Bernoulli's equation?

30. Why is it that the presence of the atmosphere reduces the maximum range of some objects (for example, tennis balls) but increases the maximum range of others (for example, Frisbees and golf balls)?

31. A discus can be thrown farther *against* a 25-mi/h wind than with it. What is the explanation? (*Hint:* Think about dynamic lift and drag.)

32. Explain why golf balls are dimpled.

33. The longer the board and the shallower the water, the farther will a surf board skim across the water. Explain. (See "The Surf Skimmer," by R. D. Edge, *American Journal of Physics,* July 1968, p. 630.)

34. When poured from a teapot, water has a tendency to run along the underside of the spout. Explain. (See "The Teapot Effect . . . a Problem," by Markus Reiner, *Physics Today,* September 1956, p. 16.)

35. Prairie dogs live in large colonies in complex interconnected burrow systems. They face the problem of maintaining a sufficient air supply to their burrows to avoid suffocation. They avoid this by building conical earth mounds about some of their many burrow openings. In terms of Bernoulli's equation, how does this air conditioning scheme work? Note that because of viscous forces the wind speed over the prairie is less close to the ground level than it is even a few inches higher up. (See *New Scientist,* January 27, 1972, p. 191.)

36. Viscosity is an example of a transport phenomenon. What property is being transported? Can you think of other transport phenomena and their corresponding properties?

37. Why do auto manufacturers recommend using "multi-viscosity" engine oil in cold weather?

38. Why is it more important to take viscosity into account for a fluid flowing in a narrow channel than in a relatively unconfined channel?

39. Viscosity can delay the onset of turbulence in fluid flow; that is, it tends to stabilize the flow. Consider syrup and water, for example, and make this plausible.

PROBLEMS

Section 18-2 Streamlines and the Equation of Continuity

1. A pipe of diameter 34.5 cm carries water moving at 2.62 m/s. How long will it take to discharge 1600 m³ of water?

2. A garden hose having an internal diameter of 0.75 in. is connected to a lawn sprinkler that consists merely of an enclosure with 24 holes, each 0.050 in. in diameter. If the water in the hose has a speed of 3.5 ft/s, at what speed does it leave the sprinkler holes?

3. Figure 27 shows the confluence of two streams to form a

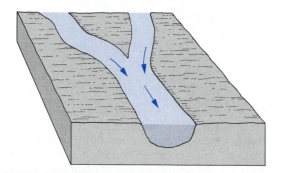

Figure 27   Problem 3.

river. One stream has a width of 8.2 m, depth of 3.4 m, and current speed of 2.3 m/s. The other stream is 6.8 m wide, 3.2 m deep, and flows at 2.6 m/s. The width of the river is 10.7 m and the current speed is 2.9 m/s. What is its depth?

4. Water is pumped steadily out of a flooded basement at a speed of 5.30 m/s through a uniform hose of radius 9.70 mm. The hose passes out through a window 2.90 m above the water line. How much power is supplied by the pump?

5. A river 21 m wide and 4.3 m deep drains a 8500-km² land area in which the average precipitation is 48 cm/y. One-fourth of this subsequently returns to the atmosphere by evaporation, but the remainder ultimately drains into the river. What is the average speed of the river current?

6. Tidal currents in narrow channels connecting coastal bays with the ocean can be very swift. Water must flow into the bay as the tide rises and back out to the sea as the tide falls. Consider the rectangular bay shown in Fig. 28*a*. The bay is connected to the sea by a channel 190 m wide and 6.5 m deep at mean sea level. The graph (Fig. 28*b*) shows the diurnal variation of the water level in the bay. Calculate the average speed of the tidal current in the channel.

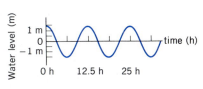

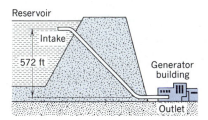

Figure 28   Problem 6.

Section 18-3 Bernoulli's Equation

7. How much work is done by pressure in forcing 1.4 m³ of water through a 13-mm internal diameter pipe if the difference in pressure at the two ends of the pipe is 1.2 atm?

8. A water intake at a storage reservoir (see Fig. 29) has a cross-sectional area of 7.60 ft². The water flows in at a speed of 1.33 ft/s. At the generator building 572 ft below the intake point, the water flows out at 31.0 ft/s. (a) Find the difference in pressure, in lb/in.², between inlet and outlet. (b) Find the area of the outlet pipe. The weight density of water is 62.4 lb/ft³.

Figure 29   Problem 8.

9. Models of torpedoes are sometimes tested in a horizontal pipe of flowing water, much as a wind tunnel is used to test model airplanes. Consider a circular pipe of internal diameter 25.5 cm and a torpedo model, aligned along the axis of the pipe, with a diameter of 4.80 cm. The torpedo is to be tested with water flowing past it at 2.76 m/s. (a) With what speed must the water flow in the unconstricted part of the pipe? (b) Find the pressure difference between the constricted and unconstricted parts of the pipe.

10. Water is moving with a speed of 5.18 m/s through a pipe with a cross-sectional area of 4.20 cm². The water gradually descends 9.66 m as the pipe increases in area to 7.60 cm². (a) What is the speed of flow at the lower level? (b) The pressure at the upper level is 152 kPa; find the pressure at the lower level.

11. Suppose that two tanks, 1 and 2, each with a large opening at the top, contain different liquids. A small hole is made in the side of each tank at the same depth h below the liquid surface, but the hole in tank 1 has half the cross-sectional area of the hole in tank 2. (a) What is the ratio ρ_1/ρ_2 of the densities of the fluids if it is observed that the mass flux is the same for the two holes? (b) What is the ratio of the flow rates (volume flux) from the two tanks? (c) It is desired to equalize the two flow rates by adding or draining fluid in tank 2. What should

be the new height of the fluid above the hole in tank 2 to make the flow rate in tank 2 equal to that of tank 1?

12. In a hurricane, the air (density 1.2 kg/m³) is blowing over the roof of a house at a speed of 110 km/h. (a) What is the pressure difference between inside and outside that tends to lift the roof? (b) What would be the lifting force on a roof of area 93 m²?

13. The windows in an office building are 4.26 m by 5.26 m. On a stormy day, air is blowing at 28.0 m/s past a window on the 53rd floor. Calculate the net force on the window. The density of the air is 1.23 kg/m³.

14. A liquid flows through a horizontal pipe whose inner radius is 2.52 cm. The pipe bends upward through a height of 11.5 m where it widens and joins another horizontal pipe of inner radius 6.14 cm. What must the volume flux be if the pressure in the two horizontal pipes is the same?

15. Figure 30 shows liquid discharging from an orifice in a large tank at a distance h below the liquid surface. The tank is open at the top. (a) Apply Bernoulli's equation to a streamline connecting points 1, 2, and 3, and show that the speed of efflux is

$$v = \sqrt{2gh}.$$

This is known as *Torricelli's law*. (b) If the orifice were curved directly upward, how high would the liquid stream rise? (c) How would viscosity or turbulence affect the analysis?

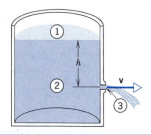

Figure 30   Problem 15.

16. A tank is filled with water to a height H. A hole is punched in one of the walls at a depth h below the water surface (Fig. 31). (a) Show that the distance x from the foot of the wall at which the stream strikes the floor is given by $x = 2\sqrt{h(H-h)}$. (b) Could a hole be punched at another depth so that this second stream would have the same range? If so, at what depth? (c) At what depth should the hole be

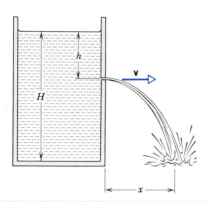

Figure 31   Problem 16.

placed to make the emerging stream strike the ground at the maximum distance from the base of the tank? What is this maximum distance?

17. A sniper fires a rifle bullet into a gasoline tank, making a hole 53.0 m below the surface of the gasoline. The tank was sealed and is under 3.10-atm absolute pressure, as shown in Fig. 32. The stored gasoline has a density of 660 kg/m³. At what speed does the gasoline begin to shoot out of the hole?

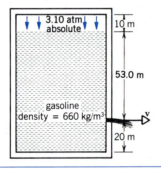

Figure 32   Problem 17.

18. Consider a uniform U-tube with a diaphragm at the bottom and filled with a liquid to different heights in each arm (see Fig. 33). Now imagine that the diaphragm is punctured so that the liquid flows from left to right. (a) Show that the application of Bernoulli's equation to points 1 and 3 leads to a contradiction. (b) Explain why Bernoulli's equation is not applicable here. (*Hint:* Is the flow steady?)

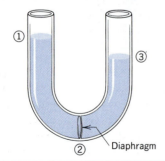

Figure 33   Problem 18.

19. If a person blows air with a speed of 15.0 m/s across the top of one side of a U-tube containing water, what will be the

difference between the water levels on the two sides? Assume the density of air is 1.20 kg/m³.

20. The fresh water behind a reservoir dam is 15.2 m deep. A horizontal pipe 4.30 cm in diameter passes through the dam 6.15 m below the water surface, as shown in Fig. 34. A plug secures the pipe opening. (a) Find the frictional force between plug and pipe wall. (b) The plug is removed. What volume of water flows out of the pipe in 3.00 h?

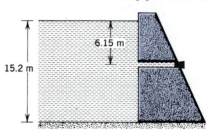

Figure 34   Problem 20.

21. A siphon is a device for removing liquid from a container that is not to be tipped. It operates as shown in Fig. 35. The tube must initially be filled, but once this has been done the liquid will flow until its level drops below the tube opening at *A*. The liquid has density ρ and negligible viscosity. (a) With what speed does the liquid emerge from the tube at *C*? (b) What is the pressure in the liquid at the topmost point *B*? (c) What is the greatest possible height h_1 that a siphon may lift water?

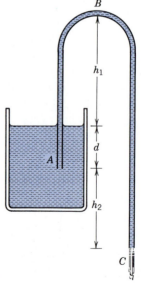

Figure 35   Problem 21.

22. (a) Consider a stream of fluid of density ρ with speed v_1 passing *abruptly* from a cylindrical pipe of cross-sectional area a_1 into a wider cylindrical pipe of cross-sectional area a_2 (see Fig. 36). The jet will mix with the surrounding fluid and, after the mixing, will flow on almost uniformly with an average speed v_2. Without referring to the details of the mixing, use momentum ideas to show that the increase in pressure due to the mixing is approximately

$$p_2 - p_1 = \rho v_2(v_1 - v_2).$$

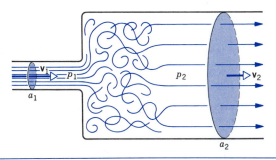

Figure 36   Problem 22.

(b) Show from Bernoulli's equation that in a *gradually* widening pipe we would get

$$p_2 - p_1 = \tfrac{1}{2}\rho(v_1^2 - v_2^2).$$

(c) Find the loss of pressure due to the abrupt enlargement of the pipe. Can you draw an analogy with elastic and inelastic collisions in particle mechanics?

23. A jug contains 15 glasses of orange juice. When you open the tap at the bottom it takes 12.0 s to fill a glass with juice. If you leave the tap open, how long will it take to fill the remaining 14 glasses and thus empty the jug?

Section 18-4 Applications of Bernoulli's Equation and the Equation of Continuity

24. A Pitot tube is mounted on an airplane wing to determine the speed of the plane relative to the air, which has a density of 1.03 kg/m³. The tube contains alcohol and indicates a level difference of 26.2 cm. What is the plane's speed relative to the air? The density of alcohol is 810 kg/m³.

25. A hollow tube has a disk *DD* attached to its end (Fig. 37). When air of density ρ is blown through the tube, the disk attracts the card *CC*. Let the area of the card be *A* and let *v* be the average air speed between the card and the disk. Calculate the resultant upward force on *CC*. Neglect the card's weight; assume that $v_0 \ll v$, where v_0 is the air speed in the hollow tube.

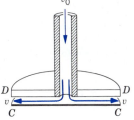

Figure 37   Problem 25.

26. A square plate with edge length 9.10 cm and mass 488 g is hinged along one side. If air is blown over the upper surface only, what speed must the air have to hold the plate horizontal? The air has density 1.21 kg/m³.

27. Air flows over the top of an airplane wing, area *A*, with speed v_t and past the underside of the wing with speed v_u. Show that Bernoulli's equation predicts that the upward lift force *L* on the wing will be

$$L = \tfrac{1}{2}\rho A(v_t^2 - v_u^2),$$

where ρ is the density of the air. (*Hint:* Apply Bernoulli's equation to a streamline passing just over the upper wing surface and to a streamline passing just beneath the lower wing surface. Can you justify setting the constants for the two streamlines equal?)

28. An airplane has a wing area (each wing) of 12.5 m². At a certain air speed, air flows over the upper wing surface at 49.8 m/s and over the lower wing surface at 38.2 m/s. (a) Find the mass of the plane. Assume that the plane travels with constant velocity and that lift effects associated with the fuselage and tail assembly are small. Discuss the lift if the airplane, flying at the same air speed, is (b) in level flight, (c) climbing at 15°, and (d) descending at 15°. The air density is 1.17 kg/m³. See Problem 27.

29. Consider the stagnant air at the front edge of a wing and the air rushing over the wing surface at a speed *v*. Assume pressure at the leading edge to be approximately atmospheric and find the greatest value possible for *v* in streamline flow; assume air is incompressible and use Bernoulli's equation. Take the density of air to be 1.2 kg/m³. How does this compare with the speed of sound under these conditions (340 m/s)? Can you explain the difference? Why should there be any connection between these quantities?

30. A Venturi tube has a pipe diameter of 25.4 cm and a throat diameter of 11.3 cm. The water pressure in the pipe is 57.1 kPa and in the throat is 32.6 kPa. Calculate the volume flux of water through the tube.

31. Consider the Venturi meter of Fig. 9. By applying Bernoulli's equation to points 1 and 2, and the equation of continuity (Eq. 3), verify Eq. 11 for the speed of flow at point 1.

32. Consider the Venturi meter of Fig. 9, containing water, without the manometer. Let $A = 4.75a$. Suppose that the pressure at point 1 is 2.12 atm. (a) Compute the values of *v* at point 1 and v' at point 2 that would make the pressure p' at point 2 equal to zero. (b) Compute the corresponding volume flow rate if the diameter at point 1 is 5.20 cm. The phenomenon at point 2 when p' falls to nearly zero is known as *cavitation*. The water vaporizes into small bubbles.

Section 18-5 Fields of Flow

33. Show that the constant in Bernoulli's equation is the same for *all* streamlines in the case of the steady, irrotational flow of Fig. 14.

34. A force field is conservative if $\oint \mathbf{F} \cdot d\mathbf{s} = 0$. The circle on the integration sign means that the integration is to be taken along a closed curve (a round trip) in the field. A flow is a potential flow (hence irrotational) if $\oint \mathbf{v} \cdot d\mathbf{s} = 0$ for every closed path in the field. Using this criterion, show that the fields of (a) Fig. 14 and (b) Fig. 17 are fields of potential flow.

35. In flows that are sharply curved, centrifugal effects are appreciable. Consider an element of fluid that is moving with speed *v* along a streamline of a curved flow in a horizontal plane (Fig. 38). (a) Show that $dp/dr = \rho v^2/r$, so that the pressure increases by an amount $\rho v^2/r$ per unit distance perpendicular to the streamline as we go from the concave to the convex side of the streamline. (b) Then use Bernoulli's equation and this result to show that *vr* equals a constant, so that speeds increase toward the center of curvature. Hence streamlines that are uniformly spaced in a straight pipe will

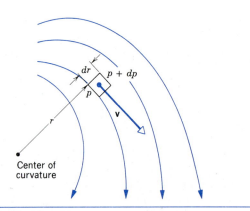

Figure 38 Problem 35.

be crowded toward the inner wall of a curved passage and widely spaced toward the outer wall. This problem should be compared to Problem 29 of Chapter 17 in which the curved motion is produced by rotating a container. There the speed varied directly with r, but here it varies inversely. (*c*) Show that this flow is irrotational.

36. Before Newton proposed his theory of gravitation, a model of planetary motion proposed by René Descartes was widely accepted. In Descartes' model the planets were caught in and dragged along by a whirlpool of ether particles centered around the Sun. Newton showed that this vortex scheme contradicted observations because: (*a*) the speed of an ether particle in the vortex varies inversely as its distance from the Sun; (*b*) the period of revolution of such a particle varies directly as the square of its distance from the Sun; and (*c*) this result contradicts Kepler's third law. Prove (*a*), (*b*), and (*c*).

Section 18-6 Viscosity, Turbulence, and Chaotic Flow

37. Figure 39 shows a cross section of the upper layers of the Earth. The surface of the Earth is broken into several rigid blocks, called plates, that slide (slowly!) over a "slushy" lower layer called the asthenosphere. See the figure for typical dimensions. Suppose that the speed of the rigid plate shown is $v_0 = 48$ mm/y, and that the base of the asthenosphere does not move. Calculate the shear stress on the base of the plate. The viscosity of the asthenosphere material is 4.0×10^{19} Pa·s. Ignore the curvature of the Earth.

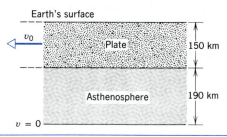

Figure 39 Problem 37.

38. Calculate the greatest speed at which blood, at 37°C, can flow through an artery of diameter 3.8 mm if the flow is to remain laminar.

39. Liquid mercury (viscosity = 1.55×10^{-3} N·s/m²) flows through a horizontal pipe of internal radius 1.88 cm and length 1.26 m. The volume flux is 5.35×10^{-2} L/min. (*a*) Show that the flow is laminar. (*b*) Calculate the difference in pressure between the two ends of the pipe.

40. The streamlines of the Poiseuille field of flow are shown in Fig. 40. The spacing of the streamlines indicates that although the motion is rectilinear, there is a velocity gradient in the transverse direction. Show that the Poiseuille flow is rotational.

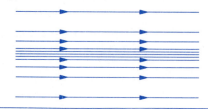

Figure 40 Problem 40.

41. A fluid of viscosity η flows steadily through a horizontal cylindrical pipe of radius R and length L, as shown in Fig. 41. (*a*) Consider an arbitrary cylinder of fluid of radius r. Show that the viscous force F due to the neighboring layer is $F = -\eta(2\pi rL)dv/dr$. (*b*) Show that the force F' pushing that cylinder of fluid through the pipe is $F' = (\pi r^2)\Delta p$. (*c*) Use the equilibrium condition to obtain an expression for dv in terms of dr. Integrate the expression to obtain Eq. 18.

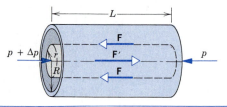

Figure 41 Problems 41 and 42.

42. Consider once again the fluid flowing through the pipe described in Problem 41 and illustrated in Fig. 41. Find an expression for the mass flux through the annular ring between radii r and $r + dr$; then integrate this result to find the total mass flux through the pipe, thereby verifying Eq. 20.

43. A soap bubble of radius 38.2 mm is blown on the end of a narrow tube of length 11.2 cm and internal diameter 1.08 mm. The other end of the tube is exposed to the atmosphere. Find the time taken for the bubble radius to fall to 21.6 mm. Assume Poiseuille flow in the tube. (For the surface tension of the soap solution use 2.50×10^{-2} N/m; the viscosity of air is 1.80×10^{-5} N·s/m².)

CHAPTER 19

WAVE MOTION

Wave motion appears in almost every branch of physics. Surface waves on bodies of water are commonly observed. Sound waves and light waves are essential to our perception of the environment, because we have evolved receptors (eyes and ears) capable of their detection. In the past century, we have learned how to produce and use radio waves. We can understand the structure of atoms and subatomic systems based on the wavelike properties of their constituent particles. The similarity of the physical and mathematical descriptions of these different kinds of waves indicates that wave motion is one of the unifying themes of physics.

In this chapter and the next we develop the verbal and mathematical descriptions of waves. We use the example of mechanical waves, *in part because we have already developed the laws of mechanics in this text. Later in the text we develop the laws that govern other types of waves (light and other electromagnetic waves, for example). For simplicity, we concentrate on the study of harmonic waves (that is, those that can be represented by sine and cosine functions), but the principles that we develop apply to more complex waveforms as well.*

19-1 MECHANICAL WAVES

Ocean waves travel thousands of miles across the ocean, but the particles of water do not make that journey. When you shout at a friend, a sound wave may travel across a room through the air, but the air molecules do not travel that distance. We are familiar with energy and momentum being transported from one place to another through the motion of particles; wave motion provides an alternative way for energy and momentum to move from one place to another without material particles making that journey.

Water waves and sound waves are examples of *mechanical waves*, which are waves that travel through a deformable or elastic medium. They originate when some portion of the medium is displaced from its normal position and released. Because of the elastic properties of the medium, the disturbance propagates through the medium. On the microscopic level, such mechanical properties as the forces between atoms are responsible for the propagation of mechanical waves.

In this chapter, we concentrate on the study of mechanical waves. One simple type of mechanical wave, involving the oscillation of a stretched string such as might be found on a guitar, is chosen as an example to illustrate some general properties of waves.

As a wave reaches a particle in the medium, it sets that particle into motion and displaces it, thus transferring both kinetic and potential energy to it. Not only energy but also information about the nature of the wave source can be transmitted over considerable distances by wave motion. We can regard the particles of the medium as moving by only small distances about their previous positions as the wave passes, without undergoing any net overall displacement in the direction of travel of the wave. For example, small floating objects such as leaves or corks show that the actual motion of the water in a passing wave is mostly up and down and perhaps slightly back and forth; after the wave passes, the object is more or less where it started before the wave passed. This fact was already realized in the 15th century by Leonardo da Vinci, who wrote of water waves: "It often happens that the wave flees the place of its creation, while the water does not; like the waves made in a field of grain by the wind, where we see the waves running across the field while the grain remains in place."

19-2   TYPES OF WAVES

In listing water waves, light waves, and sound waves as examples of wave motion, we are classifying waves according to their broad physical properties. Waves can also be classified in other ways.

We can distinguish different kinds of mechanical waves by considering how the direction of motion of the particles of matter is related to the direction of propagation of the wave. If the motion of the particles is perpendicular to the direction of propagation of the wave itself, we have a *transverse* wave. For example, when a string under tension is set oscillating back and forth at one end, a transverse wave travels along the string; the disturbance moves along the string but the string particles vibrate at right angles to the direction of propagation of the disturbance (Fig. 1*a*). Light waves, although they are not mechanical waves, are also transverse waves.

If, however, the motion of the particles in a mechanical wave is back and forth along the direction of propagation, we have a *longitudinal* wave. For example, when a spring under tension is set oscillating back and forth at one end, a longitudinal wave travels along the spring; the coils vibrate back and forth parallel to the direction in which the disturbance travels along the spring (Fig. 1*b*). Sound waves in a gas are longitudinal waves. We discuss them in greater detail in Chapter 20.

Some waves are neither purely longitudinal nor purely transverse. For example, in waves on the surface of water the particles of water move both up and down and back and forth, tracing out elliptical paths as the water waves move by.

Waves can also be classified as one-, two-, and three-dimensional, according to the number of dimensions in which they propagate energy. Waves moving along the string or spring of Fig. 1 are one-dimensional. Surface waves or ripples on water, caused by dropping a pebble into a quiet pond, are two-dimensional. Sound waves and light waves traveling radially outward from a small source are three-dimensional.

Waves may be classified further according to how the particles of the medium move in time. For example, we can produce a *pulse* traveling down a stretched string by applying a single sidewise movement at its end (Fig. 1*c*). Each particle remains at rest until the pulse reaches it, then it moves during a short time, and then it again remains at rest. If we continue to move the end of the string back and forth (Fig. 1*a*), we produce a *train of waves* traveling along the string. If our motion is periodic, we produce a *periodic train of waves* in which each particle of the string has a periodic motion. The simplest special case of a periodic wave is a *harmonic wave,* in which each particle undergoes simple harmonic motion.

Imagine a stone dropped in a still lake. Circular ripples spread outward from the point where the stone entered the water (Fig. 2). Along a given circular ripple, all points are in the same state of motion. Those points define a surface called a *wavefront.* If the medium is of uniform density, the direction of motion of the waves is at right

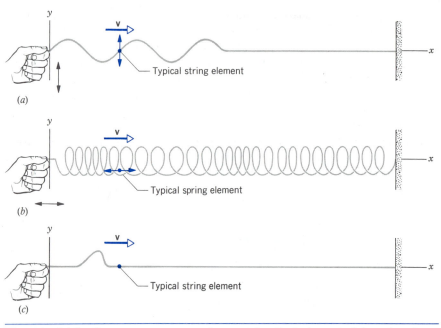

Figure 1   (*a*) Sending a transverse wave along a string. Each element of the string vibrates at right angles to the direction of propagation of the wave. (*b*) Sending a longitudinal wave along a spring. Each element of the spring vibrates parallel to the direction of propagation of the wave. (*c*) Sending a single transverse pulse along a string.

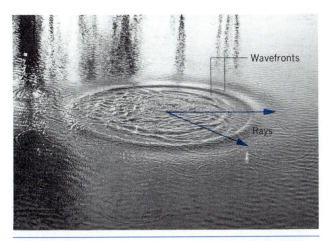

Figure 2   Waves on the surface of a lake. The circular ripples represent wavefronts. The rays, which are perpendicular to the wavefronts, indicate the direction of motion of the wave.

angles to the wavefront. A line normal to the wavefronts, indicating the direction of motion of the waves, is called a *ray*.

Wavefronts can have many shapes. A point source at the surface of water produces two-dimensional waves with circular wavefronts and rays that radiate outward from the point of the disturbance (as in Fig. 2). On the other hand, a very long stick dropped horizontally into the water would produce (near its center) disturbances that travel as straight lines, in which the rays are parallel lines. The three-dimensional analogy, in which the disturbances travel in a single direction, is the *plane wave*. At a given instant, conditions are the same everywhere on any plane perpendicular to the direction of propagation. The wavefronts are planes, and the rays are parallel straight lines (Fig. 3*a*). The three-dimensional analogy of circular waves is spherical waves. Here the disturbance is propagated outward in all directions from a point source of waves. The wavefronts are spheres, and the rays are radial lines leaving the point source in all directions (Fig. 3*b*). Far from the source the spherical wavefronts have very

small curvature, and over a limited region they can often be regarded as planes. Of course, there are many other possible shapes for wavefronts.

19-3   TRAVELING WAVES

As an example of the behavior of mechanical waves, we consider a transverse waveform that travels on a long stretched string. We assume an "ideal" string, in which the disturbance, whether it is a pulse or a train of waves, keeps its form as it travels. For this to occur, frictional losses and other means of energy dissipation must be negligibly small. The disturbance lies in the xy plane and travels in the x direction.

Figure 4*a* shows an arbitrary waveform at $t = 0$; we can consider this to be a snapshot of the pulse traveling along the string shown in Fig. 1*c*. Let the pulse move in the positive x direction with speed v. At a later time t, the pulse has moved a distance vt, as shown in Fig. 4*b*. Note that the waveform is the same at $t = 0$ as it is at later times.

The coordinate y indicates the transverse displacement of a particular point on the string. This coordinate depends on both the position x and the time t. We indicate this dependence on two variables as $y(x,t)$.

We can represent the waveform of Fig. 4*a* as

$$y(x,0) = f(x), \qquad (1)$$

where f is a function that describes the shape of the wave. At time t, the waveform must still be described by the same function f, because we have assumed that the shape does not change as the wave travels. Relative to the origin O' of a reference frame that travels with the pulse, the shape is described by the function $f(x')$, as indicated in Fig. 4*b*. The relationship between the x coordinates in the two reference frames is $x' = x - vt$, as you can see from Fig. 4*b*. Thus, at time t, the wave is described by

$$y(x,t) = f(x') = f(x - vt). \qquad (2)$$

That is, the function $f(x - vt)$ has the same shape relative

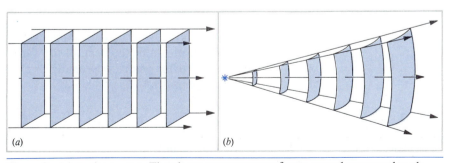

Figure 3   (*a*) A plane wave. The planes represent wavefronts spaced one wavelength apart, and the arrows represent rays. (*b*) A spherical wave. The wavefronts, spaced one wavelength apart, are spherical surfaces, and the rays are in the radial direction.

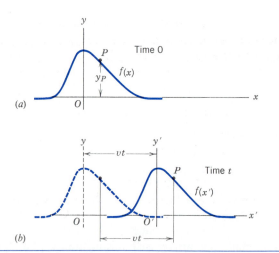

(a)

(b)

Figure 4   (a) A transverse pulse, shown as a snapshot at time $t = 0$. The point P represents a particular location on the phase of the pulse, *not* a particular point of the medium (the string, for instance). (b) At a time t later, the pulse has moved a distance vt in the positive x direction. The point P on the phase has also moved a distance vt. The peak of the pulse defines the origin of the x' coordinate.

to the point $x = vt$ at time t that the function $f(x)$ has relative to the point $x = 0$ at time $t = 0$.

To describe the wave completely, we must specify the function f. Later we shall consider harmonic waves, for which f is a sine or cosine function.

Equations 1 and 2 together indicate that we can change a function of any shape into a wave traveling in the positive x direction by merely substituting the quantity $x - vt$ for x everywhere that it appears in $f(x)$. For example, if $f(x) = x^2$, then $f(x - vt) = (x - vt)^2$. Furthermore, a wave traveling in the positive x direction must depend on x and t *only* in the combination $x - vt$; thus $x^2 - (vt)^2$ does not represent such a traveling wave.

Let us follow the motion of a particular part (or *phase*) of the wave, such as that of location P of the waveform of Fig. 4. If the wave is to keep its shape as it travels, then the y coordinate y_P of P must not change. We see from Eq. 2 that the only way this can happen is for the x coordinate of P to increase as t increases in such a way that the quantity $x - vt$ keeps a fixed value. That is, evaluating the quantity $x - vt$ gives the same result at P in Fig. 4b and at P in Fig. 4a. This remains true for any location on the waveform and for all times t. Thus for the motion of any particular phase of the wave we must have

$$x - vt = \text{constant}. \tag{3}$$

We can verify that Eq. 3 characterizes the motion of the phase of the waveform by differentiating with respect to time, which gives

$$\frac{dx}{dt} - v = 0 \quad \text{or} \quad \frac{dx}{dt} = v. \tag{4}$$

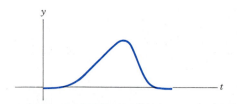

Figure 5   An observer stationed at a particular point on the x axis would record this y displacement as a function of time as the pulse of Fig. 4 passes. Note that the form appears to be reversed, because the leading edge of the traveling pulse arrives at the observer at the earliest times. That is, the displacements recorded by the observer at earlier times are closer to the origin here.

The velocity dx/dt describes the motion of the phase of the wave, and so it is known as the *phase velocity*. We take v to be a positive constant, independent of any property of the wave but possibly (as we shall see) depending on properties of the medium.

If the wave moves in the *negative* x direction, all we need do is replace v by $-v$. In this case we would obtain

$$y(x,t) = f(x + vt), \tag{5}$$

where once again $f(x)$ represents the shape at $t = 0$. That is, substituting in $f(x)$ the quantity $x + vt$ in place of x gives a wave that would move to the left in Fig. 4. The motion of any phase of the wave would then be characterized by the requirement that $x + vt = \text{constant}$, and by analogy with Eq. 4 we can show that $dx/dt = -v$, indicating that the x component of the phase velocity in this case is indeed negative.

The function $y(x,t)$ contains the complete description of the shape of the wave and its motion. At any particular time, say t_1, the function $y(x,t_1)$ gives y as a function of x, which defines a curve; this curve represents the actual shape of the string at that time and can be regarded as a "snapshot" of the wave. On the other hand, we can consider the motion of a particular point on the string, say at the fixed coordinate x_1. The function $y(x_1,t)$ then tells us the y coordinate of that point as a function of the time. Figure 5 shows how a point on the x axis might move with time as the pulse of Fig. 4 passes, moving in the positive x direction. At times near $t = 0$, the point is not moving at all. It then begins to move gradually, as the leading edge of the pulse of Fig. 4 arrives. After the peak of the wave passes, the displacement of the point drops rapidly back to zero as the trailing edge passes.

Sinusoidal Waves

The above description is quite general. It holds for arbitrary wave shapes, and it holds for transverse as well as longitudinal waves. Let us consider, for example, a transverse waveform having a sinusoidal shape, which has par-

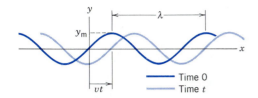

Figure 6   At $t = 0$ (darker color), the string has the sinusoidal shape given by $y = y_m \sin 2\pi x/\lambda$. At a later time t (lighter color), the wave has moved to the right a distance $x = vt$, and the string has a shape given by $y = y_m \sin 2\pi(x - vt)/\lambda$.

ticularly important applications. Suppose that at the time $t = 0$ we have a wavetrain along the string given by

$$y(x,0) = y_m \sin \frac{2\pi}{\lambda} x. \qquad (6)$$

The wave shape is shown in Fig. 6. The maximum displacement y_m is called the *amplitude* of the sine curve. The transverse displacement y has the same value at any x as it does at $x + \lambda$, $x + 2\lambda$, and so on. The symbol λ represents the *wavelength* of the wavetrain and indicates the distance between two adjacent points in the wave having the same phase. If the wave travels in the $+x$ direction with phase speed v, then the equation of the wave is

$$y(x,t) = y_m \sin \frac{2\pi}{\lambda} (x - vt). \qquad (7)$$

Note that this has the form $f(x - vt)$ required for a traveling wave (Eq. 2).

The *period T* of the wave is the time necessary for a point at any particular x coordinate to undergo one complete cycle of transverse motion. During this time T, the wave travels a distance vT that must correspond to one wavelength λ, so that

$$\lambda = vT. \qquad (8)$$

The inverse of the period is called the *frequency ν* of the wave: $\nu = 1/T$. Frequency has units of cycles per second, or hertz (Hz). Period and frequency were previously discussed in Chapter 15.

Putting Eq. 8 into Eq. 7, we obtain another expression for the wave:

$$y(x,t) = y_m \sin 2\pi \left(\frac{x}{\lambda} - \frac{t}{T} \right). \qquad (9)$$

From this form it is clear that y, at any given time, has the same value at x, $x + \lambda$, $x + 2\lambda$, and so on, and that y, at any given position, has the same value at the times t, $t + T$, $t + 2T$, and so on.

To reduce Eq. 9 to a more compact form, we introduce two quantities, the *wave number k* and the *angular frequency ω*. They are defined by

$$k = \frac{2\pi}{\lambda} \quad \text{and} \quad \omega = \frac{2\pi}{T} = 2\pi\nu. \qquad (10)$$

The wave number k is, like ω, an angular quantity, and units for both involve radians. Units for k might be, for instance, rad/m, and for ω, rad/s. In terms of these quantities, the equation of a sine wave traveling in the positive x direction (to the right in Fig. 6) is

$$y(x,t) = y_m \sin (kx - \omega t). \qquad (11)$$

The equation of a sine wave traveling in the negative x direction (to the left in Fig. 6) is

$$y(x,t) = y_m \sin (kx + \omega t). \qquad (12)$$

Comparing Eqs. 8 and 10, we see that the phase speed v of the wave is given by

$$v = \lambda\nu = \frac{\lambda}{T} = \frac{\omega}{k}. \qquad (13)$$

Phase and Phase Constant

In the traveling waves of Eqs. 11 and 12 we have assumed that the displacement y is zero at the position $x = 0$ at the time $t = 0$. This, of course, need not be the case. The general expression for a sinusoidal wave traveling in the positive x direction is

$$y(x,t) = y_m \sin (kx - \omega t - \phi). \qquad (14)$$

The quantity that appears in the argument of the sine, namely, $kx - \omega t - \phi$, is called the *phase* of the wave. Two waves with the same phase (or with phases differing by any integer multiple of 2π) are said to be "in phase"; they execute the same motion at the same time.

The angle ϕ is called the *phase constant*. The phase constant does not affect the shape of the wave; it moves the wave forward or backward in space or time. To see this, we rewrite Eq. 14 in two equivalent forms:

$$y(x,t) = y_m \sin \left[k \left(x - \frac{\phi}{k} \right) - \omega t \right] \qquad (15a)$$

or

$$y(x,t) = y_m \sin \left[kx - \omega \left(t + \frac{\phi}{\omega} \right) \right]. \qquad (15b)$$

Figure 7a shows a "snapshot" at any time t of the two waves represented by Eqs. 11 (in which $\phi = 0$) and 14. Note that any particular point on the wave described by Eq. 15a (say, a certain wave crest) is a distance ϕ/k *ahead of* the corresponding point in the wave described by Eq. 11.

Equivalently, if we were to observe the displacement at a fixed position x resulting from each of the two waves represented by Eqs. 11 and 14, we would obtain the result indicated by Fig. 7b. The wave described by Eq. 15b is similarly *ahead of* the wave having $\phi = 0$, in this case by a time difference ϕ/ω.

When the phase constant in Eq. 14 is positive, the corresponding wave is ahead of a wave described by a similar equation having $\phi = 0$. It is for this reason that we

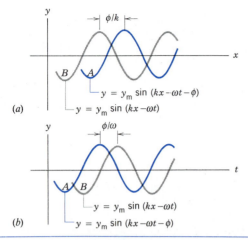

(a)

(b)

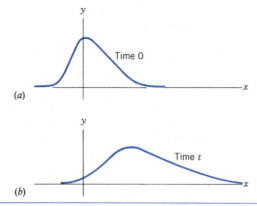

(a)

(b)

Figure 8   In a dispersive medium, the waveform changes as the wave travels.

Figure 7   (*a*) A snapshot of two sine waves traveling in the positive *x* direction. Wave *A* has phase constant ϕ, and wave *B* has $\phi = 0$. Wave *A* is a distance of ϕ/k ahead of wave *B*. (*b*) The motion of a single point in time due to the same two waves. Wave *A* is a time ϕ/ω ahead of wave *B*. Note that, in a graph of *y* versus *t*, "ahead of" means "to the left of," while in a graph of *y* versus *x*, "ahead of" means "to the right of," if the waves travel in the positive *x* direction.

introduced the phase constant with a negative sign in Eq. 14. When one wave is ahead of another in time or space, it is said to "lead." On the other hand, putting a negative phase constant into Eq. 14 moves the corresponding wave behind the one with $\phi = 0$. Such a wave is said to "lag."

If we fix our attention on a particular point of the string, say x_1, the displacement *y* at that point can be written

$$y(t) = -y_m \sin (\omega t + \phi'),$$

where we have substituted a new phase constant $\phi' = \phi - kx_1$. This expression for $y(t)$ is similar to Eq. 6 of Chapter 15 for simple harmonic motion. Hence any particular element of the string undergoes simple harmonic motion about its equilibrium position as this wavetrain travels along the string.

Group Speed and Dispersion

Pure sinusoidal waves are useful mathematical devices for helping us understand wave motion. In practice, we use other kinds of waves to transport energy and information. These waves may be periodic but nonsinusoidal, such as square waves or "sawtooth" waves, or they may be nonperiodic pulses, such as that of Fig. 4.

We have used the phase speed to describe the motion of two kinds of waves: the pulse that preserves its shape as it travels (Fig. 4) and the pure sine wave (Fig. 6). In other cases, we must use a different speed, called the *group speed,* which is the speed at which energy or information travels in a real wave.

Figure 8 shows a pulse traveling through a medium. The shape of the pulse changes as it travels; the pulse

spreads out, or *disperses.* (Dispersion is not the same as energy dissipation. The energy content of the pulse in Fig. 8 may remain constant as it travels, even though the pulse disperses. We assume that the medium is *dispersive,* but not necessarily *dissipative.*) As we see in Section 19-7, any periodic wave can be regarded as the sum or superposition of a series of sinusoidal waves of different frequencies or wavelengths. The frequencies, amplitudes, and phases of the component sinusoidal waves must be carefully chosen according to a prescribed mathematical procedure, known as *Fourier analysis,* so that the waves add to give the desired waveform. In most real media, the speed of propagation of these component waves (that is, the phase speed) depends on the frequency or wavelength of the particular component. Each component wave may travel with its own unique speed. Thus, as the wave travels, the phase relationships of the components may change, and the waveform of the sum of the components will correspondingly change as the wave travels. This is the origin of dispersion — the component waves travel at different phase speeds. There is no simple relationship between the phase speeds of the components and the group speed of the wave; the relationship depends on the dispersion of the medium.

Some real media are approximately nondispersive, in which case the wave keeps its shape, and all component waves travel with the same speed. An example is sound waves in air. If air were strongly dispersive for sound waves, conversation would be impossible, because the waveform produced by your friend's vocal cords would be jumbled by the time it reached your ears. Furthermore, the care taken by the players in an orchestra to play precisely at the same time would be to no avail, because (if air were dispersive for sound) the notes of high frequency would travel to the listener's ear at a speed different from that of the notes of low frequency, and the listener would hear the sounds at different times. Fortunately, this does not occur for sound waves. Light waves in vacuum are perfectly nondispersive; the dispersion of light waves in real media is responsible for such effects as the spectrum of colors in rainbows.

In a nondispersive medium, all the component waves in a complex waveform travel at the same phase speed, and the group speed of the waveform is equal to that common value of the phase speed. Only in this case can we speak of the phase speed of the entire waveform. In this chapter, we assume that we are dealing with mechanical waves that propagate in a nondispersive medium.

19-4 WAVE SPEED

The wave speed, which we take here to mean the phase speed of a sinusoidal wave or the group speed of a pulse in a nondispersive medium, does not depend on the frequency or wavelength. It is possible to calculate the speed of a mechanical wave from the properties of the medium by applying the basic principles of Newtonian mechanics. In this section we continue to focus our attention on transverse waves in a stretched string, and in the next section we show how to calculate the speed of such waves in the most general way. Calculations of the speed of other waves — for instance, sound waves in air — follow similar methods.

Here we consider two approaches: a treatment based on dimensional analysis and a somewhat less general mechanical analysis in which we compute the speed of a transverse pulse along a stretched string.

Dimensional Analysis

The speed of waves on the string depends on the mass of an element of the string and on the force between neighboring elements, which is the tension F under which the string is stretched. If we increase the tension (such as by tightening the tuning pegs on a guitar string), the force between neighboring elements will increase, and we expect that the wave speed should increase. We characterize the mass of an element of the string in terms of the *linear mass density* μ, the mass per unit length of the string. Assuming that the wave speed v depends only on F and μ, we can use the method of dimensional analysis (see Section 1-7) and write

$$v \propto F^a \mu^b,$$

where a and b are exponents to be determined from the dimensional analysis. In terms of the dimensions of mass M, length L, and time T, this can be written

$$[v] = [F^a][\mu^b]$$

$$LT^{-1} = (MLT^{-2})^a (ML^{-1})^b,$$

and solving by equating corresponding powers of M, L, and T gives $a = \frac{1}{2}$ and $b = -\frac{1}{2}$. Thus $v \propto \sqrt{F/\mu}$, or, introducing a constant of proportionality C,

$$v = C \sqrt{\frac{F}{\mu}}. \tag{16}$$

The most we can say from this analysis is that the wave speed is equal to a dimensionless constant times $\sqrt{F/\mu}$. The value of the constant can be obtained from a mechanical analysis of the problem or from experiment. These methods show that the constant is equal to unity.

Mechanical Analysis

Now let us derive an expression for the speed of a pulse in a stretched string by a mechanical analysis. In Fig. 9 we show a "snapshot" of a wave pulse that is moving from left to right in the string with a speed v. We can imagine instead that the entire string is moved from right to left with this same speed so that the wave pulse remains fixed in space (perhaps by pulling the string through a frictionless tube having the desired shape of the pulse). This simply means that, instead of taking our reference frame to be the walls between which the string is stretched, we choose a reference frame that is in uniform motion with respect to that one. In effect, we observe the pulse while running along the string at the same speed as the pulse. Because Newton's laws involve only accelerations, which are the same in both frames, we can use them in either frame. We just happen to choose a more convenient frame.

We consider a small section of the pulse of length δl, as shown in Fig. 9. This section approximately forms an arc of a circle of radius R. The mass δm of this element is $\mu\, \delta l$, where μ is the mass per unit length of the string. The tension F in the string is a tangential pull at each end of this small segment of the string. The horizontal components of F cancel, and the vertical components are each equal to $F \sin\theta$. Hence the total vertical force $F_\perp$ is $2F \sin\theta$. Because θ is small, we can take $\sin\theta \approx \theta$. From Fig. 9, we see that $2\theta = \delta l/R$, and so we obtain

$$F_\perp = 2F \sin\theta \approx 2F\theta = F\frac{\delta l}{R}. \tag{17}$$

This gives the force supplying the centripetal acceleration of the string particles directed toward O. The centripetal force acting on a mass $\delta m\ (= \mu\, \delta l)$ moving in a circle of radius R with speed v is $\delta m\, v^2/R$. Note that the tangential velocity v of this mass element along the top of the arc is

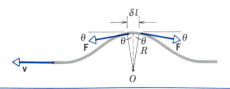

Figure 9 A pulse moving to the right on a stationary string is equivalent to a pulse in a fixed position on a string that is moving to the left. We consider the forces on a section of string of length δl on the "fixed" pulse.

horizontal and is in magnitude equal to the wave speed. Equating the net vertical force on the element, Eq. 17, to the needed centripetal force, we obtain

$$F_\perp = \frac{\delta m \, v^2}{R}$$

or

$$F \frac{\delta l}{R} = \frac{\mu \, \delta l \, v^2}{R}$$

and thus

$$v = \sqrt{\frac{F}{\mu}}. \tag{18}$$

Equation 18 shows from a mechanical analysis that the constant C in Eq. 16 has the value 1.

If the amplitude of the pulse were very large compared to the length of the string, we would have been unable to use the approximation $\sin\theta \approx \theta$. Furthermore, the tension F in the string would be changed by the presence of the pulse, whereas we assumed F to be unchanged from the original tension in the stretched string. Therefore our result holds only for relatively small transverse displacements of the string, a case that is widely applicable in practice.

A periodic wave that enters a medium usually results from an external influence that disturbs the medium at a certain frequency. The wave that travels through that medium will have the same frequency as the source of the wave. The velocity of the wave is determined by the properties of the medium. Given the frequency ν of the wave and its velocity v in the medium, the wavelength of the periodic wave *in that medium* is determined from Eq. 13, $\lambda = v/\nu$. When a wave passes from one medium to another medium of different wave speed (for example, two strings of different linear mass densities), the frequency in one medium must be the same as the frequency in the other. (Otherwise there would be a discontinuity at the point where the two strings are joined.) The wavelengths, however, will differ from one another. The relationship between the wavelengths follows from the equality of the frequencies ν_1 and ν_2 in the two media; that is, $\nu_1 = \nu_2$ gives

$$\frac{v_1}{\lambda_1} = \frac{v_2}{\lambda_2}. \tag{19}$$

Transverse Velocity of a Particle

The motion of a particle in a transverse wave such as that of Fig. 6 is in the y direction. The wave speed describes the motion of the wave along the direction of travel (the x direction). The wave speed does *not* characterize the transverse motion of the particles of the string.

To find the transverse velocity of a particle of the string, we need the change in the y coordinate with time. We focus our attention on a single particle of the string, that is, on a certain coordinate x. We therefore need the derivative of y with respect to t at constant x. This is represented

by the symbol $\partial y/\partial t$, which indicates the *partial derivative* of y with respect to t, holding constant all other variables on which y may depend. We represent the particle velocity, which varies with x (the location of the particle) as well as with t, as $u(x,t)$. Assuming we are dealing with a sinusoidal wave of the form of Eq. 14, we then have

$$u(x,t) = \frac{\partial y}{\partial t} = \frac{\partial}{\partial t}\,[y_{\mathrm{m}} \sin\,(kx - \omega t - \phi)]$$
$$= -y_{\mathrm{m}}\omega \cos\,(kx - \omega t - \phi). \tag{20}$$

Continuing in this way, we can find the transverse acceleration of the particle at this location of x according to

$$a(x,t) = \frac{\partial^2 y}{\partial t^2} = \frac{\partial u}{\partial t} = -y_{\mathrm{m}}\omega^2 \sin\,(kx - \omega t - \phi)$$
$$= -\omega^2 y. \tag{21}$$

Equation 21 has the same form as Eq. 5 of Chapter 15; the transverse acceleration of any point is proportional to its transverse displacement, but oppositely directed. This shows that each particle of the string undergoes transverse simple harmonic motion as the sinusoidal wave passes.

Sample Problem 1   A transverse sinusoidal wave is generated at one end of a long horizontal string by a bar that moves the end up and down through a distance of 1.30 cm. The motion is continuous and is repeated regularly 125 times per second. (a) If the string has a linear density of 0.251 kg/m and is kept under a tension of 96 N, find the amplitude, frequency, speed, and wavelength of the wave motion. (b) Assuming the wave moves in the $+x$ direction and that, at $t = 0$, the element of the string at $x = 0$ is at its equilibrium position $y = 0$ and moving downward, find the equation of the wave.

Solution   (a) As the bar moves a total of 1.30 cm, the end of the string moves $\frac{1}{2}(1.30\text{ cm}) = 0.65$ cm away from the equilibrium position, first above it, then below it; therefore the amplitude y_{m} is 0.65 cm.

The entire motion is repeated 125 times each second, and thus the frequency is 125 vibrations per second, or $\nu = 125$ Hz.

The wave speed is given by Eq. 18,

$$v = \sqrt{\frac{F}{\mu}} = \sqrt{\frac{96\text{ N}}{0.251\text{ kg/m}}} = 19.6\text{ m/s}.$$

The wavelength is given by $\lambda = v/\nu$, so that

$$\lambda = \frac{19.6\text{ m/s}}{125\text{ Hz}} = 0.156\text{ m} = 15.6\text{ cm}.$$

(b) The general expression for a transverse sinusoidal wave moving in the $+x$ direction is given by Eq. 14,

$$y(x,t) = y_{\mathrm{m}} \sin\,(kx - \omega t - \phi).$$

Imposing the given initial conditions ($y = 0$ and $\partial y/\partial t < 0$ for $x = 0$ and $t = 0$) yields

$$y_{\mathrm{m}} \sin\,(-\phi) = 0 \quad \text{and} \quad -y_{\mathrm{m}}\omega \cos\,(-\phi) < 0,$$

which means that the phase constant ϕ may be taken to be zero (or any integer multiple of 2π). Hence, for this wave

$$y(x,t) = y_m \sin (kx - \omega t),$$

and with the values just found,

$$y_m = 0.65 \text{ cm},$$

$$k = \frac{2\pi}{\lambda} = \frac{2\pi}{0.156 \text{ m}} = 40.3 \text{ rad/m} = 0.403 \text{ rad/cm},$$

$$\omega = vk = (19.6 \text{ m/s})(40.3 \text{ rad/m}) = 789 \text{ rad/s},$$

we obtain as the equation for the wave

$$y(x,t) = 0.65 \sin (0.403x - 789t),$$

where x and y are in centimeters and t is in seconds.

Sample Problem 2   As the wave of Sample Problem 1 passes along the string, each particle of the string moves up and down at right angles to the direction of the wave motion. (*a*) Find expressions for the velocity and acceleration of a particle P located at $x_P = 0.245$ m. (*b*) Evaluate the transverse displacement, velocity, and acceleration of this particle at $t = 1.5$ s.

Solution   (*a*) For a particle at $x_p = 0.245$ m $= 24.5$ cm in the wave of Sample Problem 1, we obtain, using Eq. 20 with $\phi = 0$,

$$u(x_P,t) = -(0.65)(789) \cos [(0.403)(24.5) - 789t]$$
$$= -513 \cos (9.87 - 789t),$$

where u is in cm/s and t is in seconds. Similarly, using Eq. 21, we find the acceleration to be

$$a(x_P,t) = -(0.65)(789)^2 \sin (9.87 - 789t)$$
$$= -(4.05 \times 10^5) \sin (9.87 - 789t),$$

where a is in cm/s^2.

(*b*) At $t = 1.5$ s, we evaluate the expressions for y, u, and a to give

$$y = +0.63 \text{ cm}, \quad u = -125 \text{ cm/s}, \quad a = -3.93 \times 10^5 \text{ cm/s}^2.$$

That is, the particle is close to its maximum positive displacement, it is moving in the negative y direction (away from that maximum), and it is accelerating in the negative y direction (its velocity is increasing in magnitude as the particle moves toward its equilibrium position).

19-5   THE WAVE EQUATION   *(Optional)*

In Chapter 15 we discussed the commonly encountered phenomenon of oscillation. One reason that this phenomenon is so common is that the basic equation that describes an oscillating system [$x = x_m \cos (\omega t + \phi)$, Eq. 6 of Chapter 15] is a solution of Eq. 5 of Chapter 15,

$$\frac{d^2x}{dt^2} = -\left(\frac{k}{m}\right) x,$$

which is an equation of a general form that can be derived from a mechanical analysis of a variety of physical situations, some of which were discussed in Section 15-5.

The situation is similar in the case of wave motion. As we

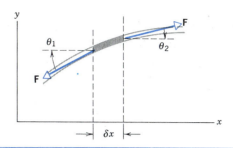

Figure 10   A small element of length δx of a long string under tension F. The figure represents a snapshot of the element at a particular time during the passage of a wave.

demonstrate in this section, the mechanical analysis gives an equation of another commonly encountered form, the solution of which is a wave of the form of Eq. 2 or 5.

Figure 10 shows an element of a long string that is under tension F. A passing wave has caused the element to be displaced from its equilibrium position at $y = 0$. We consider the element of the string of length δx, and we apply Newton's second law to analyze *how* this element is made to move.

The element is acted on by two forces, exerted by the portions of the string on either side of the element. These forces have equal magnitudes, because the tension is evenly distributed along the string, but they have slightly different directions, because they act tangent to the string at the endpoints of the element. The net force in the y direction is

$$F_y = F \sin \theta_2 - F \sin \theta_1.$$

We consider only small displacements from equilibrium, so that the angles θ_1 and θ_2 are small, and we can write $\sin \theta \approx \tan \theta$, which gives

$$F_y \approx F \tan \theta_2 - F \tan \theta_1 = F \delta(\tan \theta), \quad (22)$$

where $\delta(\tan \theta) = \tan \theta_2 - \tan \theta_1$. This resultant force must be equal to the mass of the element, $\delta m = \mu \, \delta x$, times the y component of the acceleration. Ignoring frictional or other dissipative forces, we find that Newton's second law gives

$$F_y = \delta m \, a_y$$

$$F \delta(\tan \theta) = \mu \, \delta x \, a_y$$

$$\frac{\delta(\tan \theta)}{\delta x} = \frac{\mu}{F} a_y.$$

For the y component of the acceleration a_y, we use the transverse acceleration for a particle, $\partial^2 y/\partial t^2$. We also replace $\tan \theta$, which is the slope of the string, by the equivalent partial derivative $\partial y/\partial x$. Making these substitutions, we obtain

$$\frac{\delta(\partial y/\partial x)}{\delta x} = \frac{\mu}{F} \frac{\partial^2 y}{\partial t^2}. \quad (23)$$

We now take the limit of Eq. 23 as the mass element becomes very small. The left side is in the standard form for expressing the derivative with respect to x as a limit:

$$\lim_{\delta x \to 0} \frac{\delta(\partial y/\partial x)}{\delta x} = \frac{\partial}{\partial x}\left(\frac{\partial y}{\partial x}\right) = \frac{\partial^2 y}{\partial x^2},$$

and the final result is

$$\frac{\partial^2 y}{\partial x^2} = \frac{\mu}{F}\frac{\partial^2 y}{\partial t^2} . \qquad (24)$$

Replacing μ/F with $1/v^2$, we obtain

$$\frac{\partial^2 y}{\partial x^2} = \frac{1}{v^2}\frac{\partial^2 y}{\partial t^2} . \qquad (25)$$

Equation 25 is the general form of equation that describes waves: the second derivative of the wave displacement y with respect to the coordinate x in the direction of propagation is equal to $1/v^2$ times the second derivative with respect to time. This general form of equation is called the *wave equation*. It arises not only in mechanics but in other situations as well. For example, as we discuss in Chapter 41, if we use the equations of electromagnetism instead of the equations of mechanics (Newton's laws), we obtain an equation of exactly the same form as Eq. 25, except that the displacement y is replaced by the strength of an electric or magnetic field. The speed of propagation v for electromagnetic waves traveling in a vacuum becomes the speed of light c.

Let us see how our general formula for a traveling wave, $y(x,t) = f(x \pm vt)$, is the solution of Eq. 25. We make a simple change of variable and let z represent $x \pm vt$, so that $y = f(z)$. Then, repeatedly using the chain rule of calculus,

$$\frac{\partial y}{\partial x} = \frac{df}{dz}\frac{\partial z}{\partial x} = \frac{df}{dz}$$

$$\frac{\partial^2 y}{\partial x^2} = \frac{d}{dz}\left(\frac{df}{dz}\right)\frac{\partial z}{\partial x} = \frac{d^2 f}{dz^2}$$

$$\frac{\partial y}{\partial t} = \frac{df}{dz}\frac{\partial z}{\partial t} = \pm v\frac{df}{dz}$$

$$\frac{\partial^2 y}{\partial t^2} = \frac{d}{dz}\left(\pm v\frac{df}{dz}\right)\frac{\partial z}{\partial t} = (\pm v)^2 \frac{d^2 f}{dz^2} = v^2\frac{d^2 f}{dz^2} .$$

Thus

$$\frac{d^2 f}{dz^2} = \frac{\partial^2 y}{\partial x^2} = \frac{1}{v^2}\frac{\partial^2 y}{\partial t^2} ,$$

and Eq. 25 is satisfied. It can be shown that *only* the combinations $x \pm vt$ in f satisfy the wave equation, so that all traveling waves must be in the form of Eq. 2 or 5.

To express these results in another way, Eq. 24, which was derived from Newton's laws, represents a traveling wave only when $\mu/F = 1/v^2$. This discussion thus provides an independent derivation of Eq. 18 for the velocity of propagation of waves along a stretched string. ∎

19-6   POWER AND INTENSITY IN WAVE MOTION

If, as suggested by Fig. 1, you were shaking (and thus doing work on) one end of a string, a friend at the other end could extract the resulting energy (which is transported along the string in the form of the potential energy and kinetic energy of its elements) and use the energy to do work on another system. Such transport of energy (and momentum) is in fact one of the purposes for which we produce waves. In this section we consider the rate at which the string transports energy.

Figure 11 shows a snapshot of the wave at times t and $t + dt$. A point on the string at coordinate x has at time t a transverse velocity $\mathbf{u}$, which has only a y component. This velocity, as we discussed in Section 19-4, is *not* related to the phase speed of the wave but instead has magnitude given by Eq. 20 with $\phi = 0$,

$$u = \frac{\partial y}{\partial t} = -\omega y_{\mathrm{m}}\cos(kx - \omega t)$$

for a sinusoidal wave of the form of Eq. 11.

The force exerted on an element of the string by the element to its left is also shown in Fig. 11. The force transmits energy at a rate given by Eq. 23 of Chapter 7, $P = \mathbf{u} \cdot \mathbf{F} = uF_y$. Only the component F_y of $\mathbf{F}$ along $\mathbf{u}$ con-

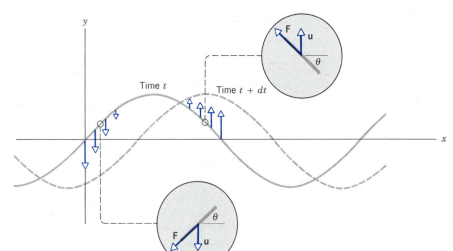

Figure 11   The vectors in the y direction show the value of the instantaneous velocity $\mathbf{u}$ of different points of the string as the sine wave travels. The dashed line shows the wave at a later time, when the particles have moved in the direction given by their velocity vectors. The insets show the force on two different elements of the string exerted by the element to its left. Note that the instantaneous power $\mathbf{u} \cdot \mathbf{F}$ is positive, no matter where we are on the phase of the wave.

tributes to the power; this component is $F \sin \theta$, which for small displacements can be approximated as $F \tan \theta = F \,\partial y/\partial x$, where $\partial y/\partial x$ is the slope of the string at the coordinate x.

Note that the y component of **F** is parallel to **u**, no matter whether that element of string happens to be moving up or down. Thus $uF_y \geq 0$, and so the power transmitted is never negative during the cycle of oscillation. *There is a continuous net flow of energy in the positive x direction (the direction of propagation of the wave).*

Substituting for the y component of the force, we obtain

$$P = uF_y = \left(\frac{\partial y}{\partial t}\right)\left(F \frac{\partial y}{\partial x}\right)$$

$$= F[-\omega y_m \cos(kx - \omega t)][-k y_m \cos(kx - \omega t)]$$

$$= y_m^2 k\omega F \cos^2(kx - \omega t)$$

or

$$P = y_m^2 \mu v \omega^2 \cos^2(kx - \omega t), \qquad (26)$$

where we have used $v = F/\mu$ and $v = \omega/k$.

Note that the power or rate of flow of energy is not constant. This is because the input power oscillates: the work done by the hand that is moving the end of the string varies with the transverse displacement of that point. As energy is transported along the string, it is stored in each element of the string as a combination of kinetic energy and potential energy of deformation. This is similar to the case of a simple harmonic oscillator.

The power input to the string is often taken to be the average over one period of motion. The average power delivered is

$$\overline{P} = \frac{1}{T} \int_t^{t+T} P \, dt, \qquad (27)$$

where T is the period. The average value of $\sin^2 \theta$ or $\cos^2 \theta$ over one cycle is $\frac{1}{2}$, and so we obtain, using Eq. 26,

$$\overline{P} = \tfrac{1}{2} y_m^2 \mu v \omega^2, \qquad (28)$$

a result that does not depend on x or t. The dependence of the rate of transfer of energy on the *square* of the wave amplitude and *square* of the wave frequency is true in general for all types of waves.

In a three-dimensional wave, such as a light wave or a sound wave from a point source, it is often more useful to specify the *intensity* of the wave. The intensity I is defined as the *average power per unit area transmitted across an area A normal to the direction in which the wave is traveling,* or

$$I = \frac{\overline{P}}{A}. \qquad (29)$$

Just as with power in the wave traveling along the string, the intensity of any wave is always proportional to the square of the amplitude. (However, for circular or spherical waves, the amplitude is not constant as a wavefront travels; see Sample Problem 3.)

As a wave progresses through space, its energy may be

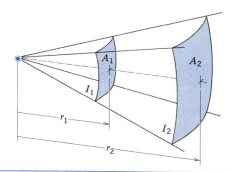

Figure 12   Sample Problem 3.

dissipated. Through internal friction or other viscous effects, the mechanical energy of the wave can be converted into internal energy of the string or heat energy transferred to the surroundings. In this chapter we neglect such energy transformations and assume that no mechanical energy is lost.

Sample Problem 3   Spherical waves travel from a source of waves whose power output, assumed constant, is P; see Fig. 12. How does the wave intensity depend on the distance from the source?

Solution   We assume that the medium is isotropic and that the source radiates uniformly in all directions, that is, its emission is spherically symmetrical.

The intensity of a wave is given by Eq. 29. The power is distributed uniformly over any spherical surface of area $A = 4\pi r^2$, and thus

$$I = \frac{P}{A} = \frac{P}{4\pi r^2}.$$

The wave intensity varies inversely as the square of its distance from the source. Since the intensity is proportional to the square of the amplitude, the amplitude of the wave must vary inversely as the distance from the source. Thus, for example, in doubling the distance from a source, the amplitude of a spherical wave decreases by half, and the intensity is only one-quarter as large.

19-7   THE PRINCIPLE OF SUPERPOSITION

We often observe two or more waves to travel simultaneously through the same region of space independently of one another. For example, the sound reaching our ears from a symphony orchestra is very complex, but we can pick out the sound made by individual instruments. The electrons in the antennas of our radio and TV sets are set into motion by a whole array of signals from different broadcasting centers, but we can nevertheless tune to any particular station, and the signal we receive from that station is in principle the same as that which we would receive if all other stations were to stop broadcasting.

The above examples illustrate the *principle of superposition,* which asserts that, when several waves combine at a point, the displacement of any particle at any given time is simply the vector sum of the displacements that each individual wave acting alone would give it. For example, suppose that two waves travel simultaneously along the same stretched string. Let $y_1(x,t)$ and $y_2(x,t)$ be the displacements that the string would experience if each wave acted alone. The displacement of the string when both waves act is then

$$y(x,t) = y_1(x,t) + y_2(x,t), \qquad (30)$$

the sum in this case being an algebraic one.

For mechanical waves in elastic media, the superposition principle holds whenever the restoring force varies linearly with the displacement. For electromagnetic waves, the superposition principle holds because the electric and magnetic fields are linearly related.

Figure 13 shows a time sequence of "snapshots" of two pulses traveling in opposite directions in the same stretched string. When the pulses overlap, the displacement of the string is the algebraic sum of the individual displacements of the string caused by each of the two

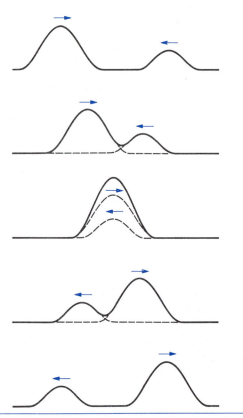

Figure 13 Two pulses travel in opposite directions along a stretched string. The superposition principle applies as they move through each other.

pulses alone, as Eq. 30 requires. The pulses simply move through one another, each moving along as if the other were not present.

The superposition principle may seem to be an obvious result, but there are instances in which it does not hold. Suppose, for instance, that one of the waves has such a large amplitude that the elastic limit of the medium is exceeded. The restoring force is no longer directly proportional to the displacement of a particle in the medium. Then, no matter what the amplitude of the second wave (even if it is very small), its effect at a point is not a linear function of its amplitude. Furthermore, the second wave will be changed by passing through the nonlinear region, and its subsequent behavior will be altered. This situation arises only very rarely, and in most circumstances the principle of superposition is valid (as we assume throughout this text).

Complex Waves

When two (or more) different waves, possibly having different amplitudes and wavelengths, are simultaneously present in a medium, we can apply the principle of superposition at every point and obtain a complex wave pattern $y(x,t)$ that may not look at all like the component waves. Nevertheless, it is an acceptable traveling waveform.

Figure 14*a* shows an example in the case of two sine waves of equal amplitude whose wavelengths are in the ratio of 3 : 1. The waves are traveling in the same direction with the same phase speed. They are in phase at $x = 0$. The darker curve shows the resultant waveform that can be calculated using Eq. 30. Note that it is not a sine wave. In Fig. 14*b*, the two combining waves are identical with those of Fig. 14*a*, except they are 180° out of phase at $x = 0$. The resultant waveform is quite different from that of Fig. 14*a*.

By changing the label on the horizontal axis in Fig. 14 from x to t, we would have a representation of the superposition of two waves as a function of time at a particular point. Such a graph might represent, for instance, the motion in time of a particular point on a string in response to the combination of two waves.

Fourier Analysis *(Optional)*

The importance of the superposition principle physically is that, where it holds, it makes it possible to analyze a complicated wave motion as a combination of simple waves. In fact, as was shown by the French mathematician J. Fourier (1768–1830), all that we need to build up the most general form of periodic wave are simple harmonic waves. Fourier showed that any periodic motion of a particle can be represented as a combination of simple harmonic motions. For example, if $y(x)$ represents the waveform (at a particular time) of a source of waves having a wavelength λ, we can analyze $y(x)$ as follows:

$$y(x) = A_0 + A_1 \sin kx + A_2 \sin 2kx + A_3 \sin 3kx + \cdots$$
$$+ B_1 \cos kx + B_2 \cos 2kx + B_3 \cos 3kx + \cdots, \quad (31)$$

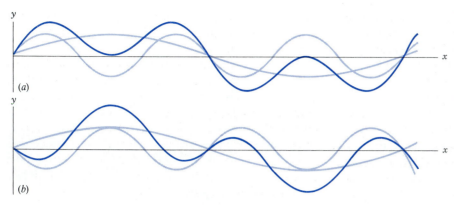

Figure 14   The addition of two waves with a wavelength ratio of 3:1 (lighter color) yields a wave whose shape (darker color) depends on the phase relationship of the two waves. Compare (a) and (b), which show different phase relationships of the added waves.

where $k = 2\pi/\lambda$. This expression is called a Fourier series. The coefficients A_i and B_i have definite values for any particular periodic motion $y(x)$. For example, the so-called sawtooth wave of Fig. 15 can be written

$$y(x) = -\frac{1}{\pi} \sin kx - \frac{1}{2\pi} \sin 2kx - \frac{1}{3\pi} \sin 3kx - \cdots.$$

If the motion is not periodic, as in the case of a pulse, the sum is replaced by an integral—the Fourier integral. Hence any motion (pulsed or continuous) of a source of waves can be represented in terms of a superposition of simple harmonic motions, and any waveform so generated can be analyzed as a combination of components that are individually simple harmonic waves. This once again illustrates the importance of harmonic motion and harmonic waves.

Only in the case of a nondispersive medium will the waveform

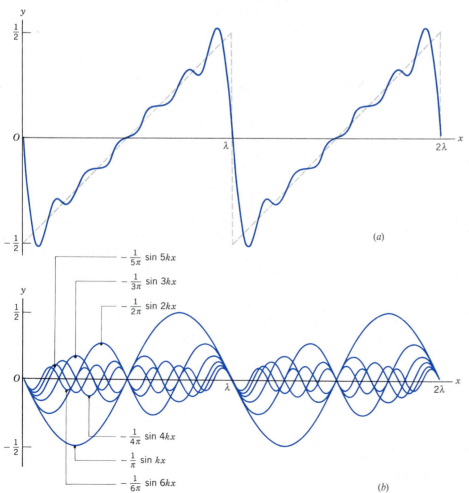

Figure 15   (a) The dashed line is a sawtooth wave commonly encountered in electronics. It can be represented as a Fourier series of sine waves. (b) The first six sine waves of the Fourier series that represents the sawtooth wave are shown, and their sum is shown as the solid curve in part (a). As more terms are included, the Fourier series becomes a better approximation of the wave.

maintain its shape as it travels. In a dispersive medium, the waveforms of the sinusoidal component waves do not change, but each may travel at a different speed. In this case, the combined waveform changes as the phase relationship between the components is altered. The wave can also change its shape if it loses mechanical energy to the medium, such as by air resistance, viscosity, or internal friction. Such dissipative forces often depend on the velocity, and so the Fourier components most strongly affected are those with higher particle velocities (that is, those with high frequencies, according to Eq. 20 in which u is seen to depend on ω). Here again the wave shape may change, as the higher frequency components lose amplitude more quickly. The decay with time of the sound of piano strings is an example of this phenomenon. The vibrational motion of a piano string, immediately after it is struck by the hammer, includes a wide range of frequencies, which give it its characteristic tone. The higher frequency components of this complex motion dissipate their energy more rapidly than the lower frequency components, and thus the character of a sustained tone may change with time. ■

Figure 16 Two wave trains, in this case circular ripples from two different disturbances, interfere where they overlap at particular points. The displacement at any point is the superposition of the individual displacements due to each of the two waves.

19-8 INTERFERENCE OF WAVES

When two or more waves combine at a particular point, they are said to *interfere,* and the phenomenon is called *interference.* As we shall see, the resultant waveform is strongly dependent on the relative phases of the interfering waves. Figure 16 shows an example of interfering waves.

Let us first consider two transverse sinusoidal waves of equal amplitude and wavelength, which travel in the x direction with the same speed. We take the phase constant of one wave to be ϕ, while the other has $\phi = 0$. Figure 17 shows the combined waveform at a particular time for the two cases of ϕ nearly 0 (the waves are nearly in phase) and

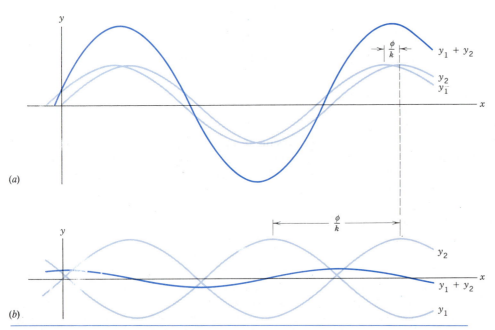

Figure 17 (*a*) The superposition of two waves of equal wavelength and amplitude that are almost in phase results in a wave of almost twice the amplitude of either component. (*b*) The superposition of two waves of equal wavelength and amplitude that are almost 180° out of phase results in a wave whose amplitude is nearly zero. Note that the wavelength of the resultant is unchanged in either case.

ϕ nearly 180° (the waves are nearly out of phase). You can see by merely adding the individual displacements at each x that in the first case there is nearly complete reinforcement of the two waves and the resultant has nearly double the amplitude of the individual components, while in the second case there is nearly complete cancellation at every point and the resultant amplitude is close to zero. These cases are known, respectively, as *constructive* interference and *destructive* interference.

Let us see how interference arises from the equations for the waves. We consider a general case in which the two waves have phase constants ϕ_1 and ϕ_2, respectively. The equations of the two waves are

$$y_1(x,t) = y_m \sin (kx - \omega t - \phi_1) \qquad (32)$$

and

$$y_2(x,t) = y_m \sin (kx - \omega t - \phi_2). \qquad (33)$$

Now let us find the resultant wave. Using the principle of superposition, we take the sum of Eqs. 32 and 33, which gives

$$y(x,t) = y_1(x,t) + y_2(x,t)$$
$$= y_m[\sin (kx - \omega t - \phi_1) \\ + \sin (kx - \omega t - \phi_2)]. \qquad (34)$$

From the trigonometric identity for the sum of the sines of two angles,

$$\sin B + \sin C = 2 \sin \tfrac{1}{2}(B + C) \cos \tfrac{1}{2}(B - C), \quad (35)$$

we obtain, after some rearrangement,

$$y(x,t) = [2y_m \cos (\Delta\phi/2)] \sin (kx - \omega t - \phi'), \quad (36)$$

where $\phi' = (\phi_1 + \phi_2)/2$. The quantity $\Delta\phi = (\phi_2 - \phi_1)$ is called the *phase difference* between the two waves.

This resultant wave corresponds to a new wave having the same frequency but with an amplitude $2y_m|\cos (\Delta\phi/2)|$. If $\Delta\phi$ is very small (compared to 180°), the resultant amplitude is nearly $2y_m$ (as shown in Fig. 17a). When $\Delta\phi$ is zero, the two waves have the same phase everywhere. The crest of one falls on the crest of the other and likewise for the valleys, which gives total constructive interference. The resultant amplitude is just twice that of either wave alone. If $\Delta\phi$ is close to 180°, on the other hand, the resultant amplitude is nearly zero (as shown in Fig. 17b). When $\Delta\phi$ is exactly 180°, the crest of one wave falls exactly on the valley of the other. The resultant amplitude is zero, corresponding to total destructive interference.

Notice that Eq. 36 always has the form of a sinusoidal wave. Thus adding two sine waves of the same wavelength and amplitude always gives a sine wave of the identical wavelength. We can also add components that have the same wavelength but different amplitudes. In this case the resultant again is a sine wave with the identical wavelength, but the resultant amplitude does not have the simple form given by Eq. 36. If the individual amplitudes are y_{1m} and y_{2m}, then if the waves are in phase ($\Delta\phi = 0$)

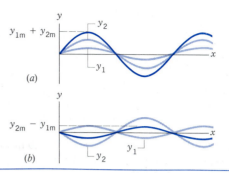

Figure 18   The addition of two waves of the same wavelength and phase but differing amplitudes (lighter color) gives a resultant of the same wavelength and phase. (*a*) The amplitudes add if the waves are in phase, and (*b*) they subtract if the waves are 180° out of phase.

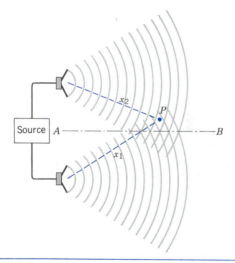

Figure 19   Two loudspeakers, driven by a common source, send signals to point *P*, where the signals interfere.

the resultant amplitude is $y_{1m} + y_{2m}$ (Fig. 18), while if they are out of phase ($\phi = 180°$) the resultant amplitude is $|y_{1m} - y_{2m}|$. There can be no complete destructive interference in this case, although there is partial destructive interference.

Figure 19 shows an example of the occurrence of interference effects. The loudspeakers are driven from the same source. At points equidistant from the speakers (on the line *AB*, which represents the entire midplane), there is complete constructive interference if the speakers are driven in phase ($\Delta\phi = 0$). There are also other points *P* where the waves arrive in phase and interfere constructively. That is, we can shift one of the waves in Fig. 18 by a phase constant of any integer multiple of 2π (or by a distance of any whole number of wavelengths), and the combined waveform is unchanged. These other points of constructive interference are located wherever the difference in distance to the two speakers is a whole number of wavelengths:

$$|x_1 - x_2| = \lambda, 2\lambda, 3\lambda, \ldots . \quad (37)$$

At other points P, the differing distances x_1 and x_2 result in the waves possibly arriving at P out of phase, even if they started out in phase at the speakers. The listening environment might therefore have "dead spots" where partial or complete destructive interference occurs for a particular wavelength λ. Maximal destructive interference occurs at points where

$$|x_1 - x_2| = \frac{\lambda}{2}, 3\frac{\lambda}{2}, 5\frac{\lambda}{2}, \ldots , \quad (38)$$

corresponding to a phase difference of 180°, 540°, 900°, and so on.

Of course, if the speakers emit a mixture of many different wavelengths, some points P might show destructive interference for one wavelength and constructive interference for another. The critical factor in determining the locations of the maxima and minima of sound intensity is the *path difference* $|x_1 - x_2|$. At points not on the midplane represented by the line AB, the two components arrive with different amplitudes (because the distances from the speakers are not the same; see Sample Problem 3). There will thus be no complete destructive interference. (In certain geometries, it is possible for the sound radiated from the *back* of a speaker to interfere with the sound radiated from the front. These two waves are 180° out of phase, and their interference can reduce the sound intensity at locations in front of the speaker. Loudspeaker enclosures are designed to eliminate this effect.)

Sample Problem 4   Two waves travel in the same direction along a string and interfere. The waves have the same wavelength and travel with the same speed. The amplitude of each wave is 9.7 mm, and there is a phase difference of 110° between them. (*a*) What is the amplitude of the combined wave resulting from the interference of the two waves? (*b*) To what value should the phase difference be changed so that the combined wave will have an amplitude equal to that of one of the original waves?

Solution   (*a*) The amplitude of the combined wave was given in Eq. 36:

$$2y_m|\cos(\Delta\phi/2)| = 2(9.7 \text{ mm})|\cos(110°/2)| = 11.1 \text{ mm}.$$

(*b*) If the quantity $2y_m|\cos(\Delta\phi/2)|$ is to equal y_m, then we must have

$$2|\cos(\Delta\phi/2)| = 1,$$

or

$$\Delta\phi = 2\cos^{-1}(\tfrac{1}{2}) = 120° \quad \text{or} \quad -120°.$$

Either wave can be leading the other by 120° (plus or minus any integer multiple of 360°) to produce the desired combination wave.

Sample Problem 5   In the geometry of Fig. 19, a listener is seated at a point a distance of 1.2 m directly in front of one speaker. The two speakers, which are separated by a distance D of 2.3 m, emit pure tones of wavelength λ. The waves are in phase when they leave the speakers. For what wavelengths will the listener hear a minimum in the sound intensity?

Solution   The minimum sound intensity occurs when the waves from the two speakers interfere destructively, according to the criteria of Eq. 38. If the listener is seated in front of speaker 2, then $x_2 = 1.2$ m, and x_1 can be found from the Pythagorean formula,

$$x_1 = \sqrt{x_2^2 + D^2} = \sqrt{(1.2 \text{ m})^2 + (2.3 \text{ m})^2} = 2.6 \text{ m}.$$

Thus $x_1 - x_2 = 2.6$ m $- 1.2$ m $= 1.4$ m, and, according to Eq. 38, we have

$$1.4 \text{ m} = \lambda/2, 3\lambda/2, 5\lambda/2, \ldots ,$$

corresponding to

$$\lambda = 2.8 \text{ m}, 0.93 \text{ m}, 0.56 \text{ m}, \ldots .$$

Complete destructive interference will not occur at this location, because the two waves arriving at the observation point have different amplitudes, if they leave the speakers with equal amplitudes.

19-9   STANDING WAVES

In the previous section we considered the effect of superposing two component waves of equal amplitude and frequency moving in the same direction on a string. What is the effect if the waves are moving along the string in *opposite* directions?

Figure 20 is a graphical indication of the effect of adding the component waveforms to obtain the resultant. Two traveling waves are shown in the figure, one moving to the left and the other to the right. "Snapshots" are shown of the two component waves and their resultant at intervals of $\frac{1}{4}$ period.

One particular feature results from this superposition: there are certain points along the string, called *nodes*, at which the displacement is zero *at all times*. (Figure 18 also showed some points in which the resultant had zero displacement, but that figure represented a snapshot of traveling waves *at a particular time*. If we took another snapshot an instant later, we would find that those points no longer had zero displacement, because the wave is traveling. In Fig. 20*c*, the zeros remain zeros at all times.) Between the nodes are the *antinodes*, where the displacement oscillates with the largest amplitude. Such a pattern of nodes and antinodes is known as a *standing wave*.

To analyze the standing wave mathematically, we represent the two waves by

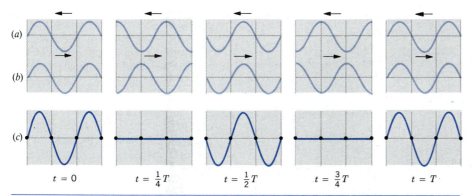

Figure 20   (*a,b*) Two traveling waves of the same wavelength and amplitude, moving in opposite directions. (*c*) The superposition of the two waves at different instants of time. The nodes in the standing wave pattern are indicated by dots. Note that the traveling waves have no nodes.

$$y_1(x,t) = y_m \sin (kx - \omega t),$$
$$y_2(x,t) = y_m \sin (kx + \omega t).$$

Hence the resultant may be written

$$y(x,t) = y_1(x,t) + y_2(x,t)$$
$$= y_m \sin (kx - \omega t) + y_m \sin (kx + \omega t) \quad (39)$$

or, making use of the trigonometric relation of Eq. 35,

$$y(x,t) = [2y_m \sin kx] \cos \omega t. \quad (40)$$

Equation 40 is the equation of a standing wave. It cannot represent a traveling wave, because x and t do *not* appear in the combination $x - vt$ or $x + vt$ required for a traveling wave.

Note that a particle at any particular location x executes simple harmonic motion as time goes on, and that all particles vibrate with the same angular frequency ω. In a traveling wave each particle of the string vibrates with the same amplitude. In a standing wave, however, *the amplitude is not the same for different particles but varies with the location x of the particle.* In fact, the amplitude, $|2y_m \sin kx|$, has a *maximum* value of $2y_m$ at positions where

$$kx = \frac{\pi}{2}, \frac{3\pi}{2}, \frac{5\pi}{2}, \dots$$

or

$$x = \frac{\lambda}{4}, \frac{3\lambda}{4}, \frac{5\lambda}{4}, \dots. \quad (41)$$

These points are the antinodes and are spaced one-half wavelength apart. The amplitude has a *minimum* value of zero at positions where

$$kx = \pi, 2\pi, 3\pi, \dots$$

or

$$x = \frac{\lambda}{2}, \lambda, \frac{3\lambda}{2}, \dots. \quad (42)$$

These points are the nodes and are also spaced one-half

wavelength apart. The separation between a node and an adjacent antinode is one-quarter wavelength.

It is clear that energy is not transported along the string to the right or to the left, for energy cannot flow past the nodes in the string, which are permanently at rest. Hence the energy remains "standing" in the string, although it alternates between vibrational kinetic energy and elastic potential energy. When the antinodes are all at their maximum displacements, the energy is stored entirely as potential energy, in particular as the elastic potential energy associated with the stretching of the string. When all parts of the string are simultaneously passing through equilibrium (as in the second and fourth snapshots of Fig. 20), the energy is stored entirely as kinetic energy. Figure 21 shows a more detailed description of the shifting of energy between kinetic and potential forms during one cycle of oscillation. Compare Fig. 21 with Fig. 6 of Chapter 8 for the oscillating block–spring system. How are these systems similar?

We can equally well regard the motion as an oscillation of the string as a whole, each particle undergoing simple harmonic motion of angular frequency ω and with an amplitude that depends on its location. Each small part of the string has inertia and elasticity, and the string as a whole can be thought of as a collection of coupled oscillators. Hence the vibrating string is the same in principle as the block–spring system, except that the block–spring system has only one natural frequency, and a vibrating string has a large number of natural frequencies (see Section 19-10).

An easy way to achieve a standing wave is to superpose a wave traveling down a string with its reflection traveling in the opposite direction. Let us now consider the process of reflection of a wave more closely. Suppose a pulse travels down a stretched string that is fixed at one end, as shown in Fig. 22*a*. When the pulse arrives at that end, it exerts an upward force on the support. The support is rigid, however, and does not move. By Newton's third law

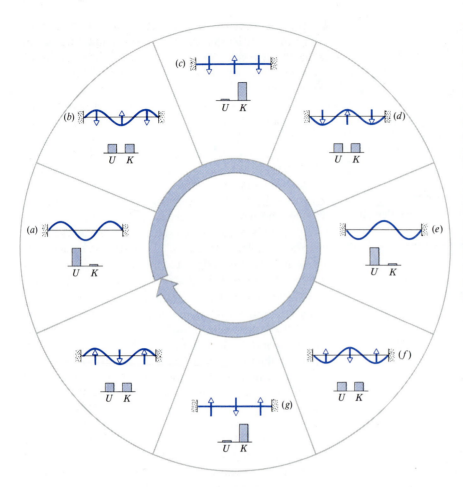

Figure 21 A standing wave on a stretched string, showing one cycle of oscillation. At (*a*) the string is momentarily at rest with the antinodes at their maximum displacement. The energy of the string is all elastic potential energy. (*b*) One-eighth of a cycle later, the displacement is reduced and the energy is partly potential and partly kinetic. The vectors show the instantaneous velocities of particles of the string at certain locations. (*c*) The displacement is zero; there is no potential energy, and the kinetic energy is maximum. The particles of the string have their maximum velocities. (*d–h*) The motion continues through the remainder of the cycle, with the energy being continually exchanged between potential and kinetic forms.

the support exerts an equal but oppositely directed force on the string. This reaction force generates a pulse at the support, which travels back along the string in a direction opposite to that of the incident pulse. We say that the incident pulse has been *reflected* at the fixed endpoint of the string. Note that the reflected pulse returns with its transverse displacement reversed. If a wavetrain is incident on the fixed endpoint, a reflected wavetrain is generated at that point in the same way. The displacement of any point along the string is the sum of the displacements caused by the incident and reflected waves. Since the endpoint is fixed, these two waves must always interfere destructively at that point so as to give zero displacement there. Hence the reflected wave is always 180° out of phase with the incident wave at a fixed boundary. *On reflection from a fixed end, a transverse wave undergoes a phase change of 180°.*

The reflection of a pulse at a free end of a stretched string, that is, at an end that is free to move transversely, is represented in Fig. 22*b*. The end of the string is attached to a very light ring free to slide without friction along a transverse rod. When the pulse arrives at the free end, it exerts a force on the element of string there. This element is accelerated, and (as in the case of a pendulum) its motion carries it past the equilibrium point; it "overshoots" and exerts a reaction force on the string. This generates a pulse that travels back along the string in a direction opposite to that of the incident pulse. Once again we get reflection, but now at a free end. The free end will obviously suffer the maximum displacement of the particles on the string; an incident and a reflected wavetrain must interfere constructively at that point if we are to have a maximum there. Hence the reflected wave is always in phase with the incident wave at that point. *At a free end, a transverse wave is reflected without change of phase.*

Figure 23 shows time exposures of the standing wave patterns that can be obtained by shaking a stretched string that is fixed at one end.

So far we have assumed that the wave reflects at the boundary with no loss of intensity. In practice, we always find that at any boundary between two media there is partial reflection and partial transmission; for example, looking at a piece of ordinary window glass, you can see some light reflected back toward you and some transmitted through the glass. We can demonstrate this effect with transverse waves on strings by tying together two strings of different mass densities. When a wave traveling along one of the strings reaches the point where the strings are

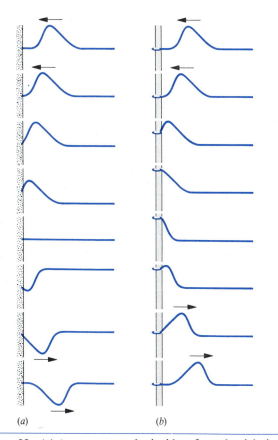

Figure 22 (*a*) A transverse pulse incident from the right is reflected by a rigid wall. Note that the phase of the reflected pulse is inverted, or changed by 180°. (*b*) Here the end of the string is free to move, the string being attached to a loop that can slide freely along the rod. The phase of the reflected pulse is unchanged.

joined, part of the wave energy is transmitted to the other string and part is reflected back. The amplitude of the reflected wave is less than the amplitude of the original incident wave, because the wave transmitted to the second string carries away some of the incident energy.

If the second string has a greater mass density than the first, the wave reflected back into the first string still suffers a phase shift of 180° on reflection. But because its amplitude is less than the incident wave, the boundary point is not a node and moves. Thus a net energy transfer occurs along the first string into the second. If the second string has a smaller mass density than the first, partial reflection occurs without change of phase, but once again energy is transmitted to the second string. In practice, the best way to realize a "free end" for a string is to attach it to a long and very much lighter string. The energy transmitted is negligible, and the second string serves to maintain the tension in the first one.

Note that the transmitted wave travels with a speed different from that of the incident and reflected waves. The wave speed is determined by the relation $v = \sqrt{F/\mu}$; the tension is the same in both strings, but their densities are different. Hence the wave travels more slowly in the denser string. The frequency of the transmitted wave is the same as that of the incident and reflected waves. (If this were not true, there would be a discontinuity at the point where the strings are joined.) Waves having the same frequency but traveling with different speeds have different wavelengths. From the relation $\lambda = v/\nu$, we conclude that in the denser string, where v is smaller, the wavelength is shorter. This phenomenon of change of

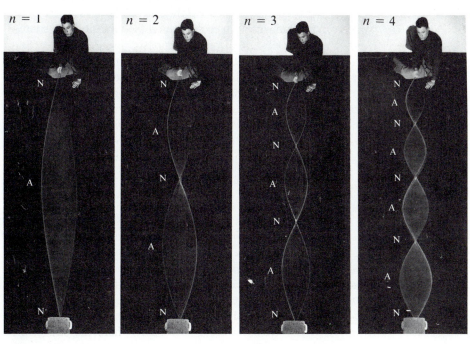

Figure 23 A student shakes a stretched string (actually a rubber tube) at four resonant frequencies, producing four different patterns of standing waves.

wavelength as a wave passes from one medium to another will be encountered frequently in our study of light waves. It also occurs for sound waves: a string, such as on a guitar, vibrates with a certain frequency and wavelength; the wave transmitted to the air has the same frequency as that of the string, but a different wavelength, because the speed of waves on the string differs from their speed in air.

19-10   RESONANCE

Look again at the standing wave patterns of Fig. 23. You will see that it shows four different standing waves that can occur. The spacing between the nodes differs in the four patterns, and since the wavelength is twice the distance between adjacent nodes, the wavelength differs as well. The phase velocity, on the other hand, is the same in all four situations, being determined only by the tension in the string. The relationship $v = \lambda v$ then tells us that if v is constant and λ changes, the frequency v must certainly be different for the different standing waves. The student must therefore be shaking the string at certain different but well-defined frequencies in the four photographs.

The photos in Fig. 23 seem to show a system with nodes at both ends. (If the student is wiggling the string at one end, he is doing it at very small amplitude so that the end is approximately a node.) The spacing between nodes is always one-half wavelength, so the condition for a standing wave to be set up in the string is that the length L of the string be equal to an integral number n of half-wavelengths:

$$L = n\frac{\lambda}{2} \qquad (n = 1,2,3, \ . \ . \ .)$$

or

$$\lambda_n = \frac{2L}{n} \qquad (n = 1,2,3, \ . \ . \ .). \qquad (43)$$

In terms of the frequency, we can write Eq. 43 as

$$v_n = \frac{v}{\lambda_n} = n\frac{v}{2L} \qquad (n = 1,2,3, \ . \ . \ .). \qquad (44)$$

That is, the student must shake the string at these particular frequencies (corresponding to $n = 1, 2, 3$, and 4) to produce the standing waves.

We can consider the frequencies of Eq. 44 to be the *natural frequencies* of the oscillating system (the string). When the frequency of the driving force (the student's hand) matches one of the allowed natural frequencies, a standing wave is produced and the system begins to move at large amplitude. This is the condition of *resonance* we previously discussed in Section 15-9.

A block on a spring is also capable of resonating, but only at a single frequency. Why then does the stretched string have an infinite number of resonant frequencies? In the block–spring system, the inertia (the block) is concentrated ("lumped") in one part of the system while the elasticity (the spring) is concentrated in another. Such a resonant system is said to have *lumped* elements. The stretched string, on the other hand, is said to have *distributed* elements, because every part of the string has both inertial and elastic properties. There are many possible ways for the string to store its kinetic and potential energies, in contrast to only a single way in the block–spring system. A lumped system of N objects has N natural frequencies, each of which corresponds to a different pattern of oscillation (Fig. 24). The limit as N tends to infinity leads us to the completely distributed system of the stretched string, with its infinite number of resonant frequencies.

If the vibrating string of Fig. 23 were set into motion and left alone, the vibrations would gradually die out. The motion of the string is damped by dissipation of energy through the supports at the ends and by resistance of the air to the motion. To maintain the vibration, the student

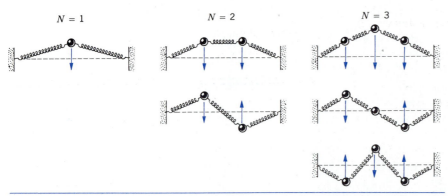

Figure 24   Some patterns of oscillation of an oscillator having lumped elements, in this case oscillating bodies connected by springs of negligible mass. Each different pattern of motion has a different natural frequency, the number of natural frequencies being equal to the number of oscillating bodies.

must pump energy into the system by applying a driving force. When the driving frequency is very different from one of the resonant frequencies, the reflected wave causes the string to do work on the student's hand; energy is lost from the string in this manner in addition to that lost through damping. At resonance, the motion of the student's hand is in phase with that of the string, and no energy is lost by the string through work done on the student's hand. All the energy supplied by the student, less that lost to damping, is stored in the oscillation, and the result is motion at a large amplitude. Eventually, a stable situation is reached, in which the energy supplied by the driving force exactly balances the losses due to damping.

This motion is analogous to that of the damped harmonic oscillator discussed in Section 15-9. The resonant frequency is almost, but not quite, a natural frequency of the string. The apparent nodes are not true nodes, because energy must be flowing past them along the string to compensate for the losses due to damping. If there were no damping, the resonant frequency would be exactly a natural frequency, and the amplitude would increase without limit as energy continued to be supplied to the string. Eventually, the elastic limit would be exceeded and the string would break. (The elastic limit can be exceeded even if damping is present, as Fig. 21 of Chapter 15 showed.)

If the student wiggles the string at a frequency that differs from one of the natural frequencies of the system, the reflected wave returns to the student's hand out of phase with the motion of the hand. In this case, the string does work on the hand, in addition to the hand doing work on the string. No fixed standing wave pattern is produced. The amplitude of the resulting motion is small and not much different from that of the motion of the student's hand. This situation is analogous to the erratic motion of a swing being pushed with a frequency other than its natural one; the resulting displacement of the swing is rather small.

At resonance, the string absorbs as much energy as it can from the student's hand. This is true for any vibrating system. In tuning a radio, the natural frequency of an electronic circuit is changed until it matches a particular frequency of the radio waves that are broadcast from the station. At this point the circuit resonates with the signal and absorbs as much energy from the signal as it can. Other similar resonance conditions occur in sound, electromagnetism, optics, and atomic and nuclear physics.

In the next chapter, we consider in greater detail the importance of resonance in understanding the properties of different musical instruments and the way they produce their characteristic sounds. Although we have used the vibrating string in this section as our example of a vibrating system, the principles discussed here apply to all vibrating systems that can sustain wave motion.

Sample Problem 6 In the arrangement of Fig. 25, a vibrator sets the string into motion at a frequency of 120 Hz. The string has a length of $L = 1.2$ m, and its linear mass density is 1.6 g/m. To what value must the tension be adjusted (by increasing the hanging weight) to obtain the pattern of motion having four loops?

Solution To find the tension, we can substitute Eq. 18 into Eq. 44 and obtain

$$F = \frac{4L^2 v^2 \mu}{n^2}.$$

The tension corresponding to $n = 4$ (for 4 loops) is found to be

$$F = \frac{4(1.2 \text{ m})^2(120 \text{ Hz})^2(0.0016 \text{ kg/m})}{4^2} = 8.3 \text{ N}.$$

This corresponds to a hanging weight of about 2 lb.

Sample Problem 7 A violin string tuned to concert A (440 Hz) has a length of 0.34 m. (a) What are the three longest wavelengths of the resonances of the string? (b) What are the corresponding wavelengths that reach the ear of the listener?

Solution (a) The resonant wavelengths of a string of length $L = 0.34$ m can be found directly from Eq. 43:

$$\lambda_1 = 2L/1 = 2(0.34 \text{ m}) = 0.68 \text{ m},$$

$$\lambda_2 = 2L/2 = 0.34 \text{ m},$$

$$\lambda_3 = 2L/3 = 0.23 \text{ m}.$$

(b) When a wave passes from one medium (the string) to another (the air) of differing wave speed, the frequency remains the same, but the wavelength changes. Equation 19 gives the relationship between the wavelengths. To find the wave speed on the string, we note that in the lowest resonant mode $\nu = 440$ Hz and $\lambda = 0.68$ m, so that

$$v = \nu\lambda = (440 \text{ Hz})(0.68 \text{ m}) = 299 \text{ m/s}.$$

In air, we take the wave speed to be 343 m/s, and from Eq. 19 we obtain

$$\lambda_{\text{air}} = \lambda_{\text{string}} \frac{v_{\text{air}}}{v_{\text{string}}} = \lambda_{\text{string}} \frac{343 \text{ m/s}}{299 \text{ m/s}} = 1.15\lambda_{\text{string}}.$$

We thus find the wavelengths in air:

$$\lambda_1 = 0.78 \text{ m}, \qquad \lambda_2 = 0.39 \text{ m}, \qquad \lambda_3 = 0.26 \text{ m}.$$

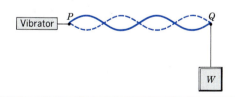

Figure 25 Sample Problem 6. A string under tension is connected to a vibrator. For a fixed vibrator frequency, standing wave patterns will occur for certain discrete values of the tension in the string.

QUESTIONS

1. How could you prove experimentally that energy is associated with a wave?

2. Energy can be transferred by particles as well as by waves. How can we experimentally distinguish between these methods of energy transfer?

3. Can a wave motion be generated in which the particles of the medium vibrate with angular simple harmonic motion? If so, explain how and describe the wave.

4. In analyzing the motion of an elastic wave through a material medium, we often ignore the molecular structure of matter. When is this justified and when isn't it?

5. How do the amplitude and the intensity of surface water waves vary with the distance from the source?

6. How can one create plane waves? Spherical waves?

7. A passing motor boat creates a wake that causes waves to wash ashore. As time goes on, the period of the arriving waves grows shorter and shorter. Why?

8. The following functions in which A is a constant are of the form $y = f(x \pm vt)$:

$$y = A(x - vt), \qquad y = A(x + vt)^2,$$
$$y = A\sqrt{x - vt}, \qquad y = A \ln(x + vt).$$

Explain why these functions are not useful in wave motion.

9. Can one produce on a string a waveform that has a discontinuity in slope at a point, that is, a sharp corner? Explain.

10. The inverse-square law does not apply exactly to the decrease in intensity of sounds with distance. Why not?

11. When two waves interfere, does one alter the progress of the other?

12. When waves interfere, is there a loss of energy? Explain your answer.

13. Why don't we observe interference effects between the light beams emitted from two flashlights or between the sound waves emitted by two violins?

14. As Fig. 20 shows, twice during the cycle the configuration of standing waves in a stretched string is a straight line, exactly what it would be if the string were not vibrating at all. Discuss from the point of view of energy conservation.

15. Two waves of the same amplitude and frequency are travel-

ing on the same string. At a certain instant the string looks like a straight line. Are the two waves necessarily traveling in the same direction? What is the phase relationship between the two waves?

16. If two waves differ only in amplitude and are propagated in opposite directions through a medium, will they produce standing waves? Is energy transported? Are there any nodes?

17. The partial reflection of wave energy by discontinuities in the path of transmission is usually wasteful and can be minimized by insertion of "impedance matching" devices between sections of the path bordering on the discontinuity. For example, a megaphone helps match the air column of mouth and throat to the air outside the mouth. Give other examples and explain qualitatively how such devices minimize reflection losses.

18. Consider the standing waves in a string to be a superposition of traveling waves and explain, using superposition ideas, why there are no true nodes in the resonating string of Fig. 25, even at the "fixed" end. (*Hint:* Consider damping effects.)

19. Standing waves in a string are demonstrated by an arrangement such as that of Fig. 25. The string is illuminated by a fluorescent light and the vibrator is driven by the same electric outlet that powers the light. The string exhibits a curious color variation in the transverse direction. Explain.

20. In the discussion of transverse waves on a string, we have dealt only with displacements in a single plane, the xy plane. If all displacements lie in one plane, the wave is said to be *plane polarized.* Can there be displacements in a plane other than the plane dealt with? If so, can two different plane polarized waves be combined? What appearance would such a combined wave have?

21. A wave transmits energy. Does it transfer momentum? Can it transfer angular momentum? (See "Energy and Momentum Transport in String Waves," by D. W. Juenker, *American Journal of Physics,* January 1976, p. 94.)

22. In the Mexico City earthquake of September 19, 1985, areas with high damage alternated with areas of low damage. Also, buildings between 5 and 15 stories high sustained the most damage. Discuss these effects in terms of standing waves and resonance.

PROBLEMS

Section 19-3 Traveling Waves

1. A wave has a wave speed of 243 m/s and a wavelength of 3.27 cm. Calculate (*a*) the frequency and (*b*) the period of the wave.

2. By rocking a boat, a child produces surface water waves on a previously quiet lake. It is observed that the boat performs 12 oscillations in 30 s and also that a given wave crest reaches shore 15 m away in 5.0 s. Find (*a*) the frequency, (*b*) the speed, and (*c*) the wavelength of the waves.

3. A sinusoidal wave travels along a string. The time for a particular point to move from maximum displacement to zero displacement is 178 ms. The wavelength of the wave is 1.38 m. Find (*a*) the period, (*b*) the frequency, and (*c*) the speed of the wave.

4. Write an expression describing a transverse wave traveling along a cord in the $+x$ direction with wavelength 11.4 cm, frequency 385 Hz, and amplitude 2.13 cm.

5. Write the equation for a wave traveling in the negative direction along the x axis and having an amplitude of 1.12 cm, a frequency of 548 Hz, and a speed of 326 m/s.

6. A wave of frequency 493 Hz has a speed of 353 m/s. (*a*) How far apart are two points differing in phase by 55.0°? (*b*) Find the difference in phase between two displacements at the same point but at times differing by 1.12 ms.

Section 19-4 Wave Speed

7. Show (*a*) that the maximum transverse speed of a particle in a string owing to a traveling wave is given by $u_{max} = \omega y_m$, and (*b*) that the maximum transverse acceleration is $a_{max} = \omega^2 y_m$.

8. The equation of a transverse wave traveling along a string is given by

$$y = (2.30 \times 10^{-3}) \sin (18.2x - 588t),$$

where x and y are in meters and t is in seconds. Find (*a*) the amplitude, (*b*) the frequency, (*c*) the velocity, (*d*) the wavelength of the wave, and (*e*) the maximum transverse speed of a particle in the string.

9. The equation of a transverse wave traveling along a very long string is given by $y = 6.0 \sin (0.020\pi x + 4.0\pi t)$, where x and y are expressed in centimeters and t in seconds. Calculate (*a*) the amplitude, (*b*) the wavelength, (*c*) the frequency, (*d*) the speed, (*e*) the direction of propagation of the wave, and (*f*) the maximum transverse speed of a particle in the string.

10. Calculate the speed of a transverse wave in a cord of length 2.15 m and mass 62.5 g under a tension of 487 N.

11. The speed of a wave on a string is 172 m/s when the tension is 123 N. To what value must the tension be increased in order to raise the wave speed to 180 m/s?

12. Show that, in terms of the tensile stress S and mass density ρ, the speed v of transverse waves in a wire is given by $v = (S/\rho)^{1/2}$.

13. The equation of a particular transverse wave on a string is $y = 1.8 \sin (23.8x + 317t)$, where x is in meters, y is in millimeters, and t is in seconds. The string is under a tension of 16.3 N. Find the linear mass density of the string.

14. A continuous sinusoidal wave is traveling on a string with speed 82.6 cm/s. The displacement of the particles of the string at $x = 9.60$ cm is found to vary with time according to the equation $y = 5.12 \sin (1.16 - 4.08t)$, where y is in centimeters and t is in seconds. The linear mass density of the string is 3.86 g/cm. (*a*) Find the frequency of the wave. (*b*) Find the wavelength of the wave. (*c*) Write the general equation giving the transverse displacement of the particles of the string as a function of position and time. (*d*) Calculate the tension in the string.

15. A simple harmonic transverse wave is propagating along a string toward the left (or $-x$) direction. Figure 26 shows a plot of the displacement as a function of position at time $t = 0$. The string tension is 3.6 N and its linear density is 25 g/m. Calculate (*a*) the amplitude, (*b*) the wavelength, (*c*) the wave speed, (*d*) the period, and (*e*) the maximum speed of a particle in the string. (*f*) Write an equation describing the traveling wave.

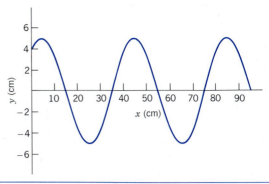

Figure 26 Problem 15.

16. Prove that the slope of a string at any point is numerically equal to the ratio of the particle speed to the wave speed at that point.

17. For a wave on a stretched cord, find the ratio of the maximum particle speed (the maximum speed with which a single particle in the cord moves transverse to the wave) to the wave speed. If a wave having a certain frequency and amplitude is imposed on a cord, would this speed ratio depend on the material of which the cord is made, such as wire or nylon?

18. In Fig. 27*a*, string #1 has a linear mass density of 3.31 g/m, and string #2 has a linear mass density of 4.87 g/m. They are under tension owing to the hanging block of mass $M = 511$ g. (*a*) Calculate the wave speed in each string. (*b*) The block is now divided into two blocks (with $M_1 + M_2 = M$) and the apparatus is rearranged as shown in Fig. 27*b*. Find M_1 and M_2 such that the wave speeds in the two strings are equal.

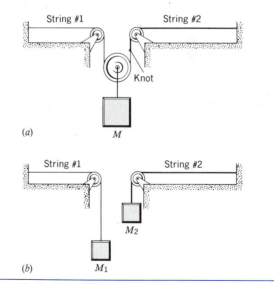

Figure 27 Problem 18.

19. A wire 10.3 m long and having a mass of 97.8 g is stretched under a tension of 248 N. If two pulses, separated in time by 29.6 ms, are generated one at each end of the wire, where will the pulses meet?

20. Find the speed of the fastest transverse wave that can be sent

along a steel wire. Allowing for a reasonable safety factor, the maximum tensile stress to which steel wires should be subject is 720 MPa. The density of steel is 7.80 g/cm³. Show that your answer does not depend on the diameter of the wire.

21. The type of rubber band used inside some baseballs and golfballs obeys Hooke's law over a wide range of elongation of the band. A segment of this material has an unstretched length L and a mass m. When a force F is applied, the band stretches an additional length ΔL. (a) What is the speed (in terms of m, ΔL, and the force constant k) of transverse waves on this rubber band? (b) Using your answer to (a), show that the time required for a transverse pulse to travel the length of the rubber band is proportional to $1/\sqrt{\Delta L}$ if $\Delta L \ll L$ and is constant if $\Delta L \gg L$.

22. A uniform rope of mass m and length L hangs from a ceiling. (a) Show that the speed of a transverse wave in the rope is a function of y, the distance from the lower end, and is given by $v = \sqrt{gy}$. (b) Show that the time it takes a transverse wave to travel the length of the rope is given by $t = 2\sqrt{L/g}$. (c) Does the actual mass of the rope affect the results of (a) and (b)?

23. A nonuniform wire of length L and mass M has a variable linear mass density given by $\mu = kx$, where x is the distance from one end of the wire and k is a constant. (a) Show that $M = kL^2/2$. (b) Show that the time t required for a pulse generated at one end of the wire to travel to the other end is given by $t = \sqrt{8ML/9F}$, where F is the tension in the wire.

24. A uniform circular hoop of string is rotating clockwise in the absence of gravity (see Fig. 28). The tangential speed is v_0. Find the speed of waves on this string. (*Remark:* The answer is independent of the radius of the hoop and the linear mass density of the string!)

Figure 28   Problem 24.

Section 19-6   Power and Intensity in Wave Motion

25. A string 2.72 m long has a mass of 263 g. The tension in the string is 36.1 N. What must be the frequency of traveling waves of amplitude 7.70 mm in order that the average transmitted power be 85.5 W?

26. A line source emits a cylindrical expanding wave. Assuming the medium absorbs no energy, find how (a) the intensity and (b) the amplitude of the wave depend on the distance from the source.

27. A wave travels out uniformly in all directions from a point source. (a) Justify the following expression for the displacement y of the medium at any distance r from the source:

$$y = \frac{Y}{r} \sin k(r - vt).$$

Consider the speed, direction of propagation, periodicity, and intensity of the wave. (b) What are the dimensions of the constant Y?

28. An observer measures an intensity of 1.13 W/m² at an unknown distance from a source of spherical waves whose power output is also unknown. The observer walks 5.30 m closer to the source and measures an intensity of 2.41 W/m² at this new location. Calculate the power output of the source.

29. (a) Show that the intensity I is the product of the energy density u (energy per unit volume) and the speed of propagation v of a wave disturbance; that is, show that $I = uv$. (b) Calculate the energy density in a sound wave 4.82 km from a 47.5-kW siren, assuming the waves to be spherical, the propagation isotropic with no atmospheric absorption, and the speed of sound to be 343 m/s.

30. A transverse sinusoidal wave is generated at one end of a long, horizontal string by a bar that moves up and down through a distance of 1.12 cm. The motion is continuous and is repeated regularly 120 times per second. The string has linear density 117 g/m and is kept under a tension of 91.4 N. Find (a) the maximum value of the transverse speed u and (b) the maximum value of the transverse component of the tension. (c) Show that the two maximum values calculated above occur at the same phase values for the wave. What is the transverse displacement y of the string at these phases? (d) What is the maximum power transferred along the string? (e) What is the transverse displacement y for conditions under which this maximum power transfer occurs? (f) What is the minimum power transfer along the string? (g) What is the transverse displacement y for conditions under which this minimum power transfer occurs?

Section 19-8   Interference of Waves

31. What phase difference between two otherwise identical traveling waves, moving in the same direction along a stretched string, will result in the combined wave having an amplitude 1.65 times that of the common amplitude of the two combining waves? Express your answer in both degrees and radians.

32. Determine the amplitude of the resultant wave when two sinusoidal waves having the same frequency and traveling in the same direction are combined, if their amplitudes are 3.20 cm and 4.19 cm and they differ in phase by $\pi/2$ rad.

33. Two pulses are traveling along a string in opposite directions, as shown in Fig. 29. (a) If the wave speed is 2.0 m/s and the pulses are 6.0 cm apart, sketch the patterns after 5.0, 10, 15, 20, and 25 ms. (b) What has happened to the energy at $t = 15$ ms?

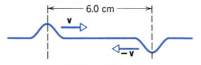

Figure 29   Problem 33.

34. Three sinusoidal waves travel in the positive x direction along the same string. All three waves have the same frequency. Their amplitudes are in the ratio $1 : \frac{1}{2} : \frac{1}{3}$ and their phase angles are 0, $\pi/2$, and π, respectively. Plot the resultant waveform and discuss its behavior as t increases.

35. Four sinusoidal waves travel in the positive x direction

along the same string. Their frequencies are in the ratio
$1:2:3:4$ and their amplitudes are in the ratio $1:\frac{1}{2}:\frac{1}{3}:\frac{1}{4}$, respectively. When $t = 0$, at $x = 0$, the first and third waves are
$180°$ out of phase with the second and fourth. Plot the resultant waveform when $t = 0$ and discuss its behavior as t increases.

36. Consider two point sources S_1 and S_2 in Fig. 30, which emit
waves of the same frequency and amplitude. The waves start
in the same phase, and this phase relation at the sources is
maintained throughout time. Consider points P at which r_1
is nearly equal to r_2. (*a*) Show that the superposition of these
two waves gives a wave whose amplitude y_m varies with the
position P approximately according to

$$y_m = \frac{2Y}{r} \cos \frac{k}{2}(r_1 - r_2),$$

in which $r = (r_1 + r_2)/2$. (*b*) Then show that total cancellation occurs when $r_1 - r_2 = (n + \frac{1}{2})\lambda$, n being any integer, and
that total re-enforcement occurs when $r_1 - r_2 = n\lambda$. The
locus of points whose difference in distance from two fixed
points is a constant is a hyperbola, the fixed points being the
foci. Hence each value of n gives a hyperbolic line of constructive interference and a hyperbolic line of destructive
interference. At points at which r_1 and r_2 are not approximately equal (as near the sources), the amplitudes of the
waves from S_1 and S_2 differ and the cancellations are only
partial. (This is the basis of the OMEGA navigation system.)

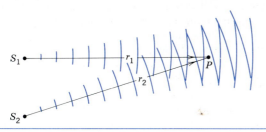

Figure 30 Problem 36.

37. A source S and a detector D of high-frequency waves are a
distance d apart on the ground. The direct wave from S is
found to be in phase at D with the wave from S that is
reflected from a horizontal layer at an altitude H (Fig. 31).
The incident and reflected rays make the same angle with
the reflecting layer. When the layer rises a distance h, no
signal is detected at D. Neglect absorption in the atmosphere
and find the relation between d, h, H, and the wavelength λ
of the waves.

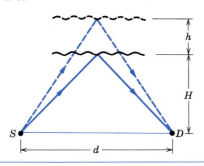

Figure 31 Problems 37 and 38.

38. Refer to Problem 37 and Fig. 31. Suppose that $d = 230$ km
and $H = 510$ km. The waves are 13.0 MHz radio waves
($v = 3.00 \times 10^8$ m/s). At the detector D the combined signal
strength varies from a maximum to zero and back to a
maximum again six times in 1 min. At what vertical speed is
the reflecting layer moving? (The layer is moving slowly, so
that the vertical distance moved in 1 min is small compared
to H and d.)

Section 19-9 Standing Waves

39. A string fixed at both ends is 8.36 m long and has a mass of
122 g. It is subjected to a tension of 96.7 N and set vibrating.
(*a*) What is the speed of the waves in the string? (*b*) What is
the wavelength of the longest possible standing wave?
(*c*) Give the frequency of that wave.

40. A nylon guitar string has a linear mass density of 7.16 g/m
and is under a tension of 152 N. The fixed supports are
89.4 cm apart. The string is vibrating in the standing wave
pattern shown in Fig. 32. Calculate the (*a*) speed, (*b*) wavelength, and (*c*) frequency of the component waves whose
superposition gives rise to this vibration.

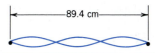

Figure 32 Problem 40.

41. The equation of a transverse wave traveling in a string is
given by

$$y = 0.15 \sin (0.79x - 13t),$$

in which x and y are expressed in meters and t is in seconds.
(*a*) What is the displacement at $x = 2.3$ m, $t = 0.16$ s?
(*b*) Write down the equation of a wave that, when added to
the given one, would produce standing waves on the string.
(*c*) What is the displacement of the resultant standing wave
at $x = 2.3$ m, $t = 0.16$ s?

42. A string vibrates according to the equation

$$y = 0.520 \sin (1.14x) \cos (137t),$$

where x and y are in centimeters and t is in seconds. (*a*) What
are the amplitude and speed of the component waves whose
superposition can give rise to this vibration? (*b*) Find the
distance between nodes. (*c*) What is the velocity of a particle
of the string at the position $x = 1.47$ cm at time $t = 1.36$ s?

43. Vibrations from a 622-Hz tuning fork set up standing waves
in a string clamped at both ends. The wave speed for the
string is 388 m/s. The standing wave has four loops and an
amplitude of 1.90 mm. (*a*) What is the length of the string?
(*b*) Write an equation for the displacement of the string as a
function of position and time.

44. Consider a standing wave that is the sum of two waves traveling in opposite directions but otherwise identical. Show that
the maximum kinetic energy in each loop of the standing
wave is $2\pi^2 \mu y_m^2 \nu v$.

45. An incident traveling wave, amplitude A_i, is only partially
reflected from a boundary, with the amplitude of the reflected wave being A_r. The resulting superposition of two
waves with different amplitudes and traveling in opposite

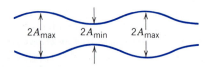

Figure 33   Problems 45 and 46.

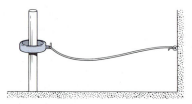

Figure 34   Problem 51.

directions gives a standing wave pattern of waves whose envelope is shown in Fig. 33. The *standing wave ratio* (SWR) is defined as $(A_i + A_r)/(A_i - A_r) = A_{max}/A_{min}$, and the percent reflection is defined as the ratio of the average power in the reflected wave to the average power in the incident wave, times 100. (*a*) Show that for 100% reflection SWR $= \infty$ and that for no reflection SWR $= 1$. (*b*) Show that a measurement of the SWR just before the boundary reveals the percent reflection occurring at the boundary according to the formula

$$\% \text{ reflection} = [(SWR - 1)^2/(SWR + 1)^2](100).$$

46. Estimate (*a*) the SWR (standing wave ratio) and (*b*) the percent reflection at the boundary for the envelope of the standing wave pattern shown in Fig. 33.

47. Two strings of linear mass density μ_1 and μ_2 are knotted together at $x = 0$ and stretched to a tension F. A wave $y = A \sin k_1(x - v_1 t)$ in the string of density μ_1 reaches the junction between the two strings, at which it is partly transmitted into the string of density μ_2 and partly reflected. Call these waves $B \sin k_2(x - v_2 t)$ and $C \sin k_1(x + v_1 t)$, respectively. (*a*) Assuming that $k_2 v_2 = k_1 v_1 = \omega$ and that the displacement of the knot arising from the incident and reflected waves is the same as that arising from the transmitted wave, show that $A = B + C$. (*b*) If it is assumed that both strings near the knot have the same slope (why?)—that is, dy/dx in string 1 $= dy/dx$ in string 2—show that

$$C = A\, \frac{k_2 - k_1}{k_2 + k_1} = A\, \frac{v_1 - v_2}{v_1 + v_2}.$$

Under what conditions is C negative?

Section 19-10   Resonance

48. A 15.0-cm violin string, fixed at both ends, is vibrating in its $n = 1$ mode. The speed of waves in this wire is 250 m/s, and the speed of sound in air is 348 m/s. What are (*a*) the frequency and (*b*) the wavelength of the emitted sound wave?

49. What are the three lowest frequencies for standing waves on a wire 9.88 m long having a mass of 0.107 kg, which is stretched under a tension of 236 N?

50. A 1.48-m-long wire has a mass of 8.62 g and is held under a tension of 122 N. The wire is held rigidly at both ends and set into vibration. Calculate (*a*) the speed of waves on the wire, (*b*) the wavelengths of the waves that produce one- and two-loop standing waves on the wire, and (*c*) the frequencies of the waves in (*b*).

51. One end of a 120-cm string is held fixed. The other end is attached to a weightless ring that can slide along a frictionless rod as shown in Fig. 34. What are the three longest possible wavelengths for standing waves in this string? Sketch the corresponding standing waves.

52. A 75.6-cm string is stretched between fixed supports. It is observed to have resonant frequencies of 420 and 315 Hz,

and no other resonant frequencies between these two. (*a*) What is the lowest resonant frequency for this string? (*b*) What is the wave speed for this string?

53. In an experiment on standing waves, a string 92.4 cm long is attached to the prong of an electrically driven tuning fork, which vibrates perpendicular to the length of the string at a frequency of 60.0 Hz. The mass of the string is 44.2 g. How much tension must the string be under (weights are attached to the other end) if it is to vibrate with four loops?

54. An aluminum wire of length $L_1 = 60.0$ cm and of cross-sectional area 1.00×10^{-2} cm^2 is connected to a steel wire of the same cross-sectional area. The compound wire, loaded with a block m of mass 10.0 kg, is arranged as shown in Fig. 35 so that the distance L_2 from the joint to the supporting pulley is 86.6 cm. Transverse waves are set up in the wire by using an external source of variable frequency. (*a*) Find the lowest frequency of excitation for which standing waves are observed such that the joint in the wire is a node. (*b*) What is the total number of nodes observed at this frequency, excluding the two at the ends of the wire? The density of aluminum is 2.60 g/cm^3 and that of steel is 7.80 g/cm^3.

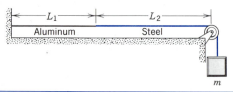

Figure 35   Problem 54.

55. A piano wire 1.4 m long is made of steel with density 7.8 g/cm^3 and Young's modulus 220 MPa. The tension in the wire produces a strain of 1.0%. Calculate the lowest resonant frequency of the wire.

Computer Projects

56. (*a*) Initially a taut string has a shape given by $f_1(x) = 0.02e^{-(x-5)^2/9}$, where f and x are in meters. Suppose the pulse moves with speed $v = 25$ m/s in the positive x direction, so the displacement of the string at coordinate x and time t is given by $y_1(x, t) = f(x - vt) = 0.02e^{-(x-vt-5)^2/9}$. Use a computer program or spreadsheet to plot $y_1(x, t)$ as a function of x from $x = 0$ to $x = 50$ m for $t = 0$, 0.5, 1.0, and 1.5 s. Preferably plot the graphs on a monitor screen and design the program so you can easily change the value of t and replot. Note the position of the pulse maximum on each graph and verify that the graphs depict a pulse that travels in the positive x direction, has a speed of 25 m/s, and moves without change in shape. (*b*) A second pulse has the form $f_2(x) = 0.02e^{-(x-45)^2/9}$ at $t = 0$ and moves in the negative x

direction with a speed of 25 m/s. Use your program to plot $y_2(x, t) = f_2(x + vt)$ from $x = 0$ to $x = 50$ m for $t = 0, 0.5, 0.8, 1.0,$ and 1.5 s. Verify that the graphs depict a pulse moving in the negative x direction. (*c*) Suppose both pulses are on the string at the same time. Use your program to plot $y_1(x, t) + y_2(x, t)$ from $x = 0$ to $x = 50$ m for $t = 0, 0.5, 1.0,$ and 1.5 s. Verify that the graphs depict the pulses moving toward each other and that when they meet the string displacement is large in the region of overlap. The pulses then move away from each other without change in shape. (*d*) Suppose the second wave has the form $f_2(x) = -0.02e^{-(x-45)^2/9}$ at $t = 0$ and travels in the negative x direction with a speed of 25 m/s. Use your program to plot $y_1(x, t) + y_2(x, t)$ from $x = 0$ to $x = 50$ m for $t = 0, 0.5, 0.8, 1.0,$ and 1.5 s. When the two pulses meet, the action of one tends to nullify the action of the other. For one value of the time, the displacement of the string is zero everywhere. The pulses then continue on their ways without change in shape.

57. Waves can be generated on a taut string by moving one end. Suppose the string is extremely long and let $g(t)$ be the displacement of the end being moved, presumed to be at $x = 0$. If the string stretches along the positive x axis, at time t the displacement at the point at x is the same as the displacement at the end but at the earlier time $t - x/v$, where v is the wave speed. Thus the displacement at x is given by $y(x, t) = g(t - x/v)$. (*a*) Suppose that, starting at $t = 0$ and continuing for 0.20 s, the string at $x = 0$ is pulled upward in the positive y direction with a constant speed of 0.15 m/s. It is then held at its final displacement. Thus $g(t) = 0$ for $t < 0$, $g(t) = 0.15t$ for $0 < t < 0.20$ s, and $g(t) = 0.15 \times 0.20 = 0.030$ m for $t > 0.20$ s. Take the wave speed to be 5.0 m/s and use a computer program to make separate graphs of $y(x, t)$ from $x = 0$ to $x = 20$ m for $t = 0, 0.1, 0.2, 1.0, 2.0,$ and 3.0 s. To do this, have the computer calculate $u = x - vt$ for each selected value of x, then set $y = 0$ if $u < 0$, set $y = 0.15u$ if $0 < u < 0.20$, and set $y = 0.03$ if $u > 0.20$. (*b*) Take the wave speed to be 15 m/s and plot $y(x, t)$ from $x = 0$ to $x = 20$ m for $t = 0, 0.1, 0.2, 0.5, 0.75, 1.0,$ and 1.25 s. (*c*) What determines the slope of the string as the pulse moves along? If the string end is raised at a greater rate does the string slope increase or decrease? If the wave speed is increased does the slope increase or decrease?

58. Starting at time $t = 0$ and continuing for 0.40 s, the end of a taut string is jiggled up and down in simple harmonic motion. Its displacement is given by $g(t) = 0.020 \sin (31.4t)$, where g is in meters and t is in seconds. Use a computer to make separate graphs of the string displacement $y(x, t)$ from $x = 0$ to $x = 20$ m for each of the times $t = 0, 0.1, 0.2, 0.3, 0.4, 0.5, 1.0, 1.5, 2.0,$ and 2.5 s. See the previous computer project for some hints.

59. A taut string is initially distorted into the shape given by $f(x) = 0.02e^{-(x-5)^2/9}$, where f and x are in meters. The pulse travels at 5.0 m/s in the positive x direction along the string until it gets to the fixed end at $x = 20$ m, where it is reflected. The displacement of the string is given by $y(x, t) = y_1(x, t) + y_2(x, t)$, where y_1 is the incident pulse and y_2 is the reflected pulse. The incident pulse, of course, is given by

$y_1(x, t) = f(x - vt) = 0.02e^{-(x-vt-5)^2/9}$. Show that the reflected pulse is given by $y_2(x, t) = -f(2L - x - vt) = -0.02e^{-(2L-x-vt-5)^2/9}$, where L is the coordinate of the fixed point. This is the only function of $x + vt$ such that $y_1(L, t) + y_2(L, t) = 0$. Use a computer program or spreadsheet to make separate graphs of the string displacement from $x = 0$ to $x = 20$ m for $t = 0, 1.0, 2.0, 2.5, 2.75, 3.0, 3.25, 3.5, 4.0,$ and 5.0 s. The function to plot is $y(x, t) = 0.020e^{-(x-vt-5)^2/9} - 0.020e^{-(2L-x-vt-5)^2/9}$.

60. A taut string carrying a wave has energy: kinetic energy because it is moving and potential energy because it is distorted. If μ is the linear mass density, then the kinetic energy in an infinitesimal length dx is given by $\frac{1}{2}\mu(\partial y/\partial t)^2 \, dx$. If F is the tension in the string, then the potential energy in an infinitesimal length is given by $\frac{1}{2}F(\partial y/\partial x)^2 \, dx$. Since $y(x, t) = f(x \pm vt)$ and $v = \sqrt{F/\mu}$, these two quantities are exactly equal for the same string length. Thus the total mechanical energy in the string from x to $x + \Delta x$ is given by

$$E = \mu \int_{x}^{x+\Delta x} (\partial y/\partial t)^2 \, dx.$$

You can use the numerical integration program described in the computer projects of Chapter 8 to evaluate integrals of this form.

(*a*) The tension in a string with a linear mass density of 0.080 kg/m is 2.0 N. At time $t = 0$ the string is distorted so it has the shape given by $f(x) = 0.02e^{-(x-5)^2/9}$, where f and x are in meters. Assume the pulse moves in the positive x direction. Show that

$$E = (0.04/9)^2\mu v^2 \int_{x}^{x+\Delta x} (x - vt - 5)^2 e^{-(x-vt-5)^2/4.5} \, dx.$$

(*b*) Use numerical integration to calculate the total energy in the string segment from $x = 0$ to $x = 20$ m at $t = 1$ s. This segment includes all of the pulse except for the very small tails. Using 200 intervals should produce 4 significant figure accuracy. (*c*) Use numerical integration to calculate the total energy in the string segment from $x = 30$ m to $x = 50$ m at $t = 7$ s. The result should be the same as that of part (*b*) and should indicate to you that the energy has moved from the region around $x = 10$ m to the region around $x = 40$ m. This makes sense because the wave speed is 5.0 m/s and the wave traveled 30 m in the intervening 6 s. (*d*) The rate at which energy passes the point at x is given by $P = -F(\partial y/\partial x)(\partial y/\partial t)$, so in the time interval from t to $t + \Delta t$ the energy passing x is given by

$$E = \int_{t}^{t+\Delta t} P \, dt = -F \int_{t}^{t+\Delta t} (\partial y/\partial x)(\partial y/\partial t) \, dt.$$

For the pulse described above show that

$$E = (0.04/9)^2 Fv \int_{t}^{t+\Delta t} (x - vt - 5)^2 e^{-(x-vt-5)^2/4.5} \, dt.$$

Use numerical integration to calculate the energy that passed the point at $x = 25$ m from $t = 1$ s to $t = 7$ s. The result is again the same as before, indicating that all of the energy around $x = 10$ m at $t = 1$ s passed $x = 25$ m on its way to the region around $x = 40$ m.

CHAPTER 20

SOUND WAVES

In Chapter 19 we considered transverse mechanical waves, such as the vibrations of a stretched string. In a longitudinal *mechanical wave, the material particles that transmit the wave vibrate in the direction of propagation of the wave. The most familiar longitudinal mechanical wave is a sound wave. Humans can detect these waves in the frequency range from about 20 Hz to about 20,000 Hz, which is called the* audible *range. Longitudinal mechanical waves of higher frequency are called* ultrasonic *and are used in locating underwater objects and in medical imaging; those of lower frequency are called* infrasonic, *for example, seismic pressure waves produced in earthquakes.*

Sound waves can travel through solids, liquids, and gases; we discuss mainly the propagation of sound through air. A vibrating system (for example, a guitar string, a friend's vocal cords, or a drum head) sets into motion the air in its immediate vicinity. That disturbance propagates through the air and eventually reaches our eardrum, where an amazingly delicate receptor and amplifier convert the mechanical disturbance into an electrical signal that travels to the brain.

In this chapter we discuss the properties of sound waves, their propagation, and their production by vibrating systems.

20-1 THE SPEED OF SOUND

Although sound waves normally travel in three dimensions, we simplify our discussion somewhat by considering a one-dimensional system. Figure 1 shows a tube fitted at one end with a moving piston, which represents, for instance, the moving cone of a loudspeaker. We assume the tube to be filled with a compressible medium such as air and to be very long, so that we need not consider reflections from the far end. As the piston moves back and forth, it alternately compresses and rarefies the medium. We can regard the compressions and rarefactions as (respectively) increases and decreases in the local density relative to its average value in the medium, or else as increases and decreases in the local pressure relative to its average value. These two descriptions convey the same information but have different mathematical forms, as we discuss in Section 20-2.

As a result of the internal mechanical forces of the medium, the compressions and rarefactions travel along the tube. As is the case for all mechanical waves, the velocity of propagation depends on the ratio between an elastic

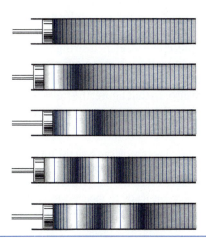

Figure 1 Sound waves generated in a tube by a moving piston, which might represent the moving cone of a loudspeaker. The vertical lines divide the compressible medium in the tube into layers of equal mass.

property of the medium (the tension, in the case of transverse waves on a string) and an inertial property of the medium (the linear mass density, in the case of the string).

For longitudinal waves, the elastic property describes how the medium responds to changes in pressure with a change in volume; this is the bulk modulus* introduced in Eq. 5 of Chapter 17:

$$B = -\frac{\Delta p}{\Delta V/V} \ , \tag{1}$$

where Δp is the change in pressure, and ΔV is the change in the volume V. The minus sign ensures that an increase in pressure ($\Delta p > 0$) causes a decrease in volume ($\Delta V < 0$).

The inertial property of the medium must be given by its density ρ. We can carry out a dimensional analysis to determine the dependence of the velocity on B and ρ using the same procedure we used in Section 19-4, and the result is

$$v = C\sqrt{\frac{B}{\rho}}, \tag{2}$$

where once again the dimensionless constant C cannot be determined from this method of analysis. To complete the derivation we turn, as we did in Section 19-4, to a mechanical analysis based on Newton's laws.

Mechanical Analysis

The spirit of this derivation follows very closely that of Section 19-4. We consider for simplicity a single compressional pulse, such as might be produced by a single stroke of the piston in Fig. 1.

Suppose that a compressional pulse travels through the tube from left to right with speed v. For simplicity we assume that the pulse has sharply defined leading and trailing faces and that it has a uniform pressure and density in its interior. When we analyzed the motion of a transverse pulse in a stretched string in Section 19-4, we found it convenient to choose a reference frame in which the pulse remained stationary. As indicated in Fig. 2, we do this here also. In that figure, the pulse (called the "compressional zone") remains stationary in our reference

---

* A given change in pressure can give rise to different changes in the volume of a compressible medium, depending on the circumstances under which the pressure is changed. For example, since a compression tends to increase the temperature of the medium, we might allow heat to leave so that the temperature remains constant. In such a case, an example being the static fluid processes considered in Chapter 17, we would observe the *isothermal* (constant temperature) bulk modulus. However, the ability of a gas to conduct heat (its *thermal conductivity;* see Section 25-7) is too small for heat to be able to flow between the warmer compressions and the cooler rarefactions at audio frequencies. In this case we need the *adiabatic* (no heat transfer) bulk modulus. For typical gases, the adiabatic bulk modulus is about 1.4 times the isothermal bulk modulus. Isothermal and adiabatic processes are considered in greater detail in Chapter 25.

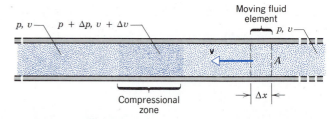

Figure 2   A single pulse (a compression) is sent down a long tube. The reference frame of this figure is chosen to be that of the pulse, so that the fluid is streaming through from right to left. A slice of fluid of width Δx is moving toward the compressional zone with speed v.

frame while the fluid moves through it from right to left with speed v.

Let us follow the motion of the moving fluid element contained between the vertical lines in Fig. 2. This element moves to the left at speed v until it strikes the compressional zone. The left edge of the fluid element enters the compressional zone at time t, and the right edge enters at time $t + \Delta t$. The time interval Δt depends on the width Δx of the element according to $\Delta t = \Delta x/v$.

During the interval Δt, while the element is entering this zone, there is a pressure $p + \Delta p$ on the leading face of the fluid element and a pressure p on the trailing face. As a result of the pressure difference Δp across the fluid element, it is compressed and *decelerated*. Inside the zone, the element moves with a lower speed $v + \Delta v$, the quantity Δv being negative. The element eventually emerges from the left face of the zone, where it expands to its original volume and is accelerated back to its original speed v as a result of the pressure differential Δp.

Let us apply Newton's laws to the fluid element during the time interval Δt in which it enters the zone. The resultant force acting during this interval is

$$F = pA - (p + \Delta p)A = -\Delta p\, A, \tag{3}$$

in which A is the cross-sectional area of the tube. Here we have taken the positive direction to be that of the velocity, that is, to the left in Fig. 2. The original volume V of the element is $A\, \Delta x = Av\, \Delta t$, and its mass is $\rho vA\, \Delta t$, where ρ is the undisturbed density of the fluid outside the compressional zone. The acceleration a is $\Delta v/\Delta t$, and since Δv is negative, a is negative. Newton's second law then gives

$$F = ma$$
$$-\Delta p\, A = (\rho vA\, \Delta t)\frac{\Delta v}{\Delta t} \ ,$$

which we can write as

$$\rho v^2 = \frac{-\Delta p}{\Delta v/v} \ . \tag{4}$$

During the interval Δt, the leading edge of the fluid element is moving at speed $v + \Delta v$, and so it moves a dis-

TABLE 1   THE SPEED OF SOUND[a]

Medium	Speed (m/s)
Gases	
Air (0°C)	331
Air (20°C)	343
Helium	965
Hydrogen	1284
Liquids	
Water (0°C)	1402
Water (20°C)	1482
Seawater[b]	1522
Solids	
Aluminum	6420
Steel	5941
Granite	6000

[a] At 0°C and 1 atm pressure, unless indicated otherwise.
[b] At 20°C and 3.5% salinity.

tance of $(v + \Delta v)\Delta t$. In that same time, the trailing edge moves a distance of $v\,\Delta t$. The width of the fluid element thus changes in that interval by a negative amount $\Delta v\,\Delta t$, and the volume changes correspondingly by the amount $\Delta V = A\,\Delta v\,\Delta t$. Hence

$$\frac{\Delta V}{V} = \frac{A\,\Delta v\,\Delta t}{Av\,\Delta t} = \frac{\Delta v}{v}$$

and we obtain, using Eq. 1,

$$\rho v^2 = \frac{-\Delta p}{\Delta V/V} = B. \qquad (5)$$

Thus

$$v = \sqrt{B/\rho}, \qquad (6)$$

which shows that the constant C in Eq. 2 has the value 1.

If the medium through which the pulse travels is a thin, solid rod rather than a fluid, the bulk modulus B in Eq. 6 must be replaced by Young's modulus (see Section 14-5). If the solid is extended, we must allow for the fact that a solid offers elastic resistance to tangential or shearing forces, and the speed of longitudinal waves depends on the shear modulus as well as the bulk modulus. (Both longitudinal and transverse waves can propagate in an extended solid. Here we consider only longitudinal waves.) Table 1 gives some representative values for the speed of sound in various media.

20-2 TRAVELING LONGITUDINAL WAVES

Consider a continuous train of compressions and rarefactions traveling along a fluid-filled tube, as in Fig. 3. If we station ourselves at some fixed position along the tube, there are two ways we can view this traveling wave. (1) We can focus attention on the back-and-forth oscillatory *displacement* of a fluid element at our location as the wave passes through it. (2) Alternatively, we can focus on the periodic variations in *pressure* that occur at our point of observation. In this section, we explore the connection between these descriptions of a sound wave as a displacement wave and a pressure wave.

As the wave advances along the tube, each small volume element of fluid oscillates about its equilibrium position. The displacement is to the right or left along the direction of propagation of the wave, which we take to be the positive x direction. We represent the displacement of

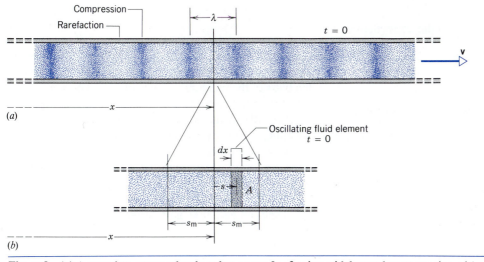

Figure 3 (*a*) A snapshot, assumed to be taken at $t = 0$, of a sinusoidal sound wave moving with speed v through a long tube filled with fluid. (*b*) An expanded view of a region near position x. A fluid element oscillates about its equilibrium position as the wave passes through. At the moment shown, the central plane of the element is displaced a distance s from its equilibrium position.

the volume element from its equilibrium position at x (our observation location) by $s(x, t)$. This function is analogous to the transverse displacement $y(x, t)$ considered in Chapter 19, with one important exception: the displacement s is *along the direction of propagation* for a longitudinal wave, whereas for a transverse wave the displacement y is at *right angles to the direction of propagation*. In the case of a sinusoidal wave, we can therefore write the equation for the longitudinal displacement as

$$s(x, t) = s_m \cos (kx - \omega t), \qquad (7)$$

where we have assumed the wave to be traveling in the positive x direction. We have also assumed a particular choice of phase constant for the displacement wave, which allows us to write it in terms of a cosine function. The amplitude s_m is quite small for sound waves; see Sample Problem 1.

It is usually more convenient to deal with pressure variations in a sound wave than with the actual displacements of the particles. Let us therefore write the equation of the wave in terms of the pressure variation rather than in terms of the displacement.

From Eq. 1, we can write

$$\Delta p = -B \frac{\Delta V}{V}.$$

Just as we let s represent the displacement from the equilibrium position, so we now let Δp represent the *change* from the undisturbed pressure p_0. We seek an expression for the change in pressure Δp as a function of location x and time t, that is, $\Delta p(x, t)$. The actual pressure at any point will then be $p_0 + \Delta p(x, t)$, which might be greater or smaller than p_0 depending on whether Δp is positive or negative at that point and at that time.

A layer of fluid at pressure p_0 with a thickness Δx and a cross-sectional area A has a volume $V = A \Delta x$. When the pressure changes, the volume changes by $A \Delta s$, where Δs is the amount by which the thickness of the layer changes during compression or rarefaction. Hence

$$\Delta p = -B \frac{\Delta V}{V} = -B \frac{A \Delta s}{A \Delta x}.$$

As we let $\Delta x \to 0$ so that the fluid layer shrinks to infinitesimal thickness, we obtain

$$\Delta p = -B \frac{\partial s}{\partial x}. \qquad (8)$$

We have used partial derivative notation because s is a function of both x and t. If the particle displacement is sinusoidal, then, from Eq. 7, we obtain

$$\frac{\partial s}{\partial x} = -k s_m \sin (kx - \omega t),$$

and from Eq. 8

$$\Delta p(x, t) = B k s_m \sin (kx - \omega t). \qquad (9)$$

Hence the pressure variation at each position x is also sinusoidal.

Because $v = \sqrt{B/\rho}$, we can write Eq. 9 more conveniently as

$$\Delta p(x, t) = [k \rho v^2 s_m] \sin (kx - \omega t). \qquad (10)$$

Recall that Δp represents the change from the undisturbed pressure p_0. The term in brackets represents the *maximum* change in pressure and is called the *pressure amplitude*. If we denote this by Δp_m, then

$$\Delta p(x, t) = \Delta p_m \sin (kx - \omega t), \qquad (11)$$

where

$$\Delta p_m = k \rho v^2 s_m. \qquad (12)$$

Hence a sound wave may be considered as either a displacement wave or a pressure wave. If the former is written as a cosine function, the other will be a sine function. The displacement wave is thus 90° out of phase with the pressure wave. That is, when the displacement from equilibrium at a point is a maximum or a minimum, the excess pressure there is zero; when the displacement at a point is zero, the excess or deficiency of pressure there is a maximum. Equation 12 gives the relation between the pressure amplitude (maximum variation of pressure from equilibrium) and the displacement amplitude (maximum variation of position from equilibrium). You should check the dimensions of each side of Eq. 12 for consistency.

Although we have described a sound wave in terms of either a pressure wave or a displacement wave, the two descriptions are in general not equivalent. Only when a single longitudinal wave is propagating in a single direction can we easily shift back and forth between the two descriptions. When we consider the reflection of a sound wave at the end of a tube, or when we superimpose two sound waves that interfere at a point, using the displacement wave description can lead to serious errors.* For example, consider two sound waves from different sources (two loudspeakers, for instance) that travel along different directions and interfere at a point, such that one wave gives a pressure change Δp and the other $-\Delta p$. On the basis of the pressure description, we expect complete destructive interference at that point, because the pressures add like scalars. However, the displacements (which are along the directions of travel of the two waves) do not add to zero, because they are vectors in different directions. *It is usually preferable to describe a sound wave as a pressure wave* to avoid such difficulties. Moreover, as we shall see in the next section, it is the pressure change, not the displacement, that is detected by ears and microphones.

---

* For a careful discussion of this point, see "Pressure and Displacement in Sound Waves," by C. T. Tindle, *American Journal of Physics,* September 1984, p. 749.

Finally, we note that in this section we have treated the fluid as a continuous medium. In a gas, however, the spaces between molecules are large (compared with the size of the molecules), and the molecules move with a random thermal motion. The oscillations produced by a sound wave are superimposed on these random thermal motions. An impulse given to one molecule is passed on to another molecule only after the first has moved through the empty space between them and collided with the second. There is thus a close connection between the average molecular speed in a fluid and the speed of sound in that fluid. In particular, as we increase the temperature, the average molecular speed and the speed of sound in a gas increase in exactly the same manner.

Sample Problem 1 The maximum pressure variation Δp_m that the ear can tolerate in loud sounds is about 28 Pa at 1000 Hz. The faintest sound that can be heard at 1000 Hz has a pressure amplitude of about 2.8×10^{-5} Pa. Find the corresponding displacement amplitudes.

Solution From Table 1, $v = 343$ m/s in air at room temperature, so that

$$k = \frac{2\pi}{\lambda} = \frac{2\pi v}{v} = \frac{2\pi \times 10^3 \text{ Hz}}{343 \text{ m/s}} = 18.3 \text{ rad/m}.$$

The density of air under these conditions is 1.21 kg/m³. Hence for $\Delta p_m = 28$ Pa, we obtain, using Eq. 12,

$$s_m = \frac{\Delta p_m}{k\rho v^2} = \frac{28 \text{ Pa}}{(18.3 \text{ rad/m})(1.21 \text{ kg/m}^3)(343 \text{ m/s})^2}$$
$$= 1.1 \times 10^{-5} \text{ m}.$$

The displacement amplitudes for the loudest sounds are about 10^{-5} m, a very small value indeed. For the faintest sounds, we have similarly

$$s_m = \frac{2.8 \times 10^{-5} \text{ Pa}}{(18.3 \text{ rad/m})(1.21 \text{ kg/m}^3)(343 \text{ m/s})^2} = 1.1 \times 10^{-11} \text{ m}.$$

This is about one-tenth of the radius of a typical atom and suggests how sensitive the ear must be to detect vibrations of such a small amplitude.

20-3 POWER AND INTENSITY OF SOUND WAVES

We follow the methods of Chapter 19 to calculate the power delivered by a sound wave, the major difference being that now the particle velocity **u** is along the direction of the wave. As the pressure wave travels, each element of fluid exerts a force on the element ahead of it; the magnitude of the net force is $F = A \Delta p$, where A is the cross-sectional area of the element of fluid. Using Eq. 11 for Δp, we find that the force is

$$F = A \Delta p_m \sin(kx - \omega t). \tag{13}$$

The velocity of the thin slice of fluid, as indicated in Fig. 3, is

$$u = \frac{\partial s}{\partial t} = -\omega s_m [-\sin(kx - \omega t)]. \tag{14}$$

The power delivered to the fluid element is

$$P = uF = A\omega \Delta p_m s_m \sin^2(kx - \omega t). \tag{15}$$

Using Eq. 12, we can write this as

$$P = \frac{A(\Delta p_m)^2}{\rho v} \sin^2(kx - \omega t). \tag{16}$$

As we did in Chapter 19 for the case of a transverse wave traveling along a string, we average the power over one cycle; since the average value of $\sin^2 \theta$ is $\frac{1}{2}$, the average power is

$$\bar{P} = \frac{A(\Delta p_m)^2}{2\rho v}. \tag{17}$$

As in the case of the transverse wave, the power depends on the *square* of the amplitude, in this case the pressure amplitude. Note also that the frequency does not appear explicitly in Eq. 17 (although it would appear if we instead expressed the average power in terms of the displacement amplitude). Hence, by measuring pressure amplitudes, we can directly compare the intensities of sounds having *different* frequencies. For this reason, instruments that measure pressure changes are preferable to those that measure displacements; moreover, as we learned from Sample Problem 1, displacements from the weakest audible sounds are very small and would be difficult to measure directly.

When we are comparing different sounds, it is more useful to use the *intensity* (average power per unit area) of the wave. From Eq. 17, we can immediately obtain the intensity I:

$$I = \frac{\bar{P}}{A} = \frac{(\Delta p_m)^2}{2\rho v}. \tag{18}$$

Because the ear is so sensitive (it is capable of responding to intensities ranging over 12 orders of magnitude), we introduce a logarithmic scale of intensity called the *sound level SL*:

$$SL = 10 \log \frac{I}{I_0}. \tag{19}$$

The *SL* is defined with respect to a reference intensity I_0, which is chosen to be 10^{-12} W/m² (a typical value for the threshold of human hearing). Sound levels defined in this way are measured in units of *decibels* (dB). A sound of intensity I_0 has a sound level of 0 dB, while sound at the upper range of human hearing, called the threshold of pain, has an intensity of 1 W/m² and a *SL* of 120 dB. Each increase in intensity I by a multiplicative factor of 10 corresponds to adding 10 dB to the *SL*.

We can also use dB as a *relative* measure to compare different sounds with one another, rather than with the reference intensity. Suppose we wish to compare two sounds of intensities I_1 and I_2:

$$SL_1 - SL_2 = 10 \log \frac{I_1}{I_0} - 10 \log \frac{I_2}{I_0}$$

$$= 10 \log \frac{I_1}{I_2}. \qquad (20)$$

For example, two sounds whose intensity ratio is 2 differ in *SL* by 10 log 2 = 3 dB.

The sensitivity of the human ear varies with frequency. The threshold of 10^{-12} W/m² applies only to the mid-range frequencies around 1000 Hz. At the higher frequencies, say 10,000 Hz, the threshold rises to about 10 dB (10^{-11} W/m²), while at a lower frequency of 100 Hz the threshold is about 30 dB (10^{-9} W/m²). It takes 1000 times the sound intensity at 100 Hz to produce the same physiological response as a given sound intensity at 1000 Hz. Figure 4 shows the variation with frequency of the thresholds of hearing and of pain, and Table 2 shows some representative sound levels and their corresponding intensities.

Sample Problem 2 Spherical sound waves are emitted uniformly in all directions from a point source, the radiated power *P* being 25 W. What are the intensity and the sound level of the sound wave a distance $r = 2.5$ m from the source?

Solution All the radiated power *P* must pass through a sphere of radius *r* centered on the source. Thus

$$I = \frac{P}{4\pi r^2}.$$

We see that the intensity of the sound drops off as the inverse square of the distance from the source. Numerically, we have

$$I = \frac{25 \text{ W}}{(4\pi)(2.5 \text{ m})^2} = 0.32 \text{ W/m}^2$$

and

$$SL = 10 \log \frac{I}{I_0}$$

$$= 10 \log \frac{0.32 \text{ W/m}^2}{10^{-12} \text{ W/m}^2} = 115 \text{ dB}.$$

A comparison of this result with Table 2 raises questions about the wisdom of buying 100-W amplifiers for home use.

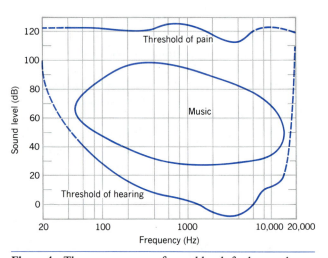

Figure 4 The average range of sound levels for human hearing. Note the dependence of the threshold levels on frequency. A sound that we can just hear at 100 Hz must have 1000 times the acoustic power (30 dB greater sound level) than one we can just hear at 1000 Hz, because our ear is that much less sensitive at 100 Hz.

20-4 STANDING LONGITUDINAL WAVES

We now consider what happens when a sound wave such as that shown in Fig. 1 reaches the end of the tube. In analogy with the transverse wave on the string (see Fig. 22

TABLE 2 SOME INTENSITIES AND SOUND LEVELS

Sound	Intensity (W/m²)	Relative Intensity (I/I_0)	Sound Level (dB)
Threshold of hearing	1×10^{-12}	10^0	0
Rustle of leaves	1×10^{-11}	10^1	10
Whisper (at 1 m)	1×10^{-10}	10^2	20
City street, no traffic	1×10^{-9}	10^3	30
Office, classroom	1×10^{-7}	10^5	50
Normal conversation (at 1 m)	1×10^{-6}	10^6	60
Jackhammer (at 1 m)	1×10^{-3}	10^9	90
Rock group	1×10^{-1}	10^{11}	110
Threshold of pain	1	10^{12}	120
Jet engine (at 50 m)	10	10^{13}	130
Saturn rocket (at 50 m)	1×10^8	10^{20}	200

of Chapter 19), a reflection occurs, and the reflected wave travels back down the tube in the opposite direction. The behavior of the wave at the reflecting end depends on whether the end of the tube is open or closed.

Let us first consider a tube that is closed at the end. As the wave travels down the tube and reaches the end, it can compress the layers of air at the closed end against the fixed barrier. At that end, the pressure can therefore vary with its maximum amplitude, and *the closed end is a pressure antinode.* A pressure wave reflects from a closed end in a manner similar to the reflection of a transverse displacement wave at the free end of a string (Fig. 22*b* of Chapter 19). If a compression, for example, is incident on the closed end, it reflects back along the tube as a compression. In analogy with our discussion of transverse waves on strings, we say that *a longitudinal pressure wave is reflected from a closed end with no change of phase.* The same effect occurs in the case of a longitudinal wave traveling on a spring such as a Slinky toy and reflecting from a *fixed* end: a compression is reflected as a compression.

Now consider what happens if the end of the tube is open. The pressure at the end of the open tube is the same as the ambient pressure p_0 in the surrounding room. We cannot change the pressure at that end of the tube unless we change the pressure in the entire room. The pressure at the open end therefore remains at the value p_0, and *the open end is a pressure node.* Comparison with Fig. 22 of Chapter 19 shows that this case is analogous to the transverse displacement wave reflecting from the fixed end of the string. The attempt by the wave incident on the open end to compress the air at that end causes a rarefaction, which travels back down the tube in the opposite direction. Thus *a longitudinal pressure wave is reflected from an open end with a phase change of 180°.* The same effect can again be observed with a coiled spring: a compression is reflected as a rarefaction.

Let us now assume we have a train of sinusoidal waves traveling down the tube. The waves are reflected at the end, which will behave either as a pressure node (if the end is open) or a pressure antinode (if the end is closed). We assume the source of the wavetrain is a speaker at the opposite end. The movement of the speaker sends a compressional wave down the tube, and the superposition of the original and reflected waves gives a pattern of standing waves, just as was the case for the transverse waves on the string. Within the tube will be a pattern of pressure nodes and antinodes (which are not points, as in the case of transverse waves on a string, but planes).

If the frequency (or wavelength) of the source of the waves is selected to be a particular value that depends on the length of the tube, then a standing wave pattern is established along the entire tube, in analogy with the case of the standing wave patterns shown in Fig. 23 of Chapter 19. If there is a node of pressure at the speaker end, then little energy is given back to the speaker from the standing wave pattern in the tube, and we have a condition of

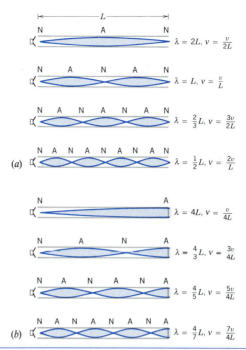

Figure 5   (*a*) The pressure waves of the first four resonant modes of a tube driven by a speaker and open at the other end. There is a pressure node N at each end, and antinodes A are located between the nodes. The curves suggest the sinusoidal variation of pressure within the tube. (*b*) The pressure waves of the first four resonant modes of a tube that is closed at one end. The closed end is a pressure antinode. Note the differences in vibrational patterns and wavelengths between the open and closed tubes.

resonance. The driving frequency must be equal to one of the natural frequencies of the system, which are determined by the length of the tube.

Figure 5*a* shows a tube that is driven by a speaker at one end and is open at the other end. As we have discussed previously, the speaker end is a pressure node at resonance and the open end is likewise a pressure node. In Fig. 5*a* are shown the resulting variations in the pressure amplitude of the standing waves.* These patterns look very similar to those of Fig. 23 of Chapter 19. In the first mode of oscillation, the length L of the tube is equal to $\lambda/2$, where λ is the wavelength of the wave produced by the speaker for that particular resonant condition. The wavelength is therefore $2L$, and the corresponding frequency is $\nu = v/\lambda = v/2L$. The other resonances shown in Fig. 5*a*

---

* A beautiful demonstration of the locations of the pressure nodes and antinodes can be obtained with a Rubens flame-tube. See "Rubens Flame-tube Demonstration," by George W. Ficken and Francis C. Stephenson, *The Physics Teacher,* May 1979, p. 306.

have successively smaller wavelengths, which can be written in general as

$$\lambda_n = \frac{2L}{n}, \qquad n = 1, 2, 3, \ldots . \qquad (21)$$

The corresponding resonant frequencies, determined using the expression $v = v/\lambda$ with the above wavelengths, are

$$v_n = n\frac{v}{2L}, \quad n = 1, 2, 3, \ldots \quad \text{(open tube)}. \quad (22)$$

Here v represents the speed of the wave in the medium that fills the tube, usually air.

Figure 5*b* shows the case in which the tube is closed at one end and open at the other end. In this case, the closed end must be a pressure antinode. In the first resonant mode, the length L of the tube is $\frac{1}{4}\lambda$, and so the source must be producing a wave whose wavelength is $4L$. In the next mode, the wavelength changes so that now L is $\frac{3}{4}\lambda$, and thus $\lambda = \frac{4}{3}L$. Continuing the series, we see that in this case the general expression for the wavelengths of the resonant modes is

$$\lambda_n = \frac{4L}{n}, \qquad n = 1, 3, 5, \ldots . \qquad (23)$$

Note that only odd values of the integer n appear in this case. The corresponding resonant frequencies are

$$v_n = n\frac{v}{4L}, \quad n = 1, 3, 5, \ldots \quad \text{(closed tube)}. \quad (24)$$

As we discuss in the next section, the resonant frequencies given by Eq. 22 or 24 determine the musical notes played by the wind instruments.

The actual location of the pressure node at an open end is not *exactly* at the end of the tube. The wave extends slightly into the medium beyond the tube, so the true length of the tube is a bit greater and the resonant frequencies are a bit smaller. For narrow tubes of cylindrical shape, the length correction is roughly equal to $0.6R$, where R is the radius of the tube. For a tube open at both ends, the length correction must be applied at each end. For a tube of length 0.6 m and radius 1 cm (typical values for the smaller wind instruments such as the clarinet or flute), the lowest frequency without the end correction would be 286 Hz if the tube were open and 143 Hz if the tube were closed. With the end correction, the corresponding values would be 280 Hz and 142 Hz. The corrections are small, but nevertheless quite important.

Sample Problem 3   Figure 6 shows an apparatus that can be used to measure the speed of sound in air by using the resonance condition. A small speaker is held above a cylindrical tube partially filled with water. By adjusting the water level, the length of the air column can be changed until the tube is in resonance, at which point an increase in the sound intensity can be heard. In

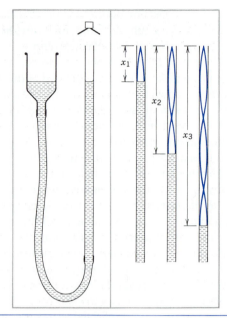

Figure 6   Sample Problem 3. An apparatus for measuring the speed of sound in air. The water level can be adjusted by raising and lowering the reservoir on the left, which is connected through a hose to the tube. At right are shown the pressure waveforms of the first three resonant modes for a fixed wavelength.

an experiment, the speaker is driven at a fixed frequency of 1080 Hz, and three resonances are observed when the water level is at distances of $x_1 = 6.5$ cm, $x_2 = 22.2$ cm, and $x_3 = 37.7$ cm below the top of the tube. Find the value of the speed of sound from these data.

Solution   The air column acts like a tube of variable length closed at one end. The pattern of standing waves shows a pressure node near the speaker and a pressure antinode at the surface of the water. Since we don't know the end correction, we can't use the given data directly to find the speed of sound from Eq. 24. However, we note from the resonance conditions shown in Fig. 5*b* that the distance between adjacent pressure nodes is $\frac{1}{2}\lambda$; the same is true for the distance between adjacent antinodes. From the data given, we therefore conclude from the first two resonances that

$$\tfrac{1}{2}\lambda = x_2 - x_1 = 22.2 \text{ cm} - 6.5 \text{ cm} = 15.7 \text{ cm},$$

and similarly, from the second and third resonances,

$$\tfrac{1}{2}\lambda = x_3 - x_2 = 37.7 \text{ cm} - 22.2 \text{ cm} = 15.5 \text{ cm}.$$

The average of these two values, which we take as the best value from this measurement, is 15.6 cm, corresponding to a wavelength of 2(15.6 cm) = 31.2 cm = 0.312 m. We therefore deduce the speed of sound to be

$$v = \lambda v = (0.312 \text{ m})(1080 \text{ Hz}) = 337 \text{ m/s}.$$

Other than the end correction, what physical factors in this experiment (including the properties of the air) might influence the measured value?

20-5 VIBRATING SYSTEMS AND SOURCES OF SOUND*

A vibrating system transmits a wave through the air to the ears of the listener. This is the basic principle of the production of sound by voice or by musical instruments. We have already studied the propagation of the sound wave, and now to understand the nature of the sound we must study the vibrating system that produces it.

As we discussed in Section 19-10 in the case of the vibrating string and in the previous section in the case of a column of air, a distributed system has a large (perhaps infinite) number of natural or resonant vibrational frequencies. These are the frequencies at which it *can* vibrate. Which of the frequencies *will* be present in the vibration depends on how the system is set into vibration.

Suppose the system is capable of vibrating at a number of frequencies $v_1, v_2, v_3, \ldots$. We write these in ascending order, so that $v_1 < v_2 < v_3 < \cdots$. The lowest frequency, v_1, is called the *fundamental* frequency, and the corresponding mode of oscillation is called the fundamental mode. The higher frequencies are called *overtones,* with v_2 being the first overtone, v_3 the second overtone, and so on.

In certain systems, the overtones are all integer multiples of the fundamental:

$$v_n = n v_1, \qquad (25)$$

where n is an integer. In such a case, the overtones are called *harmonics.* The first member of a harmonic sequence is the fundamental, the second harmonic is the first overtone, and so on.

Why do some vibrating systems produce pleasant sounds while others produce harsh or discordant sounds? When several frequencies are heard simultaneously, a pleasant sensation results if the frequencies are in the ratio of small, whole numbers, such as $3:2$ or $5:4$. If a system produces overtones that are harmonics, its vibrations will include frequencies that have these ratios, and it would produce a pleasing sound. If the overtones are not harmonics, it is likely that the sound will be discordant. Much of the effort in the design of musical instruments is devoted to the production of harmonic sequences of overtones. Some instruments, such as those based on vibrating strings, produce overtones that are automatically harmonics when the vibrations have small amplitude. In other cases, the shape of the instrument must be carefully designed to make it harmonic; a bell is an example of such an instrument. The harmonics that an instrument pro-

duces give it its richness and diversity of tone, and they are critical to the beauty of the sound of the instrument. If instruments produced only fundamentals, they would all sound exactly alike.

We can classify musical instruments into three categories: those based on vibrating strings, those based on vibrating columns of air, and more complex systems including vibrating plates, rods, and membranes.

Vibrating Strings

These instruments include the bowed strings (violins, for example), plucked strings (guitar, harpsichord), and struck strings (piano).

If a string fixed at both ends is bowed, struck, or plucked, transverse vibrations travel along the string; these disturbances are reflected at the fixed ends, and a standing wave pattern is formed. The natural modes of vibration of the string are excited, and these vibrations give rise to longitudinal waves in the surrounding air, which transmits them to our ears as a musical sound.

We have seen (Section 19-10) that a string of length L, fixed at both ends, can resonate at frequencies given by

$$v_n = n\frac{v}{2L}, \qquad n = 1, 2, 3, \ldots . \qquad (26)$$

Here v is the speed in the string of the transverse traveling waves whose superposition can be thought of as giving rise to the vibrations; the speed $v \; (= \sqrt{F/\mu})$ is the same for all frequencies. (Note that v is *not* the speed of sound in air; even though Eq. 26 looks exactly like Eq. 22, v stands for different quantities in the two equations.) At any one of these frequencies the string contains a whole number n of loops between its ends; it has nodes at each end and $n - 1$ additional nodes equally spaced along its length (Fig. 7).

If the string is initially distorted so that its shape is the

---

* For a listing of references on the physics of musical instruments and related topics, see "Resource Letter MA-2: Musical Acoustics," by Thomas D. Rossing, *American Journal of Physics,* July 1987, p. 589.

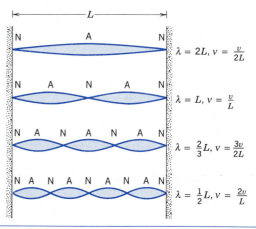

Figure 7 The first four resonant modes of a vibrating string fixed at both ends. Nodes and antinodes of displacement are denoted by N and A.

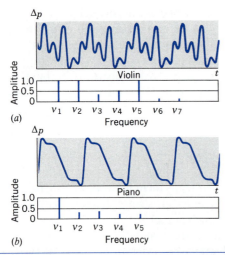

Figure 8   Waveforms and sound spectra for two stringed instruments, (*a*) violin and (*b*) piano, each playing a note of fundamental frequency $v_1 = 440$ Hz (concert A). The sound spectrum below each waveform shows the harmonics that are present in the complex tone and their relative amplitudes.

same as *any one* of the possible harmonics, it vibrates only at the frequency of that particular harmonic. The initial conditions usually arise from striking or bowing the string, however, and in such cases not only the fundamental but many of the overtones are present in the resulting vibration. We have a superposition of several natural modes of oscillation. The actual displacement is the sum of the several harmonics with various amplitudes. The impulses that are sent through the air to the ear and brain give rise to one net effect, which is characteristic of the particular stringed instrument. The quality of the sound of a particular note (fundamental frequency) played by an instrument is determined by the number of overtones present and their respective intensities. Figure 8 shows the sound spectra and corresponding waveforms for the violin and piano.

Vibrating Air Columns

An organ pipe is a simple example of sound originating in a vibrating air column. If both ends of a pipe are open and a stream of air is directed against an edge at one end, standing longitudinal waves can be set up in the tube. The air column then resonates at its natural frequencies of vibration, given by Eq. 22. As with the bowed string, the fundamental and the overtones (which are harmonics) are produced at the same time. If one end of the tube is closed, the fundamental frequency is reduced by half, relative to its value for an open tube of the same length, and only the odd harmonics are present, which changes the quality of the sound. That is, an open pipe produces the same fundamental tone as a closed pipe of half the length, but because the mixture of harmonics is different in the two pipes, the quality of the tones differs.

Reed instruments, such as the clarinet, produce their tones differently. Air is forced through a narrow opening, one side of which is covered by a reed that has elastic properties. According to Bernoulli's equation, the high-speed air passing through the narrow opening causes a local region of low pressure inside the mouthpiece. The outside pressure exceeds the inside pressure, which forces the reed inward so that it covers the opening. As soon as the opening is covered, the air flow is interrupted, the dynamic low-pressure region is eliminated, and the reed pops open, allowing the air flow to start again. This repeated opening and closing of the air passage causes maximum variations in pressure at the mouthpiece end of the instrument, which therefore behaves like an antinode of pressure. In a clarinet, the other end of the instrument is open, and therefore the resonances of the instrument are those given by Eq. 24 for a tube closed at one end and open at the other. Some wind instruments, such as the flute, use a method similar to the organ pipe to produce the tone, such that the mouthpiece behaves like an open end; their resonant frequencies are given by Eq. 22. Still others (oboe, saxophone), which use a reed to produce their tone, have a conical (that is, tapered) rather than a cylindrical bore, which gives them overtones that are approximately harmonics, odd as well as even. The brass instruments (trumpet or trombone, for example) are also called *lip reed* instruments, because the player's lip acts like a reed, but again the bore is slightly tapered, and as a result the overtones contain all the harmonics. Figure 9 shows waveforms of some wind instruments.

Other Vibrating Systems

Vibrating rods, plates, and stretched membranes also give rise to sound waves. Consider a stretched flexible membrane, such as a drumhead. If it is struck a blow, a two-dimensional pulse travels outward from the struck point and is reflected again and again at the boundary of the membrane. If some point of the membrane is forced to vibrate periodically, continuous trains of waves travel out along the membrane. Just as in the one-dimensional case of the string, so here too standing waves can be set up in the two-dimensional membrane. Each of these standing waves has a certain frequency natural to (or characteristic of) the membrane. Again the lowest frequency is called the fundamental, and the others are overtones. Generally, many overtones are present along with the fundamental when the membrane is vibrating. These vibrations may excite sound waves of the same frequency.

The nodes of a vibrating membrane are lines rather than points (as in a vibrating string) or planes (as in a tube). Since the boundary of the membrane is fixed, it must be a nodal line. For a circular membrane fixed at its edge, possible modes of vibration together with their nodal lines are shown in Fig. 10. The natural frequency of each mode is given in terms of the fundamental v_1. The frequencies of the overtones are not harmonics; that is,

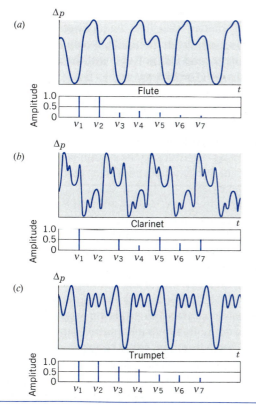

Figure 9 Waveforms of some wind instruments: (*a*) flute, (*b*) clarinet, and (*c*) trumpet, and their sound spectra, as in Fig. 8. Note that the clarinet spectrum shows mostly the odd harmonics, while the flute and trumpet have odd as well as even harmonics.

they are not integral multiples of v_1. Vibrating rods also have a nonharmonic set of natural frequencies. Rods and plates have limited use as musical instruments for this reason. In instruments such as the xylophone and the marimba, small bars of wood or metal are set into vibration by striking. The shape of the bars is carefully modified, making them thinner in the center, in such a way that the overtones become approximately harmonics.

20-6 BEATS

We have previously considered the effect of waves that are superimposed to give regions of maximum and minimum (zero) intensity, such as in the case of a standing wave in a tube. This illustrates a type of interference that we can call *interference in space.*

The same principle of superposition leads us to another type of interference, which we can call *interference in time.* In this case we examine the superposition of two waves at a given point as a function of the time. This superposition, which in general can result in quite complex waveforms, takes a particularly simple form when

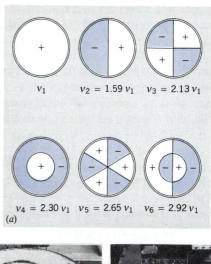

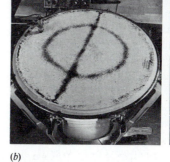

Figure 10 (*a*) The lowest six resonant modes of a circular drumhead clamped at its rim. The lines represent nodes; the rim is also a nodal line. The + or − signs indicate that a particular region is moving up out of the page or down into the page. In this case the overtones are not integral multiples of the fundamental and are thus not harmonics. (*b*) The vibrational patterns of a kettledrum in the modes numbered 4, 5, and 6, and one additional mode not illustrated in (*a*). They are made visible by sprinkling dark powder on the drumhead and setting it into vibration at the proper frequency using a mechanical vibrator. As the drumhead vibrates, the powder is shaken and eventually settles on the nodal lines, where there is no motion.

the two waves have nearly the same frequency. With sound such a condition exists when, for example, two instruments or two guitar strings are being tuned to one another.

Consider a point in space through which the waves are

passing. Figure 11*a* shows the pressure produced at that point by the two waves separately as a function of time. For simplicity we have assumed that the two waves have equal amplitude, although this is not necessary. The resultant pressure at that point as a function of time is the sum of the individual pressures and is plotted in Fig. 11*b*. We see that the amplitude of the resultant wave is not constant but varies with time. In the case of sound the varying amplitude gives rise to variations in loudness, which are called *beats*.

Let us represent the variation in pressure with time (for constant *x*) produced by one wave as

$$\Delta p_1(t) = \Delta p_m \sin \omega_1 t,$$

where we have chosen the phase constant to enable us to write the wave in this simple form. The pressure variation at the same point produced by the other wave of equal amplitude is represented as

$$\Delta p_2(t) = \Delta p_m \sin \omega_2 t.$$

By the superposition principle, the resultant pressure is

$$\Delta p(t) = \Delta p_1(t) + \Delta p_2(t)$$
$$= \Delta p_m(\sin \omega_1 t + \sin \omega_2 t). \quad (27)$$

Using the trigonometric identity

$$\sin A + \sin B = 2 \cos \frac{A - B}{2} \sin \frac{A + B}{2},$$

Eq. 27 can be written

$$\Delta p(t) = \left[2\Delta p_m \cos \left(\frac{\omega_1 - \omega_2}{2} \right) t \right] \sin \left(\frac{\omega_1 + \omega_2}{2} \right) t. \quad (28)$$

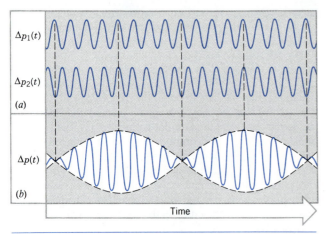

(a)

(b)

Time

Figure 11 (*a*) Two sinusoidal waveforms of nearly equal frequencies. (*b*) The superposition of the two waveforms. Note that the two waves in part (*a*) go from being in phase, giving a resultant of large amplitude, to being out of phase, giving a resultant of zero amplitude. The dashed curves show the sinusoidal variation of the modulating envelope with angular frequency ω_{amp}.

So far everything we have done applies to any two waves, no matter what their frequencies. When the frequencies are nearly the same, Eq. 28 can be simplified by writing the second factor in terms of the average angular frequency $\overline{\omega}$ of the two waves,

$$\overline{\omega} = \frac{\omega_1 + \omega_2}{2}. \quad (29)$$

The first factor, contained in the brackets of Eq. 28, gives a time-varying amplitude to the sinusoidal variation of the second factor. This amplitude factor varies with an angular frequency

$$\omega_{amp} = \frac{|\omega_1 - \omega_2|}{2}. \quad (30)$$

In terms of $\overline{\omega}$ and ω_{amp}, we can write Eq. 28 as

$$\Delta p(t) = [2\Delta p_m \cos \omega_{amp} t] \sin \overline{\omega} t. \quad (31)$$

If ω_1 and ω_2 are nearly equal, the amplitude frequency ω_{amp} is small, and the amplitude fluctuates slowly. Figure 11 shows the superposition of the two waves according to Eq. 28. Notice that in the case of nearly equal frequencies, the rapid variation of the resultant wave occurs with a frequency that is approximately that of either of the two added waves. The overall amplitude of the resultant varies slowly with the amplitude frequency ω_{amp}, which defines an "envelope" within which the more rapid variation occurs. This phenomenon is a form of *amplitude modulation,* which has a counterpart (side bands) in AM radio receivers.

In the case shown in Fig. 11*b*, the ear would perceive a tone at a frequency v $(= \overline{\omega}/2\pi)$, which is approximately the same as the frequencies v_1 $(= \omega_1/2\pi)$ or v_2 $(= \omega_2/2\pi)$ of the two component waves. The tone grows alternately loud and soft as the amplitude of the resultant varies with time, having maxima and minima as shown in Fig. 11*b*.

A beat, that is, a maximum of intensity, occurs whenever $\cos \omega_{amp} t$ equals $+1$ or -1, since the intensity depends on the *square* of the amplitude. Each of these values occurs once in each cycle of the envelope (see Fig. 11), so the number of beats per second is twice the number of cycles per second of the envelope. The beat angular frequency ω_{beat} is then

$$\omega_{beat} = 2\omega_{amp} = |\omega_1 - \omega_2|. \quad (32)$$

Using $\omega = 2\pi v$, we can rewrite this expression as

$$v_{beat} = |v_1 - v_2|. \quad (33)$$

Hence *the number of beats per second equals the difference of the frequencies of the component waves.* Beats between two tones can be detected by the ear up to a frequency of about 15 Hz. At higher frequencies individual beats cannot be distinguished in the sound produced. Musicians often listen for beats when tuning certain instruments. The tuning is changed until the beat frequency decreases and the beats disappear.

Sample Problem 4   A violin string that should be tuned to concert A (440 Hz) is slightly mistuned. When the violin string is played in its fundamental mode along with a concert A tuning fork, 3 beats per second are heard. (*a*) What are the possible values of the fundamental frequency of the string? (*b*) Suppose the string were played in its first overtone simultaneously with a tuning fork one octave above concert A (880 Hz). How many beats per second would be heard? (*c*) When the tension of the string is increased slightly, the number of beats per second in the fundamental mode increases. What was the original frequency of the fundamental?

Solution   (*a*) From Eq. 33, we know that the frequency ν_1 of the string differs by the beat frequency (3 Hz) from the frequency ν_2 of the tuning fork (440 Hz), but from the number of beats per second alone, we cannot tell whether the string has a higher or lower frequency. Thus the possible frequencies are

$$\nu_1 = 440 \text{ Hz} \pm 3 \text{ Hz} = 443 \text{ Hz} \quad \text{or} \quad 437 \text{ Hz}.$$

(*b*) In the first overtone, the frequency of the string is twice its fundamental frequency, and thus either 886 Hz or 874 Hz. When played against a 880-Hz tuning fork, the frequency difference in either case is 6 Hz, and thus 6 beats per second would be heard.

(*c*) Increasing the tension in the string raises the velocity of transverse waves and therefore raises the fundamental frequency (see Eq. 26). Since we are given that this increases the beat frequency, we conclude that the frequency of the fundamental mode was previously greater than 440 Hz, since increasing the frequency made the difference from 440 Hz even greater. Thus the string was originally tuned to 443 Hz, and to bring it into proper tuning the tension must be reduced.

20-7   THE DOPPLER EFFECT

When a listener is in motion toward a stationary source of sound, the pitch (frequency) of the sound heard is higher than when the listener is at rest. If the listener is in motion away from the stationary source, a lower pitch is heard. We obtain similar results when the source is in motion toward or away from a stationary listener. The pitch of the whistle of a locomotive or the siren of a fire engine is higher when the source is approaching the hearer than when it has passed and is receding.

In a paper written in 1842, Christian Johann Doppler (1803–1853, Austrian) called attention to the fact that the color of a luminous body must be changed by relative motion of the body and the observer. This *Doppler effect,* as it is called, applies to waves in general. Doppler himself mentions the application of his principle to sound waves. An experimental test was carried out in Holland in 1845 by Buys Ballot, "using a locomotive drawing an open car with several trumpeters."

Moving Observer, Source at Rest

We now consider the Doppler effect for sound waves, treating only the special case in which the source and observer move along the line joining them. Let us adopt a reference frame at rest in the medium through which the sound travels. Figure 12 shows a source of sound S at rest in this frame and an observer O moving toward the source at a speed v_O. The circles represent wavefronts, spaced one wavelength apart, traveling through the medium. An observer at rest in the medium would receive vt/λ waves in time t, where v is the speed of sound in the medium and λ is the wavelength. Because of the motion toward the source, however, the observer receives $v_O t/\lambda$ additional waves in this same time t. The frequency ν' that is actually heard is the number of waves received per unit time, or

$$\nu' = \frac{vt/\lambda + v_O t/\lambda}{t} = \frac{v + v_O}{\lambda} = \frac{v + v_O}{v/\nu}.$$

That is,

$$\nu' = \nu\frac{v + v_O}{v} = \nu\left(1 + \frac{v_O}{v}\right). \qquad (34)$$

The frequency ν' heard by the observer is the frequency ν heard at rest plus the increase $\nu(v_O/v)$ arising from the motion of the observer. When the observer is in motion *away from* the stationary source, there is a *decrease* in frequency $\nu(v_O/v)$ corresponding to the waves that do not reach the observer in each unit of time because of the receding motion. Then

$$\nu' = \nu\frac{v - v_O}{v} = \nu\left(1 - \frac{v_O}{v}\right). \qquad (35)$$

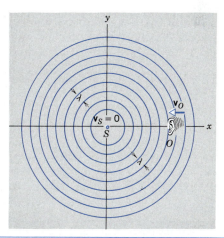

Figure 12   A stationary source of sound S emits spherical wavefronts, shown one wavelength apart. An observer O, represented by the ear, moves with speed v_O toward the source. The moving observer encounters more waves per second than an observer at rest and therefore measures a higher frequency. The observer would measure a *lower* frequency for motion away from the source.

Hence the general relation holding when the *source is at rest* with respect to the medium but the *observer is moving* through it is

$$v' = v \frac{v \pm v_O}{v}, \tag{36}$$

where the plus sign holds for motion toward the source and the minus sign holds for motion away from the source. Note that the change in frequency occurs because the observer intercepts more or fewer waves each second as a result of the motion through the medium.

Moving Source, Observer at Rest

When the source is in motion *toward* a stationary observer, the effect is a shortening of the wavelength (see Fig. 13), for the source is following after the approaching waves, and the crests therefore come closer together. If the frequency of the source is v and its speed is v_S, then during each vibration it travels a distance v_S/v, and each wavelength is shortened by this amount. Hence the wavelength of the sound arriving at the observer is not $\lambda = v/v$ but $\lambda' = v/v - v_S/v$. The frequency of the sound heard by the observer is increased and is given by

$$v' = \frac{v}{\lambda'} = \frac{v}{(v - v_S)/v} = v \frac{v}{v - v_S}. \tag{37}$$

If the source moves *away from* the observer, the wavelength emitted is v_S/v greater than λ, so that the observer hears a decreased frequency, namely,

$$v' = \frac{v}{(v + v_S)/v} = v \frac{v}{v + v_S}. \tag{38}$$

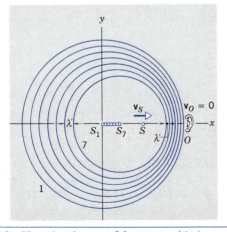

Figure 13   Here the observer O is at rest, with the source moving toward it at speed v_S. Wavefront 1 was emitted when the source was at S_1, wavefront 7 when the source was at S_7, and so on. At the instant of this drawing, the source is at S. The observer measures a shorter wavelength because of the "bunching up" of the wavefronts along the motion. An observer on the negative x axis, from whom the source would be moving away, would measure a longer wavelength.

Hence the general relation holding when the *observer is at rest* with respect to the medium but the *source is moving* through it is

$$v' = v \frac{v}{v \pm v_S}, \tag{39}$$

where the minus sign holds for motion toward the observer and the plus sign holds for motion away from the observer. Note that the change here is the shortening or increasing of the wavelength transmitted through the medium due to the motion of the source through the medium.

If *both source and observer move* through the transmitting medium, you should be able to show that the observer hears a frequency

$$v' = v \frac{v \pm v_O}{v \mp v_S}, \tag{40}$$

where the upper signs (+ numerator, − denominator) correspond to the source and observer moving along the line joining the two in the direction toward the other, and the lower signs in the direction away from the other. Equation 40 incorporates all four different possibilities, as Sample Problem 5 shows. Note that Eq. 40 reduces to Eq. 36 when $v_S = 0$ and to Eq. 39 when $v_O = 0$, as it must.

If a source of sound is moved away from an observer and toward a wall, the observer hears two notes of different frequency. The note heard directly from the receding source is lowered in pitch by the motion. The other note is due to the waves reflected from the wall, and this is raised in pitch (because the source is moving *toward* the wall, and the wall "hears" the higher frequency). The superposition of these two wave trains produces beats. A similar effect occurs if a wave from a stationary source is reflected from a moving object. The beat frequency can be used to deduce the speed of the object. This is the basic principle of radar speed monitors, and it is also used to track satellites.

The discussion in this section applies to the Doppler shift for sound waves and other similar mechanical waves. Light waves also show the Doppler effect; however, because there is no medium of propagation for light, the formulas developed in this section do not apply. See Chapters 21 and 42 for a discussion of the Doppler effect for light waves.

Sample Problem 5   The siren of a police car emits a pure tone at a frequency of 1125 Hz. Find the frequency that you would perceive in your car under the following circumstances: (*a*) your car at rest, police car moving toward you at 29 m/s (65 mi/h); (*b*) police car at rest, your car moving toward it at 29 m/s; (*c*) you and the police car moving toward one another at 14.5 m/s; (*d*) you moving at 9 m/s, police car chasing behind you at 38 m/s.

Solution   All four parts of this problem can be solved using Eq.

40. (*a*) Here $v_O = 0$ (your car is at rest) and $v_S = 29$ m/s. We choose the upper (minus) sign in the denominator of Eq. 40, because the police car is moving toward you. We thus obtain, using $v = 343$ m/s for the speed of sound in still air,

$$v' = v \frac{v}{v - v_S} = (1125 \text{ Hz}) \frac{343 \text{ m/s}}{343 \text{ m/s} - 29 \text{ m/s}} = 1229 \text{ Hz}.$$

(*b*) In this case $v_S = 0$ (the police car is at rest) and $v_O = 29$ m/s. We choose the upper (plus) sign in the numerator of Eq. 40, because you are moving toward the police car, and we find

$$v' = v \frac{v + v_O}{v} = (1125 \text{ Hz}) \frac{343 \text{ m/s} + 29 \text{ m/s}}{343 \text{ m/s}} = 1220 \text{ Hz}.$$

(*c*) In this case $v_S = 14.5$ m/s and $v_O = 14.5$ m/s. We choose the upper signs in both the numerator and denominator of Eq. 40, because you and the police car are each moving toward the other. We thus obtain

$$v' = v \frac{v + v_O}{v - v_S} = (1125 \text{ Hz}) \frac{343 \text{ m/s} + 14.5 \text{ m/s}}{343 \text{ m/s} - 14.5 \text{ m/s}} = 1224 \text{ Hz}.$$

(*d*) Here $v_O = 9$ m/s and $v_S = 38$ m/s. You are moving away from the police car, so we choose the lower (minus) sign in the numerator, but the police car is moving *toward* you, so we choose the upper (minus) sign in the denominator. The result is

$$v' = v \frac{v - v_O}{v - v_S} = (1125 \text{ Hz}) \frac{343 \text{ m/s} - 9 \text{ m/s}}{343 \text{ m/s} - 38 \text{ m/s}} = 1232 \text{ Hz}.$$

Note that in all four cases in this sample problem, the relative speed between you and the police car is the same, namely, 29 m/s, but the perceived frequencies are different in the four cases. The Doppler shift for sound is determined not only by the relative speed between source and observer, but also by both of their speeds relative to the medium that carries the sound.

Effects at High Speed   *(Optional)*

When v_O or v_S becomes comparable in magnitude to v, the formulas just given for the Doppler effect usually must be modified. One modification is required because the linear relation between restoring force and displacement assumed up until now may no longer hold in the medium. The speed of wave propagation is then no longer the normal phase velocity, and the wave shapes change in time. Components of the motion at right angles to the line joining source and observer also contribute to the Doppler effect at these high speeds. When v_O or v_S exceeds v, the Doppler formula does not apply; for example, if $v_S > v$, the source gets ahead of the wave in one direction; if $v_O > v$ and the observer moves away from the source, the wave never catches up with the observer.

There are many instances in which the source moves through a medium at a speed greater than the phase speed of the wave in that medium. In such cases the wavefront takes the shape of a cone with the moving body at its apex. Some examples are the bow wave from a speedboat on the water and the "shock wave" from an airplane or projectile moving through the air at a speed greater than the speed of sound in that medium (supersonic speeds). Another example is the so-called Cerenkov radiation, which consists of light waves emitted by charged particles that move through a medium with a speed greater than the phase

speed of light in that medium. The blue glow of the water that often surrounds the core of a nuclear reactor is one type of Cerenkov radiation.

In Fig. 14*a* we show the present positions of the spherical waves that originated at various positions of a source during its motion. The radius of each sphere at this time is the product of the wave speed v and the time t that has elapsed since the source was at its center. The envelope of these waves is a cone whose surface makes an angle θ with the direction of motion of the source. From the figure we obtain the result

$$\sin \theta = \frac{v}{v_S}. \tag{41}$$

For surface water waves the cone reduces to a pair of intersecting lines. In aerodynamics the ratio v_S/v is called the *Mach number*. An aircraft flying at supersonic speed generates a *Mach cone* similar to that shown in Fig. 14. When the edge of that cone intercepts the ground below, we hear a "sonic boom," which (contrary to common belief) is *not* associated with an aircraft "breaking the sound barrier." The sonic boom is merely the total

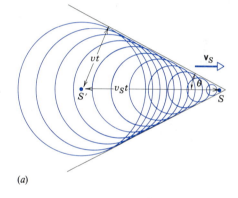

(*a*)

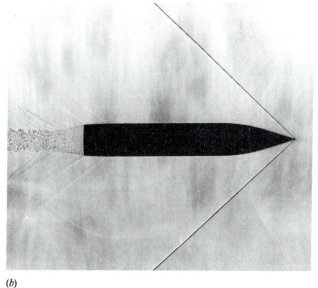

(*b*)

Figure 14   (*a*) Wavefronts of a source moving at supersonic speed. The wavefronts are spherical and their envelope is a cone. Compare this figure with Fig. 13. (*b*) A photograph of a projectile fired from a gun at Mach 2. Note the Mach cone.

effect of the concentration on one surface of the aircraft's radiated sound energy, which would normally radiate in all directions at subsonic speeds. As the photograph of Fig. 14*b* shows, it might be possible to hear two sonic booms from the same aircraft, one from the leading edge and another from the trailing edge. (Note also that the Mach cone never intercepts the projectile itself; thus the aircraft's passengers do not hear the sonic boom.) ∎

QUESTIONS

1. Why will sound not travel through a vacuum?

2. List some sources of infrasonic waves and of ultrasonic waves.

3. Ultrasonic waves can be used to reveal internal structures of the body. They can, for example, distinguish between liquid and soft human tissues far better than can x rays. How? Why do we still use x rays?

4. What experimental evidence is there for assuming that the speed of sound in air is the same for all wavelengths?

5. Give a qualitative explanation why the speed of sound in lead is less than that in copper.

6. Transverse waves on a string can be plane polarized. Can sound waves be polarized?

7. Bells frequently sound less pleasant than pianos or violins. Why?

8. A bell is rung for a short time in a school. After a while its sound is inaudible. Trace the sound waves and the energy they transfer from the time of emission until they become inaudible.

9. The pitch of the wind instruments rises and that of the string instruments falls as an orchestra warms up. Explain why.

10. Explain how a stringed instrument is tuned.

11. Is resonance a desirable feature of every musical instrument? Give examples.

12. When you strike one prong of a tuning fork, the other prong also vibrates, even if the bottom end of the fork is clamped firmly in a vise. How can this happen? That is, how does the second prong "get the word" that somebody has struck the first prong?

13. How can a sound wave travel down an organ pipe and be reflected at its open end? It would seem that there is nothing there to reflect it.

14. How can we experimentally locate the positions of nodes and antinodes on a string, in an air column, and on a vibrating surface?

15. Explain how a note is produced when you blow across the top of a test tube. What would be the effect of blowing harder? Of raising the temperature of the air in the test tube?

16. How might you go about reducing the noise level in a machine shop?

17. Foghorns emit sounds of very low pitch. For what purpose?

18. Are longitudinal waves in air always audible as sound, regardless of frequency or intensity? What frequencies would give a person the greatest sensitivity, the greatest tolerance, and the greatest range?

19. What is the common purpose of the valves of a cornet and the slide of a trombone? The bugle has no valves. How then can we sound different notes on it? To what notes is the bugler limited? Why?

20. Explain how bowing a violin string gets it to vibrate.

21. What is the meaning of zero decibels? Could the reference intensity for audible sound be set so as to permit negative sound levels in decibels? If so, how?

22. Discuss the factors that determine the range of frequencies in your voice and the quality of your voice.

23. Explain the origin of the sound in ordinary whistling.

24. What physical properties of a sound wave correspond to the human sensations of pitch, of loudness, and of tone quality?

25. What is the difference between a violin note and the same note sung by a human voice that enables us to distinguish between them?

26. Does your singing really sound better in a shower? If so, what are the physical reasons?

27. Explain the audible sound produced by drawing a wet finger around the rim of a wine glass.

28. Would a plucked violin string oscillate for a longer or shorter time if the violin had no sounding board? Explain.

29. Is a bowed violin string an example of forced damped oscillations? How would the string sound if it were not damped?

30. A tube can act like an acoustic filter, discriminating against the passage through it of sounds of frequencies different from the natural frequencies of the tube. The muffler of an automobile is an example. (*a*) Explain how such a filter works. (*b*) How can we determine the cut-off frequency, below which sound is not transmitted?

31. Discuss factors that improve the acoustics in music halls.

32. What is the effect of using a megaphone or cupping your hands in front of your mouth to project your voice over a distance?

33. A lightning flash dissipates an enormous amount of energy and is essentially instantaneous. How is that energy transformed into the sound waves of thunder? (See "Thunder," by Arthur A. Few, *Scientific American,* July 1975, p. 80.)

34. Sound waves can be used to measure the speed at which blood flows in arteries and veins. Explain how.

35. Suppose that George blows a whistle and Gloria hears it. She will hear an increased frequency whether she is running toward George or George is running toward her. Are the increases in frequency the same in each case? Assume the same speeds of running.

36. Suppose that, in the Doppler effect for sound, the source and receiver are at rest in some reference frame but the transmitting medium (air) is moving with respect to this frame. Will there be a change in wavelength, or in frequency, received?

37. You are standing in the middle of the road and a bus is coming toward you at constant speed, with its horn sounding. Because of the Doppler effect is the pitch of the horn rising, falling, or constant?

38. How might the Doppler effect be used in an instrument to detect the fetal heart beat? (Such measurements are rou-

tinely made; see "Ultrasound in Medical Diagnosis," by Gilbert B. Devey and Peter N. T. Wells, *Scientific American,* May 1978, p. 98.)

39. Bats can examine the characteristics of objects—such as size, shape, distance, direction, and motion—by sensing the way the high-frequency sounds they emit are reflected off the objects back to the bat. Discuss qualitatively how each of these features affects the reflected sound waves. (See "Information Content of Bat Sonar Echoes," by J. A. Simmons, D. J. Howell, and N. Suga, *American Scientist,* March–April 1975, p. 204.)

40. Assume that you can detect an object by bouncing waves off it (such as in sonar or radar, for instance) as long as the object is larger than the wavelength of the waves. Then consider that bats and porpoises each can emit sound waves of frequency 100 kHz; however, bats can detect objects as small as insects but porpoises only small fish. Why the difference?

41. Is there a Doppler effect for sound when the observer or the source moves at right angles to the line joining them? How then can we determine the Doppler effect when the motion has a component at right angles to this line?

42. Two ships with steam whistles of the same pitch sound off in the harbor. Would you expect this to produce an interference pattern with regions of high and low intensity? If not, why not?

43. A satellite emits radio waves of constant frequency. These waves are picked up on the ground and made to beat against some standard frequency. The beat frequency is then sent through a loudspeaker and one "hears" the satellite signals. Describe how the sound changes as the satellite approaches, passes overhead, and recedes from the detector on the ground.

44. How and why do the Doppler effects for light and for sound differ? In what ways are they the same?

PROBLEMS

Where needed in the problems, use speed of sound in air = 343 m/s and density of air = 1.21 kg/m³ unless otherwise specified.

Section 20-1 The Speed of Sound

1. Diagnostic ultrasound of frequency 4.50 MHz is used to examine tumors in soft tissue. (*a*) What is the wavelength in air of such a sound wave? (*b*) If the speed of sound in tissue is 1500 m/s, what is the wavelength of this wave in tissue?

2. If the wavelength of sound is large, by a factor of about 10, relative to the mean free path of the molecules, then sound waves can propagate through a gas. For air at room temperature the mean free path is about 0.1 μm. Calculate the frequency above which sound waves could not propagate.

3. Figure 15 shows a remarkably detailed image, of a transistor in a microelectronic circuit, formed by an acoustic microscope. The sound waves have a frequency of 4.2 GHz. The speed of such waves in the liquid helium in which the speci-

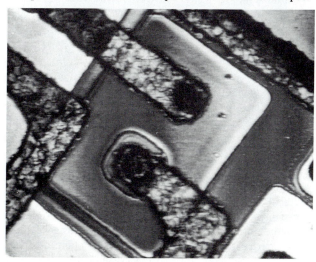

men is immersed is 240 m/s. (*a*) What is the wavelength of these ultrahigh-frequency acoustic waves? (*b*) The ribbon-like conductors in the figure are ~ 2 μm wide. To how many wavelengths does this correspond?

4. (*a*) A rule for finding your distance from a lightning flash is to count seconds from the time you see the flash until you hear the thunder and then divide the count by 5. The result is supposed to give the distance in miles. Explain this rule and determine the percent error in it at 0°C and 1 atm pressure. (*b*) Devise a similar rule for obtaining the distance in kilometers.

5. A column of soldiers, marching at 120 paces per minute, keeps in step with the music of a band at the head of the column. It is observed that the men at the rear of the column are striding forward with the left foot when those in the band are advancing with the right. What is the length of the column approximately?

6. You are at a large outdoor concert, seated 300 m from the stage microphone. The concert is also being broadcast live, in stereo, around the world via satellite. Consider a listener 5000 km away. Who hears the music first and by what time difference?

7. The speed of sound in a certain metal is *V*. One end of a long pipe of that metal of length *L* is struck a hard blow. A listener at the other end hears two sounds, one from the wave that has traveled along the pipe and the other from the wave that has traveled through the air. (*a*) If *v* is the speed of sound in air, what time interval *t* elapses between the arrival of the two sounds? (*b*) A hammer strikes a long aluminum rod at one end. A listener, whose ear is close to the other end of the rod, hears the sound of the blow twice, with a 120-ms interval between. How long is the rod?

8. Earthquakes generate sound waves in the Earth. Unlike in a gas, there are both transverse (S) and longitudinal (P) sound waves in a solid. Typically, the speed of S waves is about 4.5 km/s and that of P waves 8.2 km/s. A seismograph records P and S waves from an earthquake. The first P waves

Figure 15   Problem 3.

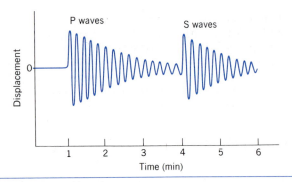

Figure 16   Problem 8.

arrive 3 min before the first S waves; see Fig. 16. How far away did the earthquake occur?

9. A stone is dropped into a well. The sound of the splash is heard 3.00 s later. What is the depth of the well?

Section 20-2 Traveling Longitudinal Waves

10. A continuous sinusoidal longitudinal wave is sent along a coiled spring from a vibrating source attached to it. The frequency of the source is 25 Hz, and the distance between successive rarefactions in the spring is 24 cm. (*a*) Find the wave speed. (*b*) If the maximum longitudinal displacement of a particle in the spring is 0.30 cm and the wave moves in the −*x* direction, write the equation for the wave. Let the source be at *x* = 0 and the displacement *s* = 0 at the source when *t* = 0.

11. The pressure in a traveling sound wave is given by the equation

$$\Delta p = (1.48 \text{ Pa}) \sin (1.07\pi x - 334\pi t),$$

where *x* is in meters and *t* is in seconds. Find (*a*) the pressure amplitude, (*b*) the frequency, (*c*) the wavelength, and (*d*) the speed of the wave.

Section 20-3 Power and Intensity of Sound Waves

12. Show that the sound wave intensity *I* can be written in terms of the frequency *v* and displacement amplitude s_m in the form

$$I = 2\pi^2 \rho v v^2 s_m^2.$$

13. A source emits spherical waves isotropically (that is, with equal intensity in all directions). The intensity of the wave 42.5 m from the source is 197 μW/m². Find the power output of the source.

14. A note of frequency 313 Hz has an intensity of 1.13 μW/m². What is the amplitude of the air vibrations caused by this sound?

15. A sound wave of intensity 1.60 μW/cm² passes through a surface of area 4.70 cm². How much energy passes through the surface in 1 h?

16. Find the intensity ratio of two sounds whose sound levels differ by 1.00 dB.

17. A certain sound level is increased by an additional 30 dB. Show that (*a*) its intensity increases by a factor of 1000 and (*b*) its pressure amplitude increases by a factor of 32.

18. A salesperson claimed that a stereo system would deliver 110 W of audio power. Testing the system with several speakers set up so as to simulate a point source, the consumer noted that she could get as close as 1.3 m with the volume full on before the sound hurt her ears. Should she report the firm to the Consumer Protection Agency?

19. A certain loudspeaker produces a sound with a frequency of 2.09 kHz and an intensity of 962 μW/m² at a distance of 6.11 m. Presume that there are no reflections and that the loudspeaker emits the same in all directions. (*a*) Find the intensity at 28.5 m. (*b*) Find the displacement amplitude at 6.11 m. (*c*) Calculate the pressure amplitude at 6.11 m.

20. (*a*) If two sound waves, one in air and one in water, are equal in intensity, what is the ratio of the pressure amplitude of the wave in water to that of the wave in air? (*b*) If the pressure amplitudes are equal instead, what is the ratio of the intensities of the waves? Assume the water is at 20°C.

21. Find the energy density in a sound wave 4.82 km from a 5.20-kW nuclear emergency siren (see Fig. 17), assuming the waves to be spherical and the propagation isotropic with no atmospheric absorption.

Figure 17   Problem 21.

22. A line source (for instance, a long freight train on a straight track) emits a cylindrical expanding wave. Assuming that the air absorbs no energy, find how (*a*) the intensity and (*b*) the amplitude of the wave depend on the distance from the source. Ignore reflections and consider points near the center of the train.

23. In Fig. 18 we show an acoustic interferometer, used to demonstrate the interference of sound waves. *S* is a source of sound (a loudspeaker, for instance), and *D* is a sound detector, such as the ear or a microphone. Path *SBD* can be varied in length, but path *SAD* is fixed. The interferometer contains air, and it is found that the sound intensity has a minimum value of $10 \, \mu\text{W/cm}^2$ at one position of *B* and continuously climbs to a maximum value of $90 \, \mu\text{W/cm}^2$ at a second position 1.65 cm from the first. Find (*a*) the frequency of the sound emitted from the source and (*b*) the relative amplitudes of the waves arriving at the detector for each of the two positions of *B*. (*c*) How can it happen that these waves have different amplitudes, considering that they originate at the same source?

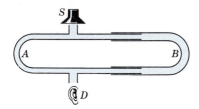

Figure 18   Problem 23.

24. You are standing at a distance *D* from an isotropic source of sound waves. You walk 51.4 m toward the source and observe that the intensity of these waves has doubled. Calculate the distance *D*.

25. Estimate the maximum possible sound level in decibels of sound waves in air. (*Hint:* Set the pressure amplitude equal to 1 atm.)

26. Suppose that the average sound level of human speech is 65 dB. How many persons in a room speaking at the same time each at 65 dB are needed to produce a sound level of 80 dB?

27. Suppose that a rustling leaf generates 8.4 dB of sound. Find the sound level from a tree with 2.71×10^5 rustling leaves.

28. In a test, a subsonic jet flies overhead at an altitude of 115 m. The sound level on the ground as the jet passes overhead is 150 dB. At what altitude should the plane fly so that the ground noise is no greater than 120 dB, the threshold of pain? Ignore the finite time required for the sound to reach the ground.

29. A certain loudspeaker (assumed to be a point source) emits 31.6 W of acoustic power. A small microphone of effective cross-sectional area $75.2 \, \text{mm}^2$ is located 194 m from the loudspeaker. Calculate (*a*) the sound intensity at the microphone, (*b*) the power incident on the microphone, and (*c*) the amount of energy that impinges on the microphone in 25.0 min.

30. A sound wave of 42.0-cm wavelength enters the tube shown in Fig. 19. What must be the smallest radius *r* such that a minimum will be heard at the detector?

Figure 19   Problem 30.

31. Two stereo loudspeakers are separated by a distance of 2.12 m. Assume the amplitude of the sound from each speaker is approximately the same at the position of a listener, who is 3.75 m directly in front of one of the speakers; see Fig. 20. (*a*) For what frequencies in the audible range (20–20,000 Hz) will there be a minimum signal? (*b*) For what frequencies is the sound a maximum?

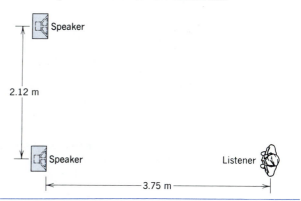

Figure 20   Problem 31.

32. A spherical sound source is placed at P_1 near a reflecting wall *AB* and a microphone is located at point P_2, as shown in Fig. 21. The frequency of the sound source is variable. Find the two lowest frequencies for which the sound intensity, as observed at P_2, will be a maximum. There is no phase change on reflection; the angle of incidence equals the angle of reflection.

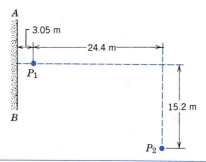

Figure 21   Problem 32.

33. Two sources of sound are separated by a distance of 5.00 m. They both emit sound at the same amplitude and frequency, 300 Hz, but they are 180° out of phase. At what points along the line connecting them will the sound intensity be the largest?

34. The *reverberation time* of an auditorium or concert hall is the time required for the sound intensity (in W/m²) to decrease by a factor of 10^6. The reverberation time depends on the frequency of the sound. Suppose that in a particular concert hall, the reverberation time for a note of a certain frequency is 2.6 s. If the note is sounded at a sound level of 87 dB, how long will it take for the sound level to fall to 0 dB (the threshold of human hearing)?

35. A large parabolic reflector having a circular opening of radius 0.50 m is used to focus sound. If the energy is delivered from the focus to the ear of a listening detective through a tube of diameter 1.0 cm with 12% efficiency, how far away can a whispered conversation be understood? (Assume that the sound level of a whisper is 20 dB at 1.0 m from the source, considered to be a point, and that the threshold for hearing is 0 dB.)

Section 20-4 Standing Longitudinal Waves

36. The strings of a cello have a length L. (a) By what length ΔL must they be shortened by fingering to change the pitch by a frequency ratio r? (b) Find ΔL, if $L = 80.0$ cm and $r = \frac{6}{5}, \frac{5}{4}, \frac{4}{3}$, and $\frac{3}{2}$.

37. A sound wave in a fluid medium is reflected at a barrier so that a standing wave is formed. The distance between nodes is 3.84 cm and the speed of propagation is 1520 m/s. Find the frequency.

38. A well with vertical sides and water at the bottom resonates at 7.20 Hz and at no lower frequency. The air in the well has a density of 1.21 kg/m³ and a bulk modulus of 1.41×10^5 Pa. How deep is the well?

39. S in Fig. 22 is a small loudspeaker driven by an audio oscillator and amplifier, adjustable in frequency from 1000 to 2000 Hz only. D is a piece of cylindrical sheetmetal pipe 45.7 cm long and open at both ends. (a) At what frequencies will resonance occur when the frequency emitted by the speaker is varied from 1000 to 2000 Hz? (b) Sketch the displacement nodes for each resonance. Neglect end effects.

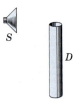

Figure 22   Problem 39.

40. The width of the terraces in an amphitheater in Los Angeles, Fig. 23, is 36 in. (=0.914 m). A single hand-clap occurring at the center of the stage will reflect back to the stage as a tone of what frequency?

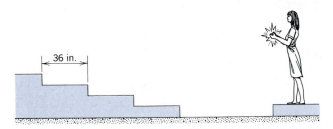

Figure 23   Problem 40.

41. A tunnel leading straight through a hill greatly amplifies tones at 135 and 138 Hz. Find the shortest length the tunnel could have.

42. The period of a pulsating variable star may be estimated by considering the star to be executing radial longitudinal pulsations in the fundamental standing wave mode; that is, the

radius varies periodically with the time, with a displacement antinode at the surface. (a) Would you expect the center of the star to be a displacement node or antinode? (b) By analogy with the open organ pipe, show that the period of pulsation T is given by

$$T = \frac{4R}{v_S},$$

where R is the equilibrium radius of the star and v_S is the average sound speed. (c) Typical white dwarf stars are composed of material with a bulk modulus of 1.33×10^{22} Pa and a density of 1.0×10^{10} kg/m³. They have radii equal to 0.009 solar radius. What is the approximate pulsation period of a white dwarf? (See "Pulsating Stars," by John R. Percy, *Scientific American,* June 1975, p. 66.)

43. In Fig. 24, a rod R is clamped at its center; a disk D at its end projects into a glass tube that has cork filings spread over its interior. A plunger P is provided at the other end of the tube. The rod is set into longitudinal vibration and the plunger is moved until the filings form a pattern of nodes and antinodes (the filings form well-defined ridges at the pressure antinodes). If we know the frequency v of the longitudinal vibrations in the rod, a measurement of the average distance d between successive antinodes determines the speed of sound v in the gas in the tube. Show that

$$v = 2vd.$$

This is Kundt's method for determining the speed of sound in various gases.

Figure 24   Problem 43.

Section 20-5 Vibrating Systems and Sources of Sound

44. (a) Find the speed of waves on an 820-mg violin string 22.0 cm long if the frequency of the fundamental is 920 Hz. (b) Calculate the tension in the string.

45. If a violin string is tuned to a certain note, by what factor must the tension in the string be increased if it is to emit a note of double the original frequency (that is, a note one octave higher in pitch)?

46. A certain violin string is 30 cm long between its fixed ends and has a mass of 2.0 g. The string sounds an A note (440 Hz) when played without fingering. Where must one put one's finger to play a C (528 Hz)?

47. An open organ pipe has a fundamental frequency of 291 Hz. The first overtone ($n = 3$) of a closed organ pipe has the same frequency as the second harmonic of the open pipe. How long is each pipe?

48. A tube 1.18 m long is closed at one end. A stretched wire is placed near the open end. The wire is 33.2 cm long and has a mass of 9.57 g. It is fixed at both ends and vibrates in its fundamental mode. It sets the air column in the tube into vibration at its fundamental frequency by resonance. Find (a) the frequency of oscillation of the air column and (b) the tension in the wire.

49. A 30.0-cm violin string with linear mass density 0.652 g/m is placed near a loudspeaker that is fed by an audio oscillator

of variable frequency. It is found that the string is set into oscillation only at the frequencies 880 and 1320 Hz as the frequency of the oscillator is varied continuously over the range 500–1500 Hz. What is the tension in the string?

Section 20-6 Beats

50. A tuning fork of unknown frequency makes three beats per second with a standard fork of frequency 384 Hz. The beat frequency decreases when a small piece of wax is put on a prong of the first fork. What is the frequency of this fork?

51. The A string of a violin is a little too taut. Four beats per second are heard when it is sounded together with a tuning fork that is vibrating accurately at the pitch of concert A (440 Hz). What is the period of the violin string vibration?

52. You are given four tuning forks. The fork with the lowest frequency vibrates at 500 Hz. By using two tuning forks at a time, the following beat frequencies are heard: 1, 2, 3, 5, 7, and 8 Hz. What are the possible frequencies of the other three tuning forks?

53. You are given five tuning forks, each of which has a different frequency. By trying every pair of tuning forks, (a) what maximum number of *different* beat frequencies might be obtained? (b) What minimum number of *different* beat frequencies might be obtained?

Section 20-7 The Doppler Effect

54. A source *S* generates circular waves on the surface of a lake, the pattern of wave crests being shown in Fig. 25. The speed of the waves is 5.5 m/s and the crest-to-crest separation is 2.3 m. You are in a small boat heading directly toward *S* at a constant speed of 3.3 m/s with respect to the shore. What frequency of the waves do you observe?

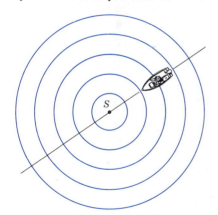

Figure 25   Problem 54.

55. The 15.8-kHz whine of the turbines in the jet engines of an aircraft moving with speed 193 m/s is heard at what frequency by the pilot of a second craft trying to overtake the first at a speed of 246 m/s?

56. An ambulance emitting a whine at 1602 Hz overtakes and passes a cyclist pedaling a bike at 2.63 m/s. After being passed, the cyclist hears a frequency of 1590 Hz. How fast is the ambulance moving?

57. A whistle of frequency 538 Hz moves in a circle of radius

71.2 cm at an angular speed of 14.7 rad/s. What are (a) the lowest and (b) the highest frequencies heard by a listener a long distance away at rest with respect to the center of the circle?

58. In 1845, Buys Ballot first tested the Doppler effect for sound. He put a trumpet player on a flatcar drawn by a locomotive and another player near the tracks. If each player blows a 440-Hz note, and if there are 4.0 beats/s as they approach each other, what is the speed of the flatcar?

59. A bullet is fired with a speed of 2200 ft/s. Find the angle made by the shock cone with the line of motion of the bullet.

60. Estimate the speed of the projectile illustrated in the photograph in Fig. 14. Assume the speed of sound in the medium through which the projectile is traveling to be 380 m/s.

61. The speed of light in water is 2.25×10^8 m/s (about three-fourths the speed in a vacuum). A beam of high-speed electrons from a betatron emits Cerenkov radiation in water, the wavefront being a cone of angle 58.0°. Find the speed of the electrons in the water.

62. Two identical tuning forks oscillate at 442 Hz. A person is located somewhere on the line between them. Calculate the beat frequency as measured by this individual if (a) she is standing still and the tuning forks both move to the right at 31.3 m/s, and (b) the tuning forks are stationary and the listener moves to the right at 31.3 m/s.

63. A plane flies at 396 m/s at constant altitude. The sonic boom reaches an observer on the ground 12.0 s after the plane flies overhead. Find the altitude of the plane. Assume the speed of sound to be 330 m/s.

64. A jet plane passes overhead at a height of 5140 m and a speed of Mach 1.52 (1.52 times the speed of sound). (a) Find the angle made by the shock wave with the line of motion of the jet. (b) How long after the jet has passed directly overhead will the shock wave reach the ground? Use 331 m/s for the speed of sound.

65. Figure 26 shows a transmitter and receiver of waves contained in a single instrument. It is used to measure the speed *V* of a target object (idealized as a flat plate) that is moving directly toward the unit, by analyzing the waves reflected from it. (a) Apply the Doppler equations twice, first with the target as observer and then with the target as a source, and show that the frequency v_r of the reflected waves at the receiver is related to their source frequency v_s by

$$v_r = v_s \left(\frac{v + V}{v - V} \right),$$

where *v* is the speed of the waves. (b) In a great many practical situations, $V \ll v$. In this case, show that the equation above becomes

$$\frac{v_r - v_s}{v_s} \approx \frac{2V}{v}.$$

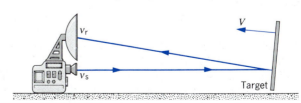

Figure 26   Problem 65.

66. A sonar device sends 148-kHz sound waves from a hiding police car toward a truck approaching at a speed of 44.7 m/s. Calculate the frequency of the reflected waves detected at the police car.

67. An acoustic burglar alarm consists of a source emitting waves of frequency 28.3 kHz. What will be the beat frequency of waves reflected from an intruder walking at 0.95 m/s directly away from the alarm?

68. A siren emitting a sound of frequency 1000 Hz moves away from you toward a cliff at a speed of 10.0 m/s. (a) What is the frequency of the sound you hear coming directly from the siren? (b) What is the frequency of the sound you hear reflected off the cliff? (c) Find the beat frequency. Could you hear the beats? Take the speed of sound in air as 330 m/s.

69. A person in a car blows a trumpet sounding at 438 Hz. The car is moving toward a wall at 19.3 m/s. Calculate (a) the frequency of the sound as received at the wall and (b) the frequency of the reflected sound arriving back at the source.

70. Two submarines are on a head-on collision course during maneuvers in the North Atlantic. The first sub is moving at 20.2 km/h and the second sub at 94.6 km/h. The first submarine sends out a sonar signal (sound wave in water) at 1030 Hz. Sonar waves travel at 5470 km/h. (a) The second sub picks up the signal. What frequency does the second sonar detector hear? (b) The first sub picks up the reflected signal. What frequency does the first sonar detector hear? See Fig. 27. The ocean is calm; assume no currents.

20.2 km/h   94.6 km/h

Figure 27 Problem 70.

71. A police car sounding its siren is moving at 27 m/s and approaching a stationary pedestrian. The police in the car

hear the siren at 12.6 kHz but the pedestrian hears the siren at 13.7 kHz. Find the air temperature. (Assume that the speed of sound increases linearly with temperature between 0°C and 20°C; see Table 1.)

72. In a discussion of Doppler shifts of ultrasonic (high-frequency) waves used in medical diagnosis, the authors remark: "For every millimeter per second that a structure in the body moves, the frequency of the incident ultrasonic wave is shifted approximately 1.3 Hz/MHz." What speed of the ultrasonic waves in tissue do you deduce from this statement?

73. A bat is flittering about in a cave, navigating very effectively by the use of ultrasonic bleeps (short emissions of high-frequency sound lasting a millisecond or less and repeated several times a second). Assume that the sound emission frequency of the bat is 39.2 kHz. During one fast swoop directly toward a flat wall surface, the bat is moving at 8.58 m/s. Calculate the frequency of the sound the bat hears reflected off the wall.

74. A submarine moving north with a speed of 75.2 km/h with respect to the ocean floor emits a sonar signal (sound waves in water used in ways similar to radar; see Table 1) of frequency 989 Hz. If the ocean at that point has a current moving north at 30.5 km/h relative to the land, what frequency is observed by a ship drifting with the current north of the submarine? (*Hint:* All speeds in the Doppler equations must be taken with respect to the medium.)

75. A 2000-Hz siren and a civil defense official are both at rest with respect to the Earth. What frequency does the official hear if the wind is blowing at 12 m/s (a) from source to observer and (b) from observer to source?

76. Two trains on parallel tracks are traveling toward each other at 34.2 m/s relative to the ground. One train is blowing a whistle at 525 Hz. (a) What frequency will be heard on the other train in still air? (b) What frequency will be heard on the other train if the wind is blowing at 15.3 m/s parallel to the tracks and toward the whistle? (c) What frequency will be heard if the wind direction reverses?

CHAPTER 21

THE SPECIAL THEORY OF RELATIVITY*

The special theory of relativity has an undeserved reputation as a difficult subject. It is not mathematically complicated; most of its details can be understood using techniques well known to readers of this text. Perhaps the most challenging aspect of special relativity is its insistence that we replace some of our ideas about space and time, which we have acquired through years of "common-sense" experiences, with new ideas.

The essential ideas of special relativity were formally presented in a paper written by Albert Einstein and published in 1905.[†] In this chapter we present the basic postulates of Einstein's theory and their consequences, introduce the mathematical techniques that allow measurements made in one frame of reference to be transformed into another, and study some of the consequences for both kinematics and dynamics.

Previously in this text we introduced some details of special relativity and contrasted them with the corresponding results of classical physics. Before starting this chapter, you should review those discussions: Section 4-6, Relative Motion; Section 7-7, Kinetic Energy at High Speed; Section 8-7, Mass and Energy; and Section 9-4, Linear Momentum of a Particle.

21-1 TROUBLES WITH CLASSICAL PHYSICS

The kinematics developed by Galileo and the mechanics developed by Newton, which form the basis of what we call *classical physics,* had many triumphs. Particularly noteworthy are the understanding of the motion of the planets and the use of kinetic theory to explain certain observed properties of gases. However, a number of exper-

---

* Some instructors may wish to delay covering relativity until after the treatment of electromagnetic waves in Chapter 41. Relativistic effects in wave motion are discussed in Chapter 42.

† In that year he also published his papers on Brownian motion and on the photoelectric effect. It was for this latter paper (and not specifically for his theory of relativity) that he was awarded the 1921 Nobel prize in physics. Einstein also proposed a *general* theory of relativity in 1917. The general theory deals with the effect of gravity on space and time, some consequences of which were discussed in Section 16-10. In this chapter we consider only the *special* theory, in which gravity plays no role.

---

imental phenomena cannot be understood with these otherwise successful classical theories. Let us consider a few of these difficulties.

Troubles with Our Ideas About Time

The pion (π^+ or π^-) is a particle that can be created in a high-energy particle accelerator. It is a very unstable particle; pions created at rest are observed to decay (to other particles) with an average lifetime of only 26.0 ns (26.0×10^{-9} s). In one particular experiment, pions were created in motion at a speed of $v = 0.913c$ (where c is the speed of light). In this case they were observed to travel in the laboratory an average distance of $D = 17.4$ m before decaying, from which we conclude that they decay in a time given by $D/v = 63.7$ ns, much larger than the lifetime measured for pions at rest (26.0 ns). This effect, called *time dilation,* suggests that something about the relative motion between the pion and the laboratory has stretched the measured time interval by a factor of about 2.5. Such an effect cannot be explained by Newtonian physics, in which time is a universal coordinate having identical values for all observers.

Troubles with Our Ideas About Length

Suppose an observer in the above laboratory placed one marker at the location of the pion's formation and another at the location of its decay. The distance between the markers is measured to be 17.4 m. Now consider the situation according to a different observer who is traveling along with the pion at a speed of $u = 0.913c$. This observer, to whom the pion appears to be at rest, measures its lifetime to be 26.0 ns, characteristic of pions at rest. To this observer, the distance between the markers showing the formation and decay of the pion is $(0.913c)(26.0 \times 10^{-9}\,\text{s}) = 7.1$ m. Thus two observers who are in relative motion measure different values for the same length interval. This is likewise inconsistent with Newtonian physics, in which spatial coordinates are absolute and give identical readings for all observers.

Troubles with Our Ideas About Velocity

Figure 1 shows a game between A and B, as seen by an observer O. All three observers are at rest in this reference frame. A throws a ball at superluminal (faster than light) speed toward B, who catches it. The light signal carrying the view of A throwing the ball travels to observer O, as does the light signal carrying the view of B catching the ball. Both light signals travel at speed c, which is less than the speed of the ball thrown by A. At the location of observer O, as shown in Fig. 1, the light signal from B arrives before the light signal from A. Therefore, according to O, B catches the ball before A throws it! Newtonian physics permits us to accelerate projectiles to unlimited velocities and therefore allows such apparent violations of cause and effect to be observed.

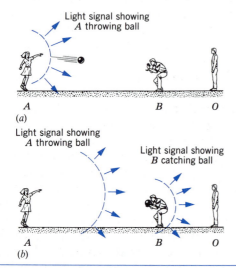

Figure 1   (*a*) A throws a ball to B. The ball moves faster than light and so is ahead of the light signal that shows A throwing the ball. (*b*) The light signal showing B catching the ball will reach the observer O before the light signal showing A throwing the ball. Such logical inconsistencies argue against the possibility of accelerating particles to speeds faster than light.

Troubles with Our Ideas About Light

Maxwell's theory of electromagnetism (which we discuss later in this text) was one of the greatest triumphs of 19th century physics. One of the deductions of this theory was that light could be described as an electromagnetic wave. Just as a mechanical wave (discussed in Chapter 19) can be analyzed in terms of oscillating particles in a medium, an electromagnetic wave can be analyzed in terms of oscillating electric and magnetic fields. It therefore became a goal of experimental physicists in the late 19th century to detect the medium in which these fields oscillate as light propagates and to measure the speed with which the Earth moves through this medium, which was called the *ether*.

Beginning in 1881, A. A. Michelson (the first U.S. recipient of a Nobel prize in physics) and E. W. Morley carried out a series of delicate optical experiments (described in Section 45-7) to measure the speed with which the Earth moves through the postulated ether. To their surprise, they found that, within their small experimental error, the result was zero! More recent experiments with lasers have improved the precision of this result by many orders of magnitude, and the value remains consistent with zero.*

Given the complex motion of the Earth (rotating about its axis and orbiting about the Sun as the Sun itself revolves about the center of the galaxy), it seems inconceivable that the ether could remain firmly attached to the moving Earth. Yet great theoretical efforts were made toward the end of the 19th century to try to explain how this could occur. Einstein's brilliant contribution to the understanding of space and time was to show that the concepts of the ether and of the medium of propagation of light were useless and unnecessary.

Einstein's Paradox

Einstein proposed his special theory of relativity in 1905, not out of any attempt to explain the outcome of the Michelson–Morley experiment but based on a thought experiment that he had devised. As a 16-year-old student, Einstein had learned Maxwell's theory of electromagnetism and had thought about a paradox: If you were to move at the speed of light parallel to a light beam traveling in empty space, you would observe "static" electric and magnetic field patterns. (In a similar way, we showed in Fig. 9 of Chapter 19 a "static" disturbance on a string, which would be seen by an observer moving along the string at the same speed as waves on the string.) However, Einstein knew that such static electric and magnetic field patterns in empty space violated Maxwell's theory.

Einstein was faced with two choices to resolve this paradox: either Maxwell's theory was wrong or else the classi-

---

* For more details about the Michelson–Morley experiment, one of the landmark experiments in the history of physics, see *Basic Concepts in Relativity and Early Quantum Theory,* by Robert Resnick and David Halliday (Wiley, 1985).

cal kinematics that permits an observer to travel along with a light beam was wrong. With the intuition that was perhaps his greatest attribute, Einstein put his faith in Maxwell's theory and sought an alternative to the kinematics of Galileo and Newton. Later in this chapter we show how this new kinematics, which forms the basis of special relativity, prevents any observer from catching a light beam. We also show how it solves the other problems with time, length, and velocity discussed previously.

The critical test of any theory is of course how well it agrees with experiment. Einstein's special theory of relativity has been subjected to exhaustive tests over the past 85 years and has passed every one. Where classical physics and relativity theory predict different results, experiment has always been found to agree with relativity theory.

21-2 THE POSTULATES OF SPECIAL RELATIVITY

A scientific theory usually begins with general statements called *postulates,* which attempt to provide a basis for the theory. From these postulates we can obtain a set of mathematical laws in the form of equations that relate physical variables. Finally, we test the predictions of the equations in the laboratory. The theory stands until contradicted by experiment, after which the postulates may be modified or replaced, and the cycle is repeated.

For about two centuries, the mechanics of Galileo and Newton withstood all experimental tests. In this case the postulates concern the absolute nature of space and time. Based on his thought experiment about catching a light beam, Einstein realized the need to replace the Galilean laws of relative motion. In his 1905 paper, entitled "On the Electrodynamics of Moving Bodies," Einstein offered two postulates that form the basis of his special theory of relativity. We can rephrase his postulates as follows:

The principle of relativity: *The laws of physics are the same in all inertial reference frames.*

The principle of the constancy of the speed of light: *The speed of light in free space has the same value c in all inertial reference frames.*

The first postulate declares that the laws of physics are absolute, universal, and the same for all inertial observers. Laws that hold for one inertial observer cannot be violated for *any* inertial observer.

The second postulate is much more difficult to accept, because it violates our "common sense," which is firmly grounded in the Galilean kinematics that we have learned from everyday experiences. Consider three observers A, B, and C, each of whom is at rest in a different inertial reference frame. A flash of light is emitted by observer A, who observes the light to travel at speed c. The frame of

observer B is moving away from A at a speed of c/4; Galilean kinematics predicts that B measures the value $c - c/4 = 3c/4$ for the speed of the light emitted by A. Observer C is in a frame that is moving *toward* A with speed c/4; according to Galileo, observer C measures a speed of $c + c/4 = 5c/4$ for the speed of the light emitted by A. Einstein's second postulate, on the other hand, asserts that *all three observers measure the same speed c for the light pulse!*

This is of course not the way ordinary objects behave. A projectile fired from a moving car has a velocity relative to the ground determined by the vector sum of the velocity of the projectile relative to the car and the velocity of the car relative to the ground. However, the velocities of light waves and particles moving at speeds close to c do not behave in this way. We discuss the relativistic law for velocity addition in Section 21-6 and show that it reduces to the "common-sense" Galilean law at low speeds.

Einstein put forth these postulates at a time when experimental tests were difficult or impossible. During the following decades, the development of high-energy particle accelerators made possible the study of the motions of particles at speeds close to c. In 1964, for example, an experiment was performed at CERN, the European high-energy particle physics laboratory near Geneva, Switzerland. The proton accelerator at CERN was used to produce a beam of particles called neutral pions (π^0), which decay rapidly (with an average lifetime of about 10^{-16} s) to two gamma rays:

$$\pi^0 \rightarrow \gamma + \gamma.$$

Gamma rays are electromagnetic radiations that travel at the speed of light. The experimenters measured directly the speed of the gamma rays emitted by the decaying pions, which were moving at a speed of 0.99975c. According to Galileo, gamma rays emitted in the direction of motion of the pions should have a speed of $c + 0.99975c = 1.99975c$ in the laboratory frame of reference. According to Einstein, they should have a speed of c. The measured speed was 2.9977×10^8 m/s, equal to c to within 1 part in 10^4, thus providing direct verification of the second postulate.

The two postulates taken together have another consequence: they imply that *it is impossible to accelerate a particle to a speed greater than c, no matter how much kinetic energy we give it.* This is also a prediction that can be tested in the laboratory, and one that brings out another difference between the postulates of relativity and those of classical physics. Classical physics places no upper limit on the speed that an object may attain; relativity does impose such a limiting speed, which, by the first postulate, must be the same for all frames of reference.

In another experiment done in 1964, electrons were accelerated by a large voltage difference (up to 15 million volts), and the speed of the electrons was directly determined. Figure 2 shows the measured speeds as a function of the kinetic energy acquired by the electrons. No matter

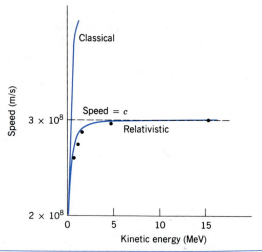

Figure 2   The points represent measurements of the speed of electrons accelerated through a large voltage difference to a known kinetic energy. The measurements show that, no matter how great the kinetic energy, the speed of the electrons does not exceed c. (See "Speed and Kinetic Energy of Relativistic Electrons," by William Bertozzi, *American Journal of Physics,* May 1964, p. 551.)

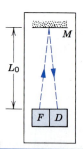

Figure 3   The clock ticks at intervals Δt_0 determined by the time necessary for a light flash to travel the distance $2L_0$ from the flashing bulb F to the mirror M and back to the detector D. (The lateral distance between F and D is assumed to be negligible in comparison with L_0.)

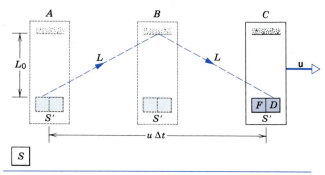

Figure 4   In the frame of reference of S, the clock carried by S' on the train moves with speed u. The dashed line, of length $2L$, shows the path of the light beam according to S.

how much the accelerating voltage is increased, the speed never quite reaches or exceeds c. Once again, experiments at high speeds are inconsistent with predictions based on the kinematics of Galileo and Newton but instead confirm the postulates of special relativity.

21-3   CONSEQUENCES OF EINSTEIN'S POSTULATES

In Section 21-1, we discussed difficulties in interpreting certain measurements of time, length, and velocity based on classical physics. Let us see how Einstein's postulates can resolve those difficulties.

The Relativity of Time

We consider two observers: S is at rest on the ground, and S' is in a train moving on a long, straight track at constant speed u relative to S. The observers carry identical timing devices, illustrated in Fig. 3, consisting of a flashing light bulb F attached to a detector D and separated by a distance L_0 from a mirror M. The bulb emits a flash of light that travels to the mirror. When the reflected light returns to D, the clock ticks and another flash is triggered. The time interval Δt_0 between ticks is just the distance $2L_0$ traveled by the light divided by the speed of light c:

$$\Delta t_0 = 2L_0/c. \qquad (1)$$

The interval Δt_0 is observed by either S or S' when the clock is at rest with respect to that observer.

We now consider the situation when one observer looks at a clock carried by the other. Figure 4 shows a representation of the sequence of events that S observes* on the clock carried by S' on the moving train. According to S, the flash is emitted at A, reflected at B, and detected at C. In this interval Δt, according to S the clock moves forward a horizontal distance of $u \Delta t$ from the location where the flash was emitted.

According to S, the light beam travels a distance $2L$, where $L = \sqrt{L_0^2 + (u \Delta t/2)^2}$, as shown in Fig. 4. The time interval measured by S for the light to travel this distance at a speed c (the same speed measured by S'!) is

$$\Delta t = \frac{2L}{c} = \frac{2\sqrt{L_0^2 + (u \Delta t/2)^2}}{c}. \qquad (2)$$

Substituting for L_0 from Eq. 1 and solving Eq. 2 for Δt gives

$$\Delta t = \frac{\Delta t_0}{\sqrt{1 - u^2/c^2}}. \qquad (3)$$

---

* We assume that S has a row of synchronized clocks, which S can use to make time measurements at points A, B, and C. Establishment of a synchronized array of clocks is discussed in Section 21-5.

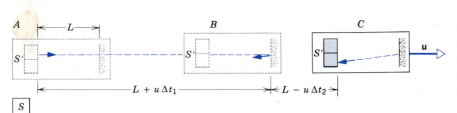

Figure 5 Here the clock carried by S' on the train emits its light flash in the direction of motion of the train. The figure at C has been displaced to the right for clarity.

The factor in the denominator of Eq. 3 is always less than or equal to 1, and thus $\Delta t \geq \Delta t_0$. That is, the observer relative to whom the clock is in motion (observer S) measures a greater interval between ticks. This effect is called *time dilation.* The time interval Δt_0 measured by an observer (S' in this case) relative to whom the clock is at rest is called the *proper time.* The proper time interval between events is the smallest interval between them that any observer can measure; all observers in motion relative to the clock measure *longer* intervals.

Equation 3 enables us to understand the difficulty with the pion decay experiments discussed in Section 21-1. A pion at rest decays in a time interval of 26.0 ns; this interval is a proper time interval and is designated as Δt_0. (The pion is in effect a clock, and the interval from formation to decay of the pion can be regarded as a tick of the clock.) An observer in the laboratory, relative to whom the pion is in motion at a speed of $u = 0.913c$, would be expected to measure a time interval of

$$\Delta t = \frac{\Delta t_0}{\sqrt{1 - u^2/c^2}} = \frac{26.0 \text{ ns}}{\sqrt{1 - (0.913)^2}} = 63.7 \text{ ns},$$

in agreement with the measured value.

Equation 3, which is deduced from Einstein's postulates, gives the relationship between time intervals according to special relativity for observers in relative motion. Note that the factor in the denominator differs appreciably from 1 only at speeds that approach the speed of light. Even at a speed of $0.1c$, Eq. 3 gives $\Delta t = 1.005\Delta t_0$. At ordinary speeds, we can take $\Delta t = \Delta t_0$ to a very high precision. This is the classical result (which is obtained directly from Eq. 3 in the limit $u \ll c$) and is in accord with our "common-sense" experience.

Equation 3 is valid for any direction of the relative motion of S and S'. It is also valid for any type of clock, not just the special one we used in its derivation. It has been verified experimentally not only with decaying elementary particles (such as the pion) moving at high speed, but also with precise atomic clocks moving relative to one another at ordinary (jet airliner) speeds. Even biological clocks such as human aging are expected to be affected by time dilation. An interesting aspect of this effect, called the *twin paradox,* is discussed later in this chapter.

The Relativity of Length

We now consider the effect of Einstein's postulates on the measurement of length intervals. Suppose that S' turns the clock on the train sideways, so that the light now

travels along the direction of motion of the train. Figure 5 shows the sequence of events as observed by S for the moving clock. According to S the length of the clock is L; as we shall see, this length is different from the length L_0 measured by S', relative to whom the clock is at rest.

A flash of light is emitted at position A in Fig. 5 and reaches the mirror (position B) a time Δt_1 later. The total distance traveled by the light in this interval is $c\,\Delta t_1$, which can also be written as the length L of the clock plus the additional distance $u\,\Delta t_1$ that the mirror moves forward in this interval due to the motion of the train. That is,

$$c\,\Delta t_1 = L + u\,\Delta t_1. \qquad (4)$$

During the return trip from the mirror to the detector (position C in Fig. 5), which takes an interval Δt_2 according to S, the light travels a distance $c\,\Delta t_2$, which must equal the length L less the distance $u\,\Delta t_2$ that the train moves forward in this interval, or

$$c\,\Delta t_2 = L - u\,\Delta t_2. \qquad (5)$$

After solving Eqs. 4 and 5 for Δt_1 and Δt_2, we add to find the total time interval Δt, which gives

$$\Delta t = \Delta t_1 + \Delta t_2 = \frac{L}{c - u} + \frac{L}{c + u}$$
$$= \frac{2L}{c}\frac{1}{1 - u^2/c^2}. \qquad (6)$$

From Eq. 3,

$$\Delta t = \frac{\Delta t_0}{\sqrt{1 - u^2/c^2}} = \frac{2L_0}{c}\frac{1}{\sqrt{1 - u^2/c^2}}. \qquad (7)$$

Setting Eqs. 6 and 7 equal to one another and solving, we obtain

$$L = L_0\sqrt{1 - u^2/c^2}. \qquad (8)$$

Equation 8 summarizes the effect known as *length contraction.* The length L_0 measured by an observer (such as S') who is at rest with respect to the object being measured is called the *rest length* (also known as the *proper length,* in analogy with the proper time). All observers in motion relative to S' measure a shorter length, but only for dimensions along the direction of motion; length measurements transverse to the direction of motion are unaffected. In the situation shown in Fig. 4, the length L_0 is unaffected by the relative motion.

Equation 8 can help us resolve the difficulties with the classical concept of length discussed in Section 21-1. The two markers placed in the laboratory at the locations of the formation and decay of the pion are separated by a distance of 17.4 m. Since the markers are at rest in the laboratory, the distance between them is a rest length. To an observer traveling with the pion, the entire laboratory is in motion at $u = 0.913c$, and the distance between the markers is measured, according to Eq. 8, to have a contracted length

$$L = (17.4 \text{ m})\sqrt{1 - (0.913)^2} = 7.1 \text{ m},$$

consistent with the discussion of Section 21-1.

Under ordinary circumstances, $u \ll c$ and the effects of length contraction are far too small to be observed. For example, a rocket of length 100 m launched from Earth with the high speed sufficient to escape the Earth's gravity ($u = 11.2$ km/s) would be measured to contract, according to an observer on the Earth, by an amount roughly equivalent to only 2 atomic diameters!

Length contraction suggests that objects in motion are measured to have a shorter length than they do at rest. No actual shrinkage is implied, merely a difference in measured results, just as two observers in relative motion measure a different frequency for the same source of sound (the Doppler effect).

The Relativistic Addition of Velocities

Let us now modify our timing device, as shown in Fig. 6. The flashing bulb F is moved to the mirror end and is replaced by a device P that emits particles at a speed v_0, as measured by an observer at rest with respect to the device. The bulb is triggered to flash when it is struck by a particle, and a light beam makes the return trip to the detector D. Thus the time interval Δt_0, measured by an observer (such as S') who is at rest with respect to the device, consists of two parts: one due to the particle traveling the distance L_0 at speed v_0 and another due to the light beam traveling the same distance at speed c:

$$\Delta t_0 = L_0/v_0 + L_0/c. \tag{9}$$

The sequence of events observed by S as the timing device is carried by S' on the train is identical with that of Fig. 5. The emitted particle, which travels at speed v according to S, reaches F after an interval Δt_1, during which time it travels a distance $v\, \Delta t_1$, which is equal to the

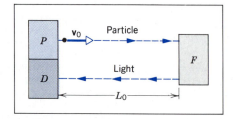

Figure 6 In this timing device, a particle is emitted by P at a speed v_0. When the particle reaches F, it triggers the emission of a flash of light that travels to the detector D.

(contracted) length L plus the additional distance $u\, \Delta t_1$ moved by the train in that interval:

$$v\, \Delta t_1 = L + u\, \Delta t_1. \tag{10}$$

In the interval Δt_2, the light beam travels a distance $c\, \Delta t_2$ equal to the length L less the distance $u\, \Delta t_2$ moved forward by the train in that interval:

$$c\, \Delta t_2 = L - u\, \Delta t_2. \tag{11}$$

Solving Eqs. 10 and 11 for Δt_1 and Δt_2, we can then find the total time interval $\Delta t = \Delta t_1 + \Delta t_2$ between ticks according to S, and we substitute that result along with Eq. 9 into Eq. 3, which gives (after using Eq. 8 to relate L_0 and L)

$$v = \frac{v_0 + u}{1 + v_0 u/c^2}. \tag{12}$$

Equation 12 gives one form of the velocity addition law consistent with Einstein's postulates; here we are concerned only with adding velocities that are along the direction of relative motion (the direction of **u**). Later in this chapter we derive more general results.

According to Galileo and Newton, a projectile fired forward at speed v_0 in a train that is moving at speed u should have a speed $v_0 + u$ relative to an observer on the ground. This clearly permits speeds in excess of c to be realized. The difference between the classical result and the relativistic result is the denominator of Eq. 12, which can certainly be replaced by 1 in ordinary circumstances when the speeds are much smaller than c. This important factor, as we see in Sample Problem 2, prevents the relative speed from ever exceeding c.

If the projectile is a light beam ($v_0 = c$ according to S'), then Eq. 12 immediately gives $v = c$ for all observers, no matter what their speed relative to S' (that is, independent of u). Thus Eq. 12 is consistent with Einstein's second postulate.

Sample Problem 1 Muons are elementary particles with a (proper) lifetime of 2.2 μs. They are produced with very high speeds in the upper atmosphere when cosmic rays (high-energy particles from space) collide with air molecules. Take the height L_0 of the atmosphere (its rest length) to be 100 km in the reference frame of the Earth, and find the minimum speed that will enable the muons to survive the journey to the surface of the Earth. Solve this problem in two ways: (*a*) in the Earth's frame of reference and (*b*) in the muon's frame of reference.

Solution (*a*) In the Earth's frame of reference (Fig. 7*a*), the decay of the moving muon is slowed by the time dilation effect. If the muon is moving at a speed that is very close to c, the time necessary for it to travel from the top of the atmosphere to the Earth is

$$\Delta t = \frac{L_0}{c} = \frac{100 \text{ km}}{3.00 \times 10^8 \text{ m/s}} = 333 \ \mu\text{s}.$$

The muon must survive for at least 333 μs in the Earth's frame of

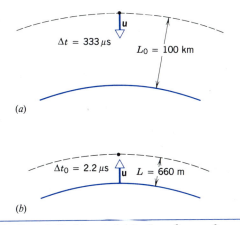

(a)

(b)

Figure 7   Sample Problem 1. (*a*) In the reference frame of the Earth, a muon takes 333 μs to travel a distance of 100 km through the atmosphere. (*b*) In the reference frame of the muon, the atmosphere is only 660 m high, and the journey takes 2.2 μs.

reference. We now find the speed that dilates the lifetime from its proper value Δt_0 ($=2.2$ μs) to this value, according to the time dilation formula (Eq. 3):

$$333 \ \mu\text{s} = \frac{2.2 \ \mu\text{s}}{\sqrt{1 - u^2/c^2}} \ .$$

Solving, we find

$$u = 0.999978c.$$

(*b*) In the muon's frame of reference, the atmosphere is rushing by at high speed. In this frame of reference the entire atmosphere must rush by in a time equal to the (proper) lifetime of the muon, and thus the height of the atmosphere can be no greater than

$$L = c \, \Delta t_0 = (3.00 \times 10^8 \text{ m/s})(2.2 \times 10^{-6} \text{ s}) = 660 \text{ m}.$$

This is of course the measured contracted length in the muon's frame of reference (see Fig. 7*b*). The relationship between the rest length L_0 ($=100$ km), measured in the Earth's frame of reference, and the contracted length, measured in the muon's frame of reference, is given by Eq. 8, and so

$$660 \text{ m} = (100 \text{ km}) \sqrt{1 - u^2/c^2}.$$

Solving for the speed u, we obtain the same result given in part (*a*).

Note that a *time dilation* in one frame of reference can be observed as a *length contraction* in another. This interrelationship of time and space is a fundamental part of special relativity.

Sample Problem 2   A spaceship is moving away from the Earth at a speed of $0.80c$ when it fires a missile parallel to the direction of motion of the ship. The missile moves at a speed of $0.60c$ relative to the ship (Fig. 8). What would be the speed of the missile as measured by an observer on the Earth? Compare with the predictions of Galilean kinematics.

Solution   This problem is similar to that of the observer and the train. Here S' is on the ship and S is on Earth, and S' moves with

Figure 8   Sample Problem 2. A spaceship moves away from Earth at a speed of $0.80c$. An observer S' on the spaceship fires a missile and measures its speed to be $0.60c$ relative to the ship.

a speed of $u = 0.80c$ relative to S. The missile moves at speed $v_0 = 0.60c$ relative to S', and we seek its speed v relative to S. Using Eq. 12, we obtain

$$v = \frac{v_0 + u}{1 + v_0 u/c^2} = \frac{0.60c + 0.80c}{1 + (0.60c)(0.80c)/c^2}$$
$$= \frac{1.40c}{1.48} = 0.95c.$$

According to classical kinematics (the numerator of Eq. 12), an observer on the Earth would see the missile moving at $0.60c + 0.80c = 1.40c$, thereby exceeding the maximum relative speed of c permitted by relativity. You can see how Eq. 12 brings about this speed limit. Even if v_0 were $0.9999 \cdots c$ and u were $0.9999 \cdots c$, the relative speed v measured by S would remain less than c.

21-4   THE LORENTZ TRANSFORMATION

Einstein's postulates provide a first step in the resolution of the difficulties we presented in Section 21-1, but a more formal mathematical basis is needed to give the theory its full power to calculate the expected outcomes of a wider variety of physical processes. For example, we might wish to compute how the results of measurements of an energy or a magnetic field strength differ for observers in relative motion.

We require a set of relationships called *transformation equations* that relate observations of a single event by two different observers. The transformation equations have three ingredients: (1) an observer S at rest in one inertial frame, (2) another observer S' at rest in a different inertial frame that is in motion at constant velocity with respect to S, and (3) a single event that is observed by *both* S and S'. The event occurs, according to each observer, at a particular set of coordinates in three-dimensional space *and* at a particular time. Knowing the relative velocity of S and S', we wish to calculate the coordinates x', y', z', t' of an event as observed by S' from the coordinates x, y, z, t of the *same event* according to S. We simplify this problem somewhat, without losing generality, by always choosing the x and x' axes to be along the direction of **u** (see Fig. 9).

This problem can be solved using the classical kine-

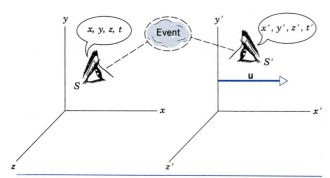

Figure 9   Two observers, whose frames of reference are represented by S and S', observe the same event. S' moves relative to S with velocity **u** along the common xx' direction. S measures the coordinates x, y, z, t of the event, while S' measures the coordinates x', y', z', t' of the same event.

matics of Galileo, and the resulting *Galilean transformation* equations are

$$x' = x - ut,$$
$$y' = y,$$
$$z' = z,$$ (13)
$$t' = t.$$

The first of these equations is consistent with our "common-sense" experience. For instance, suppose S is at rest on the ground and measures the location x of a fencepost. S', who is in a car moving at speed u relative to S, does indeed find the fencepost at the location $x' = x - ut$ (Fig. 10). The fourth equation, $t' = t$, was simply taken for granted in classical physics (as exemplified by Newton's universal time coordinate).

The relativistic relationships we seek have come to be known as the *Lorentz transformation equations.* They are named for the Dutch physicist H. A. Lorentz, who proposed them (before Einstein) for quite a different reason and who was not fully aware of their implications about the nature of space and time. The equations can be derived directly from Einstein's postulates, if we invoke certain reasonable assumptions about the symmetry and the homogeneity of space and time. As an example of this

latter property, consider an observer S who measures the length of a rod held by observer S' in a different inertial frame. The result of the measurement carried out by S should not depend on where S' is located in the reference frame or on the time of day at which S makes the measurement.

The Lorentz transformation equations, derived on these assumptions, are*

$$x' = \frac{x - ut}{\sqrt{1 - u^2/c^2}} = \gamma(x - ut),$$
$$y' = y,$$ (14)
$$z' = z,$$
$$t' = \frac{t - ux/c^2}{\sqrt{1 - u^2/c^2}} = \gamma(t - ux/c^2).$$

Note that an object located initially at the origin according to S (that is, $x = 0$ at $t = 0$) is also initially located at the origin according to S' (that is, $x' = 0$ and $t' = 0$).

In these equations, we have used the *Lorentz factor γ*, defined as

$$\gamma = \frac{1}{\sqrt{1 - u^2/c^2}}.$$ (15)

It is also convenient in relativity equations to introduce the *speed parameter β*, defined as the ratio between the relative speed u of the two coordinate systems and the speed of light:

$$\beta = u/c.$$ (16)

Some sample values of β and γ are given in Table 1, and the relationship between β and γ is shown in Fig. 11. The range of γ is from 1 (at low speed, where $u \ll c$ or $\beta \ll 1$) to ∞ (at high speed, where $u \to c$ or $\beta \to 1$).

Note that the Lorentz transformation equations reduce to those of the Galilean transformation (Eqs. 13) when $u \ll c$. One convenient way to show this is to let $c \to \infty$, so that $u/c \to 0$. In this case, as you should prove, the relativistic Eqs. 14 reduce directly to the classical Eqs. 13. All classical results derived in previous chapters agree

---

* See *Introduction to Special Relativity,* by Robert Resnick (Wiley, 1968), Section 2.2, for a derivation of these equations.

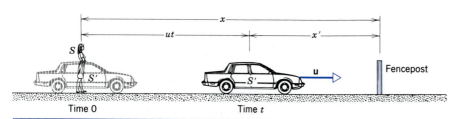

Figure 10   According to S, the fencepost is at the coordinate x. According to S', who is at coordinate ut relative to S at time t, the fencepost is at the coordinate $x' = x - ut$. Note that the origins of S and S' coincide at $t = 0$.

TABLE 1 SAMPLE VALUES OF THE SPEED PARAMETER AND THE LORENTZ FACTOR

β	γ	β	γ
0.00	1.000	0.90	2.29
0.10	1.005	0.99	7.09
0.30	1.048	0.999	22.4
0.60	1.25	0.9999	70.7

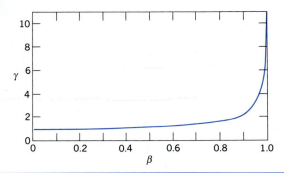

Figure 11 The Lorentz factor γ as a function of the speed parameter β.

with experiment when $u \ll c$. Only at high speeds must we take relativistic effects into account.

Equations 14 permit us to find the space and time coordinates in S' if we know those in S. Suppose, however, that we wish to know the coordinates in S, given those in S'. From the point of view of S' in Fig. 9, S appears to move in the *negative* x (or x') direction. We can obtain the *inverse Lorentz transformation* by merely switching primed and unprimed coordinates in Eqs. 14 and substituting $-u$ for u. This gives

$$
\begin{aligned}
x &= \gamma(x' + ut'), \\
y &= y', \\
z &= z', \\
t &= \gamma(t' + ux'/c^2).
\end{aligned} \tag{17}
$$

We can use a different method of inverting the Lorentz transformation (see Problem 18) by solving Eqs. 14 algebraically for x and t (treating the first and last equations as

a system of two equations in two unknowns). When we do, we obtain exactly the inverse transformation given by Eqs. 17, which were obtained directly from a symmetry argument.

Table 2 summarizes the equations of the Lorentz transformation when the relative velocity between the coordinate systems is in the common xx' direction. The equations are shown in four forms: the Lorentz transformation (Eqs. 14), the inverse Lorentz transformation (Eqs. 17), and both corresponding *interval* transformations, which are useful when we wish to transform not a coordinate but a space or time interval, such as $\Delta x' = x'_2 - x'_1$ (the distance between two events, as measured by S') or $\Delta t' = t'_2 - t'_1$ (the time between two events, as measured by S').

Sample Problem 3 In inertial frame S, a red light and a blue light are separated by a distance $\Delta x = 2.45$ km, with the red light at the larger value of x. The blue light flashes, and 5.35 μs later the red light flashes. Frame S' is moving in the direction of increasing x with a speed of $u = 0.855c$. What is the distance between the two flashes and the time between them as measured in S'?

Solution The Lorentz parameter is

$$
\gamma = \frac{1}{\sqrt{1 - u^2/c^2}} = \frac{1}{\sqrt{1 - (0.855)^2}} = 1.928.
$$

We are given the intervals in S as $\Delta x = 2450$ m and $\Delta t = 5.35 \times 10^{-6}$ s. From Table 2, we have the interval transformations

$$
\begin{aligned}
\Delta x' &= \gamma(\Delta x - u\,\Delta t) \\
&= 1.928[2450 \text{ m} - (0.855)(3.00 \times 10^8 \text{ m/s})(5.35 \times 10^{-6} \text{ s})] \\
&= 2078 \text{ m} = 2.08 \text{ km}
\end{aligned}
$$

and

$$
\begin{aligned}
\Delta t' &= \gamma(\Delta t - u\,\Delta x/c^2) \\
&= 1.928[5.35 \times 10^{-6} \text{ s} - (0.855)(2450 \text{ m})/(3.00 \times 10^8 \text{ m/s})] \\
&= -3.147 \times 10^{-6} \text{ s} = -3.15 \ \mu\text{s}.
\end{aligned}
$$

In S', the red flash is also located at the more distant coordinate, but the distance is 2.08 km rather than 2.45 km. Also, in S' the red flash comes *before* the blue flash (in contrast to what is observed in S); the time between flashes is 3.15 μs according to S'.

TABLE 2 THE LORENTZ TRANSFORMATION EQUATIONS[a]

Lorentz Transformation	Inverse Transformation	Interval Transformation	Inverse Interval Transformation
$x' = \gamma(x - ut)$	$x = \gamma(x' + ut')$	$\Delta x' = \gamma(\Delta x - u\,\Delta t)$	$\Delta x = \gamma(\Delta x' + u\,\Delta t')$
$y' = y$	$y = y'$	$\Delta y' = \Delta y$	$\Delta y = \Delta y'$
$z' = z$	$z = z'$	$\Delta z' = \Delta z$	$\Delta z = \Delta z'$
$t' = \gamma(t - ux/c^2)$	$t = \gamma(t' + ux'/c^2)$	$\Delta t' = \gamma(\Delta t - u\,\Delta x/c^2)$	$\Delta t = \gamma(\Delta t' + u\,\Delta x'/c^2)$

[a] Apply these equations only in the case of relative motion in the xx' direction. The Lorentz factor is $\gamma = 1/\sqrt{1 - u^2/c^2}$.

21-5   MEASURING THE SPACE–TIME COORDINATES OF AN EVENT

So far, we have said little about how observers S and S' go about measuring the coordinates x, y, z, t and x', y', z', t' of an event, as in the case of the light flashes in Sample Problem 3. The procedure we now describe forms a conceptual foundation on which actual laboratory procedures can be based.

We assume that S has a large team of assistants available to help in the setting up of a coordinate system. Each assistant is given a clock and a measuring rod of a certain length. For example, three assistants have measuring rods 1 m in length. They are instructed to lay out their rods, each along one of the three coordinate axes, and to wait at the position determined by the end of the rod until they see a flash of light at the origin, at which time they are to start their clocks at the preset reading of 3.33×10^{-9} s (3.33 ns, the time necessary for light to travel the distance of 1 m from the origin to the assistant's location). Three other assistants, who are similarly each assigned one of the coordinate axes, are given rods of length 2 m and are instructed to start their clocks, when they see the flash of light, at the preset time of 6.67 ns (the time for light to travel 2 m). Each assistant is sent to a post with a rod of some length L and a clock preset at $t = L/c$.

When all the assistants are at their posts, S sets off a flash of light at the origin and simultaneously starts the clock at the origin, which is preset to zero. As the light signal reaches the other clocks, each is started in turn at the preset reading. Thus the clock on the x axis at $x = 1$ m is started at the preset reading of 3.33 ns when the clock at the origin reads 3.33 ns; the clock on the x axis at $x = 2$ m starts at the preset reading of 6.67 ns when the clock at the origin and the clock at $x = 1$ m both read 6.67 ns; and so on for all the clocks in the coordinate system. All clocks in the entire system are thus perfectly synchronized. The resulting system of rods and clocks is represented in Fig. 12.

Suppose that S wishes to chart the progress of a particle as it moves through the coordinate system. All that S and the assistants must do is watch the particle as it travels and write down as it passes each point the coordinate and the reading on the clock at that coordinate.

Of course, this calibration holds only for observer S. Observer S' and all other inertial observers must carry out a similar procedure to define a coordinate system and synchronize its clocks. With such a scheme the measuring rods and clocks of each observer (which of course are at rest in the frame of that observer) are unique to that inertial frame and are independent of the rods and clocks of observers in other inertial frames.

This procedure suggests that space and time are not independent coordinates, but that the description of an event must include its coordinates in both space and time.

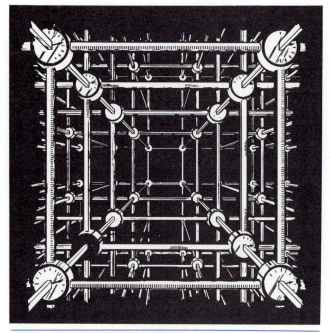

Figure 12   A framework of measuring rods and clocks that might be used by an observer in a particular reference frame to determine the space–time coordinates of an event.

(That is, we cannot use a clock at one location to record the passage of a particle through another location.) For this reason, special relativity usually is formulated in terms of combined *space–time coordinates* x, y, z, t. Space and time are treated as equivalent coordinates in special relativity.

21-6   THE TRANSFORMATION OF VELOCITIES

In this section we use the equations of the Lorentz transformation to relate the velocity $\mathbf{v}$ of a particle measured by an observer in the S frame to the velocity $\mathbf{v}'$ of the same particle measured by an observer in the S' frame, who is in turn moving with velocity $\mathbf{u}$ relative to S. In this discussion, it is important to keep in mind the meanings of these three velocities.

Suppose observer S finds the particle to move from coordinates x_1, y_1, z_1, t_1 to x_2, y_2, z_2, t_2. Observer S', on the other hand, records the observations of the initial and final coordinates of the same particle as x_1', y_1', z_1', t_1' and x_2', y_2', z_2', t_2'.

Let us calculate v_x' $(= \Delta x'/\Delta t')$, the x' component of the velocity measured by S'. From Table 2, we obtain the transformation equations for the intervals $\Delta x'$ and $\Delta t'$. Dividing these two equations, we obtain

$$v_x' = \frac{\Delta x'}{\Delta t'} = \frac{\gamma(\Delta x - u\,\Delta t)}{\gamma(\Delta t - u\,\Delta x/c^2)} = \frac{\Delta x/\Delta t - u}{1 - u(\Delta x/\Delta t)/c^2},$$

or, replacing $\Delta x / \Delta t$ by v_x,

$$v'_x = \frac{v_x - u}{1 - uv_x/c^2} . \qquad (18)$$

In similar fashion, we obtain the transformation equations for the y and z components of the velocities:

$$v'_y = \frac{v_y}{\gamma(1 - uv_x/c^2)} \quad \text{and} \quad v'_z = \frac{v_z}{\gamma(1 - uv_x/c^2)} . \qquad (19)$$

Note that $v'_y \neq v_y$, even though $\Delta y = \Delta y'$, because $\Delta t \neq \Delta t'$. Similar considerations hold for v'_z. This is another example of the difference between the way the Galilean and Lorentz transformations deal with the time coordinate. Be sure to note that the denominators of all three equations include the factor v_x.

Equations 18 and 19 give the *Lorentz velocity transformation*. They are analogous to the equations of the Lorentz coordinate transformation: they relate observations in one coordinate frame to observations in another. Table 3 summarizes these equations, along with the corresponding inverse velocity transformation. Note that the inverse transformation equation for v_x is identical with Eq. 12, which we derived in quite a different way. In Eq. 12, the speed v_0 is the same as the speed v'_x measured by S'.

Let us examine Eqs. 18 and 19 in the nonrelativistic limit. Do they reduce to the classical Galilean transformation when $u \ll c$ (or equivalently, when $c \rightarrow \infty$)? In this case Eqs. 18 and 19 reduce to

$$v'_x = v_x - u, \quad v'_y = v_y, \quad \text{and} \quad v'_z = v_z, \qquad (20)$$

which are indeed the Galilean results, as given by Eq. 43 of Chapter 4 or by differentiating Eq. 13, the Galilean coordinate transformation.

We now show directly that the Lorentz velocity transformation gives the result demanded by Einstein's second postulate (the constancy of the speed of light): a speed of c measured by one observer must also be measured to be c by any other observer. Suppose that the common event being observed by S and S' is the passage of a light beam along the x direction. Observer S measures $v_x = c$ and $v_y = v_z = 0$. What velocity does observer S' measure? Using Eqs. 18 and 19, we find the velocity components measured by S' to be

$$v'_x = \frac{v_x - u}{1 - uv_x/c^2} = \frac{c - u}{1 - uc/c^2} = \frac{c - u}{(c - u)/c} = c,$$

$$v'_y = v'_z = 0.$$

Note that this result follows *independent of the relative speed u between S and S'*. A speed of c measured in one inertial reference frame leads to a speed of c measured in *all* frames. Thus the speed of light is indeed the same for all observers. The same conclusion holds for any direction of travel of the light beam; see Problem 19.

Sample Problem 4   A particle is accelerated from rest in the laboratory until its velocity is $0.60c$. As viewed from a frame that is moving with the particle at a speed of $0.60c$ relative to the laboratory, the particle is then given an additional increment of velocity amounting to $0.60c$. Find the final velocity of the particle as measured in the laboratory frame.

Solution   Once again, the problem becomes a direct application of the Lorentz velocity transformation once we have clearly specified the reference frames S and S' and the system being observed. Clearly the particle is the system being observed, and if we seek its velocity measured in the laboratory frame it is natural to associate the laboratory with the S frame. The S' frame is then the inertial reference frame occupied by the particle after the first acceleration and before the second (see Fig. 13). Relative to this frame, the velocity of the particle after the second acceleration is $v'_x = 0.60c$. The velocity of the S' frame with respect to the S frame is $u = 0.60c$. We know v'_x and u, and we seek v_x, which is given by the inverse velocity transformation from Table 3:

$$v_x = \frac{v'_x + u}{1 + uv'_x/c^2} = \frac{0.60c + 0.60c}{1 + (0.60c)(0.60c)/c^2} = \frac{1.20c}{1.36} = 0.88c.$$

The speed is less than c, in contradiction to the prediction of the Galilean transformation, which gives $v_x = 1.20c$.

Suppose we now let the S' frame be that of the particle after the second acceleration, so that $u = 0.88c$ relative to the original S frame (the laboratory). Let there now be a *third* acceleration, so that, relative to the new S' frame, the particle again moves with velocity $v'_x = 0.60c$. By repeating the above procedure, you

---

TABLE 3   THE LORENTZ VELOCITY TRANSFORMATION

Velocity Transformation	*Inverse Velocity Transformation*
$v'_x = \dfrac{v_x - u}{1 - uv_x/c^2}$	$v_x = \dfrac{v'_x + u}{1 + uv'_x/c^2}$
$v'_y = \dfrac{v_y}{\gamma(1 - uv_x/c^2)}$	$v_y = \dfrac{v'_y}{\gamma(1 + uv'_x/c^2)}$
$v'_z = \dfrac{v_z}{\gamma(1 - uv_x/c^2)}$	$v_z = \dfrac{v'_z}{\gamma(1 + uv'_x/c^2)}$

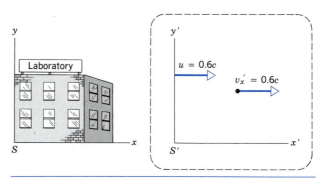

Figure 13   Sample Problem 4. S', the frame of reference of the particle after the first acceleration, moves with speed $u = 0.60c$ relative to the laboratory (frame S). Relative to S', the particle moves at speed $v'_x = 0.60c$ after its second acceleration.

should show that an observer in the laboratory (S) frame will measure a speed of $v'_x = 0.97c$ in this case.

No matter how many times we accelerate the particle in a reference frame moving with the particle, its velocity measured in the original laboratory frame (or in any other frame) will never exceed c.

21-7 CONSEQUENCES OF THE LORENTZ TRANSFORMATION

We have already shown that some unexpected consequences follow from applying Einstein's postulates to physical situations. Now we use the more mathematical basis of the Lorentz transformation to show that these same consequences and others can be obtained.

The Relativity of Time

In Section 21-3, we showed that the time dilation effect follows directly from applying Einstein's postulates to measurements of time intervals by two observers in motion relative to one another. Figure 14 shows a different view of the time dilation effect. Clock C' is at rest in the frame of S', who moves at speed u relative to S. S' measures the time interval $\Delta t' = t'_2 - t'_1$ in which the hand of the clock moves between two marks, passing the first mark at time t'_1 and the second at time t'_2.

The hand of clock C' passing the two marks can be regarded as two events, which occur at the same location x'_0 according to S' (because clock C' is at rest in that frame). However, S (whose reference frame contains a stationary set of synchronized clocks such as that described in Section 21-5) observes the hand of clock C' to pass the first mark at the location x_1 (where the local stationary clock reads time t_1) and to pass the second mark at location x_2 (where a *different* stationary clock

reads the time t_2). We can find the relationship between the time intervals Δt and $\Delta t'$ directly from the inverse Lorentz transformation. From Table 2, we have

$$\Delta t = \gamma(\Delta t' + u\,\Delta x'/c^2). \tag{21}$$

This general expression gives the time interval Δt measured by S corresponding to the time interval $\Delta t'$ measured by S' for events that are separated by a distance $\Delta x'$. According to S', relative to whom clock C' is at rest, the two events (the hand passing the two marks) take place at the same location x'_0, so $\Delta x' = 0$. Because S' is at rest relative to clock C', the time interval $\Delta t'$ measured by S' is a proper time interval, which we represent as Δt_0. Substituting $\Delta x' = 0$ and $\Delta t' = \Delta t_0$ into Eq. 21, we obtain

$$\Delta t = \gamma\,\Delta t_0 = \frac{\Delta t_0}{\sqrt{1 - u^2/c^2}},$$

which is identical to Eq. 3, the time dilation equation.

The time dilation effect is completely symmetric. If a clock C at rest in S is observed by S', then S' concludes that clock C is running slow. Each observer believes that the other's clock is running slower than the ones at rest in the reference frame of the observer. Time dilation is often summarized by the phrase, "moving clocks run slow." It is useful to remember this phrase, but do so with caution. The phrase indicates that a clock moving relative to a frame containing an array of synchronized clocks will be found to run slow *when timed by those clocks*. That is, only in the sense of comparing a single moving clock with two separated synchronized stationary clocks can we declare that "moving clocks run slow."

Consider three other consequences of the Lorentz transformation that are related to measurements of time:

1. *The relativity of simultaneity.* Suppose S' has two clocks at rest, located at x'_1 and x'_2, and separated by the interval $\Delta x' = x'_2 - x'_1$. A flash of light emitted from a point midway between the clocks reaches the two clocks simultaneously, according to S' (see Fig. 15a). That is, a

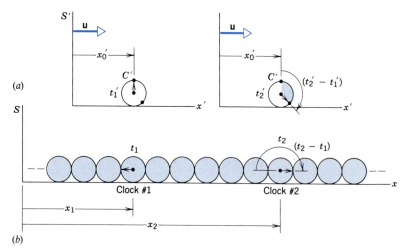

(a)

(b)

Figure 14 Clock C' is fixed at position x'_0 in reference frame S'. Observer S, relative to whom clock C' is in motion at velocity **u**, compares the reading of C' with two different stationary clocks from the array of synchronized clocks (numbered 1 and 2) established in the frame of S. As shown, the interval $t_2 - t_1$ measured by S is greater than the interval $t'_2 - t'_1$. Observer S therefore declares that, by comparison with the clocks in S, the moving clock is running slow.

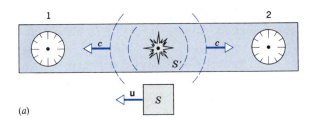

(a)

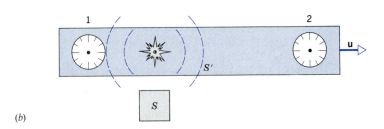

(b)

Figure 15 (a) In the frame of reference of S', a flash of light at a point midway between two clocks reaches the clocks at the same instant. (b) In the frame of reference of S, the flash of light reaches clock 1 before it reaches clock 2.

measurement by S' of the interval between the arrival of the light signals at the two clocks gives $\Delta t' = 0$. Now consider the situation from the point of view of S, relative to whom the frame of S' (including the clocks) moves with speed u (Fig. 15b). Clearly the light signal reaches clock 1 before it reaches clock 2, and thus the arrival of the light signals at the locations of the two clocks is *not* simultaneous to S. We therefore reach the following conclusion:

If two observers are in relative motion, in general they do not agree on whether two events at different locations are simultaneous. If one observer finds the two events to be simultaneous, the other does not.

This conclusion also follows directly from Eq. 21: if $\Delta t' = 0$ and $\Delta x' \neq 0$, then $\Delta t \neq 0$. Note that this occurs only when the two events occur at different locations according to S'. If the two events occur at the same location and are simultaneous according to S', they are simultaneous to S as well.

2. *The Doppler shift.* In Section 20-7 we considered the Doppler effect for sound waves, in which the motion of a source or an observer of waves relative to the medium carrying the waves causes a change in the frequency measured by the observer.

In the case of light waves, "motion relative to the medium" is not a valid concept. Special relativity gives a Doppler shift for light that depends only on the relative speed between the source and the observer; in contrast to the case of sound waves, in which we used different formulas to account for source motion and observer motion, in the case of light waves one formula, involving only the *relative* motion, is sufficient. The relativistic Doppler formula is thus simpler to apply than the classical one.

Another aspect of the Doppler effect in special relativity has no classical counterpart. This is the *transverse Doppler effect,* which, in contrast to the cases considered

in Section 20-7, occurs when the source or the observer moves perpendicular to the line connecting them. The transverse Doppler effect is actually another result of time dilation, and the precise measurements of the transverse Doppler effect provide some of our most sensitive experimental tests of time dilation. We consider the Doppler effect for light in more detail in Chapter 42.

3. *The twin paradox.* Time dilation applies not only to elementary particles but to all naturally occurring time intervals including pulse rates and human lifetimes. This fact has been used to propose an apparent puzzle that has become known as the *twin paradox.**

Suppose two twins, Fred and Ethel, are on a platform coasting in space. Ethel embarks on a journey in a high-speed spaceship to a distant star while Fred remains on the platform. During this journey, Fred is able to monitor Ethel's heart beat and average respiration rate, and he finds them to be slower by the time dilation effect; thus Ethel's entire aging process has been slowed. Fred therefore expects that, upon Ethel's return to the platform after the journey to the star, she will be younger than he.

The paradox seemingly occurs when we analyze the situation from the frame of reference of Ethel, thereby regarding Fred and the platform as the ones making the journey. According to this analysis, Fred is the traveling twin and should be the younger at the end of the journey. Here is the paradox: When they get together at the end of the journey, it cannot be true that Ethel is younger than Fred and also that Fred is younger than Ethel.

The resolution of the paradox comes when we realize that Fred and Ethel are not really in symmetric situations. For the two twins to get back together again, one of them must decelerate and reverse directions, resulting in an

---

* For more details about the twin paradox, see *Basic Concepts in Relativity and Early Quantum Theory,* 2nd edition, by Robert Resnick and David Halliday (Wiley, 1985), p. 281.

easily measurable acceleration of one of them. Put another way, Ethel must change from one inertial reference frame (the one moving away from Fred) to another (one moving toward Fred). Fred, on the other hand, experiences no acceleration and remains in the same inertial reference frame for the entire duration of the journey. It is indeed Ethel who is the traveler and who will be younger upon her return.

Although we have not yet been able to do an experiment of this sort with actual twins, it has been done with atomic clocks.* Two identical clocks were carefully calibrated; one was then flown on a commercial airliner around the world and compared with its stay-at-home twin upon its return. The speed during such a trip was of course far less than c, but atomic clocks are capable of sufficient precision that the small resulting asymmetry in the "aging" of the two clocks, amounting to about 10^{-7} s, could easily and precisely be determined. It was found that the clock in the airliner, which was the one subject to an acceleration and therefore the true traveler, was indeed "younger" (that is, running slower) after the journey.

The reading of the clock in the airliner must also be corrected for the time it spends at a different gravitational potential, an effect of *general* relativity. Corrections for special and general relativity are thus of important practical concern when such precise clocks are transported from one location to another.

The Relativity of Length

Length contraction, which we discussed in Section 21-3, also follows directly from the equations of the Lorentz transformation. Let us first realize that to measure the length of an object we must make a *simultaneous* determination of the coordinates of the ends of the object (see Fig. 16). *It does no good to measure the coordinate of one end of a moving object at one time and the coordinate of the other end at a different time.*

Suppose (see Fig. 17) that a measuring rod of rest length L_0 is carried by S'. Observer S wishes to measure its length. According to S', in whose frame of reference the rod is at rest, the ends of the rod are at coordinates x'_2 and x'_1, such that $\Delta x' = x'_2 - x'_1 = L_0$, the rest length of the rod. Observer S, using the calibrated and synchronized coordinates established according to the procedure described in Section 21-5, makes a simultaneous determination of the coordinates x_2 and x_1 of the ends of the rod. The interval $\Delta x = x_2 - x_1$ gives the length L of the rod according to S. From the interval equation in Table 2, we have

$$\Delta x' = \gamma(\Delta x - u\, \Delta t). \quad (22)$$

Putting $\Delta t = 0$ (because S made a *simultaneous* determination of x_2 and x_1), we solve for $\Delta x (= L)$ and obtain

$$L = \Delta x = \frac{\Delta x'}{\gamma} = L_0 \sqrt{1 - u^2/c^2} \, ,$$

which is identical to Eq. 8.

We have deduced time dilation and length contraction both from the postulates (Section 21-3) and from the Lorentz transformation (this section). These are not independent derivations, however, because the Lorentz transformation itself is obtained from the postulates. Ultimately, all of special relativity follows directly from Einstein's postulates.

Like time dilation, length contraction is an effect that holds for all observers in relative motion. Questions such as "Does a moving measuring rod *really* shrink?" have

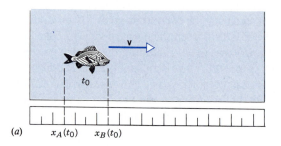

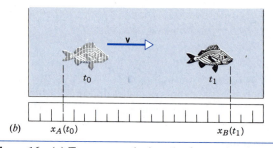

Figure 16 (*a*) To measure the length of a moving fish, you must determine *simultaneously* the positions of its head and tail. (*b*) If the determination is not simultaneous, the measurement does not give the length.

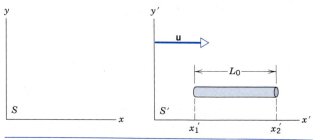

Figure 17 The ends of a measuring rod are determined to be at coordinates x'_1 and x'_2 according to S', relative to whom the rod is at rest. To determine the length of the rod, S must make a *simultaneous* determination of the coordinates x_1 and x_2 of its endpoints.

meaning only in the sense that they refer to measurements by observers in relative motion. The essence of relativity is that results of measurements of length and time are subject to the state of motion of the observer relative to the event being measured and refer only to measurements by a particular observer in a particular frame of reference. If different observers were to bring the rod to rest in their individual inertial frames, each would measure the same value for the length of the rod. In this respect, special relativity is a theory of measurement that simply says "motion affects measurement."

Sample Problem 5   An observer S is standing on a platform of length $D_0 = 65$ m on a space station. A rocket passes at a relative speed of $0.80c$ moving parallel to the edge of the platform. The observer S notes that the front and back of the rocket simultaneously line up with the ends of the platform at a particular instant (Fig. 18a). (a) According to S, what is the time necessary for the rocket to pass a particular point on the platform? (b) What is the rest length L_0 of the rocket? (c) According to an observer S' on the rocket, what is the length D of the platform? (d) According to S', how long does it take for observer S to pass the entire length of the rocket? (e) According to S, the ends of the rocket simultaneously line up with the ends of the platform. Are these events simultaneous to S'?

Solution   (a) According to S, the length L of the rocket matches the length D_0 of the platform. The time for the rocket to pass a particular point is measured by S to be

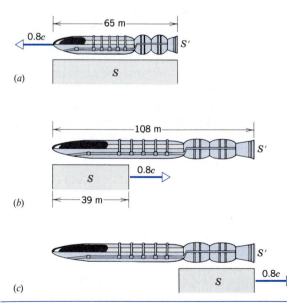

(a)

(b)

(c)

Figure 18   Sample Problem 5. (a) From the reference frame of S at rest on the platform, the passing rocket lines up simultaneously with the front and the back of the platform. (b,c) From the reference frame of the rocket, the passing platform lines up first with the front of the rocket and later with the rear. Note the differing effects of length contraction in the two reference frames.

$$\Delta t_0 = \frac{L}{0.80c} = \frac{65\ \text{m}}{2.40 \times 10^8\ \text{m/s}} = 0.27\ \mu\text{s}.$$

This is a proper time interval, because S is measuring the time interval between two events that occur at the same point in the frame of reference of S (the front of the rocket passes a point, and then the back of the rocket passes the same point).

(b) S measures the contracted length L of the rocket. We can find its rest length L_0 using Eq. 8:

$$L_0 = \frac{L}{\sqrt{1 - u^2/c^2}} = \frac{65\ \text{m}}{\sqrt{1 - (0.80)^2}} = 108\ \text{m}.$$

(c) According to S the platform is at rest, so 65 m is its rest length D_0. According to S', the contracted length of the platform is therefore

$$D = D_0\sqrt{1 - u^2/c^2} = (65\ \text{m})\sqrt{1 - (0.80)^2} = 39\ \text{m}.$$

(d) For S to pass the entire length of the rocket, S' concludes that S must move a distance equal to its rest length, or 108 m. The time needed to do this is

$$\Delta t' = \frac{108\ \text{m}}{0.80c} = 0.45\ \mu\text{s}.$$

Note that this is *not* a proper time interval for S', who determines this time interval using one clock at the front of the rocket to measure the time at which S passes the front of the rocket, and another clock on the rear of the rocket to measure the time at which S passes the rear of the rocket. The two events therefore occur at different points in S' and so cannot be separated by a proper time in S'. The corresponding time interval measured by S for the same two events, which we calculated in part (a), is a proper time interval for S, because the two events *do* occur at the same point in S. The time intervals measured by S and S' should be related by the time dilation formula:

$$\Delta t' = \gamma\,\Delta t = \frac{0.27\ \mu\text{s}}{\sqrt{1 - (0.80)^2}} = 0.45\ \mu\text{s},$$

in agreement with the value calculated above from the proper length of the rocket in S'.

(e) According to S', the rocket has a rest length of $L_0 = 108$ m and the platform has a contracted length of $D = 39$ m. There is thus no way that S' could observe the two ends of both to align simultaneously. The sequence of events according to S' is illustrated in Figs. 18b and 18c. The time interval $\Delta t'$ in S' between the two events that are simultaneous in S can be calculated from the interval equation for $\Delta t'$ in Table 2 with $\Delta t = 0$, which gives

$$\Delta t' = -\gamma u\,\Delta x/c^2 = \frac{-(0.80c)(-65\ \text{m})}{c^2\sqrt{1 - (0.80)^2}} = 0.29\ \mu\text{s}.$$

We can check this result by noting that, according to S', the time interval between the situations shown in Figs. 18b and 18c must be that necessary for the platform to move a distance of 108 m − 39 m = 69 m, which takes a time

$$\Delta t' = \frac{69\ \text{m}}{0.80c} = 0.29\ \mu\text{s},$$

in agreement with the value calculated from the interval transformation. This last result illustrates the relativity of simultaneity: two events that are simultaneous to S (the lining up of the two ends of the rocket with the two ends of the platform) *cannot* be simultaneous to S'.

21-8   RELATIVISTIC MOMENTUM

So far we have investigated the effect of Einstein's two postulates on the kinematical variables time, displacement, and velocity as viewed from two different inertial frames. In this section and the next, we broaden our efforts to include the dynamical variables momentum and energy. Here we discuss the relativistic view of linear momentum.

Consider the collision shown in Fig. 19a, viewed from the S frame of reference. Two particles, each of mass m, move with equal and opposite velocities v and −v along the x axis. They collide at the origin, and the distance between their lines of approach has been adjusted so that after the collision the particles move along the y axis with equal and opposite final velocities (Fig. 19b). We assume the collision to be perfectly elastic, so that no kinetic energy is lost. The final velocities must then be v and −v.

Using the classical formula ($\mathbf{p} = m\mathbf{v}$), the components of the momentum of the two-particle system in the S frame are

$$\text{Initial:}\quad p_{xi} = mv + m(-v) = 0,$$
$$p_{yi} = 0.$$
$$\text{Final:}\quad p_{xf} = 0,$$
$$p_{yf} = mv + m(-v) = 0.$$

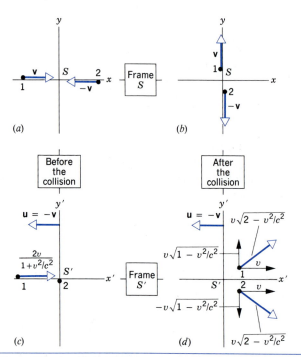

(a)   (b)

(c)   (d)

Figure 19 A collision between two particles of the same mass is shown (a) before the collision in the reference frame of S, (b) after the collision in the reference frame of S, (c) before the collision in the reference frame of S′, and (d) after the collision in the reference frame of S′.

Thus $p_{xi} = p_{xf}$ and $p_{yi} = p_{yf}$; the initial (vector) momentum is equal to the final momentum, and momentum is conserved in the S frame.

Let us now view the same collision from the S′ frame, which moves relative to the S frame with speed $u = -v$ (Fig. 19c). Note that in the S′ frame, particle 2 is at rest before the collision. We use the Lorentz velocity transformation, Eqs. 18 and 19, to find the transformed x′ and y′ components of the initial and final velocities, as they would be observed by S′. These values, which you should calculate, are shown in Figs. 19c and 19d.

We now use those velocities to find the components of the momentum of the system in the S′ frame:

$$p'_{xi} = m\left(\frac{2v}{1 + v^2/c^2}\right) + m(0) = \frac{2mv}{1 + v^2/c^2},$$
$$p'_{yi} = 0,$$
$$p'_{xf} = mv + mv = 2mv,$$
$$p'_{yf} = mv\sqrt{1 - v^2/c^2} + m(-v\sqrt{1 - v^2/c^2}) = 0.$$

We see that p'_{xi} is *not* equal to p'_{xf}, and S′ will conclude that momentum is *not* conserved.

It is clear from the above calculation that the law of conservation of linear momentum, which we have found useful in a variety of applications, does not satisfy Einstein's first postulate (the law must be the same in all inertial frames) if we calculate momentum as $p = mv$. Therefore, *if we are to retain the conservation of momentum as a general law consistent with Einstein's first postulate, we must find a new definition of momentum.* This new definition of momentum must have two properties. (1) It must yield a law of conservation of momentum that satisfies the principle of relativity; that is, if momentum is conserved according to an observer in one inertial frame, then it is conserved according to observers in *all* inertial frames. (2) At low speeds, the new definition must reduce to $\mathbf{p} = m\mathbf{v}$, which we know works perfectly well in the nonrelativistic case.

The relativistic formula for the momentum of a particle of mass m moving with velocity $\mathbf{v}$ is

$$\mathbf{p} = \frac{m\mathbf{v}}{\sqrt{1 - v^2/c^2}}, \qquad (23)$$

which we have already introduced in Eq. 22 of Chapter 9. In terms of components, we can write Eq. 23 as

$$p_x = \frac{mv_x}{\sqrt{1 - v^2/c^2}} \quad \text{and} \quad p_y = \frac{mv_y}{\sqrt{1 - v^2/c^2}}. \qquad (24)$$

The velocity v that appears in the denominator of these expressions is *always* the velocity of the particle as measured in a particular inertial frame. It is *not* the velocity of an inertial frame. The velocity in the numerator can be any of the components of the velocity vector.

Let us see how this new definition restores conservation of momentum in the collision we considered. In the S frame, the velocities before and after are equal and oppo-

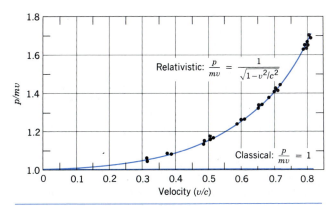

Figure 20 The ratio p/mv is plotted for electrons of various speeds. According to classical physics, $p = mv$, and thus the classical equations predict $p/mv = 1$. The data clearly agree with the relativistic result and not with the classical result. At low speeds, the classical and relativistic predictions are indistinguishable.

site, and thus Eq. 23 again gives zero for the initial and final momenta. In the S' frame, we can use the magnitudes of the velocities as given in Figs. 19c and 19d to obtain, as you should verify,

$$p'_{xi} = p'_{xf} = \frac{2mv}{1 - v^2/c^2},$$
$$p'_{yi} = p'_{yf} = 0. \tag{25}$$

Thus the initial and final momenta are equal in the S' frame. Momentum is conserved in both the S and S' frames. In fact, the definition of momentum given in Eq. 23 gives conservation of momentum in *all* inertial frames, as required by the principle of relativity.

Note also that, in the limit of low speeds, the denominator of Eq. 23 is nearly equal to 1; at low speeds Eq. 23 reduces to the familiar classical formula $\mathbf{p} = m\mathbf{v}$. Equation 23 thus also satisfies this necessary criterion of relativistic formulas.

Of course, the ultimate test is agreement with experiment. Figure 20 shows a collection of data, based on independent determinations of the momentum and velocity of electrons. The data are plotted as p/mv, which should have the constant value 1 according to classical physics. The results agree with the relativistic equation and not with the classical one. Note that the classical and relativistic predictions agree for low speeds, and in fact the difference between the two is not at all apparent until the speed exceeds $0.1c$, which accounts for our failure to observe the relativistic corrections in experiments with ordinary laboratory objects.

Sample Problem 6 What is the momentum of a proton moving at a speed of $v = 0.86c$?

Solution Using Eq. 23, we obtain

$$p = \frac{mv}{\sqrt{1 - v^2/c^2}}$$
$$= \frac{(1.67 \times 10^{-27} \text{ kg})(0.86)(3.00 \times 10^8 \text{ m/s})}{\sqrt{1 - (0.86)^2}}$$
$$= 8.44 \times 10^{-19} \text{ kg} \cdot \text{m/s}.$$

The units of kg·m/s are generally not convenient in solving problems of this type. Instead, we manipulate Eq. 23 to obtain

$$pc = \frac{mcv}{\sqrt{1 - v^2/c^2}} = \frac{mc^2(v/c)}{\sqrt{1 - v^2/c^2}} = \frac{(938 \text{ MeV})(0.86)}{\sqrt{1 - (0.86)^2}}$$
$$= 1580 \text{ MeV}.$$

Here we have used the proton's *rest energy* mc^2, a concept we introduced in Section 8-7. The momentum is obtained from this result by dividing by the symbol c (not its numerical value), which gives

$$p = 1580 \text{ MeV}/c.$$

The units of MeV/c for momentum are often used in relativistic calculations because, as we show in the next section, the quantity pc often appears in these calculations. You should be able to convert MeV/c to kg·m/s and show that the two results obtained for p are equivalent.

21-9 RELATIVISTIC ENERGY

The interplay of mass and energy from the relativistic viewpoint was previously discussed in Section 8-7. You may find it useful to review that discussion before continuing to read this section.

In analogy with our discussion of momentum in the previous section, special relativity gives us a different approach to kinetic energy. Let us first indicate the difficulty by reconsidering the collision shown in Fig. 19. If we use the classical expression $\frac{1}{2}mv^2$, the collision does not conserve kinetic energy in the S' frame. (We chose the final velocities in the S frame so that kinetic energy would be conserved.) Using the velocities shown in Figs. 19c and 19d, you can show (see Problem 46) that, with $K = \frac{1}{2}mv^2$,

$$K'_i = \frac{2mv^2}{(1 + v^2/c^2)^2},$$
$$K'_f = mv^2(2 - v^2/c^2). \tag{26}$$

Thus K'_i is not equal to K'_f, and the elastic collision apparently does not conserve kinetic energy in S'. This situation violates the relativity postulate; the type of collision (elastic versus inelastic) should depend on the properties of the colliding objects and not on the particular reference frame from which we happen to be viewing the collision. As was the case with momentum, we require a new definition of kinetic energy if we are to preserve the law of conservation of energy and the relativity postulate.

The classical expression for kinetic energy also violates the second relativity postulate by allowing speeds in excess of the speed of light. There is no limit (in either classical or relativistic dynamics) to the energy we can give to a particle. Yet, if we allow the kinetic energy to increase without limit, the classical expression $K = \frac{1}{2}mv^2$ implies that the velocity must correspondingly increase without limit, thereby violating the second postulate. We must therefore find a way to redefine kinetic energy, so that the kinetic energy of a particle can be increased without limit while its speed remains less than c.

The relativistic expression for the kinetic energy of a particle can be derived using essentially the same procedure we used to derive the classical expression, starting with the particle form of the work–energy theorem (see Problem 54). The result of this calculation is

$$K = \frac{mc^2}{\sqrt{1 - v^2/c^2}} - mc^2. \tag{27}$$

Equation 27 looks very different from the classical result $\frac{1}{2}mv^2$, but, as we showed in Section 7-7, Eq. 27 reduces to the classical expression in the limit of low speeds ($v \ll c$). You can also see from the first term of Eq. 27 that $K \rightarrow \infty$ as $v \rightarrow c$. Thus we can increase the kinetic energy of a particle without limit, and its speed will not exceed c.

We can also express Eq. 27 as

$$K = E - E_0, \tag{28}$$

where the *total relativistic energy E* is defined as

$$E = \frac{mc^2}{\sqrt{1 - v^2/c^2}}, \tag{29}$$

and the *rest energy E_0* is defined as

$$E_0 = mc^2. \tag{30}$$

The rest energy is in effect the total relativistic energy of a particle measured in a frame of reference in which the particle is at rest.

The total relativistic energy is given by Eq. 28 as

$$E = K + E_0. \tag{31}$$

In interactions of particles at relativistic speeds, we can replace our previous principle of conservation of energy with one based on the total relativistic energy:

In an isolated system of particles, the total relativistic energy remains constant.

This principle is a special case of the result previously expressed in the form of Eq. 36 of Chapter 8 ($\Delta E_0 + \Delta K = W$) in which $W = 0$ (that is, the system is isolated—no external work is done by its environment).

Using the relativistic form of kinetic energy given by Eq. 27, we can show that kinetic energy is conserved in the S' frame of the collision of Fig. 19 (see Problem 47). Because the rest energies of the initial and final particles

are equal in this collision, conservation of total relativistic energy is equivalent to conservation of kinetic energy. In general, collisions of particles at high energies can result in the production of new particles, and thus the final rest energy may not be equal to the initial rest energy (see Sample Problem 9). Such collisions must be analyzed using conservation of total relativistic energy E; kinetic energy is not conserved when the rest energy changes in a collision.

Often m in Eq. 30 is called the *rest mass m_0* and is distinguished from the "relativistic mass," which is defined as $m_0/\sqrt{1 - v^2/c^2}$. We choose not to use relativistic mass, because it can be a misleading concept. Whenever we refer to mass, we always mean rest mass.

Manipulation of Eqs. 23 and 29 gives a useful relationship among the total energy, momentum, and rest energy:

$$E = \sqrt{(pc)^2 + (mc^2)^2}. \tag{32}$$

Figure 21 shows a useful mnemonic device for remembering this relationship, which has the form of the Pythagorean theorem for the sides of a right triangle.

The relationships between kinetic energy and velocity and between kinetic energy and momentum can be tested in the relativistic regime by accelerating particles to high speeds or by using high-speed particles (namely, electrons) emitted in certain radioactive decay processes. Figure 2 showed electrons given a known kinetic energy (using a high-voltage electrical terminal) whose resulting velocities were measured. The experimental data are in perfect agreement with the relativistic expression and in disagreement with the classical expression. Similar results are obtained indirectly today at every large accelerator facility in the world. Particles are accelerated to speeds very close to c, and the design parameters of the accelerators must be based on relativistic dynamics. Thus every modern accelerator is in effect a laboratory for testing special relativity. Needless to say, the success of these accelerators is a dramatic confirmation of special relativity.

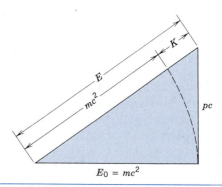

Figure 21   A useful mnemonic device for recalling the relationships between E_0, p, K, and E. Note that to put all variables in energy units, the quantity pc must be used.

equated to the initial total energy of 823 MeV. Thus we have one equation in the two unknowns p_1 and p_2.

To find a second equation in the two unknowns we apply conservation of momentum. The final momentum of the two-pion system along the beam direction is $p_1 + p_2$, and setting this equal to the initial momentum p_K gives

$$p_1c + p_2c = p_Kc = 655 \text{ MeV}. \qquad (36)$$

We now have two equations (Eqs. 35 and 36) in the two unknowns p_1 and p_2. Solving Eq. 36 for p_2c and substituting this result into Eq. 35, we obtain (after some algebraic manipulation) a quadratic equation for p_1c, which can be solved by standard algebraic techniques to give

$$p_1c = 668 \text{ MeV or } -13 \text{ MeV}.$$

Since the labels 1 and 2 of the two pions are arbitrary, the solution gives one pion traveling parallel to the beam with momentum $p_1 = 668$ MeV/c, while the other pion travels in the opposite direction with momentum $p_2 = -13$ MeV/c. The corresponding kinetic energies are found using Eqs. 28 and 32, which give

$$K = \sqrt{(pc)^2 + (m_\pi c^2)^2} - m_\pi c^2$$
$$K_1 = \sqrt{(668 \text{ MeV})^2 + (140 \text{ MeV})^2} - 140 \text{ MeV} = 543 \text{ MeV},$$
$$K_2 = \sqrt{(-13 \text{ MeV})^2 + (140 \text{ MeV})^2} - 140 \text{ MeV} = 0.6 \text{ MeV}.$$

This problem can also be solved in a different way by making a Lorentz transformation to a reference frame in which the kaons are at rest. The two pions are emitted in this frame in opposite directions (because the total momentum must be zero), and so they share the decay energy equally. Transforming back to the lab frame then gives the solution for the momenta and energies (see Problem 57). The next sample problem demonstrates another application of this technique.

Sample Problem 9   The discovery of the antiproton $\bar{p}$ (a particle with the same rest energy as a proton, 938 MeV, but with the opposite electric charge) took place in 1956 at Berkeley through the following reaction:

$$p + p \rightarrow p + p + p + \bar{p},$$

in which accelerated protons were incident on a target of protons at rest in the laboratory. The minimum incident kinetic energy needed to produce the reaction is called the *threshold* kinetic energy, for which the final particles move together as if they were a single unit. Find the threshold kinetic energy to produce antiprotons in this reaction.

Solution   This problem is conceptually the reverse case of the previous sample problem. Here particles are coming together to form a composite. We demonstrate an alternate method by solving in the center-of-mass reference frame, in which the two protons come together with equal and opposite momenta to form a new particle at rest (Fig. 22).

The final total relativistic energy in the center-of-mass frame S' is the rest energy of the products, which are produced at rest in this frame, so

$$E'_f = 4m_p c^2.$$

Sample Problem 7   In the Stanford Linear Collider* electrons are accelerated to a kinetic energy of 50 GeV. Find the speed of such an electron as (*a*) a fraction of c and (*b*) a difference from c. The rest energy of the electron is 0.511 MeV = 0.511×10^{-3} GeV.

Solution   (*a*) First we solve Eq. 27 for v, obtaining

$$v = c\sqrt{1 - \frac{1}{(1 + K/mc^2)^2}}, \qquad (33)$$

and thus

$$v = c\sqrt{1 - \frac{1}{(1 + 50 \text{ GeV}/0.511 \times 10^{-3} \text{ GeV})^2}}$$
$$= 0.999\,999\,999\,948c.$$

Calculators cannot be trusted to 12 significant digits. Here is a way to avoid this difficulty. We can write Eq. 33 as $v = c(1 + x)^{1/2}$, where $x = -1/(1 + K/mc^2)^2$. Because $K \gg mc^2$, we have $x \ll 1$, and we can use the binomial expansion to write $v \approx c(1 + \frac{1}{2}x)$, or

$$v \approx c\left[1 - \frac{1}{2(1 + K/mc^2)^2}\right], \qquad (34)$$

which gives

$$v = c(1 - 5.2 \times 10^{-11}).$$

This leads to the value of v given above.

(*b*) From the above result, we have

$$c - v = 5.2 \times 10^{-11}c = 0.016 \text{ m/s} = 1.6 \text{ cm/s}.$$

Sample Problem 8   A certain accelerator produces a beam of neutral kaons ($m_K c^2 = 498$ MeV) with kinetic energy 325 MeV. Consider a kaon that decays in flight into two pions ($m_\pi c^2 = 140$ MeV). Find the kinetic energy of each pion in the special case in which the pions travel parallel or antiparallel to the direction of the kaon beam.

Solution   The energy of the particles that remain after the decay can be found by applying principles of conservation of total relativistic energy and momentum. The initial total relativistic energy is, from Eq. 31,

$$E_K = K + m_K c^2 = 325 \text{ MeV} + 498 \text{ MeV} = 823 \text{ MeV}.$$

The initial momentum can be found from Eq. 32:

$$p_K c = \sqrt{E_K^2 - (m_K c^2)^2} = \sqrt{(823 \text{ MeV})^2 - (498 \text{ MeV})^2}$$
$$= 655 \text{ MeV}.$$

The total energy of the final system consisting of the two pions is

$$E = E_1 + E_2 = \sqrt{(p_1 c)^2 + (m_\pi c^2)^2} + \sqrt{(p_2 c)^2 + (m_\pi c^2)^2}$$
$$= 823 \text{ MeV}, \qquad (35)$$

which, by conservation of total relativistic energy, we have

---

* See "The Stanford Linear Collider," by John R. Rees, *Scientific American,* October 1989, p. 58.

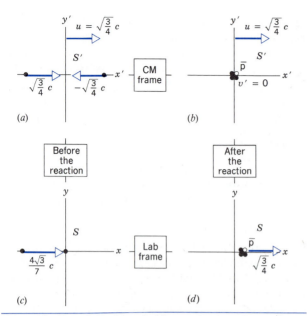

Figure 22 Sample Problem 9. The production of an antiproton, viewed from (*a,b*) the center-of-mass frame and (*c,d*) the laboratory frame. Compare with Fig. 19.

The initial energy is just the sum of the total energies of the two original reacting protons:

$$E_i' = E_1' + E_2'.$$

Conservation of energy requires $E_i' = E_f'$, and since the energies E_1' and E_2' are equal in the S' frame, we have

$$E_1' = E_2' = 2m_p c^2.$$

The corresponding magnitude of the velocity of either reacting proton in the S' frame is found by solving Eq. 29 for v/c, which gives

$$\frac{v_1'}{c} = \sqrt{1 - \left(\frac{m_p c^2}{E_1'}\right)^2} = \sqrt{1 - \left(\frac{1}{2}\right)^2} = \sqrt{\frac{3}{4}}.$$

We now make a Lorentz transformation back to the laboratory using this as the transformation speed, which brings one of the protons to rest and gives the other a velocity v that can be found from the inverse velocity transformation expression for v_x from Table 3. Using $v' = c\sqrt{3/4}$ and $u = c\sqrt{3/4}$, and dropping the x subscript, we have

$$v = \frac{v' + u}{1 + uv'/c^2} = \frac{2c\sqrt{3/4}}{1 + (\sqrt{3/4})^2} = \frac{4\sqrt{3}}{7}c.$$

This is the speed of the incident proton in the laboratory frame. Its total energy can be found from Eq. 29:

$$E = \frac{mc^2}{\sqrt{1 - v^2/c^2}} = \frac{m_p c^2}{\sqrt{1 - (4\sqrt{3}/7)^2}} = 7m_p c^2,$$

and the threshold kinetic energy is

$$K = E - m_p c^2 = 6m_p c^2 = 6(938 \text{ MeV})$$
$$= 5628 \text{ MeV} = 5.628 \text{ GeV}.$$

The Bevatron accelerator at Berkeley was designed with this experiment in mind, so that it could produce a beam of protons whose energy exceeded 5.6 GeV. The discovery of the antiproton in this reaction was honored with the award of the 1959 Nobel prize to the experimenters, Emilio Segrè and Owen Chamberlain.

21-10   THE COMMON SENSE OF SPECIAL RELATIVITY

We have reached a point where we can look back at our presentation of special relativity and think about its common sense. We must first of all note that relativity affects every aspect of physics; we have concentrated in this chapter on mechanics, and later in this text we consider the effect of relativity on electromagnetism. Indeed, we must carefully reexamine every subfield of physics from the perspective of special relativity, verifying that each is consistent with the two postulates. We must also note that relativity has passed every experimental test without the slightest discrepancy. It is a theory that is of great aesthetic value, providing us with a view more satisfying than that of classical physics about the validity of different perspectives and symmetries. It is also a theory of great practical value, providing engineers with the proper guidance to construct large particle accelerators and providing those concerned about maintaining standards with the proper procedures for correcting the readings of atomic clocks when they are transported from one location to another.

The first postulate of relativity is really an outgrowth of Newton's first law, the law of inertia, which defined the concept of inertial frames and gave us the first notion that inertial observers would draw idential conclusions from observing an experiment in which no net force acts. It is not too great a leap to extend that view to assert that inertial observers should also draw identical conclusions from observing an experiment in which there *is* a net force. Finally, why should we single out the laws of mechanics for this equivalence? By extending it to an equivalence for inertial observers of *all* the laws of physics, we arrive at the first postulate.

The second postulate is also a reasonable one. It seems unrealistic to be able to transmit a signal at an infinite speed, thereby providing instantaneous communication throughout the universe. Moreover, experiments on the relativity of time show that such instant communication between distant points is not consistent with observation. If there is a limiting speed, then surely (by the first postulate) it must be the same for all observers, regardless of their state of motion.

For some, the first exposure to the relativity of simultaneity, the apparent shrinking of moving rods, and the slowing down of time may be disturbing. However, a bit of thought will persuade you that the classical alternatives are even more disturbing. For example, a classical rigid rod of definite length is not a concept that is consistent with relativity; a signal (say, a quick movement) at one end cannot be transmitted instantly to the other end. We must give up the idea of all observers being able to use the *same* measuring rod. We replace this idea with one that gives each observer a measuring rod and permits that observer to use that rod to make measurements within a particular frame of reference. No observer's measuring instruments or results are preferred over any other's. Finally, relativity gives us a wonderful symmetry between these observers; it doesn't assert the *reality* of slowing clocks, but that, from their two differing perspectives, two observers in relative motion each observe that the other's clocks are slow. There is no necessity to grant preferred status to either of them, or to any other inertial observer.

According to classical physics, space and time are absolute. This leads to the result that the laws of physics must be different for different observers. Relativity, on the other hand, tells us that the laws of physics must be the same for all observers, and as a consequence space and time become relative concepts. Clearly, relativity is "more absolute" than classical physics. The arbitrary and complex physical world of classical physics, in which each observer must use a different set of physical laws, becomes the more uniform and simple physical world of relativity.

Relativity broadens our view of the universe by placing us among the many inertial observers of that universe. It brings together concepts that, according to classical physics, were treated separately: for instance, space and time into space–time, or mass and energy into rest energy. It points the way toward a single unifying theory that includes all possible interactions between particles: electricity and magnetism into electromagnetism; electromagnetism and the so-called weak forces (those responsible for certain radioactive decay processes) into the electroweak interaction; the electroweak and the strong nuclear interactions into one of the proposed Grand Unified Theories (GUTs); and finally GUTs and gravity into the hypothetical Theory of Everything. Einstein, who knew about only the first of these unifications, would surely be very pleased at these developments.

QUESTIONS

1. The speed of light in a vacuum is a true constant of nature, independent of the wavelength of the light or the choice of an (inertial) reference frame. Is there any sense, then, in which Einstein's second postulate can be viewed as contained within the scope of his first postulate?

2. Discuss the problem that young Einstein grappled with; that is, what would be the appearance of an electromagnetic wave to a person running along with it at speed c?

3. Is the concept of an incompressible fluid valid in relativity? What about perfectly rigid bodies?

4. A quasar (*quasi-stellar object*) travels away from the Earth at half the speed of light. What is the speed, with respect to the Earth, of the light we detect coming from it?

5. Quasars are the most intrinsically luminous objects in the universe. Many of them fluctuate in brightness, often on a time scale of a day or so. How can the rapidity of these brightness changes be used to estimate an upper limit to the size of these objects? (*Hint:* Separated points cannot change in a coordinated way unless information is sent from one to the other.)

6. The sweep rate of the tail of a comet can exceed the speed of light. Explain this phenomenon and show that there is no contradiction with relativity.

7. Consider a spherical light wavefront spreading out from a point source. As seen by an observer at the source, what is the *difference in velocity* of portions of the wavefront traveling in opposite directions? What is the *relative velocity* of one of these portions of the wavefront with respect to the other?

8. Borrowing two phrases from Herman Bondi, we can catch the spirit of Einstein's two postulates by labeling them: (1) the principle of "the irrelevance of velocity" and (2) the principle of "the uniqueness of light." In what senses are velocity irrelevant and light unique in these two statements?

9. A beam from a laser falls at right angles on a plane mirror and reflects from it. What is the speed of the reflected beam if the mirror is (*a*) fixed in the laboratory and (*b*) moving directly toward the laser with speed v?

10. Give an example from classical physics in which the motion of a clock affects its rate, that is, the way it runs. (The magnitude of the effect may depend on the detailed nature of the clock.)

11. Although in relativity (where motion is relative and not absolute) we find that "moving clocks run slow," this effect has nothing to do with the motion altering the way a clock works. With what does it have to do?

12. We have seen that if several observers watch two events, labeled A and B, one of them may say that event A occurred first but another may claim that it was event B that did so. What would you say to a friend who asked you which event *really did* occur first?

13. Two events occur at the same place and at the same time for one observer. Will they be simultaneous for all other observers? Will they also occur at the same place for all other observers?

14. Two events are simultaneous but separated in space in one inertial reference frame. Will they be simultaneous in any other frame? Will their spatial separation be the same in any other frame?

15. Let event A be the departure of an airplane from San Fran-

cisco and event *B* be its arrival in New York. Is it possible to find two observers who disagree about the time order of these events? Explain.

16. Two observers, one at rest in *S* and one at rest in *S'*, each carry a meter stick oriented parallel to their relative motion. *Each* observer finds upon measurement that the *other* observer's meter stick is the shorter of the two sticks. Does this seem like a paradox to you? Explain. (*Hint:* Compare with the following situation. Harry waves goodbye to Walter who is in the rear of a station wagon driving away from Harry. Harry says that Walter gets smaller. Walter says that Harry gets smaller. Are they measuring the same thing?)

17. How does the concept of simultaneity enter into the measurement of the length of an object?

18. In relativity the time and space coordinates are intertwined and treated on a more or less equivalent basis. Are time and space fundamentally of the same nature, or is there some essential difference between them that is preserved even in relativity?

19. In the "twin paradox," explain (in terms of heartbeats, physical and mental activities, and so on) why the younger returning twin has not lived any longer than her own proper time even though her stay-at-home brother may say that she has. Hence explain the remark: "You age according to your own proper time."

20. Can we simply substitute γm for *m* in classical equations to obtain the correct relativistic equations? Give examples.

21. If zero-mass particles have a speed *c* in one reference frame, can they be found at rest in any other frame? Can such particles have any speed other than *c*?

22. A particle with zero mass (a neutrino, possibly) can transport momentum. But, by Eq. 23, $\mathbf{p} = m\mathbf{v}/\sqrt{1 - v^2/c^2}$, the momentum is directly proportional to the mass and therefore should be zero if the mass is zero. Explain.

23. How many relativistic expressions can you think of in which the Lorentz factor γ enters as a simple multiplier?

24. Is the mass of a stable, composite particle (a gold nucleus, for example) greater than, equal to, or less than the sum of the masses of its constituents? Explain.

25. "The mass of the electron is 0.511 MeV." Exactly what does this statement mean?

26. "The relation $E_0 = mc^2$ is essential to the operation of a power plant based on nuclear fission but has only a negligible relevance for a fossil-fuel plant." Is this a true statement? Explain why or why not.

27. A hydroelectric plant generates electricity because water falls under gravity through a turbine, thereby turning the shaft of a generator. According to the mass–energy concept, must the appearance of energy (the electricity) be identified with a mass decrease somewhere? If so, where?

28. Some say that relativity complicates things. Give examples to the contrary, wherein relativity simplifies matters.

PROBLEMS

Section 21-3 Consequences of Einstein's Postulates

1. Quite apart from effects due to the Earth's rotational and orbital motions, a laboratory frame is not strictly an inertial frame because a particle placed at rest there will not, in general, remain at rest; it will fall under gravity. Often, however, events happen so quickly that we can ignore free fall and treat the frame as inertial. Consider, for example, a 1.0-MeV electron (for which $v = 0.941c$) projected horizontally into a laboratory test chamber and moving through a distance of 20 cm. (*a*) How long would it take, and (*b*) how far would the electron fall during this interval? What can you conclude about the suitability of the laboratory as an inertial frame in this case?

2. A 100-MeV electron, for which $v = 0.999987c$, moves along the axis of an evacuated tube that has a length of 2.86 m as measured by a laboratory observer *S* with respect to whom the tube is at rest. An observer *S'* moving with the electron, however, would see this tube moving past with speed *v*. What length would this observer measure for the tube?

3. A rod lies parallel to the *x* axis of reference frame *S*, moving along this axis at a speed of $0.632c$. Its rest length is 1.68 m. What will be its measured length in frame *S*?

4. The mean lifetime of muons stopped in a lead block in the laboratory is measured to be 2.20 μs. The mean lifetime of high-speed muons in a burst of cosmic rays observed from the Earth is measured to be 16.0 μs. Find the speed of these cosmic ray muons.

5. An unstable high-energy particle enters a detector and leaves a track 1.05 mm long before it decays. Its speed relative to the detector was $0.992c$. What is its proper lifetime? That is, how long would it have lasted before decay had it been at rest with respect to the detector?

6. The length of a spaceship is measured to be exactly half its rest length. (*a*) What is the speed of the spaceship relative to the observer's frame? (*b*) By what factor do the spaceship's clocks run slow, compared to clocks in the observer's frame?

7. A particle moves along the *x'* axis of frame *S'* with a speed of $0.413c$. Frame *S'* moves with a speed of $0.587c$ with respect to frame *S*. What is the measured speed of the particle in frame *S*?

8. Frame *S'* moves relative to frame *S* at $0.620c$ in the direction of increasing *x*. In frame *S'* a particle is measured to have a velocity of $0.470c$ in the direction of increasing *x'*. (*a*) What is the velocity of the particle with respect to frame *S*? (*b*) What would be the velocity of the particle with respect to *S* if it moved (at $0.470c$) in the direction of *decreasing x'* in

the S' frame? In each case, compare your answers with the predictions of the classical velocity transformation equation.

9. A spaceship of rest length 130 m drifts past a timing station at a speed of $0.740c$. (*a*) What is the length of the spaceship as measured by the timing station? (*b*) What time interval between the passage of the front and back end of the ship will the station monitor record?

10. A pion is created in the higher reaches of the Earth's atmosphere when an incoming high-energy cosmic-ray particle collides with an atomic nucleus. A pion so formed descends toward Earth with a speed of $0.99c$. In a reference frame in which they are at rest, pions have a lifetime of 26 ns. As measured in a frame fixed with respect to the Earth, how far will such a typical pion move through the atmosphere before it decays?

11. To circle the Earth in low orbit a satellite must have a speed of about 7.91 km/s. Suppose that two such satellites orbit the Earth in opposite directions. (*a*) What is their relative speed as they pass? Evaluate using the classical Galilean velocity transformation equation. (*b*) What fractional error was made because the (correct) relativistic transformation equation was not used?

Section 21-4 The Lorentz Transformation

12. What must be the value of the speed parameter β if the Lorentz factor γ is to be (*a*) 1.01? (*b*) 10.0? (*c*) 100? (*d*) 1000?

13. Find the speed parameter of a particle that takes 2 years longer than light to travel a distance of 6.0 ly.

14. Observer S assigns to an event the coordinates $x = 100$ km, $t = 200$ μs. Find the coordinates of this event in frame S', which moves in the direction of increasing x with speed $0.950c$. Assume $x = x'$ at $t = t' = 0$.

15. Observer S reports that an event occurred on the x axis at $x = 3.20 \times 10^8$ m at a time $t = 2.50$ s. (*a*) Observer S' is moving in the direction of increasing x at a speed of $0.380c$. What coordinates would S' report for the event? (*b*) What coordinates would S'' report if S'' were moving in the direction of *decreasing* x at this same speed?

16. Inertial frame S' moves at a speed of $0.60c$ with respect to frame S in the direction of increasing x. In frame S, event 1 occurs at the origin at $t = 0$ and event 2 occurs on the x axis at $x = 3.0$ km and at $t = 4.0$ μs. What times of occurrence does observer S' record for these same events? Explain the reversal of the time order.

17. An experimenter arranges to trigger two flashbulbs simultaneously, a blue flash located at the origin of his reference frame and a red flash at $x = 30.4$ km. A second observer, moving at a speed of $0.247c$ in the direction of increasing x, also views the flashes. (*a*) What time interval between them does she find? (*b*) Which flash does she say occurs first?

18. Derive Eqs. 17 for the inverse Lorentz transformation by algebraically inverting the equations for the Lorentz transformation, Eqs. 14.

Section 21-6 The Transformation of Velocities

19. Suppose observer S fires a light beam in the y direction ($v_x = 0$, $v_y = c$). Observer S' is moving at speed u in the x direction. (*a*) Find the components v'_x and v'_y of the velocity

of the light beam according to S', and (*b*) show that S' measures a speed of c for the light beam.

20. One cosmic-ray proton approaches the Earth along its axis with a velocity of $0.787c$ toward the North Pole and another, with velocity $0.612c$, toward the South Pole. See Fig. 23. Find the relative speed of approach of one particle with respect to the other. (*Hint:* It is useful to consider the Earth and one of the particles as the two inertial reference frames.)

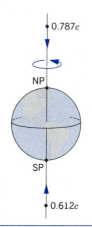

Figure 23 Problem 20.

21. Galaxy A is reported to be receding from us with a speed of $0.347c$. Galaxy B, located in precisely the opposite direction, is also found to be receding from us at this same speed. What recessional speed would an observer on galaxy A find (*a*) for our galaxy and (*b*) for galaxy B?

22. It is concluded from measurements of the red shift of the emitted light that quasar Q_1 is moving away from us at a speed of $0.788c$. Quasar Q_2, which lies in the same direction in space but is closer to us, is moving away from us at speed $0.413c$. What velocity for Q_2 would be measured by an observer on Q_1?

23. A spaceship, at rest in a certain reference frame S, is given a speed increment of $0.500c$. It is then given a further $0.500c$ increment in this new frame, and this process is continued until its speed with respect to its original frame S exceeds $0.999c$. How many increments does it require?

24. A radioactive nucleus moves with a constant speed of $0.240c$ along the x axis of a reference frame S fixed with respect to the laboratory. It decays by emitting an electron whose speed, measured in a reference frame S' moving with the nucleus, is $0.780c$. Consider first the cases in which the emitted electron travels (*a*) along the common xx' axis and (*b*) along the y' axis and find, for each case, its velocity (magnitude and direction) as measured in frame S. (*c*) Suppose, however, that the emitted electron, viewed now from frame S, travels along the y axis of that frame with a speed of $0.780c$. What is its velocity (magnitude and direction) as measured in frame S'?

25. In Fig. 24, A and B are trains on perpendicular tracks, shown radiating from station S. The velocities are in the station frame (S frame). (*a*) Find $\mathbf{v}_{AB}$, the velocity of train B with respect to train A. (*b*) Find $\mathbf{v}_{BA}$, the velocity of train A with

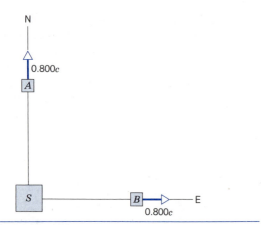

Figure 24 Problem 25.

respect to train *B*. (*c*) Comment on the fact that these two relative velocities do not point in opposite directions.

Section 21-7 Consequences of the Lorentz Transformation

26. An electron is moving at a speed such that it could circumnavigate the Earth at the equator in 1 s. (*a*) What is its speed, in terms of the speed of light? (*b*) Its kinetic energy *K*? (*c*) What percent error do you make if you use the classical formula to calculate *K*?

27. The rest radius of the Earth is 6370 km and its orbital speed about the Sun is 29.8 km/s. By how much would the Earth's diameter appear to be shortened to an observer stationed so as to be able to watch the Earth move past at this speed?

28. An airplane whose rest length is 42.4 m is moving with respect to the Earth at a constant speed of 522 m/s. (*a*) By what fraction of its rest length will it appear to be shortened to an observer on Earth? (*b*) How long would it take by Earth clocks for the airplane's clock to fall behind by 1 μs? (Assume that only special relativity applies.)

29. A spaceship whose rest length is 358 m has a speed of 0.728*c* with respect to a certain reference frame. A micrometeorite, with a speed of 0.817*c* in this frame, passes the spaceship on an antiparallel track. How long does it take this micrometeorite to pass the spaceship?

30. A clock moves along the *x* axis at a speed of 0.622*c* and reads zero as it passes the origin. (*a*) Calculate the Lorentz factor. (*b*) What time does the clock read as it passes *x* = 183 m?

31. An observer *S* sees a flash of red light 1210 m away and a flash of blue light 730 m closer and on the same straight line. *S* measures the time interval between the occurrence of the flashes to be 4.96 μs, the red flash occurring first. (*a*) Find the relative velocity, magnitude and direction, of a second observer *S'* who would record these flashes as occurring at the same place. (*b*) From the point of view of *S'*, which flash occurs first and what is the measured time interval between the flashes?

32. Consider the previous problem. Suppose now that observer *S* sees the two flashes in the same positions as in that problem but occurring closer together in time. How close together in time can they be and still have it possible to find a frame *S'* in which they occur at the same place?

33. A space traveler takes off from Earth and moves at speed 0.988*c* toward the star Vega, which is 26.0 ly distant. How much time will have elapsed by Earth clocks (*a*) when the traveler reaches Vega and (*b*) when the Earth observers receive word from him that he has arrived? (*c*) How much older will the Earth observers calculate the traveler to be when he reaches Vega than he was when he started the trip?

34. You wish to make a round trip from Earth in a spaceship, traveling at constant speed in a straight line for 6 months and then returning at the same constant speed. You wish further, on your return, to find the Earth as it will be 1000 years in the future. (*a*) How fast must you travel? (*b*) Does it matter whether or not you travel in a straight line on your journey? If, for example, you traveled in a circle for 1 year, would you still find that 1000 years had elapsed by Earth clocks when you returned?

35. Observers *S* and *S'* stand at the origins of their respective frames, which are moving relative to each other with a speed 0.600*c*. Each has a standard clock, which, as usual, they set to zero when the two origins coincide. Observer *S* keeps the *S'* clock visually in sight. (*a*) What time will the *S'* clock record when the *S* clock records 5.00 μs? (*b*) What time will observer *S* *actually read* on the *S'* clock when the *S* clock reads 5.00 μs?

36. (*a*) Can a person, in principle, travel from Earth to the galactic center (which is about 23,000 ly distant) in a normal lifetime? Explain, using either time-dilation or length-contraction arguments. (*b*) What constant speed would be needed to make the trip in 30 y (proper time)?

Section 21-8 Relativistic Momentum

37. Show that 1 kg·m/s = 1.875×10^{21} MeV/*c*.

38. A particle has a momentum equal to *mc*. Calculate its speed.

39. Calculate the speed parameter of a particle with a momentum of 12.5 MeV/*c* if the particle is (*a*) an electron and (*b*) a proton.

Section 21-9 Relativistic Energy

40. Find the speed parameter *β* and the Lorentz factor *γ* for an electron whose kinetic energy is (*a*) 1.0 keV, (*b*) 1.0 MeV, and (*c*) 1.0 GeV.

41. Find the speed parameter *β* and the Lorentz factor *γ* for a particle whose kinetic energy is 10 MeV if the particle is (*a*) an electron, (*b*) a proton, and (*c*) an alpha particle.

42. A particle has a speed of 0.990*c* in a laboratory reference frame. What are its kinetic energy, its total energy, and its momentum if the particle is (*a*) a proton or (*b*) an electron?

43. Quasars are thought to be the nuclei of active galaxies in the early stages of their formation. A typical quasar radiates energy at the rate of 1.20×10^{41} W. At what rate is the mass of this quasar being reduced to supply this energy? Express your answer in solar mass units per year, where one solar mass unit (smu) is the mass of our Sun.

44. Calculate the speed of a particle (*a*) whose kinetic energy is equal to twice its rest energy and (*b*) whose total energy is equal to twice its rest energy.

45. Find the momentum of a particle of mass *m* in order that its total energy be three times its rest energy.

46. Use the velocities given in Fig. 19 in the S' frame and show that, according to S', the kinetic energies before and after the collision, computed classically, are given by Eqs. 26.

47. Reconsider the collision shown in Fig. 19. Using Eq. 27 for the relativistic kinetic energy, calculate the initial and final kinetic energies in frame S' and thereby show that kinetic energy is conserved in this frame as in frame S.

48. Consider the following, all moving in free space: a 2.0-eV photon, a 0.40-MeV electron, and a 10-MeV proton. (*a*) Which is moving the fastest? (*b*) The slowest? (*c*) Which has the greatest momentum? (*d*) The least? (*Note:* A photon is a light-particle of zero mass.)

49. (*a*) If the kinetic energy K and the momentum p of a particle can be measured, it should be possible to find its mass m and thus identify the particle. Show that

$$m = \frac{(pc)^2 - K^2}{2Kc^2}.$$

(*b*) What does this expression reduce to as $v/c \to 0$, in which v is the speed of the particle? (*c*) Find the mass of a particle whose kinetic energy is 55.0 MeV and whose momentum is 121 MeV/c; express your answer in terms of the mass m_e of the electron.

50. In a high-energy collision of a primary cosmic-ray particle near the top of the Earth's atmosphere, 120 km above sea level, a pion is created with a total energy of 135 GeV, traveling vertically downward. In its proper frame this pion decays 35.0 ns after its creation. At what altitude above sea level does the decay occur? The rest energy of a pion is 139.6 MeV.

51. How much work must be done to increase the speed of an electron from (*a*) $0.18c$ to $0.19c$ and (*b*) $0.98c$ to $0.99c$? Note that the speed increase (= $0.01c$) is the same in each case.

52. Two identical particles, each of mass 1.30 mg, moving with equal but opposite velocities of $0.580c$ in the laboratory reference frame, collide and stick together. Find the mass of the resulting particle.

53. A particle of mass m traveling at a relativistic speed makes a completely inelastic collision with an identical particle that is initially at rest. Find (*a*) the speed of the resulting single particle and (*b*) its mass. Express your answers in terms of the Lorentz factor γ of the incident particle.

54. (*a*) Suppose we have a particle accelerated from rest by the action of a force F. Assuming that Newton's second law for a particle, $F = dp/dt$, is valid in relativity, show that the final kinetic energy K can be written, using the work–energy theorem, as $K = \int v \, dp$. (*b*) Using Eq. 23 for the relativistic momentum, show that carrying out the integration in (*a*) leads to Eq. 27 for the relativistic kinetic energy.

55. (*a*) In modern experimental high-energy physics, energetic particles are made to circulate in opposite directions in so-

called storage rings and permitted to collide head-on. In this arrangement each particle has the same kinetic energy K in the laboratory. The collisions may be viewed as totally inelastic, in that the rest energy of the two colliding particles, plus all available kinetic energy, can be used to generate new particles and to endow them with kinetic energy. Show that the available energy in this arrangement can be written in the form

$$E_{\text{new}} = 2mc^2 \left(1 + \frac{K}{mc^2} \right)$$

where m is the mass of the colliding particles. (*b*) How much energy is made available when 100-GeV protons are used in this fashion? (*c*) What proton energy would be required to make 100 GeV available? (*Note:* Compare your answers with those in Problem 56, which describes another—less energy-effective—bombarding arrangement.)

56. (*a*) A proton, mass m, accelerated in a proton synchrotron to a kinetic energy K strikes a second (target) proton at rest in the laboratory. The collision is entirely inelastic in that the rest energy of the two protons, plus all the kinetic energy consistent with the law of conservation of momentum, is available to generate new particles and to endow them with kinetic energy. Show that the energy available for this purpose is given by

$$E_{\text{new}} = 2mc^2 \sqrt{1 + \left(\frac{K}{2mc^2} \right)}.$$

(*b*) How much energy is made available when 100-GeV protons are used in this fashion? (*c*) What proton energy would be required to make 100 GeV available? (*Note:* Compare with Problem 55.)

57. (*a*) Consider the decay of the kaon described in Sample Problem 8, but use a frame of reference (the center-of-mass frame) in which the kaons are initially at rest. Show that the two pions emitted in the decay travel in opposite directions with equal speeds of $0.827c$. (*b*) What is the velocity of the original kaons as observed in the laboratory frame? (*c*) Assume the two pions are emitted in the center-of-mass frame with velocities of $v_x' = +0.827c$ and $v_x' = -0.827c$. By calculating the corresponding velocities in the laboratory frame, show that the kinetic energies in the laboratory frame are identical with those found in the solution to Sample Problem 8.

58. An alpha particle with kinetic energy 7.70 MeV strikes a ^{14}N nucleus at rest. An ^{17}O nucleus and a proton are produced, the proton emitted at 90° to the direction of the incident alpha particle and carrying kinetic energy 4.44 MeV. The rest energies of the various particles are: alpha particle, 3730.4 MeV; ^{14}N, 13,051 MeV; proton, 939.29 MeV; ^{17}O, 15,843 MeV. (*a*) Find the kinetic energy of the ^{17}O nucleus. (*b*) At what angle with respect to the direction of the incident alpha particle does the ^{17}O nucleus move?

CHAPTER 22

TEMPERATURE

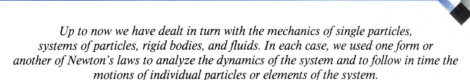

Up to now we have dealt in turn with the mechanics of single particles, systems of particles, rigid bodies, and fluids. In each case, we used one form or another of Newton's laws to analyze the dynamics of the system and to follow in time the motions of individual particles or elements of the system.

Beginning with this chapter, we broaden our perspective to deal with systems that are too complex to treat in terms of the motion of individual particles. These systems usually appear disordered because of the large number of particles involved and the many different ways they can share the energy available to the system. To analyze such systems we use the principles of thermodynamics. *In our study of thermodynamics we define a new set of physical variables to describe the state of a system, and we deduce a new set of laws that govern the behavior of systems. We also show how it is possible to understand these new laws on the basis of our previous laws of mechanics.*

A central concept of thermodynamics is temperature. *In this chapter we define temperature and discuss its measurement.*

22-1 MACROSCOPIC AND MICROSCOPIC DESCRIPTIONS

A liter of gas contains about 3×10^{22} molecules. Let us take the simplest possible case and treat the gas molecules as point particles that collide elastically with one another and with the walls of the container. If we specify the initial position and velocity of every particle, we could then apply Newton's laws and deduce the position and velocity of each particle at any future time. Given that information, we could calculate certain measurable properties of the system, such as the net impulsive force exerted on an element of area of the container. We call this the *microscopic* description of the system. Because the number of particles is so large, it is advantageous to treat the system using average values of the microscopic quantities. This approach is called *statistical mechanics* and is discussed in Chapter 24.

A different approach is based on the following question: Can we describe the system, including its interactions with its environment, in terms of a small number of overall properties that are measurable by relatively easily

performed laboratory operations? In the case of a gas confined to a container, we can indeed obtain such a description in terms of the *macroscopic* quantities—pressure, volume, temperature, quantity of matter, and internal energy, among others. For systems other than a gas, we can define and measure different macroscopic variables. For instance, in a ferromagnet such as iron, the particles interact not by impulsive forces in collisions but by magnetic forces; in the macroscopic description of a ferromagnet, the magnetization must be included among the macroscopic quantities.

Macroscopic properties can usually be measured directly in the laboratory, for example, the pressure of a confined gas or the magnetization of a piece of iron. We can also easily measure the variation of any such property with the temperature and derive an *equation of state* that describes the dependence of the macroscopic variables on one another.

For any system the macroscopic and the microscopic quantities must be related because they are simply different ways of describing the same situation. In particular, we should be able to express one in terms of the other. The pressure of a gas, a macroscopic quantity, is measured operationally using a manometer. Microscopically, pressure is related to the average rate per unit area at which the molecules of the gas deliver momentum to the manometer fluid as they strike its surface. In Section 23-3 we quantify this microscopic definition of pressure. Similarly (see Section 23-4), the temperature of a gas (also a macroscopic quantity) is related to the average kinetic energy of translation of the molecules.

If the macroscopic quantities can be expressed in terms of the microscopic quantities, then the laws of thermodynamics can be quantitatively expressed in terms of statistical mechanics. This accomplishment is one of the landmark achievements in the development of physics. As we proceed through our study of thermodynamics, this theme of the relationship between macroscopic and microscopic variables will arise frequently.

22-2   TEMPERATURE AND THERMAL EQUILIBRIUM

Consider the two systems *A* and *B* illustrated in Fig. 1*a*. They are "isolated" from one another and from the environment. By "isolated" we mean that neither energy nor matter can enter or leave either system. For example, the systems might be surrounded by walls made of thick slabs of Styrofoam, presumed to be both rigid and impermeable. The walls in this case are said to be *adiabatic.* (The word "adiabatic" comes from the Greek for "cannot be crossed." You can think of "adiabatic" as meaning "insulating.") Changes in the properties of one system have no effect on the other system.

We can replace the adiabatic wall separating *A* and *B* with one that permits the flow of energy (Fig. 1*b*) in a form that we shall come to know as heat. A thin but rigid sheet

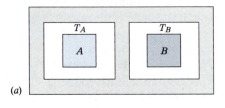

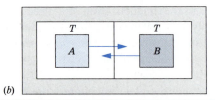

Figure 1   (*a*) Systems *A* and *B* are separated by an adiabatic wall. The systems have different temperatures T_A and T_B. (*b*) Systems *A* and *B* are separated by a diathermic wall. The systems, having come to thermal equilibrium, have the same temperature T.

of copper might be an example. Such a wall is called *diathermic.* (The word "diathermic," loosely translated from the Greek, means "heat passing through." You can think of "diathermic" as meaning "conducting.")

When two systems are placed into contact through a diathermic wall, the exchange of energy causes the macroscopic properties of the two systems to change. If the systems are confined gases, for example, the pressure might be one of the macroscopic quantities that change. The changes are relatively rapid at first, but become slower and slower as time goes on, until finally the macroscopic properties approach constant values. When this occurs, we say that the two systems are in *thermal equilibrium* with each other.

One way of testing whether bodies are in thermal equilibrium is to bring them into contact through a diathermic wall and to observe whether the macroscopic properties of the systems change with time after they are brought into contact. If no changes in the macroscopic properties are observed with time, the systems were originally in thermal equilibrium. It might, however, be inconvenient or even impossible to move two systems so that they would be in contact with one another. (The systems might be too bulky to move easily, or they might be separated by a very large distance.) We therefore generalize the concept of thermal equilibrium so that the systems need not necessarily be brought into contact with each other. The separated bodies can be said to be in thermal equilibrium if they are in states such that, if they *were* connected, they would be in thermal equilibrium.

The way to test whether such separated systems are in thermal equilibrium is to use a third system *C*. By placing *C* into contact with *A* and then with *B*, we could discover whether *A* and *B* are in thermal equilibrium without ever bringing *A* and *B* into direct contact. This is summarized as a postulate called the *zeroth law of thermodynamics,* which is often stated as follows:

If systems A and B are each in thermal equilibrium with a third system C, then A and B are in thermal equilibrium with each other.

This law may seem simple, but it is not at all obvious. If *A*, *B*, and *C* were people, it might be true that *A* and *C* may each know *B* but not know each other. If *A* and *C* are unmagnetized pieces of iron and *B* is a magnet, then *A* and *C* are each attracted to *B* without being attracted to each other.

The zeroth law has been called a logical afterthought. It came to light in the 1930s, long after the first and second laws of thermodynamics had been proposed and accepted. As we discuss later, the zeroth law in effect defines the concept of temperature, which is fundamental to the first and second laws of thermodynamics. The law that establishes the temperature should have a lower number, so it is called the zeroth law.

Temperature

When two systems are in thermal equilibrium, we say that they have the same *temperature.* Conversely, temperature is that property of a system which equals that of another system when the two systems are in thermal equilibrium. For example, suppose the systems are two gases that initially have different temperatures, pressures, and volumes. After we place them into contact and wait a sufficiently long time for them to reach thermal equilibrium, their pressures will in general *not* be equal, nor will their volumes; their temperatures, however, will always be equal in thermal equilibrium. *It is only through this argument based on thermal equilibrium that the notion of temperature can be introduced into thermodynamics.*

Although temperature in its everyday use is familiar to all of us, it is necessary to give it a precise meaning if it is to be of value as a scientific measure. Our subjective notion of temperature is not at all reliable. For example, suppose you are sitting indoors in a chair that is made partly of cloth, wood, and metal. Touch the various parts of the chair and decide which is "coldest," that is, which is at the lowest temperature. You will probably decide that the metal parts are coldest. However, we expect that all parts of the chair have been in the room long enough to come into thermal equilibrium with the air and should all therefore be at the same temperature as the air. What you are testing when you touch the metal is not only its temperature but also its ability to conduct heat away from your (presumably warmer) hand. In this case your hand is giving a subjective and incorrect measure of temperature. Furthermore, that subjective judgment will change with time if you hold your hand on the metal, as your hand and the metal approach thermal equilibrium with one another.

You can also test your subjectivity by soaking one hand in cold water and another in warm water. When you then grasp an object of intermediate temperature, you will find that the first hand senses a higher temperature than the second. You can be somewhat more objective in comparing two different samples of the same material at different temperatures by touching each with the same hand, which may distinguish the "hotter" from the "cooler." This procedure might reveal which object is at the higher temperature, but it is hardly sufficient to be quantitative about the difference. It is therefore necessary that we carefully specify an objective way of measuring temperature, which is our goal in this chapter.

In practical use of the zeroth law, we wish to identify system *C* as a thermometer. If the thermometer comes separately into thermal equilibrium with systems *A* and *B* and indicates the same temperature, then we may conclude that *A* and *B* are in thermal equilibrium and thus do indeed have the same temperature.

Another statement of the zeroth law, more formal and more fundamental, is the following:

There exists a scalar quantity called temperature, which is a property of all thermodynamic systems in equilibrium. Two systems are in thermal equilibrium if and only if their temperatures are equal.

The zeroth law thus defines the concept of temperature and specifies it as the one macroscopic property of a system that will be equal to that of another system when they are in thermal equilibrium. The zeroth law permits us to build and use thermometers to measure the temperature of a system, for we now know that a thermometer in thermal contact with a system will reach a common temperature with the system.

22-3 MEASURING TEMPERATURE

In Chapter 1 we described a two-step procedure for establishing a measuring standard for a physical quantity: we define a base unit, and we then specify a procedure for making comparisons with the base unit. For instance, in the case of time, we defined the base unit in terms of the frequency of light of a certain wavelength emitted by cesium atoms. To make 1 second takes 9,192,631.770 of those vibrations. We can (at least in principle) use this scale to measure a human lifetime or even the age of the universe by counting the corresponding number of vibrations.

Temperature is one of the seven base units (see Table 1 of Chapter 1), and we might therefore attempt to treat temperature as we treated other base units in the SI system: establish a standard and relate all other scales to the standard. However, temperature has a nature different from that of other SI base units, and so this scheme will not work in quite that simple a form. For instance, if we define one period of vibration of the light emitted by a cesium atom as a standard of time, then two such vibrations last for twice the time, and any arbitrary time interval can in effect be measured in terms of the number of vibrations. But even if we define a standard of temperature, such as that of water boiling under certain conditions, we have no procedure to determine a temperature twice as large. Two pots of boiling water, after all, have the same temperature as one pot. There is no apparent way using only this standard that we can relate the temperature of boiling water to that, for example, of boiling oil; no amount of boiling water will ever be in thermal equilibrium with boiling oil.

To establish a measuring scale for temperature we adopt the following procedure, which differs from the usual procedure for the SI base units: we find a substance that has a property that varies with temperature, and we measure that property. The substance we choose is called the *thermometric substance,* and the property that depends on temperature is called the *thermometric property.*

Examples might be the volume of a liquid (as in the common glass-bulb mercury thermometer), the pressure of a gas kept at constant volume, the electrical resistance of a wire, the length of a strip of metal, or the color of a lamp filament, all of which vary with temperature. *The choice of one of these substances leads to an individual temperature scale that is defined only for that substance and that does not necessarily agree with other independently defined temperature scales.* Removing this disagreement requires the adoption of standards for the choice of a particular thermometric substance, a particular thermometric property, and a particular relationship between that property and a universally accepted temperature scale. Each individual temperature scale can thus be calibrated against the universal scale. We describe the accepted universal scale in Sections 22-4 and 26-5.

Let us assume that our particular thermometer is based on a system in which we measure the value of the thermometric property X. The temperature T is some function of X, $T(X)$. We choose the simplest possible relationship between T and X, the linear function given by

$$T(X) = aX + b, \qquad (1)$$

where the constants a and b must be determined. This linear scale means that every interval of temperature ΔT corresponds to the same change ΔX in the value of the thermometric property. To determine a temperature on this scale, we choose two calibration points, arbitrarily define the temperatures T_1 and T_2 at those points, and measure the corresponding values X_1 and X_2 of the thermometric property.

The most familiar examples of this type of scale are the Celsius and Fahrenheit scales used in common thermometers, in which the thermometric substance may be mercury and the thermometric property may be its volume, observed by means of the length of the mercury column in a thin glass tube. The linear behavior in this case means that the intervals between degree markings on the glass tube of a thermometer are of uniform size.

The Celsius and Fahrenheit Scales*

In nearly all countries of the world, the Celsius scale (formerly called the centigrade scale) is used for all popular and commercial and most scientific measurements. The Celsius scale was originally based on two calibration

points: the normal freezing point of water, defined to be 0°C, and the normal boiling point of water, defined to be 100°C. These two points were used to calibrate thermometers, and other temperatures were then deduced by interpolation or extrapolation. Note that the degree symbol (°) is used to express temperatures on the Celsius scale.

The Fahrenheit scale, used in the United States, employs a smaller degree than the Celsius scale, and its zero is set to a different temperature. It was also originally based on two fixed points, the interval between which was set to 100 degrees: the freezing point of a mixture of ice and salt, and the normal human body temperature. On this scale, the normal freezing and boiling points of water turn out to be, respectively, 32°F and 212°F. The relation between the Celsius and the Fahrenheit scales is

$$T_F = \tfrac{9}{5}T_C + 32. \qquad (2)$$

The degree symbol is used in expressing temperatures on the Fahrenheit scale, as, for example, 98.6°F (normal human body temperature).

Transferring between the Fahrenheit and Celsius scales is easily done by remembering a few corresponding points, such as the normal freezing point (0°C = 32°F) and boiling point (100°C = 212°F) of water, and by making use of the equality between an *interval* of 9 degrees on the Fahrenheit scale and an *interval* of 5 degrees on the Celsius scale, which we express as

$$9F° = 5C°. \qquad (3)$$

Note that these *intervals* are expressed as F° and C°, *not* as °F and °C. Readings on the temperature scale are given in °F or °C (degrees Fahrenheit or degrees Celsius); *differences* in readings are given in F° or C° (Fahrenheit degrees or Celsius degrees).

The Kelvin Scale*

On the Kelvin scale, one of the calibration points is defined to be at a temperature of zero, where the thermometric property also has a value of zero; in effect, the constant b in Eq. 1 is set to zero, in which case

$$T(X) = aX. \qquad (4)$$

To determine a temperature on this scale, we need only one calibration point P. At that point, the temperature is defined to be T_P and the thermometric property has the measured value X_P. In this case

---

* Anders Celsius (1701–1744) was a Swedish astronomer who, in addition to developing the temperature scale named for him, made measurements of the length of the arc of a meridian that verified Newton's theory of the flattening of the Earth at the poles. Daniel Fahrenheit (1686–1736), a contemporary of Celsius, was a German physicist who invented both the alcohol and mercury liquid thermometers and used them to study the boiling and freezing points of liquids.

---

* Lord Kelvin (William Thomson, 1824–1907) was a Scottish physicist and engineer who made fundamental contributions to a wide variety of subjects, including not only thermodynamics but the law of conservation of energy, electricity and magnetism, acoustics, and hydrodynamics. His scientific contributions were so highly regarded that he was accorded the honor of burial in Westminster Abbey in London.

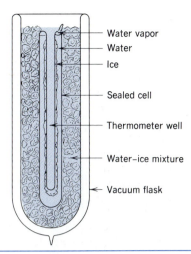

Figure 2 The National Institute of Standards and Technology (formerly National Bureau of Standards) triple-point cell. The U-shaped inner cell contains pure water and is sealed after all the air has been removed. It is immersed in a water–ice bath. The system is at the triple point when ice, water, and water vapor are all present, and in equilibrium, inside the cell. The thermometer to be calibrated is inserted into the central well.

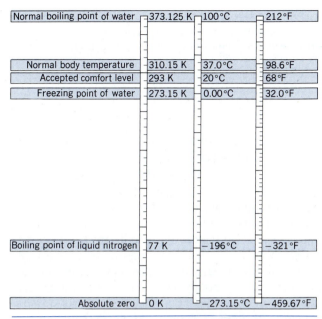

Figure 3 The Kelvin, Celsius, and Fahrenheit temperature scales compared.

$$a = \frac{T_P}{X_P} \qquad (5)$$

and so

$$T(X) = T_P \frac{X}{X_P}. \qquad (6)$$

By general agreement, we choose for our calibration the temperature at which ice, liquid water, and water vapor coexist in equilibrium. This point, which is very close to the normal freezing point of water, is called the *triple point* of water (Fig. 2). The temperature at the triple point has been set by international agreement to be

$$T_{tr} = 273.16 \text{ K},$$

where K (= kelvin) is the SI base unit of temperature on the absolute scale, which is identical with the ideal gas temperature scale discussed in the next section. The kelvin is thus defined as 1/273.16 of the temperature of the triple point of water. With this choice of calibration point, Eq. 6 becomes

$$T(X) = (273.16 \text{ K}) \frac{X}{X_{tr}}, \qquad (7)$$

where X_{tr} is the value of the thermometric property at the triple point.

A temperature determined from Eq. 7 is valid only for that particular thermometric property; other thermometric properties and thermometric substances might give different temperature readings (see Sample Problem 1). To eliminate this confusion between the readings of different thermometers, we choose as an accepted standard one type of thermometer in which the temperature

can be determined independently of the nature of the thermometric substance. This choice is discussed in the next section.

The size of the degree is the same on the Celsius and the Kelvin scales, but the zero of the Celsius scale is shifted to a more convenient value. Today we no longer use two fixed points to define the Celsius scale; instead, the Kelvin scale is defined, and the relationship between the Celsius temperature T_C and the Kelvin temperature T is now set as

$$T_C = T - 273.15. \qquad (8)$$

The freezing and boiling points of water are now measured on the Kelvin scale and then converted to Celsius using Eq. 8. The experimental values are, respectively, $0.00°C$ and $99.975°C$. Figure 3 compares the Fahrenheit, Celsius, and Kelvin scales.

Sample Problem 1 The resistance of a certain platinum wire increases by a factor of 1.392 between the triple point of water and the normal boiling point of water. Find the platinum resistance temperature of boiling water.

Solution We use Eq. 7, with the resistance R as the thermometric property X. We are not given the value of R_{tr}, but we do know that at the boiling point of water, $R = 1.392R_{tr}$. Thus

$$T(R) = T_{tr} \frac{R}{R_{tr}} = (273.16 \text{ K})(1.392) = 380.2 \text{ K}.$$

This value gives the "platinum resistance temperature" of boiling water. Other thermometers will give different values; for example, the temperature of boiling water according to a copper–constantan thermocouple is 440 K. Each of these readings is a temperature determined on a "private" scale, valid only for that device. The accepted temperature of the normal boiling point of water is 373.125 K, which is determined using the constant-volume gas thermometer described in the next section.

22-4 THE IDEAL GAS TEMPERATURE SCALE

The temperature of a system should have a well-defined value, independent of the particular means used to measure it. According to Eq. 7, different thermometric substances all give the same temperature at the triple point, but (as we have seen in Sample Problem 1) their readings at other points may differ. We might imagine doing a series of measurements in which we simultaneously use different thermometric properties to determine the temperature of a system. Results of such a test would show that the thermometers all give different readings. We might continue by choosing a particular thermometric property, such as the resistance of a wire, and measuring the temperature of the system using different kinds of wires, made of differing materials. Again we would find a wide variation in the measurements.

To obtain a definite temperature scale, we must select one particular kind of thermometer as the standard. The choice will be made, not on the basis of experimental convenience, but by inquiring whether the temperature scale defined by a particular thermometer proves to be useful in formulating the laws of physics. The smallest variation in readings is found among *constant-volume gas thermometers* using different gases, which suggests that we choose a gas as the standard thermometric substance. It turns out that as we reduce the amount of gas and therefore its pressure, the variation in readings between gas thermometers using different kinds of gas is reduced also. Hence there seems to be something fundamental about the behavior of a constant-volume thermometer containing a gas at low pressure. Let us therefore consider the properties of the constant-volume gas thermometer.

If the volume of a gas is kept constant, its pressure depends on the temperature and increases linearly with rising temperature. The constant-volume gas thermometer uses the pressure of a gas at constant volume as the thermometric property.

Figure 4 shows a diagram of the thermometer. It consists of a bulb of glass, porcelain, quartz, platinum, or platinum–iridium (depending on the temperature range over which it is to be used), connected by a capillary tube to a mercury manometer. The bulb *B* containing some gas is put into the bath or environment whose temperature *T* is to be measured; by raising or lowering the mercury reservoir *R*, the mercury in the left branch of the U-tube

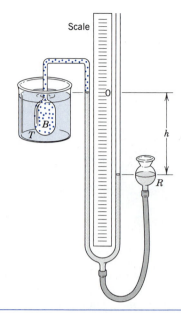

Figure 4 A constant-volume gas thermometer. The bulb *B* is immersed in a bath whose temperature *T* is to be measured. The difference between the pressure of the gas in the bulb and atmospheric pressure is determined by the height *h* of the column of mercury.

can be made to coincide with a fixed reference mark, thus keeping the confined gas at a constant volume. The difference between the pressure *p* of the confined gas on the left branch of the tube and the pressure p_0 of the atmosphere on the right branch of the tube is indicated by the height *h* of the column of mercury, and thus

$$p = p_0 - \rho g h, \qquad (9)$$

where ρ is the density of the mercury in the manometer.

In practice the apparatus is very elaborate, and we must make many corrections, for example, (1) to allow for the small volume change owing to slight contraction or expansion of the bulb and (2) to allow for the fact that not all the confined gas (such as that in the capillary) has been immersed in the bath. Let us assume that all corrections have been made, and that *p* is the corrected value of the absolute pressure at the temperature of the bath. Then the temperature is given provisionally by

$$T(p) = (273.16 \text{ K}) \frac{p}{p_{tr}} \quad \text{(constant } V). \qquad (10)$$

Let a certain amount of gas, nitrogen, for instance, be put into the bulb so that when the bulb is surrounded by water at the triple point the pressure p_{tr} is equal to a definite value, say, 80 cm Hg. Now we immerse the bulb in the system whose temperature *T* we wish to measure, and with the volume kept constant at its previous value, we measure the gas pressure *p*, as in Eq. 9, and calculate the provisional temperature *T* of the system using Eq. 10. The

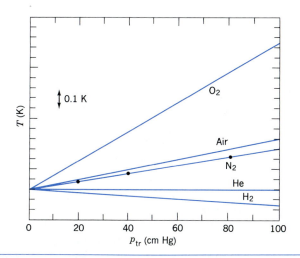

Figure 5   As the pressure of the nitrogen gas in a constant-volume gas thermometer is reduced from 80 cm Hg to 40 and then to 20, the temperature deduced for the system approaches a limit corresponding to a pressure of 0. Other gases approach the same limit, which is the ideal gas temperature T of the system. The full range of the vertical scale is about 1 K for typical conditions.

TABLE 1   TEMPERATURES OF SELECTED SYSTEMS

System	Temperature (K)
Plasma in fusion test reactor	10^8
Center of Sun	10^7
Surface of Sun	6×10^3
Melting point of tungsten	3.6×10^3
Freezing point of water	2.7×10^2
Normal boiling point of N_2	77
Normal boiling point of ^{4}He	4.2
Mean temperature of universe	2.7
^{3}He–^{4}He dilution refrigerator	5×10^{-3}
Adiabatic demagnetization of paramagnetic salt	10^{-3}
Nuclear spin cooling	2×10^{-8}

result of this measurement is plotted as a point in Fig. 5. Now we return the thermometer to the triple-point cell and remove some of the gas, so that p_{tr} has a smaller value, say, 40 cm Hg. Then we return the thermometer to the unknown system, measure the new value of p, and calculate another provisional temperature T, also plotted in Fig. 5. We continue this same procedure, reducing the amount of gas in the bulb and at each new lower value of p_{tr} calculating the temperature T. If we plot the values of T against p_{tr}, we can extrapolate the resulting curve to the intersection with the axis where $p_{tr} = 0$. The data points for N_2 and the resulting straight-line extrapolation are shown in Fig. 5.

We repeat this procedure with gases other than nitrogen in the thermometer, obtaining results shown in Fig. 5. The lines show that the temperature readings of a constant-volume gas thermometer depend on the gas used at ordinary values of the reference pressure. However, as the reference pressure is decreased, the temperature readings of constant-volume gas thermometers using different gases approach the same value T, which we can regard as the temperature of the system. *The extrapolated value of the temperature depends only on the general properties of gases and not on any particular gas.* We therefore define the *ideal gas temperature scale:*

$$T = (273.16 \text{ K}) \lim_{p_{tr} \to 0} \frac{p}{p_{tr}} \quad \text{(constant } V\text{)}. \quad (11)$$

Our standard thermometer is chosen to be a constant-volume gas thermometer using a temperature scale defined by Eq. 11.

If temperature is to be a truly fundamental physical quantity, one in which the laws of thermodynamics may be expressed, it is absolutely necessary that its definition be independent of the properties of specific materials. It would not do, for example, to have such a basic quantity as temperature depend on the expansivity of mercury, the electrical resistivity of platinum, or any other such "handbook" property. We chose the gas thermometer as our standard instrument precisely because no such specific properties of materials are involved in its operation. You can use any gas and you always get the same answer.

Although our temperature scale is independent of the properties of any one particular gas, it does depend on the properties of gases in general (that is, on the properties of a so-called ideal gas). The lowest temperature that can be measured with a gas thermometer is about 1 K. To obtain this temperature we must use low-pressure helium, which remains a gas at lower temperatures than any other gas. We cannot give experimental meaning to temperatures below about 1 K by means of a gas thermometer.

We would like to define a temperature scale in a way that is independent of the properties of any particular substance. We show in Section 26-5 that the absolute thermodynamic temperature scale, called the Kelvin scale, is such a scale. We also show that the ideal gas scale and the Kelvin scale are identical in the range of temperatures in which a gas thermometer may be used. For this reason we can use units of kelvins for the ideal gas temperature, as we have already done in Eq. 11. Table 1 lists the temperatures in kelvins of various bodies and processes.

We also show in Section 26-5 that the Kelvin scale has an *absolute zero* of 0 K and that it is impossible to cool a system below 0 K. The absolute zero of temperature has defied all attempts to reach it experimentally, but temperatures within a small range (10^{-8} K) of absolute zero have been achieved.

Although there is a direct connection, as we show in Chapter 23, between the microscopic motion of molecules and the macroscopic temperature, it is *not* true that all molecular motion ceases at the absolute zero of temperature. The connection between temperature and molecular kinetic energy is based on classical concepts, while the quantum theory tells us that there is a nonzero lower limit to the molecular kinetic energy, even at absolute zero. This *zero-point energy* cannot be inferred from classical calculations.

The International Temperature Scale

Precise measurement of a temperature with a gas thermometer is a difficult task, requiring many months of painstaking laboratory work. In practice, the gas thermometer is used only to establish certain fixed points that can then be used to calibrate other more convenient secondary thermometers.

For practical use, as in the calibration of industrial or scientific thermometers, the International Temperature Scale has been adopted. This scale consists of a set of procedures for providing in practice the best possible approximations to the Kelvin scale. The adopted scale consists of a set of fixed points, along with the instruments to be used for interpolating between these fixed points and for extrapolating beyond the highest fixed point. A new scale has been adopted by the International Committee of Weights and Measures about every 20 years; the fixed points of the most recent one (1990) are shown in Table 2.

TABLE 2 PRIMARY FIXED POINTS ON THE 1990 INTERNATIONAL TEMPERATURE SCALE[a]

Substance	State	Temperature (K)
Helium	Boiling point	$3 - 5^c$
Hydrogen	Triple point	13.8033
Hydrogen	Boiling point[b]	$17.025 - 17.045^c$
Hydrogen	Boiling point	$20.26 - 20.28^c$
Neon	Triple point	24.5561
Oxygen	Triple point	54.3584
Argon	Triple point	83.8058
Mercury	Triple point	234.3156
Water	Triple point	273.16
Gallium	Melting point	302.9146
Indium	Freezing point	429.7485
Tin	Freezing point	505.078
Zinc	Freezing point	692.677
Aluminum	Freezing point	933.473
Silver	Freezing point	1234.93
Gold	Freezing point	1337.33
Copper	Freezing point	1357.77

[a] See "The International Temperature Scale of 1990 (ITS-90)," by H. Preston-Thomas, *Metrologia,* 27 (1990), p. 3.

[b] This boiling point is for a pressure of $\frac{1}{3}$ atm. All other boiling points, melting points, or freezing points are for a pressure of 1 atm.

[c] The temperature of the boiling point varies somewhat with the pressure of the gas above the liquid. The temperature scale gives the relationship between T and p that can be used to calculate T for a given p.

22-5 THERMAL EXPANSION

You can often loosen a tight metal jar lid by holding it under a stream of hot water. As its temperature rises, the metal lid expands slightly relative to the glass jar. Thermal expansion is not always desirable, as Fig. 6 suggests. We have all seen expansion slots in the roadways of bridges. Pipes at refineries often include an expansion loop, so that the pipe will not buckle as the temperature rises. Materials used for dental fillings have expansion properties similar to those of tooth enamel. In aircraft manufacture, rivets and other fasteners are often designed so that they are to be cooled in dry ice before insertion and then allowed to expand to a tight fit. Thermometers and thermostats may be based on the differences in expansion between the components of a bimetallic strip; see Fig. 7. In a thermometer of a familiar type, the bimetallic strip is coiled into a helix that winds and unwinds as the temperature changes; see Fig. 8. The familiar liquid-in-glass thermometers are based on the fact that liquids such as mercury or alcohol expand to a different (greater) extent than do their glass containers.

We can understand this expansion by considering a simple model of the structure of a crystalline solid. The atoms are held together in a regular array by electrical forces, which are like those that would be exerted by a set of springs connecting the atoms. We can thus visualize the solid body as a microscopic bedspring (Fig. 9). These "springs" are quite stiff and not at all ideal (see Problem 3

Figure 6 Railroad tracks distorted because of thermal expansion on a very hot day. Expansion joints between sections of track can prevent this distortion.

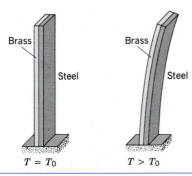

Figure 7 A bimetallic strip, consisting of a strip of brass and a strip of steel welded together, at temperature T_0. At temperatures higher than T_0, the strip bends as shown; at lower temperatures it bends the other way. Many thermostats operate on this principle, using the motion of the end of the strip to make or break an electrical contact.

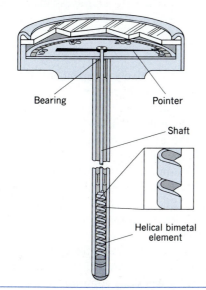

Figure 8 A thermometer based on a bimetallic strip. The strip is formed into a helix, which coils or uncoils as the temperature is changed.

of Chapter 15), and there are about 10^{23} of them per cubic centimeter. At any temperature the atoms of the solid are vibrating. The amplitude of vibration is about 10^{-9} cm, about one-tenth of an atomic diameter, and the frequency is about 10^{13} Hz.

When the temperature is increased, the atoms vibrate at larger amplitude, and the average distance between atoms increases. (See the discussion of the microscopic basis of thermal expansion at the end of this section.) This leads to an expansion of the whole solid body. The change in *any* linear dimension of the solid, such as its length, width, or thickness, is called a *linear expansion*. If the length of this linear dimension is L, the change in temperature ΔT causes a change in length ΔL. We find from experiment that, if ΔT is small enough, this change in length ΔL is

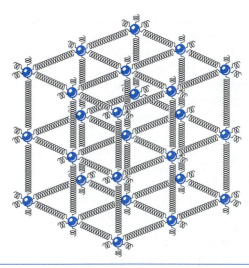

Figure 9 A solid behaves in many ways as if it were a collection of atoms joined by elastic forces (here represented by springs).

proportional to the temperature change ΔT and to the original length L. Hence we can write

$$\Delta L = \alpha L \, \Delta T, \tag{12}$$

where α, called the *coefficient of linear expansion,* has different values for different materials. Rewriting this formula, we obtain

$$\alpha = \frac{\Delta L/L}{\Delta T}, \tag{13}$$

so that α has the meaning of a fractional change in length per degree temperature change.

Strictly speaking, the value of α depends on the actual temperature and the reference temperature chosen to determine L (see Problem 23). However, its variation is usually negligible compared to the accuracy with which measurements need to be made. It is often sufficient to choose an average value that can be treated as a constant over a certain temperature range. In Table 3 we list the

TABLE 3   SOME AVERAGE COEFFICIENTS OF LINEAR EXPANSION[a]

Substance	$\alpha(10^{-6}$ per C°)
Ice	51
Lead	29
Aluminum	23
Brass	19
Copper	17
Steel	11
Glass (ordinary)	9
Glass (Pyrex)	3.2
Invar alloy	0.7
Quartz (fused)	0.5

[a] Typical average values in the temperature range 0°C to 100°C are shown, except for ice in which the range is −10°C to 0°C.

experimental values for the average coefficient of linear expansion of several common solids. For all the substances listed, the change in size consists of an expansion as the temperature rises, because α is positive. The order of magnitude of the expansion is about 1 millimeter per meter length per 100 Celsius degrees. (Note the use of C°, not °C, to express temperature changes here.)

Sample Problem 2   A steel metric scale is to be ruled so that the millimeter intervals are accurate to within about 5×10^{-5} mm at a certain temperature. What is the maximum temperature variation allowable during the ruling?

Solution   From Eq. 12, we have

$$\Delta T = \frac{\Delta L}{\alpha L} = \frac{5 \times 10^{-5} \text{ mm}}{(11 \times 10^{-6}/\text{C}°)(1.0 \text{ mm})} = 4.5 \text{ C}°,$$

where we have used the value of α for steel from Table 3. The temperature during the ruling must be kept constant to within about 5 C°, and the scale must be used within that same interval of temperature at which it was made.

Note that if the alloy invar were used instead of steel, we could achieve the same precision over a temperature interval of about 75 C°; or, equivalently, if we could maintain the same temperature variation (5 C°), we could achieve an accuracy due to temperature changes of about 3×10^{-6} mm.

For many solids, called *isotropic,* the percent change in length for a given temperature change is the same for all lines in the solid. The expansion is quite analogous to a photographic enlargement, except that a solid is three-dimensional. Thus, if you have a flat plate with a hole punched in it, $\Delta L/L$ $(=\alpha \Delta T)$ for a given ΔT is the same for the length, the thickness, the face diagonal, the body diagonal, and the hole diameter. Every line, whether straight or curved, lengthens in the ratio α per degree temperature rise. If you scratch your name on the plate, the line representing your name has the same fractional change in length as any other line. The analogy to a photographic enlargement is shown in Fig. 10.

With these ideas in mind, you should be able to show (see Problems 30 and 31) that to a high degree of accuracy the fractional change in area A per degree temperature change for an isotropic solid is 2α, that is,

$$\Delta A = 2\alpha A \, \Delta T, \qquad (14)$$

and the fractional change in volume V per degree temperature change for an isotropic solid is 3α, that is,

$$\Delta V = 3\alpha V \, \Delta T. \qquad (15)$$

Because the shape of a fluid is not definite, only the change in volume with temperature is significant. Gases respond strongly to temperature or pressure changes, whereas the change in volume of liquids with changes in

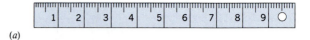

(a)

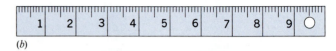

(b)

Figure 10   A steel rule at two different temperatures. The expansion increases in proportion in all dimensions: the scale, the numbers, the hole, and the thickness are all increased by the same factor. (The expansion shown is greatly exaggerated; to obtain such an expansion would require a temperature increase of about 20,000 C°!)

temperature or pressure is very much smaller. If we let β represent the coefficient of volume expansion for a liquid, so that

$$\beta = \frac{\Delta V/V}{\Delta T}, \qquad (16)$$

we find that β is relatively independent of the temperature. Liquids typically expand with increasing temperature, their volume expansion being generally about 10 times greater than that of solids.

However, the most common liquid, water, does not behave like most other liquids. In Fig. 11 we show the volume expansion curve for water. Note that above 4°C water expands as the temperature rises, although not linearly. (That is, β is not constant over these large temperature intervals.) As the temperature is lowered from 4°C to 0°C, however, water expands instead of contracting, which is the reason that lakes freeze first at their upper surface. Such an expansion with decreasing temperature is not observed in any other common liquid; it is observed in rubberlike substances and in certain crystalline solids over limited temperature intervals. The density of water is a maximum at 3.98°C, where its value is 999.973 kg/m³. (The standard kilogram and meter were originally supposed to correspond to a maximum density for water of 1000 kg/m³ or 1 g/cm³. Accurate measurements show, however, that the international standards do not correspond exactly to this value.)

Microscopic Basis of Thermal Expansion   *(Optional)*
On the microscopic level, thermal expansion of a solid suggests an increase in the average separation between the atoms in the solid. The potential energy curve for two adjacent atoms in a crystalline solid as a function of their internuclear separation is an asymmetric curve like that of Fig. 12. As the atoms move close together, their separation decreasing from the equilibrium value r_0, strong repulsive forces come into play, and the potential energy rises steeply $(F = -dU/dr)$; as the atoms move farther apart, their separation increasing from the equilibrium value, somewhat weaker attractive forces take over and the potential energy rises more slowly. At a given vibrational energy the separation of the atoms changes periodically from a minimum to a

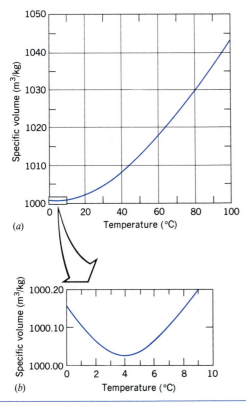

(a)

(b)

Figure 11 (*a*) The specific volume (the volume occupied by a particular mass) of water as a function of its temperature. The specific volume is the inverse of the density (the mass per unit volume). (*b*) An enlargement of the region near 4°C, showing a minimum specific volume (or a maximum density).

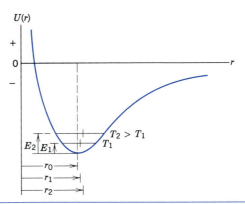

Figure 12 Potential energy curve for two adjacent atoms in a solid as a function of their internuclear separation distance. The equilibrium separation is r_0. Because the curve is asymmetric, the average separation (r_1, r_2) increases as the temperature (T_1, T_2) and the vibrational energy (E_1, E_2) increase.

maximum value, the average separation being greater than the equilibrium separation because of the asymmetric nature of the potential energy curve. At still higher vibrational energy the average separation is even greater. The effect is enhanced because, as suggested by Fig. 12, the kinetic energy is smaller at larger separations; thus the particles move slower and spend more time at large separations, which then contribute a larger share to the time average. Because the vibrational energy in-

creases as the temperature rises, the average separation between atoms increases with temperature, and the entire solid expands.

Note that if the potential energy curve were symmetric about the equilibrium separation, then the average separation would equal the equilibrium separation, no matter how large the amplitude of the vibration. Hence thermal expansion is a direct consequence of the deviation from symmetry of the characteristic potential energy curve of solids.

Some crystalline solids, in certain temperature regions, may contract as the temperature rises. The above analysis remains valid if one assumes that only compressional (longitudinal) modes of vibration exist or that these modes predominate. However, solids may vibrate in shearlike (transverse) modes as well, and these modes of vibration allow the solid to contract as the temperature rises, the average separation of the planes of atoms decreasing. For certain types of crystalline structure and in certain temperature regions, these transverse modes of vibration may predominate over the longitudinal ones, giving a net negative coefficient of thermal expansion.

It should be emphasized that the microscopic models presented here are oversimplifications of a complex phenomenon that can be treated with greater insight using statistical mechanics and quantum theory. ∎

QUESTIONS

1. Is temperature a microscopic or macroscopic concept?

2. Can we define temperature as a derived quantity, in terms of length, mass, and time? Think of a pendulum, for example.

3. Absolute zero is a minimum temperature. Is there a maximum temperature?

4. Can one object be hotter than another if they are at the same temperature? Explain.

5. Are there physical quantities other than temperature that tend to equalize if two different systems are joined?

6. A piece of ice and a warmer thermometer are suspended in an insulated evacuated enclosure so that they are not in

contact. Why does the thermometer reading decrease for a time?

7. What qualities make a particular thermometric property suitable for use in a practical thermometer?

8. What difficulties would arise if you defined temperature in terms of the density of water?

9. Let p_3 be the pressure in the bulb of a constant-volume gas thermometer when the bulb is at the triple-point temperature of 273.16 K and let p be the pressure when the bulb is at room temperature. Given three constant-volume gas thermometers: for A the gas is oxygen and $p_3 = 20$ cm Hg; for B

the gas is also oxygen but $p_3 = 40$ cm Hg; for C the gas is hydrogen and $p_3 = 30$ cm Hg. The measured values of p for the three thermometers are p_A, p_B, and p_C. (a) An approximate value of the room temperature T can be obtained with each of the thermometers using

$$T_A = (273.16 \text{ K})(p_A/20 \text{ cm Hg}),$$

$$T_B = (273.16 \text{ K})(p_B/40 \text{ cm Hg}),$$

$$T_C = (273.16 \text{ K})(p_C/30 \text{ cm Hg}).$$

Mark each of the following statements true or false: (1) With the method described, all three thermometers will give the same value of T. (2) The two oxygen thermometers will agree with each other but not with the hydrogen thermometer. (3) Each of the three will give a different value of T. (b) In the event that there is a disagreement among the three thermometers, explain how you would change the method of using them to cause all three to give the same value of T.

10. The editor-in-chief of a well-known business magazine, discussing possible warming effects associated with the increasing concentration of carbon dioxide in the Earth's atmosphere (greenhouse effect), wrote: "The polar regions might be three times warmer than now, ..." What do you suppose he meant, and what did he say literally? (See "Warmth and Temperature: A Comedy of Errors," by Albert A. Bartlett, *The Physics Teacher*, November 1984, p. 517.)

11. Although the absolute zero of temperature seems to be experimentally unattainable, temperatures as low as 0.00000002 K have been achieved in the laboratory. Why would physicists strive, as indeed they do, to obtain still lower temperatures? Isn't this low enough for all practical purposes?

12. You put two uncovered pails of water, one containing hot water and one containing cold water, outside in below-freezing weather. The pail with the hot water will usually begin to freeze first. Why? What would happen if you covered the pails?

13. Can a temperature be assigned to a vacuum?

14. Does our "temperature sense" have a built-in sense of direction; that is, does hotter necessarily mean higher temperature, or is this just an arbitrary convention? Celsius, by the way, originally chose the steam point as 0°C and the ice point as 100°C.

15. Many medicine labels inform the user to store below 86°F. Why 86? (*Hint:* Change to Celsius.) (See *The Science Almanac*, 1985–1986, p. 430.)

16. How would you suggest measuring the temperature of (a) the Sun, (b) the Earth's upper atmosphere, (c) an insect, (d) the Moon, (e) the ocean floor, and (f) liquid helium?

17. Considering the Celsius, Fahrenheit, and Kelvin scales, does any one stand out as "nature's scale"? Discuss.

18. Is one gas any better than another for purposes of a standard constant–volume gas thermometer? What properties are desirable in a gas for such purposes?

19. State some objections to using water-in-glass as a thermometer. Is mercury-in-glass an improvement? If so, explain why.

20. Explain why the column of mercury first descends and then rises when a mercury-in-glass thermometer is put in a flame.

21. What are the dimensions of α, the coefficient of linear expansion? Does the value of α depend on the unit of length used? When Fahrenheit degrees are used instead of Celsius degrees as the unit of temperature change, does the numerical value of α change? If so, how? If not, prove it.

22. A metal ball can pass through a metal ring. When the ball is heated, however, it gets stuck in the ring. What would happen if the ring, rather than the ball, were heated?

23. A bimetallic strip, consisting of two different metal strips riveted together, is used as a control element in the common thermostat. Explain how it works.

24. Two strips, one of iron and one of zinc, are riveted together side by side to form a straight bar that curves when heated. Why is the iron on the inside of the curve?

25. Explain how the period of a pendulum clock can be kept constant with temperature by attaching vertical tubes of mercury to the bottom of the pendulum.

26. Why should a chimney be freestanding, that is, not part of the structural support of the house?

27. Water expands when it freezes. Can we define a coefficient of volume expansion for the freezing process?

28. Explain why the apparent expansion of a liquid in a glass bulb does not give the true expansion of the liquid.

29. Does the change in volume of an object when its temperature is raised depend on whether the object has cavities inside, other things being equal?

30. Why is it much more difficult to make a precise determination of the coefficient of expansion of a liquid than of a solid?

31. A common model of a solid assumes the atoms to be points executing simple harmonic motion about mean lattice positions. What would be the coefficient of linear expansion of such a lattice?

32. Explain the fact that the temperature of the ocean at great depths is very constant the year round, at a temperature of about 4°C.

33. Explain why lakes freeze first at the surface.

34. What causes water pipes to burst in the winter?

35. What can you conclude about how the melting point of ice depends on pressure from the fact that ice floats on water?

PROBLEMS

Section 22-3 Measuring Temperature

1. A *resistance thermometer* is a thermometer in which the electrical resistance changes with temperature. We are free to define temperatures measured by such a thermometer in kelvins (K) to be directly proportional to the resistance R,

measured in ohms (Ω). A certain resistance thermometer is found to have a resistance R of 90.35 Ω when its bulb is placed in water at the triple-point temperature (273.16 K). What temperature is indicated by the thermometer if the bulb is placed in an environment such that its resistance is 96.28 Ω?

2. A thermocouple is formed from two different metals, joined at two points in such a way that a small voltage is produced when the two junctions are at different temperatures. In a particular iron–constantan thermocouple, with one junction held at 0°C, the output voltage varies linearly from 0 to 28.0 mV as the temperature of the other junction is raised from 0 to 510°C. Find the temperature of the variable junction when the thermocouple output is 10.2 mV.

3. The amplification or *gain* of a transistor amplifier may depend on the temperature. The gain for a certain amplifier at room temperature (20.0°C) is 30.0, whereas at 55.0°C it is 35.2. What would the gain be at 28.0°C if the gain depends linearly on temperature over this limited range?

4. Absolute zero is −273.15°C. Find absolute zero on the Fahrenheit scale.

5. If your doctor tells you that your temperature is 310 kelvins above absolute zero, should you worry? Explain your answer.

6. (*a*) The temperature of the surface of the Sun is about 6000 K. Express this on the Fahrenheit scale. (*b*) Express normal human body temperature, 98.6°F, on the Celsius scale. (*c*) In the continental United States, the lowest officially recorded temperature is −70°F at Rogers Pass, Montana. Express this on the Celsius scale. (*d*) Express the normal boiling point of oxygen, −183°C, on the Fahrenheit scale. (*e*) At what Celsius temperature would you find a room to be uncomfortably warm?

7. At what temperature, if any, do the following pairs of scales give the same reading: (*a*) Fahrenheit and Celsius, (*b*) Fahrenheit and Kelvin, and (*c*) Celsius and Kelvin?

8. At what temperature is the Fahrenheit scale reading equal to (*a*) twice that of the Celsius and (*b*) half that of the Celsius?

9. It is an everyday observation that hot and cold objects cool down or warm up to the temperature of their surroundings. If the temperature difference ΔT between an object and its surroundings ($\Delta T = T_{\text{obj}} - T_{\text{sur}}$) is not too great, the rate of cooling or warming of the object is proportional, approximately, to this temperature difference; that is,

$$\frac{d\,\Delta T}{dt} = -A(\Delta T),$$

where A is a constant. The minus sign appears because ΔT decreases with time if ΔT is positive and increases if ΔT is negative. This is known as *Newton's law of cooling*. (*a*) On what factors does A depend? What are its dimensions? (*b*) If at some instant $t = 0$ the temperature difference is ΔT_0, show that it is

$$\Delta T = \Delta T_0 e^{-At}$$

at a time t later.

10. Early in the morning the heater of a house breaks down. The outside temperature is −7.0°C. As a result, the inside temperature drops from 22 to 18°C in 45 min. How much longer will it take for the inside temperature to fall by an-

other 4.0 C°? Assume that the outside temperature does not change and that Newton's law of cooling applies; see the previous problem.

Section 22-4 The Ideal Gas Temperature Scale

11. If the gas temperature at the steam point is 373.15 K, what is the limiting value of the ratio of the pressures of a gas at the steam point and at the triple point of water when the gas is kept at constant volume?

12. A particular gas thermometer is constructed of two gas-containing bulbs, each of which is put into a water bath, as shown in Fig. 13. The pressure difference between the two bulbs is measured by a mercury manometer as shown in the figure. Appropriate reservoirs, not shown in the diagram, maintain constant gas volume in the two bulbs. There is no difference in pressure when both baths are at the triple point of water. The pressure difference is 120 mm Hg when one bath is at the triple point and the other is at the boiling point of water. Finally, the pressure difference is 90.0 mm Hg when one bath is at the triple point and the other is at an unknown temperature to be measured. Find the unknown temperature.

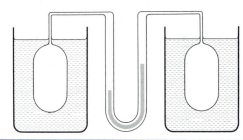

Figure 13 Problem 12.

13. Two constant-volume gas thermometers are assembled, one using nitrogen as the working gas and the other using helium. Both contain enough gas so that $p_{\text{tr}} = 100$ cm Hg. What is the difference between the pressures in the two thermometers if both are inserted into a water bath at the boiling point? Which pressure is the higher of the two? See Fig. 5.

Section 22-5 Thermal Expansion

14. An aluminum flagpole is 33 m high. By how much does its length increase as the temperature increases by 15 C°?

15. The Pyrex® glass mirror in the telescope at the Mount Palomar Observatory (the Hale telescope) has a diameter of 200 in. The most extreme temperatures ever recorded on Palomar Mountain are −10°C and 50°C. Determine the maximum change in the diameter of the mirror.

16. A circular hole in an aluminum plate is 2.725 cm in diameter at 12°C. What is its diameter when the temperature of the plate is raised to 140°C?

17. Steel railroad tracks are laid when the temperature is −5.0°C. A standard section of rail is then 12.0 m long. What gap should be left between rail sections so that there is no compression when the temperature gets as high as 42°C?

18. A glass window is 200 cm by 300 cm at 10°C. By how much has its area increased when its temperature is 40°C? Assume that the glass is free to expand.

19. A brass cube has an edge length of 33.2 cm at 20.0°C. Find (*a*) the increase in surface area and (*b*) the increase in volume when it is heated to 75.0°C.

20. What is the volume of a lead ball at −12°C if its volume at 160°C is 530 cm³?

21. Show that when the temperature of a liquid in a barometer changes by ΔT, and the pressure is constant, the height h changes by $\Delta h = \beta h \, \Delta T$, where β is the coefficient of volume expansion of the liquid. Neglect the expansion of the glass tube.

22. In a certain experiment, it was necessary to be able to move a small radioactive source at selected, extremely slow speeds. This was accomplished by fastening the source to one end of an aluminum rod and heating the central section of the rod in a controlled way. If the effective heated section of the rod in Fig. 14 is 1.8 cm, at what constant rate must the temperature of the rod be made to change if the source is to move at a constant speed of 96 nm/s?

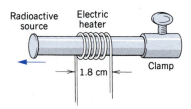

Figure 14   Problem 22.

23. Show that if α is dependent on the temperature T, then

$$L \cong L_0 \left[1 + \int_{T_0}^{T} \alpha(T) dT \right],$$

where L_0 is the length at a reference temperature T_0.

24. Soon after the Earth formed, heat released by the decay of radioactive elements raised the average internal temperature from 300 to 3000 K, at about which value it remains today. Assuming an average coefficient of volume expansion of 3.2×10^{-5} K⁻¹, by how much has the radius of the Earth increased since its formation?

25. A rod is measured to be 20.05 cm long using a steel ruler at a room temperature of 20°C. Both the rod and the ruler are placed in an oven at 270°C, where the rod now measures 20.11 cm using the same ruler. Calculate the coefficient of thermal expansion for the material of which the rod is made.

26. Consider a mercury-in-glass thermometer. Assume that the cross section of the capillary is constant at A, and that V is the volume of the bulb of mercury at 0.00°C. Suppose that the mercury just fills the bulb at 0.00°C. Show that the length L of the mercury column in the capillary at a temperature T, in °C, is

$$L = \frac{V}{A} (\beta - 3\alpha)T,$$

that is, proportional to the temperature, where β is the coeffi-

cient of volume expansion of mercury and α is the coefficient of linear expansion of glass.

27. (*a*) Show that if the lengths of two rods of different solids are inversely proportional to their respective coefficients of linear expansion at the same initial temperature, the difference in length between them will be constant at all temperatures. (*b*) What should be the lengths of a steel and a brass rod at 0°C so that at all temperatures their difference in length is 0.30 m?

28. As a result of a temperature rise of 32°C, a bar with a crack at its center buckles upward, as shown in Fig. 15. If the fixed distance $L_0 = 3.77$ m and the coefficient of linear expansion is 25×10^{-6}/C°, find x, the distance to which the center rises.

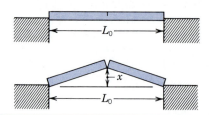

Figure 15   Problem 28.

29. A steel rod is 3.000 cm in diameter at 25°C. A brass ring has an interior diameter of 2.992 cm at 25°C. At what common temperature will the ring just slide onto the rod?

30. The area A of a rectangular plate is ab. Its coefficient of linear expansion is α. After a temperature rise ΔT, side a is longer by Δa and side b is longer by Δb. Show that if we neglect the small quantity $\Delta a \, \Delta b / ab$ (see Fig. 16), then $\Delta A = 2\alpha A \, \Delta T$, verifying Eq. 14.

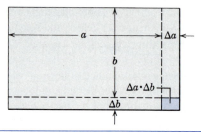

Figure 16   Problem 30.

31. Prove that, if we neglect extremely small quantities, the change in volume of a solid upon expansion through a temperature rise ΔT is given by $\Delta V = 3\alpha V \, \Delta T$, where α is the coefficient of linear expansion. See Eq. 15.

32. When the temperature of a copper penny (which is not pure copper) is raised by 100 C°, its diameter increases by 0.18%. Find the percent increase in (*a*) the area of a face, (*b*) the thickness, (*c*) the volume, and (*d*) the mass of the penny. (*e*) Calculate its coefficient of linear expansion.

33. Density is mass divided by volume. If the volume V is temperature dependent, so is the density ρ. Show that the change in density $\Delta \rho$ with change in temperature ΔT is given by

$$\Delta\rho = -\beta\rho\,\Delta T,$$

where β is the coefficient of volume expansion. Explain the minus sign.

34. When the temperature of a metal cylinder is raised from 60 to 100°C, its length increases by 0.092%. (a) Find the percent change in density. (b) Identify the metal.

35. At 100°C a glass flask is completely filled by 891 g of mercury. What mass of mercury is needed to fill the flask at -35°C? (The coefficient of linear expansion of glass is 9.0×10^{-6}/C°; the coefficient of volume expansion of mercury is 1.8×10^{-4}/C°.)

36. Figure 17 shows the variation of the coefficient of volume expansion of water between 4°C and 20°C. The density of water at 4°C is 1000 kg/m³. Calculate the density of water at 20°C.

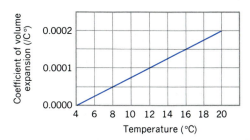

Figure 17 Problem 36.

37. A composite bar of length $L = L_1 + L_2$ is made from a bar of material 1 and length L_1 attached to a bar of material 2 and length L_2, as shown in Fig. 18. (a) Show that the effective coefficient of linear expansion α for this bar is given by $\alpha = (\alpha_1 L_1 + \alpha_2 L_2)/L$. (b) Using steel and brass, design such a composite bar whose length is 52.4 cm and whose effective coefficient of linear expansion is 13×10^{-6}/C°.

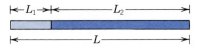

Figure 18 Problem 37.

38. (a) Prove that the change in rotational inertia I with temperature of a solid object is given by $\Delta I = 2\alpha I\,\Delta T$. (b) A thin uniform brass rod, spinning freely at 230 rev/s about an axis perpendicular to it at its center, is heated without mechanical contact until its temperature increases by 170 C°. Calculate the change in angular velocity.

39. A cylinder placed in frictionless bearings is set rotating about its axis. The cylinder is then heated, without mechanical contact, until its radius is increased by 0.18%. What is the percent change in the cylinder's (a) angular momentum, (b) angular velocity, and (c) rotational energy?

40. (a) Prove that the change in period P of a physical pendulum with temperature is given by $\Delta P = \frac{1}{2}\alpha P\,\Delta T$. (b) A clock pendulum made of invar has a period of 0.500 s and is accurate at 20°C. If the clock is used in a climate where the temperature averages 30°C, what approximate correction to the time given by the clock is necessary at the end of 30 days?

41. A pendulum clock with a pendulum made of brass is designed to keep accurate time at 20°C. How much will the error be, in seconds per hour, if the clock operates at 0°C?

42. An aluminum cup of 110 cm³ capacity is filled with glycerin at 22°C. How much glycerin, if any, will spill out of the cup if the temperature of the cup and glycerin is raised to 28°C? (The coefficient of volume expansion of glycerin is 5.1×10^{-4}/C°.)

43. A 1.28-m-long vertical glass tube is half-filled with a liquid at 20.0°C. How much will the height of the liquid column change when the tube is heated to 33.0°C? Assume that $\alpha_{glass} = 1.1 \times 10^{-5}$/C° and $\beta_{liquid} = 4.2 \times 10^{-5}$/C°.

44. A steel rod at 24°C is bolted securely at both ends and then cooled. At what temperature will it begin to yield? See Table 1, Chapter 14.

45. Three equal-length straight rods, of aluminum, invar, and steel, all at 20°C, form an equilateral triangle with hinge pins at the vertices. At what temperature will the angle opposite the invar rod be 59.95°? See Appendix H for needed trigonometric formulas.

46. Two rods of different materials but having the same lengths L and cross-sectional areas A are arranged end-to-end between fixed, rigid supports, as shown in Fig. 19a. The temperature is T and there is no initial stress. The rods are heated, so that their temperature increases by ΔT. (a) Show that the rod interface is displaced upon heating by an amount given by

$$\Delta L = \left(\frac{\alpha_1 E_1 - \alpha_2 E_2}{E_1 + E_2}\right) L\,\Delta T,$$

where α_1, α_2 are the coefficients of linear expansion and E_1, E_2 are Young's moduli of the materials. Ignore changes in cross-sectional areas; see Fig. 19b. (b) Find the stress at the interface after heating.

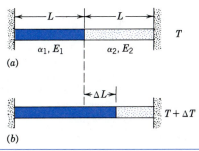

Figure 19 Problem 46.

47. An aluminum cube 20 cm on an edge floats on mercury. How much farther will the block sink when the temperature rises from 270 to 320 K? (The coefficient of volume expansion of mercury is 1.8×10^{-4}/C°.)

48. A glass tube nearly filled with mercury is attached in tandem to the bottom of an iron pendulum rod 100 cm long. How high must the mercury be in the glass tube so that the center of mass of this pendulum will not rise or fall with changes in temperature? (The cross-sectional area of the tube is equal

to that of the iron rod. Neglect the mass of the glass. Iron has a density of 7.87×10^3 kg/m³ and a coefficient of linear expansion equal to 12×10^{-6}/C°. The coefficient of volume expansion of mercury is 18×10^{-5}/C°.)

49. The distance between the towers of the main span of the Golden Gate Bridge near San Francisco is 4200 ft (Fig. 20). The sag of the cable halfway between the towers at 50°F is 470 ft. Take $\alpha = 6.5 \times 10^{-6}$/F° for the cable and compute (a) the change in length of the cable and (b) the change in sag for a temperature change from 10 to 90°F. Assume no bending or separation of the towers and a parabolic shape for the cable.

Figure 20   Problem 49.

CHAPTER 23

KINETIC THEORY AND THE IDEAL GAS

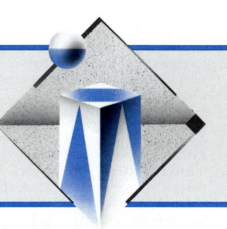

*The basic laws of thermodynamics deal with the relationships between
macroscopic properties, such as the pressure, temperature, volume, and internal
energy of an ideal gas. The laws say nothing about the fact that matter is made up of
particles (atoms or molecules). Owing to the large number of particles involved, it is not
practical to apply the laws of mechanics to find the motion of every particle in a gas.
Instead, we use averaging techniques to express the thermodynamic properties as averages
of molecular properties. If the number of particles is very large, such averages give sharply
defined quantities.*

In this chapter, we consider an approach to averaging called kinetic theory, *in which we
follow the motion of representative particles in a gas and then average this behavior over all
particles. Kinetic theory was developed in the 17th to 19th centuries by Boyle, D. Bernoulli,
Joule, Kronig, Clausius, and Maxwell, among others. Another approach to averaging is*
statistical mechanics, *in which laws of probability are applied to statistical distributions of
molecular properties. This approach is considered in Chapter 24.*

23-1 MACROSCOPIC PROPERTIES OF A GAS AND THE IDEAL GAS LAW

Figure 1 shows a gas confined to a cylinder fitted with a
movable piston. We wish to carry out a series of measurements of the *macroscopic* properties of the gas: the type
and amount of gas and its pressure, volume, and absolute
(Kelvin) temperature. We assume that we have attached
to the cylinder suitable devices for measuring these properties. We also assume that we have at our disposal the
means to change any of these properties. For example, we
suppose the gas to be in contact with an idealized device
called a *thermal reservoir,* which we can regard as a body
maintained at a temperature *T*, such that the temperature
of the reservoir does not change when our gas cylinder
comes into thermal equilibrium with it. We assume that
we can easily change the temperature of the reservoir,
thereby changing the temperature of the gas. If we wish to
change the pressure *p*, we add or subtract weight on the
piston. (The space above the piston is assumed to be evacuated, so that there is no air pressure pushing *down* on the
piston.) The volume *V* can be changed simply by chang-

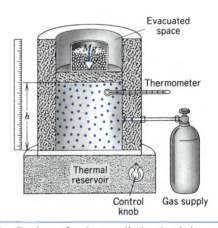

Figure 1 Gas is confined to a cylinder that is in contact with
a thermal reservoir at the (adjustable) temperature *T*. The piston exerts a total downward force *Mg* on the gas, which in
equilibrium is balanced by the upward force due to the gas
pressure. The volume of the gas can be determined from a
measurement of the height *h* of the piston above the bottom
of the cylinder, and the temperature of the gas is measured
with a suitable thermometer. A gas supply permits additional
gas to be added to the cylinder; we assume that a mechanism
is also provided for removing gas and for changing the supply
to admit different kinds of gas.

ing the position of the piston, and the amount of gas might be changed by allowing gas to enter the chamber, thereby changing the number of molecules N. After each change, we allow enough time for the gas to reach thermal equilibrium and to acquire a new set of macroscopic thermodynamic variables.

Let us now conduct the following experiments on the gas.

1.   *Dependence of V on N.* Keeping the temperature and pressure constant (that is, the gas is in contact with the thermal reservoir at a particular temperature T, and the weight on the piston is constant), we allow gas to enter or leave the chamber, and we measure the resulting volume V by observing the height of the piston. (We assume that we know the mass of each molecule and the total mass of gas that is present in the cylinder. Thus we can determine N, the total number of molecules.) Figure 2 shows typical results of such experiments. The data points appear to follow a straight line, and we conclude that, to a sufficiently good approximation, there is a direct proportion between V and N; that is, the volume increases linearly with the number of particles. Furthermore, by replacing the gas in the cylinder with an equal number of molecules of a different gas at the same pressure and temperature, we find that the new gas occupies the same volume. Thus we would conclude that the volume occupied by a gas at a particular pressure and temperature is independent of the type of gas or of the size or mass of its molecules; the volume depends only on the *number* of molecules. Mathematically, $V \propto N$, or

$$V = CN \qquad (p,T \text{ constant}). \qquad (1)$$

Here C is a constant, equal to the slope of the line in Fig. 2 and determined by the values of p and T. If we repeated this experiment with different constant values of p and T, we would still find Eq. 1 to hold, but with a different value of the constant C.

Equation 1 is sometimes known as Avogadro's law. It holds to a very good approximation for all gases, especially at low density, where the molecules are very far apart and the volume occupied by the molecules themselves is indeed a negligibly small fraction of the volume of the container to which the gas is confined. We can generalize from the behavior of these *real gases* to that of an *ideal gas* that does follow Eq. 1 exactly. In the next section we consider the microscopic properties of an ideal gas.

2.   *Dependence of V on p.* Keeping the number of particles N and the temperature T constant, we change the pressure (by changing the weight on the piston) and measure the resulting volume. The result is shown in Fig. 3a, which suggests an *inverse* relationship: as the pressure p increases, the volume V decreases. To check this, we instead plot p versus V^{-1}, as in Fig. 3b, which confirms a linear relationship. We therefore conclude $p \propto V^{-1}$, or

$$p = \frac{C'}{V} \qquad (N,T \text{ constant}). \qquad (2)$$

Here C' represents another constant, which would have a different value if we had chosen different values of N and T. Equation 2 is called *Boyle's law* and like Eq. 1 is a somewhat ideal generalization. As we discuss in Section 23-8, real gases deviate somewhat from this ideal behavior.

3.   *Dependence of V on T.* Keeping p and N constant, we vary the temperature T (by changing the temperature of the thermal reservoir), and we measure the resulting volume V. We find (Fig. 4) a direct relationship: the volume increases as the temperature increases; thus $V \propto T$, or

$$V = C''T \qquad (p,N \text{ constant}), \qquad (3)$$

where C'' is yet another constant. Equation 3 is called *Charles' law* or *Gay-Lussac's law*. Like Eqs. 1 and 2, it is an idealization of the behavior of real gases.

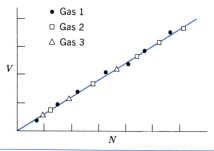

Figure 2   The volume V occupied by the gas in Fig. 1 depends on the number of molecules N. At a given temperature and pressure, different gases follow the same linear relationship.

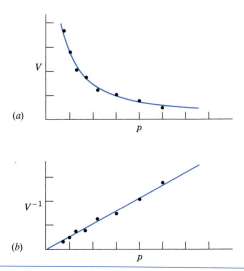

Figure 3   (*a*) The volume V occupied by the gas appears to depend inversely on the pressure p, with the temperature and the number of particles held constant. (*b*) Plotting V^{-1} against p shows that the relationship is indeed an inverse linear one.

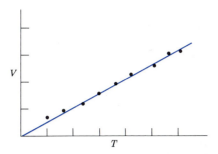

Figure 4 The volume V occupied by the gas varies linearly with the temperature T, when the pressure and the number of molecules are held constant.

Equation of State

Equations 1, 2, and 3 summarize experimental results strictly valid only for our hypothetical ideal gas but to a high degree approximately valid for most real gases. We can combine the three equations into a single equation that includes all three of the observed relationships, as follows:

$$\frac{pV}{NT} = k, \qquad (4)$$

where k is a constant. By rewriting Eq. 4 we can show that it is consistent with Eqs. 1–3:

$$V = \left(\frac{kT}{p}\right)N = CN \qquad (p, T \text{ constant}), \qquad (5a)$$

$$p = \frac{(kNT)}{V} = \frac{C'}{V} \qquad (N, T \text{ constant}), \qquad (5b)$$

$$V = \left(\frac{kN}{p}\right)T = C''T \qquad (p, N \text{ constant}). \qquad (5c)$$

The constant k in Eq. 4 is called the *Boltzmann constant*. It is a universal constant with a value determined by experiment to be

$$k = 1.38066 \times 10^{-23} \text{ J/K}.$$

It is more common to write Eq. 4 in a slightly different form. We express the quantity of gas not in terms of the number of molecules N but in terms of the *number of moles n*. The mole was defined in Section 1-5. In terms of the Avogadro constant N_A, the number of moles is

$$n = \frac{N}{N_A}, \qquad (6)$$

and we can rewrite Eq. 4 as

$$\frac{pV}{nT} = kN_A,$$

or

$$pV = nRT, \qquad (7)$$

where

$$R = N_A k$$
$$= 8.3145 \text{ J/mol}\cdot\text{K}. \qquad (8)$$

Equation 7 is called the *ideal gas law* or *ideal gas equation of state*. An equation of state of a system gives a fundamental mathematical relationship among macroscopic thermodynamic quantities. Experiments reveal that, at low enough densities, all real gases approach the ideal gas abstraction described in Eq. 7. This is the same limit that we discussed in connection with the ideal gas temperature scale in Section 22-4. The constant R has the same value for all gases and is called the *universal gas constant*.

Sample Problem 1 An insulated cylinder fitted with a piston (Fig. 1) contains oxygen at a temperature of 20°C and a pressure of 15 atm in a volume of 22 liters. The piston is lowered, decreasing the volume of the gas to 16 liters, and simultaneously the temperature is raised to 25°C. Assuming oxygen to behave like an ideal gas under these conditions, what is the final pressure of the gas?

Solution From Eq. 7, since the quantity of gas remains unchanged, we have

$$\frac{p_i V_i}{T_i} = \frac{p_f V_f}{T_f},$$

or

$$p_f = p_i \left(\frac{T_f}{T_i}\right)\left(\frac{V_i}{V_f}\right).$$

Because this is in the form of a ratio, we need not convert p and V into SI units, but *we must express T in absolute (Kelvin) temperature units*. Thus

$$p_f = (15 \text{ atm}) \left(\frac{273 + 25 \text{ K}}{273 + 20 \text{ K}}\right)\left(\frac{22 \text{ L}}{16 \text{ L}}\right) = 21 \text{ atm}.$$

23-2 THE IDEAL GAS: A MODEL

When physicists want to understand a complex system, they often invent a *model*. A model is a simplified version of the system that permits calculations to be made but still yields physical insight. A model might begin with a set of simplifying assumptions that permit the system to be analyzed using an existing set of laws, for example, Newtonian mechanics. The analysis might then lead to an equation or set of equations describing the original physical system. Because the model is a simplification of nature, the final result is generally not a true or complete description of nature, but if we have been clever at forming the model, the final result may prove to be a very good approximation of the behavior of the system. What is more important, the final result may give us a way of studying the system in the laboratory and gaining still more insight. Previously in this text, we have used a model (without calling it one) to describe the motion of a complicated object as a point particle under certain circumstances. We have also sometimes modeled the force between atoms in a molecule, or between atoms in a solid, in terms of the spring force, $\mathbf{F} = -k\mathbf{r}$, which is itself based on a kind of model that simplifies (under certain elastic conditions) the complicated internal processes in a solid under stress.

A gas confined in a container is an example of a complex system that is difficult to analyze using Newton's laws. The molecules can collide inelastically, and the energy of the collision can be absorbed by the molecules as internal energy in a variety of ways. Keeping track of these processes for all the molecules would be a project of hopeless complexity. We simplify this problem by inventing a model that describes the microscopic properties of the real gas. This model, which we call the *ideal gas model,* proves to be entirely consistent with the concept of the ideal gas that we developed experimentally in Section 23-1. In that section we saw that, especially at low density, the macroscopic properties of real gases approximately follow a general result, the ideal gas law of Eq. 7.

From the microscopic point of view our model of an ideal gas includes the following assumptions. Based on these assumptions, we use Newton's laws to analyze the mechanics of the ideal gas; this procedure forms the basis of *kinetic theory.* Later we relate this microscopic description to a macroscopic one.

1. *A gas consists of particles, called molecules.* Depending on the gas, each molecule may consist of one atom or a group of atoms. If the gas is an element or a compound and is in a stable state, we consider all its molecules to be identical.

2. *The molecules are in random motion and obey Newton's laws of motion.* The molecules move in all directions and with a range of velocities. In describing the motion, we assume that Newtonian mechanics is valid at the microscopic level.

3. *The total number of molecules is large.* The velocity (magnitude and direction) of any one molecule may change abruptly on collision with the wall or another molecule. Any particular molecule will follow a zigzag path because of these collisions. However, because there are so many molecules we assume that the resulting large number of collisions maintains the overall distribution of molecular velocities and the randomness of the motion.

4. *The volume of the molecules is a negligibly small fraction of the volume occupied by the gas.* Even though there are many molecules, they are extremely small. We know that the volume occupied by a gas can be changed through a large range of values with little difficulty, and that when a gas condenses the volume occupied by the liquid may be thousands of times smaller than that of the gas. Hence our assumption is plausible. Later we shall investigate the actual size of molecules and see whether we need to modify this assumption.

5. *No appreciable forces act on the molecules except during a collision.* That is, we assume that the range of molecular forces is comparable to the molecular size and much smaller than the typical distance between molecules. To the extent that this is true a molecule moves with constant velocity between collisions. Therefore the motion of a particular molecule is a zigzag path consisting mostly of segments with constant velocity changed by impulsive forces.

6. *Collisions are elastic and of negligible duration.* Collisions of one molecule with another or with the walls of the container conserve momentum and (we assume) kinetic energy. Molecules are not true point particles and do have internal structure; thus some kinetic energy may be converted into internal energy during the collision. We assume that the molecule does not retain this internal energy, which is then available again as kinetic energy after such a brief time (the time between collisions) that we can ignore this exchange entirely.

23-3   KINETIC CALCULATION OF THE PRESSURE

Let us now calculate the pressure of an ideal gas from kinetic theory. For simplicity, we consider a gas in a cubical container of edge length L whose walls are perfectly elastic. Call the faces normal to the x axis (Fig. 5) A_1 and A_2, each of area L^2. Consider a molecule of mass m with velocity $\mathbf{v}$, which we resolve into components v_x, v_y, and v_z. When this particle collides with A_1, it rebounds with its x component of velocity reversed; that is, $v_x \rightarrow -v_x$. There is no effect on v_y or v_z, so that the change in the particle's momentum has only an x component, given by

final momentum − initial momentum =
$$-mv_x - (mv_x) = -2mv_x. \quad (9)$$

Because the total momentum is conserved in the collision, the momentum imparted to A_1 is $+2mv_x$.

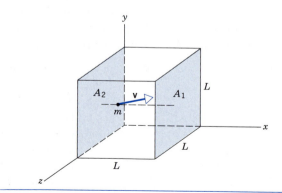

Figure 5   A cubical box of edge L containing an ideal gas. A molecule of the gas is shown moving with velocity $\mathbf{v}$ toward side A_1.

Suppose that this particle reaches A_2 without striking any other particle on the way. The time required to cross the cube is L/v_x. (If the molecule strikes one of the other faces of the box on the way to A_2, the x component of its velocity does not change, nor does the transit time.) At A_2 it again has its x component of velocity reversed and returns to A_1. Assuming no collisions with other molecules, the round trip takes a time $2L/v_x$, which is the time between collisions with A_1. The average impulsive force exerted by this molecule on A_1 is the transferred momentum divided by the time interval between transfers, or

$$F_x = \frac{2mv_x}{2L/v_x} = \frac{mv_x^2}{L}. \tag{10}$$

To obtain the *total* force on A_1, that is, the rate at which momentum is imparted to A_1 by *all* the gas molecules, we must sum the quantity mv_x^2/L for all the particles. Then, to find the pressure, we divide this force by the area of A_1, namely, L^2. The pressure is therefore

$$\begin{aligned}
p &= \frac{1}{L^2} \frac{mv_{x1}^2 + mv_{x2}^2 + \cdots}{L} \\
&= \frac{m}{L^3}(v_{x1}^2 + v_{x2}^2 + \cdots),
\end{aligned} \tag{11}$$

where v_{x1} is the x component of the velocity of particle 1, v_{x2} is that of particle 2, and so on. If N is the total number of particles in the container, then Nm is the total mass and Nm/L^3 is the density ρ. Thus $m/L^3 = \rho/N$, and

$$p = \rho \left(\frac{v_{x1}^2 + v_{x2}^2 + \cdots}{N} \right). \tag{12}$$

The quantity in parentheses in Eq. 12 is the average value of v_x^2 for all the particles in the container, which we represent by $\overline{v_x^2}$. Then

$$p = \rho \overline{v_x^2}. \tag{13}$$

For any particle, $v^2 = v_x^2 + v_y^2 + v_z^2$. Because we have many particles and because they are moving entirely at random, the average values of v_x^2, v_y^2, and v_z^2 are equal, and the value of each is exactly one-third the average value of v^2. There is no preference among the molecules for motion along any one of the three axes. Hence $\overline{v_x^2} = \frac{1}{3}\overline{v^2}$, so that Eq. 13 becomes

$$p = \frac{1}{3}\rho \overline{v^2}. \tag{14}$$

Although we derived this result by neglecting collisions between particles, the result is true even when we consider collisions. Because of the exchange of velocities in an elastic collision between identical particles, there will always be a molecule that collides with A_2 with momentum mv_x corresponding to the molecule that left A_1 with this same momentum. Equation 14 holds even if the box contains a mixture of molecules of different masses, because momentum is conserved in collisions, and the wall must receive the same impulse regardless of which molecules strike it. Also, the time spent during collisions is

negligible compared to the time spent between collisions. Hence our neglect of collisions is merely a convenient device for calculation. Similarly, we could have chosen a container of any shape: the cube merely simplifies the calculation. Although we have calculated the pressure exerted only on the side A_1, it follows from Pascal's law that the pressure is the same on all sides and everywhere in the interior. (This is true only if the density of the gas is uniform. In a large sample of gas, gravitational effects might be significant, and we should take into account the varying density. See Section 17-3 and Problem 6 of this chapter.)

The square root of $\overline{v^2}$ is called the *root-mean-square* speed of the molecules and is a kind of average molecular speed. (We consider this average in more detail in Section 24-3.) Using Eq. 14, we can calculate the root-mean-square speed from measured values of the pressure and density of the gas. Thus

$$v_{\text{rms}} \equiv \sqrt{\overline{v^2}} = \sqrt{\frac{3p}{\rho}}. \tag{15}$$

In Eq. 14 we relate a macroscopic quantity (the pressure p) to an average value of a microscopic quantity (that is, to $\overline{v^2}$ or v_{rms}^2). However, averages can be taken over short times or over long times, over small regions of space or large regions of space. The average computed in a small region for a short time might depend on the time or region chosen, so that the values obtained in this way may fluctuate. This could happen in a gas of very low density, for example. We can ignore fluctuations, however, when the number of particles in the system is large enough.

Sample Problem 2   Calculate the root-mean-square speed of hydrogen molecules at 0.00°C and 1.00 atm pressure, assuming hydrogen to be an ideal gas. Under these conditions hydrogen has a density ρ of 8.99×10^{-2} kg/m³.

Solution   Since $p = 1.00$ atm $= 1.01 \times 10^5$ Pa,

$$v_{\text{rms}} = \sqrt{\frac{3p}{\rho}} = \sqrt{\frac{3(1.01 \times 10^5 \text{ Pa})}{8.99 \times 10^{-2} \text{ kg/m}^3}} = 1840 \text{ m/s}.$$

This is of the order of a mile per second, or 3600 mi/h.

Table 1 gives the results of similar calculations for some gases at room temperature. These molecular speeds are roughly of the same order as the speed of sound at the same temperature. For example, in air at 0°C, $v_{\text{rms}} = 485$ m/s and the speed of sound is 331 m/s; in hydrogen $v_{\text{rms}} = 1838$ m/s and sound travels at 1286 m/s. These results are to be expected in terms of our model of a gas; see Problem 38. The energy of the sound wave is carried as kinetic energy from one molecule to the next one with which it collides. We might therefore expect sound waves to propagate with a speed that is roughly the same as the

TABLE 1   SOME MOLECULAR SPEEDS AT ROOM TEMPERATURE (300 K)

Gas	Molar mass M^a (g/mol)	v_{rms} (m/s)	Translational Kinetic Energy per Mole (J/mol)
Hydrogen	2.0	1920	3720
Helium	4.0	1370	3750
Water vapor	18.0	645	3740
Nitrogen	28.0	517	3740
Oxygen	32.0	483	3730
Carbon dioxide	44.0	412	3730
Sulfur dioxide	64.1	342	3750

[a] The molar mass, sometimes also known as the molecular weight, is given here for convenience in g/mol; its SI unit is kg/mol.

characteristic speed of molecular motion, which is in fact what we observe. The molecules themselves, in spite of their high speeds, do not move very far during a period of the sound vibration; they are confined to a rather small space by the effects of a large number of collisions. This explains why there is a time lag between opening an ammonia bottle at one end of a room and smelling it at the other end. Although molecular speeds are high, the large number of collisions restrains the advance of the ammonia molecules. They diffuse through the air at speeds that are very much less than molecular speeds.

Sample Problem 3   Assuming that the speed of sound in a gas is the same as the root-mean-square speed of the molecules, show how the speed of sound for an ideal gas would depend on the temperature.

Solution   The density of a gas is

$$\rho = \frac{nM}{V},$$

in which M is the molar mass (the mass of 1 mole) and n is the number of moles. Combining this with the ideal gas law $pV = nRT$ yields

$$\frac{p}{\rho} = \frac{RT}{M}.$$

We obtain from Eq. 15

$$v_{rms} = \sqrt{\frac{3p}{\rho}} = \sqrt{\frac{3RT}{M}}, \qquad (16)$$

so that the speed of sound v_1 at a temperature T_1 is related to the speed of sound v_2 in the same gas at a temperature T_2 by

$$\frac{v_1}{v_2} = \sqrt{\frac{T_1}{T_2}}.$$

For example, if the speed of sound at 273 K is 331 m/s in air, its speed in air at 300 K is

$$(331 \text{ m/s}) \sqrt{\frac{300 \text{ K}}{273 \text{ K}}} = 347 \text{ m/s}.$$

Note that the absolute (Kelvin) temperature is used here. Why?

Our initial assumption, that the speed of sound in a gas is the same as the root-mean-square speed of the molecules, is only crudely correct. In reality, the speed of sound is proportional to v_{rms}. Does this change the conclusions of this sample problem regarding the dependence of the speed of sound on the temperature? See Sample Problem 6 for a derivation of the speed of sound in a gas.

23-4   KINETIC INTERPRETATION OF THE TEMPERATURE

If we multiply each side of Eq. 14 by the volume V, we obtain

$$pV = \tfrac{1}{3}\rho V \overline{v^2},$$

where ρV is the total mass of gas, ρ being the density. We can also write the mass of gas as nM, in which n is the number of moles and M is the molar mass. Making this substitution yields

$$pV = \tfrac{1}{3}nM\overline{v^2}. \qquad (17)$$

The total translational kinetic energy of the gas is

$$\tfrac{1}{2}m(v_1^2 + v_2^2 + \cdots + v_N^2) = \tfrac{1}{2}m(N\overline{v^2}),$$

where N is the total number of molecules. The total mass of the gas can be written as $mN = nM$. The right side of Eq. 17 is therefore two-thirds of the total translational kinetic energy. We can write Eq. 17 as

$$pV = \tfrac{2}{3}(\tfrac{1}{2}nM\overline{v^2}).$$

Combining this with the equation of state of an ideal gas ($pV = nRT$), we obtain

$$\tfrac{1}{2}M\overline{v^2} = \tfrac{3}{2}RT. \qquad (18)$$

That is, the *average translational kinetic energy per mole of an ideal gas is proportional to the temperature.* This result connects the kinetic theory with the equation of state of an ideal gas. Equivalently, we may consider Eq. 18 as a connection between a macroscopic property, temperature, and a microscopic property, the kinetic energy of a molecule. Either way, we gain some insight into the meaning of temperature for gases.

The temperature of a gas is related to the average translational kinetic energy measured with respect to the center of mass of the gas. The kinetic energy associated with the motion of the center of mass of the gas has no bearing on the gas temperature. In Section 23-2 we assumed random motion as part of our statistical definition of an ideal gas and in Section 23-3 we calculated $\overline{v^2}$ on this basis. For a distribution of molecular velocities having random directions, the center of mass would be at rest. Thus, to calculate $\overline{v^2}$, we must use a reference frame in which the center of mass of the gas is at rest. In all other frames the molecules each have velocities greater by $\mathbf{u}$ (the velocity of the center of mass in that frame) than in the center-of-mass frame; hence the motions will no longer be random, and we obtain different values for $\overline{v^2}$. The temperature of a gas in a container does not increase when we put the container in a moving car!

Let us now divide each side of Eq. 18 by the Avogadro constant N_A, which is the number of molecules per mole of a gas. Thus $M/N_A = m$, the mass of a single molecule, and we have

$$\tfrac{1}{2}(M/N_A)\overline{v^2} = \tfrac{1}{2}m\overline{v^2} = \tfrac{3}{2}(R/N_A)T. \qquad (19)$$

Now $\tfrac{1}{2}m\overline{v^2}$ is the average translational kinetic energy per molecule. The ratio R/N_A is, from Eq. 8, the Boltzmann constant k, which plays the role of the gas constant per molecule. We then have

$$\tfrac{1}{2}m\overline{v^2} = \tfrac{3}{2}kT. \qquad (20)$$

Equation 20 is the molecular analogue of Eq. 18, which dealt with molar quantities. Here we see that the average translational kinetic energy of a molecule is determined by the temperature.

In the last column of Table 1 we list calculated values of $\tfrac{1}{2}Mv_{rms}^2$. As Eq. 18 predicts for an ideal gas, this quantity (the translational kinetic energy per mole) has nearly the same value for real gases at a given temperature (300 K in this case). From Eq. 20 we conclude that at a particular temperature T the ratio of the root-mean-square speeds of molecules of two different gases is equal to the square root of the inverse ratio of their masses. That is, from

$$T = \frac{2}{3k}\frac{m_1\overline{v_1^2}}{2} = \frac{2}{3k}\frac{m_2\overline{v_2^2}}{2}$$

we obtain

$$\sqrt{\frac{\overline{v_1^2}}{\overline{v_2^2}}} = \frac{v_{1,rms}}{v_{2,rms}} = \sqrt{\frac{m_2}{m_1}}. \qquad (21)$$

We can apply Eq. 21 to the diffusion of two different gases in a container with porous walls placed in an evacuated space. The lighter gas, whose molecules move more rapidly on the average, will escape faster than the heavier one. The ratio of the number of molecules of the two gases that pass through the porous walls in a short time interval, which is called the *separation factor* α, is equal to the ratio of their rms speeds, and thus according to Eq. 21 to the square root of the inverse ratio of their molecular masses or, equivalently, their molar masses:

$$\alpha = \sqrt{m_2/m_1} = \sqrt{M_2/M_1}. \qquad (22)$$

The diffusion process through porous walls is one method used to separate the atoms of an element by mass into its different isotopes.

Sample Problem 4   Natural uranium consists primarily of two isotopes, ^{235}U (0.7% abundance) and ^{238}U (99.3% abundance). Only ^{235}U is easily fissionable. In a sample of the gas UF_6 (uranium hexafluoride), it is desired to increase the abundance of ^{235}U from 0.7% to 3% by forcing the gas n times through a porous barrier. Find n.

Solution   The molar mass M of $^{235}UF_6$ is 0.349 kg/mol and that of $^{238}UF_6$ is 0.352 kg/mol. Thus after passage through a porous barrier the gas will be enriched in ^{235}U by the separation factor α, given by Eq. 22:

$$\alpha = \sqrt{\frac{M_2}{M_1}} = \sqrt{\frac{0.352 \text{ kg/mol}}{0.349 \text{ kg/mol}}} = 1.0043.$$

Each successive passage through a porous wall increases the relative fraction of ^{235}U by a factor of α. After n such passages, the relative concentration of ^{235}U will increase by α^n. To increase the concentration of ^{235}U from 0.7%, characteristic of natural uranium, to 3%, an enrichment commonly used in power reactors, the number n of porous barriers that must be passed is determined from

$$\alpha^n \left(\frac{0.007}{0.993}\right) = \left(\frac{0.03}{0.97}\right).$$

Solving, we obtain $n = 350$. In practice, this is accomplished through successive stages, in which a portion of the gas that passes most easily through a barrier (and thus is slightly enriched in ^{235}U) advances to the next stage, while the remainder (now slightly depleted of ^{235}U) is returned to feed the previous lower stage. To obtain nearly pure ^{235}U, such as is required for nuclear weapons, might require several thousand steps.

23-5   WORK DONE ON AN IDEAL GAS

If we raise the temperature of the gas in the cylinder of Fig. 1, the gas expands and raises the weight against gravity; the gas does (positive) work on the weight. The upward force exerted by the gas due to its pressure p is given by pA, where A is the area of the piston. By Newton's third law, the force exerted *by* the piston *on* the gas is equal and opposite to the force exerted *by* the gas *on* the piston. Using Eq. 7 of Chapter 7, we can therefore write the work W done on the gas as

$$W = \int F\,dx = \int (-pA)dx. \qquad (23)$$

Here dx represents the displacement of the piston, and the minus sign enters because the force exerted by the piston on the gas is in a direction opposite to the displacement of the piston. If we *reduce* the temperature of the gas, it contracts instead of expanding; the work done on the gas in that case is positive. We assume that the process described by Eq. 23 is carried out slowly, so that the gas can be considered to be in equilibrium at all intermediate stages. Otherwise, the pressure would not be clearly defined during the process, and the integral in Eq. 23 could not easily be evaluated.

We can write Eq. 23 in a more general form that turns out to be very useful. If the piston moves through a distance dx, then the volume of the gas changes by an

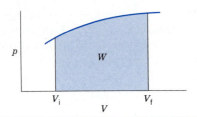

Figure 6 The magnitude of the work W done on a gas by a process of arbitrarily varying pressure is equal to the area under the pressure curve on a pV diagram between the initial volume V_i and the final volume V_f.

amount $dV = A\,dx$. Thus the work done on the gas can be written

$$W = -\int p\,dV. \qquad (24)$$

The integral is carried out between the initial volume V_i and the final volume V_f.

Equation 24 is the most general result for the work done on a gas. It makes no reference to the outside agent that does the work; it states simply that the work done on the gas can be calculated from the pressure and volume of the gas itself. Note that the algebraic sign of the work is implicitly contained in Eq. 24: if the gas expands, dV is positive and W is negative, p being a scalar quantity having only positive values. Conversely, if the gas contracts, dV is negative and the work done on the gas is positive.

Equation 24 is analogous to the general result for the work done on a system by a variable force F. You will recall from Fig. 7 of Chapter 7 that if we plot F against x, the work done by F is just the area under the curve between x_i and x_f. Figure 6 shows the similar situation for the work done on the gas. A graph in the form of Fig. 6 is called a *pV diagram*, with p plotted on the vertical axis (like F) and V plotted on the horizontal axis (like x). *The magnitude of the work done on the gas is equal to the area under the pressure curve on a pV diagram.* The sign of W is determined according to whether $V_f > V_i$ (in which case W is negative, as in Fig. 6), or $V_f < V_i$ (in which case W is positive). Once again, the work done *on* the gas is negative if the process increases the volume of the gas and positive if the process reduces the volume of the gas.

The pressure force is clearly nonconservative, as Fig. 7 demonstrates. Let us suppose we wish to take our ideal gas from the initial conditions V_i and p_i (point A) to the final conditions V_f and p_f (point D). There are many different paths we can take between A and D, of which two are shown in Fig. 7. Along path 1 (ABD), we first increase the pressure from p_i to p_f at constant volume. (We might accomplish this by turning up the control knob on the thermal reservoir, increasing the temperature of the gas, while we simultaneously add just the right amount of additional weight to the piston to keep it from moving.) We then follow path BD by increasing the temperature

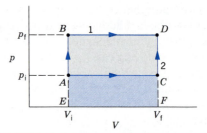

Figure 7 A gas is taken from the pressure and volume at point A to the pressure and volume at point D along two different paths, ABD and ACD. Along path 1 (ABD) the work is equal to the area of the rectangle $BDFE$, while along path 2 (ACD) the work is equal to the area of the rectangle $ACFE$.

but adding no additional weight to the piston, so that the pressure remains constant at the value p_f while the volume increases from V_i to V_f. The work done in this entire procedure is the area of the rectangle $BDFE$ (the area below the line BD).

We can find W_1, the work done on the gas along path 1, by considering the work done along the two segments AB and BD:

$$W_1 = W_{AB} + W_{BD}.$$

Because the volume is constant along AB, it follows from Eq. 24 that $W_{AB} = 0$. Along BD, the pressure is constant (at the value p_f) and comes out of the integral. The result is

$$W_1 = W_{AB} + W_{BD}$$
$$= 0 - \int p\,dV = -p_f \int_{V_i}^{V_f} dV = -p_f(V_f - V_i).$$

To follow path 2 (ACD), we first increase the temperature while holding the pressure constant at p_i (that is, adding no additional weight to the piston), so that the volume increases from V_i to V_f. We then increase the pressure from p_i to p_f at the constant volume V_f by increasing the temperature and adding weight to the piston to keep it from moving. The work done in this case is the area under the line AC or the rectangle $ACFE$. We can compute this as

$$W_2 = W_{AC} + W_{CD}$$
$$= -\int p\,dV + 0 = -p_i \int_{V_i}^{V_f} dV = -p_i(V_f - V_i).$$

Clearly $W_1 \neq W_2$, and the work depends on the path.

We can perform a variety of operations on the gas and evaluate the work done in each case.

Work Done at Constant Volume

The work is zero for any process in which the volume remains constant (as in segments AB and CD in Fig. 7):

$$W = 0 \qquad (\text{constant } V). \qquad (25)$$

We deduce directly from Eq. 24 that $W = 0$ if V is constant. Note that it is not sufficient that the process start and end with the same volume; the volume must be constant throughout the process for the work to vanish. For example, consider process $ACDB$ in Fig. 7. The volume starts and ends at V_i, but the work is certainly not zero. The work is zero only for vertical paths such as AB, representing a process at constant volume.

Work Done at Constant Pressure

Here we can easily apply Eq. 24, because the constant p comes out of the integral:

$$W = -p \int dV$$
$$= -p(V_f - V_i) \quad \text{(constant } p). \qquad (26)$$

Examples are the segments AC and BD in Fig. 7. Note that the work done on the gas is negative for both of these segments, because the volume increases in both processes.

Work Done at Constant Temperature

If the gas expands or contracts at constant temperature, the relationship between p and V, given by the ideal gas law, is

$$pV = \text{constant}.$$

On a pV diagram, the plot of the equation $pV = \text{constant}$ is exactly like a plot of the equation $xy = \text{constant}$ on an xy coordinate system: it is a hyperbola, as shown in Fig. 8.

A process done at constant temperature is called an *isothermal* process, and the corresponding hyperbolic curve on the pV diagram is called an *isotherm*. To find the work done on a gas during an isothermal process, we use Eq. 24, but we must find a way of carrying out the integral when p varies. To do this we use the ideal gas equation of state to write $p = nRT/V$, and thus

$$W = -\int_{V_i}^{V_f} p \, dV = -\int_{V_i}^{V_f} \frac{nRT}{V} \, dV = -nRT \int_{V_i}^{V_f} \frac{dV}{V},$$

where the last step can be made because we are taking T to be a constant. Carrying out the integral, we find

$$W = -nRT \ln \frac{V_f}{V_i} \quad \text{(constant } T). \qquad (27)$$

Note that this is also negative whenever $V_f > V_i$ (ln x is positive for $x > 1$) and positive whenever $V_f < V_i$.

Work Done in Thermal Isolation

Let us remove the gas cylinder in Fig. 1 from contact with the thermal reservoir and rest it on a slab of insulating material. The gas will then be in complete thermal isolation from its surroundings; if we do work on it, its temperature will change, in contrast to its behavior when it was in contact with the thermal reservoir. A process carried out in thermal isolation is called an *adiabatic* process.

If we allow the gas to expand with no other constraints, the path it will follow is represented by the hyperbola-like curve

$$pV^\gamma = \text{constant}, \qquad (28)$$

as shown in Fig. 9. The parameter γ, called the *ratio of specific heats,* must be determined empirically for any particular gas. Its values are typically in the range 1.1 – 1.8. (In Section 25-4 we discuss the specific heats of gases, and we derive Eq. 28 in Section 25-6.) Because γ is greater than 1, the curve $pV^\gamma = \text{constant}$ is a bit steeper than the curve $pV = \text{constant}$, and hence the work done in this process will be somewhat smaller in magnitude than the work done in expanding from V_i to V_f at constant T, as can be seen from Fig. 9.

The constant in Eq. 28 is determined from the pressure

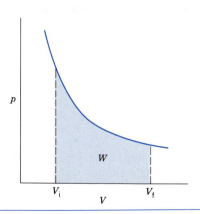

Figure 8 A process done at constant temperature (isothermal process) is represented by a hyperbola on a pV diagram. The work done in changing the volume is equal to the area under the curve between V_i and V_f.

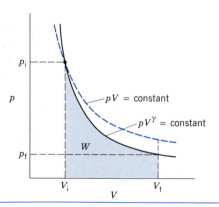

Figure 9 An adiabatic process is represented on a pV diagram by the hyperbola-like curve $pV^\gamma = \text{constant}$. The work done in changing the volume is equal to the area under the curve between V_i and V_f. Because $\gamma > 1$, the adiabatic curve has a steeper negative slope than the isothermal curve $pV = \text{constant}$.

518 *Chapter 23 Kinetic Theory and the Ideal Gas*

and volume at any particular point on the curve. Let us choose the initial point p_i, V_i in Fig. 9, and so

$$pV^\gamma = p_iV_i^\gamma$$

or

$$p = \frac{p_iV_i^\gamma}{V^\gamma}. \tag{29}$$

We can now find the adiabatic work:

$$W = -\int_{V_i}^{V_f} p\,dV = -\int_{V_i}^{V_f}\frac{p_iV_i^\gamma}{V^\gamma}\,dV = -p_iV_i^\gamma\int_{V_i}^{V_f}\frac{dV}{V^\gamma}$$
$$= -\frac{p_iV_i^\gamma}{\gamma-1}(V_i^{1-\gamma} - V_f^{1-\gamma}).$$

First by bringing a factor of $V_i^{\gamma-1}$ inside the parentheses, and second by using $p_iV_i^\gamma = p_fV_f^\gamma$, we can write the adiabatic work as

$$W = \frac{p_iV_i}{\gamma-1}\left[\left(\frac{V_i}{V_f}\right)^{\gamma-1} - 1\right]$$
$$= \frac{1}{\gamma-1}(p_fV_f - p_iV_i) \quad \text{(adiabatic).} \tag{30}$$

If the gas expands, then $V_i/V_f < 1$, and since a number less than 1 raised to any positive power remains less than 1, the work again is negative.

Sample Problem 5 A sample of gas consisting of 0.11 mol is compressed from a volume of 4.0 m³ to 1.0 m³ while its pressure increases from 10 to 40 Pa. Compare the work done along the three different paths shown in Fig. 10.

Solution Path 1 consists of two processes, one at constant pressure followed by another at constant volume. The work done at constant pressure is found from Eq. 26,

$$W = -p(V_f - V_i) = -(10\text{ Pa})(1.0\text{ m}^3 - 4.0\text{ m}^3) = 30\text{ J}.$$

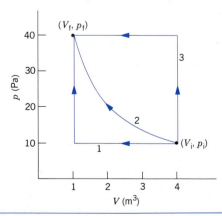

Figure 10 Sample Problem 5. A gas is taken from initial point i to final point f along three different paths. Path 2 is an isotherm.

The work done at constant volume is zero (see Eq. 25), so the total work for path 1 is

$$W_1 = 30\text{ J} + 0 = 30\text{ J}.$$

Path 2 represents an isothermal process, along which $T = $ constant. Thus $p_iV_i = p_fV_f = nRT$. The work done during the isothermal process can be found using Eq. 27, substituting p_iV_i for nRT, which gives

$$W_2 = -p_iV_i\ln\frac{V_f}{V_i} = -(10\text{ Pa})(4.0\text{ m}^3)\ln\frac{1.0\text{ m}^3}{4.0\text{ m}^3} = 55\text{ J}.$$

Path 3 consists of a process at constant volume, for which the work is again zero, followed by a process at constant pressure, and so the total work for path 3 is

$$W_3 = 0 - p_f(V_f - V_i) = -(40\text{ Pa})(1.0\text{ m}^3 - 4.0\text{ m}^3) = 120\text{ J}.$$

Note that the work is positive for all three processes, and that the magnitudes increase according to the area under each path on the pV diagram.

Sample Problem 6 (a) Find the bulk modulus B for an adiabatic process involving an ideal gas. (b) Use the adiabatic bulk modulus to find the speed of sound in the gas as a function of temperature. Evaluate for air at room temperature (20°C).

Solution (a) In the differential limit, the bulk modulus (see Eq. 5 of Chapter 17) can be written

$$B = -V\frac{dp}{dV}.$$

For an adiabatic process, Eq. 28 ($pV^\gamma = $ constant) gives, taking the derivative with respect to V,

$$\frac{d(pV^\gamma)}{dV} = \left(\frac{dp}{dV}\right)V^\gamma + p(\gamma V^{\gamma-1}) = 0,$$

or

$$V\frac{dp}{dV} = -\gamma p.$$

Thus

$$B = \gamma p$$

for an adiabatic process involving an ideal gas.

(b) In Section 20-1, we determined that the speed of sound in a gas can be written

$$v = \sqrt{B/\rho},$$

where B is the bulk modulus and ρ is the density of the gas. Using the result of part (a) and the ideal gas equation of state (Eq. 7), we obtain

$$v = \sqrt{\frac{\gamma p}{\rho}} = \sqrt{\frac{\gamma(nRT/V)}{\rho}} = \sqrt{\frac{\gamma nRT}{\rho V}}.$$

The quantity ρV is the total mass of the gas, which can also be written nM, where n is the number of moles and M is the molar mass. Making this substitution, we have

$$v = \sqrt{\frac{\gamma RT}{M}}.$$

Thus the speed of sound in a gas depends on the square root of the temperature, as we inferred in Sample Problem 3.

For air, the average molar mass is about 0.0290 kg/mol, and the parameter γ is about 1.4. Thus for $T = 20°C = 293$ K,

$$v = \sqrt{\frac{(1.4)(8.31 \text{ J/mol} \cdot \text{K})(293 \text{ K})}{0.0290 \text{ kg/mol}}} = 343 \text{ m/s}.$$

23-6   THE INTERNAL ENERGY OF AN IDEAL GAS

Our model of the ideal gas is based on molecules that are considered to be point particles. The temperature, as we have seen, depends on the *translational* kinetic energy of the molecules. For point particles, there is no other form for the internal energy E_{int} to take. There is no molecular potential energy, nor is there any internal energy associated with the rotation or the vibration of the molecule. *For an ideal gas, the internal energy can only be translational kinetic energy.* If we have n moles of an ideal gas at temperature T, then

$$E_{int} = n(\tfrac{1}{2}M\overline{v^2}) = \tfrac{3}{2}nRT \tag{31}$$

using Eq. 18. *The internal energy of an ideal gas depends only on the temperature.* It does not depend, for example, on the pressure or the volume of the gas.

One way to change the internal energy of an ideal gas is to do work on it (or to allow the gas to do work on its environment). Suppose the gas in the cylinder shown in Fig. 1 is isolated from the thermal reservoir. Let the environment do work W on the gas. The generalized law of conservation of energy (see Eq. 28 of Chapter 8) then gives

$$\Delta E_{int} = W \tag{32}$$

because internal energy is the only way the gas can store energy, and the work gives the only contribution to the change in internal energy of the gas.

Suppose the environment does work on the gas, so that W is positive in Eq. 32. It then follows that ΔE_{int} must be positive, and using Eq. 31 we can write

$$\Delta E_{int} = \tfrac{3}{2}nR\,\Delta T, \tag{33}$$

so that the temperature change is also positive.

If the piston moves upward, the environment does negative work on the gas, and by Eq. 32 the change in internal energy is negative. According to Eq. 33 the change in temperature is also negative.

Let us now modify one of the basic assumptions in our model of the ideal gas. Instead of considering a molecule to be represented as a point particle, let it be considered as two point particles separated by a given distance. This model gives a better description of *diatomic gases,* those with two atoms in each molecule, including such common gases as O_2, N_2, or CO (carbon monoxide). Such a molecule can acquire kinetic energy by rotating about its center of mass, and it is therefore necessary to consider in the internal energy the contributions of rotational kinetic energy as well as translational kinetic energy.

The rotational kinetic energy of a diatomic molecule, illustrated in Fig. 11, can be written

$$K_{rot} = \tfrac{1}{2}I_{x'}\omega_{x'}^2 + \tfrac{1}{2}I_{y'}\omega_{y'}^2$$

where I is the rotational inertia of the molecule for rotations about a particular axis. The $x'y'z'$ coordinate system is fixed to the center of mass of the molecule. For point masses, there is no kinetic energy associated with rotation about the z' axis, because $I_{z'} = 0$. The total kinetic energy of the molecule is the sum of the translational and rotational parts:

$$K = \tfrac{1}{2}mv_x^2 + \tfrac{1}{2}mv_y^2 + \tfrac{1}{2}mv_z^2 + \tfrac{1}{2}I_{x'}\omega_{x'}^2 + \tfrac{1}{2}I_{y'}\omega_{y'}^2. \tag{34}$$

Because kinetic energy is the only type of energy the molecule can have, Eq. 34 also represents the contribution of one molecule to the internal energy of the gas. To find the total internal energy of the gas, we must find the sum of expressions such as Eq. 34 over all N molecules. A simpler way is to evaluate the average energy per molecule, and then multiply by the number of molecules, N.

Suppose we do work W on the gas, increasing its internal energy. How much of this increase appears as translational kinetic energy and how much as rotational kinetic energy? This determination is very important for understanding the macroscopic properties of the gas, because *only the average translational kinetic energy of a gas contributes to its temperature.* That is, two gases with the same average translational kinetic energy have the same temperature, even if one has greater rotational energy and thus greater internal energy.

To determine the relative contributions of translational and rotational kinetic energy (and possibly other forms as

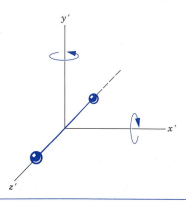

Figure 11   A diatomic molecule, consisting of two atoms considered to be point particles, is shown with its axis along the z' axis of a coordinate system. In this orientation, the rotational inertia for rotations about the z' axis is zero, and thus there is no term in the kinetic energy corresponding to such rotations. The rotational inertias for rotations about the x' and y' axes are not zero, and thus there are kinetic energy terms for such rotations.

well) to the internal energy, it is necessary to consider the average value of each different term in the expression for the internal energy of a gas, such as the five terms in Eq. 34, which is based on the assumption of a rigid diatomic molecule. For other gases, we might need to include a third rotational term, and for nonrigid molecules it is necessary to include terms in the energy corresponding to the vibrational motion (see Section 15-10). From classical statistical mechanics, which we consider in Chapter 24, we can show that, *when the number of particles is large and Newtonian mechanics holds, each of these independent terms has the same average energy of $\frac{1}{2}kT$.* In other words, the available energy depends only on the temperature and is distributed in equal shares to each of the independent ways that a molecule can store energy. This theorem, deduced by Maxwell, is called the *equipartition of energy.*

Each independent form that a system's energy can take, as, for example, the five terms of Eq. 34, is called a *degree of freedom.* A monatomic gas has only *three* degrees of freedom per molecule, since it has only translational kinetic energy ($E_{int} = \frac{1}{2}mv_x^2 + \frac{1}{2}mv_y^2 + \frac{1}{2}mv_z^2$). A diatomic gas has *five* degrees of freedom per molecule, if the molecule is rigid.

Let us use the equipartition of energy theorem to write an expression for the internal energy of a monatomic ideal gas. The average internal energy per molecule is $\frac{3}{2}kT$ (3 degrees of freedom $\times \frac{1}{2}kT$ per degree of freedom), and the total internal energy of the N molecules is

$$E_{int} = N(\tfrac{3}{2}kT) = \tfrac{3}{2}nRT \quad \text{(monatomic gas),} \quad (35)$$

where we have used Eqs. 6 and 8. Equation 35 is identical with Eq. 31.

For a diatomic gas, with 5 degrees of freedom, the result is

$$E_{int} = N(\tfrac{5}{2}kT) = \tfrac{5}{2}nRT \quad \text{(diatomic gas).} \quad (36)$$

A polyatomic gas (more than two atoms per molecule) generally has three possible axes of rotation (unless the three atoms lie in a straight line, as in CO_2). The internal kinetic energy per molecule could then have a sixth term, $\frac{1}{2}I_{z'}\omega_{z'}^2$. For 6 degrees of freedom, the internal energy is

$$E_{int} = N(\tfrac{6}{2}kT) = 3nRT \quad \text{(polyatomic gas).} \quad (37)$$

So far we have considered only the contributions of the overall translational or rotational kinetic energy to the internal energy of a gas. Other kinds of energy may also contribute. For example, a diatomic molecule that is free to vibrate (imagine the two atoms to be connected by a spring) has two additional contributions to the energy: the potential energy of the spring and the vibrational kinetic energy of the atoms. Thus a diatomic molecule free to translate, rotate, and vibrate would have 7 ($= 3 + 2 + 2$) degrees of freedom. For polyatomic molecules, the number of vibrational terms in the energy can be greater than

two. The vibrational modes in the internal energy are usually apparent only at high temperature, where the more violent collisions can cause the molecule to vibrate.

In Section 25-4, we show that the results derived in this section give a very good description of the relationship between the internal energy and the temperature of real gases. We also see that, as the temperature of a gas is lowered, the vibrational and rotational motions can be "frozen," so that at a low enough temperature only the 3 translational degrees of freedom are present. The most serious shortcoming of this ideal gas model is its failure to account for the quantum effects inherent in atomic and molecular structure. Experiments with gas collisions provided early evidence that the internal energy of an atom is quantized. We can thus say that the seeds of quantum theory lay in the kinetic theory of gases.*

Sample Problem 7   Consider once again the situation of Sample Problem 5, in which the gas begins at the initial point with volume $V_i = 4.0 \text{ m}^3$ and pressure $p_i = 10 \text{ Pa}$. Let the cylinder be removed from the thermal reservoir, and let us compress the gas adiabatically until its volume is $V_f = 1.0 \text{ m}^3$. Find the change in internal energy of the gas, assuming it to be helium (a monatomic gas with $\gamma = 1.66$).

Solution   To find the change in internal energy, we can use Eq. 33 if we know the change in temperature. We can find the initial temperature using the ideal gas law (since p_i and V_i are known), and we can find the final temperature if we know the pressure and volume of the final point. The final pressure can be found using the adiabatic relationship of Eq. 29:

$$p_f = \frac{p_i V_i^{\gamma}}{V_f^{\gamma}} = \frac{(10 \text{ Pa})(4.0 \text{ m}^3)^{1.66}}{(1.0 \text{ m}^3)^{1.66}} = 100 \text{ Pa}.$$

On the pV diagram of Fig. 10, the final point reached in the adiabatic process lies vertically far above the final point reached in the isothermal process (40 Pa). This is consistent with the adiabatic curves being steeper than the isothermal curves, as shown in Fig. 9.

We can now proceed to find the initial and final temperatures and then the change in internal energy:

$$T_i = \frac{p_i V_i}{nR} = \frac{(10 \text{ Pa})(4.0 \text{ m}^3)}{(0.11 \text{ mol})(8.31 \text{ J/mol·K})} = 44 \text{ K}.$$

$$T_f = \frac{p_f V_f}{nR} = \frac{(100 \text{ Pa})(1.0 \text{ m}^3)}{(0.11 \text{ mol})(8.31 \text{ J/mol·K})} = 109 \text{ K}.$$

$$\Delta E_{int} = \tfrac{3}{2}nR\Delta T$$
$$= \tfrac{3}{2}(0.11 \text{ mol})(8.31 \text{ J/mol·K})(109 \text{ K} - 44 \text{ K}) = 89 \text{ J}.$$

The change in internal energy is positive, consistent with Eq. 32 for this adiabatic process, because the work done in compressing the gas is similarly positive.

---

* See "On Teaching Quantum Phenomena," by Sir N. F. Mott, *Contemporary Physics,* August 1964, p. 401.

23-7   INTERMOLECULAR FORCES   *(Optional)*

Forces between molecules are of electromagnetic origin. All molecules contain electric charges in motion. These molecules are electrically neutral in the sense that the negative charge of the electrons is equal and opposite to the positive charge of the nuclei. This does not mean, however, that molecules do not interact electrically. For example, when two molecules approach each other, the charges on each are disturbed and depart slightly from their usual positions in such a way that the average distance between opposite charges in the two molecules is a little smaller than that between like charges. Hence an attractive intermolecular force results. This internal rearrangement takes place only when molecules are fairly close together, so that these forces act only over short distances; they are short-range forces. If the molecules come very close together, so that their outer charges begin to overlap, the intermolecular force becomes repulsive. The molecules repel each other because there is no way for a molecule to rearrange itself internally to prevent repulsion of the adjacent external electrons. It is this repulsion on contact that accounts for the billiard-ball character of molecular collisions in gases. If it were not for this repulsion, molecules would move right through each other instead of rebounding on collision.

Let us assume that molecules are approximately spherically symmetrical. Then we can describe intermolecular forces graphically by plotting the mutual potential energy of two molecules, U, as a function of distance r between their centers. The force F acting on each molecule is related to the potential energy U by $F = -dU/dr$. In Fig. 12a we plot a typical $U(r)$. Here we can imagine one molecule to be fixed at O. Then the other molecule is repelled from O when the slope of U is negative and is attracted to O when the slope is positive. At r_0 no force acts between the molecules; the slope is zero there. In Fig. 12b we plot the mutual force $F(r)$ corresponding to this potential energy function. The line E in Fig. 12a represents the mechanical energy of the colliding molecules. The intersection of $U(r)$ with this line is a "turning point" of the motion (see Section 8-4). The separation of the centers of two molecules at the turning point is the distance of closest approach. The separation distance at which the mutual potential energy is zero may be taken as the approximate distance of closest approach in a low-energy collision and hence as the diameter of the molecule. For simple molecules the diameter is about 2.5×10^{-10} m. The distance r_0 at which the potential is a minimum (the equilibrium point) is about 3.5×10^{-10} m for simple molecules, and the force and potential energy approach zero as r increases to about 10^{-9} m, or about 4 diameters. The molecular force thus has a very short range. Of course, different molecules have different sizes and internal arrangement of charges so that intermolecular forces vary from one molecule to another. However, they always show the qualitative behavior indicated in Fig. 12.

In a solid, molecules vibrate about the equilibrium position r_0. Their total energy E is negative, that is, lying below the horizontal axis in Fig. 12a. The molecules do not have enough energy to escape from the potential valley (that is, from the attractive binding force). The centers of vibration O are more or less fixed in a solid. In a liquid the molecules have greater vibrational energy about centers that are free to move but that remain about the same distance from one another. Molecules have their great-

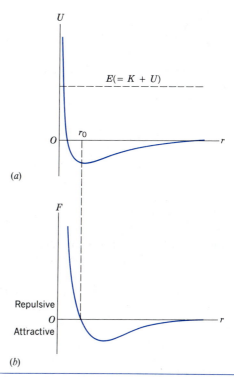

Figure 12   (a) The mutual potential energy U of two molecules as a function of their separation distance r. The mechanical energy E is indicated by the horizontal line. (b) The radial force between the molecules, given by $-dU/dr$, corresponding to this potential energy. The potential energy is a minimum at the equilibrium separation r_0, at which point the force is zero.

est kinetic energy in the gaseous state. In a gas the average distance between the molecules is considerably greater than the effective range of intermolecular forces, and the molecules move in straight lines between collisions. Maxwell discusses the relation between the kinetic theory model of a gas and the intermolecular forces as follows: "Instead of saying that the particles are hard, spherical, and elastic, we may if we please say that the particles are centers of force, of which the action is insensible except at a certain small distance, when it suddenly appears as a repulsive force of very great intensity. It is evident that either assumption will lead to the same results."

It is interesting to compare the measured intermolecular forces with the gravitational force of attraction between molecules. If we choose a separation distance of 4×10^{-10} m, for example, the force between two helium atoms is about 6×10^{-13} N. The gravitational force at that separation is about 7×10^{-42} N, smaller than the intermolecular force by a factor of 10^{29}! This is a typical result and shows that gravitation is negligible in intermolecular forces.

Although the intermolecular forces appear to be small by ordinary standards, we must remember that the mass of a molecule is so small (about 10^{-26} kg) that these forces can impart instantaneous accelerations of the order of 10^{15} m/s² (10^{14} g). These accelerations may last for only a very short time, of course, because one molecule can very quickly move out of the range of influence of the other.   ■

23-8 THE VAN DER WAALS EQUATION OF STATE *(Optional)*

Kinetic theory provides the microscopic description of the behavior of an ideal gas, but certain of the assumptions of our model of the ideal gas are not valid when applied to real gases. Many modifications to the equation of state of the ideal gas have been suggested to correct for these deficiencies. In the previous section, we showed that a realistic way of looking at the intermolecular force leads us to conclude that molecules have a small but certainly nonzero diameter (which may contradict assumption 4 of the ideal gas model) and that the range of the force may extend beyond the "collision diameter" (which contradicts assumption 5). In this section we develop a modified equation of state that takes these factors into account.

To consider the effect of the finite size of the molecules, let us regard each molecule as a hard sphere of diameter d. Two molecules are not permitted to approach one another so close that the distance between their centers would be less than d (Fig. 13). The "free volume" available for one molecule is therefore decreased by the volume of a hemisphere of *radius d* centered on the other molecule. Let b represent the decrease in the available volume due to the molecules in 1 mole of a gas. The total volume available to the entire collection of molecules in n moles is thus the volume V of the container less an amount nb that represents the volume occupied by the molecules. If we take the estimate from the previous section of $d = 2.5 \times 10^{-10}$ m, then we estimate b as

$$b = \tfrac{1}{2}N_A(\tfrac{4}{3}\pi d^3) = 2 \times 10^{-5} \text{ m}^3/\text{mol}$$
$$= 2 \times 10^{-3} \text{ L/mol}.$$

(The factor of $\tfrac{1}{2}$ comes about because, as two molecules approach one another, the volume within which they interact is not a full sphere but the hemisphere facing the direction of approach.) Under normal conditions, 1 mole of a gas has a volume of 22.4 L, and thus the correction b is normally small (0.01–0.1%), but it can become much more significant if we study a gas at high density.

The "free" volume available to the gas is thus $V - nb$, and we can modify the equation of state accordingly:

$$p(V - nb) = nRT. \tag{38}$$

Solving for p, we obtain

$$p = \frac{nRT}{V - nb}. \tag{39}$$

Equation 39 indicates that the pressure of a real gas is increased relative to that of an ideal gas under the same conditions. In effect, the reduced volume available to the molecules means that they make more collisions with the walls and thereby increase the pressure.

To account for the effect of the range of the force between molecules, let us consider a region of the gas within a distance d of one of the walls of the container (Fig. 14). We choose d to correspond to the range of the force between molecules, and we focus our attention on a particular molecule C that is about to strike the wall. When it strikes the wall, the impulse–momentum theorem, $\Delta \mathbf{p} = \int \mathbf{F}\, dt$, can be used to relate the change in momentum of the molecule to the impulse of the net force $\mathbf{F}$ that acts on it during the collision. In the ideal gas model, the molecules exert forces on one another only during collisions; thus the only force that acts on a molecule colliding with the wall is exerted by the wall. This force, by Newton's third law, is equal to the force exerted on the wall by the molecule and thus is responsible for the pressure that the gas exerts on the walls of the container, as we discussed in Section 23-3.

Now suppose that molecule C also experiences forces from the attraction of other nearby molecules (those lying within a hemisphere of radius d, the range of the force). For a molecule near the wall, the sum of all the intermolecular forces gives a resultant that acts away from the wall. (Molecules near the surface of a liquid experience a similar inward force, which is responsible for surface tension; see Section 17-6.) Thus during the collision the component of the force acting away from the wall

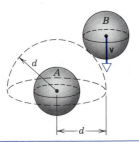

Figure 13 If molecules of a gas are considered to behave like hard spheres, then the center of molecule B is not permitted to move within the hemisphere of radius d centered on molecule A. Here d is the diameter of a molecule. The free volume available for molecule B is reduced by the volume of such a hemisphere centered on each molecule of the gas.

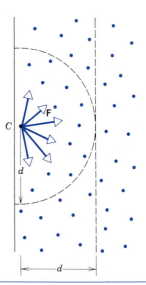

Figure 14 A gas molecule C (here considered to be a point) near the wall of the container experiences a net force away from the wall due to the attraction of the surrounding molecules within the range d of the force between molecules. The net pressure on the walls of the container is reduced by all such molecules within a distance d of the walls.

has two contributions: one from the wall and another from the surrounding molecules. For a given change in momentum from a collision with the wall, the force exerted by the wall during the collision is therefore smaller, the reaction force exerted by the molecule is smaller, and the pressure exerted by the gas is likewise smaller.

This reduction in pressure owing to the collision of molecule C with the wall is proportional to the number of molecules in the hemisphere of radius d surrounding molecule C and thus to the number of molecules per unit volume of the gas, N/V. The net effect due to *all* the molecules like C in the surface layer of thickness d is proportional to the number of molecules in that layer, which is also proportional to the number of molecules per unit volume of the gas. The total reduction in pressure resulting from the force between molecules is thus proportional to $(N/V)^2$.

That is, if we triple the number of molecules but keep the volume of the container constant, our imaginary hemisphere will have three times as many molecules and hence molecule C will suffer three times the force pulling it away from the wall. In the entire gas there will be three times as many of these molecules, each suffering the same effect. The overall effect thus increases ninefold.

The net effect of the intermolecular force is to introduce a correction to the pressure, proportional to $(N/V)^2$. Instead of writing this correction in terms of the number of molecules N, we write it in terms of the number of moles n, so that the corrected pressure becomes

$$p = \frac{nRT}{V - nb} - a\left(\frac{n}{V}\right)^2,$$ (40)

where a is a proportionality constant. The modified equation of state can be written

$$\left(p + a\frac{n^2}{V^2}\right)(V - nb) = nRT.$$ (41)

This expression, first deduced by J. D. van der Waals (1837–1923), is called the *van der Waals equation of state*. Note that Eq. 41 reduces to the ideal gas equation of state (Eq. 7) when the gas occupies a large volume (that is, the molecules are very far apart and the gas density is small).

The values of the constants a and b must be determined by experiment, which makes the equation empirical in this respect. Like the ideal gas equation of state, it is also based on a model with oversimplifying assumptions. No simple formula can be applied to all gases under all conditions, and only through experiments can we learn whether one equation is superior to another in its description of reality over a certain set of conditions.

Figure 15 compares isotherms for an ideal gas with those calculated for CO_2 using the van der Waals equation of state. Note that the deviation from ideal behavior occurs primarily at high pressure and at low temperature. For CO_2 at temperatures below 304 K, the isotherms begin to curve downward, indicating that as we decrease the volume, the pressure likewise decreases. Since this behavior is contrary to expectations for a *gas*, it suggests that some of the CO_2 is condensing into a *liquid*, leaving less of it in the gaseous state. The van der Waals model thus suggests the existence of mixtures of different phases, which the ideal gas model cannot do. If we were to compress a sample of CO_2, we would find that the actual $T = 264$ K isotherm would not dip downward as the van der Waals equation predicts, but instead would follow the horizontal segment AB in Fig. 15, as the gas condenses into a liquid at constant pressure. The van der Waals model gives an improvement over the ideal gas model, but no simple model is able to account for the behavior of the gas under all possible circumstances.

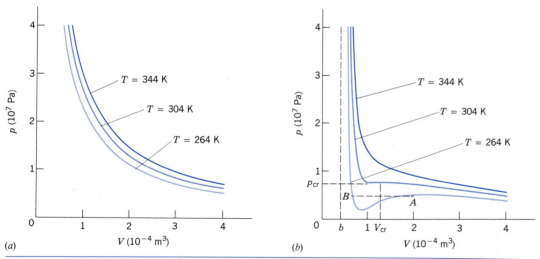

Figure 15   (a) Isotherms for 1 mole of an ideal gas. (b) Isotherms for 1 mole of CO_2 determined from the van der Waals equation. Note that at large volume, the ideal and van der Waals isotherms behave similarly. As the temperature is raised, the van der Waals isotherms behave more like those of the ideal gas. Note also that, as the pressure becomes very large, the volume approaches the value of b, as Eq. 40 requires, rather than the value of zero, as the ideal gas equation of state would predict. The dashed line AB shows a more realistic representation of the $T = 264$ K isotherm. As the gas is compressed along this isotherm, some of the gas condenses into a liquid, and the pressure remains constant.

We also find that our other results for the ideal gas are only approximately correct in their application to real gases. For example, the internal energy of a real gas depends on the volume as well as on the temperature. If there are attractive forces between the molecules, then the internal potential energy increases as we increase the average distance between the molecules. We therefore expect the internal energy of a gas to increase slightly with volume, and this expectation is consistent with experiment in most gases. If the state of the gas is such that repulsive forces are more significant than attractive forces, then increasing the distance between the molecules decreases the potential energy. For some gases (hydrogen and helium at ordinary temperatures, for instance) the internal energy is observed to decrease as the volume increases. In either case, the internal energy is not simply a function of temperature but depends on the volume as well.

Sample Problem 8   The isotherm drawn in Fig. 15*b* for CO_2 at a temperature of $T = 304$ K is called the *critical* isotherm. It is distinguished by having a minimum and a point of inflection (the point where the curvature changes from downward to upward) that coincide at a single point. Using this information along with the value of the critical pressure p_{cr}, estimate the values of the van der Waals constants a and b for CO_2.

Solution   The minimum of a curve on a pV diagram is determined by the point at which the slope dp/dV is zero, and in calculus we learn that at a point of inflection the second derivative is zero. We can find the derivatives when the van der Waals equation of state is written in the form of Eq. 40:

$$\frac{dp}{dV} = \frac{-nRT}{(V - nb)^2} + \frac{2an^2}{V^3},$$

$$\frac{d^2p}{dV^2} = \frac{2nRT}{(V - nb)^3} - \frac{6an^2}{V^4}.$$

In taking both derivatives we assume constant T, as is appropriate for an isotherm.

Setting both derivatives equal to zero and solving these equations simultaneously for a and b, we find

$$a = \frac{27R^2T_{cr}^2}{64p_{cr}},$$

$$b = \frac{RT_{cr}}{8p_{cr}}.$$

Reading $p_{cr} = 0.75 \times 10^7$ Pa from Fig. 15*b*, we then calculate

$$a = 0.364 \text{ J} \cdot \text{m}^3/\text{mol}^2 \quad \text{and} \quad b = 4.27 \times 10^{-5} \text{ m}^3/\text{mol}.$$

Although the van der Waals model gives a much more realistic description than the ideal gas model of the behavior of a real gas such as CO_2, it still represents only an approximation of the actual behavior. In the case of CO_2, for instance, the above calculation gives $V_{cr} = 3nb = 1.28 \times 10^{-4}$ m^3 for the volume of 1 mole at the critical point. The measured value, however, is 0.96×10^{-4} m^3. Nevertheless, it is a successful first step in improving the ideal gas model in cases in which the molecules are sufficiently close together that the basic assumptions of the ideal gas model do not hold, and it even suggests condensation due to the force between molecules, which the ideal gas model is completely unable to do. ∎

QUESTIONS

1. In discussing the fact that it is impossible to apply the laws of mechanics individually to atoms in a macroscopic system, Mayer and Mayer state: "The very complexity of the problem (that is, the fact that the number of atoms is large) is the secret of its solution." Discuss this sentence.

2. In kinetic theory we assume that there is a large number of molecules in a gas. Real gases behave like an ideal gas at low densities. Are these statements contradictory? If not, what conclusion can you draw from them?

3. We have assumed that the walls of the container are elastic for molecular collisions. Actually, the walls may be inelastic. Why does this make no difference as long as the walls are at the same temperature as the gas?

4. On a humid day, some say that the air is "heavy." How does the density of humid air compare with that of dry air at the same temperature and pressure?

5. Where does the root-mean-square speed of molecules in still air at room temperature fit into this sequence: 0; 2 m/s (walking speed); 30 m/s (fast car); 500 m/s (supersonic airplane); 1.1×10^4 m/s (escape speed from Earth); 3×10^8 m/s (speed of light)?

6. Two equal-size rooms communicate through an open doorway. However, the average temperatures in the two rooms are maintained at different values. In which room is there more air?

7. Molecular motions are maintained by no outside force, yet continue indefinitely with no sign of diminishing speed. What is the reason that friction does not bring these tiny particles to rest, as it does other moving particles?

8. What justification is there in neglecting the changes in gravitational potential energy of molecules in a gas?

9. We have assumed that the force exerted by molecules on the wall of a container is steady in time. How is this justified?

10. It is found that the weight of an empty flat thin plastic bag is not changed when the bag is filled with air. Why not?

11. We know that a stone will fall to the ground if we release it. We put no constraint on molecules of the air, yet they don't all fall to the ground. Why not?

12. Justify the fact that the pressure of a gas depends on the *square* of the speed of its particles by explaining the dependence of pressure on the collision frequency and the momentum transfer of the particles.

13. How is the speed of sound related to the gas variables in the kinetic theory model?

14. Consider a hot, stationary golf ball sitting on a tee and a cold golf ball just moving off the tee after being hit. The total kinetic energy of the molecules' motion relative to the tee can be the same in the two cases. Explain how. What is the difference between the two cases?

15. Far above the Earth's surface the gas kinetic temperature is reported to be on the order of 1000 K. However, a person placed in such an environment would freeze to death rather than vaporize. Explain.

16. Why doesn't the Earth's atmosphere leak away? At the top of the atmosphere atoms will occasionally be headed out with a speed exceeding the escape speed. Isn't it just a matter of time?

17. Titan, one of Saturn's many moons, has an atmosphere, but our own Moon does not. What is the explanation?

18. How, if at all, would you expect the composition of the air to change with altitude?

19. Explain why the temperature decreases with height in the lower atmosphere.

20. In large-scale inelastic collisions mechanical energy is lost through internal friction resulting in a rise of temperature owing to increased internal molecular agitation. Is there a loss of mechanical energy to heat in an inelastic collision between molecules?

21. By considering quantities that must be conserved in an elastic collision, show that in general molecules of a gas cannot have the same speeds after a collision as they had before. Is it possible, then, for a gas to consist of molecules that all have the same speed?

22. We often say that we see the steam emerging from the spout of a kettle in which water is boiling. However, steam itself is a colorless gas. What is it that you really see?

23. Why does smoke rise, rather than fall, from a lighted candle?

24. Would a gas whose molecules were true geometric points obey the ideal gas law?

25. Why do molecules not travel in perfectly straight lines between collisions and what effect, easily observable in the laboratory, occurs as a result?

26. Why must the time allowed for diffusion separation be relatively short?

27. Suppose we want to obtain ^{238}U instead of ^{235}U as the end product of a diffusion process. Would we use the same pro-

cess? If not, explain how the separation process would have to be modified.

28. Considering the diffusion of gases into each other, can you draw an analogy to a large jostling crowd with many "collisions" on a large inclined plane with a slope of a few degrees?

29. Can you describe a centrifugal device for gaseous separation? Is a centrifuge better than a diffusion chamber for separation of gases?

30. Do the pressure and volume of air in a house change when the furnace raises the temperature significantly? If not, is the ideal gas law violated?

31. Would you expect real molecules to be spherically symmetrical? If not, how would the potential energy function of Fig. 12a change?

32. Explain why the temperature of a gas drops in an adiabatic expansion.

33. If hot air rises, why is it cooler at the top of a mountain than near sea level?

34. Comment on this statement: "There are two ways to carry out an adiabatic process. One is to do it quickly and the other is to do it in an insulated box."

35. A sealed rubber balloon contains a very light gas. The balloon is released and it rises high into the atmosphere. Describe and explain the thermal and mechanical behavior of the balloon.

36. Although real gases can be liquified, an ideal gas cannot be. Explain.

37. Show that as the volume per mole of a gas increases, the van der Waals equation tends to the equation of state of an ideal gas.

38. *Extensive* quantities have values that depend on what the system's boundaries are, whereas *intensive* quantities are independent of the choice of boundaries. That is, extensive quantities are necessarily defined for a whole system, whereas intensive quantities apply uniformly to any small part of the system. Of the following quantities, determine which are extensive and which are intensive: pressure, volume, temperature, density, mass, internal energy.

PROBLEMS

Section 23-1 Macroscopic Properties of a Gas and the Ideal Gas Law

1. (a) Calculate the volume occupied by 1.00 mol of an ideal gas at standard conditions, that is, pressure of 1.00 atm ($= 1.01 \times 10^5$ Pa) and temperature of 0°C ($= 273$ K). (b) Show that the number of molecules per cubic centimeter (the Loschmidt number) at standard conditions is 2.68×10^{19}.

2. The best vacuum that can be attained in the laboratory corresponds to a pressure of about 10^{-18} atm, or 1.01×10^{-13} Pa. How many molecules are there per cubic centimeter in such a vacuum at 22°C?

3. A quantity of ideal gas at 12.0°C and a pressure of 108 kPa occupies a volume of 2.47 m^3. (a) How many moles of the gas are present? (b) If the pressure is now raised to 316 kPa

and the temperature is raised to 31.0°C, how much volume will the gas now occupy? Assume no leaks.

4. Oxygen gas having a volume of 1130 cm^3 at 42.0°C and a pressure of 101 kPa expands until its volume is 1530 cm^3 and its pressure is 106 kPa. Find (a) the number of moles of oxygen in the system and (b) its final temperature.

5. A weather balloon is loosely inflated with helium at a pressure of 1.00 atm ($= 76.0$ cm Hg) and a temperature of 22.0°C. The gas volume is 3.47 m^3. At an elevation of 6.50 km, the atmospheric pressure is down to 36.0 cm Hg and the helium has expanded, being under no restraint from the confining bag. At this elevation the gas temperature is -48.0°C. What is the gas volume now?

6. The variation in pressure in the Earth's atmosphere, assumed to be at a uniform temperature, is given by

$p = p_0 e^{-Mgy/RT}$, where M is the molar mass of the air. (See Section 17-3.) Show that $n_V = n_{V0} e^{-Mgy/RT}$, where n_V is the number of molecules per unit volume.

7. Consider a given mass of ideal gas. Compare curves representing constant pressure, constant volume, and isothermal (constant temperature) processes on (*a*) a pV diagram, (*b*) a pT diagram, and (*c*) a VT diagram. (*d*) How do these curves depend on the mass of gas chosen?

8. Estimate the mass of the Earth's atmosphere. Express your estimate as a fraction of the mass of the Earth. Recall that atmospheric pressure equals 101 kPa.

9. An automobile tire has a volume of 988 in.³ and contains air at a gauge pressure of 24.2 lb/in.² when the temperature is −2.60°C. Find the gauge pressure of the air in the tire when its temperature rises to 25.6°C and its volume increases to 1020 in.³. (*Hint:* It is not necessary to convert from British to SI units. Why? Use $p_{atm} = 14.7$ lb/in.².)

10. (*a*) Consider 1.00 mol of an ideal gas at 285 K and 1.00 atm pressure. Imagine that the molecules are for the most part evenly spaced at the centers of identical cubes. Using Avogadro's constant and taking the diameter of a molecule to be 3.00×10^{-8} cm, find the length of an edge of such a cube and calculate the ratio of this length to the diameter of a molecule. The edge length is an estimate of the distance between molecules in the gas. (*b*) Now consider a mole of water having a volume of 18 cm³. Again imagine the molecules to be evenly spaced at the centers of identical cubes and repeat the calculation in (*a*).

11. An air bubble of 19.4 cm³ volume is at the bottom of a lake 41.5 m deep where the temperature is 3.80°C. The bubble rises to the surface, which is at a temperature of 22.6°C. Take the temperature of the bubble to be the same as that of the surrounding water and find its volume just before it reaches the surface.

12. An open–closed pipe of length $L = 25.0$ m contains air at atmospheric pressure. It is thrust vertically into a freshwater lake until the water rises halfway up in the pipe, as shown in Fig. 16. What is the depth h of the lower end of the pipe? Assume that the temperature is the same everywhere and does not change.

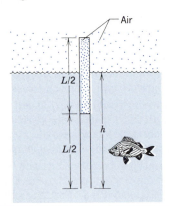

Figure 16   Problem 12.

13. Container A contains an ideal gas at a pressure of 5.0×10^5 Pa and at a temperature of 300 K. It is connected by a thin

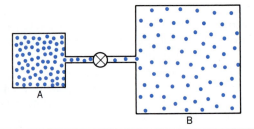

Figure 17   Problem 13.

tube to container B with four times the volume of A; see Fig. 17. B contains the same ideal gas at a pressure of 1.0×10^5 Pa and at a temperature of 400 K. The connecting valve is opened, and equilibrium is achieved at a common pressure while the temperature of each container is kept constant at its initial value. What is the final pressure in the system?

14. Two vessels of volumes 1.22 L and 3.18 L contain krypton gas and are connected by a thin tube. Initially, the vessels are at the same temperature, 16.0°C, and the same pressure, 1.44 atm. The larger vessel is then heated to 108°C while the smaller one remains at 16.0°C. Calculate the final pressure. (*Hint:* There are no leaks.)

15. Consider a sample of argon gas at 35.0°C and 1.22 atm pressure. Suppose that the radius of a (spherical) argon atom is 0.710×10^{-10} m. Calculate the fraction of the container volume actually occupied by atoms.

16. A mercury-filled manometer with two unequal-length arms of the same cross-sectional area is sealed off with the same pressure p_0 in the two arms, as in Fig. 18. With the temperature constant, an additional 10.0 cm³ of mercury is admitted through the stopcock at the bottom. The level on the left increases 6.00 cm and that on the right increases 4.00 cm. Find the pressure p_0.

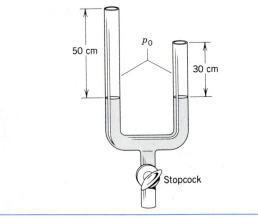

Figure 18   Problem 16.

Section 23-3 Kinetic Calculation of the Pressure

17. The temperature in interstellar space is 2.7 K. Find the root-mean-square speed of hydrogen molecules at this temperature. (See Table 1.)

18. Calculate the root-mean-square speed of ammonia (NH_3) molecules at 56.0°C. An atom of nitrogen has a mass of

2.33×10^{-26} kg and an atom of hydrogen has a mass of 1.67×10^{-27} kg.

19. At 0°C and 1.000 atm pressure the densities of air, oxygen, and nitrogen are, respectively, 1.293 kg/m³, 1.429 kg/m³, and 1.250 kg/m³. Calculate the fraction by mass of nitrogen in the air from these data, assuming only these two gases to be present.

20. The mass of the H_2 molecule is 3.3×10^{-24} g. If 1.6×10^{23} hydrogen molecules per second strike 2.0 cm² of wall at an angle of 55° with the normal when moving with a speed of 1.0×10^5 cm/s, what pressure do they exert on the wall?

21. At 44.0°C and 1.23×10^{-2} atm the density of a gas is 1.32×10^{-5} g/cm³. (*a*) Find v_{rms} for the gas molecules. (*b*) Find the molar mass of the gas and identify it.

22. *Dalton's law* states that when mixtures of gases having no chemical interaction are present together in a vessel, the pressure exerted by each constituent at a given temperature is the same as it would exert if it alone filled the whole vessel, and that the total pressure is equal to the sum of the partial pressures of each gas. Derive this law from kinetic theory, using Eq. 14.

23. A container encloses two ideal gases. Two moles of the first gas are present, with molar mass M_1. Molecules of the second gas have a molar mass $M_2 = 3M_1$, and 0.5 mol of this gas is present. What fraction of the total pressure on the container wall is attributable to the second gas? (*Hint*: See Problem 22.)

Section 23-4 Kinetic Interpretation of the Temperature

24. The Sun is a huge ball of hot ideal gas. The glow surrounding the Sun in the x-ray photo shown in Fig. 19 is the corona—the atmosphere of the Sun. Its temperature and pressure are 2.0×10^6 K and 0.030 Pa. Calculate the rms speed of free electrons in the corona.

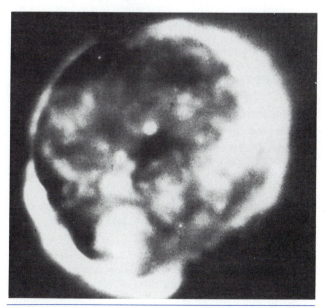

Figure 19 Problem 24.

25. (*a*) Calculate the average value in electron-volts of the translational kinetic energy of the particles of an ideal gas at 0°C

and at 100°C. (*b*) Find the translational kinetic energy per mole of an ideal gas at these temperatures, in joules.

26. At what temperature is the average translational kinetic energy of a molecule in an ideal gas equal to 1.00 eV?

27. Oxygen (O_2) gas at 15°C and 1.0 atm pressure is confined to a cubical box 25 cm on a side. Calculate the ratio of the change in gravitational potential energy of a mole of oxygen molecules falling the height of the box to the total translational kinetic energy of the molecules.

28. Gold has a molar (atomic) mass of 197 g/mol. Consider a 2.56-g sample of pure gold vapor. (*a*) Calculate the number of moles of gold present. (*b*) How many atoms of gold are present?

29. Find the average translational kinetic energy of individual nitrogen molecules at 1600 K (*a*) in joules and (*b*) in electron-volts.

30. (*a*) Find the number of molecules in 1.00 m³ of air at 20.0°C and at a pressure of 1.00 atm. (*b*) What is the mass of this volume of air? Assume that 75% of the molecules are nitrogen (N_2) and 25% are oxygen (O_2).

31. Consider a gas at temperature T occupying a volume V to consist of a mixture of atoms, namely, N_a atoms of mass m_a each having an rms speed v_a, and N_b atoms of mass m_b each having an rms speed v_b. (*a*) Give an expression for the total pressure exerted by the gas. (*b*) Suppose now that $N_a = N_b$ and that the different atoms combine at constant volume to form molecules of mass $m_a + m_b$. Once the temperature returns to its original value, what would be the ratio of the pressure after combination to the pressure before?

32. A steel tank contains 315 g of ammonia gas (NH_3) at an absolute pressure of 1.35×10^6 Pa and temperature 77.0°C. (*a*) What is the volume of the tank? (*b*) The tank is checked later when the temperature has dropped to 22.0°C and the absolute pressure has fallen to 8.68×10^5 Pa. How many grams of gas leaked out of the tank?

33. (*a*) Compute the temperatures at which the rms speed is equal to the speed of escape from the surface of the Earth for molecular hydrogen and for molecular oxygen. (*b*) Do the same for the Moon, assuming the gravitational acceleration on its surface to be 0.16*g*. (*c*) The temperature high in the Earth's upper atmosphere is about 1000 K. Would you expect to find much hydrogen there? Much oxygen?

34. At what temperature do the atoms of helium gas have the same rms speed as the molecules of hydrogen gas at 26.0°C?

35. The envelope and basket of a hot-air balloon have a combined mass of 249 kg, and the envelope has a capacity of 2180 m³. When fully inflated, what should be the temperature of the enclosed air to give the balloon a lifting capacity of 272 kg (in addition to its own mass)? Assume that the surrounding air, at 18.0°C, has a density of 1.22 kg/m³.

Section 23-5 Work Done on an Ideal Gas

36. A sample of gas expands from 1.0 to 5.0 m³ while its pressure decreases from 15 to 5.0 Pa. How much work is done on the gas if its pressure changes with volume according to each of the three processes shown in the *pV* diagram in Fig. 20?

37. Suppose that a sample of gas expands from 2.0 to 8.0 m³ along the diagonal path in the *pV* diagram shown in Fig. 21. It is then compressed back to 2.0 m³ along either path 1 or

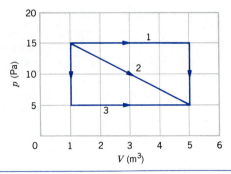

Figure 20 Problem 36.

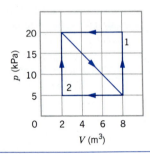

Figure 21 Problem 37.

path 2. Compute the net work done on the gas for the complete cycle in each case.

38. The speed of sound in different gases at the same temperature depends on the molar mass of the gas. Show that $v_1/v_2 = \sqrt{M_2/M_1}$ (constant T), where v_1 is the speed of sound in the gas of molar mass M_1 and v_2 is the speed of sound in the gas of molar mass M_2.

39. Air at 0.00°C and 1.00 atm pressure has a density of 1.291×10^{-3} g/cm³, and the speed of sound is 331 m/s at that temperature. Compute (a) the value of γ of air and (b) the effective molar mass of air.

40. Air that occupies 0.142 m³ at 103 kPa gauge pressure is expanded isothermally to zero gauge pressure and then cooled at constant pressure until it reaches its initial volume. Compute the work done on the gas.

41. Calculate the work done by an external agent in compressing 1.12 mol of oxygen from a volume of 22.4 L and 1.32 atm pressure to 15.3 L at the same temperature.

42. Use the result of Sample Problem 6 to show that the speed of sound in air increases about 0.59 m/s for each Celsius degree rise in temperature near 20°C.

43. Gas occupies a volume of 4.33 L at a pressure of 1.17 atm and a temperature of 310 K. It is compressed adiabatically to a volume of 1.06 L. Determine (a) the final pressure and (b) the final temperature, assuming the gas to be an ideal gas for which $\gamma = 1.40$. (c) How much work was done on the gas?

44. (a) One liter of gas with $\gamma = 1.32$ is at 273 K and 1.00 atm pressure. It is suddenly (adiabatically) compressed to half its

original volume. Find its final pressure and temperature. (b) The gas is now cooled back to 273 K at constant pressure. Find the final volume. (c) Find the total work done on the gas.

45. The gas in a cloud chamber at a temperature of 292 K undergoes a rapid expansion. Assuming the process is adiabatic, calculate the final temperature if $\gamma = 1.40$ and the volume expansion ratio is 1.28.

46. An air compressor takes air at 18.0°C and 1.00 atm pressure and delivers compressed air at 2.30 atm pressure. The compressor operates at 230 W of useful power. Assume that the compressor operates adiabatically. (a) Find the temperature of the compressed air. (b) How much compressed air, in liters, is delivered each second?

47. A thin tube, sealed at both ends, is 1.00 m long. It lies horizontally, the middle 10.0 cm containing mercury and the two equal ends containing air at standard atmospheric pressure. If the tube is now turned to a vertical position, by what amount will the mercury be displaced? Assume that the process is (a) isothermal and (b) adiabatic. (For air, $\gamma = 1.40$.) Which assumption is more reasonable?

Section 23-6 The Internal Energy of an Ideal Gas

48. Calculate the internal energy of 1 mole of an ideal gas at 25.0°C.

49. Calculate the total rotational kinetic energy of all the molecules in 1 mole of air at 25.0°C.

50. A cosmic-ray particle with energy 1.34 TeV is stopped in a detecting tube that contains 0.120 mol of neon gas. Once this energy is distributed among all the atoms, by how much is the temperature of the neon increased?

51. An ideal gas experiences an adiabatic compression from $p = 122$ kPa, $V = 10.7$ m³, $T = -23.0$°C to $p = 1450$ kPa, $V = 1.36$ m³. (a) Calculate the value of γ. (b) Find the final temperature. (c) How many moles of gas are present? (d) What is the total translational kinetic energy per mole before and after the compression? (e) Calculate the ratio of the rms speed before to that after the compression.

Section 23-8 The Van der Waals Equation of State

52. Van der Waals b for oxygen is 32 cm³/mol. Compute the diameter of an O_2 molecule.

53. Using the values of a and b for CO_2 found in Sample Problem 8, calculate the pressure at 16.0°C of 2.55 mol of CO_2 gas occupying a volume of 14.2 L. Assume (a) that the van der Waals equation is correct, then (b) that CO_2 behaves as an ideal gas.

54. Calculate the work done on n moles of a van der Waals gas in an isothermal expansion from volume V_i to V_f.

55. Show that $V_{cr} = 3nb$.

56. The constants a and b in the van der Waals equation are different for different substances. Show, however, that if we take V_{cr}, p_{cr}, and T_{cr} as the units of volume, pressure, and temperature, the van der Waals equation becomes identical for all substances.

CHAPTER 24

STATISTICAL MECHANICS

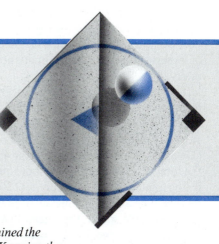

In Section 23-4, which dealt with kinetic theory, we determined the average translational kinetic energy of the molecules of a gas. Knowing the average, however, tells us nothing about how the speeds of individual molecules are distributed about the average. In some cases, the average might provide sufficient information about the properties of the gas, such as its temperature. In other cases, it might be necessary to have more information about the distribution of speeds.

If you are designing commercial passenger aircraft, you must know the average weight of the passengers and their baggage to estimate the lift required for the plane to fly safely. The number of overweight or underweight passengers is of little interest. On the other hand, if your job is to order suits for a clothing store, you must have information on the distribution of sizes; knowing the average size of the customers is of little help.

*In this chapter, we are concerned with distributions of molecular speeds and energies, and their use in computing the macroscopic properties of collections of molecules. This approach to thermodynamics is called **statistical mechanics**. Its classical formulations were first worked out in the 19th century by Maxwell, Gibbs, and Boltzmann. In the 20th century, many of these same techniques were applied by Einstein, Planck, Fermi, and others to systems governed by the laws of quantum mechanics.*

24-1 STATISTICAL DISTRIBUTIONS AND MEAN VALUES

A highway engineer might want to have some information on the distribution of speeds of the automobiles that use a certain section of roadway. By setting sensors a known distance apart, the engineer can determine the time necessary for an automobile to cross that distance and so determine its speed. After accumulating and storing such data for a period of several weeks, the engineer wishes to analyze the data to study the need for improvements in the road. How can the information be displayed in a way that permits this analysis?

A simple list of the speeds provides too much information and is unmanageable. Instead, the engineer sorts the speeds into groups. How many cars have speeds between 0 and 5 mph? Between 5 and 10 mph? The result of such a sorting might be similar to that shown in Fig. 1, which gives the *statistical distribution* of speeds in the form

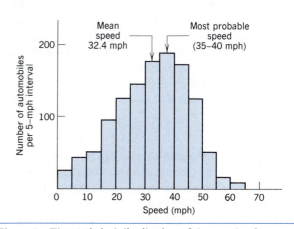

Figure 1 The statistical distribution of the speeds of automobiles traveling a particular section of roadway. The data are sorted into bins of width 5 mph. The height of each bin shows the number of automobiles with speeds in that 5-mph range. The data can be characterized by the mean speed and the most probable speed.

known as a *histogram.* Each rectangular area has a width equal to the size of the sorting interval (which need not necessarily all be equal) and a height equal to the number of observations or relative frequency of values in that interval.

The distribution gives the engineer all the essential information about the traffic on this road, on which the speed limit is 45 mph. The total number of cars in the survey is just the sum of the heights of all the rectangles (1205). The fraction that exceed the 45-mph speed limit is the total of heights of the last four intervals (194) divided by the total number in the survey (1205), which gives 0.16 or 16%. The average or mean speed is 32.4 mph, but the most probable speed (the interval with the largest number) is in the range 35–40 mph. The engineer can now decide whether road improvements might be needed to increase the mean speed or whether better enforcement might be needed to decrease the number of speeders.

Figure 1 is a statistical distribution based purely on empirical data. There is no "theory of automobile speeds" that can be used to derive a mathematical formula that predicts the shape of the distribution. Nevertheless, it provides essential information about the statistical behavior of automobiles in this situation. It is of no interest to the engineer whether a particular car has a speed of, say, 32.46 mph or 33.14 mph. Even though the speed is a continuous variable, we sort the possible speeds into discrete bins, and the relative numbers in the bins help us to understand a physical situation.

The average or mean value of the speed, $\bar{v}$, can be computed from the statistical distribution of Fig. 1. Let us assume we have a total of B bins or intervals into which the data have been sorted; in Fig. 1, $B = 13$. The intervals are labeled by the index i, where $i = 1, 2, \ldots, B$. We choose a typical or representative value v_i of the speed in each interval; that value might, for example, be the speed at the center of each interval. Each interval has a width δv (which for simplicity we assume to be the same for all intervals, although in general it need not be) and a height $n(v_i)$, the number of observations for that interval corresponding to the representative speed v_i. The total number of observations N is then given by the total of the numbers of observations in each bin, or

$$N = \sum_i n(v_i), \tag{1}$$

the sum being carried out over all the bins.

To find $\bar{v}$, we find the sum of all the observations (which is equivalent to the sum of the products of the representative speed in each interval with the number of observations in that interval), and we divide that sum by the total number of observations:

$$\bar{v} = \frac{\sum_i v_i n(v_i)}{\sum_i n(v_i)}. \tag{2}$$

Equation 2 resembles the formula for calculating the

center of mass of a system of particles (Eq. 11 of Chapter 9). We can regard the center of mass as a sort of "average" location of the particles in the system.

We can also define the relative frequency or probability of any value v_i as

$$f(v_i) = \frac{n(v_i)}{\sum_i n(v_i)} = \frac{n(v_i)}{N}. \tag{3}$$

For example, in Fig. 1, $n(v_1)$, the number in the range 0–5 mph, is 23; thus the relative probability of cars having speeds in that interval is $f(v_i) = 23/1205 = 0.019$ or 1.9%. In terms of the probability $f(v_i)$, we can write the average value of Eq. 2 as

$$\bar{v} = \sum_i v_i f(v_i). \tag{4}$$

Sample Problem 1   Figure 1 is based on the observation of a total of 1205 cars, whose speed distribution is as follows:

i	*Speed Interval* (mph)	$n(v_i)$
1	0–5	23
2	5–10	41
3	10–15	54
4	15–20	95
5	20–25	123
6	25–30	142
7	30–35	177
8	35–40	186
9	40–45	170
10	45–50	122
11	50–55	50
12	55–60	15
13	60–65	7

Find the mean value of the speed for this distribution, taking the speed at the middle of each interval as representing the entire interval.

Solution   Using Eq. 2 we have

$$\bar{v} = \frac{(2.5 \text{ mph})(23) + (7.5 \text{ mph})(41) + \cdots + (62.5 \text{ mph})(7)}{23 + 41 + \cdots + 7}$$

$$= 32.4 \text{ mph}.$$

Figure 2 shows another kind of statistical distribution, the results of a set of 1000 observations of the number of times $n(h)$, in a deal of 13 cards from a deck of 52 cards, that you would be dealt a hand of h cards of a specified suit (say, hearts). This distribution looks somewhat like that of Fig. 1: it has a mean value $\bar{h}$ (which should be roughly 3.25, equal to one-quarter of the total number of hearts, as expected since we selected one-quarter of the total number of cards from the deck). Above the mean, the probability decreases rapidly, becoming exceedingly small as h becomes larger. This case differs from that of Fig. 1 in two respects: (1) the variable *h* is discrete (that is, just the numbers 1, 2, 3, . . . , 13), rather than continuous, and (2) we can calculate *n(h)* from basic principles. We make one important assumption: any particular selection of 13 cards is as likely to occur as any other selection. From this assumption it can be shown that the relative probability *f(h)* of deals giving *h* hearts is

$$f(h) = \frac{n(h)}{N} = \frac{(39!)^2(13!)^2}{52![(13-h)!]^2(26+h)!h!} , \qquad (5)$$

where *k*! (*k* factorial) means $1 \times 2 \times 3 \times 4 \times \cdots \times k$, the product of all the integers from 1 through *k*. (See Problem 1 for a derivation of Eq. 5 and Problem 35 for a computer exercise testing Eq. 5.)

For a particular set of *N* trials (such as $N = 1000$ used in Fig. 2) we can compare the observed distribution with the predicted one. If we repeat the experiment with another 1000 trials, we would not expect to observe *exactly* the same distribution (that is, instead of 12 hands with zero hearts we might find 11 or 13), but it should have pretty much the same shape. Equation 5 gives the probabilities when the number of observations is infinite; distributions based on a finite number of observations differ somewhat from predictions based on Eq. 5, but the larger the number of observations, the smaller the deviation.

Statistical distributions, such as those of Figs. 1 and 2, give the frequency or probability to observe values of a variable within a particular range. Calculating mean or average values is one use of these distributions. In some cases, as in Fig. 1, the distribution may be purely empirical, with no underlying theory. In other cases, as in Fig. 2, the observed distribution may be compared with calculations based on particular theory.

Knowledge of the statistical distribution can help in formulating a theory in cases in which the underlying theory is unknown. For example, precise data on the wavelength distribution of thermal radiation (such as that emitted by glowing objects) led to the development of the quantum theory in the early 1900s by Planck and Einstein. We consider the distribution of thermal radiation in Chapter 49 of the extended text.

Sample Problem 2 In a deal of 13 cards, what is the probability to be dealt a hand with 10 hearts?

Solution Using Eq. 5 with $h = 10$, we find

$$f(h) = \frac{(39!)^2(13!)^2}{52!(3!)^2 36! 10!} = 4.1 \times 10^{-6},$$

or about four chances out of one million.

Many scientific calculators have a factorial key, which simplifies the evaluation of such expressions as this. Even so, some care must be taken; it may not be possible to evaluate the entire numerator, which exceeds 10^{99}, the largest number that many calculators can store. In such a case, we can evaluate the expression by alternating factors from the numerator and denominator, keeping the resulting quantity within the range of the calculator. For example, we might start with 39!, divide by 52!, multiply by 39!, divide by 36!, and so on.

If your calculator does not have a factorial key, you need not compute each factorial product to evaluate this expression. We can simplify the expression by writing $39!/36! = 39 \times 38 \times 37$ and $39!/52! = 1/40 \times 41 \times 42 \times \cdots \times 52$, reducing the number of factors that must be multiplied.

24-2 MEAN FREE PATH

Let us suppose we could follow the motion of one molecule in a gas as it zigzags around due to collisions with neighboring molecules (Fig. 3). In particular we measure the distance it travels between collisions, and we repeat the experiment many times. We wish to determine the

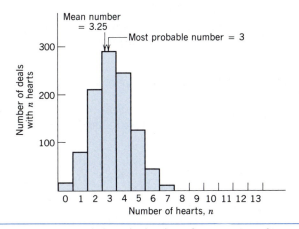

Figure 2 The statistical distribution of the number of hearts received in a deal of 13 cards from a deck of 52 cards. This sample represents one possible set of outcomes from 1000 different deals.

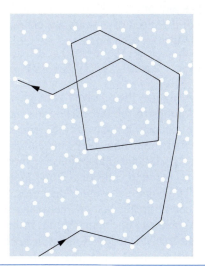

Figure 3 A molecule traveling through a gas, colliding with other molecules in its path. Of course, the other molecules are themselves moving and experiencing collisions.

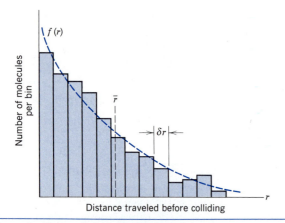

Figure 4   The statistical distribution of the distance traveled by a molecule between collisions. The mean value $\bar{r}$ gives the mean free path of the molecules.

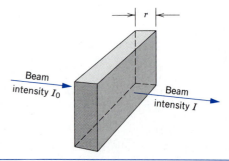

Figure 5   A beam of molecules is incident on a thin layer of gas of thickness r. The intensity I of the beam passing through the layer is a measure of the number of molecules of the beam that experience no collisions over the distance r.

average distance the molecule travels between collisions, and we hope to understand that value on the basis of a microscopic view of the gas. We are thus continuing our efforts to relate macroscopic and microscopic properties of the gas.

Let us first assume that we have no underlying theory of molecular collisions. We form a distribution based only on the results of our many measurements of the distance traveled by a molecule between collisions. This approach is similar to the experiment to study the mean speed of cars described in the previous section. After making a large number N of measurements of the distance traveled by the molecule, we sort those distances into B bins of equal widths δr. We choose a representative value of the distance r_i for each bin, and we plot the number $n(r_i)$ of molecules traveling distances within the width δr of that bin. That is, we have $n(r_1)$ observations in which the distance between collisions is between 0 and δr, $n(r_2)$ in which the distance is between δr and $2\delta r$, and so on. Figure 4 shows a plot of the resulting distribution. As we did in the case of the speed of the cars, we define an average distance $\bar{r}$ in analogy with Eq. 4:

$$\bar{r} = \sum_i r_i f(r_i), \qquad (6)$$

where $f(r_i) = n(r_i)/N$, the relative frequency or probability of observations in the bin corresponding to the distance r_i. The distance $\bar{r}$ is known as the *mean free path* and is represented by the symbol λ. The mean free path represents an average distance that a molecule moves (at constant velocity) between collisions. Some molecules may collide after moving much smaller distances, while some may move a much greater distance between collisions. As we shall see, the mean free path depends inversely on the size of the molecules (the larger the molecules, the more likely they are to collide) and on the number of molecules per unit volume (the greater the number of molecules, the more likely they are to collide).

Let us now see if we can understand the shape of the

statistical distribution of distances in Fig. 4. That is, can we find a continuous function $f(r)$ that gives the probability for a molecule to travel a distance r before having a collision? To attempt to answer this question, let us do a different experiment. We cause a beam of molecules to be incident at a rate or intensity I_0 on a thin layer of gas, and we measure the rate or intensity I at which the molecules emerge after passing through a thickness r of the gas (Fig. 5). Those molecules in the beam that have collisions are scattered into different directions, and thus the emerging beam intensity I is determined by the number of molecules that pass through the distance r without collision. We assume that no molecule in the beam is scattered more than once by a target molecule; this assumption is justified if the density of target molecules is not too great. Let us now increase the thickness of the layer by an amount dr and find the resulting change dI in the intensity I. We expect that the decrease in the intensity is proportional to I (the more molecules in the beam, the more that are scattered) and to the additional thickness dr of the gas layer. Thus

$$dI = -cI\,dr, \qquad (7)$$

where c is a (positive) constant of proportionality to be determined. Note that dI is negative (the beam intensity decreases with the addition of the thickness dr). Rewriting Eq. 7 gives a form that can be integrated:

$$\frac{dI}{I} = -c\,dr, \qquad (8)$$

and integrating between the limits I_0, corresponding to $r = 0$ (no gas layer), and I, corresponding to a thickness r, we obtain

$$\int_{I_0}^{I} \frac{dI}{I} = \int_{0}^{r} (-c\,dr)$$

$$\ln \frac{I}{I_0} = -cr$$

$$I(r) = I_0 e^{-cr}. \qquad (9)$$

That is, the beam intensity decreases exponentially with the thickness r of the gas layer.

This calculation suggests an exponential form for $f(r)$, and we choose

$$f(r) = Ae^{-cr}, \qquad (10)$$

where A is another constant to be determined. This function is in effect the dashed curve of Fig. 4, since it gives the height $n(r_i)$ of any bin when evaluated at the distance $r = r_i$, according to $n(r_i) = Nf(r_i)$. The total number of measurements N is the total of the heights of all the bins, $N = \Sigma \, n(r_i)$. We can therefore rewrite Eq. 6, replacing $\bar{r}$ by λ, as

$$\lambda = \sum_i r_i f(r_i) = \frac{\sum_i r_i n(r_i)}{N} = \frac{\sum_i r_i n(r_i)}{\sum_i n(r_i)}$$

$$= \frac{\sum_i r_i f(r_i)\delta r}{\sum_i f(r_i)\delta r}, \qquad (11)$$

where in the last step we have multiplied numerator and denominator by δr so that we can easily convert the sums to integrals. Let us now make the widths δr very small, so that we can write Eq. 11 in terms of integrals:

$$\lambda = \frac{\int_0^\infty rf(r)\,dr}{\int_0^\infty f(r)\,dr}. \qquad (12)$$

Substituting $f(r) = Ae^{-cr}$, we carry out the integrals (see Problem 2) and obtain

$$\lambda = \frac{1}{c}. \qquad (13)$$

Thus the exponential probability distribution should be written

$$f(r) = Ae^{-r/\lambda}. \qquad (14)$$

Equation 14 shows how the mean free path enters into the calculation of the probability for the molecules to travel a distance r before a collision. Note that the constant A cancels from the ratio in Eq. 12 and thus does not affect the calculation of the mean free path. We can determine A by demanding that the total of all $n(r_i)$ is N (see Problem 3).

Microscopic Calculation of the Mean Free Path

Let us now return to our agenda in thermodynamics and see how we can understand the mean free path (a macroscopic quantity) from the microscopic properties of the molecules.

If molecules were points, they would not collide at all and the mean free path would be infinite. Molecules, however, are not points and hence collisions occur. If the molecules were so numerous that they completely filled the space available to them, leaving no room for translational motion, the mean free path would be zero. Thus the

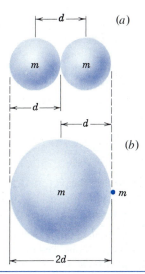

Figure 6   (a) A collision occurs when the centers of two molecules come within a distance d of each other, where d is the molecular diameter. (b) An equivalent but more convenient representation is to think of the moving molecule as having a diameter $2d$, all other molecules being points.

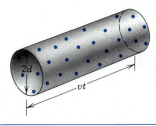

Figure 7   A molecule with an equivalent diameter $2d$ (as in Fig. 6b) traveling with speed v sweeps out a cylinder of base area πd^2 and length vt in a time t. The number of collisions suffered by the molecule in this time is equal to the number of molecules (regarded as points) that lie within the cylinder. In actuality, this cylinder would be bent many times as the direction of the molecule's path is changed by collisions; for convenience that path has been straightened.

mean free path is related to the size of the molecules and to their number per unit volume.

Consider the molecules of a gas to be spheres of diameter d. A collision will take place when the centers of two molecules approach within a distance d of one another. An equivalent description of collisions made by any one molecule is to regard that molecule as having a diameter $2d$ and all other molecules as point particles (see Fig. 6).

Imagine a typical molecule of equivalent diameter $2d$ moving with speed v through a gas of equivalent point particles, and let us temporarily assume that the molecule and the point particles exert no forces on each other. In time t our molecule sweeps out a cylinder of cross-sectional area πd^2 and length vt. If $\rho_n \, (= N/V)$ is the number of molecules per unit volume, the cylinder contains $(\pi d^2 vt)\rho_n$ particles (see Fig. 7). Since our molecule and the point particles *do* exert forces on each other, this is also the number of collisions experienced by the molecule in time t. The cylinder of Fig. 7 is, in fact, a broken one, changing direction with every collision.

The mean free path λ is the average distance between successive collisions. Hence λ is the total distance vt covered in time t divided by the number of collisions that take place in this time, or

$$\lambda = \frac{vt}{\pi d^2 \rho_n vt} = \frac{1}{\pi d^2 \rho_n}. \quad (15)$$

This equation is based on the picture of a molecule hitting stationary targets. Actually the molecule hits moving targets. When the target molecules are moving, the two v's in Eq. 15 are not the same. The one in the numerator $(= \bar{v})$ is the mean molecular speed measured with respect to the container. The one in the denominator $(= \bar{v}_{rel})$ is the mean *relative* speed with respect to other molecules; it is this relative speed that determines the collision rate.

We can see qualitatively that $\bar{v}_{rel} > \bar{v}$ as follows. Two molecules of speed v moving toward each other have a relative speed of $2v$ $(> v)$; two molecules with speed v moving at right angles on a collision course have a relative speed of $\sqrt{2}v$ (also $> v$); two molecules moving with speed v in the same direction have a relative speed of zero $(< v)$. Thus molecules arriving from *all of the forward hemisphere* and *part of the backward hemisphere* have $\bar{v}_{rel} > \bar{v}$. The molecules arriving from the rest of the backward hemisphere have $\bar{v}_{rel} < \bar{v}$ but, since there are fewer of them, the overall average gives $\bar{v}_{rel} > \bar{v}$. A quantitative calculation, taking into account the actual speed distribution of the molecules, gives $\bar{v}_{rel} = \sqrt{2}\,\bar{v}$. As a result, Eq. 15 becomes

$$\lambda = \frac{1}{\sqrt{2}\pi d^2 \rho_n}. \quad (16)$$

Taking the motion of the target molecules into account, we see that the collision frequency is increased and the mean free path is reduced, relative to their values for stationary target molecules.

The mean free path of air molecules at sea level is about 0.1 μm. At an altitude of 100 km, the density of air has dropped to such an extent that the mean free path rises to about 16 cm. At 300 km, the mean free path is about 20 km. A problem faced by those who study the physics and the chemistry of the upper atmosphere in the laboratory is the fact that no available containers are large enough to permit the contained gas samples to simulate upper atmospheric conditions. Studies of the concentrations of Freon, carbon dioxide, and ozone in the upper atmosphere are of vital public concern.

At a pressure of 10^{-7} mm Hg (about 10^{-10} atm), a reasonably good laboratory vacuum, the density of molecules (which can be found from the molecular form of the ideal gas law, $pV = NkT$) is $\rho_n = N/V = p/kT = 3 \times 10^{15}$ m^{-3} at room temperature (300 K). Equation 16 then gives about 1 km for the mean free path. That is, in an ordinary laboratory vacuum chamber of dimension typically 1 m

or less, the molecules hardly ever collide with one another; only collisions with the walls are likely to occur.

Such considerations can be important in the design of particle accelerators, in which a beam of particles must travel many times through the accelerator and must have a negligibly small probability of making collisions with any air molecules that may be present in the evacuated accelerator tube. Even though a mean free path of 1 km seems large by ordinary laboratory standards, a particle making hundreds or thousands of revolutions through a circular accelerator can travel much longer distances. In the case of proton accelerators, the diameter of a proton is of the order of 10^{-15} m, far smaller than the diameter of a molecule (10^{-10} m), and the resulting increase in λ is sufficient to reduce the probability of collisions to an acceptably small value.

Sample Problem 3   The molecular diameters of different kinds of gas molecules can be found experimentally by measuring the rates at which different gases diffuse into each other. For nitrogen, $d = 3.15 \times 10^{-10}$ m has been reported. What are the mean free path and the average rate of collision for nitrogen at room temperature ($T = 300$ K) and at atmospheric pressure?

Solution   Let us first find $\rho_n (= N/V)$, the number of molecules per unit volume under these conditions. From the ideal gas law ($pV = NkT$), we obtain

$$\rho_n = \frac{N}{V} = \frac{p}{kT} = \frac{(1 \text{ atm})(1.01 \times 10^5 \text{ Pa/atm})}{(1.38 \times 10^{-23} \text{ J/K})(300 \text{ K})}$$

$$= 2.44 \times 10^{25} \text{ molecules/m}^3.$$

Equation 16 then gives the mean free path

$$\lambda = \frac{1}{\sqrt{2}\pi \rho_n d^2} = \frac{1}{\sqrt{2}\pi(2.44 \times 10^{25} \text{ m}^{-3})(3.15 \times 10^{-10} \text{ m})^2}$$

$$= 9.3 \times 10^{-8} \text{ m}.$$

This is about 300 molecular diameters. On average, the molecules in such a gas are about $\rho_n^{-1/3} = 3.45 \times 10^{-9}$ m or 11 molecular diameters apart. Thus a molecule will typically pass about 30 other molecules before experiencing a collision.

To find the average rate of collision, we first find the average speed, for which we use v_{rms} given by Eq. 16 of Chapter 23:

$$v_{rms} = \sqrt{\frac{3RT}{M}} = \sqrt{\frac{3(8.31 \text{ J/mol} \cdot \text{K})(300 \text{ K})}{0.028 \text{ kg/mol}}} = 517 \text{ m/s}.$$

We find the average collision rate by dividing this average speed by the mean free path, or

$$\text{rate} = \frac{v}{\lambda} = \frac{517 \text{ m/s}}{9.3 \times 10^{-8} \text{ m}} = 5.6 \times 10^9 \text{ s}^{-1}.$$

On the average, every nitrogen molecule makes more than 5 billion collisions per second! Nitrogen is the principal constituent of air, and therefore the results of this sample problem can be taken to be representative of air under normal conditions of temperature and pressure.

24-3 THE DISTRIBUTION OF MOLECULAR SPEEDS

Figure 8 shows a small number of gas molecules confined to a box. We assume the gas is in thermal equilibrium at temperature T. The molecules move randomly and make collisions with the walls and with one another. As we learned in Section 23-4, the temperature determines the mean kinetic energy per molecule, but we wish to learn about the entire distribution of molecular speeds, not just the mean value. It is unlikely that all molecules could have the same speed, because collisions between molecules would soon upset this situation. However, it is similarly unlikely that too many of the molecules have speeds that are very different from the mean value; speeds close to zero or speeds much greater than v_{rms} would require a sequence of preferential collisions that would be very improbable in a condition of equilibrium.

Let us suppose we had a small probe that we could place in the box to determine the number of molecules passing through it with speeds between v and $v + \delta v$. We then sort the measured speeds into bins of width δv and plot the resulting statistical distribution. Figure 9 shows a set of results that might be obtained from this experiment.

Figure 8   The molecules of a gas confined to a chamber have a distribution of speeds.

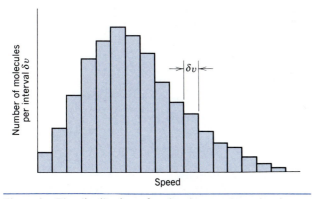

Figure 9   The distribution of molecular speeds. As in Fig. 1, the measured speeds have been sorted into bins of width δv. The height of each bin gives the number of molecules with speeds between v and $v + \delta v$.

The distribution has a clearly defined mean and falls to zero both at low speed and at high speed. Even if we introduced all the molecules into the box at the same speed, the randomizing effect of the collisions will eventually produce a distribution of the form of Fig. 9. Note the similarity between Fig. 9 and Fig. 1.

Maxwell first solved the problem of the distribution of speeds in a gas containing a large number of molecules. The Maxwellian speed distribution, for a sample of gas at temperature T containing N molecules each of mass m, is

$$n(v) = 4\pi N \left(\frac{m}{2\pi kT} \right)^{3/2} v^2 e^{-mv^2/2kT}. \qquad (17)$$

In this equation $n(v)$, which has dimensions of v^{-1}, is the number of molecules per unit speed interval having speeds between v and $v + dv$; equivalently, $n(v)dv$ is the dimensionless number of molecules in the gas sample having speeds between v and $v + dv$. Note that for a given gas the speed distribution depends only on the temperature.

We find N, the total number of molecules in the sample, by adding up (that is, by integrating) the number present in each differential speed interval from zero to infinity, or

$$N = \int_0^\infty n(v)\, dv. \qquad (18)$$

Figure 10 shows the Maxwell distribution of speeds for molecules of oxygen at room temperature. Comparison of Figs. 9 and 10 shows that the Maxwell speed distribution does indeed have the same form as the measured distribution.

Although $n(v)$ is a well-defined mathematical function, the *physically* meaningful quantity is $n(v)dv$, the number of molecules with speeds between v and $v + dv$. We cannot speak of "the number of particles with speed v," because there is a finite number of molecules but an infinite number of possible speeds. The probability that a particle

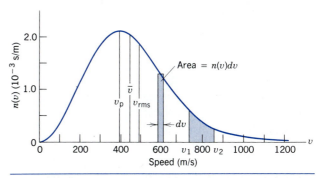

Figure 10   The Maxwell speed distribution for the molecules of a gas. The plotted curve is characteristic of oxygen molecules at $T = 300$ K. The number of molecules with speeds in any interval dv is $n(v)dv$, indicated by the narrow shaded strip. The number with speeds between any limits v_1 and v_2 is given by the area under the curve between those limits.

has a precisely stated speed, such as 279.343267. . . m/s, is exactly zero. However, we can divide the range of speeds into intervals, and the probability that a particle has a speed somewhere in a given interval dv (such as 279 m/s to 280 m/s) has a definite nonzero value. The number of molecules with speeds between any two limits, such as v_1 and v_2, is equal to the area under the $n(v)$ curve between those limits. The *total* area under the curve equals the total number of molecules N, as shown by Eq. 18.

The distribution $n(v)$ can be characterized by the most probable speed v_p [the speed where $n(v)$ has its maximum], the mean speed $\bar{v}$, and the root-mean-square speed v_{rms}. Figure 10 shows the relationship of these speeds, and Sample Problem 6 shows how to calculate them. The distribution curve is not symmetrical about the most probable speed because the lowest speed must be zero while there is no classical limit to the highest speed that a molecule can have. The mean speed is therefore larger than the most probable speed. The root-mean-square speed, which involves the mean of the squares, is still larger.

As the temperature increases, the root-mean-square speed v_{rms} (and $\bar{v}$ and v_p as well) increases, in accord with our microscopic interpretation of temperature. At higher temperatures, the range of typical speeds is greater, and the distribution is broader. Since the area under the distribution curve (which is the total number of molecules in the sample) remains the same, the distribution must also flatten as the temperature rises. Hence the number of molecules that have speeds greater than some given speed increases as the temperature increases (see Fig. 11). This explains many phenomena, such as the increase in the

rates of chemical reactions or certain nuclear reactions with rising temperature.

The distribution of speeds of molecules in a liquid also resembles the curves of Fig. 11. The speed needed for a molecule to escape from the surface of the liquid would be far out on the tail of the distribution of Fig. 11; only a very small number of molecules have speeds above this threshold. Even though the temperature of the liquid is well below its normal boiling point, these few molecules can overcome the attraction of other molecules in the surface and escape by evaporation. The escape of these energetic molecules reduces the average kinetic energy of the remaining molecules, leaving the liquid at a lower temperature. This explains why evaporation is a cooling process. If the liquid is isolated from its surroundings, it will indeed cool, and the rate of evaporation will decrease. If the liquid is not isolated, then energy from the surroundings will flow into the liquid, keeping constant the fraction of molecules with speeds above the evaporation threshold, and eventually all molecules will acquire enough energy to escape into the vapor.

From Eq. 17 we see that the distribution of molecular speeds depends on the mass of the molecule as well as on the temperature. The smaller the mass, the larger the proportion of high-speed molecules at any given temperature. Hence hydrogen is more likely to escape from the atmosphere at high altitudes than oxygen or nitrogen. The Moon has a tenuous atmosphere. For the molecules in this atmosphere to have small probability of escaping from the weak gravitational pull of the Moon, we would expect them to be molecules or atoms of the heavier elements, such as the heavy inert gases krypton and xenon, which were produced largely by radioactive decay early in the Moon's history. The atmospheric pressure on the Moon is about 10^{-13} of the Earth's atmospheric pressure.

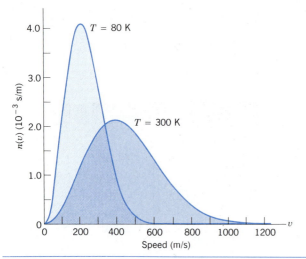

Figure 11 A comparison of the Maxwell speed distribution for oxygen molecules at two different temperatures. The molecules in general have lower average speeds at the lower temperatures, although both distributions cover the entire range of speeds. The areas of the two distributions are equal, because the total number of molecules is the same in both cases.

Sample Problem 4   The speeds of ten particles in m/s are 0, 1.0, 2.0, 3.0, 3.0, 3.0, 4.0, 4.0, 5.0, and 6.0. Find (*a*) the average speed, (*b*) the root-mean-square speed, and (*c*) the most probable speed of these particles.

Solution   (*a*) The average speed is

$$\bar{v} = \frac{0 + 1.0 + 2.0 + 3.0 + 3.0 + 3.0 + 4.0 + 4.0 + 5.0 + 6.0}{10}$$

$$= 3.1 \text{ m/s.}$$

(*b*) The mean-square speed is

$$\bar{v^2} = \frac{1}{10}[0 + (1.0)^2 + (2.0)^2 + (3.0)^2 + (3.0)^2$$
$$+ (3.0)^2 + (4.0)^2 + (4.0)^2 + (5.0)^2 + (6.0)^2]$$

$$= 12.5 \text{ m}^2/\text{s}^2,$$

and the root-mean-square speed is

$$v_{rms} = \sqrt{12.5 \text{ m}^2/\text{s}^2} = 3.5 \text{ m/s.}$$

(*c*) Of the ten particles, three have speeds of 3.0 m/s, two have speeds of 4.0 m/s, and each of the other five has a different speed. Hence the most probable speed v_p of a particle is

$$v_p = 3.0 \text{ m/s}.$$

Sample Problem 5 A container filled with N molecules of oxygen gas is maintained at 300 K. What fraction of the molecules has speeds in the range 599–601 m/s? The molar mass M of oxygen is 0.032 kg/mol.

Solution This speed interval δv (= 2 m/s) is so small that we can treat it as a differential dv. The number of molecules in this interval is $n(v)dv$, and the fraction in that interval is $f = n(v)dv/N$, where $n(v)$ is to be evaluated at $v = 600$ m/s, the midpoint of the range; see the narrow shaded strip in Fig. 10. Using Eq. 17 with the substitution $m/k = M/R$, we find the fraction

$$f = \frac{n(v)\, dv}{N} = 4\pi \left(\frac{M}{2\pi RT} \right)^{3/2} v^2 e^{-Mv^2/2RT}\, dv.$$

Substituting the given numerical values yields

$$f = 2.6 \times 10^{-3} \text{ or } 0.26\%.$$

At room temperature, 0.26% of the oxygen molecules have speeds that lie in the narrow range between 599 and 601 m/s. If the shaded strip of Fig. 10 were drawn to the scale of this problem, it would be a very thin strip indeed.

Sample Problem 6 What are the average speed $\bar{v}$, the root-mean-square speed v_{rms}, and the most probable speed v_p of oxygen molecules at $T = 300$ K? The molar mass M of oxygen is 0.032 kg/mol.

Solution To find the average speed, we sum all the measured speeds, which is done most simply by summing the products of the speed v in each interval and the number in that interval, $n(v)dv$. This sum is then divided by the total number of measurements N, which gives, in the limit of infinitesimal intervals in which the sum becomes an integral,

$$\bar{v} = \frac{1}{N} \int_0^\infty v n(v)\, dv. \tag{19}$$

The next step is to substitute for $n(v)$ from Eq. 17 and evaluate the integral. The result is

$$\bar{v} = \sqrt{\frac{8kT}{\pi m}} = \sqrt{\frac{8RT}{\pi M}}$$
$$= 1.60 \sqrt{\frac{RT}{M}} \quad \text{(average speed).} \tag{20}$$

Substituting numerical values yields

$$\bar{v} = 1.60 \sqrt{\frac{(8.31 \text{ J/mol·K})(300 \text{ K})}{0.032 \text{ kg/mol}}} = 445 \text{ m/s}.$$

To find the root-mean-square speed v_{rms} of the oxygen molecules, we proceed as above except that we find the average value of v^2 by multiplying v^2 (rather than simply v) by the numerical factor $n(v)dv$. This leads, after another integration (see Appendix H), to

$$\bar{v^2} = \frac{1}{N} \int_0^\infty v^2 n(v)\, dv = \frac{3kT}{m}. \tag{21}$$

The rms speed is the square root of this quantity, or

$$v_{rms} = \sqrt{\bar{v^2}} = \sqrt{\frac{3kT}{m}} = \sqrt{\frac{3RT}{M}}$$
$$= 1.73 \sqrt{\frac{RT}{M}} \quad \text{(rms speed).} \tag{22}$$

Equation 22 is identical to Eq. 16 of Chapter 23. The numerical calculation gives

$$v_{rms} = 1.73 \sqrt{\frac{(8.31 \text{ J/mol·K})(300 \text{ K})}{0.032 \text{ kg/mol}}} = 483 \text{ m/s}.$$

The most probable speed is the speed at which $n(v)$ of Eq. 17 has its maximum value. We find it by requiring that $dn/dv = 0$ and solving for v. Doing so yields (as you should show)

$$v_p = \sqrt{\frac{2kT}{m}} = \sqrt{\frac{2RT}{M}}$$
$$= 1.41 \sqrt{\frac{RT}{M}} \quad \text{(most probable speed).} \tag{23}$$

Numerically, this gives

$$v_p = 1.41 \sqrt{\frac{(8.31 \text{ J/mol·K})(300 \text{ K})}{0.032 \text{ kg/mol}}} = 395 \text{ m/s}.$$

Experimental Confirmation of the Maxwellian Distribution

Maxwell derived his distribution law for molecular speeds (Eq. 17) in 1859. At that early date it was not possible to check this law by direct measurement and, indeed, it was not until 1920 that Stern made the first serious attempt to do so. Techniques improved rapidly in the hands of various workers until 1955 when a high-precision experimental verification of the law (for gas molecules) was provided by Miller and Kusch of Columbia University.

Their apparatus is illustrated in Fig. 12. The walls of oven O, containing some thallium metal, were heated, in one set of experiments, to a uniform temperature of 870 ± 4 K. At this temperature thallium vapor, at a pres-

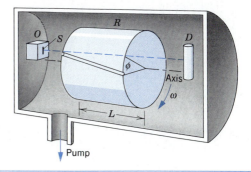

Figure 12 Apparatus used to verify the Maxwell speed distribution. A beam of thallium molecules leaves the oven O through the slit S, travels through the helical groove in the rotating cylinder R, and strikes the detector D. The angular velocity ω of the cylinder can be varied so that molecules of differing speeds will pass through the cylinder.

sure of 3.2×10^{-3} mm Hg, fills the oven. Some molecules of thallium vapor escape from slit S into the highly evacuated space outside the oven, falling on the rotating cylinder R. This cylinder, of length L, has a number of helical grooves cut into it, only one of which is shown in Fig. 12. For a given angular speed ω of the cylinder, only molecules of a sharply defined speed v can pass along the grooves without striking the walls. The speed v can be found from

$$\text{time of travel along the groove} = \frac{L}{v} = \frac{\phi}{\omega}$$

or

$$v = \frac{L\omega}{\phi}$$

in which ϕ (see Fig. 12) is the angular displacement between the entrance and the exit of a helical groove. Thus the rotating cylinder is a *velocity selector*, in which the speed is selected by the (controllable) angular speed ω. The beam intensity is recorded by detector D as a function of the selected speed v. Figure 13 shows the remarkable agreement between theory (the solid line) and experiment (the open and filled circles) for thallium vapor.

The distribution of speeds in the *beam* (as distinguished from the distribution of speeds in the *oven*) is not proportional to $v^2 e^{-mv^2/2kT}$, as in Eq. 17, but to $v^3 e^{-mv^2/2kT}$. Consider a group of molecules in the oven whose speeds lie within a certain small range v_1 to $v_1 + \delta v$, where v_1 is less than the most probable speed v_p. We can always find another equal speed interval δv, extending from v_2 to $v_2 + \delta v$, where v_2, which will be greater than v_p, is chosen

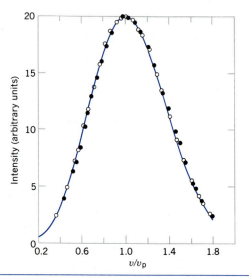

Figure 13   The results of the experiment to verify the Maxwell speed distribution. The open circles show data taken with the oven temperature at $T = 870$ K, and the filled circles show data at $T = 944$ K. When the distributions are plotted against v/v_p, the two distributions should be identical. The solid curve is the Maxwell distribution. The data agree remarkably well with the curve.

so that the two speed intervals contain the same number of molecules. However, more molecules in the higher interval than in the lower will escape from slit S to form the beam, because molecules in the higher interval "bombard" the slit with a greater frequency, by precisely the factor v_2/v_1. Thus, other things being equal, fast molecules are favored in escaping from the oven, just in proportion to their speeds, and the molecules in the beam have a v^3 rather than a v^2 distribution. This effect is included in the theoretical curve of Fig. 13.

Although the Maxwell speed distribution for gases agrees remarkably well with experiment under ordinary circumstances, it fails at high densities, where the basic assumptions of classical kinetic theory are not valid. Under these circumstances, we must use distributions based on the principles of quantum mechanics, which we discuss in Section 24-6. The quantum distributions, which are correct in all circumstances, reduce to the Maxwell distribution in the classical region (low density). Thus it is perfectly acceptable for us to use the Maxwell distribution for gases at low density, as long as we remember that the theory, like most theories, is limited in its applicability.

24-4   THE DISTRIBUTION OF ENERGIES

An alternative description of the motion of molecules can be obtained if we look for the distribution in *energy* rather than in speed. That is, we seek the distribution $n(E)$, such that $n(E)dE$ gives the number of molecules with energies between E and $E + dE$.

This problem was first solved by Maxwell. We derive the result, called the *Maxwell–Boltzmann energy distribution*, in the special case that translational kinetic energy is the only form of energy that a molecule can have.

Let us consider again the situation of Sample Problem 5, in which we obtained the fraction of oxygen molecules having speeds between 599 and 601 m/s. We found that 0.26% of the molecules in a container at a temperature of 300 K have speeds in that range. An oxygen molecule with a speed of 599 m/s has a kinetic energy of 9.54×10^{-21} J, and one with a speed of 601 m/s has a kinetic energy of 9.60×10^{-21} J. What fraction of the oxygen molecules has kinetic energies in the range of 9.54×10^{-21} to 9.60×10^{-21} J?

A bit of thought should convince you that this fraction must also be 0.26%. It makes no difference whether we count the molecules by their speeds or by their kinetic energies; as long as we set the lower and upper limits of the interval to have corresponding speeds and kinetic energies, we count the same number of molecules between the limits. That is, the number with kinetic energies between E and $E + dE$ is the same as the number with speeds between v and $v + dv$. Mathematically, we express this conclusion as

$$n(E)\, dE = n(v)\, dv, \qquad (24)$$

or

$$n(E) = n(v)\, \frac{dv}{dE}. \qquad (25)$$

Since the energy is only kinetic, we must have $E = \frac{1}{2}mv^2$ or $v = \sqrt{2E/m}$, and thus

$$\frac{dv}{dE} = \sqrt{\frac{2}{m}}\,(\tfrac{1}{2}E^{-1/2}). \qquad (26)$$

Substituting Eqs. 17 and 26 into Eq. 25, we obtain

$$n(E) = \frac{2N}{\sqrt{\pi}}\,\frac{1}{(kT)^{3/2}}\,E^{1/2}\,e^{-E/kT}. \qquad (27)$$

Equation 27 is the *Maxwell–Boltzmann energy distribution*. In deriving this result, we have assumed that the gas molecules can have only translational kinetic energy. This distribution therefore applies only to a monatomic gas. In the case of gases with more complex molecules, other factors will be present in Eq. 27. The factor $e^{-E/kT}$, however, is a general feature of the Maxwell–Boltzmann energy distribution that is present no matter what the form of the energy E. This factor is often taken as a rough estimate of the relative probability for a particle to have an energy E in a collection of particles characterized by a temperature T.

Using Eq. 27, we can calculate the fraction of the gas molecules having energies between E and $E + dE$, which is given by $n(E)dE/N$. As before, N is the total number of molecules, which is determined from

$$N = \int_0^\infty n(E)\, dE. \qquad (28)$$

One interesting feature of the Maxwell–Boltzmann energy distribution is that it is precisely the same for any gas at a given temperature, no matter what the mass of the molecules (in contrast to the Maxwell speed distribution, Eq. 17, in which the mass appears explicitly). Even a "gas" of electrons, to the extent they can be treated as classical particles, has the same energy distribution as a gas of heavy atoms. The effect of increasing the mass m by some factor is to reduce v^2 by the same factor, so that the product mv^2, and thus the kinetic energy, remains the same.

The Internal Energy of an Ideal Gas

We gain a measure of confidence in this statistical approach to thermodynamics by showing that it gives results identical with calculations based on kinetic theory. Let us therefore obtain the statistical result for the internal energy of an ideal gas and compare it with our previous result.

Since there are $n(E)dE$ molecules with energy between E and $E + dE$, their contribution to the internal energy of

this gas is $En(E)dE$. The total of all such contributions gives the internal energy of the gas:

$$E_{\text{int}} = \int_0^\infty En(E)\, dE,$$

and substituting from Eq. 27 for $n(E)$, we obtain

$$E_{\text{int}} = \frac{2N}{\sqrt{\pi}}\,\frac{1}{(kT)^{3/2}}\int_0^\infty E^{3/2}e^{-E/kT}\, dE.$$

Integrals of forms similar to this one occur often in statistical mechanics. We put it in standard form by substituting $u = E/kT$, which gives

$$E_{\text{int}} = \frac{2N}{\sqrt{\pi}}\,kT\int_0^\infty u^{3/2}e^{-u}\, du.$$

The integral can be evaluated (substitute $u = x^2$ and use a definite integral from Appendix H) to be $\frac{3}{4}\sqrt{\pi}$ and so

$$E_{\text{int}} = \frac{2N}{\sqrt{\pi}}\,kT\,(\tfrac{3}{4}\sqrt{\pi}) = \tfrac{3}{2}NkT,$$

in agreement with Eq. 31 of Chapter 23. Thus the Maxwell–Boltzmann energy distribution is entirely consistent with our previous results derived from kinetic theory.

Sample Problem 7 Find (*a*) the mean energy and (*b*) the most probable energy of a gas in thermal equilibrium at temperature T.

Solution (*a*) The mean energy $\overline{E}$ can be written, in analogy with Eq. 19 of Sample Problem 6,

$$\overline{E} = \frac{1}{N}\int_0^\infty En(E)\, dE. \qquad (29)$$

Substituting for $n(E)$ from Eq. 27 and carrying out the integral, we find

$$\overline{E} = \tfrac{3}{2}kT. \qquad (30)$$

Is this an expected result?

(*b*) To find the most probable energy, we differentiate Eq. 27, set the result equal to zero, and solve for the energy. The result, which you should derive, is

$$E_{\text{p}} = \tfrac{1}{2}kT. \qquad (31)$$

Note that this is *not* equal to $\frac{1}{2}mv_{\text{p}}^2$, which gives an energy of kT. Can you explain why the energy corresponding to the most probable speed is not the most probable energy?

24-5 BROWNIAN MOTION*

The acceptance of atomic and molecular theory during the last quarter of the 19th century was not shared by all scientists. In spite of the many quantitative agreements

---

* See "Brownian Motion," by Bernard H. Lavenda, *Scientific American*, February 1985, p. 70.

between kinetic theory and the behavior of gases, no proof of the separate existence of atoms and molecules had been obtained, nor had any observation been made that could really demonstrate the motions of the molecules. Ernst Mach (1838–1916) saw no point to "thinking of the world as a mosaic, since we cannot examine its individual pieces of stone." Early in the development of kinetic theory it had been established that an atom should be about 10^{-7} or 10^{-8} cm in diameter, so no one actually expected to see an atom or detect the effect of a single atom.

The leader of the opposition to the atomic theory was Wilhelm Ostwald, generally regarded as "the father of physical chemistry." He was a strong believer in the principle of conservation of energy and regarded energy as the ultimate reality. Ostwald argued that with a thermodynamical treatment of a process we know all that is essential about the process and that further mechanical assumptions about the mechanism of the reactions are unproved hypotheses. He abandoned the atomic and molecular theories and fought to free science "from hypothetical conceptions which lead to no immediate experimentally verifiable conclusions." Other prominent scientists were reluctant to accept the existence of atoms as an established scientific fact.

Ludwig Boltzmann objected to this attitude in an article in 1897, stressing the indispensability of atomism in natural science. The progress of science is often guided by the analogies of nature's processes that occur in the minds of investigators. Kinetic theory was such a mechanical analogy. As with most analogies, it suggests experiments to test the validity of our mental pictures and leads to further investigations and clearer knowledge.

As is always true in such controversies in science, the decision rests with experiment. The earliest and most direct experimental evidence for the reality of atoms was the proof of the atomic kinetic theory provided by the quantitative studies of Brownian motion. These observations convinced both Mach and Ostwald of the validity of the kinetic theory and the atomic description of matter on which it rests. The atomic theory gained unquestioned acceptance in later years when a wide variety of experiments all led to the same values of the fundamental atomic constants.

Brownian motion is named after the English botanist Robert Brown, who discovered in 1827 that a grain of pollen suspended in water shows a continuous random motion when viewed under a microscope. At first these motions were considered a form of life, but it was soon found that small inorganic particles behave similarly. There was no quantitative explanation of this phenomenon until the development of kinetic theory. Then, in 1905, Albert Einstein developed a theory of Brownian motion. (Einstein's theory appeared as an article in the same volume of the *Annalen der Physik* that contained his famous paper on the theory of relativity and also his paper

on the theory of the photoelectric effect.) In his *Autobiographical Notes*, Einstein writes: "My major aim in this was to find facts which would guarantee as much as possible the existence of atoms of definite size. In the midst of this I discovered that, according to atomistic theory, there would have to be a movement of suspended microscopic particles open to observation, without knowing that observations concerning the Brownian motion were already long familiar."

Einstein's basic assumption was that particles suspended in a liquid or a gas share in the thermal motions of the medium, and that on the average the translational kinetic energy of each particle is $\frac{3}{2}kT$, in accordance with the principle of equipartition of energy. In this view the Brownian motions result from impacts by molecules of the fluid, and the suspended particles acquire the same mean kinetic energy as the molecules of the fluid. (Recall that the Maxwell–Boltzmann energy distribution is independent of the mass of the particles and is determined only by the temperature.)

The suspended particles are extremely large compared to the molecules of the fluid and are being continually bombarded on all sides by them. If the particles are sufficiently large and the number of molecules is sufficiently great, equal numbers of molecules strike the particles on all sides at each instant. For smaller particles and fewer molecules the number of molecules striking various sides of the particle at any instant, being merely a matter of chance, may not be equal; that is, fluctuations occur. Hence the particle at each instant suffers an unbalanced force, causing it to move this way or that. The particles therefore act just like very large molecules in the fluid, and their motions should be qualitatively the same as the motion of the fluid molecules. If the Avogadro constant were infinite, there would be no statistical imbalance (fluctuations) and no Brownian motion. If the Avogadro constant were very small, the Brownian motion would be very large. Hence we should be able to deduce the value of the Avogadro constant from observations of the Brownian motion. Deeply ingrained in this picture is the idea of molecular motion and the smallness of molecules. The Brownian motion therefore offers a striking experimental test of the kinetic theory hypotheses.

The suspended particles are under the influence of gravity and would settle to the bottom of the fluid were it not for the molecular bombardment opposing this tendency. Since the suspended particles behave like gas molecules, we are not surprised to learn that, as for molecules in the atmosphere, their density drops off exponentially with respect to height in the fluid; they form a "miniature atmosphere" (see Section 17-3; Problem 6, Chapter 23; and Problem 32, this chapter). Jean Perrin, a French physical chemist, confirmed this prediction in 1908 by determining the number of small particles of gum resin suspended at different heights in a liquid drop (Fig. 14a). From his data he deduced a value of the Avogadro

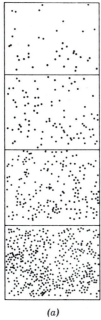

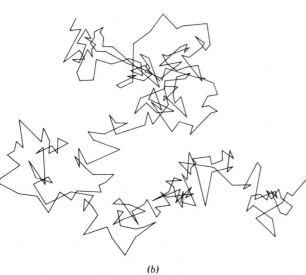

(b)

Figure 14 (*a*) A representation of particles in gum resin observed by Perrin in 1909. The horizontal lines represent layers 0.01 cm apart, and the particles have diameter 0.6×10^{-3} cm. (*b*) The motion of a tiny particle, suspended in water and viewed through a microscope. The short line segments connect its position at 30-s intervals. The path of the particle is an example of a *fractal*, a curve for which any small section resembles the curve as a whole. For example, if we take any short 30-s segment and view it in smaller intervals, perhaps 0.1 s, the plot of the motion in that 30-s segment would be similar to this entire figure.

constant $N_A = 6 \times 10^{23}$ particles/mol. Perrin also made measurements of the displacements of Brownian particles during many equal time intervals and found that they have the statistical distribution required by kinetic theory and the root-mean-square displacement predicted by

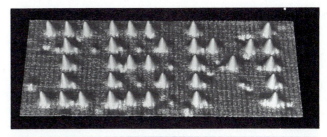

Figure 15 Using a scanning tunneling microscope, physicists at IBM in 1990 spelled their company's name using individual atoms of xenon.

Einstein (Fig. 14*b*). For this work, which was so compelling a confirmation of the existence of atoms, Perrin was awarded the 1926 Nobel Prize in physics. Today, we have more direct photographic evidence of the existence of atoms (Fig. 15).

24-6   QUANTUM STATISTICAL DISTRIBUTIONS   *(Optional)*

So far our study of kinetic theory and statistical mechanics has involved the application of classical principles (Newton's laws) to classical particles. When we consider the effects of quantum mechanics, however, statistical distributions differ greatly from their classical counterparts and lead to experimental effects that have no classical analogues. There are two principles of quantum mechanics that affect the statistical distributions of particles.

1. *In quantum mechanics, identical particles must be treated as indistinguishable.* In the ideal gas model that forms the basis of kinetic theory, we assumed all atoms or molecules of the gas to be identical. Nevertheless, we were able (see, for example, our use of kinetic theory to determine the pressure in Section 23-3) to follow the motion of a *single molecule* as it made its way across the chamber. That is, we assumed that molecule to be somehow *distinguishable* from its neighbors, even though it is identical to them in its structure and properties. In quantum mechanics, on the other hand, this is not possible. We must treat the particles as physically and mathematically *indistinguishable*; in quantum theory, we cannot consider the motion of a particular particle in a system without simultaneously considering all the other identical particles. For example, after a collision between two identical particles such as gas molecules, from the classical standpoint we are able to distinguish which was the incident particle and which the struck particle; from the quantum standpoint, either particle leaving the collision zone has a 50% chance of having been the incident particle and a 50% chance of having been the struck particle.

2. *Within the quantum domain, particles in a system can be put into two classes: those for which there is no limit on the number that can be in a particular state of motion, and those for which the limit is precisely one per state of motion.* We can regard this rule as an experimental one, based on observations of the properties of the two kinds of particles. However, there is also a deep underlying theoretical significance to this rule, based on a result that can be derived from quantum field theory: particles of the first kind, for which the number in any state of motion is unlimited, all have an intrinsic angular momentum or "spin" (see Section 13-6) that is measured to be an integral multiple of $h/2\pi$, where h is the Planck constant. Such particles, of which photons (quanta of electromagnetic radiation) and atoms of the most common isotope of helium gas are examples, are called *bosons,* and they follow a statistical distribution called the *Bose–Einstein distribution.*

The second kind of quantum particle, which is restricted to one per state of motion, is found to have an intrinsic angular momentum measured in units of half-integral $(\frac{1}{2}, \frac{3}{2}, \ldots)$ multiples of $h/2\pi$. Particles in this category, which include electrons, neutrons, and protons, are called *fermions,* and they follow a statistical distribution called the *Fermi–Dirac distribution.*

The restriction of one fermion per state of motion is also known as the *Pauli exclusion principle* and is responsible for the observed properties of atoms. The permitted states of motion of electrons in an atom form a set of discrete energy levels (see Section 8-8). Were it not for the Pauli exclusion principle, all electrons in an atom would eventually drop into the lowest energy state, and all atoms would have more-or-less similar physical and chemical properties. The diversity of our world and the very existence of chemistry and biology are direct consequences of the Pauli exclusion principle.

Sample Problem 8 A system consists of two particles, labeled 1 and 2. The particles may be in any one of three possible states of motion, labeled A, B, and C. Find the number of ways that the particles can be distributed in this system if the particles are (*a*) classical and distinguishable; (*b*) indistinguishable bosons; and (*c*) indistinguishable fermions. (*d*) Assuming all allowed arrangements are equally probable, what is the probability to find one particle in state B and the other in state C?

Solution (*a*) The chart below shows how the classical particles can be distributed among the three states:

Arrangement	State A	State B	State C
1	1, 2		
2		1, 2	
3			1, 2
4	1	2	
5	1		2
6	2	1	
7	2		1
8		1	2
9		2	1

Nine different arrangements are possible.

(*b*) If the particles are treated as indistinguishable quantum particles, then the labels 1 and 2 no longer have any meaning. Therefore it is not possible to list arrangements 4 and 6 as separate possibilities; we must combine these into a single arrangement in which one particle is in state A and the other is in state B. Similarly, arrangements 5 and 7 must be counted as only one possibility, and likewise arrangements 8 and 9. In this case, only six different arrangements are possible.

(*c*) If the particles are fermions, no more than one particle can be in any state of motion. Therefore arrangements 1, 2, and 3 are

not permitted, and as a result only three arrangements are possible (4 + 6, 5 + 7, and 8 + 9).

(*d*) For classical particles, there are two arrangements (out of the nine) that have one particle each in states B and C. The probability is therefore 2/9. For bosons, there are six arrangements, only one of which (the combination 8 + 9) has one particle in each of the states B and C. The probability is then 1/6. For fermions, only three arrangements are permitted, and the probability is therefore 1/3.

The above sample problem illustrates the differences in the statistical distributions of classical particles and the two classes of quantum particles; it also shows that the outcome of a measurement (the probability of an observation) can be directly affected by these distinctions.

The Maxwell–Boltzmann distribution function was derived from classical principles and is therefore appropriate only in describing classical particles. The characteristic part of that distribution is the exponential function

$$f_{\text{MB}}(E) = e^{-E/kT}. \tag{32}$$

From statistical considerations, the corresponding characteristic part of the Bose–Einstein distribution can be shown to be

$$f_{\text{BE}}(E) = \frac{1}{e^{(E-E_0)/kT} - 1}, \tag{33}$$

where the energy E_0 is a parameter that depends on the density of the particles and on other aspects of their nature. The Fermi–Dirac distribution is characterized by the function

$$f_{\text{FD}}(E) = \frac{1}{e^{(E-E_0)/kT} + 1}. \tag{34}$$

Note that the seemingly minor difference of a sign in the denominators of Eqs. 33 and 34 makes a drastic difference in the behavior of the particles described by these two functions. Figures 16 and 17 illustrate this difference. The function f_{BE} goes to infinity as the energy approaches E_0; there is thus nothing that prevents all the particles in a system of bosons from dropping into a single energy level. In this condition, all particles have in effect the same state of motion, and the entire substance behaves cooperatively. An example of such an effect occurs in the case of liquid helium (^{4}He), in which the cooperative motion of all the atoms is responsible for its ability to flow with absolutely no viscosity even through the narrowest of constrictions. This effect is called *superfluidity* and occurs when helium is cooled below a temperature of 2.2 K. (By way of contrast, the other stable isotope of helium, ^{3}He, is a fermion and does not show superfluidity until it is cooled below 0.002 K, at which temperature two atoms of ^{3}He can join to form a molecular system that behaves like a boson.) The superfluid is also characterized by an infinite thermal conductivity, so that boiling ceases below 2.2 K (Fig. 18), and by a cooperative effect called "film flow," in which the helium can flow over the walls of its container and be seen dripping from the bottom (Fig. 19). Other applications in which the Bose–Einstein distribution gives results different from those of the classical Maxwell–Boltzmann distribution include the calculation of the spectrum of thermal radiation emitted by objects (see Section 49-2 of the extended text) and of the heat capacities of solids (see Section 25-3); in the latter case it is the

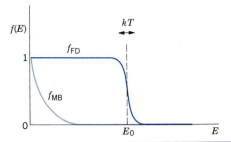

Figure 16 The Fermi–Dirac characteristic function compared with the Maxwell–Boltzmann function. The Fermi–Dirac function differs from 1 at low energy and 0 at high energy only in a narrow range of width roughly kT in the vicinity of the energy E_0.

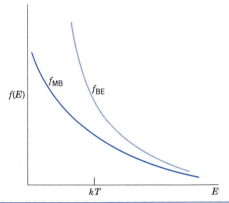

Figure 17 The Bose–Einstein characteristic function compared with the Maxwell–Boltzmann function. The two agree at high energy, where quantum effects become negligible, but disagree substantially at low energy, where f_{BE} becomes infinite as E approaches E_0, which we assume to be zero here.

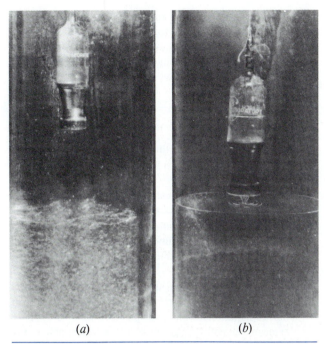

Figure 18 (a) Above 2.2 K liquid helium boils normally. (b) Below 2.2 K, the liquid suddenly stops boiling. Evaporation and cooling may continue, but they occur without boiling. Helium in this condition is in its superfluid state.

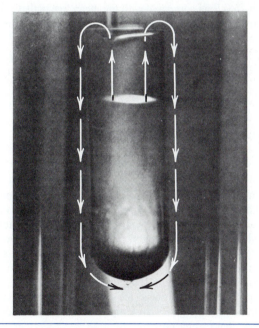

Figure 19 An illustration of the cooperative behavior of helium in the superfluid state is the film flow, in which the liquid in the container spontaneously creeps over the walls and can be seen dripping from the bottom.

vibrational modes of the atoms of the solid that behave cooperatively and are subject to the distribution.

Note from comparing Figs. 16 and 17 that the Fermi–Dirac distribution has a very different behavior at low energy than the Bose–Einstein distribution. Instead of allowing all the particles to fall into the lowest energy state, the Fermi–Dirac distribution permits at most one particle to be in any energy state. This is exactly the requirement imposed by the Pauli exclusion principle. Examples of effects that result from the application of the Fermi–Dirac distribution include electrical conduction in solids (see Chapter 53 of the extended version of this text) and the properties of neutron stars.

Superconductivity (the ability of certain materials to conduct electricity without resistance; see Section 32-8) is another example of the effect of differing quantum statistical distributions. Ordinarily, the conduction electrons in a solid obey the Fermi–Dirac distribution, which prevents cooperative behavior by permitting no more than one electron per state of motion. In a superconductor, two electrons join to form a pair, and the paired electrons behave like a boson instead of a fermion. They are thus subject to Bose–Einstein statistics, which permits the cooperative behavior that gives rise to superconductivity.

When the density of particles is very low or the temperature is very high, the exponential term in the denominators of Eqs. 33

and 34 becomes much greater than 1, and the 1 can be neglected. In this case, both f_{BE} and f_{FD} depend on E as $e^{-E/kT}$, which is the same energy dependence as f_{MB}. Thus the quantum distributions reduce to the classical one in the low-density or high-temperature limit. The low-density limit is applicable in the case of a gas;

as we discovered in Sample Problem 3, the average distance between molecules in a gas is typically more than 10 molecular diameters, and a molecule may travel several hundred diameters before encountering another molecule. The indistinguishability of molecules therefore does not affect our ability to study gas molecules, and the classical distribution gives a very accurate representation of the properties of the gas. In liquids and solids, however, the average spacing is more like 1 molecular diameter, and the classical distribution is usually not applicable. ∎

QUESTIONS

1. Consider the case in which the mean free path is greater than the longest straight line in a vessel. Is this a perfect vacuum for a molecule in this vessel?

2. List effective ways of increasing the number of molecular collisions per unit time in a gas.

3. Give a qualitative explanation of the connection between the mean free path of ammonia molecules in air and the time it takes to smell the ammonia when a bottle is opened across the room.

4. Consider Archimedes' principle applied to a gas. Isn't it true that once we accept a kinetic theory model of a gas, we need a new explanation for this principle? For example, suppose the mean free path of a gas molecule is comparable to the depth of the body immersed in the gas, or greater. What is the origin of the buoyant force then? (See "Archimedes' Principle in Gases," by Alan J. Walton, *Contemporary Physics,* March 1969, p. 181.)

5. A gas can transmit only those sound waves whose wavelength is long compared with the mean free path. Can you explain this? Describe a situation for which this limitation would be important.

6. If molecules are not spherical, what meaning can we give to *d* in Eq. 16 for the mean free path? In which gases would the molecules act most nearly like rigid spheres?

7. In what sense is the mean free path a macroscopic property of a gas rather than a microscopic one?

8. Suppose we dispense with the hypothesis of elastic collisions in kinetic theory and consider the molecules as centers of force acting at a distance. Does the concept of mean free path have any meaning under these circumstances?

9. Since the actual force between molecules depends on the distance between them, forces can cause deflections even when molecules are far from "contact" with one another. Furthermore, the deflection caused should depend on how long a time these forces act and hence on the relative speed of the molecules. (*a*) Would you then expect the measured mean free path to depend on temperature, even though the density remains constant? (*b*) If so, would you expect λ to increase or decrease with temperature? (*c*) How does this dependence enter into Eq. 16?

10. When a can of mixed nuts is shaken, why does the largest nut generally end up on the surface, even if it is denser than the others?

11. In a fixed amount of gas, how would the mean free path be affected if (*a*) the density of the gas is doubled, (*b*) the mean molecular speed is doubled, and (*c*) both the density and mean molecular speed are doubled?

12. Justify qualitatively the statement that, in a mixture of molecules of different kinds in complete equilibrium, each kind of molecule has the same Maxwellian distribution in speed that it would have if the other kinds were not present.

13. A gas consists of *N* particles. Explain why $v_{rms} \geq \bar{v}$ regardless of the distribution of speeds.

14. What observation is good evidence that not all molecules of a body are moving with the same speed at a given temperature?

15. The fraction of molecules within a given range Δv of the rms speed decreases as the temperature of a gas rises. Explain.

16. (*a*) Do half the molecules in a gas in thermal equilibrium have speeds greater than v_p? Than $\bar{v}$? Than v_{rms}? (*b*) Which speed—v_p, $\bar{v}$, or v_{rms}—corresponds to a molecule having average kinetic energy?

17. Consider the distribution of speeds shown in Fig. 20. (*a*) List v_{rms}, $\bar{v}$, and v_p in the order of increasing speed. (*b*) How does this compare with the Maxwellian distribution?

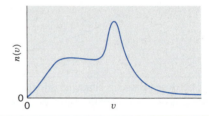

Figure 20   Question 17.

18. Figure 21 shows the distribution of the *x* component of the velocities of the molecules in a container at a fixed temperature. (*a*) The distribution is symmetrical about $v_x = 0$; make this plausible. (*b*) What does the total area under the curve represent? (*c*) How would the distribution change with an increase in temperature? (*d*) What is the most probable value of v_x? (*e*) Is the most probable speed equal to zero? Explain.

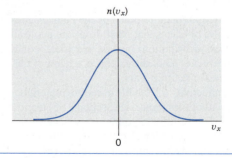

Figure 21   Question 18.

19. The slit system in Fig. 12 selects only those molecules moving in the +*x* direction. Does this destroy the validity of the experiment as a measure of the distribution of speeds of molecules moving in all directions?

20. List examples of the Brownian motion in physical phenomena.

21. Would Brownian motion occur in gravity-free space?

22. A golf ball is suspended from the ceiling by a long thread. Explain in detail why its Brownian motion is not readily apparent.

23. We have defined ρ_n to be the number of molecules per unit volume in a gas. If we define ρ_n for a very small volume in a gas, say, one equal to 10 times the volume of an atom, then ρ_n fluctuates with time through the range of values zero to some maximum value. How then can we justify a statement that ρ_n has a definite value at every point in a gas?

PROBLEMS

Section 24-1 Statistical Distributions and Mean Values

1. Suppose you have a collection of N objects. The number of ways you can select n of them, if the order isn't important, is $N!/n! \, (N - n)!$. Using this formula, derive Eq. 5. (*Hint*: Divide a pack of cards into 13 hearts and 39 other cards. In how many ways can you select h hearts from the 13? In how many ways can you select $13 - h$ cards from the 39 non-hearts? How many ways are there to select *any* 13 cards from the deck?)

Section 24-2 Mean Free Path

2. Substitute Eq. 10 into Eq. 12 and evaluate the integrals to obtain Eq. 13.

3. The denominator of Eq. 12 is equal to the total number of particles N. Use this to show that the constant A in Eq. 14 is given by $A = N/\lambda$.

4. At 2500 km above the Earth's surface the density is about 1.0 molecule/cm³. (*a*) What mean free path is predicted by Eq. 16 and (*b*) what is its significance under these conditions? Assume a molecular diameter of 2.0×10^{-8} cm.

5. At standard temperature and pressure (0°C and 1.00 atm) the mean free path in helium gas is 285 nm. Determine (*a*) the number of molecules per cubic meter and (*b*) the effective diameter of the helium atoms.

6. In a certain particle accelerator the protons travel around a circular path of diameter 23.5 m in a chamber of 1.10×10^{-6} mm Hg pressure and 295 K temperature. (*a*) Calculate the number of gas molecules per cubic meter at this pressure. (*b*) What is the mean free path of the gas molecules under these conditions if the molecular diameter is 2.20×10^{-8} cm?

7. The mean free path λ of the molecules of a gas may be determined from measurements (for example, from measurement of the viscosity of the gas). At 20.0°C and 75.0 cm Hg pressure such measurements yield values of λ_{Ar} (argon) $= 9.90 \times 10^{-6}$ cm and λ_{N_2} (nitrogen) $= 27.5 \times 10^{-6}$ cm. (*a*) Find the ratio of the effective cross-section diameters of argon to nitrogen. (*b*) What would the value be of the mean free path of argon at 20.0°C and 15.0 cm Hg? (*c*) What would the value be of the mean free path of argon at -40.0°C and 75.0 cm Hg?

8. Calculate the mean free path for 35 spherical jelly beans in a jar that is vigorously shaken. The volume of the jar is 1.0 L and the diameter of a jelly bean is 1.0 cm.

9. At what frequency would the wavelength of sound be on the order of the mean free path in nitrogen at 1.02 atm pressure and 18.0°C? Take the diameter of the nitrogen molecule to be 315 pm. See Problem 42 of Chapter 23.

10. (*a*) Find the probability that a molecule will travel a distance at least equal to the mean free path before its next collision. (*b*) After what distance of travel since the last collision is the probability of having suffered the next collision equal to $\frac{1}{2}$? Give the answer in terms of the mean free path.

11. In Sample Problem 3, at what temperature is the average rate of collision equal to $6.0 \times 10^9 \, s^{-1}$? The pressure remains unchanged.

Section 24-3 The Distribution of Molecular Speeds

12. The speeds of a group of ten molecules are 2, 3, 4, . . . , 11 km/s. (*a*) Find the average speed of the group. (*b*) Calculate the root-mean-square speed of the group.

13. (*a*) Ten particles are moving with the following speeds: four at 200 m/s, two at 500 m/s, and four at 600 m/s. Calculate the average and root-mean-square speeds. Is $v_{rms} > \bar{v}$? (*b*) Make up your own speed distribution for the 10 particles and show that $v_{rms} \geq \bar{v}$ for your distribution. (*c*) Under what condition (if any) does $v_{rms} = \bar{v}$?

14. You are given the following group of particles (n_i represents the number of particles that have a speed v_i):

n_i	v_i (km/s)
2	1
4	2
6	3
8	4
2	5

(*a*) Compute the average speed $\bar{v}$. (*b*) Compute the root-mean-square speed v_{rms}. (*c*) Among the five speeds shown, which is the most probable speed v_p for the entire group?

15. In the apparatus of Miller and Kusch (see Fig. 12) the length L of the rotating cylinder is 20.4 cm and the angle ϕ is 0.0841 rad. What rotational speed corresponds to a selected speed v of 212 m/s?

16. It is found that the most probable speed of molecules in a gas at temperature T_2 is the same as the rms speed of the molecules in this gas when its temperature is T_1. Calculate T_2/T_1.

17. Two containers are at the same temperature. The first contains gas at pressure p_1 whose molecules have mass m_1 with a root-mean-square speed v_{rms1}. The second contains molecules of mass m_2 at pressure $2p_1$ that have an average speed $\bar{v}_2 = 2v_{rms1}$. Find the ratio m_1/m_2 of the masses of their molecules.

18. A gas, not necessarily in thermal equilibrium, consists of N particles. The speed distribution is not necessarily Maxwellian. (*a*) Show that $v_{rms} \geq \bar{v}$ regardless of the distribution of speeds. (*b*) When would the equality hold?

19. Show that, for atoms of mass m emerging as a beam from a

small opening in an oven of temperature T, the most probable speed is $v_p = \sqrt{3kT/m}$.

20. An atom of germanium (diameter = 246 pm) escapes from a furnace ($T = 4220$ K) with the root-mean-square speed into a chamber containing atoms of cold argon (diameter = 300 pm) at a density of 4.13×10^{19} atoms/cm³. (a) What is the speed of the germanium atom? (b) If the germanium atom and an argon atom collide, what is the closest distance between their centers, considering each as spherical? (c) Find the initial collision frequency experienced by the germanium atom.

21. For the hypothetical gas speed distribution of N particles shown in Fig. 22 [$n(v) = Cv^2, 0 < v < v_0; n(v) = 0, v > v_0$], find (a) an expression for C in terms of N and v_0, (b) the average speed of the particles, and (c) the rms speed of the particles.

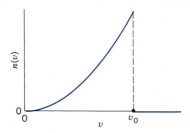

Figure 22   Problem 21.

22. A hypothetical gas of N particles has the speed distribution shown in Fig. 23 [$n(v) = 0$ for $v > 2v_0$]. (a) Express a in terms of N and v_0. (b) How many of the particles have speeds between $1.50v_0$ and $2.00v_0$? (c) Express the average speed of the particles in terms of v_0. (d) Find v_{rms}.

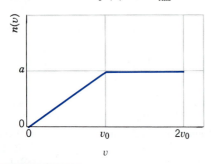

Figure 23   Problem 22.

23. For a gas in which all molecules travel with the same speed $\bar{v}$, show that $\bar{v}_{rel} = \frac{4}{3}\bar{v}$ rather than $\sqrt{2}\,\bar{v}$ (which is the result obtained when we consider the actual distribution of molecular speeds). See Eq. 16.

Section 24-4 The Distribution of Energies

24. Calculate the fraction of particles in a gas moving with translational kinetic energy between $0.01kT$ and $0.03kT$. (*Hint:* For $E \ll kT$, the term $e^{-E/kT}$ in Eq. 27 can be replaced with $1 - E/kT$. Why?)

25. (a) Find E_{rms} using the energy distribution of Eq. 27. (b) Why is $E_{rms} \neq \frac{1}{2}mv_{rms}^2$, where v_{rms} is given by Eq. 22?

26. Find the fraction of particles in a gas having translational kinetic energies within a range $0.02kT$ centered on the most probable energy E_p. (*Hint:* In this region, $n(E) \approx$ constant. Why?)

Section 24-5 Brownian Motion

27. The root-mean-square speed of hydrogen molecules at $0°C$ is 1840 m/s. Compute the speed of colloidal particles of "molar mass" 3.2×10^6 g/mol.

28. Particles of mass 6.2×10^{-14} g are suspended in a liquid at $26°C$ and are observed to have a root-mean-square speed of 1.4 cm/s. Calculate Avogadro's number from the equipartition theorem and these data.

29. Calculate the root-mean-square speed of smoke particles of mass 5.2×10^{-14} g in air at $14°C$ and 1.07 atm pressure.

30. A delicate spring balance has a force constant of 0.44 N/m. It vibrates randomly due to the bombardment by molecules in the air on the supported object (thermal fluctuations). (a) Find the rms displacement of the object from its equilibrium position at $31°C$. (*Hint:* Take the average energy of the random motion to be kT. Why?) (b) Determine the uncertainty in a weight determination due to this effect.

31. Very small solid particles, called grains, exist in interstellar space. They are continually bombarded by hydrogen atoms of the surrounding interstellar gas. As a result of these collisions, the grains execute Brownian movement in both translation and rotation. Assume the grains are uniform spheres of diameter 4.0×10^{-6} cm and density 1.0 g/cm³, and that the temperature of the gas is 100 K. Find (a) the root-mean-square speed of the grains between collisions and (b) the approximate rate (rev/s) at which the grains are spinning.

32. Colloidal particles in solution are buoyed up by the liquid in which they are suspended. Let ρ' be the density of liquid and ρ the density of the particles. If V is the volume of a particle, show that the number of particles per unit volume in the liquid varies with height as

$$n = n_0 \exp\left[-\frac{N_A}{RT}V(\rho - \rho')gh\right],$$

where n_0 is the number density of particles at height $h = 0$. This equation was tested by Perrin in his Brownian motion studies.

Section 24-6 Quantum Statistical Distributions

33. In how many ways can three (a) distinguishable classical particles and (b) indistinguishable bosons be distributed in two states?

34. (a) At what energy E is $f_{FD}(E) = \frac{1}{2}$? (b) What is the value of $f_{BE}(E)$ at the energy found in (a)?

Computer Project

35. Many computers have a random number generator, which you can use to "deal" a hand of 13 cards. For example, by assigning each card a numerical value from 1 through 52 and then by generating a random number in that interval, you can "deal" a card at random from a deck of 52. Using this technique, deal a hand of 13 cards and determine the number of hearts. Repeat 1000 times and draw a histogram similar to Fig. 2. Compare your results with the predictions of Eq. 5.

CHAPTER 25

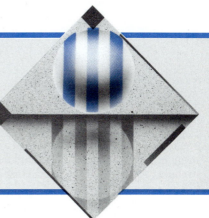

HEAT AND THE FIRST LAW OF THERMODYNAMICS

*In Chapter 22, we introduced the notion of systems in thermal contact.
If their temperatures are initially different, the systems will exchange energy
until thermal equilibrium is reached. In this chapter we deal with this energy flow from one
body to another, which we call* heat. *We also describe the effects of transferring heat to a
body, which may include increasing its temperature or changing its state, such as from solid
to liquid or liquid to vapor. Finally, we tie together the concepts of heat, internal energy,
and work through the* first law of thermodynamics, *a statement of the conservation of
energy. In essence, this topic continues and extends the program we began in Chapters 7
and 8, where we first introduced the concept of energy.*

25-1 HEAT: ENERGY IN TRANSIT

It is a common observation that if you place a hot object
(say, a cup of coffee) or a cold object (a glass of ice water)
in an environment at ordinary room temperature, the
object will tend toward thermal equilibrium with its environment. That is, the coffee gets colder and the ice water
gets warmer; the temperature of each approaches the temperature of the room.

It seems clear that such approaches to thermal equilibrium must involve some sort of exchange of energy between the system and its environment.* We define *heat*
(symbol Q) to be the energy that is transferred, such as
from the coffee to the room or from the room to the ice
water. More generally, we adopt the following definition:

*Heat is energy that flows between a system and its
environment by virtue of a temperature difference
between them.*

Figure 1 summarizes this view. If the temperature T_S of a

---

* This view has not always been held by scientists. In the 18th
century, it was believed that a material fluid, called *caloric*, was
exchanged between bodies at different temperatures. Experiments in the 19th century performed by Benjamin Thompson
(later known as Count Rumford of Bavaria) showed conclusively that mechanical work could produce heat, which resulted
in the identification of heat as a form of energy and led to the
development of the law of conservation of energy.

system is less than the temperature T_E of the environment, heat flows into the system. We choose our sign
convention so that Q is positive in this case; you can think
of this as being a process by which the internal energy of
the system is increased. Conversely, when $T_S > T_E$, heat
flows out of the system, and we take Q for the system to be
negative.

Because heat is a form of energy, its units are those of
energy, namely, the joule (J) in the SI system. Before it
was recognized that heat is a form of energy, other units
were assigned to it. In some cases these units, specifically
the *calorie* (cal) and the *British thermal unit* (Btu), are still
in use today. They are related to the joule according to

$$1 \text{ cal} = 4.186 \text{ J} \quad \text{and} \quad 1 \text{ Btu} = 1055 \text{ J}.$$

The "calorie" in common use as a measure of nutrition
(Cal) is in reality a kilocalorie; that is,

$$1 \text{ Cal} = 1000 \text{ cal} = 4186 \text{ J}.$$

The Btu is still commonly found as a measure of the
ability of an air conditioner to transfer energy (as heat)
from a room to the outside environment. A typical room
air conditioner rated at 10,000 Btu/h can therefore remove about 10^7 J from a room every hour and transfer it
to the outside environment.

Misconceptions About Heat

Heat is similar to work in that both represent a means for
the *transfer* of energy. Neither heat nor work is an intrinsic property of a system; that is, we cannot say that a

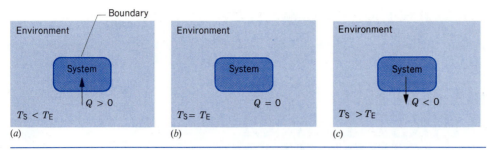

Figure 1   (*a*) If the temperature T_S of a system is less than the temperature T_E of its environment, heat flows into the system until thermal equilibrium is established, as in (*b*). (*c*) If the temperature of a system is greater than that of its environment, heat flows out of the system.

system "contains" a certain amount of heat or work. Instead, we say that it can transfer a certain amount of energy as heat or work under certain specified conditions.

Some of the confusion about the precise meaning of heat results from the popular usage of the term. Often heat is used when what is really meant is temperature or perhaps internal energy. When we hear about heat in relation to weather, or when cooking instructions indicate "heat at 300 degrees," it is *temperature* that is being discussed. On the other hand, we also hear references to the "heat generated" by the brake linings of an automobile or by briskly rubbing the palms of your hands together. In this case, as we shall see, it is usually *internal energy* that is meant. A clue to the proper usage comes from the definition of heat: when you rub your hands together, they do work on one another, thereby increasing their internal energy and raising their temperature. This excess energy can then be transferred to the environment as heat, because the hands are at a higher temperature than the environment.

The Mechanical Equivalent of Heat

In the past, when the calorie was independently defined as a unit of heat, it was necessary to determine an empirical relationship between the calorie and the joule. This was first done by James Joule in 1850 in an experiment to determine *the mechanical equivalent of heat*. A diagram of Joule's apparatus is shown in Fig. 2. Basically, the mechanical work W done by the falling weights (measured in joules) produces a measurable temperature rise of the water. The calorie was originally defined as the quantity of heat necessary to raise the temperature of 1 g of water from 14.5 to 15.5°C. From the measured temperature increase of the water, Joule was able to deduce the number of calories of heat Q that, if transferred from some external source to an equal quantity of water at the same initial temperature, would have produced the same temperature increase. The work W done on the water by the falling weights (in joules) therefore produced a temperature rise equivalent to the absorption by the water of a certain heat Q (in calories), and from this equivalence it is possible to determine the relationship between the calorie

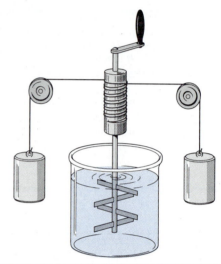

Figure 2   Joule's arrangement for measuring the mechanical equivalent of heat. The falling weights turn paddles that stir the water in the container, thus raising its temperature.

and the joule. The result of Joule's experiment, and others that followed, provided for nearly 100 years a conversion between the joule and the calorie. Today, following the adoption in 1948 of the joule as the SI unit for heat, we express all energy-related quantities, such as heat and work, in J, and so this conversion factor has lost the importance it had in Joule's time. Nevertheless, Joule's work is still noteworthy for the skill and ingenuity of his experiments, for its precision (Joule's results differ by only 1% from the SI-defined relationship between the joule and the calorie), and for the direction it provided in showing that heat, like work, could properly be regarded as a means of transferring energy.

25-2   HEAT CAPACITY AND SPECIFIC HEAT

We can change the state of a body by transferring energy to or from it in the form of heat, or by doing work on the body. One property of a body that may change in such a process is its temperature T. The change in temperature

ΔT that corresponds to the transfer of a particular quantity of heat energy Q will depends on the circumstances under which the heat was transferred. For example, in the case of a gas confined to a cylinder with a movable piston, we can add heat and keep the piston fixed (thus keeping the volume constant), or we can add heat and allow to piston to move but keep the force on the piston constant (thus keeping the gas under constant pressure). We can even change the temperature by doing work on a system, such as by rubbing together two objects that exert frictional forces on one another; in this case, no heat transfer need occur.

It is convenient to define the *heat capacity C′* of a body as the ratio of the amount of heat Q supplied to a body in any process to its corresponding temperature change ΔT; that is,

$$C' = \frac{Q}{\Delta T}. \tag{1}$$

The word "capacity" may be misleading because it suggests the essentially meaningless statement "the amount of heat a body can hold," whereas what is meant is simply the energy per degree of temperature change that is transferred as heat when the temperature of the body changes.

The heat capacity per unit mass of a body, called *specific heat capacity* or usually just *specific heat,* is characteristic of the material of which the body is composed:

$$c = \frac{C'}{m} = \frac{Q}{m\,\Delta T}. \tag{2}$$

The heat capacity is characteristic of a particular object, but the specific heat characterizes a substance. Thus we speak, on one hand, of the heat capacity of a penny but, on the other, of the specific heat of copper.

Neither the heat capacity of a body nor the specific heat of a material is constant; both depend on the temperature (and possibly on other variables as well, such as the pressure). The previous equations give only average values for these quantities in the temperature range of ΔT. In the limit, as $\Delta T \to 0$, we can speak of the specific heat at a particular temperature T.

We can find the heat that must be given to a body of mass m, whose material has a specific heat c, to increase its temperature from initial temperature T_i to final temperature T_f by dividing the temperature change into N small intervals ΔT_n, assuming that c_n is constant in each small interval, and summing the contributions to the total heat transfer from all intervals $n = 1, 2, \ldots, N$. This gives

$$Q = \sum_{n=1}^{N} mc_n \, \Delta T_n. \tag{3}$$

In the differential limit this becomes

$$Q = m \int_{T_i}^{T_f} c \, dT, \tag{4}$$

where c may be a function of the temperature. At ordinary

TABLE 1 HEAT CAPACITIES OF SOME SUBSTANCES[a]

Substance	Specific Heat Capacity (J/kg·K)	Molar Heat Capacity (J/mol·K)
Elemental solids		
Lead	129	26.7
Tungsten	135	24.8
Silver	236	25.5
Copper	387	24.6
Carbon	502	6.02
Aluminum	900	24.3
Other solids		
Brass	380	
Granite	790	
Glass	840	
Ice (−10°C)	2220	
Liquids		
Mercury	139	
Ethyl alcohol	2430	
Seawater	3900	
Water	4190	

[a] Measured at room temperature and atmospheric pressure, except where noted.

temperatures and over ordinary temperature intervals, specific heats can be considered to be constants. For example, the specific heat of water varies by less than 1% over the interval from 0°C to 100°C. We can therefore write Eq. 4 in the more generally useful form

$$Q = mc(T_f - T_i). \tag{5}$$

Equation 2 does not define specific heat uniquely. We must also specify the conditions under which the heat Q is added to the material. One common condition is that the specimen remain at normal (constant) atmospheric pressure while we add the heat, but there are many other possibilities, each leading, in general, to a different value for c. To obtain a unique value for c we must indicate the conditions, such as specific heat at constant pressure c_p, specific heat at constant volume c_v, and so on.

Table 1 shows values for the specific heat capacities of a number of common substances, measured under conditions of constant pressure. Although the units are expressed in terms of K, we can also work with temperatures in °C, because a temperature *difference* in C° is equal to the same temperature difference in K.

Sample Problem 1 A copper sample whose mass m_c is 75 g is heated in a laboratory oven to a temperature T of 312°C. The copper is then dropped into a glass beaker containing a mass m_w (=220 g) of water. The effective heat capacity C'_b of the beaker is 190 J/K. The initial temperature T_i of the water and the beaker is 12.0°C. What is the common final temperature T_f of the copper, the beaker, and the water?

Solution   Let us take as our system the water + beaker + copper. No heat enters or leaves this system so that the algebraic sum of the internal heat transfers that occur must be zero. There are three such transfers:

heat flow into the water: $Q_w = m_w c_w (T_f - T_i)$,

heat flow into the beaker: $Q_b = C_b'(T_f - T_i)$,

heat flow into the copper: $Q_c = m_c c_c (T_f - T)$.

The temperature difference is written—in all three cases—as the final temperature minus the initial temperature. We see by inspection that Q_w and Q_b are positive (indicating that heat has been transferred to the water and to the beaker) and that Q_c is negative (indicating that heat has been transferred from the copper).

Since all the energy that leaves one object in this isolated system enters another object, conservation of (heat) energy requires

$$\sum Q = 0 \qquad (6)$$

or

$$Q_w + Q_b + Q_c = 0.$$

Substituting the heat transfer expressions above yields

$$m_w c_w (T_f - T_i) + C_b'(T_f - T_i) + m_c c_c (T_f - T) = 0.$$

Solving for T_f and substituting, we have

$$T_f = \frac{m_w c_w T_i + C_b' T_i + m_c c_c T}{m_w c_w + C_b' + m_c c_c}$$

$$= \frac{(0.220 \text{ kg})(4190 \text{ J/kg·K})(12°\text{C}) + (190 \text{ J/K})(12°\text{C}) + (0.075 \text{ kg})(386 \text{ J/kg·K})(312°\text{C})}{(0.220 \text{ kg})(4190 \text{ J/kg·K}) + 190 \text{ J/K} + (0.075 \text{ kg})(386 \text{ J/kg·K})}$$

$$= 19.6°\text{C}.$$

Note that, because all temperatures were part of temperature *differences,* we can use °C in this expression. In most thermodynamic expressions, however, only Kelvin temperatures can be used.

From the given data you can show that

$$Q_w = 7010 \text{ J}, \quad Q_b = 1440 \text{ J}, \quad \text{and} \quad Q_c = -8450 \text{ J}.$$

The algebraic sum of these three heat transfers is indeed zero, as Eq. 6 requires.

Heats of Transformation

When heat enters a solid or liquid, the temperature of the sample does not necessarily rise. Instead, the sample may change from one *phase* or *state* (that is, solid, liquid, or gas) to another. Thus ice melts and water boils, absorbing heat in each case without a temperature change. In the reverse processes (water freezes, steam condenses), heat is released by the sample, again at a constant temperature.

The amount of heat per unit mass transferred during a phase change is called the *heat of transformation* or *latent heat* (symbol L) for the process. The total heat transferred in a phase change is then

$$Q = Lm, \qquad (7)$$

where m is the mass of the sample that changes phase. The

TABLE 2   SOME HEATS OF TRANSFORMATION

Substance[a]	Melting Point (K)	Heat of Fusion (kJ/kg)	Boiling Point (K)	Heat of Vaporization (kJ/kg)
Hydrogen	14.0	58.6	20.3	452
Oxygen	54.8	13.8	90.2	213
Mercury	234	11.3	630	296
Water	273	333	373	2256
Lead	601	24.7	2013	858
Silver	1235	105	2485	2336
Copper	1356	205	2840	4730

[a] Substances are listed in order of increasing melting points.

heat transferred during melting or freezing is called the *heat of fusion* (symbol L_f), and the heat transferred during boiling or condensing is called the *heat of vaporization* (symbol L_v). Table 2 shows the heats of transformation of some substances.

Knowledge of heat capacities and heats of transformation is important because we can measure a heat transfer by determining the temperature change of a material of known heat capacity or the amount of a substance of known heat of transformation converted from one phase to another. For example, in low-temperature systems involving liquid helium at 4 K, the rate at which helium gas boils from the liquid gives a measure of the rate at which heat enters the system.

25-3   HEAT CAPACITIES OF SOLIDS

From Table 1 we conclude that the specific heats of solids vary widely from one material to another. However, quite a different story emerges if we compare samples of materials that contain the same number of atoms rather than samples that have the same mass. We can do this by finding the *molar heat capacity* of the substance, defined in analogy with Eq. 2 as

$$C = \frac{C'}{n} = \frac{Q}{n \, \Delta T}, \qquad (8)$$

where n is the number of moles of the substance having heat capacity C'. Just as the specific heat capacity (symbol c, unit J/kg·K) represents the heat capacity per unit mass of a substance, the molar heat capacity (symbol C, unit J/mol·K) represents the heat capacity per mole.

In 1819 Dulong and Petit pointed out that the molar heat capacities of elemental solids, with few exceptions (see carbon in Table 1), have values close to 25 J/mol·K. The molar heat capacity, listed in the last column of Table 1, is found by multiplying the specific heat by the molar mass (the mass of 1 mole, which contains 6.02×10^{23} atoms) of the element. We see the amount of

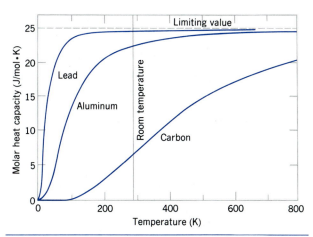

Figure 3   The molar heat capacity of three elements as a function of temperature. At high temperatures, the heat capacities of all solids approach the same limiting value. For lead and aluminum, that value is nearly reached at room temperature; for carbon it is not.

heat required *per atom* to raise the temperature of a solid by a given amount seems to be about the same for almost all materials shown in the table. This is striking evidence for the atomic theory of matter.

Actually molar heat capacities vary with temperature, approaching zero at $T \to 0$ K and approaching the Dulong–Petit value as T becomes larger. Figure 3 shows the variation for lead, aluminum, and carbon. You can see that the apparently anomalous value of Table 1 for carbon occurs because carbon has not yet achieved its limiting value at room temperature. Since the number of atoms rather than the kind of atom seems to be important in determining the heat required to increase the temperature of a body by a given amount, we are led to expect that the molar heat capacities of different substances vary with temperature in much the same way. Figure 4 shows that, indeed, the molar heat capacities of various substances can be made to fall on the same curve by a simple, empirical adjustment in the temperature scale. The horizontal scale in Fig. 4 is the dimensionless ratio T/T_D, where T

is the Kelvin temperature and T_D is a characteristic temperature, called the *Debye temperature,* that has a particular constant value for each material. For lead, T_D has the empirical value of 88 K and for carbon, $T_D = 1860$ K. Note that the molar heat capacity reaches about 80% of its limiting value at $T = 0.5 T_D$ and about 90% at $T = T_D$. Thus T_D can be taken to be a characteristic temperature of the approach to the limit. It is therefore not surprising that, at room temperature, lead ($T/T_D = 3.4$) has reached its limiting value but carbon ($T/T_D = 0.16$) has not.

We can understand the molar heat capacities of solids using results from statistical mechanics found in Chapters 23 and 24. In the high-temperature limit, we consider the atoms of a solid as free to vibrate in three directions. In Section 23-6, we introduced the concept of the number of *degrees of freedom* of a system — essentially, the number of different forms in which a system can have energy. An object that vibrates in one dimension has two degrees of freedom, one for the potential energy and one for the kinetic energy. That is, you can give an oscillator any amount of potential energy you choose, and you can also start it moving with any amount of kinetic energy. The motion of each oscillating atom in a solid can be regarded as a combination of 3 one-dimensional vibrations, each with two degrees of freedom; thus there are six degrees of freedom per atom of the solid. According to the equipartition theorem, each atom has an average energy of $\frac{1}{2}kT$ per degree of freedom, so the internal energy per mole is

$$E_{\text{int}} = 6N_A(\tfrac{1}{2}kT) = 3RT. \qquad (9)$$

If we raise the temperature of a sample of material through ΔT by transferring heat to it, no work being done in the process, the increase in internal energy per mole will be $\Delta E_{\text{int}} = 3R\, \Delta T$. Equating this increase in internal energy per mole to the heat added per mole to achieve the temperature increase, we find, using Eq. 8,

$$C_V = \frac{Q/n}{\Delta T} = \frac{\Delta E_{\text{int}}}{\Delta T} = 3R = 24.9 \text{ J/mol·K}.$$

This is the "classical" Dulong and Petit value of the molar

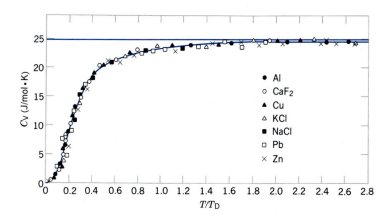

Figure 4   Some selected values of the molar heat capacities of several solids. The solid horizontal line is the Dulong–Petit limit, and the curve is the result of the theory of Debye.

heat capacity,* which according to Table 1 and Fig. 4 is in excellent agreement with the observed values in the high-temperature region. For $T < T_D$, however, the classical theory fails. In this region, the effects of quantum theory become important, and we must use a theory developed first by Einstein and later by Debye (1912). According to quantum theory (see Section 24-6), we must take into account the cooperative nature of the oscillations by using one of the quantum statistical distributions. The solid curve in Fig. 4 is obtained from Debye's calculation, and its excellent agreement with the data is a triumph of quantum physics.†

The data plotted in Fig. 4 vary smoothly and characterize materials that do not change their state in that temperature range. That is, they do not melt or change from one crystalline form to another. Often we can detect such changes by measuring the specific heat of a substance. Figure 5 shows the specific heats of tantalum and brass. In the case of tantalum, you can see that at a temperature of about 4.4 K a sudden change occurs. Below 4.4 K, tantalum is a superconductor: it offers no resistance to the flow of electric current. Such a specific heat "anomaly" does not always indicate the transition from a normal conductor to a superconductor, but it does indicate a change of some sort in the properties of the material. In the case of brass, a change in the crystal structure occurs at about 460°C from a very ordered structure below that temperature to a rather disordered structure above it. The change in structure at 460°C is clearly indicated by a sudden change in the specific heat of brass (Fig. 5b).

25-4   HEAT CAPACITIES OF AN IDEAL GAS

In calculating the heat capacities of an ideal gas, we use results from the kinetic theory of an ideal gas discussed in Chapter 23. You may find it helpful to review Sections 23-5 and 23-6.

---

* The data plotted in Figs. 3 and 4 are the molar heat capacities at constant volume, C_V, while the values given in Table 1 are C_p, the molar heat capacities at constant pressure. It is easier to measure C_p, because the thermal expansion need not be taken into account, but it is easier to calculate C_V. The two are related by the formula

$$C_p = C_V + T\beta^2 B/\rho,$$

in which β is the thermal coefficient of volume expansion, $B(= -V\,\Delta p/\Delta V)$ is the isothermal bulk modulus, and ρ is the density. At room temperature, the difference between C_p and C_V is about 5%.

† Details of Einstein's calculation, which is somewhat simpler but less applicable than Debye's, can be found in *Modern Physics*, by Kenneth S. Krane (Wiley, 1983), Chapter 12.

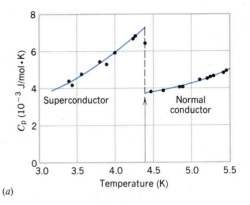

(a)

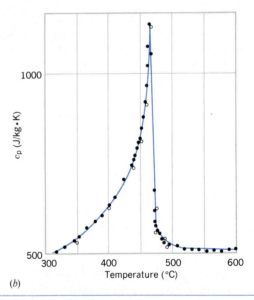

(b)

Figure 5   (a) The specific heat of tantalum near its superconducting transition temperature. (b) The specific heat of brass.

Heat Capacity at Constant Volume

Let us introduce a certain amount of energy as heat Q into a gas that is confined inside a cylinder fitted with a piston. The gas can then either (1) store the energy as the random kinetic energy of its molecules (internal energy) or (2) use the energy to do work on the environment (such as by raising a weight on the piston). Let us first consider the case in which the piston is fixed, so that the volume of the gas remains constant, and no external work is done. In this case all the heat energy goes into internal energy:

$$Q = \Delta E_{\text{int}}. \tag{10}$$

We let C_V represent the *molar heat capacity at constant volume*, so that Eq. 8 gives

$$C_V = \frac{Q}{n\,\Delta T} = \frac{\Delta E_{\text{int}}}{n\,\Delta T}. \tag{11}$$

From Eq. 35 of Chapter 23 for a monatomic ideal gas, $\Delta E_{\text{int}} = \frac{3}{2}nR \, \Delta T$, and so

$$C_{\text{V}} = \tfrac{3}{2}R = 12.5 \text{ J/mol}\cdot\text{K} \quad \text{(monatomic gas)}. \quad (12)$$

Repeating this derivation using Eqs. 36 and 37 of Chapter 23 for the diatomic and polyatomic gases, we find

$$C_{\text{V}} = \tfrac{5}{2}R = 20.8 \text{ J/mol}\cdot\text{K} \quad \text{(diatomic gas)}, \quad (13)$$

$$C_{\text{V}} = 3R = 24.1 \text{ J/mol}\cdot\text{K} \quad \text{(polyatomic gas)}. \quad (14)$$

Heat Capacity at Constant Pressure

Figure 6 shows two ideal gas isotherms differing in temperature by ΔT. Path ab is the constant-volume process considered previously. Path ac is a constant-pressure process that connects the same two isotherms. In Section 23-6, we established that *the internal energy of an ideal gas depends only on the temperature.* For all paths connecting the two isotherms of Fig. 6, the change in internal energy has the same value, because all paths correspond to the same change in temperature. In particular, the change in internal energy is the same for paths ab and ac:

$$\Delta E_{\text{int}, ab} = \Delta E_{\text{int}, ac}. \quad (15)$$

There are two contributions to the change in internal energy along path ac—the heat Q transferred to the gas and the work W done on the gas:

$$\Delta E_{\text{int}, ac} = Q + W. \quad (16)$$

Note the sign conventions that are implicit in Eq. 16. Heat transferred *from* the environment is considered to be positive and tends to increase the internal energy. If the volume decreases, the work done on the gas by the environment is positive, which tends to increase the internal energy. If the volume increases ($W < 0$), we regard the gas as doing work on the environment, which tends to decrease the supply of internal energy of the gas.

Using Eq. 8, the heat transferred in a constant-pressure process can be written

$$Q = nC_{\text{p}} \, \Delta T, \quad (17)$$

where C_{p} is the *molar heat capacity at constant pressure.* Equation 26 of Chapter 23 gives the work along path ac as $W = -p \, \Delta V$, which can be written for this constant-pressure process using the ideal gas law as

$$W = -p \, \Delta V = -nR \, \Delta T. \quad (18)$$

Using Eq. 11 to obtain the change in internal energy along path ab, we can substitute into Eq. 16 to find

$$nC_{\text{V}} \, \Delta T = nC_{\text{p}} \, \Delta T - nR \, \Delta T$$

or

$$C_{\text{p}} = C_{\text{V}} + R. \quad (19)$$

From Eqs. 12–14 we then find the molar heat capacities at constant pressure:

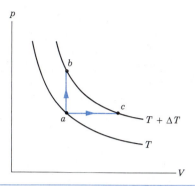

Figure 6   Two ideal-gas isotherms differing in temperature by ΔT are connected by the constant-volume process ab and the constant-pressure process ac.

$$C_{\text{p}} = \tfrac{5}{2}R = 20.8 \text{ J/mol}\cdot\text{K} \quad \text{(monatomic gas)}, \quad (20)$$

$$C_{\text{p}} = \tfrac{7}{2}R = 29.1 \text{ J/mol}\cdot\text{K} \quad \text{(diatomic gas)}, \quad (21)$$

$$C_{\text{p}} = 4R = 33.3 \text{ J/mol}\cdot\text{K} \quad \text{(polyatomic gas)}. \quad (22)$$

Another parameter of interest, which can be directly measured independently of the values of C_{p} and C_{V}, is the *ratio of molar heat capacities* γ, defined as

$$\gamma = \frac{C_{\text{p}}}{C_{\text{V}}}. \quad (23)$$

Because the specific heat capacity is related to the molar heat capacity by $c = C/M$, where M is the molar mass of the substance, we can also express γ as $c_{\text{p}}/c_{\text{V}}$. For this reason γ is often called the *ratio of specific heats* or *specific heat ratio.* We used γ previously in the expression for the speed of sound in a gas (Sample Problem 6 of Chapter 23) and in the relationship between pressure and volume in an adiabatic process (Eq. 28 of Chapter 23).

Using Eqs. 20–22 for C_{p} and Eqs. 12–14 for C_{V}, we obtain

$$\gamma = \tfrac{5}{3} = 1.67 \quad \text{(monatomic gas)}, \quad (24)$$

$$\gamma = \tfrac{7}{5} = 1.40 \quad \text{(diatomic gas)}, \quad (25)$$

$$\gamma = \tfrac{4}{3} = 1.33 \quad \text{(polyatomic gas)}. \quad (26)$$

Table 3 shows a comparison of observed values with the predictions of the ideal gas model. The agreement is excellent.

Sample Problem 2   A family enters a winter vacation cabin that has been unheated for such a long time that the interior temperature is the same as the outside temperature (0°C). The cabin consists of a single room of floor area 6 m by 4 m and height 3 m. The room contains one 2-kW electric heater. Assuming that the room is perfectly airtight and that all the heat from the electric heater is absorbed by the air, none escaping through the walls or being absorbed by the furnishings, how long after the heater is turned on will the air temperature reach the comfort level of 21°C (=70°F)?

TABLE 3 MOLAR HEAT CAPACITIES OF GASES

Gas	C_p (J/mol·K)	C_v (J/mol·K)	$C_p - C_v$ (J/mol·K)	γ
Monatomic				
Ideal	20.8	12.5	8.3	1.67
He	20.8	12.5	8.3	1.66
Ar	20.8	12.5	8.3	1.67
Diatomic				
Ideal	29.1	20.8	8.3	1.40
H_2	28.8	20.4	8.4	1.41
N_2	29.1	20.8	8.3	1.40
O_2	29.4	21.1	8.3	1.40
Polyatomic				
Ideal	33.3	24.9	8.3	1.33
CO_2	37.0	28.5	8.5	1.30
NH_3	36.8	27.8	9.0	1.31

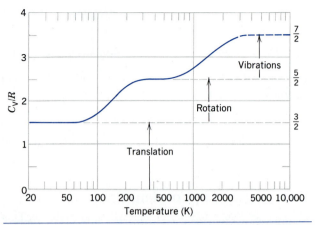

Figure 7 The ratio C_v/R for hydrogen as a function of temperature. Because the rotational and vibrational motions occur at quantized energies, only translational motion occurs at low temperatures. As the temperature increases, rotational motion can be excited during collisions. At still higher temperatures, vibrational motion can occur.

Solution Let us assume the air in the room (which is mostly nitrogen and oxygen) behaves like an ideal diatomic gas. The volume of the room is

$$V = (6\text{ m})(4\text{ m})(3\text{ m}) = 72\text{ m}^3 = 72{,}000\text{ L}.$$

Since 1 mole of an ideal gas occupies 22.4 L at 0°C and 1 atm, the number of moles is

$$n = (72{,}000\text{ L})/(22.4\text{ L/mol}) = 3.2 \times 10^3\text{ mol}.$$

If the room is airtight (see the discussion below), we can regard the absorption of heat to take place at constant volume, for which

$$Q = nC_v \Delta T = (3.2 \times 10^3\text{ mol})(20.8\text{ J/mol·K})(21\text{ K})$$
$$= 1.4 \times 10^6\text{ J}.$$

The heater delivers a power P of 2 kW and can provide this energy in a time of

$$t = \frac{Q}{P} = \frac{1.4 \times 10^6\text{ J}}{2 \times 10^3\text{ W}} = 700\text{ s}$$

or about 12 min.

This problem contained some very unphysical assumptions about the absorption of heat in this room. Try to estimate the heat capacity of some pieces of furniture to see if neglecting their effect on the heat absorption (and thus on the time to bring the room to comfort level) was reasonable. The heat loss through the walls of the room, which we consider in Section 25-7, also will have a considerable effect on this problem.

Is the assumption about the room being airtight reasonable? If the air in the cabin were originally at a pressure of 1 atm when the temperature was 0°C, what will be the interior pressure at 21°C? What will be the resulting outward force on the roof and walls? A more reasonable assumption might be that the room is not quite airtight, but that as the temperature rises some air will escape, thereby keeping the pressure constant. See Problem 30 for a calculation based on this assumption.

Effect of Quantum Theory *(Optional)*
The values shown in Table 3 are characteristic of room temperature, and it is therefore logical to ask whether the molar heat capacity of a gas shows any effect of temperature. Figure 7 shows

the molar heat capacity of hydrogen as a function of temperature. At low temperatures $C_v = \frac{3}{2}R$, characteristic of a gas with only translational degrees of freedom. From about 200 to 600 K, hydrogen has $C_v = \frac{5}{2}R$ as we expect for a diatomic gas with two rotational degrees of freedom; above about 2000 K, C_v appears to approach the value $\frac{7}{2}R$ that would be characteristic of two additional degrees of freedom associated with the vibrational motion.

The key to understanding these features lies with quantum theory. Let us first look at the rotational energy. According to Eq. 23 of Chapter 13, the smallest possible change in the angular momentum of a rotating system is $\Delta L = h/2\pi$, where h is the Planck constant. The rotational kinetic energy E_R is related to the angular momentum L through $E_R = \frac{1}{2}I\omega^2 = \frac{1}{2}L^2/I$, where I is the rotational inertia of a H_2 molecule about an axis through the center of mass and perpendicular to the line connecting the two H atoms, given by

$$I = m\left(\frac{R}{2}\right)^2 + m\left(\frac{R}{2}\right)^2 = \frac{1}{2}mR^2,$$

where m is the mass of a hydrogen atom and $R\ (= 0.074\text{ nm})$ is the equilibrium separation of the two atoms. Putting in the numerical values and computing the rotational kinetic energy corresponding to the smallest permitted change in the angular momentum $(h/2\pi)$, we estimate

$$E_R = 3.8 \times 10^{-3}\text{ eV}.$$

According to the equipartition theorem, an energy of $\frac{1}{2}kT$ is allocated to this rotation, but this value of E_R is the *minimum* energy of rotation. If T is so small that $\frac{1}{2}kT < E_R$, there is (on the average) not enough thermal energy available to provide the minimum rotational kinetic energy, and rotations cannot occur. Let us find this temperature threshold:

$$\frac{1}{2}kT = E_R$$
$$T = \frac{2E_R}{k} = \frac{2(3.8 \times 10^{-3}\text{ eV})}{8.6 \times 10^{-5}\text{ eV/K}} = 88\text{ K}.$$

The value is entirely consistent with the data plotted in Fig. 7: effects of rotation do not appear until temperatures above about 88 K.

A similar situation occurs for the vibrational energy. The vibrational frequency can be found from the effective "spring constant," which can in turn be estimated from treating the potential energy of the diatomic molecule as approximately parabolic near its minimum (see Fig. 10 of Chapter 8). For H_2, the frequency turns out to be $\nu = 1.3 \times 10^{14}$ Hz, and the quantized vibrational energy E_v (see Eq. 38 of Chapter 8) is

$$E_v = h\nu = 0.54 \text{ eV}.$$

The equipartition theorem allows a total energy of kT for the two vibrational degrees of freedom, but the molecule will not vibrate unless it has at least 0.54 eV of thermal energy available. Thus the vibrational theshold is determined by

$$kT = E_v$$

$$T = \frac{E_v}{k} = \frac{0.54 \text{ eV}}{8.6 \times 10^{-5} \text{ eV/K}} = 6300 \text{ K}.$$

This rough estimate is consistent with the data of Fig. 7.

The description we have given of the structure of hydrogen gives us insight into the behavior of molecules, but you should keep in mind that it does contradict principles of classical kinetic theory that we developed in Chapter 23. Kinetic theory is based on the application of Newtonian mechanics to a gas of particles, and the equipartition of energy (Section 23-6) follows directly from classical statistical mechanics. However, if equipartition of energy holds, then the molar heat capacity of hydrogen should be independent of temperature. Classical physics does not permit one mode of motion, such as the vibrational or rotational motion of H_2, to be "frozen" below a certain threshold of temperature, nor does it allow energy to be added to only one mode of motion at a time. Classical physics is thus in obvious disagreement with the experimental results shown in Fig. 7.

Our study of kinetic theory has indicated the inadequacy of classical mechanics and suggested the need for a new theory, quantum mechanics. Just as Newtonian mechanics must be replaced by relativity theory to describe motion at high speed (near the speed of light), so Newtonian mechanics must be replaced by quantum mechanics to describe the behavior of physical systems of small (subatomic) dimensions. Fortunately, quantum mechanics reduces directly to Newtonian mechanics in the limit of ordinary-sized objects, and so we can continue to apply classical thermodynamics with confidence to systems in which the subatomic structure is not evident.

kinetic theory, that molecules can be regarded as having no internal structure, holds true at ordinary temperatures. Only at temperatures high enough to give molecules an average translational kinetic energy comparable to the energy difference between the ground state and lowest excited state will the internal structure of the molecule change as collisions become inelastic. Conversely, the failure of classical kinetic theory in gases at high temperatures may be said to provide evidence for the quantized internal structure of atoms. ∎

25-5   THE FIRST LAW OF THERMODYNAMICS

Figure 8 shows a system consisting of two gases separated by a diathermic wall in a container that is otherwise isolated from the environment. The system has no moving parts, so no work is done. Let us assume that the gases are originally at temperatures T_1 and T_2, and that after a sufficient time in thermal contact the system comes to equilibrium at some intermediate temperature T. From the techniques already discussed in this chapter, we know how to find this temperature based on one assumption: the energy lost as heat by the hotter gas (Q_1, a negative quantity) is equal in magnitude to the energy gained as heat by the cooler gas (Q_2, a positive quantity). In effect, this is nothing more than a statement of the conservation of energy: $|Q_1| = |Q_2|$. Another way to state this is: $Q_1 + Q_2 = 0$; that is, since no heat is transferred between this combined system and its environment, the total energy of the two gases remains constant.

Instead of taking the combination as our system, let us instead choose gas 2, which absorbs heat Q_2. After this energy is absorbed, there is no change in the system other than an increase of its temperature from T_2 to T. For an ideal gas we can calculate the corresponding change in internal energy $\Delta E_{\text{int},2}$. The only source for this change in internal energy is the absorbed heat, and so $\Delta E_{\text{int},2} = Q_2$, both quantities being positive. This is a statement of the conservation of energy applied to gas 2. We can also write a similar statement of conservation of energy applied to

---

Sample Problem 3   The internal structure of hydrogen shows a series of discrete excited states, the first such state being at an energy of $E = 10.2$ eV above the lowest (ground) state. At what temperature would the average translational kinetic energy be equal to the energy of the excited state?

Solution   We require

$$\tfrac{3}{2}kT = E,$$

or

$$T = \frac{2E}{3k} = \frac{2(10.2 \text{ eV})}{3(8.6 \times 10^{-5} \text{ eV/K})} = 7.9 \times 10^4 \text{ K}.$$

Based on this calculation, we see why the basic assumption of

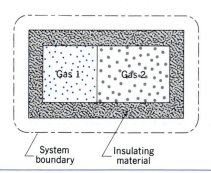

Figure 8   Two gases separated by a diathermic (heat conducting) wall in an insulated container.

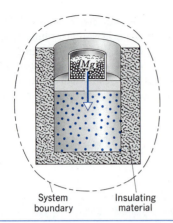

Figure 9   A gas in an insulated container has external work done on it by gravity.

gas 1 : $\Delta E_{\text{int},1} = Q_1$, in which both quantities are negative. Assuming we have taken sufficient care to get the signs correct, we can write a general equation that describes how conservation of energy may be applied to either gas, in the absence of external work:

$$\Delta E_{\text{int}} = Q. \qquad (27)$$

Now consider the familiar situation illustrated in Fig. 9. Let the system (now *including* the weight) be isolated from the environment, so no heat enters or leaves. Suppose that the load on the piston is gradually increased, so that the weighted piston descends through a certain distance. Gravity (an external force exerted by the environment) does a certain amount of (positive) work W on the system. (See Eq. 30 of Chapter 23 for an expression for the work done in this adiabatic process.) The temperature increases in this process, and the system therefore experiences a positive change in internal energy. Since no heat transfer is involved, the internal energy of the gas increases by the work done on it, or

$$\Delta E_{\text{int}} = W, \qquad (28)$$

both quantities being positive in this equation. Equation 28 is another expression of the conservation of energy applied to the system.

These two examples, one involving a change in internal

energy due to heat transfer and another due to external work, are represented schematically in Fig. 10. We consider a general thermodynamic system, and we have been careful to draw a definite boundary between the system and its environment. The system starts out in an initial equilibrium state i in Fig. 10*a*, in which the thermodynamic variables have certain values. We then subject the system to an interaction with its environment, as in Fig. 10*b*, so that work W may be done and heat Q may be exchanged. Finally, as in Fig. 10*c*, the system is in a final equilibrium state f.

We now take the system from state i to state f along a variety of different paths (see, for example, Fig. 11 and Problem 38). We know from our previous considerations that both W and Q depend on the path. Guided by our previous discussion of internal energy, we evaluate the quantity $Q + W$ for each path. *We find that in every case the quantity $Q + W$ has the same value. Even though Q and W individually depend on the path taken, the quantity $Q + W$ is independent of the path,* depending only on the initial and final equilibrium states i and f of the gas.

You will recall from our discussion of conservative forces in Chapter 8 that, when the work done on an object depends only on the initial and final states and is independent of path, we can define a function (the potential energy in Chapter 8) that depends only on the values of the initial and final coordinates, such that the work done in displacing the object is equal to the difference between the final and initial values of this function. We have a similar situation here in thermodynamics, in which the quantity $Q + W$ depends *only* on the initial and final coordinates and *not at all* on the path taken between i and f. We make a similar conclusion: *there must be a function of the thermodynamic coordinates, whose final value minus initial value equals the value of $Q + W$ in the process.*

This function is the *internal energy* of the system. We have already represented the internal energy of an ideal gas E_{int} as the sum of the translational and possibly rotational or vibrational energy of its molecules. Here we are referring to a more general internal energy function applicable to *any* thermodynamic system. In the case of a real gas, it might include (in addition to the translational and rotational kinetic energies) the potential energy between the atoms in a molecule as well as the potential energy

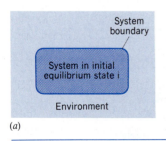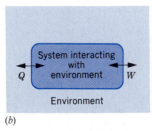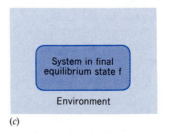

Figure 10   (*a*) A system in an initial state in equilibrium with its surroundings. (*b*) A thermodynamic process during which the system may exchange heat Q or work W with its environment. (*c*) A final equilibrium state reached as a result of the process.

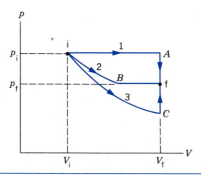

Figure 11 A system in an initial equilibrium state i is brought to a final equilibrium state f along three different paths: (1) a constant-pressure process to A, followed by a constant-volume process to f; (2) an isothermal process to B, followed by a constant-pressure process to f; (3) an adiabatic process to C, followed by a constant-volume process to f. The heat Q transferred and work W done are different for each path, but the sum $Q + W$ has the same value for every path between i and f.

between different molecules. It can even include the internal excitations of the electrons in the atoms. If our thermodynamic system is a solid, the internal energy might include the binding energy of atoms in the lattice.

The change in internal energy between equilibrium states i and f is

$$\Delta E_{int} = E_{int, f} - E_{int, i}. \qquad (29)$$

The value of $E_{int, i}$ depends only on the coordinates of the state i (perhaps on its temperature, volume, and pressure in the case of a gas, although only on temperature for an *ideal* gas). Similarly, $E_{int, f}$ depends only on the coordinates of the point f. Such a function is called a *state function*: it depends only on the state of the system and not at all on how the system arrived at that state.

This brings us to the *first law of thermodynamics*, which can be stated as follows:

In any thermodynamic process between equilibrium states i *and* f, *the quantity* $Q + W$ *has the same value for any path between* i *and* f. *This quantity is equal to the change in the value of a state function called the internal energy.*

Mathematically, the first law is*

$$\Delta E_{int} = Q + W. \qquad (30)$$

---

* The form in which the first law is written depends on the sign convention chosen for the work W, which we defined as $-\int p\,dV$ (work done *on* a system). Alternatively, the work can be defined as $\int p\,dV$ (work done *by* the system), in which case the first law would be written $\Delta E_{int} = Q - W$. We have chosen to write the first law in this form so that thermodynamic work will have the same sign convention that we used in Chapter 7 for mechanical work; that is, work done *on* a system is positive.

Implicit in the first law are three features: (1) the existence of the internal energy E_{int} (much as the zeroth law includes the existence of the temperature); (2) the mathematical relationship between E_{int}, Q, and W; and (3) the sign conventions necessary to apply the first law ($Q > 0$ when heat *enters* a system, which tends to increase E_{int}; $W > 0$ when work is done *on* the system, which also tends to increase E_{int}).

Equation 30 expresses the first law in a form that is directly related to our previous general result for energy conservation in a system of particles, Eq. 28 of Chapter 8: $\Delta U + \Delta K + \Delta E_{int} = W$. In that chapter, we did not consider energy changes by means of heat transfer, so in one sense the first law is a somewhat *more* general statement of energy conservation. On the other hand, it is at the same time somewhat less general, in that thermodynamics usually does not deal with center-of-mass motion, and so $\Delta K = 0$, nor does it deal with cases in which an external force gives a potential energy, so $\Delta U = 0$. Based on our experience with thermodynamics, we would be tempted to write a more general equation of energy conservation as $\Delta U + \Delta K + \Delta E_{int} = Q + W$, and it is appropriate to view the first law as an expression of energy conservation valid under these special circumstances.

If the system undergoes only an infinitesimal change in state, only an infinitesimal amount of heat dQ is absorbed, and only an infinitesimal amount of work dW is done. In such a case the first law is written in differential form:

$$dE_{int} = dQ + dW. \qquad (31)$$

Because Q and W are not actual functions of the state of a system, they cannot be treated as exact differentials in the mathematical sense; that is, there is no function of coordinates Q or W whose differential is dQ or dW. In Eq. 31 their meaning is that of a very small quantity. However, dE_{int} is an exact differential, because E_{int} *is* a function of the coordinates of the system.

The first law of thermodynamics is a general result that is thought to apply to every process in nature that proceeds between equilibrium states. It is not necessary that every step in the process be an equilibrium state, only the initial and final states. For example, we can apply the first law to the explosion of a firecracker in an insulated steel drum. We can account for the energy balance before the explosion and after the system has returned to equilibrium, and for this calculation we need not worry that the intermediate condition is turbulent and that pressure and temperature are not well defined.

Because of its generality, the first law is somewhat incomplete as a description of nature. It tells us that energy must be conserved in every process, but it does not tell us whether any particular process that conserves energy can actually occur. The explosion of the firecracker, for example, releases chemical energy stored in the gunpowder that eventually raises the temperature of the gas in the drum. We can imagine the hot gas giving its thermal energy back to the combustion products, turning them once again into gunpowder and reassembling the firecracker, but this is never observed. Conservation of energy works either way, but nature seems to go in a preferred direction. To provide this distinction, we need the second law of thermodynamics, to be discussed in Chapter 26.

25-6   APPLICATIONS OF THE FIRST LAW

Adiabatic Processes

In an adiabatic process, the system is well insulated so that no heat enters or leaves, in which case $Q = 0$. The first law becomes, in this case,

$$\Delta E_{int} = W \quad \text{(adiabatic process).} \quad (32)$$

Let us derive the relationship between p and V for an adiabatic process carried out on an ideal gas, which we used in Section 23-5. We assume the process to be carried out slowly, so that the pressure is always well defined. For an ideal gas, we can write Eq. 11 as

$$dE_{int} = nC_V \, dT.$$

Thus

$$p \, dV = -dW = -dE_{int} = -nC_V \, dT. \quad (33)$$

The equation of state of the gas can be written in differential form as

$$d(pV) = d(nRT)$$

$$p \, dV + V \, dp = nR \, dT. \quad (34)$$

But $p \, dV$ is just $-dW$, which is equal to $-dE_{int}$ (since Eq. 32 can be written in differential form as $dE_{int} = dW$). Solving Eq. 34 for $V \, dp$ and substituting Eq. 33, we have

$$V \, dp = nC_V \, dT + nR \, dT = nC_p \, dT, \quad (35)$$

where the last result has been obtained using Eq. 19, $C_p = C_V + R$. We now take the ratio between Eqs. 35 and 33, which gives

$$\frac{V \, dp}{p \, dV} = \frac{nC_p \, dT}{-nC_V \, dT} = -\frac{C_p}{C_V} = -\gamma,$$

using Eq. 23 for the ratio of molar heat capacities γ. Rewriting, we find

$$\frac{dp}{p} = -\gamma \frac{dV}{V},$$

which we can integrate between initial state i and final state f

$$\int_{p_i}^{p_f} \frac{dp}{p} = -\gamma \int_{V_i}^{V_f} \frac{dV}{V}$$

$$\ln \frac{p_f}{p_i} = -\gamma \ln \frac{V_f}{V_i}$$

which can be written

$$p_i V_i^{\gamma} = p_f V_f^{\gamma}. \quad (36)$$

Since i and f are arbitrary points, we can write this equation as

$$pV^{\gamma} = \text{constant.} \quad (37)$$

Equations 36 and 37 give the relationship between the pressure and volume of an ideal gas that undergoes an adiabatic process. Given the values of the pressure and volume at the initial point, the adiabatic process will proceed through final points whose pressure and volume can be calculated from Eq. 36. Equivalently, Eq. 37 defines a family of curves on a pV diagram. Every adiabatic process can be represented by a segment of one of these curves (Fig. 12).

We can rewrite these results in terms of temperature, using the ideal gas equation of state:

$$(pV)V^{\gamma-1} = \text{constant}$$

$$TV^{\gamma-1} = \text{constant.} \quad (38)$$

The constant in Eq. 38 is not the same as that in Eq. 37. Equivalently, we can write Eq. 38 as

$$T_i V_i^{\gamma-1} = T_f V_f^{\gamma-1}$$

$$T_f = T_i \left(\frac{V_i}{V_f} \right)^{\gamma-1} \quad (39)$$

From our basic definition of thermodynamic work, $W = -\int p \, dV$, we can show (see Section 23-5) that, for an adiabatic process,

$$W = \frac{1}{\gamma - 1} (p_f V_f - p_i V_i). \quad (40)$$

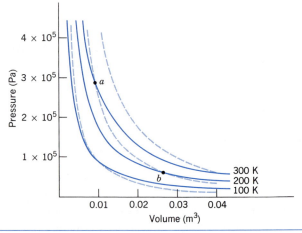

Figure 12   Isothermal processes (solid lines) and adiabatic processes (dashed lines) carried out on 1 mole of a diatomic ideal gas. Note that an adiabatic *increase* in volume (for example, the segment *ab*) is always accompanied by a *decrease* in temperature.

Suppose we compress a gas in an adiabatic process (as illustrated in Fig. 9). Then $V_i > V_f$, and Eq. 39 then requires that $T_f > T_i$. The temperature of the gas rises as it is compressed, as we frequently observe from the warming of a bicycle pump. Conversely, the temperature falls when a gas expands, which is often used as a means to achieve low temperatures in the laboratory (see Fig. 12).

Sound waves in air can be represented in terms of adiabatic processes. At audio frequencies, air is a poor conductor of heat. There is an increase in temperature in the compression zones of a sound wave, but due to the poor conduction there is no appreciable flow of heat to the neighboring cooler rarefactions; the process is thus adiabatic. The compressions and expansions of steam in a steam engine, or of the hot gases in the cylinders of an internal combustion engine, are also essentially adiabatic, because there is insufficient time for heat to flow.

Isothermal Processes

In an isothermal process, the temperature remains constant. If the system is an ideal gas, then the internal energy must therefore also remain constant. With $\Delta E_{\text{int}} = 0$, the first law gives

$$Q + W = 0 \quad \text{(isothermal process; ideal gas).} \quad (41)$$

If an amount of (positive) work W is done on the gas, an equivalent amount of heat $Q = -W$ is released by the gas to the environment. None of the work done on the gas remains with the gas as stored internal energy.

Figure 12 compares isothermal and adiabatic processes for 1 mole of a monatomic ideal gas.

Constant-Volume Processes

If the volume of a gas remains constant, it can do no work. Thus $W = 0$, and the first law gives

$$\Delta E_{\text{int}} = Q \quad \text{(constant-volume process).} \quad (42)$$

In this case all the heat that enters the gas ($Q > 0$) is stored as internal energy ($\Delta E_{\text{int}} > 0$).

Cyclical Processes

In a cyclical process, we carry out a sequence of operations that eventually restores the system to its initial state, as, for example, the three-step process illustrated in Fig. 13. Because the process starts and finishes at the point A, the internal energy change for the cycle is zero. Thus, according to the first law,

$$Q + W = 0 \quad \text{(cyclical process),} \quad (43)$$

where Q and W represent the totals for the cycle. In Fig. 13, the total work is positive, because there is more positive area under the curve representing step 3 than there is negative area under the line representing step 2. Thus

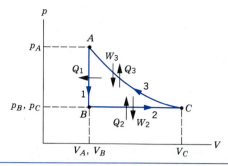

Figure 13 A gas undergoes a cyclical process starting at point A and consisting of (1) a constant-volume process AB, (2) a constant-pressure process BC, and (3) an isothermal process CA.

$W > 0$ and it follows from Eq. 43 that $Q < 0$. In fact, for any cycle that is done in a counterclockwise direction, we must have $W > 0$ (and thus $Q < 0$), while cycles performed in the clockwise direction have $W < 0$ and $Q > 0$.

Free Expansion

Figure 14 represents the process known as *free expansion*. The gas is initially in only one side of the container, and when the stopcock is opened, the gas expands into the previously evacuated half. No weights can be raised in this process, so no work is done. The container is insulated, so the process is adiabatic. Hence, with $W = 0$ and $Q = 0$, the first law gives

$$\Delta E_{\text{int}} = 0 \quad \text{(free expansion).} \quad (44)$$

Thus the internal energy of an ideal gas undergoing a free expansion remains constant, and because the internal energy of an ideal gas depends only on the temperature, its temperature must similarly remain constant.

The free expansion is a good example of a *nonequilibrium* process. If a gas has a well-defined pressure and volume (and therefore temperature), we can show the state of the gas as a point on a pV diagram. The assignment of a temperature to the gas means that it must be in thermal equilibrium; each point on a pV diagram therefore represents a system in equilibrium. In the case of a

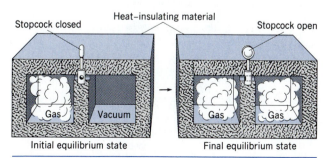

Figure 14 Free expansion. Opening the stopcock allows gas to flow from one side of the insulated container to the other. No work is done, and no heat is transferred to the environment.

TABLE 4   APPLICATIONS OF THE FIRST LAW

Process	Restriction	First Law	Other Results
All	None	$\Delta E_{int} = Q + W$	$\Delta E_{int} = nC_V \Delta T$, $W = -\int p \, dV$
Adiabatic	$Q = 0$	$\Delta E_{int} = W$	$\overline{W = (p_f V_f - p_i V_i)/(\gamma - 1)}$
Constant volume	$W = 0$	$\Delta E_{int} = Q$	$\overline{Q = nC_V \Delta T}$
Constant pressure	$\Delta p = 0$	$\Delta E_{int} = Q + W$	$W = -p \Delta V$, $Q = nC_p \Delta T$
Isothermal	$\overline{\Delta E_{int} = 0}$	$Q = -W$	$\overline{W = -nRT \ln(V_f/V_i)}$
Cycle	$\overline{\Delta E_{int} = 0}$	$Q = -W$	
Free expansion	$Q = W = 0$	$\Delta E_{int} = 0$	$\underline{\Delta T = 0}$

Items underlined apply only to ideal gases; all other items apply in general.

free expansion, the initial state (all gas on one side) is an equilibrium state, as is the final state; but at intermediate times, as the gas rushes from one side to the other, the temperature and the pressure do not have unique values, and we cannot plot this process on a pV diagram. Only the initial and final points appear on the graph. Nevertheless, we can still use the first law to analyze this process, because the change in internal energy depends only on the initial and final points.

Table 4 summarizes the processes we have considered and their energy transfers.

Sample Problem 4   Let 1.00 kg of liquid water be converted to steam by boiling at standard atmospheric pressure; see Fig. 15. The volume changes from an initial value of 1.00×10^{-3} m³ as a liquid to 1.671 m³ as steam. For this process, find (*a*) the work done on the system, (*b*) the heat added to the system, and (*c*) the change in the internal energy of the system.

Solution   (*a*) The work done on the gas during this constant-pressure process is given by Eq. 26 of Chapter 23:

$$W = -p(V_f - V_i)$$
$$= -(1.01 \times 10^5 \text{ Pa})(1.671 \text{ m}^3 - 1 \times 10^{-3} \text{ m}^3)$$
$$= -1.69 \times 10^5 \text{ J} = -169 \text{ kJ}.$$

The work done on the system is negative; equivalently, positive

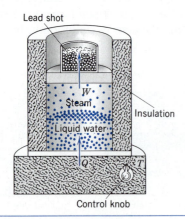

Figure 15   Sample Problem 4. Water is boiling at constant pressure. Heat flows from the reservoir until the water has changed completely into steam. Work is done by the expanding gas as it lifts the piston.

work is done *by* the system on its environment in lifting the weighted piston of Fig. 15.

(*b*) From Eq. 7 we have

$$Q = Lm = (2260 \text{ kJ/kg})(1.00 \text{ kg}) = 2260 \text{ kJ}.$$

This quantity is positive, as is appropriate for a process in which heat is added to the system.

(*c*) We find the change in internal energy from the first law:

$$\Delta E_{int} = Q + W = 2260 \text{ kJ} + (-169 \text{ kJ}) = 2090 \text{ kJ}.$$

This quantity is positive, indicating that the internal energy of the system has increased during the boiling process. This energy represents the internal work done in overcoming the strong attraction that the H_2O molecules have for each other in the liquid state.

We see that, when water boils, about 7.5% (169 kJ/2260 kJ = 0.075) of the added heat goes into external work in pushing back the atmosphere. The rest goes into internal energy that is added to the system.

Sample Problem 5   The cycle shown in Fig. 13 consists of three processes, starting at point *A*: a reduction in pressure at constant volume from point *A* to point *B*; an increase in volume at constant pressure from point *B* to point *C*; an isothermal compression (decrease in volume) from point *C* back to point *A*. Let the cycle be carried out on 0.75 mol of a diatomic ideal gas, with $p_A = 3.2 \times 10^3$ Pa, $V_A = 0.21$ m³, and $p_B = 1.2 \times 10^3$ Pa. For each of the three processes and for the cycle, find Q, W, and ΔE_{int}.

Solution   The first step is to find the values of p, V, and T at each point. At point *A*, we are given p_A and V_A, and we can solve for T_A from the ideal gas law:

$$T_A = \frac{p_A V_A}{nR} = \frac{(3.2 \times 10^3 \text{ Pa})(0.21 \text{ m}^3)}{(0.75 \text{ mol})(8.31 \text{ J/mol} \cdot \text{K})} = 108 \text{ K}.$$

At point *B*, we are given p_B and $V_B (= V_A)$, and we can similarly find T_B:

$$T_B = \frac{p_B V_B}{nR} = \frac{(1.2 \times 10^3 \text{ Pa})(0.21 \text{ m}^3)}{(0.75 \text{ mol})(8.31 \text{ J/mol} \cdot \text{K})} = 40 \text{ K}.$$

At point *C*, we know $p_C (= p_B)$ and $T_C (= T_A$, because process *CA* is an isotherm). We can then find V_C:

$$V_C = \frac{nRT_C}{p_C} = \frac{(0.75 \text{ mol})(8.31 \text{ J/mol} \cdot \text{K})(108 \text{ K})}{1.2 \times 10^3 \text{ Pa}} = 0.56 \text{ m}^3.$$

With this information, we can now calculate the heat transfer, work done, and change in internal energy for each process. For process 1 (*AB*), we have

$$Q_1 = nC_V(T_B - T_A)$$
$$= (0.75 \text{ mol})(20.8 \text{ J/mol} \cdot \text{K})(40 \text{ K} - 108 \text{ K}) = -1060 \text{ J},$$

$$W_1 = 0 \quad \text{(constant-volume process)},$$

$$\Delta E_{\text{int},1} = Q_1 + W_1 = -1060 \text{ J} + 0 = -1060 \text{ J}.$$

The system transfers energy to the environment as heat during process 1, and its temperature falls, corresponding to a negative change in internal energy.

For the constant-pressure process 2 (*BC*), we obtain

$$Q_2 = nC_p(T_C - T_B)$$
$$= (0.75 \text{ mol})(29.1 \text{ J/mol} \cdot \text{K})(108 \text{ K} - 40 \text{ K}) = 1480 \text{ J},$$

$$W_2 = -p(V_C - V_B)$$
$$= -(1.2 \times 10^3 \text{ Pa})(0.56 \text{ m}^3 - 0.21 \text{ m}^3) = -420 \text{ J},$$

$$\Delta E_{\text{int},2} = Q_2 + W_2 = 1480 \text{ J} + (-420 \text{ J}) = 1060 \text{ J}.$$

Energy is transferred to the gas as heat during process 2, and in expanding the gas does work on its environment (the environment does negative work on the gas).

Along the isotherm (*CA*), the work is given by Eq. 27 of Chapter 23:

$$W_3 = -nRT_C \ln \frac{V_A}{V_C}$$
$$= -(0.75 \text{ mol})(8.31 \text{ J/mol} \cdot \text{K})(108 \text{ K}) \ln \frac{0.21 \text{ m}^3}{0.56 \text{ m}^3}$$
$$= 660 \text{ J},$$

$$\Delta E_{\text{int},3} = 0 \quad \text{(isothermal process)},$$

$$Q_3 = \Delta E_{\text{int},3} - W_3 = 0 - 660 \text{ J} = -660 \text{ J}.$$

For the cycle, we have

$$Q = Q_1 + Q_2 + Q_3 = -1060 \text{ J} + 1480 \text{ J} + (-660 \text{ J})$$
$$= -240 \text{ J},$$

$$W = W_1 + W_2 + W_3 = 0 + (-420 \text{ J}) + 660 \text{ J} = 240 \text{ J},$$

$$\Delta E_{\text{int}} = \Delta E_{\text{int},1} + \Delta E_{\text{int},2} + \Delta E_{\text{int},3} = -1060 \text{ J} + 1060 \text{ J} + 0 = 0.$$

Note that, as expected for the cycle, $\Delta E_{\text{int}} = 0$ and $Q = -W$. The total work for the cycle is positive, as we expect for a cycle that is done in the counterclockwise direction.

25-7   THE TRANSFER OF HEAT

We have discussed the transfer of heat between a system and its environment, but we have not yet described how that transfer takes place. There are three mechanisms: conduction, convection, and radiation. We discuss each in turn.

Conduction

If you leave a metal poker in a fire for any length of time, its handle will become hot. Energy is transferred from the fire to the handle by *conduction* along the length of the

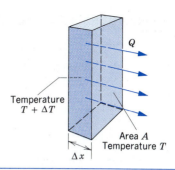

Figure 16   Heat Q flows through a rectangular slab of material of thickness Δx and area A.

metal shaft. The atoms at the hot end, by virtue of the high temperature at that end, are vibrating with large amplitude. These large vibrational amplitudes are passed along the shaft, from atom to atom, by interactions between adjacent atoms. In this way a region of rising temperature travels along the shaft to your hand.

Consider a thin slab of homogeneous material of thickness Δx and cross-sectional area A (Fig. 16). The temperature is $T + \Delta T$ on one face and T on the other. By experiment, we learn several features about the rate $H = Q/\Delta t$ at which a small quantity of heat Q will be transferred through the slab in a time Δt. The rate of heat flow through the slab is (1) directly proportional to A—the more area available, the more heat can flow per unit time; (2) inversely proportional to Δx—the thicker the slab, the less heat can flow per unit time; and (3) directly proportional to ΔT—the larger the temperature difference, the more heat can flow per unit time. [These experimental results provide clues for minimizing the loss of heat from your house in winter: make the surface area smaller (a two-story house is more efficient than a one-story house of the same total floor area); use thick walls filled with insulation; and, perhaps most important, move to a warmer climate.]

Mathematically, we can summarize these experimental results as

$$H = \frac{Q}{\Delta t} \propto A \frac{\Delta T}{\Delta x}.$$

Introducing a proportionality constant k, called the *thermal conductivity* (not to be confused with the Boltzmann constant), we can write

$$H = \frac{Q}{\Delta t} = kA \frac{\Delta T}{\Delta x}. \tag{45}$$

A substance with a large value of k is a good heat conductor; one with a small value of k is a poor conductor or a good insulator. In the case of solids, the properties of materials that make them good *electrical* conductors (namely, the ability of electrons to move relatively easily throughout the bulk of the material) also make them good *thermal* conductors. Table 5 shows some representative

TABLE 5 SOME THERMAL CONDUCTIVITIES AND R-VALUES[a]

Material	Conductivity, k (W/m·K)	R-Value (ft²·F°·h/Btu)
Metals		
Stainless steel	14	0.010
Lead	35	0.0041
Aluminum	235	0.00061
Copper	401	0.00036
Silver	428	0.00034
Gases		
Air (dry)	0.026	5.5
Helium	0.15	0.96
Hydrogen	0.18	0.80
Building materials		
Polyurethane foam	0.024	5.9
Rock wool	0.043	3.3
Fiberglass	0.048	3.0
White pine	0.11	1.3
Window glass	1.0	0.14

[a] Values are for room temperature. Note that values of k are given in SI units and those of R in the customary British units. The R-values are for a 1-in. slab.

values of k. Over the range of temperatures we normally encounter, we can regard k as a constant, but over wide temperature ranges it does show a slight variation with T. Note that solids, even those we commonly regard as insulators, have higher conductivities than gases.

Let us consider two limits of applicability of Eq. 45. We first take the case of a long rod of length L and uniform cross section A, in which one end is maintained at the high temperature T_H and the other end at the low temperature T_L (Fig. 17). That is, the ends of the rods are immersed in thermal reservoirs, so that one can supply an unlimited amount of heat and still maintain the temperature T_H, while the other can absorb an unlimited amount of heat and still maintain the temperature T_L. (A reservoir can be

a material of considerably more bulk than the rod, and thus with such a large heat capacity that the heat flowing to or from the rod is negligible, or else the reservoir can be connected to a heat engine such as a furnace or a refrigerator that can supply or absorb heat continuously at a cost of external work. A mixture of water and ice at 0°C or a mixture of steam and water at 100°C can also be considered as a thermal reservoir.)

We call this a *steady state* situation: the temperatures and the rate of heat transfer are constant in time. In this situation, every increment of heat Q that enters the rod at the hot end leaves it at the cold end. Put another way, through any cross section along the length of the rod, we would measure the same rate of heat transfer.

For this case, we can write Eq. 45 as

$$H = kA \frac{T_H - T_L}{L}. \qquad (46)$$

Here L is the thickness of the material in the direction of heat transfer. The rate of heat flow H is a constant, and the temperature decreases in linear fashion between the ends of the rod (Fig. 17b).

In choosing building materials, one often finds them rated in terms of the *thermal resistance* or R-value, defined by

$$R = \frac{L}{k}. \qquad (47)$$

Thus the lower the conductivity, the higher the R-value: good insulators have high R-values. Numerically, the R-value is evaluated according to Eq. 47 expressed in the British units of ft²·F°·h/Btu. The R-value is determined for a certain thickness of material. For example, a 1-in. thickness of fiberglass has $R = 3$, while a 1-in. thickness of wood has $R = 1$ (and therefore conducts heat at three times the rate of fiberglass). One inch of air has $R = 5$, but air is a poor thermal insulator because it can transfer more heat by convection, and the thermal conductivity is thus not a good measure of the insulating value of air. Table 5 shows the R-values of some materials.

We now consider the case in which the slab has infinitesimal thickness dx and temperature difference dT across its thickness. In this limit, we obtain

$$H = -kA \frac{dT}{dx}. \qquad (48)$$

The derivative dT/dx is often known as the *temperature gradient*, the word "gradient" being a general mathematical term for the derivative of a scalar variable with respect to a specified coordinate. As a variable, x indicates the direction of heat flow, and so we introduce a minus sign into Eq. 48 as a reminder that heat flows in the direction of *decreasing dT/dx*.

Equation 48 has particular applicability in cases when the cross section of the material is not uniform. We use this differential form in the solution to Sample Problem 7.

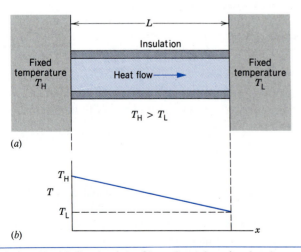

Figure 17 (*a*) Conduction of heat through an insulated conducting rod. (*b*) The variation of temperature along the rod.

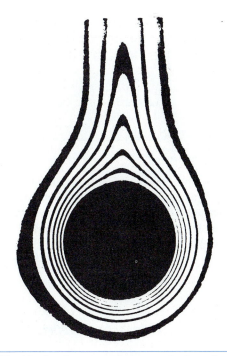

Figure 18 Air rises by convection around a heated cylinder. The dark areas represent regions of uniform temperature.

Convection

If you look at the flame of a candle or a match, you are watching heat energy being transported upward by *convection.* Heat transfer by convection occurs when a fluid, such as air or water, is in contact with an object whose temperature is higher than that of its surroundings. The temperature of the fluid that is in contact with the hot object increases, and (in most cases) the fluid expands. Being less dense than the surrounding cooler fluid, it rises because of buoyant forces; see Fig. 18. The surrounding cooler fluid falls to take the place of the rising warmer fluid, and a convective circulation is set up.

Atmospheric convection plays a fundamental role in determining the global climate patterns and in our daily weather variations. Glider pilots and condors alike seek the convective thermals that, rising from the warmer Earth beneath, keep them aloft. Huge energy transfers take place within the oceans by the same process. The outer region of the Sun, called the *photosphere,* contains a vast array of convection cells that transport energy to the solar surface and give the surface a granulated appearance.

We have been describing *free* or *natural* convection. Convection can also be forced, as when a furnace blower causes air circulation to heat the rooms of a house.

Radiation

Energy is carried from the Sun to us by electromagnetic waves that travel freely through the near vacuum of the

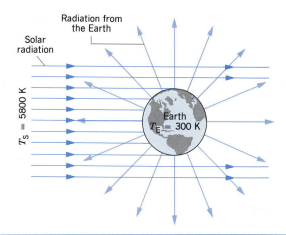

Figure 19 Solar radiation is intercepted by the Earth and is (mostly) absorbed. The temperature T_E of the Earth adjusts itself to a value at which the Earth's heat loss by radiation is just equal to the solar heat that it absorbs.

intervening space. If you stand near a bonfire or an open fireplace, you are warmed by the same process. All objects emit such electromagnetic radiation because of their temperature and also absorb some of the radiation that falls on them from other objects. The higher the temperature of an object, the more it radiates. We shall see in Chapter 49 of the extended version of this text that the energy radiated by an object is proportional to the fourth power of its temperature. The average temperature of our Earth, for example, levels off at about 300 K because at that temperature the Earth radiates energy into space at the same rate that it receives it from the Sun; see Fig. 19.

Sample Problem 6 Consider a compound slab, consisting of two materials having different thicknesses, L_1 and L_2, and different thermal conductivities, k_1 and k_2. If the temperatures of the outer surfaces are T_2 and T_1, find the rate of heat transfer through the compound slab (Fig. 20) in a steady state.

Solution Let T_x be the temperature at the interface between the two materials. Then the rate of heat transfer through slab 2 is

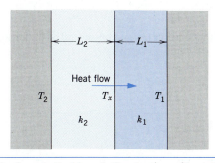

Figure 20 Sample Problem 6. Conduction of heat through two layers of matter with different thermal conductivities.

$$H_2 = \frac{k_2 A(T_2 - T_x)}{L_2}$$

and that through slab 1 is

$$H_1 = \frac{k_1 A(T_x - T_1)}{L_1}.$$

In a steady state $H_2 = H_1$ so that

$$\frac{k_2 A(T_2 - T_x)}{L_2} = \frac{k_1 A(T_x - T_1)}{L_1}.$$

Let H be the rate of heat transfer (the same for all sections). Then, solving for T_x and substituting into either of the equations for H_1 or H_2, we obtain

$$H = \frac{A(T_2 - T_1)}{(L_1/k_1) + (L_2/k_2)} = \frac{A(T_2 - T_1)}{R_1 + R_2}.$$

The extension to any number of sections in series is

$$H = \frac{A(T_2 - T_1)}{\Sigma(L_i/k_i)} = \frac{A(T_2 - T_1)}{\Sigma R_i}. \qquad (49)$$

Sample Problem 7   A thin cylindrical metal pipe is carrying steam at a temperature of $T_S = 100\,°C$. The pipe has a diameter of 5.4 cm and is wrapped with a thickness of 5.2 cm of fiberglass insulation. A length $D = 6.2$ m of the pipe passes through a room in which the temperature is $T_R = 11\,°C$. (*a*) How much heat is lost through the insulation? (*b*) How much additional insulation must be added to reduce the heat loss by half?

Solution   (*a*) Figure 21 illustrates the geometry appropriate to the calculation. In the steady state, the same constant rate of heat flow H will cross every thin cylindrical shell, such as the one indicated by the dashed lines in Fig. 21. We can regard this shell as a slab of material, having a thickness dr and an area of $2\pi r D$. Applying Eq. 48 to this geometry, we have

$$H = -kA\frac{dT}{dr} = -k(2\pi r D)\frac{dT}{dr}$$

or

$$H\frac{dr}{r} = -2\pi k D\, dT.$$

We assume that the thin metal pipe is at the temperature of the steam, so it doesn't enter into the calculation. We integrate from the inner radius r_1 of the insulation at temperature T_S to the outer radius r_2 at temperature T_R:

$$\int_{r_1}^{r_2} H\frac{dr}{r} = -2\pi k D \int_{T_S}^{T_R} dT.$$

Removing the constant H from the integral on the left and carrying out the integrations, we obtain

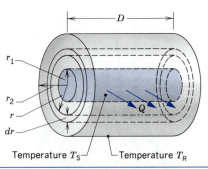

Figure 21   Sample Problem 7. The inner surface (radius r_1) of the insulation on a cylindrical pipe is at the temperature T_S and the outer surface (radius r_2) is at T_R. The same heat Q flows through every cylindrical shell of insulation, such as the intermediate one of thickness dr and radius r shown by the dashed lines.

$$H \ln\frac{r_2}{r_1} = -2\pi k D(T_R - T_S) = 2\pi k D(T_S - T_R).$$

Solving for H and inserting the numerical values, we find

$$
\begin{aligned}
H &= \frac{2\pi k D(T_S - T_R)}{\ln(r_2/r_1)} \\
&= \frac{2\pi(0.048\text{ W/m}\cdot\text{K})(6.2\text{ m})(89\text{ K})}{\ln(7.9\text{ cm}/2.7\text{ cm})} = 155\text{ W}.
\end{aligned}
$$

(*b*) To reduce the heat loss by half, we must increase r_2 to the value r_2' such that the denominator in the above expression for H becomes twice as large; that is,

$$\frac{\ln(r_2'/r_1)}{\ln(r_2/r_1)} = 2.$$

Solving for r_2', we find

$$r_2' = \frac{r_2^2}{r_1} = \frac{(7.9\text{ cm})^2}{2.7\text{ cm}} = 23\text{ cm}.$$

Thus we need nearly four times the thickness of insulation to reduce the heat transfer by half! This effect is due to the increase in the area, and therefore in the mass, contained in each thin slab as we increase the radius in the cylindrical geometry. There is more material available to conduct heat at the outer radii, and we must therefore supply an increasing amount of insulation as r grows larger. This differs from the linear geometry, in which the heat transferred decreases linearly as the insulation thickness increases. In the spherical geometry (which might be appropriate to calculating the heat lost from the Earth's core to its surface), the calculation is still different; see Problem 61.

QUESTIONS

1. Temperature and heat are often confused, as in "outside temperatures will be cold today." By example, distinguish between these two concepts as carefully as you can.

2. Give an example of a process in which no heat is transferred to or from the system but the temperature of the system changes.

3. Is the mechanical equivalent of heat a physical concept or merely a conversion factor for converting energy from heat units to mechanical units and vice versa?

4. Can heat be considered a form of stored (or potential) energy? Would such an interpretation contradict the concept of heat as energy in the process of transfer because of a temperature difference?

5. Can heat be added to a substance without causing the temperature of the substance to rise? If so, does this contradict the concept of heat as energy in the process of transfer because of a temperature difference?

6. Why must heat energy be supplied to melt ice—after all, the temperature doesn't change?

7. Explain the fact that the presence of a large body of water nearby, such as a sea or ocean, tends to moderate the temperature extremes of the climate on adjacent land.

8. As ice is heated it melts, forming a liquid, and then it boils. However, as solid carbon dioxide is heated it goes directly to the vapor state—we say it *sublimes*—without passing through a liquid state. How could liquid carbon dioxide be produced?

9. Pails of hot and cold water are set out in freezing weather. Explain (*a*) if the pails have lids, the cold water will freeze first but (*b*) if the pails do not have lids, it is possible for the hot water to freeze first.

10. Increasing pressure will always encourage condensation and generally encourage solidification. Explain these change of state tendencies in microscopic terms.

11. Why does the boiling temperature of a liquid increase with pressure?

12. A block of wood and a block of metal are at the *same* temperature. When the blocks feel cold, the metal feels colder than the wood; when the blocks feel hot, the metal feels hotter than the wood. Explain. At what temperature will the blocks feel equally hot or cold?

13. How can you best use a spoon to cool a cup of coffee? Stirring—which involves doing work—would seem to heat the coffee rather than cool it.

14. How does a layer of snow protect plants during cold weather? During freezing spells, citrus growers in Florida often spray their fruit with water, hoping it will freeze. How does that help?

15. Explain the wind-chill effect.

16. A thick glass is more likely to crack than a thin glass when you pour hot water into it. Why?

17. You put your hand in a hot oven to remove a casserole and burn your fingers on the hot dish. However, the air in the oven is at the same temperature as the casserole dish but it does not burn your fingers. Why not?

18. Metal workers have observed that they can dip a hand very briefly into hot molten metal without ill effect. Explain.

19. Why is thicker insulation used in an attic than in the walls of a house?

20. Is ice always at 0°C? Can it be colder? Can it be warmer? What about an ice–water mixture?

21. (*a*) Can ice be heated to a temperature above 0°C without its melting? Explain. (*b*) Can water be cooled to a temperature below 0°C without its freezing? Explain. (See "The Under-cooling of Liquids," by David Turnbull, *Scientific American,* January 1965, p. 38.)

22. Explain why your finger sticks to a metal ice tray just taken from your refrigerator.

23. The water in a kettle makes quite a bubbling noise while it is being heated to boiling. Once it starts boiling, however, it does so quietly. What is the explanation? (*Hint*: Think of the fate of a bubble of vapor rising from the bottom of the kettle before the water is uniformly heated.)

24. It is difficult to "boil" eggs in water at the top of a high mountain because water boils there at a relatively low temperature. What is a simple, practical way of overcoming this difficulty?

25. Will a 3-minute egg cook any faster if the water is boiling furiously than if it is simmering quietly?

26. Water is a much better coolant than most liquids. Why? Would there be instances in which another liquid might be preferred?

27. Explain why the latent heat of vaporization of a substance might be expected to be considerably greater than its latent heat of fusion.

28. Explain why the specific heat at constant pressure is greater than the specific heat at constant volume.

29. Why is the difference between C_p and C_v often neglected for solids?

30. Real gases always cool when making a free expansion, whereas an ideal gas does not. Explain.

31. Discuss the similarities and especially the distinctions between heat, work, and internal energy.

32. Discuss the process of the freezing of water from the point of view of the first law of thermodynamics. Remember that ice occupies a greater volume than an equal mass of water.

33. A thermos bottle contains coffee. The thermos bottle is vigorously shaken. Consider the coffee as the system. (*a*) Does its temperature rise? (*b*) Has heat been added to it? (*c*) Has work been done on it? (*d*) Has its internal energy changed?

34. Is the temperature of an isolated system (no interaction with the environment) conserved? Explain.

35. Is heat the same as internal energy? If not, give an example in which a system's internal energy changes without a flow of heat across the system's boundary.

36. Can you tell whether the internal energy of a body was acquired by heat transfer or by the performance of work?

37. If the pressure and volume of a system are given, is the temperature always uniquely determined?

38. Keeping in mind that the internal energy of a body consists of kinetic energy and potential energy of its particles, how would you distinguish between the internal energy of a body and its temperature?

39. The gases in two identical containers are at 1 atm of pressure and at room temperature. One contains helium gas (monatomic, molar mass = 4 g/mol) and the other contains an equal number of moles of argon gas (monatomic, molar mass = 40 g/mol). If 1 J of heat added to the helium gas increases its temperature a given amount, what amount of heat must be added to the argon gas to increase its temperature by the same amount?

40. Explain how we might keep a gas at a constant temperature during a thermodynamic process.

41. Why is it more common to excite radiation from gaseous atoms by use of electrical discharge than by thermal methods?

42. We have seen that "energy conservation" is a universal law of nature. At the same time national leaders urge "energy conservation" upon us (for example, driving slower). Explain the two quite different meanings of these words.

43. On a winter day the temperature on the inside surface of a wall is much lower than room temperature and that of the outside surface is much higher than the outdoor temperature. Explain.

44. The physiological mechanisms that maintain a person's internal temperature operate in a limited range of external temperature. Explain how this range can be extended at each extreme by the use of clothes. (See "Heat, Cold, and

Clothing," by James B. Kelley, *Scientific American,* February 1956, p. 109.)

45. What requirements for thermal conductivity, specific heat capacity, and coefficient of expansion would you want a material to be used in a cooking utensil to satisfy?

46. Both heat conduction and wave propagation involve the transfer of energy. Is there any difference in principle between these two phenomena?

47. Can heat energy be transferred through matter by radiation? If so, give an example. If not, explain why.

48. Why does stainless steel cookware often have a layer of copper or aluminum on the bottom?

49. Consider that heat can be transferred by convection and radiation, as well as by conduction, and explain why a thermos bottle is double-walled, evacuated, and silvered.

50. A lake freezes first at its upper surface. Is convection involved? What about conduction and radiation?

PROBLEMS

Section 25-2 Heat Capacity and Specific Heat

1. In a certain solar house, energy from the Sun is stored in barrels filled with water. In a particular winter stretch of five cloudy days, 5.22 GJ are needed to maintain the inside of the house at 22.0°C. Assuming that the water in the barrels is at 50.0°C, what volume of water is required?

2. Icebergs in the North Atlantic present hazards to shipping (see Fig. 22), causing the length of shipping routes to increase by about 30% during the iceberg season. Attempts to destroy icebergs include planting explosives, bombing, torpedoing, shelling, ramming, and painting with lampblack. Suppose that direct melting of the iceberg, by placing heat sources in the ice, is tried. How much heat is required to melt 10% of a 210,000-metric-ton iceberg? (One metric ton = 1000 kg.)

Figure 22 Problem 2.

3. How much water remains unfrozen after 50.4 kJ of heat have been extracted from 258 g of liquid water initially at 0°C?

4. An object of mass 6.50 kg falls through a height of 50.0 m and, by means of a mechanical linkage, rotates a paddle wheel that stirs 520 g of water. The water is initially at 15°C. What is the maximum possible temperature rise?

5. (a) Compute the possible increase in temperature for water going over Niagara Falls, 49.4 m high. (b) What factors would tend to prevent this possible rise?

6. A small electric immersion heater is used to boil 136 g of water for a cup of instant coffee. The heater is labeled 220 watts. Calculate the time required to bring this water from 23.5°C to the boiling point, ignoring any heat losses.

7. A 146-g copper bowl contains 223 g of water; both bowl and water are at 21.0°C. A very hot 314-g copper cylinder is dropped into the water. This causes the water to boil, with 4.70 g being converted to steam, and the final temperature of the entire system is 100°C. (a) How much heat was transferred to the water? (b) How much to the bowl? (c) What was the original temperature of the cylinder?

8. An athlete needs to lose weight and decides to do it by "pumping iron." (a) How many times must an 80.0-kg weight be lifted a distance of 1.30 m in order to burn off 1 lb of fat, assuming that it takes 3500 Cal to do this? (b) If the weight is lifted once every 4 s, how long does it take?

9. Calculate the minimum amount of heat required to completely melt 130 g of silver initially at 16.0°C. Assume that the specific heat does not change with temperature. See Tables 1 and 2.

10. A thermometer of mass 0.055 kg and heat capacity 46.1 J/K reads 15.0°C. It is then completely immersed in 0.300 kg of water and it comes to the same final temperature as the water. If the thermometer reads 44.4°C, what was the temperature of the water before insertion of the thermometer, neglecting other heat losses?

11. A chef, awaking one morning to find the stove out of order, decides to boil water for coffee by shaking it in a thermos flask. Suppose the chef uses 560 cm³ of tap water at 59°F, and the water falls 35 cm each shake, the chef making 30 shakes each minute. Neglecting any loss of energy, how long must the flask be shaken before the water boils?

12. In a solar water heater, energy from the Sun is gathered by rooftop collectors, which circulate water through tubes in the collector. The solar radiation enters the collector through a transparent cover and warms the water in the tubes; this water is pumped into a holding tank. Assuming that the efficiency of the overall system is 20% (that is, 80% of the incident solar energy is lost from the system), what

collector area is necessary to take water from a 200-L tank and raise its temperature from 20 to 40°C in 1.0 h? The intensity of incident sunlight is 700 W/m².

13. An aluminum electric kettle of mass 0.560 kg contains a 2.40-kW heating element. It is filled with 0.640 L of water at 12.0°C. How long will it take (*a*) for boiling to begin and (*b*) for the kettle to boil dry? (Assume that the temperature of the kettle does not exceed 100°C at any time.)

14. A *flow calorimeter* is used to measure the specific heat of a liquid. Heat is added at a known rate to a stream of the liquid as it passes through the calorimeter at a known rate. Then a measurement of the resulting temperature difference between the inflow and the outflow points of the liquid stream enables us to compute the specific heat of the liquid. A liquid of density 0.85 g/cm³ flows through a calorimeter at the rate of 8.2 cm³/s. Heat is added by means of a 250-W electric heating coil, and a temperature difference of 15 C° is established in steady-state conditions between the inflow and the outflow points. Find the specific heat of the liquid.

15. Water standing in the open at 32°C evaporates because of the escape of some of the surface molecules. The heat of vaporization is approximately equal to ϵn, where ϵ is the average energy of the escaping molecules and n is the number of molecules per kilogram. (*a*) Find ϵ. (*b*) What is the ratio of ϵ to the average kinetic energy of H₂O molecules, assuming that the kinetic energy is related to temperature in the same way as it is for gases?

16. What mass of steam at 100°C must be mixed with 150 g of ice at 0°C, in a thermally insulated container, to produce liquid water at 50°C?

17. A person makes a quantity of iced tea by mixing 520 g of the hot tea (essentially water) with an equal mass of ice at 0°C. What are the final temperature and mass of ice remaining if the initial hot tea is at a temperature of (*a*) 90.0°C and (*b*) 70.0°C?

18. (*a*) Two 50-g ice cubes are dropped into 200 g of water in a glass. If the water were initially at a temperature of 25°C, and if the ice came directly from a freezer at −15°C, what is the final temperature of the drink? (*b*) If only one ice cube had been used in (*a*), what would be the final temperature of the drink? Neglect the heat capacity of the glass.

19. A 21.6-g copper ring has a diameter of 2.54000 cm at its temperature of 0°C. An aluminum sphere has a diameter of 2.54533 cm at its temperature of 100°C. The sphere is placed on top of the ring (Fig. 23), and the two are allowed to come to thermal equilibrium, no heat being lost to the surroundings. The sphere just passes through the ring at the equilibrium temperature. Find the mass of the sphere.

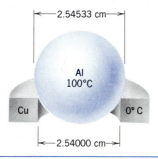

Figure 23 Problem 19.

Section 25-3 Heat Capacities of Solids

20. A certain substance has a molar mass of 51.4 g/mol. When 320 J of heat are added to a 37.1-g sample of this material, its temperature rises from 26.1 to 42.0°C. (*a*) Find the specific heat of the substance. (*b*) How many moles of the substance are present? (*c*) Calculate the molar heat capacity of the substance.

21. Near absolute zero, the molar heat capacity of aluminum varies with the absolute temperature T and is given by $C = (3.16 \times 10^{-5})T^3$, in J/mol·K. How much heat is needed to raise the temperature of 1.2 g of aluminum from 6.6 to 15 K?

22. The molar heat capacity of silver, measured at atmospheric pressure, is found to vary with temperature between 50 and 100 K by the empirical equation

$$C = 0.318T - 0.00109T^2 - 0.628,$$

where C is in J/mol·K and T is in K. Calculate the quantity of heat required to raise 316 g of silver from 50.0 to 90.0 K. The molar mass of silver is 107.87 g/mol.

23. From Fig. 3, estimate the amount of heat needed to raise the temperature of 0.45 mol of carbon from 200 to 500 K. (*Hint*: Approximate the actual curve in this region with a straight-line segment.)

Section 25-4 Heat Capacities of an Ideal Gas

24. The mass of a helium atom is 6.66×10^{-27} kg. Compute the specific heat at constant volume for helium gas (in J/kg·K) from the molar heat capacity at constant volume.

25. In an experiment, 1.35 mol of oxygen (O₂) are heated at constant pressure starting at 11.0°C. How much heat must be added to the gas to double its volume?

26. Twelve grams of nitrogen (N₂) in a steel tank are heated from 25.0 to 125°C. (*a*) How many moles of nitrogen are present? (*b*) How much heat is transferred to the nitrogen?

27. A 4.34-mol sample of an ideal diatomic gas experiences a temperature increase of 62.4 K under constant-pressure conditions. (*a*) How much heat was added to the gas? (*b*) By how much did the internal energy of the gas increase? See Eq. 36 of Chapter 23. (*c*) By how much did the internal translational kinetic energy of the gas increase?

28. A container holds a mixture of three nonreacting gases: n_1 moles of the first gas with molar specific heat at constant volume C_1, and so on. Find the molar specific heat at constant volume of the mixture, in terms of the molar specific heats and quantities of the three separate gases.

29. The molar atomic mass of iodine is 127 g. A standing wave in a tube filled with iodine gas at 400 K has nodes that are 6.77 cm apart when the frequency is 1000 Hz. Determine from these data whether iodine gas is monatomic or diatomic.

30. A room of volume V is filled with diatomic ideal gas (air) at temperature T_1 and pressure p_0. The air is heated to a higher temperature T_2, the pressure remaining constant at p_0 because the walls of the room are not airtight. Show that the internal energy content of the air remaining in the room is the same at T_1 and T_2 and that the energy supplied by the furnace to heat the air has all gone to heat the air *outside* the room. If we add no energy to the air, why bother to light the furnace? (Ignore the furnace energy used to raise the temperature of the walls, and consider only the energy used to raise the air temperature.)

Section 25-6 Applications of the First Law

31. A sample of *n* moles of an ideal gas undergoes an isothermal expansion. Find the heat flow into the gas in terms of the initial and final volumes and the temperature.

32. Gas within a chamber passes through the cycle shown in Fig. 24. Determine the net heat added to the gas during process *CA* if $Q_{AB} = 20$ J, $Q_{BC} = 0$, and $W_{BCA} = -15$ J.

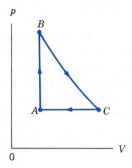

Figure 24   Problem 32.

33. Consider that 214 J of work are done on a system, and 293 J of heat are extracted from the system. In the sense of the first law of thermodynamics, what are the values (including algebraic signs) of (*a*) *W*, (*b*) *Q*, and (*c*) ΔE_{int}?

34. Figure 25*a* shows a cylinder containing gas and closed by a movable piston. The cylinder is submerged in an ice–water mixture. The piston is *quickly* pushed down from position 1 to position 2. The piston is held at position 2 until the gas is again at 0°C and then is *slowly* raised back to position 1. Figure 25*b* is a *pV* diagram for the process. If 122 g of ice are melted during the cycle, how much work has been done *on* the gas?

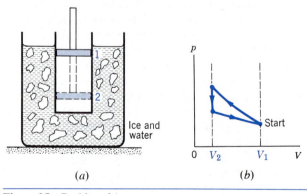

Figure 25   Problem 34.

35. (*a*) A monatomic ideal gas initially at 19.0°C is suddenly compressed to one-tenth its original volume. What is its temperature after compression? (*b*) Make the same calculation for a diatomic gas.

36. A quantity of ideal gas occupies an initial volume V_0 at a pressure p_0 and a temperature T_0. It expands to volume V_1 (*a*) at constant pressure, (*b*) at constant temperature, (*c*) adiabatically. Graph each case on a *pV* diagram. In which case is *Q* greatest? Least? In which case is *W* greatest? Least? In which case is ΔE_{int} greatest? Least?

37. A quantity of ideal monatomic gas consists of *n* moles initially at temperature T_1. The pressure and volume are then slowly doubled in such a manner as to trace out a straight line on the *pV* diagram. In terms of *n*, *R*, and T_1, find (*a*) *W*, (*b*) ΔE_{int}, and (*c*) *Q*. (*d*) If one were to define an equivalent specific heat for this process, what would be its value?

38. In Fig. 11, assume the following values: $p_i = 2.20 \times 10^5$ Pa, $V_i = 0.0120$ m³, $p_f = 1.60 \times 10^5$ Pa, $V_f = 0.0270$ m³. For each of the three paths shown, find the value of *Q*, *W*, and $Q + W$. (*Hint*: Find *p*, *V*, *T* at points *A*, *B*, *C*. Assume an ideal monatomic gas.)

39. When a system is taken from state *i* to state *f* along the path *iaf* in Fig. 26, it is found that $Q = 50$ J and $W = -20$ J. Along the path *ibf*, $Q = 36$ J. (*a*) What is *W* along the path *ibf*? (*b*) If $W = +13$ J for the curved return path *fi*, what is *Q* for this path? (*c*) Take $E_{int,i} = 10$ J. What is $E_{int,f}$? (*d*) If $E_{int,b} = 22$ J, find *Q* for process *ib* and process *bf*.

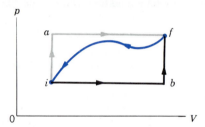

Figure 26   Problem 39.

40. Gas within a chamber undergoes the processes shown in the *pV* diagram of Fig. 27. Calculate the net heat added to the system during one complete cycle.

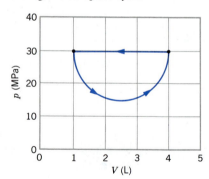

Figure 27   Problem 40.

41. Let 20.9 J of heat be added to a particular ideal gas. As a result, its volume changes from 63.0 to 113 cm³ while the pressure remains constant at 1.00 atm. (*a*) By how much did the internal energy of the gas change? (*b*) If the quantity of gas present is 2.00×10^{-3} mol, find the molar heat capacity at constant pressure. (*c*) Find the molar heat capacity at constant volume.

42. The temperature of 3.15 mol of an ideal polyatomic gas is raised 52.0 K by each of three different processes: at constant volume, at constant pressure, and by an adiabatic compression. Complete a table, showing for each process the heat added, the work done on the gas, the change in internal energy of the gas, and the change in total translational kinetic energy of the gas molecules.

43. An engine carries 1.00 mol of an ideal monatomic gas around the cycle shown in Fig. 28. Process AB takes place at constant volume, process BC is adiabatic, and process CA takes place at a constant pressure. (*a*) Compute the heat Q, the change in internal energy ΔE_{int}, and the work W for each of the three processes and for the cycle as a whole. (*b*) If the initial pressure at point A is 1.00 atm, find the pressure and the volume at points B and C. Use 1 atm = 1.013×10^5 Pa and $R = 8.314$ J/mol·K.

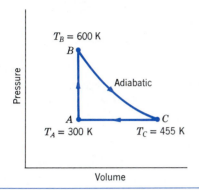

Figure 28   Problem 43.

44. A cylinder has a well-fitted 2.0-kg metal piston whose cross-sectional area is 2.0 cm² (Fig. 29). The cylinder contains water and steam at constant temperature. The piston is observed to fall slowly at a rate of 0.30 cm/s because heat flows out of the cylinder through the cylinder walls. As this happens, some steam condenses in the chamber. The density of the steam inside the chamber is 6.0×10^{-4} g/cm³ and the atmospheric pressure is 1.0 atm. (*a*) Calculate the rate of condensation of steam. (*b*) At what rate is heat leaving the chamber? (*c*) What is the rate of change of internal energy of the steam and water inside the chamber?

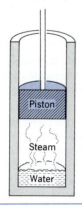

Figure 29   Problem 44.

45. In a motorcycle engine, after combustion occurs in the top of the cylinder, the piston is forced down as the mixture of gaseous products undergoes an adiabatic expansion. Find the average power involved in this expansion when the engine is running at 4000 rpm, assuming that the gauge pressure immediately after combustion is 15.0 atm, the initial volume is 50.0 cm³, and the volume of the mixture at the bottom of the stroke is 250 cm³. Assume that the gases are diatomic and that the time involved in the expansion is one-half that of the total cycle.

Section 25-7   *The Transfer of Heat*

46. Calculate the rate at which heat would be lost on a very cold winter day through a 6.2 m × 3.8 m brick wall 32 cm thick. The inside temperature is 26°C and the outside temperature is −18°C; assume that the thermal conductivity of the brick is 0.74 W/m·K.

47. The average rate at which heat flows out through the surface of the Earth in North America is 54 mW/m², and the average thermal conductivity of the near surface rocks is 2.5 W/m·K. Assuming a surface temperature of 10°C, what should be the temperature at a depth of 33 km (near the base of the crust)? Ignore the heat generated by radioactive elements; the curvature of the Earth can also be ignored.

48. (*a*) Calculate the rate at which body heat flows out through the clothing of a skier, given the following data: the body surface area is 1.8 m² and the clothing is 1.2 cm thick; skin surface temperature is 33°C, whereas the outer surface of the clothing is at 1.0°C; the thermal conductivity of the clothing is 0.040 W/m·K. (*b*) How would the answer change if, after a fall, the skier's clothes become soaked with water? Assume that the thermal conductivity of water is 0.60 W/m·K.

49. Consider the slab shown in Fig. 16. Suppose that $\Delta x = 24.9$ cm, $A = 1.80$ m², and the material is copper. If $T = -12.0$°C, $\Delta T = 136$ C°, and a steady state is reached, find (*a*) the temperature gradient, (*b*) the rate of heat transfer, and (*c*) the temperature at a point in the rod 11.0 cm from the high-temperature end.

50. A cylindrical silver rod of length 1.17 m and cross-sectional area 4.76 cm² is insulated to prevent heat loss through its surface. The ends are maintained at a temperature difference of 100 C° by having one end in a water–ice mixture and the other in boiling water and steam. (*a*) Find the rate at which heat is transferred along the rod. (*b*) Calculate the rate at which ice melts at the cold end.

51. Four square pieces of insulation of two different materials, all with the same thickness and area A, are available to cover an opening of area $2A$. This can be done in either of the two ways shown in Fig. 30. Which arrangement, (*a*) or (*b*), would give the lower heat flow if $k_2 \neq k_1$?

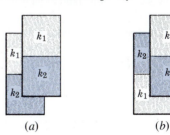

(*a*)   (*b*)

Figure 30   Problem 51.

52. Show that the temperature T_x at the interface of a compound slab (see Sample Problem 6) is given by

$$T_x = \frac{R_1 T_1 + R_2 T_2}{R_1 + R_2}.$$

53. A long tungsten heater wire is rated at 3.08 kW/m and is 0.520 mm in diameter. It is embedded along the axis of a ceramic cylinder of diameter 12.4 cm. When operating at the rated power, the wire is at 1480°C; the outside of the cylinder is at 22.0°C. Calculate the thermal conductivity of the ceramic.

54. Two identical rectangular rods of metal are welded end to end as shown in Fig. 31a, and 10 J of heat flows through the rods in 2.0 min. How long would it take for 30 J to flow through the rods if they are welded as shown in Fig. 31b?

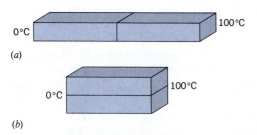

(a)

(b)

Figure 31   Problem 54.

55. (a) Calculate the rate of heat loss through a glass window of area 1.4 m² and thickness 3.0 mm if the outside temperature is −20°F and the inside temperature is +72°F. (b) A storm window is installed having the same thickness of glass but with an air gap of 7.5 cm between the two windows. What will be the corresponding rate of heat loss presuming that conduction is the only important heat-loss mechanism?

56. Compute the rate of heat flow through two storm doors 1.96 m high and 0.770 m wide. (a) One door is made with aluminum panels 1.50 mm thick and a 3.10-mm glass pane that covers 75.0% of its surface (the structural frame is considered to have negligible area). (b) The second door is made entirely of white pine averaging 2.55 cm in thickness. Take the temperature drop through the doors to be 33.0 C° (=59.4 F°). See Table 5.

57. An idealized representation of the air temperature as a function of distance from a single-pane window on a calm, winter day is shown in Fig. 32. The window dimensions are 60 cm × 60 cm × 0.50 cm. (a) At what rate does heat flow out through the window? (*Hint*: The temperature drop across the glass is very small.) (b) Estimate the difference in temperature between the inner and outer glass surfaces.

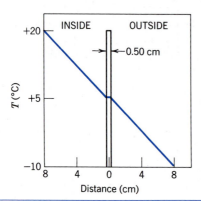

Figure 32   Problem 57.

58. A container of water has been outdoors in cold weather until a 5.0-cm-thick slab of ice has formed on its surface (Fig. 33). The air above the ice is at −10°C. Calculate the rate of formation of ice (in centimeters per hour) on the bottom

surface of the ice slab. Take the thermal conductivity and density of ice to be 1.7 W/m·K and 0.92 g/cm³. Assume that no heat flows through the walls of the tank.

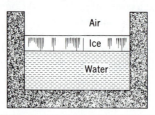

Figure 33   Problem 58.

59. Ice has formed on a shallow pond and a steady state has been reached with the air above the ice at −5.20°C and the bottom of the pond at 3.98°C. If the total depth of ice + water is 1.42 m, how thick is the ice? (Assume that the thermal conductivities of ice and water are 1.67 and 0.502 W/m·K, respectively.)

60. A wall assembly consists of a 20 ft × 12 ft frame made of 16 two-by-four vertical studs, each 12 ft long and set with their center lines 16 in. apart. The outside of the wall is faced with ¼-in. plywood sheet (R = 0.30) and ¾-in. white pine siding (R = 0.98). The inside is faced with ¼-in. plasterboard (R = 0.47), and the space between the studs is filled with polyurethane foam (R = 5.9 for a 1-in. layer). A "two-by-four" is actually 1.75 in. × 3.75 in. in size. Assume that they are made of wood for which R = 1.3 for a 1-in. slab. (a) At what rate does heat flow through this wall for a 30 F° temperature difference? (b) What is the R-value for the assembled wall? (c) What fraction of the wall area contains studs, as opposed to foam? (d) What fraction of the heat flow is through the studs, as opposed to the foam?

61. Assuming k is constant, show that the radial rate of flow of heat in a substance between two concentric spheres is given by

$$H = \frac{(T_1 - T_2)4\pi k r_1 r_2}{r_2 - r_1},$$

where the inner sphere has a radius r_1 and temperature T_1, and the outer sphere has a radius r_2 and temperature T_2.

62. (a) Use data in Problem 47 to calculate the rate at which heat flows out through the surface of the Earth. (b) Suppose that this heat flux is due to the presence of a hot core in the Earth and that this core has a radius of 3470 km. Assume also that the material lying between the core and the surface of the Earth contains no sources of heat and has an average thermal conductivity of 4.2 W/m·K. Use the result of Problem 61 to calculate the temperature of the core. (Assume that the Earth's surface is at 0°C.) The answer obtained is too high by a factor of about 10. Why?

63. At low temperatures (below about 50 K), the thermal conductivity of a metal is proportional to the absolute temperature; that is, k = aT, where a is a constant with a numerical value that depends on the particular metal. Show that the rate of heat flow through a rod of length L and cross-sectional area A whose ends are at temperatures T_1 and T_2 is given by

$$H = \frac{aA}{2L}(T_1^2 - T_2^2).$$

(Ignore heat loss from the surface.)

CHAPTER 26

ENTROPY AND THE SECOND LAW OF THERMODYNAMICS

*We can imagine many processes that conserve energy (and thus satisfy
the first law), but that are never observed. For instance, a hot cup of coffee gives
up some internal energy to rotational energy and spontaneously begins rotating; a block and
tabletop convert some of their internal energy to start the block moving; a glass of cool water
is transformed into an ice cube in a glass of warmer water. In each of these cases, however,
the reverse process is commonly observed. As we discuss in this chapter, the* second law of
thermodynamics *deals with whether such processes will occur in nature. It is often said that
the second law gives a preferred direction to the "arrow of time," telling us that systems
naturally evolve with time in one direction but not the other.*

*In this chapter, we use the second law to analyze engines that convert heat into useful
work, and we show that there is an upper limit to the efficiency at which an engine can
operate. The second law leads to a new concept,* entropy, *just as the zeroth law led to
temperature and the first law to internal energy. We conclude our study of thermodynamics
by showing how the relationship between entropy (a macroscopic quantity) and its
corresponding microscopic quantity (the statistical probability of different arrangements of a
system) strengthens the connection between thermodynamics and statistical mechanics, the
goal we first set in Section 22-1.*

26-1 REVERSIBLE AND IRREVERSIBLE PROCESSES

Consider a typical system in thermodynamic equilibrium, say *n* moles of a (real) gas confined in a cylinder–piston arrangement of volume *V*, the gas having a pressure *p* and a temperature *T*. In an equilibrium state these thermodynamic variables remain constant with time. Suppose that the cylinder, whose walls are insulating but whose base conducts heat, is placed on a large reservoir maintained at this same temperature *T*, as in Fig. 1. Now let us take the system to another equilibrium state in which the temperature *T* is the same but the volume *V* is reduced by one-half. Of the many ways in which we could do this, we discuss two extreme cases.

1. We depress the piston very rapidly; we then wait for equilibrium with the reservoir to be re-established. During this process the gas is turbulent, and its pressure and temperature are not well defined. We cannot plot the process as a continuous line on a *pV* diagram because we would not know what value of pressure (or temperature)

to associate with a given volume. The system passes from one equilibrium state *i* to another *f* through a series of nonequilibrium states (Fig. 1*a*).

2. We depress the piston (assumed to be frictionless) exceedingly slowly—perhaps by gradually adding sand to the top of the piston—so that the pressure, volume, and temperature of the gas are, at all times, well-defined quantities. We first drop a few grains of sand on the piston. This will reduce the volume of the system a little, and the temperature will tend to rise; the system will depart from equilibrium, but only slightly. A small amount of heat will be transferred to the reservoir, and in a short time the system will reach a new equilibrium state, its temperature again being that of the reservoir. Then we drop a few more grains of sand on the piston, reducing the volume further. Again we wait for a new equilibrium state to be established, and so on. By many repetitions of this procedure we finally reduce the volume by one-half. During this entire process the system is always in a state differing only slightly from an equilibrium state. If we imagine carrying out this procedure with still smaller successive increases in pressure, the intermediate states will depart from equilibrium even less. By indefinitely increasing the number

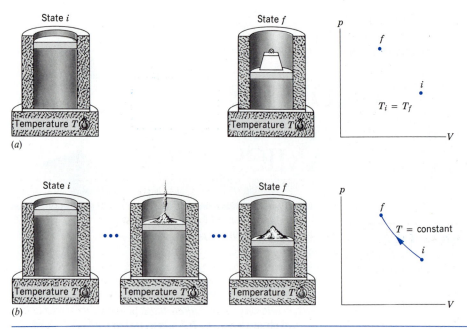

Figure 1   A real gas is made to go from initial state i (characterized by pressure p_i, volume V_i, and temperature T_i) to final state f (characterized by p_f, V_f, and T_f). The process can be carried out (*a*) irreversibly, by suddenly dropping a weight on the piston, or (*b*) reversibly, by adding sand to the piston a few grains at a time.

of changes and correspondingly decreasing the size of each change, we arrive at an ideal process in which the system passes through a continuous succession of equilibrium states, which we can plot as a continuous line on a pV diagram (Fig. 1*b*). During this process a certain amount of heat Q is tranferred from the system to the reservoir.

Processes of type 1 are called *irreversible,* and those of type 2 are called *reversible. A reversible process is one that, by a differential change in the environment, can be made to retrace its path.* That is, if we add a few grains of sand to the piston when the system is in a particular state A, the volume decreases by dV and a small amount of heat is transferred *to* the reservoir. If we next *remove* those few grains of sand (a differential change in the environment), the volume *increases* by dV and an equal amount of heat is tranferred *from* the reservoir, thereby returning both the system and the environment to the original state A. In practice all processes are irreversible, but we can approach reversibility arbitrarily closely by making appropriate experimental refinements. The strictly reversible process is a simple and useful abstraction that bears a similar relation to real processes that the ideal gas abstraction does to real gases.

Not all processes carried out very slowly are reversible. For example, if the piston in our example exerted a frictional force on the cylinder walls, it would not return to its previous state when we removed a few grains of sand. If we added sand slowly to the piston, the system would still evolve through a series of equilibrium states, but it would

not do so reversibly. The word *quasi-static* is used to describe processes that are carried out slowly enough so that the system passes through a continuous sequence of equilibrium states; a quasi-static process may or may not be reversible.

The process described in 2 is not only reversible but *isothermal,* because we have assumed that the temperature of the gas differs at all times by only a differential amount dT from the (constant) temperature of the reservoir on which the cylinder rests.

We could also reduce the volume *adiabatically* by removing the cylinder from the thermal reservoir and putting it on a nonconducting stand. In an adiabatic process no heat is allowed to enter or to leave the system. An adiabatic process can be either reversible or irreversible; the definition does not exclude either. In a reversible adiabatic process we move the piston exceedingly slowly, perhaps using the sand-loading technique; in an irreversible adiabatic process we shove the piston down quickly.

The temperature of the gas will rise during an adiabatic compression because, from the first law with $Q = 0$, the work W done on the system by the environment in pushing down the piston must appear as an increase ΔE_{int} in the internal energy of the system. The work W has different values for different rates of pushing down the piston, being given by $-\int p \, dV$ (that is, by the area under a curve on a pV diagram) only for reversible processes, for which p has a well-defined value. Thus ΔE_{int} and the corresponding temperature change ΔT are not the same for reversible and irreversible adiabatic processes.

On the other hand, for a transformation from a given initial point *i* to a given final point *f*, the internal energy change depends only on the thermodynamic coordinates (*p*, *V*, and *T*, perhaps) of *i* and *f*. *Although W and Q depend on the path, ΔE_{int} does not. In particular, if we are able to calculate ΔE_{int} for one particular reversible path, it has the same value for all other paths, including irreversible ones.* Entropy, as we shall see, is also a state variable like E_{int} whose change in any irreversible process can be found from a suitably chosen reversible process connecting the same initial and final states.

26-2   HEAT ENGINES AND THE SECOND LAW

A *heat engine* is a device for converting heat into useful work. That is, energy flows into a system in the form of heat, and some of this energy leaves the system in the form of work done on the environment. The reverse process, converting work into heat, also occurs: frictional forces can convert work into internal energy, as in the warming of two surfaces that rub together, and this energy can then be transferred to other objects in the environment as heat. For another example, mechanical work done by an electrical generator sends current to your home, where an electric heater converts the work into internal energy, which then flows as heat.

The ideal gas cylinder standing on the thermal reservoir at temperature *T* can serve as a representative example of a heat engine. If we remove a small amount of weight from the piston, the gas expands (isothermally). Heat enters the gas from the reservoir, and work is done by the gas. Because the temperature is constant, the internal energy is constant, and we thus have a device for converting heat into work.

This heat engine would not be very useful over the long term, because it could not operate indefinitely: we would either run out of weight to remove from the piston or the piston would come to the top of the cylinder. A more useful engine would be one that operated in a cycle, returning to its starting point after each unit work *W* was done, and retracing its steps continually. Figure 2 shows an example of a cyclical process that might form the basis for a heat engine. The cycle consists of several steps, all of which can be accomplished in small increments and therefore reversibly. We assume that the gas cylinder is resting on a thermal reservoir, the temperature of which can be easily adjusted.

In analyzing the steps of the cycle, it is helpful to keep in mind the sign conventions that we are using for heat and work:

Heat entering the system is considered to be positive; heat leaving a system is considered to be negative.

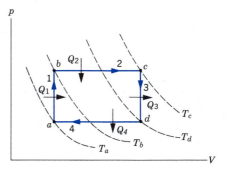

Figure 2   A cyclical process, consisting of four steps, two (*ab* and *cd*) at constant volume and two (*bc* and *da*) at constant pressure. The dashed lines show the isotherms corresponding to the temperatures T_a, T_b, T_c, and T_d.

Work done on a system, corresponding to a decrease in volume, is considered to be positive; work done by a system is the negative of work done on a system.

Work done in a cyclical process is negative if the cycle is done in a clockwise sense on a pV diagram and positive if the cycle is done counterclockwise.

To remember these sign conventions, you can relate the effect of heat transferred or work done to the change in the internal energy of the system.

The four steps in our cycle are the following:

Step 1 (*ab*). We increase the temperature of the reservoir, and we simultaneously add some additional weight to the piston, so that the pressure increases but the volume remains constant.

Step 2 (*bc*). We increase the temperature of the reservoir and allow the gas to expand at constant pressure. The piston does negative work on the gas.

Step 3 (*cd*). We decrease the temperature of the reservoir and simultaneously remove some weight from the piston, so that the volume remains constant.

Step 4 (*da*). We continue to decrease the temperature of the reservoir, but we keep the load on the piston constant, so the pressure remains constant while the volume decreases to its original value.

Note from the isotherms shown in Fig. 2 that the temperature increases for steps 1 and 2 and decreases for steps 3 and 4. Thus heat enters the system during steps 1 and 2 ($Q_1 > 0$ and $Q_2 > 0$), and heat leaves the system during steps 3 and 4 ($Q_3 < 0$ and $Q_4 < 0$). Furthermore, note that for the entire cycle $W < 0$ because the cycle is done in the clockwise direction. In step 1, we added weight when the piston was at the lower position, and in step 3 we removed weight when the piston was at the higher position. The overall effect of the cycle on the environment is thus to raise some weight *mg* through the distance *h* that the piston rises in step 2; the magnitude of the work done by the gas on the environment is equal to *mgh*.

Let us now examine the energy transfers during the cycle. The total heat Q_{in} that enters the system is $Q_1 + Q_2$, and the total heat Q_{out} that leaves the system is $Q_3 + Q_4$. The net heat transfer Q for the cycle is $Q_{in} + Q_{out}$; to remind ourselves that Q_{in} is positive and Q_{out} is negative, we write, using absolute magnitudes,

$$|Q| = |Q_{in}| - |Q_{out}|. \tag{1}$$

The change in internal energy for the cycle must be zero, because the cycle starts and finishes at the same point. The first law then gives $W = -Q$ or, again using absolute magnitudes,

$$|W| = |Q| = |Q_{in}| - |Q_{out}|. \tag{2}$$

For this cycle, $|Q_{in}| > |Q_{out}|$, so that the right-hand sides of Eqs. 1 and 2 are positive, as is necessary when we write those equations in terms of magnitudes.

We define the *efficiency e* of any cycle to be the net amount of work done *on the environment* during the cycle, divided by the *heat input Q_{in}*:

$$e = \frac{|W|}{|Q_{in}|}. \tag{3}$$

The work done on the environment, which is how we measure the useful output of an engine, is the negative of the work done by the environment on the gas. We write the ratio in Eq. 3 in this way so that both the numerator and denominator are positive quantities.

We can also write the efficiency as

$$e = \frac{|Q_{in}| - |Q_{out}|}{|Q_{in}|} = 1 - \frac{|Q_{out}|}{|Q_{in}|}. \tag{4}$$

We could make a perfectly efficient heat engine ($e = 1.00$ or 100%) if we could design a cycle to reduce $|Q_{out}|$, the heat discharged, to zero; otherwise, the efficiency is always less than 100%. One form of the *second law of thermodynamics* asserts that this goal of making a perfectly efficient heat engine is impossible:

It is not possible in a cyclical process to convert heat entirely into work, with no other change taking place.

In our example, the "other change taking place" is the exhaust heat $|Q_{out}|$, and thus the second law says that it is impossible to reduce $|Q_{out}|$ to zero. Equation 4 implies that the efficiency of the heat engine can never quite reach 100%. This form of the second law, which is sometimes called the *Kelvin–Planck* form, states that *there are no perfect heat engines*.

Figure 3 shows a schematic representation of a simplified perfect engine, which converts heat Q entirely into work, and a real engine, which obtains heat Q_H from a reservoir at the high temperature T_H and discharges heat

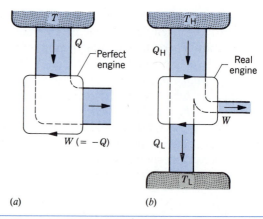

(a)   (b)

Figure 3 An engine is represented by the clockwise arrows around the central block. (*a*) In a perfect engine, all the heat extracted from a high-temperature reservoir is converted into work. (*b*) In a real engine, heat Q_H extracted from the high-temperature reservoir is converted partially into work and partially into heat Q_L exhausted to a low-temperature reservoir.

Q_L to the reservoir at the low temperature T_L. In this generalized engine, the input heat, which may be transferred in several steps, is represented simply as Q_H, and the exhaust heat is similarly represented as Q_L. The cycle involves a series of operations performed on a *working substance*; in our case, the series of operations shown in Fig. 2 was performed on an ideal gas, but in practice we can design a heat engine using any one of a great variety of working substances. In a power plant, for example, water is often the working substance, absorbing heat Q_H when it is converted into steam and discharging heat Q_L when the steam condenses back into water. (Do not confuse the fuel in an engine with the working substance; the fuel merely maintains the temperature of the hot reservoir T_H.) The low-temperature reservoir, to which the heat Q_L is exhausted, might be the atmosphere, a cooling pond, or a river.

Sample Problem 1 An automobile engine, whose thermal efficiency *e* is 22%, operates at 95 cycles per second and does work at the rate of 120 hp. (*a*) How much work per cycle is done on the system by the environment? (*b*) How much heat enters and leaves the engine in each cycle?

Solution (*a*) The work per cycle done on the system, a negative quantity, is

$$W = -\frac{(120 \text{ hp})[746 \text{ (J/s)/hp}]}{95 \text{ s}^{-1}} = -942 \text{ J}.$$

Put another way, the engine does +942 J of work per cycle on the environment.

(*b*) To find the input heat Q_H absorbed from the high-temperature reservoir (the exploding fuel mixture), we use Eq. 3:

$$|Q_H| = \frac{|W|}{e} = \frac{942 \text{ J}}{0.22} = 4.3 \times 10^3 \text{ J}.$$

From Eq. 2, we find the output heat, which is discharged to the low-temperature reservoir (the environment):

$$|Q_L| = |Q_H| - |W| = 4.3 \times 10^3 \text{ J} - 942 \text{ J} = 3.4 \times 10^3 \text{ J}.$$

Heat discharged from the engine carries a negative sign so that

$$Q_L = -3.4 \times 10^3 \text{ J}.$$

We see that this engine obtains 4.3×10^3 J of heat per cycle, which must be paid for at the gas pump, does 942 J of work, and transfers 3.4×10^3 J of exhaust heat. The engine discards 3.6 times more energy than it converts to useful purposes. Some engines can put this exhaust heat to useful purpose. For example, heat from the steam exhausted in a power plant can be transferred to commercial buildings to keep them warm in cold weather.

26-3 REFRIGERATORS AND THE SECOND LAW

A refrigerator is basically a heat engine run in reverse. Like a heat engine, a refrigerator is considered to operate in a cyclical process, and running the cycle of Fig. 2 in reverse would represent the operation of one type of refrigerator. A more general refrigerator can be represented by the engine of Fig. 3 run in reverse. Heat Q_L is extracted from the low-temperature reservoir at temperature T_L, and heat Q_H is discharged to the high-temperature reservoir at temperature T_H. Figure 4 shows a schematic representation of a perfect refrigerator, in which $|Q_L| = |Q_H|$, and a real refrigerator, in which $|Q_L| < |Q_H|$ and external work W must be supplied.

As in the case of the heat engine, there is no change of internal energy in a complete cycle, and therefore $|W| = |Q|$, or

$$|W| = |Q_H| - |Q_L|. \tag{5}$$

In the refrigerator, heat is *input* from the low-temperature reservoir, so $Q_L > 0$, and is *output* to the high-temperature reservoir, so $Q_H < 0$, as Fig. 4 suggests. Also, $W > 0$ since the environment does work on the working substance.

In analogy with the efficiency of a heat engine, we evaluate a refrigerator in terms of the *coefficient of performance* K, defined by

$$K = \frac{|Q_L|}{|W|} = \frac{|Q_L|}{|Q_H| - |Q_L|}. \tag{6}$$

In a perfect refrigerator, $W = 0$ (thus $|Q_H| = |Q_L|$), and the coefficient of performance is infinite.

An alternative statement of the second law of thermodynamics deals with the performance of a refrigerator:

It is not possible in a cyclical process for heat to flow from one body to another body at higher temperature, with no other change taking place.

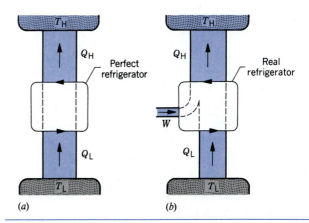

Figure 4 A refrigerator is represented by arrows pointing counterclockwise around the central block. (*a*) In a perfect refrigerator, no work is required. (*b*) In a real refrigerator, heat is extracted from a low-temperature reservoir through the performance of some external work, and the energy equivalent of the extracted heat and the work is discharged as heat to a high-temperature reservoir.

In this statement, the "other change taking place" means that during the cycle external work must be done to cause the heat to move in this way, since it would prefer on its own to flow the other way. This statement of the second law is often called the *Clausius* form, and in effect it says that *there are no perfect refrigerators.*

In an ordinary household refrigerator, the working substance is a liquid (Freon) that circulates within the system. The low-temperature reservoir is the cold chamber in which food is stored, and the high-temperature reservoir is the room in which the unit is kept. The external work is provided by a motor that drives the unit. Typical refrigerators have coefficients of performance around 5.

Equivalence of the Clausius and Kelvin–Planck Statements

The two statements we have presented for the second law are not independent and are, in fact, entirely equivalent. To show this, let us consider what would happen if the Kelvin–Planck form were incorrect, and that we could build a perfect engine, converting heat Q_H entirely into work W. Let us use this work W to drive a real refrigerator, as shown in Fig. 5a. This refrigerator takes heat $|Q'_L|$ from the low-temperature reservoir and pumps heat $|Q'_H| = |Q'_L| + |W|$ to the high-temperature reservoir.

Let us regard the combination of the perfect engine and the real refrigerator as a single device, as indicated in Fig. 5b. The work W is an internal feature of this device and does not enter into any exchanges of energy with the environment. This device takes heat $|Q'_L|$ from the low-temperature reservoir, and it transfers to the high-temperature reservoir a net amount of heat equal to $|Q'_H| - |Q_H|$. But $|Q_H| = |W|$, and so

$$|Q'_H| - |Q_H| = |Q'_H| - |W| = |Q'_L|.$$

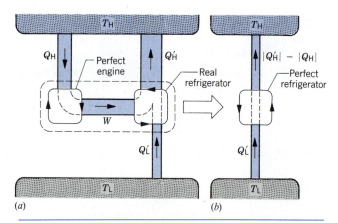

Figure 5   (*a*) A real refrigerator, driven by a perfect engine, is equivalent to (*b*) a perfect refrigerator.

Thus our combined device acts like a perfect refrigerator, taking heat $|Q'_L|$ from the low-temperature reservoir and pumping heat $|Q'_L|$ to the high-temperature reservoir, with no external work performed.

This example shows that, if we can build a perfect engine, then we can build a perfect refrigerator. That is, a violation of the Kelvin–Planck statement of the second law implies a violation of the Clausius statement. In a similar manner, a perfect refrigerator allows us to turn a real heat engine into a perfect heat engine. Thus a violation of the Clausius statement implies a violation of the Kelvin–Planck statement. Because a violation of either statement implies a violation of the other, the two statements are logically equivalent.

Sample Problem 2   A household refrigerator, whose coefficient of performance K is 4.7, extracts heat from the cooling chamber at the rate of 250 J per cycle. (*a*) How much work per cycle is required to operate the refrigerator? (*b*) How much heat per cycle is discharged to the room, which forms the high-temperature reservoir of the refrigerator?

Solution   (*a*) From Eq. 6, $K = |Q_L|/|W|$, we have

$$|W| = \frac{|Q_L|}{K} = \frac{250 \text{ J}}{4.7} = 53 \text{ J}.$$

W represents work done *on* the system, so it is a positive quantity.

(*b*) To find the heat Q_H discharged to the room (which serves as the high-temperature reservoir), we use Eq. 5, which is the first law of thermodynamics for a cyclic device and holds for refrigerators as well as for engines. We then have

$$|Q_H| = |W| + |Q_L| = 53 \text{ J} + 250 \text{ J} = 303 \text{ J}.$$

A refrigerator is also an efficient room heater! By paying for 53 J of work (done by the motor), you get 303 J of heat delivered to the room from the condenser coils at the back of the unit. (See Sample Problem 4, which deals with the operation of a *heat pump*, a device similar to a refrigerator that can heat your home.)

If you heated the room with an electric heater, you would get at most 53 J of heat for every 53 J of work that you pay for. Think about the wisdom (?) of trying to cool the kitchen on a hot day by leaving the refrigerator door open! Of course, a complete calculation of the relative efficiency of various heating systems must take into account the thermodynamic efficiency of the production of electric power at the generating station.

26-4   THE CARNOT CYCLE

We have seen that the second law of thermodynamics prevents us from building perfect heat engines and refrigerators. It is then logical to ask whether we can come as close to perfection as we like, or whether there is some other fundamental limitation on the performance of heat engines and refrigerators. It turns out that there *is* a fundamental limit, and to consider it we discuss an engine that operates on a particular cycle, called the *Carnot cycle*.*

In the Carnot cycle, an ideal gas in our usual cylinder is the working substance. We use two thermal reservoirs, one at the high temperature T_H and another at the low temperature T_L. The cycle consists of four reversible processes, two isothermal and two adiabatic. The sequence, indicated schematically in Fig. 6 and plotted on a pV diagram in Fig. 7, proceeds as follows:

Step 1 (*ab*).   Put the cylinder on the high-temperature reservoir, with the gas in a state represented by point *a* in Fig. 7. Gradually, remove some weight from the piston, allowing the gas to expand slowly to point *b*. During this process, heat $Q_1 = |Q_H|$ is absorbed by the gas from the high-temperature reservoir. Because this process is isothermal, the internal energy of the gas does not change ($\Delta E_{int} = 0$), and all the (positive) added heat appears as the (negative) work done on the gas as the weight on the piston rises.

Step 2 (*bc*).   Insulate the cylinder from the reservoir and, by incrementally removing more weight from the piston, allow the gas slowly to expand further to point *c* in Fig. 7. This expansion is adiabatic because no heat enters or leaves the system ($Q_2 = 0$). The piston does (negative) work W_2 on the gas. The temperature of the gas drops to T_L, because the energy to do the work must come from the internal energy of the gas.

Step 3 (*cd*).   Put the cylinder on the low-temperature reservoir and, by gradually adding weight to the piston, compress the gas slowly to point *d* in Fig. 7. During this process, heat $Q_3 = -|Q_L|$ is transferred from the gas to the

---

* Named for the French engineer and scientist N. L. Sadi Carnot (1796–1832), who proposed the concept in 1824.

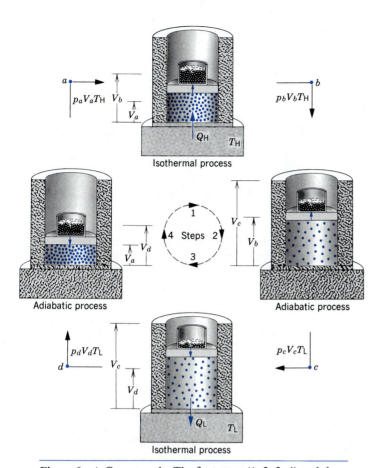

Figure 6   A Carnot cycle. The four steps (1, 2, 3, 4) and the four endpoints (*a, b, c, d*) correspond to those in Fig. 7. The cylinder–piston arrangement is shown at intermediate points, during the performance of each process.

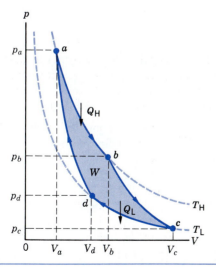

Figure 7   A *pV* diagram for the Carnot cycle illustrated in Fig. 6. The working substance is assumed to be an ideal gas.

reservoir. The compression is isothermal at temperature T_L, and (positive) work is done on the gas by the descending piston and its load.

Step 4 (*da*). Insulate the cylinder from the reservoir and, by adding still more weight, compress the gas slowly back to its initial point *a* of Fig. 7, thus completing the cycle. The compression is adiabatic because no heat enters or leaves the system. Work W_4 is done on the gas, and its temperature rises to T_H.

The energy transfers during the cycle can be summarized as follows:

	Q	W	ΔE_{int}
Step 1	>0	<0	0
Step 2	0	<0	<0
Step 3	<0	>0	0
Step 4	0	>0	>0
Cycle	>0	<0	0

Let us now calculate the efficiency of a heat engine operating on the Carnot cycle. Along the isothermal path *ab* in Fig. 7, the temperature remains constant. Because the gas is ideal, its internal energy, which depends only on the temperature, also remains constant. With $\Delta E_{int} = 0$, the first law requires that the heat Q_H transferred from the high-temperature reservoir must equal the magnitude of the work W done on the expanding gas. From Eq. 27 of Chapter 23 we then have

$$|Q_H| = |W_1| = nRT_H \ln \frac{V_b}{V_a}.$$

Similarly, for the isothermal process *cd* in Fig. 7, we can write

$$|Q_L| = |W_3| = nRT_L \ln \frac{V_c}{V_d}.$$

Dividing these two equations yields

$$\frac{|Q_H|}{|Q_L|} = \frac{T_H}{T_L} \frac{\ln (V_b/V_a)}{\ln (V_c/V_d)}. \qquad (7)$$

Equation 38 of Chapter 25 allows us to write, for the two adiabatic processes *bc* and *da*,

$$T_H V_b^{\gamma-1} = T_L V_c^{\gamma-1} \quad \text{and} \quad T_H V_a^{\gamma-1} = T_L V_d^{\gamma-1}.$$

Dividing these two equations results in

$$\frac{V_b^{\gamma-1}}{V_a^{\gamma-1}} = \frac{V_c^{\gamma-1}}{V_d^{\gamma-1}}$$

or

$$\frac{V_b}{V_a} = \frac{V_c}{V_d}. \qquad (8)$$

Combining Eqs. 7 and 8 yields

$$\frac{|Q_H|}{|Q_L|} = \frac{T_H}{T_L}. \qquad (9)$$

Equation 9 is an important and fundamental result for the Carnot cycle. We shall need this result again later in this chapter when we discuss entropy.

Using Eq. 4 with $Q_{in} = Q_H$ and $Q_{out} = Q_L$ and substituting Eq. 9, we obtain the efficiency of a heat engine operating on a Carnot cycle:

$$e = 1 - \frac{T_L}{T_H} = \frac{T_H - T_L}{T_H}. \tag{10}$$

The efficiency of a Carnot engine depends only on the temperature of the two reservoirs between which it operates. Note that the efficiency increases as T_L decreases, approaching 1 as T_L approaches 0. Since T_L can never reach 0, the efficiency must be less than 100%.

A Carnot cycle, because it is reversible, can be run backward to make a refrigerator. It is left as an exercise (see Problem 19) to show that the coefficient of performance of a Carnot refrigerator is

$$K = \frac{T_L}{T_H - T_L}. \tag{11}$$

We have used an ideal gas as an example of a working substance. The working substance can be anything at all, although the pV diagrams for other substances would be different. Common heat engines use steam or a mixture of fuel and air or fuel and oxygen as their working substance. Heat may be obtained from the combustion of a fuel such as gasoline or coal, or from the release of nuclear energy in fission reactors. Heat may be discharged at the exhaust or to a condenser. Although real heat engines do not operate on a reversible cycle, the Carnot cycle, which is reversible, gives useful information about the behavior of any heat engine. It is especially important because, as we shall see later, it sets an upper limit on the performance of real engines and thereby gives us a goal to work toward.

Carnot's Theorem and the Second Law

Based on his ideal reversible heat engine, Carnot developed a general theorem applicable to *all* heat engines:

The efficiency of any heat engine operating between two specified temperatures can never exceed the efficiency of a Carnot engine operating between the same two temperatures.

That is, the Carnot efficiency (Eq. 10) is the upper limit for the performance of a heat engine. Clausius and Kelvin showed that the Carnot theorem was a necessary consequence of the second law of thermodynamics, but it is remarkable that Carnot's work was complete long before Clausius and Kelvin developed the statements of the second law. (Carnot's work on heat engines was published in 1824, the year of Kelvin's birth and two years before Clausius was born!)

To show that violating Carnot's theorem is also a violation of the second law, let us suppose we have an engine,

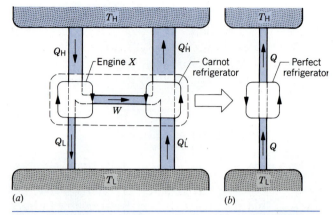

Figure 8 (*a*) Engine X drives a Carnot refrigerator. If engine X were more efficient than a Carnot engine, then the combination would be equivalent to the perfect refrigerator shown in (*b*).

which we call engine X, whose efficiency e_X exceeds the Carnot efficiency e. Let us couple engine X to a Carnot engine operating backward as a refrigerator, as in Fig. 8. Engine X extracts heat Q_H from the high-temperature reservoir and discharges heat Q_L to the low-temperature reservoir, doing work W in the process. Let this work W drive the Carnot refrigerator, which extracts heat Q'_L from the low-temperature reservoir and discharges heat Q'_H to the high-temperature reservoir.

The net heat that flows from the low-temperature reservoir due to the combination of the two devices is $|Q'_L| - |Q_L|$, and the net heat delivered to the high-temperature reservoir is $|Q'_H| - |Q_H|$. Applying the first law to each device separately, we have $|W| = |Q_H| - |Q_L|$ for engine X, and $|W| = |Q'_H| - |Q'_L|$ for the Carnot refrigerator. Equating these two expressions, we find

$$|W| = |Q_H| - |Q_L| = |Q'_H| - |Q'_L|,$$

or, defining Q to be the difference between $|Q'_H|$ and $|Q_H|$,

$$Q = |Q'_H| - |Q_H| = |Q'_L| - |Q_L|. \tag{12}$$

Our hypothesis is that the efficiency of engine X can exceed the Carnot efficiency; that is,

$$e_X > e \quad \text{(hypothesis)}. \tag{13}$$

Let us investigate the consequences of this hypothesis. From Eq. 3 for the basic definition of efficiency, our hypothesis is equivalent to

$$\frac{|W|}{|Q_H|} > \frac{|W|}{|Q'_H|}$$

or

$$|Q'_H| > |Q_H| \quad \text{(consequence of hypothesis)}. \tag{14}$$

Comparing Eqs. 12 and 14, we see that $Q > 0$ as a direct consequence of our hypothesis that we can build an engine that violates the Carnot theorem. Thus the combination of engine X and the Carnot refrigerator is equivalent to the perfect refrigerator shown in Fig. 8*b*, in which heat

Q is transferred from the low-temperature reservoir at T_L to the high-temperature reservoir at T_H without external work. This clearly violates the Clausius form of the second law, and so our original hypothesis (Eq. 13) must be false. The Carnot theorem is therefore a necessary consequence of the second law.

How would this argument differ if X were a real engine? If $e_X < e$, then Eq. 14 would change to

$$|Q'_H| < |Q_H|,$$

and from Eq. 12 we would deduce $Q < 0$. In this case, we would reverse the directions of the arrows in Fig. 8b, which would then no longer be a refrigerator. Instead, heat Q would be flowing from the high-temperature reservoir to the low-temperature reservoir, which is a natural process and violates no basic law.

If engine X operates on a cycle composed entirely of reversible processes, then its efficiency is equal to the Carnot efficiency. If the cycle is in part irreversible, then in effect a portion of the energy transferred in each cycle is lost, perhaps to friction, and cannot be regained as useful work. In Fig. 8, for example, it would not be true that all the work W produced by a partially irreversible engine X would be available to run the refrigerator; some would be lost to friction or to another cause. We can thus summarize Carnot's theorem, applied to the efficiency e of any engine, as follows:

$$\begin{aligned} e &= e_{Carnot} \quad \text{(reversible)}, \\ e &< e_{Carnot} \quad \text{(irreversible)}. \end{aligned} \tag{15}$$

Sample Problem 3   The turbine in a steam power plant takes steam from a boiler at 520°C and exhausts it into a condenser at 100°C. What is its maximum possible efficiency?

Solution   The maximum efficiency is the efficiency of a Carnot engine operating between the same two temperatures. From Eq. 10 then,

$$\begin{aligned} e_{max} &= \frac{T_H - T_L}{T_H} = \frac{793\ K - 373\ K}{793\ K} \\ &= 0.53 \quad \text{or} \quad 53\%. \end{aligned}$$

Note that the temperatures in this equation must be expressed on the Kelvin scale. Because of friction, turbulence, and unwanted thermal losses, actual efficiencies of about 40% may be realized for such a steam turbine. Note that the theoretical maximum efficiency depends only on the two temperatures involved, not on the pressures or other factors.

The theoretical efficiency of an ordinary automobile engine is about 56%, but practical considerations reduce this to about 25%.

Sample Problem 4   A heat pump (see Fig. 9) is a device that — acting as a refrigerator — can heat a house by drawing heat from the outside, doing some work, and discharging heat inside the house. The outside temperature is −10°C, and the interior is to be kept at 22°C. It is necessary to deliver heat to the interior at

Figure 9   Sample Problem 4. A heat pump.

the rate of 16 kW to make up for the normal heat losses. At what minimum rate must energy be supplied to the heat pump?

Solution   The low-temperature reservoir is the outdoors at $T_L = 273 - 10 = 263$ K, and the high-temperature reservoir is the interior at $T_H = 273 + 22 = 295$ K. From Eq. 11, the maximum coefficient of performance of the heat pump, acting as a refrigerator, is

$$K = \frac{T_L}{T_H - T_L} = \frac{263\ K}{295\ K - 263\ K} = 8.22.$$

We can recast Eq. 6 as

$$K = \frac{|Q_L|}{|W|} = \frac{|Q_H| - |W|}{|W|}.$$

Solving for $|W|$ and dividing by time to express the result in terms of power, we obtain

$$\frac{|W|}{t} = \frac{|Q_H/t|}{K + 1} = \frac{16\ kW}{8.22 + 1} = 1.7\ kW.$$

Herein lies the "magic" of the heat pump. By using the heat pump as a refrigerator to cool the great outdoors, you can deliver 16 kW to the interior of the house but you need pay for only the 1.7 kW it takes to run the pump. Actually, the 1.7 kW is a theoretical minimum requirement because it is based on an ideal performance. In practice, a greater power input would be required but there would still be a very considerable saving over, say, heating the house directly with electric heaters. In that case, you would have to pay directly for every kilowatt of heat transfer. When the outside temperature is greater than the inside temperature, the heat pump can be used as an air conditioner. Still operating as a refrigerator, it now pumps heat from inside the house to outside. Again, work must be done (and paid for) but the energy removed as heat from the house interior exceeds the energy equivalent of the work done. Another thermodynamic bargain! (See also Sample Problem 2, for another indication that a refrigerator is an efficient heater.)

26-5 THE THERMODYNAMIC TEMPERATURE SCALE

The efficiency of a reversible engine is independent of the working substance and depends only on the two temperatures between which the engine works. Since $e = 1 - |Q_L|/|Q_H|$, then $|Q_L|/|Q_H|$ can depend only on the temperatures. This led Kelvin to suggest a new scale of temperature. If we let θ_L and θ_H represent these two temperatures, his defining equation is

$$\frac{\theta_L}{\theta_H} = \frac{|Q_L|}{|Q_H|}.$$

That is, two temperatures on this scale have the same ratio as the heats absorbed and rejected, respectively, by a Carnot engine operating between these temperatures. Such a temperature scale is called the *thermodynamic* (or *Kelvin*) temperature scale.

To complete the definition of the thermodynamic scale, we assign the standard value of 273.16 to the temperature of the triple point of water. Hence, $\theta_{tr} = 273.16$ K. Then for a Carnot engine operating between reservoirs at the temperatures θ and θ_{tr}, we have

$$\frac{\theta}{\theta_{tr}} = \frac{|Q|}{|Q_{tr}|}$$

or

$$\theta = 273.16 \text{ K } \frac{|Q|}{|Q_{tr}|}. \tag{16}$$

If we compare this with Eq. 7 of Chapter 22,

$$T = 273.16 \text{ K } \frac{X}{X_{tr}},$$

we see that on the thermodynamic scale $|Q|$ plays the role of a thermometric property. However, $|Q|$ does not depend on the characteristics of any substance because *the efficiency of a Carnot engine is independent of the nature of the working substance.* Therefore we obtain a scale of temperature that is free of the objection we can raise to the ideal gas scale of Chapter 22, and in fact we arrive at a fundamental definition of temperature.

The definition of thermodynamic temperature enables us to rewrite the equation for the efficiency of a reversible engine as

$$e = \frac{|Q_H| - |Q_L|}{|Q_H|} = \frac{\theta_H - \theta_L}{\theta_H}. \tag{17}$$

But we have shown that the efficiency of a Carnot engine using an ideal gas as working substance is

$$e = \frac{|Q_H| - |Q_L|}{|Q_H|} = \frac{T_H - T_L}{T_H}, \tag{18}$$

where T is the temperature given by the constant-volume thermometer containing the ideal gas. Comparing Eqs. 17 and 18, we see that $|Q_H|/|Q_L| = T_H/T_L$ and $|Q_H|/|Q_L| =$

θ_H/θ_L. Since $\theta_{tr} = T_{tr} = 273.16$ and $\theta/\theta_{tr} = T/T_{tr}$, it follows that $\theta = T$. Hence *if an ideal gas were available for use in a constant-volume thermometer, the thermometer would yield the thermodynamic (or Kelvin) temperature.* We have seen that, although an ideal gas is not available, measurements made using the limiting process with real gases correspond to ideal gas behavior. We shall treat the ideal gas scale and the thermodynamic scale as identical, and we shall use the designation K interchangeably for each, as in fact we have already done.

Absolute Zero and Negative Temperatures *(Optional)*

In practice, we cannot have a gas below 1 K, and therefore we cannot measure temperatures below 1 K using a constant-volume gas thermometer. Fortunately, it is possible to measure temperatures below 1 K using the thermodynamic scale directly. Suppose we have a system at a temperature T_2 we wish to measure. We can take the system around a Carnot cycle (Fig. 10), first doing adiabatic work on it to raise its temperature to T_1, which is presumably known on the ideal gas scale, then transferring known heat $|Q_1|$ isothermally, then doing adiabatic work to reduce its temperature back to T_2, and finally transferring heat $|Q_2|$ necessary to return the system to its original condition. From the above arguments, we conclude

$$T_2 = T_1 \frac{|Q_2|}{|Q_1|}. \tag{19}$$

Thus knowing T_1 and measuring $|Q_1|$ and $|Q_2|$ enables us to determine the thermodynamic temperature T_2 directly. Regarding T_2 as a known temperature, we can take the system around another Carnot cycle to determine a still lower temperature T_3. In principle, we could continue this process to the absolute zero of temperature; however, the smaller the temperature, the smaller the heat $|Q|$ transferred in an isothermal process between two given adiabatic processes (Fig. 10). At the limit of the absolute zero of the thermodynamic temperature scale, the system could undergo an isothermal process with no transfer of heat.

The fundamental feature of all cooling processes is that the

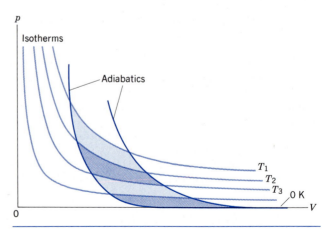

Figure 10 A series of Carnot cycles tending toward the absolute zero of temperature. The difference in slope between isothermal and adiabatic processes has been exaggerated for clarity.

lower the temperature, the more difficult it is to go still lower. This experience has led to the formulation of the *third law of thermodynamics,* which can be stated in one form as follows: *It is impossible by any procedure, no matter how idealized, to reduce any system to the absolute zero of temperature in a finite number of operations.* Hence, because we cannot obtain a reservoir at absolute zero, a heat engine with 100% efficiency is a practical impossibility.

As an alternative to the cyclical process, we can use certain *absolute* thermometers to determine the thermodynamic temperature directly. These thermometers are based on the dependence on temperature of basic results from statistical mechanics. One type of thermometer, called a *noise thermometer,* uses the Brownian motion of the electrons in a solid. In analogy with our discussion of the mean square velocity in Section 24-3, it can be shown that the mean square current of these electrons is proportional to the temperature. The average value of these fluctuating currents can be measured with sensitive probes, and the temperature can be determined directly without the use of a calibration. Temperatures in the millikelvin (0.001 K) range can be determined in this way. Another useful device in this temperature region makes use of the Maxwell–Boltzmann energy distribution to determine temperature. Suppose we have a system (an atom or a nucleus) in which the energy can take two values E_1 and $E_2 = E_1 + \Delta E$ (Fig. 11). If we have a large number of those atoms or nuclei in thermal equilibrium at a temperature T, then a rough estimate of the relative numbers of atoms or nuclei with energies E_1 and E_2 is given by the exponential part of the Maxwell–Boltzmann distribution (see Eq. 32 of Chapter 24):

$$\frac{f_{\text{MB}}(E_2)}{f_{\text{MB}}(E_1)} = \frac{e^{-E_2/kT}}{e^{-E_1/kT}} = e^{-\Delta E/kT}. \qquad (20)$$

There is a variety of ways of directly measuring the ratio $f_{\text{MB}}(E_2)/f_{\text{MB}}(E_1)$, such as by observing the electromagnetic radiation emitted by the atoms or nuclei, and thus once again we can determine the thermodynamic temperature.

In the limit of very low temperature, most of the atoms or nuclei in the scheme of Fig. 11 would be in the lower energy state, because Eq. 20 suggests $f_{\text{MB}}(E_2) \to 0$ as $T \to 0$. In the limit of high temperature, Eq. 20 indicates that $f_{\text{MB}}(E_2) = f_{\text{MB}}(E_1)$; that is, there are equal numbers of atoms or nuclei with the two energy values. Thus the increase in temperature from very low to very high is accompanied by an increase in the relative number

in the upper energy state from near 0 to 50%. Is it possible for the number to exceed 50%? We can artificially "pump" systems from the lower energy state to the upper state, such as by causing them to absorb radiation of the proper energy ΔE. (Such a procedure is basic to the operation of a laser, in which we must have a "population inversion" with more atoms in the upper energy state.) If we try to use Eq. 20 to describe a system with $f_{\text{MB}}(E_2) > f_{\text{MB}}(E_1)$, the result is a *negative* temperature. It is thus possible to have negative temperatures, but in contrast to the usual meaning of negative numbers, negative temperatures are not below zero —they are above infinity!* ∎

26-6 ENTROPY: REVERSIBLE PROCESSES

The zeroth law of thermodynamics is related to the concept of temperature T, and the first law is related to the concept of internal energy E_{int}. In this and the following sections we show that the second law of thermodynamics is related to a thermodynamic variable called *entropy*, S, and that we can express the second law quantitatively in terms of this variable.

We start by considering a Carnot cycle. For such a cycle we can write Eq. 9 as

$$\frac{|Q_H|}{T_H} = \frac{|Q_L|}{T_L}.$$

We now discard the absolute value notation, recognizing in the process that whether the Carnot cycle is carried out clockwise, as an engine, or counterclockwise, as a refrigerator, Q_H and Q_L always have opposite signs. We can therefore write

$$\frac{Q_H}{T_H} + \frac{Q_L}{T_L} = 0. \qquad (21)$$

This equation states that the sum of the algebraic quantities Q/T is zero for a Carnot cycle.

As a next step, we want to generalize Eq. 21 to any reversible cycle, not just a Carnot cycle. To do this, we approximate any reversible cycle as an assembly of Carnot cycles. Figure 12*a* shows an arbitrary reversible cycle superimposed on a family of isotherms. We can approximate the actual cycle by connecting the isotherms by suitably chosen short segments of adiabatic lines (Fig. 12*b*), thus forming an assembly of thin Carnot cycles. You should convince yourself that traversing the individual Carnot cycles in Fig. 12*b* in sequence is exactly equivalent, in terms of heat transferred and work done, to traversing the jagged sequence of isotherms and adiabatic lines that approximates the actual cycle. This is so because adjacent Carnot cycles have a common isotherm, and the

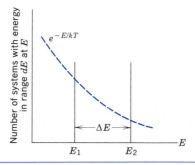

Figure 11 In a system consisting of a large number of atoms or nuclei with two discrete energy states, the relative numbers occupying each energy state can be found from the Maxwell–Boltzmann distribution (dashed line).

---

* See "Negative Absolute Temperatures," by Warren G. Proctor, *Scientific American,* August 1978, p. 90.

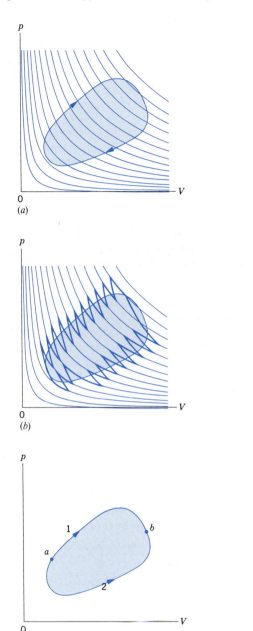

Figure 12 (*a*) A reversible cycle superimposed on a family of isotherms. (*b*) The isotherms are connected by adiabatic lines, forming an assembly of Carnot cycles that approximates the given cycle. (*c*) *a* and *b* are two arbitrary points on the cycle, and 1 and 2 are reversible paths connecting them.

two traversals, in opposite directions, cancel each other in the region of overlap as far as heat transfer and work done are concerned. By making the temperature interval between the isotherms in Fig. 12*b* small enough, we can approximate the actual cycle as closely as we wish by an alternating sequence of isotherms and adiabatic lines.

We can then write for the isothermal–adiabatic sequence of lines in Fig. 12*b*,

$$\sum \frac{Q}{T} = 0,$$

or, in the limit of infinitesimal temperature differences between the isotherms of Fig. 12*b*,

$$\oint \frac{dQ}{T} = 0, \tag{22}$$

in which $\oint$ indicates that the integral is evaluated for a complete traversal of the cycle, starting and ending at the same arbitrary point of the cycle. (Keep in mind that dQ in Eq. 22 is not an exact differential, as we previously pointed out in connection with Eq. 31 of Chapter 25. That is, there is no *function Q*, of which dQ would be the differential. We use dQ here to mean a small quantity of heat, not a true differential. In the case of Fig. 12*b*, dQ means the small quantity of heat that enters or leaves the system along a short element of the path.)

As we have seen already in Section 8-1 in the case of potential energy, if the integral of a variable around any closed path in a coordinate system is zero, then the value of that variable at a point depends only on the coordinates of the point and not at all on the path by which we arrived at that point. Such a variable is often called a *state variable*, meaning that it has a value that is characteristic only of the state of the system, regardless of how that state was arrived at. Equation 22 is such an integral, and therefore dQ/T must be a differential change in a state variable. We call this new variable the *entropy S,* such that

$$dS = \frac{dQ}{T}, \tag{23}$$

and thus Eq. 22 becomes

$$\oint dS = 0. \tag{24}$$

The SI unit for entropy is J/K.

The essential point of Eqs. 22–24 is that, although dQ is not an exact differential, dQ/T is. Gravitational potential energy U_g, internal energy E_{int}, pressure p, and temperature T are other state variables, and equations of the form $\oint dX = 0$ hold for each of them, where X is replaced by the appropriate symbol. Heat Q and work W are *not* state variables; we know that, in general, $\oint dQ \neq 0$ and $\oint dW \neq 0$, as you can easily show for the special case of a Carnot cycle.

The property of a state variable expressed by $\oint dX = 0$ can also be expressed by saying that $\int dX$ between any two equilibrium states has the same value for all (reversible) paths connecting those states. Let us prove this for the state variable entropy. We can write Eq. 24 (see Fig. 12*c*) as

$$\int_a^b dS + \int_b^a dS = 0, \tag{25}$$
$$\text{path 1} \quad \text{path 2}$$

where *a* and *b* are arbitrary points and 1 and 2 describe the paths connecting these points. Since the cycle is reversible, we can traverse path 2 in the opposite direction (that is, from *a* to *b* rather than from *b* to *a*), in which case we can write Eq. 25 as

$$\int_a^b dS - \int_a^b dS = 0$$
$$\text{path 1} \qquad \text{path 2}$$

or

$$\int_a^b dS = \int_a^b dS. \qquad (26)$$
$$\text{path 1} \qquad \text{path 2}$$

Note that changing the order of the limits in the second integral of Eq. 25 requires that we also change the sign of the integral. This gives Eq. 26, which tells us that the quantity $\int_a^b dS$ between any two equilibrium states of the system, such as *a* and *b*, is independent of the path connecting those states, for 1 and 2 are quite arbitrary paths. Recall our almost identical discussion in Section 8-1, where we introduced the concept of a conservative force.

The change in entropy between any two states *i* and *f* is then

$$\Delta S = S_f - S_i = \int_i^f dS$$

$$= \int_i^f \frac{dQ}{T} \qquad \text{(reversible process)}, \qquad (27)$$

where the integral is evaluated over *any reversible path* connecting these two states.

Sample Problem 5 A lump of ice whose mass *m* is 235 g melts (reversibly) to water, the temperature remaining at 0°C throughout the process. What is the entropy change for the ice? The heat of fusion of ice is 333 kJ/kg.

Solution The requirement that we melt the ice reversibly means that we must put the ice in contact with a heat reservoir whose temperature exceeds 0°C by only a differential amount, thereby melting only a small bit of the ice. (If we then lower the reservoir temperature by the same differential amount, the melted ice would freeze; thus the process is reversible.) For such a reversible process, we can use Eq. 27, or, since the temperature is constant,

$$S_{\text{water}} - S_{\text{ice}} = \int \frac{dQ}{T} = \frac{1}{T} \int dQ = \frac{Q}{T}.$$

Here *dQ* means the small elements of heat energy that enter the ice from the heat reservoir, and the total of all these elements is just the total heat absorbed by the ice, or

$$Q = mL = (0.235 \text{ kg})(333 \text{ kJ/kg}) = 7.83 \times 10^4 \text{ J}.$$

Thus

$$S_{\text{water}} - S_{\text{ice}} = \frac{Q}{T} = \frac{7.83 \times 10^4 \text{ J}}{273 \text{ K}} = 287 \text{ J/K}.$$

The above answer completes our analysis of the entropy change of the *system,* but let us carry the problem a bit further and consider the entropy change of the environment as well. In this case, the environment is the heat reservoir from which the heat required to melt the ice is drawn. Every unit of heat that

enters the ice must have left the reservoir, the temperature of both ice and reservoir being the same. Therefore the entropy change of the reservoir is equal in magnitude but opposite in sign to that of the ice, or

$$\Delta S_{\text{reservoir}} = -287 \text{ J/K}.$$

The entropy change for the ice + reservoir, taken together, is thus zero. This is true for any *reversible* process, because any increment of heat $+dQ$ that enters the system must originate from an equal increment $-dQ$ that leaves the reservoir.

In practice, the melting of ice is likely to be irreversible, as when you toss an ice cube into a glass of water at room temperature. The temperature difference between the ice and the reservoir (the water) in this case is not a differential amount but is about 20°C. The process proceeds in only one direction—the ice melts—and cannot be reversed at any stage by making only a differential change in the water temperature. You cannot use Eq. 27 in such a case, and the calculations of this problem are not valid. We see how to handle a calculation of this type in the next section.

26-7 ENTROPY: IRREVERSIBLE PROCESSES

Equation 27 describes the calculation of the change in entropy for a reversible process. However, there are no absolutely reversible processes in nature. Friction and unwanted heat transfers are always present, and we can seldom perform real processes in infinitesimal steps. Every thermodynamic process is therefore to some extent irreversible.

To calculate the entropy change for an irreversible process, we take advantage of the fact that *entropy is a state variable.* The difference in entropy between states *i* and *f* is independent of the path we choose from *i* to *f*. Even though nature may have chosen an irreversible path between *i* and *f* for the actual process, we can choose any convenient reversible path for the calculation of the entropy change.

To find the entropy change for an irreversible path between two equilibrium states, find a reversible process connecting the same states, and calculate the entropy change using Eq. 27.

We consider two examples.

1. *Free expansion.* As in Section 25-6 (see Fig. 14 of Chapter 25), let an ideal gas double its volume by expanding into an evacuated space. No work is done against the vacuum, so $W = 0$, and the gas is confined to an insulating container, so $Q = 0$. From the first law, we must therefore have $\Delta E_{\text{int}} = 0$. For an ideal gas, whose internal energy depends only on temperature, it follows that $T_i = T_f$.

The free expansion is certainly irreversible, because we

lose control of the system once we open the valve that separates the two compartments. There is an entropy difference between the initial and final states, but we cannot calculate it using Eq. 27, which applies only to reversible processes. Clearly, Eq. 27 must not be used directly, because the temperature is not defined for the intermediate nonequilibrium states through which the system evolves once the gas begins flowing. Furthermore, $Q = 0$, which presents a further difficulty in using Eq. 27.

To find the entropy change, we choose a reversible path from i to f for which we can do this calculation. A convenient choice is an isothermal expansion that would take an ideal gas from the same initial point (p_i, V_i, T_i) to the same final point (p_f, V_f, T_f). It represents a procedure very different from that of the free expansion, but it connects the same pair of equilibrium states. From Eq. 27, we then have

$$\Delta S = S_f - S_i = \int_i^f \frac{dQ}{T} = \frac{1}{T} \int_i^f dQ = \frac{Q}{T} = \frac{-W}{T},$$

where the last step can be made because $\Delta E_{int} = 0$ in an isothermal process, and thus $-W = Q$. Using Eq. 27 of Chapter 23 for W, we obtain

$$\Delta S = \frac{-W}{T} = nR \ln \frac{V_f}{V_i}$$
$$= nR \ln 2. \qquad (28)$$

This is equal to the entropy change for the irreversible free expansion. Note that ΔS is positive for the system.

Because there is no energy transfer of any kind to the environment in the free expansion, the entropy change of the environment is zero. Thus the total entropy of system + environment increases during a free expansion.

2.   *Irreversible heat transfer.* Figure 13a shows two blocks whose initial temperatures are T_1 and T_2. For simplicity, we assume the blocks have the same mass m and specific heat c. We remove the insulating barrier that separates the blocks and bring them into thermal contact, as shown in Fig. 13b. Eventually, they reach the common equilibrium temperature T_e. Like the free expansion, this process is totally irreversible, because we lose control once we place the blocks in thermal contact with one another.

To find the entropy change in this irreversible process, we once again choose a reversible path that leads us to the same final state. Consider first block 1 at initial temperature T_1. (Assume this to be the lower initial temperature.) Imagine a series of thermal reservoirs at temperatures T_1, $T_1 + dT$, $T_1 + 2dT$, . . . , $T_e - dT$, T_e. Start with block 1 in contact with the first reservoir, and then move it one step at a time along the sequence. At each step, an infinitesimal amount of heat dQ enters the block. The process is clearly reversible; at any point we can move the block back to the previous lower step, and the same amount of heat dQ will flow from the block back into the reservoir. Each reversible heat transfer dQ can be expressed as

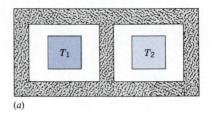

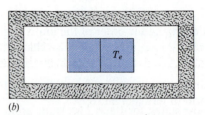

Figure 13   (*a*) The initial state: two blocks are at different temperatures in individual insulating enclosures. (*b*) The final state: the insulating wall between the blocks is removed, and they are allowed to come to equilibrium at the intermediate temperature T_e.

$mc\, dT$, and thus we can use Eq. 27 to find the entropy change for block 1:

$$\Delta S_1 = \int_i^f \frac{dQ}{T} = mc \int_{T_1}^{T_e} \frac{dT}{T} = mc \ln \frac{T_e}{T_1}. \qquad (29)$$

We can similarly construct a descending series of reservoirs for block 2 between temperatures T_2 and T_e, and we find the entropy change for block 2:

$$\Delta S_2 = mc \int_{T_2}^{T_e} \frac{dT}{T} = mc \ln \frac{T_e}{T_2}. \qquad (30)$$

The total entropy change is

$$\Delta S = \Delta S_1 + \Delta S_2 = mc \ln \frac{T_e}{T_1} + mc \ln \frac{T_e}{T_2}$$
$$= mc \ln \frac{T_e^2}{T_1 T_2}. \qquad (31)$$

If T_1 is the lower temperature, then $\Delta S_1 > 0$ and $\Delta S_2 < 0$. We can show that the total entropy change ΔS is always positive, for which we need to show that $T_e^2/T_1 T_2 > 1$. We first find the equilibrium temperature by requiring that the total heat transferred be zero:

$$Q_1 + Q_2 = mc(T_e - T_1) + mc(T_e - T_2) = 0,$$

or $T_e = (T_1 + T_2)/2$. We can therefore write the quantity $T_e^2/T_1 T_2$ as

$$\frac{T_e^2}{T_1 T_2} = \frac{(T_1 + T_2)^2}{4 T_1 T_2} = \frac{4 T_1 T_2 + (T_1 - T_2)^2}{4 T_1 T_2}$$
$$= 1 + \frac{(T_1 - T_2)^2}{4 T_1 T_2}.$$

Clearly this is greater than one (the last quantity is always positive), so the logarithm in Eq. 31 is greater than zero, and the entropy change is positive.

Placing the two blocks in thermal contact produces no change at all in the environment, so $\Delta S = 0$ for the environment. The total entropy of system + environment therefore increases in this irreversible heat transfer.

26-8 ENTROPY AND THE SECOND LAW

We are now ready to express the second law in its most general form in terms of entropy:

In any thermodynamic process that proceeds from one equilibrium state to another, the entropy of the system + environment either remains unchanged or increases.

For reversible processes, as we have seen in Section 26-6, the entropy remains unchanged. As we saw in the case of Sample Problem 5, the entropy change of the system was positive and that of the environment was negative and of equal magnitude, so that the total was zero.

For irreversible processes (that is, for all natural processes), the total entropy of system + environment must increase. It is possible that the entropy of the system might decrease, but the entropy of the environment always shows an increase of greater magnitude, so that the total change in entropy is always positive. *No natural process can ever show a decrease in the total entropy of system + environment.*

As was the case for the zeroth and first laws, implicit in this form of the second law is a statement about the existence and usefulness of a new thermodynamic variable, in this case the entropy.

The second law, like the zeroth and first laws, is a generalization from experience. It cannot be proved, but we can test it in a variety of circumstances. We can show that it is consistent with observation, in that it forbids processes that might seem to satisfy all other known laws, but that are not observed. Let us consider this statement of the second law with respect to some of the principles we have already established in this chapter.

Free Compression

Let us imagine performing the free expansion with a removable partition that separates the two halves of the container. When we remove the partition, the gas molecules that were originally moving to the right in Fig. 14 of Chapter 25 find no partition with which to collide, and so they spread into the formerly empty half of the container, eventually striking the far wall. When they recoil from

that wall, they do not all find their way back to the other half, because they are likely to collide with other molecules along the way. Eventually the collisions tend to randomize the motions of the molecules, and they fill the entire container.

What keeps us from finding all the molecules back in one half at a later time? We might call this process a free compression, the reverse of a free expansion. Equation 28 shows that a free compression, in which $V_f < V_i$, would have a negative change in entropy for the system (with no change in entropy of the environment, as in the free expansion). This statement of the second law in terms of entropy thus forbids the free compression, and so it is unlikely that you will find all the air rushing to the opposite side of the room in which you are sitting. (We give another interpretation of this non-event in the next section.)

The Kelvin–Planck Form of the Second Law

Because all engines operate in cycles, the entropy change for the system (the working substance) must be zero for one complete cycle of operation. In a perfect engine, the environment (see Fig. 3a) releases heat Q at temperature T, and its entropy change is Q/T, a negative quantity. The total entropy change of system + environment is therefore negative in a perfect engine. The existence of a perfect engine would thus violate the entropy statement of the second law.

The Clausius Form of the Second Law

In a perfect refrigerator, the system again has no entropy change in a complete cycle, but the environment releases heat $-Q$ at the temperature T_L and absorbs heat Q at the temperature T_H. The total change in entropy of the environment is therefore

$$\Delta S = \frac{Q}{T_H} - \frac{Q}{T_L} = Q\left(\frac{1}{T_H} - \frac{1}{T_L}\right).$$

Because $T_H > T_L$, this entropy change is negative. A perfect refrigerator would thus violate the entropy statement of the second law.

The Arrow of Time

It is the change in entropy that ultimately provides us with the answer to why systems will naturally evolve in one direction with time and not the other: systems always evolve in time in such a way that the total entropy of system + environment increases. If you observe a system in which the entropy appears to decrease, you can be sure that somewhere there is a change in the entropy of the environment large enough to make the total entropy change positive.

Sample Problem 6   A piece of ice of mass $m_i = 0.012$ kg is initially at a temperature $T_i = -15°C$. It is dropped into an insulated container of negligible heat capacity containing a mass $m_w = 0.056$ kg of water at a temperature $T_w = 23°C$. The system comes to equilibrium at temperature T_e. Calculate the total entropy change of the system + environment. Use the following specific heat capacities and heat of fusion: $c_i = 2220$ J/kg·K, $c_w = 4190$ J/kg·K, $L = 333$ kJ/kg.

Solution   Dropping the ice into the water is clearly an irreversible process; it is not done in infinitesimal steps, and we cannot restore the system to its original state by reversing the process. To calculate the entropy change for the system, we first must find the final equilibrium temperature. To do so, we assume that the final temperature is greater than 0°C and that all the ice melts, eventually becoming water at the equilibrium temperature. We can check this assumption for consistency later. Requiring that the total heat transferred among all the objects be zero, we can find the equilibrium temperature:

$$m_i c_i(0°C - T_i) + m_i L + m_i c_w(T_e - 0°C)$$
$$+ m_w c_w(T_e - T_w) = 0,$$

and inserting the given values and solving we find

$$T_e = 276.6 \text{ K} = 3.5°C.$$

This is certainly consistent with all the ice melting. (If, on the other hand, we would have obtained a final temperature of 0°C or below, we would suspect that our original assumption was incorrect, and we would change the solution accordingly.)

We can now find the entropy changes. First for the ice, we separate the procedure into three steps: warming the ice to 0°C, melting the ice, and then warming the resulting water to T_e. We use Eq. 29 for the (reversible) entropy change associated with a change in the temperature and use the result of Sample Problem 5 for the entropy change of melting. The result for the entropy change ΔS_i of the ice is

$$\Delta S_i = m_i c_i \ln \frac{273 \text{ K}}{T_i} + \frac{m_i L}{273 \text{ K}} + m_i c_w \ln \frac{T_e}{273 \text{ K}}$$

$$= (0.012 \text{ kg})(2220 \text{ J/kg·K}) \ln \frac{273 \text{ K}}{258 \text{ K}}$$

$$+ \frac{(0.012 \text{ kg})(333 \text{ kJ/kg})}{273 \text{ K}}$$

$$+ (0.012 \text{ kg})(4190 \text{ J/kg·K}) \ln \frac{276.6 \text{ K}}{273 \text{ K}}$$

$$= 16.7 \text{ J/K}.$$

For the water, we similarly obtain its (reversible) entropy change:

$$\Delta S_w = m_w c_w \ln \frac{T_e}{T_w} = (0.056 \text{ kg})(4190 \text{ J/kg·K}) \ln \frac{276.6 \text{ K}}{296 \text{ K}}$$

$$= -15.9 \text{ J/K}.$$

The entropy change of the environment is zero, since the entire procedure takes place in an insulated container. The total entropy change of system plus environment is therefore

$$\Delta S = \Delta S_i + \Delta S_w = 16.7 \text{ J/K} + (-15.9 \text{ J/K}) = 0.8 \text{ J/K},$$

and this is clearly positive, as required by the second law.

26-9   ENTROPY AND PROBABILITY

Entropy is a macroscopic variable, associated with the overall state of a system and calculable from the macroscopic quantities associated with its overall state. We have seen that all macroscopic variables in thermodynamics have a corresponding microscopic quantity (such as temperature, a macroscopic quantity, and mean molecular kinetic energy, a microscopic quantity). If we make certain assumptions about the microscopic properties of the system, we can usually find a way to relate the macroscopic and microscopic quantities. In the case of the temperature of a gas, these assumptions include a mechanical model of the molecules and their interactions, along with a statistical distribution of the molecular energies. We would therefore like to consider the microscopic calculation of the entropy of a system.

The microscopic quantity related to entropy is the relative probability of different ways of sorting the molecules of the system. Let us first consider some qualitative applications of this relationship:

1. *Free expansion.* In a free expansion the gas molecules confined to one half of a box are permitted to fill the entire box. Let us consider the entire box in its two circumstances: first, at the instant the partition is removed with the molecules all occupying one half of the box, and second, with the molecules filling the entire box. The first condition is a state of very low probability; left on its own, it would be very unlikely for the system to sort itself in this way. The second condition is one of rather high probability. We can regard the molecules in the free expansion as moving from a condition of low probability to one of high probability. That is, given all the possible ways of distributing or sorting the molecules randomly within the box, a large number of those ways show a rather uniform distribution of molecules, while a very small number show a nonuniform distribution. For instance, let us consider a box with only 10 molecules and evaluate the number of ways for a particular number n to be found in the left half of the box at any instant. Since each molecule has two possible locations in the box in this scheme (left half or right half), the total number of ways we can distribute the molecules, with two choices for each, is $2^{10} = 1024$. Of those 1024 ways, in only one will we find all the molecules in the left half ($n = 10$), while it can be shown that there are 252 ways of having a uniform distribution ($n = 5$). As the number of molecules increases, the relative probability of a uniform distribution increases dramatically. With 100 molecules, there is still only one way to sort them all into the left half, but there are about 10^{29} ways to distribute them equally between the two halves. The free expansion, in which there is an increase in entropy, can thus microscopically be regarded as a transformation from a state of very low probability to a state of very high probability.

2. *Heat conduction.* In this example two bodies of different temperatures T_1 and T_2 come to a uniform intermediate temperature T_e when they are placed in contact. This case is similar to the free expansion, except that we sort by speed rather than by location. Again we consider the entire system in two circumstances: just after contact, with the "hot" molecules (the mostly faster moving ones) on one side and the "cool" molecules (mostly slower moving) on the other, and at a much later time, when the distribution of speeds between the two halves is uniform. Once again, the sorted condition (fast molecules on one side, slow on the other) is a state of low probability, and the uniform distribution is a state of high probability. Upon contact, the system evolves spontaneously from a state of low probability to one of high probability.

3. *A stirred cup of coffee.* Suppose that you stir a cup of coffee and then remove the spoon. Over time, the circulation of the liquid dies out, and viscosity causes the energy of the rotating fluid to be dissipated as internal energy of the molecules. In the initial state there is an ordered motion of the swirling coffee. In the final equilibrium state there is random molecular motion. Once again, the ordered circulation of the molecules is a state of low probability, while the random disordered motion is a state of high probability. In this natural process, the system has gone from a state of low probability to a state of high probability.

In all three cases above, the system has gone spontaneously from a state of low probability to one of high probability. All three of these situations are irreversible natural processes that are characterized by an increase in the entropy of the system. It is therefore reasonable to conclude that there is a quantitative relationship between probability and entropy. This relationship, which was proposed by Boltzmann, is

$$S = k \ln P. \qquad (32)$$

Here k is the Boltzmann constant, and S is the entropy of the system. P, to define it loosely, is the number of different molecular arrangements that correspond to the same macroscopic state. If, for example, the circumstances are so special that only one arrangement is possible, then $P = 1$ and $S (= k \ln P) = 0$, and we have a completely ordered state. Larger values of P, such as that corresponding to the more uniform distribution of molecules in the container, give larger values of the entropy. When we say that state A of a system is more probable than state B, we simply mean that state A has the larger value of P.

An increase in entropy has also been said to be a measure of the increase in the disorder of a system, and thus entropy is in effect a measure of the disorder. The term "disorder" does not have a precise mathematical definition, but it is qualitatively related to the probability. A state of low disorder is a state in which the components of a system have been carefully sorted, such as by placing all the molecules with low speeds in one part of a system. A state of high disorder is a random state in which no sorting has occurred. The increase in entropy of a system in natural processes can thus also be regarded as an increase in the level of disorder of the system. Natural processes tend to make the universe more disordered.*

The second law of thermodynamics tells us that, if an isolated system undergoes a spontaneous process, its final state will be one in which the entropy (and also P) is maximum. There is always—in principle—the possibility that, by a statistical fluctuation, some other state might occur, even a state whose entropy is lower than that of the initial state. For systems with very small numbers of particles, such fluctuations from average behavior are indeed there to be seen, the Brownian motion (Section 24-5) being a good example. For macroscopic systems, however, the probability that the entropy will actually decrease in a spontaneous process proves to be unbelievably small. Thus you can predict with complete confidence that (1) the molecules of air in your room will *not* spontaneously congregate on one side, (2) a glass of room-temperature water will not spontaneously separate into ice cubes and warmer water, and (3) your coffee will not spontaneously begin rotating in the cup. Each of these processes can conserve energy, and thus would be consistent with the first law. They will, however, not be observed, even if you keep watch for a time as long as the present age of the universe! It has been said that calculations of probabilities of such events give operational meaning to the word "never." The area of application of the second law of thermodynamics is so broad and the chance of nature's contradicting it is so small that it has the distinction of being one of the most useful and general laws in all of science.

---

* For an account of attempts to violate the second law, including Maxwell's demon and a design for a perfect heat engine in which the working substance is one molecule of a gas, see "Demons, Engines, and the Second Law," by Charles H. Bennett, *Scientific American,* November 1987, p. 108.

QUESTIONS

1. Is a human being a heat engine? Explain.

2. Couldn't we just as well define the efficiency of an engine as $e = |W|/|Q_{out}|$ rather than as $e = |W|/|Q_{in}|$? Why don't we?

3. The efficiencies of nuclear power plants are less than those of fossil-fuel plants. Why?

4. Can a given amount of mechanical energy be converted completely into heat energy? If so, give an example.

5. An inventor suggested that a house might be heated in the following manner. A system resembling a refrigerator draws heat from the Earth and rejects heat to the house. The inventor claimed that the heat supplied to the house can exceed the work done by the engine of the system. What is your comment?

6. Comment on the statement: "A heat engine converts disordered mechanical motion into organized mechanical motion."

7. Is a heat engine operating between the warm surface water of a tropical ocean and the cooler water beneath the surface a possible concept? Is the idea practical? (See "Solar Sea Power," by Clarence Zener, *Physics Today,* January 1973, p. 48.)

8. Give a qualitative explanation of how frictional forces between moving surfaces produce internal energy. Why does the reverse process (internal energy producing relative motion of those surfaces) not occur?

9. Are any of the following phenomena reversible: (a) breaking an empty soda bottle; (b) mixing a cocktail; (c) winding a watch; (d) melting an ice cube in a glass of iced tea; (e) burning a log of firewood; (f) puncturing an automobile tire; (g) heating electrically an insulated block of metal; (h) isothermally expanding a non-ideal gas against a piston; (i) finishing the "Unfinished Symphony"; (j) writing this book?

10. Give some examples of irreversible processes in nature.

11. Are there any natural processes that are reversible?

12. Can we calculate the work done during an irreversible process in terms of an area on a pV diagram? Is any work done?

13. If a Carnot engine is independent of the working substance, then perhaps real engines should be similarly independent, to a certain extent. Why then, for real engines, are we so concerned to find suitable fuels such as coal, gasoline, or fissionable material? Why not use stones as fuel?

14. Under what conditions would an ideal heat engine be 100% efficient?

15. What factors reduce the efficiency of a heat engine from its ideal value?

16. You wish to increase the efficiency of a Carnot engine as much as possible. You can do this by increasing T_H a certain amount, keeping T_L constant, or by decreasing T_L the same amount, keeping T_H constant. Which would you do?

17. Explain why a room can be warmed by leaving open the door of an oven but cannot be cooled by leaving open the door of a kitchen refrigerator.

18. Why do you get poorer gasoline mileage from your car in winter than in summer?

19. From time to time inventors will claim to have perfected a device that does useful work but consumes no (or very little) fuel. What do you think is most likely true in such cases: (a) the claimants are right, (b) the claimants are mistaken in their measurements, or (c) the claimants are swindlers? Do you think that such a claim should be examined closely by a panel of scientists and engineers? In your opinion, would the time and effort be justified?

20. We have seen that real engines always discard substantial amounts of heat to their low-temperature reservoirs. It seems a shame to throw this heat energy away. Why not use this heat to run a second engine, the low-temperature reservoir of the first engine serving as the high-temperature reservoir of the second?

21. Give examples in which the entropy of a system decreases and explain why the second law of thermodynamics is not violated.

22. Do living things violate the second law of thermodynamics? As a chicken grows from an egg, for example, it becomes more and more ordered and organized. Increasing entropy, however, calls for disorder and decay. Is the entropy of a chicken actually decreasing as it grows?

23. Two containers of gases at different temperatures are isolated from the surroundings and separated from each other by a partition that allows heat exchange. What would have to happen if the entropy were to decrease? To increase? What is likely to happen?

24. Is there a change in entropy in purely mechanical motions?

25. Show that the total entropy increases when work is converted into heat by friction between sliding surfaces. Describe the increase in disorder.

26. Heat energy flows from the Sun to the Earth. Show that the entropy of the Earth–Sun system increases during this process.

27. Is it true that the heat energy of the universe is steadily growing less available? If so, why?

28. Consider a box containing a very small number of molecules, say five. It must sometimes happen by chance that all these molecules find themselves in the left half of the box, the right half being completely empty. This is just the reverse of free expansion, a process that we have declared to be *irreversible*. What is your explanation?

29. A rubber band feels warmer than its surroundings immediately after it is quickly stretched; it becomes noticeably cooler when it is allowed to contract suddenly. Also, a rubber band supporting a load contracts on being heated. Explain these observations using the fact that the molecules of rubber consist of intertwined and cross-linked long chains of atoms in roughly random orientation.

30. What entropy change occurs, if any, when a pack of 52 cards is shuffled into one particular arrangement? Is the concept of entropy appropriate in this case? If so, explain how one could get useful cooling by carrying out this process adiabatically.

31. Discuss the following comment of Panofsky and Phillips: "From the standpoint of formal physics there is only one concept which is asymmetric in the time, namely, entropy. But this makes it reasonable to assume that the second law

of thermodynamics can be used to ascertain the sense of time independent of any frame of reference; that is, we shall take the positive direction of time to be that of statistically increasing disorder, or increasing entropy." (See, in this connection, "The Arrow of Time," by David Layzer, *Scientific American,* December 1975, p. 56.)

32. Explain the statement: "Cosmic rays continually *decrease* the entropy of the Earth on which they fall." Why does this not contradict the second law of thermodynamics?

33. When we put cards together in a deck or put bricks together to build a house, for example, we increase the order in the physical world. Does this violate the second law of thermodynamics? Explain.

34. Can one use terrestrial thermodynamics, which is known to

apply to bounded and isolated bodies, for the whole universe? If so, is the universe bounded and from what is the universe isolated?

35. Temperature and pressure are examples of *intensive* properties of a system, their values for any sample of the system being independent of the size of the sample. However, entropy, like internal energy, is an *extensive* property, its value for any sample of a system being proportional to the size of the sample. Discuss.

36. The first, second, and third laws of thermodynamics may be paraphrased, respectively, as follows: (1) You can't win. (2) You can't even break even. (3) You can't get out of the game. Explain in what sense these are permissible restatements.

PROBLEMS

Section 26-2 Heat Engines and the Second Law

1. A heat engine absorbs 52.4 kJ of heat and exhausts 36.2 kJ of heat each cycle. Calculate (*a*) the efficiency and (*b*) the work done by the engine per cycle.

2. A car engine delivers 8.18 kJ of work per cycle. (*a*) Before a tuneup, the efficiency is 25.0%. Calculate, per cycle, the heat absorbed from the combustion of fuel and the heat exhausted to the atmosphere. (*b*) After a tuneup, the efficiency is 31.0%. What are the new values of the quantities calculated in (*a*)?

3. Calculate the efficiency of a fossil-fuel power plant that consumes 382 metric tons of coal each hour to produce useful work at the rate of 755 MW. The heat of combustion of coal is 28.0 MJ/kg.

4. Two moles of a monatomic ideal gas are caused to go through the cycle shown in Fig. 14. Process *bc* is a reversible adiabatic expansion. Also, $p_b = 10.4$ atm, $V_b = 1.22$ m³, and $V_c = 9.13$ m³. Calculate (*a*) the heat added to the gas, (*b*) the heat leaving the gas, (*c*) the net work done by the gas, and (*d*) the efficiency of the cycle.

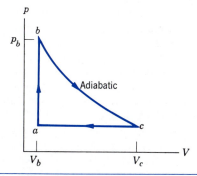

Figure 14 Problem 4.

5. One mole of a monatomic ideal gas initially at a volume of 10 L and a temperature 300 K is heated at constant volume to a temperature of 600 K, allowed to expand isothermally to its initial pressure, and finally compressed isobarically (that is, at constant pressure) to its original volume, pres-

sure, and temperature. (*a*) Compute the heat input to the system during one cycle. (*b*) What is the net work done by the gas during one cycle? (*c*) What is the efficiency of this cycle?

6. A gasoline internal combustion engine can be approximated by the cycle shown in Fig. 15. Assume an ideal diatomic gas and use a compression ratio of 4 : 1 ($V_d = 4V_a$). Assume that $p_b = 3p_a$. (*a*) Determine the pressure and temperature of each of the vertex points of the *pV* diagram in terms of p_a and T_a. (*b*) Calculate the efficiency of the cycle.

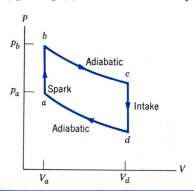

Figure 15 Problem 6.

Engine *A*, compared to engine *B*, produces, per cycle, five times the work but receives three times the heat input and exhausts twice the heat out. Determine the efficiency of each engine.

Section 26-3 Refrigerators and the Second Law

8. A refrigerator does 153 J of work to transfer 568 J of heat from its cold compartment. (*a*) Calculate the refrigerator's coefficient of performance. (*b*) How much heat is exhausted to the kitchen?

9. To make some ice, a freezer extracts 185 kJ of heat at −12.0°C. The freezer has a coefficient of performance of 5.70. The room temperature is 26.0°C. (*a*) How much heat was delivered to the room? (*b*) How much work was required to run the freezer?

Section 26-4 The Carnot Cycle

10. How much work must be done to extract 10.0 J of heat (a) from a reservoir at 7°C and transfer it to one at 27°C by means of a refrigerator using a Carnot cycle; (b) from one at −73°C to one at 27°C; (c) from one at −173°C to one at 27°C; and (d) from one at −223°C to one at 27°C?

11. In a Carnot cycle, the isothermal expansion of an ideal gas takes place at 412 K and the isothermal compression at 297 K. During the expansion, 2090 J of heat energy are transferred to the gas. Determine (a) the work performed by the gas during the isothermal expansion, (b) the heat rejected from the gas during the isothermal compression, and (c) the work done on the gas during the isothermal compression.

12. A Carnot engine has an efficiency of 22%. It operates between heat reservoirs differing in temperature by 75 C°. Find the temperatures of the reservoirs.

13. For the Carnot cycle illustrated in Fig. 7, show that the work done by the gas during process *bc* (step 2) has the same absolute value as the work done on the gas during process *da* (step 4).

14. Apparatus that liquifies helium is in a laboratory at 296 K. The helium in the apparatus is at 4.0 K. If 150 mJ of heat is transferred from the helium, find the minimum amount of heat delivered to the laboratory.

15. An air conditioner takes air from a room at 70°F and transfers it to the outdoors, which is at 95°F. For each joule of electrical energy required to run the refrigerator, how many joules of heat are transferred from the room?

16. An inventor claims to have created a heat pump that draws heat from a lake at 3.0°C and delivers heat at a rate of 20 kW to a building at 35°C, while using only 1.9 kW of electrical power. How would you judge the claim?

17. (a) A Carnot engine operates between a hot reservoir at 322 K and a cold reservoir at 258 K. If it absorbs 568 J of heat per cycle at the hot reservoir, how much work per cycle does it deliver? (b) If the same engine, working in reverse, functions as a refrigerator between the same two reservoirs, how much work per cycle must be supplied to transfer 1230 J of heat from the cold reservoir?

18. A heat pump is used to heat a building. The outside temperature is −5.0°C and the temperature inside the building is to be maintained at 22°C. The coefficient of performance is 3.8, and the pump delivers 7.6 MJ of heat to the building each hour. At what rate must work be done to run the pump?

19. If a Carnot cycle is run backward, we have an ideal refrigerator. A quantity of heat $|Q_L|$ is taken in at the lower temperature T_L and a quantity of heat $|Q_H|$ is given out at the higher temperature T_H. The difference is the work W that must be supplied to run the refrigerator. (a) Show that

$$|W| = |Q_L| \frac{T_H - T_L}{T_L}.$$

(b) The coefficient of performance K of a refrigerator is defined as the ratio of the heat extracted from the cold source to the work needed to run the refrigerator. Show that ideally

$$K = \frac{T_L}{T_H - T_L}.$$

(c) In a mechanical refrigerator the low-temperature coils

are at a temperature of −13°C and the compressed gas in the condenser has a temperature of 25°C. Find the theoretical coefficient of performance.

20. The motor in a refrigerator has a power output of 210 W. The freezing compartment is at −3.0°C and the outside air is at 26°C. Assuming that the efficiency is 85% of the ideal, calculate the amount of heat that can be extracted from the freezing compartment in 15 min.

21. Show that the efficiency of a reversible ideal heat engine is related to the coefficient of performance of the reversible refrigerator obtained by running the engine backward by the relation $e = 1/(K + 1)$.

22. (a) In a two-stage Carnot heat engine, a quantity of heat $|Q_1|$ is absorbed at a temperature T_1, work $|W_1|$ is done, and a quantity of heat $|Q_2|$ is expelled at a lower temperature T_2 by the first stage. The second stage absorbs the heat expelled by the first, does work $|W_2|$, and expels a quantity of heat $|Q_3|$ at a lower temperature T_3. Prove that the efficiency of the combination is $(T_1 - T_3)/T_1$. (b) A combination mercury–steam turbine takes saturated mercury vapor from a boiler at 469°C and exhausts it to heat a steam boiler at 238°C. The steam turbine receives steam at this temperature and exhausts it to a condenser at 37.8°C. Calculate the maximum efficiency of the combination.

23. A Carnot engine works between temperatures T_1 and T_2. It drives a Carnot refrigerator that works between two different temperatures T_3 and T_4 (see Fig. 16). Find the ratio $|Q_3|/|Q_1|$ in terms of the four temperatures.

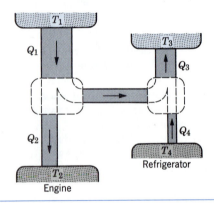

Figure 16   Problem 23.

24. An inventor claims to have invented four engines, each of which operates between heat reservoirs at 400 and 300 K. Data on each engine, per cycle of operation, are as follows: Engine A: $Q_{in} = 200$ J, $Q_{out} = -175$ J, $W = 40$ J; engine B: $Q_{in} = 500$ J, $Q_{out} = -200$ J, $W = 400$ J; engine C: $Q_{in} = 600$ J, $Q_{out} = -200$ J, $W = 400$ J; engine D: $Q_{in} = 100$ J, $Q_{out} = -90$ J, $W = 10$ J. Which of the first and second laws of thermodynamics (if either) does each engine violate?

25. In a steam locomotive, steam at a boiler pressure of 16.0 atm enters the cylinders, is expanded adiabatically to 5.60 times its original volume, and then exhausted to the atmosphere. Calculate (a) the steam pressure after expansion and (b) the greatest possible efficiency of the engine.

26. (a) Plot accurately a Carnot cycle on a *pV* diagram for 1.00

mol of an ideal gas. Let point *a* (see Fig. 7) correspond to $p = 1.00$ atm, $T = 300$ K, and let point *b* correspond to 0.500 atm, $T = 300$ K; take the low-temperature reservoir to be at 100 K. Let $\gamma = 1.67$. (*b*) Compute graphically the work done in this cycle. (*c*) Compute the work analytically.

27. One mole of an ideal monatomic gas is used as the working substance of an engine that operates on the cycle shown in Fig. 17. Calculate (*a*) the work done by the engine per cycle, (*b*) the heat added per cycle during the expansion stroke *abc*, and (*c*) the engine efficiency. (*d*) What is the Carnot efficiency of an engine operating between the highest and lowest temperatures present in the cycle? How does this compare to the efficiency calculated in (*c*)? Assume that $p_1 = 2p_0$, $V_1 = 2V_0$, $p_0 = 1.01 \times 10^5$ Pa, and $V_0 = 0.0225$ m³.

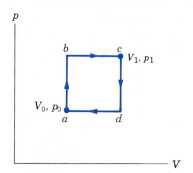

Figure 17   Problem 27.

Section 26-6 Entropy: Reversible Processes

28. In Fig. 12*c*, suppose that the change in entropy of the system in passing from state *a* to state *b* along path 1 is +0.60 J/K. What is the entropy change in passing (*a*) from state *a* to *b* along path 2 and (*b*) from state *b* to *a* along path 2?

29. An ideal gas undergoes a reversible isothermal expansion at 132°C. The entropy of the gas increases by 46.2 J/K. How much heat was absorbed?

30. Four moles of an ideal gas are caused to expand from a volume V_1 to a volume $V_2 = 3.45V_1$. (*a*) If the expansion is isothermal at the temperature $T = 410$ K, find the work done on the expanding gas. (*b*) Find the change in entropy, if any. (*c*) If the expansion were reversibly adiabatic instead of isothermal, what is the entropy change?

31. (*a*) Show that a Carnot cycle, plotted on an absolute temperature versus entropy (*TS*) diagram, graphs as a rectangle.

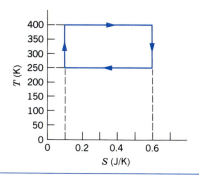

Figure 18   Problem 31.

For the Carnot cycle shown in Fig. 18, calculate (*b*) the heat that enters and (*c*) the work done on the system.

32. Find (*a*) the heat absorbed and (*b*) the change in entropy of a 1.22-kg block of copper whose temperature is increased reversibly from 25.0 to 105°C.

33. At very low temperatures, the molar specific heat of many solids is (approximately) proportional to T^3; that is, $C_V = AT^3$, where A depends on the particular substance. For aluminum, $A = 3.15 \times 10^{-5}$ J/mol·K⁴. Find the entropy change of 4.8 mol of aluminum when its temperature is raised from 5.0 to 10 K.

34. Heat can be transferred from water at 0°C and atmospheric pressure without causing the water to freeze, if done with little disturbance of the water. Suppose the water is cooled to -5.0°C before ice begins to form. Find the change in entropy occurring during the sudden freezing of 1.0 g of water that then takes place.

35. An object of constant heat capacity *C* is heated from an initial temperature T_i to a final temperature T_f by being placed in contact with a reservoir at T_f. Represent the process on a graph of C/T versus T and show graphically that the total change in entropy ΔS (object plus reservoir) is positive and (*b*) show how the use of reservoirs at intermediate temperatures would allow the process to be carried out in a way that makes ΔS as small as desired.

36. One mole of an ideal monatomic gas is caused to go through the cycle shown in Fig. 19. (*a*) How much work is done on the gas in expanding the gas from *a* to *c* along path *abc*? (*b*) What is the change in internal energy and entropy in going from *b* to *c*? (*c*) What is the change in internal energy and entropy in going through one complete cycle? Express all answers in terms of the pressure p_0 and volume V_0 at point *a* in the diagram.

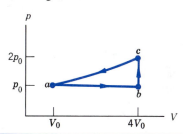

Figure 19   Problem 36.

Section 26-7 Entropy: Irreversible Processes

37. An ideal gas undergoes an isothermal expansion at 77°C, increasing its volume from 1.3 to 3.4 L. The entropy change of the gas is 24 J/K. How many moles of gas are present?

38. Suppose that the same amount of heat energy, say, 260 J, is transferred by conduction from a heat reservoir at a temperature of 400 K to another reservoir, the temperature of which is (*a*) 100 K, (*b*) 200 K, (*c*) 300 K, and (*d*) 360 K. Calculate the changes in entropy and discuss the trend.

39. A brass rod is in thermal contact with a heat reservoir at 130°C at one end and a heat reservoir at 24.0°C at the other end. (*a*) Compute the total change in the entropy arising from the process of conduction of 1200 J of heat through the rod. (*b*) Does the entropy of the rod change in the process?

40. One mole of an ideal diatomic gas is caused to pass through the cycle shown on the pV diagram in Fig. 20 where $V_2 = 3V_1$. Determine, in terms of p_1, V_1, T_1, and R: (a) p_2, p_3, and T_3 and (b) W, Q, ΔE_{int}, and ΔS for all three processes.

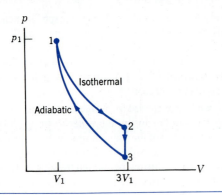

Figure 20   Problem 40.

41. One mole of a monatomic ideal gas is taken from an initial state of pressure p_0 and volume V_0 to a final state of pressure $2p_0$ and volume $2V_0$ by two different processes. (I) It expands isothermally until its volume is doubled, and then its pressure is increased at constant volume to the final state. (II) It is compressed isothermally until its pressure is doubled, and then its volume is increased at constant pressure to the final state. Show the path of each process on a pV diagram. For each process calculate in terms of p_0 and V_0: (a) the heat absorbed by the gas in each part of the process; (b) the work done on the gas in each part of the process;

(c) the change in internal energy of the gas, $E_{int,f} - E_{int,i}$; and (d) the change in entropy of the gas, $S_f - S_i$.

Section 26-8   Entropy and the Second Law

42. A 50.0-g block of copper having a temperature of 400 K is placed in an insulating box with a 100-g block of lead having a temperature of 200 K. (a) What is the equilibrium temperature of this two-block system? (b) What is the change in the internal energy of the two-block system as it changes from the initial condition to the equilibrium condition? (c) What is the change in the entropy of the two-block system? (See Table 1 in Chapter 25.)

43. A mixture of 1.78 kg of water and 262 g of ice at 0°C is, in a reversible process, brought to a final equilibrium state where the water/ice ratio, by mass, is 1 : 1 at 0°C. (a) Calculate the entropy change of the system during this process. (b) The system is then returned to the first equilibrium state, but in an irreversible way (by using a Bunsen burner, for instance). Calculate the entropy change of the system during this process. (c) Show that your answer is consistent with the second law of thermodynamics.

44. In a specific heat experiment, 196 g of aluminum at 107°C is mixed with 52.3 g of water at 18.6°C. (a) Calculate the equilibrium temperature. Find the entropy change of (b) the aluminum and (c) the water. (d) Calculate the entropy change of the system. (Hint: See Eqs. 29 and 30.)

45. A 12.6-g ice cube at −10.0°C is placed in a lake whose temperature is +15.0°C. Calculate the change in entropy of the system as the ice cube comes to thermal equilibrium with the lake. (Hint: Will the ice cube affect the temperature of the lake?)

APPENDIX A

THE INTERNATIONAL SYSTEM OF UNITS (SI)*

THE SI BASE UNITS

Quantity	Name	Symbol	Definition
length	meter	m	". . . the length of the path traveled by light in vacuum in 1/299,792,458 of a second." (1983)
mass	kilogram	kg	". . . this prototype [a certain platinum-iridium cylinder] shall henceforth be considered to be the unit of mass." (1889)
time	second	s	". . . the duration of 9,192,631,770 periods of the radiation corresponding to the transition between the two hyperfine levels of the ground state of the cesium-133 atom." (1967)
electric current	ampere	A	". . . that constant current which, if maintained in two straight parallel conductors of infinite length, of negligible circular cross section, and placed 1 meter apart in vacuum, would produce between these conductors a force equal to 2×10^{-7} newton per meter of length." (1946)
thermodynamic temperature	kelvin	K	". . . the fraction 1/273.16 of the thermodynamic temperature of the triple point of water." (1967)
amount of substance	mole	mol	". . . the amount of substance of a system which contains as many elementary entities as there are atoms in 0.012 kilogram of carbon 12." (1971)
luminous intensity	candela	cd	". . . the luminous intensity, in the perpendicular direction, of a surface of 1/600,000 square meter of a blackbody at the temperature of freezing platinum under a pressure of 101.325 newton per square meter." (1967)

* Adapted from "The International System of Units (SI)," National Bureau of Standards Special Publication 330, 1972 edition. The definitions above were adopted by the General Conference of Weights and Measures, an international body, on the dates shown. In this book we do not use the candela.

SOME SI DERIVED UNITS

Quantity	Name of Unit	Symbol	Equivalent
area	square meter	m^2	
volume	cubic meter	m^3	
frequency	hertz	Hz	s^{-1}
mass density (density)	kilogram per cubic meter	kg/m^3	
speed, velocity	meter per second	m/s	
angular velocity	radian per second	rad/s	
acceleration	meter per second squared	m/s^2	
angular acceleration	radian per second squared	rad/s^2	
force	newton	N	$kg \cdot m/s^2$
pressure	pascal	Pa	N/m^2
work, energy, quantity of heat	joule	J	$N \cdot m$
power	watt	W	J/s
quantity of electricity	coulomb	C	$A \cdot s$
potential difference, electromotive force	volt	V	$N \cdot m/C$
electric field	volt per meter	V/m	N/C
electric resistance	ohm	Ω	V/A
capacitance	farad	F	$A \cdot s/V$
magnetic flux	weber	Wb	$V \cdot s$
inductance	henry	H	$V \cdot s/A$
magnetic field	tesla	T	Wb/m^2, $N/A \cdot m$
entropy	joule per kelvin	J/K	
specific heat capacity	joule per kilogram kelvin	$J/(kg \cdot K)$	
thermal conductivity	watt per meter kelvin	$W/(m \cdot K)$	
radiant intensity	watt per steradian	W/sr	

THE SI SUPPLEMENTARY UNITS

Quantity	Name of Unit	Symbol
plane angle	radian	rad
solid angle	steradian	sr

APPENDIX B

SOME FUNDAMENTAL CONSTANTS OF PHYSICS

Constant	Symbol	Computational Value	Best (1986) value Value[a]	Uncertainty[b]
Speed of light in a vacuum	c	3.00×10^8 m/s	2.99792458	exact
Elementary charge	e	1.60×10^{-19} C	1.60217733	0.30
Electron rest mass	m_e	9.11×10^{-31} kg	9.1093897	0.59
Permittivity constant	ϵ_0	8.85×10^{-12} F/m	8.85418781762	exact
Permeability constant	μ_0	1.26×10^{-6} H/m	1.25663706143	exact
Electron rest mass[c]	m_e	5.49×10^{-4} u	5.48579902	0.023
Neutron rest mass[c]	m_n	1.0087 u	1.008664904	0.014
Hydrogen atom rest mass[c]	$m(^1H)$	1.0078 u	1.007825035	0.011
Deuterium atom rest mass[c]	$m(^2H)$	2.0141 u	2.014101779	0.012
Helium atom rest mass[c]	$m(^4He)$	4.0026 u	4.00260324	0.012
Electron charge-to-mass ratio	e/m_e	1.76×10^{11} C/kg	1.75881962	0.30
Proton rest mass	m_p	1.67×10^{-27} kg	1.6726231	0.59
Proton-to-electron mass ratio	m_p/m_e	1840	1836.152701	0.020
Neutron rest mass	m_n	1.67×10^{-27} kg	1.6749286	0.59
Muon rest mass	m_μ	1.88×10^{-28} kg	1.8835327	0.61
Planck constant	h	6.63×10^{-34} J·s	6.6260755	0.60
Electron Compton wavelength	λ_e	2.43×10^{-12} m	2.42631058	0.089
Universal gas constant	R	8.31 J/mol·K	8.314510	8.4
Avogadro constant	N_A	6.02×10^{23} mol^{-1}	6.0221367	0.59
Boltzmann constant	k	1.38×10^{-23} J/K	1.3806513	1.8
Molar volume of ideal gas at STP[d]	V_m	2.24×10^{-2} m^3/mol	2.2413992	1.7
Faraday constant	F	9.65×10^4 C/mol	9.6485309	0.30
Stefan-Boltzmann constant	σ	5.67×10^{-8} W/m^2·K^4	5.670399	6.8
Rydberg constant	R	1.10×10^7 m^{-1}	1.0973731571	0.00036
Gravitational constant	G	6.67×10^{-11} m^3/s^2·kg	6.67259	128
Bohr radius	a_0	5.29×10^{-11} m	5.29177249	0.045
Electron magnetic moment	μ_e	9.28×10^{-24} J/T	9.2847700	0.34
Proton magnetic moment	μ_p	1.41×10^{-26} J/T	1.41060761	0.34
Bohr magneton	μ_B	9.27×10^{-24} J/T	9.2740154	0.34
Nuclear magneton	μ_N	5.05×10^{-27} J/T	5.0507865	0.34
Fine structure constant	α	1/137	1/137.0359895	0.045
Magnetic flux quantum	Φ_0	2.07×10^{-15} Wb	2.06783461	0.30
Quantized Hall resistance	R_H	25800 Ω	25812.8056	0.045

[a] Same unit and power of ten as the computational value.

[b] Parts per million.

[c] Mass given in unified atomic mass units, where 1 u = $1.6605402 \times 10^{-27}$ kg.

[d] STP—standard temperature and pressure = 0°C and 1.0 bar.

APPENDIX C

SOME ASTRONOMICAL DATA

THE SUN, THE EARTH, AND THE MOON

Property	Sun[a]	Earth	Moon
Mass (kg)	1.99×10^{30}	5.98×10^{24}	7.36×10^{22}
Mean radius (m)	6.96×10^{8}	6.37×10^{6}	1.74×10^{6}
Mean density (kg/m³)	1410	5520	3340
Surface gravity (m/s²)	274	9.81	1.67
Escape velocity (km/s)	618	11.2	2.38
Period of rotation[c] (d)	$26-37^{b}$	0.997	27.3
Mean orbital radius (km)	2.6×10^{17d}	1.50×10^{8e}	3.82×10^{5f}
Orbital period	$2.4 \times 10^{8} \, \text{y}^{d}$	$1.00 \, \text{y}^{e}$	$27.3 \, \text{d}^{f}$

[a] The Sun radiates energy at the rate of 3.90×10^{26} W; just outside the Earth's atmosphere solar energy is received, assuming normal incidence, at the rate of 1380 W/m².

[b] The Sun—a ball of gas—does not rotate as a rigid body. Its rotational period varies between 26 d at the equator and 37 d at the poles.

[c] Measured with respect to the distant stars.

[d] About the galactic center.

[e] About the Sun.

[f] About the Earth.

SOME PROPERTIES OF THE PLANETS

	Mercury	Venus	Earth	Mars	Jupiter	Saturn	Uranus	Neptune	Pluto
Mean distance from Sun (10^6 km)	57.9	108	150	228	778	1,430	2,870	4,500	5,900
Period of revolution (y)	0.241	0.615	1.00	1.88	11.9	29.5	84.0	165	248
Period of rotation[a] (d)	58.7	243[b]	0.997	1.03	0.409	0.426	0.451[b]	0.658	6.39
Orbital speed (km/s)	47.9	35.0	29.8	24.1	13.1	9.64	6.81	5.43	4.74
Inclination of axis to orbit	0.0°	2.6°	23.5°	24.0°	3.08°	26.7°	82.1°	28.8°	65°
Inclination of orbit to Earth's orbit	7.00°	3.39°	—	1.85°	1.30°	2.49°	0.77°	1.77°	17.2°
Eccentricity of orbit	0.206	0.0068	0.0167	0.0934	0.0485	0.0556	0.0472	0.0086	0.250
Equatorial diameter (km)	4,880	12,100	12,800	6,790	143,000	120,000	51,800	49,500	3,400
Mass (Earth = 1)	0.0558	0.815	1.000	0.107	318	95.1	14.5	17.2	0.002
Mean density (g/cm³)	5.60	5.20	5.52	3.95	1.31	0.704	1.21	1.67	0.5(?)
Surface gravity[c] (m/s²)	3.78	8.60	9.78	3.72	22.9	9.05	7.77	11.0	0.03
Escape speed (km/s)	4.3	10.3	11.2	5.0	59.5	35.6	21.2	23.6	1.3
Known satellites	0	0	1	2	16 + rings	19 + rings	15 + rings	8 + rings	1

[a] Measured with respect to the distant stars.

[b] The sense of rotation is opposite to that of the orbital motion.

[c] Measured at the planet's equator.

APPENDIX D

PROPERTIES OF THE ELEMENTS

Element	Symbol	Atomic number, Z	Molar mass (g/mol)	Density (g/cm³) at 20°C	Melting point (°C)	Boiling point (°C)	Specific heat (J/g·C°) at 25°C
Actinium	Ac	89	(227)	—	1050	3200	0.092
Aluminum	Al	13	26.9815	2.699	660	2467	0.900
Americium	Am	95	(243)	13.7	994	2607	—
Antimony	Sb	51	121.75	6.69	630.5	1750	0.205
Argon	Ar	18	39.948	1.6626×10^{-3}	-189.2	-185.7	0.523
Arsenic	As	33	74.9216	5.72	817 (28 at.)	613	0.331
Astatine	At	85	(210)	—	302	337	—
Barium	Ba	56	137.33	3.5	725	1640	0.205
Berkelium	Bk	97	(247)	—	—	—	—
Beryllium	Be	4	9.0122	1.848	12.78	2970	1.83
Bismuth	Bi	83	208.980	9.75	271.3	1560	0.122
Boron	B	5	10.811	2.34	20.79	2550	1.11
Bromine	Br	35	79.909	3.12 (liquid)	-7.2	58	0.293
Cadmium	Cd	48	112.41	8.65	320.9	765	0.226
Calcium	Ca	20	40.08	1.55	839	1484	0.624
Californium	Cf	98	(251)	—	—	—	—
Carbon	C	6	12.011	2.25	3550	—	0.691
Cerium	Ce	58	140.12	6.768	798	3443	0.188
Cesium	Cs	55	132.905	1.873	28.40	6.69	0.243
Chlorine	Cl	17	35.453	3.214×10^{-3} (0°C)	-101	-34.6	0.486
Chromium	Cr	24	51.996	7.19	1857	2672	0.448
Cobalt	Co	27	58.9332	8.85	1495	2870	0.423
Copper	Cu	29	63.54	8.96	1083.4	2567	0.385
Curium	Cm	96	(247)	—	1340	—	—
Dysprosium	Dy	66	162.50	8.55	1412	2567	0.172
Einsteinium	Es	99	(252)	—	—	—	—
Erbium	Er	68	167.26	9.07	1529	2868	0.167
Europium	Eu	63	151.96	5.245	822	1527	0.163
Fermium	Fm	100	(257)	—	—	—	—
Fluorine	F	9	18.9984	1.696×10^{-3} (0°C)	-219.6	-188.2	0.753
Francium	Fr	87	(223)	—	(27)	(677)	—
Gadolinium	Gd	64	157.25	7.90	1313	3273	0.234
Gallium	Ga	31	69.72	5.907	29.78	2403	0.377
Germanium	Ge	32	72.61	5.323	937.4	2830	0.322
Gold	Au	79	196.967	19.32	1064.43	2808	0.131
Hafnium	Hf	72	178.49	13.31	2227	4602	0.144
Helium	He	2	4.0026	0.1664×10^{-3}	-272.2	-268.9	5.23
Holmium	Ho	67	164.930	8.79	1474	2700	0.165
Hydrogen	H	1	1.00797	0.08375×10^{-3}	-259.34	-252.87	14.4
Indium	In	49	114.82	7.31	156.6	2080	0.233
Iodine	I	53	126.9044	4.94	113.5	184.35	0.218
Iridium	Ir	77	192.2	22.5	2410	4130	0.130
Iron	Fe	26	55.847	7.87	1535	2750	0.447
Krypton	Kr	36	83.80	3.488×10^{-3}	-156.6	-152.3	0.247
Lanthanum	La	57	138.91	6.145	918	3464	0.195
Lawrencium	Lr	103	(260)	—	—	—	—

(Continued)

Element	Symbol	Atomic number, Z	Molar mass (g/mol)	Density (g/cm³) at 20°C	Melting point (°C)	Boiling point (°C)	Specific heat (J/g·C°) at 25°C
Lead	Pb	82	207.19	11.36	327.50	1740	0.129
Lithium	Li	3	6.939	0.534	180.54	1342	3.58
Lutetium	Lu	71	174.97	9.84	1663	3402	0.155
Magnesium	Mg	12	24.305	1.74	649	1090	1.03
Manganese	Mn	25	54.9380	7.43	1244	1962	0.481
Mendelevium	Md	101	(258)	—	—	—	—
Mercury	Hg	80	200.59	13.55	−38.87	357	0.138
Molybdenum	Mo	42	95.94	10.22	2617	4612	0.251
Neodymium	Nd	60	144.24	7.00	1021	3074	0.188
Neon	Ne	10	20.180	0.8387×10^{-3}	−248.67	−246.0	1.03
Neptunium	Np	93	(237)	20.25	640	3902	1.26
Nickel	Ni	28	58.69	8.902	1453	2732	0.444
Niobium	Nb	41	92.906	8.57	2468	4742	0.264
Nitrogen	N	7	14.0067	1.1649×10^{-3}	−210	−195.8	1.03
Nobelium	No	102	(259)	—	—	—	—
Osmium	Os	76	190.2	22.57	3045	5027	0.130
Oxygen	O	8	15.9994	1.3318×10^{-3}	−218.4	−183.0	0.913
Palladium	Pd	46	106.4	12.02	1554	3140	0.243
Phosphorus	P	15	30.9738	1.83	44.25	280	0.741
Platinum	Pt	78	195.09	21.45	1772	3827	0.134
Plutonium	Pu	94	(244)	19.84	641	3232	0.130
Polonium	Po	84	(209)	9.24	254	962	—
Potassium	K	19	39.098	0.86	63.25	760	0.758
Praseodymium	Pr	59	140.907	6.773	931	3520	0.197
Promethium	Pm	61	(145)	7.264	1042	(3000)	—
Protactinium	Pa	91	(231)	—	1600	—	—
Radium	Ra	88	(226)	5.0	700	1140	—
Radon	Rn	86	(222)	9.96×10^{-3} (0°C)	−71	−61.8	0.092
Rhenium	Re	75	186.2	21.04	3180	5627	0.134
Rhodium	Rh	45	102.905	12.44	1965	3727	0.243
Rubidium	Rb	37	85.47	1.53	38.89	686	0.364
Ruthenium	Ru	44	101.107	12.2	2310	3900	0.239
Samarium	Sm	62	150.35	7.49	1074	1794	0.197
Scandium	Sc	21	44.956	2.99	1541	2836	0.569
Selenium	Se	34	78.96	4.79	217	685	0.318
Silicon	Si	14	28.086	2.33	1410	2355	0.712
Silver	Ag	47	107.68	10.49	961.9	2212	0.234
Sodium	Na	11	22.9898	0.9712	97.81	882.9	1.23
Strontium	Sr	38	87.62	2.54	769	1384	0.737
Sulfur	S	16	32.066	2.07	112.8	444.6	0.707
Tantalum	Ta	73	180.948	16.6	2996	5425	0.138
Technetium	Tc	43	(98)	11.46	2172	4877	0.209
Tellurium	Te	52	127.60	6.24	449.5	990	0.201
Terbium	Tb	65	158.924	8.25	1357	3230	0.180
Thallium	Tl	81	204.38	11.85	304	1457	0.130
Thorium	Th	90	(232)	11.72	1750	(3850)	0.117
Thulium	Tm	69	168.934	9.31	1545	1950	0.159
Tin	Sn	50	118.71	7.31	231.97	2270	0.226
Titanium	Ti	22	4788	4.54	1660	3287	0.523
Tungsten	W	74	183.85	19.3	3410	5660	0.134
Uranium	U	92	(238)	19.07	1132	3818	0.117
Vanadium	V	23	50.942	6.1	1890	3380	0.490
Xenon	Xe	54	131.30	5.495×10^{-3}	−111.79	−108	0.159
Ytterbium	Yb	70	173.04	6.966	819	1196	0.155
Yttrium	Y	39	88.905	4.469	1552	5338	0.297
Zinc	Zn	30	65.37	7.133	419.58	907	0.389
Zirconium	Zr	40	91.22	6.506	1852	4377	0.276

The values in parentheses in the column of atomic masses are the mass numbers of the longest-lived isotopes of those elements that are radioactive. Melting points and boiling points in parentheses are uncertain.

All the physical properties are given for a pressure of one atmosphere except where otherwise specified.

The data for gases are valid only when these are in their usual molecular state, such as H_2, He, O_2, Ne, etc. The specific heats of the gases are the values at constant pressure.

Source: Handbook of Chemistry and Physics, 71st edition (CRC Press, 1990).

APPENDIX E

PERIODIC TABLE
OF THE
ELEMENTS

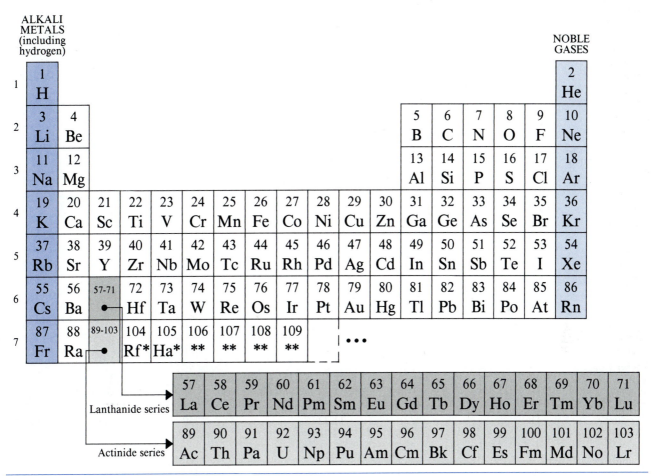

ALKALI METALS (including hydrogen)

NOBLE GASES

1 1 H																	**2** He
2 3 Li	4 Be											5 B	6 C	7 N	8 O	9 F	10 Ne
3 11 Na	12 Mg											13 Al	14 Si	15 P	16 S	17 Cl	18 Ar
4 19 K	20 Ca	21 Sc	22 Ti	23 V	24 Cr	25 Mn	26 Fe	27 Co	28 Ni	29 Cu	30 Zn	31 Ga	32 Ge	33 As	34 Se	35 Br	36 Kr
5 37 Rb	38 Sr	39 Y	40 Zr	41 Nb	42 Mo	43 Tc	44 Ru	45 Rh	46 Pd	47 Ag	48 Cd	49 In	50 Sn	51 Sb	52 Te	53 I	54 Xe
6 55 Cs	56 Ba	57-71 ●	72 Hf	73 Ta	74 W	75 Re	76 Os	77 Ir	78 Pt	79 Au	80 Hg	81 Tl	82 Pb	83 Bi	84 Po	85 At	86 Rn
7 87 Fr	88 Ra	89-103 ●	104 Rf*	105 Ha*	106 **	107 **	108 **	109 **									

● ● ●

Lanthanide series

57 La	58 Ce	59 Pr	60 Nd	61 Pm	62 Sm	63 Eu	64 Gd	65 Tb	66 Dy	67 Ho	68 Er	69 Tm	70 Yb	71 Lu

Actinide series

89 Ac	90 Th	91 Pa	92 U	93 Np	94 Pu	95 Am	96 Cm	97 Bk	98 Cf	99 Es	100 Fm	101 Md	102 No	103 Lr

* The names of these elements (Rutherfordium and Hahnium) have not been accepted because of conflicting claims of discovery. A group in the USSR has proposed the names Kurchatovium and Neilsbohrium.

** Discovery of these elements has been reported but names for them have not yet been adopted.

APPENDIX F

ELEMENTARY PARTICLES

1. THE FUNDAMENTAL PARTICLES

LEPTONS

Particle	Symbol	Anti-particle	Charge (e)	Spin ($h/2\pi$)	Rest energy (MeV)	Mean life (s)	Typical decay products
Electron	e^-	e^+	-1	1/2	0.511	∞	
Electron neutrino	ν_e	$\bar{\nu}_e$	0	1/2	<0.00002	∞	
Muon	μ^-	μ^+	-1	1/2	105.7	2.2×10^{-6}	$e^- + \bar{\nu}_e + \nu_\mu$
Muon neutrino	ν_μ	$\bar{\nu}_\mu$	0	1/2	<0.3	∞	
Tau	τ^-	τ^+	-1	1/2	1784	3.0×10^{-13}	$\mu^- + \bar{\nu}_\mu + \nu_\tau$
Tau neutrino	ν_τ	$\bar{\nu}_\tau$	0	1/2	<40	∞	

QUARKS

Flavor	Symbol	Antiparticle	Charge (e)	Spin ($h/2\pi$)	Rest energy[a] (MeV)	Other property
Up	u	$\bar{u}$	$+2/3$	1/2	300	$C = S = T = B = 0$
Down	d	$\bar{d}$	$-1/3$	1/2	300	$C = S = T = B = 0$
Charm	c	$\bar{c}$	$+2/3$	1/2	1500	Charm (C) $= +1$
Strange	s	$\bar{s}$	$-1/3$	1/2	500	Strangeness (S) $= -1$
Top[b]	t	$\bar{t}$	$+2/3$	1/2	$>40{,}000$	Topness (T) $= +1$
Bottom	b	$\bar{b}$	$-1/3$	1/2	4700	Bottomness (B) $= -1$

FIELD PARTICLES

Particle	Symbol	Interaction	Charge (e)	Spin ($h/2\pi$)	Rest energy (GeV)
Graviton[b]		Gravity	0	2	0
Weak boson	W^+, W^-	Weak	± 1	1	80.6
Weak boson	Z^0	Weak	0	1	91.2
Photon	γ	Electromagnetic	0	1	0
Gluon	g	Strong (color)	0	1	0

2. SOME COMPOSITE PARTICLES

BARYONS

Particle	Symbol	Quark content	Anti-particle	Charge (e)	Spin $(h/2\pi)$	Rest energy (MeV)	Mean life (s)	Typical decay
Proton	p	uud	$\bar{\text{p}}$	$+1$	1/2	938	$>10^{40}$	$\pi^0 + e^+$ (?)
Neutron	n	udd	$\bar{\text{n}}$	0	1/2	940	889	$p + e^- + \bar{\nu}_e$
Lambda	Λ^0	uds	$\overline{\Lambda^0}$	0	1/2	1116	2.6×10^{-10}	$p + \pi^-$
Omega	Ω^-	sss	$\overline{\Omega^-}$	-1	3/2	1673	8.2×10^{-11}	$\Lambda^0 + K^-$
Delta	Δ^{++}	uuu	$\overline{\Delta^{++}}$	$+2$	3/2	1232	5.7×10^{-24}	$p + \pi^+$
Charmed lambda	Λ_c^+	udc	$\overline{\Lambda_c^+}$	$+1$	1/2	2285	1.9×10^{-13}	$\Lambda^0 + \pi^+$

MESONS

Particle	Symbol	Quark content	Anti-particle	Charge (e)	Spin $(h/2\pi)$	Rest energy (MeV)	Mean life (s)	Typical decay
Pion	π^+	$u\bar{d}$	π^-	$+1$	0	140	2.6×10^{-8}	$\mu^+ + \nu_\mu$
Pion	π^0	$u\bar{u} + d\bar{d}$	π^0	0	0	135	8.4×10^{-17}	$\gamma + \gamma$
Kaon	K^+	$u\bar{s}$	K^-	$+1$	0	494	1.2×10^{-8}	$\mu^+ + \nu_\mu$
Kaon	K^0	$d\bar{s}$	$\overline{K^0}$	0	0	498	0.9×10^{-10}	$\pi^+ + \pi^-$
Rho	ρ^+	$u\bar{d}$	ρ^-	$+1$	1	768	4.5×10^{-24}	$\pi^+ + \pi^-$
D-meson	D^+	$c\bar{d}$	D^-	$+1$	0	1869	1.1×10^{-12}	$K^- + \pi^+ + \pi^+$
Psi	ψ	$c\bar{c}$	ψ	0	1	3097	1.0×10^{-20}	$e^+ + e^-$
B-meson	B^+	$u\bar{b}$	B^-	$+1$	0	5278	1.2×10^{-12}	$D^- + \pi^+ + \pi^+$
Upsilon	Y	$b\bar{b}$	Y	0	1	9460	1.3×10^{-20}	$e^+ + e^-$

[a] The rest energies listed for the quarks are not those associated with free quarks; since no free quarks have yet been observed, measuring their rest energies in the free state has not yet been possible. The tabulated values are effective rest energies corresponding to *constituent* quarks, those bound in composite particles.

[b] Particles expected to exist but not yet observed.

Source: "Review of Particle Properties," *Physics Letters B,* vol. 239 (April 1990).

APPENDIX G

CONVERSION FACTORS

Conversion factors may be read directly from the tables. For example, 1 degree = 2.778×10^{-3} revolutions, so $16.7° = 16.7 \times 2.778 \times 10^{-3}$ rev. The SI quantities are capitalized. Adapted in part from G. Shortley and D. Williams, *Elements of Physics,* Prentice-Hall, Englewood Cliffs, NJ, 1971.

PLANE ANGLE

	°	′	″	RADIAN	rev
1 degree =	1	60	3600	1.745×10^{-2}	2.778×10^{-3}
1 minute =	1.667×10^{-2}	1	60	2.909×10^{-4}	4.630×10^{-5}
1 second =	2.778×10^{-4}	1.667×10^{-2}	1	4.848×10^{-6}	7.716×10^{-7}
1 RADIAN =	57.30	3438	2.063×10^{5}	1	0.1592
1 revolution =	360	2.16×10^{4}	1.296×10^{6}	6.283	1

SOLID ANGLE

1 sphere = 4π steradians = 12.57 steradians

LENGTH

	cm	METER	km	in.	ft	mi
1 centimeter =	1	10^{-2}	10^{-5}	0.3937	3.281×10^{-2}	6.214×10^{-6}
1 METER =	100	1	10^{-3}	39.37	3.281	6.214×10^{-4}
1 kilometer =	10^{5}	1000	1	3.937×10^{4}	3281	0.6214
1 inch =	2.540	2.540×10^{-2}	2.540×10^{-5}	1	8.333×10^{-2}	1.578×10^{-5}
1 foot =	30.48	0.3048	3.048×10^{-4}	12	1	1.894×10^{-4}
1 mile =	1.609×10^{5}	1609	1.609	6.336×10^{4}	5280	1

1 angström = 10^{-10} m
1 nautical mile = 1852 m
  = 1.151 miles = 6076 ft
1 fermi = 10^{-15} m

1 light-year = 9.460×10^{12} km
1 parsec = 3.084×10^{13} km
1 fathom = 6 ft
1 Bohr radius = 5.292×10^{-11} m

1 yard = 3 ft
1 rod = 16.5 ft
1 mil = 10^{-3} in.
1 nm = 10^{-9} m

AREA

	METER^2	cm^2	ft^2	in.^{2}
1 SQUARE METER =	1	10^{4}	10.76	1550
1 square centimeter =	10^{-4}	1	1.076×10^{-3}	0.1550
1 square foot =	9.290×10^{-2}	929.0	1	144
1 square inch =	6.452×10^{-4}	6.452	6.944×10^{-3}	1

1 square mile = 2.788×10^{7} ft^2 = 640 acres
1 barn = 10^{-28} m^2

1 acre = 43,560 ft^2
1 hectare = 10^{4} m^2 = 2.471 acre

VOLUME

	METER^3	cm^3	L	ft^3	in.^{3}
1 CUBIC METER =	1	10^6	1000	35.31	6.102 × 10^4
1 cubic centimeter =	10^{-6}	1	1.000 × 10^{-3}	3.531 × 10^{-5}	6.102 × 10^{-2}
1 liter =	1.000 × 10^{-3}	1000	1	3.531 × 10^{-2}	61.02
1 cubic foot =	2.832 × 10^{-2}	2.832 × 10^4	28.32	1	1728
1 cubic inch =	1.639 × 10^{-5}	16.39	1.639 × 10^{-2}	5.787 × 10^{-4}	1

1 U.S. fluid gallon = 4 U.S. fluid quarts = 8 U.S. pints = 128 U.S. fluid ounces = 231 in.^{3}
1 British imperial gallon = 277.4 in^3 = 1.201 U.S. fluid gallons

MASS

	g	KILOGRAM	slug	u	oz	lb	ton
1 gram =	1	0.001	6.852 × 10^{-5}	6.022 × 10^{23}	3.527 × 10^{-2}	2.205 × 10^{-3}	1.102 × 10^{-6}
1 KILOGRAM =	1000	1	6.852 × 10^{-2}	6.022 × 10^{26}	35.27	2.205	1.102 × 10^{-3}
1 slug =	1.459 × 10^4	14.59	1	8.786 × 10^{27}	514.8	32.17	1.609 × 10^{-2}
1 u =	1.661 × 10^{-24}	1.661 × 10^{-27}	1.138 × 10^{-28}	1	5.857 × 10^{-26}	3.662 × 10^{-27}	1.830 × 10^{-30}
1 ounce =	28.35	2.835 × 10^{-2}	1.943 × 10^{-3}	1.718 × 10^{25}	1	6.250 × 10^{-2}	3.125 × 10^{-5}
1 pound =	453.6	0.4536	3.108 × 10^{-2}	2.732 × 10^{26}	16	1	0.0005
1 ton =	9.072 × 10^5	907.2	62.16	5.463 × 10^{29}	3.2 × 10^4	2000	1

1 metric ton = 1000 kg
Quantities in the colored areas are not mass units but are often used as such. When we write, for example, 1 kg "=" 2.205 lb this means that a kilogram is a *mass* that *weighs* 2.205 pounds under standard condition of gravity ($g = 9.80665$ m/s^2).

DENSITY

	slug/ft^3	KILOGRAM/METER^3	g/cm^3	lb/ft^3	lb/in.^{3}
1 slug per ft^3	1	515.4	0.5154	32.17	1.862 × 10^{-2}
1 KILOGRAM per METER^3 =	1.940 × 10^{-3}	1	0.001	6.243 × 10^{-2}	3.613 × 10^{-5}
1 gram per cm^3 =	1.940	1000	1	62.43	3.613 × 10^{-2}
1 pound per ft^3 =	3.108 × 10^{-2}	16.02	1.602 × 10^{-2}	1	5.787 × 10^{-4}
1 pound per in.^{3} =	53.71	2.768 × 10^4	27.68	1728	1

Quantities in the colored areas are weight densities and, as such, are dimensionally different from mass densities. See note for mass table.

TIME

	y	d	h	min	SECOND
1 year =	1	365.25	8.766 × 10^3	5.259 × 10^5	3.156 × 10^7
1 day =	2.738 × 10^{-3}	1	24	1440	8.640 × 10^4
1 hour =	1.141 × 10^{-4}	4.167 × 10^{-2}	1	60	3600
1 minute =	1.901 × 10^{-6}	6.944 × 10^{-4}	1.667 × 10^{-2}	1	60
1 SECOND =	3.169 × 10^{-8}	1.157 × 10^{-5}	2.778 × 10^{-4}	1.667 × 10^{-2}	1

SPEED

	ft/s	km/h	METER/SECOND	mi/h	cm/s
1 foot per second =	1	1.097	0.3048	0.6818	30.48
1 kilometer per hour =	0.9113	1	0.2778	0.6214	27.78
1 METER per SECOND =	3.281	3.6	1	2.237	100
1 mile per hour =	1.467	1.609	0.4470	1	44.70
1 centimeter per second =	3.281×10^{-2}	3.6×10^{-2}	0.01	2.237×10^{-2}	1

1 knot = 1 nautical mi/h = 1.688 ft/s   1 mi/min = 88.00 ft/s = 60.00 mi/h

FORCE

	dyne	NEWTON	lb	pdl	gf	kgf
1 dyne =	1	10^{-5}	2.248×10^{-6}	7.233×10^{-5}	1.020×10^{-3}	1.020×10^{-6}
1 NEWTON =	10^5	1	0.2248	7.233	102.0	0.1020
1 pound =	4.448×10^5	4.448	1	32.17	453.6	0.4536
1 poundal =	1.383×10^4	0.1383	3.108×10^{-2}	1	14.10	1.410×10^{-2}
1 gram-force =	980.7	9.807×10^{-3}	2.205×10^{-3}	7.093×10^{-2}	1	0.001
1 kilogram-force =	9.807×10^5	9.807	2.205	70.93	1000	1

Quantities in the colored areas are not force units but are often used as such. For instance, if we write 1 gram-force "=" 980.7 dynes, we mean that a gram-mass experiences a force of 980.7 dynes under standard conditions of gravity ($g = 9.80665$ m/s^2)

ENERGY, WORK, HEAT

	Btu	erg	ft·lb	hp·h	JOULE	cal	kW·h	eV	MeV	kg	u
1 British thermal unit =	1	1.055×10^{10}	777.9	3.929×10^{-4}	1055	252.0	2.930×10^{-4}	6.585×10^{21}	6.585×10^{15}	1.174×10^{-14}	7.070×10^{12}
1 erg =	9.481×10^{-11}	1	7.376×10^{-8}	3.725×10^{-14}	10^{-7}	2.389×10^{-8}	2.778×10^{-14}	6.242×10^{11}	6.242×10^5	1.113×10^{-24}	670.2
1 foot-pound =	1.285×10^{-3}	1.356×10^7	1	5.051×10^{-7}	1.356	0.3238	3.766×10^{-7}	8.464×10^{18}	8.464×10^{12}	1.509×10^{-17}	9.037×10^9
1 horsepower-hour =	2545	2.685×10^{13}	1.980×10^6	1	2.685×10^6	6.413×10^5	0.7457	1.676×10^{25}	1.676×10^{19}	2.988×10^{-11}	1.799×10^{16}
1 JOULE =	9.481×10^{-4}	10^7	0.7376	3.725×10^{-7}	1	0.2389	2.778×10^{-7}	6.242×10^{18}	6.242×10^{12}	1.113×10^{-17}	6.702×10^9
1 calorie =	3.969×10^{-3}	4.186×10^7	3.088	1.560×10^{-6}	4.186	1	1.163×10^{-6}	2.613×10^{19}	2.613×10^{13}	4.660×10^{-17}	2.806×10^{10}
1 kilowatt-hour =	3413	3.6×10^{13}	2.655×10^6	1.341	3.6×10^6	8.600×10^5	1	2.247×10^{25}	2.247×10^{19}	4.007×10^{-11}	2.413×10^{16}
1 electron volt =	1.519×10^{-22}	1.602×10^{-12}	1.182×10^{-19}	5.967×10^{-26}	1.602×10^{-19}	3.827×10^{-20}	4.450×10^{-26}	1	10^{-6}	1.783×10^{-36}	1.074×10^{-9}
1 million electron volts =	1.519×10^{-16}	1.602×10^{-6}	1.182×10^{-13}	5.967×10^{-20}	1.602×10^{-13}	3.827×10^{14}	4.450×10^{-20}	10^6	1	1.783×10^{-30}	1.074×10^{-3}
1 kilogram =	8.521×10^{13}	8.987×10^{23}	6.629×10^{16}	3.348×10^{10}	8.987×10^{16}	2.146×10^{16}	2.497×10^{10}	5.610×10^{35}	5.610×10^{29}	1	6.022×10^{26}
1 unified atomic mass unit =	1.415×10^{-13}	1.492×10^{-3}	1.101×10^{-10}	5.559×10^{-17}	1.492×10^{-10}	3.564×10^{-11}	4.146×10^{-17}	9.32×10^8	932.0	1.661×10^{-27}	1

Quantities in the colored areas are not properly energy units but are included for convenience. They arise from the relativistic mass-energy equivalence formula $E = mc^2$ and represent the energy equivalent of a mass of one kilogram or one unified atomic mass unit (u).

N/m^2

PRESSURE

	atm	dyne/cm²	inch of water	cm Hg	PASCAL	lb/in.²	lb/ft²
1 atmosphere =	1	1.013×10^6	406.8	76	1.013×10^5	14.70	2116
1 dyne per cm² =	9.869×10^{-7}	1	4.015×10^{-4}	7.501×10^{-5}	0.1	1.405×10^{-5}	2.089×10^{-3}
1 inch of water[a] at 4°C =	2.458×10^{-3}	2491	1	0.1868	249.1	3.613×10^{-2}	5.202
1 centimeter of mercury[a] at 0°C =	1.316×10^{-2}	1.333×10^4	5.353	1	1333	0.1934	27.85
1 PASCAL =	9.869×10^{-6}	10	4.015×10^{-3}	7.501×10^{-4}	1	1.450×10^{-4}	2.089×10^{-2}
1 pound per in.² =	6.805×10^{-2}	6.895×10^4	27.68	5.171	6.895×10^3	1	144
1 pound per ft² =	4.725×10^{-4}	478.8	0.1922	3.591×10^{-2}	47.88	6.944×10^{-3}	1

[a] Where the acceleration of gravity has the standard value 9.80665 m/s².
1 bar = 10^6 dyne/cm² = 0.1 MPa   1 millibar = 10^3 dyne/cm² = 10^2 Pa   1 torr = 1 millimeter of mercury

POWER

	Btu/h	ft·lb/s	hp	cal/s	kW	WATT
1 British thermal unit per hour =	1	0.2161	3.929×10^{-4}	6.998×10^{-2}	2.930×10^{-4}	0.2930
1 foot-pound per second =	4.628	1	1.818×10^{-3}	0.3239	1.356×10^{-3}	1.356
1 horsepower =	2545	550	1	178.1	0.7457	745.7
1 calorie per second =	14.29	3.088	5.615×10^{-3}	1	4.186×10^{-3}	4.186
1 kilowatt =	3413	737.6	1.341	238.9	1	1000
1 WATT =	3.413	0.7376	1.341×10^{-3}	0.2389	0.001	1

MAGNETIC FLUX

	maxwell	WEBER
1 maxwell =	1	10^{-8}
1 WEBER =	10^8	1

MAGNETIC FIELD

	gauss	TESLA	milligauss
1 gauss =	1	10^{-4}	1000
1 TESLA =	10^4	1	10^7
1 milligauss =	0.001	10^{-7}	1

1 tesla = 1 weber/meter²

APPENDIX H

MATHEMATICAL FORMULAS

GEOMETRY

Circle of radius r: circumference $= 2\pi r$; area $= \pi r^2$.

Sphere of radius r: area $= 4\pi r^2$; volume $= \frac{4}{3}\pi r^3$.

Right circular cylinder of radius r and height h: area $= 2\pi r^2 + 2\pi rh$; volume $= \pi r^2 h$.

Triangle of base a and altitude h: area $= \frac{1}{2}ah$.

QUADRATIC FORMULA

If $ax^2 + bx + c = 0$, then $x = \dfrac{-b \pm \sqrt{b^2 - 4ac}}{2a}$.

TRIGONOMETRIC FUNCTIONS OF ANGLE θ

$$\sin \theta = \frac{y}{r} \quad \cos \theta = \frac{x}{r}$$

$$\tan \theta = \frac{y}{x} \quad \cot \theta = \frac{x}{y}$$

$$\sec \theta = \frac{r}{x} \quad \csc \theta = \frac{r}{y}$$

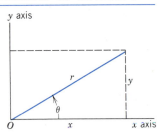

PYTHAGOREAN THEOREM

$a^2 + b^2 = c^2$

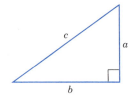

TRIANGLES

Angles A, B, C
Opposite sides a, b, c

$A + B + C = 180°$

$$\frac{\sin A}{a} = \frac{\sin B}{b} = \frac{\sin C}{c}$$

$c^2 = a^2 + b^2 - 2ab \cos C$

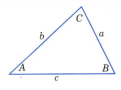

MATHEMATICAL SIGNS AND SYMBOLS

$=$ equals

$\approx$ equals approximately

$\sim$ is of the order of magnitude of

$\neq$ is not equal to

$\equiv$ is identical to, is defined as

$>$ is greater than ($\gg$ is much greater than)

$<$ is less than ($\ll$ is much less than)

$\geq$ is greater than or equal to (or, is no less than)

$\leq$ is less than or equal to (or, is no more than)

$\pm$ plus or minus ($\sqrt{4} = \pm 2$)

$\propto$ is proportional to

Σ the sum of

$\bar{x}$ the average value of x

PRODUCTS OF VECTORS

Let $\mathbf{i}$, $\mathbf{j}$, $\mathbf{k}$ be unit vectors in the x, y, z directions. Then

$$\mathbf{i} \cdot \mathbf{i} = \mathbf{j} \cdot \mathbf{j} = \mathbf{k} \cdot \mathbf{k} = 1, \quad \mathbf{i} \cdot \mathbf{j} = \mathbf{j} \cdot \mathbf{k} = \mathbf{k} \cdot \mathbf{i} = 0,$$

$$\mathbf{i} \times \mathbf{i} = \mathbf{j} \times \mathbf{j} = \mathbf{k} \times \mathbf{k} = 0,$$

$$\mathbf{i} \times \mathbf{j} = \mathbf{k}, \quad \mathbf{j} \times \mathbf{k} = \mathbf{i}, \quad \mathbf{k} \times \mathbf{i} = \mathbf{j}.$$

Any vector $\mathbf{a}$ with components a_x, a_y, a_z along the x, y, z axes can be written

$$\mathbf{a} = a_x \mathbf{i} + a_y \mathbf{j} + a_z \mathbf{k}.$$

Let $\mathbf{a}$, $\mathbf{b}$, $\mathbf{c}$ be arbitrary vectors with magnitudes a, b, c. Then

$$\mathbf{a} \times (\mathbf{b} + \mathbf{c}) = (\mathbf{a} \times \mathbf{b}) + (\mathbf{a} \times \mathbf{c})$$

$$(s\mathbf{a}) \times \mathbf{b} = \mathbf{a} \times (s\mathbf{b}) = s(\mathbf{a} \times \mathbf{b}) \quad (s = \text{a scalar}).$$

Let θ be the smaller of the two angles between $\mathbf{a}$ and $\mathbf{b}$. Then

$$\mathbf{a} \cdot \mathbf{b} = \mathbf{b} \cdot \mathbf{a} = a_x b_x + a_y b_y + a_z b_z = ab \cos \theta$$

$$\mathbf{a} \times \mathbf{b} = -\mathbf{b} \times \mathbf{a} = \begin{vmatrix} \mathbf{i} & \mathbf{j} & \mathbf{k} \\ a_x & a_y & a_z \\ b_x & b_y & b_z \end{vmatrix}$$

$$= (a_y b_z - b_y a_z)\mathbf{i} + (a_z b_x - b_z a_x)\mathbf{j} + (a_x b_y - b_x a_y)\mathbf{k}$$

$$|\mathbf{a} \times \mathbf{b}| = ab \sin \theta$$

$$\mathbf{a} \cdot (\mathbf{b} \times \mathbf{c}) = \mathbf{b} \cdot (\mathbf{c} \times \mathbf{a}) = \mathbf{c} \cdot (\mathbf{a} \times \mathbf{b})$$

$$\mathbf{a} \times (\mathbf{b} \times \mathbf{c}) = (\mathbf{a} \cdot \mathbf{c})\mathbf{b} - (\mathbf{a} \cdot \mathbf{b})\mathbf{c}$$

TRIGONOMETRIC IDENTITIES

$\sin(90° - \theta) = \cos\theta$

$\cos(90° - \theta) = \sin\theta$

$\sin\theta/\cos\theta = \tan\theta$

$\sin^2\theta + \cos^2\theta = 1 \quad \sec^2\theta - \tan^2\theta = 1 \quad \csc^2\theta - \cot^2\theta = 1$

$\sin 2\theta = 2\sin\theta\cos\theta$

$\cos 2\theta = \cos^2\theta - \sin^2\theta = 2\cos^2\theta - 1 = 1 - 2\sin^2\theta$

$\sin(\alpha \pm \beta) = \sin\alpha\cos\beta \pm \cos\alpha\sin\beta$

$\cos(\alpha \pm \beta) = \cos\alpha\cos\beta \mp \sin\alpha\sin\beta$

$\tan(\alpha \pm \beta) = \dfrac{\tan\alpha \pm \tan\beta}{1 \mp \tan\alpha\tan\beta}$

$\sin\alpha \pm \sin\beta = 2\sin\tfrac{1}{2}(\alpha \pm \beta)\cos\tfrac{1}{2}(\alpha \mp \beta)$

BINOMIAL THEOREM

$(1 \pm x)^n = 1 \pm \dfrac{nx}{1!} + \dfrac{n(n-1)x^2}{2!} + \cdots \quad (x^2 < 1)$

$(1 \pm x)^{-n} = 1 \mp \dfrac{nx}{1!} + \dfrac{n(n+1)x^2}{2!} + \cdots \quad (x^2 < 1)$

EXPONENTIAL EXPANSION

$e^x = 1 + x + \dfrac{x^2}{2!} + \dfrac{x^3}{3!} + \cdots$

LOGARITHMIC EXPANSION

$\ln(1 + x) = x - \tfrac{1}{2}x^2 + \tfrac{1}{3}x^3 - \cdots \quad (|x| < 1)$

TRIGONOMETRIC EXPANSIONS (θ in radians)

$\sin\theta = \theta - \dfrac{\theta^3}{3!} + \dfrac{\theta^5}{5!} - \cdots$

$\cos\theta = 1 - \dfrac{\theta^2}{2!} + \dfrac{\theta^4}{4!} - \cdots$

$\tan\theta = \theta + \dfrac{\theta^3}{3} + \dfrac{2\theta^5}{15} + \cdots$

DERIVATIVES AND INTEGRALS

In what follows, the letters u and v stand for any functions of x, and a and m are constants. To each of the indefinite integrals should be added an arbitrary constant of integration. The *Handbook of Chemistry and Physics* (CRC Press Inc.) gives a more extensive tabulation.

1. $\dfrac{dx}{dx} = 1$

2. $\dfrac{d}{dx}(au) = a\dfrac{du}{dx}$

3. $\dfrac{d}{dx}(u + v) = \dfrac{du}{dx} + \dfrac{dv}{dx}$

4. $\dfrac{d}{dx}x^m = mx^{m-1}$

5. $\dfrac{d}{dx}\ln x = \dfrac{1}{x}$

6. $\dfrac{d}{dx}(uv) = u\dfrac{dv}{dx} + v\dfrac{du}{dx}$

7. $\dfrac{d}{dx}e^x = e^x$

8. $\dfrac{d}{dx}\sin x = \cos x$

9. $\dfrac{d}{dx}\cos x = -\sin x$

10. $\dfrac{d}{dx}\tan x = \sec^2 x$

11. $\dfrac{d}{dx}\cot x = -\csc^2 x$

12. $\dfrac{d}{dx}\sec x = \tan x \sec x$

13. $\dfrac{d}{dx}\csc x = -\cot x \csc x$

14. $\dfrac{d}{dx}e^u = e^u\dfrac{du}{dx}$

15. $\dfrac{d}{dx}\sin u = \cos u\dfrac{du}{dx}$

16. $\dfrac{d}{dx}\cos u = -\sin u\dfrac{du}{dx}$

1. $\displaystyle\int dx = x$

2. $\displaystyle\int au\,dx = a\int u\,dx$

3. $\displaystyle\int (u + v)\,dx = \int u\,dx + \int v\,dx$

4. $\displaystyle\int x^m\,dx = \dfrac{x^{m+1}}{m+1} \quad (m \neq -1)$

5. $\displaystyle\int \dfrac{dx}{x} = \ln|x|$

6. $\displaystyle\int u\dfrac{dv}{dx}\,dx = uv - \int v\dfrac{du}{dx}\,dx$

7. $\displaystyle\int e^x\,dx = e^x$

8. $\displaystyle\int \sin x\,dx = -\cos x$

9. $\displaystyle\int \cos x\,dx = \sin x$

10. $\displaystyle\int \tan x\,dx = \ln|\sec x|$

11. $\displaystyle\int \sin^2 x\,dx = \tfrac{1}{2}x - \tfrac{1}{4}\sin 2x$

12. $\displaystyle\int e^{-ax}\,dx = -\dfrac{1}{a}e^{-ax}$

13. $\displaystyle\int xe^{-ax}\,dx = -\dfrac{1}{a^2}(ax + 1)e^{-ax}$

14. $\displaystyle\int x^2 e^{-ax}\,dx = -\dfrac{1}{a^3}(a^2x^2 + 2ax + 2)e^{-ax}$

15. $\displaystyle\int_0^\infty x^n e^{-ax}\,dx = \dfrac{n!}{a^{n+1}}$

16. $\displaystyle\int_0^\infty x^{2n} e^{-ax^2}\,dx = \dfrac{1 \cdot 3 \cdot 5 \cdots (2n-1)}{2^{n+1}a^n}\sqrt{\dfrac{\pi}{a}}$

APPENDIX I

COMPUTER PROGRAMS

Three examples are given of computer programs that have been used in the text for kinematic calculations involving nonconstant forces acting on a particle. The programs are written in the BASIC language and can easily be adapted to most personal computers. In each case, the initial velocity and position of the particle must be put into the programs at lines 40 and 50, respectively.

1. TIME-DEPENDENT FORCES

This program was used in Section 6–6 to find the position and velocity of a car whose acceleration depends on the time. The program can be used for any time-dependent acceleration by changing line 180 so that it shows the desired $a(t)$. In this case we use the example of Section 6–6, $a(t) = -2.67t$.

PROGRAM LISTING

```
10  ' BASIC KINEMATICS PROGRAM -- TIME DEPENDENT FORCES
20  ' GIVEN A(T), V0, X0;  COMPUTES V(T), X(T)
30  ' SPECIFY INITIAL VALUES
40  VO = 29.2
50  X0 = 0
60  'SPECIFY THE MAXIMUM NUMBER OF TIME UNITS
70  '     FOR WHICH THE PROGRAM SHOULD RUN
80  TMAX = 10
90  'SPECIFY THE VALUE OF ONE TIME UNIT
100 '     EXAMPLE: 0.5 FOR 0.5 SECOND
110 '     EXAMPLE: 2.0 FOR 2.0 HOUR
120 TU = .5
130 'SPECIFY THE NUMBER OF INTERVALS DT
140 '     INTO WHICH EACH TIME UNIT IS DIVIDED
150 NT = 100
160 DT = TU/NT
170 'INSERT A(T) IMMEDIATELY AFTER DEF FN IN NEXT STATEMENT
180 DEF FNA(T)=-2.67*T
190 V=V0
200 X=X0
210 PRINT "TIME          VELOCITY        POSITION"
220 LPRINT   "TIME          VELOCITY        POSITION"
230 'BEGIN ITERATION
240 FOR TIME = 1 TO TMAX
250 FOR N = 1 TO NT
260 T = (TIME-1)*TU + N*DT
270 AI=FNA(T)
280 AI1=FNA(T-DT)
290 AV=.5*(AI+AI1)
```

(Continued)

```
300 DV=AV*DT
310 V = V + DV
320 DX = .5*(V + V - DV)*DT
330 X = X+DX
340 NEXT N
350 PRINT TIME*TU,V,X
360 LPRINT TIME*TU,V,X
370 NEXT TIME
400 END
```

SAMPLE OUTPUT

```
TIME        VELOCITY        POSITION
.5          28.86625        14.54437
1           27.86499        28.75499
1.5         26.19618        42.2981
2           23.85994        54.83993
2.5         20.85615        66.04676
3           17.18486        75.5848
3.5         12.84612        83.1203
4           7.839843        88.31959
4.5         2.166084        90.84886
5          -4.175166        90.37431
```

2. VELOCITY-DEPENDENT FORCES

This program can be used as described in Section 6–7 to analyze the motion of a projectile subjected to a drag force. In this case the force is written in line 200 as $F(x) = g - bv^2$, with $g = 9.8$ and $b = 0.33$. Other expressions for velocity-dependent forces can be substituted at line 200. The output shows that the particle reaches its terminal velocity of about 5.4 m/s after a time of 1.5 s, during which it has traveled about 6 m.

PROGRAM LISTING

```
10 ' BASIC KINEMATICS PROGRAM -- VELOCITY DEPENDENT FORCES
20 ' GIVEN A(V), V0, X0;   COMPUTES V(T), X(T)
30 'SPECIFY INITIAL VALUES
40 VO = 0
50 XO = 0
60 'SPECIFY THE MAXIMUM NUMBER OF TIME UNITS
70 '     FOR WHICH THE PROGRAM SHOULD RUN
80 TMAX = 10
90 'SPECIFY THE VALUE OF ONE TIME UNIT
100 '     EXAMPLE: 0.5 FOR 0.5 SECOND
110 '     EXAMPLE: 2.0 FOR 2.0 HOUR
120 TU=.25
130 'SPECIFY THE NUMBER OF INTERVALS DT
140 '     INTO WHICH EACH TIME UNIT IS DIVIDED
150 NT=100
160 DT=TU/NT
170 V=V0
180 X=X0
190 'INSERT A(V) IMMEDIATELY AFTER DEF FN IN NEXT STATEMENT
200 DEF FNA(V)=9.8 - .33*V*V
210 PRINT "TIME          VELOCITY        POSITION"
220 LPRINT "TIME          VELOCITY        POSITION"
230 'BEGIN ITERATION
240 FOR TIME = 1 TO TMAX
250 FOR N = 1 TO NT
```

(Continued)

```
260 AV=FNA(V)
270 DV=AV*DT
280 V = V + DV
290 DX = .5*(V + V - DV)*DT
300 X = X+DX
310 NEXT N
320 PRINT TIME*TU,V,X
330 LPRINT TIME*TU,V,X
340 NEXT TIME
400 END
```

SAMPLE OUTPUT

```
TIME        VELOCITY        POSITION
 .25        2.299237        .2966358
 .5         3.905542        1.08959
 .75        4.765719        2.18636
1           5.161553        3.434
1.25        5.330923        4.748592
1.5         5.401125        6.091382
1.75        5.42984         7.445783
2           5.441519        8.804918
2.25        5.446261        10.16598
2.5         5.448183        11.52782
```

3. POSITION-DEPENDENT FORCES

This program was used in Section 8–4 to analyze the motion of an oscillating particle subject to a force $F = -kx$. The force is inserted in line 200 as $F(x) = -9.6x$, so that $k = 9.6$. The output shows that the particle oscillates with a period of 3.2 s, as expected for a particle of this mass (line 60).

PROGRAM LISTING

```
10  ' BASIC KINEMATICS PROGRAM -- POSITION DEPENDENT FORCES
20  ' GIVEN F(X), V0, X0, M;   COMPUTES V(T), X(T)
30  'SPECIFY INITIAL VALUES AND MASS OF PARTICLE
40  V0 = 0      'METERS PER SECOND
50  X0 = .05         'METERS
60  M = 2.5 'KILOGRAMS'
70  'SPECIFY THE MAXIMUM NUMBER OF TIME UNITS
80  '     FOR WHICH THE PROGRAM SHOULD RUN
90  TMAX = 40
100 'SPECIFY THE VALUE OF ONE TIME UNIT
110 '     EXAMPLE: 0.5 FOR 0.5 SECOND
120 TU=.1
130 'SPECIFY THE NUMBER OF INTERVALS DT
140 '      INTO WHICH EACH TIME UNIT IS DIVIDED
150 NT = 10
160 DT = TU/NT
170 V=V0
180 X=X0
190 'INSERT F(X) IMMEDIATELY AFTER DEF FN IN NEXT STATEMENT
200 DEF FNF(X)=-9.600001*X
210 PRINT " TIME          VELOCITY        POSITION"
220 PRINT "  (S)           (M/S)            (M) "
230 LPRINT " TIME          VELOCITY        POSITION"
240 LPRINT "  (S)           (M/S)            (M) "
250 LPRINT USING "###.##";TIME;:PRINT USING  "+##########.###";V0,X0
260 PRINT USING "###.##";TIME;:PRINT USING  "+##########.###";V0,X0
```

(Continued)

```
270  'BEGIN ITERATION
280  FOR TIME = 1 TO TMAX
290  FOR N = 1 TO NT
300  A=FNF(X)/M 'ACCELERATION IN INTERVAL
310  X = X + V*DT + .5*A*DT*DT 'POSITION AT END OF INTERVAL
320  V = V + A*DT 'VELOCITY AT END OF INTERVAL
330  NEXT N
340  PRINT USING "###.##";TIME*TU;:PRINT USING "+#########.###";V,X
350  LPRINT USING "###.##";TIME*TU;:LPRINT USING "+#########.###";V,X
360  NEXT TIME
400  END
```

SAMPLE OUTPUT

TIME (S)	VELOCITY (M/S)	POSITION (M)
0.00	+0.000	+0.050
0.10	-0.019	+0.049
0.20	-0.037	+0.046
0.30	-0.054	+0.042
0.40	-0.069	+0.035
0.50	-0.082	+0.028
0.60	-0.091	+0.019
0.70	-0.097	+0.010
0.80	-0.099	-0.000
0.90	-0.097	-0.010
1.00	-0.092	-0.019
1.10	-0.083	-0.028
1.20	-0.070	-0.036
1.30	-0.056	-0.042
1.40	-0.039	-0.047
1.50	-0.020	-0.050
1.60	-0.001	-0.051
1.70	+0.019	-0.050
1.80	+0.037	-0.047
1.90	+0.055	-0.042
2.00	+0.070	-0.036
2.10	+0.083	-0.028
2.20	+0.092	-0.020
2.30	+0.098	-0.010
2.40	+0.100	-0.000
2.50	+0.099	+0.010
2.60	+0.093	+0.019
2.70	+0.084	+0.028
2.80	+0.072	+0.036
2.90	+0.057	+0.043
3.00	+0.040	+0.047
3.10	+0.021	+0.050
3.20	+0.001	+0.052
3.30	-0.018	+0.051
3.40	-0.037	+0.048
3.50	-0.055	+0.043
3.60	-0.071	+0.037
3.70	-0.084	+0.029
3.80	-0.093	+0.020
3.90	-0.099	+0.011
4.00	-0.102	+0.001

APPENDIX J

NOBEL PRIZES IN PHYSICS*

1901	Wilhelm Konrad Röntgen	1845–1923	for the discovery of x-rays
1902	Hendrik Antoon Lorentz	1853–1928	for their researches into the influence of magnetism upon radiation
	Pieter Zeeman	1865–1943	phenomena
1903	Antoine Henri Becquerel	1852–1908	for his discovery of spontaneous radioactivity
	Pierre Curie	1859–1906	for their joint researches on the radiation phenomena discovered by
	Marie Sklowdowska-Curie	1867–1934	Professor Henri Becquerel
1904	Lord Rayleigh	1842–1919	for his investigations of the densities of the most important gases and
	(John William Strutt)		for his discovery of argon
1905	Philipp Eduard Anton von Lenard	1862–1947	for his work on cathode rays
1906	Joseph John Thomson	1856–1940	for his theoretical and experimental investigations on the conduction of electricity by gases
1907	Albert Abraham Michelson	1852–1931	for his optical precision instruments and metrological investigations carried out with their aid
1908	Gabriel Lippmann	1845–1921	for his method of reproducing colors photographically based on the phenomena of interference
1909	Guglielmo Marconi	1874–1937	for their contributions to the development of wireless
	Carl Ferdinand Braun	1850–1918	telegraphy
1910	Johannes Diderik van der Waals	1837–1932	for his work on the equation of state for gases and liquids
1911	Wilhelm Wien	1864–1928	for his discoveries regarding the laws governing the radiation of heat
1912	Nils Gustaf Dalén	1869–1937	for his invention of automatic regulators for use in conjunction with gas accumulators for illuminating lighthouses and buoys
1913	Heike Kamerlingh Onnes	1853–1926	for his investigations of the properties of matter at low temperatures which led, *inter alia,* to the production of liquid helium
1914	Max von Laue	1879–1960	for his discovery of the diffraction of Röntgen rays by crystals
1915	William Henry Bragg	1862–1942	for their services in the analysis of crystal structure by means of
	William Lawrence Bragg	1890–1971	x-rays
1917	Charles Glover Barkla	1877–1944	for his discovery of the characteristic x-rays of the elements
1918	Max Planck	1858–1947	for his discovery of energy quanta
1919	Johannes Stark	1874–1957	for his discovery of the Doppler effect in canal rays and the splitting of spectral lines in electric fields
1920	Charles-Édouard Guillaume	1861–1938	for the service he has rendered to precision measurements in Physics by his discovery of anomalies in nickel steel alloys
1921	Albert Einstein	1879–1955	for his services to Theoretical Physics, and especially for his discovery of the law of the photoelectric effect
1922	Neils Bohr	1885–1962	for the investigation of the structure of atoms, and of the radiation emanating from them
1923	Robert Andrews Millikan	1868–1953	for his work on the elementary charge of electricity and on the photoelectric effect
1924	Karl Manne Georg Siegbahn	1888–1979	for his discoveries and research in the field of x-ray spectroscopy
1925	James Franck	1882–1964	for their discovery of the laws governing the impact of an electron
	Gustav Hertz	1887–1975	upon an atom
1926	Jean Baptiste Perrin	1870–1942	for his work on the discontinuous structure of matter, and especially for his discovery of sedimentation equilibrium

* See *Nobel Lectures, Physics,* 1901–1970, Elsevier Publishing Company for biographies of the awardees and for lectures given by them on receiving the prize.

1927	Arthur Holly Compton	1892–1962	for his discovery of the effect named after him
	Charles Thomson Rees Wilson	1869–1959	for his method of making the paths of electrically charged particles visible by condensation of vapor
1928	Owen Willans Richardson	1879–1959	for his work on the thermionic phenomenon and especially for the discovery of the law named after him
1929	Prince Louis-Victor de Broglie	1892–1987	for his discovery of the wave nature of electrons
1930	Sir Chandrasekhara Venkata Raman	1888–1970	for his work on the scattering of light and for the discovery of the effect named after him
1932	Werner Heisenberg	1901–1976	for the creation of quantum mechanics, the application of which has, among other things, led to the discovery of the allotropic forms of hydrogen
1933	Erwin Schrödinger	1887–1961	for the discovery of new productive forms of atomic theory
	Paul Adrien Maurice Dirac	1902–1984	
1935	James Chadwick	1891–1974	for his discovery of the neutron
1936	Victor Franz Hess	1883–1964	for the discovery of cosmic radiation
	Carl David Anderson	1905–1991	for his discovery of the positron
1937	Clinton Joseph Davisson	1881–1958	for their experimental discovery of the diffraction of electrons
	George Paget Thomson	1892–1975	by crystals
1938	Enrico Fermi	1901–1954	for his demonstrations of the existence of new radioactive elements produced by neutron irradiation, and for his related discovery of nuclear reactions brought about by slow neutrons
1939	Ernest Orlando Lawrence	1901–1958	for the invention and development of the cyclotron and for results obtained with it, especially for artificial radioactive elements
1943	Otto Stern	1888–1969	for his contribution to the development of the molecular ray method and his discovery of the magnetic moment of the proton
1944	Isidor Isaac Rabi	1898–1988	for his resonance method for recording the magnetic properties of atomic nuclei
1945	Wolfgang Pauli	1900–1958	for the discovery of the Exclusion Principle (Pauli Principle)
1946	Percy Williams Bridgman	1882–1961	for the invention of an apparatus to produce extremely high pressures, and for the discoveries he made therewith in the field of high-pressure physics
1947	Sir Edward Victor Appleton	1892–1965	for his investigations of the physics of the upper atmosphere, especially for the discovery of the so-called Appleton layer
1948	Patrick Maynard Stuart Blackett	1897–1974	for his development of the Wilson cloud chamber method, and his discoveries therewith in nuclear physics and cosmic radiation
1949	Hideki Yukawa	1907–1981	for his prediction of the existence of mesons on the basis of theoretical work on nuclear forces
1950	Cecil Frank Powell	1903–1969	for his development of the photographic method of studying nuclear processes and his discoveries regarding mesons made with this method
1951	Sir John Douglas Cockcroft	1897–1967	for their pioneer work on the transmutation of atomic nuclei by
	Ernest Thomas Sinton Walton	1903–	artificially accelerated atomic particles
1952	Felix Bloch	1905–1983	for their development of new methods for nuclear magnetic precision
	Edward Mills Purcell	1912–	methods and discoveries in connection therewith
1953	Frits Zernike	1888–1966	for his demonstration of the phase-contrast method, especially for his invention of the phase-contrast microscope
1954	Max Born	1882–1970	for his fundamental research in quantum mechanics, especially for his statistical interpretation of the wave function
	Walther Bothe	1891–1957	for the coincidence method and his discoveries made therewith
1955	Willis Eugene Lamb	1913–	for his discoveries concerning the fine structure of the hydrogen spectrum
	Polykarp Kusch	1911–	for his precision determination of the magnetic moment of the electron
1956	William Shockley	1910–1989	for their researches on semiconductors and their discovery of the
	John Bardeen	1908–1991	transistor effect
	Walter Houser Brattain	1902–1987	
1957	Chen Ning Yang	1922–	for their penetrating investigation of the parity laws which has led to
	Tsung Dao Lee	1926–	important discoveries regarding the elementary particles
1958	Pavel Aleksejeciĉ Ĉerenkov	1904–	for the discovery and the interpretation of the Cerenkov effect
	Il' ja Michajloviĉ Frank	1908–1990	
	Igor' Evgen' eviĉ Tamm	1895–1971	
1959	Emilio Gino Segrè	1905–1989	for their discovery of the antiproton
	Owen Chamberlain	1920–	
1960	Donald Arthur Glaser	1926–	for the invention of the bubble chamber
1961	Robert Hofstadter	1915–1990	for his pioneering studies of electron scattering in atomic nuclei and for his thereby achieved discoveries concerning the structure of the nucleons
	Rudolf Ludwig Mössbauer	1929–	for his researches concerning the resonance absorption of γ-rays and his discovery in this connection of the effect which bears his name

1962	Lev Davidoviĉ Landau	1908–1968	for his pioneering theories of condensed matter, especially liquid helium
1963	Eugene P. Wigner	1902–	for his contribution to the theory of the atomic nucleus and the elementary particles, particularly through the discovery and application of fundamental symmetry principles
	Maria Goeppert Mayer	1906–1972	for their discoveries concerning nuclear shell structure
	J. Hans D. Jensen	1907–1973	
1964	Charles H. Townes	1915–	for fundamental work in the field of quantum electronics which has led to the construction of oscillators and amplifiers based on the maser-laser principle
	Nikolai G. Basov	1922–	
	Alexander M. Prochorov	1916–	
1965	Sin-itiro Tomonaga	1906–1979	for their fundamental work in quantum electrodynamics, with deep-ploughing consequences for the physics of elementary particles
	Julian Schwinger	1918–	
	Richard P. Feynman	1918–1988	
1966	Alfred Kastler	1902–1984	for the discovery and development of optical methods for studying Hertzian resonance in atoms
1967	Hans Albrecht Bethe	1906–	for his contributions to the theory of nuclear reactions, especially his discoveries concerning the energy production in stars
1968	Luis W. Alvarez	1911–1988	for his decisive contribution to elementary particle physics, in particular the discovery of a large number of resonance states, made possible through his development of the technique of using hydrogen bubble chamber and data analysis
1969	Murray Gell-Mann	1929–	for his contribution and discoveries concerning the classification of elementary particles and their interactions
1970	Hannes Alvén	1908–	for fundamental work and discoveries in magneto-hydrodynamics with fruitful applications in different parts of plasma physics
	Louis Néel	1904–	for fundamental work and discoveries concerning antiferromagnetism and ferrimagnetism which have led to important applications in solid state physics
1971	Dennis Gabor	1900–1979	for his discovery of the principles of holography
1972	John Bardeen	1908–1991	for their development of a theory of superconductivity
	Leon N. Cooper	1930–	
	J. Robert Schrieffer	1931–	
1973	Leo Esaki	1925–	for his discovery of tunneling in semiconductors
	Ivar Giaever	1929–	for his discovery of tunneling in superconductors
	Brian D. Josephson	1940–	for his theoretical prediction of the properties of a super-current through a tunnel barrier
1974	Antony Hewish	1924–	for the discovery of pulsars
	Sir Martin Ryle	1918–1984	for his pioneering work in radioastronomy
1975	Aage Bohr	1922–	for the discovery of the connection between collective motion and particle motion and the development of the theory of the structure of the atomic nucleus based on this connection
	Ben Mottelson	1926–	
	James Rainwater	1917–1986	
1976	Burton Richter	1931–	for their (independent) discovery of an important fundamental particle
	Samuel Chao Chung Ting	1936–	
1977	Philip Warren Anderson	1923–	for their fundamental theoretical investigations of the electronic structure of magnetic and disordered systems
	Nevill Francis Mott	1905–	
	John Hasbrouck Van Vleck	1899–1980	
1978	Peter L. Kapitza	1894–1984	for his basic inventions and discoveries in low-temperature physics
	Arno A. Penzias	1926–	for their discovery of cosmic microwave background radiation
	Robert Woodrow Wilson	1936–	
1979	Sheldon Lee Glashow	1932–	for their unified model of the action of the weak and electromagnetic forces and for their prediction of the existence of neutral currents
	Abdus Salam	1926–	
	Steven Weinberg	1933–	
1980	James W. Cronin	1931–	for the discovery of violations of fundamental symmetry principles in the decay of neutral K mesons
	Val L. Fitch	1923–	
1981	Nicolaas Bloembergen	1920–	for their contribution to the development of laser spectroscopy
	Arthur Leonard Schawlow	1921–	
	Kai M. Siegbahn	1918–	for his contribution of high-resolution electron spectroscopy
1982	Kenneth Geddes Wilson	1936–	for his method of analyzing the critical phenomena inherent in the changes of matter under the influence of pressure and temperature
1983	Subrehmanyan Chandrasekhar	1910–	for his theoretical studies of the structure and evolution of stars
	William A. Fowler	1911–	for his studies of the formation of the chemical elements in the universe
1984	Carlo Rubbia	1934–	for their decisive contributions to the large project, which led to the discovery of the field particles W and Z, communicators of the weak interaction
	Simon van der Meer	1925–	
1985	Klaus von Klitzing	1943–	for his discovery of the quantized Hall resistance
1986	Ernst Ruska	1906–	for his invention of the electron microscope
	Gerd Binnig	1947–	for their invention of the scanning-tunneling electron microscope
	Heinrich Rohrer	1933–	

1987	Karl Alex Müller	1927–	for their discovery of a new class of superconductors
	J. George Bednorz	1950–	
1988	Leon M. Lederman	1922–	for the first use of a neutrino beam and the discovery of the
	Melvin Schwartz	1932–	muon neutrino
	Jack Steinberger	1921–	
1989	Norman Ramsey	1915–	for their work that led to the development of atomic clocks and
	Hans Dehmelt	1922–	precision timing
	Wolfgang Paul	1913–1993	
1990	Jerome I. Friedman	1930–	for demonstrating that protons and neutrons consist of quarks
	Henry W. Kendall	1926–	
	Richard E. Taylor	1929–	
1991	Pierre de Gennes	1932–	for studies of order phenomena, such as in liquid crystals and polymers
1992	George Charpak	1924–	for his invention of fast electronic detectors for high-energy particles
1993	Joseph H. Taylor	1941–	for verifying Einstein's general relativity theory by binary pulsar
	Russell A. Hulse	1951–	observations

ANSWERS TO ODD NUMBERED PROBLEMS

CHAPTER 1

3. 52.6 min; 5.2%. **5.** -0.44%. **7.** (a) Yes. (b) 8.6 s.
9. 720 days. **11.** 55 s; about one minute. **13.** 2 d 5 h.
15. (a) 100 m; 8.56 m; 28.1 ft. (b) 1 mi; 109 m; 358 ft.
17. 1.88×10^{22} cm^3. **19.** (a) 4.00×10^4 km.
(b) 5.10×10^8 km^2. (c) 1.08×10^{12} km^3.
21. 2.86×10^{-3} ly/century.
23. (a) 4.85×10^{-6} pc; 1.58×10^{-5} ly.
(b) 9.48×10^{12} km; 3.08×10^{13} km. **25.** (a) 390.
(b) 5.9×10^7. (c) 3500 km. **27.** 5.97×10^{26}.
29. New York. **31.** 840 km. **33.** 132 kg/s.
37. 605.780211 nm. **39.** (a) 43.2 cm^2. (b) 43 cm^2.
41. $\sqrt{Gh/c^3} = 4.05 \times 10^{-35}$ m.

CHAPTER 2

1. 81 ft (24 m). **3.** 2 cm/y. **5.** 48 mi/h.
(The physicist did other moving besides this weekly trip.)
7. (a) 45.0 mi/h (72.4 km/h). (b) 42.8 mi/h (68.8 km/h).
(c) 43.9 mi/h (70.6 km/h). **9.** (a) 0, 0, -2, 0, 12 m.
(b) -2, 12 m. (c) 7, 0 m/s. **11.** (a) 5.7 ft/s. (b) 7.0 ft/s.
13. (a) 28.5 cm/s. (b) 18.0 cm/s. (c) 40.5 cm/s. (d) 28.1 cm/s.
(e) 30.4 cm/s. **15.** -2 m/s^2.
19. (a) OA: $+$, $-$; AB: 0, 0; BC: $+$, $+$; CD: $+$, 0. (b) No.
21. (e) Situations (a), (b), and (d). **23.** (a) 80 m/s.
(b) 110 m/s. (c) 20 m/s^2.
25. (b) -0.030, -0.020, -0.010, 0.0 m/s.
(c) -0.040, -0.020, 0.0, 0.020, 0.040, 0.060 m/s.
(e) 0.020, 0.020, 0.020 m/s^2. **27.** (b) 19 m/s. (c) 31 m.
29. 2.8 m/s^2 (9.4 ft/s^2). **31.** 560 ms. **33.** 1.4×10^{15} m/s^2.
35. 2.6 s. **37.** (a) 4.5×10^4 ft/s^2. (b) 5.8 ms.
39. (a) 5.71 m/s^2. (b) 3.68 s. (c) 5.78 s. (d) 95.4 m.
41. (a) 60.6 s. (b) 36.4 m/s. **43.** (a) 0.75 s. (b) 50 m.
45. (a) 82 m. (b) 19 m/s. **47.** (a) 12 ft/s^2 (3.6 m/s^2).
(b) 3.7 ft/s (1.4 m/s). **49.** (a) 0.74 s. (b) -20 ft/s^2.
51. (a) 48.5 m/s. (b) 4.95 s. (c) 34.3 m/s. (d) 3.50 s.
53. (a) 32.4 m/s. (b) 6.62 s. **55.** Mercury.
57. 1.23, 4.90, 11.0, 19.6, 30.6 cm. **59.** 3.0 m (9.8 ft).
61. (a) 350 ms. (b) 82 ms.
63. 22.2 and 88.9 cm below the nozzle. **65.** 130 m/s^2, up.
67. (a) 3.41 s. (b) 57.0 m. **69.** ≈ 0.3 s. **71.** (a) 17.1 s.
(b) 293 m. **75.** 6.8 cm.

CHAPTER 3

1. The displacements should be (a) parallel, (b) antiparallel,
(c) perpendicular. **3.** (a) 370 m, 57° east of north.
(b) Displacement magnitude = 370 m; distance walked = 420 m.

7. (a) 4.5 units, 52° north of east.
(b) 8.4 units, 25° south of east.
9. Walpole (the state prison). **11.** (a) 4.9 m. (b) 12 m.
13. 4.76 km. **15.** (a) 28 m. (b) 13 m.
17. (a) $10\mathbf{i} + 12\mathbf{j} + 14\mathbf{k}$. (b) 21 ft.
(c) It can be equal or greater, but not less. (d) 26 ft.
19. (a) $3\mathbf{i} - 2\mathbf{j} + 5\mathbf{k}$. (b) $5\mathbf{i} - 4\mathbf{j} - 3\mathbf{k}$. (c) $-5\mathbf{i} + 4\mathbf{j} + 3\mathbf{k}$.
21. (a) $1400\mathbf{i} + 2100\mathbf{j} - 48\mathbf{k}$. (b) Zero.
23. (a) $r_x = 2.50$, $r_y = 15.3$. (b) 15.5. (c) 80.7°.
27. (a) $a\mathbf{i} + a\mathbf{j} + a\mathbf{k}$, $a\mathbf{i} + a\mathbf{j} - a\mathbf{k}$, $a\mathbf{i} - a\mathbf{j} - a\mathbf{k}$, $a\mathbf{i} - a\mathbf{j} + a\mathbf{k}$.
(b) 54.7°. (c) $a\sqrt{3}$. **33.** (a) -19. (b) 27, $+z$ direction.
39. (a) -21. (b) -9. (c) $5\mathbf{i} - 11\mathbf{j} - 9\mathbf{k}$. **41.** (a) 0. (b) -16.
(c) -9. **43.** (a) 2.97. (b) $1.51\mathbf{i} - 2.67\mathbf{j} - 1.36\mathbf{k}$. (c) 48.5°.
49. 70.5°.

CHAPTER 4

1. (a) 920 mi, 63° south of east.
(b) 410 mi/h, 63° south of east. (c) 550 mi/h.
3. (a) 3.9 km/h. (b) 13°. **5.** (a) 24 ns. (b) 2.7 mm.
(c) 9.6×10^8 cm/s; 2.3×10^8 cm/s. **7.** (a) $8t\mathbf{j} + \mathbf{k}$. (b) $8\mathbf{j}$.
(c) A parabola. **9.** 60°. **11.** (a) 514 ms. (b) 9.94 ft/s.
13. (a) 18 cm. (b) 1.9 m. **15.** (a) 3.03 s. (b) 758 m.
(c) 29.7 m/s. **17.** No. **19.** (a) 1.16 s. (b) 13.0 m.
(c) 18.8 m/s; 5.56 m/s. (d) No. **21.** (b) 76.0°. **23.** (a) 99 ft.
(b) 90 ft/s. (c) 180 ft. **25.** (a) 285 km/h. (b) 33°.
27. (a) 310 ms. (b) 1.9 m and 2.9 m above the hands.
29. The third. **31.** Yes. **33.** (a) 260 m/s. (b) 45 s.
35. 23 ft/s. **37.** (a) 9.8 s. (b) 2700 ft.
39. Approximately 40 m (130 ft). **41.** (a) 20 cm.
(b) No; the ball hits the net 4.4 cm above the ground.
43. Between the angles 31° and 63° above the horizontal.
45. 115 ft/s. **47.** (a) $D = v\sqrt{(2L/g)} \sin \theta - L \cos \theta$.
(b) The projectile will pass over the observer's head if D is
positive and will fall short if D is negative.
49. 5.66 s. **51.** 8.98×10^{22} m/s^2. **53.** (a) 7.49 km/s.
(b) 8.00 m/s^2. **55.** (a) 94 cm. (b) 19 m/s. (c) 2400 m/s^2.
57. (a) 130 km/s. (b) 850 km/s^2. **61.** (a) 92. (b) 9.6.
(c) $92 = (9.6)^2$. **63.** 2.6 cm/s^2. **65.** (a) 33.6 m/s^2.
(b) 89.7 m/s^2. **67.** 36 s; no.
69. Wind blows from the west at 55 mi/h. **71.** 31 m/s.
75. (a) 5.8 m/s. (b) 17 m. (c) 67°. (d) 49°.
77. 170 km/h, 7.3° south of west. **79.** (a) 30° upstream.
(b) 69 min. (c) 80 min. (d) 80 min.
(e) Perpendicular to current; 60 min.
81. (a) Head the boat 25° upstream. (b) 0.21 h. **83.** 0.83c.
85. (b) $t = 2.16$ s, $x = 97.7$ m, $y = 22.8$ m.
(c) $t = 4.31$ s, $x = 195$ m, $v_x = 45.3$ m/s, $v_y = -21.1$ m/s.

CHAPTER 5

1. 6.3 y. **3.** (*a*) 1.0×10^{-15} N. (*b*) 8.9×10^{-30} N.
5. 8.0 cm/s^2. **7.** 6500 N. **9.** (*a*) 3.1 cm/s^2. (*b*) 1.2×10^5 km.
(*c*) 2.7 km/s. **11.** (*a*) $42\mathbf{i} + 34\mathbf{j}$, m/s. (*b*) $630\mathbf{i} + 250\mathbf{j}$, m.
13. (*a*) 1.39×10^8 N; 6.94×10^6 N. (*b*) 4.11 y; 4.19 y.
15. (*a*) 0.62 m/s^2. (*b*) 0.13 m/s^2. (*c*) 2.6 m.
17. (*a*) 44.4 slug; 1420 lb. (*b*) 412 kg; 4040 N.
19. (*a*) 12.2 N; 2.65 kg. (*b*) Zero; 2.65 kg. **21.** 1600 lb.
23. 1.19×10^6 N (133 tons). **25.** (*a*) 1.8 mN. (*b*) 3.3 mN.
27. 0.15 N. **29.** (*a*) 210 m/s^2 (710 ft/s^2). (*b*) 17 kN (4000 lb).
31. (*a*) 7.3 kg (0.50 slug). (*b*) 89 N (20 lb). **33.** (*a*) 2.1 m/s^2.
(*b*) 120 N. (*c*) 21 m/s^2. **35.** (*a*) 1.8 m/s^2. (*b*) 3.8 m/s.
(*c*) 4.0 m. (*d*) 11°. **37.** 18.4 kN. **39.** (*b*) 12 ft/s^2. (*c*) 8.9°.
41. 33 m/s. **43.** (*a*) 730 N. (*b*) 1300 N. **45.** (*a*) 3260 N.
(*b*) 2720 kg. (*c*) 1.20 m/s^2. **47.** (*a*) 5.0×10^5 N.

(*b*) 1.4×10^6 N. **49.** $2M\left(\dfrac{a}{a+g}\right)$.

51. (*a*) $g \sin \theta$, down the incline.
(*b*) $g \sin \theta$, down the incline. (*c*) $(g - a) \sin \theta$, down the incline.
(*d*) $(g + a) \sin \theta$, down the incline. (*e*) Zero.
(*f*) $m(g - a) \cos \theta$. **53.** (*a*) 6.8 m/s.
(*b*) Yes: he can climb the rope while falling.
55. (*a*) 0.97 m/s^2. (*b*) $T_1 = 1.2$ N; $T_2 = 3.5$ N.
57. (*a*) 135 N. (*b*) 45.3 N. (*c*) 75.4 N. **59.** (*a*) 0.217 m/s^2.
(*b*) 17.8 N. **61.** (*a*) 12.1 kN. (*b*) 10.5 kN.
(*c*) 1.60 kN, toward the counterweight. **63.** (*a*) 37 N.
(*b*) 55 N. (*c*) 36 m/s^2, upward.
65. (*b*) $P/(m + M)$. (*c*) $PM/(m + M)$.
(*d*) $P(m + 2M)/2(m + M)$. **67.** 230 lb.

CHAPTER 6

1. 2.3°. **3.** 9.3 m/s^2. **5.** 52 N. **7.** (*a*) 9.1 kN. (*b*) 9.0 kN.
9. (*a*) No. (*b*) A 12-lb force to the left and a 5.0-lb force up.
11. (*a*) 11.1 N. (*b*) 47.3 N. (*c*) 40.1 N. **15.** (*a*) $v_0^2/4g \sin \theta$.
(*b*) No. **17.** (*a*) 10 kg. (*b*) 2.7 m/s^2. **19.** (*a*) 61 N. (*b*) 66 N.
(*c*) 5.9 kN. **21.** (*a*) 70 lb. (*b*) 4.6 ft/s^2. **23.** (*b*) 30 MN.
25. (*a*) 1.24 m/s^2. (*b*) 13.4 N. **27.** $g(\sin \theta - \sqrt{2}\mu_k \cos \theta)$.
29. (*a*) 3.46 m/s^2. (*b*) 0.910 N, in tension.
(*c*) 3.46 m/s^2; 0.910 N in compression. **31.** (*a*) 7.6 m/s^2.
(*b*) 0.86 m/s^2. **33.** (*a*) 730 lb (3200 N). (*b*) 0.30.
35. (*a*) 0.46. (*b*) 0.92. **37.** 870 N; 17°. **39.** 0.032.
41. (*a*) 0.43. (*b*) 42 m. **43.** (*a*) 175 lb. (*b*) 50.0 lb.
45. (*a*) 30 cm/s. (*b*) 170 cm/s^2, radially inward. (*c*) 2.9 mN.
(*d*) 0.40. **47.** 2.32 km. **49.** (*a*) At the bottom of the circle.
(*b*) 31 ft/s. **51.** (*a*) 0.0337 N. (*b*) 9.77 N.
53. (*a*) $\dfrac{1}{2\pi}\sqrt{\dfrac{g(\tan \theta + \mu_s)}{r(1 - \mu_s \tan \theta)}}$. (*b*) $\dfrac{1}{2\pi}\sqrt{\dfrac{g(\tan \theta - \mu_s)}{r(1 + \mu_s \tan \theta)}}$.
55. (*a*) 235 m/s. (*b*) 107 m/s^2. (*c*) 232 N.
57. (*a*) $0.632F_0T/m$. (*b*) $0.368F_0T^2/m$. **59.** $\sqrt{mg/b}$.
61. 2.0×10^{-5} N·s/m. **63.** 1.30 m/s.
65. (*a*) $\left(\dfrac{m}{b}\right) \ln (v_i/v_f)$. (*b*) 19 s. **67.** (*b*) 370 m.

69. (*a*) 11.7 s. (*b*) 59.8 m/s. (*c*) 0.610.
71. 819, 838, 833, 805, 762 m; 30°.
73. (*a*) $t = 1.95$ s, $x = 80.4$ m, $y = 20.0$ m,
$v_x = 37.3$ m/s, $v_y = 0$, $a_x = -3.73$ m/s^2, $a_y = -9.80$ m/s^2;
(*b*) $t = 1.79$ s, $x = 68.3$ m, $y = 17.8$ m, $v_x = 31.7$ m/s, $v_y = 0$,
$a_x = -6.33$ m/s^2, $a_y = -9.80$ m/s^2; (*c*) 151 m, 121 m;
(*d*) for $b = 0.10$ s^{-1}: $v_x = 30.3$ m/s, $v_y = -18.5$ m/s;
for $b = 0.20$ s^{-1}: $v_x = 21.1$ m/s, $v_y = -16.4$ m/s

CHAPTER 7

1. (*a*) 580 J. (*b*) Zero. (*c*) Zero. **3.** (*a*) 430 J. (*b*) -400 J.
(*c*) Zero. **5.** (*a*) $-\frac{3}{4}Mgd$. (*b*) Mgd. **7.** (*a*) 2160 J.

(*b*) -1430 J. **9.** (*a*) 215 lb. (*b*) 1.01×10^4 ft·lb. (*c*) 48.0 ft.
(*d*) 1.03×10^4 ft·lb. **11.** 800 J. **13.** $(3/2)F_0x_0$.
15. (*a*) 23 mm. (*b*) 45 N. **17.** (*a*) 135 N. (*b*) 60.0 J.
19. 1200 km/s. **21.** AB: +; BC: 0; CD: −; DE: +.
23. 100 ft; no. **25.** 20.2 ft·lb (24.4 J).
27. Man, 2.41 m/s; boy, 4.82 m/s.
29. (*a*) 9.0×10^4 "megatons of TNT." (*b*) 45 km.
31. 6.55 m/s. **33.** (*a*) 304 mJ. (*b*) -1.75 J. (*c*) 3.32 m/s.
(*d*) 22.5 cm. **35.** 720 W (0.97 hp). **37.** 24 W.
39. (*a*) 2.45×10^5 ft·lb. (*b*) 0.619 hp. **41.** 90.3 kN.
43. 25 hp. **45.** (*a*) 0.77 mi. (*b*) 71 kW. **47.** 16.6 kW.
49. (*b*) mtv_f^2/t_f^2. **51.** 2.66 hp. **53.** (*b*) 1.95.
55. (*a*) 10.0 kW. (*b*) 2.97 kJ. **57.** 69 hp. **61.** (*a*) $0.13c$.
(*b*) 4.6 keV. (*c*) Low by 1.3%. **63.** (*a*) 79.1 keV.
(*b*) 3.11 MeV. (*c*) 10.9 MeV.

CHAPTER 8

1. 110 MN/m. **3.** (*a*) 7.8 MJ. (*b*) 6.2. **5.** 2.15 m/s.
7. (*a*) 27.0 kJ. (*b*) 2.94 kJ. (*c*) 158 m/s; *a, b*. **9.** (*a*) 2.56 J.
(*b*) 11.1 m/s. **11.** 830 ft. **13.** 2.75 m/s. **15.** (*a*) 1300 MW.
(*b*) 137 M\$. **19.** 4.24 m. **21.** (*a*) 34.2 ft/s. (*b*) 4.32 in.
23. $mgL/32$. **25.** 11.1 cm.
27. (*a*) 8.06mg, at 82.9° left of vertical. (*b*) $5R/2$.
29. (*a*) $U(x) = -Gm_1m_2/x$. (*b*) $Gm_1m_2\,d/x_1(x_1 + d)$.
31. (*a*) 69.2 J. (*b*) 7.99 m/s. (*c*) Conservative.
35. (*a*) 44.6 cm. (*b*) 3.47 cm. **37.** (*a*) $\sqrt{5gR}$.
(*b*) $\theta = \sin^{-1} (1/3)$. **41.** (*c*) -1.2×10^{-19} J. (*d*) 2.2×10^{-19} J.
(*e*) $\approx 1 \times 10^{-9}$ N, toward M. **45.** (*a*) $-U_0(r_0r^{-2} + r^{-1})e^{-r/r_0}$.
(*b*) 0.14, 0.0078, 6.8×10^{-6}. **47.** (*a*) 3.02 kJ. (*b*) 391 J.
(*c*) 2.63 kJ. **49.** 39 kW. **51.** 472 kJ. **53.** 4.19 m.
55. 65.1 cm/s. **57.** (*a*) 48.7 m/s. (*b*) 64.5 kJ.
59. (*a*) 24.0 ft/s. (*b*) 3.00 ft. (*c*) 9.00 ft. (*d*) 48.8 ft.
61. (*a*) 10.8 PJ. (*b*) 263,000 y. **63.** Decreases by 1.10 kg.
65. 266 times the equatorial circumference of the Earth.
67. 191. **69.** 2.21 eV. **71.** (*a*) -12.5 kJ. (*b*) 2.70 kJ.
(*c*) -9.80 kJ. (*d*) 1.70 kJ. (*e*) 100 J.
(*f*) $x = 2.95$ m, $y = -2.95$ m.
73. (*a*) 0.541, 0.541, 0.541 J. (*b*) 0.541, 1.08, 0.383 J.

CHAPTER 9

1. (*c*) $x_1 = x_{cm} - (m_2/M)(L + d_i \cos \omega t)$;
$x_2 = x_{cm} + (m_1/M)(L + d_i \cos \omega t)$; $v_1 = (m_2/M)d_i\omega \sin \omega t$;
$v_2 = -(m_1/M)d_i\omega \sin \omega t$.
3. 4640 km (1730 km beneath the Earth's surface).
5. 75.2 km/h. **7.** (*a*) Down; $mv/(m + M)$.
(*b*) Balloon again stationary. **9.** (*a*) L. (*b*) Zero.
11. (*a*) Midway between them.
(*b*) It moves 1.12 mm toward the heavier body.
(*c*) $0.00160g$, down. **13.** $g(1 - 2x/L)$. **15.** 55.2 kg.
17. $L/5$ from the heavy rod, along the symmetry axis.
19. $x_{cm} = y_{cm} = 20$ cm; $z_{cm} = 16$ cm.
21. $4R/3\pi$ away from the flat base, along the axis of symmetry.
23. (*a*) 6.94×10^4 J.
(*b*) 3.56×10^4 kg·m/s, 38.7° south of east. **25.** (*a*) 6.96 J.
(*b*) $P_i = 0.854$ kg·m/s, 27.4° above the horizontal;
$P_f = 0.854$ kg·m/s, 27.4° below the horizontal;
0.786 kg·m/s, vertically down. (*c*) 1.53 s.
29. 0.0103 ft/s, backward. **31.** $wv_{rel}/(W + w)$. **33.** 27.
35. (*a*) Rocket case: 7290 m/s; payload: 8200 m/s.
(*b*) Before: 12.71 GJ; after: 12.75 GJ.
37. (*a*) 1.4×10^{-22} kg·m/s, 150° from the electron
track and 120° from the neutrino track. (*b*) 1.0 eV.
39. (*a*) 746 m/s. (*b*) 963 m/s. **41.** Yes.

43. $\left(\dfrac{u\cos\alpha}{\sqrt{1-u\cos^2\alpha}}\right)\sqrt{2gh},\ u=\dfrac{m}{m+M}$.   **45.** 2.66 m/s.
47. (a) 1790 N. (b) 609 J.   **51.** (a) 2.72. (b) 7.39.
53. 1.33 km/s.   **55.** 60 N.   **57.** (a) 49.1 kg. (b) 141 kg.
59. (a) 23.4 kN (5260 lb). (b) 4.31 MW (5780 hp).

CHAPTER 10

1. 64 kN.   **3.** $2\mu u$.   **5.** (a) 2.40 N·s. (b) 2.40 N·s.
(c) 2.00 kN. (d) 62.7 J.   **7.** 3.29 kN (744 lb).
9. (a) 2.20 N·s, to the left. (b) 212 N, to the right.
11. (a) 1.95×10^{5} kg·m/s, for each direction of thrust.
(b) Backward: +66.1 MJ; forward: −50.9 MJ; sideways:
+7.61 MJ.   **13.** 41.7 cm/s.
17. (a) 1.03 kg·m/s. (b) 250 J. (c) 10.3 N. (d) 824 N.
19. 124 kW.   **21.** (a) 1.9 m/s, to the right. (b) Yes.
23. 4.2 m/s.   **25.** (a) 2.74 m/s. (b) 1.46 km/s.
27. ≈2 mm/y.   **29.** 1.2 kg.
31. (a) 74.4 m/s. (b) 81.5 m/s; 84.1 m/s.
33. (a) A: 4.57 m/s; B: 3.94 m/s. (b) 7.53 m/s.
35. 12.9 tons.   **37.** (a) 4.21 ft/s; 2210 ft·lb.
(b) 3.21 ft/s; 5.51 ft/s.   **39.** 41.0 N.   **41.** 35.9 cm.

43. $\sqrt{2E\left(\dfrac{M+m}{Mm}\right)}$.   **47.** (a) 4.0**i** + 5.0**j**, m/s.

(b) 700 J is gained.   **51.** (a) 26° from the incoming
proton's direction. (b) 227 m/s; 466 m/s.   **53.** $v=V/4$.
55. (a) 3.43 m/s, deflected by 17.3° to the right. (b) 954 kJ.
57. (a) 28.0°. (b) 7.44 m/s.   **61.** 2.44 m/s, to the left.
63. (a) 117 MeV. (b) K_π = 102 MeV; K_n = 15.0 MeV.
65. (a) $(-1.04\mathbf{i}+0.655\mathbf{j})\times10^{-19}$ kg·m/s. (b) 7.66 MeV.

CHAPTER 11

3. (a) 5.5×10^{15} s. (b) 26.   **5.** (a) 0.105 rad/s.
(b) 1.75×10^{-3} rad/s. (c) 1.45×10^{-4} rad/s.   **7.** 11 rad/s.
9. (a) 4.8 m/s. (b) No.   **11.** (b) 23 h 56 min.
13. (a) 8140 rev/min². (b) 425 rev.   **15.** (a) −1.28 rad/s².
(b) 248 rad. (c) 39.5 rev.   **17.** (a) 2.0 rev/s. (b) 3.8 s.
19. (a) 369 s. (b) -3.90×10^{-3} rad/s². (c) 108 s.
21. (b) -2.30×10^{-9} rad/s². (c) ≈4610. (d) 24 ms.
23. (a) 3.49 rad/s. (b) 20.6 in./s. (c) 10.1 in./s.
(d) 71.9 in./s²; 35.3 in./s².   **25.** 5.6 rad/s².
27. (a) 3.65 rad/s. (b) 38.0 m/s. (c) 6.78 m/s². (d) 139 m/s².
29. 4.56 s.   **31.** (a) −1.18 rev/min². (b) 10,300.
(c) 1.08 mm/s². (d) 30.2 m/s².
33. (a) 6.3×10^{4} ft/min (1.9×10^{4} m/min).
(b) 6.8×10^{4} ft/min (2.1×10^{4} m/min).   **35.** 16.4 s.
37. (a) $r\alpha^2t^2$. (b) $r\alpha$. (c) 44.1°.   **39.** (a) 71 rad/s.
(b) −13 rad/s². (c) 72 m.
41. (a) $x^2+y^2=R^2$; a circle of radius R;
ω is the angular speed of the object.
(b) $v_x=-\omega y$; $v_y=\omega x$; **v** is tangent to the circle; $v=\omega R$.
(c) $a=\omega^2 R$; **a** points radially inward.

CHAPTER 12

1. (a) 1305 g·cm². (b) 545 g·cm². (c) 1850 g·cm².
3. 6.75×10^{12} rad/s.   **5.** (a) 6490 kg·m². (b) 4.36 MJ.
7. 0.097 kg·m².   **9.** (b) $MR^2/4$.   **13.** (a) $dm/M=2r\,dr/R^2$.
(b) $dI=2Mr^3\,dr/R^2$. (c) $I=\frac{1}{2}MR^2$.
15. 3.66 N·m, into the page.   **17.** 12 N·m, out of the page.
19. 7.63 rad/s², out of the page.   **21.** (a) 28.2 rad/s².
(b) 338 N·m. (c) 1.36 kW.   **25.** (a) 2.57×10^{29} J.
(b) 1.94 Gy.   **27.** 690 rad/s.   **29.** (a) $2\theta/t^2$. (b) $2R\theta/T^2$.
(c) $T_1=M(g-2R\theta/t^2)$; $T_2=Mg-(2\theta/t^2)(MR+I/R)$.
31. 1.73×10^5 g·cm².   **33.** 6.11 m/s.

35. (a) −7.67 rad/s². (b) −11.7 N·m. (c) 45.8 kJ. (d) 624 rev.
(e) The energy dissipated by friction; 45.8 kJ.
37. (a) 4.82×10^{5} N. (b) 1.12×10^{4} N·m.
39. (a) 1.88×10^{12} J/s. (b) -2.67×10^{-22} rad/s².
(c) 4.06×10^{9} N.   **41.** (a) 47.9 km/h. (b) 3.65 rad/s².
(c) 8.68 kW.   **45.** (a) 56.5 rad/s. (b) −8.88 rad/s².
(c) 69.2 m.   **47.** (a) 12.5 cm/s². (b) 4.63 s. (c) 28.8 rev/s.
(d) 70.8 rev/s.   **49.** 48. m.   **51.** (a) $W/6$. (b) $2g/3$.
55. $\alpha=2F/MR;\ a=F/M$.   **57.** (a) 57.9 rad/s. (b) 4.21 m.

CHAPTER 13

5. mvd.   **11.** (a) −4.17 m/s². (b) −16.9 rad/s².
(c) −2.62 N·m.   **15.** (a) 1.49 N·m. (b) 20.8 rad. (c) −31.0 J.
(d) 20.3 W.   **17.** The center of mass moves in the direction
of the impulsive force with a speed of 2.90 m/s; the stick
rotates about its center of mass with an angular speed of 10.7
rad/s.   **21.** (b) $ML^2/(L^2+12d^2)$.   **25.** (a) 1.18 s. (b) 8.60 m.
(c) 5.18 rev. (d) 6.07 m/s.   **27.** 3.0 min.   **29.** $mv/(m+M)R$.
31. (a) 171 rev/min. (b) 0.792.   **33.** (a) 5.12 mrad/s.
(b) 1.90 cm/s.   **35.** (a) $MR^2\omega_0^2/4$; $MR^2\omega_0/2$. (b) $R^2\omega_0^2/2g$.
(c) ω_0.   **37.** $\sqrt{2gr}\sec\theta_0$.
39. (a) Each revolves in a circle of radius 1.46 m at 0.945 rad/s.
(b) 9.12 rad/s. (c) K_a = 97.5 J; K_b = 941 J.   **41.** −0.127.
43. 1.90 min.

CHAPTER 14

1. (a) Two. (b) Seven.   **5.** (a) 2.5 m. (b) 7.3°.
7. (a) Slides; 31°. (b) Tips; 34°.   **9.** 1200 lb.
11. (a) 2.78 kN. (b) 3.89 kN.
13. Left pedestal: 1.17 kN (tension);
right pedestal: 1.89 kN (compression).
15. Three-fourths the length of the beam from the worker
at the end.
17. F_muscle = 1.91 kN, up, $3W$; F_bone = 2.55 kN, down, $4W$.
19. $W\sqrt{h(2r-h)}/(r-h)$.
21. (a) $F_1=w\sin\theta_2/\sin(\theta_2-\theta_1)$;
$F_2=w\sin\theta_1/\sin(\theta_2-\theta_1)$; normal to the planes.
23. (a) 416 N. (b) 238 N; 172 N.   **25.** (a) 47.0 lb.
(b) 21.3 lb; 10.9 lb.   **27.** (a) 1460 lb. (b) 1220 lb; 1420 lb.
29. (a) $Wx/L\sin\theta$. (b) $Wx/L\tan\theta$. (c) $W(1-x/L)$.
31. (a) Lower hinge: F_h = 180 lb, F_v = 210 lb;
upper hinge: F_h = 180 lb, F_v = 60 lb.
(b) F_h = 180 lb, F_v = 60 lb, on each beam, oppositely directed.
33. (a) 47 lb. (b) F_A = 120 lb; F_E = 72 lb.   **35.** (a) 446 N.
(b) 0.500. (c) Yes; push 45° upward; 315 N.
37. (a) $L/2$, $L/4$, $L/6$. (c) $N=n$.   **41.** 75 GN/m².
43. 3.65 mm.   **45.** 201 kN.   **47.** 802 rev/min.
49. (a) 18.0 MN. (b) 14.4 MN. (c) 16.

CHAPTER 15

1. 289 ms.   **3.** 708 N/m.   **5.** (a) 495 N/m. (b) 1.57 cm.
(c) 1.55 Hz.   **7.** (a) 1250 N/cm. (b) 2.63 Hz.   **9.** 30.4 lb.
11. (b) 12.47 kg. (c) 72.85 kg.   **13.** 2.08 h.   **15.** 2.83 cm.
17. (c) $2\pi\sqrt{mb^3/a^4}$.   **19.** (a) $0.183L$. (b) Same direction.
23. (a) 1.07 Hz. (b) 4.73 cm.   **27.** (a) 6.97 MN/m.
(b) 48,500.   **29.** (a) 3.04 ms. (b) 3.84 m/s. (c) 90.7 J.
33. (a) 31.9 cm. (b) 34.4°.   **37.** (a) 5.60 J. (b) 2.80 J.
39. 24.9 cm.   **41.** 8.35 s.   **43.** (a) 436 mHz. (b) 1.31 m.
45. 906 ms.   **47.** 5.57 cm.   **49.** (a) $2\pi\sqrt{(L^2+12d^2)/12gd}$.

51. $1.22v_0$.   **55.** $\left(\dfrac{1}{2\pi}\right)\sqrt{\left(\dfrac{g}{L}\right)\sqrt{1+\left(\dfrac{v^2}{Rg}\right)^2}}$.

59. (*a*) Straight line, $y = \pm x$.
(*b*) Ellipse, $y^2 - \sqrt{3}xy + x^2 = A^2/4$. (*c*) Circle, $x^2 + y^2 = A^2$.
65. $k = 490$ N/cm; $b = 1100$ kg/s. **67.** 1.9 in.
69. 362 ms. **71.** (*a*) 8.00 u; 0.98 u; 6.86 u. (*b*) 490 N/m.
73. (*a*) 2.8 cm, 0. (*b*) 2.8 cm, 3.14 rad.
(*c*) 1.98 cm, -1.57 rad. (*d*) 1.98 cm, 1.57 rad.
(*e*) 3.43 cm, -0.615 rad. (*f*) 3.43 cm, 0.615 rad.
(*g*) 3.43 cm, 3.76 rad. (*h*) 3.43 cm, 2.53 rad.

CHAPTER 16

1. 39.2 nN. **3.** 2.60×10^5 km. **5.** 997 ms. **7.** 997 km.
9. (*a*) 1.33×10^{12} m/s². (*b*) 1.79×10^6 m/s. **11.** 9.78 m/s².
17. (*b*) 1.9 h. **21.** 7.90 km/s.
23. $\dfrac{GMm}{d^2}\left[1 - \dfrac{1}{8(1 - R/2d)^2}\right]$.

25. (*a*) 9.83 m/s². (*b*) 9.84 m/s². (*c*) 9.79 m/s².
27. (*b*) 2.0×10^8 N/m². (*c*) 180 km. **31.** 220 km/s.
37. 98.4 pJ. **39.** (*a*) 2.02 km/s. (*b*) 523 km. (*c*) 1.26 km/s.
(*d*) 4.80×10^{22} kg. **41.** (*a*) 3.34×10^7 m/s. (*b*) 5.49×10^7 m/s.
45. 6.5×10^{23} kg. **47.** 0.354 lunar months.
49. (*a*) 1.68 km/s. (*b*) 108 min. **51.** $7.20R_S$. **53.** 58.3 km/s.
55. Properties are proportional to (*a*) $r^{3/2}$; (*b*) r^{-1};
(*c*) $r^{1/2}$; (*d*) $r^{-1/2}$. **59.** 3.5 y.
61. $4\pi r^{3/2}/\sqrt{G(4M + m)}$. **63.** (*a*) 7.54 km/s. (*b*) 97.3 min.
(*c*) 405 km; 7.68 km/s; 92.3 min. (*d*) 3.18 mN.
(*e*) Of the satellite, no; of the system Earth + satellite, yes.
65. (*a*) No. (*b*) Same. (*c*) Yes.
67. (*a*) Easily; it would weigh only about 3 lb.
(*b*) Probably; you would have to be able to run at 6.9 m/s.
69. South, 35.4° above the horizon. **71.** (*a*) 7964 m/s
(7750 m/s for *A*). (*b*) 7820 m/s (7750 m/s for *A*). **73.** $\sqrt{GM/L}$.

CHAPTER 17

1. 429 kPa. **3.** 27.4 kN. **5.** 6.0 lb/in.² **7.** 0.52 m.
9. 1.29 MPa. **11.** 130 km. **13.** 0.412 cm.
15. (*a*) 600, 30, 80 tons.
(*b*) No: even though the previous answers are changed
to 3100, 280, 760 tons, atmospheric pressure acts on each
side of the walls and cancels out.
17. 809 kN. **19.** 43.5 km. **21.** 230 MPa.
23. $\frac{1}{4}\rho g A(h_2 - h_1)^2$. **25.** (*b*) *a*. **27.** (*b*) $p = \rho g h$.
31. (*a*) 35.6 kN. (*b*) Yes; decreases by 0.0851 m³.
33. 1070 g. **35.** 2.0×10^{-4}. **37.** (*a*) 38.4 kN. (*b*) 40.5 kN.
(*c*) 2.35 kN. (*d*) 2.08 kN. **39.** 4.74 MN. **41.** 56.1 cm.
43. 4. **45.** 0.190. **47.** (*a*) 1.82 m³. (*b*) 4.61 m³.
49. 740 kg/m³. **51.** 500 μJ. **53.** 61.6 μJ. **57.** (*a*) 3.25 Pa.
(*b*) 1.79 Pa. (*c*) 68.7 J. (*d*) 765 μJ.

CHAPTER 18

1. 1 h 49 min. **3.** 3.9 m. **5.** 1.1 m/s. **7.** 1.7×10^5 N·m.
9. (*a*) 2.66 m/s. (*b*) 271 Pa. **11.** (*a*) 2. (*b*) 1/2. (*c*) $h/4$, below.
13. 10.8 kN. **15.** (*b*) To height *h*. **17.** 41.0 m/s.
19. 1.38 cm. **21.** (*a*) $\sqrt{2g(h_2 + d)}$. (*b*) $p_{atm} - \rho g(h_2 + d + h_1)$.
(*c*) 10.3 m. **23.** 5 min 42 s. **25.** $\frac{1}{2}\rho v^2 A$. **29.** 410 m/s.
37. 320 kPa. **39.** (*b*) 35.5 mPa. **43.** 3630 s.

CHAPTER 19

1. (*a*) 7.43 kHz. (*b*) 135 μs. **3.** (*a*) 712 ms. (*b*) 1.40 Hz.
(*c*) 1.93 m/s.
5. $y = 0.0112 \sin(10.6x + 3440t)$, in which *x* and *y* are
in meters and *t* is in seconds.
9. (*a*) 6.0 cm. (*b*) 100 cm. (*c*) 2.0 Hz. (*d*) 200 cm/s.
(*e*) Negative *x* direction. (*f*) 75 cm/s. **11.** 135 N.

13. 91.9 g/m. **15.** (*a*) 5.0 cm. (*b*) 40 cm. (*c*) 12 m/s.
(*d*) 33 ms. (*e*) 9.4 m/s.
(*f*) $y = 5.0 \sin(0.16x + 190t + 0.93)$, with *x* and *y* in cm and
t in seconds. **17.** $2\pi y_m/\lambda$.
19. 7.54 m from the end of the wire at which the earlier pulse
originated.
21. (*a*) $\sqrt{k \, \Delta L(L + \Delta L)/m}$. **25.** 198 Hz. **27.** (*b*) Length².
29. (*b*) 474 nJ/m³. **31.** 68.8°, 1.20 rad.
37. $\lambda = 2\sqrt{4(H + h)^2 + d^2} - 2\sqrt{4H^2 + d^2}$. **39.** (*a*) 81.4 m/s.
(*b*) 16.7 m. (*c*) 4.87 Hz. **41.** (*a*) -3.9 cm.
(*b*) $y = 0.15 \sin(0.79x + 13t)$. (*c*) -14 cm. **43.** (*a*) 1.25 m.
(*b*) $y = (3.80 \times 10^{-3}) \sin 10.1x \cos 3910t$, where *x* and *y* are
in meters and *t* is in seconds.
49. 7.47, 14.9, 22.4 Hz. **51.** 480, 160, 96 cm. **53.** 36.8 N.
55. 190 Hz.

CHAPTER 20

1. (*a*) 76.2 μm. (*b*) 333 μm. **3.** (*a*) 57 nm. (*b*) ~35.
5. 170 m. **7.** (*a*) $L(V - v)/Vv$. (*b*) 43.5 m. **9.** 40.7 m.
11. (*a*) 1.48 Pa. (*b*) 167 Hz. (*c*) 1.87 m. (*d*) 312 m/s.
13. 4.47 W. **15.** 27.1 mJ. **19.** (*a*) 44.2 μW/m². (*b*) 164 nm.
(*c*) 894 mPa. **21.** 51.9 nJ/m³. **23.** (*a*) 5.20 kHz.
(*b*) $B/A = \frac{1}{2}$. **25.** 190 dB. **27.** 63 dB. **29.** (*a*) 66.8 μW/m².
(*b*) 5.02 nW. (*c*) 7.53 μJ.
31. (*a*) $v = 307n$ Hz, $n = 1, 3, \ldots$ 65.
(*b*) $v = 615n$ Hz, $n = 1, 2, \ldots$ 32.
33. At ±0.286, 0.857, 1.43, 2.00 m from the midpoint.
35. 346 m. **37.** 19.8 kHz. **39.** (*a*) 1130, 1500, 1880 Hz.
41. 57.2 m. **45.** Four.
47. Open pipe: 58.9 cm; closed pipe: 44.2 cm. **49.** 45.4 N.
51. 2.25 ms. **53.** (*a*) Ten. (*b*) Four. **55.** 17.4 kHz.
57. (*a*) 522 Hz. (*b*) 554 Hz. **59.** 31°. **61.** 2.65×10^8 m/s.
63. 7.16 km. **67.** 160 Hz. **69.** (*a*) 464 Hz. (*b*) 490 Hz.
71. 8.8°C. **73.** 41.2 kHz. **75.** (*a*) 2.0 kHz. (*b*) 2.0 kHz.

CHAPTER 21

1. (*a*) 7.1×10^{-10} s. (*b*) 2.5×10^{-18} m. **3.** 1.30 m.
5. 0.445 ps. **7.** $0.805c$. **9.** (*a*) 87.4 m. (*b*) 394 ns.
11. (*a*) 15.8 km/s. (*b*) 6.95×10^{-10}. **13.** 0.75.
15. (*a*) $x' = 3.78 \times 10^7$ m; $t' = 2.26$ s.
(*b*) 6.54×10^8 m; 3.14 s. **17.** (*a*) 25.8 μs.
(*b*) The red flash, Doppler shifted. **19.** (*a*) $-u$; $c\sqrt{1 - u^2/c^2}$.
21. (*a*) $0.347c$. (*b*) $0.619c$. **23.** Seven.
25. (*a*) $0.933c$, 31.0° east of south.
(*b*) $0.933c$, 59.0° west of north. **27.** 6.29 cm. **29.** 1.23 μs.
31. (*a*) $0.491c$, in the negative *x* direction. (*b*) 4.32 μs; red.
33. (*a*) 26.3 y. (*b*) 52.3 y. (*c*) 4.06 y. **35.** (*a*) 4.00 μs.
(*b*) 2.50 μs. **39.** (*a*) 0.999 165. (*b*) 0.0133.
41. (*a*) 0.9988; 20.6. (*b*) 0.145; 1.01. (*c*) 0.073; 1.0027.
43. 21.2 smu/y. **45.** $\sqrt{8}mc$. **49.** (*b*) $K = p^2/2m$. (*c*) $207m_e$.
51. (*a*) 996 eV. (*b*) 1.05 MeV. **53.** (*a*) $\sqrt{\dfrac{\gamma - 1}{\gamma + 1}}\,c$.
(*b*) $\sqrt{2(\gamma + 1)}m$. **55.** (*b*) 202 GeV. (*c*) 49.1 GeV.
57. (*b*) $0.796c$.

CHAPTER 22

1. 291.1 K. **3.** 31.2. **5.** No; 310 K = 98.6°F. **7.** (*a*) $-40°$.
(*b*) 575°. (*c*) None. **11.** 1.3660.
13. 0.073 cm Hg; nitrogen. **15.** 0.038 in. **17.** 6.2 mm.
19. (*a*) 13.8 cm². (*b*) 115 cm³. **25.** 2.3×10^{-5}/C°.
27. (*b*) Steel: 71 cm; brass: 41 cm. **29.** 360°C. **35.** 909 g.
37. (*b*) Use 39.3 cm of steel and 13.1 cm of brass.
39. (*a*) Zero. (*b*) -0.36%. (*c*) -0.36%. **41.** $+0.68$ s/h.

43. 0.17 mm. **45.** 66.4°C. **47.** 0.27 mm. **49.** (*a*) 2.25 ft. (*b*) 3.99 ft.

CHAPTER 23

1. (*a*) 0.0225 m³. **3.** (*a*) 113. (*b*) 0.900 m³. **5.** 5.59 m³.
9. 26.9 lb/in.² **11.** 104 cm³. **13.** 200 kPa.
15. 4.34×10^{-5}. **17.** 180 m/s. **19.** 0.76. **21.** (*a*) 531 m/s.
(*b*) 28 g/mol; N_2. **23.** 1/5. **25.** (*a*) 3.53 meV; 4.83 meV.
(*b*) 3400 J; 4650 J. **27.** 2.2×10^{-5}. **29.** (*a*) 3.31×10^{-20} J.
(*b*) 0.207 eV. **31.** (*a*) $(N_a + N_b)(kT/V)$. (*b*) $\frac{1}{2}$.
33. (*a*) 1.0×10^4 K; 1.6×10^5 K. (*b*) 440 K; 7000 K.
35. 89.0°C. **37.** 45 kJ along path 1; −45 kJ along path 2.
39. (*a*) 1.40. (*b*) 29.0 g/mol. **41.** 1.14 kJ. **43.** (*a*) 8.39 atm.
(*b*) 544 K. (*c*) 966 J. **45.** 265 K. **47.** (*a*) 2.95 cm.
(*b*) 2.11 cm. **49.** 2.48 kJ. **51.** (*a*) 1.20. (*b*) 105°C.
(*c*) 628 mol. (*d*) 1.96 MJ; 2.96 MJ. (*e*) 0.813.
53. (*a*) 423 kPa. (*b*) 431 kPa.

CHAPTER 24

5. (*a*) 2.69×10^{25}. (*b*) 0.171 nm. **7.** (*a*) 1.67.
(*b*) 49.5×10^{-6} cm. (*c*) 7.87×10^{-6} cm. **9.** 3.86 GHz.
11. −12°C. **13.** (*a*) 420 m/s; 458 m/s; yes. **15.** 13.9 rev/s.
17. 4.71. **21.** (*a*) $3N/v_0^3$. (*b*) $0.750v_0$. (*c*) $0.775v_0$.

25. (*a*) $\sqrt{\dfrac{15}{4}kT}$. **27.** 1.5 m/s. **29.** 1.5 cm/s.

31. (*a*) 35 cm/s. (*b*) 4.4×10^6 rev/s. **33.** (*a*) 8. (*b*) 4.

CHAPTER 25

1. 44.5 m³. **3.** 107 g. **5.** (*a*) 0.12 C°. **7.** (*a*) 75.4 kJ.
(*b*) 4.46 kJ. (*c*) 757°C. **9.** 42.7 kJ. **11.** 2.4 days.

13. (*a*) 117 s. (*b*) 718 s. **15.** (*a*) 6.75×10^{-20} J. (*b*) 10.7.
17. (*a*) 5.26°C; no ice left. (*b*) 0°C; 62.0 g of ice left.
19. 4.81 g. **21.** 17 mJ. **23.** 1.2 kJ. **25.** 11.3 kJ.
27. (*a*) 7880 J. (*b*) 5630 J. (*c*) 3380 J. **29.** Diatomic.
31. $nRT \ln(V_f/V_i)$. **33.** (*a*) +214 J. (*b*) −293 J.
(*c*) −79.0 J. **35.** (*a*) 1090°C. (*b*) 460°C. **37.** (*a*) $-1.5nRT_1$.
(*b*) $4.5nRT_1$. (*c*) $6nRT_1$. (*d*) 2R. **39.** (*a*) −6.0 J. (*b*) −43 J.
(*c*) 40 J. (*d*) 18 J; 18 J. **41.** (*a*) 15.9 J. (*b*) 34.4 J/mol·K.
(*c*) 26.1 J/mol·K.
43. (*a*) Q, ΔE_{int}, W: AB: 3740, 3740, 0 J;
BC: 0, −1810, −1810 J; CA: −3220, −1930, +1290 J;
cycle: 520, 0, −520 J. (*b*) $V_B = 0.0246$ m³; $p_B = 2.00$ atm;
$V_C = 0.0373$ m³; $p_C = 1.00$ atm. **45.** 12.0 kW. **47.** 720°C.
49. (*a*) 546 C°/m. (*b*) 394 kW. (*c*) 63.9°C.
51. Arrangement *b*. **53.** 1.84 W/m·K. **55.** (*a*) 24 kW.
(*b*) 25 W. **57.** (*a*) 1.8 W. (*b*) 0.025 C°. **59.** 1.15 m.

CHAPTER 26

1. (*a*) 30.9%. (*b*) 16.2 kJ. **3.** 25.4%. **5.** (*a*) 7200 J. (*b*) 960 J.
(*c*) 13%. **7.** $e_A = 33.3\%$; $e_B = 55.6\%$. **9.** (*a*) 217 kJ.
(*b*) 32.5 kJ. **11.** (*a*) 2090 J. (*b*) 1510 J. (*c*) 1510 J.
15. 21 J. **17.** (*a*) 113 J. (*b*) 305 J. **19.** (*c*) 6.8.
23. $[1 - T_2/T_1]/[1 - T_4/T_3]$. **25.** (*a*) 1.62 atm. (*b*) 43.7%.
27. (*a*) 2.27 kJ. (*b*) 14.8 kJ. (*c*) 15.3%. (*d*) 75.0%.
29. 18.7 kJ. **31.** (*b*) 200 J. (*c*) −75 J. **33.** 0.044 J/K.
37. 3.0 mol. **39.** (*a*) +1.06 J/K. (*b*) No.
41. (*a*) Path I: $Q_T = p_0V_0 \ln 2$; $Q_V = (9/2)p_0V_0$.
Path II: $Q_T = -p_0V_0 \ln 2$; $Q_p = (15/2)p_0V_0$.
(*b*) Path I: $W_T = -p_0V_0 \ln 2$; $W_V = 0$.
Path II: $W_T = p_0V_0 \ln 2$; $W_p = 3p_0V_0$.
(*c*) $(9/2)p_0V_0$ for each process. (*d*) $4R \ln 2$ for each process.
43. (*a*) −926 J/K. (*b*) 926 J/K. **45.** +0.95 J/K.

PHOTO CREDITS

CHAPTER 1
Figure 1: Courtesy National Bureau of Standards and Technology. Figure 4: Courtesy National Physical Laboratories, Teddington, England. Figure 5: Courtesy National Bureau of Standards and Technology. Figure 6: Courtesy Professor R. C. Barber, The University of Manitoba. Figure 7: Stephen Pitkin.

CHAPTER 2
Figure 21: Courtesy National Bureau of Standards and Technology. Figure 22: Courtesy Baltimore Office of Promotion and Tourism. Figure 30: NASA. Figure 31: Courtesy Marriott Marquis, N.Y.C.. Figure 33: National Basketball Association.

CHAPTER 3
Figure 21: NASA.

CHAPTER 4
Figures 6 and 7: Education Development Center, Inc. Figure 15: From *The Particle Explosion,* Oxford Press, 1987. Figure 32: Courtesy Boeing.

CHAPTER 5
Figures 11 and 23: NASA. Figure 25: Courtesy Hale Observatories. Figure 29: Courtesy Smithsonian Astrophysical Observatory. Figure 32: Courtesy A. A. Bartlett and Boeing. Figure 36: NASA. Figure 46: Courtesy USAADTA. Photo by PUT Eugenio P. Redmond.

CHAPTER 6
Figure 2: From *Friction and Lubrication of Solids,* by F. P. Bowden and S. Tabor, Clarendon Press, 1950. Figure 18: NOAA. Figure 19: NASA. Figure 23: Ira Kirschenbaum/Stock, Boston. Figure 41: EPU/Heine Pederson/Woodfin Camp.

CHAPTER 7
Figure 2: Ed Goldfarb/Black Star. Figure 21: UPI/Bettmann Archive. Figure 22: Courtesy Cunard.

CHAPTER 8
Figure 8: Courtesy Six Flags. Figure 15: From *Introduction to the Detection of Nuclear Particles in a Bubble Chamber,* Ealing Press, 1969. Courtesy Lawrence Berkeley Radiation Laboratories, University of California at Berkeley. Figure 17: Courtesy Department of Astronomy, University of Texas at Austin. Figure 45: NASA. Figure 50: Courtesy American Red Cross.

CHAPTER 9
Figure 31: United Feature Syndicate.

CHAPTER 10
Figure 1: Courtesy Harold E. Edgerton, M. I. T., Cambridge, Mass. Figure 2: PSSC, *Physics,* Haber-Scham, Cross, Dodge, and Walter, D. C. Heath and Co., Boston. Education Development Center, Newton, Mass., 1976. Figure 4: Courtesy CERN. Figure 5: Bob Kalman/The Image Works. Figure 13: Courtesy Laurence Radiation Laboratory. Figure 18: Courtesy Stanford Linear Accelerator Center. Figure 19: Courtesy Fermi National Accelerator Laboratory. Figure 21: Sylvia Johnson/Woodfin Camp and Associates. Figure 28: A registered trademark of DC Comics, Inc., copyright © 1963. Figure 31: Georg Gerster/Comstock

CHAPTER 11
Figure 1: K. Bendo. Figure 10: Reprinted with permission from *The Courier-Journal* and *The Louisville Times.* Figure 13: NASA.

CHAPTER 12
Figure 19: Education Development Center, Inc. Figure 22: Courtesy Alice Halliday. Figure 44: Courtesy Lawrence Livermore Laboratory.

CHAPTER 13
Figure 15: NASA. Figure 20: Courtesy GE Medical Systems.

CHAPTER 14
Figure 15: Courtesy Micro Measurements Division, Measurements Group, Inc., Raleigh, N. C. Figure 20: The Bettmann Archive.

CHAPTER 15
Figure 17: Courtesy Tektronix. Figure 24: NASA.

CHAPTER 16
Figures 17 and 23*a*: Courtesy Lick Observatory. Figure 23*b*: Mt. Wilson and Palomar Observatories. Figure 26: Courtesy Kitt Peak National Observatory. Figure 27: Courtesy P. J. E. Peebles, based on the Lick Observatory Catalog by C. Shane and C. Wirtanen.

CHAPTER 17
Figure 14: Mark Antman/The Image Works. Figure 16: NASA. Figure 26: Courtesy Goodyear Tire and Rubber Company.

CHAPTER 18
Figure 21: Richard Megna/Fundamental Photos. Figure 22: Imperial College, London. Figure 24: Professor Harry Swinney, University of Texas at Austin.

CHAPTER 19
Figure 2: G. Whiteby/Photo Researchers. Figure 16: Clifford Swartz. Figure 23: From PSSC, *Physics,* D. C. Heath, Lexington, Mass., 1960, with permission.

CHAPTER 20
Figure 10: Courtesy Dr. T. D. Rossing, Northern Illinois University. Figure 14: Courtesy U. S. Army Ballistic Research Laboratory. Figure 15: Courtesy John S. Foster, Stanford University. Photo by C. F. Quate. Figure 17: Courtesy Pilgrim Nuclear Power Plant/Boston Edison.

CHAPTER 22
Figure 6: AP/Wide World Photos. Figure 20: Palmer/Monkmeyer Press.

CHAPTER 23
Figure 19: NASA.

CHAPTER 24
Figure 15: Courtesy IBM. Figures 18 and 19: Mendelssohn, *The Quest for Absolute Zero.*

CHAPTER 25
Figure 18: Courtesy Soehngen.

CHAPTER 26
Figure 9: Courtesy The Bryant Day and Night and Payne Brands of Carrier Corporation.

INDEX